HIV

From Biology to Prevention and Treatment

A subject collection from *Cold Spring Harbor Perspectives in Medicine*

OTHER SUBJECT COLLECTIONS FROM *COLD SPRING HARBOR PERSPECTIVES IN MEDICINE*

The Biology of Alzheimer Disease
Angiogenesis: Biology and Pathology

SUBJECT COLLECTIONS FROM *COLD SPRING HARBOR PERSPECTIVES IN BIOLOGY*

Extracellular Matrix Biology
Protein Homeostasis
Calcium Signaling
The Golgi
Germ Cells
The Mammary Gland as an Experimental Model
The Biology of Lipids: Trafficking, Regulation, and Function
Auxin Signaling: From Synthesis to Systems Biology
The Nucleus
Neuronal Guidance: The Biology of Brain Wiring
Cell Biology of Bacteria
Cell–Cell Junctions
Generation and Interpretation of Morphogen Gradients
Immunoreceptor Signaling
NF-κB: A Network Hub Controlling Immunity, Inflammation, and Cancer
Symmetry Breaking in Biology
The Origins of Life
The p53 Family

HIV

From Biology to Prevention and Treatment

A subject collection from *Cold Spring Harbor Perspectives in Medicine*

EDITED BY

Frederic D. Bushman

University of Pennsylvania School of Medicine

Gary J. Nabel

National Institutes of Health

Ronald Swanstrom

University of North Carolina at Chapel Hill

COLD SPRING HARBOR LABORATORY PRESS
Cold Spring Harbor, New York • www.cshlpress.org

HIV: From Biology to Prevention and Treatment

A Subject Collection from *Cold Spring Harbor Perspectives in Medicine*
Articles online at www.perspectivesinmedicine.org

Executive Editor	Richard Sever
Managing Editor	Maria Smit
Production Editor	Diane Schubach
Project Manager	Barbara Acosta
Permissions Administrator	Carol Brown
Production Manager/Cover Designer	Denise Weiss
Publisher	John Inglis

Front cover artwork: The cover art is from the painting *Miracle of Hope II* by David Putnam, which depicts a battle within a single human cell between retroviral drugs (blue dots) and the protease enzyme of HIV (black dots). This painting is part of a series that, in its third panel, shows the protease enzyme 100% annihilated. In *Miracle of Hope II* the battle has reached a fever pitch and the retroviral drugs are on the verge of victory. The tide is turning in favor of the patient being treated and the impending victory is fueled not only by medical science but also the will to live, in other words, hope (http://www.daveputnamart.com).

Library of Congress Cataloging-in-Publication Data

HIV : from biology to prevention and treatment / edited by Frederic D. Bushman, Gary J. Nabel, Ronald Swanstrom.
    p. ; cm.
"A subject collection from Cold Spring Harbor perspectives in medicine"-- T.p. verso.
Includes bibliographical references and index.
ISBN 978-1-936113-40-8 (hardcover : alk. paper)
I. Bushman, Frederic D. II. Nabel, Gary J. III. Swanstrom, R. (Ronald) IV. Cold Spring Harbor perspectives in medicine.
[DNLM: 1. HIV Infections--physiopathology--Collected Works. 2. HIV Infections--therapy--Collected Works. WC 503]

LC classification not assigned
614.5'99392--dc23
                    2011030503

10 9 8 7 6 5 4 3 2 1

All World Wide Web addresses are accurate to the best of our knowledge at the time of printing.

All Cold Spring Harbor Laboratory Press publications may be ordered directly from Cold Spring Harbor Laboratory Press, 500 Sunnyside Blvd., Woodbury, New York 11797-2924. Phone: 1-800-843-4388 in Continental U.S. and Canada. All other locations: (516) 422-4100. FAX: (516) 422-4097. E-mail: cshpress@cshl.edu. For a complete catalog of all Cold Spring Harbor Laboratory Press publications, visit our website at http://www.cshlpress.org.

Contents

Contents

THERAPY AND PREVENTION

Preface

THE SCALE OF HIV RESEARCH IS STAGGERING. A search of the PubMed literature database on "HIV" in April 2011 yielded 223,877 scientific papers. In 2010, fully 13,188 papers were published on HIV, or about 36 new papers *a day*. There was once a time when a scholar could expect to read all the literature in their field, but for HIV this task is now impossible.

The breadth of the field is also vast, perhaps to an unprecedented degree. At the smallest scale, research centers on functions of the viral proteins and their interactions with small-molecule inhibitors. Some of the most creative breakthroughs in pharmacology have yielded new classes of HIV drugs, which are now approved by the FDA and widely used. The viral proteins assemble into larger structures, each of which interacts with myriad cellular proteins. The infection process involves single cells, but the disease involves the many different kinds of cells that make up complex tissues and organs. A vast effort centers on understanding the immune response against HIV and manipulating it with vaccines, representing one of the most important areas of public health research today. Research in primatology has revealed the origins of HIV and mechanisms of pathogenesis (or the lack of it) in the many combinations of simian immunodeficiency viruses and their primate hosts. Research on human behavior is another critical area, focused on reducing HIV transmission. Responding to the epidemic worldwide reaches into areas of economics and public policy. And the list goes on.

Given this scale, it is evident that comprehensive reviews of HIV research areas are critical for understanding AIDS, for the development of new treatment and prevention strategies, and for the effective education of physicians, scientists, public health officials, patients, advocates, and the public. This book is composed of 29 chapters, spanning everything from molecular mechanisms of viral replication to pathogenesis and prevention of infection. Because the HIV literature is so extensive, it is not possible to cite every publication on HIV in this book. Each chapter cites key papers and also earlier reviews, which contain large numbers of citations to earlier reports. We apologize in advance for the omission of many high-quality publications because of severe space constraints.

We have also avoided duplicating the encyclopedic *Retroviruses* book, another Cold Spring Harbor publication that is now available through the National Library of Medicine (http://www.ncbi.nlm.nih.gov/books/NBK19376/). We instead have assembled a book that allows people in specific areas of HIV research to access critical and current information about other areas. Studies of animal retroviruses and other human retroviruses are well covered in the earlier book.

Many people helped make this book possible. Barbara Acosta and Richard Sever at Cold Spring Harbor Laboratory Press worked hard to produce the final book. We are grateful to colleagues who read over chapters and helped with edits, including Troy Brady, Peter Cherepanov, Alan Engelman, Rithun Mukherjee, Karen Ocwieja, and Shannah Roth. Kushol Gupta and Fred Hunter provided essential help with movies. Caitlin Greig provided outstanding help with figures.

FREDERIC D. BUSHMAN
GARY J. NABEL[1]
RONALD SWANSTROM

---

[1]Gary J. Nabel's works as editor and author were performed outside the scope of his employment as a U.S. government employee. These works represent his personal and professional views and not necessarily those of the U.S. government.

Origins of HIV and the AIDS Pandemic

Paul M. Sharp[1] and Beatrice H. Hahn[2]

[1]Institute of Evolutionary Biology and Centre for Immunity, Infection and Evolution, University of Edinburgh, Edinburgh EH9 3JT, United Kingdom

[2]Department of Medicine, Perelman School of Medicine, University of Pennsylvania, Philadelphia, Pennsylvania 19104

Correspondence: bhahn@upenn.edu

Acquired immunodeficiency syndrome (AIDS) of humans is caused by two lentiviruses, human immunodeficiency viruses types 1 and 2 (HIV-1 and HIV-2). Here, we describe the origins and evolution of these viruses, and the circumstances that led to the AIDS pandemic. Both HIVs are the result of multiple cross-species transmissions of simian immunodeficiency viruses (SIVs) naturally infecting African primates. Most of these transfers resulted in viruses that spread in humans to only a limited extent. However, one transmission event, involving SIVcpz from chimpanzees in southeastern Cameroon, gave rise to HIV-1 group M—the principal cause of the AIDS pandemic. We discuss how host restriction factors have shaped the emergence of new SIV zoonoses by imposing adaptive hurdles to cross-species transmission and/or secondary spread. We also show that AIDS has likely afflicted chimpanzees long before the emergence of HIV. Tracing the genetic changes that occurred as SIVs crossed from monkeys to apes and from apes to humans provides a new framework to examine the requirements of successful host switches and to gauge future zoonotic risk.

Acquired Immune Deficiency Syndrome (AIDS) was first recognized as a new disease in 1981 when increasing numbers of young homosexual men succumbed to unusual opportunistic infections and rare malignancies (CDC 1981; Greene 2007). A retrovirus, now termed human immunodeficiency virus type 1 (HIV-1), was subsequently identified as the causative agent of what has since become one of the most devastating infectious diseases to have emerged in recent history (Barre-Sinoussi et al. 1983; Gallo et al. 1984; Popovic et al. 1984). HIV-1 spreads by sexual, percutaneous, and perinatal routes (Hladik and McElrath 2008; Cohen et al. 2011); however, 80% of adults acquire HIV-1 following exposure at mucosal surfaces, and AIDS is thus primarily a sexually transmitted disease (Hladik and McElrath 2008; Cohen et al. 2011). Since its first identification almost three decades ago, the pandemic form of HIV-1, also called the main (M) group, has infected at least 60 million people and caused more than 25 million deaths (Merson et al. 2008). Developing countries have experienced the greatest HIV/AIDS morbidity and mortality, with the highest prevalence rates recorded in young adults in sub-Saharan Africa (http://www.unaids.org/).

Although antiretroviral treatment has reduced the toll of AIDS- related deaths, access to therapy is not universal, and the prospects of curative treatments and an effective vaccine are uncertain (Barouch 2008; Richman et al. 2009). Thus, AIDS will continue to pose a significant public health threat for decades to come.

Ever since HIV-1 was first discovered, the reasons for its sudden emergence, epidemic spread, and unique pathogenicity have been a subject of intense study. A first clue came in 1986 when a morphologically similar but antigenically distinct virus was found to cause AIDS in patients in western Africa (Clavel et al. 1986). Curiously, this new virus, termed

human immunodeficiency virus type 2 (HIV-2), was only distantly related to HIV-1, but was closely related to a simian virus that caused immunodeficiency in captive macaques (Chakrabarti et al. 1987; Guyader et al. 1987). Soon thereafter, additional viruses, collectively termed simian immunodeficiency viruses (SIVs) with a suffix to denote their species of origin, were found in various different primates from sub-Saharan Africa, including African green monkeys, sooty mangabeys, mandrills, chimpanzees, and others (Fig. 1). Surprisingly, these viruses appeared to be largely nonpathogenic in their natural hosts, despite clustering together with the human and simian AIDS viruses in a single phylogenetic lineage

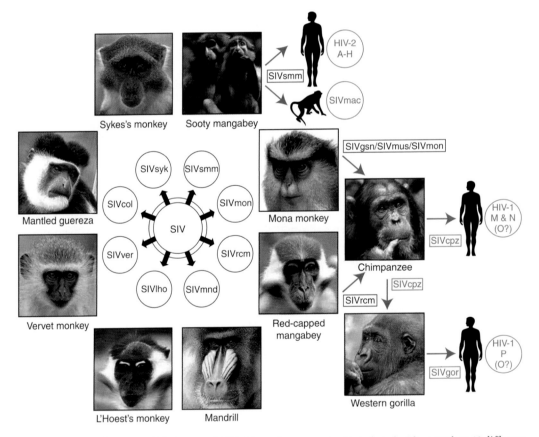

Figure 1. Origins of human AIDS viruses. Old World monkeys are naturally infected with more than 40 different lentiviruses, termed simian immunodeficiency viruses (SIVs) with a suffix to denote their primate species of origin (e.g., SIVsmm from sooty mangabeys). Several of these SIVs have crossed the species barrier to great apes and humans, generating new pathogens (see text for details). Known examples of cross-species transmissions, as well as the resulting viruses, are highlighted in red.

Cite this article as *Cold Spring Harb Perspect Med* doi: 10.1101/cshperspect.a006841

within the radiation of lentiviruses (Fig. 2). Interestingly, close simian relatives of HIV-1 and HIV-2 were found in chimpanzees (Huet et al. 1990) and sooty mangabeys (Hirsch et al. 1989), respectively. These relationships provided the first evidence that AIDS had emerged in both humans and macaques as a consequence of cross-species infections with lentiviruses from different primate species (Sharp et al. 1994). Indeed, subsequent studies confirmed that SIVmac was not a natural pathogen of macaques (which are Asian primates), but had been generated inadvertently in US primate centers by inoculating various species of macaques with blood and/or tissues from

naturally infected sooty mangabeys (Apetrei et al. 2005, 2006). Similarly, it became clear that HIV-1 and HIV-2 were the result of zoonotic transfers of viruses infecting primates in Africa (Hahn et al. 2000). In this article, we summarize what is known about the simian precursors of HIV-1 and HIV-2, and retrace the steps that led to the AIDS pandemic.

PRIMATE LENTIVIRUSES

Lentiviruses cause chronic persistent infections in various mammalian species, including bovines, horses, sheep, felines, and primates. The great majority of lentiviruses are exogenous,

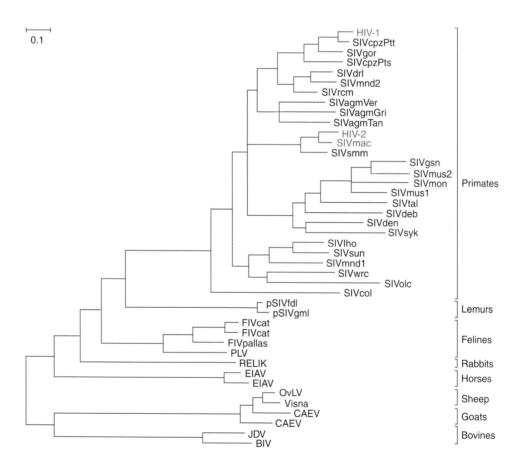

Figure 2. Phylogeny of lentiviruses. The evolutionary relationships among Pol sequences (∼ 770 amino acids) derived from various mammalian lentiviruses; host species are indicated at the right. Exogenous viruses are depicted in black, with HIV-1, HIV-2, and SIVmac highlighted in red; endogenous viruses are shown in purple. The phylogenetic tree was estimated using maximum likelihood methods (Guindon and Gascuel 2003). The scale bar represents 0.10 amino acid replacements per site.

meaning that they are transmitted horizontally between individuals. However, it has recently become clear that, on several occasions in the past, lentiviruses have infiltrated their hosts' germlines and become endogenous, vertically transmissible, genomic loci (Fig. 2). Examples include the rabbit endogenous lentivirus type K (RELIK), which became germ-line embedded approximately 12 million years ago (Katzourakis et al. 2007; van der Loo et al. 2009), and two prosimian endogenous lentiviruses, which independently invaded the germ-lines of both the grey mouse lemur (pSIVgml) and the fat-tailed dwarf lemur (pSIVfdl) about 4 million years ago (Gifford et al. 2008; Gilbert et al. 2009). These "viral fossils" are of particular interest because they provide direct evidence of the timescale of lentivirus evolution. Molecular clocks derived from extant SIV sequences suggested that ancestral SIVs existed only a few hundreds of years ago (Wertheim and Worobey 2009), but it has long been suspected that such analyses may grossly underestimate deeper evolutionary timescales (Sharp et al. 2000; Holmes 2003). Recent studies of SIV-infected monkeys on Bioko Island, Equatorial Guinea, partly substantiated this conclusion, showing that geographically isolated subspecies have been infected with the same type of SIV for at least 30,000 years and probably much longer (Worobey et al. 2010). The endogenous viruses in lemurs reveal that the span of evolutionary history of primate lentiviruses as a whole is at least two orders of magnitude greater still. Thus, it is possible that at least some SIVs, such as those infecting four closely related species of African green monkeys (*Chlorocebus* species), have coevolved with their respective hosts for an extended period of time, perhaps even before these hosts diverged from their common ancestor (Jin et al. 1994a). So far, SIV infections have only been found in African monkeys and apes, and so it seems likely that primate lentiviruses emerged in Africa sometime after the splits between lineages of African and Asian Old World monkeys, which are believed to have occurred around 6–10 million years ago (Fabre et al. 2009). However, because neither Asian nor New World primates have been sampled

exhaustively, the conclusion that SIVs are restricted to African primates must remain tentative, especially because none of these primate species has been examined for endogenous forms of SIV (Ylinen et al. 2010). Thus, our understanding of the evolutionary history of primate lentiviruses is still incomplete.

To date, serological evidence of SIV infection has been reported for over 40 primate species, and molecular data have been obtained for most of these (also see Klatt et al. 2011). The latter studies have shown that the great majority of primate species harbor a single "type" or "strain" of SIV. That is, viral sequences from members of the same species form a monophyletic clade in evolutionary trees. This host-specific clustering indicates that the great majority of transmissions occur among members of the same species; however, there are also numerous documented instances when SIVs have crossed between species. Examples range from incidental "dead-end" infections (e.g., SIVver infections of baboons) (Jin et al. 1994b; van Rensburg et al. 1998) to the generation of new SIV lineages with substantial secondary spread (e.g., SIVgor infection of gorillas) (Van Heuverswyn et al. 2006). In addition, cross-species transmissions have generated mosaic SIV lineages through superinfection and recombination in species that already harbored an SIV (e.g., SIVsab infection of sabaeus monkeys) (Jin et al. 1994a). In both mandrills (*Mandrillus sphinx*) and moustached monkeys (*Cercopithecus cephus*), such recombination events have led to the emergence of a second SIV strain that cocirculates with the original virus (Souquiere et al. 2001; Aghokeng et al. 2007). Thus, it is clear that in addition to more long-standing virus/host relationships, a number of naturally occurring SIVs have emerged more recently as a result of cross-species transmission and recombination. What remains unknown is when and how often these cross-species transfers have occurred, what impact they had on virus and host biology, and whether AIDS is a frequent consequence of SIV host switching. The prevalence of naturally occurring SIV infections varies widely, ranging from 1% in some species to over 50% in others (Aghokeng et al.

2010), and it is tempting to speculate that less ubiquitous SIVs were acquired more recently and/or may be more pathogenic.

ORIGIN AND DISTRIBUTION OF SIVcpz

Of the many primate lentiviruses that have been identified, SIVcpz has been of particular interest because of its close genetic relationship to HIV-1 (Fig. 2). However, studies of this virus have proven to be challenging because of the endangered status of chimpanzees. The first isolates of SIVcpz were all derived from animals housed in primate centers or sanctuaries, although infection was rare in these populations. Collective analyses of nearly 2,000 wild-caught or captive-born apes identified fewer than a dozen SIVcpz positive individuals (Sharp et al. 2005). Because other primate species, such as sooty mangabeys and African green monkeys, are much more commonly infected, both in captivity and in the wild (Fultz et al. 1990; Phillips-Conroy et al. 1994; Santiago et al. 2005), this finding raised doubts about whether chimpanzees represented a true SIV reservoir. To resolve this conundrum, our laboratory developed noninvasive diagnostic methods that detect SIVcpz specific antibodies and nucleic acids in chimpanzee fecal and urine samples with high sensitivity and specificity (Santiago et al. 2003; Keele et al. 2006). These technical innovations, combined with genotyping methods for species and subspecies confirmation as well as individual identification, permitted a comprehensive analysis of wild-living chimpanzee populations throughout central Africa.

Chimpanzees are classified into two species, the common chimpanzee (*Pan troglodytes*) and the bonobo (*Pan paniscus*). Common chimpanzees have traditionally been further subdivided into a number of geographically differentiated subspecies (Groves 2001). Four subspecies were defined on the basis of mitochondrial DNA sequences (Gagneux et al. 1999), namely western (*P. t. verus*), Nigeria-Cameroonian (*P. t. ellioti*, formerly termed *P. t. vellerosus*), central (*P. t. troglodytes*), and eastern (*P. t. schweinfurthii*) chimpanzees. To determine the distribution of SIVcpz among these populations, fecal (and in some cases urine) samples were collected at different field sites and tested for the presence of virus specific antibodies. Antibody positive fecal specimens were then subjected to RNA extraction and reverse transcriptase polymerase chain reaction (RT-PCR) amplification to molecularly characterize the infecting virus strain. At select field sites, mitochondrial and microsatellite analyses of host DNA were also used to confirm sample integrity and to determine the number of tested individuals. Figure 3A summarizes current molecular epidemiological data derived from the analysis of over 7,000 chimpanzee fecal samples collected at nearly 90 field sites (Santiago et al. 2002, 2003; Worobey et al. 2004; Keele et al. 2006; Van Heuverswyn et al. 2007; Li et al. 2010; Rudicell et al. 2010). These studies have identified common chimpanzees as a natural SIVcpz reservoir, but also revealed important differences between the epidemiology of SIVcpz and that of other primate lentiviruses. First, only two of the four chimpanzee subspecies were found to harbor these viruses. SIVcpz was detected at multiple sites throughout the ranges of both central and eastern chimpanzees in an area ranging from Cameroon to Tanzania, but there was no evidence of infection in western and Nigeria-Cameroonian chimpanzees, nor in bonobos, despite testing of multiple communities. In addition, SIVcpz prevalence rates among central and eastern chimpanzees varied widely, ranging from 30% to 50% in some communities to rare or absent infection in others. In contrast, other SIVs, such as those of sooty mangabeys and African green monkeys, are much more widely and evenly distributed and infect their hosts at generally higher prevalence rates (Phillips-Conroy et al. 1994; Santiago et al. 2005). Nonetheless, the puzzle of why SIVcpz was so scarce among captive chimpanzees was finally resolved: As it turned out, most of these apes were imported from West Africa and thus were members of the *P. t. verus* subspecies, which does not harbor SIVcpz (Prince et al. 2002; Switzer et al. 2005).

The absence of SIVcpz from two of the four subspecies suggested that chimpanzees had acquired this virus more recently, after their

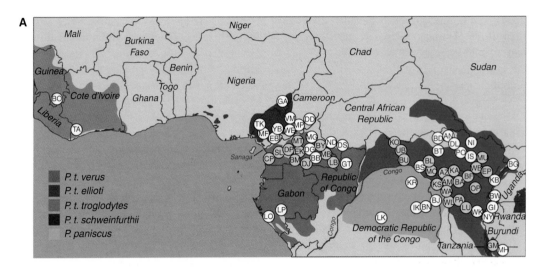

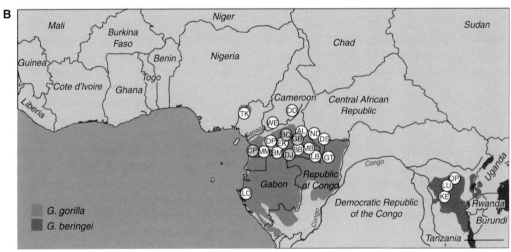

Figure 3. Geographic distribution of SIVcpz and SIVgor infections in sub-Saharan Africa. Field sites where wild-living (*A*) chimpanzees and bonobos, and (*B*) gorillas have been sampled are shown (each site is identified by a two-letter code; because of space limitations, only a subset is depicted). Sites where SIV infections were detected are highlighted in yellow. The upper panel depicts the ranges of the four subspecies of the common chimpanzee (*Pan troglodytes verus*, gray; *P. t. ellioti*, magenta; *P. t. troglodytes*, red; and *P. t. schweinfurthii*, blue) and of the bonobo (*P. paniscus*, orange). The lower panel depicts the ranges of western (*Gorilla gorilla*, green) and eastern (*G. beringei*, brown) gorillas (map courtesy of Lilian Pintea, The Jane Goodall Institute). Data were compiled from several studies (Santiago et al. 2002, 2003; Worobey et al. 2004; Keele et al. 2006; Van Heuverswyn et al. 2007; Li et al. 2010; Rudicell et al. 2010).

divergence into different subspecies. Indeed, phylogenetic analyses of full-length proviral sequences revealed that SIVcpz represents a complex mosaic, generated by recombination of two lineages of SIVs that infect monkeys (Bailes et al. 2003). In the 5′ half of the genome, as well as the *nef* gene and 3′ LTR, SIVcpz is most closely related to SIVrcm from red-capped mangabeys (*Cercocebus torquatus*); however, in the *vpu*, *tat*, *rev*, and *env* genes, SIVcpz is most closely related to a clade of SIVs infecting several *Cercopithecus* species, including greater spot-nosed (*C. nictitans*), mustached (*C. cephus*), and mona (*C. mona*) monkeys (Bailes et al.

2003). Chimpanzees are known to hunt and kill other mammals, including monkeys (Goodall 1986), suggesting that they acquired SIV in the context of predation. The current range of the central chimpanzee overlaps those of red-capped mangabeys and the various *Cercopithecus* species, and so it is likely that the cross-species transmission events that led to the emergence of SIVcpz occurred in that area, and that SIVcpz later spread to eastern chimpanzees, although it is unclear whether this occurred during or subsequent to their divergence from the central subspecies. Importantly, all of more than 30 sequenced SIVcpz strains show an identical mosaic genome structure. Moreover, there is no evidence that chimpanzees harbor any other SIV, although they, as well as bonobos, are routinely exposed to SIVs through their hunting behavior (Mitani and Watts 1999; Surbeck and Hohmann 2008; Leendertz et al. 2011).

NATURAL HISTORY OF SIVcpz INFECTION

Initially, SIVcpz was thought to be harmless for its natural host. This was because none of the few captive apes that were naturally SIVcpz infected suffered from overt immunodeficiency, although in retrospect this conclusion was based on the immunological and virological analyses of only a single naturally infected chimpanzee (Heeney et al. 2006). In addition, SIV-infected sooty mangabeys and African green monkeys showed no sign of disease despite high viral loads in blood and lymphatic tissues (Paiardini et al. 2009), leading to the belief that all naturally occurring SIV infections are nonpathogenic. However, the sporadic prevalence of SIVcpz, along with its more recent monkey origin, suggested that its natural history might differ from that of other primate lentiviruses. To address this, a prospective study was initiated in Gombe National Park, Tanzania, the only field site where SIVcpz infected chimpanzees are habituated and so can be observed in their natural habitat.

Gombe is located in northwestern Tanzania on the shores of Lake Tanganyika. The park is home to three communities, termed Kasekela, Mitumba, and Kalande, which have been studied by Goodall and colleagues since the 1960s, 1980s, and 1990s, respectively (Pusey et al. 2007). Prospective studies of SIVcpz in Gombe began in 2000 (Santiago et al. 2002). By 2009, infections were documented in all three communities, with mean biannual prevalence rates of 13%, 12%, and 46% in Mitumba, Kasekela, and Kalande, respectively (Rudicell et al. 2010). Analysis of epidemiologically linked infections revealed that SIVcpz spreads primarily through sexual routes, with an estimated transmission probability per coital act (0.0008–0.0015) that is similar to that of HIV-1 among heterosexual humans (0.0011) (Gray et al. 2001; Rudicell et al. 2010). SIVcpz also appears to be transmitted from infected mothers to their infants, and in rare cases, possibly by aggression (Keele et al. 2009). Migration of infected females constitutes a major route of virus transmission between communities (Rudicell et al. 2010).

Behavioral and virological studies also provided insight into the pathogenicity of SIVcpz. Age-corrected mortality analyses revealed that infected chimpanzees had a 10- to 16-fold increased risk of death compared to uninfected chimpanzees (Keele et al. 2009). SIVcpz-infected females were less likely to give birth and had a much higher infant mortality rate than uninfected females. Postmortem analyses revealed significant CD4$^+$ T-cell depletion in three infected individuals, but not in either of two uninfected individuals. One infected female, who died within 3 years of acquiring the virus, had histopathological findings consistent with end-stage AIDS. Taken together, these findings provided compelling evidence that SIVcpz was pathogenic in its natural host. Subsequent studies of both wild and captive chimpanzees confirmed these findings. By the end of 2010, the Kasekela and Mitumba communities had experienced three additional deaths, all SIVcpz related. One case concerned an infant born to an infected mother, whereas the other two were adult females, one of whom died with severe CD4$^+$ T cell depletion within 5 years of acquiring SIVcpz (KA Terio et al., submitted). Moreover, demographic studies revealed that the Kalande community, which showed the highest SIVcpz prevalence rates (40%–50%), had suffered a catastrophic population decline, whereas the sizes of the Mitumba and Kasekela communities, which were infected at a much lower level (12%–13%), remained stable (Rudicell et al. 2010). It has been suggested that only members of the *P. t. schweinfurthii* subspecies, or more particularly the chimpanzees of Gombe, are susceptible to SIVcpz-associated pathogenicity (Weiss and Heeney 2009; Soto et al. 2010). However, a prospective study of orphaned chimpanzees in Cameroon identified an SIVcpz infected *P. t. troglodytes* ape that suffered from progressive CD4$^+$ T cell loss, severe thrombocytopenia, and clinical AIDS (Etienne et al. 2011). Thus, it seems likely that SIVcpz has a substantial negative impact on the health, reproduction, and lifespan of all chimpanzees that harbor SIVcpz in the wild.

ORIGIN AND DISTRIBUTION OF SIVgor

Noninvasive testing also led to the unexpected finding of a new SIV lineage in wild-living gorillas (Fig. 4). Analysis of ~ 200 fecal samples from southern Cameroon identified several HIV/SIV antibody positive gorillas, and amplification of viral sequences revealed the existence of a new SIV lineage, termed SIVgor (Van Heuverswyn et al. 2006). This lineage fell within the radiation of SIVcpz, clustering with strains from *P. t. troglodytes* apes, suggesting that gorillas had acquired SIVgor by cross-species infection from sympatric chimpanzees (Fig. 4). Phylogenetic analyses of full-length SIVgor sequences confirmed this conclusion, indicating that SIVgor resulted from a single chimpanzee-to-gorilla transmission event estimated to have occurred at least 100 to 200 years ago (Takehisa et al. 2009). Subsequent screening of over 2500 fecal samples from 30 field sites across central Africa uncovered additional SIVgor infections, but only in western lowland gorillas (*Gorilla gorilla gorilla*) and not in eastern gorillas (*Gorilla beringei*). Although virus-positive apes were present at field sites more than 400 km apart, only four such sites were identified, and prevalence rates in these communities did not exceed

5% (Fig. 3B). Thus, SIVgor appears to be much less common in gorillas than SIVcpz is in chimpanzees (Neel et al. 2010). Whether SIVgor is more prevalent in communities in parts of west central Africa that have not yet been tested is not known. It is also unclear whether SIVgor is pathogenic for its host, because there has been no opportunity to study the natural history of this infection in either captive or wild-living gorillas. Finally, it remains a mystery how gorillas acquired SIVgor, because they are herbivores and do not hunt or kill other mammals. Nonetheless, gorillas and chimpanzees feed in the same forest areas, which must have led to at least one encounter that allowed transmission.

ORIGINS OF HIV-1

HIV-1 has long been suspected to be of chimpanzee origin (Gao et al. 1999); however, until recently, the perceived lack of a chimpanzee reservoir left the source of HIV-1 open to question. These uncertainties have since been resolved by noninvasive testing of wild-living ape populations. It is now well established that all naturally occurring SIVcpz strains fall into two subspecies-specific lineages, termed SIVcpz*Ptt* and SIVcpz*Pts*, respectively, that are restricted to the home ranges of their respective hosts (Figs. 3 and 4). Viruses from these two lineages are quite divergent, differing at about 30%–50% of sites in their Gag, Pol, and Env protein sequences (Vanden Haesevelde et al. 1996). Interestingly, population genetic studies have shown that central and eastern chimpanzees are barely differentiated, calling into question their status as separate subspecies (Fischer et al. 2006; Gonder et al. 2011). However, the fact that they harbor distinct SIVcpz lineages suggests that central and eastern chimpanzees have been effectively isolated for some time. In addition, molecular epidemiological studies in southern Cameroon have shown that SIVcpz*Ptt* strains show phylogeographic clustering, with viruses from particular areas forming monophyletic lineages, and the discovery of SIVgor has identified a second ape species as a potential reservoir for human infection (Van Heuverswyn et al. 2006). Collectively,

Cite this article as *Cold Spring Harb Perspect Med* doi: 10.1101/cshperspect.a006841

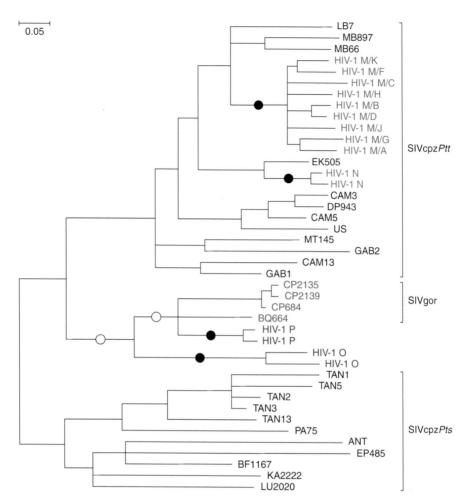

Figure 4. HIV-1 origins. The phylogenetic relationships of representative SIVcpz, HIV-1, and SIVgor strains are shown for a region of the viral *pol* gene (HIV-1/HXB2 coordinates 3887–4778). SIVcpz and SIVgor sequences are shown in black and green, respectively. The four groups of HIV-1, each of which represents an independent cross-species transmission, are shown in different colors. Black circles indicate the four branches where cross-species transmission-to-humans has occurred. White circles indicate two possible alternative branches on which chimpanzee-to-gorilla transmission occurred. Brackets at the right denote SIVcpz from *P. t. troglodytes* (SIVcpzPtt) and *P. t. schweinfurthii* (SIVcpzPts), respectively. The phylogenetic tree was estimated using maximum likelihood methods (Guindon and Gascuel 2003). The scale bar represents 0.05 nucleotide substitutions per site.

these findings have allowed the origins of HIV-1 to be unraveled (Keele et al. 2006; Van Heuverswyn et al. 2007).

HIV-1 is not just one virus, but comprises four distinct lineages, termed groups M, N, O, and P, each of which resulted from an independent cross-species transmission event. Group M was the first to be discovered and represents the pandemic form of HIV-1; it has infected

millions of people worldwide and has been found in virtually every country on the globe. Group O was discovered in 1990 and is much less prevalent than group M (De Leys et al. 1990; Gurtler et al. 1994). It represents less than 1% of global HIV-1 infections, and is largely restricted to Cameroon, Gabon, and neighboring countries (Mauclere et al. 1997; Peeters et al. 1997). Group N was identified in

1998 (Simon et al. 1998), and is even less prevalent than group O; so far, only 13 cases of group N infection have been documented, all in individuals from Cameroon (Vallari et al. 2010). Finally, group P was discovered in 2009 in a Cameroonian woman living in France (Plantier et al. 2009). Despite extensive screening, group P has thus far only been identified in one other person, also from Cameroon (Vallari et al. 2011). Although members of all of these groups are capable of causing CD4[+] T-cell depletion and AIDS, they obviously differ vastly in their distribution within the human population.

Figure 4 depicts a phylogenetic tree of representative HIV-1, SIVcpz, and SIVgor strains. It shows that all four HIV-1 groups, as well as SIVgor, cluster with SIVcpz*Ptt* from central chimpanzees, identifying this subspecies as the original reservoir of both human and gorilla infections. HIV-1 groups N and M are very closely related to SIVcpz*Ptt* strains from southern Cameroon, indicating that they are of chimpanzee origin. It has even been possible to trace their ape precursors to particular *P. t. troglodytes* communities. HIV-1 group N appears to have emerged in the vicinity of the Dja Forest in south-central Cameroon, whereas the pandemic form, group M, likely originated in an area flanked by the Boumba, Ngoko, and Sangha rivers in the southeastern corner of Cameroon (Keele et al. 2006; Van Heuverswyn et al. 2007). Existing phylogenetic data support a gorilla origin of HIV-1 group P, but too few SIVgor strains have been characterized to identify the region where this transmission might have occurred. In contrast, the immediate source of HIV-1 group O remains unknown, because there are no ape viruses that are particularly closely related to this group (Fig. 4). Thus, HIV-1 group O could either be of chimpanzee or gorilla origin. Nonetheless, the fact that group O and P viruses are more closely related to SIVcpz*Ptt* than to SIVcpz*Pts* suggests that both groups originated in west central Africa, which is consistent with their current distributions.

How humans acquired the ape precursors of HIV-1 groups M, N, O, and P is not known;

however, based on the biology of these viruses, transmission must have occurred through cutaneous or mucous membrane exposure to infected ape blood and/or body fluids. Such exposures occur most commonly in the context of bushmeat hunting (Peeters et al. 2002). Whatever the circumstances, it seems clear that human–ape encounters in west central Africa have resulted in four independent cross-species transmission events. Molecular clock analyses have dated the onset of the group M and O epidemics to the beginning of the twentieth century (Korber et al. 2000; Lemey et al. 2004; Worobey et al. 2008). In contrast, groups N and P appear to have emerged more recently, although the sequence data for these rare groups are still too limited to draw definitive conclusions.

Eastern chimpanzees are endemically infected with SIVcpz*Pts* throughout central Africa (Fig. 3A). Although prevalence rates have not been determined for all field sites, the *P. t. schweinfurthii* communities that have been studied show infection rates that are very similar to those found in *P. t. troglodytes* (Keele et al. 2006, 2009; Rudicell et al. 2010). Given that SIVcpz*Ptt* strains have been transmitted to gorillas and humans on at least five occasions, it is striking that evidence of similar transmissions from eastern chimpanzees is lacking. There are a number of possible explanations. First, the risk of human exposure to SIVcpz*Pts* may be lower, perhaps because of differences in the frequencies or types of human–ape interactions in central and east Africa. Second, SIVcpz*Pts* infections of humans may have occurred, but gone unrecognized, because of limited human sampling and a lack of lineage-specific serological tests. Finally, as discussed below, SIVcpz*Ptt* has evolved to overcome human restriction factors, such as tetherin, which may pose a barrier to cross-species transmission; because SIVcpz*Pts* is highly divergent from SIVcpz*Ptt*, viruses from this lineage may not have been able to adapt in the same way. Although SIVcpz*Ptt* and SIVcpz*Pts* strains replicate with similar kinetics in human CD4[+] T cells in vitro (Takehisa et al. 2007), such cultures are unlikely to accurately recapitulate the conditions of viral replication and transmission in vivo.

ORIGINS OF HIV-2

Since its first discovery, HIV-2 has remained largely restricted to West Africa, with its highest prevalence rates recorded in Guinea-Bissau and Senegal (de Silva et al. 2008). However, overall prevalence rates are declining, and in most West African countries HIV-2 is increasingly being replaced by HIV-1 (van der Loeff et al. 2006; Hamel et al. 2007). Viral loads tend to be lower in HIV-2 than HIV-1 infected individuals, which may explain the lower transmission rates of HIV-2 and the near complete absence of mother-to-infant transmissions (Popper et al. 2000; Berry et al. 2002). In fact, most individuals infected with HIV-2 do not progress to AIDS, although those who do, show clinical symptoms indistinguishable from HIV-1 (Rowland-Jones and Whittle 2007). Thus, it is clear that the natural history of HIV-2 infection differs considerably from that of HIV-1, which is not surprising given that HIV-2 is derived from a very different primate lentivirus.

A sooty mangabey origin of HIV-2 was first proposed in 1989 (Hirsch et al. 1989) and subsequently confirmed by demonstrating that humans in West Africa harbored HIV-2 strains that resembled locally circulating SIVsmm infections (Gao et al. 1992; Chen et al. 1996). SIVsmm was found to be highly prevalent, both in captivity and in the wild, and to be nonpathogenic in its natural host (Silvestri 2005). In a wild-living sooty mangabey community, SIVsmm was primarily found in higher-ranking females, suggesting that virus infection had no appreciable negative effect on reproductive behavior or success (Santiago et al. 2005). Finally, the fact that sooty mangabeys are frequently hunted as agricultural pests in many areas of West Africa provided plausible routes of transmission.

Since its first isolation, at least eight distinct lineages of HIV-2 have been identified, each of which appears to represent an independent host transfer (Fig. 5). By analogy with HIV-1, these lineages have been termed groups A–H, although only groups A and B have spread within humans to an appreciable degree. Group A has been found throughout western Africa (Damond et al. 2001; Peeters et al. 2003), whereas group B predominates in Cote d'Ivoire (Pieniazek et al. 1999; Ishikawa et al. 2001). All other HIV-2 "groups" were initially identified only in single individuals, suggesting that they represent incidental infection with very limited or no secondary spread. Of these, groups C, G, and H have been linked to SIVsmm strains from Cote d'Ivoire, group D is most closely related to an SIVsmm strain from Liberia, and groups E and F resemble SIVsmm strains from Sierra Leone (Gao et al. 1992; Chen et al. 1996, 1997; Santiago et al. 2005). Because of their sporadic nature, groups C–H have been assumed to represent "dead-end" transmissions. However, a second divergent HIV-2 strain has recently been placed in group F (Fig. 5). This virus was identified in an immigrant in New Jersey, who came from the same geographic area in Sierra Leone where this lineage was first discovered (Smith et al. 2008). Unlike the original index case, the newly identified group F infection was associated with reduced CD4 T cell counts and high viral loads (Smith et al. 2008). It is presently unknown whether group F has been spreading cryptically in humans, or whether the two group F viruses represents independent transmissions from sooty mangabeys.

HOST-SPECIFIC ADAPTATIONS

HIV and SIV must interact with a large number of host proteins to replicate in infected cells (Fu et al. 2009; Ortiz et al. 2009). Because the common ancestor of Old World monkeys and apes existed around 25 million years ago, the divergence of these host proteins may pose an obstacle to cross-species infection. In addition, primates (including humans) encode a number of host restriction factors, which have evolved as part of their innate immune response to protect against infection with a wide variety of viral pathogens (Malim and Emerman 2008; Neil and Bieniasz 2009; Kajaste-Rudnitski et al. 2010). Although viruses have, in turn, found ways to antagonize these restriction factors,

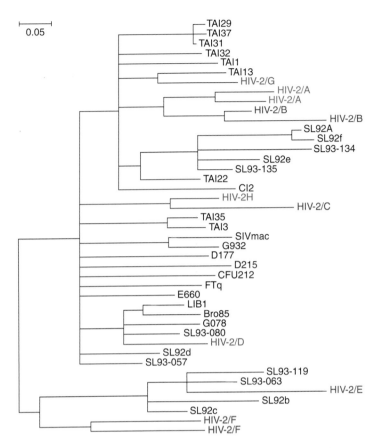

Figure 5. HIV-2 origins. The phylogenetic relationships of representative SIVsmm and HIV-2 strains are shown for a region of the viral *gag* gene (SIVmac239 coordinates 1191–1921). SIVsmm and SIVmac are shown in black; the eight groups of HIV-2, each of which represents an independent cross-species transmission, are shown in different colors. The phylogenetic tree was estimated using maximum likelihood methods (Guindon and Gascuel 2003). The scale bar represents 0.05 nucleotide substitutions per site.

these countermeasures are frequently species-specific. Thus, a number of adaptive hurdles have to be overcome before primate lentiviruses can productively infect a new species.

The first evidence of host-specific adaptation of HIV-1 came from an analysis of sites in the viral proteome that were highly conserved in the ape precursors of HIV-1, but changed— in the same way—each time these viruses crossed the species barrier to humans (Wain et al. 2007). This analysis identified one site in the viral matrix protein (Gag-30) that encoded a Met in all known strains of SIVcpz*Ptt* and SIVgor but switched to an Arg in the inferred ancestors of HIV-1 groups M, N, and O, and has subsequently been conserved as a basic amino acid (Arg or Lys) in most strains of HIV-1. The fact that the same nonconservative amino acid substitution occurred on each of the three branches involving cross-species transmission to humans suggested that this matrix residue was under strong host-specific selection pressure. This conclusion was subsequently confirmed by two additional observations. First, it was found that a reciprocal transmission, in which a chimpanzee was experimentally infected with HIV-1, led to the reversion of this host-specific signature; that is, a basic residue at Gag-30 in HIV-1 changed back to a Met on in vivo propagation in a chimpanzee (Mwaengo and Novembre 1998). Second, it was found that in chimpanzee CD4$^+$ T

Cite this article as *Cold Spring Harb Perspect Med* doi: 10.1101/cshperspect.a006841

lymphocytes, a virus with a Met at position 30 replicated more efficiently that an otherwise isogenic virus with a Lys at the same position, whereas the opposite was true in human cells (Wain et al. 2007). Interestingly, only one of the two recently discovered HIV-1 group P strains has switched from a Met to a Lys at Gag-30 (Vallari et al. 2011). Although the structure of the HIV-1 matrix protein has been determined (Hill et al. 1996), the function of the amino acid at position 30 is not known, and it remains to be determined why this site is under such strong selection pressure.

The potential of an SIV to infect a new primate species is also influenced by its ability to counteract different host restriction factors. Three classes of restriction factors have been shown to constitute barriers to SIV cross-species transmission. These include (1) APOBEC3G (apolipoprotein B mRNA editing enzyme catalytic polypeptide-like 3G), which interferes with reverse transcription (Sheehy et al. 2002); (2) TRIM5α (tripartite motif 5α protein), which interferes with viral uncoating (Stremlau et al. 2004); and (3) tetherin (also termed BST-2 and CD317) which inhibits the budding and release of virions from infected cells (Neil et al. 2008). Of these, tetherin appears to have had the greatest impact on the precursors of HIV-1 and HIV-2 (Fig. 6). Tetherin is comprised of a cytoplasmic amino-terminal region, a *trans*-membrane domain, a coiled-coiled extracellular domain, and a carboxy-terminal glycosylphosphatidylinositol (GPI) anchor (Fig. 6B). Recent studies have shown that most SIVs use their Nef protein to remove tetherin from the cell surface by targeting its cytoplasmic domain (Jia et al. 2009; Zhang et al. 2009). In contrast, HIV-1 (Neil et al. 2008; Van Damme et al. 2008) as well as SIVs from greater spot-nosed, mona, moustached, and Dent's monkeys (Sauter et al. 2009; Schmokel et al. 2011) use their Vpu protein to degrade tetherin by binding to its membrane-spanning domain (Iwabu et al. 2009; Rong et al. 2009). Still other viruses use their envelope glycoprotein to interfere with tetherin, by interacting with either its extracellular or its cytoplasmic domain (Bour et al. 1996; Gupta et al. 2009;

Le Tortorec and Neil 2009; Serra-Moreno et al. 2011). These various antitetherin responses appear to have emerged as a direct result of host-specific selection pressures following cross-species transmission (Sauter et al. 2009; Evans et al. 2010; Lim et al. 2010).

Because of the constant onslaught of viral pathogens, host restriction factors evolve rapidly (Sawyer et al. 2004, 2005; McNatt et al. 2009; Lim et al. 2010). Most notably, the human tetherin gene differs from that of other apes by a five-codon deletion in the region encoding the cytoplasmic domain (Sauter et al. 2009). Because Nef interacts with the cytoplasmic domain of tetherin, this deletion rendered the SIVcpz Nef protein inactive on transmission to humans. Gorilla tetherin does not have this deletion, and thus Nef continued to function as a tetherin antagonist on transmission of SIVcpz to gorillas (Fig. 6C). Thus, to facilitate replication in humans, SIVcpz and SIVgor had to find alternative routes to overcome tetherin. One option was to switch back to using Vpu, as in their monkey ancestors (Fig. 6C). However, this required that the SIVcpz and SIVgor Vpu proteins regained this function because neither has antitetherin activity in chimpanzee or humans cells. Not surprisingly, this was not successful in all instances (Fig. 6C). When representatives of each of the HIV-1 groups were analyzed, only the Vpu proteins of group M viruses showed potent antitetherin activity (Sauter et al. 2009). Group O and P Vpu proteins were completely inactive, whereas the group N Vpu showed only marginal activity (Sauter et al. 2009; Kirchhoff 2010). Moreover, even the latter adaptation came at a cost, because the group N Vpu lost its ability to down-modulate CD4. Thus, of the four transmitted ape viruses, only the precursor of HIV-1 group M succeeded in mounting a full antitetherin defense in human cells. It may thus not be a coincidence that only HIV-1 group M resulted in a global epidemic (Gupta and Towers 2009; Sauter et al. 2009).

Like many other SIVs, SIVsmm does not encode a *vpu* gene and uses its Nef protein to combat tetherin. Thus, on transmission to humans, SIVsmm had to overcome the same

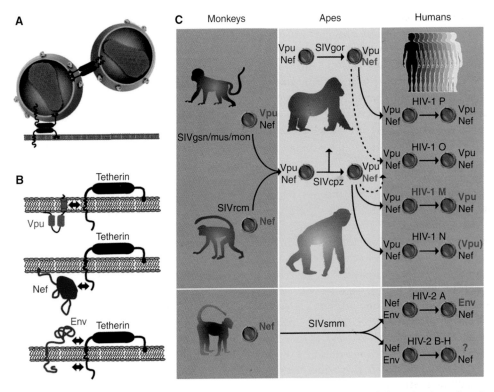

Figure 6. Tetherin function and virus specific antagonism in different primate hosts. (*A*) Mechanism of restricted virion release by tetherin; two alternative models are shown (for details, see Evans et al. 2010); (*B*) Viral antagonists of tetherin and their sites of interaction (indicated by arrows). Vpu associates with the *trans*-membrane domain of tetherin, Nef targets the cytoplasmic domain, and Env interacts either with the extracellular or the cytoplasmic domain (Kirchhoff 2010; Serra-Moreno et al. 2011). (*C*) Antitetherin function in HIV-1 and HIV-2 and their immediate simian precursors. SIVcpz acquired *vpu* and *nef* genes from different sources, the SIVgsn/mus/mon and SIVrcm lineages, respectively. During adaptation in chimpanzees, Nef (and not Vpu) evolved to become an effective tetherin antagonist. SIVgor and SIVsmm also use Nef to counteract tetherin. After transmission to humans, SIVcpz, SIVgor, and SIVsmm Nef were unable to antagonize human tetherin because of a deletion in its cytoplasmic domain. HIV-1 group M adapted by regaining Vpu-mediated antitetherin activity. The Nef and Vpu proteins of HIV-1 groups O and P remained poor tetherin antagonists. The Vpu of HIV-1 group N gained modest antitetherin activity, but lost the ability to degrade CD4. HIV-2 group A adapted by gaining Env-mediated antitetherin activity; whether HIV-2 groups B–H gained antitetherin function has not been tested. Proteins that are active against tetherin are highlighted in red, and those that are inactive are shown in gray (adapted from Kirchhoff [2010] and reprinted, with permission, from Elsevier ©2010).

hurdle of a cytoplasmic tail deleted human tetherin. In this case, the envelope glycoprotein (gp41) was recruited as an alternative tetherin antagonist (Le Tortorec and Neil 2009). However, this escape mechanism has thus far only been observed for epidemic HIV-2 group A, and it will be interesting to determine which, if any, of the other HIV-2 groups have acquired a similar antitetherin activity. It seems likely that the cytoplasmic domain deletion in human

tetherin has represented a significant barrier to SIV zoonoses, including those viruses that have succeeded in infecting humans.

ORIGIN OF THE AIDS PANDEMIC

HIV-1 evolves around one million times faster than mammalian DNA (Li et al. 1988; Lemey et al. 2006), because the HIV-1 reverse transcriptase is error prone and the viral generation time is short (Ho et al. 1995; Wei et al. 1995). This propensity for rapid genetic change has provided a unique opportunity to gain insight into when and where the AIDS pandemic had its origin. Phylogenetic and statistical analyses have dated the last common ancestor of HIV-1 group M to around 1910 to 1930, with narrow confidence intervals (Korber et al. 2000; Worobey et al. 2008). This indicates that after pandemic HIV-1 first emerged in colonial west central Africa, it spread for some 50 to 70 years before it was recognized. The probable location of the early epidemic has also been identified. Molecular epidemiological studies have indicated that most, if not all, of the early diversification of HIV-1 group M likely occurred in the area around Kinshasa, then called Leopoldville. All of the known HIV-1 group M subtypes were identified there, as well as additional lineages that have remained restricted to this area (Vidal et al. 2000). Leopoldville was also the place where the earliest strains of HIV-1 group M were discovered (Zhu et al. 1998; Worobey et al. 2008). Genetic analysis of infected blood and tissue samples collected from residents of Kinshasa in 1959 and 1960, respectively, revealed that HIV-1 had already diversified into different subtypes by that time (Worobey et al. 2008). Finally, demographic data indicate that pandemic HIV-1 emerged at a time when urban populations in west central Africa were expanding (Worobey et al. 2008). Leopoldville was the largest city in the region at that time and thus a likely destination for a newly emerging infection. Moreover, rivers, which served as major routes of travel and commerce at the time, would have provided a link between the chimpanzee reservoir of HIV-1 group M in southeastern Cameroon and Leopoldville on the banks of the Congo (Sharp and Hahn 2008). Thus, all current evidence points to Leopoldville/Kinshasa as the cradle of the AIDS pandemic.

As HIV-1 group M spread globally, its dissemination involved a number of population bottlenecks—founder events, which led to the predominance of different group M lineages, now called subtypes, in different geographic areas. HIV-1 group M is currently classified into nine subtypes (A–D, F–H, J, K), as well as more than 40 different circulating recombinant forms (CRFs), which were generated when multiple subtypes infected the same population (Taylor et al. 2008). It has been possible to trace the migration pathways of some of these subtypes and CRFs. For example, subtypes A and D originated in central Africa, but ultimately established epidemics in eastern Africa, whereas subtype C was introduced to, and predominates in, southern Africa from where it spread to India and other Asian countries. Subtype B, which accounts for the great majority of HIV-1 infections in Europe and the Americas, arose from a single African strain that appears to have first spread to Haiti in the 1960s and then onward to the US and other western countries (Gilbert et al. 2007). The recombination event that created CRF01 probably occurred in Central Africa, but this viral lineage was first noted in the late 1980s causing a heterosexual epidemic in Thailand, contemporary with subtype B viruses spreading among intravenous drug users (Taylor et al. 2008). CRF01 has gone on to dominate the AIDS epidemic in southeast Asia. Although the initial distribution of these subtypes and CRFs may have been largely caused by chance events, recent studies have suggested that viruses of different subtypes vary in their biological properties, which may influence their epidemiology (Taylor et al. 2008). For example, subtype D has been associated with greater pathogenicity (Kiwanuka et al. 2010) and an increased incidence of cognitive impairment and AIDS dementia (Sacktor et al. 2009). It thus appears that not only the genetic but also the biological diversity of HIV-1 group M subtypes and CRF is increasing.

CONCLUSIONS

Although primate lentiviruses were first identified in the late 1980s, it is only very recently that the complexities of their evolutionary origins, geographic distribution, prevalence, natural history, and pathogenesis in natural and nonnatural hosts have been appreciated. The preceding sections summarize what is known about the origins and evolution of the simian relatives of HIV-1 and HIV-2, their propensity to cross species barriers, and the host factors that govern such transmissions. Given that there are numerous additional primate species that harbor SIV, the question arises what to expect from future zoonoses? Host-specific restriction factors play a major role in preventing cross-species transmission, and they may well be responsible for the fact that only two types of SIV have thus far succeeded in colonizing humans. However, as exemplified by the various HIV-1 and HIV-2 outbreaks, they are certainly not insurmountable. Determining the entire spectrum of host restriction factors and their mechanisms of action will be required to gauge the likelihood of future zoonoses. In this regard, the role of tetherin should be examined further. Because this protein "tethers" virions to the cell surface, a lack of effective antitetherin measures may result in reduced titers of infectious virus in genital secretions. This may explain why the precursors of the rare groups of HIV-1 and HIV-2 were able to infect humans but unable to establish epidemic infections.

From the above, it is also clear that any newly introduced SIV must replicate to some extent to accumulate the necessary mutations that are required to adapt to divergent host proteins and restriction factors. Circumstances that enhance human-to-human passage would thus be expected to increase the chance of such adaptation. It has been suggested that large-scale injection campaigns conducted in west central Africa at the beginning of the twentieth century (Pepin et al. 2006, 2010; Pepin and Labbe 2008), together with the destabilization of social structures (Chitnis et al. 2000), the rapid growth of cities (Worobey et al. 2008), and an increased prevalence in sexually transmitted diseases, including genital ulcers (de Sousa et al. 2010), may have facilitated the early dissemination and adaptation of both HIV-1 and HIV-2. The fact that HIV-1 groups M and O as well as HIV-2 group A all emerged around the same time is consistent with this hypothesis (Korber et al. 2000; Lemey et al. 2003, 2004; Worobey et al. 2008; de Sousa et al. 2010). However, whether these medical interventions and/or

social factors really played a role in the emergence of HIV-1 and HIV-2, and more importantly, whether such "jump-starts" were required to spawn the AIDS pandemic, will remain unknown.

Finally, it is important to view the pathogenesis of simian and human AIDS viruses in the context of their evolution (Kirchhoff 2009). One feature of pathogenic HIV-1 and HIV-2 infection that distinguishes them from nonpathogenic SIV infections is a high level of chronic immune activation, which is a strong predictor of disease progression. In HIV-1 infection, this immune activation is fueled, at least in part, by the inability of the Nef protein to down-modulate TCR-CD3, a function that is conserved in most nonpathogenic SIV infections (Schindler et al. 2006; Arhel and Kirchhoff 2009). Lack of this Nef function is associated with increased T-cell activation and apoptosis in vitro and loss of $CD4^+$ T cells in natural SIV infection in vivo (Schindler et al. 2008). A higher state of T-cell activation is associated with enhanced levels of proviral transcription and viral replication, but also with increased expression of interferon-induced restriction factors, such as tetherin. In HIV-1, Vpu compensates for this by providing potent antitetherin activity. It has thus been proposed that to overcome the barriers of cross-species transmission, primate lentiviruses must induce an inflammatory milieu to increase their ability to replicate and accumulate mutations necessary for more adaptation (Kirchhoff 2009). If this were indeed the case, AIDS would be an inevitable consequence of SIV cross-species transmission.

ACKNOWLEDGMENTS

We thank Lilian Pintea for maps of current ape ranges, Frank Kirchhoff for unpublished results, Gerald Learn for phylogenetic tree construction, and Jamie White for artwork and manuscript preparation. This work was supported by grants from the National Institutes of Health (R01 AI50529, R01 AI58715, P30 AI 27767), and the Bristol Myers Freedom to Discover Program.

REFERENCES

*Reference is also in this collection.

Aghokeng AF, Bailes E, Loul S, Courgnaud V, Mpoudi-Ngolle E, Sharp PM, Delaporte E, Peeters M. 2007. Full-length sequence analysis of SIVmus in wild populations of mustached monkeys (*Cercopithecus cephus*) from Cameroon provides evidence for two co-circulating SIVmus lineages. *Virology* **360:** 407–418.

Aghokeng AF, Ayouba A, Mpoudi-Ngole E, Loul S, Liegeois F, Delaporte E, Peeters M. 2010. Extensive survey on the prevalence and genetic diversity of SIVs in primate bushmeat provides insights into risks for potential new cross-species transmissions. *Infect Genet Evol* **10:** 386–396.

Apetrei C, Kaur A, Lerche NW, Metzger M, Pandrea I, Hardcastle J, Falkenstein S, Bohm R, Koehler J, Traina-Dorge V, et al. 2005. Molecular epidemiology of simian immunodeficiency virus SIVsm in U.S. primate centers unravels the origin of SIVmac and SIVstm. *J Virol* **79:** 8991–9005.

Apetrei C, Lerche NW, Pandrea I, Gormus B, Silvestri G, Kaur A, Robertson DL, Hardcastle J, Lackner AA, Marx PA. 2006. Kuru experiments triggered the emergence of pathogenic SIVmac. *AIDS* **20:** 317–321.

Arhel NJ, Kirchhoff F. 2009. Implications of Nef: Host cell interactions in viral persistence and progression to AIDS. *Curr Top Microbiol Immunol* **339:** 147–175.

Bailes E, Gao F, Bibollet-Ruche F, Courgnaud V, Peeters M, Marx PA, Hahn BH, Sharp PM. 2003. Hybrid origin of SIV in chimpanzees. *Science* **300:** 1713.

Barouch DH. 2008. Challenges in the development of an HIV-1 vaccine. *Nature* **455:** 613–619.

Barre-Sinoussi F, Chermann JC, Rey F, Nugeyre MT, Chamaret S, Gruest J, Dauguet C, Axler-Blin C, Vezinet-Brun F, Rouzioux C, et al. 1983. Isolation of a T-lymphotropic retrovirus from a patient at risk for acquired immune deficiency syndrome (AIDS). *Science* **220:** 868–871.

Berry N, Jaffar S, Schim van der Loeff M, Ariyoshi K, Harding E, N'Gom PT, Dias F, Wilkins A, Ricard D, Aaby P, et al. 2002. Low level viremia and high CD4% predict normal survival in a cohort of HIV type-2-infected villagers. *AIDS Res Hum Retroviruses* **18:** 1167–1173.

Bour S, Schubert U, Peden K, Strebel K. 1996. The envelope glycoprotein of human immunodeficiency virus type 2 enhances viral particle release: A Vpu-like factor? *J Virol* **70:** 820–829.

CDC. 1981. Kaposi's sarcoma and Pneumocystis pneumonia among homosexual men—New York City and California. *MMWR Morb Mortal Wkly Rep* **30:** 305–308.

Chakrabarti L, Guyader M, Alizon M, Daniel MD, Desrosiers RC, Tiollais P, Sonigo P. 1987. Sequence of simian immunodeficiency virus from macaque and its relationship to other human and simian retroviruses. *Nature* **328:** 543–547.

Chen Z, Telfier P, Gettie A, Reed P, Zhang L, Ho DD, Marx PA. 1996. Genetic characterization of new West African simian immunodeficiency virus SIVsm: Geographic clustering of household-derived SIV strains with human immunodeficiency virus type 2 subtypes and genetically diverse viruses from a single feral sooty mangabey troop. *J Virol* **70:** 3617–3627.

Chen Z, Luckay A, Sodora DL, Telfer P, Reed P, Gettie A, Kanu JM, Sadek RF, Yee J, Ho DD, et al. 1997. Human immunodeficiency virus type 2 (HIV-2) seroprevalence and characterization of a distinct HIV-2 genetic subtype from the natural range of simian immunodeficiency virus-infected sooty mangabeys. *J Virol* **71:** 3953–3960.

Chitnis A, Rawls D, Moore J. 2000. Origin of HIV type 1 in colonial French Equatorial Africa? *AIDS Res Hum Retroviruses* **16:** 5–8.

Clavel F, Guetard D, Brun-Vezinet F, Chamaret S, Rey MA, Santos-Ferreira MO, Laurent AG, Dauguet C, Katlama C, Rouzioux C, et al. 1986. Isolation of a new human retrovirus from West African patients with AIDS. *Science* **233:** 343–346.

Cohen MS, Shaw GM, McMichael AJ, Haynes BF. 2011. Acute-HIV-1 Infection: Basic, clinical and public health perspectives. *N Engl J Med* **364:** 1943–1954.

Damond F, Descamps D, Farfara I, Telles JN, Puyeo S, Campa P, Lepretre A, Matheron S, Brun-Vezinet F, Simon F. 2001. Quantification of proviral load of human immunodeficiency virus type 2 subtypes A and B using real-time PCR. *J Clin Microbiol* **39:** 4264–4268.

De Leys R, Vanderborght B, Vanden Haesevelde M, Heyndrickx L, van Geel A, Wauters C, Bernaerts R, Saman E, Nijs P, Willems B, et al. 1990. Isolation and partial characterization of an unusual human immunodeficiency retrovirus from two persons of west-central African origin. *J Virol* **64:** 1207–1216.

de Silva TI, Cotten M, Rowland-Jones SL. 2008. HIV-2: The forgotten AIDS virus. *Trends Microbiol* **16:** 588–595.

de Sousa JD, Muller V, Lemey P, Vandamme AM. 2010. High GUD incidence in the early 20th century created a particularly permissive time window for the origin and initial spread of epidemic HIV strains. *PLoS One* **5:** e9936.

Etienne L, Nerrienet E, Lebreton M, Bibila GT, Foupouapouognigni Y, Rousset D, Nana A, Djoko CF, Tamoufe U, Aghokeng AF, et al. 2011. Characterization of a new simian immunodeficiency virus strain in a naturally infected *Pan troglodytes troglodytes* chimpanzee with AIDS related symptoms. *Retrovirology* **8:** 4.

Evans DT, Serra-Moreno R, Singh RK, Guatelli JC. 2010. BST-2/tetherin: A new component of the innate immune response to enveloped viruses. *Trends Microbiol* **18:** 388–396.

Fabre PH, Rodrigues A, Douzery EJ. 2009. Patterns of macroevolution among primates inferred from a supermatrix of mitochondrial and nuclear DNA. *Mol Phylogenet Evol* **53:** 808–825.

Fischer A, Pollack J, Thalmann O, Nickel B, Paabo S. 2006. Demographic history and genetic differentiation in apes. *Curr Biol* **16:** 1133–1138.

Fu W, Sanders-Beer BE, Katz KS, Maglott DR, Pruitt KD, Ptak RG. 2009. Human immunodeficiency virus type 1, human protein interaction database at NCBI. *Nucleic Acids Res* **37:** D417–D422.

Fultz PN, Gordon TP, Anderson DC, McClure HM. 1990. Prevalence of natural infection with simian immunodeficiency virus and simian T-cell leukemia virus type I in a breeding colony of sooty mangabey monkeys. *AIDS* **4:** 619–625.

Gagneux P, Wills C, Gerloff U, Tautz D, Morin PA, Boesch C, Fruth B, Hohmann G, Ryder OA, Woodruff DS. 1999. Mitochondrial sequences show diverse evolutionary histories of African hominoids. *Proc Natl Acad Sci* **96:** 5077–5082.

Gallo RC, Salahuddin SZ, Popovic M, Shearer GM, Kaplan M, Haynes BF, Palker TJ, Redfield R, Oleske J, Safai B, et al. 1984. Frequent detection and isolation of cytopathic retroviruses (HTLV-III) from patients with AIDS and at risk for AIDS. *Science* **224:** 500–503.

Gao F, Yue L, White AT, Pappas PG, Barchue J, Hanson AP, Greene BM, Sharp PM, Shaw GM, Hahn BH. 1992. Human infection by genetically diverse SIVsm-related HIV-2 in west Africa. *Nature* **358:** 495–499.

Gao F, Bailes E, Robertson DL, Chen Y, Rodenburg CM, Michael SF, Cummins LB, Arthur LO, Peeters M, Shaw GM, et al. 1999. Origin of HIV-1 in the chimpanzee *Pan troglodytes troglodytes*. *Nature* **397:** 436–441.

Gifford RJ, Katzourakis A, Tristem M, Pybus OG, Winters M, Shafer RW. 2008. A transitional endogenous lentivirus from the genome of a basal primate and implications for lentivirus evolution. *Proc Natl Acad Sci* **105:** 20362–20367.

Gilbert MT, Rambaut A, Wlasiuk G, Spira TJ, Pitchenik AE, Worobey M. 2007. The emergence of HIV/AIDS in the Americas and beyond. *Proc Natl Acad Sci* **104:** 18566–18570.

Gilbert C, Maxfield DG, Goodman SM, Feschotte C. 2009. Parallel germline infiltration of a lentivirus in two Malagasy lemurs. *PLoS Genet* **5:** e1000425.

Gonder MK, Locatelli S, Ghobrial L, Mitchell MW, Kujawski JT, Lankester FJ, Stewart CB, Tishkoff SA. 2011. Evidence from Cameroon reveals differences in the genetic structure and histories of chimpanzee populations. *Proc Natl Acad Sci* **108:** 4766–4771.

Goodall J. 1986. *The Chimpanzees of Gombe: Patterns of behavior.* Belknap Press, Cambridge, UK.

Gray RH, Wawer MJ, Brookmeyer R, Sewankambo NK, Serwadda D, Wabwire-Mangen F, Lutalo T, Li X, vanCott T, Quinn TC. 2001. Probability of HIV-1 transmission per coital act in monogamous, heterosexual, HIV-1-discordant couples in Rakai, Uganda. *Lancet* **357:** 1149–1153.

Greene WC. 2007. A history of AIDS: Looking back to see ahead. *Eur J Immunol* **37** (Suppl. 1): S94–S102.

Groves C. 2001. *Primate taxonomy.* Smithsonian Institution Press, Washington, DC.

Guindon S, Gascuel O. 2003. A simple, fast, and accurate algorithm to estimate large phylogenies by maximum likelihood. *Syst Biol* **52:** 696–704.

Gupta RK, Towers GJ. 2009. A tail of Tetherin: How pandemic HIV-1 conquered the world. *Cell Host Microbe* **6:** 393–395.

Gupta RK, Mlcochova P, Pelchen-Matthews A, Petit SJ, Mattiuzzo G, Pillay D, Takeuchi Y, Marsh M, Towers GJ. 2009. Simian immunodeficiency virus envelope glycoprotein counteracts tetherin/BST-2/CD317 by intracellular sequestration. *Proc Natl Acad Sci* **106:** 20889–20894.

Gurtler LG, Hauser PH, Eberle J, von Brunn A, Knapp S, Zekeng L, Tsague JM, Kaptue L. 1994. A new subtype of human immunodeficiency virus type 1 (MVP-5180) from Cameroon. *J Virol* **68:** 1581–1585.

Guyader M, Emerman M, Sonigo P, Clavel F, Montagnier L, Alizon M. 1987. Genome organization and transactivation of the human immunodeficiency virus type 2. *Nature* **326:** 662–669.

Hahn BH, Shaw GM, De Cock KM, Sharp PM. 2000. AIDS as a zoonosis: Scientific and public health implications. *Science* **287:** 607–614.

Hamel DJ, Sankale JL, Eisen G, Meloni ST, Mullins C, Gueye-Ndiaye A, Mboup S, Kanki PJ. 2007. Twenty years of prospective molecular epidemiology in Senegal: Changes in HIV diversity. *AIDS Res Hum Retroviruses* **23:** 1189–1196.

Heeney JL, Rutjens E, Verschoor EJ, Niphuis H, ten Haaft P, Rouse S, McClure H, Balla-Jhagjhoorsingh S, Bogers W, Salas M, et al. 2006. Transmission of simian immunodeficiency virus SIVcpz and the evolution of infection in the presence and absence of concurrent human immunodeficiency virus type 1 infection in chimpanzees. *J Virol* **80:** 7208–7218.

Hill CP, Worthylake D, Bancroft DP, Christensen AM, Sundquist WI. 1996. Crystal structures of the trimeric human immunodeficiency virus type 1 matrix protein: Implications for membrane association and assembly. *Proc Natl Acad Sci* **93:** 3099–3104.

Hirsch VM, Olmsted RA, Murphey-Corb M, Purcell RH, Johnson PR. 1989. An African primate lentivirus (SIVsm) closely related to HIV-2. *Nature* **339:** 389–392.

Hladik F, McElrath MJ. 2008. Setting the stage: Host invasion by HIV. *Nat Rev Immunol* **8:** 447–457.

Ho DD, Neumann AU, Perelson AS, Chen W, Leonard JM, Markowitz M. 1995. Rapid turnover of plasma virions and CD4 lymphocytes in HIV-1 infection. *Nature* **373:** 123–126.

Holmes EC. 2003. Molecular clocks and the puzzle of RNA virus origins. *J Virol* **77:** 3893–3897.

Huet T, Cheynier R, Meyerhans A, Roelants G, Wain-Hobson S. 1990. Genetic organization of a chimpanzee lentivirus related to HIV-1. *Nature* **345:** 356–359.

Ishikawa K, Janssens W, Banor JS, Shinno T, Piedade J, Sata T, Ampofo WK, Brandful JA, Koyanagi Y, Yamamoto N, et al. 2001. Genetic analysis of HIV type 2 from Ghana and Guinea-Bissau, West Africa. *AIDS Res Hum Retroviruses* **17:** 1661–1663.

Iwabu Y, Fujita H, Kinomoto M, Kaneko K, Ishizaka Y, Tanaka Y, Sata T, Tokunaga K. 2009. HIV-1 accessory protein Vpu internalizes cell-surface BST-2/tetherin through transmembrane interactions leading to lysosomes. *J Biol Chem* **284:** 35060–35072.

Jia B, Serra-Moreno R, Neidermyer W, Rahmberg A, Mackey J, Fofana IB, Johnson WE, Westmoreland S, Evans DT. 2009. Species-specific activity of SIV Nef and HIV-1 Vpu in overcoming restriction by tetherin/BST2. *PLoS Pathog* **5:** e1000429.

Jin MJ, Hui H, Robertson DL, Muller MC, Barre-Sinoussi F, Hirsch VM, Allan JS, Shaw GM, Sharp PM, Hahn BH. 1994a. Mosaic genome structure of simian immunodeficiency virus from west African green monkeys. *EMBO J* **13:** 2935–2947.

Jin MJ, Rogers J, Phillips-Conroy JE, Allan JS, Desrosiers RC, Shaw GM, Sharp PM, Hahn BH. 1994b. Infection of a yellow baboon with simian immunodeficiency virus from African green monkeys: Evidence for cross-species transmission in the wild. *J Virol* **68:** 8454–8460.

Kajaste-Rudnitski A, Pultrone C, Marzetta F, Ghezzi S, Coradin T, Vicenzi E. 2010. Restriction factors of retroviral replication: The example of Tripartite Motif (TRIM) protein 5 α and 22. *Amino Acids* **39**: 1–9.

Katzourakis A, Tristem M, Pybus OG, Gifford RJ. 2007. Discovery and analysis of the first endogenous lentivirus. *Proc Natl Acad Sci* **104**: 6261–6265.

Keele BF, Van Heuverswyn F, Li Y, Bailes E, Takehisa J, Santiago ML, Bibollet-Ruche F, Chen Y, Wain LV, Liegeois F, et al. 2006. Chimpanzee reservoirs of pandemic and nonpandemic HIV-1. *Science* **313**: 523–526.

Keele BF, Jones JH, Terio KA, Estes JD, Rudicell RS, Wilson ML, Li Y, Learn GH, Beasley TM, Schumacher-Stankey J, et al. 2009. Increased mortality and AIDS-like immunopathology in wild chimpanzees infected with SIVcpz. *Nature* **460**: 515–519.

Kirchhoff F. 2009. Is the high virulence of HIV-1 an unfortunate coincidence of primate lentiviral evolution? *Nat Rev Microbiol* **7**: 467–476.

Kirchhoff F. 2010. Immune evasion and counteraction of restriction factors by HIV-1 and other primate lentiviruses. *Cell Host Microbe* **8**: 55–67.

Kiwanuka N, Robb M, Laeyendecker O, Kigozi G, Wabwire-Mangen F, Makumbi FE, Nalugoda F, Kagaayi J, Eller M, Eller LA, et al. 2010. HIV-1 viral subtype differences in the rate of $CD4^+$ T-cell decline among HIV seroincident antiretroviral naive persons in Rakai district, Uganda. *J Acquir Immune Defic Syndr* **54**: 180–184.

Klatt NR, Silvestri G, Hirsch V. 2011. Nonpathogenic simian immunodeficiency virus infections. *Cold Spring Harb Perspect Med* doi: 10.1101/cshperspect.a007153.

Korber B, Muldoon M, Theiler J, Gao F, Gupta R, Lapedes A, Hahn BH, Wolinsky S, Bhattacharya T. 2000. Timing the ancestor of the HIV-1 pandemic strains. *Science* **288**: 1789–1796.

Leendertz SA, Locatelli S, Boesch C, Kucherer C, Formenty P, Liegeois F, Ayouba A, Peeters M, Leendertz FH. 2011. No evidence for transmission of SIVwrc from western red colobus monkeys (*Piliocolobus badius badius*) to wild West African chimpanzees (*Pan troglodytes verus*) despite high exposure through hunting. *BMC Microbiol* **11**: 24.

Lemey P, Pybus OG, Wang B, Saksena NK, Salemi M, Vandamme AM. 2003. Tracing the origin and history of the HIV-2 epidemic. *Proc Natl Acad Sci* **100**: 6588–6592.

Lemey P, Pybus OG, Rambaut A, Drummond AJ, Robertson DL, Roques P, Worobey M, Vandamme AM. 2004. The molecular population genetics of HIV-1 group O. *Genetics* **167**: 1059–1068.

Lemey P, Rambaut A, Pybus OG. 2006. HIV evolutionary dynamics within and among hosts. *AIDS Rev* **8**: 125–140.

Le Tortorec A, Neil SJ. 2009. Antagonism to and intracellular sequestration of human tetherin by the human immunodeficiency virus type 2 envelope glycoprotein. *J Virol* **83**: 11966–11978.

Li W-H, Tanimura M, Sharp PM. 1988. Rates and dates of divergence between AIDS virus nucleotide sequences. *Mol Biol Evol* **5**: 313–330.

Li Y, Ndjango J-B, Learn G, Robertson J, Takehisa J, Bibollet-Ruche F, Sharp P, Worobey M, Shaw G, Hahn B. 2010. Molecular epidemiology of simian immunodeficiency virus in eastern chimpanzees and gorillas. In *17th Conference on Retroviruses and Opportunistic Infections*. San Francisco, CA.

Lim ES, Malik HS, Emerman M. 2010. Ancient adaptive evolution of tetherin shaped the functions of Vpu and Nef in human immunodeficiency virus and primate lentiviruses. *J Virol* **84**: 7124–7134.

Malim MH, Emerman M. 2008. HIV-1 accessory proteins–ensuring viral survival in a hostile environment. *Cell Host Microbe* **3**: 388–398.

Mauclere P, Loussert-Ajaka I, Damond F, Fagot P, Souquieres S, Monny Lobe M, Mbopi Keou FX, Barre-Sinoussi F, Saragosti S, Brun-Vezinet F, et al. 1997. Serological and virological characterization of HIV-1 group O infection in Cameroon. *AIDS* **11**: 445–453.

McNatt MW, Zang T, Hatziioannou T, Bartlett M, Fofana IB, Johnson WE, Neil SJ, Bieniasz PD. 2009. Species-specific activity of HIV-1 Vpu and positive selection of tetherin transmembrane domain variants. *PLoS Pathog* **5**: e1000300.

Merson MH, O'Malley J, Serwadda D, Apisuk C. 2008. The history and challenge of HIV prevention. *Lancet* **372**: 475–488.

Mitani JC, Watts DP. 1999. Demographic influences on the hunting behavior of chimpanzees. *Am J Phys Anthropol* **109**: 439–454.

Mwaengo DM, Novembre FJ. 1998. Molecular cloning and characterization of viruses isolated from chimpanzees with pathogenic human immunodeficiency virus type 1 infections. *J Virol* **72**: 8976–8987.

Neel C, Etienne L, Li Y, Takehisa J, Rudicell RS, Bass IN, Moudindo J, Mebenga A, Esteban A, Van Heuverswyn F, et al. 2010. Molecular epidemiology of simian immunodeficiency virus infection in wild-living gorillas. *J Virol* **84**: 1464–1476.

Neil S, Bieniasz P. 2009. Human immunodeficiency virus, restriction factors, and interferon. *J Interferon Cytokine Res* **29**: 569–580.

Neil SJ, Zang T, Bieniasz PD. 2008. Tetherin inhibits retrovirus release and is antagonized by HIV-1 Vpu. *Nature* **451**: 425–430.

Ortiz M, Guex N, Patin E, Martin O, Xenarios I, Ciuffi A, Quintana-Murci L, Telenti A. 2009. Evolutionary trajectories of primate genes involved in HIV pathogenesis. *Mol Biol Evol* **26**: 2865–2875.

Paiardini M, Pandrea I, Apetrei C, Silvestri G. 2009. Lessons learned from the natural hosts of HIV-related viruses. *Annu Rev Med* **60**: 485–495.

Peeters M, Gueye A, Mboup S, Bibollet-Ruche F, Ekaza E, Mulanga C, Ouedrago R, Gandji R, Mpele P, Dibanga G, et al. 1997. Geographical distribution of HIV-1 group O viruses in Africa. *AIDS* **11**: 493–498.

Peeters M, Courgnaud V, Abela B, Auzel P, Pourrut X, Bibollet-Ruche F, Loul S, Liegeois F, Butel C, Koulagna D, et al. 2002. Risk to human health from a plethora of simian immunodeficiency viruses in primate bushmeat. *Emerg Infect Dis* **8**: 451–457.

Peeters M, Toure-Kane C, Nkengasong JN. 2003. Genetic diversity of HIV in Africa: Impact on diagnosis, treatment, vaccine development and trials. *AIDS* **17**: 2547–2560.

Pepin J, Labbe AC. 2008. Noble goals, unforeseen consequences: Control of tropical diseases in colonial Central Africa and the iatrogenic transmission of blood-borne viruses. *Trop Med Int Health* **13**: 744–753.

Pepin J, Plamondon M, Alves AC, Beaudet M, Labbe AC. 2006. Parenteral transmission during excision and treatment of tuberculosis and trypanosomiasis may be responsible for the HIV-2 epidemic in Guinea-Bissau. *AIDS* **20**: 1303–1311.

Pepin J, Labbe AC, Mamadou-Yaya F, Mbelesso P, Mbadingai S, Deslandes S, Locas MC, Frost E. 2010. Iatrogenic transmission of human T cell lymphotropic virus type 1 and hepatitis C virus through parenteral treatment and chemoprophylaxis of sleeping sickness in colonial Equatorial Africa. *Clin Infect Dis* **51**: 777–784.

Phillips-Conroy JE, Jolly CJ, Petros B, Allan JS, Desrosiers RC. 1994. Sexual transmission of SIVagm in wild grivet monkeys. *J Med Primatol* **23**: 1–7.

Pieniazek D, Ellenberger D, Janini LM, Ramos AC, Nkengasong J, Sassan-Morokro M, Hu DJ, Coulibally IM, Ekpini E, Bandea C, et al. 1999. Predominance of human immunodeficiency virus type 2 subtype B in Abidjan, Ivory Coast. *AIDS Res Hum Retroviruses* **15**: 603–608.

Plantier JC, Leoz M, Dickerson JE, De Oliveira F, Cordonnier F, Lemee V, Damond F, Robertson DL, Simon F. 2009. A new human immunodeficiency virus derived from gorillas. *Nature Med* **15**: 871–872.

Popovic M, Sarngadharan MG, Read E, Gallo RC. 1984. Detection, isolation, and continuous production of cytopathic retroviruses (HTLV-III) from patients with AIDS and pre-AIDS. *Science* **224**: 497–500.

Popper SJ, Sarr AD, Gueye-Ndiaye A, Mboup S, Essex ME, Kanki PJ. 2000. Low plasma human immunodeficiency virus type 2 viral load is independent of proviral load: Low virus production in vivo. *J Virol* **74**: 1554–1557.

Prince AM, Brotman B, Lee DH, Andrus L, Valinsky J, Marx P. 2002. Lack of evidence for HIV type 1-related SIVcpz infection in captive and wild chimpanzees (*Pan troglodytes verus*) in West Africa. *AIDS Res Hum Retroviruses* **18**: 657–660.

Pusey AE, Pintea L, Wilson ML, Kamenya S, Goodall J. 2007. The contribution of long-term research at Gombe National Park to chimpanzee conservation. *Conserv Biol* **21**: 623–634.

Richman DD, Margolis DM, Delaney M, Greene WC, Hazuda D, Pomerantz RJ. 2009. The challenge of finding a cure for HIV infection. *Science* **323**: 1304–1307.

Rong L, Zhang J, Lu J, Pan Q, Lorgeoux RP, Aloysius C, Guo F, Liu SL, Wainberg MA, Liang C. 2009. The transmembrane domain of BST-2 determines its sensitivity to down-modulation by human immunodeficiency virus type 1 Vpu. *J Virol* **83**: 7536–7546.

Rowland-Jones SL, Whittle HC. 2007. Out of Africa: What can we learn from HIV-2 about protective immunity to HIV-1? *Nat Immunol* **8**: 329–331.

Rudicell RS, Holland Jones J, Wroblewski EE, Learn GH, Li Y, Robertson JD, Greengrass E, Grossmann F, Kamenya S, Pintea L, et al. 2010. Impact of simian immunodeficiency virus infection on chimpanzee population dynamics. *PLoS Pathog* **6**: e1001116.

Sacktor N, Nakasujja N, Skolasky RL, Rezapour M, Robertson K, Musisi S, Katabira E, Ronald A, Clifford DB, Laeyendecker O, et al. 2009. HIV subtype D is associated with dementia, compared with subtype A, in immunosuppressed individuals at risk of cognitive impairment in Kampala, Uganda. *Clin Infect Dis* **49**: 780–786.

Santiago ML, Rodenburg CM, Kamenya S, Bibollet-Ruche F, Gao F, Bailes E, Meleth S, Soong SJ, Kilby JM, Moldoveanu Z, et al. 2002. SIVcpz in wild chimpanzees. *Science* **295**: 465.

Santiago ML, Lukasik M, Kamenya S, Li Y, Bibollet-Ruche F, Bailes E, Muller MN, Emery M, Goldenberg DA, Lwanga JS, et al. 2003. Foci of endemic simian immunodeficiency virus infection in wild-living eastern chimpanzees (*Pan troglodytes schweinfurthii*). *J Virol* **77**: 7545–7562.

Santiago ML, Range F, Keele BF, Li Y, Bailes E, Bibollet-Ruche F, Fruteau C, Noe R, Peeters M, Brookfield JF, et al. 2005. Simian immunodeficiency virus infection in free-ranging sooty mangabeys (*Cercocebus atys atys*) from the Tai Forest, Cote d'Ivoire: Implications for the origin of epidemic human immunodeficiency virus type 2. *J Virol* **79**: 12515–12527.

Sauter D, Schindler M, Specht A, Landford WN, Munch J, Kim KA, Votteler J, Schubert U, Bibollet-Ruche F, Keele BF, et al. 2009. Tetherin-driven adaptation of Vpu and Nef function and the evolution of pandemic and nonpandemic HIV-1 strains. *Cell Host Microbe* **6**: 409–421.

Sawyer SL, Emerman M, Malik HS. 2004. Ancient adaptive evolution of the primate antiviral DNA-editing enzyme APOBEC3G. *PLoS Biol* **2**: E275.

Sawyer SL, Wu LI, Emerman M, Malik HS. 2005. Positive selection of primate *TRIM5α* identifies a critical species-specific retroviral restriction domain. *Proc Natl Acad Sci* **102**: 2832–2837.

Schindler M, Munch J, Kutsch O, Li H, Santiago ML, Bibollet-Ruche F, Muller-Trutwin MC, Novembre FJ, Peeters M, Courgnaud V, et al. 2006. Nef-mediated suppression of T cell activation was lost in a lentiviral lineage that gave rise to HIV-1. *Cell* **125**: 1055–1067.

Schindler M, Schmokel J, Specht A, Li H, Munch J, Khalid M, Sodora DL, Hahn BH, Silvestri G, Kirchhoff F. 2008. Inefficient Nef-mediated downmodulation of CD3 and MHC-I correlates with loss of CD4⁺ T cells in natural SIV infection. *PLoS Pathog* **4**: e1000107.

Schmokel J, Sauter D, Schindler M, Leendertz FH, Bailes E, Dazza MC, Saragosti S, Bibollet-Ruche F, Peeters M, Hahn BH, et al. 2011. The presence of a *vpu* gene and the lack of Nef-mediated downmodulation of T cell receptor-CD3 are not always linked in primate lentiviruses. *J Virol* **85**: 742–752.

Serra-Moreno R, Jia B, Breed M, Alvarez X, Evans DT. 2011. Compensatory changes in the cytoplasmic tail of gp41 confer resistance to tetherin/BST-2 in a pathogenic nef-deleted SIV. *Cell Host Microbe* **9**: 46–57.

Sharp PM, Hahn BH. 2008. AIDS: Prehistory of HIV-1. *Nature* **455**: 605–606.

Sharp PM, Robertson DL, Gao F, Hahn BH. 1994. Origins and diversity of human immunodeficiency viruses. *AIDS* **8**: S27–S42.

Sharp PM, Bailes E, Gao F, Beer BE, Hirsch VM, Hahn BH. 2000. Origins and evolution of AIDS viruses: Estimating the time-scale. *Biochem Soc Trans* **28:** 275–282.

Sharp PM, Shaw GM, Hahn BH. 2005. Simian immunodeficiency virus infection of chimpanzees. *J Virol* **79:** 3891–3902.

Sheehy AM, Gaddis NC, Choi JD, Malim MH. 2002. Isolation of a human gene that inhibits HIV-1 infection and is suppressed by the viral Vif protein. *Nature* **418:** 646–650.

Silvestri G. 2005. Naturally SIV-infected sooty mangabeys: Are we closer to understanding why they do not develop AIDS? *J Med Primatol* **34:** 243–252.

Simon F, Mauclere P, Roques P, Loussert-Ajaka I, Muller-Trutwin MC, Saragosti S, Georges-Courbot MC, Barre-Sinoussi F, Brun-Vezinet F. 1998. Identification of a new human immunodeficiency virus type 1 distinct from group M and group O. *Nat Med* **4:** 1032–1037.

Smith SM, Christian D, de Lame V, Shah U, Austin L, Gautam R, Gautam A, Apetrei C, Marx PA. 2008. Isolation of a new HIV-2 group in the US. *Retrovirology* **5:** 103.

Soto PC, Stein LL, Hurtado-Ziola N, Hedrick SM, Varki A. 2010. Relative over-reactivity of human versus chimpanzee lymphocytes: Implications for the human diseases associated with immune activation. *J Immunol* **184:** 4185–4195.

Souquiere S, Bibollet-Ruche F, Robertson DL, Makuwa M, Apetrei C, Onanga R, Kornfeld C, Plantier JC, Gao F, Abernethy K, et al. 2001. Wild *Mandrillus sphinx* are carriers of two types of lentivirus. *J Virol* **75:** 7086–7096.

Stremlau M, Owens CM, Perron MJ, Kiessling M, Autissier P, Sodroski J. 2004. The cytoplasmic body component TRIM5α restricts HIV-1 infection in Old World monkeys. *Nature* **427:** 848–853.

Surbeck M, Hohmann G. 2008. Primate hunting by bonobos at LuiKotale, Salonga National Park. *Curr Biol* **18:** R906–R907.

Switzer WM, Parekh B, Shanmugam V, Bhullar V, Phillips S, Ely JJ, Heneine W. 2005. The epidemiology of simian immunodeficiency virus infection in a large number of wild- and captive-born chimpanzees: Evidence for a recent introduction following chimpanzee divergence. *AIDS Res Hum Retroviruses* **21:** 335–342.

Takehisa J, Kraus MH, Decker JM, Li Y, Keele BF, Bibollet-Ruche F, Zammit KP, Weng Z, Santiago ML, Kamenya S, et al. 2007. Generation of infectious molecular clones of simian immunodeficiency virus from fecal consensus sequences of wild chimpanzees. *J Virol* **81:** 7463–7475.

Takehisa J, Kraus MH, Ayouba A, Bailes E, Van Heuverswyn F, Decker JM, Li Y, Rudicell RS, Learn GH, Neel C, et al. 2009. Origin and biology of simian immunodeficiency virus in wild-living western gorillas. *J Virol* **83:** 1635–1648.

Taylor BS, Sobieszczyk ME, McCutchan FE, Hammer SM. 2008. The challenge of HIV-1 subtype diversity. *N Engl J Med* **358:** 1590–1602.

Vallari A, Bodelle P, Ngansop C, Makamche F, Ndembi N, Mbanya D, Kaptue L, Gurtler LG, McArthur CP, Devare SG, et al. 2010. Four new HIV-1 group N isolates from Cameroon: Prevalence continues to be low. *AIDS Res Hum Retroviruses* **26:** 109–115.

Vallari A, Holzmayer V, Harris B, Yamaguchi J, Ngansop C, Makamche F, Mbanya D, Kaptue L, Ndembi N, Gurtler L, et al. 2011. Confirmation of putative HIV-1 group P in Cameroon. *J Virol* **85:** 1403–1407.

Van Damme N, Goff D, Katsura C, Jorgenson RL, Mitchell R, Johnson MC, Stephens EB, Guatelli J. 2008. The interferon-induced protein BST-2 restricts HIV-1 release and is downregulated from the cell surface by the viral Vpu protein. *Cell Host Microbe* **3:** 245–252.

Vanden Haesevelde MM, Peeters M, Jannes G, Janssens W, van der Groen G, Sharp PM, Saman E. 1996. Sequence analysis of a highly divergent HIV-1-related lentivirus isolated from a wild captured chimpanzee. *Virology* **221:** 346–350.

van der Loeff MF, Awasana AA, Sarge-Njie R, van der Sande M, Jaye A, Sabally S, Corrah T, McConkey SJ, Whittle HC. 2006. Sixteen years of HIV surveillance in a West African research clinic reveals divergent epidemic trends of HIV-1 and HIV-2. *Int J Epidemiol* **35:** 1322–1328.

van der Loo W, Abrantes J, Esteves PJ. 2009. Sharing of endogenous lentiviral gene fragments among leporid lineages separated for more than 12 million years. *J Virol* **83:** 2386–2388.

Van Heuverswyn F, Li Y, Neel C, Bailes E, Keele BF, Liu W, Loul S, Butel C, Liegeois F, Bienvenue Y, et al. 2006. Human immunodeficiency viruses: SIV infection in wild gorillas. *Nature* **444:** 164.

Van Heuverswyn F, Li Y, Bailes E, Neel C, Lafay B, Keele BF, Shaw KS, Takehisa J, Kraus MH, Loul S, et al. 2007. Genetic diversity and phylogeographic clustering of SIVcpzPtt in wild chimpanzees in Cameroon. *Virology* **368:** 155–171.

van Rensburg EJ, Engelbrecht S, Mwenda J, Laten JD, Robson BA, Stander T, Chege GK. 1998. Simian immunodeficiency viruses (SIVs) from eastern and southern Africa: Detection of a SIVagm variant from a chacma baboon. *J Gen Virol* **79:** 1809–1814.

Vidal N, Peeters M, Mulanga-Kabeya C, Nzilambi N, Robertson D, Ilunga W, Sema H, Tshimanga K, Bongo B, Delaporte E. 2000. Unprecedented degree of human immunodeficiency virus type 1 (HIV-1) group M genetic diversity in the Democratic Republic of Congo suggests that the HIV-1 pandemic originated in Central Africa. *J Virol* **74:** 10498–10507.

Wain LV, Bailes E, Bibollet-Ruche F, Decker JM, Keele BF, Van Heuverswyn F, Li Y, Takehisa J, Ngole EM, Shaw GM, et al. 2007. Adaptation of HIV-1 to its human host. *Mol Biol Evol* **24:** 1853–1860.

Wei X, Ghosh SK, Taylor ME, Johnson VA, Emini EA, Deutsch P, Lifson JD, Bonhoeffer S, Nowak MA, Hahn BH, et al. 1995. Viral dynamics in human immunodeficiency virus type 1 infection. *Nature* **373:** 117–122.

Weiss RA, Heeney JL. 2009. Infectious diseases: An ill wind for wild chimps? *Nature* **460:** 470–471.

Wertheim JO, Worobey M. 2009. Dating the age of the SIV lineages that gave rise to HIV-1 and HIV-2. *PLoS Comput Biol* **5:** e1000377.

Worobey M, Santiago ML, Keele BF, Ndjango JB, Joy JB, Labama BL, Dhed AB, Rambaut A, Sharp PM, Shaw GM, et al. 2004. Origin of AIDS: Contaminated polio vaccine theory refuted. *Nature* **428:** 820.

Worobey M, Gemmel M, Teuwen DE, Haselkorn T, Kunstman K, Bunce M, Muyembe JJ, Kabongo JM, Kalengayi RM, Van Marck E, et al. 2008. Direct evidence of extensive diversity of HIV-1 in Kinshasa by 1960. *Nature* **455:** 661–664.

Worobey M, Telfer P, Souquiere S, Hunter M, Coleman CA, Metzger MJ, Reed P, Makuwa M, Hearn G, Honarvar S, et al. 2010. Island biogeography reveals the deep history of SIV. *Science* **329:** 1487.

Ylinen LM, Price AJ, Rasaiyaah J, Hue S, Rose NJ, Marzetta F, James LC, Towers GJ. 2010. Conformational adaptation of Asian macaque TRIMCyp directs lineage specific antiviral activity. *PLoS Pathog* **6:** e1001062.

Zhang F, Wilson SJ, Landford WC, Virgen B, Gregory D, Johnson MC, Munch J, Kirchhoff F, Bieniasz PD, Hatziioannou T. 2009. Nef proteins from simian immunodeficiency viruses are tetherin antagonists. *Cell Host Microbe* **6:** 54–67.

Zhu T, Korber BT, Nahmias AJ, Hooper E, Sharp PM, Ho DD. 1998. An African HIV-1 sequence from 1959 and implications for the origin of the epidemic. *Nature* **391:** 594–597.

HIV: Cell Binding and Entry

Craig B. Wilen[1], John C. Tilton[2], and Robert W. Doms[1]

[1]Department of Microbiology, University of Pennsylvania, Philadelphia, Pennsylvania 19104

[2]Department of General Medical Science, Center for Proteomics and Bioinformatics, Case Western Reserve University, Cleveland, Ohio 44106

Correspondence: doms@upenn.edu

The first step of the human immunodeficiency virus (HIV) replication cycle—binding and entry into the host cell—plays a major role in determining viral tropism and the ability of HIV to degrade the human immune system. HIV uses a complex series of steps to deliver its genome into the host cell cytoplasm while simultaneously evading the host immune response. To infect cells, the HIV protein envelope (Env) binds to the primary cellular receptor CD4 and then to a cellular coreceptor. This sequential binding triggers fusion of the viral and host cell membranes, initiating infection. Revealing the mechanism of HIV entry has profound implications for viral tropism, transmission, pathogenesis, and therapeutic intervention. Here, we provide an overview into the mechanism of HIV entry, provide historical context to key discoveries, discuss recent advances, and speculate on future directions in the field.

HIV ENTRY FUNDAMENTALS

HIV entry, the first phase of the viral replication cycle, begins with the adhesion of virus to the host cell and ends with the fusion of the cell and viral membranes with subsequent delivery of the viral core into the cytoplasm. The intricate series of protein–protein interactions that ultimately results in virus infection can be divided into several phases, some of which are essential and others that may serve to modulate the efficiency of the process. First, virions must bind to the target cell, with this being mediated either by the viral envelope (Env) protein or host cell membrane proteins incorporated into the virion with any one of a number of various cell attachment factors. Attachment can be relatively nonspecific, with Env interacting with negatively charged cell-surface heparan sulfate proteoglycans (Saphire et al. 2001), or can result from more specific interactions between Env and α4β7 integrin (Arthos et al. 2008; Cicala et al. 2009) or pattern recognition receptors such as dendritic cell–specific intercellular adhesion molecular 3-grabbing non-integrin (DC-SIGN) (Geijtenbeek et al. 2000; reviewed in Ugolini et al. 1999). HIV attachment to the host cell via any of these factors likely brings Env into close proximity with the viral receptor CD4 and coreceptor, increasing the efficiency of infection (Fig. 1) (Orloff et al. 1991). However, attachment factors differ from receptors in that they are not essential, and although they augment infection in vitro, their physiologic role in vivo remains unclear.

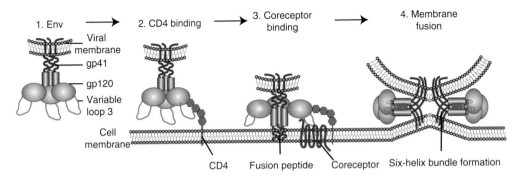

Figure 1. Overview of HIV entry. To deliver the viral payload into cells, HIV Env, comprised of gp120 and gp41 subunits (1), first attaches to the host cell, binding CD4 (2). This causes conformational changes in Env, allowing coreceptor binding, which is mediated in part by the V3 loop of Env (3). This initiates the membrane fusion process as the fusion peptide of gp41 inserts into the target membrane, followed by six-helix bundle formation and complete membrane fusion (4).

The second step of virus entry and the first absolutely required for infection entails binding of Env to its primary receptor, the host protein CD4 (Maddon et al. 1986; McDougal et al. 1986). Env is a heavily glycosylated trimer of gp120 and gp41 heterodimers. The gp120 subunit is responsible for receptor binding and gp120 contains five relatively conserved domains (C1–C5) and five variable loops (V1–V5), named for their relative genetic heterogeneity. Each of the variable regions is comprised of a loop structure formed by a disulfide bond at its base, with the exception of V5. The variable loops lie predominantly at the surface of gp120 and play critical roles in immune evasion and coreceptor binding, particularly the V3 loop (reviewed in Hartley et al. 2005). CD4 is a member of the immunoglobulin superfamily that normally functions to enhance T-cell receptor (TCR)-mediated signaling. Env interacts with the CD4 binding site (CD4bs) in gp120 (Kwong et al. 1998). Env binding to CD4 causes rearrangements of V1/V2 and subsequently V3. In addition, CD4 binding leads to formation of the bridging sheet, a four-stranded β sheet comprised of two double-stranded β sheets that are spatially separated in the unliganded state (Kwong et al. 1998; Chen et al. 2005). The bridging sheet and repositioned V3 loop play critical roles in the next step of virus entry, coreceptor engagement.

The third step of virus entry, coreceptor binding, is widely thought to be the trigger that activates the membrane fusion potential of Env. HIV strains can be broadly classified based on their coreceptor usage. Viruses that use the chemokine receptor CCR5 are termed R5 HIV, those that use CXCR4 are termed X4 HIV, and viruses that can use both coreceptors are called R5X4 HIV (Berger et al. 1998). There is no compelling evidence that coreceptors other than CCR5 and CXCR4 play important roles in supporting infection of HIV-1 in vivo. With rare exception, only R5 and R5X4 viruses are transmitted between individuals (Keele et al. 2008), likely owing to multiple imperfect but overlapping host restrictions on X4 HIV transmission (reviewed in Margolis and Shattock 2006). Interestingly, despite identification at earlier time points and despite high levels of CXCR4 expression on circulating HIV target cells, X4 or even R5X4 HIV rarely predominate until late in infection (Tersmette et al. 1989; Schuitemaker et al. 1992; Connor et al. 1997). In addition, X4 viruses are less common in clade C HIV and SIV infection (Chen et al. 1998; Ping et al. 1999; Cecilia et al. 2000; Huang et al. 2007). Several nonmutually exclusive models may explain this. First, clade B Envs may be different in their ability to adapt to CXCR4 tropism. Second, there may be differences in clade B host biology. For instance, clade B hosts may have mitigated neutralizing antibody or cytotoxic T lymphocyte responses against X4 HIV compared with R5 HIV. Finally, clade B hosts most often live in developed countries and may face different environmental stresses including fewer or different chronic coinfections, which may increase target cell CCR5 expression. Elucidating the mechanism of coreceptor switch is a critical next step because it has implications for disease progression and therapy with HIV entry inhibitors.

A fourth step of virus entry is movement of the virus particle to the site where productive membrane fusion occurs. A series of recent studies has shown that a number of viruses usurp cellular transport pathways to reach specific destinations that are either needed for infection or that make entry more efficient, and that HIV might likewise use the host cell machinery to reach sites where membrane fusion can occur (Lehmann et al. 2005; Coyne and Bergelson 2006; Sherer et al. 2010). Some viruses have been shown to "surf" along the cell surface, moving from distal sites of attachment to more proximal regions of the cell body where virus entry occurs. Retroviruses, including HIV, have been shown to use this process on some cell lines (Lehmann et al. 2005; Sherer et al. 2010). In addition, HIV may need to be internalized by the host cell's endocytic machinery for productive membrane fusion to occur, as is discussed in a later section (Miyauchi et al. 2009).

The fifth and final step of virus entry is membrane fusion mediated by Env. Coreceptor binding induces exposure of the hydrophobic gp41 fusion peptide, which inserts into the host cell membrane. This tethers the viral and host membranes, allowing the fusion peptide of each gp41 in the trimer to fold at a hinge region, bringing an amino-terminal helical region (HR-N) and a carboxy-terminal helical region (HR-C) from each gp41 subunit together to form a six-helix bundle (6HB) (Chan et al. 1997; Weissenhorn et al. 1997). Because the HR-N domain is in close proximity to the host cell membrane owing to the fusion peptide, and the HR-C domain is in close proximity to the viral membrane owing to the gp41

transmembrane domain, formation of the 6HB is the driving force that brings the opposing membranes into close apposition, resulting in the formation of a fusion pore (reviewed in Melikyan 2008). Whether one or multiple HIV Env trimers are needed for complete membrane fusion is not yet clear. In summary, coreceptor binding unlocks the potential energy of the gp41 fusion complex resulting in 6HB formation, opening and stabilization of the membrane fusion pore, and subsequent delivery of the viral contents into the host cell cytoplasm.

DISCOVERY OF THE HIV RECEPTORS

In 1981, several years before the discovery of HIV, Gottlieb and colleagues (1981) reported $CD4^+$ T-cell decline in four men who presented with pneumocystis pneumonia and mucosal candidiasis, among other opportunistic infections. Three years later, it was shown that HIV preferentially infects $CD4^+$ T cells (Klatzmann et al. 1984) and that infection is potently inhibited by CD4-specific antibodies (reviewed in Sattentau and Weiss 1988). CD4 was then shown to coimmunoprecipitate with Env (McDougal et al. 1986) and CD4 expression could rescue infection in some nonpermissive cells (Maddon et al. 1986). However, CD4 transfection into mouse cells rescued binding of virus to the cell surface but not membrane fusion or virus infection, suggesting that there were other required cofactors (Maddon et al. 1986).

Although the discovery of CD4 as the primary HIV receptor occurred shortly after the onset of the epidemic, it took more than a decade to discover the first coreceptor. In 1993, CD26 was reported as the elusive HIV coreceptor (Callebaut et al. 1993); however, this was later disproved by several groups (Lazaro et al. 1994; Stamatatos and Levy 1994). In 1995, Feng and colleagues conclusively identified CXCR4 as a major HIV coreceptor by the use of an expression cloning strategy. A critical finding of this study was that CXCR4, then termed fusin, functioned as a coreceptor for what had been termed T-cell line tropic strains of HIV but not for virus strains that could infect human macrophages but that failed to enter T-cell lines

(Feng et al. 1995, 1996). The seminal discovery of CXCR4 as a G-protein-coupled receptor (GPCR) in combination with the identification of the inhibitory effect of the β chemokines CCL3 (MIP-1α), CCL4 (MIP1β), and CCL5 (RANTES) (Cocchi et al. 1995) on some virus isolates led to the simultaneous and rapid discovery of CCR5 as the coreceptor for macrophage-tropic virus strains by five different groups (Alkhatib et al. 1996; Choe et al. 1996; Deng et al. 1996; Doranz et al. 1996; Dragic et al. 1996).

The importance of the viral coreceptors for HIV infection in vivo was shown by the discovery of a 32 base-pair deletion in *ccr5*, termed *ccr5Δ32*, which has an allelic frequency of ~10% in Caucasians (Dean et al. 1996; Liu et al. 1996; Samson et al. 1996). The Δ32 mutation results in a premature stop codon in the second extracellular loop of CCR5 and subsequent retention of the mutant protein in the endoplasmic reticulum. Homozygosity for this polymorphism results in profound resistance to HIV infection, although several Δ32 homozygotes have been infected with X4 viruses (Balotta et al. 1997; O'Brien et al. 1997; Theodorou et al. 1997). In addition, heterozygosity confers partial protection to infection (Dean et al. 1996; Samson et al. 1996) and disease progression (Dean et al. 1996; Huang et al. 1996).

Elucidating the mechanism of HIV entry has directly translated into therapeutic benefit. Currently, there are two FDA-approved entry inhibitors, enfuvirtide and maraviroc, whereas others are in various stages of development. In 2003, enfuvirtide became the first licensed entry inhibitor; it is a 36-residue-long peptide whose sequence is based on that of the HR-C in gp41. As a result, enfuvirtide behaves much like HR-C in that it binds to the HR-N prehairpin intermediate and inhibits 6HB formation and subsequent membrane fusion (Wild et al. 1992, 1993). Although enfuvirtide is a highly specific and effective membrane fusion inhibitor (Lalezari et al. 2003; Lazzarin et al. 2003), its use has been limited because it must be injected owing to its lack of oral bioavailability. Recently, protease-resistant D-peptide fusion inhibitors have been developed that also prevent 6HB

formation, which may overcome this limitation (Eckert et al. 1999; Welch et al. 2007, 2010). In addition to enfuvirtide, the CCR5 inhibitor maraviroc has been approved for clinical use. Maraviroc is a small-molecule allosteric inhibitor that binds within the CCR5 transmembrane cavity resulting in conformational changes in the extracellular loop domains of the chemokine receptor that interact with Env (Dorr et al. 2005). Similar CCR5 small-molecule inhibitors are in various stages of testing (reviewed in Tilton and Doms 2009).

KEY RECENT ADVANCES

Our understanding of the HIV entry process is derived largely from structural and in vitro studies. As the field has evolved, there is now increased emphasis on placing the now rather well-understood membrane fusion reaction in a cellular context, asking where and when virus entry takes place as well as how virus particles are transferred between cells. Increased structural detail continues to provide insight into the entry process and suggests targets for small-molecule inhibitors and neutralizing antibodies. Finally, attempts to recapitulate the *ccr5Δ32* phenotype have been developed with some being brought forward to early-stage clinical development (Perez et al. 2008).

New Structural Information

A full understanding of the HIV entry process requires detailed structural information. The structure of CD4 alone and in complex with a gp120 core fragment has been solved for HIV (Kwong et al. 1998; Huang et al. 2005) and simian immunodeficiency virus (SIV) (see online Movie 1 at www.perspectivesinmedicine.org) (Chen et al. 2005). The structure of the postfusion 6HB in gp41 has also been determined (see online Movie 2 at www.perspectivesinmedicine.org) (Chan et al. 1997; Weissenhorn et al. 1997). What has been lacking is a structure of the native Env trimer and the HIV coreceptors. However, Wu et al. (2010) recently described five independent structures of CXCR4 bound to two different small-molecule antagonists, which have

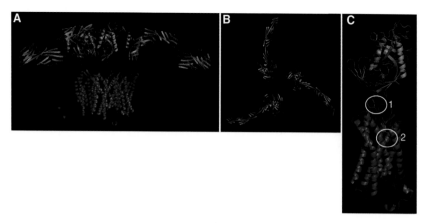

Figure 2. Model of gp120 engagement of CD4 and CXCR4. Recent structural studies have enhanced our understanding of the molecular interactions between gp120 (cyan) and its receptors. Here, CD4 (green) and CXCR4 (purple), shown as monomers for clarity, are shown simultaneously binding to gp120. (A) Lateral view. (B) Top view. However, the number of CD4 and coreceptor molecules required to interact with Env to mediate productive fusion remains unknown. (C) Gp120 has two key interactions with coreceptor. (1) The base of the V3 loop binds to the amino-terminal domain of the coreceptor, whereas the tip of the V3 loop binds to the second extracellular loop (ECL2). Although both interactions are important, viral strains differ on their dependency of each interaction. (Structural model generated by Wu et al. 2010.)

given insight into both the tertiary and quaternary structure of the native protein (Fig. 2). First, both chemokines and Env have been reported to engage CCR5 and CXCR4 in a two-site model with the chemokine receptor amino terminus as site one and the extracellular loops (ECLs), particularly ECL2, as site two. Although the orientation of the CXCR4 amino-terminal domain could not be solved owing to structural flexibility, the crystal structure provides high-resolution insight into the ECL2 binding site. Second, all five structures portray CXCR4 as a homodimer (see online Movie 3 at www.perspectivesinmedicine.org), which is consistent with biochemical studies that have suggested CXCR4 exists as an oligomer in the host cell membrane (Babcock et al. 2003). Although the implications of CXCR4 dimerization remain unclear for HIV infection, it may explain the dominant phenotype of a carboxy-terminal CXCR4 human mutation that results in WHIM syndrome, which is characterized by warts, hypogammaglobulinemia, infections, and myelokathexis (retention of neutrophils in the bone marrow) (Hernandez et al. 2003). Finally, the identified homodimer interface may represent a novel CXCR4 or potentially

CCR5 drug target, because CCR5 and CXCR4 have been reported to heterodimerize in vivo (Sohy et al. 2007, 2009). Further structural studies are needed to better define the precise interactions of Env and coreceptor and to assess the mechanisms of signaling and heterodimerization with other chemokine receptors.

Where Does Virus Entry Occur?

The entry of viruses into cells is controlled in both time and space, with these parameters being regulated by host cell factors that serve to unlock the membrane fusion potential of viral membrane proteins. Many viruses require delivery by the host cell into an acidic, intracellular compartment where low pH triggers membrane-fusion-inducing conformational changes (reviewed in Marsh and Helenius 2006). HIV entry does not require low pH; instead it is triggered by receptor engagement (Stein et al. 1987). The fact that HIV does not require low pH for cellular entry does not imply that fusion occurs at the cell surface. In fact, no spatial information is provided by the triggering mechanism. Despite this, it was often assumed that HIV fuses at the cell surface owing to several observations (reviewed in Uchil and Mothes 2009). First, Env expression on the cell surface can mediate cell-to-cell fusion, indicating not only that Env is the only viral membrane protein needed to elicit fusion but that low pH is clearly not required. Second, very early studies on HIV entry showed that lysomotropic agents, which increase endosomal pH, do not inhibit HIV infection (McClure et al. 1988). Third, inhibiting endocytosis of CD4 in cell lines by mutating its cytoplasmic domain does not affect HIV infection (Maddon et al. 1988). Together, these studies show that HIV entry is not pH dependent, but they provide no definitive information as to whether fusion occurs at the cell surface or from within endocytic vesicles, albeit in a pH-independent fashion.

The question of where HIV-membrane fusion occurs has recently been reexamined (Miyauchi et al. 2009). By combining lipid and content mixing assays with single virion fluorescent imaging, Miyauchi et al. tracked the location of virus membrane fusion in HeLa cells overexpressing CD4, CCR5, and CXCR4. They found that whereas lipid mixing can occur at the cell surface, content mixing only occurred in intracellular perinuclear compartments and thus concluded that complete fusion requires endocytosis. Whether this is always the case remains to be determined because the genetic variability of HIV and the diverse cell types it can infect make generalization difficult.

An interesting question is whether the site of entry matters; with regard to the use of entry inhibitors, probably not: Both coreceptor antagonists and fusion inhibitors block virus infection in vitro and in vivo, and neutralizing antibodies clearly function as well. However, the site of entry is more likely to have an impact on the likelihood of a productive infection actually occurring. For instance, after cellular attachment, HIV can actively surf along the cellular membrane from filopodia or microvilli to the cell body. This actin-dependent process requires receptor engagement and serves to enhance infection efficiency. Surfing toward the cell body may have several favorable consequences for the virus. First, it may facilitate endocytic HIV uptake. Second, it may bring the virus to a membrane region that has higher levels of coreceptor or important downstream signaling molecules (McDonald et al. 2003). Third, it may allow the fusion event to occur closer to the nucleus, which is the ultimate target of HIV. Thus, the site of initial HIV attachment is likely random; however, HIV hijacks the cellular machinery to traverse the cell membrane to a more favorable site of entry, be it at the plasma membrane or endosome, which ultimately serves to augment infection efficiency (Lehmann et al. 2005).

Cell–Cell Transfer and the Virological Synapse

In vitro, the rate-limiting step of virus infection is attachment to the host cell. In vivo, newly produced virions may well encounter an immediately adjoining, uninfected cell. In some cases, transfer of virus from one cell to another is a specialized process, as in the case of dendritic cells (DCs), which are professional antigen-presenting cells (APCs) that scavenge the periphery, sampling antigen. They are commonly found in the mucosa and thus may be encountered by HIV during transmission. On antigen binding, DCs migrate to the lymph nodes, process, and present the antigen to T cells to trigger an adaptive immune response. DCs are relatively resistant to productive HIV infection owing to a combination of low CD4 and coreceptor expression, host restriction factors, postintegration HIV transcription blocks, and other unknown factors (Bakri et al. 2001). However, they express a diverse range of attachment factors that facilitate the internalization and processing of pathogens before antigen presentation. HIV, along with viruses (Igakura et al. 2003; Yang et al. 2004), can take advantage of this pathway to augment infection efficiency and dissemination (Fig. 3) (reviewed in Piguet and Steinman 2007).

Cameron et al. (1992) first showed that DCs could catalyze HIV infection of cocultured CD4[+] T cells without themselves getting productively infected. Each DC can bind up to several hundred virions (McDonald et al. 2003) most likely via a C-type lectin such as DC-SIGN

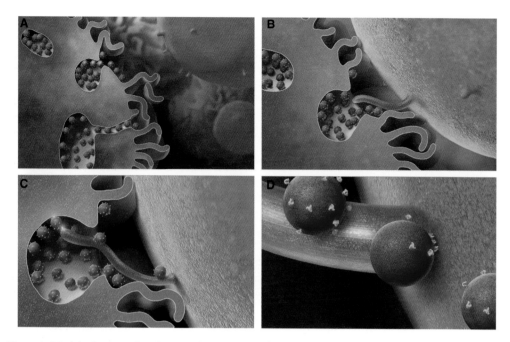

Figure 3. Model of DC-mediated transinfection of CD4$^+$ T cells. (*A*) DCs capture and concentrate virions in trypsin-resistant surface-accessible compartments. (*B*) CD4$^+$ T cells, containing membrane protrusions, bind DCs. (*C*) The CD4$^+$ T-cell protrusions invade the virus-containing compartments and efficiently bind HIV. (*D*) Virus then migrates toward the cell body to initiate infection. (Figure reproduced from Felts et al. 2010; reprinted, with permission, from *Proceedings of the National Academy of Sciences* © 2010.)

(Geijtenbeek et al. 2000; Turville et al. 2002). After binding, the virions are endocytosed into a trypsin-resistant compartment (Geijtenbeek et al. 2000), and then after DC binding to a T cell, internalized virus migrates to the DC:T-cell interface (McDonald et al. 2003) where it encounters the T-cell membrane forming the infectious synapse, analogous to the immunologic synapse that forms on MHC–TCR binding (reviewed in Vasiliver-Shamis et al. 2010). In addition to efficiently concentrating and presenting HIV at the site of T-cell contact, the infectious synapse is characterized by recruitment of CD4, CCR5, and CXCR4.

Recent advances in electron microscopy have enabled 3D (three-dimensional)-structural studies of the infectious synapse that have shed light on this mechanism (Felts et al. 2010). DCs produce membranous protrusions that engulf the surrounding extracellular environment, trapping virions in a surface-accessible but protected compartment. It remains unclear

as to whether this occurs before or after virion binding and whether it is Env induced. When CD4$^+$ T cells contact DCs, they extend filopodia, enriched for CD4 and coreceptor, into the invaginated DC compartments that contain bound virions (Fig. 3). Together, the efficient binding of HIV, relocalization to the point of CD4$^+$ T-cell contact, and the recruitment of the requisite HIV entry receptors promote HIV infection at the infectious synapse (McDonald et al. 2003; Hubner et al. 2009).

A Novel Attachment Factor: α4β7 Integrin

Although cell-to-cell transmission of HIV augments infection efficiency, the mechanism of virological synapse formation remains unclear. α4β7 integrin has been reported to bind gp120; induce activation of LFA-1 (αLβ2 integrin), which contributes to formation of the immunologic synapse (reviewed in Bromley et al. 2001); and subsequently augment infection efficiency in vitro (Arthos et al. 2008; Cicala et al. 2009).

α4β7 is a heterodimeric protein comprised of an α4 and β7 subunit that when expressed on CD4$^+$ T cells facilitates homing to the gut and other mucosal tissues. Its activation and expression are up-regulated by retinoic acid in vitro, which may also be locally secreted by mucosal DCs in vivo. The discovery of α4β7 as an attachment factor is of particular interest because HIV disrupts the integrity of the mucosal barrier and preferentially depletes gut CD4$^+$ T cells, which are more activated and express higher levels of CCR5 than peripheral CD4$^+$ T cells. α4β7 is thought to bind an LDV (Leu-Asp-Val) tripeptide motif on the second variable loop (V2) of gp120, with this resulting in lymphocyte function-associated antigen 1 (LFA-1) activation. In addition, α4β7 colocalizes with CD4 and CCR5 at the virological synapse, which may further enhance infection. Blockade of α4β7 with monoclonal antibodies or a peptide delays replication of HIV in vitro, further supporting its role in HIV infection (Cicala et al. 2009). Future work is needed to assess whether there are protective effects of inhibiting HIV-α4β7 interactions in vivo, and to validate this novel attachment factor as a therapeutic target.

FUTURE DIRECTIONS SIGNALING

Signal Transduction Mediated by HIV Env

HIV Env has the capacity to mediate signal transduction cascades through CD4 and coreceptors, although the physiological importance on viral entry, replication, and pathogenesis remains controversial. Recent discoveries have provided evidence that signaling does play an important role under certain circumstances. In this section, we briefly review what is known about Env signaling through its receptor and coreceptors and highlight some of the recent discoveries in the field.

Signal Transduction through CD4

During encounters between CD4$^+$ T cells and APCs, the CD4 molecule acts to enhance signaling through the TCR as it engages a cognate peptide bound to major histocompatibility complex class II (pMHC). Recruitment of the Src-family protein tyrosine kinase Lck to the immunological synapse through an interaction with the cytoplasmic tail of CD4 results in phosphorylation of immunoreceptor-tyrosine-based activation motifs (ITAMs) present on the CD3γ, CD3δ, CD3ε, and TCR-ζ subunits of the TCR complex and interactions with effector molecules (Love and Hayes 2010; Padhan and Varma 2010). Doubly phosphorylated ITAMs on the TCR-ζ subunit recruit the Syk-family protein tyrosine kinase ZAP-70, which in turn phosphorylates the scaffold proteins LAT and SLP-76 that recruit many additional signaling proteins, eventually resulting in T-cell activation and proliferation (reviewed in Love and Hayes 2010).

Although HIV has been shown to signal through CD4 on binding (Hivroz et al. 1993; Briant et al. 1998) and to increase the activity of Lck (Juszczak et al. 1991), it remains unclear whether signaling is essential for infectivity. Although multiple early studies with truncated forms of CD4 indicated that signaling through CD4 was not required for HIV entry (Benkirane et al. 1994; Tremblay et al. 1994), most of these assays were performed on cell lines and using cell-free virus infection. Several more recent studies have shown a potential role for Env-CD4 signaling in the context of cell–cell spread at the infectious synapse, which as noted before is a specialized junction between cells that resembles the immunological synapse between APCs and T cells. One of the key features of the immunological synapse is that T cells stop migration through the lymph node or target tissue to allow sustained interaction with the APCs. HIV gp120 was able to arrest the migration of primary activated CD4$^+$ T cells in the presence of ICAM-1 and spontaneously induce the formation of a virological synapse (Vasiliver-Shamis et al. 2008). Signal transduction through CD4 and Lck were subsequently found to be responsible for depletion of cortical F-actin underneath the virological synapse using a planar membrane model system, allowing transfer of the viral core from the plasma membrane to the nucleus (Vasiliver-Shamis et al. 2009). These studies suggest that Env signaling through CD4 may play a role in HIV entry under physiological conditions.

Signal Transduction through CCR5 and CXCR4

CCR5 and CXCR4 are both members of the seven-transmembrane-spanning family of heterotrimeric GPCRs. This family of proteins is characterized by an extracellular amino-terminal domain, seven membrane-spanning domains that form three extracellular and three intracellular loops, and a cytoplasmic tail domain. The amino terminus (site one) and three ECLs (site two) together form the binding pocket for the cognate chemokines, which appear to attach to their receptor and transmit signals in a two-site binding process (reviewed in Clark-Lewis et al. 1995). The intracellular loops and cytoplasmic tail bind to the heterotrimeric G proteins that in turn mediate effector functions.

The ability of HIV to signal through the chemokine receptors has been documented since shortly after their identification as coreceptors. Early studies tested the requirement for HIV to signal through coreceptor during entry into cells by inhibiting $G_{\alpha i}$ subunits through pertussis toxin (PTX) (Cocchi et al. 1996) or by creating chemokine mutants that abolished the ability to mobilize intracellular Ca^{2+}, a key second messenger in signal transduction pathways (Alkhatib et al. 1997; Farzan et al. 1997; Gosling et al. 1997). Neither of these interventions blocked the ability of HIV to enter cells, leading to the conclusion that signaling was dispensable for viral infection and replication in target cells. During the next decade, a multitude of studies examined the requirement for coreceptor signaling during HIV entry, with sometimes contradictory results (reviewed in Wu and Yoder 2009). Often, experiments using cell lines or activated primary cells would indicate that signaling was not required for entry, whereas experiments with resting primary cells would indicate a role for signaling. Because the majority of T cells in lymphoid tissues are in resting or nondividing states—particularly before

infection with HIV—examining the role for signaling under these conditions is particularly relevant.

Several recent studies have indicated that HIV depends on chemokine receptor signaling for efficient infection of target cells. First, Yoder and colleagues (2008) examined the ability of X4 HIV to infect resting CD4$^+$ T cells and found that HIV-mediated, $G_{\alpha i}$ signaling through CXCR4 was required for entry. This pathway triggered the activation of a cellular actin-depolymerizing molecule, cofilin, altering cortical actin dynamics near the cell surface and facilitating viral fusion. Moreover, the study suggested a role for cofilin in movement of the viral preintegration complex (PIC) toward the nucleus. In contrast to the resting cell model used in this study, most activated and cycling CD4$^+$ T cells disassemble cortical actin without the requirement of cofilin. Although this study was limited to CXCR4 viral entry, activation of cofilin may also be required for CCR5 viruses to enter resting, nondividing memory cells, such as the majority of CD4$^+$ T cells present in the lamina propria of the gut.

A second series of experiments from Harmon and colleagues (2010) showed that HIV also signals through the $G_{\alpha q}$ subunit, resulting in phospholipase C and Rac activation. Rac and the tyrosine kinase Abl then become linked to the Wave2 complex through the adapter proteins Tiam-1 and IRSp53, promoting Arp2/3-dependent actin nucleation and polymerization. Blocking activation of the Wave2 complex with small interfering RNAs (siRNAs) or Abl kinase inhibitors arrested HIV entry at the hemifusion stage. Together, these experiments suggest a critical role for envelope-coreceptor signaling-induced actin remodeling during HIV entry, particularly in the case of resting CD4$^+$ T cells.

The role of Env-mediated signaling through CD4 and coreceptor in the HIV entry process remains incompletely defined, but it appears increasingly likely that there are essential roles for signal transduction with physiologically relevant conditions and cell types. A common theme among these studies is the necessity for signaling in the reorganization of cytoskeletal actin, which may be involved in several key viral processes including "surfing," passage through the cortical actin barrier near the plasma membrane, and movement of the viral PIC into the nucleus.

CONCLUDING REMARKS

HIV entry is an active process that involves hijacking various components of the cellular machinery. Although Env engagement of CD4 and coreceptor are the most critical events of the entry process, viral surfing, endocytosis, cell-to-cell transmission, and receptor-mediated signaling likely play a role in enhancing infection efficiency by overcoming various host restrictions. Such factors likely affect the efficiency of transmission host cell tropism, and disease progression and are thus critical areas of future investigation. Although numerous questions remain, the most crucial is how can we better exploit our knowledge of HIV entry for therapeutic gain? Maraviroc and enfuvirtide have shown the efficacy of inhibiting entry, but novel therapeutic targets are needed and these likely represent the host molecules coopted by HIV.

ACKNOWLEDGMENTS

C.B.W., J.C.T., and R.W.D. were supported by grants T32 AI000632, F32 1F32AI077370, and R01 AI 040880, respectively. We thank Beili Wu, Ray Stevens, and Sriram Subramaniam for the use of figures and PDB (Protein Data Bank) files.

REFERENCES

Alkhatib G, Combadiere C, Broder CC, Feng Y, Kennedy PE, Murphy PM, Berger EA. 1996. CC CKR5: A RANTES, MIP-1α, MIP-1β receptor as a fusion cofactor for macrophage-tropic HIV-1. *Science* **272:** 1955–1958.

Alkhatib G, Locati M, Kennedy PE, Murphy PM, Berger EA. 1997. HIV-1 coreceptor activity of CCR5 and its inhibition by chemokines: Independence from G protein signaling and importance of coreceptor downmodulation. *Virology* **234:** 340–348.

Arthos J, Cicala C, Martinelli E, Macleod K, Van Ryk D, Wei D, Xiao Z, Veenstra TD, Conrad TP, Lempicki RA, et al. 2008. HIV-1 envelope protein binds to and signals

through integrin α4β7, the gut mucosal homing receptor for peripheral T cells. *Nat Immunol* **9:** 301–309.

Babcock GJ, Farzan M, Sodroski J. 2003. Ligand-independent dimerization of CXCR4, a principal HIV-1 coreceptor. *J Biol Chem* **278:** 3378–3385.

Bakri Y, Schiffer C, Zennou V, Charneau P, Kahn E, Benjouad A, Gluckman JC, Canque B. 2001. The maturation of dendritic cells results in postintegration inhibition of HIV-1 replication. *J Immunol* **166:** 3780–3788.

Balotta C, Bagnarelli P, Violin M, Ridolfo AL, Zhou D, Berlusconi A, Corvasce S, Corbellino M, Clementi M, Clerici M, et al. 1997. Homozygous Δ32 deletion of the CCR-5 chemokine receptor gene in an HIV-1-infected patient. *AIDS* **11:** F67–F71.

Benkirane M, Jeang KT, Devaux C. 1994. The cytoplasmic domain of CD4 plays a critical role during the early stages of HIV infection in T-cells. *EMBO J* **13:** 5559–5569.

Berger EA, Doms RW, Fenyo EM, Korber BT, Littman DR, Moore JP, Sattentau QJ, Schuitemaker H, Sodroski J, Weiss RA. 1998. A new classification for HIV-1. *Nature* **391:** 240.

Briant L, Robert-Hebmann V, Acquaviva C, Pelchen-Matthews A, Marsh M, Devaux C. 1998. The protein tyrosine kinase p56lck is required for triggering NF-κB activation upon interaction of human immunodeficiency virus type 1 envelope glycoprotein gp120 with cell surface CD4. *J Virol* **72:** 6207–6214.

Bromley SK, Burack WR, Johnson KG, Somersalo K, Sims TN, Sumen C, Davis MM, Shaw AS, Allen PM, Dustin ML. 2001. The immunological synapse. *Annu Rev Immunol* **19:** 375–396.

Callebaut C, Krust B, Jacotot E, Hovanessian AG. 1993. T cell activation antigen, CD26, as a cofactor for entry of HIV in CD4$^+$ cells. *Science* **262:** 2045–2050.

Cameron PU, Freudenthal PS, Barker JM, Gezelter S, Inaba K, Steinman RM. 1992. Dendritic cells exposed to human immunodeficiency virus type-1 transmit a vigorous cytopathic infection to CD4$^+$ T cells. *Science* **257:** 383–387.

Cecilia D, Kulkarni SS, Tripathy SP, Gangakhedkar RR, Paranjape RS, Gadkari DA. 2000. Absence of coreceptor switch with disease progression in human immunodeficiency virus infections in India. *Virology* **271:** 253–258.

Chan DC, Fass D, Berger JM, Kim PS. 1997. Core structure of gp41 from the HIV envelope glycoprotein. *Cell* **89:** 263–273.

Chen Z, Gettie A, Ho DD, Marx PA. 1998. Primary SIVsm isolates use the CCR5 coreceptor from sooty mangabeys naturally infected in West Africa: A comparison of coreceptor usage of primary SIVsm, HIV-2, and SIVmac. *Virology* **246:** 113–124.

Chen B, Vogan EM, Gong H, Skehel JJ, Wiley DC, Harrison SC. 2005. Structure of an unliganded simian immunodeficiency virus gp120 core. *Nature* **433:** 834–841.

Choe H, Farzan M, Sun Y, Sullivan N, Rollins B, Ponath PD, Wu L, Mackay CR, LaRosa G, Newman W, et al. 1996. The β-chemokine receptors CCR3 and CCR5 facilitate infection by primary HIV-1 isolates. *Cell* **85:** 1135–1148.

Cicala C, Martinelli E, McNally JP, Goode DJ, Gopaul R, Hiatt J, Jelicic K, Kottilil S, Macleod K, O'Shea A, et al. 2009. The integrin α4β7 forms a complex with cell-surface CD4 and defines a T-cell subset that is highly susceptible to infection by HIV-1. *Proc Natl Acad Sci* **106:** 20877–20882.

Clark-Lewis I, Kim KS, Rajarathnam K, Gong JH, Dewald B, Moser B, Baggiolini M, Sykes BD. 1995. Structure-activity relationships of chemokines. *J Leukoc Biol* **57:** 703–711.

Cocchi F, DeVico AL, Garzino-Demo A, Arya SK, Gallo RC, Lusso P. 1995. Identification of RANTES, MIP-1 α, and MIP-1 β as the major HIV-suppressive factors produced by CD8$^+$ T cells. *Science* **270:** 1811–1815.

Cocchi F, DeVico AL, Garzino-Demo A, Cara A, Gallo RC, Lusso P. 1996. The V3 domain of the HIV-1 gp120 envelope glycoprotein is critical for chemokine-mediated blockade of infection. *Nat Med* **2:** 1244–1247.

Connor RI, Sheridan KE, Ceradini D, Choe S, Landau NR. 1997. Change in coreceptor use correlates with disease progression in HIV-1–infected individuals. *J Exp Med* **185:** 621–628.

Coyne CB, Bergelson JM. 2006. Virus-induced Abl and Fyn kinase signals permit coxsackievirus entry through epithelial tight junctions. *Cell* **124:** 119–131.

Dean M, Carrington M, Winkler C, Huttley GA, Smith MW, Allikmets R, Goedert JJ, Buchbinder SP, Vittinghoff E, Gomperts E, et al. 1996. Genetic restriction of HIV-1 infection and progression to AIDS by a deletion allele of the CKR5 structural gene. Hemophilia Growth and Development Study, Multicenter AIDS Cohort Study, Multicenter Hemophilia Cohort Study, San Francisco City Cohort, ALIVE Study. *Science* **273:** 1856–1862.

Deng H, Liu R, Ellmeier W, Choe S, Unutmaz D, Burkhart M, Di Marzio P, Marmon S, Sutton RE, Hill CM, et al. 1996. Identification of a major co-receptor for primary isolates of HIV-1. *Nature* **381:** 661–666.

Doranz BJ, Rucker J, Yi Y, Smyth RJ, Samson M, Peiper SC, Parmentier M, Collman RG, Doms RW. 1996. A dual-tropic primary HIV-1 isolate that uses fusin and the β-chemokine receptors CKR5, CKR-3, and CKR-2b as fusion cofactors. *Cell* **85:** 1149–1158.

Dorr P, Westby M, Dobbs S, Griffin P, Irvine B, Macartney M, Mori J, Rickett G, Smith-Burchnell C, Napier C, et al. 2005. Maraviroc (UK-427,857), a potent, orally bioavailable, and selective small-molecule inhibitor of chemokine receptor CCR5 with broad-spectrum anti-human immunodeficiency virus type 1 activity. *Antimicrob Agents Chemother* **49:** 4721–4732.

Dragic T, Litwin V, Allaway GP, Martin SR, Huang Y, Nagashima KA, Cayanan C, Maddon PJ, Koup RA, Moore JP, et al. 1996. HIV-1 entry into CD4$^+$ cells is mediated by the chemokine receptor CC-CKR-5. *Nature* **381:** 667–673.

Eckert DM, Malashkevich VN, Hong LH, Carr PA, Kim PS. 1999. Inhibiting HIV-1 entry: Discovery of D-peptide inhibitors that target the gp41 coiled-coil pocket. *Cell* **99:** 103–115.

Farzan M, Choe H, Martin KA, Sun Y, Sidelko M, Mackay CR, Gerard NP, Sodroski J, Gerard C. 1997. HIV-1 entry and macrophage inflammatory protein-1β-mediated signaling are independent functions of the chemokine receptor CCR5. *J Biol Chem* **272:** 6854–6857.

Felts RL, Narayan K, Estes JD, Shi D, Trubey CM, Fu J, Hartnell LM, Ruthel GT, Schneider DK, Nagashima K, et al. 2010. 3D visualization of HIV transfer at the virological

synapse between dendritic cells and T cells. *Proc Natl Acad Sci* **107:** 13336–13341.

Feng Y, Zhang F, Lokey LK, Chastain JL, Lakkis L, Eberhart D, Warren ST. 1995. Translational suppression by trinucleotide repeat expansion at FMR1. *Science* **268:** 731–734.

Feng Y, Broder CC, Kennedy PE, Berger EA. 1996. HIV-1 entry cofactor: Functional cDNA cloning of a seven-transmembrane, G protein-coupled receptor. *Science* **272:** 872–877.

Geijtenbeek TB, Kwon DS, Torensma R, van Vliet SJ, van Duijnhoven GC, Middel J, Cornelissen IL, Nottet HS, KewalRamani VN, Littman DR, et al. 2000. DC-SIGN, a dendritic cell-specific HIV-1-binding protein that enhances *trans*-infection of T cells. *Cell* **100:** 587–597.

Gosling J, Monteclaro FS, Atchison RE, Arai H, Tsou CL, Goldsmith MA, Charo IF. 1997. Molecular uncoupling of C-C chemokine receptor 5-induced chemotaxis and signal transduction from HIV-1 coreceptor activity. *Proc Natl Acad Sci* **94:** 5061–5066.

Gottlieb MS, Schroff R, Schanker HM, Weisman JD, Fan PT, Wolf RA, Saxon A. 1981. *Pneumocystis carinii* pneumonia and mucosal candidiasis in previously healthy homosexual men: Evidence of a new acquired cellular immunodeficiency. *N Engl J Med* **305:** 1425–1431.

Harmon B, Campbell N, Ratner L. 2010. Role of Abl kinase and the Wave2 signaling complex in HIV-1 entry at a post-hemifusion step. *PLoS Pathog* **6:** e1000956. doi: 10.1371/journal.ppat.1000956

Hartley O, Klasse PJ, Sattentau QJ, Moore JP. 2005. V3: HIV's switch-hitter. *AIDS Res Hum Retroviruses* **21:** 171–189.

Hernandez PA, Gorlin RJ, Lukens JN, Taniuchi S, Bohinjec J, Francois F, Klotman ME, Diaz GA. 2003. Mutations in the chemokine receptor gene *CXCR4* are associated with WHIM syndrome, a combined immunodeficiency disease. *Nat Genet* **34:** 70–74.

Hivroz C, Mazerolles F, Soula M, Fagard R, Graton S, Meloche S, Sekaly RP, Fischer A. 1993. Human immunodeficiency virus gp120 and derived peptides activate protein tyrosine kinase p56lck in human CD4 T lymphocytes. *Eur J Immunol* **23:** 600–607.

Huang Y, Paxton WA, Wolinsky SM, Neumann AU, Zhang L, He T, Kang S, Ceradini D, Jin Z, Yazdanbakhsh K, et al. 1996. The role of a mutant CCR5 allele in HIV-1 transmission and disease progression. *Nat Med* **2:** 1240–1243.

Huang CC, Tang M, Zhang MY, Majeed S, Montabana E, Stanfield RL, Dimitrov DS, Korber B, Sodroski J, Wilson IA, et al. 2005. Structure of a V3-containing HIV-1 gp120 core. *Science* **310:** 1025–1028.

Huang W, Eshleman SH, Toma J, Fransen S, Stawiski E, Paxinos EE, Whitcomb JM, Young AM, Donnell D, Mmiro F, et al. 2007. Coreceptor tropism in human immunodeficiency virus type 1 subtype D: High prevalence of CXCR4 tropism and heterogeneous composition of viral populations. *J Virol* **81:** 7885–7893.

Hubner W, McNerney GP, Chen P, Dale BM, Gordon RE, Chuang FY, Li XD, Asmuth DM, Huser T, Chen BK. 2009. Quantitative 3D video microscopy of HIV transfer across T cell virological synapses. *Science* **323:** 1743–1747.

Igakura T, Stinchcombe JC, Goon PK, Taylor GP, Weber JN, Griffiths GM, Tanaka Y, Osame M, Bangham CR. 2003. Spread of HTLV-I between lymphocytes by virus-induced polarization of the cytoskeleton. *Science* 299: 1713–1716.

Juszczak RJ, Turchin H, Truneh A, Culp J, Kassis S. 1991. Effect of human immunodeficiency virus gp120 glycoprotein on the association of the protein tyrosine kinase p56lck with CD4 in human T lymphocytes. *J Biol Chem* 266: 11176–11183.

Keele BF, Giorgi EE, Salazar-Gonzalez JF, Decker JM, Pham KT, Salazar MG, Sun C, Grayson T, Wang S, Li H, et al. 2008. Identification and characterization of transmitted and early founder virus envelopes in primary HIV-1 infection. *Proc Natl Acad Sci* 105: 7552–7557.

Klatzmann D, Barre-Sinoussi F, Nugeyre MT, Danquet C, Vilmer E, Griscelli C, Brun-Veziret F, Rouzioux C, Gluckman JC, Chermann JC, et al. 1984. Selective tropism of lymphadenopathy associated virus (LAV) for helper-inducer T lymphocytes. *Science* 225: 59–63.

Kwong PD, Wyatt R, Robinson J, Sweet RW, Sodroski J, Hendrickson WA. 1998. Structure of an HIV gp120 envelope glycoprotein in complex with the CD4 receptor and a neutralizing human antibody. *Nature* 393: 648–659.

Lalezari JP, Henry K, O'Hearn M, Montaner JS, Piliero PJ, Trottier B, Walmsley S, Cohen C, Kuritzkes DR, Eron JJ, et al. 2003. Enfuvirtide, an HIV-1 fusion inhibitor, for drug-resistant HIV infection in North and South America. *N Engl J Med* 348: 2175–2185.

Lazaro I, Naniche D, Signoret N, Bernard AM, Marguet D, Klatzmann D, Dragic T, Alizon M, Sattentau Q. 1994. Factors involved in entry of the human immunodeficiency virus type 1 into permissive cells: Lack of evidence of a role for CD26. *J Virol* 68: 6535–6546.

Lazzarin A, Clotet B, Cooper D, Reynes J, Arasteh K, Nelson M, Katlama C, Stellbrink HJ, Delfraissy JF, Lange J, et al. 2003. Efficacy of enfuvirtide in patients infected with drug-resistant HIV-1 in Europe and Australia. *N Engl J Med* 348: 2186–2195.

Lehmann MJ, Sherer NM, Marks CB, Pypaert M, Mothes W. 2005. Actin- and myosin-driven movement of viruses along filopodia precedes their entry into cells. *J Cell Biol* 170: 317–325.

Liu R, Paxton WA, Choe S, Ceradini D, Martin SR, Horuk R, MacDonald ME, Stuhlmann H, Koup RA, Landau NR. 1996. Homozygous defect in HIV-1 coreceptor accounts for resistance of some multiply-exposed individuals to HIV-1 infection. *Cell* 86: 367–377.

Love PE, Hayes SM. 2010. ITAM-mediated signaling by the T-cell antigen receptor. *Cold Spring Harb Perspect Biol* doi: 10.1101/cshperspect.a002485.

Maddon PJ, Dalgleish AG, McDougal JS, Clapham PR, Weiss RA, Axel R. 1986. The *T4* gene encodes the AIDS virus receptor and is expressed in the immune system and the brain. *Cell* 47: 333–348.

Maddon PJ, McDougal JS, Clapham PR, Dalgleish AG, Jamal S, Weiss RA, Axel R. 1988. HIV infection does not require endocytosis of its receptor, CD4. *Cell* 54: 865–874.

Margolis L, Shattock R. 2006. Selective transmission of CCR5-utilizing HIV-1: The "gatekeeper" problem resolved? *Nat Rev Microbiol* 4: 312–317.

Marsh M, Helenius A. 2006. Virus entry: Open sesame. *Cell* 124: 729–740.

McClure MO, Marsh M, Weiss RA. 1988. Human immunodeficiency virus infection of CD4-bearing cells occurs by a pH-independent mechanism. *EMBO J* 7: 513–518.

McDonald D, Wu L, Bohks SM, KewalRamani VN, Unutmaz D, Hope TJ. 2003. Recruitment of HIV and its receptors to dendritic cell-T cell junctions. *Science* 300: 1295–1297.

McDougal JS, Kennedy MS, Sligh JM, Cort SP, Mawle A, Nicholson JK. 1986. Binding of HTLV-III/LAV to T4+ T cells by a complex of the 110K viral protein and the T4 molecule. *Science* 231: 382–385.

Melikyan GB. 2008. Common principles and intermediates of viral protein-mediated fusion: The HIV-1 paradigm. *Retrovirology* 5: 111.

Miyauchi K, Kim Y, Latinovic O, Morozov V, Melikyan GB. 2009. HIV enters cells via endocytosis and dynamin-dependent fusion with endosomes. *Cell* 137: 433–444.

O'Brien TR, Winkler C, Dean M, Nelson JA, Carrington M, Michael NL, White GC II. 1997. HIV-1 infection in a man homozygous for CCR5 Δ32. *Lancet* 349: 1219.

Orloff GM, Orloff SL, Kennedy MS, Maddon PJ, McDougal JS. 1991. Penetration of CD4 T cells by HIV-1. The CD4 receptor does not internalize with HIV, and CD4-related signal transduction events are not required for entry. *J Immunol* 146: 2578–2587.

Padhan K, Varma R. 2010. Immunological synapse: A multi-protein signalling cellular apparatus for controlling gene expression. *Immunology* 129: 322–328.

Perez EE, Wang J, Miller JC, Jouvenot Y, Kim KA, Liu O, Wang N, Lee G, Bartsevich VV, Lee YL, et al. 2008. Establishment of HIV-1 resistance in CD4+ T cells by genome editing using zinc-finger nucleases. *Nat Biotechnol* 26: 808–816.

Piguet V, Steinman RM. 2007. The interaction of HIV with dendritic cells: Outcomes and pathways. *Trends Immunol* 28: 503–510.

Ping LH, Nelson JA, Hoffman IF, Schock J, Lamers SL, Goodman M, Vernazza P, Kazembe P, Maida M, Zimba D, et al. 1999. Characterization of V3 sequence heterogeneity in subtype C human immunodeficiency virus type 1 isolates from Malawi: Underrepresentation of X4 variants. *J Virol* 73: 6271–6281.

Samson M, Libert F, Doranz BJ, Rucker J, Liesnard C, Farber CM, Saragosti S, Lapoumeroulie C, Cognaux J, Forceille C, et al. 1996. Resistance to HIV-1 infection in caucasian individuals bearing mutant alleles of the CCR-5 chemokine receptor gene. *Nature* 382: 722–725.

Saphire AC, Bobardt MD, Zhang Z, David G, Gallay PA. 2001. Syndecans serve as attachment receptors for human immunodeficiency virus type 1 on macrophages. *J Virol* 75: 9187–9200.

Sattentau QJ, Weiss RA. 1988. The CD4 antigen: Physiological ligand and HIV receptor. *Cell* 52: 631–633.

Schuitemaker H, Koot M, Kootstra NA, Dercksen MW, de Goede RE, van Steenwijk RP, Lange JM, Schattenkerk JK, Miedema F, Tersmette M. 1992. Biological phenotype of human immunodeficiency virus type 1 clones at different stages of infection: Progression of disease is associated with a shift from monocytotropic to T-cell-tropic virus population. *J Virol* **66:** 1354–1360.

Sherer NM, Jin J, Mothes W. 2010. Directional spread of surface-associated retroviruses regulated by differential virus-cell interactions. *J Virol* **84:** 3248–3258.

Sohy D, Parmentier M, Springael JY. 2007. Allosteric transinhibition by specific antagonists in CCR2/CXCR4 heterodimers. *J Biol Chem* **282:** 30062–30069.

Sohy D, Yano H, de Nadai P, Urizar E, Guillabert A, Javitch JA, Parmentier M, Springael JY. 2009. Hetero-oligomerization of CCR2, CCR5, and CXCR4 and the protean effects of "selective" antagonists. *J Biol Chem* **284:** 31270–31279.

Stamatatos L, Levy JA. 1994. CD26 is not involved in infection of peripheral blood mononuclear cells by HIV-1. *AIDS* **8:** 1727–1728.

Stein BS, Gowda SD, Lifson JD, Penhallow RC, Bensch KG, Engleman EG. 1987. pH-independent HIV entry into CD4-positive T cells via virus envelope fusion to the plasma membrane. *Cell* **49:** 659–668.

Tersmette M, Gruters RA, de Wolf F, de Goede RE, Lange JM, Schellekens PT, Goudsmit J, Huisman HG, Miedema F. 1989. Evidence for a role of virulent human immunodeficiency virus (HIV) variants in the pathogenesis of acquired immunodeficiency syndrome: Studies on sequential HIV isolates. *J Virol* **63:** 2118–2125.

Theodorou I, Meyer L, Magierowska M, Katlama C, Rouzioux C. 1997. HIV-1 infection in an individual homozygous for CCR5 Δ32. Seroco Study Group. *Lancet* **349:** 1219–1220.

Tilton JC, Doms RW. 2009. Entry inhibitors in the treatment of HIV-1 infection. *Antiviral Res* **85:** 91–100.

Tremblay M, Meloche S, Gratton S, Wainberg MA, Sekaly RP. 1994. Association of p56lck with the cytoplasmic domain of CD4 modulates HIV-1 expression. *EMBO J* **13:** 774–783.

Turville SG, Cameron PU, Handley A, Lin G, Pohlmann S, Doms RW, Cunningham AL. 2002. Diversity of receptors binding HIV on dendritic cell subsets. *Nat Immunol* **3:** 975–983.

Uchil PD, Mothes W. 2009. HIV entry revisited. *Cell* **137:** 402–404.

Ugolini S, Mondor I, Sattentau QJ. 1999. HIV-1 attachment: Another look. *Trends Microbiol* **7:** 144–149.

Vasiliver-Shamis G, Tuen M, Wu TW, Starr T, Cameron TO, Thomson R, Kaur G, Liu J, Visciano ML, Li H, et al. 2008. Human immunodeficiency virus type 1 envelope gp120 induces a stop signal and virological synapse formation in noninfected CD4[+] T cells. *J Virol* **82:** 9445–9457.

Vasiliver-Shamis G, Cho MW, Hioe CE, Dustin ML. 2009. Human immunodeficiency virus type 1 envelope gp120-induced partial T-cell receptor signaling creates an F-actin-depleted zone in the virological synapse. *J Virol* **83:** 11341–11355.

Vasiliver-Shamis G, Dustin ML, Hioe CE. 2010. HIV-1 virological synapse is not simply a copycat of the immunological synapse. *Viruses* **2:** 1239–1260.

Weissenhorn W, Dessen A, Harrison SC, Skehel JJ, Wiley DC. 1997. Atomic structure of the ectodomain from HIV-1 gp41. *Nature* **387:** 426–430.

Welch BD, VanDemark AP, Heroux A, Hill CP, Kay MS. 2007. Potent D-peptide inhibitors of HIV-1 entry. *Proc Natl Acad Sci* **104:** 16828–16833.

Welch BD, Francis JN, Redman JS, Paul S, Weinstock MT, Reeves JD, Lie YS, Whitby FG, Eckert DM, Hill CP, et al. 2010. Design of a potent D-peptide HIV-1 entry inhibitor with a strong barrier to resistance. *J Virol* **84:** 11235–11244.

Wild C, Oas T, McDanal C, Bolognesi D, Matthews T. 1992. A synthetic peptide inhibitor of human immunodeficiency virus replication: Correlation between solution structure and viral inhibition. *Proc Natl Acad Sci* **89:** 10537–10541.

Wild C, Greenwell T, Matthews T. 1993. A synthetic peptide from HIV-1 gp41 is a potent inhibitor of virus-mediated cell-cell fusion. *AIDS Res Hum Retroviruses* **9:** 1051–1053.

Wu Y, Yoder A. 2009. Chemokine coreceptor signaling in HIV-1 infection and pathogenesis. *PLoS Pathog* **5:** e1000520. doi: 10.1371/journal.ppat.1000520.

Wu B, Chien EY, Mol CD, Fenalti G, Liu W, Katritch V, Abagyan R, Brooun A, Wells P, Bi FC, et al. 2010. Structures of the CXCR4 chemokine GPCR with small-molecule and cyclic peptide antagonists. *Science* **330:** 1066–1071.

Yang ZY, Huang Y, Ganesh L, Leung K, Kong WP, Schwartz O, Subbarao K, Nabel GJ. 2004. pH-dependent entry of severe acute respiratory syndrome coronavirus is mediated by the spike glycoprotein and enhanced by dendritic cell transfer through DC-SIGN. *J Virol* **78:** 5642–5650.

Yoder A, Yu D, Dong L, Iyer SR, Xu X, Kelly J, Liu J, Wang W, Vorster PJ, Agulto L, et al. 2008. HIV envelope-CXCR4 signaling activates cofilin to overcome cortical actin restriction in resting CD4 T cells. *Cell* **134:** 782–792.

HIV-1 Reverse Transcription

Wei-Shau Hu[1] and Stephen H. Hughes[2]

[1]Viral Recombination Section, HIV Drug Resistance Program, National Cancer Institute, Frederick, Maryland 21702-1201

[2]Vector Design and Replication Section, HIV Drug Resistance Program, National Cancer Institute, Frederick, Maryland 21702-1201

Correspondence: hughesst@mail.nih.gov

Reverse transcription and integration are the defining features of the Retroviridae; the common name "retrovirus" derives from the fact that these viruses use a virally encoded enzyme, reverse transcriptase (RT), to convert their RNA genomes into DNA. Reverse transcription is an essential step in retroviral replication. This article presents an overview of reverse transcription, briefly describes the structure and function of RT, provides an introduction to some of the cellular and viral factors that can affect reverse transcription, and discusses fidelity and recombination, two processes in which reverse transcription plays an important role. In keeping with the theme of the collection, the emphasis is on HIV-1 and HIV-1 RT.

It has been 40 years since the discovery of reverse transcriptase (RT) was announced by Howard Temin and David Baltimore, who independently showed that retroviral virions contain an enzymatic activity that can copy RNA into DNA (Baltimore 1970; Mizutani et al. 1970). These experiments provided the crucial proof of Temin's provirus hypothesis that retroviral infections persist because the RNA genome found in the virions is converted into DNA (Temin 1964). The sequences of the genomes of eukaryotes show how pervasive reverse transcription is in nature; not only do these genomes contain large numbers of endogenous retroviruses, but also a variety of retroposons and reverse-transcribed elements. The discovery in the early 1980s, that AIDS is caused by a human retrovirus, HIV-1, invigorated retroviral research and focused attention on the viral

enzymes, which have become the primary target of anti-AIDS drugs. Not surprisingly, the focus of RT research shifted from the RTs of the murine leukemia viruses (MLV) and the avian myeloblastosis virus to HIV-1 RT. The first approved anti-HIV drug, AZT, targets RT, and of the 26 drugs currently approved to treat HIV-1 infections, 14 are RT inhibitors. In addition, RTs (primarily recombinant MLV RTs) have become extremely valuable tools that are widely used in research, in clinical/diagnostic tests, and in biotechnology. We provide here a relatively brief description of the process of reverse transcription, the structure and biochemical functions of RT, some information about how other viral and cellular factors influence reverse transcription, and briefly consider how the reverse transcription process affects both the mutations that arise during the retroviral life cycle and recombination. The focus will be HIV-1 and HIV-1 RT; however, in some cases, we will draw on insights and include information obtained with other retroviruses and other RTs. Although the issues of the inhibition of HIV-1 RT by anti-RT drugs and the mechanisms of drug resistance are of considerable importance, these issues will not be addressed in detail here; the reader is directed to Arts and Hazuda (2011). Given that the literature on RT and reverse transcription is both vast and complex, and the space allowed for this article is limited, we have had to make some difficult choices in what to present, and what to omit, both in terms of the material and the references. For the omissions, we apologize.

THE PROCESS OF REVERSE TRANSCRIPTION

When a mature HIV-1 virion infects a susceptible target cell, interactions of the envelope glycoprotein with the coreceptors on the surface of the cell brings about a fusion of the membranes of the host cell and the virion (Wilen et al. 2011). This fusion introduces the contents of the virion into the cytoplasm of the cell, setting the stage for reverse transcription. There are complexities to the early events that accompany reverse transcription in an infected cell, not all of which are well understood, which will be considered later in this article. We will begin by discussing the mechanics of the conversion of the single-stranded RNA genome found in the virion into the linear double-stranded DNA that is the substrate for the integration process. The synthesis of this linear DNA is a reasonably well-understood process; additional details and references can be found in the books *Retroviruses* (Telesnitsky and Goff 1997) and *Reverse Transcriptase* (Skalka and Goff 1993). In orthoretroviruses, including HIV-1, reverse transcription takes place in newly infected cells. There is some debate in the literature about whether reverse transcription is initiated in producer cells. Primer tagging experiments suggest that most HIV-1 virions initiate reverse transcription in newly infected cells (Whitcomb et al. 1990); however, there are claims that a small

number of nucleotides may be incorporated before the virions initiate infection of target cells (Lori et al. 1992; Trono 1992; Zhu and Cunningham 1993; Huang et al. 1997). Either way, the vast majority of the viral DNA is synthesized in newly infected cells. This is a lifestyle choice; spumaretroviruses and the more distantly related hepadna viruses carry out extensive reverse transcription in producer cells (Summers and Mason 1982; Yu et al. 1996, 1999). Although there are viral and cellular factors that assist in the process of reverse transcription (these will be discussed later) the two enzymatic activities that are necessary and sufficient to carry out reverse transcription are present in RT. These are a DNA polymerase that can copy either a RNA or a DNA template, and an RNase H that degrades RNA if, and only if, it is part of an RNA–DNA duplex.

Like many other DNA polymerases, RT needs both a primer and a template. Genomic RNA is plus-stranded (the genome and the messages are copied from the same DNA strand), and the primer for the synthesis of the first DNA strand (the minus strand) is a host tRNA whose 3' end is base paired to a complementary sequence near the 5' end of the viral RNA called the primer binding site (pbs). Different retroviruses use different host tRNAs as primers. HIV-1 uses Lys3. It would appear, based on in vitro experiments, that the addition of the first few nucleotides is slow and difficult. DNA synthesis speeds up considerably once the first five to six deoxyribonucleotides have been added to the 3' end of the tRNA primer (Isel et al. 1996; Lanchy et al. 1998). In HIV-1, the pbs is approximately 180 nucleotides from the 5' end of genomic RNA. DNA synthesis creates an RNA–DNA duplex, which is a substrate for RNase H. There are perhaps 50 RTs in an HIV-1 virion; it is unclear whether the same RT that synthesizes the DNA plays a significant role in degrading the RNA. This is not a requirement—retroviruses can replicate (at a considerably reduced efficiency) with a mixture of RTs, some of which have only polymerase activity and some that have only RNase H activity (Telesnitsky and Goff 1993; Julias et al. 2001). Moreover, in in vitro assays, little or no RNase H cleavage is detected while RT is actively synthesizing DNA; instead, cleavages occur at sites where DNA synthesis pauses (Driscoll et al. 2001; Purohit et al. 2007). Whatever the exact mechanism, RNase H degradation removes the 5′ end of the viral RNA, exposing the newly synthesized minus-strand DNA (see Fig. 1).

The ends of the viral RNA are direct repeats, called R. These repeats act as a bridge that allows the newly synthesized minus-strand DNA to be transferred to the 3′ end of the viral RNA. Retroviruses package two copies of the viral RNA

genome; the first (or minus-strand) transfer can involve the R sequence at the 3′ ends of either of the two RNAs (Panganiban and Fiore 1988; Hu and Temin 1990b; van Wamel and Berkhout 1998; Yu et al. 1998). After this transfer, minus-strand synthesis can continue along the length of the genome. As DNA synthesis proceeds, so does RNase H degradation. However, there is a purine-rich sequence in the RNA genome, called the polypurine tract, or ppt, that is resistant to RNase H cleavage and serves as the primer for the initiation of the

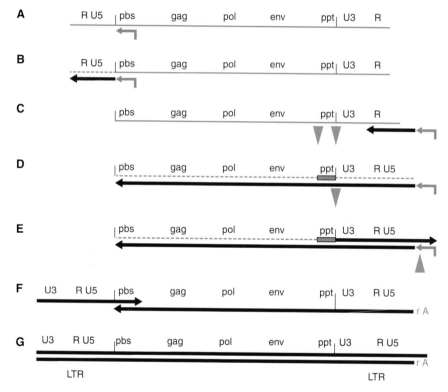

Figure 1. Conversion of the single-stranded RNA genome of a retrovirus into double-stranded DNA. (*A*) The RNA genome of a retrovirus (light blue) with a tRNA primer base paired near the 5′ end. (*B*) RT has initiated reverse transcription, generating minus-strand DNA (dark blue), and the RNase H activity of RT has degraded the RNA template (dashed line). (*C*) Minus-strand transfer has occurred between the R sequences at both ends of the genome (see text), allowing minus-strand DNA synthesis to continue (*D*), accompanied by RNA degradation. A purine-rich sequence (ppt), adjacent to U3, is resistant to RNase H cleavage and serves as the primer for the synthesis of plus-strand DNA (*E*). Plus-strand synthesis continues until the first 18 nucleotides of the tRNA are copied, allowing RNase H cleavage to remove the tRNA primer. Most retroviruses remove the entire tRNA; the RNase H of HIV-1 RT leaves the rA from the 3′ end of the tRNA attached to minus-strand DNA. Removal of the tRNA primer sets the stage for the second (plus-strand) transfer (*F*); extension of the plus and minus strands leads to the synthesis of the complete double-stranded linear viral DNA (*G*).

second (or plus) strand DNA. All retroviruses have at least one ppt. HIV-1 has two, one near the 3′ end of the RNA, the other (the central ppt) near the middle of the genome. The 3′ ppt is essential for viral replication, the central ppt probably increases the ability of the virus to complete plus-strand DNA synthesis, but is not essential (Charneau et al. 1992; Hungnes et al. 1992). When RT generates the plus-stand DNA that is initiated from the 3′ ppt, it not only copies the minus-strand DNA, but also the first 18 nucleotides of the Lys3 tRNA primer. Experiments performed with avian sarcoma-leukosis virus (ASLV) suggest that the ppt-primed plus-strand DNA synthesis stops when it encounters a modified A that RT cannot copy (Swanstrom et al. 1981). It is reasonable to expect that the same mechanism defines the portion of the HIV tRNA primer that is copied. Once the tRNA has been copied into DNA, it becomes a substrate for RNase H. Most retroviruses remove the entire tRNA; however, HIV-1 RT is the exception. It cleaves the tRNA one nucleotide from the 3′ end, leaving a single A ribonucleotide at the 5′ end of the minus strand (the specificity of RNase H cleavage is discussed at the end of this section) (Whitcomb et al. 1990; Pullen et al. 1992; Smith and Roth 1992).

In theory, minus-strand DNA synthesis can proceed along the entire length of the RNA genome; however, the genomic RNAs found in virions are often nicked. The fact that there is a second copy of the RNA genome allows minus-strand DNA synthesis to transfer to the second RNA template, thus bypassing the nick in the original template. This template switching ability contributes to efficient recombination, a topic that is considered later in this article. When minus-strand DNA synthesis nears the 5′ end of the genomic RNA, the pbs is copied, setting the stage for the second, or plus-strand transfer. The 3′ end of the plus-strand DNA contains 18 nucleotides copied from the tRNA primer, which are complementary to 18 nucleotides at the 3′ end of the minus-strand DNA that were copied from the pbs. These two complementary sequences anneal, and DNA synthesis extends both the minus and plus strands to the ends of both templates.

The synthesis of plus-strand DNA does not have to be continuous; it is clear that, in ASLV, the plus strand is made in segments (Kung et al. 1981; Hsu and Taylor 1982). It has been reported that HIV-1 plus-strand DNA is also synthesized from multiple initiation sites (Miller et al. 1995; Klarmann et al. 1997; Thomas et al. 2007); however, that raises a question about the role played by the second ppt: If plus-strand DNA is made in segments, what advantage does the second ppt give HIV-1?

The reverse transcription process creates a DNA product that is longer than the RNA genome from which it is derived: both ends of the DNA contain sequences from each end of the RNA (U3 from the 3′ end and U5 from the 5′ end). Thus, each end of the viral DNA has the same sequence, U3-R-U5; these are the long terminal repeats (LTRs) that will, after integration, be the ends of the provirus. It is important to remember that the sequences at the ends of the full-length linear viral DNA are defined, on the U5 end, by the RNase H cleavage that removes the tRNA primer, and on the U3 end, by the cleavages that generate and remove the ppt primer. Despite the fact that RNase H does not have any specific sequence recognition motifs, it cleaves these substrates with single nucleotide specificity, a specificity that appears to be based on the structures of the nucleic acid substrates when they are in a complex with RT (Pullen et al. 1993; Julias et al. 2002; Rausch et al. 2002; Dash et al. 2004; Yi-Brunozzi and Le Grice 2005). The specificity of RNase H cleavage is important because the ends of the linear viral DNA are the substrates for integration. Although DNA substrates whose ends differ modestly from the consensus sequence can be used for retroviral DNA integration, the consensus sequence is the preferred substrate (Colicelli and Goff 1985, 1988; Esposito and Craigie 1998; Oh et al. 2008).

THE GENESIS, STRUCTURE, AND ENZYMATIC FUNCTIONS OF HIV-1 RT

HIV-1 RT is produced from a Gag-Pol polyprotein by cleavage with the viral protease (PR). HIV-1 Gag-Pol is produced by a frameshift readthrough event in the p6 coding region that occurs about 5% of the time Gag is translated from an unspliced full-length viral RNA transcript. The Gag portion of Gag-Pol allows it to associate with Gag during virion assembly, ensuring that the Pol portion of Gag-Pol, which includes PR, RT, and integrase, is inside the assembled virion.

The mature form of HIV-1 RT is a heterodimer that is composed of two related subunits: the larger, p66, is 560 amino acids long; the smaller, p51, contains the first 440 amino acids of p66 (Lightfoote et al. 1986). The p66 subunit consists of two domains: polymerase and RNase H; in the mature HIV-1 RT heterodimer, p66 contains the active sites for the two enzymatic activities of RT (see online Movie 1 at www.perspectivesinmedicine.org). The polymerase domain has been compared to a human right hand and is composed of the fingers, palm, thumb, and connection subdomains (see Fig. 2) (Kohlstaedt et al. 1992). The p51 subunit corresponds closely, but not exactly, to the polymerase domain of p66, and contains the same four subdomains. However, the relative arrangement of the subdomains differs in the two subunits. The p66 domain plays the catalytic role, whereas the p51 subunit plays a structural role (Movie 1) (Kohlstaedt et al. 1992; Jacobo-Molina et al. 1993).

We are fortunate to have crystal structures that correspond to HIV-1 RT in multiple states that are important intermediates in the reverse transcription process (see Fig. 3 and online Movie 2 at www.perspectivesinmedicine.org). Some of the structures also tell us a great deal about how anti-RT drugs work, and how resistance mutations allow RT to evade the currently approved drugs (Arts and Hazuda 2011). Considering the structures in a way that corresponds to steps in reverse transcription, unliganded RT has the thumb subdomain of the p66 subunit folded over into the nucleic acid binding cleft, in a position such that the tip of thumb nearly touches the fingers (Esnouf et al. 1995; Rodgers et al. 1995; Hsiou et al. 1996). Because the thumb is in the closed configuration in unliganded RT, the thumb must move away from

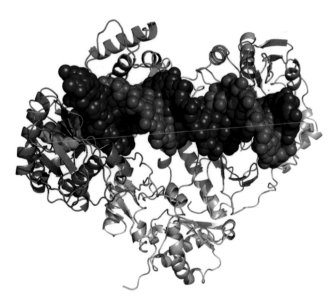

Figure 2. Structure of a ternary complex of HIV-1 RT, double-stranded DNA, and an incoming dNTP. HIV-1 RT is composed of two subunits, p51 and p66. P51 is shown in gray. The RNase H domain of p66 is gold, and the four subdomains of the polymerase domain of p66 are color-coded: fingers, blue; palm, red; thumb, green; and connection, yellow. The template strand of the DNA is brown, and the primer strand is purple. The incoming dNTP is light blue. (Figure courtesy of K. Das and E. Arnold.)

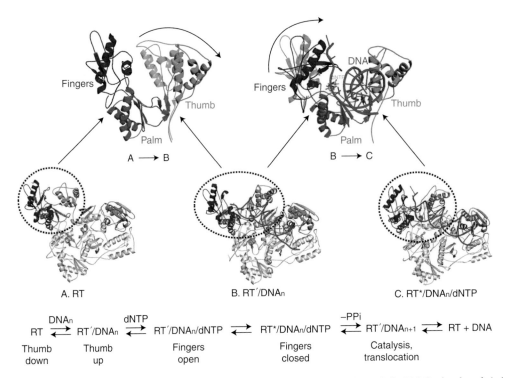

Thumb Thumb Fingers Fingers Catalysis,
down up open closed translocation

Figure 3. Structural changes in RT that occur during polymerization. In unliganded RT (*A*), the thumb is in the closed configuration. Binding a double-stranded nucleic acid substrate (*B*) is accompanied by movement of the thumb (*upper left, A,B*) that creates the nucleic acid binding site. Binding of the incoming dNTP (*C*) is accompanied by a movement of the fingers that closes the β3–β4 loop down onto the incoming dNTP (*lower left, B,C*). These movements correspond to steps in DNA synthesis (*bottom*). (Figure courtesy of K. Kirby and S. Sarafianos.)

the fingers to create the nucleic acid binding cleft for the template primer. The nucleic acid binding cleft has a structure that allows it to bind a double-stranded nucleic acid (Jacobo-Molina et al. 1993). There are modest interactions of the single-stranded 5′ extension of the template with RT; this helps to position the end of the primer at the polymerase active site, which is composed of three aspartic acid residues (D110, D185, and D186), that help position the two divalent metal ions (Mg^{2+} during viral replication) required for polymerization (Movie 2).

We do not have any structures of RT with an RNA–RNA duplex (which would correspond to minus-strand initiation); however, there is one structure with an RNA–DNA duplex and several that contain DNA–DNA duplexes

(Jacobo-Molina et al. 1993; Huang et al. 1998; Sarafianos et al. 2001, 2002; Tuske et al. 2004; Das et al. 2009; Lansdon et al. 2010; Tu et al. 2010). The sequence of the RNA in the RNA–DNA duplex structure was based on the ppt, which was chosen for this structural analysis because it is poorly cleaved by RNase H. The RNA–DNA duplex is bent approximately 40° near where it passes under the thumb. A similar bend is seen in the DNA–DNA structures (discussed below). However, the surprising thing about the RNA–DNA duplex is that it contains two unpaired and two mispaired bases that take the duplex out of, then back into, proper register (Sarafianos et al. 2001). We only have one RNA–DNA structure, so it is possible that this misalignment could be specific to the ppt, and, if it is, the misalignment could play a role in the resistance of the ppt to RNase H cleavage. Although the RNA strand is the template strand in this structure, the RNase H active site amino acids D443, E478, D498, and D549, which help position the two divalent metal ions (again, Mg^{2+} in vivo) do not make close contact with the RNA strand in the structure, which helps account for the inability of RNase H to cleave the ppt. The polymerase and RNase H active sites are 17 to 18 base pairs apart along the nucleic acid, depending on the nucleic acid substrate (Jacobo-Molina et al. 1993; Sarafianos et al. 2001). Despite the fact that the RNA template does not make close contact with the RNase H active site in the one structure we have of RT bound to an RNA–DNA duplex, the two active sites are positioned so that they should be able to simultaneously engage a nucleic acid substrate. There are several motifs that play important roles in holding and properly positioning the nucleic acid relative to the two active sites. The RNase H primer grip, which is near the RNase H active site, plays a role in positioning an RNA–DNA duplex for proper (specific) cleavage (Julias et al. 2002; Rausch et al. 2002). The primer grip and template grip, which are nearer the polymerase active site, help position nucleic acid duplexes at both the polymerase and RNase H active sites (Ghosh et al. 1996, 1997; Powell et al. 1997, 1999; Gao et al. 1998).

The complexes of RT bound to a DNA–DNA duplex are globally quite similar to the complex with the RNA–DNA duplex but with some interesting differences. None of the DNA–DNA duplexes contain unpaired bases, possibly because none of the sequences of the DNA–DNA duplexes in the RT structures have the sequence of the ppt. The DNA–DNA duplex follows a similar bent trajectory as the RNA–DNA duplex (Jacobo-Molina et al. 1993; Sarafianos et al. 2001). In the DNA–DNA duplex, the portion of the double-stranded DNA near the polymerase active site is A form. Where the DNA bends, near the thumb of p66, there is a transition, which occurs over a stretch of four base pairs, from an A-form to a B-form duplex, and the DNA beyond the thumb is B form. In contrast, the RNA–DNA duplex is neither entirely A form nor B form, being somewhat intermediate between the two (this is common for RNA–DNA duplexes; these structures have been called H form), although the region near the polymerase active site is more similar to A form and the region beyond the thumb more similar to B form.

There are also structures of HIV-1 RT with both a bound DNA–DNA duplex and an incoming dNTP or the triphosphate of a nucleoside analog (Huang et al. 1998; Tuske et al. 2004; Das et al. 2009; Lansdon et al. 2010). Overall, the structures of these ternary complexes are similar to the corresponding structures of HIV-1 RT bound to a DNA–DNA duplex, with one important difference: When there is a bound dNTP, a portion of the p66 fingers (the β3-β4 loop) closes down on the incoming triphosphate, forming part of the dNTP-binding pocket. A similar movement of the fingers has been seen with other polymerases (bacterial DNA polymerases, T7 RNA polymerase, and some viral RNA-dependent RNA polymerases) when an incoming nucleoside triphosphate is bound. In the structures of HIV-1 RT in a complex with a DNA–DNA duplex and an incoming dNTP, the last nucleotide at the 3′ end of the primer strand is a dideoxy; this prevents incorporation of the incoming dNTP. The dNTP is bound at the active site, which is also called the N, or nucleoside-binding site. The end of the primer is at the P, or priming site. During normal polymerization, the incorporation of the incoming dNTP links the α-phosphate of the dNTP to the 3′OH of the deoxyribose of the nucleotide at the 3′ end of the primer, releasing pyrophosphate. At this point, the end of the primer is still in the N site, and there are structures that correspond to this state. For polymerization to continue, the nucleic acid substrate must move (translocate) relative to RT, moving the end of the primer to the P site, so that the next incoming dNTP can bind. Release of the pyrophosphate appears to be accompanied by an opening of the fingers and it has been suggested that a movement of the conserved YMDD loop that contains two of the active site aspartates (D185 and D186) acts as a springboard, affecting translocation. In this model, the binding of the incoming dNTP causes a downward movement that loads the springboard; thus the incorporation of the nucleotide and the release of the pyrophosphate are the ultimate source of the energy that drives translocation (Sarafianos et al. 2002). Once the end of the primer is in the P site, another incoming dNTP can bind, and polymerization can continue.

The availability of the various different RT structures has guided and informed the analysis of the biochemical properties of RT. This combination of biochemical analysis and structural insights have made it possible to gain a good understanding of the roles played by structural elements like the primer grip, the template grip, the RNase H primer grip, and, in some cases, of the roles played by individual amino acids. In many cases, ideas that originated by looking at one or more of the RT structures was tested by reverse genetics: Mutations were made in specific amino acids to determine their effects on the in vitro properties of recombinant HIV-1 RT, on the replication of an HIV-1-based vector, or both. This literature is too large and complex to review here; however, a few simple ideas are worth presenting. For the most part, mutations in the structures that appear to be important in binding the substrates have phenotypes that match what the structure shows. This brings up an important point: Any mutation that causes a change in the polymerase domain of HIV-1 RT makes two changes in the mature HIV-1 RT heterodimer, one in p51 and one in p66. It is possible to express forms of recombinant RT that have changes in only one subunit. Despite the fact that both the polymerase and the RNase H active sites are in p66, in some cases the change in p51, which interacts extensively with p66 and helps form the nucleic acid binding cleft, can contribute to the behavior of the recombinant enzyme. Despite this complexity, individual mutations (e.g., active site mutations) can selectively affect one of the two activities of RT, polymerase or RNase H. However, there are a number of mutations, such as mutations that change the binding of the nucleic acid substrate,

that affect both polymerase and RNase H. For example, there are mutations in the polymerase domain that affect not only polymerase activity and the fidelity of DNA synthesis, but also RNase H cleavage (Palaniappan et al. 1997; Gao et al. 1998; Powell et al. 1999; Sevilya et al. 2001, 2003). Mutations in the RNase H primer grip can affect the specificity of RNase H cleavage, but can also have some effect on the initiation of viral DNA synthesis (Julias et al. 2002; Rausch et al. 2002). Mutations in and around the polymerase active site can profoundly affect dNTP selection and polymerization. Not surprisingly, some, but not all, of the mutations that cause resistance to nucleoside analogs are near the polymerase active site (Tantillo et al. 1994; Sarafianos et al. 2009). Mutations that do not directly impact the enzymatic activities of RT can still have important effects on reverse transcription and the viral life cycle. For example, there are mutations in HIV-1 RT that affect the stability of the heterodimer. There are also mutations that permit the degradation of RT in virions by the viral PR. Some of these mutants have been shown to have a temperature-sensitive phenotype. It would appear that the mutations allow RT to partially unfold, making it susceptible to cleavage by PR. In some cases, the mutant virions contain no detectable RT. As expected, virions in which RT is extensively degraded have little or no infectivity (Huang et al. 2003; Takehisa et al. 2007; Dunn et al. 2009; Wang et al. 2010).

REVERSE TRANSCRIPTION IN INFECTED CELLS

Although only purified RT is required to carry out DNA synthesis from an RNA template in vitro, reverse transcription in target cells is a complex process that is intimately interconnected with other early events in the viral life cycle. The reverse transcription complex (RTC), in which DNA synthesis occurs in infected cells, contains multiple proteins. At some point, late in the reverse transcription process, the RTC transitions into a preintegration complex (PIC), and the PIC is transported into the nucleus. Although the ends of the DNA are completed in the cytoplasm, the plus-strand DNA of HIV-1 comprises at least two segments before integration. Viral DNA synthesis is a highly regulated event. Mutations in other viral genes, such as CA, can have a profound effect on reverse transcription in vivo (Forshey et al. 2002).

Multiple viral proteins, including MA, CA, NC, IN, and Vpr, have been reported to be present in the RTC (Fassati and Goff 2001; Nermut and Fassati 2003; Iordanskiy et al. 2006). The role of mature MA protein in the RTC/PIC is unclear. There were claims that MA directs the nuclear import of PICs and allows HIV-1 to infect nondividing cells; however, more recent data suggest that this hypothesis is incorrect (Gallay et al. 1995a,b; Freed et al. 1997). The mature CA protein most likely provides the overall structure of the RTC. CA may play a role in the transport of the RTC and the nuclear import of the RTC/PIC, allowing HIV-1 to infect nondividing cells (Yamashita and Emerman 2004; Dismuke and Aiken 2006; Qi et al. 2008; Lee et al. 2010). NC has nucleic acid chaperone activity; it affects the reverse transcription process, both in terms of helping RT through regions of secondary structure and facilitating strand transfer (Feng et al. 1996; Zhang et al. 2002; Buckman et al. 2003; Golinelli and Hughes 2003; Houzet et al. 2008; Thomas and Gorelick 2008; Thomas et al. 2008). Vpr is present in the RTC and it has been suggested that Vpr interacts with the host enzyme uracil DNA glycosylase/uracil N-glycosylase (UNG2), a factor that could modify newly synthesized viral DNA (Selig et al. 1997; Mansky et al. 2000; Chen et al. 2004; Schrofelbauer et al. 2005). The precise role and effects of Vpr/UNG2 is not yet clear. Some lentiviruses encode a deoxyuridine triphosphatase; however, this accessory gene is not found in HIV-1 (Elder et al. 1992; Wagaman et al. 1993; Lerner et al. 1995). Other HIV-1 gene products such as Vif and Tat have been shown to affect DNA synthesis in vitro; however, their roles in in vivo DNA synthesis, if any, are unclear (Harrich et al. 1997; Kameoka et al. 2002; Liang and Wainberg 2002; Apolloni et al. 2007; Henriet

et al. 2007; Carr et al. 2008). Conversely, some, but not all, IN mutations have a profound negative effect on reverse transcription in an infected cell; however, these same IN mutations do not affect the activity of RT in viral lysates, and IN has not been shown to enhance the activity of purified RT in vitro, which suggest that IN might have an indirect role in the structure of the RTC in vivo (Engelman et al. 1995; Masuda et al. 1995; Leavitt et al. 1996; Wu et al. 1999). Certain host restriction factors, such as APOBEC3G and APOBEC3F, can be incorporated into virions and become part of the RTCs where they can cause mutations during DNA synthesis (Bishop et al. 2004b; Zheng et al. 2004); these factors are discussed briefly later in this article, and in more detail by Malim and Bieniasz (2011).

The structure of the RTC is not known. It is clear that there are changes in the structure of the core found in a mature virion (collectively called "uncoating"), which convert the core into a complex that can efficiently carry out reverse transcription. One hypothesis is that as uncoating is a continuous process, the structure of RTC changes as DNA synthesis proceeds, eventually transforming the RTC into a PIC (Forshey et al. 2002; Dismuke and Aiken 2006), which can be isolated from the cytoplasm and is capable of integrating viral DNA into a DNA target in vitro (Craigie and Bushman 2011). An alternative hypothesis is that the RTCs have a structure similar to that of the virus core, within which DNA synthesis occurs. In this proposal, the "core-like" structures are transported to the nuclear pore and converted into PICs before they enter the nucleus (Arhel et al. 2007). The RTC is a target for host restriction factors such as TRIM5α, TRIMCyp, and Fv-1; these host restriction factors interact with hexameric CA protein in the RTC, interfering with uncoating in a way that blocks reverse transcription or some later step that is essential for nuclear import and integration (Frankel et al. 1989; Best et al. 1996; Nisole et al. 2004; Sayah et al. 2004; Stremlau et al. 2004, 2006; Wu et al. 2006; Brennan et al. 2008; Newman et al. 2008; Virgen et al. 2008; Wilson et al. 2008).

Reverse transcription is initiated shortly after virus entry; viral DNA can be detected within hours of infection (Butler et al. 2001; Julias et al. 2001; Thomas et al. 2006; Mbisa et al. 2007). The rate of HIV-1 DNA synthesis had been measured in 293 T cells and activated primary human CD4$^+$ T cells: minus-strand DNA is synthesized at a rate of ~70 nucleotides per minute (Thomas et al. 2007). Plus-strand DNA synthesis is rapid, which agrees with the proposals that multiple initiation sites are used (Miller et al. 1995; Klarmann et al. 1997; Thomas et al. 2007). The minus-strand and plus-strand transfer reactions were first studied in vitro, in experiments performed with MLV and ASLV. The in vitro strand transfer reactions are slow, and the transfer intermediates, minus- and plus-strand strong-stop DNA, are easy to detect. These transfers occur more rapidly in an infected cell, where it has been estimated for HIV-1 that the transfers take ~4 and ~9 mins, respectively (Thomas et al. 2007). The rate of DNA synthesis, however, can vary depending on the nature of the target cell. For example, the rate of synthesis is expected to be slow in quiescent cells where the dNTP levels are low; it has been shown that DNA synthesis can stall in resting T cells (Zack et al. 1992). Additionally, the rate of DNA synthesis can be affected by mutations in viral genes, including RT.

DNA synthesis is often used to monitor the early stages of virus infection. The progression of reverse transcription can be determined using real-time polymerase chain reaction (PCR) and primer sets that anneal to various regions of the viral genomes (Butler et al. 2001; Julias et al. 2001; Mbisa et al. 2009). Generally, primer sets that anneal to R-U5 are used to measure the initiation of DNA synthesis, U3-R for minus-strand DNA transfer, Gag for the extension of the minus-strand DNA, and U5-5′UTR for plus-strand DNA transfer. Additionally, primers that anneal to U5-U3 can be used to measure 2-LTR circles, which are often used as a surrogate for nuclear import; lastly, primers that anneal to the LTR and human repetitive element Alu have been used to detect viral DNAs integrated in the human genome (proviruses).

GENETIC CONSEQUENCES OF REVERSE TRANSCRIPTION: MUTATION AND RECOMBINATION

Mutations and Fidelity

HIV-1 sequences vary considerably, not only between individuals, but also within an infected patient. The large variation seen in individual patients is somewhat surprising given that most patients are infected with a single virus (Keele et al. 2008); this means that the diversity of viral sequences seen in most patients arises after the patients are infected. Although it is the large numbers of infected cells and the rapid turnover of these infected cells that are the major reasons why the virus diverges so rapidly, it is the mutations that arise during the viral life cycle that are the ultimate source of viral diversity (Coffin 1995). The ability of the virus to diverge rapidly plays an important role in its ability to stay ahead of the immune system in an infected individual, and plays a key role in the ability of the virus to become resistant to all the known anti-AIDs drugs. The diversity of known viral strains also makes the daunting task of developing an effective vaccine even more difficult. As will be discussed in the next section, the fact that the virus can, and does, recombine efficiently complicates attempts to deal with these problems.

What is the cause of the mutations? A casual reading of the HIV-1 literature would suggest that mutations are caused by errors made by RT, which has no proofreading function. However, the data that speak directly to this question are quite limited. In theory, there are three ways in which mutations could arise during the HIV-1 replication cycle. Reverse transcription is one possibility. However, the RNA genome is synthesized by the host DNA-dependent RNA polymerase II (RNA pol II), another enzyme that lacks a proofreading function, and the contribution of RNA pol II to the mutation rate has not been determined. In addition, if a cell that harbors a provirus replicates, it is possible that the host DNA replication machinery could generate a mutation in the provirus. However, for exogenous retroviruses like HIV-1, in which there is a rapid turnover of infected cells, the contribution of the host DNA polymerases (which have elaborate proofreading functions) to the mutation rate is negligible. That leaves RT and RNA pol II. Unfortunately it has not been possible to separate their contributions to the overall viral mutation rate (Kim et al. 1996; O'Neil et al. 2002). It is possible to run RT fidelity assays in vitro, using purified RT, but, as will be discussed in more detail later, the in vitro data obtained with purified RT do not match the fidelity data obtained when an HIV-1 vector is used to infect cultured cells. If we could identify the errors made by RNA pol II, the errors made by RT could be identified by subtraction, but as yet there is no good way to determine which errors are made by pol II.

It has been suggested that the fidelity of HIV-1 RT is particularly low, and this accounts for the observed sequence variation. This is incorrect. The mutation rate for HIV-1 replication, which represents the combined error rate for RT and RNA pol II, is approximately 2×10^{-5} per nucleotide per replication cycle, a rate that is similar of other retroviruses (Pathak and Temin 1990a,b; Mansky and Temin 1994, 1995; Kim et al. 1996; Julias and Pathak 1998; Halvas et al. 2000; Abram et al. 2010). As has already been mentioned, the rapid variation in HIV-1 in patients is primarily because of the rapid turnover of a relatively large population of infected cells. Other retroviruses, which have a mutation rate similar to that of HIV-1, have a lower variation because the viruses replicate less rapidly in their infected hosts. In some cases (e.g., the ASLV and MLV viruses) the natural host (in this example, chickens and mice) carries closely related endogenous viruses whose proteins are expressed early enough in development to be recognized as self. This means that the exogenous viruses are sheltered, to some degree, from the host's immune system. Because the degree of protection afforded by the proteins of endogenous viruses depends on the similarity of the proteins encoded by the exogenous viruses, the immune selection tends to restrict the overall variation of these exogenous viruses. Given that the overall mutation rate for HIV-1 and other retroviruses is similar, we can infer that the error rates of their RTs are similar. It is likely that RT makes a significant contribution to the overall error rate because, in a system in which the genetic information copied by RT is supplied by RNA pol II, there can be no selection for an RT that has a fidelity higher than the enzyme that provides the template RNA copies. This idea is supported by the analysis of mutations in the LTR that must be owing to the activity of RT (Kim et al. 1996; O'Neil et al. 2002). These data leave open the possibility that RT may have a lower fidelity than RNA pol II. However, even if we assume that RT makes the majority of the errors, its fidelity could be no lower than the error rate for the viral replication cycle (2×10^{-5}). Thus, the fidelity of RT in an infected cell is at least 10 times higher than most groups have reported based on assays that involve using the purified enzyme in vitro.

Although it is convenient to calculate a specific number for the overall mutation rate, providing a single number is somewhat misleading. Errors do not arise uniformly throughout the sequence. Errors arise more frequently at some positions than at others; sites where errors occur frequently are called "hotspots." In theory, it should be possible to use the in vitro assays to understand why RT preferentially makes mistakes at certain sites. Some of the in vitro fidelity assays have been performed with a substrate (the α-complementing fragment of Lac Z) similar to the one used in the single-cycle cell culture assays. Unfortunately, none of the in vitro assays produced a pattern of hotspots that was similar to what has been seen with a viral vector in cultured cells (Mansky and Temin 1995; Abram et al. 2010). To make matters worse, the pattern of hotspots reported from the various labs that did the in vitro experiments are all different. There are several possible explanations: (1) As has already been discussed, there are a number of ancillary viral and host proteins that contribute to the efficiency of the reverse transcription process; it is possible that some of these factors also contribute to fidelity. (2) The various groups used different assay conditions, and the purified RTs used in the assays are not identical. (3) Although it is likely that RT makes a real contribution to hotspots seen in the viral vector system, some of the hotspots could be caused by RNA pol II.

One of the important underlying issues is the emergence of mutations that allow HIV-1 to evade anti-AIDS drugs and mutations that cause immune escape. It is clear that the virus almost always finds a way to evade the host's immune system and, unless the therapy completely blocks viral replication, the virus also finds ways to evade all the known anti-HIV drugs. However, the virus can use different mutations for immune escape and drug resistance. There are several possible explanations. One interesting possibility is that differences in the sequences of the various RTs could change their ability to make specific errors (mutations). As a result, these RTs (e.g., a drug-resistant RT) might, when carrying out viral DNA synthesis, generate a different spectrum of errors (mutations), thereby altering the spectrum of variants that eventually emerge in response to immunological (or drug) selection.

In addition to the errors made by RT and RNA pol II, the reverse transcription process can be affected by cellular factors, in particular, the APOBEC proteins. The APOBECs are covered in greater detail by Malim and Bieniasz (2011); however, it is important to remember that they affect the fidelity and efficiency of reverse transcription. Broadly speaking, the APOBECs are cytidine deaminases; those that affect retroviruses use a single-stranded DNA substrate (Harris et al. 2003; Mangeat et al. 2003; Bishop et al. 2004b). Although APOBEC3G is the best studied of the human APOBEC proteins, there are several human APOBECs that can affect HIV replication in cultured cells; these could have effects on HIV-1 replication in patients (Bishop et al. 2004a; Yu et al. 2004; Holmes et al. 2007). The APOBECs that affect HIV replication are packaged into virions and modify minus-strand DNA after the RNA has been degraded, but before the plus strand has been synthesized. At this stage the APOBECs can convert some of the Cs in the minus strand to Us. When the virus replicates, the C-to-U mutations in minus-strand DNA lead to the conversion of Gs in the RNA genome to As. The APOBECs are part of the host's innate defense against retroviruses, and, as might have been expected, HIV-1 has a counter, the Vif protein, which interacts with host machinery to cause the degradation of APOBEC (Mariani et al. 2003; Yu et al. 2003; Liu et al. 2004; Sawyer et al. 2004; Schrofelbauer et al. 2004; Luo et al. 2005; Fang and Landau 2007; Russell and Pathak 2007). The impact of APOBEC on HIV replication is much greater if the virus lacks a functional Vif. Moreover, the ability of the APOBECs to block HIV-1 replication does not appear to be entirely attributable to their cytidine deaminase activity; it appears that the APOBECs can have other negative effects on reverse transcription and integration, although the exact nature of these effects is not yet clear (Bishop et al. 2006; Holmes et al. 2007; Mbisa et al. 2007, 2010). There are also hints that enzymes that act as RNA adenine deaminases (ADARs) can cause mutations in HIV-1, at least in cultured cells (Abram et al. 2010) and mutations that appear to have been caused by ADARs have been reported when HIV replication is challenged with antisense RNAs (Lu et al. 2004; Mukherjee et al. 2011). However, despite the fact that viruses isolated from patients almost always have an intact Vif-coding region, it is easy to find, among the HIV-1 sequences from patients, G-to-A hypermutations that appear to be the result of APOBEC activity. A search of the same sequence databases showed no obvious indication of ADAR-induced hypermutations (Abram et al. 2010).

RECOMBINATION

The recombination rate for retroviruses is higher than for most other viruses and the recombination rate for HIV-1 is higher than other retroviruses such as MLV and spleen necrosis viruses (Hu and Temin 1990a; Anderson et al. 1998; Onafuwa et al. 2003; Rhodes et al. 2003, 2005). Mapping of HIV-1 genomes by direct sequencing shows that there is frequent recombination during DNA synthesis (Robertson et al. 1995; Jetzt et al. 2000; Zhuang et al. 2002; Dykes et al. 2004; Levy et al. 2004; Chin et al. 2008; Galli et al. 2010). During minus-strand DNA synthesis, RT can switch between the two copackaged RNAs, using portions of each RNA as a template to generate a chimeric DNA containing sequences from each of the two genomic RNAs. Template switching can occur between two copackaged RNAs with identical sequences; however, only virions that package two genetically different RNAs can generate a recombinant with a genotype distinct from that of the parents (Hu and Temin 1990a). Multiple steps are required for the generation of a novel recombinant; first, the virus producer cell needs to be infected by more than one virus, the RNAs from the two proviruses have to be copackaged into the same virion, and template switching has to occur during reverse transcription to generate a chimeric DNA copy, which needs to integrate into the genome of the target cell. Lastly, this recombinant provirus needs to be able to generate replication-competent virus for the impact of the recombination event to be observed. For these reasons, factors that affect any of these steps can influence recombination. Currently, little is known about how frequently target cells in patients are infected by more than one HIV-1 (double infection). In culture, double infection occurs more frequently than expected from random events, in both T-cell lines and primary CD4$^+$ T cells (Dang et al. 2004). This result is at least partly attributable to the fact that some cells have more receptors/coreceptors, and are, therefore, more susceptible to HIV-1 infection (Chen et al. 2005). Double infection is increased when HIV-1 is transmitted via cell-mediated events because multiple viruses are passed from the donor cell to the target cells (Dang et al. 2004).

During HIV-1 assembly, Gag packages full-length genomic RNA in a dimer form (for details, see Sundquist and Kräusslich 2011). Hence, RNA partner selection occurs before the encapsidation of the RNA genomes; a major determinant for RNA partner selection is the dimerization initiation signal (DIS) located in the loop of stem loop 1 in the 5$'$ untranslated region of HIV-1 RNA (Chin et al. 2005; Moore et al. 2007; Chen et al. 2009). Most subtype B and D variants have GCGCGC in their DIS,

whereas most subtype A, C, F, and G variants have GTGCAC, although other sequences have been found (St Louis et al. 1998; Hussein et al. 2010). It is thought that the palindromic nature of the DIS promotes an intermolecular base pairing of the two RNAs that initiates RNA dimerization. Other sequences in the viral genome can also affect the frequencies of RNA heterodimerization, albeit with a milder effect than the DIS (Chin et al. 2007).

HIV-1 recombination rates have been measured using marker genes; these results indicated that recombination rates increase proportionally with the distances that separate the two alleles when the distance between the markers is less than 0.6 kb; the maximum possible recombination rate is reached when the two alleles are separated by 1.3 kb (Rhodes et al. 2003, 2005). Although recombination has been shown to occur throughout the HIV-1 genome, RNA structure may affect the frequency of recombination of certain regions (Galetto et al. 2004). Sequence homology can affect both the recombination rate and the distribution of the crossover junctions (Baird et al. 2006); for example, there is more frequent recombination between two copackaged HIV-1 RNAs from the same subtype than there is when the two copackaged HIV-1 RNAs are from different subtypes (Galli et al. 2010).

It has been proposed that RT switches to the copackaged RNA copy where there is a break in the RNA template; this is known as the copy-choice recombination model (Fig. 4A) (Coffin 1979). The original copy-choice model has been revised and renamed the dynamic copy-choice model (Hwang et al. 2001), which proposes that a balance between polymerase activity and RNase H activity of RT determines the stability of the association between the nascent DNA and the RNA template, and that a perturbation of this balance affects the recombination rate (Fig. 4B). If the balance is shifted toward greater RNase H activity relative to the polymerase activity, there is less extensive base-pairing between nascent DNA and RNA template, which promotes dissociation of the DNA–RNA complex and template switching (Hwang et al. 2001). Indeed, RT mutants that have

A Copy-choice
recombination model

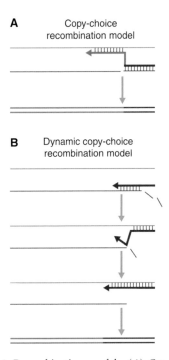

B Dynamic copy-choice
recombination model

Figure 4. Recombination models. (*A*) Copy-choice model, (*B*) dynamic copy-choice model. The *top* of *A* shows two RNA strands (thin orange and green lines). Minus-strand DNA synthesis uses the green RNA strand as a template; however, the green strand is nicked, causing DNA synthesis to switch to the orange RNA strand. This switch leads to the generation of a double-stranded DNA that is composed of sequences from both the green and the orange RNAs. In *B*, minus-strand DNA synthesis also uses the green RNA strand as a template. When the viral DNA is synthesized, RNase H degrades the green RNA strand, and the DNA that was copied from the green RNA strand can hybridize to the orange RNA (*middle* panels). This facilitates a transfer of the growing minus-strand DNA from the green to the orange strand, which results in the synthesis of a double-stranded DNA with sequences from both of the parental RNAs.

decreased RNase H activity relative to their polymerase activity show a reduction in template switching (Nikolenko et al. 2005, 2007). It has also been proposed that, for retroviruses in which plus-strand DNA synthesis is initiated at multiple sites, plus-strand DNA fragments can be annealed to a second minus-strand DNA synthesized using the copackaged RNA

as a template. After host DNA repair, a recombinant can be generated (Junghans et al. 1982). This model, which was originally proposed for recombination in avian retroviruses, requires that considerable portions of two minus-strand DNAs be synthesized. However, extensive minus-strand recombination will lead to the synthesis of a single minus-strand DNA. Currently, there is little data to suggest that plus-strand recombination occurs at a significant level during HIV-1 replication.

Last, for viral recombinants to establish themselves, they have to be able to replicate efficiently. Most template switching events use the complementarity between the nascent DNA and the acceptor template, and the recombination junctions are generally precise (Zhuang et al. 2002; Chin et al. 2008). However, because the resulting recombinants contain portions of the genomes of each parent, these sequences may or may not work together efficiently (Baird et al. 2006; Simon-Loriere et al. 2009; Galli et al. 2010). This issue is more pronounced when the two parental viruses are separated by a greater genetic distance (e.g., when the parental viruses are from different subtypes). For this reason, many newly generated intersubtype recombinants are eliminated by purifying selection during virus replication (Galli et al. 2010). The generation of a successful intersubtype recombinant faces multiple challenges: possible decreased efficiencies of RNA copackaging, relative inefficient template switching, and the impact of a decrease of replication fitness on the resulting recombinants. However, a conservative estimate suggests that >20% of the currently circulating HIV-1 variants are intersubtype recombinants (Hemelaar et al. 2006). This shows that recombination is a major force in the evolution of the HIV-1 population.

Frequent HIV-1 recombination reassorts existing mutations and increases genetic diversity in the viral population, thereby allowing the emergence of the variants that are best suited for any given environment. Recombination can combine drug-resistance mutations to produce multidrug-resistant variants (Kellam and Larder 1995; Moutouh et al. 1996); similarly, recombination can also produce

Cite this article as *Cold Spring Harb Perspect Med* doi: 10.1101/cshperspect.a006882

variants that can evade multiple challenges from the host's immune system (Streeck et al. 2008). Recombination can generate chimeras between two viruses from the same subtype, from different subtypes, or even from different groups. Therefore, recombination not only can affect the evolution of a viral population within an infected individual but can also affect HIV diversity worldwide. Given that different virus strains express different epitopes and vary in their susceptibility to antivirals, the increased diversity caused by recombination makes it more difficult to develop effective vaccines and antiviral regimens.

ANTI-RT COMPOUNDS AND RESISTANCE

Currently there are two types of antivirals targeting the reverse transcription process: nucleoside reverse transcriptase inhibitors (NRTIs), and nonnucleoside reverse transcriptase inhibitors (NNRTIs). NRTIs are given as prodrugs; after they are taken up by cells, and phosphorylated by the host cell enzymes, NRTIs can be incorporated into viral DNA by RT. NRTIs lack a 3′ hydroxyl group, thus, their incorporation blocks viral DNA synthesis. The common NRTI resistance mutations cause resistance by two general mechanisms: (1) mutations that reduce the incorporation of the NRTITP relative to the normal dNTPs, or (2) mutations that lead to a selective excision of the incorporated NRTIs by RT, unblocking the viral DNA. NNRTIs bind to RT and block the chemical step of DNA synthesis (see online Movie 3 at www.perspect ivesinmedicine.org); resistance mutations generally change the NNRTI-binding pocket in ways that make the binding of the NNRTIs less favorable. This topic will be described in more detail by Arts and Hazuda (2011).

CONCLUDING REMARKS

Reverse transcription and integration are the hallmarks of retroviruses; in this article, we provide an overview of RT and the reverse transcription process. The crystal structures of the HIV-1 RT, particularly those that reveal the structure of complexes with bound nucleic acids and incoming dNTPs, have allowed us to understand in molecular detail how the enzyme works. Structures of RT with bound anti-RT drugs have allowed us to better understand drug action and drug resistance. Complementary biochemical and genetic experiments have helped complete the picture. However, there is still much to be learned about the interactions of the various viral and cellular components that facilitate viral DNA synthesis in an infected cell and how the viral core is first converted into an RTC and then into a PIC. Reverse transcription has important genetic consequences—mutation and recombination provide the basis for the genetic diversity of HIV-1, which allows the emergence of viral strains that can escape the host's immune response and/or become resistant to drug treatment. Modern highly active antiretroviral therapy (HAART) therapy rests on the development and use of anti-AIDS drugs that target RT. However, the prevalence of drug-resistant HIV-1 strains makes it necessary to continue our efforts to develop more and better antiviral drugs. A better understanding of the complexities of the replication process, RT, and the reverse transcription pathway, can help us develop better ways to combat HIV-1.

ACKNOWLEDGMENTS

This work was supported by the Intramural Research Program of the National Institutes of Health, National Cancer Institute, Center for Cancer Research.

We thank K. Das, E. Arnold, K. Kirby, S. Sarafianos and A. Kane for help with the figures and K. Kirby and S. Sarafianos for help with movies.

REFERENCES

*Reference is also in this collection.

Abram ME, Ferris AL, Shao W, Alvord WG, Hughes SH. 2010. The nature, position and frequency of mutations made in a single-cycle of HIV-1 replication. *J Virol* **84:** 9864–9878.

Anderson JA, Bowman EH, Hu WS. 1998. Retroviral recombination rates do not increase linearly with marker distance and are limited by the size of the recombining subpopulation. *J Virol* **72:** 1195–1202.

Apolloni A, Meredith LW, Suhrbier A, Kiernan R, Harrich D. 2007. The HIV-1 Tat protein stimulates reverse transcription in vitro. *Curr HIV Res* **5:** 473–483.

Arhel NJ, Souquere-Besse S, Munier S, Souque P, Guadagnini S, Rutherford S, Prevost MC, Allen TD, Charneau P. 2007. HIV-1 DNA flap formation promotes uncoating of the pre-integration complex at the nuclear pore. *EMBO J* **26:** 3025–3037.

* Arts EJ, Hazuda DJ. 2011. HIV-1 antiretroviral drug therapy. *Cold Spring Harb Perspect Med* doi: 10.1101/cshperspect.a007161.

Baird HA, Gao Y, Galetto R, Lalonde M, Anthony RM, Giacomoni V, Abreha M, Destefano JJ, Negroni M, Arts EJ. 2006. Influence of sequence identity and unique breakpoints on the frequency of intersubtype HIV-1 recombination. *Retrovirology* **3:** 91.

Baltimore D. 1970. RNA-dependent DNA polymerase in virions of RNA tumour viruses. *Nature* **226:** 1209–1211.

Best S, Le Tissier P, Towers G, Stoye JP. 1996. Positional cloning of the mouse retrovirus restriction gene Fv1. *Nature* **382:** 826–829.

Bishop KN, Holmes RK, Sheehy AM, Davidson NO, Cho SJ, Malim MH. 2004a. Cytidine deamination of retroviral DNA by diverse APOBEC proteins. *Curr Biol* **14:** 1392–1396.

Bishop KN, Holmes RK, Sheehy AM, Malim MH. 2004b. APOBEC-mediated editing of viral RNA. *Science* **305:** 645.

Bishop KN, Holmes RK, Malim MH. 2006. Antiviral potency of APOBEC proteins does not correlate with cytidine deamination. *J Virol* **80:** 8450–8458.

Brennan G, Kozyrev Y, Hu SL. 2008. TRIMCyp expression in Old World primates *Macaca nemestrina* and *Macaca fascicularis*. *Proc Natl Acad Sci* **105:** 3569–3574.

Buckman JS, Bosche WJ, Gorelick RJ. 2003. Human immunodeficiency virus type 1 nucleocapsid Zn^{2+} fingers are required for efficient reverse transcription, initial integration processes, and protection of newly synthesized viral DNA. *J Virol* **77:** 1469–1480.

Butler SL, Hansen MS, Bushman FD. 2001. A quantitative assay for HIV DNA integration in vivo. *Nat Med* **7:** 631–634.

Carr JM, Coolen C, Davis AJ, Burrell CJ, Li P. 2008. Human immunodeficiency virus 1 (HIV-1) virion infectivity factor (Vif) is part of reverse transcription complexes and acts as an accessory factor for reverse transcription. *Virology* **372:** 147–156.

Charneau P, Alizon M, Clavel F. 1992. A second origin of DNA plus-strand synthesis is required for optimal human immunodeficiency virus replication. *J Virol* **66:** 2814–2820.

Chen R, Le Rouzic E, Kearney JA, Mansky LM, Benichou S. 2004. Vpr-mediated incorporation of UNG2 into HIV-1 particles is required to modulate the virus mutation rate and for replication in macrophages. *J Biol Chem* **279:** 28419–28425.

Chen J, Dang Q, Unutmaz D, Pathak VK, Maldarelli F, Powell D, Hu WS. 2005. Mechanisms of nonrandom human immunodeficiency virus type 1 infection and double infection: Preference in virus entry is important but is not the sole factor. *J Virol* **79:** 4140–4149.

Chen J, Nikolaitchik O, Singh J, Wright A, Bencsics CE, Coffin JM, Ni N, Lockett S, Pathak VK, Hu WS. 2009. High efficiency of HIV-1 genomic RNA packaging and heterozygote formation revealed by single virion analysis. *Proc Natl Acad Sci* **106:** 13535–13540.

Chin MP, Rhodes TD, Chen J, Fu W, Hu WS. 2005. Identification of a major restriction in HIV-1 intersubtype recombination. *Proc Natl Acad Sci* **102:** 9002–9007.

Chin MP, Chen J, Nikolaitchik OA, Hu WS. 2007. Molecular determinants of HIV-1 intersubtype recombination potential. *Virology* **363:** 437–446.

Chin MP, Lee SK, Chen J, Nikolaitchik OA, Powell DA, Fivash MJ Jr, Hu WS. 2008. Long-range recombination gradient between HIV-1 subtypes B and C variants caused by sequence differences in the dimerization initiation signal region. *J Mol Biol* **377:** 1324–1333.

Coffin JM. 1979. Structure, replication, and recombination of retrovirus genomes: Some unifying hypotheses. *J Gen Virol* **42:** 1–26.

Coffin JM. 1995. HIV population dynamics in vivo: implications for genetic variation, pathogenesis, and therapy. *Science* **267:** 483–489.

Colicelli J, Goff SP. 1985. Mutants and pseudorevertants of Moloney murine leukemia virus with alterations at the integration site. *Cell* **42:** 573–580.

Colicelli J, Goff SP. 1988. Sequence and spacing requirements of a retrovirus integration site. *J Mol Biol* **199:** 47–59.

* Craigie R, Bushman FD. 2011. HIV DNA integration. *Cold Spring Harb Perspect Med* doi: 10.1101/cshperspect.a006890.

Dang Q, Chen J, Unutmaz D, Coffin JM, Pathak VK, Powell D, KewalRamani VN, Maldarelli F, Hu WS. 2004. Nonrandom HIV-1 infection and double infection via direct and cell-mediated pathways. *Proc Natl Acad Sci* **101:** 632–637.

Das K, Bandwar RP, White KL, Feng JY, Sarafianos SG, Tuske S, Tu X, Clark AD Jr, Boyer PL, Hou X, et al. 2009. Structural basis for the role of the K65R mutation in HIV-1 reverse transcriptase polymerization, excision antagonism, and tenofovir resistance. *J Biol Chem* **284:** 35092–35100.

Dash C, Rausch JW, Le Grice SF. 2004. Using pyrrolodeoxycytosine to probe RNA/DNA hybrids containing the human immunodeficiency virus type-1 3′ polypurine tract. *Nucleic Acids Res* **32:** 1539–1547.

Dismuke DJ, Aiken C. 2006. Evidence for a functional link between uncoating of the human immunodeficiency virus type 1 core and nuclear import of the viral preintegration complex. *J Virol* **80:** 3712–3720.

Driscoll MD, Golinelli MP, Hughes SH. 2001. In vitro analysis of human immunodeficiency virus type 1 minus-strand strong-stop DNA synthesis and genomic RNA processing. *J Virol* **75:** 672–686.

Dunn LL, McWilliams MJ, Das K, Arnold E, Hughes SH. 2009. Mutations in the thumb allow human immunodeficiency virus type 1 reverse transcriptase to be cleaved by protease in virions. *J Virol* **83:** 12336–12344.

Dykes C, Balakrishnan M, Planelles V, Zhu Y, Bambara RA, Demeter LM. 2004. Identification of a preferred region for recombination and mutation in HIV-1 gag. *Virology* **326:** 262–279.

Elder JH, Lerner DL, Hasselkus-Light CS, Fontenot DJ, Hunter E, Luciw PA, Montelaro RC, Phillips TR. 1992. Distinct subsets of retroviruses encode dUTPase. *J Virol* **66:** 1791–1794.

Engelman A, Englund G, Orenstein JM, Martin MA, Craigie R. 1995. Multiple effects of mutations in human immunodeficiency virus type 1 integrase on viral replication. *J Virol* **69:** 2729–2736.

Esnouf R, Ren J, Ross C, Jones Y, Stammers D, Stuart D. 1995. Mechanism of inhibition of HIV-1 reverse transcriptase by non-nucleoside inhibitors. *Nat Struct Biol* **2:** 303–308.

Esposito D, Craigie R. 1998. Sequence specificity of viral end DNA binding by HIV-1 integrase reveals critical regions for protein-DNA interaction. *EMBO J* **17:** 5832–5843.

Fang L, Landau NR. 2007. Analysis of Vif-induced APOBEC3G degradation using an α-complementation assay. *Virology* **359:** 162–169.

Fassati A, Goff SP. 2001. Characterization of intracellular reverse transcription complexes of human immunodeficiency virus type 1. *J Virol* **75:** 3626–3635.

Feng YX, Copeland TD, Henderson LE, Gorelick RJ, Bosche WJ, Levin JG, Rein A. 1996. HIV-1 nucleocapsid protein induces "maturation" of dimeric retroviral RNA in vitro. *Proc Natl Acad Sci* **93:** 7577–7581.

Forshey BM, von Schwedler U, Sundquist WI, Aiken C. 2002. Formation of a human immunodeficiency virus type 1 core of optimal stability is crucial for viral replication. *J Virol* **76:** 5667–5677.

Frankel WN, Stoye JP, Taylor BA, Coffin JM. 1989. Genetic analysis of endogenous xenotropic murine leukemia viruses: Association with two common mouse mutations and the viral restriction locus Fv-1. *J Virol* **63:** 1763–1774.

Freed EO, Englund G, Maldarelli F, Martin MA. 1997. Phosphorylation of residue 131 of HIV-1 matrix is not required for macrophage infection. *Cell* **88:** 171–173; discussion 173–174.

Galetto R, Moumen A, Giacomoni V, Veron M, Charneau P, Negroni M. 2004. The structure of HIV-1 genomic RNA in the gp120 gene determines a recombination hot spot in vivo. *J Biol Chem* **279:** 36625–36632.

Gallay P, Swingler S, Aiken C, Trono D. 1995a. HIV-1 infection of nondividing cells: C-terminal tyrosine phosphorylation of the viral matrix protein is a key regulator. *Cell* **80:** 379–388.

Gallay P, Swingler S, Song J, Bushman F, Trono D. 1995b. HIV nuclear import is governed by the phosphotyrosine-mediated binding of matrix to the core domain of integrase. *Cell* **83:** 569–576.

Galli A, Kearney M, Nikolaitchik OA, Yu S, Chin MP, Maldarelli F, Coffin JM, Pathak VK, Hu WS. 2010. Patterns of human immunodeficiency virus type 1 recombination ex vivo provide evidence for coadaptation of distant sites, resulting in purifying selection for intersubtype recombinants during replication. *J Virol* **84:** 7651–7661.

Gao HQ, Boyer PL, Arnold E, Hughes SH. 1998. Effects of mutations in the polymerase domain on the polymerase, RNase H and strand transfer activities of human

immunodeficiency virus type 1 reverse transcriptase. *J Mol Biol* **277:** 559–572.

Ghosh M, Jacques PS, Rodgers DW, Ottman M, Darlix JL, Le Grice SF. 1996. Alterations to the primer grip of pp66 HIV-1 reverse transcriptase and their consequences for template-primer utilization. *Biochemistry* **35:** 8553–8562.

Ghosh M, Williams J, Powell MD, Levin JG, Le Grice SF. 1997. Mutating a conserved motif of the HIV-1 reverse transcriptase palm subdomain alters primer utilization. *Biochemistry* **36:** 5758–5768.

Golinelli MP, Hughes SH. 2003. Secondary structure in the nucleic acid affects the rate of HIV-1 nucleocapsid-mediated strand annealing. *Biochemistry* **42:** 8153–8162.

Halvas EK, Svarovskaia ES, Pathak VK. 2000. Role of murine leukemia virus reverse transcriptase deoxyribonucleoside triphosphate-binding site in retroviral replication and in vivo fidelity. *J Virol* **74:** 10349–10358.

Harrich D, Ulich C, Garcia-Martinez LF, Gaynor RB. 1997. Tat is required for efficient HIV-1 reverse transcription. *EMBO J* **16:** 1224–1235.

Harris RS, Bishop KN, Sheehy AM, Craig HM, Petersen-Mahrt SK, Watt IN, Neuberger MS, Malim MH. 2003. DNA deamination mediates innate immunity to retroviral infection. *Cell* **113:** 803–809.

Hemelaar J, Gouws E, Ghys PD, Osmanov S. 2006. Global and regional distribution of HIV-1 genetic subtypes and recombinants in 2004. *AIDS* **20:** W13–W23.

Henriet S, Sinck L, Bec G, Gorelick RJ, Marquet R, Paillart JC. 2007. Vif is a RNA chaperone that could temporally regulate RNA dimerization and the early steps of HIV-1 reverse transcription. *Nucleic Acids Res* **35:** 5141–5153.

Holmes RK, Koning FA, Bishop KN, Malim MH. 2007. APOBEC3F can inhibit the accumulation of HIV-1 reverse transcription products in the absence of hypermutation. Comparisons with APOBEC3G. *J Biol Chem* **282:** 2587–2595.

Houzet L, Morichaud Z, Didierlaurent L, Muriaux D, Darlix JL, Mougel M. 2008. Nucleocapsid mutations turn HIV-1 into a DNA-containing virus. *Nucleic Acids Res* **36:** 2311–2319.

Hsiou Y, Ding J, Das K, Clark AD Jr, Hughes SH, Arnold E. 1996. Structure of unliganded HIV-1 reverse transcriptase at 2.7 A resolution: Implications of conformational changes for polymerization and inhibition mechanisms. *Structure* **4:** 853–860.

Hsu TW, Taylor JM. 1982. Single-stranded regions on unintegrated avian retrovirus DNA. *J Virol* **44:** 47–53.

Hu WS, Temin HM. 1990a. Genetic consequences of packaging two RNA genomes in one retroviral particle: Pseudodiploidy and high rate of genetic recombination. *Proc Natl Acad Sci* **87:** 1556–1560.

Hu WS, Temin HM. 1990b. Retroviral recombination and reverse transcription. *Science* **250:** 1227–1233.

Huang Y, Wang J, Shalom A, Li Z, Khorchid A, Wainberg MA, Kleiman L. 1997. Primer tRNA3Lys on the viral genome exists in unextended and two-base extended forms within mature human immunodeficiency virus type 1. *J Virol* **71:** 726–728.

Huang H, Chopra R, Verdine GL, Harrison SC. 1998. Structure of a covalently trapped catalytic complex of HIV-1 reverse transcriptase: Implications for drug resistance. *Science* 282: 1669–1675.

Huang W, Gamarnik A, Limoli K, Petropoulos CJ, Whitcomb JM. 2003. Amino acid substitutions at position 190 of human immunodeficiency virus type 1 reverse transcriptase increase susceptibility to delavirdine and impair virus replication. *J Virol* 77: 1512–1523.

Hungnes O, Tjotta E, Grinde B. 1992. Mutations in the central polypurine tract of HIV-1 result in delayed replication. *Virology* 190: 440–442.

Hussein IT, Ni N, Galli A, Chen J, Moore MD, Hu WS. 2010. Delineation of the preferences and requirements of the human immunodeficiency virus type 1 dimerization initiation signal by using an in vivo cell-based selection approach. *J Virol* 84: 6866–6875.

Hwang CK, Svarovskaia ES, Pathak VK. 2001. Dynamic copy choice: Steady state between murine leukemia virus polymerase and polymerase-dependent RNase H activity determines frequency of in vivo template switching. *Proc Natl Acad Sci* 98: 12209–12214.

Iordanskiy S, Berro R, Altieri M, Kashanchi F, Bukrinsky M. 2006. Intracytoplasmic maturation of the human immunodeficiency virus type 1 reverse transcription complexes determines their capacity to integrate into chromatin. *Retrovirology* 3: 4.

Isel C, Lanchy JM, Le Grice SF, Ehresmann C, Ehresmann B, Marquet R. 1996. Specific initiation and switch to elongation of human immunodeficiency virus type 1 reverse transcription require the post-transcriptional modifications of primer tRNA3Lys. *EMBO J* 15: 917–924.

Jacobo-Molina A, Ding J, Nanni RG, Clark AD Jr, Lu X, Tantillo C, Williams RL, Kamer G, Ferris AL, Clark P, et al. 1993. Crystal structure of human immunodeficiency virus type 1 reverse transcriptase complexed with double-stranded DNA at 3.0 A resolution shows bent DNA. *Proc Natl Acad Sci* 90: 6320–6324.

Jetzt AE, Yu H, Klarmann GJ, Ron Y, Preston BD, Dougherty JP. 2000. High rate of recombination throughout the human immunodeficiency virus type 1 genome. *J Virol* 74: 1234–1240.

Julias JG, Pathak VK. 1998. Deoxyribonucleoside triphosphate pool imbalances in vivo are associated with an increased retroviral mutation rate. *J Virol* 72: 7941–7949.

Julias JG, Ferris AL, Boyer PL, Hughes SH. 2001. Replication of phenotypically mixed human immunodeficiency virus type 1 virions containing catalytically active and catalytically inactive reverse transcriptase. *J Virol* 75: 6537–6546.

Julias JG, McWilliams MJ, Sarafianos SG, Arnold E, Hughes SH. 2002. Mutations in the RNase H domain of HIV-1 reverse transcriptase affect the initiation of DNA synthesis and the specificity of RNase H cleavage in vivo. *Proc Natl Acad Sci* 99: 9515–9520.

Junghans RP, Boone LR, Skalka AM. 1982. Retroviral DNA H structures: Displacement-assimilation model of recombination. *Cell* 30: 53–62.

Kameoka M, Morgan M, Binette M, Russell RS, Rong L, Guo X, Mouland A, Kleiman L, Liang C, Wainberg MA. 2002. The Tat protein of human immunodeficiency virus type 1 (HIV-1) can promote placement of tRNA primer onto viral RNA and suppress later DNA polymerization in HIV-1 reverse transcription. *J Virol* 76: 3637–3645.

Keele BF, Giorgi EE, Salazar-Gonzalez JF, Decker JM, Pham KT, Salazar MG, Sun C, Grayson T, Wang S, Li H, et al. 2008. Identification and characterization of transmitted and early founder virus envelopes in primary HIV-1 infection. *Proc Natl Acad Sci* 105: 7552–7557.

Kellam P, Larder BA. 1995. Retroviral recombination can lead to linkage of reverse transcriptase mutations that confer increased zidovudine resistance. *J Virol* 69: 669–674.

Kim T, Mudry RA Jr, Rexrode CA 2nd, Pathak VK. 1996. Retroviral mutation rates and A-to-G hypermutations during different stages of retroviral replication. *J Virol* 70: 7594–7602.

Klarmann GJ, Yu H, Chen X, Dougherty JP, Preston BD. 1997. Discontinuous plus-strand DNA synthesis in human immunodeficiency virus type 1-infected cells and in a partially reconstituted cell-free system. *J Virol* 71: 9259–9269.

Kohlstaedt LA, Wang J, Friedman JM, Rice PA, Steitz TA. 1992. Crystal structure at 3.5 A resolution of HIV-1 reverse transcriptase complexed with an inhibitor. *Science* 256: 1783–1790.

Kung HJ, Fung YK, Majors JE, Bishop JM, Varmus HE. 1981. Synthesis of plus strands of retroviral DNA in cells infected with avian sarcoma virus and mouse mammary tumor virus. *J Virol* 37: 127–138.

Lanchy JM, Keith G, Le Grice SF, Ehresmann B, Ehresmann C, Marquet R. 1998. Contacts between reverse transcriptase and the primer strand govern the transition from initiation to elongation of HIV-1 reverse transcription. *J Biol Chem* 273: 24425–24432.

Lansdon EB, Samuel D, Lagpacan L, Brendza KM, White KL, Hung M, Liu X, Boojamra CG, Mackman RL, Cihlar T, et al. 2010. Visualizing the molecular interactions of a nucleotide analog, GS-9148, with HIV-1 reverse transcriptase-DNA complex. *J Mol Biol* 397: 967–978.

Leavitt AD, Robles G, Alesandro N, Varmus HE. 1996. Human immunodeficiency virus type 1 integrase mutants retain in vitro integrase activity yet fail to integrate viral DNA efficiently during infection. *J Virol* 70: 721–728.

Lee K, Ambrose Z, Martin TD, Oztop I, Mulky A, Julias JG, Vandegraaff N, Baumann JG, Wang R, Yuen W, et al. 2010. Flexible use of nuclear import pathways by HIV-1. *Cell Host Microbe* 7: 221–233.

Lerner DL, Wagaman PC, Phillips TR, Prospero-Garcia O, Henriksen SJ, Fox HS, Bloom FE, Elder JH. 1995. Increased mutation frequency of feline immunodeficiency virus lacking functional deoxyuridine-triphosphatase. *Proc Natl Acad Sci* 92: 7480–7484.

Levy DN, Aldrovandi GM, Kutsch O, Shaw GM. 2004. Dynamics of HIV-1 recombination in its natural target cells. *Proc Natl Acad Sci* 101: 4204–4209.

Liang C, Wainberg MA. 2002. The role of Tat in HIV-1 replication: An activator and/or a suppressor? *AIDS Rev* 4: 41–49.

Lightfoote MM, Coligan JE, Folks TM, Fauci AS, Martin MA, Venkatesan S. 1986. Structural characterization of reverse transcriptase and endonuclease polypeptides of the acquired immunodeficiency syndrome retrovirus. *J Virol* 60: 771–775.

Cite this article as *Cold Spring Harb Perspect Med* doi: 10.1101/cshperspect.a006882

Liu B, Yu X, Luo K, Yu Y, Yu XF. 2004. Influence of primate lentiviral Vif and proteasome inhibitors on human immunodeficiency virus type 1 virion packaging of APO-BEC3G. *J Virol* **78:** 2072–2081.

Lori F, di Marzo Veronese F, de Vico AL, Lusso P, Reitz MS Jr, Gallo RC. 1992. Viral DNA carried by human immunodeficiency virus type 1 virions. *J Virol* **66:** 5067–5074.

Lu X, Yu Q, Binder GK, Chen Z, Slepushkina T, Rossi J, Dropulic B. 2004. Antisense-mediated inhibition of human immunodeficiency virus (HIV) replication by use of an HIV type 1-based vector results in severely attenuated mutants incapable of developing resistance. *J Virol* **78:** 7079–7088.

Luo K, Xiao Z, Ehrlich E, Yu Y, Liu B, Zheng S, Yu XF. 2005. Primate lentiviral virion infectivity factors are substrate receptors that assemble with cullin 5-E3 ligase through a HCCH motif to suppress APOBEC3G. *Proc Natl Acad Sci* **102:** 11444–11449.

* Malim MH, Bieniasz PD. 2011. HIV restriction factors and mechanisms of evasion. *Cold Spring Harb Perspect Med* doi: 10.1101/cshperspect.a006940.

Mangeat B, Turelli P, Caron G, Friedli M, Perrin L, Trono D. 2003. Broad antiretroviral defence by human APO-BEC3G through lethal editing of nascent reverse transcripts. *Nature* **424:** 99–103.

Mansky LM, Temin HM. 1994. Lower mutation rate of bovine leukemia virus relative to that of spleen necrosis virus. *J Virol* **68:** 494–499.

Mansky LM, Temin HM. 1995. Lower in vivo mutation rate of human immunodeficiency virus type 1 than that predicted from the fidelity of purified reverse transcriptase. *J Virol* **69:** 5087–5094.

Mansky LM, Preveral S, Selig L, Benarous R, Benichou S. 2000. The interaction of vpr with uracil DNA glycosylase modulates the human immunodeficiency virus type 1 in vivo mutation rate. *J Virol* **74:** 7039–7047.

Mariani R, Chen D, Schrofelbauer B, Navarro F, Konig R, Bollman B, Munk C, Nymark-McMahon H, Landau NR. 2003. Species-specific exclusion of APOBEC3G from HIV-1 virions by Vif. *Cell* **114:** 21–31.

Masuda T, Planelles V, Krogstad P, Chen IS. 1995. Genetic analysis of human immunodeficiency virus type 1 integrase and the U3 att site: Unusual phenotype of mutants in the zinc finger-like domain. *J Virol* **69:** 6687–6696.

Mbisa JL, Barr R, Thomas JA, Vandegraaff N, Dorweiler IJ, Svarovskaia ES, Brown WL, Mansky LM, Gorelick RJ, Harris RS, et al. 2007. Human immunodeficiency virus type 1 cDNAs produced in the presence of APOBEC3G exhibit defects in plus-strand DNA transfer and integration. *J Virol* **81:** 7099–7110.

Mbisa JL, Delviks-Frankenberry KA, Thomas JA, Gorelick RJ, Pathak VK. 2009. Real-time PCR analysis of HIV-1 replication post-entry events. *Methods Mol Biol* **485:** 55–72.

Mbisa JL, Bu W, Pathak VK. 2010. APOBEC3F and APO-BEC3G inhibit HIV-1 DNA integration by different mechanisms. *J Virol* **84:** 5250–5259.

Miller MD, Wang B, Bushman FD. 1995. Human immunodeficiency virus type 1 preintegration complexes containing discontinuous plus strands are competent to integrate in vitro. *J Virol* **69:** 3938–3944.

Mizutani S, Boettiger D, Temin HM. 1970. A DNA-depenent DNA polymerase and a DNA endonuclease in virions of Rous sarcoma virus. *Nature* **228:** 424–427.

Moore MD, Fu W, Nikolaitchik O, Chen J, Ptak RG, Hu WS. 2007. Dimer initiation signal of human immunodeficiency virus type 1: Its role in partner selection during RNA copackaging and its effects on recombination. *J Virol* **81:** 4002–4011.

Moutouh L, Corbeil J, Richman DD. 1996. Recombination leads to the rapid emergence of HIV-1 dually resistant mutants under selective drug pressure. *Proc Natl Acad Sci* **93:** 6106–6111.

Mukherjee R, Plesa G, Sherrill-Mix S, Richardson MW, Riley JL, Bushman FD. HIV sequence variation associated with env antisense adoptive T-cell therapy in the hNSG mouse model. *Mol Ther* **18:** 803–811.

Nermut MV, Fassati A. 2003. Structural analyses of purified human immunodeficiency virus type 1 reverse transcription complexes. *J Virol* **77:** 8196–8206.

Newman RM, Hall L, Kirmaier A, Pozzi LA, Pery E, Farzan M, O'Neil SP, Johnson W. 2008. Evolution of a TRIM5-CypA splice isoform in old world monkeys. *PLoS Pathog* **4:** e1000003. doi: 10.1371/journal.ppat.1000003.

Nikolenko GN, Palmer S, Maldarelli F, Mellors JW, Coffin JM, Pathak VK. 2005. Mechanism for nucleoside analog-mediated abrogation of HIV-1 replication: Balance between RNase H activity and nucleotide excision. *Proc Natl Acad Sci* **102:** 2093–2098.

Nikolenko GN, Delviks-Frankenberry KA, Palmer S, Maldarelli F, Fivash MJ Jr, Coffin JM, Pathak VK. 2007. Mutations in the connection domain of HIV-1 reverse transcriptase increase 3′-azido-3′-deoxythymidine resistance. *Proc Natl Acad Sci* **104:** 317–322.

Nisole S, Lynch C, Stoye JP, Yap MW. 2004. A Trim5-cyclophilin A fusion protein found in owl monkey kidney cells can restrict HIV-1. *Proc Natl Acad Sci* **101:** 13324–13328.

Oh J, Chang KW, Wierzchoslawski R, Alvord WG, Hughes SH. 2008. Rous sarcoma virus (RSV) integration in vivo: A CA dinucleotide is not required in U3, and RSV linear DNA does not autointegrate. *J Virol* **82:** 503–512.

Onafuwa A, An W, Robson ND, Telesnitsky A. 2003. Human immunodeficiency virus type 1 genetic recombination is more frequent than that of Moloney murine leukemia virus despite similar template switching rates. *J Virol* **77:** 4577–4587.

O'Neil PK, Sun G, Yu H, Ron Y, Dougherty JP, Preston BD. 2002. Mutational analysis of HIV-1 long terminal repeats to explore the relative contribution of reverse transcriptase and RNA polymerase II to viral mutagenesis. *J Biol Chem* **277:** 38053–38061.

Palaniappan C, Wisniewski M, Jacques PS, Le Grice SF, Fay PJ, Bambara RA. 1997. Mutations within the primer grip region of HIV-1 reverse transcriptase result in loss of RNase H function. *J Biol Chem* **272:** 11157–11164.

Panganiban AT, Fiore D. 1988. Ordered interstrand and intrastrand DNA transfer during reverse transcription. *Science* **241:** 1064–1069.

Pathak VK, Temin HM. 1990a. Broad spectrum of in vivo forward mutations, hypermutations, and mutational hotspots in a retroviral shuttle vector after a single replication cycle: Deletions and deletions with insertions. *Proc Natl Acad Sci* **87**: 6024–6028.

Pathak VK, Temin HM. 1990b. Broad spectrum of in vivo forward mutations, hypermutations, and mutational hotspots in a retroviral shuttle vector after a single replication cycle: Substitutions, frameshifts, and hypermutations. *Proc Natl Acad Sci* **87**: 6019–6023.

Powell MD, Ghosh M, Jacques PS, Howard KJ, Le Grice SF, Levin JG. 1997. Alanine-scanning mutations in the "primer grip" of pp66 HIV-1 reverse transcriptase result in selective loss of RNA priming activity. *J Biol Chem* **272**: 13262–13269.

Powell MD, Beard WA, Bebenek K, Howard KJ, Le Grice SF, Darden TA, Kunkel TA, Wilson SH, Levin JG. 1999. Residues in the αH and αI helices of the HIV-1 reverse transcriptase thumb subdomain required for the specificity of RNase H-catalyzed removal of the polypurine tract primer. *J Biol Chem* **274**: 19885–19893.

Pullen KA, Ishimoto LK, Champoux JJ. 1992. Incomplete removal of the RNA primer for minus-strand DNA synthesis by human immunodeficiency virus type 1 reverse transcriptase. *J Virol* **66**: 367–373.

Pullen KA, Rattray AJ, Champoux JJ. 1993. The sequence features important for plus strand priming by human immunodeficiency virus type 1 reverse transcriptase. *J Biol Chem* **268**: 6221–6227.

Purohit V, Roques BP, Kim B, Bambara RA. 2007. Mechanisms that prevent template inactivation by HIV-1 reverse transcriptase RNase H cleavages. *J Biol Chem* **282**: 12598–12609.

Qi M, Yang R, Aiken C. 2008. Cyclophilin A-dependent restriction of human immunodeficiency virus type 1 capsid mutants for infection of nondividing cells. *J Virol* **82**: 12001–12008.

Rausch JW, Lener D, Miller JT, Julias JG, Hughes SH, Le Grice SF. 2002. Altering the RNase H primer grip of human immunodeficiency virus reverse transcriptase modifies cleavage specificity. *Biochemistry* **41**: 4856–4865.

Rhodes T, Wargo H, Hu WS. 2003. High rates of human immunodeficiency virus type 1 recombination: Near-random segregation of markers one kilobase apart in one round of viral replication. *J Virol* **77**: 11193–11200.

Rhodes TD, Nikolaitchik O, Chen J, Powell D, Hu WS. 2005. Genetic recombination of human immunodeficiency virus type 1 in one round of viral replication: Effects of genetic distance, target cells, accessory genes, and lack of high negative interference in crossover events. *J Virol* **79**: 1666–1677.

Robertson DL, Sharp PM, McCutchan FE, Hahn BH. 1995. Recombination in HIV-1. *Nature* **374**: 124–126.

Rodgers DW, Gamblin SJ, Harris BA, Ray S, Culp JS, Hellmig B, Woolf DJ, Debouck C, Harrison SC. 1995. The structure of unliganded reverse transcriptase from the human immunodeficiency virus type 1. *Proc Natl Acad Sci* **92**: 1222–1226.

Russell RA, Pathak VK. 2007. Identification of two distinct human immunodeficiency virus type 1 Vif determinants critical for interactions with human APOBEC3G and APOBEC3F. *J Virol* **81**: 8201–8210.

Sarafianos SG, Das K, Tantillo C, Clark AD Jr, Ding J, Whitcomb JM, Boyer PL, Hughes SH, Arnold E. 2001. Crystal structure of HIV-1 reverse transcriptase in complex with a polypurine tract RNA: DNA. *EMBO J* **20**: 1449–1461.

Sarafianos SG, Clark AD Jr, Das K, Tuske S, Birktoft JJ, Ilankumaran P, Ramesha AR, Sayer JM, Jerina DM, Boyer PL, et al. 2002. Structures of HIV-1 reverse transcriptase with pre- and post-translocation AZTMP-terminated DNA. *EMBO J* **21**: 6614–6624.

Sarafianos SG, Marchand B, Das K, Himmel DM, Parniak MA, Hughes SH, Arnold E. 2009. Structure and function of HIV-1 reverse transcriptase: Molecular mechanisms of polymerization and inhibition. *J Mol Biol* **385**: 693–713.

Sawyer SL, Emerman M, Malik HS. 2004. Ancient adaptive evolution of the primate antiviral DNA-editing enzyme APOBEC3G. *PLoS Biol* **2**: pE275.

Sayah DM, Sokolskaja E, Berthoux L, Luban J. 2004. Cyclophilin A retrotransposition into TRIM5 explains owl monkey resistance to HIV-1. *Nature* **430**: 569–573.

Schrofelbauer B, Chen D, Landau NR. 2004. A single amino acid of APOBEC3G controls its species-specific interaction with virion infectivity factor (Vif). *Proc Natl Acad Sci* **101**: 3927–3932.

Schrofelbauer B, Yu Q, Zeitlin SG, Landau NR. 2005. Human immunodeficiency virus type 1 Vpr induces the degradation of the UNG and SMUG uracil-DNA glycosylases. *J Virol* **79**: 10978–10987.

Selig L, Benichou S, Rogel ME, Wu LI, Vodicka MA, Sire J, Benarous R, Emerman M. 1997. Uracil DNA glycosylase specifically interacts with Vpr of both human immunodeficiency virus type 1 and simian immunodeficiency virus of sooty mangabeys, but binding does not correlate with cell cycle arrest. *J Virol* **71**: 4842–4846.

Sevilya Z, Loya S, Hughes SH, Hizi A. 2001. The ribonuclease H activity of the reverse transcriptases of human immunodeficiency viruses type 1 and type 2 is affected by the thumb subdomain of the small protein subunits. *J Mol Biol* **311**: 957–971.

Sevilya Z, Loya S, Adir N, Hizi A. 2003. The ribonuclease H activity of the reverse transcriptases of human immunodeficiency viruses type 1 and type 2 is modulated by residue 294 of the small subunit. *Nucleic Acids Res* **31**: 1481–1487.

Simon-Loriere E, Galetto R, Hamoudi M, Archer J, Lefeuvre P, Martin DP, Robertson DL, Negroni M. 2009. Molecular mechanisms of recombination restriction in the envelope gene of the human immunodeficiency virus. *PLoS Pathog* **5**: e1000418. doi: 10.1371/journal.ppat.1000418.

Skalka AM, Goff SP, ed. 1993. *Reverse transcriptase*. Cold Spring Harbor Laboratory Press, Plainview, NY.

Smith JS, Roth MJ. 1992. Specificity of human immunodeficiency virus-1 reverse transcriptase-associated ribonuclease H in removal of the minus-strand primer, tRNA(Lys3). *J Biol Chem* **267**: 15071–15079.

St Louis DC, Gotte D, Sanders-Buell E, Ritchey DW, Salminen MO, Carr JK, McCutchan FE. 1998. Infectious molecular clones with the nonhomologous dimer initiation sequences found in different subtypes of human immunodeficiency virus type 1 can recombine and initiate a spreading infection in vitro. *J Virol* **72**: 3991–3998.

Streeck H, Li B, Poon AF, Schneidewind A, Gladden AD, Power KA, Daskalakis D, Bazner S, Zuniga R, Brander C, et al. 2008. Immune-driven recombination and loss of control after HIV superinfection. *J Exp Med* **205**: 1789–1796.

Stremlau M, Owens CM, Perron MJ, Kiessling M, Autissier P, Sodroski J. 2004. The cytoplasmic body component TRIM5α restricts HIV-1 infection in Old World monkeys. *Nature* **427**: 848–853.

Stremlau M, Perron M, Lee M, Li Y, Song B, Javanbakht H, Diaz-Griffero F, Anderson DJ, Sundquist WI, Sodroski J. 2006. Specific recognition and accelerated uncoating of retroviral capsids by the TRIM5α restriction factor. *Proc Natl Acad Sci* **103**: 5514–5519.

Summers J, Mason WS. 1982. Replication of the genome of a hepatitis B–like virus by reverse transcription of an RNA intermediate. *Cell* **29**: 403–415.

* Sundquist WI, Kräusslich H-G. 2011. HIV assembly, budding, and maturation. *Cold Spring Harb Perspect Med* doi: 10.1101/cshperspect.a006924.

Swanstrom R, Varmus HE, Bishop JM. 1981. The terminal redundancy of the retrovirus genome facilitates chain elongation by reverse transcriptase. *J Biol Chem* **256**: 1115–1121.

Takehisa J, Kraus MH, Decker JM, Li Y, Keele BF, Bibollet-Ruche F, Zammit KP, Weng Z, Santiago ML, Kamenya S, et al. 2007. Generation of infectious molecular clones of simian immunodeficiency virus from fecal consensus sequences of wild chimpanzees. *J Virol* **81**: 7463–7475.

Tantillo C, Ding J, Jacobo-Molina A, Nanni RG, Boyer PL, Hughes SH, Pauwels R, Andries K, Janssen PA, Arnold E. 1994. Locations of anti-AIDS drug binding sites and resistance mutations in the three-dimensional structure of HIV-1 reverse transcriptase. Implications for mechanisms of drug inhibition and resistance. *J Mol Biol* **243**: 369–387.

Telesnitsky A, Goff SP. 1993. Two defective forms of reverse transcriptase can complement to restore retroviral infectivity. *EMBO J* **12**: 4433–4438.

Telesnitsky A, Goff G.P. 1997. Reverse transcriptase and the generation of retroviral DNA. In *Retroviruses* (ed. Coffin JM, Hughes SH, Varmus HE), pp. 121–160. Cold Spring Harbor Laboratory Press, Cold Spring Harbor, NY.

Temin HM. 1964. Homology between RNA from Rous sarcoma virus and DNA from Rous sarcoma virus-infected cells. *Proc Natl Acad Sci* **52**: 323–329.

Thomas JA, Gorelick RJ. 2008. Nucleocapsid protein function in early infection processes. *Virus Res* **134**: 39–63.

Thomas JA, Gagliardi TD, Alvord WG, Lubomirski M, Bosche WJ, Gorelick RJ. 2006. Human immunodeficiency virus type 1 nucleocapsid zinc-finger mutations cause defects in reverse transcription and integration. *Virology* **353**: 41–51.

Thomas DC, Voronin YA, Nikolenko GN, Chen J, Hu WS, Pathak VK. 2007. Determination of the ex vivo rates of human immunodeficiency virus type 1 reverse transcription by using novel strand-specific amplification analysis. *J Virol* **81**: 4798–4807.

Thomas JA, Bosche WJ, Shatzer TL, Johnson DG, Gorelick RJ. 2008. Mutations in human immunodeficiency virus type 1 nucleocapsid protein zinc fingers cause premature reverse transcription. *J Virol* **82**: 9318–9328.

Trono D. 1992. Partial reverse transcripts in virions from human immunodeficiency and murine leukemia viruses. *J Virol* **66**: 4893–4900.

Tu X, Das K, Han Q, Bauman JD, Clark AD Jr, Hou X, Frenkel YV, Gaffney BL, Jones RA, Boyer PL, et al. 2010. Structural basis of HIV-1 resistance to AZT by excision. *Nat Struct Mol Biol* **17**: 1202–1209.

Tuske S, Sarafianos SG, Clark AD Jr, Ding J, Naeger LK, White KL, Miller MD, Gibbs CS, Boyer PL, Clark P, et al. 2004. Structures of HIV-1 RT-DNA complexes before and after incorporation of the anti-AIDS drug tenofovir. *Nat Struct Mol Biol* **11**: 469–474.

van Wamel JL, Berkhout B. 1998. The first strand transfer during HIV-1 reverse transcription can occur either intramolecularly or intermolecularly. *Virology* **244**: 245–251.

Virgen CA, Kratovac Z, Bieniasz PD, Hatziioannou T. 2008. Independent genesis of chimeric TRIM5-cyclophilin proteins in two primate species. *Proc Natl Acad Sci* **105**: 3563–3568.

Wagaman PC, Hasselkus-Light CS, Henson M, Lerner DL, Phillips TR, Elder JH. 1993. Molecular cloning and characterization of deoxyuridine triphosphatase from feline immunodeficiency virus (FIV). *Virology* **196**: 451–457.

Wang J, Bambara RA, Demeter LM, Dykes C. 2010. Reduced fitness in cell culture of HIV-1 with nonnucleoside reverse transcriptase inhibitor-resistant mutations correlates with relative levels of reverse transcriptase content and RNase H activity in virions. *J Virol* **84**: 9377–9389.

Whitcomb JM, Kumar R, Hughes SH. 1990. Sequence of the circle junction of human immunodeficiency virus type 1: Implications for reverse transcription and integration. *J Virol* **64**: 4903–4906.

* Wilen CB, Tilton JC, Doms RW. 2011. HIV: Cell binding and entry. *Cold Spring Harb Perspect Med* doi: 10.1101/cshperspect.a006866.

Wilson SJ, Webb BL, Ylinen LM, Verschoor E, Heeney JL, Towers GJ. 2008. Independent evolution of an antiviral TRIMCyp in rhesus macaques. *Proc Natl Acad Sci* **105**: 3557–3562.

Wu X, Liu H, Xiao H, Conway JA, Hehl E, Kalpana GV, Prasad V, Kappes JC. 1999. Human immunodeficiency virus type 1 integrase protein promotes reverse transcription through specific interactions with the nucleoprotein reverse transcription complex. *J Virol* **73**: 2126–2135.

Wu X, Anderson JL, Campbell EM, Joseph AM, Hope TJ. 2006. Proteasome inhibitors uncouple rhesus TRIM5alpha restriction of HIV-1 reverse transcription and infection. *Proc Natl Acad Sci* **103**: 7465–7470.

Yamashita M, Emerman M. 2004. Capsid is a dominant determinant of retrovirus infectivity in nondividing cells. *J Virol* **78**: 5670–5678.

Yi-Brunozzi HY, Le Grice SF. 2005. Investigating HIV-1 polypurine tract geometry via targeted insertion of abasic lesions in the (−)-DNA template and (+)-RNA primer. *J Biol Chem* **280**: 20154–20162.

Yu SF, Baldwin DN, Gwynn SR, Yendapalli S, Linial ML. 1996. Human foamy virus replication: A pathway distinct from that of retroviruses and hepadnaviruses. *Science* **271**: 1579–1582.

Yu H, Jetzt AE, Ron Y, Preston BD, Dougherty JP. 1998. The nature of human immunodeficiency virus type 1 strand transfers. *J Biol Chem* **273:** 28384–28391.

Yu SF, Sullivan MD, Linial ML. 1999. Evidence that the human foamy virus genome is DNA. *J Virol* **73:** 1565–1572.

Yu X, Yu Y, Liu B, Luo K, Kong W, Mao P, Yu XF. 2003. Induction of APOBEC3G ubiquitination and degradation by an HIV-1 Vif-Cul5-SCF complex. *Science* **302:** 1056–1060.

Yu Q, Chen D, Konig R, Mariani R, Unutmaz D, Landau NR. 2004. APOBEC3B and APOBEC3C are potent inhibitors of simian immunodeficiency virus replication. *J Biol Chem* **279:** 53379–53386.

Zack JA, Haislip AM, Krogstad P, Chen IS. 1992. Incompletely reverse-transcribed human immunodeficiency virus type 1 genomes in quiescent cells can function as intermediates in the retroviral life cycle. *J Virol* **66:** 1717–1725.

Zhang WH, Hwang CK, Hu WS, Gorelick RJ, Pathak VK. 2002. Zinc finger domain of murine leukemia virus nucleocapsid protein enhances the rate of viral DNA synthesis in vivo. *J Virol* **76:** 7473–7484.

Zheng YH, Irwin D, Kurosu T, Tokunaga K, Sata T, Peterlin BM. 2004. Human APOBEC3F is another host factor that blocks human immunodeficiency virus type 1 replication. *J Virol* **78:** 6073–6076.

Zhu J, Cunningham JM. 1993. Minus-strand DNA is present within murine type C ecotropic retroviruses prior to infection. *J Virol* **67:** 2385–2388.

Zhuang J, Jetzt AE, Sun G, Yu H, Klarmann G, Ron Y, Preston BD, Dougherty JP. 2002. Human immunodeficiency virus type 1 recombination: Rate, fidelity, and putative hot spots. *J Virol* **76:** 11273–11282.

HIV DNA Integration

Robert Craigie[1] and Frederic D. Bushman[2]

[1]Molecular Virology Section, NIDDK, National Institutes of Health, Bethesda, Maryland 20892-0560

[2]University of Pennsylvania School of Medicine, Philadelphia, Pennsylvania 19104-6076

Correspondence: bobc@helix.nih.gov; bushman@mail.med.upenn.edu

Retroviruses are distinguished from other viruses by two characteristic steps in the viral replication cycle. The first is reverse transcription, which results in the production of a double-stranded DNA copy of the viral RNA genome, and the second is integration, which results in covalent attachment of the DNA copy to host cell DNA. The initial catalytic steps of the integration reaction are performed by the virus-encoded integrase (IN) protein. The chemistry of the IN-mediated DNA breaking and joining steps is well worked out, and structures of IN-DNA complexes have now clarified how the overall complex assembles. Methods developed during these studies were adapted for identification of IN inhibitors, which received FDA approval for use in patients in 2007. At the chromosomal level, HIV integration is strongly favored in active transcription units, which may promote efficient viral gene expression after integration. HIV IN binds to the cellular factor LEDGF/p75, which promotes efficient infection and tethers IN to favored target sites. The HIV integration machinery must also interact with many additional host factors during infection, including nuclear trafficking and pore proteins during nuclear entry, histones during initial target capture, and DNA repair proteins during completion of the DNA joining steps. Models for some of the molecular mechanisms involved have been proposed, but important details remain to be clarified.

Integration of a DNA copy of the viral genome into a host cell chromosome is an essential step in the retroviral replication cycle (Varmus et al. 1989; Coffin et al. 1997). Once integrated, the proviral DNA is replicated along with cellular DNA during cycles of cell division, as with any cellular gene. The provirus serves as the template for transcription of viral RNAs. Some viral RNAs are translated to yield the viral proteins, whereas a portion of the full-length viral RNA is recruited to serve as genomic RNA in progeny virions.

Integration is mediated by the virus-encoded IN protein, which is introduced into cells during infection along with reverse transcriptase, the viral RNA, and other proteins as a part of the viral core. After the viral DNA is synthesized by reverse transcription in the cytoplasm, it stably associates with IN and other proteins as a high-molecular-weight nucleoprotein complex that is later transported to the nucleus for subsequent integration. The mechanism of integration has been extensively studied and the basic biochemistry is quite well understood. Recently, structural studies of the active nucleoprotein complexes of IN bound to viral DNA (intasomes) have also made great progress.

Integration occurs precisely at the termini of the viral DNA but integration can take place at many locations in the host genome. Most positions in chromosomal DNA can serve as integration acceptor sites, but there are distinct regional preferences that differ among groups of retroviruses. Some of these preferences appear to involve chromatin-associated factors that also interact with IN. Understanding targeting is especially important because of the application of retroviral insertion in gene therapy, where adverse events have been associated with integration of retroviral vectors near proto-oncogenes (Howe et al. 2008; Hacein-Bey-Abina et al. 2010). The integration step is also the target of FDA-approved inhibitors (discussed by Arts and Hazuda 2011).

This review takes advantage of data from both HIV and model retroviruses because important advances came from both. First we review the evolution of models for retroviral integration, then our present picture of the biochemical steps of the integration pathway, and lastly integration in the cellular context.

HISTORY

Howard Temin's provirus hypothesis holds that the viral RNA introduced into cells during infection becomes converted to DNA by reverse transcription. Viral DNA is then integrated into the host genome. This explained how cells transformed by Rous sarcoma virus (RSV) could stably maintain the transformed state in the absence of viral replication (Temin 1976). Temin's hypothesis was vindicated by the discovery of reverse transcriptase (Baltimore 1970; Temin and Mizutani 1970) and the physical characterization of integrated viral DNA in the genome of infected cells (Weiss et al. 1984; Coffin et al. 1997).

How does the viral DNA become integrated? The first sighting of the protein we now call IN was as a nuclease activity associated with cores of avian retroviral particles (Grandgenett et al. 1978). Genetic studies later defined the IN coding region and the ends of the viral DNA to be important for integration (Donehower and Varmus 1984; Panganiban and Temin 1984;

Schwartzberg et al. 1984). In these studies, cloned copies of the viral DNA were modified in vitro, and the modified DNA was then introduced into cells by transfection, allowing production of viral particles. The phenotype of the mutant viral stocks could then be characterized by infecting fresh cells. Mutation of the $3'$ region of the pol gene resulted in viruses that were able to enter cells and carry out reverse transcription normally, but failed to integrate the reverse-transcribed DNA. This region of pol encodes a protein, now called IN. Another group of mutants with an essentially identical phenotype mapped to the ends of the viral DNA at the sites that become joined to host DNA on integration (Panganiban and Temin 1983; Murphy and Goff 1992; Murphy et al. 1993; Du et al. 1997). Studies of the synthesis of the viral proteins showed that IN is cleaved from the gag-pol polyprotein precursor by the virus-encoded protease to yield an independent protein. These studies indicated that IN likely acts on the ends of the viral DNA but could not reveal the mechanisms involved. The presence of circular forms of viral DNA in infected cells initially suggested that retroviral integration might proceed via a circular DNA intermediate, as was known to be the case for bacteriophage λ, but this idea was later refuted by biochemical studies, as described below.

KEY ADVANCES

Biochemical Studies of the Integration Mechanism

Demonstration that PICs Isolated from Infected Cells Are Competent for Integration In Vitro

The first biochemical studies of the integration reaction used viral replication intermediates purified from infected cells as a source of the integration machinery. Studies were initially performed using Moloney murine leukemia virus as a model system (Brown et al. 1987), soon followed by similar experiments with HIV (Ellison et al. 1990; Farnet and Haseltine 1990). The viral DNA made by reverse transcription within the cytoplasm was found to be part of a large nucleoprotein complex, the preintegration complex (PIC), which is derived from the core of the infecting virion (Bowerman et al. 1989). On incubation in vitro with a target DNA in the presence of a Mg^{2+} ion, the viral DNA within PICs efficiently integrates into the target DNA.

Analysis of the structure of the integration intermediates produced in these reactions (Fujiwara and Mizuuchi 1988; Brown et al. 1989) revealed the DNA cutting and joining steps of integration (Fig. 1). In the first step (3' end processing), two nucleotides are in most cases removed from each 3' end of the blunt-ended linear viral DNA. The resulting 3' ends of the viral DNA in all cases terminate with the conserved CA-3' sequence. In the second step (DNA-strand transfer), these 3' ends attack a pair of phosphodiester bonds on opposite strands of the target DNA, across the major groove. In the resulting integration intermediate the 3' ends of the viral DNA are covalently joined to the target DNA. The single-strand gaps and the two-nucleotide overhang at the 5' ends of the viral DNA must be repaired by cellular enzymes to complete integration. The sites

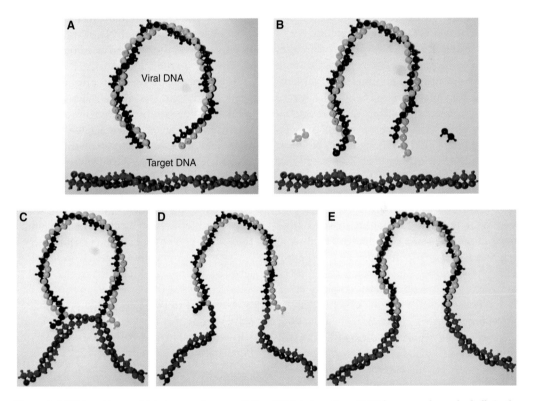

Figure 1. DNA breaking and joining reactions mediating DNA integration. DNA bases are shown by balls in the snap-together models, although the HIV DNA (10 kb) and the cellular chromosome (megabases) are not shown to scale. (A) The linear blunt-ended viral DNA (green and yellow) and target DNA (blue and red). (B) 3' end processing. Two nucleotides are in most cases removed from each 3' end of the viral DNA. (C) The 3' ends generated by 3' processing attack a pair of phosphodiester bonds in the target DNA. The sites of attack on the two target DNA strands are separated by five nucleotides in the case of HIV-1. The 3' ends of the viral DNA are joined to the 5' ends of the target DNA at the site of integration. The 5' ends of the viral DNA are not joined to target DNA in the intermediate. (D) Completion of provirus formation requires removal of the two unpaired bases at the 5' ends of the viral DNA, filling in the single-strand gaps between viral and target DNA and ligation of the 5' ends of the viral DNA to target DNA. IN catalyzes the 3' processing and DNA-strand transfer steps to form the integration intermediate. Subsequent steps are thought to be catalyzed by cellular enzymes. (E) The integrated provirus.

of joining on the two target DNA strands are separated by four nucleotides in the case of murine leukemia virus (MLV), resulting in a four-nucleotide duplication of target DNA flanking the integrated proviral DNA. For HIV, the sites are five base pairs apart, resulting in a five base-pair duplication.

Development of In Vitro Integration Assays Using Purified IN

PICs contain many viral and cellular proteins, as shown, for example, by immunoprecipitation assays (Farnet and Haseltine 1991b; Bukrinsky et al. 1993; Li et al. 2001). However, possible roles of these proteins in integration are difficult to assess, in part because the low abundance of PICs in extracts of infected cells complicates analysis. Simpler in vitro systems were therefore required to identify the minimal set of proteins required for integration. The next advance was the discovery that cell extracts containing PICs or detergent-disrupted virus particles could promote in vitro integration of linear plasmid DNA with terminal sequences that mimic the viral DNA ends (Fujiwara and Craigie 1989), albeit with low efficiency. It was later shown using the same assay that HIV-1 IN protein alone was able to support in vitro DNA integration (Bushman et al. 1990). The discovery that IN alone is sufficient to promote not only 3′ end processing (Katzman et al. 1989), but also in vitro integration (Craigie et al. 1990; Katz et al. 1990) of oligonucleotides matching the viral DNA ends opened the door to detailed biochemical studies of the integration reaction. The high efficiency of these simplified assay systems allowed the products to be directly detected by physical methods such as gel electrophoresis and provided the foundation for the development of high-throughput screens that later led to the development of IN inhibitors (Arts and Hazuda 2011).

Although highly efficient, the oligonucleotide assay system initially lacked the full fidelity of integration in vivo. Many reaction products resulted from integration of only a single viral DNA end into one strand of target DNA, rather than concerted integration of pairs of viral DNA

ends. Subsequent improvements have enabled in vitro concerted integration of both ends by HIV-1 IN, although still with somewhat low efficiency (Hindmarsh et al. 1999; Sinha et al. 2002; Li and Craigie 2005; Sinha and Grandgenett 2005). Under these conditions, IN forms a stable synaptic complex (SSC) with a pair of viral DNA ends that is an intermediate on the reaction pathway (Li et al. 2006), and in which 3′ end processing occurs on both viral DNA ends.

Integration Mechanism: Similarities to DNA Transposition

A mechanistic connection between DNA transposition and retroviral DNA integration was first suggested by the short duplication of target DNA sequences that flank integrated proviruses and integrated transposons (Ju and Skalka 1980; Shimotohno et al. 1980). In the case of DNA transposons, this duplication was known to arise by staggered cleavage of the target DNA and subsequent repair of the resulting single-strand gaps between transposon DNA and target DNA. In contrast to many DNA recombinases, these transposases splice the transposon DNA into the new target DNA by a one-step transesterification mechanism. In this reaction, the 3′ end of the viral or transposon DNA acts as a nucleophile in the attack on the target DNA backbone, breaking the target DNA and joining the viral DNA all in a single step. This is in contrast to two-step reactions involving a covalent intermediate between DNA and protein as is found in the tyrosine and serine recombinase families (reviewed in Mizuuchi 1992).

The oligonucleotide integration assay allowed the number of reaction steps to be counted in the HIV-1 DNA-strand transfer reaction using stereochemically marked phosphate atoms in target DNA substrates. The result was that integration proceeded in a single step, implicating a direct transesterification (Engelman et al. 1991), as had been previously shown for bacteriophage Mu transposase (Mizuuchi and Adzuma 1991). Reactions involving a covalent protein-DNA intermediate, in contrast, require two steps. Subsequent structural studies revealed that retroviral IN and transposases of the D,D,E family, which includes bacteriophage Mu and Tn5, share a common active site organization in which the conserved acidic amino acids bind two divalent metal atoms, confirming earlier suggestions based on mutagenesis data (Engelman and Craigie 1992; Kulkosky et al. 1992; van Gent et al. 1992; Leavitt et al. 1993).

Structural Studies of IN

In early studies, partial proteolysis of HIV IN protein revealed three structurally distinct domains (Engelman and Craigie 1992), and expression of these domains individually yielded proteins that were amenable to study. The central core domain, which from biochemical studies was expected to contain the active site (Engelman and Craigie 1992; Kulkosky et al. 1992; van Gent et al. 1992; Bushman et al. 1993; Leavitt et al. 1993), remained poorly soluble but a single amino change markedly improved solubility (Jenkins et al. 1995) and enabled crystallization. The structure (Dyda et al. 1994) confirmed the triad of acidic residues, the D,D-35-E motif identified from mutagenesis studies, to be key active site residues. The structure of the ASV IN core domain was determined soon after (Bujacz et al. 1995). The IN core structures belong to the superfamily of a functionally diverse group of polynucleotidyl transferases, which includes prokaryotic and HIV RNase H enzymes, Holiday junction resolvase RuvC, and the phage Mu transposase catalytic domain (Rice et al. 1996), collectively referred to as "RNase H superfamily."

The structures of the isolated catalytic domains provided no clue as to how IN interacts with DNA to position a pair of active sites with the correct spacing for DNA-strand transfer. Both HIV and ASV IN fragments crystallized as dimers, with similar dimer interfaces. However, the pairs of active sites were on opposite faces of the roughly spherical complexes with a separation incompatible with the five-nucleotide spacing of the sites of catalysis in the target DNA.

Structures were also solved for the amino-terminal and carboxy-terminal domains. The

amino-terminal domain, refolded in the presence of zinc, was initially solved by nuclear magnetic resonance (NMR) (Cai et al. 1997; Eijkelenboom et al. 1997). It consists of a bundle of α helices with coordination of a single Zn^{2+} ion stabilizing the structure. The carboxy-terminal domain structure was also solved by NMR, revealing an SH3-like β barrel (Eijkelenboom et al. 1995; Lodi et al. 1995). Subsequently, two-domain structures, amino-terminal plus catalytic domain or catalytic domain plus carboxy-terminal domain, were solved for several retroviral INs (reviewed in Chiu and Davies 2004; Jaskolski et al. 2009). On the basis of these partial structures, numerous models were proposed for the IN complex with viral DNA ends. However, the differences in the arrangement of domains between the various partial structures indicated that structures including viral DNA would be required to understand the organization of the active complex.

Determination of the structure of prototype foamy virus (PFV) IN in complex with viral DNA (Hare et al. 2010; Maertens et al. 2010) provided a major breakthrough in understanding IN function. The PFV intasome is comprised of a homotetramer of IN assembled on viral DNA ends (see online Movie 1 at www.perspectivesinmedicine.org). Within it, the IN domains have essentially the same structures as observed in the isolated domains previously. The tetramer has a dimer-of-dimers architecture, where the individual dimers are formed via the canonical catalytic domain dimerization interface. The tetramerization interface includes contacts between the amino-terminal domain from one monomer and the catalytic core domain of the opposing dimer. Isomorphous domain-domain interfaces have been observed in crystal structures containing two-domain constructs of HIV-1 and visna virus INs (Wang et al. 2001; Hare et al. 2009). However, the conformation of the tetramerization interface and mutual orientation of the catalytic domain dimers could not have been predicted based on existing partial structures. The tetramer is held together not only by protein–protein contacts but also protein–DNA interactions that together bury more than 10,000 Å^2

of molecular surface (Fig. 2). Two of the monomers extend across the center of symmetry in the complex, thus tightly linking the two halves. The intertwined nature of protein–protein and protein–DNA contacts is reminiscent of the structures of Tn5 and Mos1 transpososomes (Davies et al. 2000; Richardson et al. 2009). The PFV structure also confirmed that the IN active site binds a pair of divalent metal cations (Fig. 2D). A notable feature of the PFV intasome crystals is that amino- and carboxy-terminal domains of the IN subunits that do not participate in the tetramerization interface are disordered. The functions of these four domains remain to be determined.

Crystallization of the PFV intasome also provided critical new information on the mechanism of action of IN inhibitors such as raltegravir. Structures of intasome/drug complexes revealed that not only does raltegravir binding block access of target DNA to the active site, but also displaces the 3′ end of the viral DNA from the active site, thereby disrupting catalysis.

The active site region of PFV and HIV-1 IN are sufficiently similar that interactions of HIV-1 IN with inhibitors can be modeled based on the PFV structure (Krishnan et al. 2010). However, some resistance mutations map to regions that are dissimilar between the two proteins, so structures of the HIV-1 intasome will ultimately be required to fully understand the mechanism of drug resistance.

The PFV intasome presents a surface that can accommodate target DNA without any significant conformational changes, suggesting that the viral nucleoprotein complex does not undergo major structural rearrangements between 3′ processing and DNA-strand transfer. However, because the active sites of the intasome are separated by some 25 Å, a significant deformation of target DNA is required for the active sites to access the scissile phosphodiesters (Fig. 3; see online Movie 2 at www.perspectivesinmedicine.org) (Maertens et al. 2010). The target DNA-containing PFV structures explained the early observations that retroviral INs are

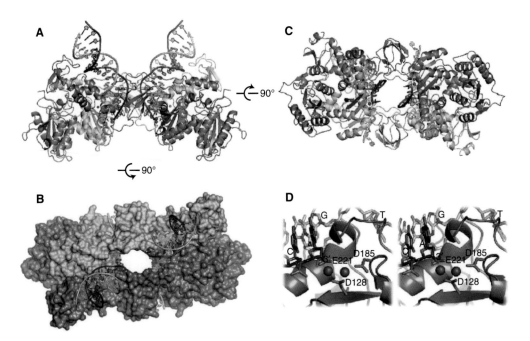

Figure 2. Structure of the complex of PFV IN and viral DNA (Hare et al. 2010). (*A,B*) Two views of the IN-DNA complex. (*C*) Top view of the complex, colored to emphasize that two monomers cross the center of symmetry and link up the two halves of the complex. (*D*) Stereo pair showing the structure of the active site, including two magnesium atoms bound to the three conserved acidic amino acids that comprise the active site.

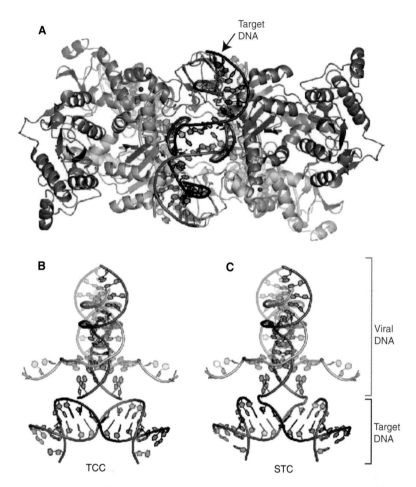

Figure 3. Target DNA capture by the complex of PFV IN and viral DNA (Maertens et al. 2010). (*A*) Overview of the complex, showing target DNA in blue. (*B,C*) Two views of the DNA only, highlighting formation of the initial covalent link between the viral DNA 3′ ends and target DNA 5′ ends.

biased toward deformed target DNA (Pryciak and Varmus 1992; Pruss et al. 1994; Katz et al. 1998).

Integration in the Cellular Context

Nuclear Localization

HIV can infect nondividing cells, implying that PICs must cross the nuclear membrane to carry out integration. The mechanism of nuclear localization has not been fully clarified (for recent reviews see Fassati 2006; Suzuki and Craigie 2007). Briefly, several known components of PICs exhibit nuclear localization prop-

erties when fused to a polypeptide that does not normally localize to the nucleus. However, mutating these determinants individually in the context of the PIC does not abolish nuclear localization. Proposed viral determinants include HIV-1 MA, CA, and IN proteins, and in addition, the three-stranded DNA flap structure generated at the central polypurine tract.

Several genome-wide screens have been performed to identify genes that when reduced in dosage using small interfering RNA (siRNA), diminish HIV infection, and several of the identified host factors have been mapped to the integration part of the viral replication cycle (Brass et al. 2008; Konig et al. 2008; Zhou et al. 2008).

Among these, several nuclear pore proteins appear to also be important not only for nuclear localization but also for efficient integration after nuclear entry, suggesting possible coupling of nuclear translocation and integration. Some cellular proteins implicated as important include transportin 3 (product of the TNPO3 gene) and Nup358 (product of the RANBP2 gene). Surprisingly, an amino acid substitution in the CA protein (N74D) was able to abrogate sensitivity to a knockdown of TNPO3, but created sensitivities to knockdown of several other nuclear pore proteins (Lee et al. 2010). These data suggest that PICs may interact with specific nuclear pore proteins during nuclear transit, and that there may be multiple redundant pathways accessible to PICs for nuclear entry.

Multiple Fates of the Viral DNA in the Nucleus

After the HIV PIC enters the nucleus, the viral DNA must become integrated into chromosomal DNA of the host for productive infection to proceed. However, the viral DNA can also undergo several circularization reactions that do not support subsequent replication and represent dead ends for the virus (Farnet and Haseltine 1991a). The two ends of the viral DNA can be ligated to each other, probably following dissociation of the PIC proteins, to yield 2-long long terminal repeat (LTR) circles. Inactivation of the host cell nonhomologous DNA end-joining (NHEJ) components Ku70/80, ligase IV, and XRCC4 blocks 2-LTR circle formation, implicating that these factors are involved in the circularization reaction (Li et al. 2001; Jeanson et al. 2002). Circles with one LTR copy can also be detected. These can be formed either by recombination between the LTRs within the nucleus, possibly involving action of the cellular MRN complex (Mre11, Rad50, and NBS1) (Kilzer et al. 2003), or as stalled products of reverse transcription that failed to complete the final steps of strand displacement synthesis (Hu and Hughes 2011). In addition, the viral DNA can use itself as an integration target, resulting in either circles with an inverted segment, or pairs of smaller circles, depending on whether each 3′ end joins initially to the same or a different DNA strand (Shoemaker et al. 1980; Farnet and Haseltine 1991a). The cellular DNA condensing protein BAF can block autointegration in vitro, suggesting that tight packaging of the viral DNA in a protein complex may protect against autointegration (Lee and Craigie 1998).

Integration Target Site Selection

Once in the nucleus, integration requires capture of host cell DNA sequences by viral PICs and completion of the chemical steps of integration. The nature of favored and disfavored target sites for retroviral integration has been the topic of close study (for reviews see Varmus et al. 1989; Coffin et al. 1997; Ciuffi and Bushman 2006). Early analysis of the DNA sequences at junctions between proviral DNA and host DNA showed that host cell sequences at the point of integration differed among proviral isolates, indicating that the integration reaction was not highly sequence specific, although close analysis subsequently showed weakly conserved sequences at target sites (Varmus et al. 1989; Stevens and Griffith 1994; Coffin et al. 1997; Holman and Coffin 2005; Wu et al. 2005; Berry et al. 2006). Early studies of gammaretroviruses suggested that integration might be favored near DNAse I hypersensitive sites in vertebrate genomes, indicating a possible association with open chromatin (Varmus et al. 1989; Coffin et al. 1997).

The development of methods for studying integration in vitro using PICs or purified IN allowed target site selection to be analyzed in reconstituted reactions. Following the idea that chromatin packing could obstruct integration, DNA templates wrapped in nucleosomes were tested as in vitro integration targets, which surprisingly showed that integration was actually favored in nucleosomal DNA (Pryciak and Varmus 1992; Pryciak et al. 1992; Pruss et al. 1994). Mapping of favored sites indicated that sharp DNA bends in the nucleosome structure were particularly favored targets (Pruss et al. 1994). Several studies have shown that DNA distortion can promote integration (Bushman and Craigie 1992; Bor et al. 1995; Katz et al. 1998), perhaps because completing the reaction cycle requires DNA distortion, so that kinking the DNA on the nucleosome may lower the activation energy. Consistent with these observations, the target DNA in the PFV target capture and strand transfer complexes is significantly deformed to position the scissile phosphates close to the two active sites (Maertens et al. 2010). The deformation is more drastic than the relatively smooth curvature of DNA on the nucleosome surface, suggesting that some remodeling of nucleosome structure may be required to facilitate integration. In addition, the presence of several sequence-specific DNA-binding proteins on integration targets was shown to obstruct integration in vitro (Pryciak and Varmus 1992; Bor et al. 1995), probably by steric occlusion.

With the completion of the draft human genome sequence in 2001 (Lander 2001; Venter 2001), it became possible to study integration target site selection genome wide using high-throughput DNA sequencing. In the first such study, 524 sites of HIV integration were mapped after acute infection of the T-cell line SupT1 and the relationship with genomic annotation analyzed (Schroder et al. 2002). This experiment revealed that HIV favored integration within transcription units quite strongly. The relationship with gene activity was then probed by transcriptional profiling analysis of the SupT1 cells, revealing that active transcription units were particularly strongly favored for integration. In the human genome, many genomic features are correlated with each other, and this complicates identifying the primary determinants of integration targeting. Active transcription units are associated with regions of high G/C content, high gene density, high CpG island density, short introns, high frequencies of Alu repeats, low frequencies of LINE repeats, and characteristic epigenetic modifications. These features are also associated with high frequencies of HIV integration (Fig. 4). Subsequently, HIV integration site selection has been studied after acute infection in many cell types (Wu et al. 2003; Mitchell et al. 2004; Barr et al. 2006; Berry et al. 2006; Ciuffi et al. 2006; Wang et al. 2007; Brady et al. 2009a), and the favoring of integration in active transcription units has been seen in all cases except in the presence of artificially engineered tethering factors (described below).

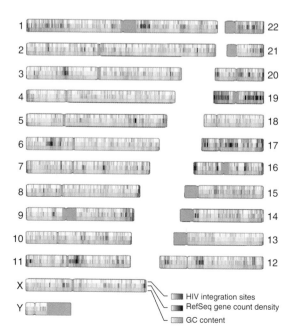

Figure 4. HIV integration site distributions on the human chromosomes (Wang et al. 2007). The human chromosomes are numbered at the sides of the diagram. HIV integration sites (20,000 total) are shown in green, gene density is shown in blue (measured as the count of RefSeq genes in a 500-kb interval), and the G/C content is shown in orange (measured in 500-kb intervals). The gray coloring indicates regions of centromeric repeats that have not been sequenced.

Why did HIV evolve to favor integration in active transcription units? Studies of many types of integrating genomic parasites show that their targeting preferences have evolved to optimize their ability to persist in their hosts and leave progeny. The yeast Ty retrotransposons, for example, must coexist with their hosts indefinitely, and they have evolved to integrate into benign genomic locations that do not harm the yeast cell (reviewed in Bushman 2001; Craig et al. 2002). For HIV, infected T cells typically have a half-life of only a day or two before cells are killed by the cellular immune system or by the toxicity of infection (Perelson et al. 1996). Thus HIV has only a limited time for the production of progeny. Recent studies show that integration within transcription units is usually favorable for efficient transcription (Jordan et al. 2001; Lewinski et al. 2005), potentially explaining the targeting preference.

Other retroviruses, however, show different favored target sites in chromosomes. The gammaretroviruses strongly favor integration at transcription start sites (Wu et al. 2003). Other retroviral genera show no strong preferences, with integration only slightly favored in transcription units or gene 5′ ends (Mitchell et al. 2004; Narezkina et al. 2004; Brady et al. 2009b).

These observations argue against the simplest version of the open chromatin model for targeting—if relative exposure of the target DNA was the only determinant of integration site selection, then how could the retroviruses be so different from each other? Considerable evidence now indicates that a simple tethering model explains HIV integration targeting. Another type of model could invoke the different modes of nuclear entry. The gammaretroviruses require mitosis for efficient infection (Roe et al. 1993), whereas HIV can infect nondividing cells, raising the possibility that PICs from the two types of retroviruses are encountering chromatin at different points in the cell cycle. If the chromatin is in different states at these stages, differences in integration targeting could result. Studies of HIV integration in two models of cells arrested at G1 showed favored integration in active transcription units for both (Barr et al. 2006; Ciuffi et al. 2006), but more

data on integration at other points in the cell cycle is needed to explore this model more fully.

An IN "swap" experiment implicated IN as a dominant determinant of integration targeting. In this study, the IN region of the gammaretrovirus MLV was substituted for that of HIV, and integration target site selection monitored (Lewinski et al. 2006). The chimeric virus showed favored integration near transcription start sites and reduced integration along the length of transcription units, thus resembling MLV and differing from HIV. Thus MLV IN was a dominant determinant of MLV-like integration in this context.

With the development of "next-generation" sequencing methods, it has become possible to generate much larger collections of integration sites for analysis. In parallel, new kinds of genome-wide annotation have become available thanks to "ChIP-seq" and other methods based on massively parallel sequencing for genome-wide mapping of bound proteins, sites of histone modification, and DNA methylation. Analysis of 40,000 sites of HIV integration in the Jurkat T-cell line has indicated that integration commonly occurs on the outer surface of DNA wrapped on nucleosomes in vivo (Wang et al. 2007), as was suggested from earlier experiments in vitro. The distributions of a wide variety of histone posttranslational modifications have been associated with distributions of HIV integration sites, indicating that marks characteristic of active transcription units are favorable (e.g., H3K4me1 and me2, H3K27me1, and H3K36me3), whereas those characteristic of intergenic regions or inactive genes are unfavorable (e.g., H3K9me2 and me3, and H3K27me2 and me3).

Extensive mapping of both HIV integration sites and genomic annotation allows extraction of the genomic features most strongly directing integration targeting. Logistic regression or machine-learning methods can be used to compare integration site distributions to random control models, and then variable selection schemes can be used to identify the most strongly associated forms of annotation. For HIV, the genomic features that best allow discrimination of HIV integration sites from random sites included the local sequence at the point of integration, G/C content, gene density, and DNAse I cleavage site density (Berry et al. 2006; Wang et al. 2007), the latter three being associated with transcriptionally active regions of the genome.

LEDGF/p75, a Host Cell Factor Affecting HIV Integration

HIV takes advantage of host cell factors to allow efficient integration and optimize integration target site selection. Of particular importance is the host cell protein LEDGF/p75 (product of the PSIP1 gene), which both boosts the efficiency of integration and mediates targeting to active transcription units.

LEDGF/p75 was first identified as a transcriptional mediator protein that promoted activator-dependent transcription in vitro (Ge et al. 1998). Subsequently, several groups used affinity-based screens to identify cellular proteins that bound to HIV IN and thereby identified LEDGF/p75 as a tight binder (Cherepanov et al. 2003; Turlure et al. 2004; Emiliani et al. 2005). Analysis of the LEDGF/p75 protein showed that it contains a PWWP chromatin-binding domain at the amino terminus, an A/T hook domain likely involved in DNA binding, a nuclear localization signal, and a carboxy-terminal domain that bound tightly to IN. Imaging studies showed that LEDGF/p75 could be visualized bound to condensed chromosomes at mitosis, and in the presence of LEDGF/p75, IN would accumulate on chromatin as well, suggesting that LEDGF/p75 could tether IN to chromatin (Maertens et al. 2003).

Initially it was unclear whether LEDGF/p75 was important for efficient HIV infection, but experiments eventually showed that even trace amounts of LEDGF/p75 were sufficient to promote HIV replication. Once LEDGF/p75 was efficiently depleted or knocked out, a substantial reduction in infectivity was detected. Furthermore, mapping the level of the block showed inhibition selectively at the integration step (Llano et al. 2006b; Shun et al. 2007). Quantitative polymerase chain reaction (PCR)

methods, in addition to documenting reduced provirus formation, also showed that the formation of 2-LTR circles was actually increased by LEDGF/p75 depletion (Llano et al. 2006b). An increase in 2-LTR circles is also seen during infection in the presence of IN inhibitors (Arts and Hazuda 2011), likely because inhibition of integration within the nucleus provides more viral cDNA substrate to the circularization reaction, boosting circle formation. Further studies showed that still greater reductions in HIV infection efficiency could be accomplished by overexpression of the LEDFGF/p75 IN binding domain fragment by itself (De Rijck et al. 2006; Llano et al. 2006a), further supporting the importance of the LEDGF/p75-IN interaction in vivo.

Mapping of integration site distributions in the presence of LEDGF/p75 knockdowns showed that much of the targeting to transcription units was lost when LEDGF/p75 was depleted, and restoration of LEDGF/p75 rescued proper targeting, directly implicating LEDGF/p75 in the targeting mechanism (Ciuffi et al. 2005; Marshall et al. 2007; Shun et al. 2007). Binding sites for LEDGF/p75 have been mapped on chromosomes experimentally for the ENCODE regions, which comprises about 1% of the human genome. Well-supported LEDGF/p75 binding sites ("LEDGF islands") were found to lie preferentially in transcription units, paralleling the observed preference of these locations for HIV integration (De Rijck et al. 2010). Several additional aspects of HIV target site selection were also affected by the LEDGF/p75 knockdown. Integration was more favored in regions of higher G/C content in the knockdown, and integration near CpG islands was increased. Overall, the data support a simple tethering model, in which LEDGF/p75 binds to HIV IN and simultaneously to chromatin at active transcription units, thereby directing integration to these locations.

Subsequent studies of HIV integration targeting in different cell types showed that the proportion of integration sites in transcription units differed among cell types, providing another angle for studying LEDGF/p75 function (Marshall et al. 2007). LEDGF expression levels were compared over these same cell types, which revealed that higher LEDGF/p75 expression correlated with a greater proportion of integration sites in transcription units. These findings support the importance of LEDGF/p75 for integration targeting in primary cells that were not subjected to harsh manipulations such as siRNA treatment or gene deletion.

The tethering model has received strong support from studies of artificial derivatives of LEDGF/p75, in which substitution of the LEDGF/p75 chromatin-binding region, comprised of the PWWP domain and a pair of A/T hooks (Llano et al. 2006b; Turlure et al. 2006) for alternative chromatin-binding domains allowed clear-cut retargeting of integration (Ferris et al. 2010; Gijsbers et al. 2010; Silvers et al. 2010). One dramatic example took advantage of the HP1/Cbx protein, which directs binding to sites of histone H3K9 di- and trimethylation. Substitution of the HP1/Cbx binding unit for the LEDGF/p75 chromatin-binding domain yielded a fusion protein that, when expressed in cells depleted for wild-type LEDGF/p75, retargeted integration to sites of H3K9 di- and trimethylation. Because this histone modification is enriched outside of transcription units, integration was reprogrammed to favor regions outside of genes. This provided strong support for the idea that LEDGF/p75 tethers integration complexes to target sites in vivo. Use of such fusions provides a possible means of controlling integration targeting for use in gene therapy applications, where it is desirable to target integration away from cancer-related genes to avoid insertional activation.

Might the IN-LEDGF/p75 interaction provide a target for small molecule therapy to inhibit HIV infection? X-ray crystallography studies have defined the interaction surface between the IN catalytic domain dimer and the LEDGF/p75 IN binding domain, defining the target for potential inhibitors (Cherepanov et al. 2005). An early study of small molecules binding to IN identified a binding site at the dimer interface in the catalytic domain (Molteni et al. 2001), and this later turned out to be the interaction site for LEDGF/p75 (Cherepanov et al. 2005). A more recent structure-based design effort yielded potent inhibitors that bind this site, block the IN-LEDGF/p75 interaction in vitro, and inhibit HIV replication in vivo (Christ et al. 2010). Thus the IN-LEDGF/p75 interaction appears to be a promising target for therapeutic inhibitors of HIV replication.

Why Does HIV Infection Not Cause Insertional Activation of Proto-Oncogenes?

Many studies document that retroviral infection in animal models can be associated with activation of proto-oncogenes and cancer, but remarkably this is never seen with HIV. HIV infection and AIDS are associated with elevated risks for several cancers, but in no case do the transformed cells harbor integrated proviruses, ruling out the known mechanisms of insertional activation. This issue is of considerable importance because HIV-based vectors are increasingly used in gene therapy, in part because of this observation. Possible explanations include (1) HIV vpr arrests the cell cycle, (2) HIV env expression is cytotoxic, (3) HIV infects terminally differentiated cells that consequently have limited proliferative potential, (4) cells expressing HIV proteins are quickly killed off by cytotoxic T cells, and (5) integration targeting is not optimal for transformation. However, in a recent lentiviral gene therapy trial that successfully treated β thalassemia major in one human subject, a cell clone expanded that contained an integrated vector in the proto-oncogene HMGA2 (Cavazzana-Calvo et al. 2010). This finding raises questions about possible effects of HIV integration on cell growth that may be more subtle than overt transformation.

Other Host Factors Affecting HIV Integration

A variety of additional host proteins are implicated in HIV integration. Following the completion of reverse transcription, the viral cDNA is likely to become coated by DNA-binding proteins, some of which may be contributed by the host cell. Product formation in integration reactions in vitro can be increased by the addition of proteins that assist assembly of protein/DNA complexes by altering DNA conformation ("architectural" DNA-binding proteins). However, multiple proteins show such activity in integration reactions in vitro, and it is unclear which are most important biologically. Proteins with reported stimulatory activity include LEDGF/p75, BAF, HMGA, HMGB, Ini-1, YY1, and the viral NC protein. Some of these proteins might contribute to PIC function by coating and condensing the viral DNA, thereby assisting the assembly of the viral nucleoprotein complexes. Several cellular chromatin proteins have also been suggested to influence integration including Ini-1, EED, SUV39H1, HP1γ, and others (for reviews of some of this work see Greene and Peterlin 2002; Bushman et al. 2005; Suzuki and Craigie 2007).

Cellular DNA repair proteins likely support the final steps of integration, in which DNA gaps at host-virus DNA junctions are processed and repaired. Collections of well-known host cell DNA repair enzymes can process model DNA substrates containing such gaps in vitro, but so far the enzymes most relevant in vivo during infection have not been identified (Yoder and Bushman 2000). The cellular double-strand break repair proteins Ku, ligase IV, and XRCC4, responsible for forming 2-LTR circles, have also been proposed to be important in supporting infection (Daniel et al. 1999), although this has been controversial (Baekelandt et al. 2000; Li et al. 2001). In several viral and transposon systems, DNA repair enzymes have also been suggested to inhibit replication, and additional DNA repair pathways are reportedly inhibitory for HIV-1 replication (Yoder et al. 2006).

Additional Roles for IN in HIV Biology

Studies of IN mutants have implicated IN protein in viral replication functions in addition to catalyzing DNA cutting and joining reactions (see Engelman et al. 1999 for review and references). Most amino acid substitutions within the IN active site residues selectively eliminate its catalytic activities without affecting other steps of the replication cycle—these have been termed "class I" mutants. However, amino acid substitutions in other parts of the IN protein, termed "class II" mutants, can have more pleiotropic effects, including disruption of correct core assembly in viral particles and impairment of reverse transcription, implicating IN in particle assembly and structure. Most deletions of IN result in a class II phenotype, suggesting that IN may be important for proper assembly of the viral core (Bukovsky and Gottlinger 1996; Dar et al. 2009).

CONCLUSIONS AND NEW RESEARCH AREAS

Progress in the integration field provides a classic example of how science ought to work. Early studies clarified the steps of the retroviral replication cycle and identified the IN protein. After the discovery of HIV, the field established assays for purified IN protein, which then provided the basis for small molecule screens to identify lead inhibitors active against virus. An enormous effort within the pharmaceutical industry then succeeded in turning early-stage inhibitors into pharmaceutical products. Full FDA approval of the first IN inhibitor was obtained in 2007. Now efforts turn to (1) developing inhibitors active against drug-resistant viruses and obligate IN cofactors, (2) understanding structures in more detail, and (3) understanding the integration system in its full cellular context.

ACKNOWLEDGMENTS

We thank Peter Cherepanov, Alan Engelman, and members of the Bushman and Craigie laboratories for helpful comments on the manuscript.

REFERENCES

*Reference is also in this collection.

* Arts EJ, Hazuda DJ. 2011. HIV-1 antiretroviral drug therapy. *Cold Spring Harb Perspect Med* doi: 10.1101/cshperspect. a007161.

Baekelandt V, Claeys A, Cherepanov P, De Clercq E, De Strooper B, Nuttin B, Debyser Z. 2000. DNA-dependent protein kinase is not required for efficient lentivirus integration. *J Virol* **74:** 11278–11285.

Baltimore D. 1970. RNA-dependent DNA polymerase in virions of RNA tumor viruses. *Nature* **226:** 1209–1211.

Barr SD, Ciuffi A, Leipzig J, Shinn P, Ecker JR, Bushman FD. 2006. HIV integration site selection: Targeting in macrophages and the effects of different routes of viral entry. *Mol Ther* **14:** 218–225.

Berry C, Hannenhalli S, Leipzig J, Bushman FD. 2006. Selection of target sites for mobile DNA integration in the human genome. *PLoS Comput Biol* **2:** e157. doi: 10.1371/journal.pcbi.0020157.

Bor YC, Bushman F, Orgel L. 1995. In vitro integration of human immunodeficiency virus type 1 cDNA into targets containing protein-induced bends. *Proc Natl Acad Sci* **92:** 10334–10338.

Bowerman B, Brown PO, Bishop JM, Varmus HE. 1989. A nucleoprotein complex mediates the integration of retroviral DNA. *Genes Dev* **3:** 469–478.

Brady T, Agosto LM, Malani N, Berry CC, O'Doherty U, Bushman F. 2009a. HIV integration site distributions in resting and activated CD4+ T cells infected in culture. *AIDS* **23:** 1461–1471.

Brady T, Lee YN, Ronen K, Malani N, Berry CC, Bieniasz PD, Bushman FD. 2009b. Integration target site selection by a resurrected human endogenous retrovirus. *Genes Dev* **23:** 633–642.

Brass AL, Dykxhoorn DM, Benita Y, Yan N, Engelman A, Xavier RJ, Lieberman J, Elledge SJ. 2008. Identification of host proteins required for HIV infection through a functional genomic screen. *Science* **319:** 921–926.

Brown PO, Bowerman B, Varmus HE, Bishop JM. 1987. Correct integration of retroviral DNA in vitro. *Cell* **49:** 347–356.

Brown PO, Bowerman B, Varmus HE, Bishop JM. 1989. Retroviral integration: Structure of the initial covalent complex and its precursor, and a role for the viral IN protein. *Proc Natl Acad Sci* **86:** 2525–2529.

Bujacz G, Jaskolski M, Alexandratos J, Wlodawer A, Merkel G, Katz RA, Skalka AM. 1995. High-resolution structure of the catalytic domain of avian sarcoma virus integrase. *J Mol Biol* **253:** 333–346.

Bukovsky A, Gottlinger H. 1996. Lack of integrase can markedly affect human immunodeficiency virus type 1 particle production in the presence of an active viral protease. *J Virol* **70:** 6820–6825.

Bukrinsky MI, Sharova N, McDonald TL, Pushkarskaya T, Tarpley GW, Stevenson M. 1993. Association of integrase, matrix, and reverse transcriptase antigens of human immunodeficiency virus type 1 with viral nucleic acids following acute infection. *Proc Natl Acad Sci* **90:** 6125–6129.

Bushman FD. 2001. *Lateral DNA transfer: Mechanisms and consequences.* Cold Spring Harbor Laboratory Press, Cold Spring Harbor, NY.

Bushman FD, Craigie R. 1992. Integration of human immunodeficiency virus DNA: Adduct interference analysis of required DNA sites. *Proc Natl Acad Sci* **89:** 3458–3462.

Bushman FD, Fujiwara T, Craigie R. 1990. Retroviral DNA integration directed by HIV integration protein in vitro. *Science* **249:** 1555–1558.

Bushman FD, Engelman A, Palmer I, Wingfield P, Craigie R. 1993. Domains of the integrase protein of human immunodeficiency virus type 1 responsible for polynucleotidyl transfer and zinc binding. *Proc Natl Acad Sci* **90:** 3428–3432.

Bushman F, Lewinski M, Ciuffi A, Barr S, Leipzig J, Hannenhalli S, Hoffmann C. 2005. Genome-wide analysis of retroviral DNA integration. *Nat Rev Microbiol* **3:** 848–858.

Cai M, Zheng R, Caffrey M, Craigie R, Clore GM, Gronenborn AM. 1997. Solution structure of the N-terminal zinc binding domain of HIV-1 integrase. *Nat Struct Biol* **4:** 567–577.

Cavazzana-Calvo M, Payen E, Negre O, Wang G, Hehir K, Fusil F, Down J, Denaro M, Brady T, Westerman K, et al. 2010. Transfusion independence and HMGA2 activation after gene therapy of human β-thalassaemia. *Nature* **467:** 318–322.

Cherepanov P, Maertens G, Proost P, Devreese B, Van Beeumen J, Engelborghs Y, De Clercq E, Debyser Z. 2003. HIV-1 integrase forms stable tetramers and associates with LEDGF/p75 protein in human cells. *J Biol Chem* **278:** 372–381.

Cherepanov P, Ambrosio AL, Rahman S, Ellenberger T, Engelman A. 2005. Structural basis for the recognition between HIV-1 integrase and transcriptional coactivator 75. *Proc Natl Acad Sci* **102:** 17308–17313.

Chiu TK, Davies DR. 2004. Structure and function of HIV-1 integrase. *Curr Top Med Chem* **4:** 965–977.

Christ F, Voet A, Marchand A, Nicolet S, Desimmie BA, Marchand D, Bardiot D, Van der Veken NJ, Van Remoortel B, Strelkov SV, et al. 2010. Rational design of small-molecule inhibitors of the LEDGF/p75–integrase interaction and HIV replication. *Nat Chem Biol* **6:** 442–448.

Ciuffi A, Bushman FD. 2006. Retroviral DNA integration: HIV and the role of LEDGF/p75. *Trends Genet* **22:** 388–395.

Ciuffi A, Llano M, Poeschla E, Hoffmann C, Leipzig J, Shinn P, Ecker JR, Bushman F. 2005. A role for LEDGF/p75 in targeting HIV DNA integration. *Nat Med* **11:** 1287–1289.

Ciuffi A, Mitchell RS, Hoffmann C, Leipzig J, Shinn P, Ecker JR, Bushman FD. 2006. Integration site selection by HIV-based vectors in dividing and growth-arrested IMR-90 lung fibroblasts. *Mol Ther* **13:** 366–373.

Coffin JM, Hughes SH, Varmus HE. 1997. *Retroviruses.* Cold Spring Harbor Laboratory Press, Cold Spring Harbor, NY.

Craig NL, Craigie R, Gellert M, Lambowitz AM. 2002. *Mobile DNA II.* ASM Press, Washington, DC.

Craigie R, Fujiwara T, Bushman F. 1990. The IN protein of Moloney murine leukemia virus processes the viral DNA ends and accomplishes their integration in vitro. *Cell* **62:** 829–837.

Daniel R, Katz RA, Skalka AM. 1999. A role for DNA-PK in retroviral DNA integration. *Science* **284:** 644–647.

Dar MJ, Monel B, Krishnan L, Shun MC, Di Nunzio F, Helland DE, Engelman A. 2009. Biochemical and virological analysis of the 18-residue C-terminal tail of HIV-1 integrase. *Retrovirology* **6:** 94.

Davies DR, Goryshin IY, Reznikoff WS, Rayment I. 2000. Three-dimensional structure of the Tn5 synaptic complex transposition intermediate. *Science* **289:** 77–85.

De Rijck J, Vandekerckhove L, Gijsbers R, Hombrouck A, Hendrix J, Vercammen J, Engelborghs Y, Christ F, Debyser Z. 2006. Overexpression of the lens epithelium-derived growth factor/p75 integrase binding domain inhibits human immunodeficiency virus replication. *J Virol* **80:** 11498–11509.

De Rijck J, Bartholomeeusen K, Ceulemans H, Debyser Z, Gijsbers R. 2010. High-resolution profiling of the LEDGF/p75 chromatin interaction in the ENCODE region. *Nucleic Acids Res* **38:** 6135–6147.

Donehower LA, Varmus HE. 1984. A mutant murine leukemia virus with a single missense codon in pol is defective in a function affecting integration. *Proc Natl Acad Sci* **81:** 6461–6465.

Du Z, Ilyinskii PO, Lally K, Desrosiers RC, Engelman A. 1997. A mutation in integrase can compensate for mutations in the simian immunodeficiency virus att site. *J Virol* **71:** 8124–8132.

Dyda F, Hickman AB, Jenkins TM, Engelman A, Craigie R, Davies DR. 1994. Crystal structure of the catalytic domain of HIV-1 integrase: Similarity to other polynucleotidyl transferases. *Science* **266:** 1981–1986.

Eijkelenboom APAM, Puras Lutzke RA, Boelens R, Plasterk RHA, Kaptein R, Hard K. 1995. The DNA binding domain of HIV-1 integrase has an SH3-like fold. *Nature Struct Biol* **2:** 807–810.

Eijkelenboom APAM, van den Ent FMI, Vos A, Doreleijers JF, Hard K, Tullius T, Plasterk RHA, Kaptein R, Boelens R. 1997. The solution structure of the amino-terminal HHCC domain of HIV-2 integrase; A three-helix bundle stabilized by zinc. *Cur Biol* **1:** 739–746.

Ellison VH, Abrams H, Roe T, Lifson J, Brown PO. 1990. Human immunodeficiency virus integration in a cell-free system. *J Virol* **64:** 2711–2715.

Emiliani S, Mousnier A, Busschots K, Maroun M, Van Maele B, Tempe D, Vandekerckhove L, Moisant F, Ben-Slama L, Witvrouw M, et al. 2005. Integrase mutants defective for interaction with LEDGF/p75 are impaired in chromosome tethering and HIV-1 replication. *J Biol Chem* **280:** 25517–25523.

Engelman A, Craigie R. 1992. Identification of conserved amino acid residues critical for human immunodeficiency virus type 1 integrase function in vitro. *J Virol* **66:** 6361–6369.

Engelman A, Mizuuchi K, Craigie R. 1991. HIV-1 DNA integration: Mechanism of viral DNA cleavage and DNA strand transfer. *Cell* **67:** 1211–1221.

Engelman A, Maramorosch K, Murphy FA, Shatkin AJ. 1999. In vivo analysis of retroviral integrase structure and function. In *Advances in virus research.* Academic Press, San Diego.

Farnet CM, Haseltine WA. 1990. Integration of human immunodeficiency virus type 1 DNA in vitro. *Proc Natl Acad Sci* **87:** 4164–4168.

Farnet CM, Haseltine WA. 1991a. Circularization of human immunodeficiency virus type 1 DNA in vitro. *J Virol* **65:** 6942–6952.

Farnet CM, Haseltine WA. 1991b. Determination of viral proteins present in the human immunodeficiency virus type 1 preintegration complex. *J Virol* **65:** 1910–1915.

Fassati A. 2006. HIV infection of non-dividing cells: A divisive problem. *Retrovirology* **3:** 74.

Ferris AL, Wu X, Hughes CM, Stewart C, Smith SJ, Milne TA, Wang GG, Shun MC, Allis CD, Engelman A, et al. 2010. Lens epithelium-derived growth factor fusion proteins redirect HIV-1 DNA integration. *Proc Natl Acad Sci* **107:** 3135–3140.

Fujiwara T, Craigie R. 1989. Integration of mini-retroviral DNA: A cell-free reaction for biochemical analysis of retroviral integration. *Proc Natl Acad Sci* **86:** 3065–3069.

Fujiwara T, Mizuuchi K. 1988. Retroviral DNA integration: Structure of an integration intermediate. *Cell* **54:** 497–504.

Ge H, Si Y, Roeder RG. 1998. Isolation of cDNAs encoding novel transcription coactivators p52 and p75 reveals an alternate regulatory mechanism of transcriptional activation. *EMBO J* **17:** 6723–6729.

Gijsbers R, Ronen K, Vets S, Malani N, De Rijck J, McNeely M, Bushman FD, Debyser Z. 2010. LEDGF hybrids efficiently retarget lentiviral integration into heterochromatin. *Mol Ther* **18:** 552–560.

Grandgenett DP, Vora AC, Schiff RD. 1978. A 32,000-dalton nucleic acid-binding protein from avian retrovirus cores possesses DNA endonuclease activity. *Virology* **89:** 119–132.

Greene WC, Peterlin BM. 2002. Charting HIV's remarkable voyage through the cell: Basic science as a passport to future therapy. *Nat Med* **8:** 673–680.

Hacein-Bey-Abina S, Hauer J, Lim A, Picard C, Wang GP, Berry CC, Martinache C, Rieux-Laucat F, Latour S, Belohradsky BH, et al. 2010. Efficacy of gene therapy for X-linked severe combined immunodeficiency. *N Engl J Med* **363:** 355–364.

Hare S, Shun MC, Gupta SS, Valkov E, Engelman A, Cherepanov P. 2009. A novel co-crystal structure affords the design of gain-of-function lentiviral integrase mutants in the presence of modified PSIP1/LEDGF/p75. *Plos Pathogens* **5**. doi: 10.1371/journal.ppat.1000259.

Hare S, Gupta SS, Valkov E, Engelman A, Cherepanov P. 2010. Retroviral intasome assembly and inhibition of DNA strand transfer. *Nature* **464:** 232–236.

Hindmarsh P, Ridky T, Reeves R, Andrake M, Skalka AM, Leis J. 1999. HMG protein family members stimulate human immunodeficiency virus type 1 avian sarcoma virus concerted DNA integration in vitro. *J Virol* **73:** 2994–3003.

Holman AG, Coffin JM. 2005. Symmetrical base preferences surrounding HIV-1, avian sarcoma/leukosis virus, and murine leukemia virus integration sites. *Proc Natl Acad Sci* **102:** 6103–6107.

Howe SJ, Mansour MR, Schwarzwaelder K, Bartholomae C, Hubank M, Kempski H, Brugman MH, Pike-Overzet K, Chatters SJ, de Ridder D, et al. 2008. Insertional mutagenesis combined with acquired somatic mutations causes leukemogenesis following gene therapy of SCID-X1 patients. *J Clin Invest* **118:** 3143–3150.

* Hu W-S, Hughes SH. 2011. HIV reverse transcription: Virally encoded enzyme. *Cold Spring Harb Perspect Med* doi: 10.1101/cshperspect.a006882.

Jaskolski M, Alexandratos JN, Bujacz G, Wlodawer A. 2009. Piecing together the structure of retroviral integrase, an important target in AIDS therapy. *FEBS J* **276:** 2926–2946.

Jeanson L, Subra F, Vaganay S, Hervy M, Marangoni E, Bourhis J, Mouscadet JF. 2002. Effect of Ku80 depletion on the preintegrative steps of HIV-1 replication in human cells. *Virology* **300:** 100–108.

Jenkins TM, Hickman AB, Dyda F, Ghirlando R, Davies DR, Craigie R. 1995. Catalytic domain of human immunodeficiency virus type 1 integrase: Identification of a soluble mutant by systematic replacement of hydrophobic residues. *Proc Natl Acad Sci* **92:** 6057–6061.

Jordan A, Defechereux P, Verdin E. 2001. The site of HIV-1 integration in the human genome determines basal transcriptional activity and response to Tat transactivation. *EMBO J* **20:** 1726–1738.

Ju G, Skalka AM. 1980. Nucleotide sequence analysis of the long terminal repeat (LTR) of avian retroviruses: Structural similarities with transposable elements. *Cell* **22:** 379–386.

Katz RA, Merkel G, Kulkosky J, Leis J, Skalka AM. 1990. The avian retroviral IN protein is both necessary and sufficient for integrative recombination in vitro. *Cell* **63:** 87–95.

Katz RA, Gravuer K, Skalka AM. 1998. A preferred target DNA structure for retroviral integrase in vitro. *J Biol Chem* **273:** 24190–24195.

Katzman M, Katz RA, Skalka AM, Leis J. 1989. The avian retroviral integration protein cleaves the terminal sequences of linear viral DNA at the in vivo sites of integration. *J Virol* **63:** 5319–5327.

Kilzer JM, Stracker TH, Beitzel B, Meek K, Weitzman MD, Bushman FD. 2003. Roles of host cell factors in circularization of retroviral DNA. *Virology* **314:** 460–467.

Konig R, Zhou Y, Elleder D, Diamond TL, Bonamy GM, Irelan JT, Chiang CY, Tu BP, De Jesus PD, Lilley CE, et al. 2008. Global analysis of host-pathogen interactions that regulate early-stage HIV-1 replication. *Cell* **135:** 49–60.

Krishnan L, Li XA, Naraharisetty HL, Hare S, Cherepanov P, Engelman A. 2010. Structure-based modeling of the functional HIV-1 intasome and its inhibition. *Proc Natl Acad Sci* **107:** 15910–15915.

Kulkosky J, Jones KS, Katz RA, Mack JPG, Skalka AM. 1992. Residues critical for retroviral integrative recombination in a region that is highly conserved among retroviral/retrotransposon integrases and bacterial insertion sequence transposases. *Mol Cell Biol* **12:** 2331–2338.

Lander E. 2001. Initial sequencing and analysis of the human genome. *Nature* **409:** 860–921.

Leavitt AD, Shiue L, Varmus HE. 1993. Site-directed mutagenesis of HIV-1 integrase demonstrates differential effects on integrase functions in vitro. *J Biol Chem* **268:** 2113–2119.

Lee MS, Craigie R. 1998. Protection of retroviral DNA from autointegration: Involvement of a cellular factor. *Proc Natl Acad Sci* **95:** 1528–1533.

Lee K, Ambrose Z, Martin TD, Oztop I, Mulky A, Julias JG, Vandegraaff N, Baumann JG, Wang R, Yuen W, et al. 2010. Flexible use of nuclear import pathways by HIV-1. *Cell Host Microbe* **7:** 221–233.

Lewinski M, Bisgrove D, Shinn P, Chen H, Verdin E, Berry CC, Ecker JR, Bushman FD. 2005. Genome-wide analysis of chromosomal features repressing HIV transcription. *J Virol* **79:** 6610–6619.

Lewinski MK, Yamashita M, Emerman M, Ciuffi A, Marshall H, Crawford G, Collins F, Shinn P, Leipzig J, Hannenhalli S, et al. 2006. Retroviral DNA integration: Viral and cellular determinants of target-site selection. *PLoS Pathog* **2:** e60. doi: 10.1371/journal.ppat.0020060.

Li M, Craigie R. 2005. Processing of viral DNA ends channels the HIV-1 integration reaction to concerted integration. *J Biol Chem* **280:** 29334–29339.

Li L, Olvera JM, Yoder K, Mitchell RS, Butler SL, Lieber MR, Martin SL, Bushman FD. 2001. Role of the non-homologous DNA end joining pathway in retroviral infection. *EMBO J* **20:** 3272–3281.

Li M, Mizuuchi M, Burke TR Jr, Craigie R. 2006. Retroviral DNA integration: Reaction pathway and critical intermediates. *EMBO J* **25:** 1295–1304.

Llano M, Saenz DT, Meehan A, Wongthida P, Peretz M, Walker WH, Teo W, Poeschla EM. 2006a. An essential role for LEDGF/p75 in HIV integration. *Science* **314:** 461–464.

Llano M, Vanegas M, Hutchins N, Thompson D, Delgado S, Poeschla EM. 2006b. Identification and characterization of the chromatin-binding domains of the HIV-1 integrase interactor LEDGF/p75. *J Mol Biol* **360:** 760–773.

Lodi PJ, Ernst JA, Kuszewski J, Hickman AB, Engelman A, Craigie R, Clore GM, Gronenborn AM. 1995. Solution structure of the DNA binding domain of HIV-1 integrase. *Biochemistry* **34:** 9826–9833.

Maertens G, Cherepanov P, Pluymers W, Busschots K, De Clercq E, Debyser Z, Engelborghs Y. 2003. LEDGF/p75 is essential for nuclear and chromosomal targeting of HIV-1 integrase in human cells. *J Biol Chem* **278:** 33528–33539.

Maertens GN, Hare S, Cherepanov P. 2010. The mechanism of retroviral integration from X-ray structures of its key intermediates. *Nature* **468:** 326–329.

Marshall H, Ronen K, Berry C, Llano M, Sutherland H, Saenz D, Bickmore W, Poeschla E, Bushman F. 2007. Role of PSIP1/LEDGF/p75 in lentiviral infectivity and integration targeting. *PLoS One* **2:** e1340. doi: 10.1371/journal.pone.0001340.

Mitchell RS, Beitzel BF, Schroder AR, Shinn P, Chen H, Berry CC, Ecker JR, Bushman FD. 2004. Retroviral DNA integration: ASLV, HIV, and MLV show distinct target site preferences. *PLoS Biol* **2:** pE234. doi: 10.1371/journal.pbio.0020234.

Mizuuchi K. 1992. Polynucleotidyl transfer reactions in transpositional DNA recombination. *J Biol Chem* **287:** 21273–21276.

Mizuuchi K, Adzuma K. 1991. Inversion of the phosphate chirality at the target site of Mu-DNA strand transfer—Evidence for a one-step transesterification mechanism. *Cell* **66:** 129–140.

Molteni V, Greenwald J, Rhodes D, Hwang Y, Kwiatkowski W, Bushman FD, Siegel JS, Choe S. 2001. Identification of a small molecule binding site at the dimer interface of the HIV integrase catalytic domain. *Acta Crystallogr D* **57:** 536–544.

Murphy JE, Goff SP. 1992. A mutation at one end of Moloney murine leukemia virus DNA blocks cleavage of both ends by the viral integrase in vivo. *J Virol* **66:** 5092–5095.

Murphy JE, De Los Santos T, Goff SP. 1993. Mutational analysis of the sequences at the termini of the Moloney murine leukemia virus DNA required for integration. *Virology* **195:** 432–440.

Narezkina A, Taganov KD, Litwin S, Stoyanova R, Hayashi J, Seeger C, Skalka AM, Katz RA. 2004. Genome-wide analyses of avian sarcoma virus integration sites. *J Virol* **78:** 11656–11663.

Panganiban AT, Temin HM. 1983. The terminal nucleotides of retrovirus DNA are required for integration but not virus production. *Nature* **306:** 155–160.

Panganiban AT, Temin HM. 1984. The retrovirus pol gene encodes a product required for DNA integration: Identification of a retrovirus int locus. *Proc Natl Acad Sci* **81:** 7885–7889.

Perelson AS, Neumann AU, Markowitz M, Leonard JM, Ho DD. 1996. HIV-1 dynamics in vivo: Virion clearance rate, infected cell life-span, and viral generation time. *Science* **271:** 1582–1586.

Pruss D, Bushman FD, Wolffe AP. 1994. Human immunodeficiency virus integrase directs integration to sites of severe DNA distortion within the nucleosome core. *Proc Natl Acad Sci* **91:** 5913–5917.

Pryciak PM, Varmus HE. 1992. Nucleosomes, DNA-binding proteins, and DNA sequence modulate retroviral integration target site selection. *Cell* **69:** 769–780.

Pryciak PM, Sil A, Varmus HE. 1992. Retroviral integration into minichromosomes in vitro. *EMBO J* **11:** 291–303.

Rice P, Craigie R, Davies DR. 1996. Retroviral integrases and their cousins. *Curr Opin Struct Biol* **6:** 76–83.

Richardson JM, Colloms SD, Finnegan DJ, Walkinshaw MD. 2009. Molecular architecture of the Mos1 paired-end complex: The structural basis of DNA transposition in a eukaryote. *Cell* **138:** 1096–1108.

Roe T, Reynolds TC, Yu G, Brown PO. 1993. Integration of murine leukemia virus DNA depends on mitosis. *EMBO J* **12:** 2099–2108.

Schroder AR, Shinn P, Chen H, Berry C, Ecker JR, Bushman F. 2002. HIV-1 integration in the human genome favors active genes and local hotspots. *Cell* **110:** 521–529.

Schwartzberg P, Colecilli J, Goff SP. 1984. Construction and analysis of deletion mutations in the pol gene of Moloney murine leukemia virus: A new viral function required for productive infection. *Cell* **37:** 1043–1052.

Shimotohno K, Mizutani S, Temin HM. 1980. Sequence of retrovirus provirus resembles that of bacterial transposable elements. *Nature* **285:** 550–554.

Shoemaker CS, Goff S, Giboa E, Paskind M, Mitra SW, Baltimore D. 1980. Structure of a cloned circular Moloney murine leukemia virus DNA molecule containing an inverted segment: Implications for retrovirus integration. *Proc Natl Acad Sci* **77:** 3932–3936.

Shun MC, Raghavendra NK, Vandegraaff N, Daigle JE, Hughes S, Kellam P, Cherepanov P, Engelman A. 2007. LEDGF/p75 functions downstream from preintegration

complex formation to effect gene-specific HIV-1 integration. *Genes Dev* **21:** 1767–1778.

Silvers RM, Smith JA, Schowalter M, Litwin S, Liang ZH, Geary K, Daniel R. 2010. Modification of integration site preferences of an HIV-1-based vector by expression of a novel synthetic protein. *Hum Gene Ther* **21:** 337–349.

Sinha S, Pursley MH, Grandgenett DP. 2002. Efficient concerted integration by recombinant human immunodeficiency virus type 1 integrase without cellular or viral cofactors. *J Virol* **76:** 3105–3113.

Sinha S, Grandgenett DP. 2005. Recombinant human immunodeficiency virus type 1 integrase exhibits a capacity for full-site integration in vitro that is comparable to that of purified preintegration complexes from virus-infected cells. *J Virol* **79:** 8208–8216.

Stevens SW, Griffith JD. 1994. Human immunodeficiency virus type 1 may preferentially integrate into chromatin occupied by L1Hs repetitive elements. *Proc Natl Acad Sci* **91:** 5557–5561.

Suzuki Y, Craigie R. 2007. The road to chromatin—Nuclear entry of retroviruses. *Nat Rev Microbiol* **5:** 187–196.

Temin HM. 1976. The DNA provirus hypothesis. *Science* **192:** 1075–1080.

Temin H, Mizutani S. 1970. RNA-dependent DNA polymerase in virions of Rous sarcoma virus. *Nature* **226:** 1211–1213.

Turlure F, Devroe E, Silver PA, Engelman A. 2004. Human cell proteins and human immunodeficiency virus DNA integration. *Front Biosci* **9:** 3187–3208.

Turlure F, Maertens G, Rahman S, Cherepanov P, Engelman A. 2006. A tripartite DNA-binding element, comprised of the nuclear localization signal and two AT-hook motifs, mediates the association of LEDGF/p75 with chromatin in vivo. *Nucleic Acids Res* **34:** 1653–1665.

van Gent DC, Oude Groeneger AAM, Plasterk RHA. 1992. Mutational analysis of the integrase protein of human immunodeficiency virus type 2. *Proc Natl Acad Sci* **89:** 9598–9602.

Varmus HE, Brown PO, Berg DE, Howe MM. 1989. Retroviruses. In *Mobile DNA*, pp. 53–108. American Society for Microbiology, Washington, DC.

Venter JC. 2001. The sequence of the human genome. *Science* **291:** 1304–1351.

Wang JY, Ling H, Yang W, Craigie R. 2001. Structure of a two-domain fragment of HIV-1 integrase: Implications for domain organization in the intact protein. *EMBO J* **20:** 7333–7343.

Wang GP, Ciuffi A, Leipzig J, Berry CC, Bushman FD. 2007. HIV integration site selection: Analysis by massively parallel pyrosequencing reveals association with epigenetic modifications. *Genome Res* **17:** 1186–1194.

Weiss R, Teich N, Varmus H, Coffin J. 1984. *RNA tumor viruses*. Cold Spring Harbor Press, Cold Spring Harbor, NY.

Wu X, Li Y, Crise B, Burgess SM. 2003. Transcription start regions in the human genome are favored targets for MLV integration. *Science* **300:** 1749–1751.

Wu X, Li Y, Crise B, Burgess SM, Munroe DJ. 2005. Weak palindromic consensus sequences are a common feature found at the integration target sites of many retroviruses. *J Virol* **79:** 5211–5214.

Yoder K, Bushman FD. 2000. Repair of gaps in retroviral DNA integration intermediates. *J Virol* **74:** 11191–11200.

Yoder K, Sarasin A, Kraemer K, McIlhatton M, Bushman F, Fishel R. 2006. The DNA repair genes XPB and XPD defend cells from retroviral infection. *Proc Natl Acad Sci* **103:** 4622–4627.

Zhou H, Xu M, Huang Q, Gates AT, Zhang XD, Castle JC, Stec E, Ferrer M, Strulovici B, Hazuda DJ, et al. 2008. Genome-scale RNAi screen for host factors required for HIV replication. *Cell Host Microbe* **4:** 495–504.

Transcriptional and Posttranscriptional Regulation of HIV-1 Gene Expression

Jonathan Karn[1] and C. Martin Stoltzfus[2]

[1]Department of Molecular Biology and Microbiology, Case Western Reserve University, Cleveland, Ohio 44106
[2]Department of Microbiology, University of Iowa, Iowa City, Iowa 52242

Correspondence: jonathan.karn@case.edu

Control of HIV-1 gene expression depends on two viral regulatory proteins, Tat and Rev. Tat stimulates transcription elongation by directing the cellular transcriptional elongation factor P-TEFb to nascent RNA polymerases. Rev is required for the transport from the nucleus to the cytoplasm of the unspliced and incompletely spliced mRNAs that encode the structural proteins of the virus. Molecular studies of both proteins have revealed how they interact with the cellular machinery to control transcription from the viral LTR and regulate the levels of spliced and unspliced mRNAs. The regulatory feedback mechanisms driven by HIV-1 Tat and Rev ensure that HIV-1 transcription proceeds through distinct phases. In cells that are not fully activated, limiting levels of Tat and Rev act as potent blocks to premature virus production.

After integration into the host genome, the HIV-1 provirus acts as a transcription template that is regulated at the transcriptional and posttranscriptional levels. Immediately after infection, HIV-1 produces only short completely spliced mRNAs encoding the viral regulatory proteins Tat and Rev. As the infection proceeds, transcription increases sharply, and larger, incompletely spliced mRNAs are produced. These encode Env and the HIV-1 accessory genes Vif, Vpr, and Vpu. Also synthesized late are the full-length unspliced transcripts which act both as the virion genomic RNA and the mRNA for the Gag-Pol polyprotein (Kim et al. 1989; Pomerantz et al. 1990).

This complex pattern of gene expression is controlled by the regulatory proteins Tat and Rev. Tat activates viral transcription by stimulating elongation from the viral long terminal repeat (LTR). Rev transports the unspliced and incompletely spliced mRNAs encoding the structural proteins from the nucleus to the cytoplasm. In this article, we review our current understanding of how these unique regulatory proteins orchestrate HIV-1 gene expression through their interactions with the cellular transcription, RNA splicing, and RNA transport machinery.

CONTROL OF HIV-1 TRANSCRIPTION BY Tat

Discovery of Transactivation by Tat

In HIV-1, as in all retroviruses, the LTR acts as the viral promoter. The first evidence that gene expression in HIV-1 also requires viral transacting factors came from experiments by Sodroski et al. (1985a,b) who noted that the expression of reporter genes placed under the control of the viral LTR was dependent on a transactivating factor, which they named Tat. Deletion analysis of the viral LTR showed that Tat activity required the transactivation-responsive region

(TAR), a regulatory element located downstream from the initiation site for transcription between nucleotides +1 and +59 (Fig. 1A). It quickly became apparent that TAR was not a typical transcription element, because it is only functional when it is placed 3′ to the HIV-1 promoter, and in the correct orientation and position (Muesing

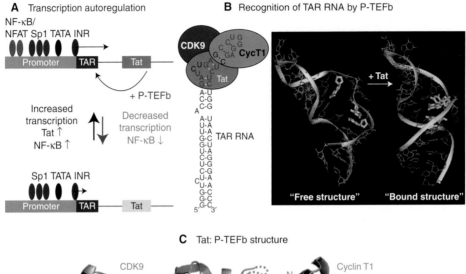

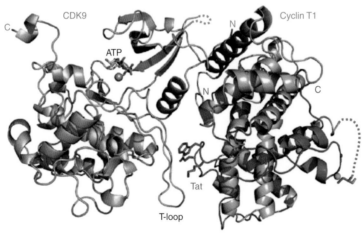

Figure 1. Tat and its interactions with P-TEFb. (*A*) Autoregulation of HIV-1 transcription by Tat. Tat binds to the TAR RNA element encoded in the HIV-1 leader sequence and recruits P-TEFb and other elongation factors to the transcription complex. Small changes in initiation efficiency, caused by epigenetic silencing or reductions in NF-κB levels in the cell, reduce Tat levels and inhibit transcription, driving the HIV-1 provirus into latency. Reinitiation by NF-κB stimulates Tat production and restores full transcription efficiency. Thus, positive feedback by Tat results in a bistable switch. (*B*) Recognition of TAR RNA by Tat and P-TEFb. The diagram on the *left* shows the bases in TAR that are recognized by Tat in the TAR bulge region and by CycT1 in the TAR loop region (red bases). The structures at *right* show the conformational changes induced by Tat binding (Aboul-ela et al. 1995). (*C*) Structure of the Tat:P-TEFb complex. Note that Tat folds on the outer surface of the CycT1 cyclin domain. The amino-terminal "activation" domain of Tat binds to the CDK9 T-loop, a region of the molecule that is essential for its enzymatic activity (Tahirov et al. 2010).

et al. 1987). Genetic evidence that TAR functions as a transcribed RNA regulatory signal came from the observation that the TAR RNA sequence forms a highly stable, nuclease-resistant, stem-loop structure; mutations that destabilize the TAR RNA structure abolish Tat-stimulated transcription (Berkhout et al. 1989; Selby et al. 1989).

The Tat/TAR RNA Interaction

Dingwall et al. (1989, 1990) showed that Tat is able to specifically recognize TAR RNA and mapped its recognition site to a U-rich bulge near the apex of the TAR RNA stem (Fig. 1B). Detailed analysis of Tat's interactions with TAR RNA by NMR subsequently revealed that Tat recognition of TAR requires conformational changes in the RNA structure (Fig. 1B). This refolding process involves displacement of the first residue in the bulge (U23) by one of the arginine side chains present in the basic binding domain of the Tat protein creating a binding pocket for the arginine side chain in the major groove together with the adjacent G26:C39 base pair (Puglisi et al. 1992; Aboul-ela et al. 1995; Brodsky and Williamson 1997; Davidson et al. 2009).

P-TEFb Is the Essential Cofactor for Tat

Although there is a strict correlation between the ability of TAR RNA to bind to Tat in vitro and the ability of these sequences to support transactivation (Churcher et al. 1993), mutations in the apical loop of the TAR element that do not interfere with Tat binding also interfere with transactivation (Feng and Holland 1988). To explain this apparent discrepancy Dingwall et al. (1990) postulated that a cellular cofactor interacts with the TAR RNA loop. This hypothesis raised further questions such as what is the role and function of the "loop factor" and what is the mechanism by which Tat stimulate gene expression after binding to TAR RNA.

The first direct evidence that Tat might be regulating HIV-1 transcriptional elongation, rather than transcriptional initiation, came from RNase protection experiments performed by Kao et al. (1987). They showed that in the ab-

sence of Tat, the majority of RNA polymerases initiating transcription stall near the promoter, whereas in the presence of Tat, there is a dramatic increase in the density of RNA polymerases found downstream from the promoter.

Rice and his colleagues (Herrmann and Rice 1995; Herrmann et al. 1996) showed that a protein kinase complex, which they called TAK (Tat-associated kinase), binds tightly and specifically to Tat. Subsequently, Zhu et al. (1997) cloned the kinase subunit of TAK. This turned out to be the CDK9 kinase, which is a component of a ubiquitous positive acting elongation factor pTEFb (Marshall and Price 1995; Marshall et al. 1996). In parallel, the search for the "loop factor" and other cofactors for Tat also pointed to P-TEFb as a critical cofactor for Tat activation of elongation. Wei et al. (1998) discovered that P-TEFb contains a cyclin component, CycT1, which can form a stable complex with CDK9, Tat, and TAR RNA (see online Movie 1 at www.perspectivesinmedicine. org). Crucially, for a putative "loop factor," complex formation between Tat, P-TEFb, and TAR requires both the Tat binding site and the loop sequence.

After these seminal biochemical observations, additional genetic and biochemical evidence showed unequivocally that P-TEFb is required for Tat-mediated transactivation. First, a set of novel CDK-9 protein kinase inhibitors were shown to be selective inhibitors of HIV-1 transcription (Mancebo et al. 1997). Second, persuasive genetic evidence showed that CycT1 is essential for Tat activity. Tat is inactive in murine cells, because the murine CycT1 sequence differs from the human sequence by a single substitution of cysteine 261 for tyrosine. Introduction of Y261 into the human CycT1 blocked HIV-1 transactivation in transfected cells whereas introduction of C261 into the murine CycT1 restored Tat-mediated transactivation (Bieniasz et al. 1998; Fujinaga et al. 1998; Garber et al. 1998; Kwak et al. 1999).

Finally, the crystal structure of a Tat:pTEFb complex was determined in 2010—the culmination of more than two decades of research on P-TEFb by David Price and his colleagues (Tahirov et al. 2010). The structure shows that Tat forms extensive contacts both with the CycT1 subunit of P-TEFb and also with the T-loop of the Cdk9 subunit (Fig. 1C).

Transactivation Mechanism

The binding of Tat to P-TEFb induces significant conformational changes in CDK9 that constitutively activate the enzyme (Wei et al. 1998; Isel and Karn 1999; Tahirov et al. 2010).

As described in Figure 2A, the transactivation mechanism involves a complex set of phosphorylation events mediated by the Tat-activated P-TEFb that modify both positive and negative cellular elongation factors.

In the absence of Tat, HIV-1 transcription elongation is highly restricted by the negative elongation factor NELF (Yamaguchi et al. 1999; Narita et al. 2003; Zhang et al. 2007). Phosphorylation of NELF-E by P-TEFb forces

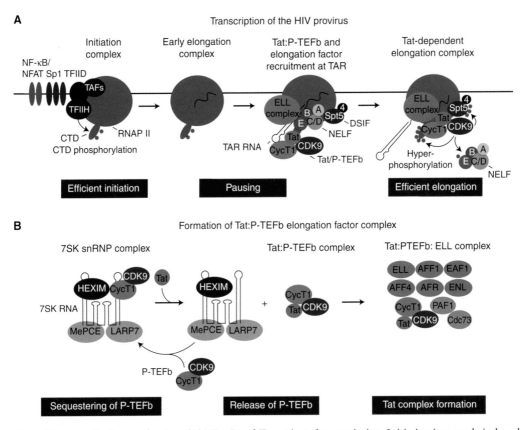

Figure 2. Transactivation mechanism. (*A*) NF-κB and Tat-activated transcription. Initiation is strongly induced by NF-κB, which acts primarily to remove chromatin restrictions near the promoter through recruitment of histone acetyltransferases. After the transcription through the TAR element, both NELF and the Tat/P-TEFb complex (including CDK9 and CycT1 and the accessory elongation factors including ELL2) are recruited to the elongation complex via binding interactions with TAR RNA. This activates the CDK9 kinase and leads to hyperphosphorylation of the CTD of RNA polymerase II, Spt5, and NELF-E. The phosphorylation of NELF-E leads to its release. The presence of hyperphosphorylated RNAP II and Spt5 allows enhanced transcription of the full HIV-1 genome. (*B*) Control of P-TEFb by 7SK and Tat. The majority of the P-TEFb in cells is found in a transcriptionally inactive snRNP complex containing 7SK RNA, HEXIM, and the RNA binding proteins MePCE and LARP7. Tat disrupts this complex by displacing HEXIM and forming a stable complex with P-TEFb. Prior to recruitment to the transcription complex, a larger complex is formed between P-TEFb and transcription elongation factors from the mixed lineage leukemia (MLL) family, including ELL2. (Figure is adapted from Karn 2011; reprinted, with permission, from Wolters Kluwer Health © 2011.)

dissociation of NELF from TAR and releases paused transcription elongation complexes (Fujinaga et al. 2004). Significantly, the NELF-E subunit is able to bind directly to TAR RNA (Yamaguchi et al. 2002; Fujinaga et al. 2004) suggesting that NELF might be recruited to the HIV-1 provirus via its interactions with TAR.

Cell-free transcription studies have shown that Tat:P-TEFb also phosphorylates the RNAP II CTD during elongation (Isel and Karn 1999; Kim et al. 2002). This reaction creates a hyperphosphorylated form of the RNA polymerase that is highly enriched for phosphorylated Ser2 residues in the CTD (Ramanathan et al. 2001; Kim et al. 2002). In addition to targeting RNAP II, P-TEFb is also able to extensively phosphorylate Spt5, a subunit of the DRB sensitivity-inducing factor (DSIF), which carries a CTD homologous to the RNAP II CTD (Ivanov et al. 2000; Bourgeois et al. 2002). Although the unmodified DSIF inhibits elongation (Yamaguchi et al. 2002), phosphorylation of Spt5 separates it from the rest of the complex and converts it into a positive elongation factor that stabilizes transcription complexes at terminator sequences (Bourgeois et al. 2002; Yamada et al. 2006). Thus, Tat and P-TEFb are able to stimulate HIV-1 transcription both through the removal of blocks to elongation imposed by NELF and DSIF and by the enhancement of RNAP II processivity through the phosphorylation of Spt5 and the RNAP II CTD.

Our picture of how Tat and P-TEFb stimulate HIV-1 elongation has recently been refined by two proteomic studies that identified large protein complexes containing Tat P-TEFb and the human transcription factors/coactivators AFF4, ENL, AF9, and ELL2 (Fig. 2B) (He et al. 2010; Sobhian et al. 2010). One of these coactivators, ELL2, an elongation factor, which was previously shown to enhance transcription elongation by preventing RNAP II backtracking, is critical both for basal HIV-1 transcription and Tat-mediated transactivation. Thus, any model for the stimulatory effects of P-TEFb on HIV-1 transcription now has to take into account the role of ELL2 and possibly several additional elongation factors.

Regulation of P-TEFb

In actively replicating cells, such as HeLa cells and Jurkat T-cells, P-TEFb activity is tightly regulated and the majority of the enzyme is sequestered into a large inactive 7SK RNP complex comprising 7SK RNA and a series of RNA-binding proteins (Fig. 2B) (Nguyen et al. 2001; Yang et al. 2001). Essential components of the 7SK RNP complex include HEXIM1 or HEXIM2, which inhibit the CDK9 kinase in a 7SK-dependent manner (Yik et al. 2003; Michels et al. 2004), and the 7SK RNA binding proteins LARP-7 (He et al. 2008; Krueger et al. 2008), and BCDIN3 (Jeronimo et al. 2007). The sequestration of P-TEFb in the 7SK RNP complex effectively prevents any basal transcriptional activation by Tat-independent recruitment of P-TEFb to the provirus. Tat overcomes this barrier by disrupting the 7SK RNP complex by competing with HEXIM for CycT1 binding (Barboric et al. 2007; Sedore et al. 2007; Krueger et al. 2010). A recent study suggests that cyclin T1 acetylation also triggers dissociation of HEXIM1 and 7SK RNA from the inactive 7SK snRNP complex and activates the transcriptional activity of P-TEFb (Cho et al. 2009).

In contrast to Jurkat T-cells both primary resting central memory T-cells (Ramakrishnan et al. 2009) and primary monocytes (Sung and Rice 2009) show highly restricted levels of CycT1. Activation of P-TEFb in these cells therefore requires multiple steps involving both the initial assembly of the 7SK RNP complex and its relocalization to nuclear speckles where it becomes accessible to Tat and the rest of the transcription machinery.

The LTR as a Promoter

The HIV-1 LTR includes multiple upstream DNA regulatory elements that serve as binding sites for cellular transcription initiation factors (Rittner et al. 1995). The core promoter is a powerful and highly optimized promoter comprised of three tandem SP1 binding sites (Jones et al. 1986), an efficient TATA element (Garcia et al. 1989), and a highly active initiator sequence (Zenzie-Gregory et al. 1993). Each of these elements participates in the cooperative binding of the initiation factor TFIID and its associated TAF cofactors to the TATA element (Rittner et al. 1995). As a result, the HIV-1 LTR is an extremely efficient promoter that is capable of supporting even higher levels of transcription than the adenovirus major late promoter or the CMV immediate early promoter.

In addition to the core promoter, HIV-1 relies on an "enhancer region" that contains two NF-κB binding motifs (Nabel and Baltimore 1987) (see online Movie 2 at www.perspectivesinmedicine.org). Members of both the NF-κB family (Liu et al. 1992) and NFAT (Kinoshita et al. 1998) can bind to the HIV-1 NF-κB motifs. Because their recognition sequences overlap, binding of these factors is mutually exclusive (Chen-Park et al. 2002; Giffin et al. 2003). Binding of NF-κB is more efficient than NFAT because it is enhanced by cooperative interactions with Sp1 (Perkins et al. 1993). Although mutation of the NF-κB sites results in only a modest inhibition of virus growth in most transformed cell lines (Chen et al. 1997), signaling through the viral enhancer is essential to reactivate latent proviruses and support virus replication in primary T-cells, regardless of whether it is stimulated by NF-κB or by NFAT (Alcami et al. 1995; Bosque and Planelles 2008).

Epigenetic Regulation of HIV-1 Transcription

When HIV-1 infects cells, it preferentially integrates into active transcription units that provide a favorable environment for viral transcription (Lewinski et al. 2006). As originally shown by Verdin et al. (1993), proviruses assemble an ordered nucleosomal structure surrounding the promoter. These nucleosomal structures play a crucial role in establishing HIV-1 latency because epigenetic modifications of the provirus restrict transcription initiation (see Siliciano and Greene 2011). Typically, transcription from latent proviruses is restricted by high levels of histone deacetylases (HDACs), deacetylated histones, methylated histones, and DNA methylation (for reviews, see Margolis 2010; Karn 2011).

Control of HIV-1 Replication by Transcriptional Feedback

Because Tat functions as part of a positive regulatory circuit, conditions that restrict transcription initiation will in turn cause a reduction in Tat levels to below threshold levels and therefore result in dramatically reduced HIV-1 transcription and eventually entry into latency (for reviews, see Karn 2011; Siliciano and Greene 2011). Insightful studies by Weinberger et al. (Weinberger et al. 2005; Weinberger and Shenk 2006) and Burnett et al. (2009) have emphasized how stochastic fluctuations in Tat gene expression can act as a molecular switch. Small changes in initiation rates, which can be experimentally mimicked by introducing mutations into the NF-κB and Sp1 binding sites, are able to reduce Tat availability and disproportionately limit HIV-1 transcription, forcing viruses into latency (Burnett et al. 2009). However, the virus remains poised to resume its replication in response to triggers that stimulate transcription initiation and restore Tat levels. This switching mechanism crucially depends on the autoregulation of Tat; when Tat is expressed *in trans* from an ectopic promoter, HIV-1 proviruses become constitutively active and are unable to enter latency (Pearson et al. 2008).

CONTROL OF HIV-1 RNA SPLICING, POLYADENYLATION, EXPORT, AND TRANSLATION

Alternative Splicing of HIV-1 mRNA

To produce the full range of mRNAs needed to encode the viral proteins, HIV-1 primary transcripts undergo extensive and complex alternative splicing in the nucleus of infected cells (Fig. 3). Most HIV-1 strains use four different splice donor or 5′ splice sites (5′ss) and eight different acceptor or 3′ splice sites (3′ss) to produce more than 40 different spliced mRNA species in infected cells. These include several incompletely spliced bicistronic mRNA species, which encode both Env and Vpu; incompletely spliced mRNAs for Vif, Vpr, and a truncated 72 aa form of Tat; and completely spliced mRNAs that encode the HIV-1 regulatory proteins Tat,

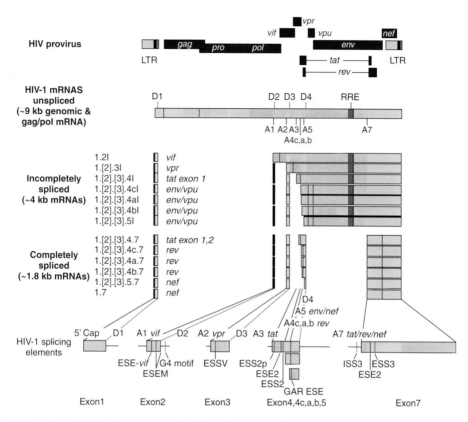

Figure 3. Locations of splice sites, exons, and splicing elements in the HIV-1 genome. (*Top*) Schematic diagram of HIV-1 genome. The dark blue rectangles indicate open reading frames and are labeled with the gene names. The LTRs are shown at each edge of the genome: U3-gray, R-black, U5-light blue. Full-length RNA transcripts begin at the 5′-end of the R region of the 5′-LTR (*left*) and 3′ processing and poly(A) addition takes place at the 3′-end of the R region in the 3′-LTR (*right*). (*Middle*) Locations of 5′ss (red bars) and 3′ss (black bars) in the HIV-1 genome. The location of the RRE is shown by the red rectangle. The exons present in the incompletely spliced ~4-kb and ~1.8-kb mRNA species corresponding to the HIV-1 genes are shown as cyan rectangles. Noncoding exon 1 is present in all spliced HIV-1 mRNA species. Either both or one of the small noncoding exons 2 and 3 shown are included in a fraction of the mRNA species. The exon compositions of the RNA species are also shown. RNA species designated by an "I" are incompletely spliced mRNA species. Brackets indicate that mRNA isoforms containing neither exon 2 nor 3, only exon 2 or 3, or both exons 2 and 3. The locations of the AUG codons used to initiate protein synthesis are shown as purple bars within the exons. (*Bottom*) Locations of the known splicing regulatory elements in HIV-1. Splicing enhancers are designated by green bars and splicing silencers are designated by red bars. (Figure is adapted from Stoltzfus 2009; reproduced, with permission, from Elsevier © 2009.)

Rev, and Nef. In each of the spliced mRNAs, the 5′ss D1 (sometimes referred to as the "major splice donor") is spliced to one of the 3′ss. As a result, all HIV-1 mRNAs include the highly structured noncoding exon 1 that extends from the 5′ cap to 5′ss D1.

Adding to the complexity of the mRNA species present in infected cells, a few viral mRNA isoforms are also produced by inclusion of exons flanked by 3′ss A1 and 5′ss D2 (exon 2)

and/or the exon flanked by 3′ss A2 and 5′ss D3 (exon 3). Exons 2 and 3 do not contain initiator AUG codons and therefore are noncoding.

RNA splicing is performed while the pre-mRNA is associated with a large complex of cellular factors referred to as the spliceosome (for recent reviews, see Wang and Burge 2008; Chen and Manley 2009). The efficiency of early splicing complex formation is determined by the intrinsic strengths of the 3′ss and downstream

5′ss, and further regulated by a number of *cis*-acting elements (Fig. 3). Control of splicing in HIV-1 involves exonic splicing enhancers (ESEs) and intronic splicing enhancers (ISEs) which facilitate splice site recognition and are selectively bound by members of the SR (Ser-Arg) protein family. In addition, there are intronic and exonic splicing silencers (ISSs and ESSs, respectively) which repress splicing and are typically bound by specific members of the heterogeneous ribonuclear protein family (hnRNPs).

Analysis of HIV-1 mRNA species in virus-infected cells showed that there are striking differences in the relative abundances of the different viral mRNA species (Purcell and Martin 1993). In general, HIV-1 3′ splice sites are relatively inefficient in comparison to constitutive cellular 3′ splice sites (for reviews of HIV-1 splicing, see Stoltzfus and Madsen 2006; Stoltzfus 2009). However, the order of intrinsic splice site strengths (Asang et al. 2008) does not correlate with the observed levels of mRNAs spliced at these 3′ splice sites, implying that the *cis*-acting splicing elements dominate the splice site selection of HIV-1 mRNAs. For example, the first *tat* coding exon contains two ESS elements (ESS2 and ESS2p) that specifically repress splicing at 3′ss A3 and reduce the levels of both incompletely and completely spliced *tat* mRNA (Jacquenet et al. 2001; Amendt et al. 1994). Similarly, splicing at 3′ss A2 is repressed by an ESS element within exon 3 (ESSV), which results in relatively low levels of *vpr* mRNA (Bilodeau et al. 2001).

By contrast, splicing at the weak 3′ss A4c, A4a, A4b, and A5 sites is greatly facilitated by a guanosine-adenosine-rich ESE (GAR) within exon 5. GAR ESE activity is able to raise the levels of incompletely spliced *env*/vpu mRNAs and completely spliced *nef* and *rev* mRNAs to the point that they become the most abundant spliced mRNA species in the HIV-1-infected cell (Purcell and Martin 1993). The GAR ESE is selectively bound by several SR proteins but the most important player in the function of the element is SF2/ASF (Caputi et al. 2004). Splicing at the relatively weak 3′ss A1, which is required for high vif mRNA expression and

inclusion of the noncoding exon 2, is facilitated by several different ESEs (ESE-Vif, ESE M1, and ESE M2) within exon 2 (Kammler et al. 2006; Exline et al. 2008). Mutations of the hnRNP A/B-dependent ESS, ESSV, activated 3′ss A2 and resulted in increased levels of mRNAs containing exon 3 and the incompletely spliced *vpr* mRNA. This excessive splicing phenotype resulted in a decrease in virus replication (Madsen and Stoltzfus 2005).

Formation of HIV-1 splicing complexes is also affected by the strengths of the various downstream 5′ss—an expected consequence of exon definition (Robberson et al. 1990; Hoffman and Grabowski 1992). Mutations of the nonconcensus D2 site that have enhanced affinity for U1 snRNP result in an excessive splicing phenotype characterized by increased inclusion of exon 2, increased levels of *vif* mRNA, reduced levels of unspliced viral RNA, and reduced virus production. Conversely, mutations of 5′ss D2 that decrease affinity for U1 snRNP result in decreased inclusion of exon 2 and decreased levels of *vif* mRNA and Vif protein (Madsen and Stoltzfus 2006; Exline et al. 2008; Mandal et al. 2008). Because of reduced levels of Vif, the replication of these virus mutants exhibit greater sensitivity than wild-type virus to inhibition by the cellular restriction factor APOBEC3G. The dramatic effects of these splicing element mutations show that maximum virus replication requires tight regulation of splicing to balance mRNA and genome RNA production.

HIV-1 Rev and the Control of RNA Export

Unspliced and incompletely spliced transcripts from cellular genes are typically degraded in the nucleus. To circumvent these surveillance mechanisms, HIV-1 and many other retroviruses, including the human T-cell leukemia viruses HTLV-1 and HTLV-II, express regulatory factors that facilitate the transport of intron-containing viral RNA out of the nucleus. The first of these factors to be discovered was the HIV-1 Rev protein which interacts with a highly structure RNA element in the *env* gene referred to as the Rev-responsive element (RRE) (Sodroski et al. 1986; Malim et al. 1989). Several other retroviruses, as originally shown for Mason-Pfizer monkey virus (MPMV), dispense with a protein factor and simply encode *cis* elements, referred to as constitutive transport elements or CTEs, that directly interact with cellular RNA export factors (for reviews of HIV-1 and MPMV RNA export, see Pollard and Malim 1998; Cullen 2003).

Initial studies showed that the ~9-kb and ~4-kb HIV-1 mRNA species, which encode the structural proteins Gag, Pol, and Env, require Rev for their transport and expression. On the other hand, the completely spliced ~1.8-kb mRNAs, which encode Tat, Rev, and Nef, are exported to the cytoplasm in the absence of Rev by an endogenous cellular pathway used by cellular mRNAs. This division of transport mechanisms is achieved because the

region of the HIV-1 env gene between 5′ss D4 and 3′ss A7 that contains the RRE is removed in the completely spliced mRNAs (Figs. 3 and 4).

Rev-regulated transport requires Rev binding to the RRE. The RRE is an elongated stem-loop structure of 351 nt (Malim et al. 1989; Mann et al. 1994; Watts et al. 2009). Rev binds initially to a high affinity site located near the apex of the RRE structure (stem IIB) (Daly et al. 1989; Heaphy et al. 1990). NMR studies have shown that, as in the case of the Tat-TAR interaction, the Rev binding to the high affinity site induces a conformational change that results in the formation of two purine–purine non-Watson-Crick base pairs (Fig. 5A). This change in the structure of the RNA helix allows binding of the Rev ARD to the major groove (Battiste et al. 1996; Daugherty et al. 2008).

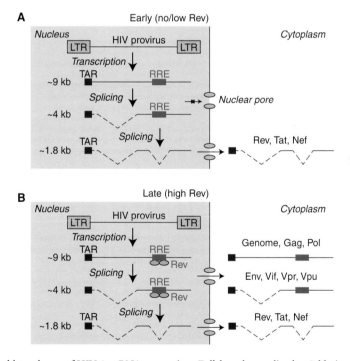

Figure 4. Early and late phases of HIV-1 mRNA expression. Full-length unspliced ~9-kb, incompletely spliced ~4-kb mRNA, and completely spliced ~1.8-kb mRNAs are expressed at both early and late times. (*A*) In the absence of Rev or when Rev is below the threshold necessary for it to function, the ~9-kb and ~4-kb mRNAs are confined to the nucleus and either spliced or degraded. Completely spliced ~1.8-kb mRNAs are constitutively exported to the cytoplasm and translated to yield Rev, Tat, and Nef. (*B*) When the levels of Rev (shown as a pink oval) in the nucleus exceed the threshold necessary for function, the ~9-kb and ~4-kb mRNAs are exported to the cytoplasm and translated. The Rev-response element (RRE) is shown as a red rectangle. (Figure adapted from Pollard and Malim 1998; reprinted, with permission, from *Annual Review of Microbiology* © 1998.)

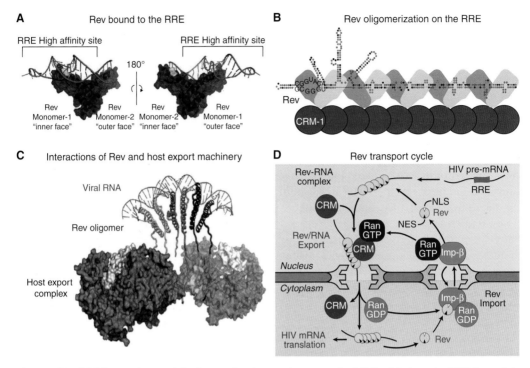

Figure 5. Rev:RRE interactions and the Rev nuclear import/export cycle. (*A*) Rev binds to the RRE through its arginine-rich domain (ARD). In this model developed by Daugherty et al. (2010), the crystal structures of a Rev dimer are combined with the NMR structures of the Rev high affinity site. Notice the distortion of the RNA helix at the site of Rev binding. (*B*) Rev oligomerizes on the RRE and forms a complex with Crm1. The full-length RRE folds into an elongated RNA-stem loop structure with the high affinity binding site for Rev at the apex (Mann et al. 1994; Watts et al. 2009). (*C*) Model for the interactions between Rev and the nuclear export complex containing CRM-1 through the Rev nuclear export sequence (NES). The NES is an extended unstructured region emerging from one face of the Rev molecule. The core arginine-rich RNA binding domains interact with the RRE (Daugherty et al. 2010). (*D*) The Rev nuclear export cycle. Rev and the nuclear export complex containing CRM-1 interacts with nuclear pore proteins and is exported through nuclear pores to the cytoplasm. Once in the cytoplasm, Ran-GTP is converted to Ran-GDP, which is mediated by RanGAP and RanBP1. Crm1 is then transported back into the nucleus and Rev is released from the RRE. Importin-β binds to Rev through the nuclear localization signal in the ARD and interacts with Ran-GDP to facilitate import through the nuclear pore into the nucleus. In the nucleus, Ran-GDP is converted to Ran-GTP in the presence of RCC1. This releases Rev, which can begin another cycle of RRE-dependent Rev export. (Figure adapted from Pollard and Malim 1998; reprinted, with permission, from *Annual Review of Microbiology* © 1998.)

Binding of Rev to the high affinity RRE site is then followed by binding of additional monomers to the complex (Malim and Cullen 1991; Zapp et al. 1991; Mann et al. 1994). The degree of oligomerization correlates with the ability of Rev to transport RNA (Fig. 5B,C) (Mann et al. 1994). Furthermore, oligomerization of Rev on the RRE is highly cooperative and results in an affinity approximately 500 times higher than Rev binding to the high affinity site alone (Daugherty et al. 2008).

Transport of HIV-1 RNA out of the nucleus, as with most cellular proteins and RNAs, occurs via the nuclear pore complexes (NPC) (Fig. 5D) (for review see Kohler and Hurt 2007). Rev bound to the RRE interacts with the karyopherin family member Crm1 (also referred to as exportin 1) through an ∼10 amino acid leucine-rich export nuclear export signal (NES) near the Rev carboxyl terminus. Crm1, like other members of the karyopherin family, binds to cargo in the presence of the GTP-bound form of Ran GTPase. After export to the cytoplasm through the NPC, the bound GTP is hydrolyzed to GDP facilitated by the proteins RanGAP (Ran GTPase-activating protein) and RanBP1. This destabilizes the Rev complex and releases factors from the RRE (Fischer et al. 1995). Rev then reenters the nucleus by binding to the nuclear import factor, importin-β (Henderson and Percipalle 1997).

In the culmination of years of effort, recent crystallographic studies have led to the solution of both the amino-terminal structure of the Rev dimer (DiMattia et al. 2010) and an intact Rev dimer (Fig. 5A) (Daugherty et al. 2010). In the Rev dimer, the arginine-rich RNA-binding helices are located at the ends of a V-shaped assembly. This allows the dimer to bind adjacent RNA sites and structurally couples dimerization and RNA recognition. A second protein–protein interface permits Rev oligomers to act as an adaptor to the host export machinery, with viral RNA bound to one face and Crm1 to another. When excess Rev is present, a defined RNP complex of three dimers bound to the RRE is formed (Fig. 5C) (Daugherty et al. 2010).

Nuclear Retention of Unspliced and Incompletely Spliced mRNAs

Rev regulation requires accumulation of a pool of unspliced and incompletely spliced mRNAs in the nucleus. Nuclear retention is achieved both because cellular factors bind to unused HIV-1 3′ss and 5′ss (Chang and Sharp 1989; Borg et al. 1997) and because of *cis*-acting repressive sequences (CRSs) or instability sequences (INSs). These elements can be introduced into heterologous expression constructs and confer Rev-regulation to RNAs produced from these constructs (Schwartz et al. 1992; Najera et al. 1999). A number of different cellular RNA-binding proteins have been implicated in the retention mediated by CRS/INS elements including poly(A)-binding protein 1 (PABP1), heterogeneous ribonuclear protein A1 (hnRNP A1), and the heterodimer of two related proteins polypyrimidine tract binding protein-associated splicing factor (PSF) and p54(nrb)

(Black et al. 1996; Afonina et al. 1997; Najera et al. 1999; Zolotukhin et al. 2003).

Guiding HIV-1 Transcripts through the Cytoplasm

In contrast to cellular mRNAs, the cytoplasmic fate of unspliced HIV-1 RNA appears to be strongly influenced by the choice of RNA export pathway. For example, HIV-1 Gag assembly in murine cells is normally very inefficient, however, altering the RNA nuclear export element used by HIV-1 *gag-pol* mRNA from the Rev response element to the constitutive transport element (CTE) restored both the trafficking of Gag to cellular membranes and efficient HIV-1 assembly (Swanson et al. 2004). Similarly, defective HIV-1 assembly occurred in human cells when export of *gag-pol* mRNA is dependent on the presence of the hepatitis B posttranscriptional element (PRE) (Jin et al. 2009).

3′ Processing and Polyadenylation of HIV-1 RNA

The 3′ processing and polyadenylation of metazoan pre-mRNAs involves recognition of the upstream AAUAAA and downstream GU-rich motifs surrounding the cleavage and poly(A) addition site. The AAUAAA signal is recognized by the cleavage/polyadenylation specificity factor (CPSF) the GU-rich motif is recognized by CstF. In addition, the cleavage reaction requires mammalian cleavage factors CF1m, CF2m, and poly(A) polymerase (for review, see Colgan and Manley 1997; Millevoi and Vagner 2010).

Like most retroviruses, HIV-1 contains a duplicated set of AAUAAA and GU-rich core elements at the ends of the R sequences found in both the 5′ and 3′ LTRs. HIV-1 uses multiple regulatory elements to direct processing to the 3′ LTR cleavage site. First, the HIV-1 U3 sequence, which is upstream of the 3′ processing signal, but not associated with the 5′ processing signal, contains upstream enhancer elements (USE) that act to facilitate binding of CPSF and enhance polyadenylation at the 3′ end of the HIV-1 transcripts (Gilmartin et al. 1995). Another USE element near the 5′ end of the

Nef gene binds the cellular SR protein 9G8 which recruits the 3′ processing factor CF1m and CPSF (Valente et al. 2009). Second, the 5′ and 3′ LTR poly(A) processing sites are imbedded in a region of secondary structure called the poly(A) hairpin located in exon 1 immediately downstream from the TAR hairpin structure. Factors binding to sequences upstream of the AAUAAA site are believed to open up the poly(A) hairpin and allow preferential use of the 3′ LTR poly(A) processing site (Das et al. 1999). Finally, the splicing factor U1 snRNP acts to inhibit the 3′ processing and poly(A) site in the 5′ LTR by binding to the adjacent 5′ss D1. Mutations of 5′ss D1 that weaken binding of U1 snRNP allow the usage of the normally silent 5′ LTR poly(A) site (Ashe et al. 1995, 2000).

HIV-1 Translation Initiation

Initiation of translation of eukaryotic mRNAs involves scanning from the 5′ cap until an initiator AUG in an appropriate Kozak consensus sequence is recognized. Because the HIV-1 exon 1 contains multiple highly structured regions including the TAR sequence, the primer binding site, the poly(A) hairpin, and RNA packaging sequences, a typical ribosomal scanning mechanism for translational initiation is precluded. Furthermore, some of the HIV-1 mRNA UTRs contain AUG sequences upstream of the authentic initiator AUG that can interfere with translation initiation at the authentic AUG. Finally, as shown in Figure 1, all HIV-1 env mRNA species are bicistronic and have an upstream vpu open reading frame overlapping the downstream env open reading frame.

Several mechanisms have been proposed to circumvent these obstacles (for a recent review, see Bolinger and Boris-Lawrie 2009). HIV-1 may include an internal ribosome binding site (IRES), similar to those as found in picornaviruses, that permits recognition of the gag initiation codon. Additionally, HIV-1 and other retroviruses contain posttranscriptional elements (PCEs) that bind to cellular RNA binding proteins that can act as enhancers to facilitate translation initiation. Gag translation can be enhanced by RHA, a DEIH helicase (Bolinger et al. 2010) as well as the RNA binding proteins SRp40 and SRp55 (Swanson et al. 2010). An additional mechanism, which is used to bypass the vpu open reading frame and permit efficient translation at the downstream env AUG, involves 5′ cap-dependent ribosome shunting in which the scanning ribosome jumps over large regions of the mRNA before recognizing the correct initiation codon (Krummheuer et al. 2007).

HIV-1 Frameshifting

In common with all other retroviruses, HIV-1 has evolved a novel mechanism of programmed frameshifting in which specific sequence and structural signals in the mRNA can specify an mRNA reading frame change during translation (for review, see Brierley and Dos Ramos 2006; Bolinger and Boris-Lawrie 2009). In HIV-1, a −1 shift in the translational reading frame is required to shift from the Gag reading frame to the pro and pol reading frame. This frameshift occurs ∼5% of the time and results in the production of about one Gag-Pro-Pol precursor for every 20 Gag precursors synthesized. Two essential cis-acting sequence elements located ∼200 nt upstream of the Gag termination codon are required for frameshifting. The first is a hexanucleotide "slippery" sequence (UUUUUUA), which is the actual site of slippage during translation. The second is a stem-loop pseudoknot structure located just 3′ to the heptanucleotide sequence that acts as a pause site that increases the time the ribosome is associated with the slippery sequence. Pausing alone appears to be insufficient for frameshifting because addition of other roadblocks to translation are unable to support frameshifting (Brierley and Dos Ramos 2006).

CONCLUSIONS

As summarized above, gene expression in HIV-1 is controlled by the RNA-binding proteins Tat and Rev, which orchestrate complex interactions with the cellular transcription, RNA splicing, and RNA transport machinery.

Cite this article as Cold Spring Harb Perspect Med doi: 10.1101/cshperspect.a006916

The mechanisms of action of both proteins, which were unprecedented at the time of their discovery, have now illuminated features of transcription elongation control, RNA splicing, and RNA export that were entirely unknown and unexpected.

Although the field has matured and expanded dramatically since the original discovery of Tat and Rev in 1985, an enormous amount still remains to be discovered about their structures and functions. On the structural side, complexes between TAR RNA and the intact P-TEFb molecule and large complexes containing P-TEFb and the RNA polymerase transcription elongation complex have yet to be tackled. The recent discovery of a whole family of additional elongation factors recruited associated with Tat certainly removes any complacency that all the cofactors required for Tat-mediated transactivation have been identified.

A particularly challenging problem is to understand the coupling that occurs between transcriptional elongation, the regulation of splicing, polyadenylation, and RNA export. Evidence that the splicing-associated c-Ski-interacting protein, SKIP, activates both Tat transactivation and HIV-1 splicing provides an intriguing insight into how these diverse events may be coordinated (Bres et al. 2005).

Understanding how the viral mRNP is remodeled during its journey from the nucleus to the cytoplasm will be essential for understanding both HIV-1 RNA translation and RNA packaging into virions. Remarkably, the export pathway used by the RNA affects viral assembly at the plasma membrane (Jin et al. 2009; Sherer et al. 2009), suggesting the site of translation influences subsequent protein function. In addition, there have been a growing number of host cell nuclear RNA binding proteins identified that bind to HIV-1 RNA or to Rev and may influence its export and cytoplasmic utilization (for reviews, see Cochrane et al. 2006; Cochrane 2009; Suhasini and Reddy 2009).

Finally, in addition to their academic interest, the HIV-1 regulatory proteins remain important targets for drug discovery, especially because Tat and Rev are both required for active viral replication and essential for the emergence of viruses from latency. Now that the structures of both proteins are known, and many of their cofactors have been identified, there is cause for optimism that renewed efforts to develop antiviral compounds will be successful.

REFERENCES

*Reference is also in this collection.

Aboul-ela F, Karn J, Varani G. 1995. The structure of the human immunodeficiency virus type 1 TAR RNA reveals principles of RNA recognition by Tat protein. *J Mol Biol* **253:** 313–332.

Afonina E, Neumann M, Pavlakis GN. 1997. Preferential binding of poly(A)-binding protein 1 to an inhibitory RNA element in the human immunodeficiency virus type 1 *gag* element. *J Biol Chem* **272:** 2307–2311.

Alcami J, de Lera TL, Folgueira L, Pedraza M-A, Jacqué J-M, Bachelerie F, Noriega AR, Hay RT, Harrich D, Gaynor RB, et al. 1995. Absolute dependence on κB responsive elements for initiation and Tat-mediated amplification of HIV transcription in blood CD4 T lymphocytes. *EMBO J* **14:** 1552–1560.

Amendt BA, Hesslein D, Chang L-J, Stoltzfus CM. 1994. Presence of negative and positive *cis*-acting RNA splicing elements within and flanking the first *tat* coding exon of the human immunodeficiency virus type 1. *Mol Cell Biol* **14:** 3960–3970.

Asang C, Hauber I, Schaal H. 2008. Insights into the selective activation of alternatively used splice acceptors by the human immunodeficiency virus type-1 bidirectional splicing enhancer. *Nucleic Acids Res* **36:** 1450–1463.

Ashe MP, Griffin P, James W, Proudfoot NJ. 1995. Poly(A) site selection in the HIV-1 provirus: Inhibition of promoter-proximal polyadenylation by the downstream major splice donor site. *Genes Dev* **9:** 3008–3025.

Ashe MP, Furger A, Proudfoot NJ. 2000. Stem-loop 1 of the U1 snRNP plays a crical role in the supression of HIV-1 polyadenylation. *RNA* **6:** 170–177.

Barboric M, Yik JH, Czudnochowski N, Yang Z, Chen R, Contreras X, Geyer M, Matija Peterlin B, Zhou Q. 2007. Tat competes with HEXIM1 to increase the active pool of P-TEFb for HIV-1 transcription. *Nucleic Acids Res* **35:** 2003–2012.

Battiste JL, Mao H, Rao NS, Tan R, Muhandiram DR, Kay LE, Frankel AD, Williamson JR. 1996. α Helix-RNA major groove recognition in an HIV-1 Rev peptide-RRE RNA complex. *Science* **273:** 1547–1551.

Berkhout B, Silverman RH, Jeang K-T. 1989. Tat *trans*-activates the human immunodeficiency virus through a nascent RNA target. *Cell* **59:** 273–282.

Bieniasz PD, Grdina TA, Bogerd HP, Cullen BR. 1998. Recruitment of a protein complex containing Tat and cyclin T1 to TAR governs the species specificity of HIV-1 Tat. *EMBO J* **17:** 7056–7065.

Bilodeau PS, Domsic JK, Mayeda A, Krainer AR, Stoltzfus CM. 2001. RNA splicing at human immunodeficiency virus type 1 3′ splice site A2 is regulated by binding of hnRNP A/B proteins to an exonic splicing silencer element. *J Virol* **75**: 8487–8497.

Black AC, Luo J, Chun S, Bakker A, Fraser JK, Rosenblatt JD. 1996. Specific binding of polypyrimidine tract binding protein and hnRNP A1 to HIV-1 CRS elements. *Virus Genes* **12**: 275–285.

Bolinger C, Boris-Lawrie K. 2009. Mechanisms employed by retroviruses to exploit host factors for translational control of a complicated proteome. *Retrovirology* **6**: 8.

Bolinger C, Sharma A, Singh D, Yu L, Boris-Lawrie K. 2010. RNA helicase A modulates translation of HIV-1 and infectivity of progeny virions. *Nucleic Acids Res* **38**: 1686–1696.

Borg KT, Favaro JP, Arrigo SJ. 1997. Involvement of human immunodeficiency virus type-1 splice sites in the cytoplasmic accumulation of viral RNA. *Virology* **236**: 95–103.

Bosque A, Planelles V. 2008. Induction of HIV-1 latency and reactivation in primary memory CD4$^+$ T cells. *Blood* **113**: 58–65.

Bourgeois CF, Kim YK, Churcher MJ, West MJ, Karn J. 2002. Spt5 cooperates with Tat by preventing premature RNA release at terminator sequences. *Mol Cell Biol* **22**: 1079–1093.

Bres V, Gomes N, Pickle L, Jones KA. 2005. A human splicing factor, SKIP, associates with P-TEFb and enhances transcription elongation by HIV-1 Tat. *Genes Dev* **19**: 1211–1226.

Brierley I, Dos Ramos FJ. 2006. Programmed ribosomal frameshifting in HIV-1 and the SARS-CoV. *Virus Res* **119**: 29–42.

Brodsky AS, Williamson JR. 1997. Solution structure of the HIV-2 TAR-argininamide complex. *J Mol Biol* **267**: 624–639.

Burnett JC, Miller-Jensen K, Shah PS, Arkin AP, Schaffer DV. 2009. Control of stochastic gene expression by host factors at the HIV promoter. *PLoS Pathog* **5**: e1000260.

Caputi M, Freund M, Kammler S, Asang C, Schaal H. 2004. A bidirectional SF2/ASF, SRp40 dependent splicing enhancer regulates HIV-1 *rev, env, vpu*, and *nef* gene expression. *J Virol* **78**: 6517–6526.

Chang DD, Sharp PA. 1989. Regulation by HIV Rev depends upon recognition of splice sites. *Cell* **59**: 789–795.

Chen M, Manley JL. 2009. Mechanisms of alternative splicing regulation: Insights from molecular and genomics approaches. *Nat Rev Mol Cell Biol* **10**: 741–754.

Chen BK, Feinberg MB, Baltimore D. 1997. The κB sites in the human immunodeficiency virus type 1 long terminal repeat enhance virus replication yet are not absolutely required for viral growth. *J Virol* **71**: 5495–5504.

Chen-Park FE, Huang DB, Noro B, Thanos D, Ghosh G. 2002. The κB DNA sequence from the HIV long terminal repeat functions as an allosteric regulator of HIV transcription. *J Biol Chem* **277**: 24701–24708.

Cho S, Schroeder S, Kaehlcke K, Kwon HS, Pedal A, Herker E, Schnoelzer M, Ott M. 2009. Acetylation of cyclin T1 regulates the equilibrium between active and inactive P-TEFb in cells. *EMBO J* **28**: 1407–1417.

Churcher M, Lamont C, Hamy F, Dingwall C, Green SM, Lowe AD, Butler PJG, Gait MJ, Karn J. 1993. High affinity binding of TAR RNA by the human immunodeficiency virus Tat protein requires amino acid residues flanking the basic domain and base pairs in the RNA stem. *J Mol Biol* **230**: 90–110.

Cochrane A. 2009. How does the journey affect the message(RNA)? *RNA Biol* **6**: 169–170.

Cochrane AW, McNally MT, Mouland AJ. 2006. The retrovirus RNA trafficking granule: From birth to maturity. *Retrovirology* **3**: 18.

Colgan DF, Manley JL. 1997. Mechanism and regulation of mRNA polyadenylation. *Genes Dev* **11**: 2755–2766.

Cullen BR. 2003. Nuclear mRNA export: Insights from virology. *Trends Biochem Sci* **28**: 419–424.

Daly TJ, Cook KS, Gary GS, Maione TE, Rusche JR. 1989. Specific binding of HIV-1 recombinant Rev protein to the Rev-responsive element in vitro. *Nature* **342**: 816–819.

Das AT, Klaver B, Berkhout B. 1999. A hairpin structure in the R region of the human immunodeficiency virus type 1 RNA genome is instrumental in polyadenylation site selection. *J Virol* **73**: 81–91.

Daugherty MD, D'Orso I, Frankel AD. 2008. A solution to limited genomic capacity: Using adaptable binding surfaces to assemble the functional HIV Rev oligomer on RNA. *Mol Cell* **31**: 824–834.

Daugherty MD, Liu B, Frankel AD. 2010. Structural basis for cooperative RNA binding and export complex assembly by HIV Rev. *Nat Struct Mol Biol* **17**: 1337–1342.

Davidson A, Leeper TC, Athanassiou Z, Patora-Komisarska K, Karn J, Robinson JA, Varani G. 2009. Simultaneous recognition of HIV-1 TAR RNA bulge and loop sequences by cyclic peptide mimics of Tat protein. *Proc Natl Acad Sci* **106**: 11931–11936.

DiMattia MA, Watts NR, Stahl SJ, Rader C, Wingfield PT, Stuart DI, Steven AC, Grimes JM. 2010. Implications of the HIV-1 Rev dimer structure at 3.2 A resolution for multimeric binding to the Rev response element. *Proc Natl Acad Sci* **107**: 5810–5814.

Dingwall C, Ernberg I, Gait MJ, Green SM, Heaphy S, Karn J, Lowe AD, Singh M, Skinner MA, Valerio R. 1989. Human immunodeficiency virus 1 Tat protein binds transactivation-responsive region (TAR) RNA in vitro. *Proc Natl Acad Sci* **86**: 6925–6929.

Dingwall C, Ernberg I, Gait MJ, Green SM, Heaphy S, Karn J, Lowe AD, Singh M, Skinner MA. 1990. HIV-1 Tat protein stimulates transcription by binding to a U-rich bulge in the stem of the TAR RNA structure. *EMBO J* **9**: 4145–4153.

Exline CM, Feng Z, Stoltzfus CM. 2008. Negative and positive mRNA splicing elements act competitively to regulate human immunodeficiency virus type 1 vif gene expression. *J Virol* **82**: 3921–3931.

Feng S, Holland EC. 1988. HIV-1 Tat transactivation requires the loop sequence within TAR. *Nature* **334**: 165–168.

Fischer U, Huber J, Boelens WC, Mattaj IW, Luhrmann R. 1995. The HIV-1 Rev activation domain is a nuclear export signal that accesses an export pathway used by specific cellular RNAs. *Cell* **82**: 475–483.

Fujinaga K, Cujec TP, Peng J, Garriga J, Price DH, Graña X, Peterlin BM. 1998. The ability of positive transcription elongation factor b to transactive human immunodeficiency virus transcription depends on a functional kinase domain, cyclin T1 and Tat. *J Virol* **72:** 7154–7159.

Fujinaga K, Irwin D, Huang Y, Taube R, Kurosu T, Peterlin BM. 2004. Dynamics of human immunodeficiency virus transcription: P-TEFb phosphorylates RD and dissociates negative effectors from the transactivation response element. *Mol Cell Biol* **24:** 787–795.

Garber ME, Wei P, KewelRamani VN, Mayall TP, Herrmann CH, Rice AP, Littman DR, Jones KA. 1998. The interaction between HIV-1 Tat and human cyclin T1 requires zinc and a critical cysteine residue that is not conserved in the murine CycT1 protein. *Genes Dev* **12:** 3512–3527.

Garcia JA, Harrich D, Soultanakis E, Wu F, Mitsuyasu R, Gaynor RB. 1989. Human immunodeficiency virus type 1 LTR TATA and TAR region sequences required for transcriptional regulation. *EMBO J* **8:** 765–778.

Giffin MJ, Stroud JC, Bates DL, von Koenig KD, Hardin J, Chen L. 2003. Structure of NFAT1 bound as a dimer to the HIV-1 LTR κB element. *Nature Struct Biol* **10:** 800–806.

Gilmartin GM, Fleming ES, Oetjen J, Graveley BR. 1995. CPSF recognition of HIV-1 mRNA 3′-processing enhancer: Multiple sequence contacts involved in poly(A) site definition. *Genes Dev* **9:** 72–83.

He N, Jahchan NS, Hong E, Li Q, Bayfield MA, Maraia RJ, Luo K, Zhou Q. 2008. A La-related protein modulates 7SK snRNP integrity to suppress P-TEFb-dependent transcriptional elongation and tumorigenesis. *Mol Cell* **29:** 588–599.

He N, Liu M, Hsu J, Xue Y, Chou S, Burlingame A, Krogan NJ, Alber T, Zhou Q. 2010. HIV-1 Tat and host AFF4 recruit two transcription elongation factors into a bifunctional complex for coordinated activation of HIV-1 transcription. *Mol Cell* **38:** 428–438.

Heaphy S, Dingwall C, Ernberg I, Gait MJ, Green SM, Karn J, Lowe AD, Singh M, Skinner MA. 1990. HIV-1 regulator of virion expression (Rev) protein binds to an RNA stem-loop structure located within the Rev-response element region. *Cell* **60:** 685–693.

Henderson BR, Percipalle P. 1997. Interactions between HIV Rev and nuclear import and export factors: The Rev nuclear localisation signal mediates specific binding to human importin-b. *J Mol Biol* **274:** 693–707.

Herrmann CH, Rice AP. 1995. Lentivirus Tat proteins specifically associate with a cellular protein kinase, TAK, that hyperphosphorylates the carboxyl-terminal domain of the large subunit of RNA polymerase II: Candidate for a Tat cofactor. *J Virol* **69:** 1612–1620.

Herrmann CH, Gold MO, Rice AP. 1996. Viral transactivators specifically target distinct cellular protein kinases that phosphorylate the RNA polymerase II C-terminal domain. *Nucleic Acids Res* **24:** 501–508.

Hoffman BE, Grabowski PJ. 1992. U1 snRNP targets an essential splicing factor, U2AF65, to the 3′ splice site by a network of interactions spanning the exon. *Genes Dev* **6:** 2554–2568.

Isel C, Karn J. 1999. Direct evidence that HIV-1 Tat activates the Tat-associated kinase (TAK) during transcriptional elongation. *J Mol Biol* **290:** 929–941.

Ivanov D, Kwak YT, Guo J, Gaynor RB. 2000. Domains in the SPT5 protein that modulate its transcriptional regulatory properties. *Mol Cell Biol* **20:** 2970–2983.

Jacquenet S, Mereau A, Bilodeau PS, Damier L, Stoltzfus CM, Branlant C. 2001. A second exon splicing silencer within human immunodeficiency virus type 1 tat exon 2 represses splicing of Tat mRNA and binds protein hnRNP H. *J Biol Chem* **276:** 40464–40475.

Jeronimo C, Forget D, Bouchard A, Li Q, Chua G, Poitras C, Therien C, Bergeron D, Bourassa S, Greenblatt J, et al. 2007. Systematic analysis of the protein interaction network for the human transcription machinery reveals the identity of the 7SK capping enzyme. *Mol Cell* **27:** 262–274.

Jin J, Sturgeon T, Weisz OA, Mothes W, Montelaro RC. 2009. HIV-1 matrix dependent membrane targeting is regulated by Gag mRNA trafficking. *PLoS One* **4:** e6551.

Jones K, Kadonaga J, Luciw P, Tjian R. 1986. Activation of the AIDS retrovirus promoter by the cellular transcription factor, Sp1. *Science* **232:** 755–759.

Kammler S, Otte M, Hauber I, Kjems J, Hauber J, Schaal H. 2006. The strength of the HIV-1 3′ splice sites affects Rev function. *Retrovirology* **3:** 89.

Kao S-Y, Calman AF, Luciw PA, Peterlin BM. 1987. Anti-termination of transcription within the long terminal repeat of HIV-1 by Tat gene product. *Nature* **330:** 489–493.

Karn J. 2011. The molecular biology of HIV latency: Breaking and restoring the Tat-dependent transcriptional circuit. *Curr Opin HIVAIDS* **6:** 4–11.

Kim S, Byrn R, Groopman J, Baltimore D. 1989. Temporal aspects of DNA and RNA synthesis during human immunodeficiency virus infection: Evidence for differential gene expression. *J Virol* **63:** 3708–3713.

Kim YK, Bourgeois CF, Isel C, Churcher MJ, Karn J. 2002. Phosphorylation of the RNA polymerase II carboxyl-terminal domain by CDK9 is directly responsible for human immunodeficiency virus type 1 Tat-activated transcriptional elongation. *Mol Cell Biol* **22:** 4622–4637.

Kinoshita S, Chen BK, Kaneshima H, Nolan GP. 1998. Host control of HIV-1 parasitism in T cells by the nuclear factor of activated T cells. *Cell* **95:** 595–604.

Kohler A, Hurt E. 2007. Exporting RNA from the nucleus to the cytoplasm. *Nat Rev Mol Cell Biol* **8:** 761–773.

Krueger BJ, Jeronimo C, Roy BB, Bouchard A, Barrandon C, Byers SA, Searcey CE, Cooper JJ, Bensaude O, Cohen EA, et al. 2008. LARP7 is a stable component of the 7SK snRNP while P-TEFb, HEXIM1 and hnRNP A1 are reversibly associated. *Nucleic Acids Res* **36:** 2219–2229.

Krueger BJ, Varzavand K, Cooper JJ, Price DH. 2010. The mechanism of release of P-TEFb and HEXIM1 from the 7SK snRNP by viral and cellular activators includes a conformational change in 7SK. *PLoS One* **5:** e12335.

Krummheuer J, Johnson AT, Hauber I, Kammler S, Anderson JL, Hauber J, Purcell DF, Schaal H. 2007. A minimal uORF within the HIV-1 vpu leader allows efficient translation initiation at the downstream env AUG. *Virology* **363:** 261–271.

Kwak YT, Ivanov D, Guo J, Nee E, Gaynor RB. 1999. Role of the human and murine cyclin T proteins in regulating HIV-1 Tat-activation. *J Mol Biol* **288:** 57–69.

Lewinski MK, Yamashita M, Emerman M, Ciuffi A, Marshall H, Crawford G, Collins F, Shinn P, Leipzig J, Hannenhalli S, et al. 2006. Retroviral DNA integration: Viral and cellular determinants of target-site selection. *Plos Pathog* **2**: e60.

Liu J, Perkins ND, Schmid RM, Nabel GJ. 1992. Specific NF-kB subunits act in concert with tat to stimulate human immunodeficiency virus type 1 transcription. *J Virol* **66**: 3883–3887.

Madsen JM, Stoltzfus CM. 2005. An exonic splicing silencer downstream of 3′ splice site A2 is required for efficient human immunodeficiency virus type 1 replication. *J Virol* **79**: 10478–10486.

Madsen JM, Stoltzfus CM. 2006. A suboptimal 5′ splice site downstream of HIV-1 splice site A1 is required for unspliced viral mRNA accumulation and efficient virus replication. *Retrovirology* **3**: 10.

Malim MH, Cullen BR. 1991. HIV-1 structural gene expression requires the binding of multiple Rev monomers to the viral RRE: Implications for HIV-1 latency. *Cell* **65**: 241–248.

Malim MH, Hauber J, Le S-Y, Maizel JV, Cullen BR. 1989. The HIV-1 *rev* trans-activator acts through a structured target sequence to activate nuclear export of unspliced viral mRNA. *Nature* **338**: 254–257.

Mancebo HSY, Lee G, Flygare J, Tomassini J, Luu P, Zhu Y, Peng J, Blau C, Hazuda D, Price D, et al. 1997. p-TEFb kinase is required for HIV Tat transcriptional activation in vivo and in vitro. *Genes Dev* **11**: 2633–2644.

Mandal D, Feng Z, Stoltzfus CM. 2008. Gag-processing defect of human immunodeficiency virus type 1 integrase E246 and G247 mutants is caused by activation of an overlapping 5′ splice site. *J Virol* **82**: 1600–1604.

Mann DA, Mikaélian I, Zemmel RW, Green SM, Lowe AD, Kimura T, Singh M, Butler PJG, Gait MJ, Karn J. 1994. A molecular rheostat: Co-operative Rev binding to Stem I of the Rev-response element modulates human immunodeficiency virus type-1 late gene expression. *J Mol Biol* **241**: 193–207.

Margolis DM. 2010. Mechanisms of HIV latency: An emerging picture of complexity. *Curr HIV/AIDS Rep* **7**: 37–43.

Marshall NF, Price DH. 1995. Purification of p-TEFb, a transcription factor required for the transition into productive elongation. *J Biol Chem* **270**: 12335–12338.

Marshall NF, Peng J, Xie Z, Price DH. 1996. Control of RNA polymerase II elongation potential by a novel carboxyl-terminal domain kinase. *J Biol Chem* **271**: 27176–27183.

Michels AA, Fraldi A, Li Q, Adamson TE, Bonnet F, Nguyen VT, Sedore SC, Price JP, Price DH, Lania L, et al. 2004. Binding of the 7SK snRNA turns the HEXIM1 protein into a P-TEFb (CDK9/cyclin T) inhibitor. *EMBO J* **23**: 2608–2619.

Millevoi S, Vagner S. 2010. Molecular mechanisms of eukaryotic pre-mRNA 3′ end processing regulation. *Nucleic Acids Res* **38**: 2757–2774.

Muesing MA, Smith DH, Capon DJ. 1987. Regulation of mRNA accumulation by a human immunodeficiency virus *trans*-activator protein. *Cell* **48**: 691–701.

Nabel G, Baltimore DA. 1987. An inducible transcription factor activates expression of human immunodeficiency virus in T cells. *Nature* **326**: 711–713.

Najera I, Krieg M, Karn J. 1999. Synergistic stimulation of HIV-1 Rev-dependent export of unspliced mRNA to the cytoplasm by HnRNP A1. *J Mol Biol* **285**: 1951–1964.

Narita T, Yamaguchi Y, Yano K, Sugimoto S, Chanarat S, Wada T, Kim DK, Hasegawa J, Omori M, Inukai N, et al. 2003. Human transcription elongation factor NELF: Identification of novel subunits and reconstitution of the functionally active complex. *Mol Cell Biol* **23**: 1863–1873.

Nguyen VT, Kiss T, Michels AA, Bensaude O. 2001. 7SK small nuclear RNA binds to and inhibits the activity of CDK9/cyclin T complexes. *Nature* **414**: 322–325.

Pearson R, Kim YK, Hokello J, Lassen K, Friedman J, Tyagi M, Karn J. 2008. Epigenetic silencing of human immunodeficiency virus (HIV) transcription by formation of restrictive chromatin structures at the viral long terminal repeat drives the progressive entry of HIV into latency. *J Virol* **82**: 12291–12303.

Perkins ND, Edwards NL, Duckett CS, Agranoff AB, Schmid RM, Nabel GJ. 1993. A cooperative interaction between NF-kB and Sp1 is required for HIV-1 enhancer activation. *EMBO J* **12**: 3551–3558.

Pollard VW, Malim MH. 1998. The HIV-1 Rev protein. *Annu Rev Microbiol* **52**: 491–532.

Pomerantz RJ, Trono D, Feinberg MB, Baltimore D. 1990. Cells nonproductively infected with HIV-1 exhibit an aberrant pattern of viral RNA expression: A molecular model for latency. *Cell* **61**: 1271–1276.

Puglisi JD, Tan R, Calnan BJ, Frankel AD, Williamson JR. 1992. Conformation of the TAR RNA-arginine complex by NMR spectroscopy. *Science* **257**: 76–80.

Purcell DFJ, Martin MA. 1993. Alternative splicing of human immunodeficiency virus type 1 mRNA modulates viral protein expression, replication and infectivity. *J Virol* **67**: 6365–6378.

Ramakrishnan R, Dow EC, Rice AP. 2009. Characterization of Cdk9 T-loop phosphorylation in resting and activated CD4$^+$ T lymphocytes. *J Leukoc Biol* **86**: 1345–1350.

Ramanathan Y, Rajpara SM, Reza SM, Lees E, Shuman S, Mathews MB, Pe'ery T. 2001. Three RNA polymerase II carboxyl-terminal domain kinases display distinct substrate preferences. *J Biol Chem* **276**: 10913–10920.

Rittner K, Churcher MJ, Gait MJ, Karn J. 1995. The human immunodeficiency virus long terminal repeat includes a specialised initiator element which is required for Tat-responsive transcription. *J Mol Biol* **248**: 562–580.

Robberson BL, Cote GJ, Berget SM. 1990. Exon definition may facilitate splice site selection in RNAs with multiple exons. *Mol Cell Biol* **10**: 84–94.

Schwartz S, Felber BK, Pavlakis GN. 1992. Distinct RNA sequences in the *gag* region of human immunodeficiency virus type 1 decrease RNA stability and inhibit expression in the absence of Rev protein. *J Virol* **66**: 150–159.

Sedore SC, Byers SA, Biglione S, Price JP, Maury WJ, Price DH. 2007. Manipulation of P-TEFb control machinery by HIV: Recruitment of P-TEFb from the large form by Tat and binding of HEXIM1 to TAR. *Nucleic Acids Res* **35**: 4347–4358.

Selby MJ, Bain ES, Luciw P, Peterlin BM. 1989. Structure, sequence and position of the stem-loop in TAR determine transcriptional elongation by Tat through the HIV-1 long terminal repeat. *Genes Dev* **3:** 547–558.

Sherer NM, Swanson CM, Papaioannou S, Malim MH. 2009. Matrix mediates the functional link between human immunodeficiency virus type 1 RNA nuclear export elements and the assembly competency of Gag in murine cells. *J Virol* **83:** 8525–8535.

* Siliciano RF, Greene WC. 2011. HIV latency. *Cold Spring Harb Perspect Med* doi: 10.1101/cshperspect.a007096.

Sobhian B, Laguette N, Yatim A, Nakamura M, Levy Y, Kiernan R, Benkirane M. 2010. HIV-1 Tat assembles a multifunctional transcription elongation complex and stably associates with the 7SK snRNP. *Mol Cell* **38:** 439–451.

Sodroski J, Patarca R, Rosen C, Wong-Staal F, Haseltine WA. 1985a. Location of the transacting region on the genome of human T-cell lymphotropic virus type III. *Science* **229:** 74–77.

Sodroski JG, Rosen CA, Wong-Staal F, Salahuddin SZ, Popovic M, Arya S, Gallo RC, Haseltine WA. 1985b. *Trans*-acting transcriptional regulation of human T-cell leukemia virus type III long terminal repeat. *Science* **227:** 171–173.

Sodroski J, Goh WC, Rosen CA, Dayton A, Terwilliger E, Haseltine WA. 1986. A second post-transcriptional activator gene required for HTLV-III replication. *Nature* **321:** 412–417.

Stoltzfus CM. 2009. Regulation of HIV-1 alternative RNA splicing and its role in virus replication. *Adv Virus Res* **74:** 1–40.

Stoltzfus CM, Madsen JM. 2006. Role of viral splicing elements and cellular RNA binding proteins in regulation of HIV-1 alternative RNA splicing. *Curr HIV Res* **4:** 43–55.

Suhasini M, Reddy TR. 2009. Cellular proteins and HIV-1 Rev function. *Curr HIV Res* **7:** 91–100.

Sung TL, Rice AP. 2009. miR-198 inhibits HIV-1 gene expression and replication in monocytes and its mechanism of action appears to involve repression of cyclin T1. *PLoS Pathog* **5:** e1000263.

Swanson CM, Puffer BA, Ahmad KM, Doms RW, Malim MH. 2004. Retroviral mRNA nuclear export elements regulate protein function and virion assembly. *EMBO J* **23:** 2632–2640.

Swanson CM, Sherer NM, Malim MH. 2010. SRp40 and SRp55 promote the translation of unspliced human immunodeficiency virus type 1 RNA. *J Virol* **84:** 6748–6759.

Tahirov TH, Babayeva ND, Varzavand K, Cooper JJ, Sedore SC, Price DH. 2010. Crystal structure of HIV-1 Tat complexed with human P-TEFb. *Nature* **465:** 747–751.

Valente ST, Gilmartin GM, Venkatarama K, Arriagada G, Goff SP. 2009. HIV-1 mRNA 3′ end processing is distinctively regulated by eIF3f, CDK11, and splice factor 9G8. *Mol Cell* **36:** 279–289.

Verdin E, Paras PJ, Van Lint C. 1993. Chromatin disruption in the promoter of human immunodeficiency virus type 1 during transcriptional activation. *EMBO J* **12:** 3249–3259.

Wang Z, Burge CB. 2008. Splicing regulation: From a parts list of regulatory elements to an integrated splicing code. *RNA* **14:** 802–813.

Watts JM, Dang KK, Gorelick RJ, Leonard CW, Bess JW Jr, Swanstrom R, Burch CL, Weeks KM. 2009. Architecture and secondary structure of an entire HIV-1 RNA genome. *Nature* **460:** 711–716.

Wei P, Garber ME, Fang S-M, Fischer WH, Jones KA. 1998. A novel cdk9-associated c-type cyclin interacts directly with HIV-1 Tat and mediates its high-affinity, loop specific binding to TAR RNA. *Cell* **92:** 451–462.

Weinberger LS, Shenk T. 2006. An HIV feedback resistor: Auto-regulatory circuit deactivator and noise buffer. *PLoS Biol* **5:** e9.

Weinberger LS, Burnett JC, Toettcher JE, Arkin AP, Schaffer DV. 2005. Stochastic gene expression in a lentiviral positive-feedback loop: HIV-1 Tat fluctuations drive phenotypic diversity. *Cell* **122:** 169–182.

Yamada T, Yamaguchi Y, Inukai N, Okamoto S, Mura T, Handa H. 2006. P-TEFb-mediated phosphorylation of hSpt5 C-terminal repeats is critical for processive transcription elongation. *Mol Cell* **21:** 227–237.

Yamaguchi Y, Takagi T, Wada T, Yano K, Furuya A, Sugimoto S, Hasegawa J, Handa H. 1999. NELF, a multisubunit complex containing RD, cooperates with DSIF to repress RNA polymerase II elongation. *Cell* **97:** 41–51.

Yamaguchi Y, Inukai N, Narita T, Wada T, Handa H. 2002. Evidence that negative elongation factor represses transcription elongation through binding to a DRB sensitivity-inducing factor/RNA polymerase II complex and RNA. *Mol Cell Biol* **22:** 2918–2927.

Yang Z, Zhu Q, Luo K, Zhou Q. 2001. The 7SK small nuclear RNA inhibits the CDK9/cyclin T1 kinase to control transcription. *Nature* **414:** 317–322.

Yik JH, Chen R, Nishimura R, Jennings JL, Link AJ, Zhou Q. 2003. Inhibition of P-TEFb (CDK9/Cyclin T) kinase and RNA polymerase II transcription by the coordinated actions of HEXIM1 and 7SK snRNA. *Mol Cell* **12:** 971–982.

Zapp ML, Hope TJ, Parslow TG, Green MR. 1991. Oligomerization and RNA binding domains of the type 1 human immunodeficiency virus Rev protein: A dual function for an arginine-rich binding motif. *Proc Natl Acad Sci* **88:** 7734–7738.

Zenzie-Gregory B, Sheridan P, Jones KA, Smale ST. 1993. HIV-1 core promoter lacks a simple initiator element but contains bipartite activator at the transcription start site. *J Biol Chem* **268:** 15823–15832.

Zhang Z, Klatt A, Gilmour DS, Henderson AJ. 2007. Negative elongation factor NELF represses human immunodeficiency virus transcription by pausing the RNA polymerase II complex. *J Biol Chem* **282:** 16981–16988.

Zhu Y, Pe'ery T, Peng J, Ramanathan Y, Marshall N, Marshall T, Amendt B, Mathews MB, Price DH. 1997. Transcription elongation factor P-TEFb is required for HIV-1 Tat transactivation in vitro. *Genes Dev* **11:** 2622–2632.

Zolotukhin AS, Michalowski D, Bear J, Smulevitch SV, Traish AM, Peng R, Patton J, Shatsky IN, Felber BK. 2003. PSF acts through the human immunodeficiency virus type 1 mRNA instability elements to regulate virus expression. *Mol Cell Biol* **23:** 6618–6630.

HIV-1 Assembly, Budding, and Maturation

Wesley I. Sundquist[1] and Hans-Georg Kräusslich[2]

[1]Department of Biochemistry, University of Utah School of Medicine, Salt Lake City, Utah 84112-5650

[2]Department of Infectious Diseases, Virology, University of Heidelberg, 69120 Heidelberg, Germany

Correspondence: wes@biochem.utah.edu; hans-georg.kraeusslich@med.uni-heidelberg.de

A defining property of retroviruses is their ability to assemble into particles that can leave producer cells and spread infection to susceptible cells and hosts. Virion morphogenesis can be divided into three stages: *assembly*, wherein the virion is created and essential components are packaged; *budding*, wherein the virion crosses the plasma membrane and obtains its lipid envelope; and *maturation*, wherein the virion changes structure and becomes infectious. All of these stages are coordinated by the Gag polyprotein and its proteolytic maturation products, which function as the major structural proteins of the virus. Here, we review our current understanding of the mechanisms of HIV-1 assembly, budding, and maturation, starting with a general overview and then providing detailed descriptions of each of the different stages of virion morphogenesis.

The assembling virion packages all of the components required for infectivity. These include two copies of the positive sense genomic viral RNA, cellular tRNA^{Lys,3} molecules to prime cDNA synthesis, the viral envelope (Env) protein, the Gag polyprotein, and the three viral enzymes: protease (PR), reverse transcriptase (RT), and integrase (IN). The viral enzymes are packaged as domains within the Gag-Pro-Pol polyprotein, which is generated when translating ribosomes shift into the -1 reading frame at a site near the $3'$ end of the *gag* open reading frame, and then go on to translate the *pol* gene.

HIV-1 virion assembly occurs at the plasma membrane, within specialized membrane microdomains. The HIV-1 Gag (and Gag-Pro-Pol) polyprotein itself mediates all of the essential events in virion assembly, including binding the plasma membrane, making the protein–protein interactions necessary to create spherical particles, concentrating the viral Env protein, and packaging the genomic RNA via direct interactions with the RNA packaging sequence (termed Ψ). These events all appear to occur simultaneously at the plasma membrane, where conformational change(s) within Gag couples membrane binding, virion assembly, and RNA packaging. Although Gag itself can bind membranes and assemble into spherical particles, the budding event that releases the virion from the plasma membrane is mediated by the host ESCRT (endosomal sorting complexes required for transport) machinery.

Folded domains within Gag are separated by flexible linker regions which contain the PR

---

cleavage sites (Fig. 1B,C). The amino-terminal Gag domain is called MA, and it functions to bind the plasma membrane and to recruit the viral Env protein. The central domain of Gag is called CA, and it mediates the protein–protein interactions required for immature virion assembly and then creates the conical shell (called the capsid) of the mature viral core. The basic Gag NC domain contains two copies of the retroviral zinc finger motif. NC captures the viral genome during assembly, and also functions as a nucleic acid "chaperone" during tRNA^{Lys,3} primer annealing and reverse transcription. Finally, the carboxy-terminal Gag p6 region contains binding sites for several other proteins, including the accessory viral protein Vpr, as well as two short sequence motifs, termed "late assembly domains," which bind the TSG101 and ALIX proteins of the cellular ESCRT pathway. Gag also contains two spacer peptides, termed SP1 and SP2, which help to regulate the conformational changes that accompany viral maturation.

The virion acquires its lipid envelope and Env protein spikes as it buds from the plasma membrane. Unlike Gag, Env is an integral membrane protein. It is inserted cotranslationally into ER membranes and then travels through the cellular secretory pathway where it is glycosylated, assembled into trimeric complexes, processed into the trans-membrane (TM; gp41) and surface (SU; gp120) subunits by the cellular protease furin, and delivered to the plasma membrane via vesicular transport.

The Gag polyprotein initially assembles into spherical immature particles, in which the membrane-bound Gag molecules project radially toward the virion interior (Fig. 1D,F). As the immature virion buds, PR is activated and cleaves Gag into its constituent MA, CA, NC, and p6 proteins, thereby also releasing the SP1 and SP2 peptides. Proteolysis is required for conversion of the immature virion into its mature infectious form (Fig. 1E,G). Like other retroviral proteases, HIV-1 PR is a dimeric aspartic protease (Fig. 1H). PR recognizes specific sites within Gag and cleaves them in an ordered fashion (Fig. 1B,C, arrowheads). Gag proteolysis triggers major changes which

include condensing and stabilizing the dimeric RNA genome, assembling the conical capsid about the genomic RNA–NC-enzyme complex, and preparing the virion to enter, replicate, and uncoat in the next host cell. Thus, viral maturation can be viewed as the switch that converts the virion from a particle that can assemble and bud from a producer cell into a particle that can enter and replicate in a new host cell. The following sections review our current understanding of HIV-1 assembly, budding, and maturation.

VIRION COMPOSITION AND RNA PACKAGING

The main constituents of HIV-1 are Gag, which makes up ~50% of the entire virion mass and the viral membrane lipids, which account for ~30% of virion mass (reviewed in Carlson 2008). Other viral and cellular proteins together contribute an additional ~20%, whereas the genomic RNA and other small RNAs amount to ~2.5% of virion mass. Gag, Gag-Pro-Pol, Env, the two copies of genomic RNA, the tRNA primer, and the lipid envelope are all essential for viral replication, whereas the relevance of virion incorporation of other cellular and viral accessory proteins, small RNA molecules, and specific lipids is generally less well understood.

All viral gene products are encoded on the genomic RNA, which also serves as mRNA for Gag and Gag-Pro-Pol, whereas singly or multiply spliced RNAs are translated to produce Env and accessory proteins, respectively. Unspliced and incompletely spliced HIV-1 RNAs are exported from the nucleus via a Rev-dependent export pathway, whereas completely spliced mRNAs exit the nucleus via the normal mRNA export route. Translation of Gag, Gag-Pro-Pol and most accessory proteins occurs on cytosolic polysomes. The two viral membrane proteins, Env and the accessory protein Vpu, which are encoded by the same mRNA, are translated on the rough ER. All virion components need to traffic from their point of synthesis to sites of assembly on the plasma membrane. The coordinated synthesis of structural proteins and enzymes as domains of the Gag and Gag-

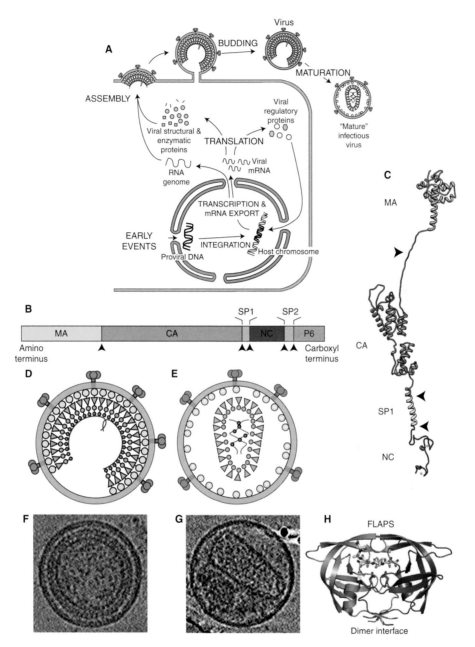

Figure 1. HIV-1 assembly, budding, and maturation. (*A*) Schematic illustration showing the different stages of HIV-1 assembly, budding, and maturation. (*B*) Domain structure of the HIV-1 Gag protein; arrows denote the five sites that are cleaved by the viral PR during maturation. (*C*) Structural model of the HIV-1 Gag protein, created by combining structures of the isolated MA-CA$_{NTD}$ (2GOL), CA$_{CTD}$ (1BAJ), and NC (1MFS) proteins, with a helical model for SP1. (*D*) Schematic model showing the organization of the immature HIV-1 virion. (*E*) Schematic model showing the organization of the mature HIV-1 virion. (*F*) Central section from a cryo-EM tomographic reconstruction of an immature HIV-1 virion. (*G*) Central section from a tomographic reconstruction of a mature HIV-1 virion. (*H*) Structure of HIV-1 protease (PR, 3D3T). The two subunits in the dimer are shown in different shades of purple, the "flap" and dimerization interfaces are labeled, positions of the active site Asp25 residues are shown in red, and a bound peptide corresponding to the SP2-p6 cleavage site is shown as a stick model, with oxygen atoms in red and nitrogen atoms in blue.

Pro-Pol polyproteins (which are produced at a ratio of ~20:1) ensures that these components are made at the proper stoichiometry (Jacks et al. 1988). Gag then binds other virion components through direct protein–protein and protein–RNA interactions, which allows the virus to assemble all of its components using a single targeting signal (Frankel and Young 1998; Freed 2001).

Protein Trafficking and Virion Incorporation

The HIV Gag and Gag-Pro-Pol polyproteins traffic from their sites of synthesis in the cytoplasm to the plasma membrane and then sort into detergent-resistant membrane microdomains (Ono and Freed 2001). Virus production is cholesterol and sphingolipid dependent, and the virus is enriched in "raft"-associated proteins and lipids (Ono 2009). Gag has been reported to interact with the cellular motor protein KIF4 (Tang et al. 1999; Martinez et al. 2008) and with various components of intracellular vesicle trafficking pathways (Batonick et al. 2005; Dong et al. 2005; Camus et al. 2007), but the role of microtubules and/or membrane trafficking for Gag membrane transport remains unconfirmed. Gag molecules do not polymerize extensively before they reach the membrane (Kutluay and Bieniasz 2010). Instead, soluble monomeric Gag proteins appear to fold into a compact, auto-inhibited conformation(s), which subsequently undergoes conformational changes that cooperatively couple MA–membrane (see online Movie 1 at www.perspectivesinmedicine.org), NC–RNA, and Gag–Gag interactions (Chukkapalli et al. 2010; Datta et al. 2011a; Jones et al. 2011). Gag molecules thus arrive at the plasma membrane as small oligomers, probably monomers or dimers, which polymerize onto nucleation sites composed of Gag–RNA complexes (Jouvenet et al. 2009). The cellular ATPase, ABCE1, has also been implicated in binding and chaperoning membrane-bound assembly intermediates (Dooher et al. 2007), although mechanistic details remain to be elucidated.

Gag membrane targeting requires myristoylation and a basic patch on the MA domain as well as the plasma membrane-specific lipid phosphatidyl inositol (4,5) bisphosphate ($PI(4,5)P_2$ [Ono et al. 2004]). Binding of the MA^{Gag} domain to $PI(4,5)P_2$ exposes the amino-terminal myristoyl group (Saad et al. 2006, 2007), and this "myristoyl switch" provides an elegant mechanism for anchoring Gag stably on the inner leaflet (Movie 1; Fig. 2). Electrostatic interactions with acidic phospholipids, which are strongly enriched in the HIV-1 lipidome (Brugger et al. 2006), probably also contribute to membrane anchoring. MA domains from other retroviruses can also bind $PI(4,5)P_2$ (Hamard-Peron et al. 2010), but the energetics of $PI(4,5)P_2$ binding and myristoyl

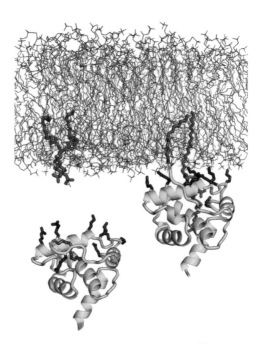

Figure 2. Myristoyl switch model for MA^{Gag} recognition of the plasma membrane (Saad et al. 2006). MA^{Gag} (yellow) proteins are shown with the aliphatic myristoyl group (brown) sequestered within the soluble protein (*left*, 1UPH), and with the myristoyl group extruded into the membrane when bound to the plasma membrane specific phosphatidyl inositide, $PI(4,5)P_2$, shown in red (*right*, 2H3F). The $PI(4,5)P_2$ inositol head group and unsaturated 2'-fatty acid bind within MA, allosterically inducing extrusion of the myristoyl group, whereas the saturated 1'-fatty acid of $PI(4,5)P_2$ remains embedded in the membrane. Basic residues on the membrane binding surface of MA^{Gag} are shown in blue.

sequestration vary, and not all retroviral MA proteins have amino-terminal myristoyl groups, suggesting that additional factors may govern membrane targeting, at least in those cases (Saad et al. 2008; Inlora et al. 2011).

The viral Env glycoproteins reach the plasma membrane independently of Gag. Genetic and biochemical analyses indicate that the long intracellular tail of TM helps sort Env into "raft"-like domains and mediates specific interactions with MA^{Gag} that promote Env virion incorporation (Yu et al. 1993; Cosson 1996; Murakami and Freed 2000; Wyma et al. 2000). Deletion of the TM cytoplasmic tail abolishes viral infectivity in most cell lines, but does not prevent Env incorporation (Einfeld 1996). Moreover, virions can be efficiently pseudotyped by heterologous glycoproteins without specific HIV-1 Gag interactions (Briggs et al. 2003a), implying that the MA–Env interaction is not absolutely essential for Env incorporation. Interestingly, pseudotyped particles seem to segregate into distinct classes, apparently displaying only one or the other glycoprotein (Leung et al. 2008). HIV-1 displays only ∼7–14 glycoprotein trimers per virion (Chertova et al. 2002; Zhu et al. 2006). This number is considerably lower than for the related simian immunodeficiency virus, which has approximately 80 Env trimers per virion (Zhu et al. 2006), and suggests that clustering of the sparsely distributed Env trimers may be important for HIV-1 entry (Sougrat et al. 2007). Other constituents of the virion are the accessory protein Vpr (incorporated at a ratio of 1:7 to Gag [Muller et al. 2000]) through a specific interaction with the p6 domain (Kondo et al. 1995) as well as a few copies of the accessory proteins Vif and Nef.

Analysis of purified virion preparations by immunoblotting or mass spectroscopy has identified many cellular proteins in HIV-1 particles (Ott 2008), but their importance for virus assembly, budding, and/or infectivity is currently not well established in most cases. These cellular proteins include plasma membrane proteins like ICAM-1 (which may mediate virus adherence to cells) and HLA-II (which may modulate immune responses), as well as cytoplasmic proteins which may be incorporated via direct or indirect Gag interactions (e.g., actin and actin-binding proteins, cyclophilin A, ubiquitin, lysyl-tRNA-synthetase, and many RNA-binding proteins [Ott 2008]).

Viral Lipid Composition

HIV-1 buds at the plasma membrane of infected cells and the viral membrane is therefore derived from the cellular plasma membrane. Aloia et al. (1988, 1993) initially reported differences between the lipid compositions of the producer cell and HIV-1 membranes, with virion enrichment of sphingomyelin (SM), phosphatidyl serine (PS), phosphatidyl ethanol (PE), and cholesterol, as well as decreased membrane fluidity. Recent advances in lipid mass spectrometry have allowed a comprehensive, quantitative analysis of the entire lipid composition (the lipidome) of purified HIV-1 including determination of side chains. These analyses revealed strong virion enrichment of the "raft lipids" SM, cholesterol, and plasmalogen-PE, with an increase in saturated fatty acids compared with the producer cell membrane. The inner leaflet of the viral membrane is enriched in PS, and the overall lipid composition of HIV-1 strongly resembles detergent-resistant membranes isolated from producer cells (Brugger et al. 2006). The native HIV-1 membrane exhibits a liquid-ordered structure (Lorizate et al. 2009), providing further evidence for its raft-like nature. The HIV-1 membrane is also enriched in $PI(4,5)P_2$, consistent with the idea that this phosphatidylinositide plays an important role in targeting Gag to the membrane (Chan et al. 2008; also see above). Membranes of the murine leukemia virus are also enriched in cholesterol, ceramide, and glycosphingolipids (Chan et al. 2008), indicating that other retroviruses also bud from raft-like membrane domains.

RNA Trafficking and Incorporation

HIV-1, like all retroviruses, selectively incorporates two copies of the capped and polyadenylated full-length RNA genome into the virion

(Johnson and Telesnitsky 2010; Lever 2007). The two RNA strands are noncovalently dimerized in their 5′UTR. RNA dimerization initiates through formation of a "kissing-loop" structure mediated by Watson–Crick base pairing of the self-complementary sequence within the loop of the dimer initiation site (DIS) (Fig. 3), which then expands into a more extended helix linkage during viral maturation. RNA dimerization is required for RNA packaging and viral infectivity (Moore and Hu 2009), and involves *cis*-acting sequences within the 5′UTR of the viral genome, which are recognized by the NC^{Gag} domain. RNA–Gag interactions also appear to convert the compact, auto-inhibited Gag conformation into its extended assembly conformation which is required for virus assembly (Rein et al. 2011). Removal of the RNA packaging signal does not abolish particle production, but the resulting particles do contain abnormally high levels of nonspecific cellular mRNAs (Rulli et al. 2007), implying that RNA facilitates virion assembly by concentrating and aligning Gag molecules at the plasma membrane.

Although NC can bind RNA nonspecifically, genome packaging requires specific recognition of the dimeric, unspliced HIV-1 RNA. Recent single virion analyses by fluorescence microscopy have confirmed that nearly all HIV-1 particles contain genomic RNA, which is dimeric in most cases (Chen et al. 2009). Efficient genome packaging depends on a structural element of approximately 150 nucleotides located in the 5′ region spanning the major splice donor and the Gag initiation codon (the Ψ-site) (D'Souza and Summers 2005). The requirements for elements located downstream from the first splice donor explain why unspliced viral RNA is selectively packaged, at least in the case of HIV-1. The Ψ-site comprises four stable stem loop structures within the highly structured 5′UTR (Fig. 3), but efficient genome packaging appears to be affected by almost the entire 5′UTR, particularly elements close to the DIS. Structural analyses of individual loops from the Ψ-site in complex with NC revealed specific interactions between the CCHC-type zinc knuckles of NC and exposed loop residues of the RNA (Fig. 3), providing insight into the

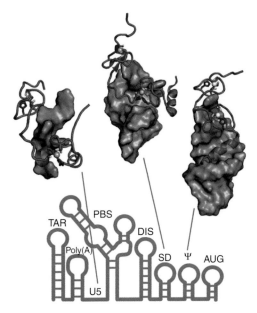

Figure 3. 5′ Untranslated region (UTR) of the HIV-1 RNA genome and its interactions with the viral NC protein. *Lower* image shows a secondary structure model for the 5′UTR, highlighting the TAR stem loop structure (which binds the viral Tat protein), the polyadenylation site, the U5 element, the primer binding site (PBS, which anneals to the tRNA^{Lys,3} primer), and four stem loops within the packaging site, which contain the dimer initiation site (DIS, stem-loop I, which forms a kissing loop structure that initiates association of the two copies of the genomic RNA), the splice donor (SD, stem loop II, which acts as the 5′ donor for splicing of subgenomic RNAs), the Psi site (ψ, stem loop III, which forms an essential part of the packaging signal), and the Gag start codon (AUG, stem loop IV, which contains the start site for Gag translation). *Upper* structures show three different complexes between the NC protein (red, with zinc atoms shown in grey and Zn-coordinating side chains shown explicitly) and viral RNAs (blue), corresponding to the U5 region (Spriggs et al. 2008), the SD stem loop (1F6U), and the ψ stem loop (1A1T).

mechanism of RNA recognition (see online Movie 2 at www.perspectivesinmedicine.org). The structural basis of dimeric RNA packaging is best understood for the genome of murine leukemia virus, where the high-affinity NC binding sites are sequestered and become exposed only upon RNA dimerization

(Miyazaki et al. 2010). Retroviruses can copy information from either of their two packaged RNA strands during reverse transcription. This "pseudodiploid" property confers the distinctive advantage of high recombination potential and helps retroviruses to overcome environmental and therapeutic pressures.

Gag polyproteins of avian and murine retroviruses and of foamy virus reportedly enter the nucleus, where they bind newly transcribed RNA and cotraffic to sites of assembly (Schliephake and Rethwilm 1994; Scheifele et al. 2002; Andrawiss et al. 2003). Nuclear import and export has also been suggested for HIV-1 Gag (Dupont et al. 1999), but productive Gag-RNA packaging interactions appear to occur in the cytoplasm in this case. A cell fusion–dependent recombination assay was used to show that HIV-1 RNA dimerization, which is essential for packaging, occurs in the cytoplasm (Moore et al. 2009), and that dimerization frequency depends on the complementarity of the DIS loop sequences (Fig. 3). Overall, current data support a model in which genomic RNA dimerizes and associates with a few Gag molecules in the cytoplasm. These RNP complexes then traffic to the plasma membrane where they nucleate assembly (Kutluay and Bieniasz 2010).

In addition to the viral genome, HIV-1 particles also package small cellular RNAs, most notably tRNAs required for the initiation of reverse transcription (Kleiman et al. 2010). Two copies of $tRNA^{Lys3}$ anneal via Watson–Crick base pairing to an 18 base-pair sequence known as the primer binding sites (PBS), which is located within the 5′LTR (Fig. 3). The $tRNA^{Lys1,2}$ isoacceptors are also selectively packaged, and it has been suggested that Gag-Pro-Pol, genomic RNA, and lysyl-tRNA synthetase are all involved in specific tRNA packaging (Kleiman et al. 2010). HIV-1 particles also contain 7SL, the RNA component of host signal recognition particle (SRP), at a sevenfold molar excess over genomic RNA (Onafuwa-Nuga et al. 2006). However, SRP protein constituents are not incorporated and the functional significance of 7SL incorporation is unknown. Other RNA constituents of the virion include the 5S, 18S, and 28S rRNAs, which may simply be present owing to their cytoplasmic abundance.

HIV ASSEMBLY AND THE IMMATURE LATTICE

Direct visualization of individual budding and release events at high temporal resolution using fluorescently labeled Gag polyproteins and live-cell microscopy has recently revealed the kinetics of HIV-1 assembly and host cell factor recruitment (Jouvenet et al. 2008; Ivanchenko et al. 2009). Furthermore, live-cell fluorescence imaging revealed cytoplasmic genomic RNA to be highly dynamic in the absence of Gag, whereas a fraction of RNA molecules became immobilized at specific sites at the plasma membrane upon coexpression of Gag (Jouvenet et al. 2009). These sites marked the position of subsequent assembly events in most cases. Gag levels increase exponentially at individual plasma membrane assembly sites until a plateau is reached and budding ensues. The mean assembly time for individual particles is ~10 min, albeit with significant variability between sites (Jouvenet et al. 2008; Ivanchenko et al. 2009). Assembling Gag molecules are largely derived from the rapidly diffusing cytoplasmic pool (Ivanchenko et al. 2009) and do not enter via lateral diffusion within the membrane or from vesicle-associated Gag transport.

Gag assembly leads to formation of the immature lattice. Immature particles have historically been studied using virions that lacked PR activity or using virus-like particles assembled in vitro from bacterially expressed Gag proteins. However, recent direct imaging of budding viruses has confirmed that the immature Gag lattice is established as the virus assembles (Fig. 4A; Carlson et al. 2010). The Gag molecules in the immature virion are extended and oriented radially, with their amino-terminal MA domains bound to the inner membrane leaflet and their carboxy-terminal p6 domains facing the interior of the particle (Fig. 4B; Fuller et al. 1997). The immature lattice is stabilized primarily by lateral protein–protein interactions involving the CA-SP1 region, with the carboxy-terminal domain of

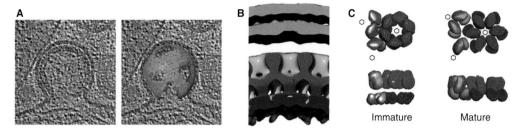

Figure 4. Assembly and structure of immature HIV-1 particles. (*A*) (*Left*) Central slice through a cryo-EM tomogram of a budding HIV-1 virion. (*Right*) Map of the Gag lattice in the budding virion. Positions of Gag hexagons are colored according to their hexagonal order, from low (brown) to high (green). (*B*) Cryo-EM tomogram of an immature HIV-1 virion, showing the structure of the immature HIV-1 Gag lattice. The surface was cut perpendicular to the membrane to reveal the two membrane leaflets, the two CA domains (orange and burnt orange), and the NC layer (red). (*C*) Schematic of the conformational rearrangements in the capsid lattice during maturation. (*Left*) Arrangement of the amino-terminal (orange) and carboxy-terminal (burnt orange) domains of CA in the immature lattice, viewed from outside the particle (*upper*), and rotated 90° around the horizontal axis (*lower*). Domains from neighboring hexamers are indicated in lighter colors, and sixfold lattice positions are marked by hexagons. (*Right*) Equivalent interactions of CA subunits within the mature HIV-1 capsid lattice.

CA (CA_{CTD}) and SP1 making particularly important interactions (Fig. 4B). MA and NC help to align and concentrate Gag molecules, but MA and p6 are dispensable for immature lattice formation, and NC can be replaced by a heterologous protein dimerization domain (Zhang et al. 1998; Accola et al. 2000). Extensive mutational analyses have identified the contributions of individual residues to immature and mature particle assembly (e.g., von Schwedler et al. 2003). These studies have generally confirmed the critical importance of the CA_{CTD} and SP1 regions for immature particle assembly, and in particular of the "major homology region" within CA_{CTD} which is conserved across retroviruses.

Over the past few years, cryoelectron tomography and image processing analyses have defined the three-dimensional structure of the immature Gag lattice in greater detail (Wright et al. 2007; Briggs et al. 2009). These studies confirmed that Gag molecules are arranged as hexamers with 8 nm spacings (Fig. 4). The CA_{NTD} forms six-membered rings with large central holes (Fig. 4B). At the current resolution (~2 nm), high-resolution structures cannot be unequivocally positioned within the density maps, but the lattice does not appear to be consistent with hexameric arrangements seen in crystallographic studies of the murine leukemia virus CA_{NTD} (Mortuza et al. 2004). Density for the CA_{CTD} domain resides beneath the hexamers, and the domain appears to make both intra- and interhexameric contacts (Fig. 4B and C). Like the full-length CA protein, CA_{CTD} constructs dimerize in solution and several different high-resolution CA_{CTD} structures have been reported, two of which appear to be used in constructing the mature capsid lattice (see below). Two other CA_{CTD} dimer structures have been suggested to play a role in stabilizing the immature lattice: a domain-swapped dimer (Ivanov et al. 2007) in which the MHR elements from each monomer associate to create a large interface, and a CA_{CTD} dimer observed in the presence of a peptide assembly inhibitor (CAI) or for certain mutations in the CAI binding pocket (Ternois et al. 2005; Bartonova et al. 2008). The CAI-induced dimer appeared to be the best fit to the reconstructed EM density, but it is currently unclear whether any of the CA_{CTD} dimer structures is actually reconstituted within the immature lattice, and higher resolution structures of immature particles are eagerly awaited.

Residues at the carboxy-terminal end of CA and in the adjacent SP1 region are also essential for immature lattice formation, and this region has been suggested to form a continuous α helix

(Accola et al. 1998). This model is consistent with tomographic reconstructions of the immature lattice, which revealed rod-like structures descending toward the NC layer along the six-fold symmetry axis below CA_{CTD} (Fig. 4B; Wright et al. 2007). These features can be modeled as six helix bundles, although this model again awaits definitive testing at higher resolution. Interestingly, the helical propensity of the CA-SP1 junction sequence is rather weak (Morellet et al. 2005), but recent evidence suggests that helix formation is strongly induced by molecular crowding (Datta et al. 2011b). Thus, helix formation could provide a switch that helps trigger immature lattice formation as the Gag molecules coalesce (Datta et al. 2011b). Subsequent proteolytic cleavage at the CA-SP1 junction would then destroy the helix during maturation, destabilize immature lattice interactions, and help drive conversion to the mature capsid lattice.

In addition to defining the structure of Gag hexamers, the cryo-EM tomographic studies revealed how the hexameric lattice curves into a spherical structure. A perfect hexagonal lattice lacks declination, and must therefore include nonhexameric defects in order to enclose space. In the mature capsid, this is achieved by including 12 pentameric defects (see below). In contrast, the immature lattice contains small, irregularly shaped defects and holes that permit it to curve (Fig. 4A; Briggs et al. 2009). Nevertheless, the Gag shell forms a contiguous lattice rather than consisting of smaller "islands" of regular Gag arrays. It does contain one large gap, however, which covers approximately one-third of the surface area of the membrane (Wright et al. 2007; Briggs et al. 2009). This gap in the Gag lattice is created when the virus buds (Fig. 4A), because Gag assembly and budding appear to be competitive processes, and immature virions typically bud before the Gag molecules have finished polymerizing into fully closed shells (Carlson et al. 2010). As a result of this gap, virions contain fewer Gag molecules than was initially calculated based on the assumption of a complete virus shell. The precise number of Gag molecules depends on the size of the individual virion

and the completeness of the Gag shell, with a virion of 130 nm diameter containing roughly 2500 Gag molecules (Carlson et al. 2008).

HIV BUDDING

Late Domains and ESCRT Pathway Recruitment

Although the viral Gag protein is responsible for cofactor packaging and virion assembly, the virus usurps the host ESCRT pathway to terminate Gag polymerization and catalyze release (Morita and Sundquist 2004; Bieniasz 2009; Carlton and Martin-Serrano 2009; Usami et al. 2009; Hurley and Hanson 2010; Peel et al. 2011). ESCRT factors also catalyze the topologically equivalent membrane fission reactions that release vesicles into endosomal multivesicular bodies (Hurley and Hanson 2010; Peel et al. 2011) and that separate daughter cells during the abscission stage of cytokinesis (Carlton and Martin-Serrano 2007, 2009; Hurley and Hanson 2010; Elia et al. 2011; Guizetti et al. 2011; Peel et al. 2011). This ability of the ESCRT machinery to facilitate membrane fission from within the necks of thin, cytoplasm-filled membrane vesicles and tubules explains why HIV-1, and many other enveloped viruses, have evolved to use the pathway to bud from cells.

HIV-1 p6^{Gag} contains two different "late domain" motifs that bind and recruit early-acting ESCRT factors (Morita and Sundquist 2004; Bieniasz 2009; Carlton and Martin-Serrano 2009; Usami et al. 2009). The primary "PTAP" late domain binds the TSG101 subunit of the heterotetrameric ESCRT-I complex (Garrus et al. 2001; Martin-Serrano et al. 2001; VerPlank et al. 2001; Demirov et al. 2002; Morita and Sundquist 2004; Bieniasz 2009). Each of the four residues (Pro-Thr/Ser-Ala-Pro) makes specific contacts within an extended groove on the amino-terminal ubiquitin E2 variant (UEV) domain of TSG101 (Pornillos et al. 2002; Im et al. 2010). PTAP motifs are also found within HRS and related proteins that recruit ESCRT to endosomal membranes (Ren and Hurley 2011). Thus, the HIV-1 p6^{Gag} PTAP late domain mimics a cellular ESCRT-I recruiting motif, and Gag and HRS can both be viewed as membrane-specific adaptors for the ESCRT pathway (Pornillos et al. 2003).

The second p6^{Gag} late domain, designated "YPXL" (Tyr-Pro-X-Leu, where "X" can vary in sequence and length), binds the ESCRT factor ALIX (Strack et al. 2003; Usami et al. 2009). The YPXL late domain contributes significantly to HIV-1 replication (Fujii et al. 2009; Eekels et al. 2011) but is less critical than the PTAP motif in most cell types. Retroviral YPXL late domains exhibit considerable sequence variation, but a conserved tyrosine binds deep within a pocket on the second arm of the ALIX V domain in all cases, and downstream hydrophobic residues contact ALIX along a shallow adjacent groove (Zhai et al. 2008, 2011). Once again, the virus is mimicking a motif used by cellular ALIX ligands in fungi (Vincent et al. 2003), although YPXL-containing binding partners for mammalian ALIX proteins remain to be characterized. The amino-terminal ALIX Bro domain also interacts with NC^{Gag}, and NC mutants can exhibit budding defects, reflecting the apparent functional importance of this interaction (Popov et al. 2008, 2009; Dussupt et al. 2009). Finally, the carboxy-terminal domain of CA^{Gag} interacts with NEDD4L, a member of the human NEDD4 ubiquitin E3 ligase protein family (Chung et al. 2008; Usami et al. 2008; Weiss et al. 2010). Although this interaction contributes only modestly to HIV-1 budding, NEDD4 family members play critical roles in the budding of other retroviruses through direct interactions with their PPXY (Pro-Pro-X-Tyr) late domains (Morita and Sundquist 2004; Bieniasz 2009; Carlton and Martin-Serrano 2009; Usami et al. 2009).

Assembly of the Core ESCRT Machinery

The human ESCRT pathway comprises more than 30 different proteins, and this complexity is expanded further by associated regulatory and ubiquitylation machinery. Essential ESCRT pathway functions and mechanisms are conserved across eukaryotes and archaea, and many basic principles have been elucidated

through biochemical and genetic analyses of the simpler yeast pathway (Saksena et al. 2007; Hurley and Hanson 2010). Recent functional studies have identified a minimal core set of human ESCRT machinery that is essential for HIV-1 budding (Fig. 5). In essence, TSG101/ ESCRT-I and ALIX both function by recruiting downstream ESCRT-III and VPS4 complexes, which in turn mediate membrane fission and ESCRT factor recycling (Morita and Sundquist 2004; Bieniasz 2009; Carlton and Martin-Serrano 2009; Hurley and Hanson 2010; Peel

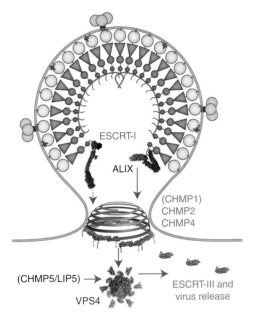

Figure 5. Summary of the essential core ESCRT machinery used in HIV-1 budding (with auxiliary factors shown in parentheses), illustrating a leading model for the budding mechanism. Late domain motifs within p6^{Gag} bind directly to the UEV domain of the TSG101 subunit of the heterotetrameric ESCRT-I complex (red, with bound ubiquitin in black, 1S1Q, 2P22) and the V domain of ALIX (blue, 2OEV). These interactions result in the recruitment of the ESCRT-III proteins of the CHMP1, CHMP2, and CHMP4 families (green, 2GD5), which apparently polymerize into a "dome" that promotes closure of the membrane neck (Peel et al. 2011). They also recruit the VPS4 ATPases (purple, 1XWI, 1YXR), which completes the membrane fission reaction and uses the energy of ATPase to release the ESCRT-III from the membrane and back into the cytoplasm. See text for details.

Cite this article as *Cold Spring Harb Perspect Med* doi: 10.1101/cshperspect.a006924

et al. 2011). Humans have 12 different ESCRT-III-like proteins (predominantly known by "CHMP" designations), which can be subdivided into seven families. These proteins share a common architecture, with an amino-terminal core domain comprising an extended four-helix bundle (Muziol et al. 2006; Bajorek et al. 2009; Xiao et al. 2009), and carboxy-terminal tails that can fold back and autoinhibit core oligomerization (Lin et al. 2005; Zamborlini et al. 2006; Lata et al. 2008a; Bajorek et al. 2009). Only the CHMP2 and CHMP4 families play critical functional roles in HIV-1 budding, although CHMP1 and CHMP3 family members may also contribute modestly (Jouvenet et al. 2011; Morita et al. 2011). The ESCRT-I interactions that lead to CHMP2/CHMP4 recruitment are not yet clear, but the ALIX branch of the pathway is better understood. As illustrated in Figure 5, the following sequence of events is consistent with current analyses of ESCRT pathway recruitment. (1) Late domain binding induces the soluble ALIX protein to undergo a conformation change that leads to dimerization and activates ALIX for membrane binding and ESCRT-III recruitment (Pires et al. 2009; Usami et al. 2009). (2) The Bro domain of the activated ALIX protein binds carboxy-terminal helices located within the tails of all three human CHMP4 proteins (McCullough et al. 2008). (3) This interaction relieves CHMP4 autoinhibition and induces the protein to polymerize into filaments within the virion neck (Hanson et al. 2008; Pires et al. 2009; Usami et al. 2009; Hurley and Hanson 2010; Peel et al. 2011). (4) CHMP4 filaments then recruit (or copolymerize with) the CHMP2 proteins (Lata et al. 2008b; Morita et al. 2011). (5) Deposition of CHMP2 exposes the protein's carboxy-terminal tail, which contains a helical sequence motif that binds the amino-terminal MIT domains of VPS4 ATPases (Obita et al. 2007; Stuchell-Brereton et al. 2007; Hurley and Yang 2008). (6) The recruited VPS4 proteins assemble into enzymatically active higher order complexes (Babst et al. 1998; Scott et al. 2005; Shestakova et al. 2010). Like most other AAA ATPases, VPS4 forms hexameric rings, and the active enzyme appears to comprise two stacked,

inequivalent hexameric rings (Yu et al. 2008) (although alternative models have been proposed). Each virus budding site recruits approximately three to five VPS4 dodecamers (Baumgartel et al. 2011), which may be linked together through bridges composed of the CHMP5/LIP5 activator complex (Yang and Hurley 2010). The entire ESCRT assembly process takes approximately 10 minutes and occurs in multiple stages, with a gradual and concomitant buildup of the Gag and ALIX proteins, followed by short (~2 min) bursts of ESCRT-III and VPS4 recruitment immediately prior to virus budding (Baumgartel et al. 2011; Jouvenet et al. 2011).

Models for Membrane Fission

The detailed mechanism of ESCRT-mediated membrane fission is an active research frontier, but several important aspects of the process have recently emerged. CHMP4 subunits, possibly in complex with CHMP2 and other ESCRT-III proteins, appear to form spiraling filaments within the neck of the budding virus (Fig. 5; Ghazi-Tabatabai et al. 2008; Hanson et al. 2008; Lata et al. 2008b; Teis et al. 2008; Saksena et al. 2009; Wollert et al. 2009; Elia et al. 2011; Guizetti et al. 2011). As the filaments spiral inward, they may create closed "domes" that constrict the opposing membranes and promote fission (Fabrikant et al. 2009; Hurley and Hanson 2010; Peel et al. 2011). VPS4 also apparently plays an active role in the membrane fission reaction because the enzyme is recruited immediately prior to virus budding (Baumgartel et al. 2011; Elia et al. 2011; Jouvenet et al. 2011), and because ESCRT-III recruitment alone is insufficient for virus release (Jouvenet et al. 2011). Possible roles for VPS4 include helping to promote ESCRT-III dome formation and/or removing ESCRT-III subunits from the dome, thereby destabilizing hemi-fission intermediates and helping to drive fission to completion (Hanson et al. 2008; Lata et al. 2008b; Baumgartel et al. 2011; Peel et al. 2011). In the final stage of the cycle, VPS4 uses the energy of ATP hydrolysis to disassemble the filaments and release the ESCRT-III subunits back into the cytoplasm as soluble, autoinhibited proteins (Babst et al. 1998; Ghazi-Tabatabai et al. 2008; Lata et al. 2008b; Wollert et al. 2009; Davies et al. 2010). Thus, the energy required for virus budding is provided by ATP hydrolysis, in the form of VPS4-mediated protein disassembly and refolding.

Blocks to HIV-1 Release

HIV-1 budding and release are essential for spreading viral infection, and it is therefore not surprising that innate immune pathways have evolved to interfere with these processes. It is now well established that the antiviral protein tetherin blocks HIV-1 dissemination by tethering newly budded viral particles to the cell surface. Tetherin is antagonized by the HIV-1 Vpu protein, and this remarkable restriction system is described in detail in Malim and Bieniasz (2011). Recently, Leis and colleagues have reported that another interferon-inducible protein, ISG-15, can inhibit HIV-1 release at the earlier budding stage by interfering with ESCRT-III protein activities. ISG-15 is a ubiquitin-like protein that can be covalently attached to the lysine side chains of ESCRT-III subunits (Skaug and Chen 2010). ISGylation of CHMP5 and other ESCRT-III subunits appears to impair VPS4 function by sequestering the LIP5-CHMP5 activator away from the enzyme and by reducing VPS4 recruitment and activity (Pincetic et al. 2010; Kuang et al. 2011). It will now be important to determine the relative contributions of these activities in inhibiting HIV-1 replication in vivo and how the virus overcomes such blocks.

HIV MATURATION

Architecture of the Mature HIV-1 Virion

Viral maturation begins concomitant with (or immediately following) budding, and is driven by viral PR cleavage of the Gag and Gag-Pro-Pol polyproteins at ten different sites, ultimately producing the fully processed MA, CA, NC, p6, PR, RT, and IN proteins (Fig. 1A–C; see online Movie 3 at www.perspectivesinmedicine.org; Swanstrom and Wills 1997; Hill et al. 2005). Over the course of maturation, these processed proteins rearrange dramatically to create the mature infectious virion, with its characteristic conical core (Fig. 1D–G). MA remains associated with the inner leaflet of the viral membrane, forming a discontinuous matrix shell that lacks long-range order. The outer capsid shell of the core particle is composed of approximately 1200 copies of CA and is typically conical, although tubes and other aberrant assemblies, including double capsids, also form at lower frequencies (Briggs et al. 2003b; Benjamin et al. 2005). The capsid approaches the matrix closely at both ends (Benjamin et al. 2005; Briggs et al. 2006), particularly at the narrow end, which may represent the nucleation site for assembly (Briggs et al. 2006). The capsid surrounds the nucleocapsid, which typically resides at the wide end of the capsid and lacks obvious long-range order (Briggs et al. 2006).

The Viral Capsid

The capsid performs essential functions during the early stages of HIV-1 replication, although these functions are not yet fully understood in mechanistic detail. In newly infected cells, the capsid interacts with both positive-acting host factors like cyclophilin A and transportin-3, and with restriction factors of the TRIM5-α family (Luban 2007; Sebastian and Luban 2007; Yamashita and Emerman 2009; Krishnan et al. 2010). CA mutations that block capsid assembly or destabilize the capsid typically inhibit reverse transcription, implying that the capsid helps to organize the replicating genome (Forshey et al. 2002). Conversely, mutations that hyperstabilize the capsid also inhibit reverse transcription, implying that the capsid must disassemble or uncoat in a timely fashion (Forshey et al. 2002). CA mutations can also inhibit or alter nuclear localization, indicating that the capsid (or at least CA subunits) probably play important roles in nuclear targeting and/or import of the preintegration complex (Dismuke and Aiken 2006; Yamashita et al. 2007; Lee et al. 2010).

HIV-1 capsids are geometric structures called "fullerene cones" (Fig. 6A), which are a family of related structures comprising conical

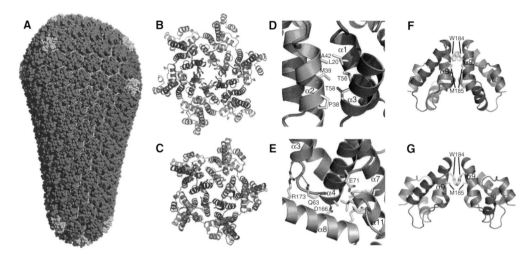

Figure 6. Fullerene cone model for the HIV-1 capsid. (*A*) Molecular model of the HIV-1 capsid, with CA hexamers in orange and pentamers in tan (adapted from Pornillos et al. 2011). (*B*) Structure of the HIV-1 CA hexamer (3H47). (*C*) Structure of the HIV-1 CA pentamer (3P05). (*D*) Detailed structure of the CA_{NTD}–CA_{NTD} interface that stabilizes the hexameric ring. Hydrophobic residues that stabilize the interaction between CA helices 1, 2, and 3 are highlighted. (*E*) Detailed structure of the CA_{NTD}(orange)–CA_{CTD}(tan) interface that forms a "girdle" around the hexameric and pentameric rings. Interface residues are highlighted, as are a series of salt bridges and hydrogen bonds that stabilize the interface while allowing it to "swivel" in response to changes in lattice curvature. (*F*,*G*) Two alternative structures of the CA_{CTD} dimer, 1A43and 2KOD. Two key interface residues (W184 and M185) are shown in each case to emphasize the fact that similar CA surfaces are used in both dimers.

hexagonal nets that close at both ends through the introduction of 12 pentagonal defects (Ganser et al. 1999; Ganser-Pornillos et al. 2008). Thus, unlike icosahedral viruses, HIV-1 capsids are best viewed as a continuum of related structures rather than as a single unique assembly. Most authentic HIV-1 capsids exhibit the 19.2° cone angle required by conical hexagonal packing, but their overall lengths and cap shapes vary owing to the insertion of pentagons at different positions in the hexagonal net (Benjamin et al. 2005; Briggs et al. 2006). Positioning of pentamers at alternate sites also accounts for the different capsid shapes seen in other retroviruses because "tubular" capsids are created when the 12 pentamers are symmetrically distributed at either end of the tube, and "spherical" capsids are created when the pentamers are distributed more evenly throughout the hexagonal net (Ganser-Pornillos et al. 2004).

A series of different intersubunit CA interactions stabilize the capsid lattice. The CA subunits are organized into hexameric and pentameric rings, and both of these structures have now been visualized crystallographically (Fig. 6B,C; see online Movie 4 at www.perspectivesinmedicine.org; Pornillos et al. 2009, 2011). The organization of subunits within the two types of rings is remarkably similar, and HIV-1 CA hexamers and pentamers are therefore an excellent example of the principle of quasi-equivalence as originally envisioned by Caspar and Klug (1962). In both cases, the rings are stabilized by interactions between the first three helices of CA_{NTD} (Fig. 6D). The CA rings are buttressed by an exterior "girdle" formed by CA_{CTD}, with each CA_{CTD} domain contacting the CA_{NTD} of a neighboring subunit in the ring (Fig. 6B,C,E). CA_{CTD} also makes important inter-ring contacts across the local two- and three-fold axes, thereby stabilizing the extended lattice (Ganser-Pornillos et al. 2007; Byeon et al. 2009). Isolated CA_{CTD} polypeptides can make two different types of twofold symmetric dimers which utilize the same basic interface but differ in their detailed packing interactions, and both of these dimers may participate in capsid assembly (Fig. 6F,G; Gamble et al. 1997; Ganser-Pornillos et al. 2007; Byeon et al. 2009).

Fullerene cones lack symmetry, which implies that all of the different CA subunits must reside in distinct local environments. The individual CA subunits must therefore have sufficient flexibility to accommodate at least three different types of heterogeneity. Both CA hexamers and pentamers can be formed because the interface between subunits in the hexameric ring is highly hydrated, and mobile water molecules are well suited for adjusting to the subtle changes required to make the pentamer interface. Alterations in lattice curvature across different regions of the cone surface are accommodated by flexibility in the NTD-CTD linker and in the NTD–CTD interface (Fig. 6E), which allows the two CA domains to move relative to one another. Finally, differences in inter-ring packing interactions may be accommodated by using both types of CTD-CTD dimers at different positions of the cone (Pornillos et al. 2011).

The Viral Protease

HIV-1 PR is one of the most extensively characterized proteins in molecular biology owing to its importance as a drug target (Wlodawer and Gustchina 2000; Louis et al. 2007; Ali et al. 2010; and see Arts and Hazuda 2011). Like other aspartic acid proteases, HIV-1 PR uses two aspartic acid side chains within a characteristic Asp-Thr-Gly "fireman's grip" motif to activate the nucleophilic water molecule that cleaves the peptide bond (Cooper 2002). Retroviral proteases are unusual, however, in that the active enzyme is a dimer of two identical subunits. The active site transverses the dimer interface, stabilized by the fireman's grip and by a four-stranded mixed β-sheet created by the amino and carboxyl termini of each subunit. Two extended flexible loops lie on the other side of the active site, and these "flaps" open to allow substrates access to the active site (Fig. 1H).

The mechanism by which PR is activated during Gag assembly and budding is still not fully understood. PR is essentially inactive within the Gag-Pro-Pol polyprotein because the mature active enzyme is a dimer, whereas unprocessed PR constructs with amino-terminal extensions dimerize only very weakly. However, such constructs can form transient dimeric "encounter" complexes. A small fraction of these encounter complexes have enzymatic activity and can accommodate insertion of the amino-terminal cleavage site into the substrate-binding cleft, which can lead to autoprocessing and formation of a stable dimer with full catalytic activity (Tang et al. 2008). Gag trafficking probably helps to regulate PR activation by preventing premature PR dimerization until the Gag molecules coalesce at the plasma membrane. Consistent with this idea, artificially dimerized PR subunits are activated prematurely, which inhibits particle production (Krausslich et al. 1991). Assembly-mediated PR dimerization is unlikely to account for the entire activation mechanism, however, because Mason-Pfizer monkey virus and other betaretroviruses preassemble into immature particles in the cytoplasm, yet their PR enzymes are not activated until they are transported and bud from the plasma membrane (Parker and Hunter 2001). Thus, PR must be activated by additional mechanisms, at least in the betaretroviruses.

During viral maturation, PR cleaves five different sites within Gag (Fig. 1B) and five different sites within Gag-Pro-Pol (not shown). Schiffer and colleagues have determined crystal structures of HIV-1 PR in complex with peptides that correspond to seven of these cleavage sites (Prabu-Jeyabalan et al. 2002, 2004). Binding of asymmetric substrates breaks the twofold symmetry, so that the two PR subunits are no longer equivalent. Substrates bind in extended, β-strand conformations and make sheet-like interactions with the flaps and the base of the PR active site. The eight side chains from four residues on either side of the scissile bond bind in a series of six different enzyme pockets (with the first and third residues on either side of the scissile bond sharing common pockets). The enzyme appears to recognize the overall shape of the substrate rather than its specific sequence, as evidenced by the fact that the different Gag and Gag-Pro-Pol cleavage sites vary considerably in amino acid sequence, yet all bind within the enzyme with very similar steric footprints (Prabu-Jeyabalan et al. 2002).

Maturation Dynamics

Important changes that occur during HIV-1 maturation include activation of the fusogenic activity of the viral Env protein (Murakami et al. 2004; Wyma et al. 2004), capsid assembly, stabilization of the genomic RNA dimer, and rearrangement of the tRNA^{Lys,3} primer–genome complex (Moore et al. 2009; Rein 2010). Maturation is a dynamic, multistep process that involves a series of conformational switches and subunit rearrangements. Temporal control of viral maturation is provided, at least in part, by the very different rates of processing at the five Gag processing sites, whose cleavage rates vary by up to 400-fold. These cleavage sites fall into three different categories: rapid (SP1/NC), intermediate (SP2/p6, MA/CA), and slow (NC/SP2, CA/SP1) (Pettit et al. 1994). Analyses of mutant virions with blockages at the different Gag sites indicate that each cleavage event performs a different function. SP1/NC cleavage activates Env (Wyma et al. 2004) and promotes condensation of the RNP particle (de Marco et al. 2010), SP2 processing frees NC to chaperone formation of the stable genomic RNA dimer (Kafaie et al. 2008; Ohishi et al. 2011), MA/CA cleavage disassembles the immature lattice and releases CA-SP1, and CA-SP1 cleavage frees CA to form the conical capsid (de Marco et al. 2010). In at least some cases, Gag proteolysis induces local conformational changes that favor alternative protein–protein interactions. For example, cleavage at the MA/CA site causes the first 13 CA residues, which are disordered in the MA-CA protein, to fold into a β-hairpin that packs above the first three CA$_{NTD}$ helices and promotes the formation of mature CA hexamers (Gitti et al. 1996; Tang et al. 2002). Cleavage at the CA-SP1 junction also appears to trigger a conformational rearrangement (Gross et al. 2000), although the molecular details of this transition are not yet well understood.

SITES OF VIRUS RELEASE AND CELL-TO-CELL TRANSMISSION

HIV-1 Assembles and Buds at the Plasma Membrane

The general view that HIV-1 assembles and buds at the plasma membrane of the infected cell was challenged by electron microscopy studies reporting budding sites and released virions in apparently intracellular compartments in macrophages and other cell lines. These compartments were reactive for late endosomal markers, which—in combination with the ESCRT requirement for virion release—led to the suggestion that HIV buds into multivesicular bodies and is subsequently released from the cell via vesicular transport and fusion at the plasma membrane (Pelchen-Matthews et al. 2003). This concept is incompatible with several more recent results, however, showing that newly synthesized Gag reaches the plasma membrane before it accumulates at late endosomes and that endosomal accumulation can be inhibited without effect on virus release (Jouvenet et al. 2006). More importantly, fluorescence microscopy analysis of HIV-1 formation in living cells (Jouvenet et al. 2008; Ivanchenko et al. 2009) only detected single particle assembly at the plasma membrane, inconsistent with vesicular release of preassembled virions from a vesicular compartment. The apparently intracellular sites of HIV-1 assembly and budding in macrophages were ultimately shown to be deep plasma membrane invaginations that were connected to the surface via narrow channels, but were nevertheless accessible to membrane-impermeant stains (Deneka et al. 2007; Welsch et al. 2007) and traceable through ion-abrasion scanning electron microscopy (Bennett et al. 2009). Thus, HIV-1 assembly and budding appears to occur predominantly at the plasma membrane in all physiologically relevant cells.

The Virological Synapse

Polarized HIV-1 release was initially observed in electron microscopy studies that reported accumulation of HIV-1 budding at sites of cell-to-cell contact (Phillips and Bourinbaiar 1992). These observations were consistent with the idea that direct cell-to-cell spread of HIV-1 via specialized cell contacts is much more efficient than infection with cell-free virus and is thus likely to represent the predominant mode of virus spread in vivo, particularly within lymphoid tissues. Microscopic studies of CD4$^+$ T-cell cultures later revealed close connections between infected donor cells and uninfected target cells, and cell-to-cell transmission of virions across these contact zones, which were termed virological synapses (VS) (Piguet and Sattentau 2004; Mothes et al. 2010; Jolly and Sattentau 2004; Haller and Fackler 2008). This name was chosen because the structures were reminiscent of immunological synapses, even though the two structures are clearly distinct in terms of protein composition, signaling, and dynamics (Vasiliver-Shamis et al. 2010). Multiple contact zones between one HIV-infected CD4$^+$ T cell and several uninfected cells were also observed and have been termed polysynapses (Rudnicka 2009). VS have been described between infected and uninfected CD4$^+$ T cells, between macrophages and CD4$^+$ T cells, and between virus-exposed dendritic cells and CD4$^+$ T cells. Although all of these contact zones are likely to serve analogous functions in enhancing virus transmission, there are some fundamental differences: In VS between T cells, polarized HIV-1 release and subsequent target cell entry occurs across the VS (Piguet and Sattentau 2004), whereas in macrophages virions that have accumulated within the invaginated compartment (see above) are brought toward the contact zone (Gousset et al. 2008). Mature DCs capture cell-free virions into a compartment that stains for late endosomal markers, with subsequent transport and release toward the cell contact zone. The ultrastructure of this contact zone has been analyzed by ion abrasion scanning electron microscopy and electron tomography, which show a close envelopment of the T cells into DC-derived sheet-like membrane extensions, sometimes containing T-cell filopodia. This close interaction provides a shielded environment that allows CD4-dependent transfer of

sequestered virions from DCs to the T-cell surface and subsequent infection (Felts et al. 2010).

Viral and cellular components that polarize on the producer cell of the T-cell VS include HIV-1 Gag and Env, cellular tetraspanins, and the microtubule organizing center (MTOC), whereas HIV-1 receptors (CD4 and coreceptors) and the actin cytoskeleton polarize on the target cell (Haller and Fackler 2008). VS formation is dependent on HIV-1 Env proteins on the producer cell and viral receptors on the target cell, and is therefore a virus-induced structure. The contact zone in the producer cell is enriched in "raft" marker proteins and the VS is destroyed by cholesterol depletion, suggesting that membrane microdomains also make important contributions. Polarization and VS formation are dependent on an intact microtubule and actin cytoskeleton, and the VS can be stabilized by interactions of the cellular adhesion molecules ICAM-1 and LFA-1 (Jolly et al. 2007). ZAP-70, a kinase that regulates T-cell signaling and immunological synapse formation, is also required for polarized localization of HIV structural proteins, VS formation, and efficient cell-to-cell spread (Sol-Foulon et al. 2007).

The intracellular tail of the TM protein appears to be essential for polarized targeting of Env and Gag as well as for efficient spread of both murine leukemia virus (Jin et al. 2009) and HIV-1 (Emerson et al. 2010). HIV-1 Gag localizes to the trailing end of polarized T cells, termed the uropod, and this localization appears to depend on the formation of higher order oligomers and requires the NC domain (Llewellyn et al. 2010).

Despite their different names, infections that occur via cell-to-cell transfer and via cell-free particles both involve production of cell-free virions and likely involve fundamentally similar molecular mechanisms of virus assembly, budding, and release. Electron microscopy and tomographic (Martin et al. 2010) studies of VS revealed HIV budding events and numerous virions within the T-cell VS, which appeared to be loosely structured and to have few points of immediate contact, suggesting that the structure is relatively permeable. Transfer of viral material through the VS was

Cite this article as *Cold Spring Harb Perspect Med* doi: 10.1101/cshperspect.a006924

observed by video microscopy (Hubner et al. 2009) and reportedly involved micrometer-sized Gag-positive structures, which are far larger than individual virions, whereas parallel EM images showed individual HIV-1 budding sites within in the contact zone. It is unclear, therefore, whether the observed transfer events corresponded to HIV-1 infections or to transfer of vesicular material destined for lysosomal destruction. A cautious interpretation of these observations is also warranted because only a low number of actual infections were observed despite the presence of numerous budding sites and virions in the contact zone.

In principle, the VS could enhance HIV-1 transmission in several different ways. Virion release within a relatively closed environment and in the immediate vicinity of a target cell surface enriched in entry receptors provides one obvious advantage over cell-free infections, which are limited by extracellular diffusion and random cell surface attachment. Cell-to-cell contacts between uninfected and infected cells may also induce signaling pathways and activate the target cell, thus making it more permissive for HIV-1 infection. Finally, several studies have suggested that the VS could shield newly produced virions from circulating neutralizing antibodies and entry inhibitors (Chen et al. 2007; Hubner et al. 2009). However, those reports have not been confirmed in more recent studies (Massanella et al. 2009; Martin et al. 2010) and also seem inconsistent with ultrastructural analyses that suggest a relatively loose contact zone. Thus, the VS appears not to confer protection from neutralization or chemotherapeutic agents.

Membrane Bridges and Cell-to-Cell Transmission

Besides VS, several types of membrane bridges have been reported to enhance cell-to-cell transmission of HIV-1 and other retroviruses (Sherer and Mothes 2008; Mothes et al. 2010). Such structures include filopodial bridges, termed cytonemes, which have been shown to support the spread of murine leukemia virus (Sherer 2007), and membrane nanotubes, which can link

$CD4^+$ T cells and support HIV transfer (Sherer and Mothes 2008; Sowinski et al. 2008). Importantly, virion transport again occurs on the outer surfaces of these membrane connections, and therefore apparently involves normal budding and receptor-mediated entry processes. Although these structures are certainly intriguing, their relative contributions to retrovirus spread in vivo are difficult to quantify.

PERSPECTIVES

Despite considerable progress, a number of key aspects of HIV-1 assembly, budding, and maturation are not yet well understood and represent important research frontiers. The most pressing issues that need to be addressed include (1) determining the structure of the immature Gag lattice at high resolution, (2) defining the precise mechanism of membrane fission during virus budding, (3) elucidating the mechanism of PR activation, (4) defining the different stages of viral maturation in molecular detail, and (5) characterizing viral capsid functions and the mechanism of capsid uncoating during the early stages of the viral life cycle. Significant progress has already been made in identifying small molecule inhibitors of capsid assembly, budding, and maturation (Adamson et al. 2009; Blair et al. 2009; Jiang et al. 2011), and these efforts will undoubtedly be aided by a greater understanding of the molecular virology, biochemistry, and structural biology of these important viral processes.

ACKNOWLEDGMENTS

We thank John Briggs, Grant Jensen, John McCullough, Owen Pornillos, Jack Skalicky, and Mike Summers for help with figures and Barbara Müller and Oliver Fackler for suggestions and critical reading of the manuscript. HIV research in the Sundquist lab is supported by NIH grants R37 AI045405, R01 AI051174, and P50 GM082545; the Kräusslich lab is supported by grants from the Deutsche Forschungsgemeinschaft (SFB544 B11, SFB638, A9, SPP1175).

REFERENCES

* Reference is also in this collection.

Accola MA, Hoglund S, Gottlinger HG. 1998. A putative alpha-helical structure which overlaps the capsid-p2 boundary in the human immunodeficiency virus type 1 Gag precursor is crucial for viral particle assembly. *J Virol* 72: 2072–2078.

Accola MA, Strack B, Gottlinger HG. 2000. Efficient particle production by minimal Gag constructs which retain the carboxy-terminal domain of human immunodeficiency virus type 1 capsid-p2 and a late assembly domain. *J Virol* 74: 5395–5402.

Adamson CS, Salzwedel K, Freed EO. 2009. Virus maturation as a new HIV-1 therapeutic target. *Expert Opin Ther Targets* 13: 895–908.

Ali A, Reddy GS, Nalam MN, Anjum SG, Cao H, Schiffer CA, Rana TM. 2010. Structure-based design, synthesis, and structure-activity relationship studies of HIV-1 protease inhibitors incorporating phenyloxazolidinones. *J Med Chem* 53: 7699–7708.

Aloia RC, Jensen FC, Curtain CC, Mobley PW, Gordon LM. 1988. Lipid composition and fluidity of the human immunodeficiency virus. *Proc Natl Acad Sci* 85: 900–904.

Aloia RC, Tian H, Jensen FC. 1993. Lipid composition and fluidity of the human immunodeficiency virus envelope and host cell plasma membranes. *Proc Natl Acad Sci* 90: 5181–5185.

Andrawiss M, Takeuchi Y, Hewlett L, Collins M. 2003. Murine leukemia virus particle assembly quantitated by fluorescence microscopy: role of Gag-Gag interactions and membrane association. *J Virol* 77: 11651–11660.

* Arts EJ, Hazuda DJ. 2011. HIV-1 antiretroviral drug therapy. *Cold Spring Harb Perspect Med* doi: 10.1101/cshperspect.a007161.

Babst M, Wendland B, Estepa EJ, Emr SD. 1998. The Vps4p AAA ATPase regulates membrane association of a Vps protein complex required for normal endosome function. *EMBO J* 17: 2982–2993.

Bajorek M, Schubert HL, McCullough J, Langelier C, Eckert DM, Stubblefield WM, Uter NT, Myszka DG, Hill CP, Sundquist WI. 2009. Structural basis for ESCRT-III protein autoinhibition. *Nat Struct Mol Biol* 16: 754–762.

Bartonova V, Igonet S, Sticht J, Glass B, Habermann A, Vaney MC, Sehr P, Lewis J, Rey FA, Kräusslich HG. 2008. Residues in the HIV-1 capsid assembly inhibitor binding site are essential for maintaining the assembly-competent quaternary structure of the capsid protein. *J Biol Chem* 283: 32024–32033.

Batonick M, Favre M, Boge M, Spearman P, Honing S, Thali M. 2005. Interaction of HIV-1 Gag with the clathrin-associated adaptor AP-2. *Virology* 342: 190–200.

Baumgartel V, Ivanchenko S, Dupont A, Sergeev M, Wiseman PW, Kräusslich HG, Brauchle C, Muller B, Lamb DC. 2011. Live-cell visualization of dynamics of HIV budding site interactions with an ESCRT component. *Nat Cell Biol* 13: 469–474.

Benjamin J, Ganser-Pornillos BK, Tivol WF, Sundquist WI, Jensen GJ. 2005. Three-dimensional structure of HIV-1 virus-like particles by electron cryotomography. *J Mol Biol* 346: 577–588.

Bennett AE, Narayan K, Shi D, Hartnell LM, Gousset K, He H, Lowekamp BC, Yoo TS, Bliss D, Freed EO, et al. 2009. Ion-abrasion scanning electron microscopy reveals surface-connected tubular conduits in HIV-infected macrophages. *PLoS Pathog* 5: e1000591.

Bieniasz PD. 2009. The cell biology of HIV-1 virion genesis. *Cell Host Microbe* 5: 550–558.

Blair WS, Cao J, Fok-Seang J, Griffin P, Isaacson J, Jackson RL, Murray E, Patick AK, Peng Q, Perros M, et al. 2009. New small-molecule inhibitor class targeting human immunodeficiency virus type 1 virion maturation. *Antimicrob Agents Chemother* 53: 5080–5087.

Briggs JA, Wilk T, Fuller SD. 2003a. Do lipid rafts mediate virus assembly and pseudotyping? *J Gen Virol* 84: 757–768.

Briggs JA, Wilk T, Welker R, Krausslich HG, Fuller SD. 2003b. Structural organization of authentic, mature HIV-1 virions and cores. *EMBO J* 22: 1707–1715.

Briggs JA, Grunewald K, Glass B, Forster F, Krausslich HG, Fuller SD. 2006. The mechanism of HIV-1 core assembly: Insights from three-dimensional reconstructions of authentic virions. *Structure* 14: 15–20.

Briggs JA, Riches JD, Glass B, Bartonova V, Zanetti G, Krausslich HG. 2009. Structure and assembly of immature HIV. *Proc Natl Acad Sci* 106: 11090–11095.

Brugger B, Glass B, Haberkant P, Leibrecht I, Wieland FT, Krausslich HG. 2006. The HIV lipidome: A raft with an unusual composition. *Proc Natl Acad Sci* 103: 2641–2646.

Byeon IJ, Meng X, Jung J, Zhao G, Yang R, Ahn J, Shi J, Concel J, Aiken C, Zhang P, et al. 2009. Structural convergence between Cryo-EM and NMR reveals intersubunit interactions critical for HIV-1 capsid function. *Cell* 139: 780–790.

Camus G, Segura-Morales C, Molle D, Lopez-Verges S, Begon-Pescia C, Cazevieille C, Schu P, Bertrand E, Berlioz-Torrent C, Basyuk E. 2007. The clathrin adaptor complex AP-1 binds HIV-1 and MLV Gag and facilitates their budding. *Mol Biol Cell* 18: 3193–3203.

Carlson LA, Briggs JA, Glass B, Riches JD, Simon MN, Johnson MC, Muller B, Grunewald K, Krausslich HG. 2008. Three-dimensional analysis of budding sites and released virus suggests a revised model for HIV-1 morphogenesis. *Cell Host Microbe* 4: 592–599.

Carlson LA, de Marco A, Oberwinkler H, Habermann A, Briggs JA, Krausslich HG, Grunewald K. 2010. Cryo electron tomography of native HIV-1 budding sites. *PLoS Pathog* 6: e1001173.

Carlton JG, Martin-Serrano J. 2007. Parallels between cytokinesis and retroviral budding: A role for the ESCRT machinery. *Science* 316: 1908–1912.

Carlton JG, Martin-Serrano J. 2009. The ESCRT machinery: New functions in viral and cellular biology. *Biochem Soc Trans* 37: 195–199.

Caspar DL, Klug A. 1962. Physical principles in the construction of regular viruses. *Cold Spring Harb Symp Quant Biol* 27: 1–24.

Chan R, Uchil PD, Jin J, Shui G, Ott DE, Mothes W, Wenk MR. 2008. Retroviruses human immunodeficiency virus and murine leukemia virus are enriched in phosphoinositides. *J Virol* 82: 11228–11238.

Chen P, Hubner W, Spinelli MA, Chen BK. 2007. Predominant mode of human immunodeficiency virus transfer between T cells is mediated by sustained Env-dependent neutralization-resistant virological synapses. *J Virol* **81:** 12582–12595.

Chen J, Nikolaitchik O, Singh J, Wright A, Bencsics CE, Coffin JM, Ni N, Lockett S, Pathak VK, Hu WS. 2009. High efficiency of HIV-1 genomic RNA packaging and heterozygote formation revealed by single virion analysis. *Proc Natl Acad Sci* **106:** 13535–13540.

Chertova E, Bess JW Jr, Crise BJ, Sowder IR, Schaden TM, Hilburn JM, Hoxie JA, Benveniste RE, Lifson JD, Henderson LE, et al. 2002. Envelope glycoprotein incorporation, not shedding of surface envelope glycoprotein (gp120/ SU), is the primary determinant of SU content of purified human immunodeficiency virus type 1 and simian immunodeficiency virus. *J Virol* **76:** 5315–5325.

Chukkapalli V, Oh SJ, Ono A. 2010. Opposing mechanisms involving RNA and lipids regulate HIV-1 Gag membrane binding through the highly basic region of the matrix domain. *Proc Natl Acad Sci* **107:** 1600–1605.

Chung HY, Morita E, von Schwedler U, Muller B, Krausslich HG, Sundquist WI. 2008. NEDD4L overexpression rescues the release and infectivity of human immunodeficiency virus type 1 constructs lacking PTAP and YPXL late domains. *J Virol* **82:** 4884–4897.

Cooper JB. 2002. Aspartic proteinases in disease: A structural perspective. *Curr Drug Targets* **3:** 155–173.

Cosson P. 1996. Direct interaction between the envelope and matrix proteins of HIV-1. *EMBO J* **15:** 5783–5788.

D'Souza V, Summers MF. 2005. How retroviruses select their genomes. *Nat Rev* **3:** 643–655.

Datta SA, Heinrich F, Raghunandan S, Krueger S, Curtis JE, Rein A, Nanda H. 2011a. HIV-1 Gag extension: Conformational changes require simultaneous interaction with membrane and nucleic acid. *J Mol Biol* **406:** 205–214.

Datta SA, Temeselew LG, Crist RM, Soheilian F, Kamata A, Mirro J, Harvin D, Nagashima K, Cachau RE, Rein A. 2011b. On the role of the SP1 domain in HIV-1 particle assembly: A molecular switch? *J Virol* **85:** 4111–4121.

Davies BA, Azmi IF, Payne J, Shestakova A, Horazdovsky BF, Babst M, Katzmann DJ. 2010. Coordination of substrate binding and ATP hydrolysis in Vps4-mediated ESCRT-III disassembly. *Mol Biol Cell* **21:** 3396–3408.

de Marco A, Muller B, Glass B, Riches JD, Krausslich HG, Briggs JA. 2010. Structural analysis of HIV-1 maturation using cryo-electron tomography. *PLoS Pathog* **6:** e1001215.

Demirov DG, Ono A, Orenstein JM, Freed EO. 2002. Overexpression of the N-terminal domain of TSG101 inhibits HIV-1 budding by blocking late domain function. *Proc Natl Acad Sci* **99:** 955–960.

Deneka M, Pelchen-Matthews A, Byland R, Ruiz-Mateos E, Marsh M. 2007. In macrophages, HIV-1 assembles into an intracellular plasma membrane domain containing the tetraspanins CD81, CD9, and CD53. *J Cell Biol* **177:** 329–341.

Dismuke DJ, Aiken C. 2006. Evidence for a functional link between uncoating of the human immunodeficiency virus type 1 core and nuclear import of the viral preintegration complex. *J Virol* **80:** 3712–3720.

Dong X, Li H, Derdowski A, Ding L, Burnett A, Chen X, Peters TR, Dermody TS, Woodruff E, Wang JJ, et al. 2005. AP-3 directs the intracellular trafficking of HIV-1 Gag and plays a key role in particle assembly. *Cell* **120:** 663–674.

Dooher JE, Schneider BL, Reed JC, Lingappa JR. 2007. Host ABCE1 is at plasma membrane HIV assembly sites and its dissociation from Gag is linked to subsequent events of virus production. *Traffic* **8:** 195–211.

Dupont S, Sharova N, DeHoratius C, Virbasius CM, Zhu X, Bukrinskaya AG, Stevenson M, Green MR. 1999. A novel nuclear export activity in HIV-1 matrix protein required for viral replication. *Nature* **402:** 681–685.

Dussupt V, Javid MP, Abou-Jaoude G, Jadwin JA, de La Cruz J, Nagashima K, Bouamr F. 2009. The nucleocapsid region of HIV-1 Gag cooperates with the PTAP and LYPXnL late domains to recruit the cellular machinery necessary for viral budding. *PLoS Pathog* **5:** e1000339.

Eekels JJ, Geerts D, Jeeninga RE, Berkhout B. 2011. Long-term inhibition of HIV-1 replication with RNA interference against cellular co-factors. *Antivir Res* **89:** 43–53.

Einfeld D. 1996. Maturation and assembly of retroviral glycoproteins. *Curr Top Microbiol Immunol* **214:** 133–176.

Elia N, Sougrat R, Spurlin TA, Hurley JH, Lippincott-Schwartz J. 2011. Dynamics of endosomal sorting complex required for transport (ESCRT) machinery during cytokinesis and its role in abscission. *Proc Natl Acad Sci* **108:** 4846–4851.

Emerson V, Haller C, Pfeiffer T, Fackler OT, Bosch V. 2010. Role of the C-terminal domain of the HIV-1 glycoprotein in cell-to-cell viral transmission between T-lymphocytes. *Retrovirology* **7:** 43.

Fabrikant G, Lata S, Riches JD, Briggs JA, Weissenhorn W, Kozlov MM. 2009. Computational model of membrane fission catalyzed by ESCRT-III. *PLoS Comput Biol* **5:** e1000575.

Felts RL, Narayan K, Estes JD, Shi D, Trubey CM, Fu J, Hartnell LM, Ruthel GT, Schneider DK, Nagashima K, et al. 2010. 3D visualization of HIV transfer at the virological synapse between dendritic cells and T cells. *Proc Natl Acad Sci* **107:** 13336–13341.

Forshey BM, von Schwedler U, Sundquist WI, Aiken C. 2002. Formation of a human immunodeficiency virus type 1 core of optimal stability is crucial for viral replication. *J Virol* **76:** 5667–5677.

Frankel AD, Young JA. 1998. HIV-1: Fifteen proteins and an RNA. *Annu Rev Biochem* **67:** 1–25.

Freed EO. 2001. HIV-1 replication. *Somat Cell Molec Gen* **26:** 13–33.

Fujii K, Munshi UM, Ablan SD, Demirov DG, Soheilian F, Nagashima K, Stephen AG, Fisher RJ, Freed EO. 2009. Functional role of Alix in HIV-1 replication. *Virology* **391:** 284–292.

Fuller SD, Wilk T, Gowen BE, Krausslich HG, Vogt VM. 1997. Cryo-electron microscopy reveals ordered domains in the immature HIV-1 particle. *Curr Biol* **7:** 729–738.

Gamble TR, Yoo S, Vajdos FF, von Schwedler UK, Worthylake DK, Wang H, McCutcheon JP, Sundquist WI, Hill CP. 1997. Structure of the carboxyl-terminal dimerization domain of the HIV-1 capsid protein. *Science* **278:** 849–853.

Ganser BK, Li S, Klishko VY, Finch JT, Sundquist WI. 1999. Assembly and analysis of conical models for the HIV-1 core. *Science* **283:** 80–83.

Ganser-Pornillos BK, von Schwedler UK, Stray KM, Aiken C, Sundquist WI. 2004. Assembly properties of the human immunodeficiency virus type 1 CA protein. *J Virol* **78:** 2545–2552.

Ganser-Pornillos BK, Cheng A, Yeager M. 2007. Structure of full-length HIV-1 CA: A model for the mature capsid lattice. *Cell* **131:** 70–79.

Ganser-Pornillos BK, Yeager M, Sundquist WI. 2008. The structural biology of HIV assembly. *Curr Opin Struct Biol* **18:** 203–217.

Garrus JE, von Schwedler UK, Pornillos OW, Morham SG, Zavitz KH, Wang HE, Wettstein DA, Stray KM, Cote M, Rich RL, et al. 2001. Tsg101 and the vacuolar protein sorting pathway are essential for HIV-1 budding. *Cell* **107:** 55–65.

Ghazi-Tabatabai S, Saksena S, Short JM, Pobbati AV, Veprintsev DB, Crowther RA, Emr SD, Egelman EH, Williams RL. 2008. Structure and disassembly of filaments formed by the ESCRT-III subunit Vps24. *Structure* **16:** 1345–1356.

Gitti RK, Lee BM, Walker J, Summers MF, Yoo S, Sundquist WI. 1996. Structure of the amino-terminal core domain of the HIV-1 capsid protein. *Science* **273:** 231–235.

Gousset K, Ablan SD, Coren LV, Ono A, Soheilian F, Nagashima K, Ott DE, Freed EO. 2008. Real-time visualization of HIV-1 GAG trafficking in infected macrophages. *PLoS Pathog* **4:** e1000015.

Gross I, Hohenberg H, Wilk T, Wiegers K, Grattinger M, Muller B, Fuller S, Krausslich HG. 2000. A conformational switch controlling HIV-1 morphogenesis. *EMBO J* **19:** 103–113.

Guizetti J, Schermelleh L, Mantler J, Maar S, Poser I, Leonhardt H, Muller-Reichert T, Gerlich DW. 2011. Cortical constriction during abscission involves helices of ESCRT-III-dependent filaments. *Science* **331:** 1616–1620.

Haller C, Fackler OT. 2008. HIV-1 at the immunological and T-lymphocytic virological synapse. *Biol Chem* **389:** 1253–1260.

Hamard-Peron E, Juillard F, Saad JS, Roy C, Roingeard P, Summers MF, Darlix JL, Picart C, Muriaux D. 2010. Targeting of murine leukemia virus gag to the plasma membrane is mediated by PI(4,5)P2/PS and a polybasic region in the matrix. *J Virol* **84:** 503–515.

Hanson PI, Roth R, Lin Y, Heuser JE. 2008. Plasma membrane deformation by circular arrays of ESCRT-III protein filaments. *J Cell Biol* **180:** 389–402.

Hill M, Tachedjian G, Mak J. 2005. The packaging and maturation of the HIV-1 Pol proteins. *Curr HIV Res* **3:** 73–85.

Hubner W, McNerney GP, Chen P, Dale BM, Gordon RE, Chuang FY, Li XD, Asmuth DM, Huser T, Chen BK. 2009. Quantitative 3D video microscopy of HIV transfer across T cell virological synapses. *Science* **323:** 1743–1747.

Hurley JH, Hanson PI. 2010. Membrane budding and scission by the ESCRT machinery: It's all in the neck. *Nat Rev Mol Cell Biol* **11:** 556–566.

Hurley JH, Yang D. 2008. MIT domainia. *Dev Cell* **14:** 6–8.

Im YJ, Kuo L, Ren X, Burgos PV, Zhao XZ, Liu F, Burke TR Jr, Bonifacino JS, Freed EO, Hurley JH. 2010. Crystallographic and functional analysis of the ESCRT-I/HIV-1 Gag PTAP interaction. *Structure* **18:** 1536–1547.

Inlora J, Chukkapalli V, Derse D, Ono A. 2011. Gag localization and virus-like particle release mediated by the matrix domain of human T-lymphotropic virus type 1 Gag are less dependent on phosphatidylinositol-(4,5)-bisphosphate than those mediated by the matrix domain of HIV-1 Gag. *J Virol* **85:** 3802–3810.

Ivanchenko S, Godinez WJ, Lampe M, Krausslich HG, Eils R, Rohr K, Brauchle C, Muller B, Lamb DC. 2009. Dynamics of HIV-1 assembly and release. *PLoS Pathog* **5:** e1000652.

Ivanov D, Tsodikov OV, Kasanov J, Ellenberger T, Wagner G, Collins T. 2007. Domain-swapped dimerization of the HIV-1 capsid C-terminal domain. *Proc Natl Acad Sci* **104:** 4353–4358.

Jacks T, Power MD, Masiarz FR, Luciw PA, Barr PJ, Varmus HE. 1988. Characterization of ribosomal frameshifting in HIV-1 *gag-pol* expression. *Nature* **331:** 280–283.

Jiang Y, Liu X, De Clercq E. 2011.New therapeutic approaches targeted at the late stages of the HIV-1 replication cycle. *Curr Med Chem* **18:** 16–28.

Jin J, Sherer NM, Heidecker G, Derse D, Mothes W. 2009. Assembly of the murine leukemia virus is directed towards sites of cell-cell contact. *PLoS Biol* **7:** e1000163.

Johnson SF, Telesnitsky A. 2010. Retroviral RNA dimerization and packaging: The what, how, when, where, and why. *PLoS Pathog* **6**.

Jolly C, Sattentau QJ. 2004. Retroviral spread by induction of virological synapses. *Traffic* **5:** 643–650.

Jolly C, Mitar I, Sattentau QJ. 2007. Adhesion molecule interactions facilitate human immunodeficiency virus type 1-induced virological synapse formation between T cells. *J Virol* **81:** 13916–13921.

Jones CP, Datta SA, Rein A, Rouzina I, Musier-Forsyth K. 2011. Matrix domain modulates HIV-1 Gag's nucleic acid chaperone activity via inositol phosphate binding. *J Virol* **85:** 1594–1603.

Jouvenet N, Neil SJ, Bess C, Johnson MC, Virgen CA, Simon SM, Bieniasz PD. 2006. Plasma membrane is the site of productive HIV-1 particle assembly. *PLoS Biol* **4:** e435.

Jouvenet N, Bieniasz PD, Simon SM. 2008. Imaging the biogenesis of individual HIV-1 virions in live cells. *Nature* **454:** 236–240.

Jouvenet N, Simon SM, Bieniasz PD. 2009. Imaging the interaction of HIV-1 genomes and Gag during assembly of individual viral particles. *Proc Natl Acad Sci* **106:** 19114–19119.

Jouvenet N, Zhadina M, Bieniasz PD, Simon SM. 2011. Dynamics of ESCRT protein recruitment during retroviral assembly. *Nat Cell Biol* **13:** 394–401.

Kafaie J, Song R, Abrahamyan L, Mouland AJ, Laughrea M. 2008. Mapping of nucleocapsid residues important for HIV-1 genomic RNA dimerization and packaging. *Virology* **375:** 592–610.

Kleiman L, Jones CP, Musier-Forsyth K. 2010. Formation of the tRNALys packaging complex in HIV-1. *FEBS Lett* **584:** 359–365.

Kondo E, Mammano F, Cohen EA, Gottlinger HG. 1995. The p6gag domain of human immunodeficiency virus type 1 is sufficient for the incorporation of Vpr into heterologous viral particles. *J Virol* **69:** 2759–2764.

Krausslich HG, Traenckner AM, Rippmann F. 1991. Expression and characterization of genetically linked homo- and hetero-dimers of HIV proteinase. *Adv Exp Med Biol* **306:** 417–428.

Krishnan L, Matreyek KA, Oztop I, Lee K, Tipper CH, Li X, Dar MJ, Kewalramani VN, Engelman A. 2010. The requirement for cellular transportin 3 (TNPO3 or TRN-SR2) during infection maps to human immunodeficiency virus type 1 capsid and not integrase. *J Virol* **84:** 397–406.

Kuang Z, Seo EJ, Leis J. 2011. The mechanism of inhibition of retrovirus release from cells by interferon induced gene ISG15. *J Virol* **85:** 7153–7161.

Kutluay SB, Bieniasz PD. 2010. Analysis of the initiating events in HIV-1 particle assembly and genome packaging. *PLoS Pathog* **6:** e1001200.

Lata S, Roessle M, Solomons J, Jamin M, Gottlinger HG, Svergun DI, Weissenhorn W. 2008a. Structural basis for autoinhibition of ESCRT-III CHMP3. *J Mol Biol* **378:** 816–825.

Lata S, Schoehn G, Jain A, Pires R, Piehler J, Gottlinger HG, Weissenhorn W. 2008b. Helical structures of ESCRT-III are disassembled by VPS4. *Science* **321:** 1354–1357.

Lee K, Ambrose Z, Martin TD, Oztop I, Mulky A, Julias JG, Vandegraaff N, Baumann JG, Wang R, Yuen W, et al. 2010. Flexible use of nuclear import pathways by HIV-1. *Cell Host Microbe* **7:** 221–233.

Leung K, Kim JO, Ganesh L, Kabat J, Schwartz O, Nabel GJ. 2008. HIV-1 assembly: Viral glycoproteins segregate quantally to lipid rafts that associate individually with HIV-1 capsids and virions. *Cell Host Microbe* **3:** 285–292.

Lever AM. 2007. HIV-1 RNA packaging. *Adv Pharmacol* **55:** 1–32.

Lin Y, Kimpler LA, Naismith TV, Lauer JM, Hanson PI. 2005. Interaction of the mammalian endosomal sorting complex required for transport (ESCRT) III protein hSnf7–1 with itself, membranes, and the AAA+ ATPase SKD1. *J Biol Chem* **280:** 12799–12809.

Llewellyn GN, Hogue IB, Grover JR, Ono A. 2010. Nucleocapsid promotes localization of HIV-1 gag to uropods that participate in virological synapses between T cells. *PLoS Pathog* **6:** e1001167.

Lorizate M, Brugger B, Akiyama H, Glass B, Muller B, Anderluh G, Wieland FT, Krausslich HG. 2009. Probing HIV-1 membrane liquid order by Laurdan staining reveals producer cell-dependent differences. *J Biol Chem* **284:** 22238–22247.

Louis JM, Ishima R, Torchia DA, Weber IT. 2007. HIV-1 protease: Structure, dynamics, and inhibition. *Adv Pharmacol* **55:** 261–298.

Luban J. 2007. Cyclophilin A, TRIM5, and resistance to human immunodeficiency virus type 1 infection. *J Virol* **81:** 1054–1061.

* Malim MH, Bieniasz PD. 2011. HIV restriction factors and mechanisms of evasion. *Cold Spring Harb Perspect Med* doi: 10.1101/cshperspect.a006940.

Martin N, Welsch S, Jolly C, Briggs JA, Vaux D, Sattentau QJ. 2010. Virological synapse-mediated spread of human immunodeficiency virus type 1 between T cells is sensitive to entry inhibition. *J Virol* **84:** 3516–3527.

Martinez NW, Xue X, Berro RG, Kreitzer G, Resh MD. 2008. Kinesin KIF4 regulates intracellular trafficking and stability of the human immunodeficiency virus type 1 Gag polyprotein. *J Virol* **82:** 9937–9950.

Martin-Serrano J, Zang T, Bieniasz PD. 2001. HIV-1 and Ebola virus encode small peptide motifs that recruit Tsg101 to sites of particle assembly to facilitate egress. *Nat Med* **7:** 1313–1319.

Massanella M, Puigdomenech I, Cabrera C, Fernandez-Figueras MT, Aucher A, Gaibelet G, Hudrisier D, Garcia E, Bofill M, Clotet B, et al. 2009. Antigp41 antibodies fail to block early events of virological synapses but inhibit HIV spread between T cells. *AIDS* **23:** 183–188.

McCullough J, Fisher RD, Whitby FG, Sundquist WI, Hill CP. 2008. ALIX-CHMP4 interactions in the human ESCRT pathway. *Proc Natl Acad Sci* **105:** 7687–7691.

Miyazaki Y, Garcia EL, King SR, Iyalla K, Loeliger K, Starck P, Syed S, Telesnitsky A, Summers MF. 2010. An RNA structural switch regulates diploid genome packaging by Moloney murine leukemia virus. *J Mol Biol* **396:** 141–152.

Moore MD, Hu WS. 2009. HIV-1 RNA dimerization: It takes two to tango. *AIDS Rev* **11:** 91–102.

Moore MD, Nikolaitchik OA, Chen J, Hammarskjold ML, Rekosh D, Hu WS. 2009. Probing the HIV-1 genomic RNA trafficking pathway and dimerization by genetic recombination and single virion analyses. *PLoS Pathog* **5:** e1000627.

Morellet N, Druillennec S, Lenoir C, Bouaziz S, Roques BP. 2005. Helical structure determined by NMR of the HIV-1 (345–392)Gag sequence, surrounding p2: Implications for particle assembly and RNA packaging. *Protein Sci* **14:** 375–386.

Morita E, Sundquist WI. 2004. Retrovirus budding. *Annu Rev Cell Dev Biol* **20:** 395–425.

Morita E, Sandrin V, McCullough J, Katsuyama A, Baci Hamilton I, Sundquist WI. 2011. ESCRT-III protein requirements for HIV-1 budding. *Cell Host Microbe* **9:** 235–242.

Mortuza GB, Haire LF, Stevens A, Smerdon SJ, Stoye JP, Taylor IA. 2004. High-resolution structure of a retroviral capsid hexameric amino-terminal domain. *Nature* **431:** 481–485.

Mothes W, Sherer NM, Jin J, Zhong P. 2010. Virus cell-to-cell transmission. *J Virol* **84:** 8360–8368.

Muller B, Tessmer U, Schubert U, Krausslich HG. 2000. Human immunodeficiency virus type 1 Vpr protein is incorporated into the virion in significantly smaller amounts than gag and is phosphorylated in infected cells. *J Virol* **74:** 9727–9731.

Murakami T, Freed EO. 2000. Genetic evidence for an interaction between human immunodeficiency virus type 1 matrix and alpha-helix 2 of the gp41 cytoplasmic tail. *J Virol* **74:** 3548–3554.

Murakami T, Ablan S, Freed EO, Tanaka Y. 2004. Regulation of human immunodeficiency virus type 1 Env-mediated membrane fusion by viral protease activity. *J Virol* **78**: 1026–1031.

Muziol T, Pineda-Molina E, Ravelli RB, Zamborlini A, Usami Y, Gottlinger H, Weissenhorn W. 2006. Structural basis for budding by the ESCRT-III factor CHMP3. *Dev Cell* **10**: 821–830.

Obita T, Saksena S, Ghazi-Tabatabai S, Gill DJ, Perisic O, Emr SD, Williams RL. 2007. Structural basis for selective recognition of ESCRT-III by the AAA ATPase Vps4. *Nature* **449**: 735–739.

Ohishi M, Nakano T, Sakuragi S, Shioda T, Sano K, Sakuragi J. 2011. The relationship between HIV-1 genome RNA dimerization, virion maturation and infectivity. *Nucleic Acids Res* **39**: 3404–3417.

Onafuwa-Nuga AA, Telesnitsky A, King SR. 2006. 7SL RNA, but not the 54-kd signal recognition particle protein, is an abundant component of both infectious HIV-1 and minimal virus-like particles. *RNA* **12**: 542–546.

Ono A. 2009. HIV-1 assembly at the plasma membrane: Gag trafficking and localization. *Future Virol* **4**: 241–257.

Ono A, Freed EO. 2001. Plasma membrane rafts play a critical role in HIV-1 assembly and release. *Proc Natl Acad Sci* **98**: 13925–13930.

Ono A, Ablan SD, Lockett SJ, Nagashima K, Freed EO. 2004. Phosphatidylinositol (4,5) bisphosphate regulates HIV-1 Gag targeting to the plasma membrane. *Proc Natl Acad Sci* **101**: 14889–14894.

Ott DE. 2008. Cellular proteins detected in HIV-1. *Rev Med Virol* **18**: 159–175.

Parker SD, Hunter E. 2001. Activation of the Mason-Pfizer monkey virus protease within immature capsids in vitro. *Proc Natl Acad Sci* **98**: 14631–14636.

Peel S, Macheboeuf P, Martinelli N, Weissenhorn W. 2011. Divergent pathways lead to ESCRT-III-catalyzed membrane fission. *Trends Biochem Sci* **36**: 199–210.

Pelchen-Matthews A, Kramer B, Marsh M. 2003. Infectious HIV-1 assembles in late endosomes in primary macrophages. *J Cell Biol* **162**: 443–455.

Pettit SC, Moody MD, Wehbie RS, Kaplan AH, Nantermet PV, Klein CA, Swanstrom R. 1994. The p2 domain of human immunodeficiency virus type 1 Gag regulates sequential proteolytic processing and is required to produce fully infectious virions. *J Virol* **68**: 8017–8027.

Phillips DM, Bourinbaiar AS. 1992. Mechanism of HIV spread from lymphocytes to epithelia. *Virology* **186**: 261–273.

Piguet V, Sattentau Q. 2004. Dangerous liaisons at the virological synapse. *J Clin Invest* **114**: 605–610.

Pincetic A, Kuang Z, Seo EJ, Leis J. 2010. The interferon-induced gene ISG15 blocks retrovirus release from cells late in the budding process. *J Virol* **84**: 4725–4736.

Pires R, Hartlieb B, Signor L, Schoehn G, Lata S, Roessle M, Moriscot C, Popov S, Hinz A, Jamin M, et al. 2009. A crescent-shaped ALIX dimer targets ESCRT-III CHMP4 filaments. *Structure* **17**: 843–856.

Popov S, Popova E, Inoue M, Gottlinger HG. 2008. Human immunodeficiency virus type 1 Gag engages the Bro1 domain of ALIX/AIP1 through the nucleocapsid. *J Virol* **82**: 1389–1398.

Popov S, Popova E, Inoue M, Gottlinger HG. 2009. Divergent Bro1 domains share the capacity to bind human immunodeficiency virus type 1 nucleocapsid and to enhance virus-like particle production. *J Virol* **83**: 7185–7193.

Pornillos O, Alam SL, Davis DR, Sundquist WI. 2002. Structure of the Tsg101 UEV domain in complex with the PTAP motif of the HIV-1 p6 protein. *Nat Struct Biol* **9**: 812–817.

Pornillos O, Higginson DS, Stray KM, Fisher RD, Garrus JE, Payne M, He GP, Wang HE, Morham SG, Sundquist WI. 2003. HIV Gag mimics the Tsg101-recruiting activity of the human Hrs protein. *J Cell Biol* **162**: 425–434.

Pornillos O, Ganser-Pornillos BK, Kelly BN, Hua Y, Whitby FG, Stout CD, Sundquist WI, Hill CP, Yeager M. 2009. X-ray structures of the hexameric building block of the HIV capsid. *Cell* **137**: 1282–1292.

Pornillos O, Ganser-Pornillos BK, Yeager M. 2011. Atomic-level modelling of the HIV capsid. *Nature* **469**: 424–427.

Prabu-Jeyabalan M, Nalivaika E, Schiffer CA. 2002. Substrate shape determines specificity of recognition for HIV-1 protease: Analysis of crystal structures of six substrate complexes. *Structure* **10**: 369–381.

Prabu-Jeyabalan M, Nalivaika EA, King NM, Schiffer CA. 2004. Structural basis for coevolution of a human immunodeficiency virus type 1 nucleocapsid-p1 cleavage site with a V82A drug-resistant mutation in viral protease. *J Virol* **78**: 12446–12454.

Rein A. 2010. Nucleic acid chaperone activity of retroviral Gag proteins. *RNA Biol* **7**: 700–705.

Rein A, Datta SA, Jones CP, Musier-Forsyth K. 2011. Diverse interactions of retroviral Gag proteins with RNAs. *Trends Biochem Sci* **36**: 373–380.

Ren X, Hurley JH. 2011. Proline-rich regions and motifs in trafficking: From ESCRT interaction to viral exploitation. *Traffic*.

Rulli SJ Jr, Hibbert CS, Mirro J, Pederson T, Biswal S, Rein A. 2007. Selective and nonselective packaging of cellular RNAs in retrovirus particles. *J Virol* **81**: 6623–6631.

Saad JS, Miller J, Tai J, Kim A, Ghanam RH, Summers MF. 2006. Structural basis for targeting HIV-1 Gag proteins to the plasma membrane for virus assembly. *Proc Natl Acad Sci* **103**: 11364–11369.

Saad JS, Loeliger E, Luncsford P, Liriano M, Tai J, Kim A, Miller J, Joshi A, Freed EO, Summers MF. 2007. Point mutations in the HIV-1 matrix protein turn off the myristyl switch. *J Mol Biol* **366**: 574–585.

Saad JS, Ablan SD, Ghanam RH, Kim A, Andrews K, Nagashima K, Soheilian F, Freed EO, Summers MF. 2008. Structure of the myristylated human immunodeficiency virus type 2 matrix protein and the role of phosphatidylinositol-(4,5)-bisphosphate in membrane targeting. *J Mol Biol* **382**: 434–447.

Saksena S, Sun J, Chu T, Emr SD. 2007. ESCRTing proteins in the endocytic pathway. *Trends Biochem Sci* **32**: 561–573.

Saksena S, Wahlman J, Teis D, Johnson AE, Emr SD. 2009. Functional reconstitution of ESCRT-III assembly and disassembly. *Cell* **136**: 97–109.

Scheifele LZ, Garbitt RA, Rhoads JD, Parent LJ. 2002. Nuclear entry and CRM1-dependent nuclear export of the Rous sarcoma virus Gag polyprotein. *Proc Natl Acad Sci* **99:** 3944–3949.

Schliephake AW, Rethwilm A. 1994. Nuclear localization of foamy virus Gag precursor protein. *J Virol* **68:** 4946–4954.

Scott A, Chung HY, Gonciarz-Swiatek M, Hill GC, Whitby FG, Gaspar J, Holton JM, Viswanathan R, Ghaffarian S, Hill CP, et al. 2005. Structural and mechanistic studies of VPS4 proteins. *EMBO J* **24:** 3658–3669.

Sebastian S, Luban J. 2007. The retroviral restriction factor TRIM5alpha. *Curr Infect Dis Rep* **9:** 167–173.

Sherer NM, Mothes W. 2008. Cytonemes and tunneling nanotubules in cell-cell communication and viral pathogenesis. *Trends Cell Biol* **18:** 414–420.

Shestakova A, Hanono A, Drosner S, Curtiss M, Davies BA, Katzmann DJ, Babst M. 2010. Assembly of the AAA ATPase Vps4 on ESCRT-III. *Mol Biol Cell* **21:** 1059–1071.

Skaug B, Chen ZJ. 2010. Emerging role of ISG15 in antiviral immunity. *Cell* **143:** 187–190.

Sol-Foulon N, Sourisseau M, Porrot F, Thoulouze MI, Trouillet C, Nobile C, Blanchet F, di Bartolo V, Noraz N, Taylor N, et al. 2007. ZAP-70 kinase regulates HIV cell-to-cell spread and virological synapse formation. *EMBO J* **26:** 516–526.

Sougrat R, Bartesaghi A, Lifson JD, Bennett AE, Bess JW, Zabransky DJ, Subramaniam S. 2007. Electron tomography of the contact between T cells and SIV/HIV-1: Implications for viral entry. *PLoS Pathog* **3:** e63.

Sowinski S, Jolly C, Berninghausen O, Purbhoo MA, Chauveau A, Kohler K, Oddos S, Eissmann P, Brodsky FM, Hopkins C, et al. 2008. Membrane nanotubes physically connect T cells over long distances presenting a novel route for HIV-1 transmission. *Nat Cell Biol* **10:** 211–219.

Spriggs S, Garyu L, Connor R, Summers MF. 2008. Potential intra- and intermolecular interactions involving the unique-5′ region of the HIV-1 5′-UTR. *Biochemistry* **47:** 13064–13073.

Strack B, Calistri A, Craig S, Popova E, Gottlinger HG. 2003. AIP1/ALIX is a binding partner for HIV-1 p6 and EIAV p9 functioning in virus budding. *Cell* **114:** 689–699.

Stuchell-Brereton MD, Skalicky JJ, Kieffer C, Karren MA, Ghaffarian S, Sundquist WI. 2007. ESCRT-III recognition by VPS4 ATPases. *Nature* **449:** 740–744.

Swanstrom R, Wills JW. 1997. Synthesis, assembly, and processing of viral proteins. In *Retroviruses* (eds. Coffin JM, Hughes SH, Varmus HE). Cold Spring Harbor Press, Cold Spring Harbor, NY.

Tang Y, Winkler U, Freed EO, Torrey TA, Kim W, Li H, Goff SP, Morse HC. 1999. *J Virol* **73:** 10508–10513.

Tang C, Ndassa Y, Summers MF. 2002. Structure of the N-terminal 283-residue fragment of the immature HIV-1 Gag polyprotein. *Nat Struct Biol* **9:** 537–543.

Tang C, Louis JM, Aniana A, Suh JY, Clore GM. 2008. Visualizing transient events in amino-terminal autoprocessing of HIV-1 protease. *Nature* **455:** 693–696.

Teis D, Saksena S, Emr SD. 2008. Ordered assembly of the ESCRT-III complex on endosomes is required to sequester cargo during MVB formation. *Dev Cell* **15:** 578–589.

Ternois F, Sticht J, Duquerroy S, Krausslich HG, Rey FA. 2005. The HIV-1 capsid protein C-terminal domain in

complex with a virus assembly inhibitor. *Nat Struct Mol Biol* **12:** 678–682.

Usami Y, Popov S, Popova E, Gottlinger HG. 2008. Efficient and specific rescue of human immunodeficiency virus type 1 budding defects by a Nedd4-like ubiquitin ligase. *J Virol* **82:** 4898–4907.

Usami Y, Popov S, Popova E, Inoue M, Weissenhorn W, Gottlinger H. 2009. The ESCRT pathway and HIV-1 budding. *Biochem Soc Trans* **37:** 181–184.

Vasiliver-Shamis G, Dustin ML, Hioe CE. 2010. HIV-1 virological synapse is not simply a copycat of the immunological synapse. *Viruses* **2:** 1239–1260.

VerPlank L, Bouamr F, LaGrassa TJ, Agresta B, Kikonyogo A, Leis J, Carter CA. 2001. Tsg101, a homologue of ubiquitin-conjugating (E2) enzymes, binds the L domain in HIV type 1 Pr55Gag. *Proc Natl Acad Sci* **98:** 7724–7729.

Vincent O, Rainbow L, Tilburn J, Arst HN Jr, Penalva MA. 2003. YPXL/I is a protein interaction motif recognized by Aspergillus PalA and its human homologue, AIP1/Alix. *Mol Cell Biol* **23:** 1647–1655.

von Schwedler U, Stray KM, Garrus JE, Sundquist WI. 2003. Functional surfaces of the human immunodeficiency virus type 1 capsid protein. *J Virol* **77:** 5439–5450.

Weiss ER, Popova E, Yamanaka H, Kim HC, Huibregtse JM, Gottlinger H. 2010. Rescue of HIV-1 release by targeting widely divergent NEDD4-type ubiquitin ligases and isolated catalytic HECT domains to Gag. *PLoS Pathog* **6:** e1001107.

Welsch S, Keppler OT, Habermann A, Allespach I, Krijnse-Locker J, Krausslich HG. 2007. HIV-1 buds predominantly at the plasma membrane of primary human macrophages. *PLoS Pathog* **3:** e36.

Wilk T, Gross I, Gowen BE, Rutten T, de Haas F, Welker R, Krausslich HG, Boulanger P, Fuller SD. 2001. Organization of immature human immunodeficiency virus type 1. *J Virol* **75:** 759–771.

Wlodawer A, Gustchina A. 2000. Structural and biochemical studies of retroviral proteases. *Biochim Biophys Acta* **1477:** 16–34.

Wollert T, Wunder C, Lippincott-Schwartz J, Hurley JH. 2009. Membrane scission by the ESCRT-III complex. *Nature* **458:** 172–177.

Wright ER, Schooler JB, Ding HJ, Kieffer C, Fillmore C, Sundquist WI, Jensen GJ. 2007. Electron cryotomography of immature HIV-1 virions reveals the structure of the CA and SP1 Gag shells. *EMBO J* **26:** 2218–2226.

Wyma DJ, Kotov A, Aiken C. 2000. Evidence for a stable interaction of gp41 with Pr55(Gag) in immature human immunodeficiency virus type 1 particles. *J Virol* **74:** 9381–9387.

Wyma DJ, Jiang J, Shi J, Zhou J, Lineberger JE, Miller MD, Aiken C. 2004. Coupling of human immunodeficiency virus type 1 fusion to virion maturation: A novel role of the gp41 cytoplasmic tail. *J Virol* **78:** 3429–3435.

Xiao J, Chen XW, Davies BA, Saltiel AR, Katzmann DJ, Xu Z. 2009. Structural basis of Ist1 function and Ist1-Did2 interaction in the multivesicular body pathway and cytokinesis. *Mol Biol Cell* **20:** 3514–3524.

Yamashita M, Emerman M. 2009. Cellular restriction targeting viral capsids perturbs human immunodeficiency virus type 1 infection of nondividing cells. *J Virol* **83:** 9835–9843.

Yamashita M, Perez O, Hope TJ, Emerman M. 2007. Evidence for direct involvement of the capsid protein in HIV infection of nondividing cells. *PLoS Pathog* **3:** 1502–1510.

Yang D, Hurley JH. 2010. Structural role of the Vps4-Vta1 interface in ESCRT-III recycling. *Structure* **18:** 976–984.

Yu X, Yuan X, McLane MF, Lee TH, Essex M. 1993. Mutations in the cytoplasmic domain of human immunodeficiency virus type 1 transmembrane protein impair the incorporation of Env proteins into mature virions. *J Virol* **67:** 213–221.

Yu Z, Gonciarz MD, Sundquist WI, Hill CP, Jensen GJ. 2008. Cryo-EM structure of dodecameric Vps4p and its 2:1 complex with Vta1p. *J Mol Biol* **377:** 364–377.

Zamborlini A, Usami Y, Radoshitzky SR, Popova E, Palu G, Gottlinger H. 2006. Release of autoinhibition converts ESCRT-III components into potent inhibitors of HIV-1 budding. *Proc Natl Acad Sci* **103:** 19140–19145.

Zhai Q, Fisher RD, Chung HY, Myszka DG, Sundquist WI, Hill CP. 2008. Structural and functional studies of ALIX interactions with YPX(n)L late domains of HIV-1 and EIAV. *Nat Struct Mol Biol* **15:** 43–49.

Zhai Q, Landesman MB, Robinson H, Sundquist WI, Hill CP. 2011. Identification and structural characterization of the ALIX-binding late domains of simian immunodeficiency virus SIVmac239 and SIVagmTan-1. *J Virol* **85:** 632–637.

Zhang Y, Qian H, Love Z, Barklis E. 1998. Analysis of the assembly function of the human immunodeficiency virus type 1 gag protein nucleocapsid domain. *J Virol* **72:** 1782–1789.

Zhu P, Liu J, Bess J Jr, Chertova E, Lifson JD, Grise H, Ofek GA, Taylor KA, Roux KH. 2006. Distribution and three-dimensional structure of AIDS virus envelope spikes. *Nature* **441:** 847–852.

HIV Restriction Factors and Mechanisms of Evasion

Michael H. Malim[1] and Paul D. Bieniasz[2]

[1]Department of Infectious Diseases, King's College London School of Medicine, Guy's Hospital, London Bridge, London SE1 9RT, United Kingdom

[2]Howard Hughes Medical Institute, Aaron Diamond AIDS Research Center, The Rockefeller University, New York, New York 10016

Correspondence: michael.malim@kcl.ac.uk

Retroviruses have long been a fertile model for discovering host–pathogen interactions and their associated biological principles and processes. These advances have not only informed fundamental concepts of viral replication and pathogenesis but have also provided novel insights into host cell biology. This is illustrated by the recent descriptions of host-encoded restriction factors that can serve as effective inhibitors of retroviral replication. Here, we review our understanding of the three restriction factors that have been widely shown to be potent inhibitors of HIV-1: namely, APOBEC3G, TRIM5α, and tetherin. In each case, we discuss how these unrelated proteins were identified, the mechanisms by which they inhibit replication, the means used by HIV-1 to evade their action, and their potential contributions to viral pathogenesis as well as inter- and intraspecies transmission.

HIV-1, in common with all viruses, requires the concerted contributions of numerous positively acting cellular factors and pathways to achieve efficient replication (Bushman et al. 2009). Conversely, mammalian cells also express a number of diverse, dominantly acting proteins that are widely expressed and function in a cell-autonomous manner to suppress virus replication. These have been termed restriction factors and/or intrinsic resistance factors, and they provide an initial (or early) line of defense against infection as a component of, or even preceding, innate antiviral responses. This work discusses the most extensively described examples of such factors, focusing on their impact on HIV-1. These are the apolipoprotein B messenger RNA (mRNA)-editing enzyme catalytic polypeptide-like 3 (APOBEC3) family of proteins (in particular, APOBEC3G), tetherin/ bone marrow stromal cell antigen 2 (BST2)/ CD317 (hereafter called tetherin), and tripartite-motif-containing 5α (TRIM5α).

A fundamental concept to the biology of restriction factors is that HIV-1 generally evades their potent inhibitory activities in human cells, thereby allowing virus replication to proceed efficiently. In contrast, the ability of HIV-1 to replicate in nonhuman cells is often severely compromised by restriction factors, thus marking these proteins as important determinants of viral host range and cross-species transmission. The mechanisms for evasion from restriction factors are virus encoded and frequently involve HIV-1's regulatory/accessory proteins, namely, Vif, Nef, Vpu, and Vpr. Indeed, the need to escape intrinsic resistance appears to have been an important driving force behind the acquisition of these viral genes. Aside from restriction factor evasion, the Vpu and Nef proteins regulate the expression and localization of a number of host proteins important during HIV-1 replication. Prominent among these interactions, Vpu and Nef both inhibit the cell-surface expression of the primary entry receptor CD4, as well as major histocompatibility class I complexes (MHC class I), whereas Nef also helps promote T-cell activation and HIV-1 particle infectivity (reviewed by Kirchhoff 2010).

HISTORY: DISCOVERY OF HIV-1 RESTRICTION FACTORS

The intellectual framework for considering restriction factors was established through studies of ecotropic murine leukemia virus (MLV). Specifically, mice encode a gene, $Fv1$, with two principal allelic forms, $Fv1^n$ and $Fv1^b$. $Fv1^n$ cells are up to 1000-fold more susceptible to infection by N-tropic strains of MLV than B-tropic strains, and $Fv1^n$ mice are correspondingly highly receptive to N-MLV induced disease. The opposite is true for $Fv1^b$ cells and animals, which are susceptible to B-rather than N-MLV. Heterozygous $Fv1^{n/b}$ cells are resistant to both N and B viruses, illustrating the general principle that a restricting phenotype is dominant over susceptibility. The sequence of the $Fv1$ gene most closely resembles that of an endogenous retrovirus gag gene (Best et al. 1996), Fv1 blocks infection by a poorly understood mechanism that operates after reverse transcription but before integration and likely requires direct recognition of infecting viral capsids, as N/B-tropism is determined by sequence differences in the capsid (CA) portion of the viral Gag protein.

Beginning in the 1990s, sporadic evidence emerged that hinted at the existence of additional restriction factors, including factors affecting HIV-1. For instance, (1) virus infectivity or the capacity of viral accessory genes to function could be profoundly affected by the animal species of the cells under experimental examination (Simon et al. 1998b; Hofmann et al. 1999); and (2) the requirements for individual accessory genes during virus replication could vary enormously between human cell lines (Gabuzda et al. 1992; Varthakavi et al. 2003). Drawing on the Fv1 analogy, but recognizing the lack of similarity among the phenotypic manifestations of these replication barriers, the concept that primate cells express a range of restriction factors that target HIV-1 and other lentiviruses gradually gained acceptance.

One experimental approach that added weight to these arguments, and parallels the resistance of $Fv1^{n/b}$ cells to N- and B-MLV infection, is illustrated in Figure 1. Here, cells that are restrictive or susceptible for a viral function of step or replication (also called nonpermissive and permissive cells, respectively) are fused in vitro to form heterokaryons that consequently express the contents of both cells. The capacity of these cell hybrids to support the viral activity in question is then assessed. A restricting phenotype points to the presence of a dominant restriction factor that is absent from susceptible cells, whereas a susceptible phenotype suggests that a positively acting cofactor has been lost from the nonpermissive cells. Cell fusion studies of this genre established that distinct restriction factor activities were apparently countered by HIV-1 Vif and Vpu, or evaded by sequence changes in CA (Madani and Kabat 1998; Simon et al. 1998a; Cowan et al. 2002; Varthakavi et al. 2003).

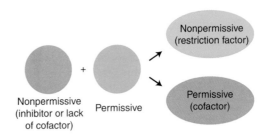

Figure 1. Cell fusion studies can illustrate restriction factor activity.

Cite this article as *Cold Spring Harb Perspect Med* doi: 10.1101/cshperspect.a006940

Two experimental strategies have been used to discover genes encoding restriction factors. First, comparative transcriptomics (e.g., gene arrays or copy DNA [cDNA] subtraction) has been used to identify genes that are preferentially expressed in restrictive cells relative to susceptible cells. Lists of candidates may be pruned further when additional characteristics, such as interferon responsiveness, are known. Candidate cDNAs are then validated functionally by asking whether ectopic expression converts susceptible cells into restrictive cells. A second more direct screening approach has exploited the expression of cDNA libraries derived from restrictive cells in susceptible cells, followed by the selection of cells that have acquired viral resistance and the isolation of the cDNA conferring resistance. The former approach was used to identify APOBEC3G and tetherin (Sheehy et al. 2002; Neil et al. 2008), which initially engage their viral substrates at postintegration stages of replication, whereas the latter scheme is well suited for finding factors that act at preintegration steps, and led to the identification of TRIM5α (Stremlau et al. 2004). In all cases, a cardinal feature that defines a restriction factor is the capacity to display potent antiviral function as single genes (i.e., without the requirement for specific cellular cofactors). This and other shared features of the known restriction factors are listed in Table 1.

Table 1. Cardinal and shared features of HIV-1 restriction factors

Germline-encoded, expressed constitutively, and interferon (IFN)-inducible.

Dominantly acting, cell-autonomous mechanisms of action.

Largely inactive against contemporary "wild-type" viruses in cells of natural hosts.

Mediate potent species-specific suppression: control of cross-species transmission?

Some (APOBEC3 and tetherin) are regulated by HIV/SIV accessory proteins.

Display hallmarks of positive genetic selection (high d_N/d_S ratio), reflecting host–pathogen coevolution.

Function and/or regulation involves the cellular ubiquitin/proteasome system.

KEY ADVANCES IN THE STUDY OF RESTRICTION FACTORS

Beyond the recognition of their existence, major advances in restriction factor biology include the identification of the proteins responsible for restriction activity, the elucidation of mechanisms of action, the recognition of specific viral countermeasures and means of evasion, and the emerging paradigm that restriction factors and their antagonists continually coevolve. For each of the factors highlighted in this work, remarkable and unanticipated biology has been uncovered.

APOBEC3 PROTEINS AND Vif

Identification

The interplay between human APOBEC3 proteins and HIV-1 was discovered through efforts to understand the function of the ~23 kDa viral protein Vif (an acronymn for virion infectivity factor) (reviewed by Malim 2009). Vif is required for HIV-1 replication in primary cell types, particularly CD4$^+$ T cells, as well as some cell lines, yet is dispensable in other lines. Cell fusion experiments attributed causation to a restriction factor, and a cDNA subtraction-based screen revealed the human gene *APOBEC3G* as being sufficient to repress the replication of *vif*-deficient HIV-1 (Sheehy et al. 2002).

APOBEC3 Proteins

APOBEC3G (A3G) is a member of a family of vertebrate proteins (humans encode 11) with polynucleotide (RNA or DNA) cytidine deaminase activity. This reaction results in the postsynthetic editing of cytidine residues to uridines, thereby altering the nucleotide sequence and, with DNA substrates, introducing an unnatural base. APOBEC3 proteins are expressed widely in human tissues and cell types, and particularly in hematopoietic cells (Koning et al. 2009; Refsland et al. 2010). All APOBEC proteins contain one or two copies of a characteristic zinc-coordinating deaminase domain (the Z domain) (LaRue et al. 2009) that comprises a platform of five β strands, flanking α helices and connecting loops, a constellation of three histidine or cysteine residues that coordinate an essential Zn^{2+} ion, and a catalytic glutamic acid residue (Chen et al. 2008). A3G has two such domains: the carboxy-terminal domain mediates deamination, whereas the amino-terminal Z domain does not have catalytic activity (for unknown reasons), mediates incorporation into HIV-1 particles (see below), and is recognized by Vif (see below) (Fig. 2) (reviewed by Malim 2009; Albin and Harris 2010). A3G also forms dimers, in an RNA-dependent manner, and this attribute is thought to be important for packaging and antiviral function (Huthoff et al. 2009). However, in the absence of structures for full length A3G, there is ongoing debate not only regarding the relative arrangement of the Z domains within a single A3G molecule, but also into the nature, determinants, and significance of dimerization.

Viral Hypermutation

In the absence of Vif, A3G is packaged into assembling HIV-1 virions through the combined action of RNA binding and interactions between the amino-terminal Z domain of A3G and the nucleocapsid (NC) region of Gag (Bogerd and Cullen 2008). A3G is transferred to target cells by ensuing virus infection where, through its association with the viral reverse transcriptase complex (RTC), it deaminates cytidine residues in nascent single-stranded (mostly) negative-strand cDNA (Harris et al. 2003; Mangeat et al. 2003; Zhang et al. 2003; Yu et al. 2004). Up to 10% of cytidines may be edited, resulting in guanosine-to-adenosine hypermutation of viral plus strand sequence and the debilitating loss of genetic integrity (this can be considered as error catastrophe) (Fig. 3). Not all cytidines are equivalent targets

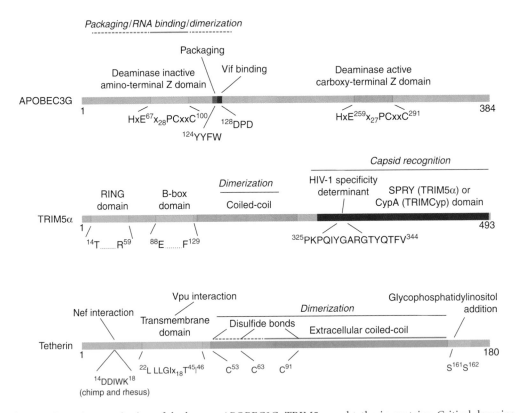

Figure 2. Domain organization of the human APOBEC3G, TRIM5α, and tetherin proteins. Critical domains and motifs are highlighted in color, their functions and attributes are indicated above, and important sequence motifs are shown below. The number of amino acids in each protein is also indicated.

Cite this article as *Cold Spring Harb Perspect Med* doi: 10.1101/cshperspect.a006940

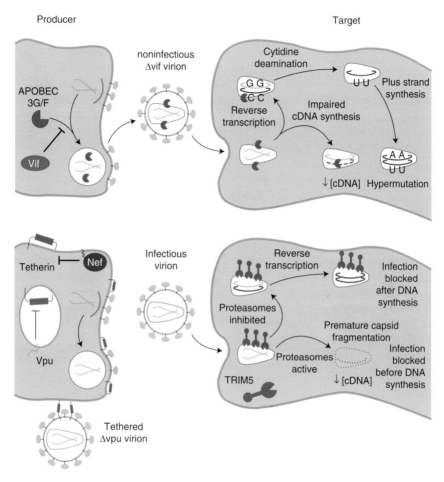

Figure 3. Mechanisms of restriction factor function and evasion. HIV-1 RNA is shown in light blue, HIV-1 cDNA in dark blue, and restriction factors in red. The sites of Vif, Vpu, and Nef antagonism are indicated. Refer to text for further details.

for A3G and there is a marked local sequence preference for 5′-CC<u>C</u>A (the deaminated cytidine is underlined) (Harris et al. 2003; Yu et al. 2004). Studies in vitro indicate that partiality for the 3′ cytidine is attributable to the 3′-to-5′ processivity of the enzyme on its DNA substrate (Chelico et al. 2006).

Mutation is not the only consequence of A3G action, as the levels of cDNA that accumulate during new HIV-1 infection are also diminished. It was attractive to believe that recognition of uridine-containing DNA by host DNA repair enzymes could initiate DNA degradation; however, this notion has been discounted because inhibition of uracil DNA

glycosidases fails to reverse A3G's effect on DNA levels (Langlois and Neuberger 2008). Rather, it appears that A3G impedes the translocation of reverse transcriptase along the viral RNA template, although the mechanistic underpinning of this effect awaits full elucidation (Fig. 3) (Iwatani et al. 2007; Bishop et al. 2008).

Vif Inhibits A3G Function

The antiviral activity of A3G is antagonized by HIV-1 Vif: Indeed, Vif is so efficient that physiologic levels of A3G have no overt effect on wild-type (Vif expressing) HIV-1 infection or replication in cultured cells. Vif's principal activity is to bind to A3G and recruit it to a cellular ubiquitin ligase complex that comprises the cullin5 scaffold protein, elongins B and C, Rbx2, and an as yet unidentified E2 conjugating enzyme (Yu et al. 2003). This results in A3G polyubiquitylation and proteasomal degradation, and therefore averts the encapsidation of A3G into nascent viral particles (Marin et al. 2003; Sheehy et al. 2003; Stopak et al. 2003; Yu et al. 2003). Recently, it has also been proposed that the central polypurine tract (cPPT) of HIV-1 helps mitigate the mutagenic effects of A3G by limiting substrate availability through reducing the length of time that minus-strand cDNA remains single stranded (Hu et al. 2010).

Extensive investigations have delineated various interactions that contribute to the assembly of the A3G-Vif-ligase complex. These are portrayed in Figure 4, although their temporal relationships with each other during complex formation are unknown (reviewed by Malim 2009; Albin and Harris 2010). Critical interactions include the binding of Vif's suppressor

of cytokine signaling (SOCS) box to the elonginB/C heterodimer (Bergeron et al. 2010), the interaction between the Zn^{2+} coordinating motif of Vif and cullin5, and the recognition of the A3G substrate by discontinuous elements within the amino-terminal region of HIV-1 Vif. This last interaction is of particular interest as it modulates the species-specific regulation of A3G by HIV and SIV Vif proteins; e.g., the African green monkey (AGM) A3G protein contains a lysine at the position corresponding to the aspartic acid at position 128 in human A3G, and this permits recognition and regulation by SIV_{AGM}, but not HIV-1, Vif (reviewed by Malim 2009; Albin and Harris 2010). Indeed, it has been argued that the capacity of the Vif proteins of ancestral viruses to counteract the APOBEC3 proteins of new hosts has played an important role in past zoonotic transmissions of SIVs into humans (Gaddis et al. 2004).

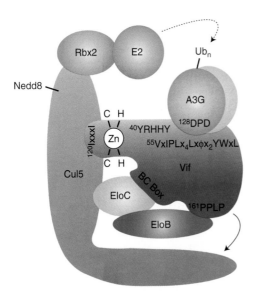

Figure 4. Components and intermolecular contacts in the cullin5-elonginBC-Vif ubiquitin ligase complex. HIV-1 Vif serves as a receptor protein that interacts with cullin5, elonginB (through a proline-rich motif), elonginC (through a BC box), and A3G. Refer to text for further details.

APOBEC3F and APOBEC3H

Of the remaining 10 human APOBEC proteins, many have been assigned HIV-1 inhibitory function in the context of overexpression studies in vitro. However, at more relevant levels of expression, the current weight of evidence indicates that only APOBEC3F (A3F) and one allelic form of APOBEC3H (A3H), haplotype II, significantly suppress HIV-1 (reviewed by Albin and Harris 2010). This can be viewed as making biological sense because these two proteins (but not other APOBEC3 proteins) are regulated by Vif, perhaps implying that Vif-mediated neutralization is only important for APOBEC3 proteins that naturally encounter HIV-1 during in vivo infections. The mechanisms of antiviral action of A3F/H mirror those of A3G, although their intrinsic potencies appear to be lower (Holmes et al. 2007; Miyagi et al. 2010), and the preferred target site for deamination is 5′-TC rather than 5′-CC. Their levels of expression are also lower (Koning et al. 2009; Refsland et al. 2010), supporting the view that A3G is the most significant family member for inhibiting HIV-1 infection.

Regulation of A3G

The most obvious form of regulation is Vif-initiated degradation during virus infection (above). There is a consensus that APOBEC3 proteins are transcriptionally induced by type 1 IFNs, particularly in myeloid cells (Koning et al. 2009; Refsland et al. 2010), a feature that characterizes many cellular proteins involved in early/innate control of viral infection. Being a DNA mutator, it is also important to consider how host cell (chromosomal) DNA might be spared. There are a number of possibilities: (1) It is not, but mutations are repaired before fixation; (2) it is partially repaired/protected and some level of mutagenesis takes place; (3) A3G is localized to the cytoplasm; and (4) A3G is sequestered in ribonucleoprotein particles (RNPs) that suppress deaminase activity (Chiu et al. 2006; Kozak et al. 2006; Gallois-Montbrun et al. 2007). Interestingly, A3G-RNPs (and A3F-RNPs) further accumulate in mRNA processing bodies (P-bodies) (Wichroski et al. 2006; Gallois-Montbrun et al. 2007), although the relevance of this to A3G function or antiviral activity is uncertain.

APOBEC3 Proteins and Natural HIV-1 Infections

A3G/F/H appear to encounter HIV-1 during in vivo infection. In addition to the sensitivity of these proteins to Vif, guanosine-to-adenosine hypermutated HIV-1 sequences with the expected local nucleotide preferences are readily recovered from infected persons. This shows that APOBEC3 proteins can escape complete inhibition by Vif, and this may be due to variation in Vif function, allelic variation in *APOBEC3* genes, excessive APOBEC3 protein expression or activity, or simply stochastic events. It has also been proposed that infrequent APOBEC3-induced mutations (as opposed to hypermutation) can (1) contribute to sequence diversification and evolution, perhaps in ways that are beneficial to the virus in terms of immune escape or drug resistance (Wood et al. 2009; Kim et al. 2010; Sadler et al. 2010); or (2) generate nonsense (or missense) mutations that result in the expression of truncated

or misfolded viral proteins, and the enhanced presentation of viral epitopes to cytotoxic T cells (Casartelli et al. 2010).

Can variation in the APOBEC3 landscape influence the course of HIV-1 infection in humans? Many publications have started to address this point, although it remains challenging to distinguish between causation and consequence. Even though there are inconsistencies among some findings, there is a discernible trend that increased levels of A3G/F expression tend to correlate with clinical benefit (reviewed by Albin and Harris 2010).

TRIM5α AND TRIMCYP

Identification

The existence of an antiretroviral protein that targets HIV-1, SIV, and other retroviral capsids was predicted by descriptions of restricted infection in a number of mammalian cell lines (Bieniasz 2003). The characteristics of these resistance phenotypes were highly reminiscent of those displayed by Fv1: Restriction was saturable, dominant in heterokaryons, independent of the route of entry, and could be encountered or avoided by manipulating retroviral CA sequence. However, resistance was apparent in nonmurine species, including humans, which lack the *Fv1* gene. A screen for rhesus macaque genes that could restrict HIV-1 infection when expressed in human cells resulted in the identification of TRIM5α (Stremlau et al. 2004), and this protein has subsequently been shown to be responsible for the majority of similar postentry restriction phenomena in a number of mammalian species (Hatziioannou et al. 2004; Keckesova et al. 2004; Perron et al. 2004; Yap et al. 2004; Johnson and Sawyer 2009).

TRIM5α and TRIMCyp Proteins

TRIM5α is an ∼500 amino acid cytoplasmic protein that acts following the entry of retroviral capsids and their contents into the cytoplasm of target cells. Its action is generally accompanied by a failure to synthesize viral cDNA (Stremlau et al. 2004). TRIM5 is one of a family of ∼70 so-called "tripartite motif" (TRIM)-containing proteins. Family members have a broadly similar domain organization (Nisole et al. 2005). The TRIM domain is composed of amino-terminal RING and B-box type 2 domains linked to a central coiled-coil domain (Fig. 2). In the case of TRIM5, the coiled coil drives the formation of dimers. The nature of the carboxy-terminal domain can vary widely among proteins of the TRIM family (Nisole et al. 2005). In the case of TRIM5α, and several other TRIM proteins, this carboxy-terminal domain is called the B30.2 or PRYSPRY domain.

The range of retroviruses that are inhibited by a particular TRIM5α varies dramatically, depending on the species of origin. For instance, the human TRIM5α protein is an effective inhibitor of N-MLV, as well as equine infectious anemia virus (EIAV) (Hatziioannou et al. 2004; Keckesova et al. 2004; Perron et al. 2004; Yu et al. 2004); however, it is virtually inactive against HIV-1. Conversely, TRIM5α proteins from Old World monkey species generally inhibit HIV-1 infection (Stremlau et al. 2004). In general, TRIM5α proteins are poor inhibitors of retroviruses that are found naturally in the same host species, but are quite often active against retroviruses that are found in other species. As such, TRIM5α can impose a quite formidable barrier to cross-species transmission of primate lentiviruses (Hatziioannou et al. 2006).

The carboxy-terminal SPRY domain contains most of the determinants that govern substrate selection for a given TRIM5α protein, and has undergone rapid evolution, as evidenced by high numbers of nonsynonymous differences in interspecies sequence comparisons compared to the genome average (Sawyer et al. 2005; Song et al. 2005; Johnson and Sawyer 2009). Three peptide segments (V1–V3) within the SPRY domain that are hypervariable in both length and sequence likely encode surface exposed loops, by analogy with SPRY domains of known structure. At least one of these variable loops (V1) can be shown experimentally to be a key determinant of antiretroviral specificity. Indeed, it is possible to make a small number of changes, even a single amino acid substitution, in V1 segments of primate TRIM5α proteins and alter

their ability to recognize HIV-1 and SIV strains (Perez-Caballero et al. 2005; Sawyer et al. 2005; Stremlau et al. 2005; Yap et al. 2005). This genetic evidence, coupled with studies showing that the PRYSPRY governs the ability of TRIM5α proteins to bind to HIV-1 capsid-like assemblies in vitro (Stremlau et al. 2006a), indicates that this TRIM5α domain has evolved to specifically recognize particular capsids in the cytoplasm of target cells.

A number of lentivirus capsids bind to the abundant host cell chaperone protein, cyclophilin A (CypA), via a peptide loop that is exposed on the surface of the assembled capsid (see Sundquist and Kraeusslich 2011). The precise role for this interaction is not completely clear, but both the sequence of this exposed loop and the CypA protein itself can affect the sensitivity of HIV-1 to TRIM5α (Berthoux et al. 2005; Keckesova et al. 2006; Stremlau et al. 2006b). In owl monkeys (Sayah et al. 2004), and independently in some macaques (Liao et al. 2007; Brennan et al. 2008; Newman et al. 2008; Virgen et al. 2008; Wilson et al. 2008), retrotransposition events have placed CypA cDNAs into the TRIM5 locus, so that the resulting chimeric gene is expressed as a TRIM5-CypA fusion protein (TRIMCyp) with a CypA protein domain replacing the PRYSPRY domain. Predictably, TRIMCyp proteins are, in general, potent inhibitors of lentiviruses whose capsids bind CypA. However, evolved modification of the capsid binding specificity of the CypA domains in TRIMPCyp proteins can occur through mutation in the CypA encoding sequence, acquired during or after its retrotransposition into the TRIM5 locus. For example, the TRIMCyp proteins found in macaques and owl monkeys have been shown to differ greatly in their ability to inhibit HIV-1 infection (Virgen et al. 2008; Price et al. 2009).

Mechanisms of Infection Inhibition by TRIM5α and TRIMCyp Proteins

The mechanisms by which TRIM5 proteins act to block retroviral infection are not completely understood (Fig. 3). This is at least partly because the processes that occur during the postentry phase of the retroviral life cycle that are perturbed by TRIM5 proteins are not easily analyzed with currently available biochemical and biophysical techniques. Nevertheless, it is clear that TRIM5α and TRIMCyp bind directly to HIV-1 capsids (Stremlau et al. 2006a) and, at least in the case of TRIMCyp, recognition of the incoming capsid must occur within 15–30 min of viral entry for inhibition to be effective (Perez-Caballero et al. 2005). Some studies have also revealed that incoming retroviral capsids lose their particulate nature on entry into the target cell cytoplasm if they encounter a TRIM5α or TRIMCyp protein (Stremlau et al. 2006a). These findings suggest a model in which TRIM5α and TRIMCyp accelerate capsid fragmentation soon after viral entry, thereby disrupting RTC architecture and blocking reverse transcription (Fig. 3).

Two zinc-binding domains (RING and B-box type 2) at the amino terminus of TRIM5α protein are important for the full antiviral activity of the protein. The B-box domain, although not required for TRIM5α dimerization, appears to constitute a second self-associating domain (Li and Sodroski 2008). Thus, the B-box contributes to the formation of higher-order multimers and the propensity of TRIM5α to assemble into preaggresomal structures, termed "cytoplasmic bodies," which are visible by fluorescent microscopy. Although cytoplasmic body formation may not be essential for antiviral activity, higher-order TRIM5α multimerization does appear to increase the efficiency with which TRIM5α interacts with the capsid lattice, and thus promotes antiviral potency (Diaz-Griffero et al. 2009). Indeed, recent cryoelectron microscopy analyses of purified TRIM5α show a propensity to assemble into hexagonal lattices that can interact in an ordered, polyvalent manner with preformed hexagonal lattices of HIV-1 CA (Ganser-Pornillos et al. 2011).

Although the RING domain of TRIM5α proteins possesses E3 ubiquitin ligase activity, ubiquitin or proteasomes appear not to be essential for antiviral function. Indeed, TRIM5α and TRIMCyp display potent anti-HIV-1 activity in the presence of proteasome inhibitors, or

in a cell line containing an inactive ubiquitin activating (E1) enzyme (Perez-Caballero et al. 2005). However, proteasome inhibition prevents TRIM5-promoted capsid disassembly and restores reverse transcription, without enabling infection (Wu et al. 2006; Diaz-Griffero et al. 2007). Thus, it appears that neither proteasome activity nor accelerated capsid fragmentation or inhibition of reverse transcription is absolutely required for TRIM5α to exert antiretroviral activity (Fig. 3). In fact, aside from capsid binding and multimerization, no activity that has been associated with TRIM5α has been definitively shown to be required for antiretroviral activity. One possible explanation for this is that TRIM5α is capable of inhibiting infection in two or more redundant ways. Alternatively, it is conceivable that perturbing the ubiquitin/proteasome system simply slows whatever process (e.g., capsid fragmentation) that is responsible for inhibiting infection, such that it is not completed until after reverse transcription.

TETHERIN

Identification

The identification of tetherin was based on the finding that the Vpu accessory protein was required for efficient virion release from some cell lines but completely dispensable in others. The requirement for Vpu was found to be dominant in heterokaryons (Varthakavi et al. 2003) and inducible in cells from which it was ordinarily absent by treatment with IFN-α (Neil et al. 2007). Other studies indicated that the absence of Vpu rendered HIV-1 sensitive to a protein-based adhesive or tethering mechanism that trapped nascent virions on the surface of infected cells (Neil et al. 2006). These accumulated findings suggested the existence of IFN-induced protein tethers and provided the basis for a microarray/candidate gene-based discovery of tetherin, a membrane protein whose expression was necessary and sufficient to impose a requirement Vpu for the efficient release of HIV-1 particles (Neil et al. 2008; Van Damme et al. 2008).

The Tetherin Protein and Mechanism of Virion Retention

In the absence of Vpu, and the presence of tetherin, HIV-1 particles are assembled normally, their lipid envelopes undergo ESCRT-protein-mediated fission from the plasma membrane, and they adopt a mature morphology. However, tetherin causes virions to remain trapped at the surface of the infected cell from which they are derived and to accumulate thereafter in endosomes following internalization (Neil et al. 2006). Tetherin is an unusual type II single-pass transmembrane protein in that it has both a transmembrane anchor close to its amino terminus and a glycophosphatidylinositol (GPI) lipid anchor at its carboxyl terminus (Fig. 2) (Kupzig et al. 2003). The entire extracellular portion of tetherin forms a single long α helix, much of which adopts a canonical dimeric coiled-coil structure (Fig. 5) (Hinz et al. 2010; Schubert et al. 2010; Yang et al. 2010). Additionally, each of three extracellular cysteines in each monomer forms disulfide bonds with a corresponding cysteine in another tetherin molecule (Figs. 2 and 5).

Several lines of evidence suggest that a simple and direct tethering mechanism underlies the ability of tetherin to cause retention of nascent virions (Fig. 5). Although tetherin's overall configuration (Fig. 2) is required for virion retention, it is surprisingly tolerant of mutations, including substitution of entire tetherin domains with protein domains that have similar predicted structures but lack sequence homology (Perez-Caballero et al. 2009). Indeed, a completely artificial tetherin-like protein, assembled from structurally similar but unrelated protein domains, can effectively mimic its activity. This fact, and the finding that tetherin can block viruses from various families whose structural proteins have no sequence or structural homology to each other, makes it unlikely that specific recognition of viral proteins is required for function (Jouvenet et al. 2009; Kaletsky et al. 2009; Mansouri et al. 2009).

Both the amino-terminal transmembrane domain and the carboxy-terminal GPI modification are essential for tethering function, and

Figure 5. A structural model of how tetherin causes virion retention. In the scenario depicted here, tetherin transmembrane domains infiltrate the virion envelope, whereas the glycophosphatidylinositol anchors remain in the plasma membrane. The tetherin coiled-coil domain thereby physically links cell and virion membranes. Refer to Perez-Caballero et al. (2009) and Fitzpatrick et al. (2010) for alternative models.

tetherin mutants lacking either are efficiently incorporated into the lipid envelope of HIV-1 particles via the remaining membrane anchor (Perez-Caballero et al. 2009). The intact tetherin protein can also be found in virions in some circumstances. A protected dimeric amino-terminal tetherin fragment can be found in the envelope of tethered virions that can be recovered from the surface of cells by protease, indicating that at least some tethered virions have dimeric tetherin amino termini inserted into their lipid envelope (Perez-Caballero et al. 2009). Thus, either or both tetherin membrane anchors seem to be inserted into the lipid envelope of budding virions as they emerge from infected cells (Fig. 3). Concordantly, both fluorescent and electron microscopic analyses reveal that tetherin colocalizes with virions on the cell surface (Neil et al. 2008; Jouvenet et al. 2009; Perez-Caballero et al. 2009; Fitzpatrick et al. 2010; Hammonds et al. 2010).

Models for tetherin action include the possible scenario that one pair of tetherin membrane anchors (e.g., the TM domains) infiltrates the lipid envelope of the assembling virion, whereas the other pair (e.g., the GPI anchors) remains in the cell membrane (Fig. 5). In this configuration, the coiled coil could promote the spatial separation of two pairs of membrane anchors, increasing the probability that one of the two pairs is incorporated into the virion envelope. Alternatively, it is possible that noncovalent interactions might mediate the adhesion of tetherin dimers that are incorporated into the virions and those that remain in the host cell membrane. Tetherin expression causes the accumulation of virions that appear to be tethered to each other as well as to the cell surface. Importantly, this is only possible if both types of TM anchor can be incorporated into virion envelopes, and can occur following the budding of a virion at the same site on the plasma membrane as that already occupied by a tethered virion.

Antagonism of Tetherin by HIV and SIV Vpu, Nef, and Env Proteins

Because the relatively invariant, host-derived lipid envelope, rather than a viral protein, is the target for tetherin action, it would seem difficult for a virus to evade tetherin by avoiding interaction with it. Faced with this problem, HIVs and SIVs have independently evolved new biological activities in the form of *trans*-acting tetherin antagonists.

The HIV-1 Vpu protein is used by HIV-1 strains as an antagonist of tetherin (Neil et al. 2008; Van Damme et al. 2008). It is ~14 kDa and is composed of a single transmembrane helix and a small cytoplasmic domain. There is some uncertainty as to how antagonism is achieved, as several different mechanisms have been reported. There is, however, general agreement that Vpu colocalizes with and can be coimmunoprecipitated with tetherin and reduces the level of tetherin at the cell surface (Van Damme et al. 2008), either by retarding its progress through the secretory pathway or by causing its internalization (Mitchell et al. 2009; Dube et al. 2010). Vpu can also reduce the overall steady-state level of tetherin in cells, at least under conditions of transient overexpression (Bartee et al. 2006). The latter effect is reversed by proteasome inhibitors (Douglas et al. 2009; Miyagi et al. 2009), but because proteasome inhibition can deplete ubiquitin and thereby affect the trafficking of some cargoes through the endosomal system, it is not yet clear whether the proteasome or the endolysosomal system is directly responsible for tetherin degradation. Some studies find that tetherin down-regulation from the cell surface and/or degradation is modest or nonexistent in cell types where Vpu appears fully functional. Thus, a clear overall picture of precisely how Vpu antagonizes tetherin is currently lacking.

Most SIVs do not encode Vpu proteins, and instead use another accessory gene product, the ~27 kDa Nef protein, to antagonize tetherin (Jia et al. 2009; Zhang et al. 2009). Nef is an amino-terminally myristoylated peripheral membrane protein that can interact with a number of cellular partners, including clathrin adapter protein (AP) complexes, several kinases, and dynamin-2. It is not currently known whether any of these Nef binding proteins play a role in tetherin antagonism or how antagonism is achieved. In some instances, primate lentivirus Env proteins can also act as tetherin antagonists, and in these cases it appears the Env proteins engage tetherin and cause it to be sequestered within intracellular compartments (Gupta et al. 2009b; Le Tortorec and Neil 2009).

Evolution of Tetherin and Viral Antagonists

The HIV-1 Vpu protein is an efficient antagonist of human tetherin but is ineffective against tetherins from other animals such as monkeys or rodents. This species-specific action of Vpu has been used to derive genetic and biochemical evidence that Vpu and tetherin interact via their transmembrane helices (Fig. 2) (Gupta et al. 2009a; McNatt et al. 2009). Similarly, although SIV Nef proteins effectively counteract monkey tetherin proteins that are found in their natural host species, they are often inactive against other tetherins. In this case, sequence variation in the tetherin cytoplasmic tail defines its sensitivity to Nef, and a five codon deletion in the cytoplasmic tail of human tetherin that renders it Nef resistant defines a particularly important target of Nef action on tetherin (Fig. 2) (Jia et al. 2009; Zhang et al. 2009).

As outlined in Sharp and Hahn (2011), primate lentiviruses have been transmitted from species to species on numerous occasions. Most recently, humans have been the recipients of these zoonoses, acquiring SIVcpz from chimpanzees and SIVsm from sooty mangabeys, resulting in the viruses we now call HIV-1 and HIV-2. Analyses of Vpu and Nef proteins from many SIVs have revealed that these proteins effectively exchanged the role of tetherin antagonist as they were passed from species to species and encountered tetherin proteins with varying TM and cytoplasmic tail sequences (Sauter et al. 2009; Lim et al. 2010). Most notably, SIVcpz uses the Nef protein as a tetherin antagonist, whereas its immediate descendent, HIV-1, uses Vpu. Similarly, SIVsm (which lacks Vpu) uses Nef to counteract tetherin, whereas its descendent, HIV-2, uses its Env protein in this role. In both cases, the acquisition of tetherin antagonist activity by HIV-1 Vpu and HIV-2 Env proteins likely occurred because human tetherin lacks a key five residue determinant of Nef sensitivity in its cytoplasmic tail (Fig. 3) (Sauter et al. 2009; Lim et al. 2010). These gain-of-function events that have occurred as a consequence of transmission to a new species illustrate the functional plasticity of HIV/SIV accessory proteins.

NEW RESEARCH AREAS

Although much has been discovered about restriction factors, there is significant scope for future discovery and exploitation. There are strong suspicions that there are more, perhaps many more, restriction factors to be uncovered. For example, the basis for the inhibitory effects of type 1 IFNs on the early steps of HIV-1 infection are largely unexplained, and may involve the action of unidentified restriction factors (Goujon and Malim 2010). Additionally, the functions of the HIV-1 accessory genes that have not been elucidated in their entirety and, based on the precedents described herein, are quite likely to include interactions with, and regulation of, restriction factors. In addition to the general approaches described above that led to the discovery of APOBEC3, TRIM5, and tetherin, as well as the MLV inhibitor ZAP (Gao et al. 2002), we predict techniques that can identify factors that interact with regulators of viral infectivity, or can monitor alterations in protein abundance and form, such as SILAC (stable isotope labeling with amino acids in cell culture), will have utility for finding new restriction factors.

Indeed, two groups very recently reported using proteomic methods to identify SAMHD1 (sterile α motif domain-, HD domain-containing protein 1) as the myeloid-specific, degradable cellular target of HIV-2/SIVsm Vpx proteins (Hrecka et al. 2011; Laguette et al. 2011). This protein is of considerable interest as its restriction of HIV and SIV infections (Goujon et al. 2008) correlates with limited viral cDNA accumulation, and mutations in the *SAMHD1* gene in humans result in a disease called Aicardi-Goutières syndrome (AGS) that is characterized by excessive IFN production and inflammation, and therefore, mimics congenital virus infection (Lee-Kirsch 2010). The mechanism of SAMHD1-mediated restriction remains to be defined, but the involvement of HD domains in nucleotide metabolism is suggestive of direct interaction with viral nucleic acids.

Therapeutic exploitation of restriction factor biology has yet to receive widespread attention. Indeed, the pharmacologic mobilization of restriction factors, possibly by blocking interactions between a viral antagonist and a host restriction factor, would appear to be an attractive approach for the development of novel antivirals. For instance, inhibitors of Vif function that have viral inhibitory activity in cell culture have recently been described (Nathans et al. 2008; Cen et al. 2010) and provide a paradigm for possible exploitation of such targets.

Host cell factors mentioned: APOBEC3G, TRIM5α, and tetherin.

REFERENCES

*Reference is also in this collection.

Albin JS, Harris RS. 2010. Interactions of host APOBEC3 restriction factors with HIV-1 in vivo: Implications for therapeutics. *Expert Rev Mol Med* **12:** e4. doi: 10.1017/S1462399409001343.

Bartee E, McCormack A, Fruh K. 2006. Quantitative membrane proteomics reveals new cellular targets of viral immune modulators. *PLoS Pathog* **2:** e107. doi: 10.1371/journal.ppat.0020107.

Bergeron JR, Huthoff H, Veselkov DA, Beavil RL, Simpson PJ, Matthews SJ, Malim MH, Sanderson MR. 2010. The SOCS-box of HIV-1 Vif interacts with ElonginBC by induced-folding to recruit its Cul5-containing ubiquitin ligase complex. *PLoS Pathog* **6:** e1000925. doi: 10.1371/journal.ppat.1000925.

Berthoux L, Sebastian S, Sokolskaja E, Luban J. 2005. Cyclophilin A is required for TRIM5α-mediated resistance to HIV-1 in Old World monkey cells. *Proc Natl Acad Sci* **102:** 14849–14853.

Best S, Le Tissier P, Towers G, Stoye JP. 1996. Positional cloning of the mouse retrovirus restriction gene Fv1. *Nature* **382:** 826–829.

Bieniasz PD. 2003. Restriction factors: A defense against retroviral infection. *Trends Microbiol* **11:** 286–291.

Bishop KN, Verma M, Kim EY, Wolinsky SM, Malim MH. 2008. APOBEC3G inhibits elongation of HIV-1 reverse transcripts. *PLoS Pathog* **4:** e1000231. doi: 10.1371/journal.ppat.1000231.

Bogerd HP, Cullen BR. 2008. Single-stranded RNA facilitates nucleocapsid: APOBEC3G complex formation. *RNA* **14:** 1228–1236.

Brennan G, Kozyrev Y, Hu SL. 2008. TRIMCyp expression in Old World primates *Macaca nemestrina* and *Macaca fascicularis*. *Proc Natl Acad Sci* **105:** 3569–3574.

Bushman FD, Malani N, Fernandes J, D'Orso I, Cagney G, Diamond TL, Zhou H, Hazuda DJ, Espeseth AS, Konig R, et al. 2009. Host cell factors in HIV replication: Meta-analysis of genome-wide studies. *PLoS Pathog* **5:** e1000437. doi: 10.1371/journal.ppat.1000437.

Casartelli N, Guivel-Benhassine F, Bouziat R, Brandler S, Schwartz O, Moris A. 2010. The antiviral factor APO-BEC3G improves CTL recognition of cultured HIV-infected T cells. *J Exp Med* **207:** 39–49.

Cen S, Peng ZG, Li XY, Li ZR, Ma J, Wang YM, Fan B, You XF, Wang YP, Liu F, et al. 2010. Small molecular compounds inhibit HIV-1 replication through specifically stabilizing APOBEC3G. *J Biol Chem* **285:** 16546–16552.

Chelico L, Pham P, Calabrese P, Goodman MF. 2006. APO-BEC3G DNA deaminase acts processively $3' \rightarrow 5'$ on single-stranded DNA. *Nat Struct Mol Biol* **13:** 392–399.

Chen KM, Harjes E, Gross PJ, Fahmy A, Lu Y, Shindo K, Harris RS, Matsuo H. 2008. Structure of the DNA deaminase domain of the HIV-1 restriction factor APO-BEC3G. *Nature* **452:** 116–119.

Chiu YL, Witkowska HE, Hall SC, Santiago M, Soros VB, Esnault C, Heidmann T, Greene WC. 2006. High-molecular-mass APOBEC3G complexes restrict Alu retrotransposition. *Proc Natl Acad Sci* **103:** 15588–15593.

Cowan S, Hatziioannou T, Cunningham T, Muesing MA, Gottlinger HG, Bieniasz PD. 2002. Cellular inhibitors with Fv1-like activity restrict human and simian immunodeficiency virus tropism. *Proc Natl Acad Sci* **99:** 11914–11919.

Diaz-Griffero F, Kar A, Perron M, Xiang SH, Javanbakht H, Li X, Sodroski J. 2007. Modulation of retroviral restriction and proteasome inhibitor-resistant turnover by changes in the TRIM5α B-box 2 domain. *J Virol* **81:** 10362–10378.

Diaz-Griffero F, Qin XR, Hayashi F, Kigawa T, Finzi A, Sarnak Z, Lienlaf M, Yokoyama S, Sodroski J. 2009. A B-box 2 surface patch important for TRIM5α self-association, capsid binding avidity, and retrovirus restriction. *J Virol* **83:** 10737–10751.

Douglas JL, Viswanathan K, McCarroll MN, Gustin JK, Fruh K, Moses AV. 2009. Vpu directs the degradation of the human immunodeficiency virus restriction factor BST-2/Tetherin via a βTrCP-dependent mechanism. *J Virol* **83:** 7931–7947.

Dube M, Roy BB, Guiot-Guillain P, Binette J, Mercier J, Chiasson A, Cohen EA. 2010. Antagonism of tetherin restriction of HIV-1 release by Vpu involves binding and sequestration of the restriction factor in a perinuclear compartment. *PLoS Pathog* **6:** e1000856. doi: 10.1371/journal.ppat.1000856.

Fitzpatrick K, Skasko M, Deerinck TJ, Crum J, Ellisman MH, Guatelli J. 2010. Direct restriction of virus release and incorporation of the interferon-induced protein BST-2 into HIV-1 particles. *PLoS Pathog* **6:** e1000701. doi: 10.1371/journal.ppat.1000701.

Gabuzda DH, Lawrence K, Langhoff E, Terwilliger E, Dorfman T, Haseltine WA, Sodroski J. 1992. Role of *vif* in replication of human immunodeficiency virus type 1 in CD4$^+$ T lymphocytes. *J Virol* **66:** 6489–6495.

Gaddis NC, Sheehy AM, Ahmad KM, Swanson CM, Bishop KN, Beer BE, Marx PA, Gao F, Bibollet-Ruche F, Hahn BH, et al. 2004. Further investigation of simian immunodeficiency virus Vif function in human cells. *J Virol* **78:** 12041–12046.

Gallois-Montbrun S, Kramer B, Swanson CM, Byers H, Lynham S, Ward M, Malim MH. 2007. Antiviral protein APOBEC3G localizes to ribonucleoprotein complexes found in P bodies and stress granules. *J Virol* **81:** 2165–2178.

Ganser-Pornillos BK, Chandrasekaran V, Pornillos O, Sodroski JG, Sundquist WI, Yeager M. 2011. Hexagonal assembly of a restricting TRIM5α protein. *Proc Natl Acad Sci* **108:** 534–539.

Gao G, Guo X, Goff SP. 2002. Inhibition of retroviral RNA production by ZAP, a CCCH-type zinc finger protein. *Science* **297:** 1703–1706.

Goujon C, Malim MH. 2010. Characterization of the alpha interferon-induced postentry block to HIV-1 infection in primary human macrophages and T cells. *J Virol* **84:** 9254–9266.

Goujon C, Arfi V, Pertel T, Luban J, Lienard J, Rigal D, Darlix JL, Cimarelli A. 2008. Characterization of simian immunodeficiency virus SIVSM/human immunodeficiency virus type 2 Vpx function in human myeloid cells. *J Virol* **82:** 12335–12345.

Gupta RK, Hue S, Schaller T, Verschoor E, Pillay D, Towers GJ. 2009a. Mutation of a single residue renders human tetherin resistant to HIV-1 Vpu-mediated depletion. *PLoS Pathog* **5:** e1000443. doi: 10.1371/journal.ppat. 1000443.

Gupta RK, Mlcochova P, Pelchen-Matthews A, Petit SJ, Mattiuzzo G, Pillay D, Takeuchi Y, Marsh M, Towers GJ. 2009b. Simian immunodeficiency virus envelope glycoprotein counteracts tetherin/BST-2/CD317 by intracellular sequestration. *Proc Natl Acad Sci* **106:** 20889–20894.

Hammonds J, Wang JJ, Yi H, Spearman P. 2010. Immunoelectron microscopic evidence for Tetherin/BST2 as the physical bridge between HIV-1 virions and the plasma membrane. *PLoS Pathog* **6:** e1000749. doi: 10.1371/journal.ppat.1000749.

Harris RS, Bishop KN, Sheehy AM, Craig HM, Petersen-Mahrt SK, Watt IN, Neuberger MS, Malim MH. 2003. DNA deamination mediates innate immunity to retroviral infection. *Cell* **113:** 803–809.

Hatziioannou T, Perez-Caballero D, Yang A, Cowan S, Bieniasz PD. 2004. Retrovirus resistance factors Ref1 and Lv1 are species-specific variants of TRIM5α. *Proc Natl Acad Sci* **101:** 10774–10779.

Hatziioannou T, Princiotta M, Piatak M Jr, Yuan F, Zhang F, Lifson JD, Bieniasz PD. 2006. Generation of simiantropic HIV-1 by restriction factor evasion. *Science* **314:** 95.

Hinz A, Miguet N, Natrajan G, Usami Y, Yamanaka H, Renesto P, Hartlieb B, McCarthy AA, Simorre JP, Gottlinger H, et al. 2010. Structural basis of HIV-1 tethering to membranes by the BST-2/tetherin ectodomain. *Cell Host Microbe* **7:** 314–323.

Hofmann W, Schubert D, LaBonte J, Munson L, Gibson S, Scammell J, Ferrigno P, Sodroski J. 1999. Species-specific, postentry barriers to primate immunodeficiency virus infection. *J Virol* **73:** 10020–10028.

Holmes RK, Koning FA, Bishop KN, Malim MH. 2007. APOBEC3F can inhibit the accumulation of HIV-1 reverse transcription products in the absence of hypermutation. Comparisons with APOBEC3G. *J Biol Chem* **282:** 2587–2595.

Hrecka K, Hao C, Gierszewska M, Swanson SK, Kesik-Brodacka M, Srivastava S, Florens L, Washburn MP, Skowronski J. 2011. Vpx relieves inhibition of HIV-1 infection of macrophages mediated by the SAMHD1 protein. *Nature* **474:** 658–661.

Hu C, Saenz DT, Fadel HJ, Walker W, Peretz M, Poeschla EM. 2010. The HIV-1 central polypurine tract functions as a second line of defense against APOBEC3G/F. *J Virol* **84:** 11981–11993.

Huthoff H, Autore F, Gallois-Montbrun S, Fraternali F, Malim MH. 2009. RNA-dependent oligomerization of APOBEC3G is required for restriction of HIV-1. *PLoS Pathog* **5:** e1000330. doi: 10.1371/journal.ppat.1000330.

Iwatani Y, Chan DS, Wang F, Maynard KS, Sugiura W, Gronenborn AM, Rouzina I, Williams MC, Musier-Forsyth K, Levin JG. 2007. Deaminase-independent inhibition of HIV-1 reverse transcription by APOBEC3G. *Nucleic Acids Res* **35:** 7096–7108.

Jia B, Serra-Moreno R, Neidermyer W, Rahmberg A, Mackey J, Fofana IB, Johnson WE, Westmoreland S, Evans DT. 2009. Species-specific activity of SIV Nef and HIV-1 Vpu in overcoming restriction by tetherin/BST2. *PLoS Pathog* **5:** e1000429. doi: 10.1371/journal.ppat.1000429.

Johnson WE, Sawyer SL. 2009. Molecular evolution of the antiretroviral TRIM5 gene. *Immunogenetics* **61:** 163–176.

Jouvenet N, Neil SJ, Zhadina M, Zang T, Kratovac Z, Lee Y, McNatt M, Hatziioannou T, Bieniasz PD. 2009. Broadspectrum inhibition of retroviral and filoviral particle release by tetherin. *J Virol* **83:** 1837–1844.

Kaletsky RL, Francica JR, Agrawal-Gamse C, Bates P. 2009. Tetherin-mediated restriction of filovirus budding is antagonized by the Ebola glycoprotein. *Proc Natl Acad Sci* **106:** 2886–2891.

Keckesova Z, Ylinen LM, Towers GJ. 2004. The human and African green monkey TRIM5α genes encode Ref1 and Lv1 retroviral restriction factor activities. *Proc Natl Acad Sci* **101:** 10780–10785.

Keckesova Z, Ylinen LM, Towers GJ. 2006. Cyclophilin A renders human immunodeficiency virus type 1 sensitive to Old World monkey but not human TRIM5α antiviral activity. *J Virol* **80:** 4683–4690.

Kim EY, Bhattacharya T, Kunstman K, Swantek P, Koning FA, Malim MH, Wolinsky SM. 2010. Human APOBEC3G-mediated editing can promote HIV-1 sequence diversification and accelerate adaptation to selective pressure. *J Virol* **84:** 10402–10405.

Kirchhoff F. 2010. Immune evasion and counteraction of restriction factors by HIV-1 and other primate lentiviruses. *Cell Host Microbe* **8:** 55–67.

Koning FA, Newman EN, Kim EY, Kunstman KJ, Wolinsky SM, Malim MH. 2009. Defining APOBEC3 expression patterns in human tissues and hematopoietic cell subsets. *J Virol* **83:** 9474–9485.

Kozak SL, Marin M, Rose KM, Bystrom C, Kabat D. 2006. The anti-HIV-1 editing enzyme APOBEC3G binds HIV-1 RNA and messenger RNAs that shuttle between polysomes and stress granules. *J Biol Chem* **281:** 29105–29119.

Kupzig S, Korolchuk V, Rollason R, Sugden A, Wilde A, Banting G. 2003. Bst-2/HM1.24 is a raft-associated apical membrane protein with an unusual topology. *Traffic* **4:** 694–709.

Laguette N, Sobhian B, Casartelli N, Ringeard M, Chable-Bessia C, Segeral E, Yatim A, Emiliani S, Schwartz O, Benkirane M. 2011. SAMHD1 is the dendritic- and myeloid-cell-specific HIV-1 restriction factor counteracted by Vpx. *Nature* **474:** 654–657.

Langlois MA, Neuberger MS. 2008. Human APOBEC3G can restrict retroviral infection in avian cells and acts independently of both UNG and SMUG1. *J Virol* **82:** 4660–4664.

LaRue RS, Andresdottir V, Blanchard Y, Conticello SG, Derse D, Emerman M, Greene WC, Jonsson SR, Landau NR, Lochelt M, et al. 2009. Guidelines for naming nonprimate APOBEC3 genes and proteins. *J Virol* **83:** 494–497.

Le Tortorec A, Neil SJ. 2009. Antagonism to and intracellular sequestration of human tetherin by the human immunodeficiency virus type 2 envelope glycoprotein. *J Virol* **83:** 11966–11978.

Lee-Kirsch MA. 2010. Nucleic acid metabolism and systemic autoimmunity revisited. *Arthritis Rheum* **62:** 1208–1212.

Li X, Sodroski J. 2008. The TRIM5α B-box 2 domain promotes cooperative binding to the retroviral capsid by mediating higher-order self-association. *J Virol* **82:** 11495–11502.

Liao CH, Kuang YQ, Liu HL, Zheng YT, Su B. 2007. A novel fusion gene, TRIM5-Cyclophilin A in the pig-tailed macaque determines its susceptibility to HIV-1 infection. *AIDS* 21 Suppl **8:** S19–S26.

Lim ES, Malik HS, Emerman M. 2010. Ancient adaptive evolution of tetherin shaped the functions of Vpu and Nef in human immunodeficiency virus and primate lentiviruses. *J Virol* **84:** 7124–7134.

Madani N, Kabat D. 1998. An endogenous inhibitor of human immunodeficiency virus in human lymphocytes is overcome by the viral Vif protein. *J Virol* **72:** 10251–10255.

Malim MH. 2009. APOBEC proteins and intrinsic resistance to HIV-1 infection. *Philos Trans R Soc Lond B Biol Sci* **364:** 675–687.

Mangeat B, Turelli P, Caron G, Friedli M, Perrin L, Trono D. 2003. Broad antiretroviral defence by human APOBEC3G through lethal editing of nascent reverse transcripts. *Nature* **424:** 99–103.

Mansouri M, Viswanathan K, Douglas JL, Hines J, Gustin J, Moses AV, Fruh K. 2009. Molecular mechanism of BST2/tetherin downregulation by K5/MIR2 of Kaposi's sarcoma-associated herpesvirus. *J Virol* **83:** 9672–9681.

Marin M, Rose KM, Kozak SL, Kabat D. 2003. HIV-1 Vif protein binds the editing enzyme APOBEC3G and induces its degradation. *Nat Med* **9:** 1398–1403.

McNatt MW, Zang T, Hatziioannou T, Bartlett M, Fofana IB, Johnson WE, Neil SJ, Bieniasz PD. 2009. Species-specific activity of HIV-1 Vpu and positive selection of tetherin transmembrane domain variants. *PLoS Pathog* **5:** e1000300. doi: 10.1371/journal.ppat.1000300.

Mitchell RS, Katsura C, Skasko MA, Fitzpatrick K, Lau D, Ruiz A, Stephens EB, Margottin-Goguet F, Benarous R, Guatelli JC. 2009. Vpu antagonizes BST-2-mediated restriction of HIV-1 release via β-TrCP and endolysosomal trafficking. *PLoS Pathog* **5:** e1000450. doi: 10.1371/journal.ppat.1000450.

Miyagi E, Andrew AJ, Kao S, Strebel K. 2009. Vpu enhances HIV-1 virus release in the absence of Bst-2 cell surface down-modulation and intracellular depletion. *Proc Natl Acad Sci* **106:** 2868–2873.

Miyagi E, Brown CR, Opi S, Khan M, Goila-Gaur R, Kao S, Walker RC Jr, Hirsch V, Strebel K. 2010. Stably expressed APOBEC3F has negligible antiviral activity. *J Virol* **84:** 11067–11075.

Nathans R, Cao H, Sharova N, Ali A, Sharkey M, Stranska R, Stevenson M, Rana TM. 2008. Small-molecule inhibition of HIV-1 Vif. *Nat Biotechnol* **26:** 1187–1192.

Neil SJ, Eastman SW, Jouvenet N, Bieniasz PD. 2006. HIV-1 Vpu promotes release and prevents endocytosis of nascent retrovirus particles from the plasma membrane. *PLoS Pathog* **2:** e39. doi: 10.1371/journal.ppat.0020039.

Neil SJ, Sandrin V, Sundquist WI, Bieniasz PD. 2007. An interferon-α-induced tethering mechanism inhibits HIV-1 and Ebola virus particle release but is counteracted by the HIV-1 Vpu protein. *Cell Host Microbe* **2:** 193–203.

Neil SJ, Zang T, Bieniasz PD. 2008. Tetherin inhibits retrovirus release and is antagonized by HIV-1 Vpu. *Nature* **451:** 425–430.

Newman RM, Hall L, Kirmaier A, Pozzi LA, Pery E, Farzan M, O'Neil SP, Johnson W. 2008. Evolution of a TRIM5-CypA splice isoform in old world monkeys. *PLoS Pathog* **4:** e1000003. doi: 10.1371/journal.ppat.1000003.

Nisole S, Stoye JP, Saib A. 2005. TRIM family proteins: Retroviral restriction and antiviral defence. *Nat Rev Microbiol* **3:** 799–808.

Perez-Caballero D, Hatziioannou T, Yang A, Cowan S, Bieniasz PD. 2005. Human tripartite motif 5α domains responsible for retrovirus restriction activity and specificity. *J Virol* **79:** 8969–8978.

Perez-Caballero D, Zang T, Ebrahimi A, McNatt MW, Gregory DA, Johnson MC, Bieniasz PD. 2009. Tetherin inhibits HIV-1 release by directly tethering virions to cells. *Cell* **139:** 499–511.

Perron MJ, Stremlau M, Song B, Ulm W, Mulligan RC, Sodroski J. 2004. TRIM5α mediates the postentry block to N-tropic murine leukemia viruses in human cells. *Proc Natl Acad Sci* **101:** 11827–11832.

Price AJ, Marzetta F, Lammers M, Ylinen LM, Schaller T, Wilson SJ, Towers GJ, James LC. 2009. Active site remodeling switches HIV specificity of antiretroviral TRIMCyp. *Nat Struct Mole Biol* **16:** 1036–1042.

Refsland EW, Stenglein MD, Shindo K, Albin JS, Brown WL, Harris RS. 2010. Quantitative profiling of the full APOBEC3 mRNA repertoire in lymphocytes and tissues: Implications for HIV-1 restriction. *Nucleic Acids Res* **38:** 4274–4284.

Sadler HA, Stenglein MD, Harris RS, Mansky LM. 2010. APOBEC3G contributes to HIV-1 variation through sublethal mutagenesis. *J Virol* **84:** 7396–7404.

Sauter D, Schindler M, Specht A, Landford WN, Munch J, Kim KA, Votteler J, Schubert U, Bibollet-Ruche F, Keele BF, et al. 2009. Tetherin-driven adaptation of Vpu and Nef function and the evolution of pandemic and nonpandemic HIV-1 strains. *Cell Host Microbe* **6:** 409–421.

Sawyer SL, Wu LI, Emerman M, Malik HS. 2005. Positive selection of primate TRIM5α identifies a critical species-specific retroviral restriction domain. *Proc Natl Acad Sci* **102:** 2832–2837.

Sayah DM, Sokolskaja E, Berthoux L, Luban J. 2004. Cyclophilin A retrotransposition into TRIM5 explains owl monkey resistance to HIV-1. *Nature* **430:** 569–573.

Schubert HL, Zhai Q, Sandrin V, Eckert DM, Garcia-Maya M, Saul L, Sundquist WI, Steiner RA, Hill CP. 2010. Structural and functional studies on the extracellular domain of BST2/tetherin in reduced and oxidized conformations. *Proc Natl Acad Sci* **107:** 17951–17956.

* Sharp PM, Hahn BH. 2011. Origins of HIV and the AIDS pandemic. *Cold Spring Harb Perspect Med* doi: 10.1101/cshperspect.a006841.

Sheehy AM, Gaddis NC, Choi JD, Malim MH. 2002. Isolation of a human gene that inhibits HIV-1 infection and is suppressed by the viral Vif protein. *Nature* **418:** 646–650.

Sheehy AM, Gaddis NC, Malim MH. 2003. The antiretroviral enzyme APOBEC3G is degraded by the proteasome in response to HIV-1 Vif. *Nat Med* **9:** 1404–1407.

Simon JH, Gaddis NC, Fouchier RA, Malim MH. 1998a. Evidence for a newly discovered cellular anti-HIV-1 phenotype. *Nat Med* **4:** 1397–1400.

Simon JH, Miller DL, Fouchier RA, Soares MA, Peden KW, Malim MH. 1998b. The regulation of primate immunodeficiency virus infectivity by Vif is cell species restricted: A role for Vif in determining virus host range and cross-species transmission. *EMBO J* **17:** 1259–1267.

Song B, Gold B, O'Huigin C, Javanbakht H, Li X, Stremlau M, Winkler C, Dean M, Sodroski J. 2005. The B30.2(SPRY) domain of the retroviral restriction factor TRIM5α exhibits lineage-specific length and sequence variation in primates. *J Virol* **79:** 6111–6121.

Stopak K, de Noronha C, Yonemoto W, Greene WC. 2003. HIV-1 Vif blocks the antiviral activity of APOBEC3G by impairing both its translation and intracellular stability. *Mole Cell* **12:** 591–601.

Stremlau M, Owens CM, Perron MJ, Kiessling M, Autissier P, Sodroski J. 2004. The cytoplasmic body component TRIM5α restricts HIV-1 infection in Old World monkeys. *Nature* **427:** 848–853.

Stremlau M, Perron M, Welikala S, Sodroski J. 2005. Species-specific variation in the B30.2(SPRY) domain of TRIM5α determines the potency of human immunodeficiency virus restriction. *J Virol* **79:** 3139–3145.

Stremlau M, Perron M, Lee M, Li Y, Song B, Javanbakht H, Diaz-Griffero F, Anderson DJ, Sundquist WI, Sodroski J. 2006a. Specific recognition and accelerated uncoating of retroviral capsids by the TRIM5α restriction factor. *Proc Natl Acad Sci* **103:** 5514–5519.

Stremlau M, Song B, Javanbakht H, Perron M, Sodroski J. 2006b. Cyclophilin A: An auxiliary but not necessary cofactor for TRIM5α restriction of HIV-1. *Virology* **351:** 112–120.

* Sundquist WI, Kräusslich H-G. 2011. HIV assembly, budding, and maturation. *Cold Spring Harb Perspect Med* doi: 10.1101/cshperspect.a06924.

Van Damme N, Goff D, Katsura C, Jorgenson RL, Mitchell R, Johnson MC, Stephens EB, Guatelli J. 2008. The interferon-induced protein BST-2 restricts HIV-1 release and is downregulated from the cell surface by the viral Vpu protein. *Cell Host Microbe* **3:** 245–252.

Varthakavi V, Smith RM, Bour SP, Strebel K, Spearman P. 2003. Viral protein U counteracts a human host cell restriction that inhibits HIV-1 particle production. *Proc Natl Acad Sci* **100:** 15154–15159.

Virgen CA, Kratovac Z, Bieniasz PD, Hatziioannou T. 2008. Independent genesis of chimeric TRIM5-cyclophilin proteins in two primate species. *Proc Natl Acad Sci* **105:** 3563–3568.

Wichroski MJ, Robb GB, Rana TM. 2006. Human retroviral host restriction factors APOBEC3G and APOBEC3F localize to mRNA processing bodies. *PLoS Pathog* **2:** e41. doi: 10.1371/journal.ppat.0020041.

Wilson SJ, Webb BL, Ylinen LM, Verschoor E, Heeney JL, Towers GJ. 2008. Independent evolution of an antiviral TRIMCyp in rhesus macaques. *Proc Natl Acad Sci* **105:** 3557–3562.

Wood N, Bhattacharya T, Keele BF, Giorgi E, Liu M, Gaschen B, Daniels M, Ferrari G, Haynes BF, McMichael A, et al. 2009. HIV evolution in early infection: Selection pressures, patterns of insertion and deletion, and the impact of APOBEC. *PLoS Pathog* **5:** e1000414. doi: 10.1371/journal.ppat.1000414.

Wu X, Anderson JL, Campbell EM, Joseph AM, Hope TJ. 2006. Proteasome inhibitors uncouple rhesus TRIM5α restriction of HIV-1 reverse transcription and infection. *Proc Natl Acad Sci* **103:** 7465–7470.

Yang H, Wang J, Jia X, McNatt MW, Zang T, Pan B, Meng W, Wang HW, Bieniasz PD, Xiong Y. 2010. Structural insight into the mechanisms of enveloped virus tethering by tetherin. *Proc Natl Acad Sci* **107:** 18428–18432.

Yap MW, Nisole S, Lynch C, Stoye JP. 2004. Trim5α protein restricts both HIV-1 and murine leukemia virus. *Proc Natl Acad Sci* **101:** 10786–10791.

Yap MW, Nisole S, Stoye JP. 2005. A single amino acid change in the SPRY domain of human Trim5α leads to HIV-1 restriction. *Curr Biol* **15:** 73–78.

Yu X, Yu Y, Liu B, Luo K, Kong W, Mao P, Yu XF. 2003. Induction of APOBEC3G ubiquitination and degradation by an HIV-1 Vif-Cul5-SCF complex. *Science* **302:** 1056–1060.

Yu Q, Konig R, Pillai S, Chiles K, Kearney M, Palmer S, Richman D, Coffin JM, Landau NR. 2004. Single-strand specificity of APOBEC3G accounts for minus-strand deamination of the HIV genome. *Nat Struct Mol Biol* **11:** 435–442.

Zhang H, Yang B, Pomerantz RJ, Zhang C, Arunachalam SC, Gao L. 2003. The cytidine deaminase CEM15 induces hypermutation in newly synthesized HIV-1 DNA. *Nature* **424:** 94–98.

Zhang F, Wilson SJ, Landford WC, Virgen B, Gregory D, Johnson MC, Munch J, Kirchhoff F, Bieniasz PD, Hatziioannou T. 2009. Nef proteins from simian immunodeficiency viruses are tetherin antagonists. *Cell Host Microbe* **6:** 54–67.

HIV Transmission

George M. Shaw[1] and Eric Hunter[2]

[1]Department of Medicine, Perelman School of Medicine, University of Pennsylvania, Philadelphia, Pennsylvania 19104

[2]Department of Pathology and Laboratory Medicine, Emory Vaccine Center, Emory University, Atlanta, Georgia 30329

Correspondence: ehunte4@emory.edu

HIV-1 is transmitted by sexual contact across mucosal surfaces, by maternal-infant exposure, and by percutaneous inoculation. For reasons that are still incompletely understood, CCR5-tropic viruses (R5 viruses) are preferentially transmitted by all routes. Transmission is followed by an orderly appearance of viral and host markers of infection in the blood plasma. In the acute phase of infection, HIV-1 replicates exponentially and diversifies randomly, allowing for an unambiguous molecular identification of transmitted/founder virus genomes and a precise characterization of the population bottleneck to virus transmission. Sexual transmission of HIV-1 most often results in productive clinical infection arising from a single virus, highlighting the extreme bottleneck and inherent inefficiency in virus transmission. It remains to be determined if HIV-1 transmission is largely a stochastic process whereby any reasonably fit R5 virus can be transmitted or if there are features of transmitted/founder viruses that facilitate their transmission in a biologically meaningful way. Human tissue explant models of HIV-1 infection and animal models of SIV/SHIV/HIV-1 transmission, coupled with new challenge virus strains that more closely reflect transmitted/founder viruses, have the potential to elucidate fundamental mechanisms in HIV-1 transmission relevant to vaccine design and other prevention strategies.

HIV-1 transmission results from virus exposure at mucosal surfaces or from percutaneous inoculation. Because such exposures in humans are inaccessible to direct analysis, our understanding of the transmission event must necessarily come from insights gleaned from studies of HIV-1 epidemiology, viral and host genetics, risk factor and behavior analyses, animal models, human explant tissues, and in vitro studies of virus-target cell interactions. In this article, we explore themes that connect these varied aspects of HIV-1 infection with the ultimate goal of understanding the molecular basis of HIV-1 transmission.

EPIDEMIOLOGY OF HIV-1: IMPLICATIONS FOR TRANSMISSION BIOLOGY

At the broadest level, HIV-1 transmission must be viewed in the context of the global pandemic. Population level phylogenetic patterns of endemic, epidemic, and pandemic strains of HIV-1 can provide insight into clinically relevant aspects of virus transmission. HIV-1 is classified phylogenetically into groups M, N, O, and P, each reflecting a separate introduction of simian immunodeficiency viruses (SIVs) from naturally infected great apes into humans (see Sharp and Hahn 2011). Of these, only group M underwent pandemic spread. Viral and host factors that may have influenced the relative transmissibility of the different HIV-1 groups are discussed elsewhere (Malim and Bieniasz 2011; Sharp and Hahn 2011). For group M viruses, mathematical modeling of virus diversification suggests a most recent common ancestor near 1910–1930 (Korber et al. 2000; Worobey et al. 2008), followed by subclinical endemic spread of the virus in human populations in West Central Africa. There, largely as a consequence of founder effects and viral population bottlenecks, HIV-1 emerged as early as the late 1950s as phylogenetically distinct subtypes, of which nine (subtypes A, B, C, D, F, G, H, J, K) are now recognized as contributing to the global pandemic (Taylor et al. 2008).

The extraordinary sequence diversity within and among these different HIV-1 group M subtypes, which can reach 25%–35% in *env*, has made possible a precise tracking of HIV-1 transmission within and between populations on a global scale (Gilbert et al. 2007; Taylor et al. 2008) and between individuals on a micro scale (Derdeyn et al. 2004; Haaland et al. 2009). A global survey of the distribution of HIV-1 subtypes and intersubtype circulating recombinant forms (CRFs) in 2004 revealed that subtype C accounted for 50% of infections worldwide, with subtypes A, B, D, and G accounting for 12%, 10%, 3%, and 6%, respectively, and subtypes F, H, J, and K together accounting for 1% (Hemelaar et al. 2006). The global distribution and genetic complexity of HIV-1 subtypes and CRFs have continued to evolve (Taylor et al. 2008), and with this has come the potential for virus evolution, adaptation, and altered transmissibility. Despite this fertile environment for mutation, there is little evidence for meaningful differences among HIV-1 subtypes in their patterns or efficiencies of transmission. The best example of differences in transmission probability is for CRF01_AE in injection drug users (IDUs) in Thailand, which

increased substantially in prevalence between 1995–1998 compared with subtype B viruses. Yet, it is unclear whether epidemiological factors, host factors, or viral properties were responsible (Hudgens et al. 2002). There is also an epidemiological observation that subtype D viruses from Uganda and neighboring regions may show R5/X4 dual tropism more commonly than do other HIV-1 subtypes, but this may have a greater effect on viral pathogenesis than on transmission per se (Church et al. 2010). Thus, unlike influenza virus in which ongoing genetic mutation dramatically impacts the frequency and patterns of virus transmission, such is not the case for pandemic HIV-1 group M viruses, which have been remarkably consistent in their transmissibility over expanses of time, geography, and target populations (Taylor et al. 2008) and as the virus moved between individuals and groups of individuals with widely different risk behavior and virus transmission routes (Kouyos et al. 2010).

TRANSMISSION ROUTES AND RISKS FOR HIV-1 INFECTION

In 2009, an estimated 2.6 million people globally became newly infected by HIV-1 (UNAIDS 2010). This represents a reduction in new infections by 21% compared with 1997 when incident infections peaked. But declining HIV-1 incidence rates have not been uniform across all regions and risk groups, highlighting the importance of different transmission routes and risk behaviors in facilitating HIV-1 transmission. The most extreme example is in Eastern Europe and Central Asia where HIV-1 prevalence tripled between 2001–2009 as a consequence of concentrated epidemics associated with sex work, drug use, and men who have sex with men (MSM) (UNAIDS 2010). Table 1 summarizes the risks for HIV-1 transmission associated with different transmission routes and their relative contributions to HIV-1 prevalence worldwide. What is evident from these estimates is that heterosexual transmission is responsible for nearly 70% of HIV-1 infections worldwide with the remainder largely attributable to MSM, maternal-infant infection, and

Cite this article as *Cold Spring Harb Perspect Med* doi: 10.1101/cshperspect.a006965

Table 1. Transmission routes and risks for HIV-1 infection

HIV invasion site	Anatomical sublocation	Type of epithelium	Transmission medium	Transmission probability per exposure event	Estimated contribution to HIV cases worldwide
Female genital tract	Vagina	Squamous, nonkeratinized	Semen; blood	1 in 200–1 in 2000	12.6 million
	Ectocervix	Squamous, nonkeratinized			
	Endocervix	Columnar, single layer			
	Other	Various			
Male genital tract	Inner foreskin	Squamous, poorly keratinized	Cervicovaginal and rectal secretions; blood	1 in 700–1 in 3000	10.2 million[a]
	Penile urethra	Columnar, stratified			
	Other	Various			
Intestinal tract	Rectum	Columnar, single layer	Semen; blood	1 in 20–1 in 300	3.9 million[b]
	Upper GI tract	Various	Semen; blood	1 in 2500	1.5 million
			Maternal blood, genital secretions (intrapartum)	1 in 5–1 in 10	960,000[c]
			Breast milk	1 in 5–1 in 10	960,000[c]
Placenta	Chorionic villi	Two-layer epithelium (cyto- and syncytiotrophoblast)	Maternal blood (intrauterine)	1 in 10–1 in 20	480,000[c]
Bloodstream			Blood products, sharps	95 in 100–1 in 150	2.6 million[d]

Adapted from the 2010 UNAIDS/WHO AIDS epidemic update and Hladik and McElrath (2008).

[a]Includes men having sex with men (MSM), bisexual men, and heterosexual men.

[b]Includes MSM, bisexual men, and women infected via anal receptive intercourse.

[c]Mother-to-child transmission.

[d]Mostly intravenous drug use, but includes infections by transfusions and health-care-related accidents. GI, gastrointestinal.

injection drug use. This is the case despite the fact that transmission probability per coital act is lowest for heterosexual exposures (1 in 200–1 in 3000) (Hladik and McElrath 2008; McElrath et al. 2008). However, two recent meta-analyses of HIV-1 incidence and prevalence data (Powers et al. 2008; Boily et al. 2009) suggest a far wider range in HIV-1 transmission risk for heterosexual exposures depending on confounding risk factors such as genital ulcer disease, male circumcision, HIV disease stage, and exposure route. For example, penile-vaginal transmission of HIV-1 was reported at a frequency as high as 1 in 10 exposures and penile-anal transmission as high as 1 in 3 depending on confounding risk cofactors (Powers et al. 2008). Thus, the commonly quoted risk estimate for heterosexual HIV-1 acquisition of 1 in 1000 exposures must be considered a lower bound.

Socioeconomic factors can also influence HIV-1 transmission indirectly. El-Sadr and colleagues have suggested that in the United States, the risk of acquiring HIV-1 infection is defined more by a person's "sexual network" than by their individual risk behaviors, with the former influenced primarily by socioeconomic factors (El-Sadr et al. 2010). On a global basis, HIV-1 transmission estimates per coital act vary significantly between high-income countries, where male to female (MTF) transmission estimates are 0.08% and female to male (FTM) rates are 0.04%, and low-income countries, where MTF and FTM transmission rates are 0.38% and 0.3%, respectively (Powers et al. 2008; Boily et al. 2009). These differences may in part reflect the relative frequency of HIV serodiscordant couples in the two settings since it has been estimated that they represent the source of a majority of adult infections in many low-income countries (Dunkle et al. 2008).

Viral load (vL) in the transmitting partner plays a major role in determining the risk of HIV-1 transmission from one individual to another. Although vL likely affects all modes of transmission, it has been best characterized in HIV-1 discordant couples, in which as much as a 2.5-fold increase in transmission was observed for every 10-fold increase in vL (Quinn 2000; Fideli et al. 2001). Moreover, although vL in genital secretions may not correlate directly with that in the blood, partners with plasma vL less than 1000 rarely transmitted to their partners (Quinn et al. 2000; Fideli et al. 2001). The recent observation (HIV Prevention Trials Network Study 052, unpubl.) that antiretroviral treatment of the positive partner of discordant couples can result in a 96% reduction in transmission is consistent with this finding (Cohen et al. 2011b). The clinical stage of infection (acute vs. middle vs. late) in the transmitting partner can also play a key role in defining the efficiency of transmission, with the risk of infection from individuals with acute or early infection being higher than that in established infection (Wawer et al. 2005; Brenner et al. 2007; Powers et al. 2008; Miller et al. 2010). This likely reflects the high vLs observed in acute infection, a lack of neutralizing antibodies that may inactivate circulating virus in established infection, and clonal amplification of highly fit viruses in acute infection that are especially suitable for initiating productive infection (Richman et al. 2003; Wei et al. 2003; Wawer et al. 2005; Keele et al. 2008; Cohen et al. 2011a). Indeed, in the Indian rhesus macaque model of SIV transmission, SIV in the plasma from animals in the acute stage of infection had a specific infectivity up to 750 times greater than that of virus in the plasma from chronically infected animals (Ma et al. 2009).

The efficiency of HIV-1 transmission can be modulated by still other factors, including sexually transmitted diseases, particularly those that result in genital inflammation and ulcers, which can elevate HIV shedding into the genital tract and can increase infection susceptibility by two- to 11-fold (Galvin and Cohen 2004); pregnancy, during which a greater than twofold increase in HIV acquisition risk has been observed (Gray et al. 2005); and circumcision, which in a series of clinical trials has been shown to decrease transmission acquisition risk in the male partner by 60% (Auvert et al. 2005; Bailey et al. 2007; Gray et al. 2007; Quinn 2007).

THE CLINICAL TRANSMISSION EVENT AND ACUTE HIV-1 INFECTION

Despite the many routes by which HIV-1 can be transmitted from one individual to another, there is generally an orderly and reproducible appearance in the blood of viral and host markers of infection following a clinically productive transmission event (Fig. 1). This sequential appearance of laboratory markers of new HIV-1 infection was systematically evaluated by Fiebig and colleagues who devised a laboratory staging system for acute and early infection (Fiebig et al. 2003). The initial period between the moment when the first cell is infected and when virus is first detectable in the blood is termed the eclipse phase. The duration of this period has been estimated to be approximately 7–21 d, based on clinical histories of high risk exposure events (Gaines et al. 1988; Clark et al. 1991; Schacker et al. 1996; Little et al. 1999; Lindback et al. 2000a,b) and

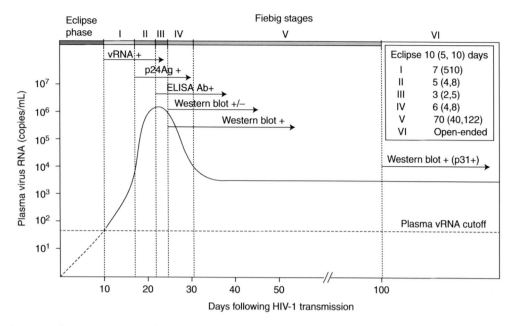

Figure 1. Laboratory staging and natural history of acute and early HIV-1 infection. The average durations and 95% confidence intervals (parentheses) of the eclipse phase and Fiebig stages (Fiebig et al. 2003) of acute infection are shown in the inset.

mathematical modeling of early HIV-1 replication and diversification (Keele et al. 2008; Lee et al. 2009). During the eclipse phase, which is clinically silent, virus is propagated in CD4$^+$ T cells in mucosa, submucosa, draining lymphatics, and perhaps to a modest extent in gut-associated lymphoid tissue (GALT) and systemic lymphatic tissues (Haase 2010). Once virus becomes detectable in blood plasma, it increases exponentially (Ribeiro et al. 2010) as a consequence of explosive replication in GALT and peripheral lymphoid tissue compartments (Veazey et al. 1998; Guadalupe et al. 2003; Brenchley et al. 2004; Mehandru et al. 2004; Li et al. 2005b; Mattapallil et al. 2005; Haase 2010). Fiebig stages I–VI reflect the ordered appearance in the plasma of HIV-1 viral RNA (Fiebig I), viral p24 antigen (Fiebig II), and virus-specific antibodies detectable first by recombinant protein-based enzyme-linked immunosorbent assay (Fiebig III), and then by Western immunoblotting (indeterminant banding pattern: Fiebig IV; diagnostic banding pattern but missing p31 reactivity: Fiebig V;

diagnostic banding pattern with p31 reactivity: Fiebig VI). The average durations of the eclipse phase and Fiebig stages along with estimated 95% confidence intervals are shown in the inset of Figure 1. Diagnosing HIV-1 infection as soon as possible after the transmission event is important clinically to ensure the safety of blood for transfusion and of body organs for transplantation. It is also important for preventing forward transmission of the virus from index patients to their contacts. Current HIV-1 nucleic acid tests based on different amplification platforms can detect viral sequences qualitatively at 1–5 copies per milliliter of plasma (Palmer et al. 2003; Nugent et al. 2009) and quantitatively at levels exceeding 50 copies per milliliter plasma (Damond et al. 2010). Alternatively, a recently approved fourth-generation combined p24 antigen-antibody test can detect HIV-1 at viral loads exceeding 10,000 virions per milliliter plasma (Eshleman et al. 2009; Cohen et al. 2011a). Thus, commercially available p24 antigen tests can detect HIV-1 infection approximately 1 wk before the first detection of anti-HIV antibodies, and vRNA tests can detect HIV-1 infection approximately 1 wk before that (Fig. 1). There are well-documented examples of HIV-1 antibody seroconversion as late as 6 mo or more after an exposure event but such cases are rare (Ridzon et al. 1997); what is more remarkable is the consistency in the timing of appearance of viral and host markers of infection (Fig. 1) despite the different cells and tissues that are involved in the transmission process at the different sites of HIV-1 invasion (Table 1).

A POPULATION BOTTLENECK TO HIV-1 TRANSMISSION

The concept that transmission of HIV involves a population bottleneck, in which a limited number of variants from a genetically diverse virus quasispecies in the transmitting partner establishes productive infection in the newly infected partner, was first established nearly two decades ago (Wolfs et al. 1992; Wolinsky et al. 1992; Zhang et al. 1993; Zhu et al. 1993). These early findings were of particular interest to investigators involved in HIV-1 vaccine research and to those interested in HIV-1 transmission biology, pathogenesis, and natural history because they raised the possibility that selective pressures could be at play in virus transmission. This idea was strengthened by the subsequent discovery of CCR5 and CXCR4 as obligate coreceptors for HIV-1 entry into cells, by the finding of a strong preference for CCR5 use by HIV-1 strains from acute and early infection, and by the observation that humans who are homozygous defective for CCR5 expression are protected from HIV-1 infection (see Berger et al. 1999; Wilen et al. 2011b). Subsequent studies continued to explore the quasispecies complexity and genetic and biologic composition of HIV-1 in acutely infected individuals in an attempt to better define the features of transmitted viruses and cofactors that enhance the risk of transmission (Poss et al. 1995; Zhu et al. 1996; Long et al. 2000; Learn et al. 2002; Derdeyn et al. 2004; Grobler et al. 2004; Ritola et al. 2004; Sagar et al. 2004, 2009). The results of all of these studies, together

with the earlier work, led to the concept that whereas chronically infected subjects were invariably infected by a genetically diverse viral quasispecies, acutely infected individuals could be grouped into those with relatively "homogeneous" infections, presumably resulting from transmission and productive clinical infection by one or few closely related viruses, and those individuals with more "heterogeneous" infections that resulted from transmission of multiple viruses with greater diversity. This distinction between homogeneous and heterogeneous virus diversity in acute HIV-1 infection was for the most part a qualitative description, which was further confounded by the suggestion that acutely infected individuals could harbor genetically diverse HIV-1 genotypes that in the weeks following transmission underwent purifying selection so as to give the appearance of a homogeneous acute infection (Learn et al. 2002). Thus, although it was clear that there was a population bottleneck to HIV-1 transmission, a quantitative and molecularly precise description of this bottleneck was not possible. This led to more controversy than consensus regarding the multiplicity of HIV-1 infection in different patient populations with distinct demographic and behavioral risk profiles as well as the genetic and biological properties of such viruses.

The stimulus to a more focused examination of the genotype and phenotypic properties of the viruses that initiate new HIV-1 infections was prompted by a report by Derdeyn and colleagues (2004), which combined an extensive phylogenetic comparison of viral *env* sequences in chronically infected subjects and their newly infected sexual partners with a phenotypic analysis of viruses pseudotyped by the encoded chronic and acute Env proteins. This study suggested that one or a few virus variants initiated infection in most of the cases examined and that such viruses differed in a consistent way from the bulk of the viruses in the chronically-infected partner's viral quasispecies. With an obvious need to further clarify the molecular and biological features of transmitted viruses, Keele and colleagues (2008) devised a novel experimental sequencing strategy to enable a more precise molecular identification and enumeration of transmitted HIV-1 genomes. This new approach was based on single-genome amplification (SGA) of endpoint-diluted plasma vRNA/cDNA or peripheral blood mononuclear cell (PBMC) DNA followed by direct population sequencing of the uncloned DNA amplicon. Sequences were then analyzed phylogenetically in the context of a mathematical model of exponential virus growth and random virus evolution so as to identify actual transmitted/founder viral genomes that were responsible for productive clinical infection (Keele et al. 2008; Salazar-Gonzalez et al. 2008; Lee et al. 2009). This SGA approach had several key advantages over previous methods: First, SGA-direct sequencing eliminates *Taq* polymerase errors in finished sequences because such base substitutions are essentially random in distribution and any one substitution is present in exceedingly low proportions in the uncloned amplified product; because the amplified product is directly sequenced, such mutations are not evident on sequence chromatograms unless they occur in the initial polymerase chain reaction (PCR) amplification cycles, in which case they are identified as "double peaks" and the sequence is disregarded. Second, SGA eliminates both template switching (recombination) between genetically distinct viral genomes and template resampling, because amplification is initiated from single genomes. Third, SGA-direct sequencing avoids misrepresentation of target sequence frequencies because of unequal cloning.

SGA-direct amplicon sequencing was thus applied in rapid succession to a large number of clinical cohorts of individuals acutely infected by HIV-1 subtypes A, B, C, D, CRF01_AE, and others. In the first large cohort study of 102 subjects with acute HIV-1 subtype B infection, SGA-direct amplicon sequences of 3449 complete *env* gp160 genes from plasma vRNA/cDNA were generated (Keele et al. 2008). A key feature of this study was the categorization of study subjects by Fiebig staging (Fig. 1), which allowed for a systematic analysis of HIV-1 sequence diversification in the earliest phases of acute infection. Sequences showed discrete low-diversity lineages representing the progeny of distinct transmitted viral genomes. Sequences generally showed a Poisson distribution of mutations and star-like phylogeny that coalesced to unambiguous transmitted/founder genomes at or near the moment of virus transmission. This was true for subjects who acquired their infection by any of several routes including vaginal, rectal, penile, or intravenous transmission. Sequences evolved over time consistent with the Poisson distribution and a model that accounted for exponential virus growth, reproductive ratio, virus generation time, and reverse transcriptase error rate estimated for HIV-1 (Lee et al. 2009). This, in turn, allowed for estimates of time to a most recent common ancestor for all sequences. By this approach, a precise molecular identification and enumeration of transmitted/founder virus genomes was obtained for 98 of 102 consecutively studied subjects. Seventy-eight of 102 subjects had evidence of productive clinical infection by a single virus, and 24 subjects had evidence of acute infection by a minimum of two to five genetically distinct viruses from a single partner. Since this study identified the sequences of full-length transmitted/founder *env* gp160 genes, these sequences could be molecularly cloned and characterized biologically. The key difference between this approach and previous strategies for analyzing cloned *env* genes is that in the SGA-direct amplicon sequencing method the cloning step is done *after* an exact and unambiguous sequence determination of the transmitted/founder *env* gene has been made. This allows the cloned version of the transmitted/founder *env* to be matched exactly with the inferred *env* sequence of the transmitted/founder virus. This same SGA approach was used in subsequent studies to clone full-length transmitted/founder HIV-1 genomes (Salazar-Gonzalez et al. 2009; Li et al. 2010). Thus, in the initial Keele study (Keele et al. 2008), 55 *env* genes whose sequences matched exactly with transmitted/founder genes were molecularly cloned, expressed in 293T cells, and used to pseudotype in *trans* Env-deficient HIV-1 viruses. All Envs were biologically functional with respect to virus entry and all were CD4 dependent. Fifty-four of 55 Envs were CCR5-tropic and one was CCR5/CXCR4 dual-tropic. Transmitted/founder Envs showed neutralization sensitivity profiles typical of tier 2 or 3 primary virus strains (Li et al. 2005a). Thus, the findings from this study provided new clarity to the quantitative and molecular aspects of HIV-1 transmission.

Based on the empirical findings described by Keele et al. (2008) and the mathematical model of early HIV-1 replication and diversification developed by Korber, Perelson, and colleagues (Keele et al. 2008; Lee et al. 2009), a conceptual framework for describing HIV-1 transmission and early diversification was developed (Fig. 2). Direct evidence in support of this transmission model and the notion that actual transmitted/founder viral genomes could be identified unambiguously by SGA-direct

amplicon sequencing came from three additional sets of studies: Indian rhesus macaque monkeys inoculated intra-rectally with low-dose SIVmac251 in which viruses in the inoculum and in newly infected animals were found to be identical in *env* gene sequences (Keele et al. 2009); human transmission pairs in which HIV-1 *env* and full-length viral genomic sequences were found to be identical in donors and recipients (Haaland et al. 2009; J. Salazar-Gonzalez and G. Shaw, unpubl.); and 454 deep sequencing of tens of thousands of plasma vRNA/cDNA HIV-1 genomes from three acutely-infected human subjects, demonstrating productive clinical infection by the same transmitted/founder viruses as identified by SGA-direct amplicon sequencing (Fischer et al. 2010). Of note, the terms "transmitted/founder sequence"

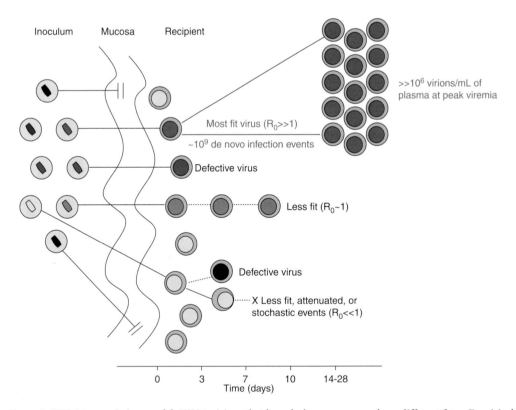

Figure 2. HIV-1 transmission model. HIV-1 virions that breach the mucosa may have different fates. Empirical measurements of virus replication and diversification, together with a mathematical model of random virus evolution, allow for a precise molecular identification of transmitted/founder viruses that are responsible for productive clinical infection (Keele et al. 2008; Lee et al. 2009). R_0 is the reproductive ratio. $R_0 > 1$ leads to productive clinical infection, whereas $R_0 < 1$ results in an extinguished infection.

and "transmitted/founder virus" are used to designate the coalescent or consensus sequence of a low-diversity virus lineage in acutely infected humans or monkeys because such sequences correspond literally to founder sequences that are responsible for establishing productive clinical infection. In most instances, the founder sequence corresponds to the actual transmitted sequence. Exceptions are likely uncommon and include founder sequences that differ from the actual transmitted sequence by one or a few nucleotides because of reverse transcriptase errors in the initial one or two replication cycles or, more substantially, because of recombination between copackaged heterozygous RNA genomes during reverse transcription in the first infected cell (Lee et al. 2009).

Additional studies of acutely infected human cohorts using SGA-direct amplicon sequencing provided further insights into the HIV-1 population bottleneck. These studies focused on heterosexual transmission in sub-Saharan Africa (Abrahams et al. 2009; Haaland et al. 2009; J Baalwa and G Shaw, unpubl.), MSM transmission in the United States (Kearney et al. 2009; Li et al. 2010), injection drug use transmission in Canada, Russia, and Thailand (Bar et al. 2010; Masharsky et al. 2010; K Bar and G Shaw, unpubl.), and most recently, maternal-infant transmission in Malawi (Russell et al. 2011). Each of these studies provided unambiguous molecular identification of transmitted/founder viral genes or genomes and precise estimates of the numbers of transmitted/founder viruses responsible for productive clinical infection. Two studies used SGA-direct amplicon sequencing to identify full-length transmitted founder genomes (Salazar-Gonzalez et al. 2009; Li et al. 2010). A summary of all of these studies, which included over 300 acutely infected subjects, indicated that approximately 80% of heterosexual subjects are productively infected by a single viral genome (Keele et al. 2008; Abrahams et al. 2009; Haaland et al. 2009) and approximately 60% and 40% of MSM and IDUs, respectively, are productively infected by single genomes. The range in transmitted/founder genomes that resulted in productive clinical infection in these studies

was 1–16, and not surprisingly, the highest numbers were in IDUs (Bar et al. 2010; Li et al. 2010; Masharsky et al. 2010; K Bar and G Shaw, unpubl.). For maternal-infant transmissions, 68% were productively infected by a single virus and this varied depending on intra-uterine versus intrapartum transmission routes (Russell et al. 2011). The observed differences in the multiplicity of viral infection associated with different routes of transmission roughly parallels the relative risk of clinical infection on a per event basis (Table 1). As predicted for transmitted/founder viruses, transmitted/founder *env* genes were invariably biologically functional and full-length genomes were invariably replication competent (Keele et al. 2008; Salazar-Gonzalez et al. 2009; Li et al. 2010; C Ochsenbauer, J Kappes, and G Shaw, unpubl.). Interestingly, in studies of primarily heterosexual transmission (Keele et al. 2008; Abrahams et al. 2009), in those individuals in which multiple viruses initiated infection, the number of infecting variants did not follow a Poisson distribution. This finding is inconsistent with each variant being transmitted independently and with low probability and suggests that transmission cofactors such as sexually transmitted infections (STIs) or hormonal contraceptives lower the barrier to transmission (Abrahams et al. 2009; Haaland et al. 2009). This interpretation is consistent with the model portrayed in Figure 2, in which virus transmission can fail at multiple steps following inoculation onto a mucosal surface. Thus, STIs including genital ulcer disease may abrogate the barrier imposed by an intact mucosa by inducing breaks in the epithelial lining, thereby allowing more virus variants to initiate infection in the mucosal tissue. Alternatively, inflammation induced by STIs could increase the availability of activated CD4 cells required to initiate a spreading infection, in this way allowing infections that would have failed because of lack of available target cells to expand. Finally, recent SGA comparisons of the genetic diversity of HIV in the genital tract of the transmitting partner to the transmitted/founder virus in the acutely infected partner argue against genital compartmentalization of specific viral variants in the donor as being responsible for origin of the population bottleneck, consistent with the model presented in Figure 2 (D Boeras, and E Hunter, unpubl.).

GENETIC SIGNATURES AND PHENOTYPES OF TRANSMITTED/FOUNDER VIRUSES

The identification and enumeration of transmitted/founder viruses in acutely infected heterosexuals, MSM, IDUs, and infants born to infected mothers provided an opportunity to explore such viruses for genetic signatures and phenotypic properties that might distinguish them from other viruses. Such studies had been undertaken previously with early but not transmitted/founder viruses, and they yielded intriguing findings, but the molecular identification of transmitted/founder viruses allowed for the first time such studies to be undertaken with far greater precision. From earlier studies, it was well established that the transmission of HIV-1 in humans selected for CCR5 tropic viruses, and that R5 viruses could be distinguished from CXCR4-tropic viruses (X4 viruses) genotypically by amino acid "signatures" primarily in the V3 region of Env (Resch et al. 2001; Jensen et al. 2003). But the important question now raised was if there were additional genetic signatures and corresponding phenotypic properties that distinguish transmitted R5 from others. Put differently, is HIV-1 transmission essentially a stochastic process in which any reasonably fit R5 virus can be transmitted, or are there critical properties of viruses that are transmitted that distinguish them from the innumerable variants that circulate in every chronically infected individual? If such signatures and phenotypes could be identified, they could conceivably represent targets of rational vaccine or drug design.

The first suggestion that HIV-1 transmission might select for traits other than coreceptor usage came from a genetic comparison of the viral *env* sequences from eight subtype C HIV-1 transmission pairs (Derdeyn et al. 2004). This study was conducted prior to reports describing SGA and thus carries the caveat that molecular clones of *env* could have

contained in vitro generated artifacts from *Taq* polymerase misincorporation or recombination. Nonetheless, in each subject pair, regardless of whether virus was transmitted MTF or FTM, the newly transmitted viruses encoded statistically shorter and less glycosylated V1–V4 regions than did *env*s from the chronically infected partner. This finding raised the possibility that more compact Env glycoproteins might interact more efficiently with relevant target cells in the genital mucosa. This observation was subsequently confirmed using SGA methods in an independent cohort of 10 subtype C transmission pairs (Haaland et al. 2009). Similar results, albeit with non-SGA methods, were obtained in studies comparing Envs from subtype A HIV-1 acutely infected sex workers to a database of matched chronic virus sequences (Chohan et al. 2005) and 13 subtype D and A transmission pairs from the Rakai district of Uganda (Sagar et al. 2009). Interestingly, such differences in acute versus chronic Envs were not seen in studies of recently transmitted subtype B HIV-1 heterosexual or MSM infections, suggesting that subtype differences in the virus or maturity of the epidemics may influence the contribution of such a phenotype to HIV-1 transmission (Chohan et al. 2005; Frost et al. 2005; Wilen et al. 2011a). Most recently, Korber and colleagues (Gnanakaran et al. 2011) compared over 7000 SGA-derived *env* gp160 sequences from 275 acutely or chronically infected subtype B subjects and observed statistically robust signatures comprising single amino acids, glycosylation motifs, and multisite patterns of clustered amino acids. These included signatures near the CCR5 coreceptor binding surface, near the CD4-binding site, in the cytoplasmic domain of gp41, and in the signal peptide. The motif with highest statistical significance was at amino acid position 12 in the signal peptide, which may affect Env expression and incorporation of Env into virions (Asmal et al. 2011). The second most significant signature was at amino acid position 413–415, which affected a glycan involved in escape from antibody neutralization. How these changes might affect HIV-1 transmission is currently unknown.

A second approach to analyzing transmitted HIV-1 genomes for distinguishing properties is by phenotypic characterization. It has been difficult to link sequence differences observed between chronic and acute Envs, even within transmission pairs, to a distinct phenotypic property that might influence transmission other than CCR5 tropism. In an analysis of Zambian subtype C transmission pairs (Derdeyn et al. 2004), it was found that Envs from the newly transmitted viruses retained sensitivity to neutralization by the partner's antibodies and in fact were more sensitive on average than Env variants derived from the transmitting partner (Derdeyn et al. 2004). This characteristic coupled with shorter variable loops in early viruses led to the hypothesis that transmissibility was linked to loss of Env modifications required for neutralization resistance in the chronically infected host but dispensable in the immunologically naïve partner. Despite this, the acute Envs were not generally more sensitive to neutralization by either pooled HIV-1–positive sera or to a majority of broadly neutralizing antibodies (Derdeyn et al. 2004; Li et al. 2006). Moreover, the acute and chronic Envs had similar requirements for high CD4 and CCR5 levels on target cells and they showed comparable utilization of CCR5 chimeric proteins and alternative coreceptors (Isaacman-Beck et al. 2009; Alexander et al. 2010). Also, following up on genetic signatures of virus transmission identified by Korber and colleagues (Gnanakaran et al. 2011), Asmal et al. (2011) characterized the phenotypic effects of the position 12 polymorphism in the Env leader sequence, which was found to be an enriched motif in transmitted/founder viruses. Experiments showed an association between a positive amino acid (histidine) at position 12 and higher Env expression, higher virion Env incorporation, and higher virion infectivity compared with control viruses.

The first biological analysis of complete HIV-1 subtype B transmitted/founder *env* genes (Keele et al. 2008), and of subtype B and C transmitted/founder full-length viral genomes (Salazar-Gonzalez et al. 2009; Li et al. 2010), showed these viruses to be uniformly CD4 dependent and CCR5 or CCR5/CXCR4 dual tropic. These findings thus revealed that CCR5 tropism is a property of the transmitted virus itself and not a phenotype that evolves in the initial days and weeks of infection. Similarly, transmitted/founder viruses were found to show neutralization sensitivity patterns typical of tier 2 or 3 primary virus strains (Li et al. 2005a) with V3 and coreceptor binding surface regions well protected from binding by neutralizing monoclonal or polyclonal HIV-1 antibodies (Keele et al. 2008). Again, these were properties of the transmitted/founder virus at or near the moment of virus transmission and were not properties that evolved in early infection. Subtype B and C transmitted/founder genomes encoded viruses that replicated efficiently in primary human CD4$^+$ T cells but much less well in monocyte-derived macrophages (Salazar-Gonzalez et al. 2009; Li et al. 2010; C Ochsenbauer, J Kappes, and G Shaw, unpubl.), consistent with results obtained with Env pseudotyped viruses (Isaacman-Beck et al. 2009). These observations were extended to subtype A transmitted/founder viruses, which replicated efficiently in primary human CD4$^+$ T cells but very poorly in monocyte-derived macrophages. Interestingly, a substantial proportion of subtype D transmitted/founder viruses replicated efficiently in both CD4$^+$ T cells and macrophages (J Baalwa and G Shaw, unpubl.), which may correlate with enhanced neuropathogenesis of subtype D viruses more generally (Sacktor et al. 2009). Again, these findings described features of transmitted/ founder viruses at or near the moment of transmission and not virus properties that evolved in the newly infected host. Thus, these findings have particular relevance to rational vaccine design efforts and studies of virus transmission in general. With respect to animal models and human tissue explant studies of HIV-1 transmission, the results suggest that tissue macrophages may not play a significant role in HIV-1 transmission and that prototypic macrophage-tropic HIV-1 strains such as BaL, ADA, and YU2 may not be best suited as challenge viruses.

Swanstrom and colleagues (Russell et al. 2011) recently examined the genetic and biological basis of the HIV-1 population bottleneck in mother-infant HIV-1 transmission in subtype C infections in Malawi. They used SGA techniques to characterize 19 transmission pairs, of which 10 involved intrauterine transmission and nine intrapartum transmission. There was a stringent transmission bottleneck in each case. Thirteen of 19 transmissions were estimated to be from a single virus. Intrapartum (but not intrauterine) transmissions were characterized by transmitted/founder Envs that were shorter and had fewer potential amino-linked glycans in V1–V5. Mother and infant viruses were similar, however, in their sensitivity to soluble CD4, a panel of neutralizing monoclonal antibodies, and to autologous and heterologous polyclonal antibody neutralization. Thus, a distinguishing transmission phenotype was not evident.

The most comprehensive and systematic assessment thus far of the biology of transmitted/founder Envs compared with chronic HIV-1 Envs has been conducted by Doms and colleagues (Wilen et al. 2011a). These investigators compared the biological activity of 24 clade B transmitted/founder Envs with that of 17 chronic controls. To increase the likelihood of an intact mucosal barrier in the acutely infected recipients and thus the likelihood of identifying phenotypic properties associated with mucosal transmission, only transmitted/founder Envs from individuals productively infected by a single virus were enrolled in the acute infection arm of the study. Env pseudotyping was used to assess envelope function in single-round infectivity assays to compare coreceptor tropism, CCR5 utilization, primary CD4^{+} T-cell subset tropism, dendritic cell *trans*-infection, Env fusion kinetics, and neutralization sensitivity between acute and chronic Envs. Transmitted/founder and chronic Envs were phenotypically equivalent in most assays, although transmitted/founder Envs were slightly more sensitive to neutralization by the CD4-binding-site antibodies b12 and VRC01 and by pooled human HIV hyperimmune immunoglobulin (HIVIG). These findings were independently validated using a panel of 14 additional chronic HIV-1 Env controls.

With a relatively large number of transmitted/founder and acute Envs now having been examined across a number of different studies and by different investigative groups, it seems that no single major genetic or phenotypic signature is required for transmission beyond the use of CCR5. Current data suggest that an array of genetic traits, including but not limited to shorter variable loops and reduced numbers of amino-linked glycosylation sites is associated with enhanced virus transmission at least for subtypes A, C, and possibly B. The structural implications of these signatures and of the modestly enhanced neutralization sensitivity of subtype B viruses are not yet understood. However, the possibility exists that relatively subtle alterations of Env structure and function in the context of the native Env trimer could provide sufficient selective advantage during the eclipse phase of HIV-1 transmission to result in preferential transmission of such viruses. The recent finding that HIV-1 gp120 binds to the CD4^{+} T-cell gut homing integrin α4β7 has raised the possibility that this capacity is important to an infecting virus by targeting cells capable of trafficking to the gut-associated lymphoid tissue (Arthos et al. 2008). It has been reported that α4β7 highly expressing CD4^{+} T cells are more susceptible to productive infection by HIV-1 than those expressing low levels of the integrin, in part because this subset is enriched for activated CD4^{+} cells, and in part because α4β7hi cells express high levels of CCR5 and low levels of CXCR4 (Cicala et al. 2009). Interestingly, a small sample of acute subtype A and C virus Envs bound α4β7 with high affinity, and in some cases, later virus strains showed significantly reduced binding (Nawaz et al. 2011), consistent with an early requirement for infection of α4β7-expressing cells that is dispensable once infection in the gut mucosa has been established. Given that HIV-1 transmission is inherently inefficient and likely represents the most vulnerable point in the natural history of HIV-1 infection, identifying unique properties and potential vulnerabilities of transmitted/founder viruses remains an important objective.

MODELS OF HIV-1 TRANSMISSION

Because the eclipse phase of infection and mucosal invasion sites invariably obscure the earliest virus-host interactions, investigators have turned to models of HIV-1 transmission to study the earliest events that initiate productive clinical infection. Figures 1 and 2 illustrate early HIV-1 replication dynamics and diversification along with initial host antibody responses for which there is now substantial direct experimental validation in humans. Figure 3 illustrates early virus-host interactions, many of which are not amenable to direct analysis in humans but which have been studied in human tissue explants and in the primate SIV model of HIV-1 transmission. Importantly, experimental data supporting all three models of HIV/SIV replication, diversification, and transmission biology are largely internally consistent and mutually reinforcing and have begun to provide a window on the HIV-1 transmission process not previously possible.

Mathematical Models

Central to the development and validation of a strategy to identify transmitted/founder virus genomes by SGA, direct amplicon sequencing, and phylogenetic analysis was the development of a mathematical model to describe early virus replication and diversification. This model was developed and refined by Korber, Perelson, and colleagues (Keele et al. 2008; Lee et al. 2009) and used previously estimated parameters of HIV-1 generation time (2 d), reproductive ratio ($R_o = 6$), and reverse transcriptase error rate (2.16×10^{-5}) and assumed that initial virus replicates exponentially infecting R_o new cells at each generation and diversifying under a model of evolution that assumes no selection. Viruses were predicted to show a Poisson distribution of mutations and a star-like phylogeny. This model led Keele and colleagues (2008) to posit that the progeny of individual transmitted/founder viruses that establish productive clinical infection could thus be identified in acute infection as distinct low-diversity genetic lineages and that the consensus sequence,

or coalescent, of each lineage would correspond to actual transmitted/founder viruses. These hypotheses were proven to be valid in both acute HIV-1 and SIV infection (Keele et al. 2008, 2009; Li et al. 2010; Liu et al. 2010; Stone et al. 2010). More recently, Perelson and coworkers developed a stochastic model of the early HIV-1 infection (Pearson et al. 2011), as opposed to deterministic models, to probe the earliest virus-host events following virus inoculation. They distinguish virus that is released from cells continuously versus in a burst and show that these different mechanisms of virus production lead to substantially different early viral dynamics and different probabilities of virus extinction. The stochastic model moves us a step closer to what is believed to be the most vulnerable point in HIV-1 infection, in the hours and days immediately following virus exposure, and provides insights relevant to vaccine and therapeutic interventions at this critical period in the transmission process of HIV-1.

Primate-SIV Infection Models

The primate-SIV infection model captures many of the essential elements of virus-host interactions illustrated in Figures 1–3. Because more data are available for the primate cervicovaginal infection model, our discussion will be limited to it, although recent work with a penile SIV infection model is promising (Ma et al. 2011; Yeh et al. 2011). Although the eclipse phase of mucosal infection by SIVmac251 or SIVsmE660 is generally shorter than the 7–21 d reported for HIV-1 infection of humans, early SIV replication kinetics in acute SIV infection of macaques is quite comparable with HIV-1 infection of humans (Liu et al. 2010; Stone et al. 2010). In fact, Barouch and colleagues showed that in a low-dose rectal mucosal infection model of Indian rhesus macaques, the eclipse phase lengthens to approximately 7 d as the multiplicity of infection approaches one virus (Liu et al. 2010). Moreover, the early diversification of mucosally transmitted SIV in macaques is virtually indistinguishable from HIV-1 in humans (Keele

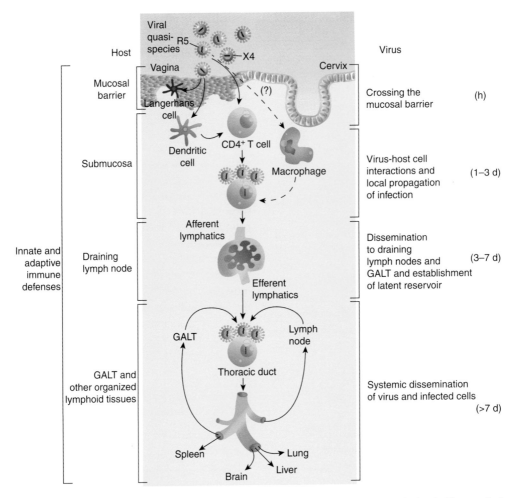

Figure 3. Model of cervicovaginal infection by HIV-1. Preferential R5 HIV-1 transmission is illustrated along with potential roles for Langerhans cells, dendritic cells, and tissue macrophages. Most HIV-1 transmitted/founder viruses replicate efficiently in CD4⁺ T cells but not in monocyte-derived macrophages (Salazar-Gonzalez et al. 2009; Li et al. 2010), raising questions about the role of macrophages in HIV-1 transmission. Virus-host cell interactions in the initial days of infection have been elucidated primarily in the SIV-Indian rhesus macaque infection model (Haase 2010) and in human tissue explants (Hladik and McElrath 2008). (Adapted from Pope and Haase 2003.)

et al. 2008, 2009; Liu et al. 2010; Stone et al. 2010). The virus-host cell interactions believed to underlie HIV-1 transmission based on human tissue explant studies and in vitro analyses of virus-host cell interactions are also generally similar to SIV infection of macaques. Figure 3 represents a mucosal infection model based on a large amount of SIV–rhesus macaque cervicovaginal transmission and infection data, human tissue explant findings, and in vitro

analysis of HIV–host cell interactions. SIV can cross mucosal epithelial surfaces within hours (Hu et al. 2000) and infects small numbers of cells productively (Zhang et al. 1999; Miller et al. 2005). In a high-dose mucosal challenge model, the numbers of viruses that cross the epithelial surfaces are great, as are the numbers of initially infected cells (Haase 2010; B Keele, J Estes, and J Lifson, unpubl.). In low-dose mucosal challenge models, these numbers are much smaller and more characteristic of human mucosal infection in which the numbers of transmitted/founder viruses are typically one or few (Keele et al. 2008, 2009; Abrahams et al. 2009; Haaland et al. 2009; Li et al. 2010; Liu et al. 2010; Stone et al. 2010). In macaques, the initial founder populations of productively infected cells undergo rapid expansion and within 1 wk or less, SIV disseminates widely to secondary lymphoid organs including GALT, in which a self-propagating systemic infection is established (Veazey et al. 1998; Miller et al. 2005). Although these events likely have close parallels in human HIV-1 infection, a key difference is that in high-dose mucosal SIV infection, every animal becomes productively infected by generally large numbers of transmitted/founder viruses following a single virus inoculation. In low-dose mucosal SIV infection, only a fraction of animals becomes productively infected following a single inoculation, and then generally by one or a few viruses. In uncomplicated human vaginal exposure to HIV-1, the virus inoculum is effectively still lower, resulting in productive clinical infection in one in every 200–2000 exposed women, and then by a single virion in 80% of cases (Table 1) (Keele et al. 2008; Abrahams et al. 2009; Haaland et al. 2009). Thus, it is unclear at the present time how closely the in situ virus-host interactions described thus far for the earliest phases of SIV infection of macaques (Haase 2010) correspond to eclipse phase events in human HIV-1 infection. Beyond these earliest events, SIV generally becomes detectable in the blood by the second week of infection and increases exponentially at a rate comparable or even faster than that of HIV-1 in humans (Lifson et al. 1997; Nowak et al. 1997; Ribeiro et al. 2010).

SIV and HIV-1 presumably invade the new host where mucosal surfaces can most easily be breached, and for HIV-1 this can be aided by microabrasions caused by the trauma of sexual intercourse or associated with sexually transmitted diseases or genital ulcerative disease (Abrahams et al. 2009; Haaland et al. 2009). It is of note that hepatitis C virus (HCV) generally circulates at even higher titers in human plasma than does HIV-1, yet it is exceedingly rare for it to be transmitted by penile-vaginal intercourse—so rare that the U.S. Department of Health and Human Services and Centers for Disease Control do not recommend barrier protection (condoms) to prevent heterosexual transmission of HCV in discordant couples. This suggests that the mechanism of transmission of HIV-1 across the vaginal and cervical mucosa is fundamentally different from a simple exchange of virus-contaminated blood or blood plasma across an abraded or otherwise traumatized mucosal surface. A single layer of columnar epithelium covers the endocervix and part of the transformation zone between endocervix and ectocervix. The endocervix and transformation zone appear to be preferential but not exclusive sites of HIV-1 entry and early replication (Li et al. 2009a). The stratified squamous epithelium of the vagina contains Langerhans cells whose extensions may reach the luminal surface. Langerhans cells are infectable by HIV-1 and can traffic to the submucosa. The submucosa of both the vagina and cervix contain dendritic cells, which also can become infected or transmit HIV-1 to CD4$^+$ T cells (Boggiano and Littman 2007; Boggiano et al. 2007; Hladik et al. 2007).

In macaques, the earliest focal collections of productively infected cells in cervical mucosa and submucosa are CD4$^+$ T cells that show a largely "resting" phenotype and low-level CCR5 expression (Zhang et al. 1999, 2004; Li et al. 2005b). In acute HIV-1 infection of humans, "resting" CD4$^+$ T cells also play a substantial role in supporting high-level virus replication (Zhang et al. 1999; Schacker et al. 2001). It is believed that such cells are memory CD4$^+$ T cells that have reverted to a largely resting phenotype but have retained sufficient CCR5 and activation pathway molecules to support SIV/HIV replication. This may be accentuated in GALT, in which $\alpha4\beta7$ expression on helper CD17$^+$/CD4$^+$ T cells may facilitate virus infection and replication (Arthos et al. 2008; Kader et al. 2009). Interestingly, tissue macrophages are not a primary target of early SIV infection, consistent with the failure of most transmitted/founder HIV-1 genomes to replicate efficiently in this cell type in vitro

(Salazar-Gonzalez et al. 2009; Li et al. 2010; C Ochsenbauer, J Kappes, and G Shaw, unpubl.).

Perelson has suggested that mucosal infections by HIV-1, and by analogy SIV, may be extinguished because of stochastic events in the transmission process (Pearson et al. 2011). This could be caused by limitations in target cell availability that may be more significant in cervicovaginal mucosa and submucosa than in rectal tissue (Edwards and Morris 1985; Ma et al. 2001; Zhang et al. 2004; Pudney et al. 2005; Miyake et al. 2006). A complex interplay of early innate host defenses may contribute to early virus containment or elimination, but paradoxically such responses can facilitate transmission by recruiting new potential target cells for infection (Li et al. 2009a; Haase 2010). Once infection disseminates to locoregional lymphatics, the GALT, and other secondary lymphoid tissues, massive depletion of memory $CD4^+$ T cells in the lamina propria ensues (Clayton et al. 1997; Veazey et al. 1998; Guadalupe et al. 2003; Brenchley et al. 2004; Mehandru et al. 2004; Mattapallil et al. 2005; Li et al. 2005b). Massive immune activation, further loss of $CD4^+$ T cells, and an acute enteropathy commonly follow (Kotler et al. 1984; Heise et al. 1994; Li et al. 2008). Defining critical steps in the mucosal transmission of SIV and HIV-1, including those most vulnerable to therapeutic or immunologic modulation or intervention, remains a high priority. Modulation of immune activation in the SIV–rhesus macaque infection model by glycerol monolaurate represents proof-of-concept of such an approach to prevention (Li et al. 2009a). Passive immunization with neutralizing monoclonal antibodies including 2F5, 4E10, 2G12, and B12 in a SHIV infection model provides another important proof-of-concept of the vulnerability of HIV-1 at this critical juncture (Mascola 2002; Hessell et al. 2007, 2009a,b).

Human Tissue Explant Models

An extensive body of literature describes human tissue explant models and their application to the analysis of HIV-1 and SIV transmission. Much of this has been summarized in recent reviews (Hladik and McElrath 2008; Wira and Veronese 2011). Although increasing attention in recent years has been paid to penile (Fischetti et al. 2009; Dinh et al. 2011) and gastrointestinal (Fletcher et al. 2006; Grivel et al. 2007; Shen et al. 2009, 2010) explant models, most information (Collins et al. 2000; Greenhead et al. 2000; Hu et al. 2004; Maher et al. 2005; Hladik et al. 2007; Saba et al. 2010; Merbah et al. 2011), can be correlated with in situ studies of primate infection by SIV (Haase 2010). Hladik and colleagues (2007) described a human vaginal organ explant culture system that involved the physical separation of vaginal epithelium from underlying stroma. This allowed for an identification of HIV-1 infection of cells exclusively in the epithelium and tracking of these cells as they exited the epithelium into surrounding culture medium. By using confocal fluorescent microscopy, virions were observed to bind and productively infect $CD4^+$ $CCR5^+$ CD45RO (memory) T cells within 2 hours of virus exposure to the intact epithelial surface. Both binding and infection of these cells was dependent on $CD4^+$ $CCR5^+$ engagement by the viral Env glycoprotein and could be blocked by receptor antibodies. HIV-1 also bound to Langerhans cells in the epithelium and localized to a perinuclear intracellular compartment but without evidence of productive infection. Antibody blocking of $CD4^+$ and $CCR5^+$, and mannan inhibition of C-type lectin binding by Env, did not inhibit Langerhans cell uptake of HIV-1, suggesting alternative pathways of virus binding and uptake by Langerhans cells. As long as 60 h posttransmission, HIV-1 could be identified in vacuolar compartments of Langerhans cells, which were frequently observed to associate with $CD4^+$ T cells. Whether Langerhans cell $CD4^+$ T-cell associations represented an "immunological synapse" active in virus transmission was not determined. This study thus showed rapid productive infection of human memory $CD4^+$ T cells consistent with in situ findings in the SIV–macaque infection model (Haase 2010), and internalization and persistence of HIV-1 in Langerhans cells in association with $CD4^+$ T cells (Fig. 3). Studies by Saba et al. (2010) in a cervicovaginal explant model also observed that CCR5$^+$ CD4$^+$ T cells with the effector memory phenotype are a primary target for infection. Human cervicovaginal tissue ex vivo was found to preferentially support productive infection by R5 HIV-1 rather than by X4 HIV-1 despite ample expression of CXCR4. Productive infection by R5 HIV-1 occurred preferentially in activated CD38$^+$ CD4$^+$ T cells in association with activation of HIV-1–uninfected (bystander) CD4$^+$ T cells that may amplify viral infection. That CXCR4-tropic HIV-1 replicated only in the few tissues that were enriched in CD27$^+$ CD28$^+$ effector memory CD4$^+$ T cells, if translatable to tissue in vivo, could in part explain the selection of R5 viruses during transmission. Still other human cervical explant studies were conducted by Shattock and colleagues (Greenhead et al. 2000). This work characterized cellular factors involved in HIV-1 entry (Hu et al. 2004) and identified potential therapeutic agents that block infection (Fletcher et al. 2005; Cummins et al. 2007; Buffa et al. 2009). In this model, blockade of CD4 or CCR5/CXCR4 prevented localized mucosal infection and trafficking by dendritic cells.

An extensive ex vivo tissue explant literature thus describes virus-host cell interactions in female and male genital mucosa that may contribute to natural HIV-1 transmission. This work includes descriptions of cell targets, cell surface molecules that mediate virus attachment or infection, and cell signaling pathways that facilitate virus infection and transmission (Hladik and McElrath 2008). However, a key question still unanswered by in vitro cell culture analyses, ex vivo tissue explant studies, or in vivo SIV-macaque infection studies is what is the contribution of Langerhans cells, dendritic cells, or macrophages, compared with CD4$^+$ CCR5$^+$ T cells alone to mucosal transmission of HIV-1 (Fig. 3). As this question is revisited, attention to the biological relevance and authenticity of the challenge viruses used in the model systems is warranted. The macrophage-tropic HIV-1 BaL, ADA, and YU2 strains, and the T-cell line-adapted NL4.3 and HXB-2 strains, have been used extensively in previous transmission model studies, yet none of these

viruses appears to faithfully reflect the properties described thus far for transmitted/founder viruses from acutely infected humans (Keele et al. 2008; Salazar-Gonzalez et al. 2009; Li et al. 2010; Wilen et al. 2011a). Future mucosal transmission studies may benefit from using virus challenge strains that more closely reflect actual transmitted/founder viruses.

CONCLUSIONS

Interrupting HIV-1 transmission by vaccination, microbicides, pre- or postexposure antiviral drug prophylaxis, or by any of a number of other prevention strategies is of paramount importance in curbing the HIV/AIDS pandemic. Elucidation of molecular viral-host interactions responsible for virus transmission and productive clinical infection can be instrumental in achieving this end. A new strategy for the molecular identification of transmitted/founder HIV-1 genomes, together with improved tissue explant and in vivo models of HIV-1 transmission, brings the possibility of elucidating critical HIV-1 transmission events and vulnerabilities within reach. Future studies can benefit from building on what has become an increasingly firm foundation in our knowledge of HIV-1 transmission.

REFERENCES

*Reference is also in this collection.

Abrahams MR, Anderson JA, Giorgi EE, Seoighe C, Mlisana K, Ping LH, Athreya GS, Treurnicht FK, Keele BF, Wood N, et al. 2009. Quantitating the multiplicity of infection with human immunodeficiency virus type 1 subtype C reveals a non-poisson distribution of transmitted variants. *J Virol* 83: 3556–3567.

Alexander M, Lynch R, Mulenga J, Allen S, Derdeyn CA, Hunter E. 2010. Donor and recipient envs from heterosexual human immunodeficiency virus subtype C transmission pairs require high receptor levels for entry. *J Virol* 84: 4100–4104.

Arthos J, Cicala C, Martinelli E, Macleod K, Van Ryk D, Wei D, Xiao Z, Veenstra TD, Conrad TP, Lempicki RA, et al. 2008. HIV-1 envelope protein binds to and signals through integrin α4β7, the gut mucosal homing receptor for peripheral T cells. *Nat Immunol* 9: 301–309.

Asmal M, Hellmann I, Liu W, Keele BF, Perelson AS, Bhattacharya T, Gnanakaran S, Daniels M, Haynes BF, Korber BT. 2011. A signature in HIV-1 envelope leader peptide associated with transition from acute to chronic infection impacts envelope processing and infectivity. *PLoS One* **6:** e23673.

Auvert B, Taljaard D, Lagarde E, Sobngwi-Tambekou J, Sitta R, Puren A. 2005. Randomized, controlled intervention trial of male circumcision for reduction of HIV infection risk: The ANRS 1265 trial. *PLoS Med* **2:** e298.

Bailey RC, Moses S, Parker CB, Agot K, Maclean I, Krieger JN, Williams CF, Campbell RT, Ndinya-Achola JO. 2007. Male circumcision for HIV prevention in young men in Kisumu, Kenya: A randomised controlled trial. *Lancet* **369:** 643–656.

Bar KJ, Li H, Chamberland A, Tremblay C, Routy JP, Grayson T, Sun C, Wang S, Learn GH, Morgan CJ, et al. 2010. Wide variation in the multiplicity of HIV-1 infection among injection drug users. *J Virol* **84:** 6241–6247.

Berger EA, Murphy PM, Farber JM. 1999. Chemokine receptors as HIV-1 coreceptors: Roles in viral entry, tropism, and disease. *Annu Rev Immunol* **17:** 657–700.

Boggiano C, Littman DR. 2007. HIV's vagina travelogue. *Immunity* **26:** 145–147.

Boggiano C, Manel N, Littman DR. 2007. Dendritic cell-mediated trans-enhancement of human immunodeficiency virus type 1 infectivity is independent of DC-SIGN. *J Virol* **81:** 2519–2523.

Boily MC, Baggaley RF, Wang L, Masse B, White RG, Hayes RJ, Alary M. 2009. Heterosexual risk of HIV-1 infection per sexual act: Systematic review and meta-analysis of observational studies. *Lancet Infect Dis* **9:** 118–129.

Brenchley JM, Schacker TW, Ruff LE, Price DA, Taylor JH, Beilman GJ, Nguyen PL, Khoruts A, Larson M, Haase AT, et al. 2004. CD4⁺ T cell depletion during all stages of HIV disease occurs predominantly in the gastrointestinal tract. *J Exp Med* **200:** 749–759.

Brenner BG, Roger M, Routy JP, Moisi D, Ntemgwa M, Matte C, Baril JG, Thomas R, Rouleau D, Bruneau J, et al. 2007. High rates of forward transmission events after acute/early HIV-1 infection. *J Infect Dis* **195:** 951–959.

Buffa V, Stieh D, Mamhood N, Hu Q, Fletcher P, Shattock RJ. 2009. Cyanovirin-N potently inhibits human immunodeficiency virus type 1 infection in cellular and cervical explant models. *J Gen Virol* **90:** 234–243.

Chohan B, Lang D, Sagar M, Korber B, Lavreys L, Richardson B, Overbaugh J. 2005. Selection for human immunodeficiency virus type 1 envelope glycosylation variants with shorter V1-V2 loop sequences occurs during transmission of certain genetic subtypes and may impact viral RNA levels. *J Virol* **79:** 6528–6531.

Church JD, Huang W, Mwatha A, Musoke P, Jackson JB, Bagenda D, Omer SB, Donnell D, Nakabiito C, Eure C, et al. 2010. Analysis of HIV tropism in Ugandan infants. *Curr HIV Res* **8:** 498–503.

Cicala C, Martinelli E, McNally JP, Goode DJ, Gopaul R, Hiatt J, Jelicic K, Kottilil S, Macleod K, O'Shea A, et al. 2009. The integrin α4β7 forms a complex with cell-surface CD4 and defines a T-cell subset that is highly susceptible to infection by HIV-1. *Proc Natl Acad Sci* **106:** 20877–20882.

Clark SJ, Saag MS, Decker WD, Campbell-Hill S, Roberson JL, Veldkamp PJ, Kappes JC, Hahn BH, Shaw GM. 1991. High titers of cytopathic virus in plasma of patients with symptomatic primary HIV-1 infection. *N Engl J Med* **324:** 954–960.

Clayton F, Snow G, Reka S, Kotler DP. 1997. Selective depletion of rectal lamina propria rather than lymphoid aggregate CD4 lymphocytes in HIV infection. *Clin Exp Immunol* **107:** 288–292.

Cohen MS, Shaw GM, McMichael AJ, Haynes BF. 2011a. Acute HIV-1 infection—basic, clinical, and public health perspectives. *N Engl J Med* **364:** 1943–1954.

Cohen MS, Chen YQ, McCauley M, Gamble T, Hosseinipour MC, Kumarasamy N, Hakim JG, Kumwenda J, Grinsztejn B, Pilotto JHS. 2011b. Prevention of HIV-1 infection with early antiretroviral therapy. *N Engl J Med* **365:** 493–505.

Collins KB, Patterson BK, Naus GJ, Landers DV, Gupta P. 2000. Development of an in vitro organ culture model to study transmission of HIV-1 in the female genital tract. *Nat Med* **6:** 475–479.

Cummins JE Jr, Guarner J, Flowers L, Guenthner PC, Bartlett J, Morken T, Grohskopf LA, Paxton L, Dezzutti CS. 2007. Preclinical testing of candidate topical microbicides for anti-human immunodeficiency virus type 1 activity and tissue toxicity in a human cervical explant culture. *Antimicrob Agents Chemother* **51:** 1770–1779.

Damond F, Avettand-Fenoel V, Collin G, Roquebert B, Plantier JC, Ganon A, Sizmann D, Babiel R, Glaubitz J, Chaix ML, et al. 2010. Evaluation of an upgraded version of the Roche Cobas AmpliPrep/Cobas TaqMan HIV-1 test for HIV-1 load quantification. *J Clin Microbiol* **48:** 1413–1416.

Derdeyn CA, Decker JM, Bibollet-Ruche F, Mokili JL, Muldoon M, Denham SA, Heil ML, Kasolo F, Musonda R, Hahn BH, et al. 2004. Envelope-constrained neutralization-sensitive HIV-1 after heterosexual transmission. *Science* **303:** 2019–2022.

Dinh MH, Fahrbach KM, Hope TJ. 2011. The role of the foreskin in male circumcision: An evidence-based review. *Am J Reprod Immunol* **65:** 279–283.

Dunkle KL, Stephenson R, Karita E, Chomba E, Kayitenkore K, Vwalika C, Greenberg L, Allen S. 2008. New heterosexually transmitted HIV infections in married or cohabiting couples in urban Zambia and Rwanda: An analysis of survey and clinical data. *Lancet* **371:** 2183–2191.

Edwards JN, Morris HB. 1985. Langerhans' cells and lymphocyte subsets in the female genital tract. *Br J Obstet Gynaecol* **92:** 974–982.

El-Sadr WM, Mayer KH, Hodder SL. 2010. AIDS in America—forgotten but not gone. *N Engl J Med* **362:** 967–970.

Eshleman SH, Khaki L, Laeyendecker O, Piwowar-Manning E, Johnson-Lewis L, Husnik M, Koblin B, Coates T, Chesney M, Vallari A, et al. 2009. Detection of individuals with acute HIV-1 infection using the ARCHITECT HIV Ag/Ab Combo assay. *J Acquir Immune Defic Syndr* **52:** 121–124.

Fideli US, Allen SA, Musonda R, Trask S, Hahn BH, Weiss H, Mulenga J, Kasolo F, Vermund SH, Aldrovandi GM. 2001. Virologic and immunologic determinants of heterosexual transmission of human immunodeficiency virus type 1 in Africa. *AIDS Res Hum Retroviruses* **17:** 901–910.

Fiebig EW, Wright DJ, Rawal BD, Garrett PE, Schumacher RT, Peddada L, Heldebrant C, Smith R, Conrad A, Kleinman SH, et al. 2003. Dynamics of HIV viremia and antibody seroconversion in plasma donors: Implications for diagnosis and staging of primary HIV infection. *AIDS* **17:** 1871–1879.

Fischer W, Ganusov VV, Giorgi EE, Hraber PT, Keele BF, Leitner T, Han CS, Gleasner CD, Green L, Lo CC, et al. 2010. Transmission of single HIV-1 genomes and dynamics of early immune escape revealed by ultra-deep sequencing. *PLoS One* **5:** e12303.

Fischetti L, Barry SM, Hope TJ, Shattock RJ. 2009. HIV-1 infection of human penile explant tissue and protection by candidate microbicides. *AIDS* **23:** 319–328.

Fletcher P, Kiselyeva Y, Wallace G, Romano J, Griffin G, Margolis L, Shattock R. 2005. The nonnucleoside reverse transcriptase inhibitor UC-781 inhibits human immunodeficiency virus type 1 infection of human cervical tissue and dissemination by migratory cells. *J Virol* **79:** 11179–11186.

Fletcher PS, Elliott J, Grivel JC, Margolis L, Anton P, McGowan I, Shattock RJ. 2006. Ex vivo culture of human colorectal tissue for the evaluation of candidate microbicides. *AIDS* **20:** 1237–1245.

Frost SD, Liu Y, Pond SL, Chappey C, Wrin T, Petropoulos CJ, Little SJ, Richman DD. 2005. Characterization of human immunodeficiency virus type 1 (HIV-1) envelope variation and neutralizing antibody responses during transmission of HIV-1 subtype B. *J Virol* **79:** 6523–6527.

Gaines H, von Sydow M, Pehrson PO, Lundbegh P. 1988. Clinical picture of primary HIV infection presenting as a glandular-fever-like illness. *BMJ* **297:** 1363–1368.

Galvin SR, Cohen MS. 2004. The role of sexually transmitted diseases in HIV transmission. *Nat Rev Microbiol* **2:** 33–42.

Gilbert MT, Rambaut A, Wlasiuk G, Spira TJ, Pitchenik AE, Worobey M. 2007. The emergence of HIV/AIDS in the Americas and beyond. *Proc Natl Acad Sci* **104:** 18566–18570.

Gnanakaran S, Bhattacharya T, Daniels M, Keele BF, Hraber PT, Lapedes AS, Shen T, Gaschen B, Krishnamoorthy M, Li H, Decker JM. 2011. Recurrent signature patterns in HIV-1 B clade envelope glycoproteins associated with either early or chronic infections. *PloS Pathog* **7:** e1002209.

Gray RH, Wawer MJ, Brookmeyer R, Sewankambo NK, Serwadda D, Wabwire-Mangen F, Lutalo T, Li X, vanCott T, Quinn TC. 2001. Probability of HIV-1 transmission per coital act in monogamous, heterosexual, HIV-1-discordant couples in Rakai, Uganda. *Lancet* **357:** 1149–1153.

Gray RH, Li X, Kigozi G, Serwadda D, Brahmbhatt H, Wabwire-Mangen F, Nalugoda F, Kiddugavu M, Sewankambo N, Quinn TC, et al. 2005. Increased risk of incident HIV during pregnancy in Rakai, Uganda: A prospective study. *Lancet* **366:** 1182–1188.

Gray RH, Kigozi G, Serwadda D, Makumbi F, Watya S, Nalugoda F, Kiwanuka N, Moulton LH, Chaudhary MA, Chen MZ, et al. 2007. Male circumcision for HIV prevention in men in Rakai, Uganda: A randomised trial. *Lancet* **369:** 657–666.

Greenhead P, Hayes P, Watts PS, Laing KG, Griffin GE, Shattock RJ. 2000. Parameters of human immunodeficiency virus infection of human cervical tissue and inhibition by vaginal virucides. *J Virol* **74:** 5577–5586.

Grivel JC, Elliott J, Lisco A, Biancoto A, Condack C, Shattock RJ, McGowan I, Margolis L, Anton P. 2007. HIV-1 pathogenesis differs in rectosigmoid and tonsillar tissues infected ex vivo with CCR5- and CXCR4-tropic HIV-1. *AIDS* **21:** 1263–1272.

Grobler J, Gray CM, Rademeyer C, Seoighe C, Ramjee G, Karim SA, Morris L, Williamson C. 2004. Incidence of HIV-1 dual infection and its association with increased viral load set point in a cohort of HIV-1 subtype C-infected female sex workers. *J Infect Dis* **190:** 1355–1359.

Guadalupe M, Reay E, Sankaran S, Prindiville T, Flamm J, McNeil A, Dandekar S. 2003. Severe CD4$^+$ T-cell depletion in gut lymphoid tissue during primary human immunodeficiency virus type 1 infection and substantial delay in restoration following highly active antiretroviral therapy. *J Virol* **77:** 11708–11717.

Haaland RE, Hawkins PA, Salazar-Gonzalez J, Johnson A, Tichacek A, Karita E, Manigart O, Mulenga J, Keele BF, Shaw GM, et al. 2009. Inflammatory genital infections mitigate a severe genetic bottleneck in heterosexual transmission of subtype A and C HIV-1. *PLoS Pathog* **5:** e1000274.

Haase AT. 2010. Targeting early infection to prevent HIV-1 mucosal transmission. *Nature* **464:** 217–223.

Heise C, Miller CJ, Lackner A, Dandekar S. 1994. Primary acute simian immunodeficiency virus infection of intestinal lymphoid tissue is associated with gastrointestinal dysfunction. *J Infect Dis* **169:** 1116–1120.

Hemelaar J, Gouws E, Ghys PD, Osmanov S. 2006. Global and regional distribution of HIV-1 genetic subtypes and recombinants in 2004. *AIDS* **20:** W13–W23.

Hessell AJ, Hangartner L, Hunter M, Havenith CE, Beurskens FJ, Bakker JM, Lanigan CM, Landucci G, Forthal DN, Parren PW, et al. 2007. Fc receptor but not complement binding is important in antibody protection against HIV. *Nature* **449:** 101–104.

Hessell AJ, Poignard P, Hunter M, Hangartner L, Tehrani DM, Bleeker WK Parren PW, Marx PA, Burton DR. 2009a. Effective, low-titer antibody protection against low-dose repeated mucosal SHIV challenge in macaques. *Nat Med* **15:** 951–954.

Hessell AJ, Rakasz EG, Poignard P, Hangartner L, Landucci G, Forthal DN, Koff WC, Watkins DI, Burton DR. 2009b. Broadly neutralizing human anti-HIV antibody 2G12 is effective in protection against mucosal SHIV challenge even at low serum neutralizing titers. *PLoS Pathog* **5:** e1000433.

Hladik F, McElrath MJ. 2008. Setting the stage: Host invasion by HIV. *Nat Rev Immunol* **8:** 447–457.

Hladik F, Sakchalathorn P, Ballweber L, Lentz G, Fialkow M, Eschenbach D, McElrath MJ. 2007. Initial events in establishing vaginal entry and infection by human immunodeficiency virus type-1. *Immunity* **26:** 257–270.

Hu J, Gardner MB, Miller CJ. 2000. Simian immunodeficiency virus rapidly penetrates the cervicovaginal mucosa after intravaginal inoculation and infects intraepithelial dendritic cells. *J Virol* **74:** 6087–6095.

Hu Q, Frank I, Williams V, Santos JJ, Watts P, Griffin GE, Moore JP, Pope M, Shattock RJ. 2004. Blockade of attachment and fusion receptors inhibits HIV-1 infection of human cervical tissue. *J Exp Med* **199:** 1065–1075.

Hudgens MG, Longini IM Jr, Vanichseni S, Hu DJ, Kitaya-porn D, Mock PA, Halloran ME, Satten GA, Choopanya K, Mastro TD. 2002. Subtype-specific transmission probabilities for human immunodeficiency virus type 1 among injecting drug users in Bangkok, Thailand. *Am J Epidemiol* **155:** 159–168.

Isaacman-Beck J, Hermann EA, Yi Y, Ratcliffe SJ, Mulenga J, Allen S, Hunter E, Derdeyn CA, Collman RG. 2009. Heterosexual transmission of human immunodeficiency virus type 1 subtype C: Macrophage tropism, alternative coreceptor use, and the molecular anatomy of CCR5 utilization. *J Virol* **83:** 8208–8220.

Jensen MA, Li FS, van 't Wout AB, Nickle DC, Shriner D, He HX, McLaughlin S, Shankarappa R, Margolick JB, Mullins JI. 2003. Improved coreceptor usage prediction and genotypic monitoring of R5-to-X4 transition by motif analysis of human immunodeficiency virus type 1 env V3 loop sequences. *J Virol* **77:** 13376–13388.

Kader M, Wang X, Piatak M, Lifson J, Roederer M, Veazey R, Mattapallil JJ. 2009. $\alpha 4^+\beta 7(hi)CD4^+$ memory T cells harbor most Th-17 cells and are preferentially infected during acute SIV infection. *Mucosal Immunol* **2:** 439–449.

Kearney M, Maldarelli F, Shao W, Margolick JB, Daar ES, Mellors JW, Rao V, Coffin JM, Palmer S. 2009. Human immunodeficiency virus type 1 population genetics and adaptation in newly infected individuals. *J Virol* **83:** 2715–2727.

Keele BF, Giorgi EE, Salazar-Gonzalez JF, Decker JM, Pham KT, Salazar MG, Sun C, Grayson T, Wang S, Li H, et al. 2008. Identification and characterization of transmitted and early founder virus envelopes in primary HIV-1 infection. *Proc Natl Acad Sci* **105:** 7552–7557.

Keele BF, Li H, Learn GH, Hraber P, Giorgi EE, Grayson T, Sun C, Chen Y, Yeh WW, Letvin NL, et al. 2009. Low-dose rectal inoculation of rhesus macaques by SIVsmE660 or SIVmac251 recapitulates human mucosal infection by HIV-1. *J Exp Med* **206:** 1117–1134.

Korber B, Muldoon M, Theiler J, Gao F, Gupta R, Lapedes A, Hahn BH, Wolinsky S, Bhattacharya T. 2000. Timing the ancestor of the HIV-1 pandemic strains. *Science* **288:** 1789–1796.

Kotler DP, Gaetz HP, Lange M, Klein EB, Holt PR. 1984. Enteropathy associated with the acquired immunodeficiency syndrome. *Ann Intern Med* **101:** 421–428.

Kouyos RD, von Wyl V, Yerly S, Boni J, Taffe P, Shah C, Burgisser P, Klimkait T, Weber R, Hirschel B, et al. 2010. Molecular epidemiology reveals long-term changes in HIV type 1 subtype B transmission in Switzerland. *J Infect Dis* **201:** 1488–1497.

Learn GH, Muthui D, Brodie SJ, Zhu T, Diem K, Mullins JI, Corey L. 2002. Virus population homogenization following acute human immunodeficiency virus type 1 infection. *J Virol* **76:** 11953–11959.

Lee HY, Giorgi EE, Keele BF, Gaschen B, Athreya GS, Salazar-Gonzalez JF, Pham KT, Goepfert PA, Kilby JM, Saag MS, et al. 2009. Modeling sequence evolution in acute HIV-1 infection. *J Theor Biol* **261:** 341–360.

Li M, Gao F, Mascola JR, Stamatatos L, Polonis VR, Koutsoukos M, Voss G, Goepfert P, Gilbert P, Greene KM, et al. 2005a. Human immunodeficiency virus type 1 env clones from acute and early subtype B infections

for standardized assessments of vaccine-elicited neutralizing antibodies. *J Virol* **79:** 10108–10125.

Li Q, Duan L, Estes JD, Ma ZM, Rourke T, Wang Y, Reilly C, Carlis J, Miller CJ, Haase AT. 2005b. Peak SIV replication in resting memory $CD4^+$ T cells depletes gut lamina propria $CD4^+$ T cells. *Nature* **434:** 1148–1152.

Li M, Salazar-Gonzalez JF, Derdeyn CA, Morris L, Williamson C, Robinson JE, Decker JM, Li Y, Salazar MG, Polonis VR, et al. 2006. Genetic and neutralization properties of subtype C human immunodeficiency virus type 1 molecular env clones from acute and early heterosexually acquired infections in Southern Africa. *J Virol* **80:** 11776–11790.

Li Q, Estes JD, Duan L, Jessurun J, Pambuccian S, Forster C, Wietgrefe S, Zupancic M, Schacker T, Reilly C, et al. 2008. Simian immunodeficiency virus-induced intestinal cell apoptosis is the underlying mechanism of the regenerative enteropathy of early infection. *J Infect Dis* **197:** 420–429.

Li Q, Estes JD, Schlievert PM, Duan L, Brosnahan AJ, Southern PJ, Reilly CS, Peterson ML, Schultz-Darken N, Brunner KG, et al. 2009a. Glycerol monolaurate prevents mucosal SIV transmission. *Nature* **458:** 1034–1038.

Li Q, Smith AJ, Schacker TW, Carlis JV, Duan L, Reilly CS, Haase AT. 2009b. Microarray analysis of lymphatic tissue reveals stage-specific, gene expression signatures in HIV-1 infection. *J Immunol* **183:** 1975–1982.

Li H, Bar KJ, Wang S, Decker JM, Chen Y, Sun C, Salazar-Gonzalez JF, Salazar MG, Learn GH, Morgan CJ, et al. 2010. High multiplicity infection by HIV-1 in men who have sex with men. *PLoS Pathog* **6:** e1000890.

Lifson JD, Nowak MA, Goldstein S, Rossio JL, Kinter A, Vasquez G, Wiltrout TA, Brown C, Schneider D, Wahl L, et al. 1997. The extent of early viral replication is a critical determinant of the natural history of simian immunodeficiency virus infection. *J Virol* **71:** 9508–9514.

Lindback S, Karlsson AC, Mittler J, Blaxhult A, Carlsson M, Briheim G, Sonnerborg A, Gaines H. 2000a. Viral dynamics in primary HIV-1 infection. Karolinska Institutet Primary HIV Infection Study Group. *AIDS* **14:** 2283–2291.

Lindback S, Thorstensson R, Karlsson AC, von Sydow M, Flamholc L, Blaxhult A, Sonnerborg A, Biberfeld G, Gaines H. 2000b. Diagnosis of primary HIV-1 infection and duration of follow-up after HIV exposure. Karolinska Institute Primary HIV Infection Study Group. *AIDS* **14:** 2333–2339.

Little SJ, McLean AR, Spina CA, Richman DD, Havlir DV. 1999. Viral dynamics of acute HIV-1 infection. *J Exp Med* **190:** 841–850.

Liu J, Keele BF, Li H, Keating S, Norris PJ, Carville A, Mansfield KG, Tomaras GD, Haynes BF, Kolodkin-Gal D, et al. 2010. Low-dose mucosal simian immunodeficiency virus infection restricts early replication kinetics and transmitted virus variants in rhesus monkeys. *J Virol* **84:** 10406–10412.

Long EM, Martin HL Jr, Kreiss JK, Rainwater SM, Lavreys L, Jackson DJ, Rakwar J, Mandaliya K, Overbaugh J. 2000. Gender differences in HIV-1 diversity at time of infection. *Nat Med* **6:** 71–75.

Ma Z, Lu FX, Torten M, Miller CJ. 2001. The number and distribution of immune cells in the cervicovaginal mucosa remain constant throughout the menstrual cycle of rhesus macaques. *Clin Immunol* **100:** 240–249.

Ma Z-M, Stone M, Piatak M Jr, Schweighardt B, Haigwood NL, Montefiori D, Lifson JD, Busch MP, Miller CJ. 2009. High specific infectivity of plasma virus from the pre-ramp-up and ramp-up stages of acute simian immunodeficiency virus infection. *J Virol* **83:** 3288–3297.

Ma Z-M, Keele BF, Qureshi H, Stone M, DeSilva V, Fritts L, Lifson JD, Miller CJ. 2011. SIVmac251 is inefficiently transmitted to rhesus macaques by penile inoculation with a single SIVenv variant found in ramp-up phase plasma. *AIDS Res Hum Retroviruses* doi: 10.1089/aid. 2011.0090.

Maher D, Wu X, Schacker T, Horbul J, Southern P. 2005. HIV binding, penetration, and primary infection in human cervicovaginal tissue. *Proc Natl Acad Sci* **102:** 11504–11509.

* Malim MH, Bieniasz PD. 2011. HIV restriction factors and mechanisms of evasion. *Cold Spring Harb Perspect Med* doi: 10.1101/cshperspect.a006940.

Mascola JR. 2002. Passive transfer studies to elucidate the role of antibody-mediated protection against HIV-1. *Vaccine* **20:** 1922–1925.

Masharsky AE, Dukhovlinova EN, Verevochkin SV, Toussova OV, Skochilov RV, Anderson JA, Hoffman I, Cohen MS, Swanstrom R, Kozlov AP. 2010. A substantial transmission bottleneck among newly and recently HIV-1-infected injection drug users in St Petersburg, Russia. *J Infect Dis* **201:** 1697–1702.

Mattapallil JJ, Douek DC, Hill B, Nishimura Y, Martin M, Roederer M. 2005. Massive infection and loss of memory CD4$^+$ T cells in multiple tissues during acute SIV infection. *Nature* **434:** 1093–1097.

McElrath MJ, De Rosa SC, Moodie Z, Dubey S, Kierstead L, Janes H, Defawe OD, Carter DK, Hural J, Akondy R, et al. 2008. HIV-1 vaccine-induced immunity in the test-of-concept step study: A case-cohort analysis. *Lancet* **372:** 1894–1905.

Mehandru S, Poles MA, Tenner-Racz K, Horowitz A, Hurley A, Hogan C, Boden D, Racz P, Markowitz M. 2004. Primary HIV-1 infection is associated with preferential depletion of CD4$^+$ T lymphocytes from effector sites in the gastrointestinal tract. *J Exp Med* **200:** 761–770.

Merbah M, Introini A, Fitzgerald W, Grivel JC, Lisco A, Vanpouille C, Margolis L. 2011. Cervico-vaginal tissue ex vivo as a model to study early events in HIV-1 infection. *Am J Reprod Immunol* **65:** 268–278.

Miller CJ, Li Q, Abel K, Kim EY, Ma ZM, Wietgrefe S, La Franco-Scheuch L, Compton L, Duan L, Shore MD, et al. 2005. Propagation and dissemination of infection after vaginal transmission of simian immunodeficiency virus. *J Virol* **79:** 9217–9227.

Miller WC, Rosenberg NE, Rutstein SE, Powers KA. 2010. Role of acute and early HIV infection in the sexual transmission of HIV. *Curr Opin HIV AIDS* **5:** 277–282.

Miyake A, Ibuki K, Enose Y, Suzuki H, Horiuchi R, Motohara M, Saito N, Nakasone T, Honda M, Watanabe T, et al. 2006. Rapid dissemination of a pathogenic simian/human immunodeficiency virus to systemic organs and active replication in lymphoid tissues following intrarectal infection. *J Gen Virol* **87:** 1311–1320.

Nawaz F, Cicala C, Van Ryk D, Block KE, Jelicic K, McNally JP, Ogundare O, Pascuccio M, Patel N, Wei D, et al. 2011. The genotype of early-transmitting HIV gp120s promotes αβ-reactivity, revealing αβCD4$^+$ T cells as key targets in mucosal transmission. *PLoS Pathog* **7:** e1001301.

Nowak MA, Lloyd AL, Vasquez GM, Wiltrout TA, Wahl LM, Bischofberger N, Williams J, Kinter A, Fauci AS, Hirsch VM, et al. 1997. Viral dynamics of primary viremia and antiretroviral therapy in simian immunodeficiency virus infection. *J Virol* **71:** 7518–7525.

Nugent CT, Dockter J, Bernardin F, Hecht R, Smith D, Delwart E, Pilcher C, Richman D, Busch M, Giachetti C. 2009. Detection of HIV-1 in alternative specimen types using the APTIMA HIV-1 RNA Qualitative Assay. *J Virol Methods* **159:** 10–14.

Palmer S, Wiegand AP, Maldarelli F, Bazmi H, Mican JM, Polis M, Dewar RL, Planta A, Liu S, Metcalf JA, et al. 2003. New real-time reverse transcriptase-initiated PCR assay with single-copy sensitivity for human immunodeficiency virus type 1 RNA in plasma. *J Clin Microbiol* **41:** 4531–4536.

Pearson JE, Krapivsky P, Perelson AS. 2011. Stochastic theory of early viral infection: Continuous versus burst production of virions. *PLoS Comput Biol* **7:** e1001058.

Pope M, Haase AT. 2003. Transmission, acute HIV-1 infection and the quest for strategies to prevent infection. *Nat Med* **9:** 847–852.

Poss M, Martin HL, Kreiss JK, Granville L, Chohan B, Nyange P, Mandaliya K, Overbaugh J. 1995. Diversity in virus populations from genital secretions and peripheral blood from women recently infected with human immunodeficiency virus type 1. *J Virol* **69:** 8118–8122.

Powers KA, Poole C, Pettifor AE, Cohen MS. 2008. Rethinking the heterosexual infectivity of HIV-1: A systematic review and meta-analysis. *Lancet Infect Dis* **8:** 553–563.

Pudney J, Quayle AJ, Anderson DJ. 2005. Immunological microenvironments in the human vagina and cervix: Mediators of cellular immunity are concentrated in the cervical transformation zone. *Biol Reprod* **73:** 1253–1263.

Quinn TC. 2007. Circumcision and HIV transmission. *Curr Opin Infect Dis* **20:** 33–38.

Quinn TC, Wawer MJ, Sewankambo N, Serwadda D, Li C, Wabwire-Mangen F, Meehan MO, Lutalo T, Gray RH. 2000. Viral load and heterosexual transmission of human immunodeficiency virus type 1. Rakai Project Study Group. *N Engl J Med* **342:** 921–929.

Resch W, Hoffman N, Swanstrom R. 2001. Improved success of phenotype prediction of the human immunodeficiency virus type 1 from envelope variable loop 3 sequence using neural networks. *Virology* **288:** 51–62.

Ribeiro RM, Qin L, Chavez LL, Li D, Self SG, Perelson AS. 2010. Estimation of the initial viral growth rate and basic reproductive number during acute HIV-1 infection. *J Virol* **84:** 6096–6102.

Richman DD, Wrin T, Little SJ, Petropoulos CJ. 2003. Rapid evolution of the neutralizing antibody response to HIV type 1 infection. *Proc Natl Acad Sci* **100:** 4144–4149.

Ridzon R, Gallagher K, Ciesielski C, Ginsberg MB, Robertson BJ, Luo CC, DeMaria A Jr, 1997. Simultaneous transmission of human immunodeficiency virus and hepatitis C virus from a needle-stick injury. *N Engl J Med* **336:** 919–922.

Ritola K, Pilcher CD, Fiscus SA, Hoffman NG, Nelson JA, Kitrinos KM, Hicks CB, Eron JJ Jr, Swanstrom R. 2004. Multiple V1/V2 env variants are frequently present during primary infection with human immunodeficiency virus type 1. *J Virol* **78:** 11208–11218.

Russell ES, Kwiek JJ, Keys J, Barton K, Mwapasa V, Montefiori DC, Meshnick SR, Swanstrom R. 2011. The genetic bottleneck in vertical transmission of subtype C HIV-1 is not driven by selection of especially neutralization-resistant virus from the maternal viral population. *J Virol* **85:** 8253–8262.

Saba E, Grivel JC, Vanpouille C, Brichacek B, Fitzgerald W, Margolis L, Lisco A. 2010. HIV-1 sexual transmission: Early events of HIV-1 infection of human cervico-vaginal tissue in an optimized ex vivo model. *Mucosal Immunol* **3:** 280–290.

Sacktor N, Nakasujja N, Skolasky RL, Rezapour M, Robertson K, Musisi S, Katabira E, Ronald A, Clifford DB, Laeyendecker O, et al. 2009. HIV subtype D is associated with dementia, compared with subtype A, in immunosuppressed individuals at risk of cognitive impairment in Kampala, Uganda. *Clin Infect Dis* **49:** 780–786.

Sagar M, Kirkegaard E, Long EM, Celum C, Buchbinder S, Daar ES, Overbaugh J. 2004. Human immunodeficiency virus type 1 (HIV-1) diversity at time of infection is not restricted to certain risk groups or specific HIV-1 subtypes. *J Virol* **78:** 7279–7283.

Sagar M, Laeyendecker O, Lee S, Gamiel J, Wawer MJ, Gray RH, Serwadda D, Sewankambo NK, Shepherd JC, Toma J, et al. 2009. Selection of HIV variants with signature genotypic characteristics during heterosexual transmission. *J Infect Dis* **199:** 580–589.

Salazar-Gonzalez JF, Bailes E, Pham KT, Salazar MG, Guffey MB, Keele BF, Derdeyn CA, Farmer P, Hunter E, Allen S, et al. 2008. Deciphering human immunodeficiency virus type 1 transmission and early envelope diversification by single-genome amplification and sequencing. *J Virol* **82:** 3952–3970.

Salazar-Gonzalez JF, Salazar MG, Keele BF, Learn GH, Giorgi EE, Li H, Decker JM, Wang S, Baalwa J, Kraus MH, et al. 2009. Genetic identity, biological phenotype, and evolutionary pathways of transmitted/founder viruses in acute and early HIV-1 infection. *J Exp Med* **206:** 1273–1289.

Schacker T, Collier AC, Hughes J, Shea T, Corey L. 1996. Clinical and epidemiologic features of primary HIV infection. *Ann Intern Med* **125:** 257–264.

Schacker T, Little S, Connick E, Gebhard K, Zhang ZQ, Krieger J, Pryor J, Havlir D, Wong JK, Schooley RT, et al. 2001. Productive infection of T cells in lymphoid tissues during primary and early human immunodeficiency virus infection. *J Infect Dis* **183:** 555–562.

* Sharp PM, Hahn BH. 2011. Origins of HIV and the AIDS pandemic. *Cold Spring Harbor Perspect Med* doi: 10.1101cshperspect.a006841.

Shen R, Richter HE, Clements RH, Novak L, Huff K, Bimczok D, Sankaran-Walters S, Dandekar S, Clapham PR, Smythies LE, et al. 2009. Macrophages in vaginal but not intestinal mucosa are monocyte-like and permissive to human immunodeficiency virus type 1 infection. *J Virol* **83:** 3258–3267.

Shen R, Smythies LE, Clements RH, Novak L, Smith PD. 2010. Dendritic cells transmit HIV-1 through human small intestinal mucosa. *J Leukoc Biol* **87:** 663–670.

Stone M, Keele BF, Ma ZM, Bailes E, Dutra J, Hahn BH, Shaw GM, Miller CJ. 2010. A limited number of simian immunodeficiency virus (SIV) env variants are transmitted to rhesus macaques vaginally inoculated with SIVmac251. *J Virol* **84:** 7083–7095.

Taylor BS, Sobieszczyk ME, McCutchan FE, Hammer SM. 2008. The challenge of HIV-1 subtype diversity. *N Engl J Med* **358:** 1590–1602.

UNAIDS. 2010. Global Report 2010, Geneva. http://www.unaids.org/globalreport/documents/20101123_Global Report_full_en.pdf.

Veazey RS, DeMaria M, Chalifoux LV, Shvetz DE, Pauley DR, Knight HL, Rosenzweig M, Johnson RP, Desrosiers RC, Lackner AA. 1998. Gastrointestinal tract as a major site of CD4+ T cell depletion and viral replication in SIV infection. *Science* **280:** 427–431.

Wawer MJ, Gray RH, Sewankambo NK, Serwadda D, Li X, Laeyendecker O, Kiwanuka N, Kigozi G, Kiddugavu M, Lutalo T, et al. 2005. Rates of HIV-1 transmission per coital act, by stage of HIV-1 infection, in Rakai, Uganda. *J Infect Dis* **191:** 1403–1409.

Wei X, Decker JM, Wang S, Hui H, Kappes JC, Wu X, Salazar-Gonzalez JF, Salazar MG, Kilby JM, Saag MS, et al. 2003. Antibody neutralization and escape by HIV-1. *Nature* **422:** 307–312.

Wilen CB, Parrish NF, Pfaff JM, Decker JM, Henning EA, Haim H, Sodroski J, Haynes BF, Montefiori DC, Tilton JC, et al. 2011a. Phenotypic and immunologic comparison of clade B transmitted/founder and chronic HIV-1 envelope glycoproteins. *J Virol* (in press).

* Wilen CB, Tilton JC, Doms RW. 2011b. Cell binding and entry. *Cold Spring Harbor Perspect Med* doi: 10.1101/cshperspect.a006866.

Wira CR, Veronese F. 2011. Sexual transmission of HIV in the 21st century. *Am J Reprod Immunol* **65:** 181–376.

Wolfs TF, Zwart G, Bakker M, Goudsmit J. 1992. HIV-1 genomic RNA diversification following sexual and parenteral virus transmission. *Virology* **189:** 103–110.

Wolinsky SM, Wike CM, Korber BT, Hutto C, Parks WP, Rosenblum LL, Kunstman KJ, Furtado MR, Munoz JL. 1992. Selective transmission of human immunodeficiency virus type-1 variants from mothers to infants. *Science* **255:** 1134–1137.

Worobey M, Gemmel M, Teuwen DE, Haselkorn T, Kunstman K, Bunce M, Muyembe JJ, Kabongo JM, Kalengayi RM, Van Marck E, et al. 2008. Direct evidence of extensive diversity of HIV-1 in Kinshasa by 1960. *Nature* **455:** 661–664.

Yeh WW, Rao SS, Lim S-Y, Zhang J, Hraber PT, Brassard LM, Luedemann C, Todd JP, Dodson A, Shen L. 2011. The TRIM5 gene modulates penile mucosal acquisition of simian immunodeficiency virus in Rhesus monkeys. *J Virol* **85:** 10389–10398.

Zhang LQ, MacKenzie P, Cleland A, Holmes EC, Brown AJ, Simmonds P. 1993. Selection for specific sequences in the external envelope protein of human immunodeficiency virus type 1 upon primary infection. *J Virol* **67:** 3345–3356.

Zhang Z, Schuler T, Zupancic M, Wietgrefe S, Staskus KA, Reimann KA, Reinhart TA, Rogan M, Cavert W, Miller CJ, et al. 1999. Sexual transmission and propagation of SIV and HIV in resting and activated CD4$^+$ T cells. *Science* **286:** 1353–1357.

Zhang ZQ, Wietgrefe SW, Li Q, Shore MD, Duan L, Reilly C, Lifson JD, Haase AT. 2004. Roles of substrate availability and infection of resting and activated CD4$^+$ T cells in transmission and acute simian immunodeficiency virus infection. *Proc Natl Acad Sci* **101:** 5640–5645.

Zhu T, Mo H, Wang N, Nam DS, Cao Y, Koup RA, Ho DD. 1993. Genotypic and phenotypic characterization of HIV-1 patients with primary infection. *Science* **261:** 1179–1181.

Zhu T, Wang N, Carr A, Nam DS, Moor-Jankowski R, Cooper DA, Ho DD. 1996. Genetic characterization of human immunodeficiency virus type 1 in blood and genital secretions: Evidence for viral compartmentalization and selection during sexual transmission. *J Virol* **70:** 3098–3107.

HIV-1 Pathogenesis: The Virus

Ronald Swanstrom[1] and John Coffin[2]

[1]Department of Biochemistry and Biophysics, University of North Carolina at Chapel Hill, Chapel Hill, North Carolina 27599

[2]Department of Molecular Biology and Microbiology, Tufts University, Boston, Massachusetts 02111

Correspondence: risunc@med.unc.edu

Transmission of HIV-1 results in the establishment of a new infection, typically starting from a single virus particle. That virion replicates to generate viremia and persistent infection in all of the lymphoid tissue in the body. HIV-1 preferentially infects T cells with high levels of CD4 and those subsets of T cells that express CCR5, particularly memory T cells. Most of the replicating virus is in the lymphoid tissue, yet most of samples studied are from blood. For the most part the tissue and blood viruses represent a well-mixed population. With the onset of immunodeficiency, the virus evolves to infect new cell types. The tropism switch involves switching from using CCR5 to CXCR4 and corresponds to an expansion of infected cells to include naïve CD4+ T cells. Similarly, the virus evolves the ability to enter cells with low levels of CD4 on the surface and this potentiates the ability to infect macrophages, although the scope of sites where infection of macrophages occurs and the link to pathogenesis is only partly known and is clear only for infection of the central nervous system. A model linking viral evolution to these two pathways has been proposed. Finally, other disease states related to immunodeficiency may be the result of viral infection of additional tissues, although the evidence for a direct role for the virus is less strong. Advancing immunodeficiency creates an environment in which viral evolution results in viral variants that can target new cell types to generate yet another class of opportunistic infections (i.e., HIV-1 with altered tropism).

The viral population from the time of initiation of infection to the time of overt immunodeficiency undergoes remarkable changes. The large viral population in an infected person is usually founded by a single infected CD4+ T cell in the mucosal tissue proximal to the site of exposure. For much of the time course of the infection, viral evolution is apparent, a result of evading the humoral and cell-mediated immune responses, while the virus continues to replicate in CD4+ T cells using CCR5 as the co-receptor. Initially, T cells in the gut associated lymphoid tissue (GALT) are massively depleted even though a majority of these cells are not in the activated state, which is preferred for HIV-1 infection in cell culture. The massive loss of GALT CD4+ T cells happens early and therefore cannot be the direct cause of immunodeficiency, which occurs late. However, the GALT is likely the source for a significant fraction of the virus in the blood, although the relationship between production of virus in lymphoid tissue and its transfer to the blood is unknown.

Important insights have been gained from examining the dynamics of both the infected cell and free virus particles, especially when the system is perturbed with antiviral drugs. These lessons are summarized by Coffin and Swanstrom (2011) and they fill out the story of virus-host interactions viewed from the perspective of the virus. In most settings, the virus turns over quickly such that changes in the production of virus are readily measured, at least for 99.9% of the virus. Most of the time virus is produced from CD4$^+$ T cells that have a short half-life. However, some cells are latently infected and present a major challenge to eradication of the virus (Siliciano and Greene 2011). Recently it has been possible to identify a variant of HIV-1 that has evolved to replicate in a new cell type with a different half-life (see below). Thus, the dynamics of virus and infected cell turnover offer important lessons into how the virus sustains itself in the host (Coffin and Swanstrom 2011).

Although the long-term persistent replication of virus leads to immunodeficiency, the damage to the host that leads to this state must be multifactorial. The early loss of most of the CD4$^+$ T cells in the GALT results in the translocation of bacterial products beyond the gut, potentially exacerbating one of the key correlates of disease progression—immune activation (Lackner et al. 2011). Loss of the capacity to make T cells and loss of the support structure to mature and regulate T cells may also contribute to the loss of immunologic capacity.

The onset of immunodeficiency sets the stage for opportunistic infections by common microbes that are otherwise controlled by the healthy host. The virus contributes to this phenomenon as shown by the appearance of variants that allow the virus to replicate in new cell types. At any one time, the virus is limited to cell types in which it can maintain a steady-state infection that is not cleared by the immune system. Growth in alternative cells likely is a challenge for the virus because replication in suboptimal cell targets would likely result in slow replication and easier containment by the immune system. With immunodeficiency the host response to replication in alternative cell types would be slower, giving the virus a chance to adapt to the new environment. However, in the clearest cases the virus stays close to home in that it always requires CD4, but can adapt to use lower levels, and it at most swaps one chemokine coreceptor for another, CXCR4 for CCR5. Whether the virus can evolve beyond these limits in an infected host is unknown but the tools for finding such viruses are in place.

In this article we follow the virus from transmission through dissemination and the adaptation to new target cells late in infection. Some of the ideas overlap other articles, which we have cited and which can be consulted for more detail. The potential for the virus to generate a localized replicating population is well established for several tissues/compartments, and it would not be surprising if other clear examples are demonstrated in the future, although such studies are difficult given the inaccessibility of the tissues of interest. Still, the potential for the virus to participate in pathogenic processes beyond immunodeficiency make such questions relevant.

TRANSMISSION

Frequency and Mechanism

Transmission is covered at length by Shaw and Hunter (2011). There are several aspects of transmission that influence viral pathogenesis. In general, transmission rates for HIV-1 are low, with the highest rates suggested to range from 10% per exposure to 0.1% (Powers et al. 2008; Boily et al. 2009). Under these circumstances, it is understandable that the infecting dose is a single virus particle; that is, limiting rates of infection result in infection with the minimal infectious dose. This inference has largely been confirmed with the molecular analysis of HIV-1 present after acute infection. In most cases, systemic infection is established by a single genetic variant in sexual transmission, transmission by intravenous drug use, or vertical transmission, although the frequencies

 Cite this article as *Cold Spring Harb Perspect Med* doi: 10.1101/cshperspect.a007443

of single variant versus multiple variant infections may vary by route (see Shaw and Hunter 2011).

Nature of the Founder Virus

There is ongoing discussion about where the genetic bottleneck associated with transmission occurs. There may be compartmentalization and restriction of the population in the donor at the site that produces the transmitted virus, although there is little evidence in support of this (see below). Alternatively, there may be physical barriers that limit the exposure in the recipient to very small amounts of virus; such barriers might include mucus and epithelium in the genital tract, which would limit access of virus to susceptible cells. Finally, multiple cells could become infected at the site of transmission, but only one variant succeed in establishing a systemic infection. There are rare examples in which the initially observed virus is not the same as the transmitted virus (Kim et al. 2005; Russell et al. 2011). However, in the setting of low frequency infection in humans, transmission of a single variant appears to be the norm (Shaw and Hunter 2011).

An early observation important for understanding viral pathogenesis was that some viruses isolated from people late in infection were able to grow and cause syncytia in transformed T-cell lines (Asjo et al. 1986), earning the name syncytium-inducing (SI). The remaining viruses were dubbed nonsyncytium-inducing (NSI). These NSI viruses were found early in infection and had some capacity to infect macrophages. This supported the initial dichotomy of less pathogenic/macrophage-tropic viruses being transmitted (van't Wout et al. 1994), which evolved into more pathogenic SI viruses. This picture has been supplanted with the understanding that SI viruses use CXCR4, which is expressed on transformed T-cell lines, in which the normal HIV-1 coreceptor CCR5 typically is absent, accounting for the apparent pathogenicity in T-cell lines of late viral isolates that have evolved to use CXCR4 (see Wilen et al. 2011). More recently,

it was discovered that only a fraction of viruses that use CCR5 can infect macrophages because of the low levels of surface CD4 (see below) and these are not the transmitted viruses. Indeed, viruses with macrophage-tropic Env proteins appear only late in infection, calling into question the role of macrophage infection in systemic infection. Thus the transmitted virus typically uses CCR5 and replicates in activated T cells as evidenced by its need for high levels of surface CD4 (Isaacman-Beck et al. 2009; Salazar-Gonzalez et al. 2009; Alexander et al. 2010).

Early Events in Replication

It is now clear that $CD4^+$ T cells in or near the epithelial layer must represent the initial target cells for mucosal transmission. In vitro, dendritic cells (DCs) can efficiently capture virus and transmit it to target $CD4^+$ T cells in the absence of viral replication (potentially by the high mannose glycans on the viral Env protein [Doores et al. 2010] binding to lectins on the surface of DCs [Wu and KewalRamani 2006]), but the importance of DCs facilitating the initial infection of $CD4^+$ T cells in vivo is still an open question (Martin et al. 2004; Boily-Larouche et al. 2009). In any case, there is little evidence to suggest that viral replication in DCs is an important part of the mechanism of transmission. The initial focal replication in both activated and resting $CD4^+$ T cells in the proximal tissue has been documented in the SIV/macaque model (Zhang et al. 2004; Miller et al. 2005), and this early replication is followed by movement of the virus to proximal lymphoid organs and then to a systemic infection (see Lackner et al. 2011).

There are several potential mechanisms for the virus to move from a localized infection to a systemic infection. The simplest model is that the concentration of virus is sufficiently high in the extracellular space to diffuse to adjacent target cells and tissues. Alternatively, virus could be transported by dendritic cells to proximal lymphoid tissue. Similarly, infected T cells could migrate to different body compartments to deliver virus. Finally, it has been suggested that the viral Env protein in the transmitted virus acts as a mimic of α4β7 integrin to target virus to CD4$^+$ T cells in the GALT (Nawaz et al. 2011); a corollary of this model is that viruses without this feature might initiate a local mucosal infection but not establish a systemic infection because of failure to traffic to the GALT. If stable infections were failing because of an absence of the ability to mimic α4β7 integrin to target GALT one might expect to see seroconversion based on transient replication in the proximal mucosal tissues, as is seen for HSV-2 infection, but this is not observed for putative abortive HIV-1 infections. It will be important to place this binding property of Env in the context of a large sampling of transmitted viruses to determine the frequency of this property.

TARGET CELLS

T-Cell Subsets

CD4 is required for natural isolates of HIV-1 to infect cells. Thus, robust infection of cells is limited to those expressing CD4. The normal function of CD4 is to act as a coreceptor along with the T-cell receptor in binding to Class II MHC, which is on antigen presenting cells and has the role of presenting heterologous peptides to the CD4$^+$ T helper cell. Other cell types can express lower levels of CD4, for example monocytes and macrophages, and it has been reported that CD4 plays an alternative role as the receptor for IL-16 (Liu et al. 1999).

CD4$^+$ T cells are heterogeneous in the expression of the CCR5 coreceptor. In the peripheral blood the memory cell subset and not the naive cell subset expresses significant levels of CCR5, whereas CXCR4 is expressed at relatively high levels on both memory and naive CD4$^+$ T cells (Bleul et al. 1997; Lee et al. 1999; Nicholson et al. 2001). This pattern of CCR5 expression is consistent with memory CD4$^+$ T cells being the predominant cell that is infected in vivo (Sleasman et al. 1996; Douek et al. 2002; Brenchley et al. 2004a), with a minor population of CD8$^+$ T cells that express a low level of CD4 also infected (Cochrane et al. 2004).

However, these studies have relied on linking the presence of viral DNA to cell surface markers, as opposed to using a marker of active viral replication. Because of down regulation of CD4 by HIV-1, cells actively producing virus are seen as CD4/CD8 double-negative T cells (Kaiser et al. 2007). In cell culture HIV-1 infects activated cells with much greater efficiency than quiescent cells (Korin and Zack 1998), with central and effector memory cells as the primary targets (Pfaff et al. 2010). However, in vivo there appears to be a type of CD4$^+$ T cell that does not express surface activation markers but supports significant levels of infection, particularly in the gut mucosa (Veazey et al. 2000; Brenchley et al. 2004b; Li et al. 2005; Mattapallil et al. 2005; Mehandru et al. 2007).

As immunodeficiency progresses the virus can evolve to enter cells using CXCR4 (Wilen et al. 2011). The appearance of CXCR4-using viruses (X4 viruses) is correlated with more rapid progression of disease, but it is still unclear if the evolution of these variants is the cause or a marker of rapid disease progression (Schuitemaker et al. 1992; Brumme et al. 2005; Hunt et al. 2006). Both may be true. X4 viruses are rarely transmitted, and it is curious why the virus primarily uses CCR5 when more CD4$^+$ T cells in the blood express CXCR4 than CCR5. However, CCR5 is generally up-regulated with infection and immune activation and CXCR4 is down-regulated (Ostrowski et al. 1998), and CCR5-expressing cells are significantly enriched in lymphoid tissue such as the GALT (Agace et al. 2000; Anton et al. 2000; Veazey et al. 2003; Brenchley et al. 2004b). Naïve CD4$^+$ T cells become more extensively infected either by infection with the X4 virus (Blaak et al. 2000; van Rij et al. 2000) or by expansion of infection into these cells by both the X4 and R5 viruses (Heeregrave et al. 2009). Although R5-to-X4 evolution is common, it is not essential for progression to disease. A large fraction of untreated HIV-infected individuals progress to AIDS and die with no evidence of X4 virus.

There are distinctive sequence features of the Env protein that allow use of CXCR4. An accumulation of basic amino acid substitutions at specific positions in the V3 loop of SU is associated with the coreceptor switch, which is presumed to increase specificity or affinity for CXCR4 (Wilen et al. 2011). There are other positions in Env but outside of V3 that also contribute to the X4 phenotype, but the specific contribution of these other sites is not clear (Hoffman et al. 2002; Pastore et al. 2006; Huang et al. 2008, 2011). There also seem to be differences among the subtypes for their propensity to evolve X4 variants, with X4 variants being more common in subtype D isolates (Tscherning et al. 1998; Huang et al. 2007). The reason for this difference is unclear, although the evolutionary distance between R5 and X4 variants of each subtype might be different. Conversely, X4 variants of subtype C appear to evolve less frequently and often include more dramatic sequence changes in V3, including deletions, compared to the sequence changes seen in subtype B (Coetzer et al. 2006, 2011).

Monocytes, Macrophages, and NK Cells

Monocytes are found in the blood and migrate to tissue where they differentiate into macrophages. This can be mimicked in cell culture by differentiating isolated blood monocytes into macrophages (monocyte-derived macrophages MDM) by exposure to cytokines. Monocytes and dendritic cells isolated from blood express very low levels of CD4 (Lee et al. 1999). Peripheral monocytes have been reported to be infected in vivo (Zhu et al. 2002; Fulcher et al. 2004; Ellery et al. 2007; Xu et al. 2008), along with complex collections of other blood myeloid cells (Centlivre et al. 2011). However, a recent analysis of viral DNA in blood cell subsets failed to detect a significant amount of viral DNA in the monocyte pool (Joseffson et al. 2011). Infection of monocytes in vitro is limited by the low levels of surface CD4 and blocks to entry (Arfi et al. 2008), reduced viral DNA synthesis (Sonza et al. 1996; Triques and Stevenson 2004; Arfi et al. 2008), and reduced viral gene expression (Dong et al. 2009).

Macrophage-tropic viruses infect cells with low levels of surface CD4 (Gorry et al. 2002; Peters et al. 2004; Martin-Garcia et al. 2006;

Thomas et al. 2007). The only place where there is clear evidence for the evolution of these viruses is in the brain, where at least a fraction of the cases of HIV-associated dementia involve the presence of macrophage-tropic virus (see Spudich and Gonzalez-Scarano 2011). There are two potential cell targets in the brain, microglia cells (macrophage-like cells in the parenchyma), and perivascular macrophages that migrate into the brain as part of an inflammatory response. It is not known which cell type supports the evolution and replication of macrophage-tropic viruses in the central nervous system (CNS). In addition to detecting these viruses in brain tissue at autopsy, it has now been possible to link the slow decay of virus in the CSF during therapy (Ellis et al. 2000; Haas et al. 2000; Cinque et al. 2001; Schnell et al. 2009), which is distinct from the rapid decay seen in the blood (Ho et al. 1995; Wei et al. 1995), with the presence of macrophage-tropic virus in the CSF, thus indicating replication in a long-lived cell (Schnell et al. 2011).

It has been possible to isolate viruses from blood that can enter macrophages (Li et al. 1999). However, these are not the viruses that evolve in the CNS, which are absent from the blood (Schnell et al. 2011) and the peripheral tissue (Peters et al. 2006). Understanding the range of infection in the body by macrophage-tropic viruses will require sampling different tissues but under conditions of more extensive disease progression to look for these late-evolving variants. In this regard, macrophages appear to be preferentially infected in macaque tissues after extensive depletion of $CD4^+$ T cells in lymphoid tissue by SIV (Veazey et al. 1998; Igarashi et al. 2001).

Both X4 viruses and macrophage-tropic viruses evolve from R5 progenitors, which require high levels of CD4, as found on activated T cells. Both of these variants appear late in the disease course and are therefore linked to increasing immunodeficiency of the host. Neither is efficiently transmitted. Thus, they both appear to be evolutionary dead-ends that evolve anew in each host, representing the extension of infection into new cell types (Fig. 1). One

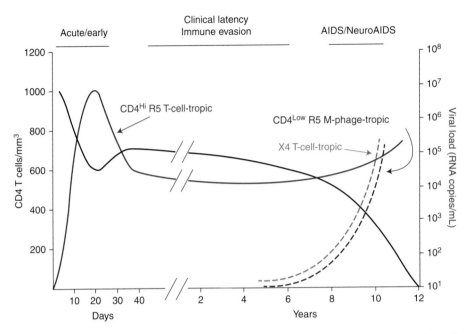

Figure 1. Time course of a typical HIV-1 infection with the appearance of host range variants late. The time course for different components of the infection are shown. There is an initial loss of CD4$^+$ T cells during acute infection followed by a partial recovery and then a slow decay during the period of clinical latency (black line). There is an initial viremia of the transmitted virus that uses CCR5 (R5) and requires high levels of CD4 to enter cells, and this virus establishes a set point during the period of clinical latency (CD4^{Hi} R5 T-cell-tropic, green line). With the loss of CD4$^+$ T cells there is increasing immunodeficiency (AIDS, including NeuroAIDS), a trend toward increasing viral load, and in a subset of subjects the appearance of host range variants that evolve to use a different coreceptor (X4 T-cell-tropic, red line) or evolve the ability to infect cells, presumably macrophages, with low levels of CD4 (CD4^{Low} R5 M-phage-tropic, blue line).

evolutionary model is based on adaptation of the Env protein structure in response to a changing host environment (Cheng-Mayer et al. 2009; Ince et al. 2010). This model suggests that in the presence of a weakened antibody response, the Env protein can evolve to be in a more open conformation rather than the closed, neutralization-resistant conformation that avoids sensitivity to antibodies targeting the coreceptor binding face. This concept is shown in Figure 2, in which the evolution of the open conformation may allow faster entry into T cells but also potentiate interaction with CXCR4 and the ability to enter cells with lower levels of CD4, setting the stage for further evolution.

NK cells can also be infected in vitro, although with a limited range of viruses (Robinson et al. 1988; Valentin et al. 2002; Harada et al.

2007; Bernstein et al. 2009). Viral DNA can also be found in this cell lineage in vivo (Valentin et al. 2002).

DISSEMINATION AND PERSISTENCE IN TARGET TISSUES

HIV-1 has been isolated from every bodily fluid and it can be assumed the virus will replicate in activated CD4$^+$ T cells virtually anywhere in the body. Thus, the virus is broadly disseminated in the body. Given this wide distribution, it is not clear what the source of virus is in the blood, from which viral samples have been most extensively studied. CD4$^+$ T cells in the blood represent just a few percent of the CD4$^+$ T cells in the body, making it likely that most viral replication is taking place in lymphoid tissues rich in CD4$^+$ T cells. It is not clear whether the virus in the

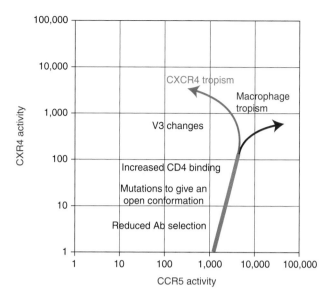

Figure 2. Evolution of host range variants. In this model the variant of HIV-1 that is replicating in memory T cells (requiring high levels of CD4 on the surface of the cell and using CCR5 as the coreceptor, shown in green) is exposed to reduced host surveillance in the form of reduced selective pressure from antibodies. This allows the virus to evolve such that the Env protein assumes a more open conformation that allows increased binding to CD4. The open conformation may also expose a latent low level tropism for CXCR4. These changes potentiate the subsequent evolution to use CXCR4 efficiently (X4 virus—red) or to use CD4 more efficiently (macrophage tropism—blue).

blood is produced by infected cells in tissue or in blood, or if all virus-producing tissues shunt virus into the blood with equal efficiency. Studies of tissues require biopsy or analysis of autopsy material, preventing careful time course studies, and the analysis of viral sequences is usually of DNA, which can include archival and defective viral DNA that may not represent the currently replicating virus. Also, free virus, the most sensitive and reliable instantaneous indicator of the state of infection, is not accessible from solid tissue samples. For this reason, more accessible fluids such as semen, cervicovaginal mucus, and cerebral spinal fluid are often used as surrogates for the corresponding tissue.

Genetic analysis of virus from tissues (or their liquid surrogates) has generated several unifying observations. In many cases the genetic diversity of the viral population in the blood overlaps the population in the tissue, suggesting a well-mixed relationship in which the virus in the blood is derived from that tissue or imported into the tissue, and if imported into the tissue the virus must undergo little replication given its similarity to the virus in the blood. A variation on the mixing includes a specific lineage of the virus in the compartment disproportionately and transiently expanding, resulting in clonal expansion or amplification of a subset of viral sequences. In a third scenario, the viral population in a tissue can be distinct from the virus in the blood, indicating an independently replicating population that is not exchanging between the compartments. Examples of these types of genetic relationships are shown in Figure 3 (comparing blood and semen, taken from Anderson et al. 2010). One unifying model is that for many tissues virus is imported by some mechanism from the blood compartment at a low level, giving the appearance of equilibrated populations. Only when local viral replication reaches a level in which it significantly increases the local viral RNA load does it become apparent that there is an independently replicating population that can be recognized as genetically distinct.

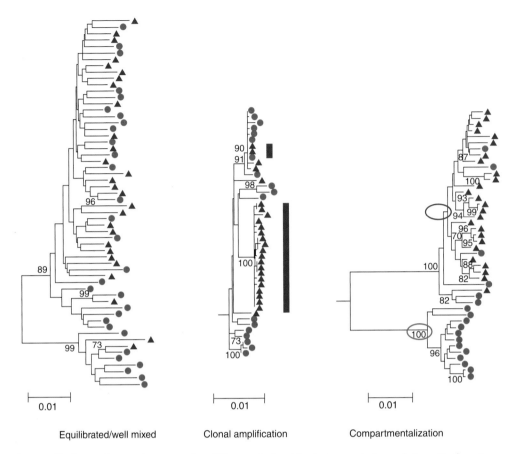

Equilibrated/well mixed Clonal amplification Compartmentalization

Figure 3. Phylogenetic trees demonstrating different relationships between viral populations. In these trees are examples of viral populations in blood (red) and semen (blue). On the *left* is an example of well-mixed populations with the sequences derived from blood and semen intermingled. In the *middle* is an example of clonal amplification (blue bars) in which a nearly homogeneous set of sequences appears only in the semen creating a population that is distinct from the blood. On the *right* is an example of compartmentalization in which the sequences in the semen are distinct from the bulk of the sequences in the blood. These two lineages are indicated by the circles. In addition, sequences present in the seminal tract appear to have migrated back into the blood compartment. (These trees were originally published in Anderson et al. 2010; reprinted with permission from the author.)

Studies of compartmentalization can be confounded by the use of a bulk PCR strategy to generate material for cloning and/or sequencing. The use of cDNA end point dilution followed by multiple PCR amplifications (also called single genome sequencing or single genome amplification) avoids both PCR resampling (which results in artifactual population homogeneity) and PCR-mediated recombination (which confounds the genetic structure of the population) and therefore provides the most reliable source of sequences for the analysis of issues dealing with viral population structure.

Blood and Lymph Nodes

Viral populations in the blood tend to be complex genetically, consistent with a large and diverse population. The rapid decay of viral loads with the initiation of therapy is consistent with 99% of the virus in the blood being produced from short lived cells, presumably activated CD4[+] T cells (Ho et al. 1995; Wei et al.

1995). Genetic variants within this population decay at similar rates, including both X4 and R5 variants, indicating that the virus-producing cells have mostly similar half-lives (Ince et al. 2009). One exception is a small percentage of infected cells that appear to be slow to integrate viral DNA, so that when an integrase inhibitor is used an even larger fraction of the viral load is accounted for in cells with a short half-life (Murray et al. 2007). A small fraction of the virus in the blood decays with much slower kinetics. The significance of these populations is discussed by Coffin and Swanstrom (2011) and Siliciano and Greene (2011).

Within the lymph node the infected $CD4^+$ T cells are found outside of the germinal centers within the paracortex (Haase 2011). In addition, there is a diffuse distribution of virus trapped on follicular dendritic cells throughout the lymph node (Lackner et al. 2011). The high efficiency of virus spread between cells has the potential to generate overlapping foci of clonal infection, resulting in local spread of a genetically homogeneous population (Gratton et al. 2000). In this view the virus creates many independent sites of replication in lymphoid tissue driven by high local concentrations of virus, with the complex population seen in the blood the sum of production from all of these independent foci. This effect could limit the chance for recombination to sites of overlapping foci, or require virus produced at distal sites to colonize new tissues to allow encounter of new recombination partners. However, the findings of recombinants in vivo (discussed in Brown et al. 2011), and the ease with which the entire population turns over during development of drug resistance, indicate there must be significant mixing of viral populations between tissues (Shriner et al. 2006). The similarity of the viral sequences at blood and tissue sites implies that they represent the same well-mixed population, consistent with the observation that CTL escape mutations sequentially move through these compartments (Vanderford et al. 2011).

Ongoing viral replication eventually results in irreversible damage in the capacity to replenish $CD4^+$ T cells. The extent of the damage is related to the length of time of infection (or more appropriately to the nadir of $CD4^+$ T-cell count). This damage is seen by an inability to completely replace the $CD4^+$ T-cell compartment after therapy in initiated (Kelley et al. 2009), which may be due in part to depletion of the host capacity to generate $CD4^+$ T cells over time, or to damage of the architecture of the lymphoid tissue that results in reduced capacity to support $CD4^+$ T-cell function (Lackner et al. 2011).

Gut-Associated Lymphoid Tissue (GALT)

The GALT is the largest lymphoid tissue in the body, containing on the order of one-half or more of the lymphoid cells. The $CD4^+$ T cells in this tissue are rapidly depleted in primary infection, more so than in the blood or other lymphoid sites. Curiously, the cells that are depleted are in more of a resting state rather than an activated state, but none-the-less able to support robust viral replication (Li et al. 1999; Veazey et al. 2000; Mehandru et al. 2007). Massive and early damage to the GALT is proposed to contribute to damage to the intestinal lining, which results in the translocation of bacterial products to the blood where they have the potential to enhance the generalized immune activation that is a central feature of HIV disease (see Lackner et al. 2011). Exposure of an epithelial cell monolayer to virus or purified gp120 can impair the integrity of the monolayer through the induction of inflammatory cytokines, as one potential mechanism of damage (Nazli et al. 2010). As noted above, there is little evidence for compartmentalization of viral sequences between blood and GALT, implying frequent exchange of virus or infected cells between these sites (Vanderford et al. 2011). Viral RNA can be detected in rectal secretions (Zuckerman et al. 2004), however, this virus has not been extensively studied in terms of compartmentalization or equilibration with the blood population of virus.

CNS

Virus can be detected in the CSF early after infection and in many subjects throughout infection (Spudich and Gonzalez-Scarano 2011).

However, it is not clear if low levels of virus can move (by an unknown mechanism) into the CSF, or if detection of virus in CSF always means there is ongoing viral replication in the CNS itself. Compartmentalized virus, distinct from that in blood, can be detected in the CSF (and in brain tissue at autopsy) late in disease associated with dementia, providing clear evidence for independent replication in the CNS. The compartmentalized virus can involve either macrophage-tropic or T-cell-tropic (R5) virus, which suggests a complex interaction between HIV-1 and CNS tissue (Schnell et al. 2011). An important question in understanding HIV-1 pathogenesis is determining when independent replication can occur in the CNS because such replication is likely to have an associated pathogenic outcome. A compartmentalized CSF population has been detected early in infection associated with a diagnosis of aseptic meningitis, indicating the potential for HIV-1 to establish an independently replicating population of virus associated with a pathogenic outcome even early after infection (Schnell et al. 2010). Compartmentalization has been detected in chronic infection in the presence or absence of neurocognitive impairment (Harrington et al. 2009). Viral replication in the CNS will remain an important consideration of viral evolution given the role of this tissue in pathogenesis.

Genital Tract

There is special interest in virus in the male genital tract (MGT) and in the female genital tract (FGT) as these are the sites involved in sexual transmission. Virus is present in both seminal plasma and FGT mucus in the absence of blood contamination, indicating there must be mechanisms for introduction of virus into genital secretions. A key question is whether genital tract virus is equilibrated with the virus in the blood or is genetically distinct, indicating local sites of viral production.

The MGT contains potential target cells at several anatomical sites (Le Tortorec et al. 2008). It has been reported that viral load does not change after vasectomy, suggesting

that virus is not produced in large amounts in the testes (Anderson et al. 1991; Krieger et al. 1998). However, this observation has not been tied to subjects who had evidence of local production of virus in the MGT so the lack of involvement of the testes is clearest for virus that is likely equilibrated with the blood. Virus can be produced locally within the MGT because in a subset of men the virus in semen can be genetically distinct from the virus in the blood (Delwart et al. 1998; Paranjpe et al. 2002; Diem et al. 2008; Anderson et al. 2010). There appear to be two related mechanisms at work in local production of virus. In one there is sustained independent replication to generate a population that is distinct from that in the blood. In the other, there is a rapid short-term expansion of a relatively homogeneous variant that becomes a significant portion of the local population. This pattern of clonal amplification has been observed in the background of virus that is otherwise similar to that in blood or of virus compartmentalized within the MGT (Anderson et al. 2010). Given the limited complexity of the clonally amplified population, it is assumed that this expansion occurs over a short period of time but is not sustained, although the longitudinal relationship between clonal amplification and the long term compartmentalization of a genetically distinct population is not clear.

Virus production in the FGT occurs in the context of a thick layer of squamous epithelium in the vagina and a single cell layer of columnar epithelium starting at the cervix. There is no obvious single tissue mass (equivalent to the prostate, for example) that could provide a source of localized virus production. Several studies have reported compartmentalization of viral sequences in the FMT within at least a subset of subjects (Poss et al. 1998; Kemal et al. 2003; Adal et al. 2005; Andreoletti et al. 2007). More recently, this compartmentalization was interpreted to be largely the result of clonal amplification (Bull et al. 2009). There is a need for both longitudinal analysis of viral populations within the genital tract, and an examination of the impact of concurrent localized bacterial infections with attendant inflammation.

 Cite this article as *Cold Spring Harb Perspect Med* doi: 10.1101/cshperspect.a007443

Other Cell and Tissue Types

Long-term infection with HIV-1 can have a variety of clinical manifestations in additional organ systems. The ongoing debate is over the extent to which they reflect infection of cells in the affected organ, dysfunction of cells in the organ caused by interaction with soluble viral proteins, or indirect pathogenesis because of concurrent systemic changes such as constitutive immune activation. As discussed above, HIV-1 evolves to infect cells using a different coreceptor (X4 tropism) and cells with low levels of CD4 (macrophage tropism). Phenotypic and genotypic evidence for viral evolution to infect additional cell types is much less convincing compared to CXCR4 tropism and macrophage tropism. It is possible that alternative cell types are fortuitously infected at a low level in vivo, but it is not clear that infection in these cells can be independently sustained, which would likely be accompanied by some type of cell-specific viral evolution. It is also possible that such evolution goes on in one or more tissue types but signatures of this process have thus far evaded detection. Similarly, the potential pathogenic role of soluble viral proteins is limited by the level of protein in vivo that reaches the target cell or tissue. However, given the difficulty is sampling tissues beyond blood, much of the work addressing these questions has been performed in cell culture systems either by infection or by exposure to purified viral proteins; in some cases this approach has been complemented with work using mice transgenic for HIV-1 genes.

Kidney: HIV-associated nephropathy can play a significant role in end-stage renal disease in infected people (Medapalli et al. 2011). Infection of renal epithelial cells has been frequently invoked to explain this condition. HIV-1 sequences have been detected in these cells (Bruggeman et al. 2000; Marras et al. 2002). The viral Nef protein has been implicated in dysregulation of cell function through interaction with cellular signaling cascades (Husain et al. 2005).

Liver: Another potential site where HIV-1 could have a pathogenic effect is the liver, in which Kupffer cells represent at least 10% of the total cells and are of the macrophage lineage. It is clear that HIV-1 infection can substantially accelerate liver disease in patients coinfected with HIV-1 and HCV (Macias et al. 2009), although it is not known if this is because of general immunodeficiency or local HIV-1 replication. Viral DNA, RNA, and protein have been detected in liver samples from HIV-1-infected subjects (Housset et al. 1990; Cao et al. 1992; van't Wout et al. 1998). In cell culture systems, HIV-1 has been reported to infect Kupffer cells (Schmitt et al. 1990) and stellate cells (Tuyama et al. 2010). HIV-1 infection of stellate cells may contribute to liver fibrosis by promoting collagen I expression and secretion of the proinflammatory cytokine monocyte chemoattractant protein-1 (Tuyama et al. 2010). Another proposed pathogenic mechanism is induction of apoptosis in hepatocytes after exposure to HCV E2 and HIV gp120 (Munshi et al. 2003; Vlahakis et al. 2003).

Lung: The possibility of local infection in the lung is of interest given the importance of the lung as the site of opportunistic infections and pulmonary TB. Both $CD4^+$ T cells and alveolar macrophages are potential target cells in the lung. The isolation of the macrophage-tropic strain Ba-L from a lavage sample (Gartner et al. 1986) is consistent with the potential for viral replication and evolution in this compartment. Initial reports suggested that independent replication can occur in the lung (Itescu et al. 1994; Nakata et al. 1995). However, more recent analysis of viral populations in blood and lung showed only modest evidence for compartmentalization (including local clonal expansion), representing a relationship that suggests mixing between the R5 T-cell-tropic viruses in the blood and the lung compartment (Heath et al. 2009). One limitation in the work to date in defining the genetics of virus in the lung is that it has not focused on later stage infections when expansion of virus into new cell types, particularly macrophages, is more likely to be occurring, such as during severe immunodeficiency (Jeffrey et al. 1991).

Hematopoietic stem cells: The potential for infection of HSC was shown many years ago (Folks et al. 1988; Stanley et al. 1992), although they are relatively difficult to infect in culture (Shen et al. 1999; Zhang et al. 2007). More recently, interest in this phenomenon has focused on potential pathogenic outcomes such as anemia (Redd et al. 2007), or as a source of long-lived virus making up part of the HIV-1 latent reservoir (Carter et al. 2010). For subtype B HIV-1 the ability to infect a multipotent cell appears to be restricted to X4 viruses (Carter et al. 2011).

Breast: Viral dissemination into breast tissue is relevant in the context of transmission during breast feeding. Recent work comparing virus in blood and breast milk has indicated that compartmentalization is largely limited to clonal amplification (Gantt et al. 2010; Salazar-Gonzalez et al. 2011). It is possible that milk production results in a flushing effect of this compartment that limits the opportunity for sustained localized replication.

MECHANISM OF CELL KILLING

The mechanisms by which an infected cell dies remain controversial. In culture, infected $CD4^+$ T cells die in a matter of days, whereas infected monocyte-derived macrophages can produce virus over a period of weeks, indicating there is nothing inherent about viral replication that leads to cell death. Indeed many retroviruses infect cells without killing them. There must be specific features of the virus and its interaction with the host cell that are responsible for cell killing. The ability to kill $CD4^+$ T cells in culture is consistent with the short half-life of virus in the blood after the initiation of therapy (Ho et al. 1995; Wei et al. 1995). That this is in large part a virologic effect is seen by the fact that the decay of virus in the blood goes on at a similar rate in the presence or absence of $CD8^+$ cytotoxic T cells (Klatt et al. 2010). In addition to the direct killing of infected cells, a number of mechanisms have been proposed that result in indirect killing of uninfected cells because of their proximity to infected cells and the general state of immune activation (Lackner et al. 2011).

Many viral proteins have been implicated as participating in cellular apoptotic (and anti-apoptotic) pathways. Expression of Env protein is toxic to the infected cell and surface expression can mediate syncytium formation or interaction with chemokine receptors on adjacent cells. However, other viral proteins have also been implicated in several cellular apoptotic pathways (Varbanov et al. 2006). Recently, a proteolytic fragment of Caspase 8, generated by the viral protease, has been shown to induce apoptosis through Caspase 9 activation (Sainski et al. 2011). Also, partial viral DNA products have been implicated in a cell killing mechanism of abortive infection that may contribute to apparent indirect killing (Zhou et al. 2008; Doitsh et al. 2010). It is clear that during acute infection massive loss of $CD4^+$ T cells is largely the result of direct killing of infected cells (Li et al. 2005; Mattapallil et al. 2005). However, the mix of direct killing versus indirect killing over the course of the infection and the range of mechanisms involved remain under discussion.

SUMMARY

Viral dissemination and evolution are central to HIV pathogenesis and the constantly changing virus-host interaction. Evidence for evolution to infect new cell types is strongest for X4 viruses and macrophage-tropic viruses, but even in these cases their contributions to viral pathogenesis are incompletely understood. Many additional mechanisms of viral pathogenesis have been proposed and will continue to be explored. The evolution of virus to form distinct replicating populations, immune escape, or distinct entry phenotypes has provided the most compelling evidence for the virus responding to a changing virus-host environment. Linking viral biology to genetic markers of viral evolution will continue to provide a strong experimental paradigm for viral pathogenesis.

ACKNOWLEDGMENTS

We would like to thank Elena Dukhovlinova and Sarah Joseph for assistance with specific sections. We also thank Elizabeth Pollom for Figure 1 and William Ince for Figure 2. The authors' own research has been supported by funding from the National Institutes of Health.

REFERENCES

Reference is also in this collection.

Adal M, Ayele W, Wolday D, Dagne K, Messele T, Tilahun T, Berkhout B, Mayaan S, Pollakis G, Dorigo-Zetsma W. 2005. Evidence of genetic variability of human immunodeficiency virus type 1 in plasma and cervicovaginal lavage in Ethiopian women seeking care for sexually transmitted infections. *AIDS Res Hum Retroviruses* **21**: 649–653.

Agace WW, Roberts AI, Wu L, Greineder C, Ebert EC, Parker CM. 2000. Human intestinal lamina propria and intraepithelial lymphocytes express receptors specific for chemokines induced by inflammation. *Eur J Immunol* **30**: 819–826.

Alexander M, Lynch R, Mulenga J, Allen S, Derdeyn CA, Hunter E. 2010. Donor and recipient envs from heterosexual human immunodeficiency virus subtype C transmission pairs require high receptor levels for entry. *J Virol* **84**: 4100–4104.

Anderson DJ, Politch JA, Martinez A, Van Voorhis BJ, Padian NS, O'Brien TR. 1991. White blood cells and HIV-1 in semen from vasectomised seropositive men. *Lancet* **338**: 573–574.

Anderson JA, Ping LH, Dibben O, Jabara CB, Arney L, Kincer L, Tang Y, Hobbs M, Hoffman I, Kazembe P, et al. 2010. HIV-1 Populations in Semen Arise through Multiple Mechanisms. *PLoS Pathog* **6**: e1001053.

Andreoletti L, Skrabal K, Perrin V, Chomont N, Saragosti S, Gresenguet G, Moret H, Jacques J, Longo Jde D, Matta M, et al. 2007. Genetic and phenotypic features of blood and genital viral populations of clinically asymptomatic and antiretroviral-treatment-naive clade a human immunodeficiency virus type 1-infected women. *J Clin Microbiol* **45**: 1838–1842.

Anton PA, Elliott J, Poles MA, McGowan IM, Matud J, Hultin LE, Grovit-Ferbas K, Mackay CR, Chen ISY, Giorgi JV. 2000. Enhanced levels of functional HIV-1 co-receptors on human mucosal T cells demonstrated using intestinal biopsy tissue. *AIDS* **14**: 1761–1765.

Arfi V, Riviere L, Jarrosson-Wuilleme L, Goujon C, Rigal D, Darlix JL, Cimarelli A. 2008. Characterization of the early steps of infection of primary blood monocytes by human immunodeficiency virus type 1. *J Virol* **82**: 6557–6565.

Asjo B, Morfeldt-Manson L, Albert J, Biberfeld G, Karlsson A, Lidman K, Fenyo EM. 1986. Replicative capacity of human immunodeficiency virus from patients with varying severity of HIV infection. *Lancet* **2**: 660–662.

Bernstein HB, Wang G, Plasterer MC, Zack JA, Ramasastry P, Mumenthaler SM, Kitchen CM. 2009. CD4$^+$ NK cells can be productively infected with HIV, leading to downregulation of CD4 expression and changes in function. *Virology* **387**: 59–66.

Blaak H, van't Wout AB, Brouwer M, Hooibrink B, Hovenkamp E, Schuitemaker H. 2000. In vivo HIV-1 infection of CD45RA$^+$CD4$^+$ T cells is established primarily by syncytium-inducing variants and correlates with the rate of CD4$^+$ T cell decline. *Proc Natl Acad Sci* **97**: 1269–1274.

Bleul CC, Wu L, Hoxie JA, Springer TA, Mackay CR. 1997. The HIV coreceptors CXCR4 and CCR5 are differentially

expressed and regulated on human T lymphocytes. *Proc Natl Acad Sci* **94**: 1925–1930.

Boily MC, Baggaley RF, Wang L, Masse B, White RG, Hayes RJ, Alary M. 2009. Heterosexual risk of HIV-1 infection per sexual act: Systematic review and meta-analysis of observational studies. *Lancet Infect Dis* **9**: 118–129.

Boily-Larouche G, Iscache AL, Zijenah LS, Humphrey JH, Mouland AJ, Ward BJ, Roger M. 2009. Functional genetic variants in DC-SIGNR are associated with mother-to-child transmission of HIV-1. *PLoS One* **4**: e7211.

Brenchley JM, Hill BJ, Ambrozak DR, Price DA, Guenaga FJ, Casazza JP, Kuruppu J, Yazdani J, Migueles SA, Connors M, et al. 2004a. T-cell subsets that harbor human immunodeficiency virus (HIV) in vivo: Implications for HIV pathogenesis. *J Virol* **78**: 1160–1168.

Brenchley JM, Schacker TW, Ruff LE, Price DA, Taylor JH, Beilman GJ, Nguyen PL, Khoruts A, Larson M, Haase AT, et al. 2004b. CD4$^+$ T cell depletion during all stages of HIV disease occurs predominantly in the gastrointestinal tract. *J Exp Med* **200**: 749–759.

Brown RJ, Peters PJ, Caron C, Gonzalez-Perez MP, Stones L, Ankghuambom C, Pondei K, McClure CP, Alemnji G, Taylor S, et al. 2011. Intercompartmental recombination of HIV-1 contributes to env intrahost diversity and modulates viral tropism and sensitivity to entry inhibitors. *J Virol* **85**: 6024–6037.

Bruggeman LA, Ross MD, Tanji N, Cara A, Dikman S, Gordon RE, Burns GC, D'Agati VD, Winston JA, Klotman ME, et al. 2000. Renal epithelium is a previously unrecognized site of HIV-1 infection. *J Am Soc Nephrol* **11**: 2079–2087.

Brumme ZL, Goodrich J, Mayer HB, Brumme CJ, Henrick BM, Wynhoven B, Asselin JJ, Cheung PK, Hogg RS, Montaner JS, et al. 2005. Molecular and clinical epidemiology of CXCR4-using HIV-1 in a large population of antiretroviral-naive individuals. *J Infect Dis* **192**: 466–474.

Bull M, Learn G, Genowati I, McKernan J, Hitti J, Lockhart D, Tapia K, Holte S, Dragavon J, Coombs R, et al. 2009. Compartmentalization of HIV-1 within the female genital tract is due to monotypic and low-diversity variants not distinct viral populations. *PLoS One* **4**: e7122.

Cao YZ, Dieterich D, Thomas PA, Huang YX, Mirabile M, Ho DD. 1992. Identification and quantitation of HIV-1 in the liver of patients with AIDS. *AIDS* **6**: 65–70.

Carter CC, Onafuwa-Nuga A, McNamara LA, Riddell JT, Bixby D, Savona MR, Collins KL. 2010. HIV-1 infects multipotent progenitor cells causing cell death and establishing latent cellular reservoirs. *Nat Med* **16**: 446–451.

Carter CC, McNamara LA, Onafuwa-Nuga A, Shackleton M, Riddell J 4th, Bixby D, Savona MR, Morrison SJ, Collins KL. 2011. HIV-1 utilizes the CXCR4 chemokine receptor to infect multipotent hematopoietic stem and progenitor cells. *Cell Host Microbe* **9**: 223–234.

Centlivre M, Legrand N, Steingrover R, van der Sluis R, Grijsen ML, Bakker M, Jurriaans S, Berkhout B, Paxton WA, Prins JM, et al. 2011. Altered dynamics and differential infection profiles of lymphoid and myeloid cell subsets during acute and chronic HIV-1 infection. *J Leukoc Biol* **89**: 785–795.

Cheng-Mayer C, Tasca S, Ho SH. 2009. Coreceptor switch in infection of nonhuman primates. *Curr HIV Res* **7**: 30–38.

Cinque P, Presi S, Bestetti A, Pierotti C, Racca S, Boeri E, Morelli P, Carrera P, Ferrari M, Lazzarin A. 2001. Effect of genotypic resistance on the virological response to highly active antiretroviral therapy in cerebrospinal fluid. *AIDS Res Hum Retroviruses* 17: 377–383.

Cochrane A, Imlach S, Leen C, Scott G, Kennedy D, Simmonds P. 2004. High levels of human immunodeficiency virus infection of CD8 lymphocytes expressing CD4 in vivo. *J Virol* 78: 9862–9871.

Coetzer M, Cilliers T, Ping LH, Swanstrom R, Morris L. 2006. Genetic characteristics of the V3 region associated with CXCR4 usage in HIV-1 subtype C isolates. *Virology* 356: 95–105.

Coetzer M, Nedellec R, Cilliers T, Meyers T, Morris L, Mosier DE. 2011. Extreme genetic divergence is required for coreceptor switching in HIV-1 subtype C. *J Acquir Immune Defic Syndr* 56: 9–15.

* Coffin L, Swanstrom R. 2011. HIV pathogenesis: Dynamics and genetics of viral populations and infected cells. *Cold Spring Harb Perspect Med* doi: 10.1101/cshperspect. a012526.

Delwart EL, Mullins JI, Gupta P, Learn GH Jr, Holodniy M, Katzenstein D, Walker BD, Singh MK. 1998. Human immunodeficiency virus type 1 populations in blood and semen. *J Virol* 72: 617–623.

Diem K, Nickle DC, Motoshige A, Fox A, Ross S, Mullins JI, Corey L, Coombs RW, Krieger JN. 2008. Male genital tract compartmentalization of human immunodeficiency virus type 1 (HIV). *AIDS Res Hum Retroviruses* 24: 561–571.

Doitsh G, Cavrois M, Lassen KG, Zepeda O, Yang Z, Santiago ML, Hebbeler AM, Greene WC. 2010. Abortive HIV infection mediates CD4 T cell depletion and inflammation in human lymphoid tissue. *Cell* 143: 789–801.

Dong C, Kwas C, Wu L. 2009. Transcriptional restriction of human immunodeficiency virus type 1 gene expression in undifferentiated primary monocytes. *J Virol* 83: 3518–3527.

Doores KJ, Bonomelli C, Harvey DJ, Vasiljevic S, Dwek RA, Burton DR, Crispin M, Scanlan CN. 2010. Envelope glycans of immunodeficiency virions are almost entirely oligomannose antigens. *Proc Natl Acad Sci* 107: 13800–13805.

Douek DC, Brenchley JM, Betts MR, Ambrozak DR, Hill BJ, Okamoto Y, Casazza JP, Kuruppu J, Kunstman K, Wolinsky S, et al. 2002. HIV preferentially infects HIV-specific CD4+ T cells. *Nature* 417: 95–98.

Ellery PJ, Tippett E, Chiu YL, Paukovics G, Cameron PU, Solomon A, Lewin SR, Gorry PR, Jaworowski A, Greene WC, et al. 2007. The CD16+ monocyte subset is more permissive to infection and preferentially harbors HIV-1 in vivo. *J Immunol* 178: 6581–6589.

Ellis RJ, Gamst AC, Capparelli E, Spector SA, Hsia K, Wolfson T, Abramson I, Grant I, McCutchan JA. 2000. Cerebrospinal fluid HIV RNA originates from both local CNS and systemic sources. *Neurology* 54: 927–936.

Folks TM, Kessler SW, Orenstein JM, Justement JS, Jaffe ES, Fauci AS. 1988. Infection and replication of HIV-1 in purified progenitor cells of normal human bone marrow. *Science* 242: 919–922.

Fulcher JA, Hwangbo Y, Zioni R, Nickle D, Lin X, Heath L, Mullins JI, Corey L, Zhu T. 2004. Compartmentalization

of human immunodeficiency virus type 1 between blood monocytes and CD4+ T cells during infection. *J Virol* 78: 7883–7893.

Gantt S, Carlsson J, Heath L, Bull ME, Shetty AK, Mutsvangwa J, Musingwini G, Woelk G, Zijenah LS, Katzenstein DA, et al. 2010. Genetic analyses of HIV-1 env sequences demonstrate limited compartmentalization in breast milk and suggest viral replication within the breast that increases with mastitis. *J Virol* 84: 10812–10819.

Gartner S, Markovits P, Markovitz DM, Kaplan MH, Gallo RC, Popovic M. 1986. The role of mononuclear phagocytes in HTLV-III/LAV infection. *Science* 233: 215–219.

Gorry PR, Taylor J, Holm GH, Mehle A, Morgan T, Cayabyab M, Farzan M, Wang H, Bell JE, Kunstman K, et al. 2002. Increased CCR5 affinity and reduced CCR5/CD4 dependence of a neurovirulent primary human immunodeficiency virus type 1 isolate. *J Virol* 76: 6277–6292.

Gratton S, Cheynier R, Dumaurier MJ, Oksenhendler E, Wain-Hobson S. 2000. Highly restricted spread of HIV-1 and multiply infected cells within splenic germinal centers. *Proc Natl Acad Sci* 97: 14566–14571.

Haas DW, Clough LA, Johnson BW, Harris VL, Spearman P, Wilkinson GR, Fletcher CV, Fiscus S, Raffanti S, Donlon R, et al. 2000. Evidence of a source of HIV type 1 within the central nervous system by ultraintensive sampling of cerebrospinal fluid and plasma. *AIDS Res Hum Retroviruses* 16: 1491–1502.

Haase AT. 2011. Early events in sexual transmission of HIV and SIV and opportunities for interventions. *Annu Rev Med* 62: 127–139.

Harada H, Goto Y, Ohno T, Suzu S, Okada S. 2007. Proliferative activation up-regulates expression of CD4 and HIV-1 co-receptors on NK cells and induces their infection with HIV-1. *Eur J Immunol* 37: 2148–2155.

Harrington PR, Schnell G, Letendre SL, Ritola K, Robertson K, Hall C, Burch CL, Jabara CB, Moore DT, Ellis RJ, et al. 2009. Cross-sectional characterization of HIV-1 env compartmentalization in cerebrospinal fluid over the full disease course. *AIDS* 23: 907–915.

Heath L, Fox A, McClure J, Diem K, van 't Wout AB, Zhao H, Park DR, Schouten JT, Twigg HL, Corey L, Mullins JI, Mittler JE. 2009. Evidence for limited genetic compartmentalization of HIV-1 between lung and blood. *PLoS One* 4: e6949.

Heeregrave EJ, Geels MJ, Brenchley JM, Baan E, Ambrozak DR, van der Sluis RM, Bennemeer R, Douek DC, Goudsmit J, Pollakis G, et al. 2009. Lack of in vivo compartmentalization among HIV-1 infected naive and memory CD4+ T cell subsets. *Virology* 393: 24–32.

Ho DD, Neumann AU, Perelson AS, Chen W, Leonard JM, Markowitz M. 1995. Rapid turnover of plasma virions and CD4 lymphocytes in HIV-1 infection. *Nature* 373: 123–126.

Hoffman NG, Seillier-Moiseiwitsch F, Ahn J, Walker JM, Swanstrom R. 2002. Variability in the human immunodeficiency virus type 1 gp120 Env protein linked to phenotype-associated changes in the V3 loop. *J Virol* 76: 3852–3864.

Housset C, Lamas E, Brechot C. 1990. Detection of HIV1 RNA and pp24 antigen in HIV1-infected human liver. *Res Virol* 141: 153–159.

Huang W, Eshleman SH, Toma J, Fransen S, Stawiski E, Paxinos EE, Whitcomb JM, Young AM, Donnell D, Mmiro F, et al. 2007. Coreceptor tropism in human immunodeficiency virus type 1 subtype D: High prevalence of CXCR4 tropism and heterogeneous composition of viral populations. *J Virol* **81:** 7885–7893.

Huang W, Toma J, Fransen S, Stawiski E, Reeves JD, Whitcomb JM, Parkin N, Petropoulos CJ. 2008. Coreceptor tropism can be influenced by amino acid substitutions in the gp41 transmembrane subunit of human immunodeficiency virus type 1 envelope protein. *J Virol* **82:** 5584–5593.

Huang W, Frantzell A, Toma J, Fransen S, Whitcomb JM, Stawiski E, Petropoulos CJ. 2011. Mutational pathways and genetic barriers to CXCR4-mediated entry by human immunodeficiency virus type 1. *Virology* **409:** 308–318.

Hunt PW, Harrigan PR, Huang W, Bates M, Williamson DW, McCune JM, Price RW, Spudich SS, Lampiris H, Hoh R, et al. 2006. Prevalence of CXCR4 tropism among antiretroviral-treated HIV-1-infected patients with detectable viremia. *J Infect Dis* **194:** 926–930.

Husain M, D'Agati VD, He JC, Klotman ME, Klotman PE. 2005. HIV-1 Nef induces dedifferentiation of podocytes in vivo: A characteristic feature of HIVAN. *AIDS* **19:** 1975–1980.

Igarashi T, Brown CR, Endo Y, Buckler-White A, Plishka R, Bischofberger N, Hirsch V, Martin MA. 2001. Macrophage are the principal reservoir and sustain high virus loads in rhesus macaques after the depletion of CD4$^+$ T cells by a highly pathogenic simian immunodeficiency virus/HIV type 1 chimera (SHIV): Implications for HIV-1 infections of humans. *Proc Natl Acad Sci* **98:** 658–663.

Ince WL, Harrington PR, Schnell GL, Patel-Chhabra M, Burch CL, Menezes P, Price RW, Eron JJ Jr, Swanstrom RI. 2009. Major coexisting human immunodeficiency virus type 1 env gene subpopulations in the peripheral blood are produced by cells with similar turnover rates and show little evidence of genetic compartmentalization. *J Virol* **83:** 4068–4080.

Ince WL, Zhang L, Jiang Q, Arrildt K, Su L, Swanstrom R. 2010. Evolution of the HIV-1 *env* gene in the Raγ2$^{-/-}$ $\gamma_C^{-/-}$ humanized mouse model. *J Virol* **84:** 2740–2752.

Isaacman-Beck J, Hermann EA, Yi Y, Ratcliffe SJ, Mulenga J, Allen S, Hunter E, Derdeyn CA, Collman RG. 2009. Heterosexual transmission of human immunodeficiency virus type 1 subtype C: Macrophage tropism, alternative coreceptor use, and the molecular anatomy of CCR5 utilization. *J Virol* **83:** 8208–8220.

Itescu S, Simonelli PF, Winchester RJ, Ginsberg HS. 1994. Human immunodeficiency virus type 1 strains in the lungs of infected individuals evolve independently from those in peripheral blood and are highly conserved in the C-terminal region of the envelope V3 loop. *Proc Natl Acad Sci* **91:** 11378–11382.

Jeffrey AA, Israel-Biet D, Andrieu JM, Even P, Venet A. 1991. HIV isolation from pulmonary cells derived from bronchoalveolar lavage. *Clin Exp Immunol* **84:** 488–492.

Josefsson L, King MS, Makitalo B, Brönnström H, Shao W, Maldarelli F, Kearney MF, Hu WS, Chen J, Gaines H, et al. 2011. Majority of CD4$^+$ T cells from peripheral blood of HIV-1-infected individuals contain only one HIV DNA molecule. *Proc Natl Acad Sci* **108:** 11199–11204.

Kaiser P, Joos B, Niederost B, Weber R, Gunthard HF, Fischer M. 2007. Productive human immunodeficiency virus type 1 infection in peripheral blood predominantly takes place in CD4/CD8 double-negative T lymphocytes. *J Virol* **81:** 9693–9706.

Kelley CF, Kitchen CM, Hunt PW, Rodriguez B, Hecht FM, Kitahata M, Crane HM, Willig J, Mugavero M, Saag M, et al. 2009. Incomplete peripheral CD4$^+$ cell count restoration in HIV-infected patients receiving long-term antiretroviral treatment. *Clin Infect Dis* **48:** 787–794.

Kemal KS, Foley B, Burger H, Anastos K, Minkoff H, Kitchen C, Philpott SM, Gao W, Robison E, Holman S, et al. 2003. HIV-1 in genital tract and plasma of women: Compartmentalization of viral sequences, coreceptor usage, and glycosylation. *Proc Natl Acad Sci* **100:** 12972–12977.

Kim EY, Busch M, Abel K, Fritts L, Bustamante P, Stanton J, Lu D, Wu S, Glowczwskie J, Rourke T, et al. 2005. Retroviral recombination in vivo: Viral replication patterns and genetic structure of simian immunodeficiency virus (SIV) populations in rhesus macaques after simultaneous or sequential intravaginal inoculation with SIVmac239δvpx/δvpr and SIVmac239δnef. *J Virol* **79:** 4886–4895.

Klatt NR, Shudo E, Ortiz AM, Engram JC, Paiardini M, Lawson B, Miller MD, Else J, Pandrea I, Estes JD, et al. 2010. CD8$^+$ lymphocytes control viral replication in SIVmac239-infected rhesus macaques without decreasing the lifespan of productively infected cells. *PLoS Pathog* **6:** e1000747.

Korin YD, Zack JA. 1998. Progression to the G1b phase of the cell cycle is required for completion of human immunodeficiency virus type 1 reverse transcription in T cells. *J Virol* **72:** 3161–3168.

Krieger JN, Nirapathpongporn A, Chaiyaporn M, Peterson G, Nikolaeva I, Akridge R, Ross SO, Coombs RW. 1998. Vasectomy and human immunodeficiency virus type 1 in semen. *J Urol* **159:** 820–825; discussion 825–826.

* Lackner AA, Lederman MM, Rodriguez B. HIV pathogenesis—The host. *Cold Spring Harb Perspect Med* doi: 10.1101/cshperspect.a007005.

Le Tortorec A, Satie AP, Denis H, Rioux-Leclercq N, Havard L, Ruffault A, Jegou B, Dejucq-Rainsford N. 2008. Human prostate supports more efficient replication of HIV-1 R5 than X4 strains ex vivo. *Retrovirology* **5:** 119.

Lee B, Sharron M, Montaner LJ, Weissman D, Doms RW. 1999. Quantification of CD4, CCR5, and CXCR4 levels on lymphocyte subsets, dendritic cells, and differentially conditioned monocyte-derived macrophages. *Proc Natl Acad Sci* **96:** 5215–5220.

Li S, Juarez J, Alali M, Dwyer D, Collman R, Cunningham A, Naif HM. 1999. Persistent CCR5 utilization and enhanced macrophage tropism by primary blood human immunodeficiency virus type 1 isolates from advanced stages of disease and comparison to tissue-derived isolates. *J Virol* **73:** 9741–9755.

Li Q, Duan L, Estes JD, Ma ZM, Rourke T, Wang Y, Reilly C, Carlis J, Miller CJ, Haase AT. 2005. Peak SIV replication in resting memory CD4$^+$ T cells depletes gut lamina propria CD4$^+$ T cells. *Nature* **434:** 1148–1152.

Liu Y, Cruikshank WW, O'Loughlin T, O'Reilly P, Center DM, Kornfeld H. 1999. Identification of a CD4 domain required for interleukin-16 binding and lymphocyte activation. *J Biol Chem* **274:** 23387–23395.

Macias J, Berenguer J, Japon MA, Giron JA, Rivero A, Lopez-Cortes LF, Moreno A, Gonzalez-Serrano M, Iribarren JA, Ortega E, et al. 2009. Fast fibrosis progression between repeated liver biopsies in patients coinfected with human immunodeficiency virus/hepatitis C virus. *Hepatology* **50:** 1056–1063.

Marras D, Bruggeman LA, Gao F, Tanji N, Mansukhani MM, Cara A, Ross MD, Gusella GL, Benson G, D'Agati VD, et al. 2002. Replication and compartmentalization of HIV-1 in kidney epithelium of patients with HIV-associated nephropathy. *Nat Med* **8:** 522–526.

Martin MP, Lederman MM, Hutcheson HB, Goedert JJ, Nelson GW, van Kooyk Y, Detels R, Buchbinder S, Hoots K, Vlahov D, et al. 2004. Association of DC-SIGN promoter polymorphism with increased risk for parenteral, but not mucosal, acquisition of human immunodeficiency virus type 1 infection. *J Virol* **78:** 14053–14056.

Martin-Garcia J, Cao W, Varela-Rohena A, Plassmeyer ML, Gonzalez-Scarano F. 2006. HIV-1 tropism for the central nervous system: Brain-derived envelope glycoproteins with lower CD4 dependence and reduced sensitivity to a fusion inhibitor. *Virology* **346:** 169–179.

Mattapallil JJ, Douek DC, Hill B, Nishimura Y, Martin M, Roederer M. 2005. Massive infection and loss of memory CD4$^+$ T cells in multiple tissues during acute SIV infection. *Nature* **434:** 1093–1097.

Medapalli RK, He JC, Klotman PE. 2011. HIV-associated nephropathy: Pathogenesis. *Curr Opin Nephrol Hypertens* **20:** 306–311.

Mehandru S, Poles MA, Tenner-Racz K, Manuelli V, Jean-Pierre P, Lopez P, Shet A, Low A, Mohri H, Boden D, et al. 2007. Mechanisms of gastrointestinal CD4$^+$ T-cell depletion during acute and early human immunodeficiency virus type 1 infection. *J Virol* **81:** 599–612.

Miller CJ, Li Q, Abel K, Kim EY, Ma ZM, Wietgrefe S, La Franco-Scheuch L, Compton L, Duan L, Shore MD, et al. 2005. Propagation and dissemination of infection after vaginal transmission of simian immunodeficiency virus. *J Virol* **79:** 9217–9227.

Munshi N, Balasubramanian A, Koziel M, Ganju RK, Groopman JE. 2003. Hepatitis C and human immunodeficiency virus envelope proteins cooperatively induce hepatocytic apoptosis via an innocent bystander mechanism. *J Infect Dis* **188:** 1192–1204.

Murray JM, Emery S, Kelleher AD, Law M, Chen J, Hazuda DJ, Nguyen BY, Teppler H, Cooper DA. 2007. Antiretroviral therapy with the integrase inhibitor raltegravir alters decay kinetics of HIV, significantly reducing the second phase. *AIDS* **21:** 2315–2321.

Nakata K, Weiden M, Harkin T, Ho D, Rom WN. 1995. Low copy number and limited variability of proviral DNA in alveolar macrophages from HIV-1-infected patients: Evidence for genetic differences in HIV-1 between lung and blood macrophage populations. *Mol Med* **1:** 744–757.

Nawaz F, Cicala C, Van Ryk D, Block KE, Jelicic K, McNally JP, Ogundare O, Pascuccio M, Patel N, Wei D, et al. 2011. The genotype of early-transmitting HIV gp120s promotes αβ-reactivity, revealing αβCD4$^+$ T cells as

key targets in mucosal transmission. *PLoS Pathog* **7:** e1001301.

Nazli A, Chan O, Dobson-Belaire WN, Ouellet M, Tremblay MJ, Gray-Owen SD, Arsenault AL, Kaushic C. 2010. Exposure to HIV-1 directly impairs mucosal epithelial barrier integrity allowing microbial translocation. *PLoS Pathog* **6:** e1000852.

Nicholson JK, Browning SW, Hengel RL, Lew E, Gallagher LE, Rimland D, McDougal JS. 2001. CCR5 and CXCR4 expression on memory and naive T cells in HIV-1 infection and response to highly active antiretroviral therapy. *J Acquir Immune Defic Syndr* **27:** 105–115.

Ostrowski MA, Justement SJ, Catanzaro A, Hallahan CA, Ehler LA, Mizell SB, Kumar PN, Mican JA, Chun TW, Fauci AS. 1998. Expression of chemokine receptors CXCR4 and CCR5 in HIV-1-infected and uninfected individuals. *J Immunol* **161:** 3195–3201.

Paranjpe S, Craigo J, Patterson B, Ding M, Barroso P, Harrison L, Montelaro R, Gupta P. 2002. Subcompartmentalization of HIV-1 quasispecies between seminal cells and seminal plasma indicates their origin in distinct genital tissues. *AIDS Res Hum Retroviruses* **18:** 1271–1280.

Pastore C, Nedellec R, Ramos A, Pontow S, Ratner L, Mosier DE. 2006. Human immunodeficiency virus type 1 coreceptor switching: V1/V2 gain-of-fitness mutations compensate for V3 loss-of-fitness mutations. *J Virol* **80:** 750–758.

Peters PJ, Bhattacharya J, Hibbitts S, Dittmar MT, Simmons G, Bell J, Simmonds P, Clapham PR. 2004. Biological analysis of human immunodeficiency virus type 1 R5 envelopes amplified from brain and lymph node tissues of AIDS patients with neuropathology reveals two distinct tropism phenotypes and identifies envelopes in the brain that confer an enhanced tropism and fusigenicity for macrophages. *J Virol* **78:** 6915–6926.

Peters PJ, Sullivan WM, Duenas-Decamp MJ, Bhattacharya J, Ankghuambom C, Brown R, Luzuriaga K, Bell J, Simmonds P, Ball J, et al. 2006. Non-macrophage-tropic human immunodeficiency virus type 1 R5 envelopes predominate in blood, lymph nodes, and semen: Implications for transmission and pathogenesis. *J Virol* **80:** 6324–6332.

Pfaff JM, Wilen CB, Harrison JE, Demarest JF, Lee B, Doms RW, Tilton JC. 2010. HIV-1 resistance to CCR5 antagonists associated with highly efficient use of CCR5 and altered tropism on primary CD4$^+$ T cells. *J Virol* **84:** 6505–6514.

Poss M, Rodrigo AG, Gosink JJ, Learn GH, de Vange Panteleeff D, Martin HL Jr, Bwayo J, Kreiss JK, Overbaugh J. 1998. Evolution of envelope sequences from the genital tract and peripheral blood of women infected with clade A human immunodeficiency virus type 1. *J Virol* **72:** 8240–8251.

Powers KA, Poole C, Pettifor AE, Cohen MS. 2008. Rethinking the heterosexual infectivity of HIV-1: A systematic review and meta-analysis. *Lancet Infect Dis* **8:** 553–563.

Redd AD, Avalos A, Essex M. 2007. Infection of hematopoietic progenitor cells by HIV-1 subtype C, and its association with anemia in southern Africa. *Blood* **110:** 3143–3149.

Robinson WE Jr, Mitchell WM, Chambers WH, Schuffman SS, Montefiori DC, Oeltmann TN. 1988. Natural killer cell infection and inactivation in vitro by the human immunodeficiency virus. *Hum Pathol* **19:** 535–540.

Russell ES, Kwiek JJ, Keys J, Barton K, Mwapasa V, Montefiori DC, Meshnick SR, Swanstrom R. 2011. The genetic bottleneck in vertical transmission of subtype C HIV-1 is not driven by selection of especially neutralization resistant virus from the maternal viral population. *J Virol* **85:** 8253–8262.

Sainski AM, Natesampillai S, Cummins NW, Bren GD, Taylor J, Saenz DT, Poeschla EM, Badley AD. 2011. The HIV-1 specific protein Casp8p41, induces death of infected cells through Bax/Bak. *J Virol* **85:** 7965–7975.

Salazar-Gonzalez JF, Salazar MG, Keele BF, Learn GH, Giorgi EE, Li H, Decker JM, Wang S, Baalwa J, Kraus MH, et al. 2009. Genetic identity, biological phenotype, and evolutionary pathways of transmitted/founder viruses in acute and early HIV-1 infection. *J Exp Med* **206:** 1273–1289.

Salazar-Gonzalez JF, Salazar MG, Learn GH, Fouda GG, Kang HH, Mahlokozera T, Wilks AB, Lovingood RV, Stacey A, Kalilani L, et al. 2011. Origin and evolution of HIV-1 in breast milk determined by single-genome amplification and sequencing. *J Virol* **85:** 2751–2763.

Schmitt MP, Gendrault JL, Schweitzer C, Steffan AM, Beyer C, Royer C, Jaeck D, Pasquali JL, Kirn A, Aubertin AM. 1990. Permissivity of primary cultures of human Kupffer cells for HIV-1. *AIDS Res Hum Retroviruses* **6:** 987–991.

Schnell G, Spudich S, Harrington P, Price RW, Swanstrom R. 2009. Compartmentalized human immunodeficiency virus type 1 originates from long-lived cells in some subjects with HIV-1-associated dementia. *PLoS Pathog* **5:** e1000395.

Schnell G, Price RW, Swanstrom R, Spudich S. 2010. Compartmentalization and clonal amplification of HIV-1 variants in the cerebrospinal fluid during primary infection. *J Virol* **84:** 2395–2407.

Schnell G, Joseph S, Spudich S, Price RW, Swanstrom R. 2011. HIV-1 replication in the central nervous system occurs in two distinct cell types. *PLoS Pathogens* (in press).

Schuitemaker H, Koot M, Kootstra NA, Dercksen MW, de Goede RE, van Steenwijk RP, Lange JM, Schattenkerk JK, Miedema F, Tersmette M. 1992. Biological phenotype of human immunodeficiency virus type 1 clones at different stages of infection: Progression of disease is associated with a shift from monocytotropic to T-cell-tropic virus population. *J Virol* **66:** 1354–1360.

* Shaw GM, Hunter E. 2011. HIV transmission. *Cold Spring Harb Perspect Med* doi: 10.1101/cshperspect.a006965.

Shen H, Cheng T, Preffer FI, Dombkowski D, Tomasson MH, Golan DE, Yang O, Hofmann W, Sodroski JG, Luster AD, et al. 1999. Intrinsic human immunodeficiency virus type 1 resistance of hematopoietic stem cells despite coreceptor expression. *J Virol* **73:** 728–737.

Shriner D, Liu Y, Nickle DC, Mullins JI. 2006. Evolution of intrahost HIV-1 genetic diversity during chronic infection. *Evolution* **60:** 1165–1176.

* Siliciano RF, Greene WC. 2001. HIV latency. *Cold Spring Harb Perspect Med* doi: 10.1101/cshperspect.a007096.

Sleasman JW, Aleixo LF, Morton A, Skoda-Smith S, Goodenow MM. 1996. CD4$^+$ memory T cells are the predominant population of HIV-1-infected lymphocytes in neonates and children. *AIDS* **10:** 1477–1484.

Sonza S, Maerz A, Deacon N, Meanger J, Mills J, Crowe S. 1996. Human immunodeficiency virus type 1 replication is blocked prior to reverse transcription and integration in freshly isolated peripheral blood monocytes. *J Virol* **70:** 3863–3869.

* Spudich S, González-Scarano F. 2011. HIV-1-related CNS disease: Current issues in pathogenesis, diagnosis, and treatment. *Cold Spring Harb Perspect Med* doi: 10.1101/cshperspect.a007120.

Stanley SK, Kessler SW, Justement JS, Schnittman SM, Greenhouse JJ, Brown CC, Musongela L, Musey K, Kapita B, Fauci AS. 1992. CD34$^+$ bone marrow cells are infected with HIV in a subset of seropositive individuals. *J Immunol* **149:** 689–697.

Thomas ER, Dunfee RL, Stanton J, Bogdan D, Taylor J, Kunstman K, Bell JE, Wolinsky SM, Gabuzda D. 2007. Macrophage entry mediated by HIV Envs from brain and lymphoid tissues is determined by the capacity to use low CD4 levels and overall efficiency of fusion. *Virology* **360:** 105–119.

Triques K, Stevenson M. 2004. Characterization of restrictions to human immunodeficiency virus type 1 infection of monocytes. *J Virol* **78:** 5523–5527.

Tscherning C, Alaeus A, Fredriksson R, Bjorndal A, Deng H, Littman DR, Fenyo EM, Albert J. 1998. Differences in chemokine coreceptor usage between genetic subtypes of HIV-1. *Virology* **241:** 181–188.

Tuyama AC, Hong F, Saiman Y, Wang C, Ozkok D, Mosoian A, Chen P, Chen BK, Klotman ME, Bansal MB. 2010. Human immunodeficiency virus (HIV)-1 infects human hepatic stellate cells and promotes collagen I and monocyte chemoattractant protein-1 expression: Implications for the pathogenesis of HIV/hepatitis C virus-induced liver fibrosis. *Hepatology* **52:** 612–622.

Valentin A, Rosati M, Patenaude DJ, Hatzakis A, Kostrikis LG, Lazanas M, Wyvill KM, Yarchoan R, Pavlakis GN. 2002. Persistent HIV-1 infection of natural killer cells in patients receiving highly active antiretroviral therapy. *Proc Natl Acad Sci* **99:** 7015–7020.

van Rij RP, Blaak H, Visser JA, Brouwer M, Rientsma R, Broersen S, de Roda Husman AM, Schuitemaker H. 2000. Differential coreceptor expression allows for independent evolution of non-syncytium-inducing and syncytium-inducing HIV-1. *J Clin Invest* **106:** 1569.

van't Wout AB, Kootstra NA, Mulder-Kampinga GA, Albrecht-van Lent N, Scherpbier HJ, Veenstra J, Boer K, Coutinho RA, Miedema F, Schuitemaker H. 1994. Macrophage-tropic variants initiate human immunodeficiency virus type 1 infection after sexual, parenteral, and vertical transmission. *J Clin Invest* **94:** 2060–2067.

van't Wout AB, Ran LJ, Kuiken CL, Kootstra NA, Pals ST, Schuitemaker H. 1998. Analysis of the temporal relationship between human immunodeficiency virus type 1 quasispecies in sequential blood samples and various organs obtained at autopsy. *J Virol* **72:** 488–496.

Vanderford TH, Bleckwehl C, Engram JC, Dunham RM, Klatt NR, Feinberg MB, Garber DA, Betts MR, Silvestri G. 2011. Viral CTL escape mutants are generated in lymph nodes and subsequently become fixed in plasma and rectal mucosa during acute SIV infection of macaques. *PLoS Pathog* **7:** e1002048.

Varbanov M, Espert L, Biard-Piechaczyk M. 2006. Mechanisms of CD4 T-cell depletion triggered by HIV-1 viral proteins. *AIDS Rev* **8:** 221–236.

Veazey RS, DeMaria M, Chalifoux LV, Shvetz DE, Pauley DR, Knight HL, Rosenzweig M, Johnson RP, Desrosiers RC, Lackner AA. 1998. Gastrointestinal tract as a major site of CD4$^+$ T cell depletion and viral replication in SIV infection. *Science* **280:** 427–431.

Veazey RS, Mansfield KG, Tham IC, Carville AC, Shvetz DE, Forand AE, Lackner AA. 2000. Dynamics of CCR5 expression by CD4$^+$ T cells in lymphoid tissues during simian immunodeficiency virus infection. *J Virol* **74:** 11001–11007.

Veazey RS, Marx PA, Lackner AA. 2003. Vaginal CD4$^+$ T cells express high levels of CCR5 and are rapidly depleted in simian immunodeficiency virus infection. *J Infect Dis* **187:** 769–776.

Vlahakis SR, Villasis-Keever A, Gomez TS, Bren GD, Paya CV. 2003. Human immunodeficiency virus-induced apoptosis of human hepatocytes via CXCR4. *J Infect Dis* **188:** 1455–1460.

Wei X, Ghosh SK, Taylor ME, Johnson VA, Emini EA, Deutsch P, Lifson JD, Bonhoeffer S, Nowak MA, Hahn BH, et al. 1995. Viral dynamics in human immunodeficiency virus type 1 infection. *Nature* **373:** 117–122.

* Wilen CB, Tilton JC, Doms RW. 2011. HIV: Cell binding and entry. *Cold Spring Harb Perspect Med* doi: 10.1101/cshperspect.a006866.

Wu L, KewalRamani VN. 2006. Dendritic-cell interactions with HIV: Infection and viral dissemination. *Nat Rev Immunol* **6:** 859–868.

Xu Y, Zhu H, Wilcox CK, van't Wout A, Andrus T, Llewellyn N, Stamatatos L, Mullins JI, Corey L, Zhu T. 2008. Blood monocytes harbor HIV type 1 strains with diversified phenotypes including macrophage-specific CCR5 virus. *J Infect Dis* **197:** 309–318.

Zhang ZQ, Wietgrefe SW, Li Q, Shore MD, Duan L, Reilly C, Lifson JD, Haase AT. 2004. Roles of substrate availability and infection of resting and activated CD4$^+$ T cells in transmission and acute simian immunodeficiency virus infection. *Proc Natl Acad Sci* **101:** 5640–5645.

Zhang J, Scadden DT, Crumpacker CS. 2007. Primitive hematopoietic cells resist HIV-1 infection via p21. *J Clin Invest* **117:** 473–481.

Zhou Y, Shen L, Yang HC, Siliciano RF. 2008. Preferential cytolysis of peripheral memory CD4$^+$ T cells by in vitro X4-tropic human immunodeficiency virus type 1 infection before the completion of reverse transcription. *J Virol* **82:** 9154–9163.

Zhu T, Muthui D, Holte S, Nickle D, Feng F, Brodie S, Hwangbo Y, Mullins JI, Corey L. 2002. Evidence for human immunodeficiency virus type 1 replication in vivo in CD14$^+$ monocytes and its potential role as a source of virus in patients on highly active antiretroviral therapy. *J Virol* **76:** 707–716.

Zuckerman RA, Whittington WL, Celum CL, Collis TK, Lucchetti AJ, Sanchez JL, Hughes JP, Coombs RW. 2004. Higher concentration of HIV RNA in rectal mucosa secretions than in blood and seminal plasma, among men who have sex with men, independent of antiretroviral therapy. *J Infect Dis* **190:** 156–161.

HIV Pathogenesis: Dynamics and Genetics of Viral Populations and Infected Cells

John Coffin[1] and Ronald Swanstrom[2]

[1]Department of Molecular Biology and Microbiology, Tufts University, Boston, Massachusetts 02111

[2]Department of Biochemistry and Biophysics, University of North Carolina, Chapel Hill, North Carolina 27599

Correspondence: john.coffin@tufts.edu

In the absence of treatment, HIV-1 infection, usually starting with a single virion, leads inexorably to a catastrophic decline in the numbers of CD4$^+$ T cells and to AIDS, characterized by numerous opportunistic infections as well as other symptoms, including dementia and wasting. In the 30 years since the AIDS pandemic came to our attention, we have learned a remarkable amount about HIV-1, the responsible virus—the molecular details about how it functions and interacts with the host cell, its evolution within the host, and the countermeasures it has evolved to overcome host defenses against viral infection. Despite these advances, we remain remarkably ignorant about how HIV-1 infection leads to disease and the death of the human host. In this brief article, we introduce and discuss important lessons that we have learned by examining the dynamics of viral populations and infected cells. These studies have revealed important features of the virus–host interaction that now form the basis of our understanding of the importance and consequence of ongoing viral replication during HIV-1 infection.

TIME COURSE OF INFECTION

As discussed elsewhere (Shaw and Hunter 2011), HIV-1 infection is usually initiated with a single virion infecting a single target cell at the portal of entry. The subsequent course of infection can be monitored in several ways: overt symptoms, such as fever, wasting, opportunistic infections, neurological symptoms, and so on; blood levels of the CD4$^+$ T-cell target and antiviral antibodies; and viremia (virus in blood), measured by infectivity, immunoassay for viral proteins, and, most accurately, by PCR for viral RNA. A typical time course of infection relating these properties to one another is shown in Figure 1. Although their timing can vary considerably from individual to individual, as can the levels of viremia, the general outline is essentially the same in virtually every infected person who does not receive effective antiviral therapy.

1. (~1–2 wk) An eclipse phase, during which the virus is freely replicating and spreading from the initial site of infection to the many tissues and organs that provide the sites for replication. Viremia is undetectable, and neither immune response nor symptoms of infection are yet visible.

2. (~2–4 wk) The acute (or primary) infection phase, characterized by relatively high levels

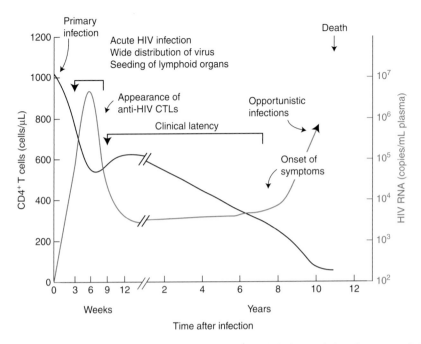

Figure 1. Time course of typical HIV infection. Patterns of CD4$^+$ T cell decline and viremia vary greatly from one patient to another. (From Fauci and Desrosiers 1997; reprinted, with permission, from Cold Spring Harbor Laboratory Press © 1997.)

of viremia (up to 10^7 or more copies of viral RNA per milliliter of blood), and large fractions of infected CD4$^+$ T cells in blood and lymph nodes. This phase is often, but not always, accompanied by "flu-like" symptoms—fever, enlarged lymph nodes, and the like. Around the time of peak viremia, the immune response begins to appear, both in the form of antibodies against all viral proteins, and a CD8$^+$ T-cell response against HIV-1 antigens expressed on infected cells. The high levels of viremia that characterize this phase most likely result from the absence of the early immune response and the generation, as part of the host response, of large numbers of activated CD4$^+$ T cells, providing a wealth of targets for viral replication. At the end of the acute phase, the level of viremia declines sharply, 100-fold or more, a result of both partial control by the immune system and exhaustion of activated target cells. This phase is also characterized by a transient decline in the numbers of CD4$^+$ T cells in the blood.

3. (~1–20 yr) Chronic infection, or "clinical latency," is characterized by a constant or slowly increasing level of viremia, usually on the order of 1–100,000 copies/mL, sometimes referred to as the "set point," and steady, near normal (~1000 cells/μL), or gradually falling levels of CD4$^+$ T cells. As a rule, patients in this phase are asymptomatic and usually unaware that they have been infected. Despite the term "latency," the viral infection is far from latent, with large numbers of CD4$^+$ T cells becoming infected and dying every day.

4. Finally, the number of CD4$^+$ T cells declines to the point (~200 cells/μL) at which immune control of adventitious infectious agents can no longer be maintained, and opportunistic infections (discussed in Lackner et al. 2011) begin to appear. Control of the HIV-1 infection itself is also lost, and the level of viremia rises during the AIDS phase, culminating in death of the infected patient. Indeed, untreated HIV-1 infection

is one of the most uniformly lethal infectious diseases known, with a mortality rate well over 95%.

SIGNIFICANCE OF VIREMIA

As clinical researchers were developing tools for diagnosis and prognosis of HIV-1 infection, it became apparent that the most powerful of them is the measurement of viremia using quantitative nucleic acid hybridization or PCR assays for HIV-1 genomic RNA in blood. RNA measurements have the virtue of an extraordinarily wide dynamic range. Routine commercial assays can now accurately measure levels of viremia as low as 50 copies/mL and as high as 10^7 or more copies/mL. More sensitive research assays have reduced the lower end to much less than 1 copy/mL with good quantitative accuracy (Palmer et al. 2003).

The first recognition of the prognostic importance of viremia came from analysis of a large prospective study of gay men at risk for AIDS, the Multicenter AIDS Cohort Study (MACS), in which volunteers provided blood samples and health information at frequent intervals over long periods of time, during which many men who were initially infected, but still healthy, progressed to AIDS and death (Mellors et al. 1996). When the levels of viremia at study entry were compared with outcome, a strong inverse correlation between the level of

viremia and the time of progression to AIDS was observed (Fig. 2). The data showed that the group with the highest level of viremia progressed to AIDS on average about fourfold faster than the lowest group. Thus, the level of virus present in the blood provides a measure of the rate at which viral infection damages the immune system of the infected host.

The second important insight made possible by the ability to accurately measure levels of viremia was in analyzing the consequences of antiviral therapy, where it was found that the ability of an antiretroviral drug to halt or even reverse progression of an infected patient to AIDS was well correlated with its ability to suppress viremia. Effective therapies, such as combinations of antiviral drugs discussed by Arts and Hazuda (2011), which can lead to sustained reductions in viremia to undetectable levels of <50 copies/mL by standard clinical assays, can forestall progression to AIDS indefinitely, although they do not cure the infection. Indeed, if therapy is suspended, even after many years, the level of viremia will rapidly return to the prior set point.

The correlation between viremia and disease has a straightforward underlying explanation (Coffin 1995). Infected cells produce virions, some fraction of which are released into the blood, from which they are rapidly removed or degraded. Although different cells may produce and release different amounts of virus into the

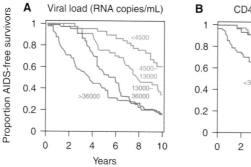
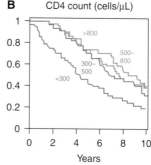

Figure 2. Relationship between viral load (viremia) and clinical progression. Shown are Kaplan–Meier plots of AIDS-free survival divided into quartiles according to virus load (*A*) or CD4 count (*B*) at the time of diagnosis. (From Fauci and Desrosiers 1997; reprinted, with permission, from Cold Spring Harbor Laboratory Press © 1997.)

blood, their numbers are very large, and, on average, the amount of virus per infected cell will be about the same from time to time in the same patient and similar from one patient to another. Also, the rate of removal of virions from the blood will be similar (and, as has been measured, quite rapid) from time to time and patient to patient. Because the level of viremia is determined by the balance between virion release into and removal from the bloodstream, assuming a constant rate of release per cell and a constant rate of removal, the level of viremia will be an instantaneous relative measure of the number of productively infected cells at any one time.

INSIGHTS FROM THERAPEUTIC DRUG STUDIES

This understanding of the significance of viremia in HIV-1 infection, combined with the new availability of potent antiretroviral drugs

(Arts and Hazuda 2011), led to an experiment that provided considerable insight into the process of HIV-1 infection and pathogenesis in vivo (Fig. 3) (Coffin 1995; Ho et al. 1995; Wei et al. 1995). The study was based on the principle that all antiviral drugs affect the ability of the virus to infect cells and do not affect the lifetime of infected cells or their ability to produce viral particles (albeit noninfectious ones in the case of protease inhibitors). Therefore, any decrease in viremia following initiation of suppressive therapy must reflect a combination of the length of time that productively infected cells live and produce virus after infection and the lifetime of the virus in the blood. Because the latter factor is very small, it can be ignored with little effect on the overall result. Thus, viremia was assayed by quantitative PCR at closely spaced time points after initiation of suppressive therapy, with striking results: After a brief lag, viremia drops rapidly, to ~1%– 10% of the set point level with a half-time of

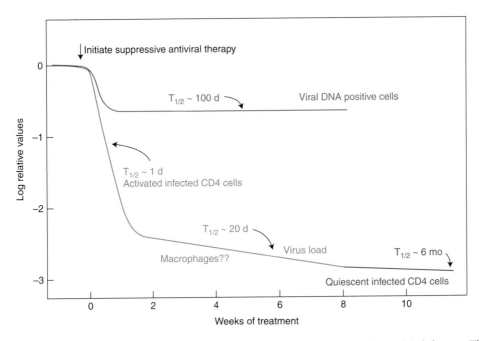

Figure 3. Decay of circulating virus and infected cells after initiation of suppressive antiviral therapy. These results imply the presence of at least four classes of HIV-infected cells: productively infected, a second class, perhaps macrophages, with a half-life of ~2 wk; latently infected resting CD4 cells, with a half-life of ~6 mo; and long-lived DNA positive cells, most of which are nonproductive. A fourth class of infected cells, inferred to have an approximately infinite half-life, is not shown. (Figure modified from Fauci and Desrosiers 1997.)

Cite this article as *Cold Spring Harb Perspect Med* doi: 10.1101/cshperspect.a012526

~1–2 d. Following this first phase of decay, there is a second phase in which most of the remaining viremia decays with a half-time of ~2 wk or so, down to the detection limit of standard assays. Two more phases, which have been defined with more sensitive assays, are discussed below.

The first phase of decay almost certainly reflects the lifetime of productively infected CD4$^+$ T cells, although the exact cells and their location have not been identified. The second phase has been speculatively attributed to macrophages (Perelson et al. 1997), but this point is far from established. Given that this phase is much less prominent when the treatment includes an integrase inhibitor, it may reflect at least in part a population of infected cells in which integration is delayed for some reason (Murray et al. 2007).

THE HIV-1 STEADY STATE

The results discussed above, combined with the fact that the characteristic level of viremia (set point) in any infected individual is essentially constant, led to the conclusion that HIV-1 exists in the infected host as a quasi-steady state, in which the level of viremia is largely determined by the number of virus-producing cells, which therefore must also be constant day in and day out throughout most of the course of infection. In any given patient, therefore, there are about the same number of infected cells today as there were yesterday and last year, but they must be different cells arising from infection of fresh target cells, which then rapidly die after producing sufficient virus, on average, to infect the same number of cells again. This concept is key to our current understanding of the HIV–host interaction. It has many important implications.

1. The replication cycle time of the virus is, on average, ~1–2 d. Thus, virus in an individual at a year after infection is 200–300 generations removed from the initial infecting virion.

2. The population size is determined by the steady-state number of productively infected cells, not by the number of virions, except that the latter must be at least large enough to ensure that the same number of cells is infected every day.

3. The steady state is remarkably robust, changing little throughout the clinically latent phase, and returning to the same value after even prolonged periods (7 yr or more) when it is undetectable on suppressive therapy, or after transient increases due to immune stimulation by vaccination or by infection with some other agent.

4. The virus exists as a replicating population that is carried forward by repeated infection of large numbers of cells with no evidence of bottlenecks or other major alterations in population size from time to time.

5. The level of viremia is determined by a combination of the number of available target cells, the infectivity of the virus, the number of virions produced per cell, the productive lifetime of the infected cell, and the clearance rate of the virus.

6. We know little about the infected cells that are responsible for producing the virus observed in the blood and can only infer their properties indirectly.

7. The extraordinary extent of genetic diversity that accumulates during HIV-1 infection, relative to other RNA viruses, is not due to its unusually high error rate, but rather to replication of a large population of virus with an average error rate repeated over and over again, accumulating thousands of generations over the course of infection of a single individual.

With these properties in mind, we can look at some specific issues.

Nature and Lifetimes of Infected Cells

The initial therapeutic studies that led to the steady-state concept implied two populations of target cells: one with a half-life of ~1–2 d and another of ~2 wk. These conclusions generated considerable optimism because the proportionality of the level of viremia to the number of productively infected cells suggested that sufficiently prolonged treatment could eventually reduce the number of infected cells to <1, effectively curing the patient of the HIV-1 infection. Unfortunately, these hopes were dashed when further studies, using more sensitive techniques, revealed the presence of at least two more phases of decay of HIV-1 viremia on fully suppressive therapy, apparently due to small populations of infected cells capable of surviving and releasing virus many years after infection (Siliciano and Green 2011). Although such cells constitute only a tiny fraction of the total number of infected cells present before therapy, they decay very slowly (with a half-life of ~6 mo by one estimate) (Zhang et al. 1999) after treatment. Also, much more sensitive PCR assays, with a limit of detection well below 1 copy/mL of viral RNA in plasma (Palmer et al. 2003), revealed that the majority patients on therapy with "undetectable" viremia for more that 1 yr in fact had an average of about 3 copies of HIV-1 RNA per milliliter, or about 1/10,000 of the level before therapy, and that this level declined further with a curve consistent with two more phases of decay: one with a half-life of ~39 wk, not significantly different from that of inducible cells, perhaps from the same source, and another with a half-life not different from infinity, suggesting the possibility of a dividing reservoir. Viremia was still detectable in all patients as long as 7 yr after the start of therapy (Palmer et al. 2008). The existence of these populations of cells, apparently infected before therapy and capable of producing virus many years later, rules out eradicating infection by antiviral therapy alone.

One more population of infected cells bears discussion. In a typical chronically infected individual, about one CD4$^+$ T cell per 100 to one per 1000 contains HIV-1 DNA, largely in the form of an integrated provirus. Very few cells contain more than one provirus (Josefsson et al. 2011). After the initiation of therapy, there is an initial decline in the numbers of these cells, reflecting the decay in viremia and implying that at least some of the virus in the blood comes from populations of cells also found in the blood. However, as can be seen in Figure 3, the level of viral DNA positive cells soon levels off, typically somewhere around 1%–10% of the starting value, implying that a significant fraction of these cells is not involved in phase 1 or 2 viral production. Furthermore, the numbers of DNA positive cells after prolonged therapy is typically 100-fold greater than the number of cells inducible to produce virus ex vivo (Chun et al. 1997), implying that most of them are incapable of ever producing infectious virus, perhaps because of mutations in the provirus, unfavorable sites of integration, or some undefined cellular factor that prevents expression of the silent provirus. Simple modeling implies that over time, because of their much longer lifetime, cells containing inactive or latent proviruses should accumulate to a considerable extent relative to productively infected cells, most of which die within a few days. Thus, only a small fraction of the population of cells containing detectable proviruses represent cells that are latently infected, that is, that contain a provirus that is not expressed but is capable of being induced to express infectious virus by some means, such as stimulation to reenter the cell cycle. For this reason, it cannot be assumed (although it often is) that the nature of cells in the so-called latent reservoir presumed responsible for the persistence of HIV-1 during years of therapy can be inferred from the properties of all provirus-containing cells (such as sites of integration, expression of transcription factors, etc.), the large majority of which must contain dead proviruses incapable of reincarnation. The physical identification and manipulation of the tiny fraction of latently infected cells in suppressed infection is a major unsolved problem in HIV-1 pathogenesis (Maldarelli 2011; Siliciano and Green 2011).

Persistence of HIV-1 Infection on Therapy

Although it is well accepted that the large majority of patients on suppressive therapy have low levels of viremia as well as latently infected cells capable of being reactivated to produce HIV-1 ex vivo, the link between the two phenomena remains somewhat controversial, with some investigators arguing that the persistent viremia observed in patients on long-term therapy could also be due to a small amount of "escape" replication of virus in anatomic sites or cells that are poorly accessible to antiviral therapy (referred to as "sanctuaries") (Sharkey and Stevenson 2001). In support of this claim, they point to the ongoing presence of two-LTR circles (Craigie and Bushman 2011) in PBMCs in a fraction of patients (Buzon et al. 2011), an observation interpreted as meaning that productive infection of new cells is continuing to occur. However, the underlying assumption that the circular forms of viral DNA are inherently unstable in cells in vivo remains unsupported, and there are several lines of evidence in favor of the idea that the low-level viremia observed in the majority of patients on long-term suppressive therapy is derived from cells infected before the start of treatment.

1. The distribution of the levels of viremia among treated patients is independent of the drugs used for treatment (Maldarelli et al. 2007). This result is not what would be expected for replication in a sanctuary because it is improbable that such sites or cells would be equally unavailable to different drugs.

2. The level of viremia is not affected by the inclusion of another drug. Several studies have been performed in which a suppressive three-drug regimen was intensified by the addition of a fourth inhibitor of a class previously unseen by the patient (Dinoso et al. 2009; Gandhi et al. 2010; McMahon et al. 2010). Despite some differences in design and inhibitors used, all studies gave the same result: There was no difference in the amount of virus during and after the intensification period as compared with before. Again, it is highly improbable that sanctuaries of replication would be equally inaccessible to all drugs, and these results are inconsistent with low-level ongoing replication.

3. As is discussed in more detail in the next section, accumulation of genetic variation due

to the rapid turnover of infected cells is a hallmark of HIV-1 infection. In the chronic infection phase, there is a slow but perceptible turnover of the viral population such that genetically distinct populations arise every few years (F Maldarelli, M Kearney, S Palmer, et al., unpubl.). In contrast, in patients on maximally suppressive therapy, evolution of viral genomes is frozen at the point at which therapy began (Ruff et al. 2002; Tobin et al. 2005), and no evidence for continuing evolution can be seen even after many years. In patients for which therapy is partially suppressive, ongoing evolution can still be observed (Tobin et al. 2005), as is also true in patients known as "elite controllers," who naturally control HIV-1 infection at levels similar to those seen in patients on suppressive antiviral therapy (Mens et al. 2010). The apparent lack of evolution on therapy is consistent with the absence of new cycles of infection.

4. As noted, the appearance of 2-LTR circles in a minority of patients undergoing intensification with an integrase inhibitor raises the possibility of a low level of infection occurring in the face of antiviral drugs. However, this result remains to be fully understood or reproduced.

5. It has recently been suggested, on the basis of mathematical and cell culture modeling, that the high multiplicity of infection that would result from cell–cell spread could greatly reduce the effectiveness of antiviral therapy and allow persistent viral replication (Sigal et al. 2011). However, this model predicts that cells infected in vivo should have multiple proviruses and that genetic variation should continue to accrue on suppressive therapy, neither of which is observed.

Regarding point 2, in the small amount of virus in the blood of some patients whose viremia has been suppressed for more than 3 yr or so, significant subpopulations of virus with genomes identical in sequence can be observed and, in some cases, can become a large fraction of the total population (Bailey et al. 2006).

These subpopulations have been called predominant plasma clones (PPCs). Their origin is unknown, but it can be speculated that they may arise from a very small fraction of infected cells that is capable of dividing and releasing virus but that do not die as a result of the expansion. Such clonal populations of virus can also be seen in the blood of patients who have stopped suppressive therapy after long-term suppression (Joos et al. 2008), although the two phenomena have not yet been firmly connected. It seems likely that the last, apparently stable, phase of viremia on therapy reflects clonal viral populations, but this point remains to be established.

It is important to keep in mind that the infected cell populations that give rise to long-lived, persistent viremia on therapy are not created by the therapy; rather, they must exist throughout the course of infection, only to be exposed as the shorter-lived cells of phases 1 and 2 die off and are not replenished and the virus produced by them disappears. The cells responsible for the production of the persistent virus must represent an extremely small fraction of the total population of productively infected cells in chronic infection—somewhere between 10^{-4} and 10^{-5} is a reasonable estimate. Given that HIV-1 infection is carried forward, day in and day out, by repeated cycles of infection, viral production, and death of target cells, these long-lived cell populations must be irrelevant to the natural history of HIV-1 infection (as well as SIV infection of the natural hosts). They are of critical importance to attempts to treat HIV-1 infection, however, because the virus they produce is capable of rekindling full-blown infection from the faintest embers remaining after prolonged completely suppressive therapy.

GENETIC VARIATION AND EVOLUTION

One of the first unusual features of HIV-1 infection to become apparent was the remarkable accumulation of genetic diversity during the course of infection of a single individual. Indeed, when first observed, it represented the fastest rate of evolution in any natural eukaryotic system; for example, the extent of HIV-1 diversity that accumulates in one infected individual exceeds that of all influenza virus isolates worldwide in any given year (Korber et al. 2001). (Actually, the record diversity no longer belongs to HIV, having since been eclipsed by hepatitis C virus.) Although it was first thought that the remarkable diversification reflected an extraordinarily high underlying error rate associated with HIV-1 replication, we now know that this rate is about average for an RNA virus and that it is the pace of replication, the duration of infection, and the size of the replicating population, allowing it to evolve rapidly in response to selective influences, that set HIV-1 (and HCV) apart from all other known viral infections. These features, in addition to the ease of generating large numbers of sequences either directly from infected individuals and populations of individuals with well-timed infections or from massive databases of such sequences (Kuiken et al. 2003), have made HIV-1 a favorite topic of study for evolutionary biologists, and they have used it effectively to develop and test new mathematical models for evolution in general, not just for HIV-1 or just for viruses. We do not present any specific models here, but the next section discusses the role of specific factors of evolution that all models must include and how they apply to HIV-1 in an infected human host.

Factors of HIV-1 Evolution

By "evolution," we mean the process by which genetic change accumulates during generations of replication. For any replicating entity, organisms or viruses, the general principles of evolution are similar, although their application can differ greatly. As a rule, models are based on consideration of four principal factors. We discuss them here in the context of HIV.

Mutation

In the case of HIV-1 and other retroviruses, mutations are primarily attributed to error-prone reverse transcription, but there are actually two other possible sources of error: transcription by host RNA polymerase II, and G-to-A hypermutation mediated by ABOBEC3G (or

F). As discussed in other articles in this collection (Hu and Hughes 2011; Karn and Stoltz-fus 2011; Malim and Bieniasz 2011), the relative contribution of these mechanisms is a matter of some debate. Whatever the source, the overall single-step point mutation rate for HIV-1 is $\sim 3 \times 10^{-5}$ mutations per base per replication cycle, and about 10-fold less for transversions than transitions (Mansky and Temin 1995). Thus, about one genome in three contains a mutation after a single round of replication. Other types of mutation, including frameshifts (insertion or deletion of small numbers of bases), as well as large-scale deletions, insertions, duplications, and inversions, are also common (Svarovskaia et al. 2003). The error rate of retrovirus replication, while much greater than that of cellular and viral DNA polymerases, is about average for RNA viruses.

Selection results from the ability of one variant to replicate more effectively than another, a property referred to as "fitness." In the case of viruses, fitness differences lead to differences in the number of infectious progeny at each replication cycle. Fitness differences can result from differences at any stage of the series of events leading from one infected cell to the next, including efficiency of entry, rate of infectious progeny production by an infected cell, and its productive lifetime, stability of the infectious virion, and so on. Note that there is no such thing as absolute fitness: The concept is meaningful only in relative terms and only in a specific context (e.g., in vitro vs. in vivo, in the presence or absence of neutralizing antibodies or antiviral drugs). It is often expressed as a selection coefficient, the difference in the relative number of progeny produced by one variant compared with another per generation. Thus, for a variant that yields 1% more infectious progeny than another, the selection coefficient is 0.01. For large populations undergoing multiple repeated cycles of infection, even small selection coefficients can have major effects on the frequency of slightly advantageous or disadvantageous mutations, with the former leading to positive selection and the latter to purifying selection. Because synonymous mutations (those that do not change the

corresponding amino acid) generally have much smaller selection coefficients than nonsynonymous ones, the ratio between the two in a population (d_N/d_S) is a useful indicator of the nature of selective forces acting at any given site or set of sites.

Drift

"Drift" is a term for the change in frequency of a variant independent of selection or mutation, resulting from sampling error that arises when replicating population sizes are small. In the case of viral populations, it is important to bear in mind that the population size is determined by the number of cells infected at each round of replication, not by the number of virions produced by them. In the case of HIV-1 infection, perhaps 10^{11} virions are produced daily; the number of cells infected in the same time span is still a matter of some debate, but is unlikely to exceed 10^9. A further important consideration is the effective population size: in simple terms, the number of infected cells multiplied by the probability that any one cell gives rise to progeny that go on to infect other target cells. If there is an infinite supply of target cells and the population of infected and target cells is well mixed, then the effective population size can be similar to the census population size (defined as the total number of infected cells). These conditions obviously do not apply in infected individuals, and the effective population size is probably much smaller. At very large effective population sizes, evolution can be considered to be deterministic, in that the frequency of a mutation at any given time (in generations) in the future can be exactly determined from its present frequency, mutation rate, and selection coefficient, as long as these factors remain constant. At smaller sizes, evolution becomes more and more stochastic, that is, subject to random factors, so that the frequency can only be predicted as an average, with the variance increasing with decreasing size. A useful relationship to keep in mind is that, after many generations of deterministic evolution, slightly deleterious mutations will reach an equilibrium frequency equal to the mutation rate divided by their selection coefficient. In the stochastic case, this calculation will give the average value over many observations but will vary considerably from one individual to the next. This consideration is of more than academic interest because it defines the probability that drug-resistance mutations, which are slightly deleterious in the absence of the drug, will already be present in the replicating viral population before initiation of antiviral therapy (the so-called genetic barrier) (see Arts and Hazuda 2011). Although transition from deterministic to stochastic evolution is continuous with decreasing population size, a useful approximate crossover point to keep in mind is a size equal to the inverse of the mutation rate, at which it becomes probable that all possible single mutations will arise at each round of replication (Rouzine and Coffin 2005).

Linkage

To complicate matters further, selection does not act on individual mutations, but rather on whole genomes, which may contain other mutations as well. In very large deterministic populations, where any mutation is always present in many genomes, this effect is unimportant; in smaller populations, it can prevent selection from operating on individual mutations, and even, in extreme cases (and at least in theory) lead to declining fitness of a population due to gradual accumulation of deleterious mutations (Muller's ratchet). Linkage can thus be visualized as increasing the role of stochastic effects on evolution, even at large population sizes. The effect of linkage on evolution can be reduced by recombination to allow mutations to be redistributed among genomes. It is important to keep in mind that recombination does not generate diversity, but rather redistributes it among replicating genomes, allowing effects of selection to be felt more strongly. As discussed elsewhere (Hu and Hughes 2011), recombination between genetically distinct genomes in retrovirus replication occurs at very high rates, but only during reverse transcription of genomes from heterozygous progeny of cells containing two or more proviruses. Although

it was originally reported that such multiply infected cells are very frequent in HIV-infected individuals (Jung et al. 2002), more recent results suggest that they are much rarer, constituting perhaps 10% of infected CD4$^+$ T cells in blood (Josefsson et al. 2011). Nevertheless, there is ample evidence for the occurrence and importance of recombination in HIV-1 infection, particularly in rare individuals coinfected with two or more genetically distinct viruses (Keele et al. 2008; Kearney et al. 2009).

Course of Evolution following Infection

Monitoring of genetic diversity of the virus following HIV-1 infection has provided a very important tool for understanding many important aspects of the virus–host interaction. For reasons discussed above, the sequences of viral genomes in the blood provide the most accurate instantaneous indicator for the genetics of the replicating viral population, particularly when single-genome sequencing (SGS) technology, which avoids many PCR artifacts (Palmer et al. 2005), is used. Figure 4 shows, in cartoon form, the important events in virus evolution during the course of a "typical" infection, starting with a clonal population and proceeding toward the highly diverse populations characteristic of chronic infection. The initial accumulation of diversity is at a rate very close to the mutation rate, with little evidence for selection, except for highly deleterious mutations; as time goes on, purifying selection takes hold, and the increase in diversity begins to level off, approaching an asymptotic value of an average pairwise distance ~1%–2% for gag, pro, and pol, which are largely subject to purifying selection, as indicated by low d_N/d_S ratios. Rates are much higher for env, where the greater, and constantly increasing, diversity and higher d_N/d_S ratios point to constant positive selection resulting from an ever-changing antibody response (Richman et al. 2003). In the case of gag, pro, and pol (as well as accessory genes, particularly nef), even though d_N/d_S ratios are low, positive selection can be discerned at sites corresponding to T-cell epitopes. Indeed, virtually all of the nonsynonymous diversity accumulating in

Cite this article as Cold Spring Harb Perspect Med doi: 10.1101/cshperspect.a012526

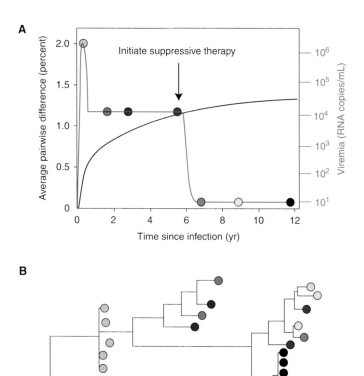

Figure 4. Typical evolution of HIV after infection. (*A*) The level of viremia (orange) and the genetic diversity, measured as the average pairwise difference (black) of the HIV population as a function of time after infection. (*B*) Phylogenetic relationship of HIV-1 sequences at various stages of infection. The colored circles correspond to sequences of plasma virus RNA sampled at the corresponding time points in *A*: At peak viremia (yellow), the viral population (in most patients) is highly uniform, diversifying gradually over the first few years of infection. Samples taken from chronic infection (green and blue) are considerably more diverse, but take 3 yr or more to diverge perceptibly (red). Following initiation of suppressive therapy, divergence ceases, even after long times (orange, gray). After years of therapy, clonal populations of virus arise in some patients (black).

these genes in untreated infection can be attributed to escape from the cytotoxic T-cell (CTL) response (Kearney et al. 2009; Walker and McMichael 2011). In *env*, escape from the antibody response is characterized not only by point mutation, particularly affecting the display of glycosylation sites, but also by gross structural changes—insertions and duplications—in the hypervariable loops V1, V2, and V4, which tend to increase in length and numbers of glycosylation sites (the "glycan shield")

during the course of infection. It is important to bear in mind that, given the conditions under which HIV populations are evolving, even weak selective forces (on the order of a few percent) can have dramatic effects on the genetics of the viral population. Thus, it cannot be inferred from these results alone that the immune response plays a significant role in controlling HIV infection. The partial control of viral replication by the host is likely the sum of effects of the immune response and the fitness cost incurred with the initial escape by the virus, a fitness cost that is in part alleviated over time through compensatory mutations.

A remarkable aspect of HIV genetic diversity during chronic infection is its stability. Except in the case of escape from antiretroviral therapy, genetic turnover of the viral population in the blood is very slow, with no evidence of a genetic or physical bottleneck (Shankarappa et al. 1999; F Maldarelli, M Kearney, and S Palmer, unpubl.). This observation is consistent with a large replicating population that can be steered by immune selection, but without ever having to recover from significant depletion. Because mutations leading to immune escape will often reduce fitness, when that immune response is not present, such as after transmission to another person, these mutations will tend to revert to the consensus sequence, although at highly variable rates. For example, mutations increasing the length of the variable loops tend to reset to shorter length following transmission (Shaw and Hunter 2011).

In addition to immune escape, resistance to antiretroviral therapy is an important consequence of genetic diversity accumulating during chronic HIV-1 infection. Both theory (discussed above) and observation are consistent with the idea that drug-resistance mutations that accumulate to low levels before therapy are an important cause of failure. Indeed, in virtually all cases, treatment of HIV-1 infection with a single agent (monotherapy) leads to rapid reappearance of viremia, rebounding to levels near the pretherapy set point. A striking example is in women treated with a single dose of the nonnucleoside RT inhibitor (NNRTI) nevirapine to prevent mother-to-child transmission (Chersich and Gray 2005). A single pill taken during labor prevents a significant fraction of infections of the newborn but leads to a very high frequency of drug-resistant virus in the mother a few weeks later. The reproducibility of this dramatic effect implies a low (and so far undetectable) frequency of NNRTI-resistant mutants in the replicating viral population combined with a relatively long lifetime of the drug, allowing their selection during the ensuing cycles of HIV-1 replication.

Unlike the immune response, antiviral therapy can impose a significant physical and genetic bottleneck on the HIV-1 population, with drug-resistant virus that rebounds after monotherapy often showing substantially less diversity than the starting population (Kitrinos et al. 2005).

Another important consequence of HIV genetic variation is the appearance of populations of virus infecting specific anatomical compartments and cell types. As discussed by Swanstrom and Coffin (2011), HIV can be found in a large number of organs and tissues. In some sites, such as the central nervous system (CNS), there is a clear genetic separation of independently evolving populations, often leading to macrophage-tropic viruses that likely replicate in either macrophages or microglial cells in the brain. Macrophage-tropic viruses can also be found late in infection at other anatomical sites, but their pathophysiological significance is much less clear. The final important consequence of HIV evolution in vivo is the appearance of virus capable of using alternative coreceptors, such as CXCR4, distinct from the CCR5 chemokine receptor used by the founder virus.

HOW DOES HIV-1 CAUSE AIDS?

As is apparent from this article and the rest of the collection, in the 25+ years since its discovery, we have learned an enormous amount about HIV, but we still cannot answer the one big question: How does HIV-1 cause AIDS? At first thought, the answer is obvious. HIV-1 infects and kills $CD4^+$ T cells. AIDS is a result of loss of $CD4^+$ T cells. *QED.* Unfortunately, it is not that simple.

For starters, although HIV-1 infection kills infected cells ex vivo, by a mechanism that is still controversial, there is still considerable debate about how infected cells die in vivo. There are three alternatives: virus-mediated killing of infected cells, as in culture systems; immune-mediated effects due to HIV-specific $CD8^+$ T cells or antibodies; or one or more of a variety of indirect effects. In support of the role of such direct killing is the short half-life of infected cells early in infection, before the appearance of antibodies or CTL, ruling out the adaptive immune response as essential for cell killing in vivo. The innate (i.e., interferon) response could still play some role at this point, however. Although the antibody response appears to have little effect on the life of the infected cell, there is evidence for an important role of the CTL response in cell killing because treatment of SIV-infected macaques with antibody directed against CD8 leads to a significant (100-fold) but transient increase in the level of viremia (Jin et al. 1999; Schmitz et al. 1999). The most straightforward explanation for this result is that CTLs shorten the productive lifetime of the infected cell, although more indirect effects have also been proposed (Davenport and Petravic 2010; Klatt et al. 2010; Wong et al. 2010). Two features of this hypothesis are worth noting. First, the productive lifetime of an infected cell is the time from the onset of virus production to cell death by CTL and is likely to be not much shorter than the lifetime defined by virus-mediated killing, perhaps only an hour or so. Second, it is likely that part of the advantage conferred on HIV-1 (and other complex viruses) by the two-step replication strategy mediated by *tat* and *rev* is the delay of the synthesis of virion proteins as long as possible in the life cycle to minimize exposure to the CTL response and maximize virion output per cell. Although cells can die from the effects of infection alone, their productive lifetime is somewhat shortened by the CTL response, that is, specifically during the time when virus is being produced. Thus, CTLs can modulate infection, but they do not eliminate it. Effects due to immune activation could include not only degradation of immune function late in infection, but also stimulation of production of activated memory $CD4^+$ T cells, increasing the number of targets available to the virus.

Even if we knew the mechanism of HIV-mediated cell killing, we would not know how HIV-1 causes $CD4^+$ T-cell decline and AIDS in humans. The observation that virus and cell turnover rates in various SIVs in their natural hosts (such as SIV_{sm} in sooty mangabeys), which do not progress to AIDS, are essentially identical to those in humans, who *do* progress, implies that cell killing alone cannot account for AIDS pathogenesis. Indeed, this result is consistent with the high natural turnover rate of activated effector memory helper T cells, the primary target for HIV-1 infection, on the order of 10^{10} cells per day, of which only a small fraction are infected after the initial primary infection phase. Clearly, in the natural host, these viruses have evolved to replicate in a population of cells that is replenished constantly and in excess over what is needed for good health. In unnatural hosts, such as HIV-1 infection of humans, or SIV_{sm} infection of macaques, something else must be happening to lead to AIDS. Two features that distinguish the pathogenic from nonpathogenic infection are chronic immune activation, discussed above, and greater levels of infection of other $CD4^+$ T-cell subsets, particularly central memory helper T cells. The former may lead eventually to immune exhaustion and damaged lymphoid architecture, the latter (which might, itself, be a result of immune activation) to loss of critical cells necessary for maintenance of a diverse response to common environmental pathogens. Sorting out the roles of these and other side effects of HIV-1 infection in the causation of AIDS will provide subject matter for HIV-1 researchers for a long time to come.

SUMMARY AND PERSPECTIVE

Quantitation of viral RNA in plasma, of viral DNA in cells, and of cells capable of being stimulated to express quiescent proviruses has led to a conceptual framework of the dynamics of HIV-1 replication and persistence in the host. The availability of suppressive antiviral therapy has brought disease progression under control, and it has also provided a means to perturb the dynamics of viral replication that has revealed additional features of viral persistence, features that show the necessity for lifelong therapy given the current treatment strategies. The dynamic changes in the sequence of the viral genome provide another critical source of information about viral replication and the selective pressures at work.

Given the ongoing generation of HIV-1 sequence information and the dynamic nature of viral populations, this virus will continue to provide the basis for studies of viral evolution and evolution in general. Deep sequencing strategies will reveal features of the viral population that have previously been hidden. There will be continuing efforts to define selective pressures in the context of sequence changes, thus providing a link between the selective pressure and the affected viral protein. Viral dynamics and evolution will continue to be central tools as our understanding of HIV-1 deepens. The lessons learned to date provide the foundation for continuing studies of virus–host interactions and for efforts to develop strategies to eradicate the virus completely.

REFERENCES

*Reference is also in this collection.

* Arts EJ, Hazuda DJ. 2011. HIV-1 antiretroviral drug therapy. *Cold Spring Harb Perspect Med* doi: 10.1101/cshperspect.a007161.

Bailey JR, Sedaghat AR, Kieffer T, Brennan T, Lee PK, Wind-Rotolo M, Haggerty CM, Kamireddi AR, Liu Y, Lee J, et al. 2006. Residual human immunodeficiency virus type 1 viremia in some patients on antiretroviral therapy is dominated by a small number of invariant clones rarely found in circulating CD4+ T cells. *J Virol* 80: 6441–6457.

Buzon MJ, Massanella M, Llibre JM, Esteve A, Dahl V, Puertas MC, Gatell JM, Domingo P, Paredes R, Sharkey M, et al. 2011. HIV-1 replication and immune dynamics are affected by raltegravir intensification of HAART-suppressed subjects. *Nat Med* 16: 460–465.

Chersich MF, Gray GE. 2005. Progress and emerging challenges in preventing mother-to-child transmission. *Curr Infect Dis Rep* 7: 393–400.

Chun TW, Carruth L, Finzi D, Shen X, DiGiuseppe JA, Taylor H, Hermankova M, Chadwick K, Margolick J, Quinn TC, et al. 1997. Quantification of latent tissue reservoirs and total body viral load in HIV-1 infection. *Nature* 387: 183–188.

Coffin JM. 1995. HIV population dynamics in vivo: Implications for genetic variation, pathogenesis, and therapy. *Science* 267: 483–488.

* Craigie R, Bushman FD. 2011. HIV DNA integration. *Cold Spring Harb Perspect Med* doi: 10.1101/cshperspect.a006890.

Davenport MP, Petravic J. 2010. CD8+ T cell control of HIV—A known unknown. *PLoS Pathog* 6: e1000728. doi: 10.1371/journal.ppat.1000728.

Dinoso JB, Kim SY, Wiegand AM, Palmer SE, Gange SJ, Cranmer L, O'Shea A, Callender M, Spivak A, Brennan T, et al. 2009. Treatment intensification does not reduce

residual HIV-1 viremia in patients on highly active antiretroviral therapy. *Proc Natl Acad Sci* 106: 9403–9408.

Fauci AS, Desrosiers RC. 1997. Pathogenesis of HIV and SIV. In *Retroviruses* (ed. Coffin JM, et al.), pp. 587–635. Cold Spring Harbor Laboratory Press, Cold Spring Harbor, NY.

Gandhi RT, Bosch RJ, Aga E, Albrecht M, Demeter LM, Dykes C, Bastow B, Para M, Lai J, Siliciano RF, et al. 2010. No evidence for decay of the latent reservoir in HIV-1-infected patients receiving intensive enfuvirtide-containing antiretroviral therapy. *J Infect Dis* 201: 293–296.

Ho DD, Neumann AU, Perelson AS, Chen W, Leonard JM, Markowitz M. 1995. Rapid turnover of plasma virions and CD4 lymphocytes in HIV infection. *Nature* 373: 123–126.

* Hu WS, Hughes SH. 2011. HIV-1 reverse transcription. *Cold Spring Harb Perspect Med* doi: 10.1101/cshperspect.a006882.

Jin X, Bauer DE, Tuttleton SE, Lewin S, Gettie A, Blanchard J, Irwin CE, Safrit JT, Mittler J, Weinberger L, et al. 1999. Dramatic rise in plasma viremia after CD8+ T cell depletion in simian immunodeficiency virus-infected macaques. *J Exp Med* 189: 991–998.

Joos B, Fischer M, Kuster H, Pillai SK, Wong JK, Boni J, Hirschel B, Weber R, Trkola A, Gunthard HF. 2008. HIV rebounds from latently infected cells, rather than from continuing low-level replication. *Proc Natl Acad Sci* 105: 16725–16730.

Josefsson L, King MS, Makitalo B, Brannstrom J, Shao W, Maldarelli F, Kearney MF, Hu WS, Chen J, Gaines H, et al. 2011. Majority of CD4+ T cells from peripheral blood of HIV-1-infected individuals contain only one HIV DNA molecule. *Proc Natl Acad Sci* 108: 11199–11204.

Jung A, Maier R, Vartanian JP, Bocharov G, Jung V, Fischer U, Meese E, Wain-Hobson S, Meyerhans A. 2002. Recombination: Multiply infected spleen cells in HIV patients. *Nature* 418: 144.

* Karn J, Stoltzfus CM. 2011. Transcriptional and post-transcriptional regulation of HIV-1 gene expression. *Cold Spring Harb Perspect Med* doi: 10.1101/cshperspect.a006916.

Kearney M, Maldarelli F, Shao W, Margolick JB, Daar ES, Mellors JW, Rao V, Coffin JM, Palmer S. 2009. Human immunodeficiency virus type 1 population genetics and adaptation in newly infected individuals. *J Virol* 83: 2715–2727.

Keele BF, Giorgi EE, Salazar-Gonzalez JF, Decker JM, Pham KT, Salazar MG, Sun C, Grayson T, Wang S, Li H, et al. 2008. Identification and characterization of transmitted and early founder virus envelopes in primary HIV-1 infection. *Proc Natl Acad Sci* 105: 7552–7557.

Kitrinos KM, Nelson JA, Resch W, Swanstrom R. 2005. Effect of a protease inhibitor-induced genetic bottleneck on human immunodeficiency virus type 1 env gene populations. *J Virol* 79: 10627–10637.

Klatt NR, Shudo E, Ortiz AM, Engram JC, Paiardini M, Lawson B, Miller MD, Else J, Pandrea U, Estes JD, et al. 2010. CD8+ lymphocytes control viral replication in SIV-mac239-infected rhesus macaques without decreasing

Cite this article as *Cold Spring Harb Perspect Med* doi: 10.1101/cshperspect.a012526

the lifespan of productively infected cells. *PLoS Pathog* **6**: e1000747.

Korber B, Gaschen B, Yusim K, Thakallapally R, Kesmir C, Detours V. 2001. Evolutionary and immunological implications of contemporary HIV-1 variation. *Br Med Bull* **58**: 19–42.

Kuiken C, Korber B, Shafer RW. 2003. HIV sequence databases. *AIDS Rev* **5**: 52–61.

* Lackner AA, Lederman MM, Rodriguez B. 2011. HIV pathogenesis—The host. *Cold Spring Harb Perspect Med* doi: 10.1101/cshperspect.a007005.

Maldarelli F. 2011. Targeting viral reservoirs: Ability of antiretroviral therapy to stop viral replication. *Curr Opin HIV AIDS* **6**: 49–56.

Maldarelli F, Palmer S, King MS, Wiegand A, Polis MA, Mican J, Kovacs JA, Davey RT, Rock-Kress D, Dewar R, et al. 2007. ART suppresses plasma HIV-1 RNA to a stable set point predicted by pretherapy viremia. *PLoS Pathog* **3**: e46. doi: 10.1371/journal.ppat.0030046.

* Malim MH, Bieniasz PD. 2011. HIV restriction factors and mechanisms of evasion. *Cold Spring Harb Perspect Med* doi: 10.1101/cshperspect.a006940.

Mansky LM, Temin HM. 1995. Lower in vivo mutation rate of human immunodeficiency virus type 1 than that predicted from the fidelity of purified reverse transcriptase. *J Virol* **69**: 5087–5094.

McMahon D, Jones J, Wiegand A, Gange SJ, Kearney M, Palmer S, McNulty S, Metcalf JA, Acosta E, Rehm C, et al. 2010. Short-course raltegravir intensification does not reduce persistent low-level viremia in patients with HIV-1 suppression during receipt of combination antiretroviral therapy. *Clin Infect Dis* **50**: 912–919.

Mellors JW, Rinaldo CRJ, Gupta P, White RM, Todd JA, Kingsley LA. 1996. Prognosis in HIV-1 infection predicted by the quantity of virus in plasma. *Science* **272**: 1167–1170.

Mens H, Kearney M, Wiegand A, Shao W, Schonning K, Gerstoft J, Obel N, Maldarelli F, Mellors JW, Benfield T, et al. 2010. HIV-1 continues to replicate and evolve in patients with natural control of HIV infection. *J Virol* **84**: 12971–12981.

Murray JM, Emery S, Kelleher AD, Law M, Chen J, Hazuda DJ, Nguyen BY, Teppler H, Cooper DA. 2007. Antiretroviral therapy with the integrase inhibitor raltegravir alters decay kinetics of HIV, significantly reducing the second phase. *AIDS* **21**: 2315–2321.

Palmer S, Wiegand AP, Maldarelli F, Bazmi H, Mican JM, Polis M, Dewar RL, Planta A, Liu S, Metcalf JA, et al. 2003. New real-time reverse transcriptase-initiated PCR assay with single-copy sensitivity for human immunodeficiency virus type 1 RNA in plasma. *J Clin Microbiol* **41**: 4531–4536.

Palmer S, Kearney M, Maldarelli F, Halvas EK, Bixby CJ, Bazmi H, Rock D, Falloon J, Davey RT Jr, Dewar RL, et al. 2005. Multiple, linked human immunodeficiency virus type 1 drug resistance mutations in treatment-experienced patients are missed by standard genotype analysis. *J Clin Microbiol* **43**: 406–413.

Palmer S, Maldarelli F, Wiegand A, Bernstein B, Hanna GJ, Brun SC, Kempf DJ, Mellors JW, Coffin JM, King MS. 2008. Low-level viremia persists for at least 7 years in patients on suppressive antiretroviral therapy. *Proc Natl Acad Sci* **105**: 3879–3884.

Perelson AS, Essunger P, Cao Y, Vesanen M, Hurley A, Saksela K, Markowitz M, Ho DD. 1997. Decay characteristics of HIV-1-infected compartments during combination therapy. *Nature* **387**: 188–191.

Richman DD, Wrin T, Little SJ, Petropoulos CJ. 2003. Rapid evolution of the neutralizing antibody response to HIV type 1 infection. *Proc Natl Acad Sci* **100**: 4144–4149.

Rouzine IM, Coffin JM. 2005. Evolution of human immunodeficiency virus under selection and weak recombination. *Genetics* **170**: 7–18.

Ruff CT, Ray SC, Kwon P, Zinn R, Pendleton A, Hutton N, Ashworth R, Gange S, Quinn TC, Siliciano RF, et al. 2002. Persistence of wild-type virus and lack of temporal structure in the latent reservoir for human immunodeficiency virus type 1 in pediatric patients with extensive antiretroviral exposure. *J Virol* **76**: 9481–9492.

Schmitz JE, Kuroda MJ, Santra S, Sasseville VG, Simon MA, Lifton MA, Racz P, Tenner-Racz K, Dalesandro M, Scallon BJ, et al. 1999. Control of viremia in simian immunodeficiency virus infection by CD8$^+$ lymphocytes. *Science* **283**: 857–860.

Shankarappa R, Margolick JB, Gange SJ, Rodrigo AG, Upchurch D, Farzadegan H, Gupta P, Rinaldo CR, Learn GH, He X, et al. 1999. Consistent viral evolutionary changes associated with the progression of human immunodeficiency virus type 1 infection. *J Virol* **7312**: 10489–10502.

Sharkey ME, Stevenson M. 2001. Two long terminal repeat circles and persistent HIV-1 replication. *Curr Opin Infect Dis* **14**: 5–11.

* Shaw GM, Hunter E. 2011. HIV transmission. *Cold Spring Harb Perspect Med* doi: 10.1101/cshperspect.a006965.

Sigal A, Kim JT, Balazs AB, Dekel E, Mayo A, Milo R, Baltimore D. 2011. Cell-to-cell spread of HIV permits ongoing replication despite antiretroviral therapy. *Nature* **477**: 95–98.

* Siliciano RF, Greene WC. 2011. HIV latency. *Cold Spring Harb Perspect Med* doi: 10.1101/cshperspect.a007096.

Svarovskaia ES, Cheslock SR, Zhang WH, Hu WS, Pathak VK. 2003. Retroviral mutation rates and reverse transcriptase fidelity. *Front Biosci* **8**: D117–D134.

* Swanstrom R, Coffin J. 2011. HIV-1 pathogenesis: The virus. *Cold Spring Harb Perspect Med* doi: 10.1101/cshperspect.a007443.

Tobin NH, Learn GH, Holte SE, Wang Y, Melvin AJ, McKernan JL, Pawluk DM, Mohan KM, Lewis PF, Mullins JI, et al. 2005. Evidence that low-level viremias during effective highly active antiretroviral therapy result from two processes: Expression of archival virus and replication of virus. *J Virol* **79**: 9625–9634.

* Walker B, McMichael A. 2011. The T-cell response to HIV. *Cold Spring Harb Perspect Med* doi: 10.1101/cshperspect.a007054.

Wei X, Ghosh S, Taylor ME, Johnson VA, Emini EA, Deutsch P, Lifson JD, Bonhoeffer S, Nowak MA, Hahn BH, et al. 1995. Viral dynamics in human immunodeficiency virus type 1 infection. *Nature* **373**: 117–122.

Wong JK, Strain MC, Porrata R, Reay E, Sankaran-Walters S, Ignacio CC, Russell T, Pillai SK, Looney DJ, Dandekar S. 2010. In vivo CD8[+] T-cell suppression of siv viremia is not mediated by CTL clearance of productively infected cells. *PLoS Pathog* **61:** e1000748.

Zhang L, Ramratnam B, Tenner-Racz K, He Y, Vesanen M, Lewin S, Talal A, Racz P, Perelson AS, Korber BT, et al. 1999. Quantifying residual HIV-1 replication in patients receiving combination antiretroviral therapy. *N Engl J Med* **340:** 1605–1613.

HIV Pathogenesis: The Host

A.A. Lackner[1], Michael M. Lederman[2], and Benigno Rodriguez[2]

[1]Tulane National Primate Research Center, Tulane University Health Science Center, Covington, Louisiana 70443

[2]Department of Molecular Biology and Microbiology, Case Western Reserve University, Cleveland, Ohio 44106

Correspondence: alackner@tulane.edu

Human immunodeficiency virus (HIV) pathogenesis has proven to be quite complex and dynamic with most of the critical events (e.g., transmission, CD4[+] T-cell destruction) occurring in mucosal tissues. In addition, although the resulting disease can progress over years, it is clear that many critical events happen within the first few weeks of infection when most patients are unaware that they are infected. These events occur predominantly in tissues other than the peripheral blood, particularly the gastrointestinal tract, where massive depletion of CD4[+] T cells occurs long before adverse consequences of HIV infection are otherwise apparent. Profound insights into these early events have been gained through the use of nonhuman primate models, which offer the opportunity to examine the early stages of infection with the simian immunodeficiency virus (SIV), a close relative of HIV that induces an indistinguishable clinical picture from AIDS in Asian primate species, but importantly, fails to cause disease in its natural African hosts, such as sooty mangabeys and African green monkeys. This article draws from data derived from both human and nonhuman primate studies.

Untreated, about half of HIV-infected persons will develop major opportunistic complications reflective of profound immune deficiency within 10 years of acquiring infection; some succumb within months, yet others remain well as long as 20 years or more after acquiring infection. Occasional variability in disease course may be owing to variations in HIV itself; rare deletions in the *nef* gene have been associated with a slower disease course (Deacon et al. 1995; Kirchhoff et al. 1995) and even less frequently, mutations in the *vpr* gene have been found in some slow progressors (Lum et al. 2003).

Host defenses undoubtedly play an important role in the course of HIV disease. The importance of diversity of T-cell recognition in disease control was suggested by the observation that homozygosity for class I human leukocyte antigen (HLA) molecules is associated with an accelerated disease course (Carrington et al. 1999; Tang et al. 1999). More specifically, certain class I HLA types (Goulder et al. 1996; Kaslow et al. 1996) are associated with a more benign disease course, whereas others (Kaslow et al. 1990) are associated with a more aggressive disease course. Genetic analyses also have implicated natural killer cells and their ligands as important genetic determinants of disease outcome (Martin et al. 2002). Other host elements appear to be important in determining the course of disease. Although persons who are homozygous for a 32-base-pair deletion in CCR5 are nearly completely protected from acquiring HIV infection (Dean et al. 1996; Liu et al. 1996; Samson et al. 1996), heterozygous individuals are infectible, yet their course tends to be less aggressive (Ioannidis et al. 2001).

A very small proportion of infected persons manage to control HIV replication in the absence of antiretroviral therapy. Although rare persons with relatively slowly progressive disease have been infected with defective viruses as indicated above, these "elite controllers" appear to be infected with viruses that are fully replication competent and lacking unique signatures (Blankson et al. 2007; Miura et al. 2008). Humoral defenses mediated by neutralizing antibodies do not appear to mediate control of viral replication (Bailey et al. 2006), but evidence to date implicates T-cell-mediated responses to HIV as important determinants of elite control. Approximately half of these elite controllers express HLA-B*57, HLA-B*5801, or HLA-B*27 (Emu et al. 2008), implicating HLA-class I–restricted recognition of HIV peptides as important in control of HIV replication. The quality of these T-cell responses also may be important as elite controllers tend to have $CD8^+$ T cells that are polyfunctional in terms of their ability to express cytokines and degranulate after HIV peptide stimulation (Migueles et al. 2002, 2008; Betts et al. 2006; Emu et al. 2008; Owen et al. 2010).

Our understanding of the pathogenesis of AIDS has evolved dramatically since its initial discovery. Although originally thought to involve a period of viral latency, it is now clear that HIV replication occurs at a high level throughout infection and that there is a highly dynamic interplay between the host immune response, attempts by the host to replenish cells that are destroyed, as well as virus and viral evolution that appear to differ among various tissue compartments (Horton et al. 2002; Paranjpe et al. 2002; Ryzhova et al. 2002; Gonzalez-Scarano and Martin-Garcia 2005). Progress in defining

both molecular and cellular viral targets of HIV infection has also led to important discoveries that allow us to better understand the various stages of infection as well as the events leading to immunodeficiency. The cellular receptors for HIV and SIV are the CD4 molecule on T cells and monocyte/macrophage lineage cells along with a chemokine receptor; most commonly CCR5 and CXCR4 (Alkhatib et al. 1996; Moore et al. 2004). In humans, infection is typically established by virus that uses CCR5 for cellular entry, but with time, viruses often emerge that are capable of using another receptor, CXCR4, for infection of target cells (Koot et al. 1999). These viruses become more prevalent with advanced disease and although it is likely that advancing immune deficiency predisposes to the emergence of these variants, CXCR4 is more broadly expressed by human CD4 T cells than is CCR5, and the emergence of CXCR4 using viruses in untreated infection is often associated with an accelerated disease course. Infection of $CD4^+$ T cells may lead to their destruction (even without productive infection for CXCR4 using viruses [Doitsh et al. 2010]). Infection of monocyte/macrophage lineage cells is also important, as these are likely major reservoirs for viral replication and persistence, and may also contribute to disease progression, immune deficiency, and AIDS-associated syndromes in nonlymphoid organs such as the brain, heart, and kidney (Shannon 2001; Ross and Klotman 2004; Hasegawa et al. 2009).

HISTORY

Clinical Manifestations of HIV Infection

Most persons who become HIV infected experience an illness characterized often by fever, sore throat, lymphadenopathy, and rash (Schacker et al. 1996). These symptoms are often severe enough that persons will seek medical attention, but as they are nonspecific and self-limited, they are often attributed to nonspecific viral infections, and testing for HIV is often not performed. In the first few weeks of infection, levels of serum antibodies to HIV proteins are typically not sufficiently elevated to permit diagnosis of infection by enzyme-linked immunosorbent assay (ELISA) and immunoblot, but high levels of HIV RNA are readily detectable in plasma. During these first few weeks of infection, there is profound destruction of CCR5$^+$ CD4$^+$ memory cells in gut tissue in both SIV and HIV infection (Veazey et al. 1998; Brenchley et al. 2004b), but interestingly, gastrointestinal symptoms are not common during this period in HIV infection (Schacker et al. 1996). High-level viremia typically diminishes as acute infection symptoms resolve and a "set point" of viremia is established that varies from rare elite controllers in whom virus levels in plasma are typically below levels of detection by commercial assays (<40 copies/mL) to levels in excess of 100,000 copies per mL. With resolution of symptoms, the HIV-infected person may be completely without signs or symptoms of disease, yet most will experience progressive depletion of CD4$^+$ T cells from circulation and from lymph nodes. Most persons will remain free of AIDS-defining illness until the circulating CD4 T-cell count falls to levels of 200 cells/μL or lower. Although persons with higher levels of virus in plasma tend to progress to the immune deficiency of AIDS more rapidly (Mellors et al. 1996, 1997), the magnitude of viremia is an incomplete predictor of the pace of disease progression (Rodriguez et al. 2006) and it appears that markers of systemic immune activation are useful predictors of disease progression risk (Giorgi et al. 1993; Liu et al. 1996; Deeks et al. 2004).

The clinical complications of advanced untreated HIV infection typically comprise infectious or malignant complications reflective of the profound impairments in T-cell-mediated immunity. Thus infections attributable to organisms such as *Pneumocystis jirovecii*, mycobacteria, cytomegalovirus, *Toxoplasma gondii*, and Cryptococcus as well as the occurrence of malignancies related to viral pathogens such as non-Hodgkins lymphoma and Kaposi's sarcoma are common. Nonetheless, the profound immune deficiency also affects humoral defenses, placing infected persons at increased risk for infection with pathogens like *Streptococcus pneumoniae* (Janoff et al. 1992; Hirschtick et al. 1995). With effective suppression of HIV replication after administration of antiretroviral therapies, immune function typically improves and risks for these life-threatening complications diminish. In the current era and where there is broad access to effective combination antiretroviral therapies, the predicted survival of the HIV-infected patient can approach that of the general population if treatment is initiated early in the course of infection (Antiretroviral Therapy Cohort Collaboration 2008; van Sighem et al. 2010). In the current era, cardiovascular disease, serious liver disease related to coinfection with hepatitis viruses, renal insufficiency, and a changing spectrum of malignant disorders are major causes of morbidity and mortality in HIV infection (Palella et al. 2006; Marin et al. 2009; Hasse et al. 2011).

Early Recognition of Key Aspects of Pathogenesis in Humans

With the first reports of AIDS, astute clinicians recognized that impairments in host defenses must underlie the opportunistic infections they were seeing (Gottlieb et al. 1981; Masur et al. 1981; Siegal et al. 1981). Thus, profound depletion of CD4 T cells was recognized immediately (Gottlieb et al. 1981; Masur et al. 1981; Siegal et al. 1981) as was dysregulation of B-cell function (Lane et al. 1983). Interestingly, early investigators also recognized that despite profound immune deficiency, immune cells also showed evidence of immune activation (Gottlieb et al. 1981; Lane et al. 1983). With the identification of HIV (Barre-Sinoussi et al. 1983; Gallo et al. 1984; Levy et al. 1984) and the recognition that infection of CD4 T cells was cytopathic (Zagury et al. 1986), a key mechanism for circulating CD4$^+$ T-cell losses was reasonably imputed. The central role of viral replication in CD4 T-cell losses and immune deficiency was further confirmed by the reliable increases in CD4 T-cell numbers and enhancement of immune function with administration of suppressive antiretroviral therapies (Autran et al. 1997; Lederman et al. 1998). Nonetheless, the precise mechanisms whereby infection with HIV resulted in progressive immune deficiency remained ill defined. Indirect mechanisms for cell loss were suggested by the relative infrequency with which circulating blood and lymph node cells could be shown to be HIV infected (Douek et al. 2002). The role and potential importance of immune activation in disease pathogenesis was suggested by epidemiologic studies in which markers of immune activation proved powerful predictors of the risk of disease progression (Giorgi et al. 1993, 1999; Liu et al. 1997, 1998; Hazenberg et al. 2003; Deeks et al. 2004; Wilson et al. 2004).

Cellular Targets for SIV/HIV

HIV and SIV use a two-receptor model for infection that requires both CD4 and a chemokine receptor that results in $CD4^+$ T cells and monocyte/macrophage lineage cells being the primary targets for infection. In addition, the activation state of $CD4^+$ T lymphocytes has a significant impact on the ability of the virus to replicate successfully. As newly produced lymphocytes emerge from the thymus, they are generally considered "naïve" in that they have never encountered their cognate antigen and are thus in a "resting" state. Naïve, resting cells are abundant in the blood and in organized lymphoid tissues (lymph nodes, intestinal Peyer's patches, etc.). Cells that have previously encountered their antigen are considered memory cells, which can be distinguished by expression of specific cell-surface antigens. In addition, memory cells can also be subdivided into short-lived, "effector memory" cells, which are actively secreting cytokines, and/or long-lived "central" memory cells, which may be "resting" or rapidly activated to mount immune responses on further exposure to the antigen. Activated $CD4^+$ T cells, identified in part by expression of CD25, CD69, HLA-DR, etc., are able to support HIV and SIV infection quite well, whereas resting $CD4^+$ T cells, and especially naïve CD4 T cells, do not (Stevenson et al. 1990; Zack et al. 1990; Chou et al. 1997). In part this may be because resting naïve $CD4^+$ T cells generally do not express CCR5 and thus are resistant to SIV and to HIV, viruses that typically use CCR5 for cellular entry. However, resting central memory cells, which express low levels of CCR5, have been shown

to be significant targets for SIV in vivo (Li et al. 2005). In addition, activated cells are generally transcribing DNA, which logically would promote more viral replication.

KEY ADVANCES

Mucosal Tissues and HIV Infection

It is now evident in SIV-infected macaques and HIV-infected humans that mucosal tissues are not only primary sites of viral transmission but also the major sites for viral replication and $CD4^+$ T-cell destruction, regardless of route of transmission. Furthermore, intestinal mucosal tissues appear to be major sites of HIV/SIV persistence even after administration of suppressive antiretroviral therapies (Chun et al. 2008). Understanding the basis of this central role of mucosal tissues in the pathogenesis of AIDS is critical for efforts to develop strategies to prevent or treat AIDS.

General features of mucosal immune system. The intestinal immune system is considered the largest single immunologic organ in the body containing upwards of 40% of all lymphocytes (Schieferdecker et al. 1992; MacDonald and Spencer 1994). Thus, when considering the rest of the mucosal immune system, including the lungs, reproductive tract, urinary tract, mammary glands, etc., it is clear that the mucosal immune system dwarfs the systemic immune system. Furthermore, in contrast to the systemic immune system, most of the $CD4^+$ T cells in the mucosal immune system are $CCR5^+$, activated memory $CD4^+$ T cells. This is of enormous importance with respect to the pathogenesis of AIDS because these represent the preferred cellular target for HIV/SIV infection (Veazey et al. 2002; Brenchley et al. 2004b; Mehandru et al. 2004). This is also reflected in the fact that productive infection of peripheral $CD4^+$ T cells is rare (0.01%–1%) (Brenchley et al. 2004a), whereas infection of mucosal CD4 cells is quite common with estimates of 60% of mucosal memory $CD4^+$ cells infected within days of infection (Mattapallil et al. 2005).

The unique challenges faced by the mucosal tissues and the immune system have resulted in a structurally and functionally distinct mucosal immune system. In the case of the intestine, this includes both inductive, organized lymphoid tissues and diffuse effector lymphoid tissues. The inductive sites in the intestine and most other mucosal sites include widely scattered but well-organized lymphoid follicles best exemplified by solitary and aggregated (Peyer's patches) lymphoid follicles. In general, these are found in the tonsils, the terminal portion of the small intestine (particularly the ileum), as well as the terminal portion of the large intestine (rectum), cecum, and appendix.

In addition to the organized lymphoid tissues of the inductive arm of the mucosal immune system, there is an even larger pool of immune cells diffusely scattered throughout mucosal tissues that serves as the "effector" arm of the mucosal immune system (Mowat and Viney 1997). The effector arm consists of large numbers of various subsets of lymphocytes, macrophages, dendritic cells, and other immune cells that are scattered diffusely throughout the lamina propria and epithelium of mucosal tissues (Mowat and Viney 1997). These cells are responsible for carrying out the major effector functions of the intestinal immune response that are "initiated" in inductive sites.

Immunophenotypic composition of the mucosal immune system. As mentioned above, mucosal tissues contain the majority of all the lymphocytes and macrophages in the body. From an anatomic perspective, the lymphocyte populations can be divided into those present in epithelium (intraepithelial lymphocytes [IEL]) and those in the underlying lamina propria (lamina propria lymphocytes [LPL]). The LPL can be further subdivided into those from inductive sites (organized lymphoid nodules) and effector sites (diffuse lamina propria). More than 90% of the IEL are $CD3^+$ T cells, $\sim$80% of which express CD8 (Mowat and Viney 1997; Veazey et al. 1997). In addition, $\sim$10% of IEL express the $\gamma\delta$ T-cell receptor (TCR) (Viney et al. 1990). In contrast, the phenotype of lymphocytes in the lamina propria is remarkably different, with most lymphoid phenotypes being represented (Mowat and Viney 1997). Most importantly, the lamina propria of mucosal tissues

contains a vast reservoir of $CD4^+$ T cells. In normal humans and nonhuman primates, the ratio of $CD4^+$ to $CD8^+$ T cells in the lamina propria is similar to that in peripheral blood and lymph nodes (Mowat and Viney 1997; Veazey et al. 1997; Veazey 2003). However, in contrast to peripheral lymphoid tissues, a much larger percentage of mucosal $CD4^+$ T cells express CCR5, have a memory phenotype, and express markers of activation particularly when examining LPL from the diffuse lamina propria separately from organized lymphoid nodules (James et al. 1987; Zeitz et al. 1988; Schieferdecker et al. 1992; Mowat and Viney 1997; Veazey et al. 1997, 1998). Furthermore, a large percentage of $CD4^+$ T cells also produce cytokines in situ, indicating that they are activated, terminally differentiated effector cells (Mowat and Viney 1997).

Combined, these data indicate that the largest pool of activated, terminally differentiated, memory $CCR5^+CD4^+$ T cells resides in mucosal tissues (particularly the diffuse lamina propria) and not in peripheral blood or lymph nodes. HIV and SIV preferentially infect these memory $CCR5^+$ $CD4^+$ T cells in immune effector sites (diffuse lamina propria), causing rapid depletion of these cells by 21 d after infection. Subsequently, most infected cells in the intestine are present in immune inductive sites represented by organized lymphoid nodules in the lamina propria (Veazey et al. 1998, 2000b, 2001b). This dramatic and rapid loss of $CD4^+$ T cells in mucosal effector sites in SIV-infected macaques is associated with subclinical opportunistic infections as well as significant alterations in intestinal structure and function (Heise et al. 1994; Stone et al. 1994). In humans, symptomatic disease of the intestine is rare in early HIV infection, yet the early damage to this important defense system may play a key role in both progressive immune deficiency and immune dysregulation.

Of additional importance is a subset of $CD4^+$ T cells known as Th17 cells because they produce IL-17 and IL-22 but not interferon γ or IL-4 (Steinman 2007). Of particular relevance for this discussion is the role these cells likely play in enterocyte homeostasis and production of antimicrobial defensins, both of which are critical for maintenance of the mucosal barrier (Kolls and Linden 2004; Liang et al. 2006). Recent evidence indicates that Th17 cells are even more profoundly depleted in the intestinal mucosa of HIV- and SIV-infected individuals than the general CD4$^+$ CCR5$^+$ T-cell population (Brenchley et al. 2008; Favre et al. 2009). Thus, the loss of these Th17 cells provides a possible direct link between CD4$^+$ T-cell destruction and dysfunction of the intestinal mucosa.

Interactions between the mucosal immune system and intestinal structure and function. Alterations in intestinal structure and function associated with HIV/SIV infection has long been recognized (Batman et al. 1989; Ullrich et al. 1989; Cummins et al. 1990; Heise et al. 1993, 1994). Histologically, villus atrophy and increased epithelial apoptosis in the villus tips was often linked to increased proliferation of crypt cells leading to crypt hyperplasia. This lesion of "crypt hyperplastic villous atrophy" had been associated with mucosal T-cell activation in vitro (MacDonald and Spencer 1988; Ferreira et al. 1990; Field 2006; Turner 2009). However, in the case of AIDS, the dominant recognized feature was one of immune suppression rather than activation, although even the earliest reports of AIDS in humans provide evidence of immune activation (Gottlieb et al. 1981; Masur et al. 1981). Over time, however, it was recognized that immune activation is a major feature of SIV and HIV infection and that intestinal immune dysfunction can result in structural changes to the intestinal mucosa and cause breakdown of the intestinal epithelial barrier (MacDonald and Spencer 1992; Clayburgh et al. 2004; Kolls and Linden 2004; Liang et al. 2006; Weber and Turner 2007; Estes et al. 2010). The molecular basis for damage to the intestinal epithelial barrier is now beginning to come into focus aided by functional genomics approaches and their frequent application to studies of AIDS pathogenesis.

Normal function of the mucosal barrier requires not only an intact epithelium joined by tight junctions, but also coordinated function of multiple cell types that occupy distinct anatomical positions and maintain reciprocal interrelationships (Traber 1997; Turner 2009). The sudden and massive destruction of activated effector memory CD4$^+$, CCR5$^+$ cells, and Th17 cells would be expected to disrupt this communication network linking epithelial cells and the intestinal immune system (Shanahan 1999). A consequence of this disruption is likely deprivation of epitheliotropic factors required for epithelial cell growth, maintenance, and renewal leading to increased epithelial cell apoptosis and death. In support of this concept, significant down-modulation of genes regulating intestinal epithelial cell growth and renewal along with increased expression of inflammation and immune activation genes, and activated caspase 3 protein expression in epithelial cells has been observed in primary HIV infection (Sankaran et al. 2005, 2008; George et al. 2008). Additionally, increases in proinflammatory cytokine production in the colon as early as 6–10 d post-SIV infection (Abel et al. 2005) and in the intestine of HIV-infected patients (Reka et al. 1994; Olsson et al. 2000; McGowan et al. 2004) may further facilitate mucosal damage by activating myosin light chain kinase (MLCK), which has been implicated as a major player in initiating damage to the intestinal epithelial barrier (Turner 2006, 2009).

Early Targets of Infection, Amplification, and Viral Dissemination

Although HIV/SIV may undergo limited replication within dendritic cells in mucosal surfaces that contain them (vagina, anus, tonsil—all lined by stratified squamous epithelium) (Spira et al. 1996; Hu et al. 2000), the primary substrate for HIV/SIV replication is memory CD4$^+$ T cells expressing CCR5 (hereafter referred to as "primary target cells"). How the virus reaches these cells, which are abundant in the lamina propria of all mucosal tissues, varies depending on the route and site of transmission.

In the case of the rectal mucosa, once the virus crosses the epithelium, either via small mucosal breaks or via M cells, the virus will encounter a high density of primary target cells to support significant levels of viral replication (amplification). It is particularly worth noting that M cells form an intraepithelial pocket containing CD4$^+$ memory cells and dendritic cells, which would greatly facilitate HIV/SIV replication (Pope et al. 1994; Neutra et al. 1996). After local replication and amplification, it is likely that virus and viral-infected cells will migrate to draining lymph nodes and from there to the rest of the body.

For transmission via the vagina (and presumably the anus), intraepithelial dendritic cells appear to play a major role. Although there are significant numbers of primary target cells in the vaginal lamina propria (Veazey et al. 2003), there are also data that indicates that dendritic cells can rapidly carry virus to regional lymph nodes (Hu et al. 2000). In this case, it appears that spread to regional lymph nodes occurs before there is significant local replication of virus in the vaginal lamina propria. This is likely because the virus is subverting normal trafficking patterns of intraepithelial dendritic cells that bring antigen to immune inductive sites (regional lymph nodes), which are lacking in the vaginal mucosa as compared with intestinal mucosa.

Although SIV can be found in draining lymph nodes within 18 h of vaginal inoculation, it is interesting to note that a delay in viremia often occurs with mucosal inoculation compared with intravenous inoculations of macaques (Ma et al. 2004; Miller et al. 2005). This likely occurs because the virus has to replicate locally for a period of time to generate sufficient progeny to cause a spreading infection or, in the case of the vagina, because of the low density of primary target cells that would be found in a regional lymph node. Although lymph nodes contain a high density of lymphocytes, unless that node is draining a site of inflammation, the vast majority of the T cells will be CCR5$^-$, resting, and naïve, and thus relatively resistant to infection. In support of this, Miller et al. (2005) have shown focal viral replication occurring in the lamina propria of cervicovaginal tissues before productive systemic infection, suggesting that local viral replication at the site of exposure was necessary to amplify virus before systemic infections could proceed.

In contrast to mucosal transmission, which provides a selective barrier based on the ability of the virus to contact target cells either directly or by using existing biological processes, intravenous transmission poses no such barrier. Thus, the virus quickly disseminates to all tissues, including those that support high levels of viral replication (mucosal tissues). Viremia can be detected as early as 2–3 d after intravenous infection in macaques, with peak viremia occurring 10–14 d after intravenous infection. At the time of peak viremia, virus can be found in lymphoid tissues throughout the body including thymus, spleen, peripheral lymphoid organs, and mucosal lymphoid tissues. In addition, virus is readily found in the central nervous system by 14 d after infection. Although virus is readily found in tissues by 14 d after infection, it is difficult to find infected cells in tissues by in situ hybridization or immunohistochemistry before that time, except in effector sites in mucosal lymphoid tissues such as the lamina propria of the intestinal tract, where significant numbers of productively infected cells have been detected within 3–4 d of intravenous inoculation (Sasseville et al. 1996). Combined, these data suggest that replication in mucosal tissues is not only important for transmission, but also critical for initial viral replication and amplification, regardless of the route of transmission.

Systemic Lymphoid Tissues

By 2 wk after intravenous inoculation of macaques with pathogenic SIV, the virus is widely distributed and easily found in all lymphoid organs. Within these tissues, evidence of productive infection is first seen in individual cells in the paracortex of the lymph nodes, periarteriolar lymphoid sheaths in the spleen (where T cells predominate), and in the thymic medulla, which contains mature lymphocytes as opposed to the thymic cortex. Recent work has shown that within these tissues at these early time points, the primary targets are memory phenotype CCR5$^+$ CD4$^+$ T cells just as they are in mucosal tissues (Veazey et al. 2000a; Mattapallil et al. 2005). The primary difference is that these cells represent a minority of the cells present in systemic lymphoid tissues.

Between 2 and 3 wk after infection, the picture changes somewhat. Although the majority of infected cells in lymphoid tissues are still CD4$^+$ T cells, infection of macrophages becomes readily apparent. This is generally thought to be a result of viral evolution, particularly in the case of cloned viruses such as SIVmac239, which does not readily infect macrophages in vitro.

In addition to infection of macrophages, diffuse labeling for viral RNA and protein over germinal centers in lymphoid organs (referred to as a "follicular" pattern) generally appears in this same time frame. This is largely owing to trapping of antigen/antibody complexes that contain intact virions on follicular dendritic cells. It has been hypothesized that this pool of virions on follicular dendritic cells may represent a major reservoir of infectious HIV-1 (Haase et al. 1996). The appearance of abundant virus on follicular dendritic cells is dependent on the generation of a humoral immune response, which probably occurs more consistently in humans than in macaques where up to 25% or more of the animals mount very poor immune responses and progress to disease quite rapidly (200 d or less) (Westmoreland et al. 1998).

Infection of the thymus is of particular interest because of its role in T-cell renewal. It is well established that dramatic thymic dysinvolution occurs in both HIV-infected humans and SIV-infected macaques. This led to the hypothesis that loss of thymic function was at least partially responsible for the decline in CD4$^+$ T cells that accompanies AIDS progression. During the first few weeks of infection, significant changes in cell proliferation, apoptosis, and percentages of T-cell precursors are observed in the thymus coincident with the presence of infected cells and primary viremia (Wykrzykowska et al. 1998). Of particular interest is the marked rebound in T-cell progenitors accompanied by increased levels of cell proliferation in the thymus. This occurred in the face of persistent high-level virus replication and provides strong evidence that the thymus has significant regenerative capacity through at least the first 2 mo of infection. However, by

24 wk of infection, morphologic evidence of severe thymic damage is evident in most SIV-infected animals (but can occur earlier in rapid progressors) (Lackner 1994). The length of this apparent window during which the thymus can regenerate is of importance when considering when to start antiretrovirals and for immune restoration strategies. Data from nonhuman primate studies imply that if combination drug therapy for HIV is not started early enough in infection, limited T-cell regeneration will occur with minimal help from the thymus, mostly as the result of clonal expansion of preexisting cells, resulting in a limited T-cell repertoire. Although much of these data would suggest that the thymus should be important in regeneration, infection studies in thymectomized macaques clearly show that the thymus has little if any role to play in disease progression or the rate of CD4$^+$ T-cell depletion in SIV-infected macaques (Arron et al. 2005). In humans with HIV infection, there is ample evidence of thymic dysfunction as characterized by diminished numbers of recent thymic emigrants and circulating naïve T cells (Douek et al. 1998; Dion et al. 2007), and these indices are linked to the outcome of infection both in the absence and presence of antiretroviral therapy (Dion et al. 2007). Yet in small numbers of humans with HIV infection who had had thymectomy, the course of infection did not appear dramatically altered (Haynes et al. 1999).

Early Immune Response

It is now clear that both HIV and SIV selectively infect and destroy memory CD4$^+$ T cells (both central and effector cells) resulting in subsequent impairment of immune responses to not only the infecting virus, but to other antigens as well. This tropism for memory CD4$^+$ T cells eventually leads to the profound immunodeficiency of AIDS and likely underlies the fact that effective immunity resulting in clearance of the infection has yet to be documented in an HIV-infected patient. This rapid and profound elimination of memory CD4$^+$ T cells in infected hosts undoubtedly affects the immune system from the onset, but understanding these consequences is confounded at least in part by the compartmentalization, dynamics, and resilience of the immune system, especially in mucosal tissues.

Acute SIV infection elicits early and relatively robust immune responses in SIV-infected macaques. Within 1–4 wk of SIV infection, marked increases in $CD8^+$ (fivefold to 10-fold) and natural killer (NK) cell (two- to threefold) proliferation are observed in the blood (Kaur et al. 2000). Interestingly, most of this proliferation appears to be nonspecific as few of the responding $CD8^+$ T cells can be shown to be specific for SIV antigens during peak viremia (Veazey et al. 2003a). Similarly, few of the $CD8^+$ T cells and even fewer $CD4^+$ T cells are demonstrably virus specific in HIV-infected patients (Betts et al. 2001). This discrepancy may be in part owing to the specificity of current assays, but is more likely a result of immune activation mediated through the destruction of infected cells or through cytokines/chemokines produced by cells directly responding to antigens or viral gene products (Grossman et al. 2006).

Mucosal tissues are also a major site for generation of virus-specific immune responses. Using tetramer technology in genetically defined macaques, strong virus-specific cytotoxic T-lymphocyte (CTL) responses are detected in mucosal sites within 14–21 d of infection (Veazey et al. 2003a; Reynolds et al. 2005). In both intravenously and rectally inoculated macaques, virus-specific CTLs appear to emerge simultaneously in blood and intestines, although the percentages of mucosal CTLs often exceed those in the blood in both early and chronic infection (Veazey et al. 2001a, 2003a; Stevceva et al. 2002). Interestingly, few virus-specific CTLs were detected in the gut of vaginally inoculated animals using similar (tetramer) techniques (Reynolds et al. 2005), which could reflect differences in CTL development or homing depending on the route of transmission, but this remains to be fully explored. In addition to cell-mediated immune responses, infection with SIV or HIV also results in generation of diverse antibody responses, although some strains of SIV are quite poor at eliciting neutralizing antibodies.

Regardless, neither robust cellular nor humoral immune responses are sufficient to clear the infection, and correlates of effective immunity to SIV and HIV remain to be determined.

Although there is consensus that early infection with SIV and HIV results in robust early immune responses, it is also apparent that the magnitude as well as quality of the immune response diminishes with time. In humans, CTL-mediated killing is more rapid in early versus chronic HIV infection (Asquith et al. 2006). Moreover, $CD4^+$ immune responses to tetanus toxoid and hepatitis C virus (in coinfected patients) also decline as the disease progresses (Harcourt et al. 2006). In the animal model of pathogenic HIV infection, studies using tetramer technology have shown that the levels of SIV-specific CTL diminish with time (Veazey et al. 2003a). Thus, both human and animal data suggest that the development of AIDS occurs gradually despite the fact that most of the memory $CD4^+$ T cells are eliminated within days of infection and never fully restored, at least in animals that progress to AIDS. This may be partly explained by the fact that, in the majority of animals, sustained increases in $CD4^+$ T-cell turnover throughout SIV infection usually result in maintenance of "threshold levels" of mucosal $CD4^+$ T cells (5%–10% of normal values), which seems sufficient to maintain immune function, although subclinical opportunistic infections are frequently found in macaques within weeks of infection (Lackner et al. 1994). Therefore, the ongoing destruction of memory $CD4^+$ T cells is likely balanced by continuous proliferation of these cells in attempts to maintain this threshold. Further evidence for this model comes from studies demonstrating that macaques that fail to maintain proliferation of memory $CD4^+$ T cells rapidly progress to AIDS (Picker et al. 2004).

Antiretroviral Therapy and the Mucosal Immune System

The potential for antiretroviral therapy (ART) to restore mucosal $CD4^+$ T cells has only begun to be examined and has been particularly difficult to assess in acute infection. Small studies in humans and SIV-infected macaques have suggested near-complete restoration of mucosal $CD4^+$ T cells when treatment is initiated very early (George et al. 2005; Guadalupe et al. 2006). In contrast, other studies have not shown a significant restoration of mucosal $CD4^+$ T cells either early or late in infection (Anton et al. 2003; Mehandru et al. 2006; Poles et al. 2006), whereas still other studies have shown significant restoration of $CD4^+$ T cells including the Th17 subset (Macal et al. 2008). In summary, although $CD4^+$ T-cell numbers in the peripheral blood often fully reconstitute in patients on ART, there is considerable controversy regarding the capacity of ART to restore intestinal $CD4^+$ T cells. Furthermore, even after suppression of detectable plasma viremia by ART, HIV can be detected and recovered in the intestinal mucosa and other tissues (Anton et al. 2003; Mehandru et al. 2006; Poles et al. 2006; Belmonte et al. 2007; Chun et al. 2008). Persistent viral replication in the intestine of SIV-infected long-term nonprogressing macaques with undetectable viremia has also been described (Ling et al. 2004). These data suggest that the intestinal immune system is an important reservoir of SIV/HIV infection and that ongoing viral replication occurs in the intestine of patients on ART, despite what appears to be nearly complete suppression of viral levels in the blood. Thus, a major challenge for antiretroviral control of HIV infection appears to be in mucosal tissues, particularly the intestine.

NEW RESEARCH AREAS

The Role of Immune Activation in HIV and SIV Disease Pathogenesis

Despite profound immune deficiency, there is evidence of profound immune activation in HIV infection. T lymphocytes, B lymphocytes, and antigen-presenting cells of the innate immune system have phenotypic and functional evidence of activation. Hyperglobulinemia and increased circulating levels of proinflammatory cytokines are characteristic, and although type 1 interferon levels are often difficult to measure in circulation, transcriptional analyses indicate

that HIV infection is associated with profound activation of interferon-responsive genes (Woelk et al. 2004; Hyrcza et al. 2007). T lymphocytes often express high levels of activation markers such as CD38 and HLA-DR (Giorgi et al. 1993). Markers of immune senescence such as CD57 (Brenchley et al. 2003) and immune exhaustion such as programmed death receptor type 1 (PD-1) (Day et al. 2006; Trautmann et al. 2006) are elevated, and cells expressing each of these markers have demonstrable impairments in response to TCR stimulation. Markers of immune activation are recognized predictors of disease outcome in HIV infection (Giorgi et al. 1993, 1999; Liu et al. 1997, 1998; Hazenberg et al. 2003; Deeks et al. 2004; Wilson et al. 2004). Expression of the activation marker CD38 on T cells is a valuable predictor of disease outcome in HIV infection (Giorgi et al. 1993, 1999; Liu et al. 1997). Likewise, plasma levels of IL-6, TNF receptors and markers of coagulation (d-dimer levels) predict mortality in treated HIV infection (Kuller et al. 2008; Kalayjian et al. 2010). One of the hallmarks of immune activation in HIV infection is a marked increase in T-cell turnover, as measured by incorporation of bromodeoxyuridine or deuterated glucose and expression of the nuclear antigen Ki-67, which indicates cell cycling (Sachsenberg et al. 1998; Douek et al. 2001; Kovacs et al. 2001; Mohri et al. 2001). This increase in cycling is seen in both CD4 and CD8 T-cell populations (Kovacs et al. 2001) and is especially striking among central memory cells in both humans and in SIV-infected macaques (Picker et al. 2004; Sieg et al. 2005). Activated cycling $CD4^+$ T cells are both more susceptible to productive HIV infection (Zack et al. 1990; Ramilo et al. 1993) and also tend to die ex vivo, likely as a result of programmed cell death (Sieg et al. 2008).

In nonhuman primate models of SIV infection, immune activation and inflammation distinguish the pathogenic models of SIV infection in rhesus macaques from the nonpathogenic outcomes of SIV infection in naturally adapted hosts that tolerate SIV replication typically with no or minimal losses of circulating CD4 T cells (Chakrabarti et al. 2000; Silvestri et al. 2003).

Several potential drivers have been postulated to account for this state of systemic immune activation in progressive HIV and SIV infection. Among these is the virus itself, which can drive activation of innate immune receptors such as TLR 7 and 8 through poly(U)-rich sequences in its genome (Beignon et al. 2005; Meier et al. 2007) as well as possibly through activation of other innate immune receptors by capsid proteins (Manel et al. 2010) or viral DNAs (Yan et al. 2010). A rapid decrease in immune activation indices is recognized with administration of suppressive antiviral drugs and it is likely that some of this decrease is a consequence of lower levels of HIV replication (Evans et al. 1998; Tilling et al. 2002). Some level of T-cell activation in HIV infection also may be mediated directly through recognition of peptides by TCRs. These peptides may be derived from HIV itself but also from opportunistic microbes (such as cytomegalovirus and other herpes viruses) that have been permitted to replicate more effectively in the setting of HIV-related immune deficiency (Hunt et al. 2011). It is also possible that some level of immune activation in HIV and pathogenic SIV infection is related to homeostatic mechanisms, that is, a need to replenish lymphocyte populations at effector sites of potential microbial invasion (Okoye et al. 2007). Finally, there is increasing evidence that in HIV and in pathogenic SIV infection, early damage to mucosal CD4$^+$ T-cell defenses permits increased translocation of microbial products from the gut to the systemic circulation (Brenchley et al. 2006) and these microbial products can drive T-cell and innate immune cell activation (Brenchley et al. 2006; Funderburg et al. 2008). These mechanisms are summarized in Figure 1.

Understanding Microbial Translocation

The human gastrointestinal mucosal surface comprises an estimated surface area of >2700 square feet designed to promote absorption of needed nutrients and fluids and to contain within the lumen the dense population of colonizing microbes and their products. Yet even in healthy subjects, microbial products such as the lipopolysaccharide components of bacterial cell walls can be found in circulation (Brenchley et al. 2006). During acute HIV infection in humans and SIV infection in both African naturally adapted hosts for SIV infection and Asian macaques that develop AIDS, there is a dramatic loss of mucosal CD4$^+$ CCR5$^+$ T cells that are critical targets for productive HIV and SIV infection (Veazey et al. 1998; Brenchley et al. 2004b). In both humans and rhesus, this is followed by an apparent breakdown in the mucosal barrier to systemic translocation of microbial products (Fig 2). Thus, in these systems, high levels of bacterial products can be found in circulation and this microbial translocation is linked to indices of immune activation. The precise mechanisms of the loss of barrier function are incompletely understood but epithelial damage (Estes et al. 2010) and relatively selective losses of Th17 CD4 cells at mucosal sites (Ferreira et al. 1990; Brenchley et al. 2008; Macal et al. 2008; Raffatellu et al. 2008) have been shown. How microbial translocation affects immune homeostasis and HIV pathogenesis is unproven, but in vitro, these microbial products can activate human T cells (Funderburg et al. 2008), and in vivo, indices of microbial translocation activation are linked both to a more aggressive course of HIV infection (Sandler et al. 2011) and inversely to the magnitude of CD4 T-cell restoration with antiviral therapy (Fig. 1) (Brenchley et al. 2006; Jiang et al. 2009). Correlation of course does not prove causality and interventional studies will be required to ascertain if a causal relationship among microbial translocation, immune activation, and HIV pathogenesis exists.

A number of strategies are in development in an effort to preserve mucosal integrity, to limit or prevent microbial translocation in HIV infection, and to test the hypothesis that microbial translocation is an important contributor to both immune activation and HIV pathogenesis. Although data are limited, there is evidence that suppressive antiretroviral therapy is associated with both improvement in mucosal integrity (Guadalupe et al. 2006; Macal et al. 2008; Sheth et al. 2008; Epple et al. 2009) and reduction in the microbial translocation that is its

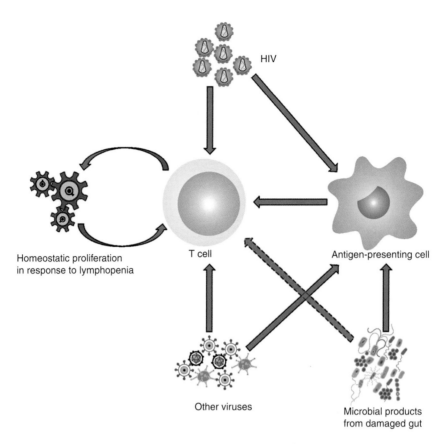

HIV

Homeostatic proliferation
in response to lymphopenia

T cell

Antigen-presenting cell

Other viruses

Microbial products
from damaged gut

Figure 1. Drivers of immune activation and cellular turnover in HIV infection. In progressive HIV and SIV infection, increased immune activation and accelerated cellular turnover are thought to play a central role in T-cell depletion and immune suppression. Whereas HIV can contribute directly to immune activation, both through TLR signaling and direct antigenic stimulation of T cells (*top center*), much of the uncontrolled immune activation is thought to occur via non-HIV-specific mechanisms. Microbial products such as lipopolysaccharide derived from intestinal bacteria enter the systemic circulation in increased quantities, owing to damage to the mucosal immune system and the integrity of the epithelial barrier function, and stimulate innate immune cells through pathogen-associated molecular pattern recognition receptors, which in turn activate adaptive T cells through proinflammatory cytokine expression (*bottom* and *center right*). Other viruses, including cytomegalovirus (CMV) and other human herpes viruses, which are more prevalent in HIV infection, are emerging as potential contributors to this process as well (*bottom center*). As T cells become depleted, decreasing numbers trigger homeostatic mechanisms that further drive existing cells into cycle and potentially contribute to further depletion as HIV infection preferentially infects and destroys activated CD4$^+$ T cells (*left*).

presumed consequence (Brenchley et al. 2006). Several targeted interventions have been designed in an effort to block systemic translocation of microbial products in HIV disease. Oral administration of bovine colostrum prepared from cows immunized with *Escherichia coli* has been tested in two clinical trials with either no effect (Purcell et al. 2011) or very modest effects on indices of immune activation (Yadavalli et al.

2011). Sevalemer is a phosphate-binding resin that also binds lipopolysaccharide (LPS) and its administration to patients with renal insufficiency has been associated with decreased plasma LPS levels. An ongoing trial in HIV infection (ACTG 5296) will test the effects of intraluminal binding of LPS by sevalemer on immune activation and CD4 T-cell homeostasis. In SIV-infected rhesus macaques, administration

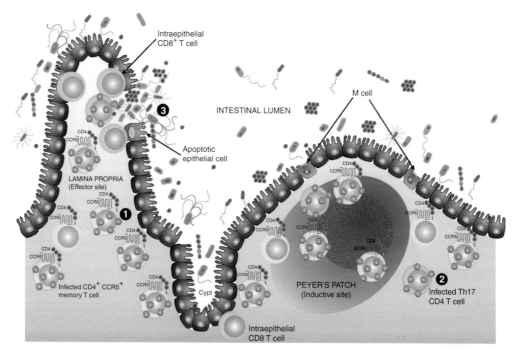

Figure 2. The intestinal immune system in HIV infection. Within days after initial HIV or SIV infection, and regardless of the route of acquisition, terminally differentiated memory CD4$^+$ CCR5$^+$ T cells in the intestinal submucosa are massively depleted, as they are preferentially targeted by HIV (1). Th17 T cells, which play a crucial role in maintaining the integrity of the mucosa and defending against luminal pathogens, are especially depleted (2), further diminishing the effectiveness of the epithelial barrier function. Additionally, tight junctions among epithelial cells are damaged and enterocytes undergo apoptosis, resulting in physical breaches of epithelial integrity, which facilitate systemic exposure to luminal pathogens (3). Antiretroviral therapy appears to correct these defects only to a minimal extent, and the resulting, ongoing exposure to microbial-derived products is thought to be partly responsible for the persistent immune activation, inflammation, and adverse events seen even in HIV-infected persons who have achieved maximal suppression of viral replication.

of antibacterial agents was associated with only transient decreases in plasma LPS levels (Brenchley et al. 2006). Nonetheless, a trial of oral administration of the nonabsorbable antibiotic rifaximin is under way in persons with chronic HIV infection (ACTG 5286).

Moving downstream, the antimalarial drugs chloroquine and hydroxychloroquine have the capability of blocking signaling after ligation of toll-like receptors (Martinson et al. 2010). In one small trial, chloroquine administration to persons with chronic HIV infection decreased indices of immune activation (CD38 and HLA-DR) on CD8 but not CD4 T cells (Murray et al. 2010). In another single-arm study, administration of hydroxychloroquine to HIV-infected persons who experienced suboptimal CD4 T-cell

gains after highly active antiretroviral therapy (HAART) resulted in decreased levels of T-cell activation and decreased levels of inflammatory cytokines IL-6 and TNF in plasma (Piconi et al. 2011). Thus, there is some indication that activation of the toll-like receptor signaling pathway plays some role in the immune activation and inflammation that characterize HIV infection.

The Role of Inflammation/Activation in the Complications of Treated HIV Infection

With widespread use of HAART, deaths attributed to AIDS have diminished rapidly (Palella et al. 2006) and in the HAART era, major AIDS-defining opportunistic infections are no longer the major cause of mortality. Instead, cardiovascular disease, liver disease, and a broadening spectrum of malignancies appear to comprise the major causes of morbidity and death in HIV-infected persons, particularly in the setting of late initiation of antiretroviral therapy (Lau et al. 2007; Marin et al. 2009; Mocroft et al. 2010). These events appear to be more common in HIV-infected persons than among the general population (Weber et al. 2006; Choi et al. 2007; Kirk et al. 2007; Triant et al. 2007; Engels et al. 2008; Joshi et al. 2011) and overall, the risk of these events appears greater in patients with lower circulating $CD4^+$ T-cell counts (Weber et al. 2005; Baker et al. 2008). The determinants of these outcomes are not entirely clear; however, several large studies have linked mortalities in the HAART era to plasma makers of inflammation and coagulation (Kuller et al. 2008; Kalayjian et al. 2010). In a very large study of intermittent versus continuous antiretroviral therapy (the SMART study), plasma levels of IL-6, C-reactive protein, and d-dimer products of thrombolysis independently predicted mortality and cardiovascular morbidities (El-Sadr et al. 2006; Kuller et al. 2008). What is driving these inflammatory and coagulation markers is not entirely clear, but in a nested case-control substudy of SMART, plasma levels of the LPS receptor (sCD14) independently predicted mortality (Sandler et al. 2011). It is likely that HIV replication plays an important role in immune activation and inflammation, as both immune activation indices and plasma inflammatory markers are elevated in untreated infection and diminish after suppressive antiretroviral therapies (Evans et al. 1998; Tilling et al. 2002). On the other hand, some persons who initiate antiretroviral therapies late in the course of disease are unable to raise their circulating CD4 T-cell counts to "normal" levels despite apparent complete suppression of HIV replication (Kelley et al. 2009). The determinants of immune failure in this setting are incompletely understood but what has

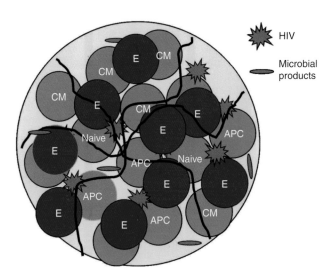

Figure 3. Activation of adaptive and innate immune mechanisms drives HIV pathogenesis in lymphoid tissue. HIV replication within lymphoid tissue promotes an increased local accumulation of HIV-reactive effector (E) T cells that are activated and expanded as a result of exposure to HIV peptides. HIV also activates antigen-presenting cells (APC)—monocytes/macrophages and dendritic cells via ligation of innate immune receptors to express inflammatory cytokines. This proinflammatory environment promotes more effector cell sequestration and also drives central memory (CM) T cells into cell cycle. Inflammation drives collagen deposition and progressive fibrosis, hindering intercellular communications and access to IL-7 that is necessary for homeostatic T-cell expansion. With further translocation of microbial products from the damaged gut, more APC are activated through innate receptors to induce a proinflammatory and procoagulant state that may underlie the increased cardiovascular risk seen in HIV infection.

been described as "fibrosis" of lymphoid tissues is associated with failure of CD4 T-cell restoration on HAART (Fig. 3) (Schacker et al. 2002, 2005). Interestingly, in these incomplete immune responders, immune activation indices are elevated as are plasma inflammatory and coagulation markers (Hunt et al. 2008; Marchetti et al. 2008; Shive et al. 2011). Markers of microbial translocation tend to be elevated in these subjects (Hunt et al. 2008; Marchetti et al. 2008; Shive et al. 2011) and the profile of T-cell activation and cycling is similar to the profile seen after in vitro exposure to microbial products (Funderburg et al. 2008; Lederman et al. 2010).

CONCLUSIONS

It is now clear that HIV and SIV prefer to infect activated memory CD4$^+$ T cells that express CCR5 and that most of the T cells of this phenotype reside in the intestine and other mucosal sites. The recognition that progressive HIV and SIV infection is linked to immune activation, which in turn is linked to a leaky gut, has only recently focused intense interest on the effects of HIV and SIV infection on the intestinal epithelial barrier. The details of how infection and loss of intestinal CD4$^+$ T cells leads to a "leaky gut" are unclear, but multiple avenues of investigation have begun to be explored. If it were possible to prevent or decrease the breakdown of the mucosal barrier through therapeutic means, it is possible that this could greatly slow AIDS disease progression, as appears to be the case in natural nonhuman primate hosts of SIV that are persistently infected, suffer acute loss of intestinal CD4$^+$ T cells, but apparently do not have a leaky gut nor chronic immune activation and rarely progress to AIDS.

REFERENCES

Abel K, Rocke DM, Chohan B, Fritts L, Miller CJ. 2005. Temporal and anatomic relationship between virus replication and cytokine gene expression after vaginal simian immunodeficiency virus infection. *J Virol* 79: 12164–12172.

Alkhatib G, Combadiere C, Broder CC, Feng Y, Kennedy PE, Murphy PM, Berger EA. 1996. CC CKR5: A RANTES,

MIP-1a, MIP-1b receptor as a fusion cofactor for macrophage-tropic HIV-1. *Science* 272: 1955–1958.

Antiretroviral Therapy Cohort Collaboration. 2008. Life expectancy of individuals on combination antiretroviral therapy in high-income countries: A collaborative analysis of 14 cohort studies. *Lancet* 372: 293–299.

Anton PA, Mitsuyasu RT, Deeks SG, Scadden DT, Wagner B, Huang C, Macken C, Richman DD, Christopherson C, Borellini F, et al. 2003. Multiple measures of HIV burden in blood and tissue are correlated with each other but not with clinical parameters in aviremic subjects. *AIDS* 17: 53–63.

Apetrei C, Kaur A, Lerche NW, Metzger M, Pandrea I, Hardcastle J, Falkenstein S, Bohm R, Koehler J, Traina-Dorge V, et al. 2005. Molecular epidemiology of simian immunodeficiency virus SIVsm in U.S. primate centers unravels the origin of SIVmac and SIVstm. *J Virol* 79: 8991–9005.

Arron ST, Ribeiro RM, Gettie A, Bohm R, Blanchard J, Yu J, Perelson AS, Ho DD, Zhang L. 2005. Impact of thymectomy on the peripheral T cell pool in rhesus macaques before and after infection with simian immunodeficiency virus. *Eur J Immunol* 35: 46–55.

Asquith B, Edwards CT, Lipsitch M, McLean AR. 2006. Inefficient cytotoxic T lymphocyte-mediated killing of HIV-1–infected cells in vivo. *PLoS Biol* 4: e90. doi: 10.1371/journal.pbio.0040090.

Autran B, Carcelain G, Li TS, Blanc C, Mathez D, Tubiana R, Katlama C, Debre P, Leibowitch J. 1997. Positive effects of combined antiretroviral therapy on CD4$^+$ T cell homeostasis and function in advanced HIV disease. *Science* 277: 112–116.

Baggaley RF, White RG, Boily MC. 2008. Systematic review of orogenital HIV-1 transmission probabilities. *Int J Epidemiol* 37: 1255–1265.

Bailey JR, Lassen KG, Yang HC, Quinn TC, Ray SC, Blankson JN, Siliciano RF. 2006. Neutralizing antibodies do not mediate suppression of human immunodeficiency virus type 1 in elite suppressors or selection of plasma virus variants in patients on highly active antiretroviral therapy. *J Virol* 80: 4758–4770.

Baker JV, Peng G, Rapkin J, Abrams DI, Silverberg MJ, MacArthur RD, Cavert WP, Henry WK, Neaton JD. 2008. CD4$^+$ count and risk of non-AIDS diseases following initial treatment for HIV infection. *AIDS* 22: 841–848.

Barre-Sinoussi F, Chermann JC, Rey F, Nugeyre MT, Chamaret S, Gruest J, Dauguet C, Axler-Blin C, Vezinet-Brun F, Rouzioux C, et al. 1983. Isolation of a T-lymphotropic retrovirus from a patient at risk for acquired immune deficiency syndrome (AIDS). *Science* 220: 868–871.

Batman PA, Miller AR, Forster SM, Harris JR, Pinching AJ, Griffin GE. 1989. Jejunal enteropathy associated with human immunodeficiency virus infection: Quantitative histology. *J Clin Pathol* 42: 275–281.

Beignon AS, McKenna K, Skoberne M, Manches O, DaSilva I, Kavanagh DG, Larsson M, Gorelick RJ, Lifson JD, Bhardwaj N. 2005. Endocytosis of HIV-1 activates plasmacytoid dendritic cells via Toll-like receptor-viral RNA interactions. *J Clin Invest* 115: 3265–3275.

Belmonte L, Olmos M, Fanin A, Parodi C, Bare P, Concetti H, Perez H, de Bracco MM, Cahn P. 2007. The intestinal mucosa as a reservoir of HIV-1 infection after successful HAART. *AIDS* 21: 2106–2108.

Betts MR, Ambrozak DR, Douek DC, Bonhoeffer S, Brenchley JM, Casazza JP, Koup RA, Picker LJ. 2001. Analysis of total human immunodeficiency virus (HIV)-specific CD4$^+$ and CD8$^+$ T-cell responses: Relationship to viral load in untreated HIV infection. *J Virol* 75: 11983–11991.

Betts MR, Nason MC, West SM, De Rosa SC, Migueles SA, Abraham J, Lederman MM, Benito JM, Goepfert PA, Connors M, et al. 2006. HIV nonprogressors preferentially maintain highly functional HIV-specific CD8$^+$ T cells. *Blood* 107: 4781–4789.

Blankson JN, Bailey JR, Thayil S, Yang HC, Lassen K, Lai J, Gandhi SK, Siliciano JD, Williams TM, Siliciano RF. 2007. Isolation and characterization of replication-competent human immunodeficiency virus type 1 from a subset of elite suppressors. *J Virol* 81: 2508–2518.

Boily MC, Baggaley RF, Wang L, Masse B, White RG, Hayes RJ, Alary M. 2009. Heterosexual risk of HIV-1 infection per sexual act: Systematic review and meta-analysis of observational studies. *Lancet Infect Dis* 9: 118–129.

Brenchley JM, Karandikar NJ, Betts MR, Ambrozak DR, Hill BJ, Crotty LE, Casazza JP, Kuruppu J, Migueles SA, Connors M, et al. 2003. Expression of CD57 defines replicative senescence and antigen-induced apoptotic death of CD8$^+$ T cells. *Blood* 101: 2711–2720.

Brenchley JM, Hill BJ, Ambrozak DR, Price DA, Guenaga FJ, Casazza JP, Kuruppu J, Yazdani J, Migueles SA, Connors M, et al. 2004a. T-cell subsets that harbor human immunodeficiency virus (HIV) in vivo: Implications for HIV pathogenesis. *J Virol* 78: 1160–1168.

Brenchley JM, Schacker TW, Ruff LE, Price DA, Taylor JH, Beilman GJ, Nguyen PL, Khoruts A, Larson M, Haase AT, et al. 2004b. CD4$^+$ T cell depletion during all stages of HIV disease occurs predominantly in the gastrointestinal tract. *J Exp Med* 200: 749–759.

Brenchley JM, Price DA, Schacker TW, Asher TE, Silvestri G, Rao S, Kazzaz Z, Bornstein E, Lambotte O, Altmann D, et al. 2006. Microbial translocation is a cause of systemic immune activation in chronic HIV infection. *Nat Med* 12: 1365–1371.

Brenchley JM, Paiardini M, Knox KS, Asher AI, Cervasi B, Asher TE, Scheinberg P, Price DA, Hage CA, Kholi LM, et al. 2008. Differential Th17 CD4 T-cell depletion in pathogenic and nonpathogenic lentiviral infections. *Blood* 112: 2826–2835.

Carrington M, Nelson GW, Martin MP, Kissner T, Vlahov D, Goedert JJ, Kaslow R, Buchbinder S, Hoots K, O'Brien SJ. 1999. HLA and HIV-1: Heterozygote advantage and *B***35-Cw***04* disadvantage. *Science* 283: 1748–1752.

Chakrabarti LA, Lewin SR, Zhang L, Gettie A, Luckay A, Martin LN, Skulsky E, Ho DD, Cheng-Mayer C, Marx PA. 2000. Normal T-cell turnover in sooty mangabeys harboring active simian immunodeficiency virus infection. *J Virol* 74: 1209–1223.

Choi AI, Rodriguez RA, Bacchetti P, Bertenthal D, Volberding PA, O'Hare AM. 2007. Racial differences in end-stage renal disease rates in HIV infection versus diabetes. *J Am Soc Nephrol* 18: 2968–2974.

Chou CS, Ramilo O, Vitetta ES. 1997. Highly purified CD25-resting T cells cannot be infected de novo with HIV-1. *Proc Natl Acad Sci* 94: 1361–1365.

Chun TW, Nickle DC, Justement JS, Meyers JH, Roby G, Hallahan CW, Kottilil S, Moir S, Mican JM, Mullins JI, et al. 2008. Persistence of HIV in gut-associated lymphoid tissue despite long-term antiretroviral therapy. *J Infect Dis* 197: 714–720.

Clayburgh DR, Shen L, Turner JR. 2004. A porous defense: The leaky epithelial barrier in intestinal disease. *Lab Invest* 84: 282–291.

Cummins AG, LaBrooy JT, Stanley DP, Rowland R, Shearman DJ. 1990. Quantitative histological study of enteropathy associated with HIV infection. *Gut* 31: 317–321.

Day CL, Kaufmann DE, Kiepiela P, Brown JA, Moodley ES, Reddy S, Mackey EW, Miller JD, Leslie AJ, DePierres C, et al. 2006. PD-1 expression on HIV-specific T cells is associated with T-cell exhaustion and disease progression. *Nature* 443: 350–354.

Deacon NJ, Tsykin A, Solomon A, Smith K, Ludford-Menting M, Hooker DJ, McPhee DA, Greenway AL, Ellett A, Chatfield C, et al. 1995. Genomic structure of an attenuated quasi species of HIV-1 from a blood transfusion donor and recipients. *Science* 270: 988–991.

Dean M, Carrington M, Winkler C, Huttley GA, Smith MW, Allikmets R, Goedert JJ, Buchbinder SP, Vittinghoff E, Gomperts E, et al. 1996. Genetic restriction of HIV-1 infection and progression to AIDS by a deletion allele of the CKR5 structural gene. Hemophilia Growth and Development Study, Multicenter AIDS Cohort Study, Multicenter Hemophilia Cohort Study, San Francisco City Cohort, ALIVE Study. *Science* 273: 1856–1862.

Deeks SG, Kitchen CM, Liu L, Guo H, Gascon R, Narvaez AB, Hunt P, Martin JN, Kahn JO, Levy J, et al. 2004. Immune activation set point during early HIV infection predicts subsequent CD4$^+$ T-cell changes independent of viral load. *Blood* 104: 942–947.

Dion ML, Bordi R, Zeidan J, Asaad R, Boulassel MR, Routy JP, Lederman MM, Sekaly RP, Cheynier R. 2007. Slow disease progression and robust therapy-mediated CD4$^+$ T-cell recovery are associated with efficient thymopoiesis during HIV-1 infection. *Blood* 109: 2912–2920.

Doitsh G, Cavrois M, Lassen KG, Zepeda O, Yang Z, Santiago ML, Hebbeler AM, Greene WC. 2010. Abortive HIV infection mediates CD4 T cell depletion and inflammation in human lymphoid tissue. *Cell* 143: 789–801.

Douek DC, McFarland RD, Keiser PH, Gage EA, Massey JM, Haynes BF, Polis MA, Haase AT, Feinberg MB, Sullivan JL, et al. 1998. Changes in thymic function with age and during the treatment of HIV infection. *Nature* 396: 690–695.

Douek DC, Betts MR, Hill BJ, Little SJ, Lempicki R, Metcalf JA, Casazza J, Yoder C, Adelsberger JW, Stevens RA, et al. 2001. Evidence for increased T cell turnover and decreased thymic output in HIV infection. *J Immunol* 167: 6663–6668.

Douek DC, Brenchley JM, Betts MR, Ambrozak DR, Hill BJ, Okamoto Y, Casazza JP, Kuruppu J, Kunstman K, Wolinsky S, et al. 2002. HIV preferentially infects HIV-specific CD4$^+$ T cells. *Nature* 417: 95–98.

El-Sadr WM, Lundgren JD, Neaton JD, Gordin F, Abrams D, Arduino RC, Babiker A, Burman W, Clumeck N, Cohen

CJ, et al. 2006. CD4$^+$ count-guided interruption of anti-retroviral treatment. *N Engl J Med* **355:** 2283–2296.

Emu B, Sinclair E, Hatano H, Ferre A, Shacklett B, Martin JN, McCune JM, Deeks SG. 2008. HLA class I-restricted T-cell responses may contribute to the control of human immunodeficiency virus infection, but such responses are not always necessary for long-term virus control. *J Virol* **82:** 5398–5407.

Engels EA, Biggar RJ, Hall HI, Cross H, Crutchfield A, Finch JL, Grigg R, Hylton T, Pawlish KS, McNeel TS, et al. 2008. Cancer risk in people infected with human immuno-deficiency virus in the United States. *Int J Cancer* **123:** 187–194.

Epple HJ, Schneider T, Troeger H, Kunkel D, Allers K, Moos V, Amasheh M, Loddenkemper C, Fromm M, Zeitz M, et al. 2009. Impairment of the intestinal barrier is evident in untreated but absent in suppressively treated HIV-infected patients. *Gut* **58:** 220–227.

Estes JD, Harris LD, Klatt NR, Tabb B, Pittaluga S, Paiardini M, Barclay GR, Smedley J, Pung R, Oliveira KM, et al. 2010. Damaged intestinal epithelial integrity linked to microbial translocation in pathogenic simian immunodeficiency virus infections. *PLoS Pathog* **6:** e1001052. doi: 10.1371/journal.ppat.1001052.

Evans TG, Bonnez W, Soucier HR, Fitzgerald T, Gibbons DC, Reichman RC. 1998. Highly active antiretroviral therapy results in a decrease in CD8$^+$ T cell activation and preferential reconstitution of the peripheral CD4$^+$ T cell population with memory rather than naive cells. *Antiviral Res* **39:** 163–173.

Favre D, Lederer S, Kanwar B, Ma ZM, Proll S, Kasakow Z, Mold J, Swainson L, Barbour JD, Baskin CR, et al. 2009. Critical loss of the balance between Th17 and T regulatory cell populations in pathogenic SIV infection. *PLoS Pathog* **5:** e1000295. doi: 10.1371/journal.ppat. 1000295.

Ferreira RC, Forsyth LE, Richman PI, Wells C, Spencer J, MacDonald TT. 1990. Changes in the rate of crypt epithelial cell proliferation and mucosal morphology induced by a T-cell-mediated response in human small intestine. *Gastroenterol* **98:** 1255–1263.

Field M. 2006. T cell activation alters intestinal structure and function. *J Clin Invest* **116:** 2580–2582.

Funderburg N, Luciano AA, Jiang W, Rodriguez B, Sieg SF, Lederman MM. 2008. Toll-like receptor ligands induce human T cell activation and death, a model for HIV pathogenesis. *PLoS One* **3:** e1915. doi: 10.1371/journal. pone.0001915.

Gallo RC, Salahuddin SZ, Popovic M, Shearer GM, Kaplan M, Haynes BF, Palker TJ, Redfield R, Oleske J, Safai B, et al. 1984. Frequent detection and isolation of cytopathic retroviruses (HTLV-III) from patients with AIDS and at risk for AIDS. *Science* **224:** 500–503.

Gao F, Yue L, White AT, Pappas PG, Barchue J, Hanson AP, Greene BM, Sharp PM, Shaw GM, Hahn BH. 1992. Human infection by genetically diverse SIVSM-related HIV-2 in west Africa. *Nature* **358:** 495–499.

Gao F, Bailes E, Robertson DL, Chen Y, Rodenburg CM, Michael SF, Cummins LB, Arthur LO, Peeters M, Shaw GM, et al. 1999. Origin of HIV-1 in the chimpanzee Pan troglodytes troglodytes. *Nature* **397:** 436–441.

George MD, Reay E, Sankaran S, Dandekar S. 2005. Early antiretroviral therapy for simian immunodeficiency virus infection leads to mucosal CD4$^+$ T-cell restoration and enhanced gene expression regulating mucosal repair and regeneration. *J Virol* **79:** 2709–2719.

George MD, Wehkamp J, Kays RJ, Leutenegger CM, Sabir S, Grishina I, Dandekar S, Bevins CL. 2008. In vivo gene expression profiling of human intestinal epithelial cells: Analysis by laser microdissection of formalin fixed tissues. *BMC Genomics* **9:** 209.

Giorgi JV, Liu Z, Hultin LE, Cumberland WG, Hennessey K, Detels R. 1993. Elevated levels of CD38$^+$ CD8$^+$ T cells in HIV infection add to the prognostic value of low CD4$^+$ T cell levels: Results of 6 years of follow-up. The Los Angeles Center, Multicenter AIDS Cohort Study. *J Acquir Immune Defic Syndr* **6:** 904–912.

Giorgi JV, Hultin LE, McKeating JA, Johnson TD, Owens B, Jacobson LP, Shih R, Lewis J, Wiley DJ, Phair JP, et al. 1999. Shorter survival in advanced human immunodeficiency virus type 1 infection is more closely associated with T lymphocyte activation than with plasma virus burden or virus chemokine coreceptor usage. *J Infect Dis* **179:** 859–870.

Goedert JJ, Biggar RJ, Weiss SH, Eyster ME, Melbye M, Wilson S, Ginzburg HM, Grossman RJ, DiGioia RA, Sanchez WC, et al. 1986. Three-year incidence of AIDS in five cohorts of HTLV-III-infected risk group members. *Science* **231:** 992–995.

Gonzalez-Scarano F, Martin-Garcia J. 2005. The neuropathogenesis of AIDS. *Nat Rev Immunol* **5:** 69–81.

Gottlieb MS, Schroff R, Schanker HM, Weisman JD, Fan PT, Wolf RA, Saxon A. 1981. *Pneumocystis carinii* pneumonia and mucosal candidiasis in previously healthy homosexual men: Evidence of a new acquired cellular immunodeficiency. *N Engl J Med* **305:** 1425–1431.

Goulder PJ, Bunce M, Krausa P, McIntyre K, Crowley S, Morgan B, Edwards A, Giangrande P, Phillips RE, McMichael AJ. 1996. Novel, cross-restricted, conserved, and immunodominant cytotoxic T lymphocyte epitopes in slow progressors in HIV type 1 infection. *AIDS Res Hum Retrov* **12:** 1691–1698.

Grossman Z, Meier-Schellersheim M, Paul WE, Picker LJ. 2006. Pathogenesis of HIV infection: What the virus spares is as important as what it destroys. *Nat Med* **12:** 289–295.

Guadalupe M, Sankaran S, George MD, Reay E, Verhoeven D, Shacklett BL, Flamm J, Wegelin J, Prindiville T, Dandekar S. 2006. Viral suppression and immune restoration in the gastrointestinal mucosa of human immunodeficiency virus type 1-infected patients initiating therapy during primary or chronic infection. *J Virol* **80:** 8236–8247.

Haase AT, Henry K, Zupancic M, Sedgewick G, Faust RA, Melroe H, Cavert W, Gebhard K, Staskus K, Zhang Z-Q, et al. 1996. Quantitative image analysis of HIV-1 infection in lymphoid tissue. *Science* **274:** 985–989.

Harcourt G, Gomperts E, Donfield S, Klenerman P. 2006. Diminished frequency of hepatitis C virus specific interferon gamma secreting CD4$^+$ T cells in human immunodeficiency virus/hepatitis C virus coinfected patients. *Gut* **55:** 1484–1487.

Hasegawa A, Liu H, Ling B, Borda JT, Alvarez X, Sugimoto C, Vinet-Oliphant H, Kim WK, Williams KC, Ribeiro RM, et al. 2009. The level of monocyte turnover predicts disease progression in the macaque model of AIDS. *Blood* 114: 2917–2925.

Hasse B, Ledergerber B, Egger M, Vernazza P, Furrer H, Battegay M, Hirschel B, Cavassini M, Bernasconi E, Weber R. 2011. Aging and (non-HIV-associated) co-morbidity in HIV-positive persons: The Swiss HIV cohort study (SHCS). In *18th Conference on Retroviruses and Opportunistic Infections, Abstract O-161.* Boston.

Haynes BF, Hale LP, Weinhold KJ, Patel DD, Liao HX, Bressler PB, Jones DM, Demarest JF, Gebhard-Mitchell K, Haase AT, et al. 1999. Analysis of the adult thymus in reconstitution of T lymphocytes in HIV-1 infection. *J Clin Invest* 103: 453–460.

Hazenberg MD, Otto SA, van Benthem BH, Roos MT, Coutinho RA, Lange JM, Hamann D, Prins M, Miedema F. 2003. Persistent immune activation in HIV-1 infection is associated with progression to AIDS. *AIDS* 17: 1881–1888.

Heise C, Vogel P, Miller CJ, Halsted CH, Dandekar S. 1993. Simian immunodeficiency virus infection of the gastrointestinal tract of rhesus macaques: Functional, pathological and morphological changes. *Am J Pathol* 142: 1759–1771.

Heise C, Miller CJ, Lackner A, Dandekar S. 1994. Primary acute simian immunodeficiency virus infection of intestinal lymphoid tissue is associated with gastrointestinal dysfunction. *J Infect Dis* 169: 1116–1120.

Hirschtick RE, Glassroth J, Jordan MC, Wilcosky TC, Wallace JM, Kvale PA, Markowitz N, Rosen MJ, Mangura BT, Hopewell PC. 1995. Bacterial pneumonia in persons infected with the human immunodeficiency virus. Pulmonary Complications of HIV Infection Study Group. *N Engl J Med* 333: 845–851.

Horton H, Vogel T, O'Connor D, Picker L, Watkins DI. 2002. Analysis of the immune response and viral evolution during the acute phase of SIV infection. *Vaccine* 20: 1927–1932.

Hu JJ, Gardner MB, Miller CJ. 2000. Simian immunodeficiency virus rapidly penetrates the cervicovaginal mucosa after intravaginal inoculation and infects intraepithelial dendritic cells. *J Virol* 74: 6087–6095.

Hunt PW, Brenchley J, Sinclair E, McCune JM, Roland M, Page-Shafer K, Hsue P, Emu B, Krone M, Lampiris H, et al. 2008. Relationship between T cell activation and CD4+ T cell count in HIV-seropositive individuals with undetectable plasma HIV RNA levels in the absence of therapy. *J Infect Dis* 197: 126–133.

Hunt P, Martin J, Sinclair E, Epling L, Teague J, Jacobson M, Tracy R, Corey L, Deeks SG. 2011. Valganciclovir reduces T cell activation in HIV-infected individuals with incomplete CD4+ T cell recovery on antiretroviral therapy. *J Infec Dis* 203: 1474–1483.

Hyrcza MD, Kovacs C, Loutfy M, Halpenny R, Heisler L, Yang S, Wilkins O, Ostrowski M, Der SD. 2007. Distinct transcriptional profiles in ex vivo CD4+ and CD8+ T cells are established early in human immunodeficiency virus type 1 infection and are characterized by a chronic interferon response as well as extensive transcriptional changes in CD8+ T cells. *J Virol* 81: 3477–3486.

Ioannidis JP, Rosenberg PS, Goedert JJ, Ashton LJ, Benfield TL, Buchbinder SP, Coutinho RA, Eugen-Olsen J, Gallart T, Katzenstein TL, et al. 2001. Effects of CCR5-Δ32, CCR2–64I, and SDF-1 3′A alleles on HIV-1 disease progression: An international meta-analysis of individual-patient data. *Ann Intern Med* 135: 782–795.

James SP, Graeff AS, Zeitz M. 1987. Predominance of helper-inducer T cells in mesenteric lymph nodes and intestinal lamina propria of normal nonhuman primates. *Cell Immunol* 107: 372–383.

Janoff EN, Breiman RF, Daley CL, Hopewell PC. 1992. Pneumococcal disease during HIV infection. Epidemiologic, clinical, and immunologic perspectives. *Ann Intern Med* 117: 314–324.

Jiang W, Lederman MM, Hunt P, Sieg SF, Haley K, Rodriguez B, Landay A, Martin J, Sinclair E, Asher AI, et al. 2009. Plasma levels of bacterial DNA correlate with immune activation and the magnitude of immune restoration in persons with antiretroviral-treated HIV infection. *J Infect Dis* 199: 1177–1185.

Joshi D, O'Grady J, Dieterich D, Gazzard B, Agarwal K. 2011. Increasing burden of liver disease in patients with HIV infection. *Lancet* 377: 1198–1209.

Kalayjian RC, Machekano RN, Rizk N, Robbins GK, Gandhi RT, Rodriguez BA, Pollard RB, Lederman MM, Landay A. 2010. Pretreatment levels of soluble cellular receptors and interleukin-6 are associated with HIV disease progression in subjects treated with highly active antiretroviral therapy. *J Infect Dis* 201: 1796–1805.

Kaslow RA, Duquesnoy R, VanRaden M, Kingsley L, Marrari M, Friedman H, Su S, Saah AJ, Detels R, Phair J, et al. 1990. A1, Cw7, B8, DR3 HLA antigen combination associated with rapid decline of T-helper lymphocytes in HIV-1 infection. A report from the Multicenter AIDS Cohort Study. *Lancet* 335: 927–930.

Kaslow RA, Carrington M, Apple R, Park L, Munoz A, Saah AJ, Goedert JJ, Winkler C, O'Brien SJ, Rinaldo C, et al. 1996. Influence of combinations of human major histocompatibility complex genes on the course of HIV-1 infection. *Nat Med* 2: 405–411.

Kaur A, Hale CL, Ramanujan S, Jain RK, Johnson RP. 2000. Differential dynamics of CD4+ and CD8+ T-lymphocyte proliferation and activation in acute simian immunodeficiency virus infection. *J Virol* 74: 8413–8424.

Kelley CF, Kitchen CM, Hunt PW, Rodriguez B, Hecht FM, Kitahata M, Crane HM, Willig J, Mugavero M, Saag M, et al. 2009. Incomplete peripheral CD4+ cell count restoration in HIV-infected patients receiving long-term antiretroviral treatment. *Clin Infect Dis* 48: 787–794.

Kirchhoff F, Greenough TC, Brettler DB, Sullivan JL, Desrosiers RC. 1995. Brief report: Absence of intact nef sequences in a long-term survivor with nonprogressive HIV-1 infection. *N Engl J Med* 332: 228–232.

Kirk GD, Merlo C, O'Driscoll P, Mehta SH, Galai N, Vlahov D, Samet J, Engels EA. 2007. HIV infection is associated with an increased risk for lung cancer, independent of smoking. *Clin Infect Dis* 45: 103–110.

Kolls JK, Linden A. 2004. Interleukin-17 family members and inflammation. *Immunity* 21: 467–476.

Koot M, van Leeuwen R, de Goede RE, Keet IP, Danner S, Eeftinck Schattenkerk JK, Reiss P, Tersmette M, Lange JM, Schuitemaker H. 1999. Conversion rate towards a syncytium-inducing (SI) phenotype during different stages of human immunodeficiency virus type 1 infection and prognostic value of SI phenotype for survival after AIDS diagnosis. *J Infect Dis* **179**: 254–258.

Kovacs JA, Lempicki RA, Sidorov IA, Adelsberger JW, Herpin B, Metcalf JA, Sereti I, Polis MA, Davey RT, Tavel J, et al. 2001. Identification of dynamically distinct subpopulations of T lymphocytes that are differentially affected by HIV. *J Exp Med* **194**: 1731–1741.

Kuller LH, Tracy R, Belloso W, De Wit S, Drummond F, Lane HC, Ledergerber B, Lundgren J, Neuhaus J, Nixon D, et al. 2008. Inflammatory and coagulation biomarkers and mortality in patients with HIV infection. *PLoS Med* **5**: e203. doi: 10.1371/journal.pmed.0050203.

Lackner AA. 1994. Pathology of simian immunodeficiency virus induced disease. In *Current topics in microbiology and immunology: Simian immunodeficiency virus* (ed. Desrosiers RC, Letvin N), pp. 35–64. Springer-Verlag, Berlin.

Lackner AA, Vogel P, Ramos RA, Kluge JD, Marthas M. 1994. Early events in tissues during infection with pathogenic (SIVmac239) and nonpathogenic (SIVmac1A11) molecular clones of simian immunodeficiency virus. *Am J Pathol* **145**: 428–439.

Lane HC, Masur H, Edgar LC, Whalen G, Rook AH, Fauci AS. 1983. Abnormalities of B-cell activation and immunoregulation in patients with the acquired immunodeficiency syndrome. *N Engl J Med* **309**: 453–458.

Lau B, Gange SJ, Moore RD. 2007. Risk of non-AIDS-related mortality may exceed risk of AIDS-related mortality among individuals enrolling into care with CD4$^+$ counts greater than 200 cells/mm^3. *J Acquir Immune Defic Syndr* **44**: 179–187.

Lederman MM, Connick E, Landay A, Kuritzkes DR, Spritzler J, St Clair M, Kotzin BL, Fox L, Chiozzi MH, Leonard JM, et al. 1998. Immunologic responses associated with 12 weeks of combination antiretroviral therapy consisting of zidovudine, lamivudine, and ritonavir: Results of AIDS clinical trials group protocol 315. *J Infect Dis* **178**: 70–79.

Lederman MM, Rodriguez B, Clagett B, Funderburg N, Medvik K, Gripshover B, Kalayjian RC, Sieg S, Calabrese L. 2010. Immune failure after suppressive ART: High level CD4 and CD8 T cell activation but only memory CD4 cells are cycling. In *18th Conference on Retroviruses and Opportunistic infections, Abstract 47LB*. Boston.

Levy JA, Hoffman AD, Kramer SM, Landis JA, Shimabukuro JM, Oshiro LS. 1984. Isolation of lymphocytopathic retroviruses from San Francisco patients with AIDS. *Science* **225**: 840–842.

Li Q, Duan L, Estes JD, Ma ZM, Rourke T, Wang Y, Reilly C, Carlis J, Miller CJ, Haase AT. 2005. Peak SIV replication in resting memory CD4$^+$ T cells depletes gut lamina propria CD4$^+$ T cells. *Nature* **434**: 1148–1152.

Liang SC, Tan XY, Luxenberg DP, Karim R, Dunussi-Joannopoulos K, Collins M, Fouser LA. 2006. Interleukin (IL)-22 and IL-17 are coexpressed by Th17 cells and cooperatively enhance expression of antimicrobial peptides. *J Exp Med* **203**: 2271–2279.

Ling B, Apetrei C, Pandrea I, Veazey RS, Lackner AA, Gormus B, Marx PA. 2004. Classic AIDS in a sooty mangabey after an 18 year natural infection. *J Virol* **78**: 8902–8908.

Liu R, Paxton WA, Choe S, Ceradini D, Martin SR, Horuk R, MacDonald ME, Stuhlmann H, Koup RA, Landau NR. 1996. Homozygous defect in HIV-1 coreceptor accounts for resistance of some multiply-exposed individuals to HIV-1 infection. *Cell* **86**: 367–377.

Liu Z, Cumberland WG, Hultin LE, Prince HE, Detels R, Giorgi JV. 1997. Elevated CD38 antigen expression on CD8$^+$ T cells is a stronger marker for the risk of chronic HIV disease progression to AIDS and death in the Multicenter AIDS Cohort Study than CD4$^+$ cell count, soluble immune activation markers, or combinations of HLA-DR and CD38 expression. *J Acquir Immune Defic Syndr Hum Retrovirol* **16**: 83–92.

Liu Z, Cumberland WG, Hultin LE, Kaplan AH, Detels R, Giorgi JV. 1998. CD8$^+$ T-lymphocyte activation in HIV-1 disease reflects an aspect of pathogenesis distinct from viral burden and immunodeficiency. *J Acquir Immune Defic Syndr Hum Retrovirol* **18**: 332–340.

Lum JJ, Cohen OJ, Nie Z, Weaver JG, Gomez TS, Yao XJ, Lynch D, Pilon AA, Hawley N, Kim JE, et al. 2003. Vpr R77Q is associated with long-term nonprogressive HIV infection and impaired induction of apoptosis. *J Clin Invest* **111**: 1547–1554.

Ma ZM, Abel K, Rourke T, Wang Y, Miller CJ. 2004. A period of transient viremia and occult infection precedes persistent viremia and antiviral immune responses during multiple low-dose intravaginal simian immunodeficiency virus inoculations. *J Virol* **78**: 14048–14052.

Macal M, Sankaran S, Chun TW, Reay E, Flamm J, Prindiville TJ, Dandekar S. 2008. Effective CD4$^+$ T-cell restoration in gut-associated lymphoid tissue of HIV-infected patients is associated with enhanced Th17 cells and polyfunctional HIV-specific T-cell responses. *Mucosal Immunol* **1**: 475–488.

MacDonald TT, Spencer J. 1988. Evidence that activated mucosal T cells play a role in the pathogenesis of enteropathy in human small intestine. *J Exp Med* **167**: 1341–1349.

MacDonald TT, Spencer J. 1992. Cell-mediated immune injury in the intestine. *Gastroenterol Clin N Am* **21**: 367–386.

MacDonald TT, Spencer J. 1994. Lymphoid cells and tissues of the gastrointestinal tract. In *Gastrointestinal and hepatic immunology* (ed. Heatley RH), pp. 1–23. Cambridge University Press, Cambridge.

Manel N, Hogstad B, Wang Y, Levy DE, Unutmaz D, Littman DR. 2010. A cryptic sensor for HIV-1 activates antiviral innate immunity in dendritic cells. *Nature* **467**: 214–217.

Marchetti G, Bellistri GM, Borghi E, Tincati C, Ferramosca S, La Francesca M, Morace G, Gori A, Monforte AD. 2008. Microbial translocation is associated with sustained failure in CD4$^+$ T-cell reconstitution in HIV-infected patients on long-term highly active antiretroviral therapy. *AIDS* **22**: 2035–2038.

Marin B, Thiebaut R, Bucher HC, Rondeau V, Costagliola D, Dorrucci M, Hamouda O, Prins M, Walker S, Porter K, et al. 2009. Non-AIDS-defining deaths and immunodeficiency in the era of combination antiretroviral therapy. *AIDS* **23**: 1743–1753.

Martin MP, Gao X, Lee JH, Nelson GW, Detels R, Goedert JJ, Buchbinder S, Hoots K, Vlahov D, Trowsdale J, et al.

2002. Epistatic interaction between KIR3DS1 and HLA-B delays the progression to AIDS. *Nat Genet* **31:** 429–434.

Martinson JA, Montoya CJ, Usuga X, Ronquillo R, Landay AL, Desai SN. 2010. Chloroquine modulates HIV-1-induced plasmacytoid dendritic cell α interferon: Implication for T-cell activation. *Antimicrob Agents Chemother* **54:** 871–881.

Marx PA, Li Y, Lerche NW, Sutjipto S, Gettie A, Yee JA, Brotman BH, Prince AM, Hanson A, Webster RG, et al. 1991. Isolation of a simian immunodeficiency virus related to human immunodeficiency virus type 2 from a West African pet sooty mangabey. *J Virol* **65:** 4480–4485.

Masur H, Michelis MA, Greene JB, Onorato I, Stouwe RA, Holzman RS, Wormser G, Brettman L, Lange M, Murray HW, et al. 1981. An outbreak of community-acquired *Pneumocystis carinii* pneumonia: Initial manifestation of cellular immune dysfunction. *N Engl J Med* **305:** 1431–1438.

Mattapallil JJ, Douek DC, Hill B, Nishimura Y, Martin M, Roederer M. 2005. Massive infection and loss of memory CD4$^+$ T cells in multiple tissues during acute SIV infection. *Nature* **434:** 1093–1097.

McGowan I, Elliott J, Fuerst M, Taing P, Boscardin J, Poles M, Anton P. 2004. Increased HIV-1 mucosal replication is associated with generalized mucosal cytokine activation. *J Acquir Immune Defic Syndr* **37:** 1228–1236.

Mehandru S, Poles MA, Tenner-Racz K, Horowitz A, Hurley A, Hogan C, Boden D, Racz P, Markowitz M. 2004. Primary HIV-1 infection is associated with preferential depletion of CD4$^+$ T lymphocytes from effector sites in the gastrointestinal tract. *J Exp Med* **200:** 761–770.

Mehandru S, Poles MA, Tenner-Racz K, Jean-Pierre P, Manuelli V, Lopez P, Shet A, Low A, Mohri H, Boden D, et al. 2006. Lack of mucosal immune reconstitution during prolonged treatment of acute and early HIV-1 infection. *PLoS Med* **3:** e484. doi: 10.1371/journal.pmed.0030484.

Meier A, Alter G, Frahm N, Sidhu H, Li B, Bagchi A, Teigen N, Streeck H, Stellbrink HJ, Hellman J, et al. 2007. MyD88-dependent immune activation mediated by human immunodeficiency virus type 1-encoded Toll-like receptor ligands. *J Virol* **81:** 8180–8191.

Mellors JW, Rinaldo CR Jr, Gupta P, White RM, Todd JA, Kingsley LA. 1996. Prognosis in HIV-1 infection predicted by the quantity of virus in plasma. *Science* **272:** 1167–1170.

Mellors JW, Munoz A, Giorgi JV, Margolick JB, Tassoni CJ, Gupta P, Kingsley LA, Todd JA, Saah AJ, Detels R, et al. 1997. Plasma viral load and CD4$^+$ lymphocytes as prognostic markers of HIV-1 infection. *Ann Intern Med* **126:** 946–954.

Migueles SA, Laborico AC, Shupert WL, Sabbaghian MS, Rabin R, Hallahan CW, Van Baarle D, Kostense S, Miedema F, McLaughlin M, et al. 2002. HIV-specific CD8$^+$ T cell proliferation is coupled to perforin expression and is maintained in nonprogressors. *Nat Immunol* **3:** 1061–1068.

Migueles SA, Osborne CM, Royce C, Compton AA, Joshi RP, Weeks KA, Rood JE, Berkley AM, Sacha JB, Cogliano-Shutta NA, et al. 2008. Lytic granule loading of CD8$^+$ T cells is required for HIV-infected cell elimination associated with immune control. *Immunity* **29:** 1009–1021.

Miller CJ, Li Q, Abel K, Kim EY, Ma ZM, Wietgrefe S, La Franco-Scheuch L, Compton L, Duan L, Shore MD, et al. 2005. Propagation and dissemination of infection after vaginal transmission of simian immunodeficiency virus. *J Virol* **79:** 9217–9227.

Miura T, Brockman MA, Brumme CJ, Brumme ZL, Carlson JM, Pereyra F, Trocha A, Addo MM, Block BL, Rothchild AC, et al. 2008. Genetic characterization of human immunodeficiency virus type 1 in elite controllers: Lack of gross genetic defects or common amino acid changes. *J Virol* **82:** 8422–8430.

Mocroft A, Reiss P, Gasiorowski J, Ledergerber B, Kowalska J, Chiesi A, Gatell J, Rakhmanova A, Johnson M, Kirk O, et al. 2010. Serious fatal and nonfatal non-AIDS-defining illnesses in Europe. *J Acquir Immune Defic Syndr* **55:** 262–270.

Mohri H, Perelson AS, Tung K, Ribeiro RM, Ramratnam B, Markowitz M, Kost R, Hurley A, Weinberger L, Cesar D, et al. 2001. Increased turnover of T lymphocytes in HIV-1 infection and its reduction by antiretroviral therapy. *J Exp Med* **194:** 1277–1287.

Moore JP, Kitchen SG, Pugach P, Zack JA. 2004. The CCR5 and CXCR4 coreceptors—Central to understanding the transmission and pathogenesis of human immunodeficiency virus type 1 infection. *AIDS Res Hum Retrov* **20:** 111–126.

Mowat AM, Viney JL. 1997. The anatomical basis of intestinal immunity. *Immunol Rev* **156:** 145–166.

Murray SM, Down CM, Boulware DR, Stauffer WM, Cavert WP, Schacker TW, Brenchley JM, Douek DC. 2010. Reduction of immune activation with chloroquine therapy during chronic HIV infection. *J Virol* **84:** 12082–12086.

Neutra MR, Frey A, Kraehenbuhl J-P. 1996. Epithelial M cells: Gateways for mucosal infection and immunization. *Cell* **86:** 345–348.

Okoye A, Meier-Schellersheim M, Brenchley JM, Hagen SI, Walker JM, Rohankhedkar M, Lum R, Edgar JB, Planer SL, Legasse A, et al. 2007. Progressive CD4$^+$ central memory T cell decline results in CD4$^+$ effector memory insufficiency and overt disease in chronic SIV infection. *J Exp Med* **204:** 2171–2185.

Olsson J, Poles M, Spetz AL, Elliott J, Hultin L, Giorgi J, Andersson J, Anton P. 2000. Human immunodeficiency virus type 1 infection is associated with significant mucosal inflammation characterized by increased expression of CCR5, CXCR4, and β-chemokines. *J Infect Dis* **182:** 1625–1635.

Owen RE, Heitman JW, Hirschkorn DF, Lanteri MC, Biswas HH, Martin JN, Krone MR, Deeks SG, Norris PJ. 2010. HIV$^+$ elite controllers have low HIV-specific T-cell activation yet maintain strong, polyfunctional T-cell responses. *AIDS* **24:** 1095–1105.

Palella FJ Jr, Baker RK, Moorman AC, Chmiel JS, Wood KC, Brooks JT, Holmberg SD. 2006. Mortality in the highly active antiretroviral therapy era: Changing causes of death and disease in the HIV outpatient study. *J Acquir Immune Defic Syndr* **43:** 27–34.

Pandrea I, Sodora DL, Silvestri G, Apetrei C. 2008. Into the wild: Simian immunodeficiency virus (SIV) infection in natural hosts. *Trends Immunol* **29:** 419–428.

Paranjpe S, Craigo J, Patterson B, Ding M, Barroso P, Harrison L, Montelaro R, Gupta P. 2002. Subcompartmentalization of HIV-1 quasispecies between seminal cells and seminal plasma indicates their origin in distinct genital tissues. *AIDS Res Hum Retrov* **18:** 1271–1280.

Picker LJ, Hagen SI, Lum R, Reed-Inderbitzin EF, Daly LM, Sylwester AW, Walker JM, Siess DC, Piatak M Jr, Wang C, et al. 2004. Insufficient production and tissue delivery of CD4+ memory T cells in rapidly progressive simian immunodeficiency virus infection. *J Exp Med* **200:** 1299–1314.

Piconi S, Parisotto S, Capetti A, Passerini S, Argenteri B, Terzi R, Meraviglia P, Rizzardini G, Clerici M, Trabattoni D. 2011. Hyperimmune bovine colostrum may decrease plasma levels of microbial products and immune activation in untreated HIV infection. In *18th Conference on Retroviruses and Opportunistic Infections, Abstract 382,* Boston.

Poles MA, Boscardin WJ, Elliott J, Taing P, Fuerst MM, McGowan I, Brown S, Anton PA. 2006. Lack of decay of HIV-1 in gut-associated lymphoid tissue reservoirs in maximally suppressed individuals. *J Acquir Immune Defic Syndr* **43:** 65–68.

Pope M, Betjes MGH, Romani N, Hirmand H, Cameron PU, Hoffman L, Gezelter S, Schuler G, Steinman RM. 1994. Conjugates of dendritic cells and memory T lymphocytes from skin facilitate productive infection with HIV-1. *Cell* **78:** 389–398.

Purcell D, Lewin SR, Byakwaga H, French M, Kelleher A, Amin J, Haskleberg H, Kelly M, Cooper D, Emery S. 2011. No correlation between microbial translocation, immune activation, and low-level HIV viremia in HIV-infected individuals with poor CD4+ T cell recovery despite suppressive ART. In *18th Conference on Retroviruses and Opportunistic Infections, Abstract 304.* Boston.

Raffatellu M, Santos RL, Verhoeven DE, George MD, Wilson RP, Winter SE, Godinez I, Sankaran S, Paixao TA, Gordon MA, et al. 2008. Simian immunodeficiency virus-induced mucosal interleukin-17 deficiency promotes *Salmonella* dissemination from the gut. *Nat Med* **14:** 421–428.

Ramilo O, Bell KD, Uhr JW, Vitetta ES. 1993. Role of CD25+ and CD25-T cells in acute HIV infection in vitro. *J Immunol* **150:** 5202–5208.

Reka S, Garro ML, Kotler DP. 1994. Variation in the expression of human immunodeficiency virus RNA and cytokine mRNA in rectal mucosa during the progression of infection. *Lymphokine Cytok Res* **13:** 391–398.

Reynolds MR, Rakasz E, Skinner PJ, White C, Abel K, Ma ZM, Compton L, Napoe G, Wilson N, Miller CJ, et al. 2005. CD8+ T-lymphocyte response to major immunodominant epitopes after vaginal exposure to simian immunodeficiency virus: Too late and too little. *J Virol* **79:** 9228–9235.

Rodriguez B, Sethi AK, Cheruvu VK, Mackay W, Bosch RJ, Kitahata M, Boswell SL, Mathews WC, Bangsberg DR, Martin J, et al. 2006. Predictive value of plasma HIV RNA level on rate of CD4 T-cell decline in untreated HIV infection. *JAMA* **296:** 1498–1506.

Ross MJ, Klotman PE. 2004. HIV-associated nephropathy. *AIDS* **18:** 1089–1099.

Ryzhova EV, Crino P, Shawyer L, Westmoreland SV, Lackner AA, Gonzalez-Scarano F. 2002. Simian immunodeficiency virus encephalitis: Analysis of envelope sequences from individual brain multinucleated giant cells and tissue samples. *Virology* **297:** 57–67.

Sachsenberg N, Perelson AS, Yerly S, Schockmel GA, Leduc D, Hirschel B, Perrin L. 1998. Turnover of CD4+ and CD8+ T lymphocytes in HIV-1 infection as measured by Ki-67 antigen. *J Exp Med* **187:** 1295–1303.

Salkowitz JR, Purvis SF, Meyerson H, Zimmerman P, O'Brien TR, Aledort L, Eyster ME, Hilgartner M, Kessler C, Konkle BA, et al. 2001. Characterization of high-risk HIV-1 seronegative hemophiliacs. *Clin Immunol* **98:** 200–211.

Samson M, Libert F, Doranz BJ, Rucker J, Liesnard C, Farber CM, Saragosti S, Lapoumeroulie C, Cognaux J, Forceille C, et al. 1996. Resistance to HIV-1 infection in caucasian individuals bearing mutant alleles of the CCR-5 chemokine receptor gene. *Nature* **382:** 722–725.

Sandler NG, Wand H, Roque A, Law M, Nason MC, Nixon DE, Pedersen C, Ruxrungtham K, Lewin SR, Emery S, et al. 2011. Plasma levels of soluble CD14 independently predict mortality in HIV infection. *J Infect Dis* **203:** 780–790.

Sankaran S, Guadalupe M, Reay E, George MD, Flamm J, Prindiville T, Dandekar S. 2005. Gut mucosal T cell responses and gene expression correlate with protection against disease in long-term HIV-1-infected nonprogressors. *Proc Natl Acad Sci* **102:** 9860–9865.

Sankaran S, George MD, Reay E, Guadalupe M, Flamm J, Prindiville T, Dandekar S. 2008. Rapid onset of intestinal epithelial barrier dysfunction in primary human immunodeficiency virus infection is driven by an imbalance between immune response and mucosal repair and regeneration. *J Virol* **82:** 538–545.

Sasseville VG, Du Z, Chalifoux LV, Pauley DR, Young HL, Sehgal PK, Desrosiers RC, Lackner AA. 1996. Induction of lymphocyte proliferation and severe gastrointestinal disease in macaques by a *nef* gene variant of SIVmac239. *Am J Pathol* **149:** 163–176.

Schacker T, Collier AC, Hughes J, Shea T, Corey L. 1996. Clinical and epidemiologic features of primary HIV infection. *Ann Intern Med* **125:** 257–264.

Schacker TW, Nguyen PL, Beilman GJ, Wolinsky S, Larson M, Reilly C, Haase AT. 2002. Collagen deposition in HIV-1 infected lymphatic tissues and T cell homeostasis. *J Clin Invest* **110:** 1133–1139.

Schacker TW, Reilly C, Beilman GJ, Taylor J, Skarda D, Krason D, Larson M, Haase AT. 2005. Amount of lymphatic tissue fibrosis in HIV infection predicts magnitude of HAART-associated change in peripheral CD4 cell count. *AIDS* **19:** 2169–2171.

Schieferdecker HL, Ullrich R, Hirseland H, Zeitz M. 1992. T cell differentiation antigens on lymphocytes in the human intestinal lamina propria. *J Immunol* **149:** 2816–2822.

Shanahan F. 1999. Intestinal lymphoepithelial communication. *Adv Exp Med Biol* **473:** 1–9.

Shannon RP. 2001. SIV cardiomyopathy in non-human primates. *Trends Cardiovasc Med* **11:** 242–246.

Sheth PM, Chege D, Shin LY, Huibner S, Yue FY, Loutfy M, Halpenny R, Persad D, Kovacs C, Chun TW, et al. 2008.

Immune reconstitution in the sigmoid colon after long-term HIV therapy. *Mucosal Immunol* **1**: 382–388.

Shive C, Lederman MM, Calabrese L, Funderburg N, Bonilla H, Gripshover B, Salata RA, McComsey GA, Clagett B, Medvik K, et al. 2011. Immunologic failure despite suppressive antiretroviral therapy (ART) is related to increased inflammation and evidence of microbial translocation. In *18th Conference on Retroviruses and Opportunistic Infections, Abstract 320*. Boston.

Sieg SF, Rodriguez B, Asaad R, Jiang W, Bazdar DA, Lederman MM. 2005. Peripheral S-phase T cells in HIV disease have a central memory phenotype and rarely have evidence of recent T cell receptor engagement. *J Infect Dis* **192**: 62–70.

Sieg SF, Bazdar DA, Lederman MM. 2008. S-phase entry leads to cell death in circulating T cells from HIV-infected persons. *J Leukocyte Biol* **83**: 1382–1387.

Siegal FP, Lopez C, Hammer GS, Brown AE, Kornfeld SJ, Gold J, Hassett J, Hirschman SZ, Cunningham-Rundles C, Adelsberg BR, et al. 1981. Severe acquired immunodeficiency in male homosexuals, manifested by chronic perianal ulcerative herpes simplex lesions. *N Engl J Med* **305**: 1439–1444.

Silvestri G, Sodora DL, Koup RA, Paiardini M, O'Neil SP, McClure HM, Staprans SI, Feinberg MB. 2003. Nonpathogenic SIV infection of sooty mangabeys is characterized by limited bystander immunopathology despite chronic high-level viremia. *Immunity* **18**: 441–452.

Spira AI, Marx PA, Patterson BK, Mahoney J, Koup RA, Wolinsky SM, Ho DD. 1996. Cellular targets of infection and route of viral dissemination after an intravaginal inoculation of simian immunodeficiency virus into rhesus macaques. *J Exp Med* **183**: 215–225.

Steinman L. 2007. A brief history of T(H)17, the first major revision in the T(H)1/T(H)2 hypothesis of T cell-mediated tissue damage. *Nat Med* **13**: 139–145.

Stevceva L, Kelsall B, Nacsa J, Moniuszko M, Hel Z, Tryniszewska E, Franchini G. 2002. Cervicovaginal lamina propria lymphocytes: Phenotypic characterization and their importance in cytotoxic T-lymphocyte responses to simian immunodeficiency virus SIVmac251. *J Virol* **76**: 9–18.

Stevenson M, Stanwick TL, Dempsey MP, Lamonica CA. 1990. HIV-1 replication is controlled at the level of T cell activation and proviral integration. *EMBO J* **9**: 1551–1560.

Stone JD, Heise CC, Miller CJ, Halsted CH, Dandekar S. 1994. Development of malabsorption and nutritional complications in simian immunodeficiency virus-infected rhesus macaques. *AIDS* **8**: 1245–1256.

Tang J, Costello C, Keet IP, Rivers C, Leblanc S, Karita E, Allen S, Kaslow RA. 1999. HLA class I homozygosity accelerates disease progression in human immunodeficiency virus type 1 infection. *AIDS Res Hum Retrov* **15**: 317–324.

Tilling R, Kinloch S, Goh LE, Cooper D, Perrin L, Lampe F, Zaunders J, Hoen B, Tsoukas C, Andersson J, et al. 2002. Parallel decline of CD8$^+$/CD38^{++} T cells and viraemia in response to quadruple highly active antiretroviral therapy in primary HIV infection. *AIDS* **16**: 589–596.

Traber PG. 1997. Epithelial cell growth and differentiation. V. Transcriptional regulation, development, and neoplasia

of the intestinal epithelium. *Am J Physiol* **273**: G979–G981.

Trautmann L, Janbazian L, Chomont N, Said EA, Gimmig S, Bessette B, Boulassel MR, Delwart E, Sepulveda H, Balderas RS, et al. 2006. Upregulation of PD-1 expression on HIV-specific CD8$^+$ T cells leads to reversible immune dysfunction. *Nat Med* **12**: 1198–1202.

Triant VA, Lee H, Hadigan C, Grinspoon SK. 2007. Increased acute myocardial infarction rates and cardiovascular risk factors among patients with human immunodeficiency virus disease. *J Clin Endocrinol Metab* **92**: 2506–2512.

Turner JR. 2006. Molecular basis of epithelial barrier regulation: From basic mechanisms to clinical application. *Am J Pathol* **169**: 1901–1909.

Turner JR. 2009. Intestinal mucosal barrier function in health and disease. *Nat Rev Immunol* **9**: 799–809.

Ullrich R, Zeitz M, Heise W, L'age M, Hoffken G, Riecken EO. 1989. Small intestinal structure and function in patients infected with human immunodeficiency virus (HIV): Evidence for HIV-induced enteropathy. *Ann Intern Med* **111**: 15–21.

van Sighem AI, Gras LA, Reiss P, Brinkman K, de Wolf F. 2010. Life expectancy of recently diagnosed asymptomatic HIV-infected patients approaches that of uninfected individuals. *AIDS* **24**: 1527–1535.

Veazey RS. 2003. Vaginal CD4$^+$ T cells express high levels of CCR5 and are rapidly depleted in simian immunodeficiency virus infection. *J Infec Dis* **187**: 769–776.

Veazey RS, Rosenzweig M, Shvetz DE, Pauley DR, DeMaria M, Chalifoux LV, Johnson RP, Lackner AA. 1997. Characterization of gut-associated lymphoid tissue (GALT) of normal rhesus macaques. *Clin Immunol Immunopathol* **82**: 230–242.

Veazey RS, DeMaria M, Chalifoux LV, Shvetz DE, Pauley DR, Knight HL, Rosenzweig M, Johnson RP, Desrosiers RC, Lackner AA. 1998. Gastrointestinal tract as a major site of CD4$^+$ T cell depletion and viral replication in SIV infection. *Science* **280**: 427–431.

Veazey RS, Mansfield KG, Tham IC, Carville AC, Shvetz DE, Forand AE, Lackner AA. 2000a. Dynamics of CCR5 expression by CD4$^+$ T cells in lymphoid tissues during SIV infection. *J Virol* **74**: 11001–11007.

Veazey RS, Tham IC, Mansfield KG, DeMaria M, Forand AE, Shvetz DE, Chalifoux LV, Sehgal PK, Lackner AA. 2000b. Identifying the target cell in primary SIV infection: Highly activated "memory" CD4$^+$ T cells are rapidly eliminated in early SIV infection in vivo. *J Virol* **74**: 57–64.

Veazey RS, Gaudin M-C, Mansfield KG, Tham IC, Altman JD, Lifson JD, Lackner AA, Johnson RP. 2001a. Emergence of simian immunodeficiency virus-specific CD8$^+$ T cells in the intestine of macaques during primary infection. *J Virol* **75**: 10515–10519.

Veazey RS, Marx PA, Lackner AA. 2001b. The mucosal immune system: Primary target for HIV infection and fundamental component of AIDS pathogenesis. *Trends Immunol* **22**: 626–633.

Veazey RS, Marx PA, Lackner AA. 2002. Importance of the state of activation and/or differentiation of CD4$^+$ T cells in AIDS pathogenesis. *Trends Immunol* **23**: 129.

Veazey RS, Lifson JD, Schmitz JE, Kuroda MJ, Piatak M, Pandrea I, Purcell J, Bohm R, Blanchard J, Williams KC, et al. 2003. Dynamics of simian immunodeficiency virus-specific cytotoxic T-cell responses in tissues. *J Med Primatol* **32:** 194–200.

Viney J, MacDonald TT, Spencer J. 1990. γ/δ T cells in the gut epithelium. *Gut* **31:** 841–844.

Weber CR, Turner JR. 2007. Inflammatory bowel disease: Is it really just another break in the wall? *Gut* **56:** 6–8.

Weber R, Friis-Moller N, Sabin C, Reiss P, Monforte A, Dabis F, El-Sadr WM, De Wit S, Morfeldt L, Law M, et al. 2005. HIV and non-HIV-related deaths and their relationship to immunodeficiency: The D:A:D study. In *12th Conference on Retroviruses and Opportunistic Infections, Abstract 595*. Boston.

Weber R, Sabin CA, Friis-Moller N, Reiss P, El-Sadr WM, Kirk O, Dabis F, Law MG, Pradier C, De Wit S, et al. 2006. Liver-related deaths in persons infected with the human immunodeficiency virus: The D:A:D study. *Arch Intern Med* **166:** 1632–1641.

Westmoreland SV, Halpern E, Lackner AA. 1998. Simian immunodeficiency virus encephalitis in rhesus macaques is associated with rapid disease progression. *J Neurovirol* **4:** 260–268.

Wilson CM, Ellenberg JH, Douglas SD, Moscicki AB, Holland CA. 2004. CD8+CD38+ T cells but not HIV type 1 RNA viral load predict CD4+ T cell loss in a predominantly minority female HIV+ adolescent population. *AIDS Res Hum Retrov* **20:** 263–269.

Woelk CH, Ottones F, Plotkin CR, Du P, Royer CD, Rought SE, Lozach J, Sasik R, Kornbluth RS, Richman DD, et al. 2004. Interferon gene expression following HIV type 1 infection of monocyte-derived macrophages. *AIDS Res Hum Retrov* **20:** 1210–1222.

Wykrzykowska JJ, Rosenzweig M, Veazey RS, Simon MA, Halvorsen K, Desrosiers RC, Johnson RP, Lackner AA. 1998. Early regeneration of thymic progenitors in rhesus macaques infected with simian immunodeficiency virus. *J Exp Med* **187:** 1767–1778.

Yadavalli GK, Lederman MM, Sieg S, Funderburg N, Clagett B, Medvik K, Salata RA, Gripshover B, Fulton S, Rodriguez B. 2011. Hyperimmune bovine colostrum may decrease plasma levels of microbial products and immune activation in untreated HIV infection. In *6th IAS Conference on HIV Pathogenesis, Treatment and Prevention, Abstract 3842*. Rome.

Yan N, Regalado-Magdos AD, Stiggelbout B, Lee-Kirsch MA, Lieberman J. 2010. The cytosolic exonuclease TREX1 inhibits the innate immune response to human immunodeficiency virus type 1. *Nat Immunol* **11:** 1005–1013.

Zack JA, Arrigo SJ, Weitsman SR, Go AS, Haislip A, Chen IS. 1990. HIV-1 entry into quiescent primary lymphocytes: Molecular analysis reveals a labile, latent viral structure. *Cell* **61:** 213–222.

Zagury D, Bernard J, Leonard R, Cheynier R, Feldman M, Sarin PS, Gallo RC. 1986. Long-term cultures of HTLV-III–infected T cells: A model of cytopathology of T-cell depletion in AIDS. *Science* **231:** 850–853.

Zeitz M, Greene WC, Peffer NJ, James SP. 1988. Lymphocytes isolated from the intestinal lamina propria of normal nonhuman primates have increased expression of genes associated with T-cell activation. *Gastroenterol* **94:** 647–655.

The Antibody Response against HIV-1

Julie Overbaugh[1] and Lynn Morris[2]

[1]Division of Human Biology, Fred Hutchinson Cancer Research Center, Seattle, Washington 98109

[2]AIDS Virus Research Unit, National Institute for Communicable Diseases, Johannesburg 2131, South Africa

Correspondence: joverbau@fhcrc.org

Neutralizing antibodies (NAbs) typically play a key role in controlling viral infections and contribute to the protective effect of many successful vaccines. In the case of HIV-1 infection, there is compelling data in experimental animal models that NAbs can prevent HIV-1 acquisition, although there is no similar data in humans and their role in controlling established infection in humans is also limited. It is clear HIV-specific NAbs drive the evolution of the HIV-1 envelope glycoprotein within an infected individual. The virus's ability to evade immune selection may be the main reason HIV-1 NAbs exert limited control during infection. The extraordinary antigenic diversity of HIV-1 also presents formidable challenges to defining NAbs that could provide broad protection against diverse circulating HIV-1 strains. Several new potent monoclonal antibodies (MAbs) have been identified, and are beginning to yield important clues into the epitopes common to diverse HIV-1 strains. In addition, antibodies can also act in concert with effector cells to kill HIV-infected cells; this could provide another mechanism for antibody-mediated control of HIV-1 replication. Understanding the impact of antibodies on HIV-1 transmission and pathogenesis is critical to helping move forward with rational HIV-1 vaccine design.

Antibodies have the potential to block HIV-1 replication through multiple pathways, and they exert immune pressure on the virus that leads to escape. Neutralizing antibodies (NAbs) bind cell-free virus and prevent the virion from infecting the host target cells, thereby disrupting subsequent rounds of replication (Fig. 1A). HIV-1 specific antibodies can also complex with the Fcγ receptor to counter HIV-1 through effector cell mechanisms—a process that has the potential to contain cell–cell HIV-1 spread (Fig. 1B,C). It is not possible to predict which of these antibody mechanisms will be most effective in containing HIV-1 because the relative contribution of cell-free versus cell–cell spread in HIV-1 transmission and pathogenesis is not well defined. Thus, the ability of antibodies to block HIV-1 infection by each of these pathways is the topic of intense study.

Current efforts to identify protective HIV-1 vaccine immunogens are focused primarily on those that elicit broadly NAb responses (Fauci et al. 2008), inspired by the results of experimental studies in macaques showing that passively transferred NAbs can protect against viruses related to HIV-1 (see Lifson and Haigwood 2011). However, as discussed below, the

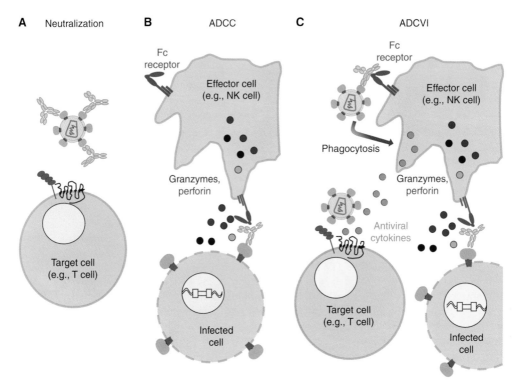

Figure 1. Schematic representation of the mechanism of action of NAbs and antibodies that act through ADCC and ADCVI. (*A*) Antibody neutralization of cell-free virus. Neutralizing antibodies bind to HIV-1 envelope glycoproteins and block the interaction of viral particles with CD4 and CCR5, essential receptors on target cells required for infection. (*B*) Antibody-dependent cellular cytotoxicity leads to the killing of infected cells. In the case of ADCC, a complex between the IgG Fab portion of antibody bound to envelope protein on the cell surface and the Fc portion to the Fc receptors on effector cells leads to lysis of the infected cell. (*C*) Antibody-dependent cell-mediated virus inhibition. ADCVI measures the effects of ADCC-mediated cell killing, which lead to reduced virus production, as well as virus inhibition by antiviral cytokines and other secondary effects of FcR-virus interactions such as phagocytosis.

ability of NAb to provide protection from HIV-1 infection and/or disease progression in humans still remains poorly defined.

Here we review what is known about the role of antibodies in driving virus escape, in controlling an established infection, and in preventing new infections. Much of the article focuses on the neutralizing function of antibodies, including recently discovered HIV-specific broadly neutralizing monoclonal antibodies (MAbs), because antibodies capable of neutralizing cell-free virus infections have garnered the most attention in the HIV-1 field. However, there is increasing interest in antibodies that act through effector mechanisms, and a summary of this topic is included. This article will focus

on studies of HIV-1 in humans but seminal studies in the nonhuman primate model that clarify important issues will be noted where relevant. A more complete review of the related studies in the nonhuman primate model can be found in Lifson and Haigwood (2011).

EARLY ANTIBODIES AND VIRAL ESCAPE

B cell responses to HIV-1 infection first develop within ~1 week of detectable viremia, and initially are detected as antigen-antibody complexes (Tomaras et al. 2008). This phase is followed by circulating anti-gp41 antibodies a few days later, with anti-gp120 antibodies delayed a further few weeks and primarily

Cite this article as *Cold Spring Harb Perspect Med* doi: 10.1101/cshperspect.a007039

targeting the V3 loop. However, these binding antibodies have no detectable effect on viremia (Tomaras et al. 2008) and apparently do not exert any selective immune pressure on the envelope (Keele et al. 2008). NAbs against the infecting strain (autologous virus) appear months later but are not able to neutralize more divergent viruses isolated from other individuals (heterologous viruses) (Moore et al. 1994; Legrand et al. 1997; Moog et al. 1997; Pilgrim et al. 1997; Richman et al. 2003; Wei et al. 2003; Frost et al. 2005; Deeks et al. 2006; Gray et al. 2007). These autologous NAbs drive neutralization escape as evidenced by the fact that contemporaneous viruses are less sensitive to autologous neutralization than earlier viruses, although later viruses remain sensitive to new NAb responses. Escape occurs through single amino acid substitutions, insertions and deletions, and through an "evolving glycan shield," in which shifting glycans prevent access of NAbs to their cognate epitopes (Overbaugh and Rudensey 1992; Chackerian et al. 1997; Herrera et al. 2003; Richman et al. 2003; Wei et al. 2003; Bunnik et al. 2008).

Recent data suggest that autologous responses in the first year of infection comprise one or two antibody specificities targeting variable regions of the HIV-1-envelope glycoprotein (Moore et al. 2009a; Rong et al. 2009). These potent but highly type-specific antibodies appear sequentially and show temporal fluctuations as escape variants emerge. It has been proposed that an immunological hierarchy of NAb responses exists, similar to what is seen for binding antibodies (Moore et al. 2009a). Alternatively, stimulation of new specificities may occur as additional epitopes become immunodominant on the escape variants. If so, this may ultimately in some instances, result in the exposure of more conserved regions that will lead to the induction of broadly cross-neutralizing antibodies in some individuals (Moore et al. 2009a).

A major target of autologous NAbs is the V1V2 region, which can serve both as a direct antibody target but also, as a result of its location and extensive glycosylation, act as a shield for other vulnerable sites (Pinter et al. 2004; Rong et al. 2007). This was first shown in the SIV/macaque model, where the earliest genetic changes were in V1V2 (Burns and Desrosiers 1991; Johnson et al. 1991; Overbaugh et al. 1991; Overbaugh and Rudensey 1992; Laird et al. 2008) and biochemical studies showed that the addition of N- and O-linked carbohydrates was a key pathway of NAb escape (Chackerian et al. 1997). In HIV-1 subtype C infection, the C3 region and specifically the α2-helix is also a major target of autologous NAbs that may be related to the more amphipathic nature of the subtype C α-2 helix in combination with a shorter V4 loop (Gnanakaran et al. 2007; Rong et al. 2009). The V1V2 as well as sites at the base of the V3 loop have been shown to be targets of autologous NAbs in HIV-1 subtype B (Fig. 2). In particular, a glycan at the base of the V3 which contributes to the 2G12 epitope was identified in two subtype B infected individuals as mediating autologous NAb escape (Tang et al. 2011). Despite the strain-specificity and their transient effect, studies on autologous NAbs are important because they identify vulnerable immunogenic regions of the HIV-1 envelope glycoprotein.

DEVELOPMENT OF BROADLY CROSS-NEUTRALIZING ANTIBODIES

NAbs tend to increase in potency over time and broadly cross neutralizing (BCN) responses, capable of recognizing heterologous HIV-1 variants, develop in a subset of individuals after primary infection (Moore et al. 1996; Moog et al. 1997; Beirnaert et al. 2000; Binley et al. 2008; Piantadosi et al. 2009; Sather et al. 2009; Simek et al. 2009; Gray et al. 2011). The reasons why some individuals develop BCN antibodies is unclear but is related to the duration of infection and viral levels suggesting that years of persistent viral stimulation are necessary for their generation (Piantadosi et al. 2009; Sather et al. 2009; Euler et al. 2010; Gray et al. 2011). However, the finding that not all individuals develop BCN antibodies points to the role of additional factors that could be viral or host or most likely an interplay between the two. There is some

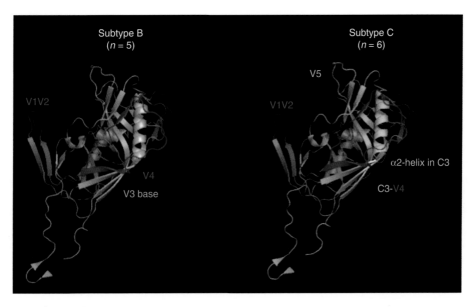

Figure 2. Targets of autologous NAbs on HIV-1 subtype B and subtype C envelopes. The figure summarizes data from studies in which the precise targets of autologous NAbs on gp120 have been defined. This includes two studies in HIV-1 subtype B infection (KJ Bar, unpubl.; Tang et al. 2011) and two in HIV-1 subtype C infection (Moore et al. 2009; Rong et al. 2009). The sites targeted are listed in color with the corresponding regions on the gp120 structures in the same color. Sites are highlighted on the crystal structure of HIV-1 JR-FL gp120 core protein containing the third variable region (V3) complexed with CD4 and the X5 antibody (PDB 2B4C) (Huang and Tang 2005).

evidence that viral genetic subtype may also be important as the breadth and potency of humoral responses is reported to be higher in subtype C and A than in subtype B infections (Bures et al. 2002; Li et al. 2006; Brown et al. 2008; Dreja et al. 2010).

A large number of studies published in the last few years have described the presence of BCN antibodies in different cohorts. These have been important studies as they indicate that such sera and antibodies are not rare (Stamatatos et al. 2009). Although there is no standard definition of breadth, all studies identified a proportion ($\sim$30%) of individuals with BCN antibodies based on their ability to neutralize a significant number of heterologous viruses usually of multiple genetic subtypes. In some cases, the specificities of the antibodies conferring breadth have been mapped and are reactive with conserved envelope regions, as exemplified by well-characterized neutralizing MAbs IgG1b12 (anti-CD4bs), 4E10 (carboxy-terminal MPER), 2F5 (amino-terminal MPER), and 2G12 (glycan array on gp120 outer domain). The ability of recombinant proteins and peptides to deplete neutralizing activity was a major step toward unraveling the components of these complex plasma/sera and the use of mutant proteins allowed for mapping to the CD4bs (Dhillon et al. 2007; Binley et al. 2008; Gray et al. 2009b; Li et al. 2009; Sather et al. 2009) or to CD4 induced epitopes (CD4i) (Gray et al. 2009b; Li et al. 2009). In other cases, neutralizing activity could be mapped to linear epitopes within the membrane proximal region (MPER) of gp41 (Binley 2009; Gray et al. 2009a,b; Li et al. 2009; Sather et al. 2009). As new NAb epitopes are discovered, mapping tools are expanding to detect these specificities. Thus, PG9/16-like antibodies are frequently being identified in BCN sera (Fig. 3). However, not all the BCN activities have been accounted for, suggesting that additional important neutralizing epitopes on the HIV-1 envelope glycoprotein are yet to be discovered. Collectively, these studies provide evidence that the human

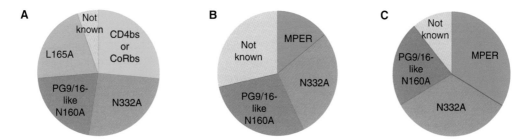

Figure 3. Specificity of antibodies in cross-neutralizing samples. Summary of the three most comprehensive mapping studies to date examining the specificities of antibodies mediating neutralization breadth. This includes (*A*) 19 samples from Walker et al. (2009), (*B*) seven samples from Gray et al. (2011), and (*C*) nine samples from Tomaras et al. (2010). PG9/16-like NAbs and an undefined epitope that involves a glycan at position 332 in the outer domain of gp120 were the most common targets in all three studies. Antibodies found to be responsible for breadth in fewer samples included those targeting the MPER (*B,C*) or CD4bs/CoRbs (*A*). In all three studies the antibody specificities in some of the broadly cross-neutralizing sera/plasma could not be identified.

immune system can generate BC NAbs against HIV-1. Intensive efforts are underway by many laboratories to isolate MAbs from these individuals (see section on MAbs later) as this may lead to the identification of new sites on the HIV envelope that could serve as targets for vaccine induced immune responses.

The question of whether breadth is conferred by single or multiple specificities has in part been addressed by some of these studies. Certainly, cases of multiple specificities have been documented, but there is also evidence that single MAbs can account for the breadth displayed by BCN plasma, most recently through the isolation of VRC01, a highly potent and broad anti-CD4bs mAb from an individual whose plasma neutralizing activity was shown to have anti-CD4bs activity (Wu et al. 2010). On the other hand, a study by Nussensweig and colleagues suggested that neutralization breadth is the result of multiple antibody specificities, each one targeting a few viral variants (Scheid et al. 2009). For vaccine design, it would likely be considerably more challenging to induce multiple rather than single specificities because different antigens and thus a more complex vaccine would be required to stimulate a polyclonal response.

Another important question is whether escape from BCN antibodies occurs as readily as it does from autologous NAbs. There is less

data on this but one might imagine that because BCN antibodies target more conserved sites, escape would be more difficult to achieve. Certainly, this is suggested in cases in which contemporaneous neutralization is seen more frequently among plasma with BCN activity (Mahalanabis et al. 2009; Bosch et al. 2010; Kirchherr et al. 2011). However, cloned viruses from the individual who was the source of VRC01, the highly potent anti-CD4bs mAb that targets one of the most conserved regions on the envelope glycoprotein, were resistant to this MAb (Wu and Wang 2010). Studies on MAbs and polyclonal plasma are difficult to compare as the lack of escape from polyclonal plasma may be the result of multiple somatic variants of a neutralizing specificity which target different variants within the quasispecies (Morris and Chen, submitted).

The development of neutralization breadth may include a requirement for extensive somatic mutational diversification that may be impeded by the B cell dysfunction that occurs in HIV-1 infection (Moir and Fauci 2009). A role for affinity maturation in the ontogeny of anti-HIV-1 neutralizing capacity is supported by Toran and coworkers (Toran et al. 1999), who showed that all anti-gp120 NAbs in a long-term nonprogressor were clonally related with considerable somatic hypermutation. Furthermore, immunogenetic analysis of existing MAbs suggests that they have undergone multiple rounds of affinity maturation to achieve BCN activity (Zhou et al. 2007; Walker et al. 2009; Xiao et al. 2009). These MAbs are characterized by unusual physical features such as heavy chain domain-swapping, sulphated tyrosines, long and hydrophobic CDR-H3's, and high levels of hypermutation (Burton et al. 2005; Kwong and Wilson 2009). Whether these events are required for antibodies to mediate cross-reactivity is not known, and longitudinal studies of immunoglobulin genes among those individuals who develop breadth are needed to understand how the immune system can generate such antibodies. If indeed long-term antigenic exposure is required, vaccination strategies will need to be revised to not only consider the immunogen used, but to incorporate the required number of booster and/or antigen exposures needed to reach the required level of antibody affinity.

ANTIBODIES AND DISEASE PROGRESSION

Several studies have examined whether HIV-1-specific NAbs contribute to viral control. The first studies focused primarily on select populations, such as long-term nonprogressors (LTNP) but these were conflicting, with some providing evidence that NAb contributed to control of infection (Cao et al. 1995; Montefiori et al. 1996; Pilgrim et al. 1997; Zhang et al. 1997; Carotenuto et al. 1998; Cecilia et al. 1999), others suggesting the opposite—that NAbs responses were low in LTNPs (Pereyra et al. 2008; Doria-Rose et al. 2009; Lambotte et al. 2009), and still others not finding consistent evidence for or against a role for NAb in controlling infection in LTNPs (Harrer et al. 1996; Barker et al. 1998; Loomis-Price et al. 1998; Bailey et al. 2006).

The first studies of more typical HIV-1-infected populations showed that individuals with greater NAb breadth had higher contemporaneous viral loads, suggesting that antigenic stimulation could be key to eliciting these NAbs (Deeks et al. 2006; Rodriguez et al. 2007; Pereyra et al. 2008; Doria-Rose et al. 2009; Sather et al. 2009). Recent studies, focused on populations followed from the time of infection, showed that the association of NAb breadth and contemporaneous viral load in chronic infection was not significant after controlling for set point viral levels, suggesting that higher viral replication was a cause, rather than a consequence of NAb breadth (Piantadosi et al. 2009). In addition, it was found that greater envelope diversity early in infection contributed to the breadth of the subsequent NAb responses, further suggesting that antigenic stimulation drives NAb breadth. Importantly, these longitudinal studies showed that the breadth of the NAb did not impact progression to AIDS (Piantadosi et al. 2009; Euler et al. 2010; Gray et al. 2011) suggesting that the presence of broad NAbs in and of itself did not lead to viral control. This finding was consistent with findings from studies of passively administered, broadly neutralizing antibodies in chronic HIV-1 infection, in which it was estimated that antibody levels that were 10- to100-fold higher than those already present in chronic HIV-1 infection would be required to see clinical benefit of monoclonal antibody therapy (Trkola et al. 2008).

Although these studies collectively suggest that the breadth of the NAb response, as measured against heterologous virus panels, does not correlate with viral control or pathogenecity, it is important to note that most have not addressed whether NAb specifically targeted to the virus in that individual provide clinical benefit. A recent case report, showing a drop in viral load coincident with the detection of NAb that recognizes the autologous virus, suggested the possibility that autologous NAb have the potential to contribute to viral control, although escape and loss of viral control is rapid (Moore et al. 2009b).

Another case report of a HIV-1-positive individual who underwent a B cell depletion protocol as part of clinical care unrelated to HIV-1 also showed a temporal relationship between changes in the autologous NAb response and viral levels, although viral load changes from other effects of B cell depletion could not be ruled out (Huang et al. 2010). These results are consistent with the results of B cell depletion studies in the SIV/rhesus macaque model (Schmitz et al. 2003; Miller et al. 2007, and see Lifson and Haigwood 2011). In addition, one cross-sectional population study suggested that, unlike heterologous NAb responses, autologous NAb inversely correlate with viral levels; however, in this study there was limited follow-up, and longer term clinical outcomes measures were not available (Deeks et al. 2006). Thus, it is possible that NAb are exerting some control on virus replication, but detailed studies of autologous NAb responses in relation to clinical outcome in longitudinal cohorts are needed to clarify whether NAbs exert any control on HIV-1 infection.

ANTIBODIES AND TRANSMISSION

Because effective protection afforded by most vaccines correlates with antibody responses (Karlsson Hedestam et al. 2008), there is considerable interest in determining whether NAbs can protect against HIV-1 transmission, and if so, in defining the NAb levels and specificities required for protection. As discussed by Lifson and Haigwood (2011), passively infused NAbs can block SIV and SHIV-1 infection in the nonhuman primate model (Prince et al. 1991; Emini et al. 1992; Conley et al. 1996; Mascola et al. 1999, 2000; Shibata et al. 1999; Baba et al. 2000; Parren et al. 2001; Mc Cann et al. 2005; Hessell et al. 2009). However, in most of the initial, proof-of-principle passive antibody studies, the antibody levels used to achieve protection were high, and typically the virus strain used for challenge was very sensitive to neutralization by the antibody tested. In more recent studies, lower NAb levels, closer to those found in HIV-1 infected humans, were shown to increase the number of exposures required to achieve infection (Hessell et al. 2009). In this study, the viral challenge strain was highly sensitive to the NAb, making it hard to extrapolate from these findings to predict an effective dose in humans, who are exposed to diverse HIV-1 strains that can differ by several orders of magnitude in their neutralization sensitivity (Blish et al. 2009; Walker et al. 2009; Seaman et al. 2010). There are limited studies from exposed humans that can inform

our understanding of the potential of HIV-specific antibodies to protect against these diverse, circulating strains of HIV-1, and these studies primarily have focused on exposed infants and individuals who become superinfected. In both of these cases, HIV-1-specific NAbs are present at the time of virus exposure.

HIV-infected high-risk individuals often continue to be exposed to HIV-positive partners and thus are at risk of being superinfected (reinfected) (reviewed in Chohan et al. 2010). Cases of superinfection have provided the chance to examine whether there are particular deficits in HIV-1 immunity in individuals who become superinfected versus those who do not despite similar continued exposure. The first study of NAbs in cases of superinfection suggested that individuals who become superinfected ($N = 3$) had relatively weak NAbs compared to controls (Smith et al. 2006). However, this study examined responses to only three viral strains, and at an early time after the first infection when NAbs tend to be weak in general. A second study focusing on six cases of superinfection, including cases that occurred several years after the first infection, showed that the neutralizing antibody breadth, defined using a panel of 16 circulating recently transmitted viral variants, was similar in individuals who became superinfected to NAb breadth in individuals who did not become superinfected (Blish et al. 2008). Furthermore, the antibodies that were present near the time of superinfection could neutralize the strain that established the second infection in most cases (Blish et al. 2008), suggesting that reinfection occurred in the face of NAbs that recognized the incoming variant. These studies suggest that the level of antibody elicited during chronic HIV-1 infection may not be adequate to protect against reinfection by diverse HIV-1 variants; however, studies to-date have been small.

The setting of mother-to-child transmission (MTCT) also offers a chance to explore whether Nab, present at the time of exposure, protects against HIV-1 infection in humans. Very few studies have focused specifically on the role of passively acquired antibodies in the exposed infant in protection, although one recent study showed that uninfected infants of HIV-1-positive mothers have HIV-1-specific NAb levels at birth that are comparable to those in an infected individual (Lynch et al. 2011). However, there was no evidence that the breadth or potency of infant antibodies, defined using a heterologous virus panel, correlated with protection from infection during the breastfeeding period (Lynch et al. 2011). Nonetheless, studies of maternal NAb do provide some support for a protective role for NAb in the mother, who in this situation is the index case. These studies have focused primarily on the study of maternal autologous virus, which is a more direct test of the potential of the antibodies to neutralize the specific virus that the infant encounters than the study of virus panels. Most of these studies suggest that mothers with more potent autologous NAbs are less likely to transmit to the infant than mothers with low autologous NAb levels (Scarlatti et al. 1993a,b; Kliks et al. 1994; Dickover et al. 2006). However, the results of studies of the effect of maternal NAb on transmission have reached divergent conclusions as to whether NAbs are important during all stages of mother-to-child transmission (in utero, intrapartum, and breastfeeding); for example, one study suggested a protective effect of NAb only for intrapartum transmission (Barin et al. 2006; Samleerat et al. 2009), whereas two others suggest NAb protection only in utero transmission (Dickover et al. 2006). Moreover, some studies have reported no association of maternal NAbs and infant infection (Husson et al. 1995; Hengel et al. 1998). These divergent findings could reflect methodological differences, such as the relative timing of the NAb and viruses studied, which is critical given the rapid evolution of both virus and antibody responses (Burton et al. 2005), whether the study examined autologous or heterologous responses and the assays used, as well as the sample size of the study, which in the earliest studies was typically small. However, to date, no specific study design differences have been identified to explain these variable findings. Although most of the larger studies do show a positive association between maternal antibodies and protection from HIV-1 infection in the

infant, this remains a critical issue to resolve fully.

In MTCT, as in other modes of HIV-1 infection, only one or a few HIV-1 variants are transmitted (reviewed in Sagar 2010 and discussed more fully in Shaw and Hunter 2011). However, there are interesting differences in neutralization sensitivity among viruses transmitted sexually compared to vertically. One detailed study of sexually transmitted viruses suggested that they are generally more neutralization sensitive compared to the population of viruses in the source partner (Derdeyn et al. 2004). In sexual transmission, immune selection pressure is not expected to play a role because HIV-1-specific NAb is not likely to be present in the exposed individual, which is in contrast to the setting of MTCT. Similar studies of individual variants isolated from mother-infant transmission pairs showed that the transmitted variants tended to be among the more neutralization-resistant viruses compared to the maternal viruses (Wu et al. 2006; Zhang et al. 2010). Although antibodies present in some transmitting mothers did neutralize the variants transmitted to their infants, the most highly neutralization-sensitive maternal variants were generally not transmitted (Wu et al. 2006; Zhang et al. 2010). Thus, the studies of MTCT support the notion that there may be some level of NAb protection, at least against the most neutralization-sensitive viruses, raising the possibility that higher levels of broadly cross-reactive NAb than those present in natural infection could prevent infection by more diverse variants.

There have also been studies suggesting a protective effect of NAbs in highly HIV-1-exposed, uninfected individuals (Hirbod et al. 2008; Hasselrot et al. 2009). These studies, although intriguing, remain controversial because there is limited precedent to suggest that antibodies are elicited by exposure in the absence of viral infection. However, the more frequent detection of this HIV-1-neutralizing activity from exposed uninfected population compared to controls has now been shown in several cohorts (Hirbod et al. 2008; Hasselrot et al. 2009). The activity was detected in the

IgA antibody fraction, but the nature of this neutralizing activity remains to be defined.

IDENTIFICATION AND CHARACTERIZATION OF MONOCLONAL ANTIBODIES

HIV-1-specific monoclonal antibodies (MAbs) have been exceptionally useful reagents for characterizing neutralization-sensitive targets on the HIV-1 envelope glycoproteins and for revealing the structural dimensions of this complex heterotrimer. One of the first MAbs to be extensively used for this purpose was IgG1b12, which was isolated in 1994 and targets the CD4 binding site (CD4bs). The CD4bs represents one of the most attractive targets for vaccine-induced immune responses because it is a highly conserved site present on all HIV-1 genetic subtypes as a result of the absolute requirement for engaging the CD4 molecule to gain entry into human cells. Recently, VRC01, which is a considerably more potent anti-CD4bs MAb, has provided additional insights into the nature of this epitope (for additional information see Kwong et al. 2011). A cluster of new antibodies have helped to define a highly structural epitope present only on the trimeric envelope that includes conserved regions of both V2 and V3. The first identified MAb directed to this target, MAb 2090, was type-specific and neutralized only SF162 because this isolate carries an unusual lysine at position 160 in V2 (Honnen et al. 2007). PG9/16 isolated from a subtype A chronically infected individual with exceptional plasma neutralization breadth binds the more common N160 variant and are hence more broadly cross-reactive (Walker et al. 2009). These types of antibodies have been isolated from other individuals with breadth (Bonsignori et al. 2010), and additional antibodies with different fine specificities await isolation (Tomaras et al. 2010; Gray et al. 2011). The 2G12 MAb, which was isolated in 1996 (Trkola et al. 1996), focused attention on glycans. The recent discovery of new MAbs that target the glycan at position 332 and compete with 2G12, but display more breadth and potency

than 2G12, suggests that interest in this potential target is likely to increase (Walker et al. 2010). MAbs against the gp41 and the MPER have also recently been described and all define new epitopes in this region (Corti et al. 2010; Sabin et al. 2010; Zhu et al. 2010; L Morris and S Chen, submitted). One of the interesting aspects of this new era in MAb isolation is that all studies have isolated multiple highly mutated clonally related antibodies (Scheid et al. 2009; Walker et al. 2009; Bonsignori et al. 2010; Wu et al. 2010), underscoring the importance of the affinity maturation process that takes place in the development of antienvelope Abs.

A few studies have shown a skewed germline gene usage among HIV-1 MAbs. Those that target CD4i frequently use the 1-69 variable heavy chain gene family (V_H1-69) (Huang et al. 2004). It has been proposed that the two hydrophobic residues at the tip of the CDR-H2 loop in the V_H1-69 germline genes allows the Ab to access hydrophobic recessed pockets such as the CD4i epitope and the MPER. In contrast, anti-V3 MAbs appear to preferentially use the V_H5-51 germline gene (Gorny et al. 2009). These studies suggest that particular epitopes select for certain characteristics in the antibody-binding site that might only be provided by certain germlines. More studies with additional MAbs are needed to determine the full extent of this and whether vaccination strategies will need to select for B cells bearing specific immunoglobulin genes.

NONNEUTRALIZING ANTIBODIES

There is renewed interest in the potential role of nonneutralizing antibodies in control of HIV-1 infection. An important mechanism by which nonneutralizing antibody can clear virus is to bind to infected cells and recruit activated effector cells, which in turn induce cytolysis or apoptosis of infected cells. Antibody-dependent cellular cytotoxicity (ADCC) is the result of the formation of a complex between the IgG Fab portion of the antibody with the viral protein on the cell surface and binding of the Fc portion to the Fc receptors (FcγRs), on effector cells (see Fig. 1B). Potential effector cells include NK cells, macrophages, dendritic cells, γδ T-cells, and neutrophils. In addition to causing cytolysis of the infected cell (Kagi et al. 1994), binding to the Fcγ receptors can lead to release of antiviral cytokines (Berke 1995; Russell and Ley 2002), depending on the nature of the receptor. Thus, Fc receptor-mediated antibody activities can inhibit virus spread in multiple ways, and it can be measured as either a cytotoxic effect on infected cells (e.g., ADCC) or as virus inhibition (antibody-dependent cell-mediated virus inhibition, ADCVI), which measures not only virus reduction because of cell killing, but also virus inhibition by antiviral cytokines and other secondary factors (Fig. 1B). ADCVI may reduce the spread of both cell-free and cell-associated virus.

There is some evidence from vaccine studies in both humans and nonhuman primate model systems that nonneutralizing antibodies may afford some protection from infection. In human vaccine studies, a correlation was observed between both antibody binding activity and ADCVI antibody activity and the incidence of HIV-1 infection in vaccine recipients in the Vaxgen Phase III efficacy trial (Gilbert et al. 2005; Forthal et al. 2007a). One important caveat to this finding is that there was no overall protective vaccine effect of the vaccine. However, several vaccine challenge studies in rhesus macaques have shown that ADDC/ADCVI activity correlates with reduced viral load postchallenge (Banks et al. 2002; Gomez-Roman et al. 2005; Florese et al. 2009; Hidajat et al. 2009; Xiao et al. 2010). Perhaps the most compelling evidence for a protective role for antibodies that work through ADCC/ADCVI comes from passive immunization showing that a HIV-specific MAb engineered with decreased FcγR binding potential provided only partial protection compared to the original MAb, despite having similar neutralizing activity against the challenge virus (Hessell et al. 2007). These studies suggested that FcγR function is important in the protective effect of the MAb tested (IgG1b12), perhaps by reducing cell–cell virus spread.

Studies of the role of Fc receptor-mediated antibody activities in HIV-1 disease progression have been relatively few compared to studies of NAbs and CD8 T cell responses. Antibodies capable of ADCC/ADCVI have been detected early in infection, and a higher magnitude of Fc-mediated NK cell responses early correlated with lower set-point viral load (Connick et al. 1996; Forthal et al. 2001a,b); however, long-term follow-up was not available to determine associations with clinical outcome. Additional indirect support for a specific role of ADCC/ADCVI in HIV-1 pathogenesis comes from studies of host genes that encode the Fc receptor proteins. Mutations in FcR impact IgG subclass binding affinity to the receptor, and thus may modulate ADCC/ADCVI responses. FcγRIIa genotype has been associated with the rate of $CD4^+$ cell count decline in infected adults (Forthal et al. 2007b). FcγRIIa genotype in infants has also been linked with risk of infection, suggesting that Fc receptor interactions with passively acquired HIV-1-specific antibodies could be playing a role in infant infection (Brouwer et al. 2004). However, to date, there have been no studies demonstrating an association between ADCC/ADCVI in infants and their risk of HIV-1 acquisition.

CONCLUSION

There have been significant advances in the last few years that have renewed enthusiasm in the field and provided new insights into the nature of neutralizing antibodies. The finding that low doses of neutralizing MAbs at levels close to those found in natural infection can protect against neutralization-sensitive challenge strains via the mucosal route in monkeys (Hessell et al. 2009) may provide clues to the amount and specificity of antibody needed for protection. The virus targets are also becoming more extensively characterized; many studies have shown that transmission is mediated in most cases by just one or a few viral particles (Mc-Nearney et al. 1992; Wolinsky et al. 1992; Zhang et al. 1993; Zhu et al. 1993; Poss et al. 1995; Long et al. 2000; Sagar et al. 2003, 2004, 2006; Derdeyn et al. 2004; Keele et al. 2008 and many subsequent references) and transmitted viruses tend to be less glycosylated (Derdeyn et al. 2004; Chohan et al. 2005). In addition, the flurry of new MAbs isolated because of significant improvements in cohort characterization and antibody mapping methods, high-throughput neutralization screening and technological advances in immunoglobulin gene rescue have identified new targets on the HIV-1 envelope vulnerable to antibody attack with the promise of more to come. A number of studies on passive immunotherapy with these new more potent MAbs are being planned to test whether these antibodies can prevent infection in humans. Most human vaccine trials that focused on envelope immunogens have so far been disappointing in terms of NAb induction. However, the modest protection observed in the vaccine efficacy trial of ALVAC and AIDSVAX in Thailand (Rerks-Ngarm et al. 2009, and see O'Connell et al. 2011), using immunogens that do not elicit strong NAbs or T cell responses, has also renewed interest in examining other functions of antibodies. These activities include ADCC and ADVCI, as discussed above, as well as antibodies that prevent transcytosis across epithelial barriers (Shen et al. 2010) or mediate complement activation (Huber et al. 2006). The potential role of HIV-1-specific antibodies at mucosal sites in protection is poorly understood and is likely to be a topic of future studies. More basic research is needed into the full array of antibody-mediated effector functions in both infected populations as well as in ongoing vaccine trials in humans to determine the role of antibodies in protection.

ACKNOWLEDGMENTS

We would like to thank Dr. Penny Moore, Dr. Elin Gray, and Constantinos Kurt Wibmer for help with the figures and for critical reading of the manuscript.

REFERENCES

*Reference is also in this collection.

Baba TW, Liska V, Hofmann-Lehmann R, Vlasak J, Xu W, Ayehunie S, Cavacini LA, Posner MR, Katinger H, Stiegler G, et al. 2000. Human neutralizing monoclonal antibodies of the IgG1 subtype protect against mucosal simian-human immunodeficiency virus infection. *Nat Med* **6:** 200–206.

Bailey JR, Lassen KG, Yang HC, Quinn TC, Ray SC, Blankson JN, Siliciano RF. 2006. Neutralizing antibodies do not mediate suppression of human immunodeficiency virus type 1 in elite suppressors or selection of plasma virus variants in patients on highly active antiretroviral therapy. *J Virol* **80:** 4758–4770.

Banks ND, Kinsey N, Clements J, Hildreth JE. 2002. Sustained antibody-dependent cell-mediated cytotoxicity (ADCC) in SIV-infected macaques correlates with delayed progression to AIDS. *AIDS Res Hum Retrov* **18:** 1197–1205.

Barin F, Jourdain G, Brunet S, Ngo-Giang-Huong N, Weerawatgoompa S, Karnchanamayul W, Ariyadej S, Hansudewechakul R, Achalapong J, Yuthavisuthi P, et al. 2006. Revisiting the role of neutralizing antibodies in mother-to-child transmission of HIV-1. *J Infect Dis* **193:** 1504–1511.

Barker E, Mackewicz CE, Reyes-Teran G, Sato A, Stranford SA, Fujimura SH, Christopherson C, Chang SY, Levy JA. 1998. Virological and immunological features of long-term human immunodeficiency virus-infected individuals who have remained asymptomatic compared with those who have progressed to acquired immunodeficiency syndrome. *Blood* **92:** 3105–3114.

Beirnaert E, Nyambi P, Willems B, Heyndrickx L, Colebunders R, Janssens W, van der Groen G. 2000. Identification and characterization of sera from HIV-infected individuals with broad cross-neutralizing activity against group M (env clade A-H) and group O primary HIV-1 isolates. *J Med Virol* **62:** 14–24.

Berke G. 1995. Unlocking the secrets of CTL and NK cells. *Immunol Today* **16:** 343–346.

Binley J. 2009. Specificities of broadly neutralizing anti-HIV-1 sera. *Curr Opin HIVAIDS* **4:** 364–372.

Binley JM, Lybarger EA, Crooks ET, Seaman MS, Gray E, Davis KL, Decker JM, Wycuff D, Harris L, Hawkins N, et al. 2008. Profiling the specificity of neutralizing antibodies in a large panel of plasmas from patients chronically infected with human immunodeficiency virus type 1 subtypes B and C. *J Virol* **82:** 11651–11668.

Blish CA, Dogan OC, Derby NR, Nguyen MA, Chohan B, Richardson BA, Overbaugh J. 2008. HIV-1 superinfection occurs despite relatively robust neutralizing antibody responses. *J Virol* **82:** 12094–12103.

Blish CA, Jalalian-Lechak Z, Rainwater S, Nguyen MA, Dogan OC, Overbaugh J. 2009. Cross-subtype neutralization sensitivity despite monoclonal antibody resistance among early subtype A, C, and D HIV-1 envelope variants. *J Virol* **83:** 7783–7788.

Bonsignori M, Hwang K. 2010. Immunoregulation of HIV-1 broadly neutralizing antibody responses: Deciphering maturation paths for antibody induction. In *AIDS Vaccine 2010*, p. A-153. Atlanta, Georgia.

Bosch KA, Rainwater S, Jaoko W, Overbaugh J. 2010. Temporal analysis of HIV envelope sequence evolution and antibody escape in a subtype A-infected individual with a broad neutralizing antibody response. *Virology* **398:** 115–124.

Brouwer KC, Lal RB, Mirel LB, Yang C, van Eijk AM, Ayisi J, Otieno J, Nahlen BL, Steketee R, Lal AA, et al. 2004. Polymorphism of Fc receptor IIa for IgG in infants is associated with susceptibility to perinatal HIV-1 infection. *Aids* 18: 1187–1194.

Brown BK, Wieczorek L, Sanders-Buell E, Rosa Borges A, Robb ML, Birx DL, Michael NL, McCutchan FE, Polonis VR. 2008. Cross-clade neutralization patterns among HIV-1 strains from the six major clades of the pandemic evaluated and compared in two different models. *Virology* 375: 529–538.

Bunnik EM, Pisas L, van Nuenen AC, Schuitemaker H. 2008. Autologous neutralizing humoral immunity and evolution of the viral envelope in the course of subtype B human immunodeficiency virus type 1 infection. *J Virol* 82: 7932–7941.

Bures R, Morris L, Williamson C, Ramjee G, Deers M, Fiscus SA, Abdool-Karim S, Montefiori DC. 2002. Regional clustering of shared neutralization determinants on primary isolates of clade C human immunodeficiency virus type 1 from South Africa. *J Virol* 76: 2233–2244.

Burns DP, Desrosiers RC. 1991. Selection of genetic variants of simian immunodeficiency virus in persistently infected rhesus monkeys. *J Virol* 65: 1843–1854.

Burton DR, Stanfield RL, Wilson IA. 2005. Antibody vs. HIV in a clash of evolutionary titans. *Proc Natl Acad Sci* 102: 14943–14948.

Cao Y, Qin L, Zhang L, Safrit J, Ho DD. 1995. Virologic and immunologic characterization of long-term survivors of human immunodeficiency virus type 1 infection [see comments]. *N Engl J Med* 332: 201–208.

Carotenuto P, Looij D, Keldermans L, de Wolf F, Goudsmit J. 1998. Neutralizing antibodies are positively associated with CD4+ T-cell counts and T-cell function in long-term AIDS-free infection. *AIDS* 12: 1591–1600.

Cecilia D, Kleeberger C, Munoz A, Giorgi JV, Zolla-Pazner S. 1999. A longitudinal study of neutralizing antibodies and disease progression in HIV-1-infected subjects. *J Infect Dis* 179: 1365–1374.

Chackerian B, Rudensey L, Overbaugh J. 1997. Specific N-linked and O-linked glycosylation additions in the envelope V1 domain of SIV variants that evolve in the host alter neutralizing antibody recognition. *J Virol* 71: 7719–7727.

Chohan B, Lang D, Sagar M, Korber B, Lavreys L, Richardson B, Overbaugh J. 2005. Selection for human immunodeficiency virus type 1 envelope glycosylation variants with shorter V1-V2 loop sequences occurs during transmission of certain genetic subtypes and may impact viral RNA levels. *J Virol* 79: 6528–6531.

Chohan BH, Piantadosi A, Overbaugh J. 2010. HIV-1 Superinfection and its implications for vaccine design. *Curr HIV Res* 8: 596–601.

Conley AJ, Kessler J-A II, Boots LJ, McKenna PM, Schleif WA, Emini EA, Mark G-E III, Katinger H, Cobb EK, Lunceford SM, et al. 1996. The consequence of passive administration of an anti-human immunodeficiency virus type 1 neutralizing monoclonal antibody before challenge of chimpanzees with a primary virus isolate. *J Virol* 70: 6751–6758.

Connick E, Marr DG, Zhang XQ, Clark SJ, Saag MS, Schooley RT, Curiel TJ. 1996. HIV-specific cellular and humoral immune responses in primary HIV infection. *AIDS Res Hum Retrov* 12: 1129–1140.

Corti D, Langedijk JP, Hinz A, Seaman MS, Vanzetta F, Fernandez-Rodriguez BM, Silacci C, Pinna D, Jarrossay D, Balla-Jhagjhoorsingh S, et al. 2010. Analysis of memory B cell responses and isolation of novel monoclonal antibodies with neutralizing breadth from HIV-1-infected individuals. *PLoS One* 5: e8805.

Deeks SG, Schweighardt B, Wrin T, Galovich J, Hoh R, Sinclair E, Hunt P, McCune JM, Martin JN, Petropoulos CJ, et al. 2006. Neutralizing antibody responses against autologous and heterologous viruses in acute versus chronic human immunodeficiency virus (HIV) infection: Evidence for a constraint on the ability of HIV to completely evade neutralizing antibody responses. *J Virol* 80: 6155–6164.

Derdeyn CA, Decker JM, Bibollet-Ruche F, Mokili JL, Muldoon M, Denham SA, Heil ML, Kasolo F, Musonda R, Hahn BH, et al. 2004. Envelope-constrained neutralization-sensitive HIV-1 after heterosexual transmission. *Science* 303: 2019–2022.

Dhillon AK, Donners H, Pantophlet R, Johnson WE, Decker JM, Shaw GM, Lee FH, Richman DD, Doms RW, Vanham G, et al. 2007. Dissecting the neutralizing antibody specificities of broadly neutralizing sera from human immunodeficiency virus type 1-infected donors. *J Virol* 81: 6548–6562.

Dickover R, Garratty E, Yusim K, Miller C, Korber B, Bryson Y. 2006. Role of maternal autologous neutralizing antibody in selective perinatal transmission of human immunodeficiency virus type 1 escape variants. *J Virol* 80: 6525–6533.

Doria-Rose NA, Klein RM, Manion MM, O'Dell S, Phogat A, Chakrabarti B, Hallahan CW, Migueles SA, Wrammert J, Ahmed R, et al. 2009. Frequency and phenotype of human immunodeficiency virus envelope-specific B cells from patients with broadly cross-neutralizing antibodies. *J Virol* 83: 188–199.

Dreja H, O'Sullivan E, Pade C, Greene KM, Gao H, Aubin K, Hand J, Isaksen A, D'Souza C, Leber W, et al. 2010. Neutralization activity in a geographically diverse East London cohort of human immunodeficiency virus type 1-infected patients: Clade C infection results in a stronger and broader humoral immune response than clade B infection. *J Gen Virol* 91: 2794–2803.

Emini EA, Schleif WA, Nunberg JH, Conley AJ, Eda Y, Tokiyoshi S, Putney SD, Matsushita S, Cobb KE, Jett CM, et al. 1992. Prevention of HIV-1 infection in chimpanzees by gp120 V3 domain-specific monoclonal antibody. *Nature* 355: 728–730.

Euler Z, van Gils MJ, Bunnik EM, Phung P, Schweighardt B, Wrin T, Schuitemaker H. 2010. Cross-reactive neutralizing humoral immunity does not protect from HIV type 1 disease progression. *J Infect Dis* 201: 1045–1053.

Fauci AS, Johnston MI, Dieffenbach CW, Burton DR, Hammer SM, Hoxie JA, Martin M, Overbaugh J, Watkins DI, Mahmoud A, et al. 2008. HIV vaccine research: The way forward. *Science* 321: 530–532.

Florese RH, Demberg T, Xiao P, Kuller L, Larsen K, Summers LE, Venzon D, Cafaro A, Ensoli B, Robert-Guroff M. 2009. Contribution of nonneutralizing vaccine-elicited antibody activities to improved protective efficacy in rhesus macaques immunized with Tat/Env compared with multigenic vaccines. *J Immunol* 182: 3718–3727.

Cite this article as *Cold Spring Harb Perspect Med* doi: 10.1101/cshperspect.a007039

Forthal DN, Landucci G, Daar ES. 2001a. Antibody from patients with acute human immunodeficiency virus (HIV) infection inhibits primary strains of HIV type 1 in the presence of natural-killer effector cells. *J Virol* **75**: 6953–6961.

Forthal DN, Landucci G, Keenan B. 2001b. Relationship between antibody-dependent cellular cytotoxicity, plasma HIV type 1 RNA, and CD4+ lymphocyte count. *AIDS Res Hum Retrov* **17**: 553–561.

Forthal DN, Gilbert PB, Landucci G, Phan T. 2007a. Recombinant gp120 vaccine-induced antibodies inhibit clinical strains of HIV-1 in the presence of Fc receptor-bearing effector cells and correlate inversely with HIV infection rate. *J Immunol* **178**: 6596–6603.

Forthal DN, Landucci G, Bream J, Jacobson LP, Phan TB, Montoya B. 2007b. FcγRIIa genotype predicts progression of HIV infection. *J Immunol* **179**: 7916–7923.

Frost SD, Wrin T, Smith DM, Kosakovsky Pond SL, Liu Y, Paxinos E, Chappey C, Galovich J, Beauchaine J, Petropoulos CJ, et al. 2005. Neutralizing antibody responses drive the evolution of human immunodeficiency virus type 1 envelope during recent HIV infection. *Proc Natl Acad Sci* **102**: 18514–18519.

Gilbert PB, Peterson ML, Follmann D, Hudgens MG, Francis DP, Gurwith M, Heyward WL, Jobes DV, Popovic V, Self SG, et al. 2005. Correlation between immunologic responses to a recombinant glycoprotein 120 vaccine and incidence of HIV-1 infection in a phase 3 HIV-1 preventive vaccine trial. *J Infect Dis* **191**: 666–677.

Gnanakaran S, Lang D, Daniels M, Bhattacharya T, Derdeyn CA, Korber B. 2007. Clade-specific differences between human immunodeficiency virus type 1 clades B and C: Diversity and correlations in C3-V4 regions of gp120. *J Virol* **81**: 4886–4891.

Gomez-Roman VR, Patterson LJ, Venzon D, Liewehr D, Aldrich K, Florese R, Robert-Guroff M. 2005. Vaccine-elicited antibodies mediate antibody-dependent cellular cytotoxicity correlated with significantly reduced acute viremia in rhesus macaques challenged with SIVmac251. *J Immunol* **174**: 2185–2189.

Gorny MK, Wang XH, Williams C, Volsky B, Revesz K, Witover B, Burda S, Urbanski M, Nyambi P, Krachmarov C, et al. 2009. Preferential use of the VH5-51 gene segment by the human immune response to code for antibodies against the V3 domain of HIV-1. *Mol Immunol* **46**: 917–926.

Gray ES, Moore PL, Choge IA, Decker JM, Bibollet-Ruche F, Li H, Leseka N, Treurnicht F, Mlisana K, Shaw GM, et al. 2007. Neutralizing antibody responses in acute human immunodeficiency virus type 1 subtype C infection. *J Virol* **81**: 6187–6196.

Gray ES, Madiga MC, Moore PL, Mlisana K, Abdool Karim SS, Binley JM, Shaw GM, Mascola JR, Morris L. 2009a. Broad neutralization of human immunodeficiency virus type 1 mediated by plasma antibodies against the gp41 membrane proximal external region. *J Virol* **83**: 11265–11274.

Gray ES, Taylor N, Wycuff D, Moore PL, Tomaras GD, Wibmer CK, Puren A, DeCamp A, Gilbert PB, Wood B, et al. 2009b. Antibody specificities associated with neutralization breadth in plasma from human

immunodeficiency virus type 1 subtype C-infected blood donors. *J Virol* **83**: 8925–8937.

Gray ES, Madiga MC, Hermanus T, Moore PL, Wibmer CK, Tumba NL, Werner L, Mlisana K, Sibeko S, Williamson C, et al. 2011. HIV-1 neutralization breadth develops incrementally over 4 years and is associated with CD4+ T cell decline and high viral load during acute infection. *J Virol* **85**: 4828–4840.

Harrer T, Harrer E, Kalams SA, Elbeik T, Staprans SI, Feinberg MB, Cao Y, Ho DD, Yilma T, Caliendo AM, et al. 1996. Strong cytotoxic T cell and weak neutralizing antibody responses in a subset of persons with stable nonprogressing HIV type 1 infection. *AIDS Res Hum Retrov* **12**: 585–592.

Hasselrot K, Saberg P, Hirbod T, Soderlund J, Ehnlund M, Bratt G, Sandstrom E, Broliden K. 2009. Oral HIV-exposure elicits mucosal HIV-neutralizing antibodies in uninfected men who have sex with men. *AIDS* **23**: 329–333.

Hengel RL, Kennedy MS, Steketee RW, Thea DM, Abrams EJ, Lambert G, McDougal JS. 1998. Neutralizing antibody and perinatal transmission of human immunodeficiency virus type 1. New York City Perinatal HIV Transmission Collaborative Study Group. *AIDS Res Hum Retrov* **14**: 475–481.

Herrera C, Spenlehauer C, Fung MS, Burton DR, Beddows S, Moore JP. 2003. Nonneutralizing antibodies to the CD4-binding site on the gp120 subunit of human immunodeficiency virus type 1 do not interfere with the activity of a neutralizing antibody against the same site. *J Virol* **77**: 1084–1091.

Hessell AJ, Hangartner L, Hunter M, Havenith CE, Beurskens FJ, Bakker JM, Lanigan CM, Landucci G, Forthal DN, Parren PW, et al. 2007. Fc receptor but not complement binding is important in antibody protection against HIV. *Nature* **449**: 101–104.

Hessell AJ, Poignard P, Hunter M, Hangartner L, Tehrani DM, Bleeker WK, Parren PW, Marx PA, Burton DR. 2009. Effective, low-titer antibody protection against low-dose repeated mucosal SHIV challenge in macaques. *Nat Med* **15**: 951–954.

Hidajat R, Xiao P, Zhou Q, Venzon D, Summers LE, Kalyanaraman VS, Montefiori DC, Robert-Guroff M. 2009. Correlation of vaccine-elicited systemic and mucosal nonneutralizing antibody activities with reduced acute viremia following intrarectal simian immunodeficiency virus SIVmac251 challenge of rhesus macaques. *J Virol* **83**: 791–801.

Hirbod T, Kaul R, Reichard C, Kimani J, Ngugi E, Bwayo JJ, Nagelkerke N, Hasselrot K, Li B, Moses S, et al. 2008. HIV-neutralizing immunoglobulin A and HIV-specific proliferation are independently associated with reduced HIV acquisition in Kenyan sex workers. *AIDS* **22**: 727–735.

Honnen WJ, Krachmarov C, Kayman SC, Gorny MK, Zolla-Pazner S, Pinter A. 2007. Type-specific epitopes targeted by monoclonal antibodies with exceptionally potent neutralizing activities for selected strains of human immunodeficiency virus type 1 map to a common region of the V2 domain of gp120 and differ only at single positions from the clade B consensus sequence. *J Virol* **81**: 1424–1432.

Huang CC, Tang M. 2005. Structure of a V3-containing HIV-1 gp120 core. *Science* **310**: 1025–1028.

Huang CC, Venturi M, Majeed S, Moore MJ, Phogat S, Zhang MY, Dimitrov DS, Hendrickson WA, Robinson J, Sodroski J, et al. 2004. Structural basis of tyrosine sulfation and VH-gene usage in antibodies that recognize the HIV type 1 coreceptor-binding site on gp120. *Proc Natl Acad Sci* **101**: 2706–2711.

Huang KH, Bonsall D, Katzourakis A, Thomson EC, Fidler SJ, Main J, Muir D, Weber JN, Frater AJ, Phillips RE, et al. 2010. B-cell depletion reveals a role for antibodies in the control of chronic HIV-1 infection. *Nat Commun* **1**: 102.

Huber M, Fischer M, Misselwitz B, Manrique A, Kuster H, Niederost B, Weber R, von Wyl V, Gunthard HF, Trkola A. 2006. Complement lysis activity in autologous plasma is associated with lower viral loads during the acute phase of HIV-1 infection. *PLoS Med* **3**: e441.

Husson RN, Lan Y, Kojima E, Venzon D, Mitsuya H, McIntosh K. 1995. Vertical transmission of human immunodeficiency virus type 1: Autologous neutralizing antibody, virus load, and virus phenotype. *J Pediatr* **126**: 865–871.

Johnson PR, Hamm TE, Goldstein S, Kitov S, Hirsch VM. 1991. The genetic fate of molecularly cloned simian immunodeficiency virus in experimentally infected macaques. *Virology* **185**: 217–228.

Kagi D, Vignaux F, Ledermann B, Burki K, Depraetere V, Nagata S, Hengartner H, Golstein P. 1994. Fas and perforin pathways as major mechanisms of T cell-mediated cytotoxicity. *Science* **265**: 528–530.

Karlsson Hedestam GB, Fouchier RA, Phogat S, Burton DR, Sodroski J, Wyatt RT. 2008. The challenges of eliciting neutralizing antibodies to HIV-1 and to influenza virus. *Nat Rev Microbiol* **6**: 143–155.

Keele BF, Giorgi EE, Salazar-Gonzalez JF, Decker JM, Pham KT, Salazar MG, Sun C, Grayson T, Wang S, Li H, et al. 2008. Identification and characterization of transmitted and early founder virus envelopes in primary HIV-1 infection. *Proc Natl Acad Sci* **105**: 7552–7557.

Kirchherr JL, Hamilton J, Lu X, Gnanakaran S, Muldoon M, Daniels M, Kasongo W, Chalwe V, Mulenga C, Mwananyanda L, et al. 2011. Identification of amino acid substitutions associated with neutralization phenotype in the human immunodeficiency virus type-1 subtype C gp120. *Virology* **409**: 163–174.

Kliks SC, Wara DW, Landers DV, Levy JA. 1994. Features of HIV-1 that could influence maternal-child transmission. *JAMA* **272**: 467–474.

Kwong PD, Wilson IA. 2009. HIV-1 and influenza antibodies: Seeing antigens in new ways. *Nat Immunol* **10**: 573–578.

* Kwong PD, Mascola JR, Nabel GJ. 2011. Rational design of vaccines to elicit broadly neutralizing antibodies to HIV-1. *Cold Spring Harb Perspect Med* doi: 10.1101/cshperspect.a007278.

Laird ME, Igarashi T, Martin MA, Desrosiers RC. 2008. Importance of the V1/V2 loop region of simian-human immunodeficiency virus envelope glycoprotein gp120 in determining the strain specificity of the neutralizing antibody response. *J Virol* **82**: 11054–11065.

Lambotte O, Ferrari G, Moog C, Yates NL, Liao HX, Parks RJ, Hicks CB, Owzar K, Tomaras GD, Montefiori DC,

et al. 2009. Heterogeneous neutralizing antibody and antibody-dependent cell cytotoxicity responses in HIV-1 elite controllers. *AIDS* **23**: 897–906.

Legrand E, Pellegrin I, Neau D, Pellegrin JL, Ragnaud JM, Dupon M, Guillemain B, Fleury HJ. 1997. Course of specific T lymphocyte cytotoxicity, plasma and cellular viral loads, and neutralizing antibody titers in 17 recently seroconverted HIV type 1-infected patients. *AIDS Res Hum Retrov* **13**: 1383–1394.

Li M, Salazar-Gonzalez JF, Derdeyn CA, Morris L, Williamson C, Robinson JE, Decker JM, Li Y, Salazar MG, Polonis VR, et al. 2006. Genetic and neutralization properties of subtype C human immunodeficiency virus type 1 molecular env clones from acute and early heterosexually acquired infections in Southern Africa. *J Virol* **80**: 11776–11790.

Li Y, Svehla K, Louder MK, Wycuff D, Phogat S, Tang M, Migueles SA, Wu X, Phogat A, Shaw GM, et al. 2009. Analysis of neutralization specificities in polyclonal sera derived from human immunodeficiency virus type 1-infected individuals. *J Virol* **83**: 1045–1059.

* Lifson JD, Haigwood NL. 2011. Lessons in nonhuman primate models for AIDS vaccine research: From minefields to milestones. *Cold Spring Harb Perspect Med* doi: 10.1101/cshperspect.a007310.

Long EM, Martin HL Jr, Kreiss JK, Rainwater SM, Lavreys L, Jackson DJ, Rakwar J, Mandaliya K, Overbaugh J. 2000. Gender differences in HIV-1 diversity at time of infection. *Nat Med* **6**: 71–75.

Loomis-Price LD, Cox JH, Mascola JR, VanCott TC, Michael NL, Fouts TR, Redfield RR, Robb ML, Wahren B, Sheppard HW, et al. 1998. Correlation between humoral responses to human immunodeficiency virus type 1 envelope and disease progression in early-stage infection. *J Infect Dis* **178**: 1306–1316.

Lynch JN, Nduati R, Blish C, Richardson BA, Mabuka J, Lechak Z, John-Stewart GC, Overbaugh J. 2011. The breadth and potency of passively acquired HIV- specific neutralizing antibodies does not correlate with risk of infant HIV infection. *J Virol* **85**: 5252–5261.

Mahalanabis M, Jayaraman P, Miura T, Pereyra F, Chester EM, Richardson B, Walker B, Haigwood NL. 2009. Continuous viral escape and selection by autologous neutralizing antibodies in drug-naive human immunodeficiency virus controllers. *J Virol* **83**: 662–672.

Mascola JR, Lewis MG, Stiegler G, Harris D, VanCott TC, Hayes D, Louder MK, Brown CR, Sapan CV, Frankel SS, et al. 1999. Protection of macaques against pathogenic simian/human immunodeficiency virus 89.6PD by passive transfer of neutralizing antibodies. *J Virol* **73**: 4009–4018.

Mascola JR, Stiegler G, VanCott TC, Katinger H, Carpenter CB, Hanson CE, Beary H, Hayes D, Frankel SS, Birx DL, et al. 2000. Protection of macaques against vaginal transmission of a pathogenic HIV-1/SIV chimeric virus by passive infusion of neutralizing antibodies. *Nat Med* **6**: 207–210.

Mc Cann CM, Song RJ, Ruprecht RM. 2005. Antibodies: Can they protect against HIV infection? *Curr Drug Targets Infect Disord* **5**: 95–111.

McNearney T, Hornickova Z, Markham R, Birdwell A, Arens M, Saah A, Ratner L. 1992. Relationship of human immunodeficiency virus type 1 sequence heterogeneity to stage of disease. *Proc Natl Acad Sci* **89:** 10247–10251.

Miller CJ, Genesca M, Abel K, Montefiori D, Forthal D, Bost K, Li J, Favre D, McCune JM. 2007. Antiviral antibodies are necessary for control of simian immunodeficiency virus replication. *J Virol* **81:** 5024–5035.

Moir S, Fauci AS. 2009. B cells in HIV infection and disease. *Nat Rev Immunol* **9:** 235–245.

Montefiori DC, Pantaleo G, Fink LM, Zhou JT, Zhou JY, Bilska M, Miralles GD, Fauci AS. 1996. Neutralizing and infection-enhancing antibody responses to human immunodeficiency virus type 1 in long-term nonprogressors. *J Infect Dis* **173:** 60–67.

Moog C, Fleury HJA, Pellegrin I, Kirn A, Aubertin AM. 1997. Autologous and heterologous neutralizing antibody responses following initial seroconversion in human immunodeficiency virus type 1-infected individuals. *J Virol* **71:** 3734–3741.

Moore JP, Cao Y, Ho DD, Koup RA. 1994. Development of the anti-gp120 antibody response during seroconversion to human immunodeficiency virus type 1. *J Virol* **68:** 5142–5155.

Moore JP, Cao Y, Leu J, Qin L, Korber B, Ho DD. 1996. Inter- and intraclade neutralization of human immunodeficiency virus type 1: Genetic clades do not correspond to neutralization serotypes but partially correspond to gp120 antigenic serotypes. *J Virol* **70:** 427–444.

Moore PL, Gray ES, Morris L. 2009a. Specificity of the autologous neutralizing antibody response. *Curr Opin HIV AIDS* **4:** 358–363.

Moore PL, Ranchobe N, Lambson BE, Gray ES, Cave E, Abrahams MR, Bandawe G, Mlisana K, Abdool Karim SS, Williamson C, et al. 2009b. Limited neutralizing antibody specificities drive neutralization escape in early HIV-1 subtype C infection. *PLoS Pathog* **5:** e1000598.

* O'Connell RJ, Kim JH, Corey L, Michael NL. 2011. Human immunodeficiency virus vaccine trials. *Cold Spring Harb Perspect Med* doi: 10.1101/cshperspect.a007351.

Overbaugh J, Rudensey LM. 1992. Alterations in potential sites for glycosylation predominate during evolution of the simian immunodeficiency virus envelope gene in macaques. *J Virol* **66:** 5937–5948.

Overbaugh J, Rudensey LM, Papenhausen MD, Benveniste RE, Morton WR. 1991. Variation in simian immunodeficiency virus *env* is confined to V1 and V4 during progression to simian AIDS. *J Virol* **65:** 7025–7031.

Parren PW, Marx PA, Hessell AJ, Luckay A, Harouse J, Cheng-Mayer C, Moore JP, Burton DR. 2001. Antibody protects macaques against vaginal challenge with a pathogenic R5 simian/human immunodeficiency virus at serum levels giving complete neutralization in vitro. *J Virol* **75:** 8340–8347.

Pereyra F, Addo MM, Kaufmann DE, Liu Y, Miura T, Rathod A, Baker B, Trocha A, Rosenberg R, Mackey E, et al. 2008. Genetic and immunologic heterogeneity among persons who control HIV infection in the absence of therapy. *J Infect Dis* **197:** 563–571.

Piantadosi A, Panteleeff D, Blish CA, Baeten JM, Jaoko W, McClelland RS, Overbaugh J. 2009. HIV-1 neutralizing antibody breadth is affected by factors early in infection, but does not influence disease progression. *J Virol* **83:** 10269–10274.

Pilgrim AK, Pantaleo G, Cohen OJ, Fink LM, Zhou JY, Zhou JT, Bolognesi DP, Fauci AS, Montefiori DC. 1997. Neutralizing antibody responses to human immunodeficiency virus type 1 in primary infection and long-term-nonprogressive infection. *J Infect Dis* **176:** 924–932.

Pinter A, Honnen WJ, He Y, Gorny MK, Zolla-Pazner S, Kayman SC. 2004. The V1/V2 domain of gp120 is a global regulator of the sensitivity of primary human immunodeficiency virus type 1 isolates to neutralization by antibodies commonly induced upon infection. *J Virol* **78:** 5205–5215.

Poss M, Martin HL, Kreiss JK, Granville L, Chohan B, Nyange P, Mandaliya K, Overbaugh J. 1995. Diversity in virus populations from genital secretions and peripheral blood from women recently infected with human immunodeficiency virus type 1. *J Virol* **69:** 8118–8122.

Prince AM, Reesink H, Pascual D, Horowitz B, Hewlett I, Murthy KK, Cobb KE, Eichberg JW. 1991. Prevention of HIV infection by passive immunization with HIV immunoglobulin. *AIDS Res Hum Retrov* **7:** 971–973.

Rerks-Ngarm S, Pitisuttithum P, Nitayaphan S, Kaewkungwal J, Chiu J, Paris R, Premsri N, Namwat C, de Souza M, Adams E, et al. 2009. Vaccination with ALVAC and AIDSVAX to prevent HIV-1 infection in Thailand. *New Engl J Med* **361:** 2209–2220.

Richman DD, Wrin T, Little SJ, Petropoulos CJ. 2003. Rapid evolution of the neutralizing antibody response to HIV type 1 infection. *Proc Natl Acad Sci* **100:** 4144–4149.

Rodriguez SK, Sarr AD, MacNeil A, Thakore-Meloni S, Gueye-Ndiaye A, Traore I, Dia MC, Mboup S, Kanki PJ. 2007. Comparison of heterologous neutralizing antibody responses of human immunodeficiency virus type 1 (HIV-1)- and HIV-2-infected Senegalese patients: Distinct patterns of breadth and magnitude distinguish HIV-1 and HIV-2 infections. *J Virol* **81:** 5331–5338.

Rong R, Bibollet-Ruche F, Mulenga J, Allen S, Blackwell JL, Derdeyn CA. 2007. Role of V1V2 and other human immunodeficiency virus type 1 envelope domains in resistance to autologous neutralization during clade C infection. *J Virol* **81:** 1350–1359.

Rong R, Li B, Lynch RM, Haaland RE, Murphy MK, Mulenga J, Allen SA, Pinter A, Shaw GM, Hunter E, et al. 2009. Escape from autologous neutralizing antibodies in acute/early subtype C HIV-1 infection requires multiple pathways. *PLoS Pathog* **5:** e1000594.

Russell JH, Ley TJ. 2002. Lymphocyte-mediated cytotoxicity. *Annu Rev Immunol* **20:** 323–370.

Sabin C, Corti D, Buzon V, Seaman MS, Lutje Hulsik D, Hinz A, Vanzetta F, Agatic G, Silacci C, Mainetti L, et al. 2010. Crystal structure and size-dependent neutralization properties of HK20, a human monoclonal antibody binding to the highly conserved heptad repeat 1 of gp41. *PLoS Pathog* **6:** e1001195.

Sagar M. 2010. HIV-1 transmission biology: Selection and characteristics of infecting viruses. *J Infect Dis* **202** (Suppl 2): S289–S296.

Sagar M, Lavreys L, Baeten JM, Richardson BA, Mandaliya K, Ndinya-Achola JO, Kreiss JK, Overbaugh J. 2003. Identification of modifiable factors that affect the genetic diversity of the transmitted HIV-1 population. *AIDS* **18:** 1–5.

Sagar M, Kirkegaard E, Long EM, Celum C, Buchbinder S, Daar ES, Overbaugh J. 2004. Human immunodeficiency virus type 1 (HIV-1) diversity at time of infection is not restricted to certain risk groups or specific HIV-1 subtypes. *J Virol* **78:** 7279–7283.

Sagar M, Kirkegaard E, Lavreys L, Overbaugh J. 2006. Diversity in HIV-1 envelope V1-V3 sequences early in infection reflects sequence diversity throughout the HIV-1 genome but does not predict the extent of sequence diversity during chronic infection. *AIDS Res Hum Retrov* **22:** 430–437.

Samleerat T, Thenin S, Jourdain G, Ngo-Giang-Huong N, Moreau A, Leechanachai P, Ithisuknanth J, Pagdi K, Wannarit P, Sangsawang S, et al. 2009. Maternal neutralizing antibodies against a CRF01_AE primary isolate are associated with a low rate of intrapartum HIV-1 transmission. *Virology* **387:** 388–394.

Sather DN, Armann J, Ching LK, Mavrantoni A, Sellhorn G, Caldwell Z, Yu X, Wood B, Self S, Kalams S, et al. 2009. Factors associated with the development of cross-reactive neutralizing antibodies during Human Immunodeficiency Virus Type 1 infection. *J Virol* **83:** 757–769.

Scarlatti G, Albert J, Rossi P, Hodara V, Biraghi P, Muggiasca L, Fenyo EM. 1993a. Mother-to-child transmission of human immunodeficiency virus type 1: Correlation with neutralizing antibodies against primary isolates. *J Infect Dis* **168:** 207–210.

Scarlatti G, Leitner T, Hodara V, Halapi E, Rossi P, Albert J, Fenyo EM. 1993b. Neutralizing antibodies and viral characteristics in mother-to-child transmission of HIV-1. *Aids* **7 (Suppl 2):** S45–S48.

Scheid JF, Mouquet H, Feldhahn N, Seaman MS, Velinzon K, Pietzsch J, Ott RG, Anthony RM, Zebroski H, Hurley A, et al. 2009. Broad diversity of neutralizing antibodies isolated from memory B cells in HIV-infected individuals. *Nature* **458:** 636–640.

Schmitz JE, Kuroda MJ, Santra S, Simon MA, Lifton MA, Lin W, Khunkhun R, Piatak M, Lifson JD, Grosschupff G, et al. 2003. Effect of humoral immune responses on controlling viremia during primary infection of rhesus monkeys with simian immunodeficiency virus. *J Virol* **77:** 2165–2173.

Seaman MS, Janes H, Hawkins N, Grandpre LE, Devoy C, Giri A, Coffey RT, Harris L, Wood B, Daniels MG, et al. 2010. Tiered categorization of a diverse panel of HIV-1 Env pseudoviruses for assessment of neutralizing antibodies. *J Virol* **84:** 1439–1452.

Shaw GM, Hunter E. 2011. HIV transmission. *Cold Spring Harb Perspect Med* doi: 10.1101/cshperspect.a006965.

Shen R, Drelichman ER, Bimczok D, Ochsenbauer C, Kappes JC, Cannon JA, Tudor D, Bomsel M, Smythies LE, Smith PD. 2010. GP41-specific antibody blocks cell-free HIV type 1 transcytosis through human rectal mucosa and model colonic epithelium. *J Immunol* **184:** 3648–3655.

Shibata R, Igarashi T, Haigwood N, Buckler-White A, Ogert R, Ross W, Willey R, Cho MW, Martin MA. 1999. Neutralizing antibody directed against the HIV-1 envelope glycoprotein can completely block HIV-1/SIV chimeric virus infections of macaque monkeys. *Nat Med* **5:** 204–210.

Simek MD, Rida W, Priddy FH, Pung P, Carrow E, Laufer DS, Lehrman JK, Boaz M, Tarragona-Fiol T, Miiro G, et al. 2009. Human immunodeficiency virus type 1 elite neutralizers: Individuals with broad and potent neutralizing activity identified by using a high-throughput neutralization assay together with an analytical selection algorithm. *J Virol* **83:** 7337–7348.

Smith DM, Strain MC, Frost SDW, Pillai SK, Wong JK, Wrin T, Petropolous CJ, Daar ES, Little SJ, Richman DD. 2006. Lack of nuetralizaing antibody response to HIV-1 predisposes to superinfection. *Virology* **355:** 1–5.

Stamatatos L, Morris L, Burton DR, Mascola JR. 2009. Neutralizing antibodies generated during natural HIV-1 infection: Good news for an HIV-1 vaccine? *Nat Med* **15:** 866–870.

Tang H, Robinson JE, Gnanakaran S, Li M, Rosenberg ES, Perez LG, Haynes BF, Liao HX, Labranche CC, Korber BT, et al. 2011. Epitopes immediately below the base of the V3 loop of gp120 as targets for the initial autologous neutralizing antibody response in two HIV-1 subtype B-infected individuals. *J Virol* **85:** 9286–9299.

Tomaras G, Binley J. 2010. Epitope specificities of elite neutralizing sera from HIV-1-infected individuals. In *Abstracts from AIDS Vaccine 2010*, p. A-24. Atlanta, Georgia.

Tomaras GD, Yates NL, Liu P, Qin L, Fouda GG, Chavez LL, Decamp AC, Parks RJ, Ashley VC, Lucas JT, et al. 2008. Initial B-cell responses to transmitted human immunodeficiency virus type 1: Virion-binding immunoglobulin M (IgM) and IgG antibodies followed by plasma anti-gp41 antibodies with ineffective control of initial viremia. *J Virol* **82:** 12449–12463.

Toran JL, Kremer L, Sanchez-Pulido L, de Alboran IM, del Real G, Llorente M, Valencia A, de Mon MA, Martinez AC. 1999. Molecular analysis of HIV-1 gp120 antibody response using isotype IgM and IgG phage display libraries from a long-term non-progressor HIV-1-infected individual. *Eur J Immunol* **29:** 2666–2675.

Trkola A, Purtscher M, Muster T, Ballaun C, Buchacher A, Sullivan N, Srinivasan K, Sodroski J, Moore JP, Katinger H. 1996. Human monoclonal antibody 2G12 defines a distinctive neutralization epitope on the gp120 glycoprotein of human immunodeficiency virus type 1. *J Virol* **70:** 1100–1108.

Trkola A, Kuster H, Rusert P, von Wyl V, Leemann C, Weber R, Stiegler G, Katinger H, Joos B, Gunthard HF. 2008. In vivo efficacy of human immunodeficiency virus neutralizing antibodies: Estimates for protective titers. *J Virol* **82:** 1591–1599.

Walker LM, Chan-Hui P. 2010. High through-put functional screening of activated B cells from 4 African elite neutralizers yields a panel of novel broadly neutralizing antibodies. In *Abstracts from AIDS Vaccine 2010*, pp. A-149–A-150. Atlanta, Georgia.

Walker LM, Phogat SK, Chan-Hui PY, Wagner D, Phung P, Goss JL, Wrin T, Simek MD, Fling S, Mitcham JL, et al. 2009. Broad and potent neutralizing antibodies from an African donor reveal a new HIV-1 vaccine target. *Science* **326:** 285–289.

Wei X, Decker JM, Wang S, Hui H, Kappes JC, Wu X, Salazar-Gonzalez JF, Salazar MG, Kilby JM, Saag MS,

The T-Cell Response to HIV

Bruce Walker[1] and Andrew McMichael[2]

[1]Ragon Institute of MGH, MIT, and Harvard Mass General Hospital-East, Charlestown, Massachusetts 02129
[2]Weatherall Institute of Molecular Medicine, Oxford University, Oxford OX3 9DS, United Kingdom

Correspondence: bwalker@partners.org

HIV is a disease in which the original clinical observations of severe opportunistic infections gave the first clues regarding the underlying pathology, namely that HIV is essentially an infection of the immune system. HIV infects and deletes CD4$^+$ T cells that normally coordinate the adaptive T- and B-cell response to defend against intracellular pathogens. The immune defect is immediate and profound: At the time of acute infection with an AIDS virus, typically more than half of the gut-associated CD4$^+$ T cells are depleted, leaving a damaged immune system to contend with a life-long infection.

The earliest studies of T-cell function among infected persons revealed aspects of HIV immunopathogenesis that are still not fully understood. Despite initial and persistent damage to CD4$^+$ T cells, and a lack of detectable HIV-specific CD4$^+$ T helper cells (Murray et al. 1984; Lane et al. 1985), the magnitude and breadth of CD8$^+$ T-cell responses to HIV in infected humans were found to be robust, with direct effector function of such a magnitude that it could be readily detected in freshly isolated lymphocytes from peripheral blood and brochoalveolar lavage in persons with AIDS (Plata et al. 1987; Walker et al. 1987; Nixon et al. 1988). HIV was already known to be an immunosuppressive disease, yet these cells were present in such robust quantity that they could be detected by assays measuring the ability of freshly isolated peripheral blood cells to lyse autologous B cells infected with recombinant vaccinia-HIV vectors or peptide pulsed targets (Walker et al. 1987; Nixon et al. 1988).

Moreover, CD8$^+$ T cells from infected persons were able to inhibit HIV replication (Walker et al. 1986), clearly showing that these cells were functional at least in vitro, but despite this, persons were progressing to AIDS.

Subsequent studies using other approaches such as interferon γ Elispot (Dalod et al. 1999), intracellular cytokine (Maecker et al. 2001), and peptide MHC tetramer assays (Altman et al. 1996) confirmed and quantified these robust CD8$^+$ T-cell responses. Most chronically infected persons target more than a dozen CD8$^+$ epitopes simultaneously (Addo et al. 2003), and in some instances up to 19% of CD8$^+$ T cells are specific for HIV (Richardson 2000; Betts et al. 2001; Papagno et al. 2002), yet control of viremia is not achieved. At the same time, HIV-specific CD4$^+$ T-cell responses were found to be severely impaired, particularly as measured by the ability of these cells to proliferate to viral antigens (Wahren et al. 1987). Thus, from the early studies there was a clear disconnect between the lack of HIV-specific $CD4^+$ T cells and the abundance of HIV-specific $CD8^+$ T cells.

Despite these conundrums, there were early signs that $CD8^+$ T cells play a role in controlling HIV disease progression. Sequencing of autologous virus from infected persons revealed evidence of immune selection pressure mediated by these responses (Phillips et al. 1991) and an association with the initial decline in peak viremia after acute infection (Borrow et al. 1994; Koup et al. 1994). The development of HLA-class I-peptide tetramers confirmed the presence of robust induction of responses to multiple epitopes (Altman et al. 1996), and as larger numbers of patients were studied, it also became clear that among the strongest associations with disease outcome was the expression of certain HLA class I alleles (Kaslow et al. 1996; Migueles et al. 2000; Gao et al. 2001), implicating class I restricted cytotoxic T lymphocytes (CTLs) as a major modulator of disease progression. The relationship between $CD8^+$ T-cell immune function and viral control was shown by experimental depletion of $CD8^+$ T cells in animal models of AIDS virus infection (Jin et al. 1999; Schmitz et al. 1999). As sensitive viral load assays became available, it also became clear that some infected persons were able to control viremia to levels below detection by the most sensitive RNA assays, and that these persons were characterized by robust HIV-specific $CD4^+$ T-cell responses (Rosenberg et al. 1997). However, now a quarter century into the epidemic the precise role of T cells in HIV control, and the precise phenotype, specificity and function of T cells that should be induced with a vaccine, remain unclear. And from the standpoint of vaccine development, it remains controversial as to how much insight is to be gained from the study of persons who have become infected, because by definition these are not protective responses.

CTL RESPONSES IN ACUTE INFECTION

Acute HIV infection is often associated with a transient febrile illness, much like infectious mononucleosis (Ho et al. 1985), which has

facilitated the identification of persons in the earliest stages of infection, who are HIV RNA positive but not yet HIV antibody positive. Important justification for the intense study of acute infection is the demonstration that early events predict subsequent disease progression (Mellors et al. 1996; Lyles et al. 2000). Acute infection is associated with a decline in peak viremia from $\sim 1-5$ million viral copies/mL to a steady state of $\sim 30,000$ copies (Lyles et al. 2000), but peak levels of viremia may be much lower in asymptomatic persons (Fiebig et al. 2003). Very few studies have actually examined $CD8^+$ T-cell responses in the earliest stages of infection (Fiebig stage I and II), but when those studies have been performed it is clear that initial immune responses are very narrowly directed (Goonetilleke et al. 2009) and remain narrowly directed as viral set point is achieved (Dalod et al. 1999; Altfeld et al. 2001). Even when responses to autologous virus are measured, the initial detectable response has usually been to only one to three epitopes (Goonetilleke et al. 2009), even though the autologous viruses may contain wild-type epitopes to which $CD8^+$ T cells are subsequently generated (Radebe et al. 2011). An important caveat is that there may be much more happening in the tissues than is sampled in the blood in these early stages. Studies comparing lymph node $CD8^+$ T-cell responses to those in peripheral blood have shown the former to be detectable well in advance of detectable responses in the periphery (Altfeld et al. 2002). Even so, there is a clear hierarchy in initial responses and immunodominance that correlates with set point viremia, which is lost during the transition to chronic infection (Streeck et al. 2009).

Despite the narrowness of the acute phase $CD8^+$ T-cell response, it is associated with a dramatic decline in initial viremia, suggesting that these early, presumably narrowly directed responses are at least partially effective. Although functional studies of early CTL responses are few (Ferrari et al. 2011), evidence of $CD8^+$ T-cell efficacy is suggested by detailed analysis of viral genomes during these early phases of infection, and by modeling studies. 80% of acute infections are established by a single founder virus, which first begins to diversify at around the time of peak infection (Salazar-Gonzalez et al. 2009; Shaw and Hunter 2011). By single genome amplification one can detect viral evolution at sites of CTL pressure as early as peak viremia (Goonetilleke et al. 2009), clear indication that there is effective pressure being applied because it can lead to a wholesale turnover in the replicating virus population. This has been modeled, suggesting that 15%–35% of infected cells can be killed by CTLs of a single specificity per day in vivo during acute infection (Goonetilleke et al. 2009). This $CD8^+$ T-cell-mediated immune pressure is also apparent in population studies, which have shown clear HLA-associated signature mutations following acute infection (Brumme et al. 2008), persisting in the chronic phase of infection (Moore et al. 2002), and transmission and reversion of CTL escape variants (Goulder et al. 2001; Goepfert et al. 2008).

The characteristics and antiviral efficacy of the acute phase $CD8^+$ T-cell responses are being progressively defined. These clearly occur in the setting of acute phase proteins and proinflammatory cytokines (Stacey et al. 2009). The initial response is narrowly directed, predominantly at epitopes in Env and Nef, regions that are among the most variable in the virus (Lichterfeld et al. 2004; Goonetilleke et al. 2009; Turnbull et al. 2009). Indeed, early responses appear to be targeted to epitopes of higher entropy for reasons that are not clear (Bansal et al. 2005). The breadth of responses increases over time, as do the number of HLA alleles that are involved in recognition of infected cells (Altfeld et al. 2006; Streeck et al. 2009). Some acute phase epitopes are not those that are targeted in chronic infection and may be novel (Goonetilleke et al. 2009), as revealed by studies using peptides representing autologous virus. Often these earliest responses fade away as soon as the epitope escape has occurred. Although these detailed findings have been made in just four patients, ongoing studies in a further 12 show very similar patterns (N Goonetilleke, M Liu, and AJ McMichael, unpubl. data).

The influence of host genetics on acute phase CTL responses is readily apparent. HLA alleles associated with protection from disease progression are preferentially targeted in acute infection, whereas coexpression of risk alleles does not induce responses, a process called immunodominantion (Altfeld et al. 2006). Clinical symptoms of acute infection have been shown to be reduced in persons expressing HLA B57, and this allele is associated with improved outcome and also dominates the early response in persons who express it (Altfeld et al. 2003a). At least one study indicates the selective loss of high-avidity CTLs in acute infection, and when these are maintained they are associated with a lower viral load (Lichterfeld et al. 2007). However, the ability of $CD8^+$ T cells to produce interferon γ in primary infection is a poor predictor of viral set point and disease progression (Gray et al. 2009), suggesting that if CTLs do influence viral set point, this assay may not discern their function. In contrast to the wealth of information regarding $CD8^+$ T-cell responses in the peripheral blood, almost nothing is known about responses in tissues in acute infection, which is the major site of HIV replication.

There has been considerable interest and some controversy over whether highly exposed but HIV seronegative people, such as sex workers in high-HIV-1-prevalence communities, make CTL responses to HIV-1 and whether these are helping them avoid infection (Rowland-Jones et al. 1995). The controversy has been partly resolved by using very sensitive highly controlled and blinded assays that show anti-HIV-specific $CD4^+$ T cells in both HIV-1-exposed and -unexposed uninfected people, with stronger and more sustained T-cell responses in the former (Ritchie et al. 2011). No $CD8^+$ T-cell responses were found in that study but the level of exposure in the exposed cohort in the study was much less than in African sex workers during the early 1990s, so the earlier findings cannot be disregarded. More likely, as shown by (Ritchie et al. 2011), there are relatively frequent $CD4^+$ T-cell responses, primed by crossreacting antigens, in unexposed people and that these are amplified by HIV-1 exposure without infection, for example, in persons homozygous for the CCR5 Δ32 mutation. With very high exposure these T-cell responses could include CD8$^+$ T cells. However, there is no evidence that either of these T-cell responses is protective.

CTL EVOLUTION FOLLOWING ACUTE INFECTION

Although a narrowly directed immune response is found at the time of maximal decline in peak viremia, responses subsequently broaden, such that during chronic infection the average person targets a median of 14 epitopes simultaneously (Addo et al. 2003), and given that these studies were performed with a reference set of peptides rather than autologous peptides, the actual number may be 20%–30% higher (Altfeld et al. 2003b; Goonetilleke et al. 2009). Immunization studies in animal models indicate that the CD8$^+$ T-cell compartment has enormous expansion capacity, without affecting the size of the naïve CD4$^+$, CD8$^+$, or B-cell populations, and while preserving memory CD8$^+$ T-cell populations to other pathogens (Vezys et al. 2009). HIV-specific CD8$^+$ T-cell responses remain detectable throughout the course of disease, and are actually broader and higher in persons with progressive infection than in those with controlled infection (Pereyra et al. 2008).

The relationship between the earliest CTL responses, viral set point, and disease progression remains controversial. Kinetics of immune responses in acute infection are associated with serial waves of responses (Turnbull et al. 2009), but detectable responses tend to be narrowly directed and of low magnitude (Dalod et al. 1999; Altfeld et al. 2001; Radebe 2011). This may be caused by localization of responses at inductive and effector sites within lymphoid and gut mucosa, respectively, and responses detected in lymph nodes precede those detectable in peripheral blood and are of higher magnitude (Altfeld et al. 2002). There is a high predictability of initial and subsequent epitopes targeted, based on the HLA type of the individual (Goulder et al. 2001; Moore et al. 2002; Yu et al. 2002; Draenert et al. 2006; Brumme et al. 2009). Remarkably, most responses detectable at the time a quasi-set point is achieved persist even in very late stage infection (Koibuchi

et al. 2005), though function may be lost, as evidenced by up-regulation of negative immunoregulatory molecules such as PD-1 (Day et al. 2006; Petrovas et al. 2006; Trautmann et al. 2006), and up-regulation of transcription factors that inhibit cell proliferation and cytokine secretion, such as BATF-1 (Quigley et al. 2010).

Specificity of responses during the chronic phase of infection repeatedly suggests that Gag targeting is associated with lower viral load (Edwards et al. 2002; Zuniga et al. 2006; Kiepiela et al. 2007). In a large study of persons with clade C virus infection, the broader the Gag-specific response the lower the viral load, and somewhat paradoxically, the broader the Env-specific response the higher the viral load (Kiepiela et al. 2007; Ngumbela et al. 2008). A similar association between targeting of Gag and viral load is also seen in children (Huang et al. 2008). It is noteworthy that the most protective HLA molecules B57, B58, B81, B14, and B27 target epitopes in Gag, often in relatively conserved regions. It may also be relevant that Gag p24 variation may often carry a fitness cost to the virus, possibly because mutants affect the complex assembly of the viral capsid (Schneidewind et al. 2008). This may make T-cell responses to these epitopes more favorable to the host because the virus is either controlled or escapes with a loss of virulence (Martinez-Picado et al. 2006). Eventually, however, the virus may restore its virulence with compensatory mutations (Schneidewind et al. 2007, 2008, 2009; Goepfert et al. 2008; Crawford et al. 2009).

In addition to targeting epitopes within the nine expressed HIV proteins, recent studies show that antisense peptides and alternative reading frame products are also targeted by CD8$^+$ T cells, as shown first in the SIV model (Maness et al. 2007, 2009) and more recently in humans infected with HIV (Bansal et al. 2010; Berger et al. 2010). Moreover, almost all studies have used peptide-pulsed target cells to define responses, and emerging evidence suggests that processing and presentation are likely critical (Allen et al. 2004; Draenert et al. 2004; Lazaro et al. 2009). Mutations surrounding epitopes are associated with altered processing and lack of presentation (Le Gall et al. 2007), and this is only detectable when the epitopes are processed and presented in infected cells. Even the infected cell type appears to influence this processing, based on digestion of longer peptides using cytosolic extracts—and perhaps importantly CD4$^+$ T cells have much lower proteolytic activity than monocytes, suggesting that antigen processing may be least effective in the cells that are the most critical to target (Lazaro et al. 2009).

CD4$^+$ T-CELL RESPONSES IN ACUTE AND CHRONIC HIV-1 INFECTION

Generally in those infected with HIV-1, the T-cell responses are dominated by CD8$^+$ T cells. These are much stronger than CD4$^+$ T-cell responses (Ramduth et al. 2005), which are damaged by the virus. Given the important role that CD4$^+$ T cells have in maintaining CD8$^+$ T-cell responses, it is remarkable that the latter are so strong (Kalams et al. 1999). In murine models in which CD4$^+$ T cells are depleted either with antibody infusion or genetically, CD8$^+$ T-cell responses are greatly impaired (Janssen et al. 2003; Shedlock and Shen 2003; Sun and Bevan 2003). On antigen stimulation, they expand rapidly to exhaustion and their IL-2-dependent progression to long term memory populations is abrogated (Kamimura and Bevan 2007). In HIV-1 infection CD4$^+$ T cells, though greatly depleted, are not entirely absent, but abnormalities in the development of CD8$^+$ T-cell responses could be consistent with partial loss of CD4$^+$ T-cell help, or impaired function of what cells remain (Pitcher et al. 1999).

In acute HIV-1 infection, memory CD4$^+$ T cells are massively depleted from the lymphoid system, particularly in the gut involving both direct targeting by the virus and bystander activation-induced cell death (Mattapallil et al. 2005; Douek et al. 2009). This applies to all memory CD4$^+$ T-cell populations but those specific for HIV may be preferentially infected and destroyed (Douek et al. 2002). However, the percentage of HIV-specific CD4$^+$ T cells that are infected, even in the presence of high level viremia, is typically only a few percent or less, suggesting that the majority of these cells

somehow escape infection despite being activated at a time of very high viremia.

Early studies showed a lack of CD4$^+$ T-cell responses, measured by antigen stimulated proliferation assays both in early and late infection (Lane et al. 1985). However, Pitcher et al. (1999) found using antigen stimulated cytokine production, that HIV-specific T cells were present at all stages of infection, though in relatively low numbers. A lack of IL-2 production by antigen-specific CD4$^+$ T cells and CD8$^+$ T cells may account for the apparent discrepancy between results with the two assays (Zimmerli et al. 2005). In very early infection however, HIV-specific CD4$^+$ T cells may be present in larger numbers. Rosenberg et al. (2000) showed that when patients were treated very early with antiretroviral drugs, that strong CD4$^+$ T-cell responses to HIV antigens could be rescued, as measured by lymphocyte proliferation assays. Recently, Gray et al. (C Riou, VV Ganusov, C Gray, et al., submitted) have supported this by showing strong CD4$^+$ T-cell responses around the time of peak viremia, but these T cells are rapidly lost as the infection progresses. This loss may reflect the damage to the whole CD4$^+$ T-cell compartment, but it should be noted that in other acute and chronic virus infections (including responses to live vaccines) CD4$^+$ T-cell responses may peak early and then decline to a relatively low level in the blood, in the absence of damage to CD4$^+$ T cells (e.g., Amyes et al. 2003; Goonetilleke et al. 2009). At this point almost nothing is known about HIV-specific CD4$^+$ T-cell responses in lymphoid tissues and the gut, a clearly needed focus for future studies.

Persons who progress very slowly to AIDS, in the absence of treatment, make stronger CD4$^+$ T-cell responses and there is a correlation between the strength of the response and slow progression (Rosenberg et al. 1997), which is also striking in the slow progression of HIV-2 infection (Duvall et al. 2006). In a subset of persons who control viremia spontaneously, the gene expression profile of CD4$^+$ cells is similar to that of uninfected persons (Vigneault et al. 2011). Cross-sectional data in chronically infected persons indicate a link between strong CD4$^+$ T-cell responses and effective CD8$^+$

T-cell responses (Kalams et al. 1999). Recent data implicate CD4$^+$ T cells that make IL21 as particularly important in maintaining CD8$^+$ responses (Chevalier et al. 2011; Williams et al. 2011).

The epitopes recognized by CD4$^+$ T cells occur in all viral proteins but there is a preponderance of T cells specific for Gag and Nef and several epitopes have been defined, though far fewer than the number defined for CD8$^+$ T-cell responses (Kaufmann et al. 2004) (LANL HIV Immunology database: http://www.hiv.lanl.gov/content/immunology/compendium.html). As is often the case for HLA class II restricted T-cell responses, the peptides can be presented by more than one, often several, different HLA class II molecules, in contrast to the generally tighter restriction by HLA class I molecules. In marked contrast to HLA class I presented peptides, mutational escape seems rare, or possibly absent, in HLA class II restricted T-cell responses (Rychert et al. 2007). This may reflect a lack of direct cytotoxic action on virus-infected cells as well as the relative lack of HLA class II on infected cells and also their propensity to interact primarily with dendritic cells, but additional studies are clearly warranted. And genetic studies in a number of cohorts suggest a relationship between expression of certain HLA class II alleles and HIV control/disease progression, again suggesting a possible role for these responses in active immune containment (Lacap et al. 2008; Julg et al. 2011).

Weak HIV-specific CD4$^+$ T-cell responses have been reported in HIV exposed but seronegative donors in a number of studies (Goh et al. 1999; Schenal et al. 2005). Remarkably around 25% of HIV unexposed persons make similar, though weaker and less sustained than in those exposed, CD4$^+$ T-cell responses to previously described HIV epitopes (Ritchie et al. 2011). These responses are probably cross-reactive with other pathogens and other antigens to which the person has been exposed. At least some of these epitopes are the same as those found in HIV infection (Kaufmann et al. 2004), so these preinfection cross-reactive responses could help explain why these particular epitopes are immunodominant after infection.

Several of the vaccine candidates that have been tested in humans were more effective at stimulating CD4$^+$ T-cell responses than CD8$^+$ T cells, particularly when plasmid DNA and pox virus vectors were used (Goonetilleke et al. 2006; Graham et al. 2006; Harari et al. 2008). Similar results have been found in macaques vaccinated with DNA and pox virus vectors (Sun et al. 2010). In the latter study, the vaccine-induced CD4$^+$ T-cell response did not protect against challenge with the SHIV 89.6P virus, although an earlier study had shown that longer survival in vaccinated monkeys challenged with SIV mac251 was associated with vaccine-induced SIV-specific CD4$^+$ T-cell responses (Sun et al. 2006). Although concern has been expressed that vaccine induction of HIV-specific CD4$^+$ T-cell responses may increase the risk of HIV-1 acquisition by providing more activated CD4$^+$ T cells to the site of infection (Douek et al. 2002; Staprans et al. 2004), this has only been observed in the absence of concomitant CD8$^+$ T-cell responses (Staprans et al. 2004).

IMMUNE SELECTION PRESSURE AND VIRAL ESCAPE

The impact of CTL selection pressure is readily observed by sequencing autologous virus, which undergoes rapid immune-driven evolution in vivo following acute infection (Fig. 1). New approaches involving single genome amplification techniques have shown clear evidence of CTL escape within 25–32 days of infection, around the time of peak viremia following acute infection, and before full seroconversion (Goonetilleke et al. 2009). Indeed, deep sequencing has revealed that the virus explores multiple pathways to escape before going to fixation, indicating that there are constraints on HIV evolution that appear to be caused by imposed fitness alterations (Fischer et al. 2010). The pathways to immune escape appear limited, as shown by studies of genetically identical twins infected by the same virus as adults through shared injection drug use, in whom the earliest targeted epitopes, kinetics of escape and earliest escape variants that arose

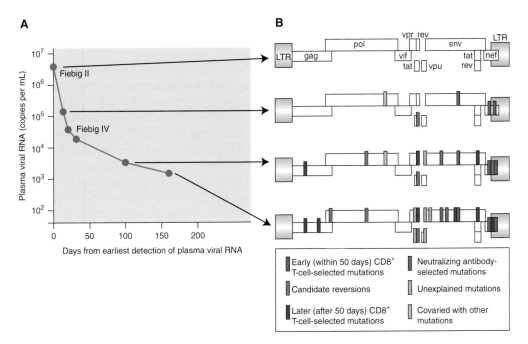

Figure 1. Evolution of the transmitted/founder virus in acute HIV-1 infection is largely driven by CD8[+] T-cell responses. (*A*) The graph shows the falling viral load from the time of peak viremia, about 3 weeks after infection, to the establishment of the viral set point 160 days later. Virus taken at the first sampling (the time of screening or "day 0") was characteristic of a single founder virus when sequenced; this is represented on the genetic map of the virus at the *top right* (*B*). At subsequent time points mutations appeared in all the viruses sequenced, shown by the bars on the genetic map to indicate the sites of the mutations. The bars are color coded to indicate rapid escape selected by CD8[+] T cells, slower or late escape selected by CD8[+] T cells, selection by neutralizing antibodies targeting the virus envelope, "reversions" from unusual sequences in the founder virus sequence to the sequence consensus in the whole database (the transmitted sequences may have been selected by CD8[+] T cells in the transmitting sexual partner of the patient). At a few sites it was not clear what was selecting the amino acid change and these are shown in gray. (Adapted from McMichael et al. 2010; reprinted, with express permission, from the author.)

were strikingly similar (Draenert et al. 2006), despite differences in T-cell receptor (TCR) usage (Yu et al. 2007).

Escape from CTL responses can occur through multiple mechanisms. Mutations in regions flanking CTL epitopes can clearly impact processing and subsequent recognition (Draenert et al. 2004; Le Gall et al. 2007; Tenzer et al. 2009), as can mutations within epitopes. Mutations in TCR contact residues have a variable impact on CTL recognition, and the ability to cross-recognize variants appears to be influenced by thymic selection (Kosmrlj et al. 2010). Those HLA alleles associated with better outcome following infection are associated with

presentation of fewer self-peptides in the thymus, resulting in a repertoire in which more clones survive negative selection, and are thus are likely to be more cross-reactive to variants that arise. This may also explain the association between protective HLA alleles in HIV infection and an association between these same alleles and autoimmunity/hypersensitivity (Kosmrlj et al. 2010).

The strongest temporal relationship between CTL escape and viral load is associated with HLA B*27 restricted recognition of an epitope in the p24 Gag protein, which requires an initial mutation within the epitope followed by a second distant mutation in order for the anchor residue to mutate and completely abrogate recognition (Schneidewind et al. 2007). However, for other mutations the effects on viral load are not as clear, likely because of competing effects of immune escape and alterations in viral fitness. Studies clearly indicate that some mutations come at the cost of a significant decrease in viral replicative capacity, suggesting that escape can be beneficial to the host (Martinez-Picado et al. 2006). In persons who control viremia without therapy, their viruses have impaired replication capacity, and this has also been shown for acute infection (Miura et al. 2010); moreover, there is a clear link between certain HLA class I alleles and viral fitness (Miura et al. 2009a), suggesting that this is immune mediated by class I restricted CD8$^+$ T cells. The evolving view is thus that CTL efficacy is not only caused by inhibiting viral replication, but is also impacted by the mutations induced by these responses. Such a view helps to reconcile the finding that the breadth of responses to Gag, a necessarily conserved protein, are associated with lower viral load, whereas breadth of responses to Env, which readily tolerates mutations without affecting viral fitness, is associated with higher viral loads (Fig. 2) (Kiepiela et al. 2007).

Some additional complexities of CTL escape have been revealed by studies of an immunodominant epitope restricted by the protective allele HLA B*57. Population studies have shown this to be one of the earliest escape mutations to arise for HLA B57 is mutation with the dominant B57-restricted epitope TW10 in Gag (Brumme et al. 2008) that causes a clear decrease in replication capacity (Martinez-Picado et al. 2006). In elite controllers, HLA B57 can be associated with additional rare mutations at other sites within and surrounding the epitope that are even more fitness impairing (Miura et al. 2009b). In acute infection a large population study showed that viruses from individuals expressing protective HLA alleles are less fit than in persons lacking these alleles, and replication capacity in the acute stage of infection correlated with the HLA B-associated Gag polymorphisms (Goepfert et al. 2008; Brockman et al. 2010; Wright et al. 2011). However,

these relationships were lost in chronic infection, but compensatory mutations that restore fitness were significantly increased. Together these data suggest that protective alleles may function in part through induction of fitness-impairing mutations. Indeed, in elite controllers, who maintain undetectable viral loads in the absence of therapy (Deeks and Walker 2007), viruses are significantly less fit (Miura et al. 2009a), and this is associated with specific HLA alleles, and more pronounced for Gag than for Env (Troyer et al. 2009). Studies of HIV controllers from the time of acute infection indicate that fitness impairing mutations are selected early or are transmitted from the donor (Miura et al. 2010), and early impairment of viral replication, including those caused by antiviral drug resistance in the donor virus, may impact long term control (Miura et al. 2010; Wright et al. 2011).

The impact of escape at a population level is becoming recognized. Transmitted escape mutations are associated with lower viral loads (Chopera et al. 2008; Goepfert et al. 2008; Crawford et al. 2009), and reversion of these mutations can be slow. Population studies show that the frequency of specific mutations increases with the expression of the selecting HLA allele in the population, and that in some geographic areas escape mutants have become the dominant circulating viruses (Kawashima et al. 2009). Because compensatory mutations can also be transmitted, the end result can be the deletion of an epitope with no associated impairment in viral fitness (Schneidewind et al. 2009). Thus, the host CD8$^+$ T-cell response is helping to drive global HIV evolution, and the geographic differences in HIV clades relates at least in part to selection of mutations by locally prevalent HLA alleles.

IMMUNOGENETICS OF HIV-SPECIFIC T-CELL RESPONSES

Population studies have repeatedly shown dramatically different outcomes following HIV infection, with some persons succumbing to AIDS within less than a year of infection, and others surviving for more than three decades

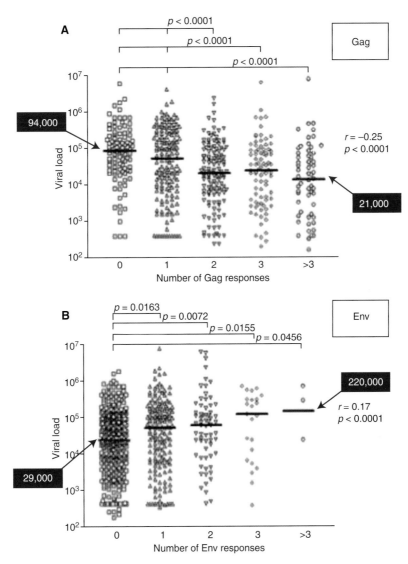

Figure 2. Breadth of protein-specific CD8[+] T-cell responses in relation to viral load. 578 persons infected with clade C virus were recruited, and comprehensive analysis of epitope targeting by PBMC was determined using a panel of overlapping peptides spanning all expressed viral proteins. Viral load of individuals in terms of the number of Gag (*A*)- and Env (*B*)-specific responses detected is shown, revealing that the broader the Gag-specific response, the lower the viral load, and the higher the Env-specific response, the higher the viral load, indicating that specificity impacts control. (Adapted from Kiepiela et al. 2007; reprinted, with permission, from Nature Publishers © 2007.)

without the need for antiviral therapy, and with no apparent adverse effects (Deeks and Walker 2007). Early on, it was noted that there was a strong association between expression of certain class I alleles and disease outcome, but the mechanistic basis for this was unclear (Migueles

et al. 2000). Indeed, whether this was a simple association or actually causally related could not be determined. Remarkably, both the positive and negative associations between viral load and HLA class I alleles were observed to be limited to HLA B alleles (Kiepiela et al. 2004), suggesting an active immune control mechanism. Although some associations with HLA C alleles were noted, these may be entirely explained by linkage disequilibrium (Kiepiela et al. 2004).

The first large scale attempt to define a genetic association with viral control was performed on a cohort of persons with acute HIV infection, in whom set point viral load data were available (Fellay et al. 2007). This study, and others that followed, revealed a SNP that is a proxy for HLA B57, as well as another that is associated with HLA C expression, to be the only significant associations with the level of viremia in the year following infection. This unbiased approach confirmed something that had already been recognized in cohort studies, namely that HLA B57 is in some way associated with better control, and also shed new light on HLA C, unique among class I alleles in that it is not down-regulated by the effects of Nef (Le Gall et al. 2000).

Recent population genetic studies comparing persons who progress rapidly to disease with controllers who maintain low viral loads in the absence of antiviral therapy have now shed additional light on these associations (Pereyra et al. 2010). At a viral load below 2000 copies the chances of disease progression drop dramatically, as does the likelihood of transmission (Quinn et al. 2000; Wawer et al. 2005). Through an international collaboration (the International HIV Controllers Study 2010) involving nearly 300 collaborators, a genome wide association study was performed on a cohort of nearly 974 HIV controllers, who maintain viral loads of less than 2000 copies/mL in the absence of therapy, and 2648 HIV progressors with viral loads in excess of 50,000 copies. Analysis of more than one million SNPs in each person revealed over 300 that were statistically associated with influencing viral load, and remarkably all of these lay within the HLA region of chromosome 6. However, after correction for multiple comparisons, only four SNPs remained significant, two of which were the same as those identified to be associated with lower set point viral load after acute infection (Fellay et al. 2007), as well as a

SNP associated with psoriasis and with a noncoding SNP in the *MICA* gene.

The above findings were still unable to reconcile whether these were simply associations, and not causal, or whether this reflected an HLA-mediated impact on HIV control. Subsequent sequence analysis within the HLA region, and stepwise regression analysis, revealed that 5 amino acids within HLA B, all of which are involved in binding of viral peptide in the HLA binding groove (Fig. 3), along with a minor effect of a SNP associated with HLA C expression, explained the entire genetic signal being detected, and revealed the major genetic influence on HIV control (Pereyra et al. 2010). The specific amino acids at these positions reconcile both the protective and risk HLA alleles, and thus provide a parsimonious explanation for the long-recognized HLA associations with HIV control, namely that it is the nature of viral peptide presentation that is the major genetic determinant of viral control, likely caused by differences in the relative efficacy of CD8[+] T-cell responses induced in the context of slightly altered T-cell-infected cell interactions. Indeed, the site with the strongest association with control is position 97 in HLA B, which sits at the base of the C pocket and depending on its allelic form, has the most impact on the nature of the binding groove and the way in which peptides are bound. Extension of these studies has now shown that the HLA C-associated SNP actually marks a polymorphism that affects microRNA binding, and thereby expression of HLA C (Kulkarni et al. 2011).

The overall impact of CD8[+] T-cell-mediated immune pressure on the virus is readily apparent from population analysis of expressed HLA types and sequencing of autologous viruses. Such selection pressure, first shown associated with mother to child transmission (Goulder et al. 2001) and revealed through examination of RT sequences and HLA alleles (Moore et al. 2002), has not only shown clear adaptation of HIV to host HLA genes, but also that some immunodominant epitopes are being eliminated over time (Kawashima et al. 2009). The extent to which CD4[+] T-cell responses are driving HIV evolution appears to

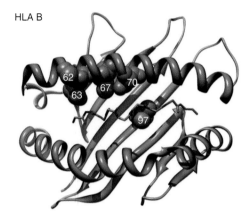

HLA B

Figure 3. Three-dimensional ribbon representation of the HLA B protein, highlighting amino acid positions 62, 63, 67, 70, and 97 lining the peptide binding pocket that are significantly associated with HIV control. The peptide backbone of the epitope is also displayed. (Adapted from the International HIV Controllers Study 2010; reprinted, with permission, from Science © 2010.)

be negligible, although mutations within $CD4^+$ T-cell epitopes have been documented in early infection (Rychert et al. 2007). And the impact of transmitted $CD8^+$ T-cell-selected mutations on viral set point and disease progression, likely caused by fitness impairing mutations, is clear (Goepfert et al. 2008; Matthews et al. 2008), and consistent with defects in viral fitness as described above.

FUNCTION OF $CD8^+$ T CELLS IN HIV INFECTION

Given the strong evidence that HIV-1-specific $CD8^+$ T cells contribute to the control of HIV-1 in the acute and chronic stages of infection, it is important to know what T-cell functions are responsible for this. Antiviral $CD8^+$ T cells were first identified as T cells that mediate lysis of virus-infected cells and are often referred to as cytotoxic T lymphocytes (Plata et al. 1975). Although most antigen-specific $CD8^+$ T cells have this activity, they can use other effector mechanisms in addition. These include production of interferon-γ, IL-2, TNF-α, MIP-1α (renamed CCL3), MIP-1β (CCL4), and RANTES (CCL5) (Betts et al. 2006; Streeck

et al. 2008). This list is probably not complete as revealed by transcriptional profiling, which also reveals subtle interplay between multiple pathways linking function, proliferation, and survival (Quigley et al. 2010). It is also clear that all $CD8^+$ T cells do not display all functions at all times (Ferrari et al. 2011). Naive T cells are quiescent and require days of antigen stimulation to show these functions (Veiga-Fernandes et al. 2000; Peixoto et al. 2007), in contrast memory $CD8^+$ T cells respond rapidly producing interferon-γ within a few hours (Lalvani et al. 1997). Production of lytic granules requires a bit longer but once activated, effector memory $CD8^+$ T cells can release perforin and granzymes within minutes (Barber et al. 2003). The delay in activating lytic functions in memory T cells probably protects the body from autoimmune attack when the TCR encounters weakly binding self antigens, but has the disadvantage that it may take many hours or days before these T cells can be recruited into lytic antipathogen activity. This may be a disadvantage for vaccines that are unlikely to maintain effector $CD8^+$ T cells in a fully primed state and therefore only kill infected cells after some hours.

The relative importance of lytic mechanisms versus cytokine or chemokine production in anti-HIV activity is much discussed. In very early HIV-1 infection $CD8^+$ T cells rapidly select virus escape mutants, often selecting a new mutant to replace all of the previous virus population in a few days (Goonetilleke et al. 2009). It is easiest to explain this by effector $CD8^+$ T cells killing HIV-infected target cells they recognize and thereby reducing their life span and capacity to generate new virus particles. However, new virus production could also be lowered by, for instance, β-chemokines (CCL3, 4, and 5) secreted by the $CD8^+$ T cells reducing infection of new target cells by direct competition for binding to CCR5 or by reducing its expression on the target cells. However, this would not shorten the life of the infected cells, which is what the mathematical model, described by Goonetilleke et al. (2009), implied. These studies suggest that lysis of infected cells is the most important antiviral function of the

CD8$^+$ T cells in acute infection. This is consistent with the finding that the T cells that select early virus escape mutants are high perforin expressors (Hersperger et al. 2010) whereas these early T cells are rather limited in their expression of cytokines and chemokines (Ferrari et al. 2011). This importance of cytolysis in HIV-1 infection contrasts with some other virus infections such as hepatitis B virus infection where, in animal models, perforin-deficient T cells are highly effective because the main antiviral activity is mediated nonlytically by interferon-γ (Yang et al. 2010). And it is important to note that there are data from in vivo analysis of SIV-infected macaques indicating that depletion of CD8$^+$ T cells 2 and 6 months after infection does not result in a measureable change in the lifespan of infected cells, suggesting that direct killing may not be not be the main mechanism of control, at least in this model.

During chronic infection, once viral set point is established, the other functions of CD8$^+$ T cells may become more important, although lytic potential may still be essential (Betts and Harari 2008). In patients who control virus well, the T cells are more quiescent than in acute infection. Many studies have shown that T cells in those who control HIV-1 well are polyfunctional, showing not only cytolytic potential but also have the capacity to produce cytokines and chemokines (Betts and Harari 2008), although it is not clear whether this is cause or effect. Prolonged antigen stimulation in the absence of excessive activation and exhaustion, as occurs in slow progressors, could favor expression of multiple functions. Production of IL-2 may be important in the long term persistence of CD8$^+$ T cells and can be provided by the CD8$^+$ T cell itself or by CD4$^+$ T cells, which survive much better in those whose disease progresses slowly (Rosenberg et al. 1997; Zimmerli et al. 2005). Similar observations have been made in HIV-2 infection in which elite controllers are relatively common (Duvall et al. 2008). These findings are entirely consistent with data in CD4$^+$ T-cell-depleted mice that show the importance of IL-2 in the maintenance of long-term CD8$^+$ T-cell memory (Williams et al. 2006).

THE INTERSECTION OF VIRAL FITNESS AND IMMUNE CONTROL

Although most of the focus on immune control and lack of control has been on CD8$^+$ T-cell function and differential induction of negative immunoregulatory molecules, an increasing body of data suggests that immune-mediated mutations within CD8$^+$ T-cell epitopes lead to reduced viral fitness. These data include assays in which replication of virus containing a B57-selected mutation is out-competed by wild-type virus (Martinez-Picado et al. 2006), evidence of reduced viral fitness in Gag-PR in persons who control virus spontaneously (Miura et al. 2009a), and evidence of compensatory mutations leading to restoration of fitness (Schneidewind et al. 2007, 2009). Importantly, mutations in Env do not lead to reduced fitness, suggesting that structural constraints are likely key to this effect (Troyer et al. 2009). More recent studies have shed further light on this, by demonstrating that there are multidimensional constraints on HIV evolution because certain combinations of mutations must occur in a coordinated manner to maintain virus viability, and thus constrain immune escape pathways (Dahirel et al. 2011). Persons who spontaneously control HIV without medications preferentially target sites that are most constrained, providing further evidence that the specific sites targeted by the immune system may have a major impact on overall control (Dahirel et al. 2011).

FUTURE DIRECTIONS: TOWARD A UNIFIED EXPLANATION FOR HIV PATHOGENESIS

A unified view of HIV pathogenesis is emerging, but there remain many unknowns, and a better understanding will undoubtedly require an integrated analysis of innate and adaptive immune responses, host genetics, and viral genetics. Although both viral load and CD4$^+$ cell count predict disease progression, we remain convinced that the adaptive immunologic response is also predictive of the subsequent course. Clear signals are there: CD8$^+$ T cells are associated with initial control, depletion of these cells in animal models of AIDS leads to an increase in viremia, virus is evolving to escape detection by CTLs, specific functions mediated by $CD8^+$ T cells have demonstrable antiviral effects in vivo, the effect of HLA far outweighs any other genetic factors, $CD8^+$ and $CD4^+$ T-cell dysfunction is associated with lack of viral control, and immune-induced mutations reduce viral fitness and likely contribute to the antiviral efficacy of the $CD8^+$ T-cell response. But important questions remain, and in particular questions regarding the actual functional profile of $CD8^+$ T cells that might lead to long-term control of HIV, or prevention of disseminated infection. And the extent to which such data will be important to vaccine design remains unclear—although animal models suggest that $CD8^+$ T cells may be able to prevent progressive systemic dissemination of infection (Hansen et al. 2009) and even reduce the level of virus to barely detectable levels (Hansen et al. 2011).

REFERENCES

*Reference is also in this collection.

Addo MM, Yu XG, Rathod A, Cohen D, Eldridge RL, Strick D, Johnston MN, Corcoran C, Wurcel AG, Fitzpatrick CA, et al. 2003. Comprehensive epitope analysis of human immunodeficiency virus type 1 (HIV-1)-specific T-cell responses directed against the entire expressed HIV-1 genome demonstrate broadly directed responses, but no correlation to viral load. J Virol 77: 2081–2092.

Allen TM, Altfeld M, Yu XG, O'Sullivan KM, Lichterfeld M, Le Gall S, John M, Mothe BR, Lee PK, Kalife ET, et al. 2004. Selection, transmission, and reversion of an antigen-processing cytotoxic T-lymphocyte escape mutation in human immunodeficiency virus type 1 infection. J Virol 78: 7069–7078.

Altfeld M, Rosenberg ES, Shankarappa R, Mukherjee JS, Hecht FM, Eldridge RL, Addo MM, Poon SH, Phillips MN, Robbins GK, et al. 2001. Cellular immune responses and viral diversity in individuals treated during acute and early HIV-1 infection. J Exp Med 193: 169–180.

Altfeld M, van Lunzen J, Frahm N, Yu XG, Schneider C, Eldridge RL, Feeney ME, Meyer-Olson D, Stellbrink HJ, Walker BD. 2002. Expansion of pre-existing, lymph node-localized CD8+ T cells during supervised treatment interruptions in chronic HIV-1 infection. J Clin Invest 109: 837–843.

Altfeld M, Addo MM, Rosenberg ES, Hecht FM, Lee PK, Vogel M, Yu XG, Draenert R, Johnston MN, Strick D, et al. 2003a. Influence of HLA-B57 on clinical presentation and viral control during acute HIV-1 infection. AIDS 17: 2581–2591.

Altfeld M, Addo MM, Shankarappa R, Lee PK, Allen TM, Yu XG, Rathod A, Harlow J, O'Sullivan K, Johnston MN, et al. 2003b. Enhanced detection of human immunodeficiency virus type 1-specific T-cell responses to highly variable regions by using peptides based on autologous virus sequences. J Virol 77: 7330–7340.

Altfeld M, Kalife ET, Qi Y, Streeck H, Lichterfeld M, Johnston MN, Burgett N, Swartz ME, Yang A, Alter G, et al. 2006. HLA alleles associated with delayed progression to AIDS contribute strongly to the initial CD8+ T cell response against HIV-1. PLoS Med 3: e403.

Altman JD, Moss PA, Goulder PJ, Barouch DH, McHeyzer-Williams MG, Bell JI, McMichael AJ, Davis MM. 1996. Phenotypic analysis of antigen-specific T lymphocytes. Science 274: 94–96.

Amyes E, Hatton C, Montamat-Sicotte D, Gudgeon N, Rickinson AB, McMichael AJ, Callan MF. 2003. Characterization of the CD4+ T cell response to Epstein-Barr virus during primary and persistent infection. J Exp Med 198: 903–911.

Bansal A, Gough E, Sabbaj S, Ritter D, Yusim K, Sfakianos G, Aldrovandi G, Kaslow RA, Wilson CM, et al. 2005. CD8 T-cell responses in early HIV-1 infection are skewed towards high entropy peptides. AIDS 19: 241–250.

Bansal A, Carlson J, Yan J, Akinsiku OT, Schaefer M, Sabbaj S, Bet A, Levy DN, Heath S, Tang J, et al. 2010. CD8 T cell response and evolutionary pressure to HIV-1 cryptic epitopes derived from antisense transcription. J Exp Med 207: 51–55.

Barber DL, Wherry EJ, Ahmed R. 2003. Cutting edge: Rapid in vivo killing by memory CD8 T cells. J Immunol 171: 27–31.

Berger CT, Carlson JM, Brumme CJ, Hartman KL, Brumme ZL, Henry LM, Rosato PC, Piechocka-Trocha A, Brockman MA, Harrigan PR, et al. 2010. Viral adaptation to immune selection pressure by HLA class I-restricted CTL responses targeting epitopes in HIV frameshift sequences. J Exp Med 207: 61–75.

Betts MR, Ambrozak DR, Douek DC, Bonhoeffer S, Brenchley JM, Casazza JP, Koup RA, Picker LJ. 2001. Analysis of total human immunodeficiency virus (HIV)-specific CD4+ and CD8+ T-cell responses: Relationship to viral load in untreated HIV infection. J Virol 75: 11983–11991.

Betts MR, Nason MC, West SM, De Rosa SC, Migueles SA, Abraham J, Lederman MM, Benito JM, Goepfert PA, Connors M, et al. 2006. HIV nonprogressors preferentially maintain highly functional HIV-specific CD8+ T cells. Blood 107: 4781–4789.

Betts MR, Harari A. 2008. Phenotype and function of protective T cell immune responses in HIV. Curr Opin HIV AIDS 3: 349–355.

Borrow P, Lewicki H, Hahn BH, Shaw GM, Oldstone MB. 1994. Virus-specific CD8+ cytotoxic T-lymphocyte activity associated with control of viremia in primary human immunodeficiency virus type 1 infection. J Virol 68: 6103–6110.

Brockman MA, Brumme ZL, Brumme CJ, Miura T, Sela J, Rosato PC, Kadie CM, Carlson JM, Markle TJ, Streeck H, et al. 2010. Early selection in Gag by protective HLA alleles contributes to reduced HIV-1 replication capacity that may be largely compensated in chronic infection. *J Virol* **84:** 11937–11949.

Brumme ZL, Brumme CJ, Carlson J, Streeck H, John M, Eichbaum Q, Block BL, Baker B, Kadie C, Markowitz M, et al. 2008. Marked epitope- and allele-specific differences in rates of mutation in human immunodeficiency type 1 (HIV-1) Gag, Pol, and Nef cytotoxic T-lymphocyte epitopes in acute/early HIV-1 infection. *J Virol* **82:** 9216–9227.

Brumme ZL, John M, Carlson JM, Brumme CJ, Chan D, Brockman MA, Swenson LC, Tao I, Szeto S, Rosato P, et al. 2009. HLA-associated immune escape pathways in HIV-1 subtype B Gag, Pol and Nef proteins. *PLoS One* **4:** e6687.

Chevalier MF, Jülg B, Pyo A, Flanders M, Ranasinghe S, Soghoian DZ, Kwon DS, Rychert J, Lian J, Muller MI, et al. 2011. HIV-1-specific interleukin-21$^+$ CD4$^+$ T cell responses contribute to durable viral control through the modulation of HIV-specific CD8$^+$ T cell function. *J Virol* **85:** 733–741.

Chopera DR, Woodman Z, Mlisana K, Mlotshwa M, Martin DP, Seoighe C, Treurnicht F, de Rosa DA, Hide W, Karim SA, et al. 2008. Transmission of HIV-1 CTL escape variants provides HLA-mismatched recipients with a survival advantage. *PLoS Pathog* **4:** e1000033.

Crawford H, Lumm W, Leslie A, Schaefer M, Boeras D, Prado JG, Tang J, Farmer P, Ndung'u T, Lakhi S, et al. 2009. Evolution of HLA-B*5703 HIV-1 escape mutations in HLA-B*5703-positive individuals and their transmission recipients. *J Exp Med* **206:** 909–921.

Dahirel V, Shekhar K, Pereyra F, Miura T, Artyomov M, Talsania S, Allen TM, Altfeld M, Carrington M, Irvine DJ, et al. 2011. Coordinate linkage of HIV evolution reveals regions of immunologic vulnerability. *Proc Natl Acad Sci* **108:** 11530–11535.

Dalod M, Dupuis M, Deschemin JC, Goujard C, Deveau C, Meyer L, Ngo N, Rouzioux C, Guillet JG, Delfraissy JF, et al. 1999. Weak anti-HIV CD8$^+$ T-cell effector activity in HIV primary infection. *J Clin Invest* **104:** 1431–1439.

Day CL, Kaufmann DE, Kiepiela P, Brown JA, Moodley ES, Reddy S, Mackey EW, Miller JD, Leslie AJ, DePierres C, et al. 2006. PD-1 expression on HIV-specific T cells is associated with T-cell exhaustion and disease progression. *Nature* **443:** 350–354.

Deeks SG, Walker BD. 2007. Human immunodeficiency virus controllers: Mechanisms of durable virus control in the absence of antiretroviral therapy. *Immunity* **27:** 406–416.

Douek DC, Brenchley JM, Betts MR, Ambrozak DR, Hill BJ, Okamoto Y, Casazza JP, Kuruppu J, Kunstman K, Wolinsky S, et al. 2002. HIV preferentially infects HIV-specific CD4$^+$ T cells. *Nature* **417:** 95–98.

Douek DC, Roederer M, Koup RA. 2009. Emerging concepts in the immunopathogenesis of AIDS. *Annu Rev Med* **60:** 471–484.

Draenert R, Le Gall S, Pfafferott KJ, Leslie AJ, Chetty P, Brander C, Holmes EC, Chang SC, Feeney ME, Addo MM, et al. 2004. Immune selection for altered antigen processing leads to cytotoxic T lymphocyte escape in chronic HIV-1 infection. *J Exp Med* **199:** 905–915.

Draenert R, Allen TM, Liu Y, Wrin T, Chappey C, Verrill CL, Sirera G, Eldridge RL, Lahaie MP, Ruiz L, et al. 2006.

Constraints on HIV-1 evolution and immunodominance revealed in monozygotic adult twins infected with the same virus. *J Exp Med* **203:** 529–539.

Duvall MG, Jaye A, Dong T, Brenchley JM, Alabi AS, Jeffries DJ, van der Sande M, Togun TO, McConkey SJ, Douek DC, et al. 2006. Maintenance of HIV-specific CD4$^+$ T cell help distinguishes HIV-2 from HIV-1 infection. *J Immunol* **176:** 6973–6981.

Duvall MG, Precopio ML, Ambrozak DA, Jaye A, McMichael AJ, Whittle HC, Roederer M, Rowland-Jones SL, Koup RA. 2008. Polyfunctional T cell responses are a hallmark of HIV-2 infection. *Eur J Immunol* **38:** 350–363.

Edwards BH, Bansal A, Sabbaj S, Bakari J, Mulligan MJ, Goepfert PA. 2002. Magnitude of functional CD8$^+$ T-cell responses to the gag protein of human immunodeficiency virus type 1 correlates inversely with viral load in plasma. *J Virol* **76:** 2298–2305.

Fellay J, Frahm N, Shianna KV, Cirulli ET, Casimiro DR, Robertson MN, Haynes BF, Geraghty DE, McElrath MJ, Goldstein DB, et al. 2007. A whole-genome association study of major determinants for host control of HIV-1. *Science* **317:** 944–947.

Ferrari G, Korber B, Goonetilleke N, Liu MK, Turnbull EL, Salazar-Gonzalez JF, Hawkins N, Self S, Watson S, Betts MR, et al. 2011. Relationship between functional profile of HIV-1 specific CD8 T cells and epitope variability with the selection of escape mutants in acute HIV-1 infection. *PLoS Pathog* **7:** e1001273.

Fiebig EW, Wright DJ, Rawal BD, Garrett PE, Schumacher RT, Peddada L, Heldebrant C, Smith R, Conrad A, Kleinman SH, et al. 2003. Dynamics of HIV viremia and antibody seroconversion in plasma donors: Implications for diagnosis and staging of primary HIV infection. *AIDS* **17:** 1871–1879.

Fischer W, Ganusov VV, Giorgi EE, Hraber PT, Keele BF, Leitner T, Han CS, Gleasner CD, Green L, Lo CC, et al. 2010. Transmission of single HIV-1 genomes and dynamics of early immune escape revealed by ultra-deep sequencing. *PLoS One* **5:** e12303.

Gao X, Nelson GW, Karacki P, Martin MP, Phair J, Kaslow R, Goedert JJ, Buchbinder S, Hoots K, Vlahov D, et al. 2001. Effect of a single amino acid change in MHC class I molecules on the rate of progression to AIDS. *N Engl J Med* **344:** 1668–1675.

Goepfert PA, Lumm W, Farmer P, Matthews P, Prendergast A, Carlson JM, Derdeyn CA, Tang J, Kaslow RA, Bansal A, et al. 2008. Transmission of HIV-1 Gag immune escape mutations is associated with reduced viral load in linked recipients. *J Exp Med* **205:** 1009–1017.

Goh WC, Markee J, Akridge RE, Meldorf M, Musey L, Karchmer T, Krone M, Collier A, Corey L, Emerman M, et al. 1999. Protection against human immunodeficiency virus type 1 infection in persons with repeated exposure: Evidence for T cell immunity in the absence of inherited CCR5 coreceptor defects. *J Infect Dis* **179:** 548–557.

Goonetilleke N, Moore S, Dally L, Winstone N, Cebere I, Mahmoud A, Pinheiro S, Gillespie G, Brown D, Loach V, et al. 2006. Induction of multifunctional human immunodeficiency virus type 1 (HIV-1)-specific T cells capable of proliferation in healthy subjects by using a prime-boost regimen of DNA- and modified vaccinia virus Ankara-vectored vaccines expressing HIV-1 Gag coupled to CD8$^+$ T-cell epitopes. *J Virol* **80:** 4717–4728.

Goonetilleke N, Liu MK, Salazar-Gonzalez JF, Ferrari G, Giorgi E, Ganusov VV, Keele BF, Learn GH, Turnbull EL, Salazar MG, et al. 2009. The first T cell response to transmitted/founder virus contributes to the control of acute viremia in HIV-1 infection. *J Exp Med* **206:** 1253–1272.

Goulder PJ, Brander C, Tang Y, Tremblay C, Colbert RA, Addo MM, Rosenberg ES, Nguyen T, Allen R, Trocha A, et al. 2001. Evolution and transmission of stable CTL escape mutations in HIV infection. *Nature* **412:** 334–338.

Graham BS, Koup RA, Roederer M, Bailer RT, Enama ME, Moodie Z, Martin JE, McCluskey MM, Chakrabarti BK, Lamoreaux L, et al. 2006. Phase 1 safety and immunogenicity evaluation of a multiclade HIV-1 DNA candidate vaccine. *J Infect Dis* **194:** 1650–1660.

Gray CM, Mlotshwa M, Riou C, Mathebula T, de Assis Rosa D, Mashishi T, Seoighe C, Ngandu N, van Loggerenberg F, Morris L, Mlisana K, et al. 2009. Human immunodeficiency virus-specific γ interferon enzyme-linked immunospot assay responses targeting specific regions of the proteome during primary subtype C infection are poor predictors of the course of viremia and set point. *J Virol.* **83:** 470–478.

Hansen SG, Vieville C, Whizin N, Coyne-Johnson L, Siess DC, Drummond DD, Legasse AW, Axthelm MK, Oswald K, Trubey CM, et al. 2009. Effector memory T cell responses are associated with protection of rhesus monkeys from mucosal simian immunodeficiency virus challenge. *Nat Med* **15:** 293–299.

Hansen SG, Ford JC, Lewis MS, Ventura AB, Hughes CM, Coyne-Johnson L, Whizin N, Oswald K, Shoemaker R, Swanson T, et al. 2011. Profound early control of highly pathogenic SIV by an effector memory T-cell vaccine. *Nature* **473:** 523–527.

Harari A, Bart PA, Stohr W, Tapia G, Garcia M, Medjitna-Rais E, Burnet S, Cellerai C, Erlwein O, Barber T, et al. 2008. An HIV-1 clade C DNA prime, NYVAC boost vaccine regimen induces reliable, polyfunctional, and long-lasting T cell responses. *J Exp Med* **205:** 63–77.

Hersperger AR, Pereyra F, Nason M, Demers K, Sheth P, Shin LY, Kovacs CM, Rodriguez B, Sieg SF, Teixeira-Johnson L, et al. 2010. Perforin expression directly ex vivo by HIV-specific CD8$^+$ T-cells is a correlate of HIV elite control. *PLoS Pathog* **6:** e1000917.

Ho DD, Sarngadharan MG, Resnick L, Dimarzoveronese F, Rota TR, Hirsch MS. 1985. Primary human T-lymphotropic virus type III infection. *Ann Intern Med* **103:** 880–883.

Huang S, Dunkley-Thompson J, Tang Y, Macklin EA, Steel-Duncan J, Singh-Minott I, Ryland EG, Smikle M, Walker BD, Christie CD, et al. 2008. Deficiency of HIV-Gag-specific T cells in early childhood correlates with poor viral containment. *J Immunol* **181:** 8103–8111.

International HIV Controllers Study, Pereyra F, Jia X, McLaren PJ, Telenti A, de Bakker PI, Walker BD, Ripke S, Brumme CJ, Pulit SL, Carrington M, et al. 2010. The major genetic determinants of HIV-1 control affect HLA class I peptide presentation. *Science* **330:** 1551–1557.

Janssen EM, Lemmens EE, Wolfe T, Christen U, von Herrath MG, Schoenberger SP. 2003. CD4$^+$ T cells are required for secondary expansion and memory in CD8$^+$ T lymphocytes. *Nature* **421:** 852–856.

Jin X, Bauer DE, Tuttleton SE, Lewin S, Gettie A, Blanchard J, Irwin CE, Safrit JT, Mittler J, Weinberger L, et al. 1999. Dramatic rise in plasma viremia after CD8$^+$ T cell depletion in simian immunodeficiency virus-infected macaques. *J Exp Med* **189:** 991–998.

Julg B, Moodley ES, Qi Y, Ramduth D, Reddy S, Mncube Z, Gao X, Goulder PJ, Detels R, Ndung'u T, et al. 2011. Possession of HLA Class II DRB1*1303 associates with reduced viral loads in chronic HIV clade C and B infection. *J Infect Dis* **203:** 803–809.

Kalams SA, Buchbinder SP, Rosenberg ES, Billingsley JM, Colbert DS, Jones NG, Shea AK, Trocha AK, Walker BD. 1999. Association between virus-specific cytotoxic T-lymphocyte and helper responses in human immunodeficiency virus type 1 infection. *J Virol* **73:** 6715–6720.

Kamimura D Bevan MJ 2007. Naive CD8$^+$ T cells differentiate into protective memory-like cells after IL-2 anti IL-2 complex treatment in vivo. *J Exp Med* **204:** 1803–1812.

Kaslow RA, Carrington M, Apple R, Park L, Munoz A, Saah AJ, Goedert JJ, Winkler C, O'Brien SJ, Rinaldo C, et al. 1996. Influence of combinations of human major histocompatibility complex genes on the course of HIV-1 infection. *Nat Med* **2:** 405–411.

Kaufmann DE, Bailey PM, Sidney J, Wagner B, Norris PJ, Johnston MN, Cosimi LA, Addo MM, Lichterfeld M, Altfeld M, et al. 2004. Comprehensive analysis of human immunodeficiency virus type 1-specific CD4 responses reveals marked immunodominance of gag and nef and the presence of broadly recognized peptides. *J Virol* **78:** 4463–4477.

Kawashima Y, Pfafferott K, Frater J, Matthews P, Payne R, Addo M, Gatanaga H, Fujiwara M, Hachiya A, Koizumi H, et al. 2009. Adaptation of HIV-1 to human leukocyte antigen class I. *Nature* **458:** 641–645.

Kiepiela P, Leslie AJ, Honeyborne I, Ramduth D, Thobakgale C, Chetty S, Rathnavalu P, Moore C, Pfafferott KJ, Hilton L, et al. 2004. Dominant influence of HLA-B in mediating the potential co-evolution of HIV and HLA. *Nature* **432:** 769–775.

Kiepiela P, Ngumbela K, Thobakgale C, Ramduth D, Honeyborne I, Moodley E, Reddy S, de Pierres C, Mncube Z, Mkhwanazi N, et al. 2007. CD8$^+$ T-cell responses to different HIV proteins have discordant associations with viral load. *Nat Med* **13:** 46–53.

Klatt NR, Shudo E, Ortiz AM, Engram JC, Paiardini M, Lawson B, Miller MD, Else J, Pandrea I, Estes JD, et al. 2010. CD8$^+$ Lymphocytes control viral replication in SIV-mac239-infected rhesus macaques without decreasing the lifespan of productively infected cells. *PLoS Pathog* **6:** e1000747.

Koibuchi T, Allen TM, Lichterfeld M, Mui SK, O'Sullivan KM, Trocha A, Kalams SA, Johnson RP, Walker BD. 2005. Limited sequence evolution within persistently targeted CD8 epitopes in chronic human immunodeficiency virus type 1 infection. *J Virol* **79:** 8171–8181.

Kosmrlj A, Read EL, Qi Y, Allen TM, Altfeld M, Deeks SG, Pereyra F, Carrington M, Walker BD, Chakraborty AK. 2010. Effects of thymic selection of the T-cell repertoire on HLA class I-associated control of HIV infection. *Nature* **465:** 350–354.

Koup RA, Safrit JT, Cao Y, Andrews CA, McLeod G, Borkowsky W, Farthing C, Ho DD. 1994. Temporal association of cellular immune responses with the initial control of viremia in primary human immunodeficiency virus type 1 syndrome. *J Virol* **68:** 4650–4655.

Kulkarni S, Savan R, Qi Y, Gao X, Yuki Y, Bass SE, Martin MP, Hunt P, Deeks SG, Telenti A, et al. 2011. Differential microRNA regulation of HLA-C expression and its association with HIV control. *Nature* **472:** 495–498.

Lacap PA, Huntington JD, Luo M, Nagelkerke NJ, Bielawny T, Kimani J, Wachihi C, Ngugi EN, Plummer FA. 2008. Associations of human leukocyte antigen DRB with resistance or susceptibility to HIV-1 infection in the Pumwani Sex Worker Cohort. *AIDS* **22:** 1029–1038.

Lalvani A, Brookes R, Hambleton S, Britton WJ, Hill AV, McMichael AJ. 1997. Rapid effector function in CD8$^+$ memory T cells. *J Exp Med* **186:** 859–865.

Lane HC, Depper JM, Greene WC, Whalen G, Waldmann TA, Fauci AS. 1985. Qualitative analysis of immune function in patients with the acquired immunodeficiency syndrome. Evidence for a selective defect in soluble antigen recognition. *N Engl J Med* **313:** 79–84.

Lazaro E, Godfrey SB, Stamegna P, Ogbechie T, Kerrigan C, Zhang M, Walker BD, Le Gall S. 2009. Differential HIV epitope processing in monocytes and CD4 T cells affects cytotoxic T lymphocyte recognition. *J Infect Dis* **200:** 236–243.

Le Gall S, Buseyne F, Trocha A, Walker BD, Heard JM, Schwartz O. 2000. Distinct trafficking pathways mediate Nef-induced and clathrin-dependent major histocompatibility complex class I down-regulation. *J Virol* **74:** 9256–9266.

Le Gall S, Stamegna P, Walker BD. 2007. Portable flanking sequences modulate CTL epitope processing. *J Clin Invest* **117:** 3563–3575.

Lichterfeld M, Yu XG, Cohen D, Addo MM, Malenfant J, Perkins B, Pae E, Johnston MN, Strick D, Allen TM, et al. 2004. HIV-1 Nef is preferentially recognized by CD8 T cells in primary HIV-1 infection despite a relatively high degree of genetic diversity. *AIDS* **18:** 1383–1392.

Lichterfeld M, Yu XG, Mui SK, Williams KL, Trocha A, Brockman MA, Allgaier RL, Waring MT, Koibuchi T, Johnston MN, et al. 2007. Selective depletion of high-avidity human immunodeficiency virus type 1 (HIV-1)-specific CD8$^+$ T cells after early HIV-1 infection. *J Virol* **81:** 4199–4214.

Lyles RH, Munoz A, Yamashita TE, Bazmi H, Detels R, Rinaldo CR, Margolick JB, Phair JP, Mellors JW. 2000. Natural history of human immunodeficiency virus type 1 viremia after seroconversion and proximal to AIDS in a large cohort of homosexual men. Multicenter AIDS Cohort Study. *J Infect Dis* **181:** 872–880.

Maecker HT, Dunn HS, Suni MA, Khatamzas E, Pitcher CJ, Bunde T, Persaud N, Trigona W, Fu TM, Sinclair E, et al. 2001. Use of overlapping peptide mixtures as antigens for cytokine flow cytometry. *J Immunol Methods* **255:** 27–40.

Maness NJ, Valentine LE, et al. 2007. AIDS virus specific CD8$^+$ T lymphocytes against an immunodominant cryptic epitope select for viral escape. *J Exp Med* **204:** 2505–2512.

Maness NJ, Sacha JB, Piaskowski SM, Weisgrau KL, Rakasz EG, May GE, Buechler MB, Walsh AD, Wilson NA, Watkins DI. 2009. Novel translation products from the SIVmac239 Env-encoding mRNA contain both Rev and cryptic T cell epitopes. *J Virol* **83:** 10280–10285.

Martinez-Picado J, Prado JG, Fry EE, Pfafferott K, Leslie A, Chetty S, Thobakgale C, Honeyborne I, Crawford H, Matthews P, et al. 2006. Fitness cost of escape mutations in p24 Gag in association with control of human immunodeficiency virus type 1. *J Virol* **80:** 3617–3623.

Mattapallil JJ, Douek DC, Hill B, Nishimura Y, Martin M, Roederer M. 2005. Massive infection and loss of memory CD4$^+$ T cells in multiple tissues during acute SIV infection. *Nature* **434:** 1093–1097.

Matthews PC, Prendergast A, Leslie A, Crawford H, Payne R, Rousseau C, Rolland M, Honeyborne I, Carlson J, Kadie C, et al. 2008. Central role of reverting mutations in HLA associations with human immunodeficiency virus set point. *J Virol* **82:** 8548–8559.

McMichael AJ, Borrow P, Tomaras GD, Goonetilleke N, Haynes BF. 2010. The immune response during acute HIV-1 infection: Clues for vaccine development. *Nat Rev Immunol* **10:** 11–23.

Mellors JW, Rinaldo CR Jr, Gupta P, White RM, Todd JA, Kingsley LA. 1996. Prognosis in HIV-1 infection predicted by the quantity of virus in plasma. *Science* **272:** 1167–1170.

Migueles SA, Sabbaghian MS, Shupert WL, Bettinotti MP, Marincola FM, Martino L, Hallahan CW, Selig SM, Schwartz D, Sullivan J, et al. 2000. HLA B*5701 is highly associated with restriction of virus replication in a subgroup of HIV-infected long term nonprogressors. *Proc Natl Acad Sci* **97:** 2709–2714.

Miura T, Brockman MA, Brumme ZL, Brumme CJ, Pereyra F, Trocha A, Block BL, Schneidewind A, Allen TM, Heckerman D, et al. 2009a. HLA-associated alterations in replication capacity of chimeric NL4-3 viruses carrying gag-protease from elite controllers of human immunodeficiency virus type 1. *J Virol* **83:** 140–149.

Miura T, Brockman MA, Schneidewind A, Lobritz M, Pereyra F, Rathod A, Block BL, Brumme ZL, Brumme CJ, Baker B, et al. 2009b. HLA-B57/B*5801 human immunodeficiency virus type 1 elite controllers select for rare gag variants associated with reduced viral replication capacity and strong cytotoxic T-lymphocyte [corrected] recognition. *J Virol* **83:** 2743–2755.

Miura T, Brumme ZL, Brockman MA, Rosato P, Sela J, Brumme CJ, Pereyra F, Kaufmann DE, Trocha A, Block BL, Daar ES, et al. 2010. Impaired replication capacity of acute/early viruses in persons who become HIV controllers. *J Virol* **84:** 7581–7591.

Moore CB, John M, James IR, Christiansen FT, Witt CS, Mallal SA. 2002. Evidence of HIV-1 adaptation to HLA-restricted immune responses at a population level. *Science* **296:** 1439–1443.

Murray HW, Rubin BY, Masur H, Roberts RB. 1984. Impaired production of lymphokines and immune (γ)

interferon in the acquired immunodeficiency syndrome. *N Engl J Med* **310:** 883–889.

Ngumbela KC, Day CL, Mncube Z, Nair K, Ramduth D, Thobakgale C, Moodley E, Reddy S, de Pierres C, Mkhwanazi N, et al. 2008. Targeting of a CD8 T cell env epitope presented by HLA-B*5802 is associated with markers of HIV disease progression and lack of selection pressure. *AIDS Res Hum Retroviruses* **24:** 72–82.

Nixon DF, Townsend AR, Elvin JG, Rizza CR, Gallwey J, McMichael AJ. 1988. HIV-1 gag-specific cytotoxic T lymphocytes defined with recombinant vaccinia virus and synthetic peptides. *Nature* **336:** 484–487.

Papagno L, Appay V, Sutton J, Rostron T, Gillespie GM, Ogg GS, King A, Makadzanhge AT, Waters A, Balotta C. et al. 2002. Comparison between HIV- and CMV-specific T cell responses in long-term HIV infected donors. *Clin Exp Immunol* **130:** 509–517.

Peixoto A, Evaristo C, Munitic I, Monteiro M, Charbit A, Rocha B, Veiga-Fernandes H. 2007. CD8 single-cell gene coexpression reveals three different effector types present at distinct phases of the immune response. *J Exp Med* **204:** 1193–1205.

Pereyra F, Addo MM, Kaufmann DE, Liu Y, Miura T, Rathod A, Baker B, Trocha A, Rosenberg R, Mackey E, Ueda P, Lu Z, et al. 2008. Genetic and immunologic heterogeneity among persons who control HIV infection in the absence of therapy. *J Infect Dis* **197:** 563–571.

Petrovas C, Casazza JP, Brenchley JM, Price DA, Gostick E, Adams WC, Precopio ML, Schacker T, Roederer M, Douek DC, et al. 2006. PD-1 is a regulator of virus-specific CD8+ T cell survival in HIV infection. *J Exp Med* **203:** 2281–2292.

Phillips RE, Rowland-Jones S, Nixon DF, Gotch FM, Edwards JP, Ogunlesi AO, Elvin JG, Rothbard JA, Bangham CR, Rizza CR. 1991. Human immunodeficiency virus genetic variation that can escape cytotoxic T cell recognition. *Nature* **354:** 453–459.

Pitcher CJ, Quittner C, Peterson DM, Connors M, Koup RA, Maino VC, Picker LJ. 1999. HIV-1-specific CD4+ T cells are detectable in most individuals with active HIV-1 infection, but decline with prolonged viral suppression. *Nat Med* **5:** 518–525.

Plata F, Cerottini JC, et al. 1975. Primary and secondary in vitro generation of cytolytic T lymphocytes in the murine sarcoma virus system. *Eur J Immunol* **5:** 227–233.

Plata F, Autran B, Martins LP, Wain-Hobson S, Raphaël M, Mayaud C, Denis M, Guillon JM, Debré P. 1987. AIDS virus-specific cytotoxic T lymphocytes in lung disorders. *Nature* **328:** 348–351.

Quigley M, Pereyra F, Nilsson B, Porichis F, Fonseca C, Eichbaum Q, Julg B, Jesneck JL, Brosnahan K, Imam S, et al. 2010. Transcriptional analysis of HIV-specific CD8+ T cells shows that PD-1 inhibits T cell function by upregulating BATF. *Nat Med* **16:** 1147–1151.

Quinn TC, Wawer MJ, et al. 2000. Viral load and heterosexual transmission of human immunodeficiency virus type 1. Rakai Project Study Group. *N Engl J Med* **342:** 921–929.

Radebe M, Nair K, Chonco F, Bishop K, Wright JK, van der Stok M, Bassett IV, Mncube Z, Altfeld M, Walker BD, Ndung'u T. 2011. Limited immunogenicity of HIV CD8+ T-cell epitopes in acute clade C virus Infection. *J Infect Dis* **204:** 768–776.

Ramduth D, Chetty P, et al. 2005. Differential immunogenicity of HIV-1 clade C proteins in eliciting CD8+ and CD4+ cell responses. *J Infect Dis* **192:** 1588–1596.

Richardson JS. 2000. Early ribbon drawings of proteins. *Nat Struct Biol* **7:** 624–625.

Ritchie AJ, Campion SL, Kopycinski J, Moodie Z, Wang ZM, Pandya K, Moore S, Liu MK, Brackenridge S, Kuldanek K, et al. 2011. Differences in HIV-specific T cell responses between HIV-exposed and -unexposed HIV-seronegative individuals. *J Virol* **85:** 3507–3516.

Rosenberg ES, Billingsley JM, Caliendo AM, Boswell SL, Sax PE, Kalams SA, Walker BD. 1997. Vigorous HIV-1-specific CD4+ T cell responses associated with control of viremia. *Science* **278:** 1447–1450.

Rosenberg ES, Altfeld M, Poon SH, Phillips MN, Wilkes BM, Eldridge RL, Robbins GK, D'Aquila RT, Goulder PJ, Walker BD. 2000. Immune control of HIV-1 after early treatment of acute infection. *Nature* **407:** 523–526.

Rowland-Jones S, Sutton J, Ariyoshi K, Dong T, Gotch F, McAdam S, Whitby D, Sabally S, Gallimore A, Corrah T. 1995. HIV-specific cytotoxic T-cells in HIV-exposed but uninfected Gambian women. *Nat Med* **1:** 59–64.

Rychert J, Saindon S, Placek S, Daskalakis D, Rosenberg E. 2007. Sequence variation occurs in CD4 epitopes during early HIV infection. *J Acquir Immune Defic Syndr* **46:** 261–267.

Salazar-Gonzalez JF, Salazar MG, Keele BF, Learn GH, Giorgi EE, Li H, Decker JM, Wang S, Baalwa J, Kraus MH, et al. 2009. Genetic identity, biological phenotype, and evolutionary pathways of transmitted/founder viruses in acute and early HIV-1 infection. *J Exp Med* **206:** 1273–1289.

Schenal M, Lo Caputo S, Fasano F, Vichi F, Saresella M, Pierotti P, Villa ML, Mazzotta F, Trabattoni D, Clerici M. 2005. Distinct patterns of HIV-specific memory T lymphocytes in HIV-exposed uninfected individuals and in HIV-infected patients. *AIDS* **19:** 653–661.

Schmitz JE, Kuroda MJ, Santra S, Sasseville VG, Simon MA, Lifton MA, Racz P, Tenner-Racz K, Dalesandro M, Scallon BJ, et al. 1999. Control of viremia in simian immunodeficiency virus infection by CD8+ lymphocytes. *Science* **283:** 857–860.

Schneidewind A, Brockman MA, Yang R, Adam RI, Li B, Le Gall S, Rinaldo CR, Craggs SL, Allgaier RL, Power KA, et al. 2007. Escape from the dominant HLA-B27-restricted cytotoxic T-lymphocyte response in Gag is associated with a dramatic reduction in human immunodeficiency virus type 1 replication. *J Virol* **81:** 12382–12393.

Schneidewind A, Brockman MA, Sidney J, Wang YE, Chen H, Suscovich TJ, Li B, Adam PI, Allgaier RL, Mothé BR, et al. 2008. Structural and functional constraints limit options for cytotoxic T-lymphocyte escape in the immunodominant HLA-B27-restricted epitope in human immunodeficiency virus type 1 capsid. *J Virol* **82:** 5594–5605.

Schneidewind A, Brumme ZL, Brumme CJ, Power KA, Reyor LL, O'Sullivan K, Gladden A, Hempel U, Kuntzen T, Wang YE, et al. 2009. Transmission and long-term stability of compensated CD8 escape mutations. *J Virol.* **83:** 3993–3997.

* Shaw GM, Hunter E. 2011. HIV transmission. *Cold Spring Harb Perspect Med* doi: 10.1101/cshperspect. a006965.

Shedlock DJ, Shen H. 2003. Requirement for CD4 T cell help in generating functional CD8 T cell memory. *Science* **300:** 337–339.

Stacey AR, Norris PJ, Qin L, Haygreen EA, Taylor E, Heitman J, Lebedeva M, DeCamp A, Li D, Grove D, et al. 2009. Induction of a striking systemic cytokine cascade prior to peak viremia in acute human immunodeficiency virus type 1 infection, in contrast to more modest and delayed responses in acute hepatitis B and C virus infections. *J Virol* **83:** 3719–3733.

Staprans SI, Barry AP, Silvestri G, Safrit JT, Kozyr N, Sumpter B, Nguyen H, McClure H, Montefiori D, Cohen JI, et al. 2004. Enhanced SIV replication and accelerated progression to AIDS in macaques primed to mount a CD4 T cell response to the SIV envelope protein. *Proc Natl Acad Sci* **101:** 13026–13031.

Streeck H, Brumme ZL, Anastario M, Cohen KW, Jolin JS, Meier A, Brumme CJ, Rosenberg ES, Alter G, Allen TM, et al. 2008. Antigen load and viral sequence diversification determine the functional profile of HIV-1-specific CD8[+] T cells. *PLoS Med* **5:** e100.

Streeck H, Jolin JS, Qi Y, Yassine-Diab B, Johnson RC, Kwon DS, Addo MM, Brumme C, Routy JP, Little S, et al. 2009. Human immunodeficiency virus type 1-specific CD8[+] T-cell responses during primary infection are major determinants of the viral set point and loss of CD4[+] T cells. *J Virol* **83:** 7641–7648.

Sun JC, Bevan MJ. 2003. Defective CD8 T cell memory following acute infection without CD4 T cell help. *Science* **300:** 339–342.

Sun Y, Schmitz JE, Buzby AP, Barker BR, Rao SS, Xu L, Yang ZY, Mascola JR, Nabel GJ, Letvin NL. 2006. Virus-specific cellular immune correlates of survival in vaccinated monkeys after simian immunodeficiency virus challenge. *J Virol* **80:** 10950–10956.

Sun Y, Santra S, Buzby AP, Mascola JR, Nabel GJ, Letvin NL. 2010. Recombinant vector-induced HIV/SIV-specific CD4[+] T lymphocyte responses in rhesus monkeys. *Virology* **406:** 48–55.

Tenzer S, Wee E, Burgevin A, Stewart-Jones G, Friis L, Lamberth K, Chang CH, Harndahl M, Weimershaus M, Gerstoft J, et al. 2009. Antigen processing influences HIV-specific cytotoxic T lymphocyte immunodominance. *Nat Immunol* **10:** 636–646.

Trautmann L, Janbazian L, Chomont N, Said EA, Gimmig S, Bessette B, Boulassel MR, Delwart E, Sepulveda H, Balderas RS, Routy JP, Haddad EK, Sekaly RP, 2006. Upregulation of PD-1 expression on HIV-specific CD8[+] T cells leads to reversible immune dysfunction. *Nat Med* **12:** 1198–1202.

Troyer RM, McNevin J, Liu Y, Zhang SC, Krizan RW, Abraha A, Tebit DM, Zhao H, Avila S, Lobritz MA, et al. 2009. Variable fitness impact of HIV-1 escape mutations to cytotoxic T lymphocyte (CTL) response. *PLoS Pathog* **5:** e1000365.

Turnbull EL, Wong M, Wang S, Wei X, Jones NA, Conrod KE, Aldam D, Turner J, Pellegrino P, Keele BF, et al.

2009. Kinetics of expansion of epitope-specific T cell responses during primary HIV-1 infection. *J Immunol* **182:** 7131–7145.

Veiga-Fernandes H, Walter U, Bourgeois C, McLean A, Rocha B. 2000. Response of naive and memory CD8[+] T cells to antigen stimulation in vivo. *Nat Immunol* **1:** 47–53.

Vezys V, Yates A, Casey KA, Lanier G, Ahmed R, Antia R, Masopust D. 2009. Memory CD8 T-cell compartment grows in size with immunological experience. *Nature* **457:** 196–199.

Vigneault F, Woods M, Buzon MJ, Li C, Pereyra F, Crosby SD, Rychert J, Church G, Martinez-Picado J, Rosenberg ES, et al. 2011. Transcriptional profiling of CD4 T cells identifies distinct subgroups of HIV-1 elite controllers. *J Virol* **85:** 3015–3019.

Wahren B, Morfeldt-Månsson L, Biberfeld G, Moberg L, Sönnerborg A, Ljungman P, Werner A, Kurth R, Gallo R, Bolognesi D. 1987. Characteristics of the specific cell-mediated immune response in human immunodeficiency virus infection. *J Virol* **61:** 2017–2023.

Walker CM, Moody DJ, Stites DP, Levy JA. 1986. CD8[+] lymphocytes can control HIV infection in vitro by suppressing virus replication. *Science* **234:** 1563–1566.

Walker BD, Chakrabarti S, Moss B, Paradis TJ, Flynn T, Durno AG, Blumberg RS, Kaplan JC, Hirsch MS, et al. 1987. HIV-specific cytotoxic T lymphocytes in seropositive individuals. *Nature* **328:** 345–348.

Wawer MJ, Gray RH, Sewankambo NK, Serwadda D, Li X, Laeyendecker O, Kiwanuka N, Kigozi G, Kiddugavu M, Lutalo T, et al. 2005. Rates of HIV-1 transmission per coital act, by stage of HIV-1 infection, in Rakai, Uganda. *J Infect Dis* **191:** 1403–1409.

Williams MA, Tyznik AJ, Bevan MJ. 2006. Interleukin-2 signals during priming are required for secondary expansion of CD8[+] memory T cells. *Nature* **441:** 890–893.

Williams LD, Bansal A, Sabbaj S, Heath SL, Song W, Tang J, Zajac AJ, Goepfert PA. 2011. Interleukin-21-producing HIV-1-specific CD8 T cells are preferentially seen in elite controllers. *J Virol* **85:** 2316–2324.

Wright JK, Novitsky V, Brockman MA, Brumme ZL, Brumme CJ, Carlson JM, Heckerman D, Wang B, Losina E, Leshwedi M, et al. 2011. Influence of gag-protease-mediated replication capacity on disease progression in individuals recently infected with HIV-1 subtype C. *J Virol* **85:** 3996–4006.

Yang PL, Althage A, Chung J, Maier H, Wieland S, Isogawa M, Chisari FV. 2010. Immune effectors required for hepatitis B virus clearance. *Proc Natl Acad Sci* **107:** 798–802.

Yu XG, Addo MM, Rosenberg ES, Rodriguez WR, Lee PK, Fitzpatrick CA, Johnston MN, Strick D, Goulder PJ, Walker BD, et al. 2002. Consistent patterns in the development and immunodominance of human immunodeficiency virus type 1 (HIV-1)-specific CD8[+] T-cell responses following acute HIV-1 infection. *J Virol* **76:** 8690–8701.

Yu XG, Lichterfeld M, Williams KL, Martinez-Picado J, Walker BD. 2007. Random T-cell receptor recruitment in human immunodeficiency virus type 1 (HIV-1)-specific CD8[+] T cells from genetically identical twins infected with the same HIV-1 strain. *J Virol* **81:** 12666–12669.

Zimmerli SC, Harari A, Cellerai C, Vallelian F, Bart PA, Pantaleo G. 2005. HIV-1-specific IFN-γ/IL-2-secreting CD8 T cells support CD4-independent proliferation of HIV-1-specific CD8 T cells. *Proc Natl Acad Sci* **102:** 7239–7244.

Zuñiga R, Lucchetti A, Galvan P, Sanchez S, Sanchez C, Hernandez A, Sanchez H, Frahm N, Linde CH, Hewitt HS, et al. 2006. Relative dominance of Gag p24–specific cytotoxic T lymphocytes is associated with human immunodeficiency virus control. *J Virol* **80:** 3122–3125.

Innate Immune Control of HIV

Mary Carrington[1,2] and Galit Alter[1]

[1]Ragon Institute of MGH, MIT, and Harvard, Charlestown, Massachusetts 02129

[2]Cancer and Inflammation Program, Laboratory of Experimental Immunology, SAIC Frederick, NCI Frederick, Frederick, Maryland 21702

Correspondence: galter@partners.org

Mounting evidence suggests a role for innate immunity in the early control of HIV infection, before the induction of adaptive immune responses. Among the early innate immune effector cells, dendritic cells (DCs) respond rapidly following infection aimed at arming the immune system, through the recognition of viral products via pattern recognition receptors. This early response results in the potent induction of a cascade of inflammatory cytokines, intimately involved in directly setting up an antiviral state, and indirectly activating other antiviral cells of the innate immune system. However, epidemiologic data strongly support a role for natural killer (NK) cells as critical innate mediators of antiviral control, through the recognition of virally infected cells through a network of receptors called the killer immunoglobulin-like receptors (KIRs). In this review, the early events in innate immune recognition of HIV, focused on defining the biology underlying KIR-mediated NK-cell control of HIV viral replication, are discussed.

Early events following HIV infection determine the course of disease progression in such a way that more robust control of viral replication in acute HIV infection, resulting in lower viral set-point levels, is associated with slower HIV disease progression (Pantaleo et al. 1997). However, reduction in viral replication during acute HIV infection often occurs before the induction of adaptive immune responses such as CD8[+] T-cell responses (Alter et al. 2007b), strongly suggesting that the innate immune system, our body's first line of defense against invading pathogens, may play an early essential role in antiviral control.

THE INNATE IMMUNE SYSTEM

The innate immune system has evolved over millennia to nonspecifically control and clear invading pathogens. Unlike the adaptive arm of the immune system, which uses antigen-specific receptors to recognize foreign antigens, the innate immune system uses an array of pattern recognition receptors to detect patterns associated with bacteria, viruses, and/or parasites. These patterns relate to carbohydrate, protein, or lipid structures that are unique to pathogens, not normally produced in human cells (Murphy et al. 2011). Three classes of pattern recognition receptors have been identified to date, including the (RIG-I)-like receptors (RLRs), the toll-like receptors (TLRs), and the nucleotide oligomerization domain (NOD)-like receptors (NLRs). Activation of different combinations of these receptors, on distinct innate immune cell subsets, results in the induction of distinct inflammatory cues that result in the creation of

Cite this article as *Cold Spring Harb Perspect Med* doi: 10.1101/cshperspect.a007070

a nonspecific antiviral environment through the release of cytokines (including interferons [IFNs]) that block viral growth, the activation and recruitment of other immune cells, and the induction of adaptive immune responses.

HIV, like other single-stranded RNA viruses, triggers innate immune receptors, including TLR7 and TLR8, resulting in the potent activation of dendritic cells (DCs) and the release of copious amounts of type 1 IFNs and tumor necrosis factor α (TNF-α), both involved in shutting down viral replication in infected cells while also promoting the activation of the immune response (Diebold et al. 2004; Heil et al. 2004; Beignon et al. 2005). Interestingly, recent data suggest that DCs from females produce higher levels of IFN-α, compared with DCs from age-matched men, on HIV RNA triggering of TLR7/8 (Meier et al. 2009). Given that women show overall lower viral set points than men, it is plausible that enhanced viral control in females may in part relate to this enhanced antiviral innate immune response. The difference in the ability of DCs from women and men to respond to TLR7/8 triggering likely reflects a hormonal sensitization of DCs, specifically promoting TLR-induced IFN-α, but not TNF-α, production in women. However, whether enhanced antiviral control reflects the direct activity of IFN-α alone, or its added effects on activating other innate immune cells (including natural killer [NK] cells), or in the induction of a more potent adaptive immune response is yet to be defined.

In addition to TLR7/8 recognition of HIV, TLR2, TLR4, and TLR9 have been implicated in recognition and modulation of HIV viral replication. Both TLR2 and TLR4 triggering on DCs has been associated with increased and reduced transmission of HIV, respectively, owing to differential induction of type 1 IFNs (Thibault et al. 2009). Furthermore, recent evidence also points to a direct role for gp120 binding to TLR9, resulting in pDC activation, type 1 IFN secretion, and activation of NK cells that may promote early antiviral control (Martinelli et al. 2007). However, the overall role of individual or combined TLR sensing in early recognition and control of HIV has not been fully elucidated.

The early HIV-mediated triggering of DCs, and other TLR expressing innate immune cells, is associated with the induction of a robust cytokine storm (Stacey et al. 2009). This early response is marked by the rapid induction of IFN-α, interleukin-15 (IL-15), and inducible protein-10 (IP-10), followed by a slower increase in proinflammatory factors, associated finally with a sustained increase in immunoregulatory cytokines. Interestingly, the acute cytokine cascade is strikingly more pronounced following HIV infection compared with hepatitis B and C infections. Thus, although the dramatic increase in immunomodulators may be geared toward the priming of a robust immune response against the incoming pathogen, it is plausible that the intensity and magnitude of this cascade may also contribute in part to the observed immunopathology associated with early HIV disease.

INNATE IMMUNE CELLS

An array of cell subsets, all derived from the bone marrow, forms the arsenal of the innate immune system that responds to the acute cytokine cascade, each expressing distinct sets of innate immune receptors, endowing them with a unique capacity to respond to incoming pathogens. These cells include phagocytes (monocytes, macrophages, DCs) primed for antigen clearance, cytolytic cells (NK cells and neutrophils) geared toward the direct destruction of the pathogen or pathogen-infected cells, and professional antigen-presenting cells (DCs) aimed at capturing foreign antigens to present to the adaptive immune response for the induction of immunological memory. These cells persistently patrol peripheral tissues, primed to respond to foreign antigens on receptor engagement without the need for antigen sensitization. Thus, the innate immune response is not only responsible for early pathogen containment, but also plays a central role in shaping the quality of the ensuing adaptive immune response through the release of potent inflammatory cues and the qualitative modulation of DCs.

Among the innate immune cells involved in early antiviral control of HIV, epidemiologic evidence strongly points to a central role for NK cells in antiviral containment. Most convincingly, the coexpression of particular NK-cell receptors (the killer immunoglobulin receptors) in conjunction with their ligands (major histocompatibility complex [MHC] class I alleles) is associated with slower HIV disease progression and early viral control of viremia (Martin et al. 2002, 2007). These data strongly support a role for these cytolytic effector cells early in infection, whereas the adaptive immune response is just developing. However, whether NK cells mediate their antiviral control strictly through cytolytic removal of infected cells or through the editing of particular DC populations resulting in more potent adaptive immune responses is unknown.

NK CELLS

Unlike CD8$^+$ T cells, NK cells are a subset of large granular lymphocytes that do not express an antigen-specific receptor, but rather express a variety of inhibitory and activating receptors on their surface that are involved in sensing changes in their ligands on the surface of the body's cells (Lanier 1998). As such, these cells are classified as cells of the innate immune system, as they are able to sense viral infection before antigen sensitization. Given that these cells are loaded with cytolytic granules that can cause a great deal of immunopathology, the activation of these cells is under tight regulation by a network of inhibitory and activating self-reactive receptor/ligand interactions. NK cells survey the body for MHC class I expression, using a network of receptors called the killer immunoglobulin-like receptors (KIRs), and are inhibited on interaction with MHC class I. However, lack of engagement of inhibitory receptors alone is not sufficient to activate an NK cell to kill a target cell, but rather an NK cell must receive an additional activating signal through recognition of ligand to induce cytolytic elimination of the target cell (Fig. 1) (Karre et al. 1986; Ljunggren and Karre 1990; Moretta et al. 1993). Alternatively, target cells that up-regulate activating NK receptor ligands to levels that outcompete the dominant inhibitory

signals delivered through normal MHC recognition by KIRs can also result in NK cell activation (Cerwenka and Lanier 2001a). Ultimately, NK-cell activation hinges on the delicate balance between inhibition and activation delivered through a variety of NK-cell receptors, including KIRs, that fine-tune their lytic activity. This concept has refined the "missing self" model of NK recognition to include two basic steps: (1) loss of self, which may occur following infection or tumor transformation, as a first signal to alert NK cells that a cell is aberrant, and (2) an activating signal that is required to fully unleash the cytolytic activity of NK cells. Furthermore, over the past decade, accumulating evidence suggests that NK cells may not be as innate as once believed, but that individual NK-cell clones may show some target cell specificity (Malnati et al. 1995; Peruzzi et al. 1996), allowing them to play a critical early role in early antiviral control following infection with HIV.

NK CELLS IN HIV

The first immunomodulators in the acute cytokine storm (IFN-α and IL-15) (Stacey et al. 2009) are centrally involved in rapidly arming and activating NK cells following infection (Biron 1999). Thus, as anticipated, NK cells expand rapidly following acute infection, specifically in the acute seronegative window, with a preferential expansion of the cytolytic CD56^{dim} NK-cell subpopulation (Alter et al. 2007b; Alter et al. 2009). However, to compensate for this early burst of innate cytolytic effector cells, HIV has devised multiple strategies to evade NK-cell recognition, indicating that these cells are able to place pressure on the virus.

HIV EVASION OF NK CELLS THROUGH Nef

Viruses have evolved multiple strategies to evade the immune system, including NK-cell recognition, suggesting a role for these cells in the early response to infection (Lodoen and Lanier 2005). Many viruses have specifically evolved strategies to down-regulate MHC class I from the surface of infected cells in an effort to avoid

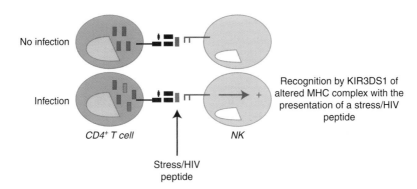

Figure 1. A model of KIR3DS1$^+$ natural killer (NK)-cell recognition of an HIV-infected target cell. Accumulating evidence suggests that specific amino acid changes in the peptides presented by major histocompatibility complex (MHC) class I can have a profound impact on KIR recognition of peptide/MHC complexes. Along these lines, it is plausible that a viral or stress peptide generated during infection presented by Bw4-80I may alter the affinity of the activating KIR3DS1 receptor expressed on NK cells for its putative ligand, resulting in the potent activation of NK cells and rapid elimination of virally infected cells.

CD8$^+$ T-cell recognition. However, this loss of MHC class I renders infected cells vulnerable to NK-cell-mediated recognition through inhibitory NK-cell receptors. Viruses such as cytomegalovirus (CMV) have evolved a compensatory repertoire of MHC class I homologs aimed at providing inhibitory signals to NK cells (Cerwenka and Lanier 2001b; Arase et al. 2002). Whereas CMV is a large DNA virus that has the opportunity to accommodate multiple genes for the evasion of both innate and adaptive immune responses, HIV is a small RNA virus that encodes only nine genes. Yet a number of studies have shown that HIV uses a single nonstructural gene, Nef, to evade both the innate and adaptive immune response.

Most notably, HIV-1 Nef protein triggers the accelerated endocytosis or retention of MHC class I molecules in the Golgi, resulting in reduced MHC class I expression on the surface of infected cells (Schwartz et al. 1996), thereby preventing recognition by HIV-specific CD8$^+$ T cells. However, reduced MHC class I expression may alert NK cells of a possible infection. Interestingly, Nef may overcome both CD8- and NK-cell-mediated recognition by down-regulating the dominant T-cell receptor ligands HLA-A and -B molecules, while sparing the dominant inhibitory KIR2D ligands, HLA-C (Le Gall et al. 1998; Cohen et al. 1999).

However, HLA-A appear to be down-regulated robustly, as compared with HLA-B (Cohen et al. 1999). These data strongly suggest that Nef has evolved a means to spare some KIR ligands, allowing it to strike a balance between T- and NK-cell evasion.

Loss of MHC class I expression is not sufficient to trigger NK-cell destruction of an HIV-infected cell, but requires a second activating signal. Viral infection often results in the up-regulation of the stress-inducible ligands for the activating c-type lectin NK-cell receptor NKG2D (Raulet 2003). These NKG2D-stress ligands, the MHC class I-related chain-A and -B (MIC-A/B) or UL-16 binding proteins-1, -2, and -3 (ULBP-1/2/3), are homologs of MHC class I alleles that are typically expressed following tumor transformation or infection (Raulet 2003). Recent studies reveal that the expression of MIC and ULBP on human tumor cells is sufficient to overcome the inhibitory effects of MHC class I expression (Zhang et al. 2005). To circumvent this activity, the HIV Nef protein has evolved the capacity to prevent the expression of some NKG2D ligands, such as MIC A, ULBP-1, and -2, at the surface of infected cells (Cerboni et al. 2007). It appears then that Nef regulation of host protein expression targets two host defense mechanisms, one involving KIRs and the other NKG2D.

A ROLE FOR KIR IN MODULATING HIV DISEASE PROGRESSION

KIRs can be divided into four groups based on two features: the number of extracellular domains (two domain [2D] or three domain [3D] and the length of the cytoplasmic tail (long [L] or short [S]). The length of the cytoplasmic domains dictates whether the receptor is activating or inhibitory, as long-tail KIRs contain immunoreceptor tyrosine-based inhibition motifs (ITIMs) that deliver strong inhibitory signals, whereas the short cytoplasmic tails associate with molecules that contain immunoreceptor tyrosine-based activation motifs (ITAMs) (Lanier et al. 1998). In addition to differences in gene content, most KIR genes show allelic polymorphism as well (Shilling et al. 2002; Carrington and Norman 2003).

Both epidemiological data and genome-wide association studies (GWASs) have pointed to a central role for particular MHC class I alleles in modulating the rate of disease progression (Carrington and O'Brien 2003; Fellay et al. 2007), the majority of which are encoded by the MHC class 1-B locus. Most of the protective HLA-B alleles express the Bw4 epitope, the primary ligands for KIR3DL1. Given the remarkable homology between alleles of KIR3DL1 and its activating counterpart KIR3DS1, epidemiological studies aimed at defining whether these three-domain KIR had any role in modulating disease progression were tested (Martin et al. 2002, 2007). Interestingly, both the activating and a subset of inhibitory variants of this KIR gene had a profound impact on modulating HIV disease progression in the context of their putative MHC ligands. Furthermore, duplications and deletions within the 3DL1/S1 segment have been observed (Martin et al. 2003), resulting in KIR haplotypes that can have zero or two copies of the KIR3DL1/S1 gene, and increasing doses of KIR3DS1 in the presence of KIR3DL1 and its putative ligand are associated with more robust control of HIV viremia in early disease (K Pelak and DG Goldstein, pers. comm.). These results suggest that NK cells may contribute to control through KIRs through at least two different mechanisms,

one modulated by inhibitory receptors and a second mediated by an activating receptor, and that the activating and inhibitory receptors may interact to promote enhanced control of HIV viral replication.

KIR3DS1-MEDIATED CONTROL OF HIV

A number of studies have highlighted the impact of particular KIR/MHC combinations on HIV-1 disease outcome (Martin et al. 2002, 2007; Jennes et al. 2006). Martin et al. showed that subjects that coexpressed the activating KIR3DS1 allele in conjunction with its putative MHC class I ligand, Bw4 alleles with an isoleucine at position 80 of the peptide-binding groove (Bw4-80I) (Barber et al. 1997), progressed significantly more slowly toward AIDS than individuals that do not have this compound genotype (Martin et al. 2002). Although the physical interaction between KIR3DS1 and HLA-Bw4-80I molecules has yet to be shown, this genetic epistasis suggests that this KIR/MHC interaction confers some antiviral signal to NK cells to allow them to control HIV infection more effectively.

Functional data support the interaction between KIR3DS1 and Bw4-80I, as KIR3DS1$^+$ NK cells degranulated more potently in response to HIV-infected Bw4-80I$^+$ CD4$^+$ T cells and suppressed viral replication in a Bw4-80I-dependent manner (Alter et al. 2007a). Additionally, these KIR3DS1$^+$ NK cells expanded robustly following acute HIV infection (Alter et al. 2009), but only in subjects that coexpressed Bw4-80I, further suggesting that KIR3DS1 may receive proliferative signals from its putative ligand early on following infection, allowing NK cells expressing this receptor to expand robustly to help contain early viral replication. Moreover, NK cells derived from individuals that encoded for KIR3DS1 responded more potently to HLA-class I negative target cells than NK cells from KIR3DS1^{neg} subjects (Long et al. 2008). Although KIR3DS1 alone was sufficient to confer elevated NK-cell responsiveness to class I devoid targets, NK-cell responses were strongest among individuals that coexpressed KIR3DS1 and Bw4-80I (Long et al. 2008). Finally, elevated KIR3DS1 transcripts were identified in persistently negative but highly exposed individuals, suggesting that KIR3DS1 may also be involved in protection from infection (Ravet et al. 2007). Taken together, these epidemiological and functional data support a role for KIR3DS1$^+$ NK cells in restricting HIV infection in a "specific" manner in individuals that coexpress its putative ligand Bw4-80I.

Although a physical interaction has yet to be observed between KIR3DS1 and Bw4-80I, epidemiological and functional evidence strongly support that these two molecules are likely to interact either directly or indirectly to activate NK cells during HIV infection. Several potential scenarios may underlie this enigmatic interaction, including the possibility that a viral or stress peptide generated during infection presented by Bw4-80I may alter the affinity of the activating 3DS1 for its putative ligand (Fig. 2). Although data exist demonstrating that amino acid variation within a peptide, particularly at positions 7 and 8, can dramatically alter inhibitory KIR recognition of MHC class I complexes on a target cell, little is known about the particular changes in the MHC class I bound peptide that may alter activating KIR binding and activation. However, recent data now suggest that KIR3DS1 may in fact recognize discrete amino acids within the HIV proteome, as distinct footprints have now been identified that emerge preferentially in individuals that express this activating KIR (G Alter, unpubl.). Like the escape mutations that emerge in CD8$^+$ T-cell-restricted epitopes, it is plausible that KIR-associated footprints may also reflect NK-restricted antiviral pressure. Alternatively, data from the murine model of Ly49p-mediated protection in murine cytomegalovirus (MCMV) infection suggest that instead of a peptide, the activating Ly49p NK-cell receptor interacts with its putative ligand, H2-Dk, only in the presence of a third, undefined protein (Lee et al. 2001).

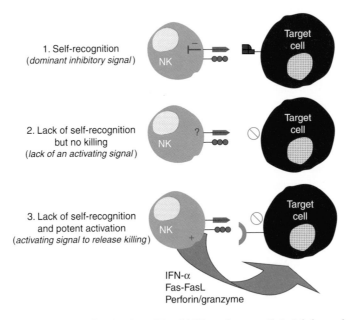

Figure 2. A model of two-step NK-cell activation. NK-cell killing of target cells is tightly regulated by a balance of activating and inhibitory signals delivered through the arsenal of NK-cell receptors expressed on the surface of a given NK-cell clone. NK cells survey the body's cells for normal MHC class-I expression, delivering a potent inhibitory signal to NK cells through inhibitory KIRs (1). Although the missing self-hypothesis states that the loss of MHC class I should trigger NK-cell killing of a target cell, this loss of inhibition is not sufficient to release the cytolytic activity of NK cells (2). Instead, an activating signal (including a stress ligand), to tip the balance toward activation, releases the full cytolytic power of a given NK-cell clone.

Overall, these data suggest that the affinity of 3DS1 for Bw4-80I may be altered during HIV infection, either by a stress/viral peptide or coactivating protein, resulting in potent NK-cell activation. The attraction of the latter possibility is that it implies nonspecificity of 3DS1 for HIV, which is what is expected for these innate immune receptors.

3DL1-MEDIATED CONTROL OF HIV

In addition to 3DS1, epidemiological studies later showed that additional inhibitory allotypes of 3DL1 are also associated with slower HIV disease progression (Martin et al. 2007). Distinct 3DL1 allotypes are expressed at variable levels on the surface of NK cells (Yawata et al. 2006), resulting in differing NK-cell functional potencies. Among the 3DL1 allotypes, three subclassifications have been defined: (1) high-expressing alleles that are associated with potent NK-cell effector functions in the presence of MHC-devoid target cells (3DL1*001, *002, *005, *008, *015, and *020), (2) low-expressing alleles that are associated with weaker NK-cell responsiveness to the same target cells (3DL1*005, *007, and *009), and (3) a nonexpressing allotype with unknown functional properties (3DL1*004). Interestingly, 3DL1 alleles expressed at high levels or not expressed at all were associated with slower HIV-1 disease progression, when coexpressed with Bw4-80I alleles (Martin et al. 2007).

Although the role of the nonexpressed 3DL1*004 allele remains an enigma, an explanation has been proposed for the high-expressed 3DL1 alleles. In 2005, a breakthrough was achieved in our understanding of the influence of KIR on NK cell function. In addition to the role of inhibitory KIR in monitoring for normal expression of MHC class I on the surface of cells ("missing self" hypothesis), a series of reports indicated that both Ly49 in mice and KIR in humans regulate NK-cell function by recognition of self-MHC class I providing signals for functional competence of the NK cell during development, a process called "licensing" (Fernandez et al. 2005; Kim et al. 2005; Anfossi et al. 2006; Kim et al. 2008). These studies

suggest that NK cells undergo a self-MHC class I-dependent maturation process that delivers a positive signal resulting in the ability of NK cells to distinguish self from autologous target cells that have lost MHC class I (Kim et al. 2005; Anfossi et al. 2006). This model helped explain the fraction of NK cells in the periphery that are hyporesponsive, which are the subgroup of NK cells that lack inhibitory KIR for self, and are not educated to respond against aberrant targets (Anfossi et al. 2006). Additionally, more detailed models termed "arming" or "tuning" helped to refine the licensing model, taking into account the balance between activating MHC-binding receptors that are sometimes expressed in the absence or lower levels of inhibitory self-binding receptors. In these models, the investigators proposed that the presence of a dominant inhibitory signal during development helps to "arm" an NK cell, whereas lack of inhibition and/or excessive activation leads to disarming (Fernandez et al. 2005) or tuning (Salcedo et al. 1998) of NK-cell responsiveness, resulting in the accumulation of a subset of hyporesponsive cells.

Thus, KIR3DL1 protection may be related to NK-cell education, where higher expression of KIR3DL1 on a developing NK cell in the presence of its ligand may result in the generation of a larger pool of functionally competent cytolytic cells, which on infection may respond more aggressively (Fig. 3). This possibility relates to the "missing self" hypothesis in that cells expressing higher levels of 3DL1 are expected to require a greater number of KIR/MHC interactions to inhibit such a cell, so they may be more sensitive to small losses of MHC class I following infection, responding vigorously to the target.

A POTENTIAL ROLE FOR TWO-DOMAIN KIRs IN CONTROL OF HIV?

GWASs in large cohorts of HIV-infected individuals identified a number of single nucleotide polymorphisms (SNPs) associated with slower HIV disease progression, all of which mapped to a single region of the human genome on chromosome 6 located within the MHC.

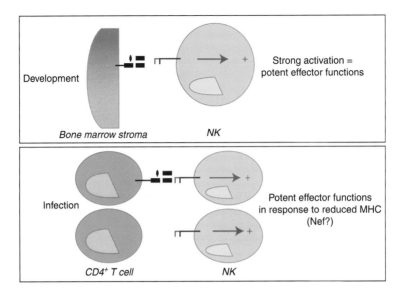

Figure 3. A model of KIR3DL1$^+$ NK-cell recognition of an HIV-infected target cell. Given that inhibitory KIRs have been recently implicated in NK-cell education, in such a way that inhibitory KIRs expressed at higher levels are associated with the generation of more functional NK-cell clones, it is possible that the expression of KIR3DL1 at higher levels on a developing NK cell in the presence of its ligand may result in the generation of a larger pool of functionally competent cytolytic cells. These more functionally competent cells may then respond more aggressively on HIV infection to cells that have lost MHC class I ligands, that are down-regulated by the HIV Nef protein.

Among these SNPs, the GWASs confirmed previous epidemiological data demonstrating a protective role for the HLA-B allele B*57 (Carrington and O'Brien 2003), but also identified a number of additional SNPs, including one located 35 kb upstream of HLA-C (Thomas et al. 2009). The protective variant is associated with increased HLA-C expression on the surface of CD3$^+$ T cells (Fellay et al. 2007; Thomas et al. 2009). Interestingly, the protective effect of this SNP could not be assigned to a specific HLA-C allele or phylogenetically related subgroup (Thomas et al. 2009), suggesting a potential non-CD8-dependent protective mechanism. As HLA-C alleles serve as ligands for KIR2D receptors (Vitale et al. 1995; Stewart et al. 2005), several groups have now begun to speculate that this protective effect in HIV infection is NK-cell-dependent through the interaction of KIR2D with its ligand. Based on the NK-cell education models, HLA-C alleles expressed at higher levels on the surface of a cell during development may generate more potent cytolytic NK

cells (Kim et al. 2008; Brodin et al. 2009), but this possibility remains to be answered.

KIRs DRIVE VIRAL EVOLUTION

Most recently, efforts to define the mechanism by which NK cells may contribute to HIV viral control have sought to determine whether NK cells may recognize and place pressure on the virus directly in vivo. Historically, the identification of "footprints," amino acid substitutions in the viral proteome that accumulate specifically in the presence of specific HLA-class I alleles, have been regarded as a marker of CD8$^+$ T-cell pressure (Allen et al. 2000). Likewise, recent data have shown that similar footprints arise in the HIV proteome in the presence of distinct KIR genes (Alter et al. 2011). These data suggest that like T cells, NK cells may also recognize specific regions of the HIV virus, placing pressure on the virus.

How can KIRs see specific regions of the HIV proteome? Several lines of evidence suggest that affinity changes between KIRs and histocompatibility leukocyte antigen (HLA) class I may be induced by the peptide bound in the MHC class I binding groove. Crystal structures of KIR/MHC class I complexes show that KIR interacts with the α1 and α2 helix of MHC class I and makes direct contact with the carboxy-terminal portion of the bound peptide (Boyington et al. 2000; Fan et al. 2001). The impact of the bound peptide on KIR/MHC interactions has further been examined in a number of studies demonstrating that particular amino acid changes in the peptide, particularly at positions 7 or 8, results in the abrogation of inhibition through KIR, resulting in target cell lysis (Correa and Raulet 1995; Malnati et al. 1995; Peruzzi et al. 1996; Rajagopalan and Long 1997; Zappacosta et al. 1997; Fadda et al. 2010). Thus, it is possible that whereas self-peptides bound to MHC class I provide a strong inhibitory signal to the inhibitory KIR, particular viral peptides produced during infection may bind differentially to KIR, whereby decreased binding to an inhibitory KIR may trigger "missing self" NK-cell activation, or increased binding to an activating KIR may activate NK-cell cytotoxicity. This direct KIR-mediated antiviral pressure may drive the virus to incorporate "escape mutations" aimed at evading this form of innate recognition (Alter et al. 2011). However, the overall impact of this specific innate immune response has yet to be defined.

CONCLUSIONS

Over the past two decades, significant advances have been achieved in our basic understanding of the role of innate immunity in the control of viral infections. Moreover, we have come to appreciate that this arm of the immune response may directly contribute to antiviral control but may also play a significant role in modulating the quality of the ensuing adaptive immune response. In the context of HIV infection, mounting epidemiologic data strongly implicate a role for NK cells in antiviral control, underscored by the fact that these innate immune cells expand robustly in response to TLR-induced DC-secreted cytokines and have

now been shown to specifically place pressure on HIV in vivo. The failure of recent HIV-1 vaccine trials to induce protective immunity in humans has highlighted our lack of understanding of the correlates of immune protection in HIV-1 infection. Therefore, new therapeutic strategies aimed at harnessing the power of the innate immune response, and particular NK cells, may provide a new approach aimed at enhancing the quality of immune control induced via vaccination.

ACKNOWLEDGMENTS

This project was funded, in whole or in part, by the National Cancer Institute, National Institutes of Health (NIH), contract no. HHSN261200800001E. The content of this publication does not necessarily reflect the views or policies of the Department of Health and Human Services, nor does mention of trade names, commercial products, or organizations imply endorsement by the U.S. government. This research was also supported, in part, by the Intramural Research Program of the NIH, National Cancer Institute, Center for Cancer Research.

REFERENCES

Allen TM, O'Connor DH, Jing P, Dzuris JL, Mothe BR, Vogel TU, Dunphy E, Liebl ME, Emerson C, Wilson N, et al. 2000. Tat-specific cytotoxic T lymphocytes select for SIV escape variants during resolution of primary viraemia. *Nature* **407:** 386–390.

Alter G, Martin MP, Teigen N, Carr WH, Suscovich TJ, Schneidewind A, Streeck H, Waring M, Meier A, Brander C, et al. 2007a. Differential natural killer cell-mediated inhibition of HIV-1 replication based on distinct KIR/HLA subtypes. *J Exp Med* **204:** 3027–3036.

Alter G, Teigen N, Ahern R, Streeck H, Meier A, Rosenberg ES, Altfeld M. 2007b. Evolution of innate and adaptive effector cell functions during acute HIV-1 infection. *J Infect Dis* **195:** 1452–1460.

Alter G, Rihn S, Walter K, Nolting A, Martin M, Rosenberg ES, Miller JS, Carrington M, Altfeld M. 2009. HLA class I subtype-dependent expansion of KIR3DS1+ and KIR3DL1+ NK cells during acute human immunodeficiency virus type 1 infection. *J Virol* **83:** 6798–6805.

Alter G, Heckerman D, Schneidewind A, Fadda L, Kadie CM, Carlson JM, Oniangue-Ndza C, Martin M, Li B, Khakoo SI, et al. 2011. HIV-1 adaptation to NK-cell-mediated immune pressure. *Nature* **476:** 96–100.

Anfossi N, Andre P, Guia S, Falk CS, Roetynck S, Stewart CA, Breso V, Frassati C, Reviron D, Middleton D, et al. 2006. Human NK cell education by inhibitory receptors for MHC class I. *Immunity* **25**: 331–342.

Arase H, Mocarski ES, Campbell AE, Hill AB, Lanier LL. 2002. Direct recognition of cytomegalovirus by activating and inhibitory NK cell receptors. *Science* **296**: 1323–1326.

Barber LD, Percival L, Arnett KL, Gumperz JE, Chen L, Parham P. 1997. Polymorphism in the α 1 helix of the HLA-B heavy chain can have an overriding influence on peptide-binding specificity. *J Immunol* **158**: 1660–1669.

Bashirova AA, Martin MP, McVicar DW, Carrington M. 2006. The killer immunoglobulin-like receptor gene cluster: Tuning the genome for defense. *Annu Rev Genomics Hum Genet* **7**: 277–300.

Beignon AS, McKenna K, Skoberne M, Manches O, Dasilva I, Kavanagh DG, Larsson M, Gorelick RJ, Lifson JD, Bhardwaj N. 2005. Endocytosis of HIV-1 activates plasmacytoid dendritic cells via Toll-like receptor-viral RNA interactions. *J Clin Invest* **115**: 3265–3275.

Biron CA. 1999. Initial and innate responses to viral infections—Pattern setting in immunity or disease. *Curr Opin Microbiol* **2**: 374–381.

Boyington JC, Motyka SA, Schuck P, Brooks AG, Sun PD. 2000. Crystal structure of an NK cell immunoglobulin-like receptor in complex with its class I MHC ligand. *Nature* **405**: 537–543.

Brennan J, Mager D, Jefferies W, Takei F. 1994. Expression of different members of the Ly-49 gene family defines distinct natural killer cell subsets and cell adhesion properties. *J Exp Med* **180**: 2287–2295.

Brodin P, Lakshmikanth T, Johansson S, Karre K, Hoglund P. 2009. The strength of inhibitory input during education quantitatively tunes the functional responsiveness of individual natural killer cells. *Blood* **113**: 2434–2441.

Carrington M, Norman P. 2003. *The KIR gene cluster*. NCBI, Bethesda, MD.

Carrington M, O'Brien SJ. 2003. The influence of HLA genotype on AIDS. *Annu Rev Med* **54**: 535–551.

Cella M, Longo A, Ferrara GB, Strominger JL, Colonna M. 1994. NK3-specific natural killer cells are selectively inhibited by Bw4-positive HLA alleles with isoleucine 80. *J Exp Med* **180**: 1235–1242.

Cerboni C, Neri F, Casartelli N, Zingoni A, Cosman D, Rossi P, Santoni A, Doria M. 2007. Human immunodeficiency virus 1 Nef protein downmodulates the ligands of the activating receptor NKG2D and inhibits natural killer cell-mediated cytotoxicity. *J Gen Virol* **88**: 242–250.

Cerwenka A, Lanier LL. 2001a. Ligands for natural killer cell receptors: Redundancy or specificity. *Immunol Rev* **181**: 158–169.

Cerwenka A, Lanier LL. 2001b. Natural killer cells, viruses and cancer. *Nat Rev Immunol* **1**: 41–49.

Ciccone E, Pende D, Viale O, Di Donato C, Tripodi G, Orengo AM, Guardiola J, Moretta A, Moretta L. 1992. Evidence of a natural killer (NK) cell repertoire for (allo) antigen recognition: Definition of five distinct NK-determined allospecificities in humans. *J Exp Med* **175**: 709–718.

Cohen GB, Gandhi RT, Davis DM, Mandelboim O, Chen BK, Strominger JL, Baltimore D. 1999. The selective downregulation of class I major histocompatibility complex proteins by HIV-1 protects HIV-infected cells from NK cells. *Immunity* **10**: 661–671.

Correa I, Raulet DH. 1995. Binding of diverse peptides to MHC class I molecules inhibits target cell lysis by activated natural killer cells. *Immunity* **2**: 61–71.

Diebold SS, Kaisho T, Hemmi H, Akira S, Reis e Sousa C. 2004. Innate antiviral responses by means of TLR7-mediated recognition of single-stranded RNA. *Science* **303**: 1529–1531.

Dorfman JR, Raulet DH. 1996. Major histocompatibility complex genes determine natural killer cell tolerance. *Eur J Immunol* **26**: 151–155.

Dorfman JR, Raulet DH. 1998. Acquisition of Ly49 receptor expression by developing natural killer cells. *J Exp Med* **187**: 609–618.

Fadda L, Borhis G, Ahmed P, Cheent K, Pageon SV, Cazaly A, Stathopoulos S, Middleton D, Mulder A, Claas FH, et al. 2010. Peptide antagonism as a mechanism for NK cell activation. *Proc Natl Acad Sci* **107**: 10160–10165.

Fan QR, Long EO, Wiley DC. 2001. Crystal structure of the human natural killer cell inhibitory receptor KIR2DL1-HLA-Cw4 complex. *Nat Immunol* **2**: 452–460.

Fellay J, Shianna KV, Ge D, Colombo S, Ledergerber B, Weale M, Zhang K, Gumbs C, Castagna A, Cossarizza A, et al. 2007. A whole-genome association study of major determinants for host control of HIV-1. *Science* **317**: 944–947.

Fernandez NC, Treiner E, Vance RE, Jamieson AM, Lemieux S, Raulet DH. 2005. A subset of natural killer cells achieves self-tolerance without expressing inhibitory receptors specific for self-MHC molecules. *Blood* **105**: 4416–4423.

Gardiner CM, Guethlein LA, Shilling HG, Pando M, Carr WH, Rajalingam R, Vilches C, Parham P. 2001. Different NK cell surface phenotypes defined by the DX9 antibody are due to KIR3DL1 gene polymorphism. *J Immunol* **166**: 2992–3001.

Hanke T, Raulet DH. 2001. Cumulative inhibition of NK cells and T cells resulting from engagement of multiple inhibitory Ly49 receptors. *J Immunol* **166**: 3002–3007.

Hanke T, Takizawa H, McMahon CW, Busch DH, Pamer EG, Miller JD, Altman JD, Liu Y, Cado D, Lemonnier FA, et al. 1999. Direct assessment of MHC class I binding by seven Ly49 inhibitory NK cell receptors. *Immunity* **11**: 67–77.

Hanke T, Takizawa H, Raulet DH. 2001. MHC-dependent shaping of the inhibitory Ly49 receptor repertoire on NK cells: Evidence for a regulated sequential model. *Eur J Immunol* **31**: 3370–3379.

Heil F, Hemmi H, Hochrein H, Ampenberger F, Kirschning C, Akira S, Lipford G, Wagner H, Bauer S. 2004. Species-specific recognition of single-stranded RNA via toll-like receptor 7 and 8. *Science* **303**: 1526–1529.

Held W, Dorfman JR, Wu MF, Raulet DH. 1996. Major histocompatibility complex class I-dependent skewing of the natural killer cell Ly49 receptor repertoire. *Eur J Immunol* **26**: 2286–2292.

Hsu KC, Liu XR, Selvakumar A, Mickelson E, O'Reilly RJ, Dupont B. 2002. Killer Ig-like receptor haplotype analysis by gene content: Evidence for genomic diversity with a minimum of six basic framework haplotypes, each with multiple subsets. *J Immunol* **169:** 5118–5129.

Jennes W, Verheyden S, Demanet C, Adje-Toure CA, Vuylsteke B, Nkengasong JN, Kestens L. 2006. Cutting edge: Resistance to HIV-1 infection among African female sex workers is associated with inhibitory KIR in the absence of their HLA ligands. *J Immunol* **177:** 6588–6592.

Karre K. 2002. NK cells, MHC class I molecules and the missing self. *Scand J Immunol* **55:** 221–228.

Karre K, Ljunggren HG, Piontek G, Kiessling R. 1986. Selective rejection of H-2-deficient lymphoma variants suggests alternative immune defence strategy. *Nature* **319:** 675–678.

Kim S, Poursine-Laurent J, Truscott SM, Lybarger L, Song YJ, Yang L, French AR, Sunwoo JB, Lemieux S, Hansen TH, et al. 2005. Licensing of natural killer cells by host major histocompatibility complex class I molecules. *Nature* **436:** 709–713.

Kim S, Sunwoo JB, Yang L, Choi T, Song YJ, French AR, Vlahiotis A, Piccirillo JF, Cella M, Colonna M, et al. 2008. HLA alleles determine differences in human natural killer cell responsiveness and potency. *Proc Natl Acad Sci* **105:** 3053–3058.

Lanier LL. 1998. NK cell receptors. *Annu Rev Immunol* **16:** 359–393.

Lanier LL, Corliss BC, Wu J, Leong C, Phillips JH. 1998. Immunoreceptor DAP12 bearing a tyrosine-based activation motif is involved in activating NK cells. *Nature* **391:** 703–707.

Lee SH, Girard S, Macina D, Busa M, Zafer A, Belouchi A, Gros P, Vidal SM. 2001. Susceptibility to mouse cytomegalovirus is associated with deletion of an activating natural killer cell receptor of the C-type lectin superfamily. *Nat Genet* **28:** 42–45.

Le Gall S, Erdtmann L, Benichou S, Berlioz-Torrent C, Liu L, Benarous R, Heard JM, Schwartz O. 1998. Nef interacts with the mu subunit of clathrin adaptor complexes and reveals a cryptic sorting signal in MHC I molecules. *Immunity* **8:** 483–495.

Ljunggren HG, Karre K. 1990. In search of the "missing self": MHC molecules and NK cell recognition. *Immunol Today* **11:** 237–244.

Ljunggren HG, Van Kaer L, Ploegh HL, Tonegawa S. 1994. Altered natural killer cell repertoire in Tap-1 mutant mice. *Proc Natl Acad Sci* **91:** 6520–6524.

Lodoen MB, Lanier LL. 2005. Viral modulation of NK cell immunity. *Nat Rev Microbiol* **3:** 59–69.

Long BR, Ndhlovu LC, Oksenberg JR, Lanier LL, Hecht FM, Nixon DF, Barbour JD. 2008. Conferral of enhanced natural killer cell function by KIR3DS1 in early human immunodeficiency virus type 1 infection. *J Virol* **82:** 4785–4792.

Maenaka K, Juji T, Nakayama T, Wyer JR, Gao GF, Maenaka T, Zaccai NR, Kikuchi A, Yabe T, Tokunaga K, et al. 1999. Killer cell immunoglobulin receptors and T cell receptors bind peptide-major histocompatibility complex class I with distinct thermodynamic and kinetic properties. *J Biol Chem* **274:** 28329–28334.

Malnati MS, Peruzzi M, Parker KC, Biddison WE, Ciccone E, Moretta A, Long EO. 1995. Peptide specificity in the

recognition of MHC class I by natural killer cell clones. *Science* **267:** 1016–1018.

Martin MP, Gao X, Lee JH, Nelson GW, Detels R, Goedert JJ, Buchbinder S, Hoots K, Vlahov D, Trowsdale J, et al. 2002. Epistatic interaction between KIR3DS1 and HLA-B delays the progression to AIDS. *Nat Genet* **31:** 429–434.

Martin MP, Bashirova A, Traherne J, Trowsdale J, Carrington M. 2003. Cutting edge: Expansion of the KIR locus by unequal crossing over. *J Immunol* **171:** 2192–2195.

Martin MP, Qi Y, Gao X, Yamada E, Martin JN, Pereyra F, Colombo S, Brown EE, Shupert WL, Phair J, et al. 2007. Innate partnership of HLA-B and KIR3DL1 subtypes against HIV-1. *Nat Genet* **39:** 733–740.

Martinelli E, Cicala C, Van Ryk D, Goode DJ, Macleod K, Arthos J, Fauci AS. 2007. HIV-1 gp120 inhibits TLR9-mediated activation and IFN-α secretion in plasmacytoid dendritic cells. *Proc Natl Acad Sci* **104:** 3396–3401.

Maxwell LD, Wallace A, Middleton D, Curran MD. 2002. A common KIR2DS4 deletion variant in the human that predicts a soluble KIR molecule analogous to the KIR1D molecule observed in the rhesus monkey. *Tissue Antigens* **60:** 254–258.

Meier A, Chang JJ, Chan ES, Pollard RB, Sidhu HK, Kulkarni S, Wen TF, Lindsay RJ, Orellana L, Mildvan D, et al. 2009. Sex differences in the Toll-like receptor-mediated response of plasmacytoid dendritic cells to HIV-1. *Nat Med* **15:** 955–959.

Moesta AK, Abi-Rached L, Norman PJ, Parham P. 2009. Chimpanzees use more varied receptors and ligands than humans for inhibitory killer cell Ig-like receptor recognition of the MHC-C1 and MHC-C2 epitopes. *J Immunol* **182:** 3628–3637.

Moesta AK, Graef T, Abi-Rached L, Older Aguilar AM, Guethlein LA, Parham P. 2010. Humans differ from other hominids in lacking an activating NK cell receptor that recognizes the C1 epitope of MHC class I. *J Immunol* **185:** 4233–4237.

Moretta A, Vitale M, Bottino C, Orengo AM, Morelli L, Augugliaro R, Barbaresi M, Ciccone E, Moretta L. 1993. P58 molecules as putative receptors for major histocompatibility complex (MHC) class I molecules in human natural killer (NK) cells. Anti-p58 antibodies reconstitute lysis of MHC class I-protected cells in NK clones displaying different specificities. *J Exp Med* **178:** 597–604.

Moretta A, Bottino C, Mingari MC, Biassoni R, Moretta L. 2002. What is a natural killer cell? *Nat Immunol* **3:** 6–8.

Murphy K, Travers P, Walport M. 2011. *Janeway's immunobiology,* 8th ed. Garland Science, New York.

Norman PJ, Stephens HA, Verity DH, Chandanayingyong D, Vaughan RW. 2001. Distribution of natural killer cell immunoglobulin-like receptor sequences in three ethnic groups. *Immunogenetics* **52:** 195–205.

Pantaleo G, Demarest JF, Schacker T, Vaccarezza M, Cohen OJ, Daucher M, Graziosi C, Schnittman SS, Quinn TC, Shaw GM, et al. 1997. The qualitative nature of the primary immune response to HIV infection is a prognosticator of disease progression independent of the initial level of plasma viremia. *Proc Natl Acad Sci* **94:** 254–258.

Peruzzi M, Parker KC, Long EO, Malnati MS. 1996. Peptide sequence requirements for the recognition of HLA-

B*2705 by specific natural killer cells. *J Immunol* **157**: 3350–3356.

Pereyra F, Jia X, McLaren PJ, Telenti A, de Bakker PI, Walker BD, Ripke S, Brumme CJ, Pulit SL, Carrington M, et al. 2010. The major genetic determinants of HIV-1 control affect HLA class I peptide presentation. *Science* **330**: 1551–1557.

Rajagopalan S, Long EO. 1997. The direct binding of a p58 killer cell inhibitory receptor to human histocompatibility leukocyte antigen (HLA)-Cw4 exhibits peptide selectivity. *J Exp Med* **185**: 1523–1528.

Raulet DH. 2003. Roles of the NKG2D immunoreceptor and its ligands. *Nat Rev Immunol* **3**: 781–790.

Raulet DH, Held W, Correa I, Dorfman JR, Wu MF, Corral L. 1997. Specificity, tolerance and developmental regulation of natural killer cells defined by expression of class I-specific Ly49 receptors. *Immunol Rev* **155**: 41–52.

Ravet S, Scott-Algara D, Bonnet E, Tran HK, Tran T, Nguyen N, Truong LX, Theodorou I, Barre-Sinoussi F, Pancino G, et al. 2007. Distinctive NK-cell receptor repertoires sustain high-level constitutive NK-cell activation in HIV-exposed uninfected individuals. *Blood* **109**: 4296–4305.

Salcedo M, Andersson M, Lemieux S, Van Kaer L, Chambers BJ, Ljunggren HG. 1998. Fine tuning of natural killer cell specificity and maintenance of self tolerance in MHC class I-deficient mice. *Eur J Immunol* **28**: 1315–1321.

Schwartz O, Marechal V, Le Gall S, Lemonnier F, Heard JM. 1996. Endocytosis of major histocompatibility complex class I molecules is induced by the HIV-1 Nef protein. *Nat Med* **2**: 338–342.

Shilling HG, Guethlein LA, Cheng NW, Gardiner CM, Rodriguez R, Tyan D, Parham P. 2002. Allelic polymorphism synergizes with variable gene content to individualize human KIR genotype. *J Immunol* **168**: 2307–2315.

Stacey AR, Norris PJ, Qin L, Haygreen EA, Taylor E, Heitman J, Lebedeva M, DeCamp A, Li D, Grove D, et al. 2009. Induction of a striking systemic cytokine cascade prior to peak viremia in acute human immunodeficiency virus type 1 infection, in contrast to more modest and delayed responses in acute hepatitis B and C virus infections. *J Virol* **83**: 3719–3733.

Stewart CA, Laugier-Anfossi F, Vely F, Saulquin X, Riedmuller J, Tisserant A, Gauthier L, Romagne F, Ferracci G, Arosa FA, et al. 2005. Recognition of peptide-MHC class I complexes by activating killer immunoglobulin-like receptors. *Proc Natl Acad Sci* **102**: 13224–13229.

Storkus WJ, Alexander J, Payne JA, Cresswell P, Dawson JR. 1989a. The α1/α2 domains of class I HLA molecules confer resistance to natural killing. *J Immunol* **143**: 3853–3857.

Storkus WJ, Alexander J, Payne JA, Dawson JR, Cresswell P. 1989b. Reversal of natural killing susceptibility in target cells expressing transfected class I HLA genes. *Proc Natl Acad Sci* **86**: 2361–2364.

Thibault S, Fromentin R, Tardif MR, Tremblay MJ. 2009. TLR2 and TLR4 triggering exerts contrasting effects with regard to HIV-1 infection of human dendritic cells and subsequent virus transfer to CD4⁺ T cells. *Retrovirology* **6**: 42.

Thomas R, Apps R, Qi Y, Gao X, Male V, O'hUigin C, O'Connor G, Ge D, Fellay J, Martin JN, et al. 2009. HLA-C cell surface expression and control of HIV/AIDS correlate with a variant upstream of HLA-C. *Nat Genet* **41**: 1290–1294.

Uhrberg M, Valiante NM, Shum BP, Shilling HG, Lienert-Weidenbach K, Corliss B, Tyan D, Lanier LL, Parham P. 1997. Human diversity in killer cell inhibitory receptor genes. *Immunity* **7**: 753–763.

Vales-Gomez M, Reyburn H, Strominger J. 2000. Interaction between the human NK receptors and their ligands. *Crit Rev Immunol* **20**: 223–244.

Valiante NM, Uhrberg M, Shilling HG, Lienert-Weidenbach K, Arnett KL, D'Andrea A, Phillips JH, Lanier LL, Parham P. 1997. Functionally and structurally distinct NK cell receptor repertoires in the peripheral blood of two human donors. *Immunity* **7**: 739–751.

Vilches C, Parham P. 2002. KIR: Diverse, rapidly evolving receptors of innate and adaptive immunity. *Annu Rev Immunol* **20**: 217–251.

Vitale M, Sivori S, Pende D, Moretta L, Moretta A. 1995. Coexpression of two functionally independent p58 inhibitory receptors in human natural killer cell clones results in the inability to kill all normal allogeneic target cells. *Proc Natl Acad Sci* **92**: 3536–3540.

Wende H, Colonna M, Ziegler A, Volz A. 1999. Organization of the leukocyte receptor cluster (LRC) on human chromosome 19q13.4. *Mamm Genome* **10**: 154–160.

Wilson MJ, Torkar M, Haude A, Milne S, Jones T, Sheer D, Beck S, Trowsdale J. 2000. Plasticity in the organization and sequences of human KIR/ILT gene families. *Proc Natl Acad Sci* **97**: 4778–4783.

Witt CS, Dewing C, Sayer DC, Uhrberg M, Parham P, Christiansen FT. 1999. Population frequencies and putative haplotypes of the killer cell immunoglobulin-like receptor sequences and evidence for recombination. *Transplantation* **68**: 1784–1789.

Yawata M, Yawata N, Draghi M, Little AM, Partheniou F, Parham P. 2006. Roles for HLA and KIR polymorphisms in natural killer cell repertoire selection and modulation of effector function. *J Exp Med* **203**: 633–645.

Yokoyama WM. 2002. The search for the missing "missing-self" receptor on natural killer cells. *Scand J Immunol* **55**: 233–237.

Yokoyama WM, Kehn PJ, Cohen DI, Shevach EM. 1990. Chromosomal location of the Ly-49 (A1, YE1/48) multigene family. Genetic association with the NK 1.1 antigen. *J Immunol* **145**: 2353–2358.

Yu J, Heller G, Chewning J, Kim S, Yokoyama WM, Hsu KC. 2007. Hierarchy of the human natural killer cell response is determined by class and quantity of inhibitory receptors for self-HLA-B and HLA-C ligands. *J Immunol* **179**: 5977–5989.

Zappacosta F, Borrego F, Brooks AG, Parker KC, Coligan JE. 1997. Peptides isolated from HLA-Cw*0304 confer different degrees of protection from natural killer cell-mediated lysis. *Proc Natl Acad Sci* **94**: 6313–6318.

Zhang C, Zhang J, Wei H, Tian Z. 2005. Imbalance of NKG2D and its inhibitory counterparts: How does tumor escape from innate immunity? *Int Immunopharmacol* **5**: 1099–1111.

HIV Latency

Robert F. Siliciano[1] and Warner C. Greene[2]

[1]Department of Medicine, Johns Hopkins University School of Medicine, Howard Hughes Medical Institute, Baltimore, Maryland 21205

[2]Gladstone Institute of Virology and Immunology, Department of Medicine, Microbiology and Immunology, University of California, San Francisco, California 94158

Correspondence: rsiliciano@jhmi.edu; wgreene@gladstone.ucsf.edu

HIV-1 can establish a state of latent infection at the level of individual T cells. Latently infected cells are rare in vivo and appear to arise when activated CD4$^+$ T cells, the major targets cells for HIV-1, become infected and survive long enough to revert back to a resting memory state, which is nonpermissive for viral gene expression. Because latent virus resides in memory T cells, it persists indefinitely even in patients on potent antiretroviral therapy. This latent reservoir is recognized as a major barrier to curing HIV-1 infection. The molecular mechanisms of latency are complex and include the absence in resting CD4$^+$ T cells of nuclear forms of key host transcription factors (e.g., NFκB and NFAT), the absence of Tat and associated host factors that promote efficient transcriptional elongation, epigenetic changes inhibiting HIV-1 gene expression, and transcriptional interference. The presence of a latent reservoir for HIV-1 helps explain the presence of very low levels of viremia in patients on antiretroviral therapy. These viruses are released from latently infected cells that have become activated and perhaps from other stable reservoirs but are blocked from additional rounds of replication by the drugs. Several approaches are under exploration for reactivating latent virus with the hope that this will allow elimination of the latent reservoir.

HISTORY AND DEFINITIONS

Viral latency is a state of reversibly nonproductive infection of individual cells. For some viruses, notably select members of the Herpes virus family, latency provides an important mechanism for viral persistence and escape from immune recognition (Perng and Jones 2010). Several Herpes viruses have elaborate genetic programs that allow persistence of viral genomes with minimal viral gene expression. For retroviruses, stable integration of reverse transcribed viral cDNA into host cell chromosomes is an essential step in the life cycle that allows persistence of viral genomes for the lifespan of infected cells. Some retroviruses also establish a state of latent infection. In early studies of sheep infected with the visna lentivirus, a restricted pattern of gene expression from proviral DNA was observed that suggested latency (Brahic et al. 1981). For HIV-1, the term latency was initially used in the clinical sense to describe the long asymptomatic period between initial infection and the development of AIDS. However, with the advent of sensitive RT-PCR assays for viremia (Piatak et al. 1993), it became clear that HIV-1 replicates actively throughout the course of the infection, even during the asymptomatic period. The major mechanism by which HIV-1 evades immune responses is not latency but rather through rapid evolution of escape mutations that abrogate recognition by neutralizing antibodies and cytolytic T lymphocytes (Bailey et al. 2004). Nevertheless, it has become clear that HIV-1 can establish a state of latent infection at the level of individual T cells.

Initial evidence suggesting HIV-1 latency came from in vitro infections of transformed cells lines. Surviving infected cells showed low or absent HIV-1 gene expression that could be up-regulated by various stimuli, including those causing T-cell activation (Folks et al. 1986). Molecular studies had established that transcription from the HIV-1 LTR was stimulated by inducible host transcription factors, such as NFκB. This factor rapidly translocates into the nucleus in response to T-cell-activating stimuli because of stimulus-coupled degradation of its cytoplasmic inhibitor, IκBα (Nabel and Baltimore 1987; Siekevitz et al. 1987; Bohnlein et al. 1988; Duh et al. 1989). Thus, a connection between latent infection and resting but not activated CD4$^+$ T cells seemed quite likely.

Interestingly, HIV-1 does not efficiently establish productive infection in resting CD4$^+$ T cells (Zack et al. 1990; Bukrinsky et al. 1991; Zhou et al. 2005). How then can a stable state of latent infection develop in these resting cells? One plausible hypothesis is based on the normal physiology of CD4$^+$ T cells (Fig. 1A). In response to antigen, resting CD4$^+$ T cells undergo a burst of cellular proliferation and differentiation, giving rise to effector cells. Most effector cells die quickly, but a subset survives and reverts to a resting G$_0$ state. They persist as memory cells, with an altered pattern of gene expression enabling long-term survival and rapid responses after reexposure to antigen. Activated CD4$^+$ T cells are highly susceptible to HIV-1 infection and typically die quickly as a result of the cytopathic effects of the virus or host immune responses (Ho et al. 1995; Wei

et al. 1995). However, some activated CD4$^+$ T cells may become infected and then survive long enough to revert back to a resting state (Fig. 1B). Because HIV-1 gene expression is dependent on inducible host transcription factors that are only transiently activated after exposure to antigen, HIV-1 gene expression may be extinguished as the cells revert to a resting memory state. The result is a stably integrated but transcriptionally silent form of the virus in a cell whose function it is to survive for long periods of time. With time, additional epigenetic mechanisms discussed below may enforce latency. In the latent state, the virus persists simply as information (in the form of 10 kb of integrated HIV-1 DNA), and it is thus unaffected by antiretroviral drugs or immune responses. However, if the host cell becomes activated by an encounter with antigen or other activating stimuli, latency may be reversed, and the cell may begin to produce a virus. Therefore, by this hypothesis, HIV-1 latency inadvertently exploits the most fundamental characteristic of the immune system, the immunological memory that resides in long-lived resting lymphocytes.

A prediction of this model is that a stable latent form of HIV-1 should reside in the memory subset of resting CD4$^+$ T cells (Fig. 1). Proving that latently infected cells were present in vivo required the development of methods for isolating extremely pure populations of resting CD4$^+$ T cells and demonstrating that within these populations were cells carrying the HIV-1 genome stably integrated into host chromosomes. In addition, it was necessary to show that replication-competent virus could be rescued from resting CD4$^+$ T cells by cellular activation. This was particularly important because the mere presence of viral nucleic acids in resting CD4$^+$ T cells could also be compatible with the presence of defective viruses. In 1995, the presence of integrated HIV-1 DNA in highly purified populations of resting CD4$^+$ T cells was definitively shown by inverse PCR (Chun et al. 1995). Replication-competent virus could be rescued by cellular activation (Chun et al. 1995). This study established that latently infected cells were present in infected

 Cite this article as *Cold Spring Harb Perspect Med* doi: 10.1101/cshperspect.a007096

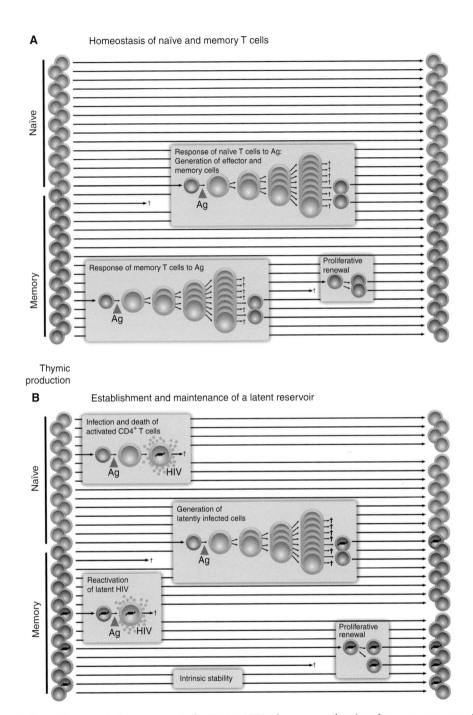

Figure 1. Establishment of a latent reservoir for HIV-1. HIV-1 latency may be viewed as a consequence of the tropism of the virus for activated CD4$^+$ T cells. (*A*) Generation of memory CD4$^+$ T cells. A fraction of the CD4$^+$ T cells that respond to a given antigen survive and revert back to a resting state as long-lived memory T cells. (*B*) Generation of latently infected cells. Latency is established when activated CD4$^+$ T cells become infected and survive long enough to revert back to a resting memory state that is nonpermissive for viral gene expression. The resulting latent reservoir is intrinsically stable because memory cells have a long lifespan and can undergo a process of proliferative renewal through homeostatic proliferation.

individuals. Interestingly, the frequency of latently infected cells was low. A critical further advance was the development of a culture assay that allowed quantitation of cells harboring replication-competent virus (Chun et al. 1997a). In this assay, serial dilutions of purified resting CD4$^+$ T cells are stimulated with a mitogen that induces 100% of the cells to undergo blast transformation. This cellular activation reverses latency, allowing virus production. Released viruses are expanded in CD4$^+$ T lymphoblasts from normal donors that are added to the culture. Virus growth is detected by ELISA of HIV-1 p24 antigen in the supernatant, and frequencies are calculated using limiting dilution statistics. With this approach, resting CD4$^+$ T cells harboring replication-competent virus were detected at low frequency ($1/10^6$) in the blood and lymph nodes of all infected individuals studied. Although several mechanisms have been proposed for the origin of latently infected cells, the finding that latent HIV-1 resides predominantly in the memory subset of resting CD4$^+$ T cells (Chun et al. 1997a; Chomont et al. 2009) is consistent with the hypothesis presented above. This model is further supported by recent studies describing the in vitro production of latently infected resting CD4$^+$ T cells from activated primary CD4$^+$ T cells that are infected and then allowed to return to a quiescent state (reviewed in Yang 2011).

Shortly after the in vivo detection of latently infected cells was first reported, combinations of antiretroviral drugs known as highly active antiretroviral therapy, or HAART, were shown to reduce plasma virus levels to below the limit of detection, raising hopes for viral eradication (Gulick et al. 1997; Hammer et al. 1997; Perelson et al. 1997). The question immediately became whether the pool of latently infected cells would persist in patients who had suppression of viremia to undetectable levels on HAART. Using the culture assay approach described above, three groups simultaneously showed in 1997 that latently infected cells persisted in patients who were responding well to HAART (Chun et al. 1997b; Finzi et al. 1997; Wong et al. 1997). Subsequent longitudinal studies showed that the decay rate of the pool

of latently infected cells was extremely slow, with a half-life of 44 months (Finzi et al. 1999; Siliciano et al. 2003). At this rate, over 70 years of treatment would be required to eradicate the latent reservoir. These studies led to the proposal that the latent reservoir in resting memory CD4$^+$ T cells guarantees lifetime persistence of the virus, even in patients on suppressive HAART regimens (Finzi et al. 1999; Siliciano et al. 2003; Strain et al. 2003). Because of the existence of stable viral reservoirs, a rebound in viremia inevitably occurs within weeks after interruption of therapy (Davey et al. 1999). Although multiple reservoirs may exist, replication-competent HIV-1 can be isolated from resting CD4$^+$ T cells of all patients on HAART, regardless of the duration of treatment, making this reservoir the best-established barrier to HIV-1 eradication.

MOLECULAR MECHANISMS OF HIV-1 LATENCY

As discussed, a leading theory for how HIV-1 latency is initially established involves infection of activated CD4$^+$ T cells as they are returning to a resting state to form long-lived memory T cells (Fig. 1B). Although this transition may give rise to proviral latency, it remains unclear how the latent state is maintained within these cells. Many studies have been performed exploring the mechanistic basis for latency. All of these studies share the ultimate aim of improving our understanding of this process sufficiently to allow a successful rational attack on the latent reservoir.

Chromatin and HIV-1 Latency

One possible mechanism for maintaining the latent state involves epigenetic changes in chromatin that suppress HIV-1 gene expression. Expression of host genes is dynamically regulated through changes in chromatin structure, chiefly reflecting altered states of chromosomal DNA compaction (Felsenfeld and Groudine 2003). DNA compaction can be more than four orders of magnitude greater in a condensed metaphase chromosome than in an actively transcribed gene. Transcribed genes are usually found within "relaxed" chromosomal DNA, termed euchromatin; nonexpressed genes commonly reside within more condensed chromosomal DNA, termed heterochromatin (Tamaru 2010). Heterochromatin formation impairs gene expression by impeding transcription factor access to the underlying DNA (Fig. 2). In an informative cell line model of HIV-1 latency called J-Lat (Jordan et al. 2001), integrated latent HIV-1 proviruses were identified within heterochromatic regions, including centromeric alphoid repeats (Jordan et al. 2003). Nevertheless, these latent viruses could be successfully activated when cells were exposed to various stimuli, including phorbol esters or tumor necrosis factor α. Further studies with this system revealed that inducible HIV-1 gene expression with integration into gene deserts,

centromeric heterochromatin or highly expressed cellular genes (Lewinski et al. 2005). Although the notion of integration within heterochromatic sites was interesting from a mechanistic perspective, extensive studies of HIV-1 integration sites in cell lines infected in vitro with HIV-1 had established that HIV-1 generally integrates within cellular genes (Schroder et al. 2002). In vivo studies of HIV-1 integrations sites in resting CD4[+] T cells from patients on HAART showed that most of the integrated HIV-1 DNA was within cellular genes that are actively transcribed in resting CD4[+] T cells (Han et al. 2004). These findings prompted a general reconsideration of the underlying mechanisms of HIV-1 latency and raised the notion that the mechanisms might in fact be multifactoral (Lassen et al. 2004).

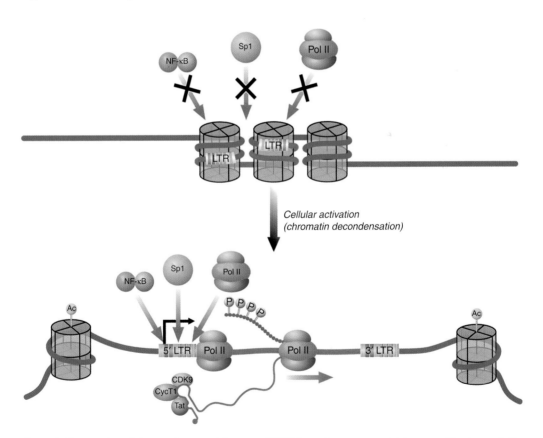

Figure 2. Compaction of chromatin around integrated HIV-1 proviruses may restrict access of key transcription factors to the 5′ long terminal repeat. However, this condensed chromatin state is reversed when cells are activated. Activation leads to transcription factor access and effective RNA Pol II elongation, giving rise to high-level virus production when HIV-1 Tat is produced.

Although integration within heterochromatin does not appear to be a major mechanism of HIV-1 latency, other aspects of chromatin biology are clearly involved. Nucleosomes are the fundamental structural unit of chromatin (Bai and Morozov 2010). A single nucleosome comprises 147 base pairs of DNA wrapped in 1.67 left-handed superhelical turns around an octet of histone proteins that includes H2A, H2B, H3, and H4. Nucleosomes are separated by ∼80 nucleotides of linker DNA, where histone H1 binds. Nucleosomes are regulated by both SWI/SNF remodeling complexes (Liu et al. 2011) and posttranslational modifications of the histone tails, including acetylation, methylation, ubiquitylation, phosphorylation, sumoylation, and poly ADP-ribosylation (Goldberg et al. 2007). Unique combinations of these modifications have been proposed to generate a "histone code" that may provide a more universal set of operating instructions governing chromatin assembly and gene expression (Fischle et al. 2003).

Strikingly, two nucleosomes, nuc-0 and nuc-1, consistently form within the 5′ long terminal repeat (LTR) of HIV even when the provirus is integrated within euchromatin (Verdin et al. 1993). These nucleosomes regulate the basal transcriptional activity of the 5′ LTR because they overlap with the binding sites for many key transcription factors that drive HIV-1 gene expression. Reactivation of latent HIV-1 proviruses by external stimuli is consistently associated with remodeling of nuc-1. Maintenance of nuc-1 is highly dependent on the state of histone acetylation based on the finding that histone deacetylase (HDAC) inhibitors promote effective remodeling of nuc-1 and transcriptional activation of HIV-1 (Laughlin et al. 1993; Van Lint et al. 1996). By analogy, it seems likely that one or more HDACs are active on the LTR of HIV-1 proviruses where nuc-1 is maintained. HDACs do not directly bind to DNA; rather these enzymes assemble with other proteins that bind DNA. The resulting protein complexes exert repressive effects on transcription. Subsequent studies revealed that HDAC-1 is effectively recruited to the HIV-1 LTR through its interactions with multiple host factors, including p50 homodimers of NF-κB, LSF1, YY1, thyroid hormone receptor, and other DNA binding partners (Coull et al. 2000; Hsia and Shi 2002; Williams et al. 2006). Importantly, addition of HDAC inhibitors not only promotes effective remodeling of nuc-1 but also increases the effective recruitment of RNA Pol II, leading to the commencement of RNA synthesis (Williams et al. 2006). Administration of valproic acid, a weakly active HDAC inhibitor, to HIV-1-infected patients receiving antiretroviral therapy was initially reported to result in modest decreases in the latent reservoir (Lehrman et al. 2005). Disappointingly, in subsequent studies, this agent did not produce durable effects on the latent reservoir (Siliciano et al. 2007; Archin et al. 2010).

DNA Methylation and HIV Latency

DNA methylation is another mechanism that reinforces HIV-1 latency. The HIV-1 transcription start site is flanked by two CpG islands that are methylated in J-Lat T cells and a primary CD4$^+$ T cells model of HIV-1 latency (Blazkova et al. 2009; Kauder et al. 2009). Methyl CpG binding domain protein-2 (MBD2) and HDAC2 are detectable at one of these CpG islands during latency, and inhibition of methylation on deoxycytosines by addition of 5-aza-2′- deoxycytidine (aza-CdR) prevents recruitment of MBD2 and HDAC2 (Kauder et al. 2009). Of note, synergistic reactivation of latent HIV-1 gene expression occurs when cells are cocultured with aza-CdR and inducers of NFκB (e.g., prostratin or tumor necrosis factor α). CpG methylation may, in fact, correspond to a relatively late silencing event in a sequential program of modifications that serve to reinforce latency. Consistent with this model, proviruses in memory CD4$^+$ T cells of some patients on HAART contain a high proportion of nonmethylated cytosines within the 5′ LTR (Blazkova et al. 2009). The renewed appreciation of the potential regulatory role of DNA methylation will likely prompt new studies of DNA methylation inhibitors in combination with other latency antagonists.

Cite this article as *Cold Spring Harb Perspect Med* doi: 10.1101/cshperspect.a007096

Transcriptional Interference as a Potential Driver of HIV Latency

In general, HIV-1 favors integration into actively transcribed genes (Schroder et al. 2002), likely because the host factor LEDGF/p75 (Meehan et al. 2009) binds to integrase and directs integration to intronic regions within these highly expressed genes. Nevertheless, the finding that latent proviruses are frequently integrated within highly expressed genes in both J-Lat cells (Lewinski et al 2005) and primary CD4[+] T cells (Han et al. 2004) was quite surprising. These findings suggest that some forms of latency mechanistically involve transcriptional interference (Fig. 3). Different forms of transcriptional interference occur, depending on the orientation of the HIV-1 provirus relative to the host gene. If both share the same polarity, promoter occlusion can occur in which read-through by an upstream elongating RNA Pol II displaces key transcription factors on HIV-1 LTR, thereby reinforcing viral latency (Greger et al. 1998; Lenasi et al. 2008). In a primary cell model of HIV-1 latency, a modest preference for integration with the same polarity was observed (Shan et al. 2011). Conversely, if the cellular gene and provirus are arranged in opposite polarity, RNA Pol II collisions may occur, leading to premature termination of transcription at one or both promoters (Han et al. 2008). This form of transcriptional interference could also generate double-stranded viral RNAs that might participate in RNA interference, lead to RNA-mediated methylation, or result in biologically active antisense RNAs. Despite the mechanism of transcriptional interference, activation of the HIV-1 LTR converts this promoter into a very strong transcription unit possibly because

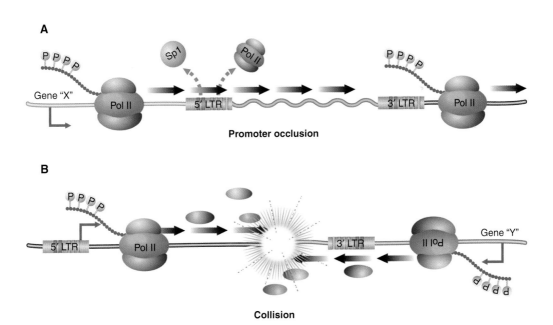

Figure 3. Two scenarios in which transcriptional interference may promote HIV-1 latency. In the first case, the HIV-1 provirus is integrated in the same polarity as an upstream gene within an intron of this gene. Read through by RNA Pol II initiating at the upstream promoter occludes the 5′LTR and displaces key transcription factors thereby promoting viral latency. Alternatively, the HIV-1 provirus and the cellular gene may be arranged in opposite polarity. In this situation, initiating polymerases collide, leading to decreased expression of one or both transcription units. The fact that latent HIV-1 proviruses are commonly found in actively transcribed genes suggests that transcriptional interference may be important in the maintenance of proviral latency.

of the tight binding affinity of NF-κB that overcomes interference and results in effective proviral reactivation (De Marco et al. 2008).

RNA Pol II Initiation

Initiation of HIV-1 gene expression is critically dependent on a range of host transcription factors whose activities are induced by extracellular stimuli, including T-cell receptor ligation or the action of various cytokines (Colin and Van Lint 2009). Key inducible transcription factors include NF-κB, NFAT, and AP-1, which interact with their cognate enhancers within the HIV-1 LTR. In resting cells, both NF-κB and NFAT are sequestered in the cytoplasm and thus are not available in the nucleus to promote the activation of latent HIV-1 proviruses. Sp1 also plays a key role by binding at three sites in the core promoter. Elimination of one or more of these sites greatly impairs both basal and Tat-dependent activation of the HIV-1 LTR (Ross et al. 1991; Sune and Garcia-Blanco 1995). Sp1 may act, at least in part, through its associations with TATA-box-binding protein (TBP), TBP-associated factor (TAF) 250, and TAF55 (Emili et al. 1994; Chiang and Roeder 1995; Shao et al. 1995). The juxtaposition of the Sp1 and κB sites also appears to be critically important for full LTR activation (Perkins et al. 1993). Of note, both NF-κB and NFAT bind to the κB sites (Kinoshita et al. 1998). Which of these factors is more important for activation of latent HIV-1 proviruses continues to be hotly debated in the field, but it seems likely that cellular context is critical. In contrast to the key roles of the κB enhancers and Sp1 sites in the reactivation of latent provirus, binding sites for other described transcription factors (e.g., LEF-1, COUP-TF, YY1, Ets-1, and USF) likely play more supporting roles (Rohr et al. 2003).

Members of the NF-κB/Rel family of transcription factors exert both activating and inhibitory effects on the HIV-1 LTR. For example, the binding of homodimers of p50 recruits HDAC1 and leads to transcriptional inhibition because of histone deacetylation and chromatin condensation (Williams et al. 2006). Conversely, p50/RelA heterodimers

(the prototypical NF-κB complex) readily displace p50 homodimers and promote strong transcriptional activation. These effects of NF-κB, in part, involve the recruitment of p300/CBP (Gerritsen et al. 1997), which mediates acetylation of histone tails and leads to increased RNA Pol II and TFIIH/CDK7 binding. The latter complex helps mediate promoter clearance and phosphorylates the carboxy-terminal domain (CTD) of RNA Pol II on serine-5 residues. Finally, RelA has also been implicated in the recruitment of P-TEFb to the HIV LTR (Barboric et al. 2001), although BRD4 also may play a role (Yang et al. 2005). The PTEFb complex, comprising cyclin T1 and CDK9, phosphorylates serine 2 in the CTD of RNA Pol II, thereby promoting effective Pol II elongation. This P-TEFb complex is hijacked by HIV Tat to drive high-level Pol II elongation (Fig. 4) (Wei et al. 1998). Like the negative regulation of NF-κB and NFAT, P-TEFb is also partially sequestered in an inactive ribonucleoprotein complex composed of 7SK RNA and HEXIM-1 (7SK snRNP complex) (Yang et al. 2001).

Just as posttranslational modification of histone tails plays a key role in regulating HIV-1 gene expression, reversible acetylation of nonhistone substrates, notably the NF-κB RelA subunit, also regulates HIV-1 LTR activation (Chen and Greene 2004). The RelA subunit is targeted by p300/CBP for acetylation at multiple sites, including lysines 218, 221, and 310. These modifications regulate distinct NF-κB functions (Chen et al. 2001, 2002), including DNA binding, assembly with IκBα, and the overall transcriptional activity of NF-κB. Acetylation of RelA is reversed by the actions of two deacetylases, HDAC3 (Chen et al. 2001), and SIRT1 (Yeung et al. 2004). Of note, SIRT1-mediated deacetylation of RelA at lysine 310 is blocked by HIV Tat. Because acetylation of lysine 310 in RelA greatly enhances transcriptional activity, this inhibitory effect of Tat could markedly boost NF-κB action and increase viral gene expression as well as expression of genes involved in inflammation and immune activation (Kwon et al. 2008). Importantly, HDAC3-mediated deacetylation of lysine 221 restores the ability of newly synthesized IκBα to bind to RelA (Sun et al. 1993),

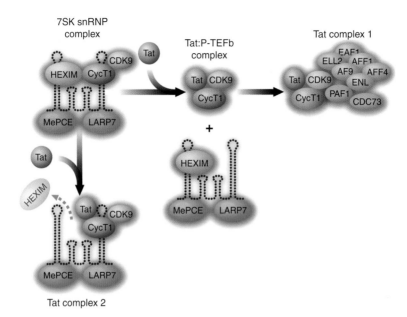

Figure 4. HIV-1 Tat effectively antagonizes HIV-1 latency by liberating P-TEFb from an inactive 7SK snRNP complex. In addition to the Tat-Cyclin T1-CDK9 complex that binds to TAR and promotes serine-2 phosphorylation of the carboxy-terminal domain of RNA Pol II, thereby eliminating promoter proximal pausing, two additional Tat complexes have recently been discovered. Tat complex 1 corresponds to the assembly of Tat-Cyclin T1-CDK9 with AFF1, AFF4, AFR, ENL, EAF1, and ELL2. This complex has been termed a "super-elongation complex." ELL2 blocks backtracking by RNA Pol II. AFR, ENL, and AF9 correspond to transcription factors/coactivators; AF9 potentiates CDK9 kinase activity. Tat complex 2 corresponds to the 7SK snRNP complex lacking HEXIM. This RNP complex may correspond to an additional reservoir of Tat-CyclinT1-CDk9 within cells. However, its precise function remains to be delineated.

leading in turn to nuclear export of NF-κB, thereby terminating its transcriptional activity.

These studies highlight how dynamic changes in the acetylation state of RelA serve as an intranuclear molecular switch regulating both the strength and duration of the NF-κB transcriptional response and consequently the activation of its target genes, including integrated latent HIV-1 proviruses. Many agents being considered for purging of the latent reservoir correspond to inducers of the NF-κB pathway. Unfortunately, because NF-κB is a "master regulator" of so many key responses in mammals, side effects will almost certainly occur.

RNA Pol II Elongation

HIV-1 transcription can be viewed as occurring in two phases: an early Tat-independent phase that promotes at best low-level expression of viral genes but importantly allows production of Tat, and a later Tat-dependent phase that mediates high-level viral transcription (Barboric and Peterlin 2005). The relative absence of Tat may serve as a central driver of HIV-1 latency (Karn 2011). Tat acts by binding to the transactivation-responsive element (TAR), a rather simple RNA stem-loop structure located at the 5′ end of all viral transcripts (Fig. 2). The Tat-TAR complex in turn binds P-TEFb, which mediates serine-2 phosphorylation of the CTD of Pol II (Parada and Roeder 1996; Kim et al. 2002) and phosphorylation of Spt5, a factor that prevents premature Pol II release at terminator sequences (Bourgeois et al. 2002). These events result in highly effective RNA Pol II elongation. In the absence of Tat, the elongating Pol II complex transcripts are naturally obstructed because of binding of two factors, termed negative elongation factor (NELF) and

DRB sensitivity-inducing factor (DSIF). This obstruction results in the production of short viral transcripts of ~60 nucleotides in length. Phosphorylation of NELF by P-TEFb (Fujinaga et al. 2004) and the promoter-clearing action of TFIIH/CDK7 remove these inhibitors and, when coupled with the positive effects of Pol II and Spt5 phosphorylation, set the stage for unimpeded high-level Pol II elongation. In resting cells, cyclin T1 levels are often low, resulting in decreased P-TEFb activity (Liou et al. 2002). This could also contribute to latency in resting cells.

Recently, the first crystal structure for Tat bound to P-TEFb was reported (Tahirov et al. 2010). This structure, resolved at 2.1 angstroms, reveals intimate contacts between Tat and the T loop of CDK9 and marked interactions of Tat with cyclin T1. Tat-induced changes in the conformation of P-TEFb likely explain how this viral protein liberates this complex from HEXIM-1 and 7SK RNA (Fig. 4). This structure could lead to the rational design of a new class of HIV-1 inhibitors targeting Tat. Because Tat plays such a pivotal role in controlling viral gene expression, it is understandable how small stochastic changes in its expression can determine whether viral latency or productive infection ensues (Weinberger et al. 2005).

As noted, Tat cannot produce its effects in the absence of P-TEFb, a kinase complex that is normally bound in a ribonucleoprotein complex of 7SK RNA and HEXIM-1 or HEXIM-2 (Fig. 4). Tat effectively liberates P-TEFb from this complex. However, the Tat-P-TEFb complex might not be as simple as originally thought. Rather than just three proteins, a much larger complex of the coactivators AFF4, ENL, and AF9 and the elongation factor ELL2 has recently been detected by two groups (He et al. 2010; Sobhian et al. 2010) (Fig. 4, Tat complex 1). ELL2 promotes elongation by curbing RNA Pol II "backtracking," and as such, its presence in the Tat complex is quite intriguing. A second complex of Tat, P-TEFb, and 7SK RNA but lacking Hexim-1 has also been isolated (Sobhian et al. 2010) (Fig. 4, Tat complex 2). The function of this complex is not yet clear, but it could

correspond to a partially active pool of P-TEFb (Karn 2011).

RNA Splicing and Export

Considerable attention has focused on the role of chromatin and transcription in the control of HIV-1 latency. However, another rather unexpected mechanism has emerged. Specifically, resting CD4$^+$ T cells from HIV-1-infected patients on antiretroviral therapy retain both Tat and Rev transcripts in their nuclei (Lassen et al. 2006). This finding stands in sharp contrast to the effective export of these multiply spliced viral transcripts in activated CD4$^+$ T cells. This defect in viral RNA export can be rescued by ectopic expression of polypyrimidine tract binding protein (PTB), an RNA export protein. Of note, a short form of PTB is present in resting CD4$^+$ T cells, whereas full-length PTB is virtually undetectable. Whether this export difference is rooted in changes in the splicing of the export protein or a selective degradation of PTB in resting CD4$^+$ T cells remains to be determined.

CLINICAL SIGNIFICANCE OF HIV-1 LATENCY

The existence of a latent reservoir for HIV-1 has implications for eradication efforts and also for understanding the clinical response to antiretroviral therapy, in particular the decay in viremia after initiation of treatment. In 1995, George Shaw and David Ho showed that treatment with a potent antiretroviral drug caused plasma viral levels to fall exponentially (Fig. 5) (Ho et al. 1995; Wei et al. 1995). They reasoned that the newer antiretroviral drugs were powerful enough to stop all new infection of susceptible cells. Importantly, all antiretroviral drugs in clinical use block new infection of susceptible cells but not the release of virions by cells that contain integrated proviruses. If the drugs completely block de novo infection, virus appearing in the plasma must come from cells infected before therapy. Because the decay rate of free virus is very fast, the decrease in plasma virus levels after initiation of effective

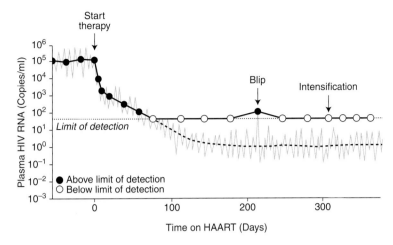

Figure 5. Dynamics of viral decay in patients on HAART medications. After initiation of HAART, levels of plasma virus fall rapidly, reflecting the exponential decay of the activated CD4$^+$ T cells that produce most of the plasma virus and the slower decay of a second population of infected cells that have a half-life of about 2 weeks. The second phase decay brings the level of plasma virus down to a new steady state that is below the limit of detection of clinical assays (50 copies/ml). The average level of residual viremia is around 1 copy/ml. Residual viremia appears to reflect release of virus from stable reservoirs and is not reduced further by treatment intensification. Biological and statistical fluctuations in the level of this residual viremia are occasionally captured in clinical measurements as "blips," but these transient elevations do not reflect the evolution of resistant virus.

therapy is determined mainly by the decay rate of previously infected cells. The rapid initial drop in plasma virus level directly reflects the short half-life ($\sim$1 day) of the CD4$^+$ T lymphoblasts that produce most of the plasma virus (Fig. 5). If the initial therapy is monotherapy, viremia rebounds because of the outgrowth of preexistent resistant variants (Wei et al. 1995). Conversely, when appropriate combinations of three antiretroviral drugs are started, plasma virus levels almost always fall to below the limit of detection (Gulick et al. 1997; Hammer et al. 1997; Perelson et al. 1997). After the rapid initial decay during the first 1–2 weeks of treatment, plasma virus levels decline at a slower rate, reflecting the release of virus from another population of infected cells with a slower decay rate (Fig. 5). Interestingly, the nature of the cells responsible for this second-phase decay is still unclear (Spivak et al. 2011). Initially, it was hoped that the second-phase decay would continue, eventually leading to eradication (Perelson et al. 1997). However, studies with especially sensitive assays for plasma HIV-1 revealed at least one more phase in the decay

curve (Fig. 5). In many patients, viremia appears to level off at around 1 copy/ml and remain at that level indefinitely (Dornadula et al. 1999; Maldarelli et al. 2007; Palmer et al. 2008). This persistent viremia at levels below the limit of detection of the standard clinical assay (50 copies/ml) is termed residual viremia.

Initially, residual viremia was thought to result from ongoing rounds of replication despite drug treatment (Dornadula et al. 1999). Such a scenario raises the possibility of the eventual emergence of drug-resistant virus and viral rebound. However, viral rebound is not seen in adherent patients, and as discussed below, direct sequence analysis of residual viremia suggests another explanation, namely, the release of virus from stable reservoirs, such as that found in resting CD4$^+$ T cells. Every day, a small fraction of the cells in this latent reservoir may become activated by encounter with antigen or cytokines. Although the virus produced by these activated cells will not infect other cells if antiretroviral treatment completely blocks new infection events, the released virus can still be detected in the plasma by sensitive RT-PCR assays.

Evidence that this residual viremia is derived at least in part from the reactivation of latent virus in resting CD4$^+$ T cells comes from studies in which the rare viruses in the plasma of patients on suppressive HAART regimens have been directly sequenced. These viruses lack new resistance mutations (Hermankova et al. 2001; Kieffer et al. 2004; Persaud et al. 2004; Nettles et al. 2005; Bailey et al. 2006), even during the blips or transient elevations in viremia that many patients experience (Fig. 5). When phylogenetic analyses are performed, sequences from the plasma virus and those from viruses present in the latent reservoir in resting CD4$^+$ T cells are intermingled and, in some cases, identical. These findings indicate that at least some of the residual viremia may be derived from latently infected cells that have become activated (Tobin et al. 2005; Bailey et al. 2006). However, several recent studies suggest that the relationship between the free virus particles detected in the plasma of patients on HAART and the virus archived in the latent reservoir in resting CD4$^+$ T cells is more complex (see below). In any event, direct analysis of residual viremia has provided little evidence for the notion that these viruses are derived from ongoing productive rounds of viral replication. Rather, these viruses are largely derived from stable reservoirs.

This conclusion is strongly supported by the results of intensification studies. Dinoso et al. (2009a,b) showed that intensification of optimal three-drug HAART regimens by addition of a fourth drug, either the potent reverse transcriptase inhibitor efavirenz or boosted forms of the protease inhibitors atazanavir or lopinavir, had no effect on the level of residual viremia (Dinoso et al. 2009a). Subsequent studies showed that intensification with the integrase inhibitor raltegravir similarly did not reduce residual viremia (Gandhi et al. 2010a; McMahon et al. 2010). Intensification also fails to reduce the size of the latent reservoir (Gandhi et al. 2010b). These findings are consistent with the idea that residual viremia reflects release of virus from stable reservoirs rather than ongoing viral replication. The highly cooperative dose-response curves of some classes of

antiretroviral drugs allow extraordinarily high levels (10 logs) of inhibition of viral replication (Shen et al. 2008), providing a pharmacodynamic explanation for the ability of combinations of antiretroviral drugs to completely block viral replication.

From a clinical standpoint, the latent reservoir for HIV-1 in resting CD4$^+$ T cells is also important as a repository for drug-resistant viruses (Persaud et al. 2000; Noe et al. 2005). During periods of active viral replication, new sequences are continuously deposited in the latent reservoir. In patients who are failing therapy with drug-resistant virus, these resistant viruses are deposited in the latent reservoir and can potentially reemerge if treatment with the failed regimen is reinstituted. Interestingly, the original wild-type viruses also persist in the latent reservoir and can reemerge if all treatment is stopped in patients who are failing therapy (Deeks et al. 2001; Ruff et al. 2002).

OTHER VIRAL RESERVOIRS

Since the original description of the latent reservoir in resting CD4$^+$ T cells, several cell types and anatomical sites have been proposed as additional reservoirs. Unfortunately, the terms latency and reservoir have been used rather loosely. An accepted definition of latency is given above. A practical definition for an HIV-1 reservoir is a cell type or anatomical site that allows persistence of replication-competent HIV-1 for long periods of time in patients on optimal HAART regimens. In a practical sense, reservoirs are barriers to eradication. Eradication strategies will likely be attempted only in patients who have had suppression of active viral replication with HAART for prolonged periods (greater than 1 year, for example). Thus, the only reservoirs that are clinically significant are those that allow persistence on a time scale of years. To date, only the latent reservoir in resting CD4$^+$ T cells has met this criterion (Finzi et al. 1999; Siliciano et al. 2003; Strain et al. 2003).

Mullins and colleagues proposed an elegant phylogenetic approach for defining reservoirs. A reservoir allows long-term persistence of viral genomes, and thus, sequences obtained from a reservoir show a distinct lack of temporal structure in that sequences isolated at later time points do not show greater divergence from the most recent common ancestor. The latent reservoir in resting CD4$^+$ T cells meets this criteria (Ruff et al. 2002). For example, wild-type sequences are preserved in the latent reservoir even in patients failing therapy with drug-resistant viruses (Ruff et al. 2002).

As discussed above, another approach to identifying reservoirs is to examine the sequences from the residual viremia in patients on HAART. Because the half-life of virions in the plasma is on the order of minutes, the continued presence of free virus in the plasma indicates ongoing virus production. As discussed above, this does not appear to reflect new cycles of replication because residual viremia is not reduced by intensification of HAART. Rather it reflects release of virus from stable reservoirs, and therefore, analysis of residual viremia might provide important clues to the nature of these reservoirs. In general, the residual viremia resembles viral sequences in resting CD4$^+$ T cells (Bailey et al. 2006). However, in some patients, the residual viremia is dominated by a small number of viral clones that are not well represented in circulating CD4$^+$ T cells (Bailey et al. 2006). In addition, analysis of viral sequences in resting CD4$^+$ T cell and in the residual viremia with phylogenetic methods designed to detect population structure revealed that some of the viruses in the plasma represent a distinct population not found in resting CD4$^+$ T cells in the blood (Ruff et al. 2002; Nickle et al. 2003). These studies suggest that, in addition to the latent reservoir in resting CD4$^+$ T cells, an additional source of residual viremia may exist in patients on HAART.

Cells of the monocyte-macrophage lineage are clearly infected by HIV-1 (Gartner et al. 1986; Igarashi et al. 2001). The role of macrophages in HIV-1 pathogenesis is described in Brennan et al. (2009), Sahu et al. (2009), and Swanstrom and Coffin (2011). Infected macrophages may constitute a particularly important source of virus late in the course of disease when CD4$^+$ T cells have been largely depleted

(Igarashi et al. 2001). It is less clear whether a true state of latency is established in macrophages in vivo; the infection is typically productive in these cells. There is some evidence for infection of circulating monocytes (Zhu et al. 2002; Arfi et al. 2008), but these cells cannot be considered a reservoir because they circulate only for about 1 day before entering the tissues and differentiating into macrophages. Importantly, monocytes can cross the blood–brain barrier and differentiate into macrophages and microglial cells. Thus, infection of monocytes may provide a mechanism for entry of virus into the central nervous system (CNS) (Gras and Kaul 2010). Infected macrophages and microglial cells clearly contribute to viral persistence and CNS pathology in HIV-1 infection (Schnell et al. 2009), and there is evidence in the SIV model for latent infection (Clements et al. 2002). In determining the importance of these cells as long-term reservoirs for HIV-1 in patients on HAART, it is important to understand their turnover rates. These cells are continually replaced by monocytes, but their half-lives are poorly understood. There is considerable controversy over the question of whether progenitor cells in the monocyte-macrophage lineage or hematopoietic stem cells (HSC) are infected in vivo. Interest in this issue has been heightened by recent evidence that HIV-1 can infect hematopoietic progenitor cells and establish a state of latent infection (Carter et al. 2010). The extent to which these cells survive in patients on HAART remains unclear.

ERADICATION STRATEGIES

There is now great interest in finding a way to eliminate the latent reservoir in resting CD4$^+$ T cells (Richman et al. 2009). Some investigators have suggested that the stability of the latent reservoir is a reflection of continued reseeding of the reservoir by new rounds of infection despite the presence of potent antiviral drugs (Chun et al. 2005). If this were true, then intensification of HAART might reduce residual viremia and accelerate the decay of the latent reservoir. However, as discussed above, intensification does not affect residual viremia. The remarkable stability of the reservoir is more likely because of the fact that it resides in long-lived resting memory CD4[+] T cells (Chun et al. 1997a). Recent results suggest that the low level of homeostatic proliferation that maintains the pool of memory T cells may contribute to the stability of the reservoir (Chomont et al. 2009).

Because intensification of HAART is unlikely to lead to eradication, current efforts have focused on finding ways to reactivate latent HIV-1 in patients on HAART. The presumption is that infected cells in which latency is reversed will die from viral cytopathic effects or be killed by HIV-1-specific cytolytic T lymphocytes. It is also presumed that HAART will prevent new rounds of infections by virus released in the process. Initial efforts to reverse latency involved the cytokine IL-2. Although treated patients appeared to have reduced numbers of latently infected cells, rapid rebound in viremia was observed when treatment was interrupted (Davey et al. 1999). Attempts to induce global T-cell activation with anti-CD3 antibodies were also unsuccessful. There is now interest in the cytokine IL-7 that is more effective in reversing latency than IL-2 (Brooks et al. 2003). Unfortunately, IL-7 also drives homeostatic proliferation of memory T cells and could lead to expansion of the latent reservoir (Chomont et al. 2009). More recent efforts have taken advantage of advances in our understanding of the molecular mechanisms regulating HIV-1 latency. Agonists of protein kinase C (Kulkosky et al. 2001; Korin et al. 2002; Brooks et al. 2003; Williams et al. 2004) and inhibitors of histone deacetylases (Ylisastigui et al. 2004; Lehrman et al. 2005; Williams et al. 2006; Archin et al. 2009) have shown promise in in vitro models. These approaches are discussed more fully in Hoxie and June (2011). A major problem with all of these approaches is their lack of specificity for infected cells.

Appropriate experimental models are likely to be critical to the development of approaches for eliminating the latent reservoir. Recently, several laboratories developed primary cell models for HIV-1 latency (reviewed in Gulick et al. 1997). These models appear to recapitulate the biology of the quiescent G_0 cells that harbor

latent HIV-1 in vivo better than models based on continuously proliferating cell lines. However, it is not yet clear what models most accurately reflect the behavior of latently infected cells in vivo. Another major advance has been the development of SIV/macaque models for HIV-1 latency. SIV establishes a state of latent infection in resting CD4[+] T cells in the same way that HIV-1 does (Shen et al. 2003). Several groups recently developed SIV or SHIV models that are sensitive to HAART (Shen et al. 2003; Dinoso et al. 2009b; North et al. 2010). In these models, agents that might effectively purge virus from the latent reservoir can be evaluated.

CONCLUDING PERSPECTIVE

A subset of resting memory CD4[+] T cells harboring integrated but transcriptionally silent HIV-1 proviruses currently poses an insurmountable barrier to viral eradication in infected subjects. Achieving the lofty goal of viral eradication will certainly not be easy because of both the intrinsic biological properties of latent proviruses and the nature of their cellular hosts. However, a better understanding of the mechanisms that underlie viral latency and a fuller appreciation for the range of cells participating in meaningful cellular reservoirs could result in a rational attack on latent HIV-1 reservoirs. Achieving the goal of complete viral eradication or alternatively a functional cure, in which subjects continue to harbor virus but do not require antiretroviral therapy, could be critical for the millions of infected individuals in the developing world where the availability of lifelong antiretroviral therapy is uncertain at best.

REFERENCES

Reference is also in this collection.

Archin NM, Espeseth A, Parker D, Cheema M, Hazuda D, Margolis DM. 2009. Expression of latent HIV induced by the potent HDAC inhibitor suberoylanilide hydroxamic acid. *AIDS Res Hum Retrov* 25: 207–212.

Archin NM, Cheema M, Parker D, Wiegand A, Bosch RJ, Coffin JM, Eron J, Cohen M, Margolis DM. 2010. Antiretroviral intensification and valproic acid lack sustained effect on residual HIV-1 viremia or resting CD4[+] cell infection. *PLoS One* 5: e9390.

Arfi V, Riviere L, Jarrosson-Wuilleme L, Goujon C, Rigal D, Darlix JL, Cimarelli A. 2008. Characterization of the early steps of infection of primary blood monocytes by human immunodeficiency virus type 1. *J Virol* **82:** 6557–6565.

Bai L, Morozov AV. 2010. Gene regulation by nucleosome positioning. *Trends Genet* **26:** 476–483.

Bailey J, Blankson JN, Wind-Rotolo M, Siliciano RF. 2004. Mechanisms of HIV-1 escape from immune responses and antiretroviral drugs. *Curr Opin Immunol* **16:** 470–476.

Bailey JR, Sedaghat AR, Kieffer T, Brennan T, Lee PK, Wind-Rotolo M, Haggerty CM, Kamireddi AR, Liu Y, Lee J, et al. 2006. Residual human immunodeficiency virus type 1 viremia in some patients on antiretroviral therapy is dominated by a small number of invariant clones rarely found in circulating CD4$^+$ T cells. *J Virol* **80:** 6441–6457.

Barboric M, Peterlin BM. 2005. A new paradigm in eukaryotic biology: HIV Tat and the control of transcriptional elongation. *PLoS Biol* **3:** e76.

Barboric M, Nissen RM, Kanazawa S, Jabrane-Ferrat N, Peterlin BM. 2001. NF-κB binds P-TEFb to stimulate transcriptional elongation by RNA polymerase II. *Mol Cell* **8:** 327–337.

Blazkova J, Trejbalova K, Gondois-Rey F, Halfon P, Philibert P, Guiguen A, Verdin E, Olive D, Van Lint C, Hejnar J, et al. 2009. CpG methylation controls reactivation of HIV from latency. *PLoS Pathog* **5:** e1000554.

Bohnlein E, Lowenthal JW, Siekevitz M, Ballard DW, Franza BR, Greene WC. 1988. The same inducible nuclear proteins regulates mitogen activation of both the interleukin-2 receptor-α gene and type 1 HIV. *Cell* **53:** 827–836.

Bourgeois CF, Kim YK, Churcher MJ, West MJ, Karn J. 2002. Spt5 cooperates with human immunodeficiency virus type 1 Tat by preventing premature RNA release at terminator sequences. *Mol Cell Biol* **22:** 1079–1093.

Brahic M, Stowring L, Ventura P, Haase AT. 1981. Gene expression in visna virus infection in sheep. *Nature* **292:** 240–242.

Brennan TP, Woods JO, Sedaghat AR, Siliciano JD, Siliciano RF, Wilke CO. 2009. Analysis of human immunodeficiency virus type 1 viremia and provirus in resting CD4$^+$ T cells reveals a novel source of residual viremia in patients on antiretroviral therapy. *J Virol* **83:** 8470–8481.

Brooks DG, Hamer DH, Arlen PA, Gao L, Bristol G, Kitchen CM, Berger EA, Zack JA. 2003. Molecular characterization, reactivation, and depletion of latent HIV. *Immunity* **19:** 413–423.

Bukrinsky MI, Stanwick TL, Dempsey MP, Stevenson M. 1991. Quiescent T lymphocytes as an inducible virus reservoir in HIV-1 infection. *Science* **254:** 423–427.

Carter CC, Onafuwa-Nuga A, McNamara LA, Riddell J IV, Bixby D, Savona MR, Collins KL. 2010. HIV-1 infects multipotent progenitor cells causing cell death and establishing latent cellular reservoirs. *Nat Med* **16:** 446–451.

Chen LF, Greene WC. 2004. Shaping the nuclear action of NF-κB. *Nat Rev Mol Cell Biol* **5:** 392–401.

Chen LF, Mu Y, Greene WC. 2002. Acetylation of RelA at discrete sites regulates distinct nuclear functions of NF-κB. *EMBO J* **21:** 6539–6548.

Chen L, Fischle W, Verdin E, Greene WC. 2001. Duration of nuclear NF-κB action regulated by reversible acetylation. *Science* **293:** 1653–1657.

Chiang CM, Roeder RG. 1995. Cloning of an intrinsic human TFIID subunit that interacts with multiple transcriptional activators. *Science* **267:** 531–536.

Chomont N, El-Far M, Ancuta P, Trautmann L, Procopio FA, Yassine-Diab B, Boucher G, Boulassel MR, Ghattas G, Brenchley JM, et al. 2009. HIV reservoir size and persistence are driven by T cell survival and homeostatic proliferation. *Nat Med* **15:** 893–900.

Chun TW, Finzi D, Margolick J, Chadwick K, Schwartz D, Siliciano RF. 1995. In vivo fate of HIV-1-infected T cells: Quantitative analysis of the transition to stable latency. *Nat Med* **1:** 1284–1290.

Chun TW, Carruth L, Finzi D, Shen X, Digiuseppe JA, Taylor H, Hermankova M, Chadwick K, Margolick J, Quinn TC, et al. 1997a. Quantitation of latent tissue reservoirs and total body load in HIV-1 infection. *Nature* **387:** 183–188.

Chun TW, Stuyver L, Mizell SB, Ehler LA, Mican JA, Baseler M, Lloyd AL, Nowak MA, Fauci AS. 1997b. Presence of an inducible HIV-1 latent reservoir during highly active antiretroviral therapy. *Proc Natl Acad Sci* **94:** 13193–13197.

Chun TW, Nickle DC, Justement JS, Large D, Semerjian A, Curlin ME, O'Shea MA, Hallahan CW, Daucher M, Ward DJ, et al. 2005. HIV-infected individuals receiving effective antiviral therapy for extended periods of time continually replenish their viral reservoir. *J Clin Invest* **115:** 3250–3255.

Clements JE, Babas T, Mankowski JL, Suryanarayana K, Piatak M Jr, Tarwater PM, Lifson JD, Zink MC. 2002. The central nervous system as a reservoir for simian immunodeficiency virus (SIV): Steady-state levels of SIV DNA in brain from acute through asymptomatic infection. *J Infect Dis* **186:** 905–913.

Colin L, Van Lint C. 2009. Molecular control of HIV-1 postintegration latency: Implications for the development of new therapeutic strategies. *Retrovirology* **6:** 111.

Coull JJ, Romerio F, Sun JM, Volker JL, Galvin KM, Davie JR, Shi Y, Hansen U, Margolis DM. 2000. The human factors YY1 and LSF repress the human immunodeficiency virus type 1 long terminal repeat via recruitment of histone deacetylase 1. *J Virol* **74:** 6790–6799.

Davey RT Jr, Bhat N, Yoder C, Chun TW, Metcalf JA, Dewar R, Natarajan V, Lempicki RA, Adelsberger JW, Miller KD, et al. 1999. HIV-1 and T cell dynamics after interruption of highly active antiretroviral therapy (HAART) in patients with a history of sustained viral suppression. *Proc Natl Acad Sci* **96:** 15109–15114.

Deeks SG, Wrin T, Liegler T, Hoh R, Hayden M, Barbour JD, Hellmann NS, Petropoulos CJ, McCune JM, Hellerstein MK, et al. 2001. Virologic and immunologic consequences of discontinuing combination antiretroviral-drug therapy in HIV-infected patients with detectable viremia. *N Engl J Med* **344:** 472–480.

De Marco A, Biancotto C, Knezevich A, Maiuri P, Vardabasso C, Marcello A. 2008. Intragenic transcriptional *cis*-activation of the human immunodeficiency virus 1 does not result in allele-specific inhibition of the endogenous gene. *Retrovirology* **5:** 98.

Dinoso JB, Kim SY, Wiegand AM, Palmer SE, Gange SJ, Cranmer L, O'Shea A, Callender M, Spivak A, Brennan T, et al. 2009a. Treatment intensification does not reduce residual HIV-1 viremia in patients on highly active antiretroviral therapy. *Proc Natl Acad Sci* **106:** 9403–9408.

Dinoso JB, Rabi SA, Blankson JN, Gama L, Mankowski JL, Siliciano RF, Zink MC, Clements JE. 2009b. A simian immunodeficiency virus-infected macaque model to study viral reservoirs that persist during highly active antiretroviral therapy. *J Virol* **83:** 9247–9257.

Dornadula G, Zhang H, VanUitert B, Stern J, Livornese L Jr, Ingerman MJ, Witek J, Kedanis RJ, Natkin J, DeSimone J, et al. 1999. Residual HIV-1 RNA in blood plasma of patients taking suppressive highly active antiretroviral therapy. *JAMA* **282:** 1627–1632.

Duh EJ, Maury WJ, Folks TM, Fauci AS, Rabson AB. 1989. Tumor necrosis factor α activates human immunodeficiency virus type 1 through induction of nuclear factor binding to the NF-κB sites in the long terminal repeat. *Proc Natl Acad Sci* **86:** 5974–5978.

Emili A, Greenblatt J, Ingles CJ. 1994. Species-specific interaction of the glutamine-rich activation domains of Sp1 with the TATA box-binding protein. *Mol Cell Biol* **14:** 1582–1593.

Felsenfeld G, Groudine M. 2003. Controlling the double helix. *Nature* **421:** 448–453.

Finzi D, Hermankova M, Pierson T, Carruth LM, Buck C, Chaisson RE, Quinn TC, Chadwick K, Margolick J, Brookmeyer R, et al. 1997. Identification of a reservoir for HIV-1 in patients on highly active antiretroviral therapy. *Science* **278:** 1295–1300.

Finzi D, Blankson J, Siliciano JD, Margolick JB, Chadwick K, Pierson T, Smith K, Lisziewicz J, Lori F, Flexner C, et al. 1999. Latent infection of CD4⁺ T cells provides a mechanism for lifelong persistence of HIV-1, even in patients on effective combination therapy. *Nat Med* **5:** 512–517.

Fischle W, Wang Y, Allis CD. 2003. Binary switches and modification cassettes in histone biology and beyond. *Nature* **425:** 475–479.

Folks T, Powell DM, Lightfoote MM, Benn S, Martin MA, Fauci AS. 1986. Induction of HTLV-III/LAV from a nonvirus-producing T-cell line: Implications for latency. *Science* **231:** 600–602.

Fujinaga K, Irwin D, Huang Y, Taube R, Kurosu T, Peterlin BM. 2004. Dynamics of human immunodeficiency virus transcription: P-TEFb phosphorylates RD and dissociates negative effectors from the transactivation response element. *Mol Cell Biol* **24:** 787–795.

Gandhi RT, Zheng L, Bosch RJ, Chan ES, Margolis DM, Read S, Kallungal B, Palmer S, Medvik K, Lederman MM, et al. 2010a. The effect of raltegravir intensification on low-level residual viremia in HIV-infected patients on antiretroviral therapy: A randomized controlled trial. *PLoS Med* **7:** 1000321.

Gandhi RT, Bosch RJ, Aga E, Albrecht M, Demeter LM, Dykes C, Bastow B, Para M, Lai J, Siliciano RF, et al. 2010b. No evidence for decay of the latent reservoir in HIV-1-infected patients receiving intensive enfuvirtide-containing antiretroviral therapy. *J Infect Dis* **201:** 293–296.

Gartner S, Markovits P, Markovitz DM, Kaplan MH, Gallo RC, Popovic M. 1986. The role of mononuclear phagocytes in HTLV-III/LAV infection. *Science* **233:** 215–219.

Gerritsen ME, Williams AJ, Neish AS, Moore S, Shi Y, Collins T. 1997. CREB-binding protein/p300 are transcriptional coactivators of 65. *Proc Natl Acad Sci* **94:** 2927–2932.

Goldberg AD, Allis CD, Bernstein E. 2007. Epigenetics: A landscape takes shape. *Cell* **128:** 635–638.

Gras G, Kaul M. 2010. Molecular mechanisms of neuroinvasion by monocytes-macrophages in HIV-1 infection. *Retrovirology* **7:** 30.

Greger IH, Demarchi F, Giacca M, Proudfoot NJ. 1998. Transcriptional interference perturbs the binding of Sp1 to the HIV-1 promoter. *Nucl Acids Res* **26:** 1294–1301.

Gulick RM, Mellors JW, Havlir D, Eron JJ, Gonzalez C, McMahon D, Richman DD, Valentine FT, Jonas L, Meibohm A, et al. 1997. Treatment with indinavir, zidovudine, and lamivudine in adults with human immunodeficiency virus infection and prior antiretroviral therapy. *N Engl J Med* **337:** 734–739.

Hammer SM, Squires KE, Hughes MD, Grimes JM, Demeter LM, Currier JS, Eron JJ Jr, Feinberg JE, Balfour HH Jr, Deyton LR, et al. 1997. A controlled trial of two nucleoside analogues plus indinavir in persons with human immunodeficiency virus infection and CD4 cell counts of 200 per cubic millimeter or less. AIDS Clinical Trials Group 320 Study Team. *N Engl J Med* **337:** 725–733.

Han Y, Lassen K, Monie D, Sedaghat AR, Shimoji S, Liu X, Pierson TC, Margolick JB, Siliciano RF, Siliciano JD. 2004. Resting CD4⁺ T cells from human immunodeficiency virus type 1 (HIV-1)-infected individuals carry integrated HIV-1 genomes within actively transcribed host genes. *J Virol* **78:** 6122–6133.

Han Y, Lin YB, An W, Xu J, Yang HC, O'Connell K, Dordai D, Boeke JD, Siliciano JD, Siliciano RF. 2008. Orientation-dependent regulation of integrated HIV-1 expression by host gene transcriptional readthrough. *Cell Host Microbe* **4:** 134–146.

He N, Liu M, Hsu J, Xue Y, Chou S, Burlingame A, Krogan NJ, Alber T, Zhou Q. 2010. HIV-1 Tat and host AFF4 recruit two transcription elongation factors into a bifunctional complex for coordinated activation of HIV-1 transcription. *Mol Cell* **38:** 428–438.

Hermankova M, Ray SC, Ruff C, Powell-Davis M, Ingersoll R, D'Aquila RT, Quinn TC, Siliciano JD, Siliciano RF, Persaud D. 2001. HIV-1 drug resistance profiles in children and adults with viral load of <50 copies/ml receiving combination therapy. *JAMA* **286:** 196–207.

Ho DD, Neumann AU, Perelson AS, Chen W, Leonard JM, Markowitz M. 1995. Rapid turnover of plasma virions and CD4 lymphocytes in HIV-1 infection. *Nature* **373:** 123–126.

* Hoxie JA, June CH. 2011. Novel cell and gene therapies for HIV. *Cold Spring Harb Perspect Med* doi: 10.1101/cshperspect.a007179.

Hsia SC, Shi YB. 2002. Chromatin disruption and histone acetylation in regulation of the human immunodeficiency virus type 1 long terminal repeat by thyroid hormone receptor. *Mol Cell Biol* **22:** 4043–4052.

Igarashi T, Brown CR, Endo Y, Buckler-White A, Plishka R, Bischofberger N, Hirsch V, Martin MA. 2001. Macrophage are the principal reservoir and sustain high virus loads in rhesus macaques after the depletion of CD4$^+$ T cells by a highly pathogenic simian immunodeficiency virus/HIV type 1 chimera (SHIV): Implications for HIV-1 infections of humans. *Proc Natl Acad Sci* **98:** 658–663.

Jordan A, Defechereux P, Verdin E. 2001. The site of HIV-1 integration in the human genome determines basal transcriptional activity and response to Tat transactivation. *EMBO J* **20:** 1726–1738.

Jordan A, Bisgrove D, Verdin E. 2003. HIV reproducibly establishes a latent infection after acute infection of T cells in vitro. *EMBO J* **22:** 1868–1877.

Karn J. 2011. The molecular biology of HIV latency: Breaking and restoring the Tat-dependent transcriptional circuit. *Curr Opin HIV AIDS* **6:** 4–11.

Kauder SE, Bosque A, Lindqvist A, Planelles V, Verdin E. 2009. Epigenetic regulation of HIV-1 latency by cytosine methylation. *PLoS Pathog* **5:** e1000495.

Kieffer TL, Finucane MM, Nettles RE, Quinn TC, Broman KW, Ray SC, Persaud D, Siliciano RF. 2004. Genotypic analysis of HIV-1 drug resistance at the limit of detection: Virus production without evolution in treated adults with undetectable HIV loads. *J Infect Dis* **189:** 1452–1465.

Kim YK, Bourgeois CF, Isel C, Churcher MJ, Karn J. 2002. Phosphorylation of the RNA polymerase II carboxyl-terminal domain by CDK9 is directly responsible for human immunodeficiency virus type 1 Tat-activated transcriptional elongation. *Mol Cell Biol* **22:** 4622–4637.

Kinoshita S, Chen BK, Kaneshima H, Nolan GP. 1998. Host control of HIV-1 parasitism in T cells by the nuclear factor of activated T cells. *Cell* **95:** 595–604.

Korin YD, Brooks DG, Brown S, Korotzer A, Zack JA. 2002. Effects of prostratin on T-cell activation and human immunodeficiency virus latency. *J Virol* **76:** 8118–8123.

Kulkosky J, Culnan DM, Roman J, Dornadula G, Schnell M, Boyd MR, Pomerantz RJ. 2001. Prostratin: Activation of latent HIV-1 expression suggests a potential inductive adjuvant therapy for HAART. *Blood* **98:** 3006–3015.

Kwon HS, Brent MM, Getachew R, Jayakumar P, Chen LF, Schnolzer M, McBurney MW, Marmorstein R, Greene WC, Ott M. 2008. Human immunodeficiency virus type 1 Tat protein inhibits the SIRT1 deacetylase and induces T cell hyperactivation. *Cell Host Microbe* **3:** 158–167.

Lassen K, Han Y, Zhou Y, Siliciano J, Siliciano RF. 2004. The multifactorial nature of HIV-1 latency. *Trends Mol Med* **10:** 525–531.

Lassen KG, Ramyar KX, Bailey JR, Zhou Y, Siliciano RF. 2006. Nuclear retention of multiply spliced HIV-1 RNA in resting CD4$^+$ T cells. *PLoS Pathog* **2:** e68.

Laughlin MA, Zeichner S, Kolson D, Alwine JC, Seshamma T, Pomerantz RJ, Gonzalez-Scarano F. 1993. Sodium butyrate treatment of cells latently infected with HIV-1 results in the expression of unspliced viral RNA. *Virology* **196:** 496–505.

Lehrman G, Hogue IB, Palmer S, Jennings C, Spina CA, Wiegand A, Landay AL, Coombs RW, Richman DD, Mellors JW, et al. 2005. Depletion of latent HIV-1 infection in vivo: A proof-of-concept study. *Lancet* **366:** 549–555.

Lenasi T, Contreras X, Peterlin BM. 2008. Transcriptional interference antagonizes proviral gene expression to promote HIV latency. *Cell Host Microbe* **4:** 123–133.

Lewinski MK, Bisgrove D, Shinn P, Chen H, Hoffmann C, Hannenhalli S, Verdin E, Berry CC, Ecker JR, Bushman FD. 2005. Genome-wide analysis of chromosomal features repressing human immunodeficiency virus transcription. *J Virol* **79:** 6610–6619.

Liou LY, Herrmann CH, Rice AP. 2002. Transient induction of cyclin T1 during human macrophage differentiation regulates human immunodeficiency virus type 1 Tat transactivation function. *J Virol* **76:** 10579–10587.

Liu N, Balliano A, Hayes JJ. 2011. Mechanism(s) of SWI/SNF-induced nucleosome mobilization. *Chembiochem* **12:** 196–204.

Maldarelli F, Palmer S, King MS, Wiegand A, Polis MA, Mican J, Kovacs JA, Davey RT, Rock-Kress D, Dewar R, et al. 2007. ART suppresses plasma HIV-1 RNA to a stable set point predicted by pretherapy viremia. *PLoS Pathog* **3:** e46.

McMahon D, Jones J, Wiegand A, Gange SJ, Kearney M, Palmer S, McNulty S, Metcalf JA, Acosta E, Rehm C, et al. 2010. Short-course raltegravir intensification does not reduce persistent low-level viremia in patients with HIV-1 suppression during receipt of combination antiretroviral therapy. *Clin Infect Dis* **50:** 912–919.

Meehan AM, Saenz DT, Morrison JH, Garcia-Rivera JA, Peretz M, Llano M, Poeschla EM. 2009. LEDGF/p75 proteins with alternative chromatin tethers are functional HIV-1 cofactors. *PLoS Pathog* **5:** e1000522.

Nabel G, Baltimore D. 1987. An inducible transcription factor activates expression of human immunodeficiency virus in T cells. *Nature* **326:** 711–713.

Nettles RE, Kieffer TL, Kwon P, Monie D, Han Y, Parsons T, Cofrancesco J Jr, Gallant JE, Quinn TC, Jackson B, et al. 2005. Intermittent HIV-1 viremia (Blips) and drug resistance in patients receiving HAART. *JAMA* **293:** 817–829.

Nickle DC, Jensen MA, Shriner D, Brodie SJ, Frenkel LM, Mittler JE, Mullins JI. 2003. Evolutionary indicators of human immunodeficiency virus type 1 reservoirs and compartments. *J Virol* **77:** 5540–5546.

Noe A, Plum J, Verhofstede C. 2005. The latent HIV-1 reservoir in patients undergoing HAART: An archive of pre-HAART drug resistance. *J Antimicrob Chemother* **55:** 410–412.

North TW, Higgins J, Deere JD, Hayes TL, Villalobos A, Adamson L, Shacklett BL, Schinazi RF, Luciw PA. 2010. Viral sanctuaries during highly active antiretroviral therapy in a nonhuman primate model for AIDS. *J Virol* **84:** 2913–2922.

Palmer S, Maldarelli F, Wiegand A, Bernstein B, Hanna GJ, Brun SC, Kempf DJ, Mellors JW, Coffin JM, King MS. 2008. Low-level viremia persists for at least 7 years in patients on suppressive antiretroviral therapy. *Proc Natl Acad Sci* **105:** 3879–3884.

Parada CA, Roeder RG. 1996. Enhanced processivity of RNA polymerase II triggered by Tat-induced phosphorylation of its carboxy-terminal domain. *Nature* **384:** 375–378.

Perelson AS, Essunger P, Cao Y, Vesanen M, Hurley A, Saksela K, Markowitz M, Ho DD. 1997. Decay characteristics of HIV-1-infected compartments during combination therapy. *Nature* **387:** 188–191.

Perkins ND, Edwards NL, Duckett CS, Agranoff AB, Schmid RM, Nabel GJ. 1993. A cooperative interaction between NF-κ B and Sp1 is required for HIV-1 enhancer activation. *EMBO J* **12:** 3551–3558.

Perng GC, Jones C. 2010. Towards an understanding of the herpes simplex virus type 1 latency-reactivation cycle. *Interdiscip Perspect Infect Dis* **2010:** 262415.

Persaud D, Pierson T, Ruff C, Finzi D, Chadwick KR, Margolick JB, Ruff A, Hutton N, Ray S, Siliciano RF. 2000. A stable latent reservoir for HIV-1 in resting CD4$^+$ T lymphocytes in infected children. *J Clin Invest* **105:** 995–1003.

Persaud D, Siberry GK, Ahonkhai A, Kajdas J, Monie D, Hutton N, Watson DC, Quinn TC, Ray SC, Siliciano RF. 2004. Continued production of drug-sensitive human immunodeficiency virus type 1 in children on combination antiretroviral therapy who have undetectable viral loads. *J Virol* **78:** 968–979.

Piatak M Jr, Saag MS, Yang LC, Clark SJ, Kappes JC, Luk KC, Hahn BH, Shaw GM, Lifson JD. 1993. High levels of HIV-1 in plasma during all stages of infection determined by competitive PCR. *Science* **259:** 1749–1754.

Richman DD, Margolis DM, Delaney M, Greene WC, Hazuda D, Pomerantz RJ. 2009. The challenge of finding a cure for HIV infection. *Science* **323:** 1304–1307.

Rohr O, Marban C, Aunis D, Schaeffer E. 2003. Regulation of HIV-1 gene transcription: From lymphocytes to microglial cells. *J Leukocyte Biol* **74:** 736–749.

Ross EK, Buckler-White AJ, Rabson AB, Englund G, Martin MA. 1991. Contribution of NF-κB and Sp1 binding motifs to the replicative capacity of human immunodeficiency virus type 1: Distinct patterns of viral growth are determined by T-cell types. *J Virol* **65:** 4350–4358.

Ruff CT, Ray SC, Kwon P, Zinn R, Pendleton A, Hutton N, Ashworth R, Gange S, Quinn TC, Siliciano RF, et al. 2002. Persistence of wild-type virus and lack of temporal structure in the latent reservoir for human immunodeficiency virus type 1 in pediatric patients with extensive antiretroviral exposure. *J Virol* **76:** 9481–9492.

Sahu GK, Paar D, Frost SD, Smith MM, Weaver S, Cloyd MW. 2009. Low-level plasma HIVs in patients on prolonged suppressive highly active antiretroviral therapy are produced mostly by cells other than CD4 T-cells. *J Med Virol* **81:** 9–15.

Schnell G, Spudich S, Harrington P, Price RW, Swanstrom R. 2009. Compartmentalized human immunodeficiency virus type 1 originates from long-lived cells in some subjects with HIV-1-associated dementia. *PLoS Pathog* **5:** e1000395.

Schroder AR, Shinn P, Chen H, Berry C, Ecker JR, Bushman F. 2002. HIV-1 integration in the human genome favors active genes and local hotspots. *Cell* **110:** 521–529.

Shan L, Yang HC, Rabi SA, Bravo HC, Irizarry RA, Zhang H, Margolick JB, Siliciano JD, Siliciano RF. 2011. Influence of host gene transcription level and orientation on HIV-1 latency in a primary cell model. *J Virol* doi: 101128/JVI02536-10.

Shao Z, Ruppert S, Robbins PD. 1995. The retinoblastoma-susceptibility gene product binds directly to the human TATA-binding protein-associated factor TAFII250. *Proc Natl Acad Sci* **92:** 3115–3119.

Shen A, Zink MC, Mankowski JL, Chadwick K, Margolick JB, Carruth LM, Li M, Clements JE, Siliciano RF. 2003. Resting CD4$^+$ T lymphocytes but not thymocytes provide a latent viral reservoir in a simian immunodeficiency virus-Macaca nemestrina model of human immunodeficiency virus type 1-infected patients on highly active antiretroviral therapy. *J Virol* **77:** 4938–4949.

Shen L, Peterson S, Sedaghat AR, McMahon MA, Callender M, Zhang H, Zhou Y, Pitt E, Anderson KS, Acosta EP, et al. 2008. Dose-response curve slope sets class-specific limits on inhibitory potential of anti-HIV drugs. *Nat Med* **14:** 762–766.

Siekevitz M, Josephs SF, Dukovich M, Peffer N, Wong-Staal F, Greene WC. 1987. Activation of the HIV-1 LTR by T cell mitogens and the trans-activator protein of HTLV-I. *Science* **238:** 1575–1578.

Siliciano JD, Kajdas J, Finzi D, Quinn TC, Chadwick K, Margolick JB, Kovacs C, Gange SJ, Siliciano RF. 2003. Long-term follow-up studies confirm the stability of the latent reservoir for HIV-1 in resting CD4$^+$ T cells. *Nat Med* **9:** 727–728.

Siliciano JD, Lai J, Callender M, Pitt E, Zhang H, Margolick JB, Gallant JE, Cofrancesco J Jr, Moore RD, Gange SJ, et al. 2007. Stability of the latent reservoir for HIV-1 in patients receiving valproic acid. *J Infect Dis* **195:** 833–836.

Sobhian B, Laguette N, Yatim A, Nakamura M, Levy Y, Kiernan R, Benkirane M. 2010. HIV-1 Tat assembles a multifunctional transcription elongation complex and stably associates with the 7SK snRNP. *Mol Cell* **38:** 439–451.

Spivak AM, Rabi SA, McMahon MA, Shan L, Sedaghat AR, Wilke CO, Siliciano RF. 2011. Dynamic constraints on the second phase compartment of HIV-infected cells. *AIDS Res Hum Retrov* doi: 10.1089/aid.2010.0199.

Strain MC, Gunthard HF, Havlir DV, Ignacio CC, Smith DM, Leigh-Brown AJ, Macaranas TR, Lam RY, Daly OA, Fischer M, et al. 2003. Heterogeneous clearance rates of long-lived lymphocytes infected with HIV: Intrinsic stability predicts lifelong persistence. *Proc Natl Acad Sci* **100:** 4819–4824.

Sun SC, Ganchi PA, Ballard DW, Greene WC. 1993. NF-κB controls expression of inhibitor I κ B α: Evidence for an inducible autoregulatory pathway. *Science* **259:** 1912–1915.

Sune C, Garcia-Blanco MA. 1995. Sp1 transcription factor is required for in vitro basal and Tat-activated transcription from the human immunodeficiency virus type 1 long terminal repeat. *J Virol* **69:** 6572–6576.

* Swanstorm R, Coffin J. 2011. HIV-1 pathogenesis: The virus. *Cold Spring Harb Perspect Med* doi: 10.1101/cshperspect.a007443.

Tahirov TH, Babayeva ND, Varzavand K, Cooper JJ, Sedore SC, Price DH. 2010. Crystal structure of HIV-1 Tat complexed with human P-TEFb. *Nature* **465:** 747–751.

Tamaru H. 2010. Confining euchromatin/heterochromatin territory: Jumonji crosses the line. *Genes Dev* **24:** 1465–1478.

Tobin NH, Learn GH, Holte SE, Wang Y, Melvin AJ, McKernan JL, Pawluk DM, Mohan KM, Lewis PF, Mullins JI, et al. 2005. Evidence that low-level viremias during effective highly active antiretroviral therapy result from two processes: Expression of archival virus and replication of virus. *J Virol* **79:** 9625–9634.

Van Lint C, Emiliani S, Ott M, Verdin E. 1996. Transcriptional activation and chromatin remodeling of the HIV-1 promoter in response to histone acetylation. *EMBO J* **15:** 1112–1120.

Verdin E, Paras P Jr, Van Lint C. 1993. Chromatin disruption in the promoter of human immunodeficiency virus type 1 during transcriptional activation. *EMBO J* **12:** 3249–3259.

Wei X, Ghosh SK, Taylor ME, Johnson VA, Emini EA, Deutsch P, Lifson JD, Bonhoeffer S, Nowak MA, Hahn BH, et al. 1995. Viral dynamics in human immunodeficiency virus type 1 infection. *Nature* **373:** 117–122.

Wei P, Garber ME, Fang SM, Fischer WH, Jones KA. 1998. A novel CDK9-associated C-type cyclin interacts directly with HIV-1 Tat and mediates its high-affinity, loop-specific binding to TAR RNA. *Cell* **92:** 451–462.

Weinberger LS, Burnett JC, Toettcher JE, Arkin AP, Schaffer DV. 2005. Stochastic gene expression in a lentiviral positive-feedback loop: HIV-1 Tat fluctuations drive phenotypic diversity. *Cell* **122:** 169–182.

Williams SA, Chen LF, Kwon H, Fenard D, Bisgrove D, Verdin E, Greene WC. 2004. Prostratin antagonizes HIV latency by activating NF-κB. *J Biol Chem* **279:** 42008–42017.

Williams SA, Chen LF, Kwon H, Ruiz-Jarabo CM, Verdin E, Greene WC. 2006. NF-κB pp50 promotes HIV latency through HDAC recruitment and repression of transcriptional initiation. *EMBO J* **25:** 139–149.

Wong JK, Hezareh M, Gunthard HF, Havlir DV, Ignacio CC, Spina CA, Richman DD. 1997. Recovery of replication-competent HIV despite prolonged suppression of plasma viremia. *Science* **278:** 1291–1295.

Yang HC. 2011. Primary cell models of HIV latency. *Curr Opin HIV AIDS* **6:** 62–67.

Yang Z, Zhu Q, Luo K, Zhou Q. 2001. The 7SK small nuclear RNA inhibits the CDK9/cyclin T1 kinase to control transcription. *Nature* **414:** 317–322.

Yang Z, Yik JH, Chen R, He N, Jang MK, Ozato K, Zhou Q. 2005. Recruitment of P-TEFb for stimulation of transcriptional elongation by the bromodomain protein Brd4. *Mol Cell* **19:** 535–545.

Yeung F, Hoberg JE, Ramsey CS, Keller MD, Jones DR, Frye RA, Mayo MW. 2004. Modulation of NF-κB-dependent transcription and cell survival by the SIRT1 deacetylase. *EMBO J* **23:** 2369–2380.

Ylisastigui L, Archin NM, Lehrman G, Bosch RJ, Margolis DM. 2004. Coaxing HIV-1 from resting CD4 T cells: Histone deacetylase inhibition allows latent viral expression. *AIDS* **18:** 1101–1108.

Zack JA, Arrigo SJ, Weitsman SR, Go AS, Haislip A, Chen IS. 1990. HIV-1 entry into quiescent primary lymphocytes: Molecular analysis reveals a labile, latent viral structure. *Cell* **61:** 213–222.

Zhou Y, Zhang H, Siliciano JD, Siliciano RF. 2005. Kinetics of human immunodeficiency virus type 1 decay following entry into resting CD4+ T cells. *J Virol* **79:** 2199–2210.

Zhu T, Muthui D, Holte S, Nickle D, Feng F, Brodie S, Hwangbo Y, Mullins JI, Corey L. 2002. Evidence for human immunodeficiency virus type 1 replication in vivo in CD14+ monocytes and its potential role as a source of virus in patients on highly active antiretroviral therapy. *J Virol* **76:** 707–716.

HIV-1-Related Central Nervous System Disease: Current Issues in Pathogenesis, Diagnosis, and Treatment

Serena Spudich[1] and Francisco González-Scarano[2]

[1]Department of Neurology, Yale University School of Medicine, New Haven, Connecticut 06520

[2]Department of Neurology, The University of Texas School of Medicine at San Antonio, San Antonio, Texas 78229

Correspondence: scarano@mail.med.upenn.edu

HIV-associated central nervous system (CNS) injury continues to be clinically significant in the modern era of HIV infection and therapy. A substantial proportion of patients with suppressed HIV infection on optimal antiretroviral therapy have impaired performance on neuropsychological testing, suggesting persistence of neurological abnormalities despite treatment and projected long-term survival. In the underresourced setting, limited accessibility to antiretroviral medications means that CNS complications of later-stage HIV infection continue to be a major concern. This article reviews key recent advances in our understanding of the neuropathogenesis of HIV, focusing on basic and clinical studies that reveal viral and host features associated with viral neuroinvasion, persistence, and immunopathogenesis in the CNS, as well as issues related to monitoring and treatment of HIV-associated CNS injury in the current era.

HIV-1 infects the nervous system in virtually all patients with systemic infection and frequently causes central nervous system (CNS) and peripheral nervous system (PNS) disorders. Until the introduction of combination antiretroviral therapy (cART) in the mid-1990s, HIV-1-associated dementia (HAD) and related cognitive and motor disorders affected 20%–30% of patients with advanced immunosuppression or AIDS. The incidence of overt HAD in countries where effective combination antiretroviral medications are widely available is now markedly diminished. However, in the setting of chronic, apparently systemically suppressive treatment, there appears to be a continued prevalence of mild-moderate neurocognitive impairment in a significant proportion or even a majority of patients. This disquieting finding, combined with the staggering numbers of patients who continue to be newly infected with HIV worldwide, and the limited availability of optimal antiretroviral treatment in many of the persons affected with this condition, make understanding and effectively preventing HIV-1-related neurological injury a continued key area of investigation. To encompass this more complex range of disorders seen in patients treated with cART, most investigators now refer to HIV-associated neurocognitive disorders (HAND) rather than HAD as the principal primary CNS complication of HIV infection.

HISTORY

A dementing illness characterized by attention and memory deficits, motor impairment, and personality changes was recognized in a significant proportion of patients with advanced AIDS within the first years of the HIV epidemic (Navia et al. 1986b). Further investigation of this disorder revealed that these complications were a direct result of HIV-1 infection and attendant inflammation in the CNS. The neuropathology was characterized by diffuse brain atrophy with large ventricles, widespread low-grade inflammation with microglial nodules, perivascular lymphocyte cuffing, multinucleated cells expressing HIV p24 and other antigens, and patchy demyelination and white matter gliosis (Gabuzda et al. 1986; Navia et al. 1986a). Although inexorably progressive to severe disability and death in the absence of disease-modifying HIV therapy, the course of this clinical disorder has been altered considerably by treatment with antiretroviral therapy and especially cART. Originally defined as the AIDS-dementia complex (ADC) based on motor, cognitive, and behavioral symptoms and signs, current research nosology defines a broader spectrum now called "HIV-associated neurocognitive disorder," with graded classifications based on abnormal performance on neuropsychological testing, and the presence or absence of a patient's perception of functional limitation related to cognitive impairment (Antinori et al. 2007). Changes in the severity of neurological disease in the current era may also be accompanied by alterations in the underlying etiology of neurological morbidity in the setting of long-term survival with HIV, including the consequences of possible ongoing low-grade viral replication and inflammation within the CNS, cumulative exposure to antiretroviral and other medications, chronic systemic inflammation leading to accelerated vascular disease, and the effects of comorbidities and neurodegeneration that occur with aging. Additionally, because cART appears to be beneficial in the amelioration and prevention of the most severe forms of HAND, newfound attention has been focused on the possible long-term cognitive benefits of initiation of cART in early stages of HIV infection.

KEY ADVANCES IN THE AREA

Viral Entry and Maintenance of Infection in the Nervous System

As with some other viruses that circulate in the bloodstream, HIV entry into the CNS is largely mediated through blood lymphocytes and monocytes that enter the perivascular spaces either in the course of their natural surveillance, or because they are attracted by chemokines to sites of inflammation. Viral strains isolated from the brain are more commonly CCR5-tropic and replicate effectively in cultured macrophages, suggesting that monocytes may predominate as "Trojan horses" in the process of CNS entry, as described years ago for classically described lentiviruses such as visna virus of sheep (Haase 1986). Alternatively, HIV may be brought into the CNS by lymphocytes, which can harbor viruses that replicate in macrophages (Collman et al. 1992), or conceivably as free virions, where the means of entry would be through endothelial cells. Regardless of the mechanism of entry, cells of the macrophage lineage are the only cells in the brain that are routinely found to harbor HIV antigens or RNA by conventional methods such as immunohistochemistry or in situ hybridization for viral RNA, although other cell types such as astrocytes may harbor HIV sequences without robust expression of RNA (or proteins) (Wiley et al. 1986). Detection of such infection requires other methodology such as in situ polymerase chain reaction (PCR) amplification or laser capture microdissection followed by PCR (Churchill et al. 2009). Recent studies have also concentrated on determining whether a subset of monocytes is particularly important in either delivering virus to the CNS or in amplifying a local inflammatory process, whether such a subset is increased or enhanced by systemic inflammation mediated by circulating bacterial products, and which cytokines and chemokines increase the recruitment of monocytes and lymphocytes into the CNS and by extension are more likely to deliver virus into the brain (Fig. 1).

Although there is general agreement that the predominant infected cells are macrophage-like, there is controversy regarding which of the several subtypes of CNS macrophages are harboring HIV (macrophage subtypes are reviewed by Perry et al. 2010). Most investigators agree that perivascular macrophages, which are mostly derived from the circulating monocytes, are highly infected in the brains of HIV-infected persons or in macaques (rhesus or pig-tailed) that are infected experimentally with SIV (Kim et al. 2006); these may be labeled by CD163, a marker for this subtype that also appears to be increased in circulating monocytes in macaques with CNS infection (Borda et al. 2008). The life span of these perivascular macrophages was previously thought to be days or weeks; recent studies in rhesus macaques have indicated that they are probably longer-lived (Soulas et al. 2009). Multinucleated giant and other CNS lesions containing SIV in an experimental model or HIV in

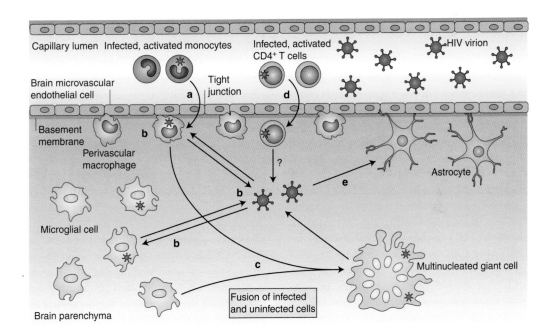

Figure 1. Potential models for HIV neuroinvasion and infection of the central nervous system (CNS). (a) HIV-infected monocytes with an activated phenotype may transport HIV into the nervous system via migration across the blood–brain barrier. (b) Infected monocytes likely differentiate into perivascular tissue macrophages and proceed to produce HIV within the CNS. This macrophage infection and replication allows for release of free virions and may facilitate infection of microglial cells. (c) Cell-to-cell fusion involving cells that express CD4 and HIV coreceptors results in formation of multinucleated giant cells within the brain, a hallmark of HIV-related brain pathology. (d) Infected CD4+ T lymphocytes may also serve as a mechanism of entry of HIV into the brain. There is varied evidence regarding the relative contribution of CD4+ T lymphocytes versus cells of the monocyte/macrophage in initiating and sustaining HIV infection within the CNS. (e) Although astrocytes might harbor HIV and also contribute to HIV-related brain disease through mechanisms of astrogliosis induced by local chemokines and cytokines, astrocytes infection is not thought to support ongoing replication within the CNS. (Adapted from Gonzalez-Scarano et al. 2005; with permission, from Macmillan Publishers Ltd. © 2005.)

autopsy samples are composed of cells expressing different macrophage markers, including CD163, CD68, and CD387; some cells appear productively infected; others do not express viral antigens (Soulas et al. 2011). Nevertheless, although it has previously been suggested that because of their rapid turnover perivascular macrophages could not contribute to the long-term presence of HIV/SIV within the CNS, which could then serve as a reservoir when systemic infection was cured, these more recent discoveries would propose that indeed this perivascular population could harbor virus for long periods of time. Parenthetically, the astrocytes found to be infected using sensitive methodologies are often close to the perivascular spaces.

Microglia, which are parenchymal, or deeper within the brain, are known to be long-lived and are replaced only infrequently during an individual's lifetime. Recent studies in mice have indicated that they may represent a separate ontogeny within bone marrow cells (Ginhoux et al. 2010), although hematopoietic circulating cells could potentially also give rise to morphologically appearing microglial cells, albeit a small minority. However, there is some controversy as to whether true microglia are commonly infected by HIV, and some investigators believe that they are never infected. Multinucleated giant cells, the pathological hallmark of HIV in the CNS, arise from macrophage-type cells, and although their frequent perivascular location would suggest they are from that perivascular macrophage population, there are no markers that can reliably confirm this. Furthermore, the life span of multinucleated giant cells is completely unknown, although one can speculate based on their presence in pathological specimens that they last at least days.

Studies performed well over a decade ago first proposed that a subset of monocytes, characterized by expression of the markers CD14 and CD69, were particularly prominent in patients with HIV neurological complications and specifically HAD (Pulliam et al. 1997). More recently, investigators from the same group related these original findings to a less impressive but still measurable increase in CD14/CD69 positive cells in patients on cART with dementia in comparison to those without (Kusdra et al. 2002). Similarly, in the rhesus macaque model of SIV encephalitis (SIVE) increased monocyte turnover and expression of the CD163 marker are associated with brain penetration and encephalitic changes (Burdo et al. 2010).

While these observations were developing, independent evidence that systemic inflammation driven by depletion in the gut immunological system and consequent microbial translocation had a role in HIV pathogenesis arose from several areas, including animal models and human observations (see Lackner et al. 2011 for details). This led to a series of discoveries suggesting that microbial translocation and concomitant immunological activation are associated with the presence of neurological complications in HIV infections (Ancuta et al. 2008). Furthermore, this finding may explain why there has been a strong correlation between inflammatory activity, as characterized by the presence of macrophages expressing activation markers in the CNS, and the development of HAND or HAD. This correlation may be as strong as that of the presence of viral proteins and other evidence of HIV replication.

A model that incorporates current concepts of systemic pathogenesis and the CNS-specific observations would then propose that infection of the CNS is driven by systemic activation of monocytes—at least partly owing to microbial translocation from a depleted gut immune system—which then are more likely to invade the brain perivascular space. As some of these cells are infected, they bring in virus that spreads locally and sets up a nidus of replication independent from the systemic circulation (see next section).

Concomitantly, the perivascular inflammation results in the secretion of cytokines and chemokines that in turn amplify the reaction, attracting in addition other circulating monocytes and infected $CD4^+$ T lymphocytes that can also add to the CNS viral burden (Xing et al. 2009). Chief among the chemokines associated with HIV infection of the CNS is MCP1 (CCL2), which is present in easily measurable concentrations in the cerebrospinal fluid (CSF) and is associated with dementia, but also IP10 and others (see next sections).

CNS Compartmentalization

Studies in acute HIV infection have shown that virus is present in the CSF at early points during HIV infection, including in some patients with primary infection (Schnell et al. 2010); whether such an early seeding forms the basis for independent replication in the brain, or whether it is cleared and virus penetrates at other points in the course of the infection, has been the subject of many excellent studies without a clear consensus (Caragounis et al. 2008; Harrington et al. 2009). It is likely that different scenarios take place depending on the host, the route of infection, and possibly the individual isolates. Nevertheless, although the details may vary, long-term HIV infection leads to genetically isolated populations in the CNS, as evidenced by studies using *pol*, *env*, *nef*, and other genes (Thomas et al. 2007; Brown et al. 2011; Cowley et al. 2011; Gray et al. 2011). The most recent studies (Brown et al. 2011) have used single genome analysis (SGA) to overcome potential PCR artifacts and confirmed the conclusions derived with bulk amplification before the development of SGA. Additionally, studies in experimental infection of rhesus macaques with a cloned SIV isolate suggested that virus is not only compartmentalized in the CNS, but that different regions have potentially different *env* genotypes, setting the stage for independent entry events, some potentially early in the course of infection, some perhaps much later (Chen et al. 2006; Reeve et al. 2010).

Important yet still unresolved questions are whether the genetic compartmentalization observed in the CNS is the result of a founder effect with concomitant independent divergence, or whether there are specific selective pressures that promote selection, and what those selective pressures might be. Most likely the end result observed in cross-sectional analyses such as the ones cited previously are attributable to a combination of factors. There have been comparatively few studies designed to

differentiate between genetic divergence owing to a founder effect and adaptive evolution; even fewer have estimated the timing of divergence in the CNS. Most investigators who have compared the rate of nonsynonymous to synonymous changes have concluded that evolution in *pol*, *env*, or *nef* in brain isolates is adaptive (Huang et al. 2002; Thomas et al. 2007b; Gray et al. 2011). What is less clear is which specific pressures are driving the adaptation. Theoretically, enhanced replication in macrophages—elaborated below—response to antiretrovirals, or the peculiarities of the immune response within the CNS could be playing a role.

Chief among the potential selective pressures are the requirement for robust replication in macrophages. Macrophages express CD4 at lower levels than CD4$^+$ lymphocytes (Lee et al. 1999). Accordingly, HIV isolates from the CNS tend to have an increased capacity to use reduced levels of CD4 for entry and infection (Peters et al. 2004; Martín-Garcia et al. 2006; Thomas et al. 2007a); however, many of these isolates have been obtained postmortem, and could reflect an end-stage phenotype. Similarly, the CNS is an environment with a relatively low penetration of antibodies, and configurations that might promote neutralization are better "tolerated" under these circumstances of "immunological privilege." In fact, isolates from the brain have been shown to be sensitive to neutralizing sera, and particularly to a monoclonal antibody (b12) that overlaps the CD4 binding site (van Marle et al. 2002; Martín-Garcia et al 2005; Dunfee et al. 2007).

Antiretroviral use and its penetration into the CNS is another potential selective mechanism for HIV strains in this regard, and a few studies have shown discordance between the resistance phenotypes in the blood and CSF-derived strains (Haas 2004; Smit et al. 2004). However, the turnover of virus-infected cells can be different between the plasma and CSF compartments (Schnell et al. 2009) consistent with the predominantly infected cell type in each compartment—lymphocytes in the circulation, longer-lived macrophages in the CNS—and discordance in the sensitivity to antiretrovirals may be attributable to viral genetic information being obtained as a "snap-shot" rather than representing a true biological phenomenon (Haas 2004). For example, resistance genotypes detected in the circulation at one point may potentially not appear in the CSF until later because of slower replication cycles.

Mechanisms of CNS Injury

The pathophysiology of HAD and HAND must eventually involve neurons, the principal effector cells of the nervous system. To understand the apparent paradox of cognitive and motor symptoms in what is principally a macrophage and microglial infection, investigators have developed a number of in vitro and in vivo models, many using combinations of infected monocyte-derived macrophages with mammalian (often rat, but also human) neurons. A somewhat simplistic summary of the vast literature on this subject divides the putative neurotoxic molecules into those that are the direct result of virus in the extracellular fluid, and those that propose that neurotoxicity is the end result of macrophage and microglial reaction to a chronic infection with HIV.

Among the viral proteins that have been implicated in neurotoxicity are gp120, Tat, Vpu, and Vpr. Of these, gp120 and Tat have received the widest attention, and unquestionably they can cause neurotoxicity in vitro. More complicated is the question of whether the concentrations of proteins that can be achieved in the in vivo extracellular fluid ever approach the concentrations required to affect neurons. For gp120, this may be the case, but extracellular concentrations of Tat in the nanogram range are difficult to visualize.

A second pathway of neurotoxicity involves the production of potentially neurotoxic factors in association with macrophage infection. Among the factors implicated in neurotoxicity are quinolinic acid, tumor necrosis factor, platelet activating factor, and arachidonic acid metabolites. Some of these have been detected in the CSF, and related to neurocognitive functioning, whereas others have been primarily tested in vitro. Many investigators have

proposed that a common end pathway of toxicity is through excitation of N-methyl-D-aspartate (NMDA)-subtype glutamate receptors, which has the potential for mediating apoptosis. Evidence supporting this model includes experiments that show that decreasing glutamate secretion by infected macrophages is partially protective in an in vitro model of macrophage-mediated neurotoxicity (O'Donnell et al. 2006), that toxicity associated with viral proteins is dependent on expression of these receptors, and that the areas that are most affected are associated with a concentration of NMDA receptors.

An intriguing recent finding, seemingly unrelated to HIV, comes from the work of Shau-Kwaun Chen and colleagues. They described a mouse mutant in the *Hoxb8* gene that has a phenotype of excessive grooming (Chen et al. 2010). However, in the relevant CNS regions, this gene is normally expressed in microglia only, and the phenotype could be rescued by bone marrow transplantation. This article raises the possibility that microglia can affect complex behaviors even in the absence of neurodegeneration, although it is still quite possible that the effects are attributable to aberrant secretion of cytokines. As such, it opens the door for a mechanism of HIV effect, potentially reversible, that does not depend on the classic findings of neuronal dropout, and could occur without such histopathological changes.

Strain-Specific Neuropathogenesis

A number of studies have related specific HIV sequences, primarily in gp120, to the development of HAND. Among these, those that associated either tropism in macrophages/microglia with specific genotypes are the most worthy of note. For example, position 306 influenced M tropism and CCR5 binding in a subset of brain-derived isolates (Dunfee et al. 2007). Similarly, a variant at position 283 (N283) was associated with brain infection and also had enhanced tropism for macrophages. These collective findings suggest that changes that enhance M tropism are associated with brain infection, but also that specific mutations are often context-dependent and that a universal brain signature is unlikely.

Similarly, in view of the considerable burden of HIV infection in Africa as well as other developing countries, many investigators have examined the prevalence of neurological disorders in individuals who are infected with HIV clades other than the clade B that is predominant in the developed world. Those studies are difficult, because neurological and psychiatric care is suboptimal in developing countries, and because most if not all of the more sophisticated instruments that are used to determine CNS involvement depend on cultural context. Nevertheless, it is clear that HIV strains from clades other than B are associated with HAND. For example, Mahadevan et al. (2007) studied the brains of patients with clade C infection and found evidence of p24 antigen expression in macrophage-like cells in patients who had opportunistic infections such as toxoplasmosis. The pattern was similar to that of clade B CNS infiltration, but there were no multinucleated giant cells. In addition, Sacktor and coworkers in Uganda (Sacktor et al. 2009) showed that dementia occurs in patients with clade D infection, possibly in greater proportion than in patients with clade A infection in the same region.

Biomarkers of CNS Disease

As the brain and spinal cord are relatively inaccessible for assessment, surrogate biomarkers of CNS disease may provide some insight into ongoing processes relevant to HIV infection. However, a major problem has been identifying markers that are specific enough and also measurable with assays routinely available.

Measurement of HIV RNA in the CSF is the most practical means of assessing CNS "viral load." CSF HIV RNA is ubiquitous during chronic untreated HIV infection, with levels that trend as the levels in blood but typically are 10-fold lower in absolute terms (Spudich et al. 2005; Marra et al. 2007). In untreated patients, HIV RNA levels may be higher in patients with active neurological disease as compared with asymptomatic individuals (Brew et al.

1997; Robertson et al. 1998). However, the CSF HIV RNA may arise from sources other than the CNS, and in addition to the brain, may also reflect virus in the systemic circulation that has been transported or "leaked" to the nervous system. Importantly, genetic compartmentalization of virus and detection of divergent viral quasispecies between these tissues indicates that CSF HIV is not entirely a spillover from that present in blood (see previous sections for references).

With the recognition that markers of cellular activation and inflammation were useful indicators of disease activity in systemic HIV infection, attention turned to the utility of following such measures in the CNS compartment. Soluble CSF markers of macrophage activation (neopterin), chemokines stimulating ingress of macrophages and lymphocytes across the blood–brain barrier (CCL2/MCP1 and CXCL10/IP10), and molecules involved at various stages in the pathways for cell turnover and activation within the nervous system compartment are used to monitor processes that are thought to serve as the substrate for neuropathology in HAND (for review, see Cinque et al. 2007). In one small study, moderately elevated CSF neopterin predicted subsequent progression to HAD (Brew et al. 1996). However, although such markers have been correlated to disease activity, they have not been clinically used for diagnosis or monitoring of Neuro-AIDS owing to lack of specificity for active neurological disease in the setting of the immune activation characterizing HIV infection (Gisslén et al. 2009). Recently, attention has turned to plasma markers related to immunopathogenesis of systemic HIV, including soluble CD14 and lipopolysaccharide (Ancuta et al. 2008; Sun et al. 2010). Direct markers of neurological injury assayed in CSF, including neurofilament light chain protein (NFL), tau protein, and precursors and products of amyloid protein (amyloid precursor proteins and Aβ1-42) may be more valuable as measures of active neurodegeneration or injury (Hagberg et al. 2000; Gisslen et al. 2007, 2009; Clifford et al. 2009).

Imaging of the brain has been extensively investigated and used; overt HAD may be characterized by cerebral atrophy with or without periventricular white matter hyperintensities, which are diffuse, largely symmetric, and not characterized by edema or mass effect (Price et al. 1991). However, these findings are neither specific for nor ubiquitous in HAD, especially in its earlier stages, and, conversely, brain atrophy is noted in many neuroasymptomatic HIV-infected patients. Magnetic resonance spectroscopy (MRS), which detects cellular and biochemical processes based on diffusion of molecules through cerebral tissues, has yielded more specific insight into the inflammatory and neuronal processes occurring in the nervous system throughout the course of HIV infection. Overt HAD is associated with reduced relative levels of N-acetylaspartate, indicating decreased neuronal function, and elevated levels of choline, associated with brain inflammation and membrane turnover (Meyerhoff et al. 1994; Chang 1995). Similar, although less severe, patterns are seen in asymptomatic, untreated HIV infection, indicating that MRS may be a valuable preclinical marker of active CNS disease (Meyerhoff et al. 1999). Although changes in cerebral metabolites may indicate regional inflammation and neuronal injury, more subtle and potentially more neuropathologically relevant information may be obtained by systematic evaluation of white matter tracts or white matter morphometry in the brain by diffusion tensor imaging (DTI). Some early work in this area suggests that in neuroasymptomatic HIV, there are reductions in major white matter tracts in a number of brain regions (Pomara et al. 2001; Thurnher et al. 2005; Chang et al. 2008). More global sophisticated brain morphometry measurement may be used to detect focal atrophy of gray or white matter structures (Wang et al. 2009). Functional magnetic resonance imaging (fMRI), which takes advantage of the fact that hemodynamics in the brain are closely linked to neural activity, uses techniques that measure cerebral blood flow and blood oxygen level dependence (BOLD) signals. Early studies in this area show reduced baseline cerebral blood flow and increased functional demand in the brain parenchyma in HIV-infected patients

(Ances et al. 2008, 2010). Detection of abnormalities in neuroasymptomatic patients underscores the potential utility of functional MRI as assessment before development of overt neurological disease. Similar pathology of impaired blood flow may be detected by simpler cerebral perfusion imaging, which may additionally have a role in the assessment and monitoring of HIV-related CNS disease (Ances et al. 2009).

Beneficial Effects of Antiretroviral Therapy in HIV-Associated CNS Disease

Combination antiretroviral therapy has had a dramatic beneficial impact on the incidence and prevalence of severe forms of HAND or HAD. The Euro-SIDA cohort study clearly showed a decline in incidence of severe dementia (then termed AIDS-dementia complex, or ADC), related to introduction of protease inhibitors and use of cART (d'Arminio Monforte et al. 2004). More recent evidence from the CHARTER study indicates a greatly reduced prevalence (2% overall) of severe HAD in a cohort of HIV-infected individuals in the current era (recruited between 2003 and 2007) (Heaton et al. 2010). This improvement in the prevalence and incidence of severe HAND with cART reflects the generally beneficial effect of initiation of cART on neurocognitive performance witnessed in studies of initiation of antiretroviral therapy (Marra et al. 2003; Robertson et al. 2004; Cysique et al. 2006).

What are the biological underpinnings of this improvement? Blood and CSF viral burden are clearly reduced by cART (Marra et al. 2003; Spudich et al. 2005, 2006), and the initiation of cART is associated with sequential reduction in HIV RNA levels in both compartments over time (Ellis et al. 2000). CNS inflammation, the putative substrate of ongoing CNS injury in the setting of HIV, is also partly ameliorated by cART. Treatment is associated with reduced levels of markers of intrathecal inflammation, such as cellular markers of T-cell activation, CSF white blood cell (WBC), CSF neopterin, and β-2 microglobulin (Yilmaz et al. 2004; Spudich et al. 2006; Sinclair et al. 2008). Finally, markers of active neural injury in the CSF, including CSF NFL and tau protein are reduced in the setting of cART and have been observed to decay over time in subjects initiating therapy (Mellgren et al. 2007).

Persistent Evidence of HAND despite Antiretroviral Therapy

HAND is a clinical diagnosis, currently defined based on abnormal cognitive and motor performance on neuropsychological tests according to criteria that denote three levels of HIV-associated neurological disease: asymptomatic neurocognitive impairment (ANI), mild neurocognitive disorder (MND), and HAD (Antinori et al. 2007, see Table 1). A number of recent studies have documented persistence of neurocognitive abnormalities, predominantly along the milder spectrum of HAND, in the setting of cART, with a prevalence ranging between 18% and 52% in varied settings (Robertson et al. 2007; Heaton et al. 2010; Cysique and Brew 2011). Studies including subjects with comorbidities that potentially confound the diagnosis of HAND, including current or past substance abuse, mental health disorders, head trauma, low education level, and coinfection with hepatitis, find a higher prevalence (up to 83% HAND in the highest "comorbidity" group in CHARTER), whereas those excluding subjects with significant relevant comorbidities and with low CD4 nadirs had

lower, but not negligible, frequency of impairment (Table 1) (Heaton et al. 2010).

The dramatic change in the severity and perhaps phenotype of the clinical disorders associated with HIV suggests that the etiology of Neuro-AIDS may have altered during this time. A small number of studies have focused on brain pathology in the era since the introduction of cART. Examination of 589 brains obtained mainly from subjects on antiretroviral therapy available through the National Neuro-AIDS Tissue Consortium (NNTC) revealed a significant reduction in the proportion of subjects with typical HIV-related brain pathology (including encephalitis, microglial nodules, and leukoencephalopathy) compared with the pre-antiretroviral era (Everall et al. 2009; Heaton et al. 2010). However, 78% overall had neuropathological abnormalities including vascular pathologies, Alzheimer type II gliosis, and other infectious and noninfectious pathologies. Premorbid HAND diagnosis (in 82%) in this cohort did not correlate with HIV-related brain pathology.

Although classical HIV-related brain pathology may no longer be the only substrate for HAND, HIV-driven mechanisms may still be important CNS abnormalities in the setting of cART. Real-time PCR quantification of HIV RNA in autopsy specimens from the NNTC has revealed detectable HIV-1 RNA in brain in a majority of cART-treated subjects (Kumar

Table 1. Diagnostic research criteria for HAND

Diagnostic entity	Cognitive performance	Functional status
Normal	Normal	Normal
Asymptomatic neurocognitive impairment (ANI)	Acquired impairment in at least two cognitive domains (<1 SD)	No perceived impact on daily function
Mild neurocognitive disorder (MND)	Acquired impairment in at least two cognitive domains (<1 SD)	Perceived interference with daily function to at least a mild degree (work inefficiency, reduced mental acuity)
HIV-associated dementia (HAD)	Acquired impairment in at least two domains, typically in multiple domains with at least two domains with severe impairment (<2 SD)	Marked impact on daily function

Adapted from Antinori et al. 2007; with permission, from Wolters Kluwer Health © 2007.

et al. 2007), and abnormal levels of microglial activation in the CNS were found to persist in neuroasymptomatic subjects with successful plasma viral suppression on cART in another recent neuropathological study (Anthony 2005). These studies are corroborated by evidence that intrathecal inflammation as detected in CSF persists in the setting of long-term, systemically effective cART (Eden et al. 2007; Yilmaz et al. 2008). Although even in the setting of therapy that fails to successfully suppress plasma HIV RNA (Spudich et al. 2006), cART usually successfully suppresses CSF HIV RNA below detectable levels, viral "escape" in the CSF may occur in up to 10% of individuals on current regimens (Eden et al. 2010), and in rare cases dramatic CSF "escape" has been associated with clinically progressive neurological disease (Canestri et al. 2010).

CNS Penetration of cART

The CNS is separated from the systemic circulation by blood–brain and blood–CSF barriers, thus potentially allowing for a "sanctuary" of infection that is only partially reached by some antiretroviral medications which, owing to molecular size or hydrophilicity, do not readily cross the blood–brain barrier. Based on structural composition and effect on CSF HIV RNA, antiretrovirals have been ranked according to their estimated CNS penetration effectiveness (CPE), with a combination of drugs in a regimen assigned a combined "CPE" score based on the sum of their individual rankings (Table 2) (Letendre et al. 2008). Given that treatment with cART benefits the nervous system in terms of both the detection and magnitude of HIV infection and attendant inflammation within the CNS, more potent activity of antiretroviral medications within the CNS might provide additional benefit in the setting of treatment of HAND. Despite predominant concurrence between numerous studies that regimens with higher CPE scores tend to lead to more successful suppression of HIV RNA levels in the CNS (Letendre et al. 2008; Marra et al. 2009), the evidence that enhanced CPE scores are related to improved neurocognitive outcomes in subjects with HIV infection is less definitive. Whereas some observational studies show a cognitive benefit of initiation of regimens with higher CPE scores in the setting of HIV-related CNS disease (Letendre et al. 2004; Tozzi et al. 2009), others show poorer neurocognitive performance in subjects treated with

Table 2. Antiretroviral central nervous system penetration effectiveness (CPE) scoring system used in recent clinical studies

Increasing CNS penetration →	0	0.5	1
Nucleoside reverse transcriptase inhibitors	Didanosine Tenofovir Zalcitabine Adefovir	Emtricitabine Lamivudine Stavudine	Abacavir Zidovudine
Nonnucleoside reverse transcriptase inhibitors		Efavirenz	Delavirdine Nevirapine
Protease inhibitors	Nelfinavir Ritonavir Saquinavir Saquinavir/r Tipranavir/r	Amprenavir Atazanavir Fosamprenavir Indinavir	Amprenavir/r Atazanavir/r Fosamprenavir/r Indinavir/r Lopinavir/r
Entry inhibitors	Enfuvirtide T-1249		Maraviroc Vicriviroc
Integrase inhibitors		Raltegravir Elvitegravir	

Adapted from Smurzynski et al. 2011; with permission, from Wolters Kluwer Health © 2011.

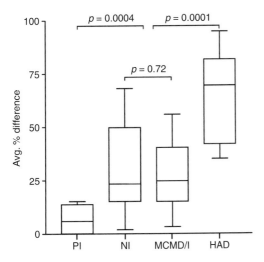

Figure 2. Heterduplex tracking assay (HTA) was used to compare V1/V2 and V4/V5 *env* populations in blood plasma (BP) and CSF from subjects with distinct stages of HIV infection and neurological status (PI, primary infection; NI, not impaired; MCMD/I, minor cognitive and motor disorder; HAD, HIV-associated dementia). The % difference for BP/CSF HTA band patterns was determined for each patient and the data compiled. A higher percentage difference indicates more discordant BP/CSF viral genetic populations for the particular region of *env* analyzed. The mean of V1/V2 and V4/V5 percent difference results was determined for each patient to reflect global *env* compartmentalization between BP and CSF, and results were compiled for comparison between the different disease categories. *p* values shown were determined by Wilcoxon rank-sum test. (Adapted from Harrington et al. 2009; with permission, from Wolters Kluwer Health © 2009.)

regimens with higher CPE scores (Marra et al. 2009). A variety of issues may lead to the discrepant results of such studies. High CPE regimens may be especially important in conditions of robust brain and meningeal infection with HIV, which typically occur in later stages of chronic infection, when targeted therapy to CNS tissues may be key for control of compartmentalized viral replication (Fig. 2). Furthermore, retrospective, observational studies may be biased; for example, subjects with more profound neurological deficits may be started on regimens with higher CNS penetration. Such deficits may not be entirely reversible and thus

persist despite high CPE regimens. A recent study did find a modest neurocognitive benefit of higher CPE scores (although only between patients with more than three drugs in their regimens and higher or lower CPE scores) in the context of randomized assignment of antiretroviral regimens (Smurzynski et al. 2011); additional randomized studies are warranted to directly examine this issue.

New Research Areas

The Relationship among Aging, New Comorbidities, and HAND

In resource-rich regions where life expectancy after HIV diagnosis in a young adult is estimated to be approximately 12 years below that of a noninfected individual (Lohse et al. 2007), longer duration of survival with chronic HIV-1 infection may change the scope and etiology of disorders affecting the CNS. A great deal of recent attention has been paid to the question of whether aging and HIV-1 will have a synergistic effect on neurodegeneration within the brain, "accelerating" injury triggered earlier by neuropathological processes associated with HIV. Systemic immunologic changes noted in HIV infection are in part characterized by markers of immunosenesence, and chronic infection with HIV is associated with the earlier onset of cancers and vascular disorders seen in HIV-uninfected persons at more advanced age (Desai and Landay 2010). However, the evidence that aging and HIV have an enhanced combined deleterious effect in the brain is mixed. A large cohort study enrolling subjects with HIV and advanced age, the Hawaii aging cohort, found a higher prevalence of cognitive deficits in older HIV-infected patients, related to CD4 nadir, insulin resistance, and presence of Apo E alleles (Valcour et al. 2004), and a recent neuroimaging study found that HIV and aging caused independent reductions in cerebral blood flow (Ances et al. 2010). However, the Multicenter AIDS Cohort Study (MACS) found no difference between older HIV$^+$ individuals and older HIV$^-$ individuals in rate of change in performance on longitudinal testing, suggesting the absence of an acceleration of neurological disease with aging (Becker et al. 2009). Two additional recent studies using well-matched HIV-uninfected comparison groups in analyses of age effects on neurocognitive performance showed no combined effects of HIV and age on cognitive function (Cysique et al. 2011; Valcour et al. 2011). Other conditions are emerging in the setting of long-term survival with chronic HIV infection that may influence the integrity of the CNS and alter the substrate for cognitive and neurological impairment in HIV. In particular, cardiovascular risk factors such as hyperlipidemia, hypertension, carotid intima-media thickness, and past history of cardiovascular disease have been identified recently as associated with reduced neurocognitive performance in HIV-infected subjects (Becker et al. 2009; Foley et al. 2010; Wright et al 2010). It is unclear whether these risk factors are non-HIV-related cofactors that influence cognitive outcomes, HIV- or cART-related systemic effects that parallel processes of CNS injury, or conditions that are directly involved in the mechanisms of HAND in the current era. Further efforts to investigate the association between vascular changes and clinical, pathological, and imaging changes in the setting of HIV are warranted to expand possible treatment approaches to HAND.

Importance of Acute/Early Infection

Although severe HAD clearly is a condition associated with long-term chronic HIV infection and immunosuppression, recent evidence that acute and early HIV are crucial for systemic disease pathogenesis has raised questions about whether early stages of HIV might also be important for neuropathogenesis. It has long been known that HIV may enter the nervous system within the first weeks after initial systemic infection (Schacker et al. 1996; Pilcher et al. 2001). It is now clear that acute infection also initiates a cascade of neuroinflammation, providing conditions for inflammation-mediated injury within the CNS. Analysis of CSF from 96 antiretroviral naïve subjects at a median

less than three months after HIV transmission revealed elevations in CSF WBC counts, neopterin, and CXCL10 equal to those in subjects with chronic HIV infection (Fig. 3) (Spudich et al. 2011).

Furthermore, neuroimaging studies in recently HIV-infected humans reveal lower n-acetylacetate in the frontal cortex during early infection, suggestive of neuronal dysfunction or injury during this early period (Lentz et al. 2009). Follow-up longitudinal studies in this group reveal dynamic patterns of cerebral metabolites over the first year of infection (Lentz et al. 2011). Given that early HIV infection is characterized by neuroinflammation and evolving metabolite changes within the CNS, it is possible that either antiretroviral or anti-inflammatory treatment initiated during early stages of infection may ameliorate injury sustained in the CNS during the early years of infection before immune systemic immunosuppression. Finally, the fact that CSF compartmentalization of HIV species may begin within the first year of HIV infection in some individuals (Schnell et al. 2010) suggests that treatment and eradication efforts may need to consider the CNS as a potential independent site of replication and mutation, beginning in the early stages of HIV.

CONCLUSIONS

There has been substantial progress in the recognition and treatment of the most severe forms of CNS HIV infection, and although the mechanisms leading to neurological dysfunction are still under investigation, the central role of macrophages and microglia is well established. Less well understood are the less severe forms of CNS disease now seen in the developed countries—where the use of cART is common—and the role of virus, inflammation, and CNS penetrance of antiretrovirals are areas of potential new discoveries. Similarly, there has not been any successful "adjuvant" therapy: that is, one designed to treat the CNS specifically rather than the virus. Such adjuvant treatment may be important in preventing or ameliorating HAND in the setting of cART.

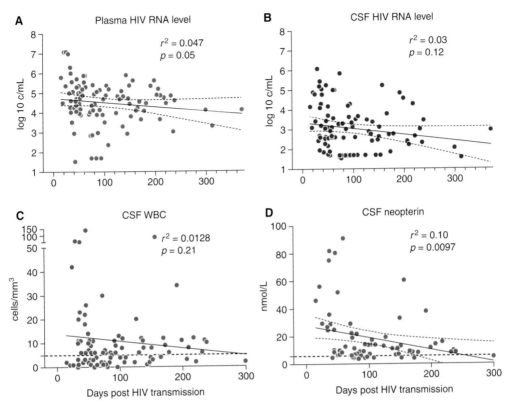

Figure 3. Baseline levels of markers of HIV infection and inflammation in a primary infection cohort are shown according to estimated days after transmission at blood and CSF sampling. Simple linear regression between the number of days post-estimated HIV transmission and (*A*) blood plasma HIV RNA levels, (*B*) CSF HIV RNA levels, (*C*) CSF WBC counts, and (*D*) CSF neopterin levels. Regression lines (solid) and 95% confidence intervals (dotted) are indicated; upper limit of normal values for CSF WBC and CSF neopterin are indicated on each graph by a dotted horizontal line. (Adapted from Spudich et al. 2011; with permission, from Oxford University Press © 2011.)

ACKNOWLEDGMENTS

We thank Dr. Anna Aldovini for her helpful review of this article. We also thank the National Institutes of Health for funding support (P50NS027405 and R01MH081772).

REFERENCES

*Reference is also in this collection.

Ancuta P, Kamat A, Kunstman KJ, Kim EY, Autissier P, Wurcel A, Zaman T, Stone D, Mefford M, Morgello S, et al. 2008. Microbial translocation is associated with increased monocyte activation and dementia in AIDS patients. *PLoS One* **3:** e2516. doi: 10.1371/journal.pone.0002516.

Ances BM, Roc AC, Korczykowski M, Wolf RL, Kolson DL. 2008. Combination antiretroviral therapy modulates the blood oxygen level-dependent amplitude in human immunodeficiency virus-seropositive patients. *J Neurovirol* **14:** 418–424.

Ances BM, Sisti D, Vaida F, Liang CL, Leontiev O, Perthen JE, Buxton RB, Benson D, Smith DM, Little SJ, et al. 2009. Resting cerebral blood flow: A potential biomarker of the effects of HIV in the brain. *Neurology* **73:** 702–708.

Ances BM, Vaida F, Yeh MJ, Liang CL, Buxton RB, Letendre S, McCutchan JA, Ellis RJ. 2010. HIV infection and aging independently affect brain function as measured by functional magnetic resonance imaging. *J Infect Dis* **201:** 336–340.

Anthony IC, Ramage SN, Carnie FW, Simmonds P, Bell JE. 2005. Influence of HAART on HIV-related CNS disease and neuroinflammation. *J Neuropathol Exp Neurol* **64:** 529–536.

Antinori A, Arendt G, Becker JT, Brew BJ, Byrd DA, Cherner M, Clifford DB, Cinque P, Epstein LG, Goodkin K, et al. 2007. Updated research nosology for HIV-associated neurocognitive disorders. *Neurology* **69:** 1789–1799.

Becker JT, Kingsley L, Mullen J, Cohen B, Martin E, Miller EN, Ragin A, Sacktor N, Selnes OA, Visscher BR. 2009. Vascular risk factors, HIV serostatus, and cognitive dysfunction in gay and bisexual men. *Neurology* **73:** 1292–1299.

Borda JT, Alvarez X, Mohan M, Hasegawa A, Bernardino A, Jean S, Aye P, Lackner AA. 2008. CD163, a marker of perivascular macrophages, is up-regulated by microglia in simian immunodeficiency virus encephalitis after haptoglobin-hemoglobin complex stimulation and is suggestive of breakdown of the blood-brain barrier. *Am J Pathol* **172:** 725–737.

Brew BJ, Dunbar N, Pemberton L, Kaldor J. 1996. Predictive markers of AIDS dementia complex: CD4 cell count and cerebrospinal fluid concentrations of β 2-microglobulin and neopterin. *J Infect Dis* **174:** 294–298.

Brew BJ, Pemberton L, Cunningham P, Law MG. 1997. Levels of human immunodeficiency virus type 1 RNA in cerebrospinal fluid correlate with AIDS dementia stage. *J Infect Dis* **175:** 963–966.

Brown RJ, Peters PJ, Caron C, Gonzalez-Perez MP, Stones L, Ankghuambom C, Pondei K, McClure CP, Alemnji G, Taylor S, et al. 2011. Inter-compartment recombination of HIV-1 contributes to env intra-host diversity and modulates viral tropism and senstivity to entry inhibitors. *J Virol* **85:** 6024–6037.

Burdo TH, Soulas C, Orzechowski K, Button J, Krishnan A, Sugimoto C, Alvarez X, Kuroda MJ, Williams KC. 2010. Increased monocyted turnover from bone marrow correlates with severity of SIV encephalitis and CD163 levels in plasma. *PLoS Pathog* **6:** e1000842. doi: 10.1371/journal.ppat.1000842.

Canestri A, Lescure FX, Jaureguiberry S, Moulignier A, Amiel C, Marcelin AG, Peytavin G, Tubiana R, Pialoux G, Katlama C. 2010. Discordance between cerebral spinal fluid and plasma HIV replication in patients with neurological symptoms who are receiving suppressive antiretroviral therapy. *Clin Infect Dis* **50:** 773–778.

Caragounis EC, Gisslén M, Lindh M, Nordborg C, Westergren S, Hagberg L, Svennerholm B. 2008. Comparison of HIV-1 pol and env sequences of blood, CSF, brain and spleen isolates collected ante-mortem and post-mortem. *Acta Neurol Scand* **117:** 108–116.

Chang L. 1995. In vivo magnetic resonance spectroscopy in HIV and HIV-related brain diseases. *Rev Neurosci* **6:** 365–378.

Chang L, Wong V, Nakama H, Watters M, Ramones D, Miller EN, Cloak C, Ernst T. 2008. Greater than age-related changes in brain diffusion of HIV patients after 1 year. *J Neuroimmune Pharmacol* **3:** 265–274.

Chen MF, Westmoreland S, Ryzhova EV, Martín-García J, Soldan SS, Lackner A, González-Scarano F. 2006. Simian immunodeficiency virus envelope compartmentalizes in brain regions independent of neuropathology. *J Neurovirol* **12:** 73–89.

Chen SK, Tvrdik P, Peden E, Cho S, Wu S, Spangrude G, Capecchi MR. 2010. Hematopoietic origin of pathological grooming in Hoxb8 mutant mice. *Cell* **141:** 775–785.

Churchill MJ, Wesslingh SL, Cowley D, Pardo CA, McArthur JC, Brew BJ, Gorry PR. 2009. Extensive astrocyte infection is prominent in HIV-associated dementia. *Ann Neurology* **66:** 253–258.

Cinque P, Brew BJ, Gisslen M, Hagberg L, Price RW. 2007. Cerebrospinal fluid markers in central nervous system HIV infection and AIDS dementia complex. *Handb Clin Neurol* **85:** 261–300.

Clifford DB, Fagan AM, Holtzman DM, Morris JC, Teshome M, Shah AR, Kauwe JS. 2009. CSF biomarkers of Alzheimer disease in HIV-associated neurologic disease. *Neurology* **73:** 1982–1987.

Collman R, Balliet JW, Gregory SA, Friedman H, Kolson DL, Nathanson N, Srinivasan A. 1992. An infectious molecular clone of an unusual macrophage-tropic and highly cytopathic strain of human immunodeficiency virus type 1. *J Virol* **66:** 7517–7521.

Cowley D, Gray LR, Wesselingh SL, Gorry PR, Churchill MJ. 2011. Genetic and functional heterogeneity of CNS-derived tat alleles from patients with HIV-associated dementia. *J Neurovirol* **17:** 70–81.

Cysique LA, Brew BJ. 2011. Prevalence of non-confounded HIV-associated neurocognitive impairment in the context of plasma HIV RNA suppression. *J Neurovirol* **17:** 176–183.

Cysique LA, Maruff P, Brew BJ. 2006. Variable benefit in neuropsychological function in HIV-infected HAART-treated patients. *Neurology* **66:** 1447–1450.

Cysique LA, Maruff P, Bain MP, Wright E, Brew BJ. 2011. HIV and age do not substantially interact in HIV-associated neurocognitive impairment. *J Neuropsych Clin Neurosci* **23:** 83–89.

d'Arminio Monforte A, Cinque P, Mocroft A, Goebel FD, Antunes F, Katlama C, Justesen US, Vella S, Kirk O, Lundgren J. 2004. Changing incidence of central nervous system diseases in the EuroSIDA cohort. *Ann Neurol* **55:** 320–328.

Desai S, Landay A. 2010. Early immune senescence in HIV disease. *Curr HIV/AIDS Rep* **7:** 4–10.

Dunfee RL, Thomas ER, Wang J, Kunstman K, Wolinsky SM, Gabuzda D. 2007. Loss of the N-linked glycosylation site at position 386 in the HIV envelope V4 region enhances macrophage tropism and is associated with dementia. *Virology* **367:** 222–234.

Dunfee RL, Thomas ER, Gabuzda D. 2009. Enhanced macrophage tropism of HIV in brain and lymphoid tissues is associated with sensitivity to the broadly neutralizing CD4 binding site antibody b12. *Retrovirology* **6:** 69.

Eden A, Price RW, Spudich S, Fuchs D, Hagberg L, Gisslen M. 2007. Immune activation of the central nervous system is still present after >4 years of effective highly active antiretroviral therapy. *J Infect Dis* **196:** 1779–1783.

Eden A, Fuchs D, Hagberg L, Nilsson S, Spudich S, Svennerholm B, Price RW, Gisslen M. 2010. HIV-1 viral escape in cerebrospinal fluid of subjects on suppressive antiretroviral treatment. *J Infect Dis* **202:** 1819–1825.

Ellis RJ, Gamst AC, Capparelli E, Spector SA, Hsia K, Wolfson T, Abramson I, Grant I, McCutchan JA. 2000. Cerebrospinal fluid HIV RNA originates from both local CNS and systemic sources. *Neurology* **54:** 927–936.

Everall I, Vaida F, Khanlou N, Lazzaretto D, Achim C, Letendre S, Moore D, Ellis R, Cherne M, Gelman B, et al. 2009. Cliniconeuropathologic correlates of human immunodeficiency virus in the era of antiretroviral therapy. *J Neurovirol* **15:** 1–11.

Foley J, Ettenhofer M, Wright MJ, Siddiqi I, Choi M, Thames AD, Mason K, Castellon S, Hinkin CH. Neurocognitive functioning in HIV-1 infection: Effects of cerebrovascular risk factors and age. *Clin Neuropsychol* 24: 265–285.

Gabuzda DH, Ho DD, de la Monte SM, Hirsch MS, Rota TR, Sobel RA. 1986. Immunohistochemical identification of HTLV-III antigen in brains of patients with AIDS. *Ann Neurol* 20: 289–295.

Ginhoux F, Greter M, Leboeuf M, Nandi S, See P, Gokhan S, Mehler MF, Conway SJ, Ng LG, Stanley ER, Merad M, et al. 2010. Fate mapping reveals that adult microglia derive from primitive macrophages. *Science* 330: 841–845.

Gisslen M, Hagberg L, Brew BJ, Cinque P, Price RW, Rosengren L. 2007. Elevated cerebrospinal fluid neurofilament light protein concentrations predict the development of AIDS dementia complex. *J Infect Dis* 195: 1774–1778.

Gisslén M, Hagberg L, Cinque P, Brew B, Price R. 2008. CSF markers in the management of CNS HIV infection and the AIDS dementia complex. In *The spectrum of neuro-AIDS disorders: Pathophysiology, diagnosis, and treatment* (ed. Goodkin K, Shapshak P, Vrma A), pp. 173–179. American Society for Microbiology, Washington, DC.

Gisslen M, Krut J, Andreasson U, Blennow K, Cinque P, Brew BJ, Spudich S, Hagberg L, Rosengren L, Price RW, et al. 2009. Amyloid and tau cerebrospinal fluid biomarkers in HIV infection. *BMC Neurol* 9: 63.

Gray LR, Gabuzda D, Cowley D, Ellett A, Chiavaroli L, Wesselingh SL, Churchill MJ, Gorry PR. 2011. CD4 and MHC class 1 down-modulation activities of nef alleles from brain- and lymphoid tissue-derived primary HIV-1 isolates. *J Neurovirol* 17: 82–91.

Gonzalez-Scarano F, Martin-Garcia J. 2005. The neuropathogenesis of AIDS. *Nat Rev Immunol* 5: 69–81.

Haas DW. 2004. Sequence heterogeneity and viral dynamics in cerebrospinal fluid and plasma during antiretroviral therapy. *J Neurovirol* 10 Suppl 1: 33–37.

Haase AT. 1986. Pathogenesis of lentivirus infections. *Nature* 322: 130–136.

Hagberg L, Fuchs D, Rosengren L, Gisslen M. 2000. Intrathecal immune activation is associated with cerebrospinal fluid markers of neuronal destruction in AIDS patients. *J Neuroimmunol* 102: 51–55.

Harrington PR, Schnell G, Letendre SL, Ritola K, Robertson K, Hall C, Burch CL, Jabara CB, Moore DT, Ellis RJ, et al. 2009. Cross-sectional characterization of HIV-1 env compartmentalization in cerebrospinal fluid over the full disease course. *AIDS* 23: 907–915.

Heaton RK, Clifford DB, Franklin DR Jr, Woods SP, Ake C, Vaida F, Ellis RJ, Letendre SL, Marcotte TD, Atkinson JH, et al. 2010. HIV-associated neurocognitive disorders persist in the era of potent antiretroviral therapy: CHARTER study. *Neurology* 75: 2087–2096.

Huang KJ, Alter GM, Wooley DP. 2002. The reverse transcriptase sequence of human immunodeficiency virus type 1 is under positive evolutionary selection within the central nervous system. *J Neurovirol* 8: 281–94.

Hughes ES, Bell JE, Simmonds P. 1997. Investigation of the dynamics of the spread of HIV to brain and other tissues by evolutionary analysis of sequences from the p17gag and env genes. *J Virol* 71: 1272–1280.

Kim W-K, Alvarez X, Fisher J, Bronfin B, Westmoreland W, McLaurin J, Williams K. 2006. CD163 identifies perivascular macrophages in normal and viral encephalitic brains and potential precursors to perivascular macrophages in blood. *Am J Pathol* 168: 822–834.

Kumar AM, Borodowsky I, Fernandez B, Gonzalez L, Kumar M. 2007. Human immunodeficiency virus type 1 RNA levels in different regions of human brain: Quantification using real-time reverse transcriptase-polymerase chain reaction. *J Neurovirol* 13: 210–224.

Kusdra L, McGuire D, Pulliam L. 2002. Changes in monocyte/macrophage neurotoxicity in the era of HAART: Implications for HIV-associated dementia. *AIDS* 16: 31–38.

* Lackner AA, Lederman MM, Rodriguez B. 2011. HIV pathogenesis—The host. *Cold Spring Harb Perspect Med* doi: 10.1101/cshperspect.a007005.

Lee B, Sharron M, Montaner LJ, Weissman D, Doms RW. 1999. Quantification of CD4, CCR5, and CXCR4 levels on lymphocyte subsets, dendritic cells, and differentially conditioned monocyte-derived macrophages. *Proc Natl Acad Sci* 96: 5215–5220.

Lentz MR, Kim WK, Lee V, Bazner S, Halpern EF, Venna N, Williams K, Rosenberg ES, Gonzalez RG. 2009. Changes in MRS neuronal markers and T cell phenotypes observed during early HIV infection. *Neurology* 72: 1465–1472.

Lentz MR, Kim WK, Kim H, Soulas C, Lee V, Venna N, Halpern EF, Rosenberg ES, Williams K, Gonzalez RG. 2011. Alterations in brain metabolism during the first year of HIV infection. *J Neurovirol* 17: 220–229

Letendre SL, McCutchan JA, Childers ME, Woods SP, Lazzaretto D, Heaton RK, Grant I, Ellis RJ. 2004. Enhancing antiretroviral therapy for human immunodeficiency virus cognitive disorders. *Ann Neurol* 56: 416–423.

Letendre S, Marquie-Beck J, Capparelli E, Best B, Clifford D, Collier AC, Gelman BB, McArthur JC, McCutchan JA, Morgello S, et al. 2008. Validation of the CNS penetration-effectiveness rank for quantifying antiretroviral penetration into the central nervous system. *Arch Neurol* 65: 65–70.

Lohse N, Hansen AB, Pedersen G, Kronborg G, Gerstoft J, Sorensen HT, Vaeth M, Obel N. 2007. Survival of persons with and without HIV infection in Denmark, 1995–2005. *Ann Intern Med* 146: 87–95.

Mahadevan A, Shankar SK, Parthasarathy S, Ranga U, Chickabasaviah YT, Santosh V, Vasanthapuram R, Pardo CA, Nath A, Zink MC. 2007. Characterization of HIV infected cells in infiltrates associated with CNS opportunistic infections in patients with HIV clade C infection. *J Neuropathol Exp Neurol* 66: 799–808.

Marra CM, Lockhart D, Zunt JR, Perrin M, Coombs RW, Collier AC. 2003. Changes in CSF and plasma HIV-1 RNA and cognition after starting potent antiretroviral therapy. *Neurology* 60: 1388–1390.

Marra CM, Maxwell CL, Collier AC, Robertson KR, Imrie A. 2007. Interpreting cerebrospinal fluid pleocytosis in HIV in the era of potent antiretroviral therapy. *BMC Infect Dis* 7: 37.

Marra CM, Zhao Y, Clifford DB, Letendre S, Evans S, Henry K, Ellis RJ, Rodriguez B, Coombs RW, Schifitto G, et al. 2009. Impact of combination antiretroviral therapy on cerebrospinal fluid HIV RNA and neurocognitive performance. *AIDS* 23: 1359–1366.

Martín-García J, Cocklin S, Chaiken IM, González-Scarano F. 2005. Interaction with CD4 and antibodies to CD4-induced epitopes of the envelope gp120 from a microglia-adapted human immunodeficiency virus type 1 isolate. *J Virol* 79, 6703–6713.

Martín-García J, Cao W, Varela-Rohena A, Plassmeyer ML, González-Scarano F. 2006. HIV-1 tropism for the central nervous system: Brain-derived envelope glycoproteins with lower CD4-dependence and reduced sensitivity to a fusion inhibitor. *Virology* 346: 169–179.

Mellgren A, Price RW, Hagberg L, Rosengren L, Brew BJ, Gisslen M. 2007. Antiretroviral treatment reduces increased CSF neurofilament protein (NFL) in HIV-1 infection. *Neurology* 69: 1536–1541.

Meyerhoff DJ, MacKay S, Poole N, Dillon WP, Weiner MW, Fein G. 1994. N-Acetylaspartate reductions measured by 1H MRSI in cognitively impaired HIV-seropositive individuals. *Magn Reson Imaging* 12: 653–659.

Meyerhoff DJ, Bloomer C, Cardenas V, Norman D, Weiner MW, Fein G. 1999. Elevated subcortical choline metabolites in cognitively and clinically asymptomatic HIV+ patients. *Neurology* 52: 995–1003.

Navia BA, Cho ES, Petito CK, Price RW. 1986a. The AIDS dementia complex: II. Neuropathology. *Ann Neurol* 19: 525–535.

Navia BA, Jordan BD, Price RW. 1986b. The AIDS dementia complex: I. Clinical features. *Ann Neurol* 19: 517–524.

O'Donnell LA, Agrawal A, Jordan-Sciutto KL, Dichter MA, Lynch DR, Kolson DL. 2006. Human immunodeficiency virus (HIV)-induced neurotoxicity: Roles for the NMDA receptor subtypes. *J Neurosci* 26: 981–990.

Perry H, Nicoll JAR, Holmes C. 2010. Microglia in neurodegenerative disease. *Nat Rev Neurol* 6: 193–201.

Peters PJ, Bhattacharya J, Hibbitts S, Dittmar MT, Simmons G, Bell J, Simmonds P, Clapham PR. 2004. Biological analysis of human immunodeficiency virus type 1 R5 envelopes amplified from brain and lymph node tissues of AIDS patients with neuropathology reveals two distinct tropism phenotypes and identifies envelopes in the brain that confer an enhanced tropism and fusigenicity for macrophages. *J Virol* 78: 6915–6926.

Pilcher CD, Shugars DC, Fiscus SA, Miller WC, Menezes P, Giner J, Dean B, Robertson K, Hart CE, Lennox JL, et al. 2001. HIV in body fluids during primary HIV infection: Implications for pathogenesis, treatment and public health. *Aids* 15: 837–845.

Pillai SK, Pond SL, Liu Y, Good BM, Strain MC, Ellis RJ, Letendre S, Smith DM, Günthard HF, Grant I, et al. 2006. Genetic attributes of cerebrospinal fluid-derived HIV-1 env. *Brain* 129 (Pt 7): 1872–1883.

Pomara N, Crandall DT, Choi SJ, Johnson G, Lim KO. 2001. White matter abnormalities in HIV-1 infection: A diffusion tensor imaging study. *Psychiatry Res* 106: 15–24.

Price RW, Sidtis JJ, Brew BJ. 1991. AIDS dementia complex and HIV-1 infection: A view from the clinic. *Brain Pathol* 1: 155–162.

Pulliam L, Gascon R, Stubblebine M, McGuire D, McGrath MS. 1997. Unique monocyte subset in patients with AIDS dementia. *Lancet* 349: 692–695.

Reeve AB, Pearce NC, Patel K, Augustus KV, Novembre FJ. 2010. Neuropathogenic SIVsmmFGb genetic diversity and selection-induced tissue-specific compartmentalization during chronic infection and temporal evolution of viral genes in lymphoid tissues and regions of the central nervous system. *AIDS Res Hum Retroviruses* 26: 663–679.

Robertson K, Fiskus S, Kapoor C, Robertson W, Schneider G, Shepard R, Howe L, Silva S, Hall C. 1998. CSF, plasma viral load and HIV associated dementia. *J Neurovirol* 4: 90–94.

Robertson KR, Robertson WT, Ford S, Watson D, Fiscus S, Harp AG, Hall CD. 2004. Highly active antiretroviral therapy improves neurocognitive functioning. *J Acquir Immune Defic Syndr* 36: 562–566.

Robertson KR, Smurzynski M, Parsons TD, Wu K, Bosch RJ, Wu J, McArthur JC, Collier AC, Evans SR, Ellis RJ. 2007. The prevalence and incidence of neurocognitive impairment in the HAART era. *AIDS* 21: 1915–1921.

Sacktor N, Nakasujja N, Skolasky RL, Rezapour M, Robertson K, Musisi S, Katabira E, Ronald A, Clifford DB, Laeyendecker O, et al. 2009. HIV subtype D is associated with dementia, compared with subtype A, in immunosuppressed individuals at risk of cognitive impairment in Kampala, Uganda. *Clinical Infect Dis* 49: 780–786.

Shau-Kwaun C, Tvrdik P, Peden E, Cho S, Wu S, Spangrude G, Capecchi MR. 2010. Hematopoietic origin of pathological grooming in Hoxb8 mutant mice. *Cell* 141: 775–785.

Schacker T, Collier AC, Hughes J, Shea T, Corey L. 1996. Clinical and epidemiologic features of primary HIV infection. *Ann Intern Med* 125: 257–264.

Schnell G, Spudich S, Harrington P, Price RW, Swanstrom R. 2009. Compartmentalized HIV-1 originates from long-lived cells in some subjects with HIV-1-associated dementia. *PLoS Pathog* 5: e1000395. doi: 10.1371/journal.ppat.1000395.

Schnell G, Price RW, Swanstrom R, Spudich S. 2010. Compartmentalization and clonal amplification of HIV-1 variants in the cerebrospinal fluid during primary infection. *J Virol* 84: 2395–2407.

Shieh JTC, Martín J, Baltuch G, Malim MH, González-Scarano F. 2000. Determinants of syncytia–formation in microglia by the human immunodeficiency virus type 1 (HIV-1): Role of the V1/V2 domains. *J Virol* 74: 693–701.

Sinclair E, Ronquillo R, Lollo N, Deeks SG, Hunt P, Yiannoutsos CT, Spudich S, Price RW. 2008. Antiretroviral treatment effect on immune activation reduces cerebrospinal fluid HIV-1 infection. *J Acquir Immune Defic Syndr* 47: 544–552.

Smit TK, Brew BJ, Tourtellotte W, Morgello S, Gelman BB, Saksena NK. 2004. Independent evolution of human immunodeficiency virus (HIV) drug resistance mutations in diverse areas of the brain in HIV-infected patients, with and without dementia, on antiretroviral treatment. *J Virol* 78: 10133–10148.

Smurzynski M, Wu K, Letendre S, Robertson K, Bosch RJ, Clifford DB, Evans S, Collier AC, Taylor M, Ellis R. 2011. Effects of central nervous system antiretroviral penetration on cognitive functioning in the ALLRT cohort. *AIDS* 25: 357–365.

Soulas C, Donahue RE, Dunbar CE, Persons DA, Alvarez X, Williams KC. 2009. Genetically modified CD34$^+$ hematopoietic stem cells contribute to turnover of brain perivascular macrophages in long-term repopulated primates. *Am J Pathol* **174:** 1808–1817.

Soulas C, Conerly C, Kim W-K, Burdo TH, Alvarez X, Lackner AA, Williams KD. 2011. Recently infiltrating MAC387$^+$ monocytes/macrophages. A third macrophage population involved in SIV and HIV encephalitic lesion formation. *Am. J Pathol* **178:** 2121–2135.

Spudich S, Nilsson A, Lollo N, Liegler T, Petropoulos C, Deeks S, Paxinos E, Price R. 2005. Cerebrospinal fluid HIV infection and pleocytosis: Relation to systemic infection and antiretroviral treatment. *BMC Infect Dis* **5:** 98.

Spudich S, Lollo N, Liegler T, Deeks SG, Price RW. 2006. Treatment benefit on cerebrospinal fluid HIV-1 levels in the setting of systemic virological suppression and failure. *J Infect Dis* **194:** 1686–1696.

Spudich S, Gisslén M, Hagberg L, Lee E, Liegler T, Brew B, Fuchs D, Tambussi G, Cinque P, Hecht F, et al. 2011. Central nervous system immune activation characterizes primary HIV-1 infection even in subjects with minimal cerebrospinal fluid viral burden. *J Infect Dis* **204:** 753–760.

Sun B, Abadjian L, Rempel H, Calosing C, Rothlind J, Pulliam L. 2010. Peripheral biomarkers do not correlate with cognitive impairment in highly active antiretroviral therapy-treated subjects with human immunodeficiency virus type 1 infection. *J Neurovirol* **16:** 115–124.

Thomas ER, Dunfee RL, Stanton J, Bogdan D, Taylor J, Kunstman K, Bell JE, Wolinsky SM, Gabuzda D. 2007a. Macrophage entry mediated by HIV Envs from brain and lymphoid tissues is determined by the capacity to use low CD4 levels and overall efficiency of fusion. *Virology* **360:** 105–119.

Thomas ER, Dunfee RL, Stanton J, Bogdan D, Kunstman K, Wolinsky SM, Gabuzda D. 2007b. High frequency of defective vpu compared with tat and rev genes in brain from patients with HIV type 1-associated dementia. *AIDS Res Hum Retroviruses* **23:** 575–580.

Thurnher MM, Castillo M, Stadler A, Rieger A, Schmid B, Sundgren PC. 2005. Diffusion-tensor MR imaging of the brain in human immunodeficiency virus-positive patients. *AJNR Am J Neuroradiol* **26:** 2275–2281.

Tozzi V, Balestra P, Salvatori MF, Vlassi C, Liuzzi G, Giancola ML, Giulianelli M, Narciso P, Antinori A. 2009. Changes in cognition during antiretroviral therapy: Comparison of 2 different ranking systems to measure antiretroviral drug efficacy on HIV-associated neurocognitive disorders. *J Acquir Immune Defic Syndr* **52:** 56–63.

Valcour VGS, Cecilia M, Watters RM, Sacktor NC. 2004. Cognitive impairment in older HIV-1-seropositive individuals: Prevalence and potential mechanisms. *AIDS* **18** Suppl 1: S79–S86.

Valcour V, Paul R, Neuhaus J, Shikuma C. 2011. The effects of age and HIV on neuropsychological performance. *J Int Neuropsychol Soc* **17:** 190–195.

van Marle G, Rourke SB, Zhang K, Silva C, Ethier J, Gill MJ, Power C. 2002. HIV dementia patients exhibit reduced viral neutralization and increased envelope sequence diversity in blood and brain. *AIDS* **16:** 1905–1914.

Wang Y, Zhang J, Gutman B, Chan TF, Becker JT, Aizenstein HJ, Lopez OL, Tamburo RJ, Toga AW, Thompson PM. 2009. Multivariate tensor-based morphometry on surfaces: Application to mapping ventricular abnormalities in HIV/AIDS. *Neuroimage* **49:** 2141–2157.

Wiley CA, Schrier RD, Nelson JA, Lampert PW, Oldstone MB. 1986. Cellular localization of human immunodeficiency virus infection within the brains of acquired immune deficiency syndrome patients. *Proc Natl Acad Sci* **83:** 7089–7093.

Wright EJ, Grund B, Robertson K, Brew BJ, Roediger M, Bain MP, Drummond F, Vjecha MJ, Hoy J, Miller C, et al. Cardiovascular risk factors associated with lower baseline cognitive performance in HIV-positive persons. *Neurology* **75:** 864–873.

Xing HQ, Hayakawa H, Izumo K, Kubota R, Gelpi E, Budka H, Izumo S. 2009. In vivo expression of proinflammatory cytokines in HIV encephalitis: An analysis of 11 autopsy cases. *Neuropathology* **29:** 433–442.

Yilmaz A, Stahle L, Hagberg L, Svennerholm B, Fuchs D, Gisslen M. 2004. Cerebrospinal fluid and plasma HIV-1 RNA levels and lopinavir concentrations following lopinavir/ritonavir regimen. *Scand J Infect Dis* **36:** 823–828.

Yilmaz A, Price RW, Spudich S, Fuchs D, Hagberg L, Gisslen M. 2008. Persistent intrathecal immune activation in HIV-1-infected individuals on antiretroviral therapy. *J Acquir Immune Defic Syndr* **47:** 168–173.

Nonpathogenic Simian Immunodeficiency Virus Infections

Nichole R. Klatt[1], Guido Silvestri[2], and Vanessa Hirsch[1]

[1]Laboratory of Molecular Microbiology, NIAID, NIH, Bethesda, Maryland 20892

[2]Department of Pathology and Laboratory Medicine, Emory University School of Medicine, and Yerkes National Primate Research Center, Atlanta, Georgia 30322

Correspondence: vhirsch@niaid.nih.gov

The simian immunodeficiency viruses (SIVs) are a diverse group of viruses that naturally infect a wide range of African primates, including African green monkeys (AGMs) and sooty mangabey monkeys (SMs). Although natural infection is widespread in feral populations of AGMs and SMs, this infection generally does not result in immunodeficiency. However, experimental inoculation of Asian macaques results in an immunodeficiency syndrome remarkably similar to human AIDS. Thus, natural nonprogressive SIV infections appear to represent an evolutionary adaptation between these animals and their primate lentiviruses. Curiously, these animals maintain robust virus replication but have evolved strategies to avoid disease progression. Adaptations observed in these primates include phenotypic changes to CD4$^+$ T cells, limited chronic immune activation, and altered mucosal immunity. It is probable that these animals have achieved a unique balance between T-cell renewal and proliferation and loss through activation-induced apoptosis, and virus-induced cell death. A clearer understanding of the mechanisms underlying the lack of disease progression in natural hosts for SIV infection should therefore yield insights into the pathogenesis of AIDS and may inform vaccine design.

The simian immunodeficiency viruses (SIVs) are a genetically diverse group of viruses that naturally infect a wide range of African nonhuman primates and are the source of the human immunodeficiency viruses (HIV-1 and HIV-2). The origins of HIV-1 and HIV-2, as well as the SIVcpz and SIVgor strains, are discussed in greater detail in Sharp and Hahn (2011). SIVs first came to the attention of AIDS researchers with the occurrence of immunodeficiency in macaques in the California, New England, and Washington primate centers

(Apetrei et al. 2005). Almost simultaneously, transfer of tissues from sooty mangabeys to macaques resulted in a similar disease at the Tulane Primate Center (Murphey-Corb et al. 1986). The virus isolated from these macaques originated from sooty mangabey monkeys, either by experimental infection at Tulane or through cohousing of African and Asian monkeys earlier in the history of the primate centers (Apetrei et al. 2005). This event represented the birth of a new animal model for HIV infection based on the use of SIVmac and SIVsmm infection of rhesus macaques (Johnson and Hirsch 1992). Subsequent studies have determined that SIV infection of macaques, although more rapid than HIV infection in humans, is remarkably similar in terms of pathogenesis, and the model has been used extensively for vaccine development (Haigwood 2009). However, the use of the sooty mangabey–derived SIV as an experimental macaque model for AIDS is only a small aspect of the overall scientific interest in the SIV infection of natural African host species. First, these primate viruses are the source of the HIV-1 (cross-species transmission of SIVcpz) and HIV-2 (cross-species transmission of SIVsmm) epidemics in humans (Sharp and Hahn 2010). Second, these animals present a fascinating enigma: lack of progression to AIDS in the face of active viral replication (Hirsch 2004; Paiardini et al. 2009a; Pandrea and Apetrei 2010). It is hoped that the study of the mechanism(s) underlying their resistance to AIDS will help us understanding the pathogenesis of HIV in humans and to design vaccine and therapeutic strategies (Sodora et al. 2009). This article will focus on the natural hosts of SIVs and their viruses and the lessons they teach us about the pathogenesis of AIDS.

THE PRIMATE LENTIVIRUSES

Origins and Phylogeny of SIVs

There is a wide variety of SIVs in African nonhuman primates, although only a fraction of them have been molecularly characterized (see Table 1). To date, serological evidence of SIV infection has been reported in 36 different primate species, and partial or full-length viral sequences have been characterized from 30 of these (Apetrei et al. 2004). Primate lentiviruses have been detected in most of the African monkeys of the genus *Cercopithecus*, African green monkeys (*Chlorocebus*), mandrills and drills (*Mandrillus*), the mangabeys (*Cercocebus*), a variety of colobus monkeys (*Colobus, Pilocolbus*), and within the great apes, two subspecies of chimpanzees (*Pan*) and gorillas (*Gorilla*) (Table 1). However, interestingly, infection of Asian monkeys such as macaques (genus

Macaca) and the Asian great apes, orangutans, has not been detected in the wild. This restriction of primate lentiviruses to African monkeys suggests that this is an ancient virus that has coevolved with its primate host. The observation of similar viruses in related species such as the multiple species of African green monkeys (AGMs) that are infected, despite geographic separation, suggests that the ancestor of the current day SIVs may date back to the time of phylogenetic divergence and geographical separation of African and Asian monkeys. A conservative estimate of the age of natural SIV infections based on the phylogeny of SIVs isolated from nonhuman primates at the Bioko island places these infections at least 32,000 years ago, although less conservative analysis suggests a much longer time (Worobey et al. 2010). In contrast, HIV-1 and HIV-2 infections of humans represent relatively recent introductions into human populations by cross-species transmission (Sharp and Hahn 2010).

As detailed in Table 1 and Figure 1, there are at least seven distinct lineages of primate lentiviruses: (1) SIVsm from sooty mangabeys (Hirsch et al. 1989), including HIV-2; (2) SIVagm from the four different species of AGMs (Allan et al. 1991; Hirsch 2004); (3) SIV from monkeys of the genus *Cercopithecus*, commonly called guenons (e.g., SIVgsn, SIVdeb, SIVmus) (Courgnaud et al. 2002, 2003); (4) SIVcpz from two species of chimpanzees (Keele et al. 2006) and SIVgor from gorillas; (5) SIVlho and SIVsun from the related L'Hoest and suntailed monkeys (Beer et al. 1999, 2000); (6) SIVcol from black and white colobus monkeys (Courgnaud et al. 2001); and (7) SIVrcm from red capped mangabeys (Beer et al. 2001) and SIVmnd and drl from mandrills and drill monkeys (Clewley et al. 1998; Hu et al. 2003). Each of the various lineages are approximately equidistant from any other lineage, sharing 40%–50% identity in the most conserved Gag and Pol proteins (Hirsch et al. 1995a). The phylogenetic relationship between many of these various SIVs, with distinct species-specific lineages, is shown in Figure 1. The primate lentiviruses all share a common genomic organization encoding the structural and enzymatic proteins Gag, Pol,

Table 1. Natural host species harboring simian immunodeficiency viruses (SIVs)

Genus	Common name	Species	SIV
Pan	Chimpanzees	*troglodytes troglodytes*	SIVcpzPtt
		troglodytes schweinfurthi	SIVcpzPts
Gorilla	Gorilla	*gorilla*	SIVgor
Cercocebus	Sooty mangabey	*atys*	SIVsmm
	Redcapped mangabey	*torquatus*	SIVrcm
Mandrillus	Mandrill	*sphinx*	SIVmnd-1, mnd-2
	Drill	*leucophaeus*	SIVdrl
Miopithecus	Talapoin	*talapoin*	SIVtal
Colobus	Black and white colobus	*guerza*	SIVcol
Procolobus	Olive colobus	*verus*	SIVolc
Pilocolobus	Western red colobus	*badius*	SIVwrc
Chlorocebus	African green monkeys		
	Sabaeus	*sabaeus*	SIVagmSab
	Vervet	*pygerythrus*	SIVagmVer
	Tantalus	*tantalus*	SIVagmTan
	Grivet	*aethiops*	SIVagmGri
Cercopithecus	The guenons		
	Greater spot-nosed	*nictitans*	SIVgsn
	Blue monkey	*mitis*	SIVblu
	Mona monkey	*mona*	SIVmon
	Dents' monkey	*denti*	SIVden
	Moustached monkey	*cephus*	SIVmus
	Red-eared monkey	*erythrotis*	SIVery
	Redtailed monkey	*ascanius*	SIVasc
	DeBrazza	*neglectus*	SIVdeb
	Sykes monkey	*albogularis*	SIVsyk
	L'Hoests monkey	*l'hoesti*	SIVlho
	Suntailed monkey	*solatus*	SIVsun
	Preussis monkey	*preussi*	SIVpre

and Env but also a variety of accessory proteins. All SIV and HIV strains share open reading frames for *tat, rev, vif, vpr,* and *nef* genes. However, the *vpu* gene is unique to HIV-1, SIVcpz, and a variety of SIV strains from *Cercopithecus* monkeys including SIVgsn (greater spot nosed monkey) (Courgnaud et al. 2003), consistent with an ancestral relationship between these viruses. In contrast, the *vpx* gene is unique to SIVs from mangabeys, SIVsmm and SIVrcm; in these two viruses, some of the various functions of the *vpr* gene have been segregated to *vpx*. Recent studies suggest that some of the functions attributed to *vpu* gene were acquired by the *nef* or the *env* genes in viruses that lack *vpu* (Sauter et al. 2009) suggesting that they played a role in overcoming intrinsic host

restriction factors on cross-species transmission as discussed in Sharp and Hahn (2010).

This family of primate lentiviruses has also been the source of two separate epidemics in humans. HIV-1 arose from multiple cross-species transmissions events with HIV-1 Groups M and N arising from SIVcpzPtt and SIVcpzPts, and HIV-1 Group O from SIVgor (Fig. 1, red lines) (reviewed in Sharp and Hahn 2010). Similarly, SIVsmm (Hirsch et al. 1989) was the source of the HIV-2 epidemic in West Africa (Fig. 1, red lines). Although SIVs are generally not pathogenic in their natural host species, the infection of humans by these animals, and evolution as HIV-1 and -2, was associated with the acquisition of virulence.

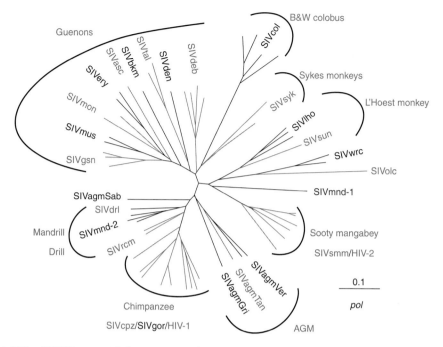

Figure 1. SIV and HIV lineages. Phylogenetic tree of a region of polymerase showing the relationship between the various primate lentiviruses. HIV strains are shown in red. The nonhuman primate natural hosts for are shown for each cluster. (Adapted from Peeters et al. 2008 and reprinted with permission from EDP Science © 2008.)

Epidemiology, Cross-Species Transmission, and Recombination

Serologic surveys and evaluation of bushmeat in Africa reveal that infection of nonhuman primates is widespread, although the prevalence may vary, depending on species and geographic location (Aghokeng et al. 2006, 2010). SIV infection has been detected in nearly all species of African nonhuman primates with the possible exception of one species of chimpanzee (*Pan troglodytes vellerosus*), baboons, and patas monkeys. In some species of monkeys, there is serological evidence of infection, but viruses have not yet been isolated or characterized. The prevalence of infection increases with age, with fairly uniform lack of infection in infants and juveniles suggesting that sexual routes as well as aggression may play a role in transmission. Furthermore, mother to infant transmission appears to be somewhat rare based on the lack of infection in infants and juveniles. This notion is also supported by studies of the colony

of sooty mangabeys housed at the Yerkes primate center, in which only ~5% of infants born to SIV-infected mothers appear to have contracted the infection vertically (Chahroudi et al. 2011). However, the route of transmission in wild populations has not been clearly defined.

SIVs generally appear to be species-specific, and with each species showing only one virus except for mandrills that are infected with two viruses (SIVmnd-1 and SIVmnd-2). This pattern is consistent with coevolution during the speciation and migration of the different primate species throughout Africa. For example, SIVagm strains are present within all species of AGM (vervet, grivet, tantalus, and sabaeus) throughout sub-Saharan Africa, but each of the species harbor a distinct, but related virus (70% identity). In at least two cases, the species of AGM have been geographically isolated from one another for thousands of years, ruling out contemporaneous spread of the virus (Muller and Barre-Sinoussi 2003; Hirsch 2004).

However, with more extensive characterization, the picture has becomes increasingly complex with evidence of multiple cross-species transmissions and recombination events. There is evidence for present-day natural transmission occurring in the wild, for example SIVagm infection of baboons and patas monkeys (Jin et al. 1994a; Bibollet-Ruche et al. 1996). However, there is also evidence of a long history of cross-species transmission, coinfections, and apparent recombination. Many of the SIV strains appear to be recombinant and often the parental strains are difficult to define because of genetic divergence from representative strains. The most notable recombinant is HIV-1 and its ancestor SIVcpz. SIVcpz appears to be a recombinant between ancestral viruses that gave rise to SIVrcm from red-capped mangabeys, and the clade of viruses found that infect species of *Cercopithecus* monkeys or guenons that includes greater spot-nosed, mona, and mustached monkeys (Bailes et al. 2003; Courgnaud et al. 2003). Other obvious recombinants are (1) SIVagmSab from the West African sabaeus species of AGM that is a recombinant of an ancestral forms of SIVagm and SIVrcm (Jin et al. 1994b); (2) SIVrcm from red capped mangabeys (Beer et al. 2001); and (3) SIVdrl and SIVmnd-2 from drills and mandrills, respectively, that are recombinants between SIVrcm and SIVmnd-1 (Hu et al. 2003). These latter viruses share a common breakpoint, suggesting a common origin and another example of cross-species transmission. Presumably, there has been a long history of cross-species transmission events and recombination within the primate lentiviruses. However, it is still evident from the species specificity of many of these strains that these viruses are ancient and have coevolved with their host species over long periods of time.

PATHOGENESIS STUDIES IN ANIMAL MODELS

Experimental Animal Models

Despite the wide range of SIV-infected African nonhuman primates in the wild, there are only three available models for experimental manipulation. Essential elements for such studies are a molecularly characterized SIV strain that can reproduce the kinetics of viral replication seen in natural infection and availability of the correct species from which this virus was initially derived. Animal models that satisfy these criteria are (1) SIVsmm infection of sooty mangabeys (SMs); (2) SIVagm infection of AGMs; and (3) SIVmnd infection of mandrills (Pandrea et al. 2003; Onanga et al. 2006). Two lineages of SIVagm have been evaluated in vivo, SIVagmVer and SIVagmSab from vervet and sabaeus AGM, respectively (Goldstein et al. 2006; Pandrea et al. 2006a,b). Initial studies with SIVagmSab were performed using sabaeus AGMs of African origin but a model has now been established using the same species of AGM of Caribbean origin, as these animals were imported from Africa more than 300 years earlier and are more readily available for experimental manipulations (Pandrea et al. 2006b). Initial studies of various strains of SIVagm in different species of AGMs revealed that these viruses are adapted for their specific host species (Goldstein et al. 2006). Thus, SIVagmVer strains are restricted in terms of replication in sabaeus AGMs relative to replication in their matched host, vervet AGMs. Because of cost and ethical issues, SIVcpz infection of chimpanzees has only been studied on a small scale or in terms of its impact in wild, habituated chimpanzee populations (Keele et al. 2009) and is discussed in more detail in article by Hahn and Sharp (2011). Infection of mandrills with SIVmnd has also been studied to a limited degree, because of restricted availability of these animals in captivity (Pandrea et al. 2003; Onanga et al. 2006). Therefore, much of what we know about natural hosts has been gleaned from studies of SMs and AGMs.

General Characteristics of Natural Infection

The lack of virulence of SIV isolates for their homologous natural host species is intriguing when contrasted with their effect in Asian macaques, and with the typically pathogenic effect of HIV-1 in humans. Natural SIV hosts, such as SMs infected with SIVsmm or AGMs infected with SIVagm, generally show no evidence of immunodeficiency. There have been sporadic reports of the development of immunodeficiency, as defined by opportunistic infection or neoplasms normally associated with AIDS. Indeed, CD4[+] T-cell depletion was observed in one naturally infected mandrill, and immunodeficiency was observed in a SM that had been naturally SIV-infected for more than 18 years (Ling et al. 2004; Pandrea et al. 2009). AIDS has also been observed in at least one chimpanzee inoculated with HIV-1 and progressive infection seen in a subset of HIV-infected chimpanzees (Novembre et al. 1997; O'Neil et al. 2000). Although HIV-1 is only a close relative of SIVcpz, this study suggests that these African primates are not immune to the pathogenic effects of primate lentiviruses under specific circumstances. Indeed, recent studies in wild habituated chimpanzee populations show a significantly greater mortality rate associated with SIVcpz infection (Keele et al. 2009). This is in agreement with the idea that SIVcpz is perhaps less adapted to chimpanzees than SIV in such hosts as SMs, consistent with a more recent introduction into chimpanzees from the monkeys on which they prey. The conclusion is that SIV infection of most natural host species is generally asymptomatic within the time frame of the lifespan of the animal. This lack of pathogenicity has been postulated to be the result of an evolutionary adaptation that, in AGMs and SMs, allows for a mutual coexistence between the host and the virus.

Despite their general lack of pathogenicity in their matched host species, SIVs clearly do not not lack the intrinsic potential to cause AIDS, revealed by the either accidental or experimental introduction of SIVsmm into rhesus macaques (RMs) (Apetrei et al. 2005). In addition, experimental infection of macaque species with SIVsmm, SIVagm, and SIVlho results in a syndrome remarkably similar in pathogenesis of AIDS in humans. Interestingly, uncloned/unpassaged SIVsmm in RMs results in levels of virus replication that are lower than in SIV-infected SMs (Bosinger et al. 2009); however, when SIVsmm becomes adapted to RM cells through in vitro and/or in vivo passage,

the level of virus replication become even higher than in SMs (Johnson et al. 1990; Hirsch et al. 1995b). In the case of SIVagm and SIVlho, infection of pigtail macaques (*Macaca nemstrina*) was required to achieve efficient replication and subsequent disease (Hirsch et al. 1995b; Beer et al. 2005). The majority of pathogenesis and vaccine studies have focused on SIVmac infection of RMs because of its more uniform course of infection. Interestingly, recent studies revealed that allelic variation in the rhesus macaque TRIM5α gene results in differences in susceptibility to infection and viral replication in the early stages of cross-species transmission of SIVsmm and that emergence of pathogenic SIVmac in RMs required adaptations in the viral capsid protein (CA) to overcome suppression by two distinct types of TRIM5α allele (Kirmaier et al. 2010). Presumably similar types of adaptations occurred for both HIV-1 and HIV-2.

SIV infections of SMs, AGMs, and mandrills share many similar features with pathogenic infections such as SIVmac infection of macaques and HIV infection of humans (Johnson and Hirsch 1991; Paiardini et al. 2009b). The general features of SIV infection of AGM and SM are compared with pathogenic infection in with differences highlighted (Table 2). Common features include the kinetics of primary viremia with robust peak viremia and persistence of viremia into the chronic phase of infection in both models (Pandrea et al. 2006a). Infection is associated with the development of adaptive and innate immune responses that are similar or lower in kinetics and magnitude, and fail to control virus replication. Based on the rapidity of viral clearance following treatment with antiviral drugs, SIV infection targets cells whose lifespan is short (i.e., 1–2 d) (Gordon et al. 2008; Pandrea et al. 2008a) and even acute depletion/loss of mucosal CD4[+] T cells, once thought to be pathognomonic for pathogenic primate lentivirus infections, is observed in both models. The most obvious difference is the clinical course of disease in natural hosts, which is typically nonprogressive. Other distinguishing features of natural host infections are (1) the maintenance of peripheral CD4[+] T cells

Table 2. Comparisons of the characteristics of primate lentivirus infection of natural SIV hosts versus macaques and humans

	Host species	
Characteristics	Sooty mangabeys and African green monkeys	Humans and Asian macaques
Clinical outcome	Depletion of peripheral CD4$^+$ T cells rare	Progressive depletion of CD4$^+$ T cells
	Progression to AIDS is very rare.	Progression to AIDS is common.
	Survival for up to 20–30 yr	Median survival 1.5 yr
	Generally no evidence of AIDS	Opportunistic infections, neoplasia, encephalitis
	Vertical transmission is rare.	Vertical transmission is more common.
Viremia	High peak viremia in 1–2 wk ($10^6 - 10^8$/mL)	High peak viremia in 1–2 wk ($10^6 - 10^8$/mL)
	Post-peak decline of viral load	Variable post peak decline in viral load
	High to moderate set point virus load ($10^3 - 10^6$/mL)	High set point of virus replication ($10^5 - 10^8$/mL)
	Stable setpoint	Increasing viral load over time
Immune responses	Adaptive immune responses	Adaptive immune responses
	Marked acute type I interferon response	Marked acute type I interferon response
	Resolution of type I interferon responses	Persistence of type I interferon responses
	Variable bystander T-cell activation and apoptosis	High levels of bystander T-cell activation and apoptosis
	Establishment of an anti-inflammatory milieu	Establishment of a proinflammatory milieu
	Limited immune activation and T-cell apoptosis	Generalized immune activation and significant T-cell apoptosis
	Preservation of mucosal TH17 cells	Preferential loss of mucosal TH17 cells
	Absence of microbial translocation	Microbial translocation is prominent.
Target cells	Short-lived, activated CD4$^+$ T cells	Short-lived, activated CD4$^+$ T cells
	Early depletion of mucosal CD4$^+$ T cells	Early depletion of mucosal CD4$^+$ T cells
	Healthy CD4$^+$ T-cell counts in most animals	Progressive depletion of circulating and mucosal CD4$^+$ T cells
	CD4 depletion is not associated with AIDS.	CD4 depletion of <100/μL is associated with AIDS.

levels in the majority of animals; (2) the lack of chronic immune activation following resolution of primary infection; (3) the absence of microbial translocation; and (4) the preferential sparing of CD4$^+$ central memory T cells from infection.

Targets Cells for SIV Replication

Similar to pathogenic models of HIV/SIV infection, SIV expression is primarily observed in lymphoid tissues and the gastrointestinal tract by polymerase chain reaction or in situ hybridization (Goldstein et al. 2006). The vast majority (i.e., >90%) of SIVsmm replication in naturally SIV-infected SMs and AGMs occurs in short-lived cells (Perelson et al. 1993; Ho et al. 1995; Wei et al. 1995; Nowak et al. 1997), suggesting that activated CD4$^+$ T cells are the major site for viral replication (Gordon et al. 2008; Pandrea et al. 2008a). This finding was shown by treating SMs and AGMs with reverse transcriptase inhibitors, and the lifespan of productively infected cells was calculated based on the slope of the decline of SIV plasma viremia after initiation of ART using a widely accepted mathematical model (Perelson et al. 1997; Gordon et al. 2008; Pandrea et al. 2008a). In addition, in situ hybridization studies have shown that SIVsmm and SIVagm colocalizes with CD3$^+$ lymphocytes in lymph nodes and mucosal tissues of SMs and AGMs, respectively (Pandrea et al. 2008a; Sodora et al. 2009). Further evidence that SIVsmm infected activated CD4$^+$ T cells in vivo was shown by depletion of CD4$^+$ T cells in SIV-infected SMs; the subsequent levels of viremia correlated directly with the number of activated CD4$^+$ T cells (Klatt et al. 2008). The rapid depletion of mucosal CD4$^+$ T cells during acute SIVsmm and SIVagm infection also suggested that CD4$^+$ T cells are the main targets of SIV replication in SM and AGM (Gordon et al. 2007; Pandrea et al. 2007a). However, despite destruction of mucosal CD4$^+$ T cells during acute infection and concomitant development of chronic viremia, these animals maintain relatively normal mucosal immune function, with preserved levels of Th17 cells and a lack of microbial translocation (Brenchley et al. 2006, 2008; Sumpter et al. 2007). Although maintenance of peripheral CD4$^+$ T cells in natural hosts is a striking feature of natural SIV infection, this model also shows that loss of CD4$^+$ T cells alone is not sufficient to cause AIDS (Kosub et al. 2008). Indeed, naturally and experimentally SIV-infected SMs exist that are depleted of peripheral CD4$^+$ T cells during infection, but remain AIDS-free (Sumpter et al. 2007; Milush et al. 2007). Furthermore, experimental CD4$^+$ lymphocyte depletion in SIV-infected SMs does not result in AIDS (Klatt et al. 2008), nor does CD4$^+$ depletion in either uninfected SMs or RMs result in an AIDS-like phenomenon (Engram et al. 2010). Thus, preservation of peripheral CD4$^+$ T cells during natural SIV infection does not by itself explain the lack of disease progression in natural hosts.

Adaptive Immune Responses

The fact that natural hosts maintain high viral load (Rey-Cuille et al. 1998; Chakrabarti et al. 2000; Silvestri et al. 2003) indicates that the disease resistance of these animals is unlikely to be because of particularly effective SIV-specific immune responses. This hypothesis is supported by the observations of lower levels of virus-specific T-cell responses in SIV-infected SMs than HIV-infected individuals (Dunham et al. 2006). In addition, depletion of CD8$^+$ T cells results in minimal increase in virus replication in either chronically SIV-infected SMs or during primary infection of AGMs (Schmitz et al. 1999; Barry et al. 2007). In contrast, CD8$^+$ lymphocyte depletion in SIV-infected rhesus macaques or pigtail macaques results in increased virus replication and rapid disease progression (Schmitz et al. 1999, 2009; Klatt et al. 2010). Furthermore, lack of disease progression cannot be accounted for by humoral responses. Depletion of CD20$^+$ B cells in AGMs significantly delays seroconversion but does not result in significant changes in viremia (Schmitz et al. 1999; Gaufin et al. 2009). Moreover, autologous neutralizing antibody levels in SIV infection of SMs are much lower than those observed in HIV-infected humans (Li et al. 2010). Thus, immune control during chronic SIV infection of natural hosts likely does not account for the nonpathogenic nature of the infection.

Innate Immune Responses

Despite the lack of immune control of SIV replication in natural hosts, there is a rapid and robust innate immune response to SIV during acute infection. Similar to pathogenic SIV infection (Table 2), acute SIV infection of SMs and AGMs results in a rapid increase in proliferating T cells (Bosinger et al. 2009; Jacquelin et al. 2009). Activation of an innate immune response was observed as an induction and massive up-regulation of interferon responsive genes measured by gene expression during acute SIV infection of both SMs and AGMs (Bosinger et al. 2009; Jacquelin et al. 2009; Lederer et al. 2009). This is associated with the production of type I interferons by plasmacytoid dendritic cells (pDCs) as measured by immunohistochemical staining in tissues during acute SIVsmm infection (Harris et al. 2010). Moreover, during acute SIV infection, in vitro production of type I interferons by SIV-stimulated pDCs was enhanced in AGMs compared to RMs (Jacquelin et al. 2009). However, in stark contrast to pathogenic SIV infection of RMs or HIV infection of humans, this robust innate immune response to the virus is rapidly attenuated after acute SIV infection of both SMs and AGMs (Bosinger et al. 2009; Jacquelin et al. 2009; Harris et al. 2010). Consistent with this, a similar phenomena is observed in a minor subset of HIV-infected individuals who are highly viremic but maintain high CD4$^+$ T cell counts (Rotger et al. 2011). The genetic profile of T cells isolated from these viremic nonprogressors is similar to that of natural hosts, and, furthermore, interferon-stimulated gene expression was decreased compared to progressive infection, but similar to chronically infected natural hosts (Rotger et al. 2011). Indeed, many reports indicate that a striking and consistent feature of SIV infection of natural hosts is the lack of chronic, generalized immune activation (Silvestri et al. 2003; Paiardini et al. 2006; Sumpter et al. 2007), which is one of the major correlates of disease progression in pathogenic models (McCune 2001; Picker 2006; Fauci 2008).

Chronic Immune Activation and Microbial Translocation

Systemic immune activation, characterized by increased cell proliferation, high rates of lymphocyte apoptosis, cell cycle dysregulation, and increased levels of proinflammatory cytokines (Paiardini et al. 2004, 2006; Hurtrel et al. 2005; Hunt et al. 2008) is a very strong predictor of disease progression during pathogenic HIV/SIV infections. Massive infection of CD4$^+$ T cells in MALT early in HIV/SIV infections is proposed to be associated with breakdown of mucosal integrity, which allows microbial products to translocate from the lumen of the gastrointestinal (GI) tract into peripheral circulation (Brenchley et al. 2006). Translocation of microbial products during pathogenic HIV/SIV infections, can be shown by an increase in plasma lipopolysacharide (LPS) and bacterial DNA levels and is significantly correlated with systemic immune activation (Brenchley et al. 2006). A consistent feature of natural infection

is the absence of generalized chronic immune activation that is characteristically associated with disease progression in pathogenic SIV and HIV infection. Indeed, SIV-infected SMs have low levels of immune activation, T-cell turnover and cell cycle perturbation as compared to SIV-infected RMs or HIV-infected humans, and more comparable levels in uninfected animals (Silvestri et al. 2003; Paiardini et al. 2006; Sumpter et al. 2007). Moreover, SIV infection of natural hosts does not result in microbial translocation, as shown by lack of LPS or sCD14 in the plasma of SIV-infected RMs or AGM (Brenchley et al. 2006; Pandrea et al. 2007a). Furthermore, experimentally induced immune activation with LPS in natural hosts results in significantly increased virus replication and CD4$^+$ T-cell depletion (Pandrea et al. 2008b), which indicates that lack of chronic, systemic immune activation, and microbial translocation may play a role in the lack of disease progression observed in natural hosts.

POTENTIAL MECHANISMS UNDERLYING LACK OF DISEASE PROGRESSION

The key to understanding the benign nature of SIV infection in natural hosts likely lies in understanding the differences between pathogenic and nonpathogenic infections. The most notable differences between SIV-infected natural hosts and SIV-infected macaques and HIV-infected humans are the lack of CD4$^+$ T-cell depletion and attenuation of immune activation.

Attenuated Chronic Immune Activation

As mentioned above, natural hosts have the ability to dampen acute innate immune responses to SIV after a few weeks of infection. This feature contrasts with pathogenic SIV and HIV infections in which immune activation persists throughout the course of infection (Fig. 2). The precise mechanisms underlying resolution of acute immune activation in SIV-infected SMs and AGMs remain poorly understood and are likely quite complex. Several hypotheses have been proposed, including (1) rapid up-regulation of the membrane

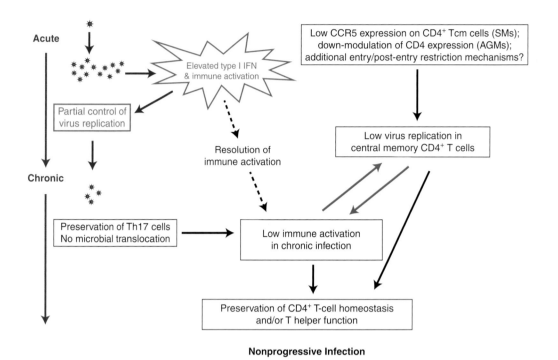

Figure 2. Pathophysiology of natural SIV infections.

receptor programmed death 1 (PD-1); (2) lack of up-regulation of genes such as TRAIL (tumor necrosis factor–related apoptosis-inducing ligand/Apo-2 ligand) and other associated death receptors that trigger apoptosis after pathogenic HIV/SIV infections of human and RMs (Kim et al. 2007); (3) early enhanced Treg responses; (4) reduced response to TLR ligands by pDCs during chronic infection (Mandl et al. 2008); and (5) the ability of Nef alleles to down-modulate the CD3-TCR complex from the surface of infected cells. Additional studies addressing genetic regulation of immune activation after acute SIV infection of natural hosts compared to nonnatural hosts will be crucial in determining a precise mechanism in which natural hosts resolve acute immune activation.

Lack of Microbial Translocation

One potentially important mechanism that underlies the lack of immune activation in SIV-infected natural hosts is absence of significant microbial translocation in these animals (Fig. 2). During SIV infection of RMs, damage to the mucosal barrier of the GI tract is associated with microbial translocation and ensuing immune activation (Estes et al. 2010). In natural infection, preservation of mucosal immune function and of the tight epithelial barrier of the GI tract appears to prevent microbial translocation from occurring. The absence of microbial translocation is associated with preservation of cell subsets integral to mucosal health, including γ-δ T cells and Th17 cells. γ-δ T cells, which are important for mucosal immunity and response to bacterial antigens, are dysregulated during HIV infection but preserved during SIV infection of SMs (Kosub et al. 2008). Th17 cells are specialized CD4$^+$ T cells that produce IL-17 as a signature cytokine in response to bacterial and fungal antigens, that are preferentially depleted from mucosal tissues during pathogenic HIV/SIV infections (Brenchley et al. 2008; Klatt and Brenchley 2010) but are maintained in natural hosts (Klatt and Brenchley 2010). Thus, despite loss of mucosal CD4$^+$ T cells after SIV infection of natural hosts, particular subsets important for mucosal immunity such as γ-δ and Th17 cells are maintained. The retention of such cells may underlie the lack of damage to the GI tract and ensuing microbial translocation and immune activation in natural SIV hosts (Brenchley et al. 2008; Favre et al. 2009; Paiardini 2010).

Target Cell Restriction

Low levels of mucosal $CCR5^+CD4^+$ T-cell targets appears to be a common characteristic of natural hosts (Pandrea et al. 2007b). Indeed, recent studies suggest additional mechanisms used by these hosts to restrict access of SIV to crucial central memory (TCM) $CD4^+$ T cells, "the target restriction" hypothesis (Fig. 2) (Brenchley et al. 2010). $CD4^+$ TCMs are a subset of antigen-experienced T cells that serve as a self-renewing source, whereas $CD4^+$ effector memory (TEM) are more "expendable" and activated than $CD4^+$ TCM cells, and produce the most virus (Grossman et al. 2006). These cells play a central role in AIDS pathogenesis as shown in the pathogenic model of SIV infection of rhesus macaques (RMs), in which depletion of $CD4^+$ TCM cells is crucial for progression to AIDS and, conversely, preservation of these cells is a correlate of vaccine efficacy (Letvin et al. 2006; Mattapallil et al. 2006; Okoye et al. 2007). Protection of $CD4^+$ TCM in natural SIV hosts appears to occur, at least in part, at the entry level. In AGMs, protection of memory $CD4^+$ T cells from SIVagm infection is achieved through down-regulation of the expression of the CD4 molecule as these cells enter the memory pool (Beaumier et al. 2009). AGM "helper" T cells that have down-modulated CD4 expression maintain functions that are typical of $CD4^+$ T cells, including production of IL-2 and IL-17, expression of FOX-P3 and CD40 ligand, and restriction by major histocompatibility complex class II molecules (Beaumier et al. 2009). In the case of SIV-infected SMs, expression of CD4 is maintained as these cells enter the memory pool. However, $CD4^+$ TCMs of SM express significantly lower levels of CCR5 both while resting and when undergoing in vivo and in vitro activation (Pandrea et al. 2007b; Paiardini et al. 2011). Thus, purified $CD4^+$ TCM show on average >10-fold lower levels of cell-associated SIV-DNA when compared to either $CD4^+$ TEM of SMs or $CD4^+$ TCM of SIV-infected rhesus macaques, RMs (Paiardini et al. 2011). It is not clear whether this is a common mechanism observed in all natural hosts of SIV because samples are not readily available for this analysis. Although regulation of CCR5 expression is central in determining the level of SIV infection of $CD4^+$ TCM in SMs, other mechanisms must also play a role. For example, 6% of SIV-infected SMs are homozygous for a 2 bp deletion in CCR5 (Δ2) that abrogates surface expression of CCR5 (Riddick et al. 2010). These animals show ~0.5 log lower viral load than CCR5 wild-type SMs and slightly elevated levels of $CD4^+$ T cells. In addition, SMs infected with an CXCR4-tropic SIVsmm that depletes most $CD4^+$ TCM may be protected by $CD3^+$ $CD4^-CD8^-$ cells that produce "helper" cytokines and show a TCM-like phenotype (Milush et al. 2011).

In terms of pathophysiology, sparing of the $CD4^+$ TCM subset may limit the homeostatic strain on the pool of $CD4^+$ TCMs and may help preserve a normal total pool of $CD4^+$ T cells that maintains the low level of immune activation observed in natural hosts. Taken together with other mechanisms of attenuated immune activation, including down-regulation of acute immune responses, lack of microbial translocation, and preservation of mucosal immune cells, natural hosts have evolved efficient mechanisms by which they remain free of disease after SIV infection.

CONCLUSION

Although these hypotheses are quite intriguing, more work needs to be performed to ascertain whether the degree to which lack of microbial translocation, low immune activation, and protection of $CD4^+$ TCM are essential to determine the nonprogressive course of infection observed in natural SIV hosts. Future research goals include the identification of the cellular and molecular mechanisms responsible for the rapid down-modulation of the immune activation in natural SIV hosts as well as the discovery of major cellular and viral factors that may protect CD4$^+$ TCM from SIV in natural hosts. Furthermore, a more complete understanding of the pathophysiologic link between microbial translocation, virus replication in CD4$^+$ TCM cells and immune activation is needed. Indeed, a more complete understanding of other potential mechanisms by which natural hosts attenuate immune activation would help to define the basis for reduced pathogenicity of these natural infections. Elucidation of how natural hosts prevent microbial translocation will also benefit from a definition of the mechanism of mucosal protection that preserves mucosal barrier function and prevents damage to the gut epithelium and microbial translocation during nonpathogenic SIV infection. Furthermore, an assessment of the degree of pathogenicity in natural and nonnatural hosts of SIV molecular clones that express accessory gene products (i.e., Nef, Vpu, etc.) that have lost specific functions will establish the role of specific viral factors in the pathophysiology of natural SIV infections. Finally, the development of a model in which AIDS is induced in natural hosts by increasing their immune activation and/or expanding their target cell tropism, or, conversely, SIV-infected macaques are rendered AIDS-free by reducing the immune activation in the chronic phase of infection and/or by protecting their CD4$^+$ TCM resistant from infection would provide substantial insight into HIV infection.

There are clearly major lessons to be learned from the natural hosts of SIV, who have spent thousands of years coevolving with the virus, and in fact it is quite possible that a full understanding of the reasons why HIV causes AIDS in humans will not be possible until the mechanisms by which SIVs do not cause disease in natural hosts are fully clarified. In this view, studies of natural SIV hosts have a tremendous impact in AIDS research as a better understanding of HIV pathogenesis will likely result in novel therapeutic and vaccination strategies to delay or prevent HIV transmission and/or disease progression in humans.

REFERENCES

*Reference is also in this collection.

Aghokeng AF, Liu W, Bibollet-Ruche F, Loul S, Mpoudi-Ngole E, Laurent C, Mwenda JM, Langat DK, Chege GK, McClure HM, et al. 2006. Widely varying SIV prevalence rates in naturally infected primate species from Cameroon. *Virology* **345:** 174–189.

Aghokeng AF, Ayouba A, Mpoudi-Ngole E, Loul S, Liegeois F, Delaporte E, Peeters M. 2010. Extensive survey on the prevalence and genetic diversity of SIVs in primate bushmeat provides insights into risks for potential new cross-species transmissions. *Infect Genet Evol* **10:** 386–396.

Allan JS, Short M, Taylor ME, Su S, Hirsch VM, Johnson PR, Shaw GM, Hahn BH. 1991. Species-specific diversity among simian immunodeficiency viruses from African green monkeys. *J Virol* **65:** 2816–2828.

Apetrei C, Robertson DL, Marx PA. 2004. The history of SIVS and AIDS: Epidemiology, phylogeny and biology of isolates from naturally SIV infected non-human primates (NHP) in Africa. *Front Biosci* **9:** 225–254.

Apetrei C, Kaur A, Lerche NW, Metzger M, Pandrea I, Hardcastle J, Falkenstein S, Bohm R, Koehler J, Traina-Dorge V, et al. 2005. Molecular epidemiology of simian immunodeficiency virus SIVsm in U.S. primate centers unravels the origin of SIVmac and SIVstm. *J Virol* **79:** 8991–9005.

Bailes E, Gao F, Bibollet-Ruche F, Courgnaud V, Peeters M, Marx PA, Hahn BH, Sharp PM. 2003. Hybrid origin of SIV in chimpanzees. *Science* **300:** 1713.

Barry AP, Silvestri G, Safrit JT, Sumpter B, Kozyr N, McClure HM, Staprans SI, Feinberg MB. 2007. Depletion of CD8$^+$ cells in sooty mangabey monkeys naturally infected with simian immunodeficiency virus reveals limited role for immune control of virus replication in a natural host species. *J Immunol* **178:** 8002–8012.

Beaumier CM, Harris LD, Goldstein S, Klatt NR, Whitted S, McGinty J, Apetrei C, Pandrea I, Hirsch VM, Brenchley JM. 2009. CD4 downregulation by memory CD4$^+$ T cells in vivo renders African green monkeys resistant to progressive SIVagm infection. *Nat Med* **15:** 879–885.

Beer BE, Bailes E, Goeken R, Dapolito G, Coulibaly C, Norley SG, Kurth R, Gautier JP, Gautier-Hion A, Vallet D, et al. 1999. Simian immunodeficiency virus (SIV) from sun-tailed monkeys (*Cercopithecus solatus*): Evidence for host-dependent evolution of SIV within the *C. lhoesti* superspecies. *J Virol* **73:** 7734–7744.

Beer BE, Bailes E, Dapolito G, Campbell BJ, Goeken RM, Axthelm MK, Markham PD, Bernard J, Zagury D, Franchini G, et al. 2000. Patterns of genomic sequence diversity among their simian immunodeficiency viruses suggest that L'Hoest monkeys (*Cercopithecus lhoesti*) are a natural lentivirus reservoir. *J Virol* **74:** 3892–3898.

Beer BE, Foley BT, Kuiken CL, Tooze Z, Goeken RM, Brown CR, Hu J, St Claire M, Korber BT, Hirsch VM. 2001. Characterization of novel simian immunodeficiency viruses from red-capped mangabeys from Nigeria (SIVrcmNG409 and -NG411). *J Virol* **75:** 12014–12027.

Beer BE, Brown CR, Whitted S, Goldstein S, Goeken R, Plishka R, Buckler-White A, Hirsch VM. 2005. Immunodeficiency in the absence of high viral load in pig-tailed macaques infected with simian immunodeficiency virus SIVsun or SIVlhoest. *J Virol* **79:** 14044–14056.

Bibollet-Ruche F, Galat-Luong A, Cuny G, Sarni-Manchado P, Galat G, Durand JP, Pourrut X, Veas F. 1996. Simian immunodeficiency virus infection in a patas monkey (*Erythrocebus patas*): Evidence for cross-species transmission from African green monkeys (*Cercopithecus aethiops sabaeus*) in the wild. *J Gen Virol* **77:** 773–781.

Bosinger SE, Li Q, Gordon SN, Klatt NR, Duan L, Xu L, Francella N, Sidahmed A, Smith AJ, Cramer EM, et al. 2009. Global genomic analysis reveals rapid control of a robust innate response in SIV-infected sooty mangabeys. *J Clin Invest* **119:** 3556–3572.

Brenchley JM, Price DA, Schacker TW, Asher TE, Silvestri G, Rao S, Kazzaz Z, Bornstein E, Lambotte O, Altmann D, et al. 2006. Microbial translocation is a cause of systemic immune activation in chronic HIV infection. *Nat Med* **12:** 1365–1371.

Brenchley JM, Paiardini M, Knox KS, Asher AI, Cervasi B, Asher TE, Scheinberg P, Price DA, Hage CA, Kholi LM, et al. 2008. Differential Th17 CD4 T-cell depletion in pathogenic and nonpathogenic lentiviral infections. *Blood* **112:** 2826–2835.

Brenchley JM, Silvestri G, Douek DC. 2010. Nonprogressive and progressive primate immunodeficiency lentivirus infections. *Immunity* **32:** 737–742.

Chahroudi A, Meeker T, Lawson B, Ratcliffe S, Else J, Silvestri G. 2011. Mother-to-infant transmission of simian immunodeficiency virus is rare in sooty mangabeys and is associated with low viremia. *J Virol* **85:** 5757–S763.

Chakrabarti LA, Lewin SR, Zhang L, Gettie A, Luckay A, Martin LN, Skulsky E, Ho DD, Cheng-Mayer C, Marx PA. 2000. Normal T-cell turnover in sooty mangabeys harboring active simian immunodeficiency virus infection. *J Virol* **74:** 1209–1223.

Clewley JP, Lewis JC, Brown DW, Gadsby EL. 1998. A novel simian immunodeficiency virus (SIVdrl) pol sequence from the drill monkey, *Mandrillus leucophaeus*. *J Virol* **72:** 10305–10309.

Courgnaud V, Pourrut X, Bibollet-Ruche F, Mpoudi-Ngole E, Bourgeois A, Delaporte E, Peeters M. 2001. Characterization of a novel simian immunodeficiency virus from guereza colobus monkeys (*Colobus guereza*) in Cameroon: A new lineage in the nonhuman primate lentivirus family. *J Virol* **75:** 857–866.

Courgnaud V, Salemi M, Pourrut X, Mpoudi-Ngole E, Abela B, Auzel P, Bibollet-Ruche F, Hahn B, Vandamme AM, Delaporte E, et al. 2002. Characterization of a novel simian immunodeficiency virus with a vpu gene from greater spot-nosed monkeys (*Cercopithecus nictitans*) provides new insights into simian/human immunodeficiency virus phylogeny. *J Virol* **76:** 8298–8309.

Courgnaud V, Abela B, Pourrut X, Mpoudi-Ngole E, Loul S, Delaporte E, Peeters M. 2003. Identification of a new simian immunodeficiency virus lineage with a *vpu* gene present among different *Cercopithecus* monkeys (*C. mona*, *C. cephus*, and *C. nictitans*) from Cameroon. *J Virol* **77:** 12523–12534.

Dunham R, Pagliardini P, Gordon S, Sumpter B, Engram J, Moanna A, Paiardini M, Mandl JN, Lawson B, Garg S, et al. 2006. The AIDS resistance of naturally SIV-infected

sooty mangabeys is independent of cellular immunity to the virus. *Blood* **108:** 209–217.

Engram JC, Cervasi B, Borghans JA, Klatt NR, Gordon SN, Chahroudi A, Else JG, Mittler RS, Sodora DL, de Boer RJ, et al. 2010. Lineage-specific T-cell reconstitution following in vivo CD4+ and CD8+ lymphocyte depletion in nonhuman primates. *Blood* **116:** 748–758.

Estes JD, Harris LD, Klatt NR, Tabb B, Pittaluga S, Paiardini M, Barclay GR, Smedley J, Pung R, Oliveira KM, et al. 2010. Damaged intestinal epithelial integrity linked to microbial translocation in pathogenic simian immunodeficiency virus infections. *PLoS Pathog* **6:** e1001052.

Fauci AS. 2008. 25 years of HIV. *Nature* **453:** 289–290.

Favre D, Lederer S, Kanwar B, Ma ZM, Proll S, Kasakow Z, Mold J, Swainson L, Barbour JD, Baskin CR, et al. 2009. Critical loss of the balance between Th17 and T regulatory cell populations in pathogenic SIV infection. *PLoS Pathog* **5:** e1000295.

Gaufin T, Pattison M, Gautam R, Stoulig C, Dufour J, MacFarland J, Mandell D, Tatum C, Marx MH, Ribeiro RM, et al. 2009. Effect of B-cell depletion on viral replication and clinical outcome of simian immunodeficiency virus infection in a natural host. *J Virol* **83:** 10347–10357.

Goldstein S, Brown CR, Ourmanov I, Pandrea I, Buckler-White A, Erb C, Nandi JS, Foster GJ, Autissier P, Schmitz JE, et al. 2006. Comparison of simian immunodeficiency virus SIVagmVer replication and CD4+ T-cell dynamics in vervet and sabaeus African green monkeys. *J Virol* **80:** 4868–4877.

Gordon SN, Klatt NR, Bosinger SE, Brenchley JM, Milush JM, Engram JC, Dunham RM, Paiardini M, Klucking S, Danesh A, et al. 2007. Severe depletion of mucosal CD4+ T cells in AIDS-free simian immunodeficiency virus-infected sooty mangabeys. *J Immunol* **179:** 3026–3034.

Gordon SN, Dunham RM, Engram JC, Estes J, Wang Z, Klatt NR, Paiardini M, Pandrea IV, Apetrei C, Sodora DL, et al. 2008. Short-lived infected cells support virus replication in sooty mangabeys naturally infected with simian immunodeficiency virus: Implications for AIDS pathogenesis. *J Virol* **82:** 3725–3735.

Grossman Z, Meier-Schellersheim M, Paul WE, Picker LJ. 2006. Pathogenesis of HIV infection: What the virus spares is as important as what it destroys. *Nat Med* **12:** 289–295.

Haigwood NL. 2009. Update on animal models for HIV research. *Eur J Immunol* **39:** 1994–1999.

Harris LD, Tabb B, Sodora DL, Paiardini M, Klatt NR, Douek DC, Silvestri G, Müller-Trutwin M, Vasile-Pandrea I, Apetrei C, et al. 2010. Downregulation of robust acute type I interferon responses distinguishes nonpathogenic simian immunodeficiency virus (SIV) infection of natural hosts from pathogenic SIV infection of rhesus macaques. *J Virol* **84:** 7886–7891.

Hirsch VM. 2004. What can natural infection of African monkeys with simian immunodeficiency virus tell us about the pathogenesis of AIDS? *AIDS Rev* **6:** 40–53.

Hirsch VM, Olmsted RA, Murphey-Corb M, Purcell RH, Johnson PR. 1989. An African primate lentivirus (SIVsm) closely related to HIV-2. *Nature* **339:** 389–392.

Hirsch VM, Dapolito G, Goeken R, Campbell BJ. 1995a. Phylogeny and natural history of the primate lentiviruses, SIV and HIV. *Curr Opin Genet Dev* **5:** 798–806.

Hirsch VM, Dapolito G, Johnson PR, Elkins WR, London WT, Montali RJ, Goldstein S, Brown C. 1995b. Induction of AIDS by simian immunodeficiency virus from an African green monkey: Species-specific variation in pathogenicity correlates with the extent of in vivo replication. *J Virol* **69:** 955–967.

Ho DD, Neumann AU, Perelson AS, Chen W, Leonard JM, Markowitz M. 1995. Rapid turnover of plasma virions and CD4 lymphocytes in HIV-1 infection. *Nature* **373:** 123–126.

Hu J, Switzer WM, Foley BT, Robertson DL, Goeken RM, Korber BT, Hirsch VM, Beer BE. 2003. Characterization and comparison of recombinant simian immunodeficiency virus from drill (*Mandrillus leucophaeus*) and mandrill (*Mandrillus sphinx*) isolates. *J Virol* **77:** 4867–4880.

Hunt PW, Brenchley J, Sinclair E, McCune JM, Roland M, Page-Shafer K, Hsue P, Emu B, Krone M, Lampiris H, et al. 2008. Relationship between T cell activation and CD4$^+$ T cell count in HIV-seropositive individuals with undetectable plasma HIV RNA levels in the absence of therapy. *J Infect Dis* **197:** 126–133.

Hurtrel B, Petit F, Arnoult D, Müller-Trutwin M, Silvestri G, Estaquier J. 2005. Apoptosis in SIV infection. *Cell Death Differ* **12:** 979–990.

Jacquelin B, Mayau V, Targat B, Liovat AS, Kunkel D, Petit-jean G, Dillies MA, Roques P, Butor C, Silvestri G, et al. 2009. Nonpathogenic SIV infection of African green monkeys induces a strong but rapidly controlled type I IFN response. *J Clin Invest* **119:** 3544–3555.

Jin MJ, Rogers J, Phillips-Conroy JE, Allan JS, Desrosiers RC, Shaw GM, Sharp PM, Hahn BH. 1994a. Infection of a yellow baboon with simian immunodeficiency virus from African green monkeys: Evidence for cross-species transmission in the wild. *J Virol* **68:** 8454–8460.

Jin MJ, Hui H, Robertson DL, Müller MC, Barré-Sinoussi F, Hirsch VM, Allan JS, Shaw GM, Sharp PM, Hahn BH. 1994b. Mosaic genome structure of simian immunodeficiency virus from west African green monkeys. *EMBO J* **13:** 2935–2947.

Johnson PR, Hirsch VM. 1991. Pathogenesis of AIDS: The non-human primate model. *Aids* **5:** S43–S48.

Johnson PR, Hirsch VM. 1992. SIV infection of macaques as a model for AIDS pathogenesis. *Int Rev Immunol* **8:** 55–63.

Johnson PR, Goldstein S, London WT, Fomsgaard A, Hirsch VM. 1990. Molecular clones of SIVsm and SIVagm: Experimental infection of macaques and African green monkeys. *J Med Primatol* **19:** 279–286.

Keele BF, Van Heuverswyn F, Li Y, Bailes E, Takehisa J, Santiago ML, Bibollet-Ruche F, Chen Y, Wain LV, Liegeois F, et al. 2006. Chimpanzee reservoirs of pandemic and nonpandemic HIV-1. *Science* **313:** 523–526.

Keele BF, Jones JH, Terio KA, Estes JD, Rudicell RS, Wilson ML, Li Y, Learn GH, Beasley TM, Schumacher-Stankey J, et al. 2009. Increased mortality and AIDS-like immunopathology in wild chimpanzees infected with SIVcpz. *Nature* **460:** 515–519.

Kim N, Dabrowska A, Jenner RG, Aldovini A. 2007. Human and simian immunodeficiency virus-mediated upregulation of the apoptotic factor TRAIL occurs in antigen-presenting cells from AIDS-susceptible but not from AIDS-resistant species. *J Virol* **81:** 7584–7597.

Kirmaier A, Wu F, Newman RM, Hall LR, Morgan JS, O'Connor S, Marx PA, Meythaler M, Goldstein S, Buckler-White A, et al. 2010. TRIM5 suppresses cross-species transmission of a primate immunodeficiency virus and selects for emergence of resistant variants in the new species. *PLoS Biol* **8:** e1000462.

Klatt NR, Brenchley JM. 2010. Th17 cell dynamics in HIV infection. *Curr Opin HIVAIDS* **5:** 135–140.

Klatt NR, Villinger F, Bostik P, Gordon SN, Pereira L, Engram JC, Mayne A, Dunham RM, Lawson B, Ratcliffe SJ, et al. 2008. Availability of activated CD4$^+$ T cells dictates the level of viremia in naturally SIV-infected sooty mangabeys. *J Clin Invest* **118:** 2039–2049.

Klatt NR, Shudo E, Ortiz AM, Engram JC, Paiardini M, Lawson B, Miller MD, Else J, Pandrea I, Estes JD, et al. 2010. CD8$^+$ lymphocytes control viral replication in SIVmac239-infected rhesus macaques without decreasing the lifespan of productively infected cells. *PLoS Pathog* **6:** e1000747.

Kosub DA, Lehrman G, Milush JM, Zhou D, Chacko E, Leone A, Gordon S, Silvestri G, Else JG, Keiser P, et al. 2008. Gamma/delta-cell functional responses differ after pathogenic human immunodeficiency virus and nonpathogenic simian immunodeficiency virus infections. *J Virol* **82:** 1155–1165.

Lederer S, Favre D, Walters KA, Proll S, Kanwar B, Kasakow Z, Baskin CR, Palermo R, McCune JM, Katze MG. 2009. Transcriptional profiling in pathogenic and non-pathogenic SIV infections reveals significant distinctions in kinetics and tissue compartmentalization. *PLoS Pathog* **5:** e1000296.

Letvin NL, Mascola JR, Sun Y, Gorgone DA, Buzby AP, Xu L, Yang ZY, Chakrabarti B, Rao SS, Schmitz JE, et al. 2006. Preserved CD4$^+$ central memory T cells and survival in vaccinated SIV-challenged monkeys. *Science* **312:** 1530–1533.

Li B, Stefano-Cole K, Kuhrt DM, Gordon SN, Else JG, Mulenga J, Allen S, Sodora DL, Silvestri G, Derdeyn CA. 2010. Nonpathogenic simian immunodeficiency virus infection of sooty mangabeys is not associated with high levels of autologous neutralizing antibodies. *J Virol* **84:** 6248–6253.

Ling B, Apetrei C, Pandrea I, Veazey RS, Lackner AA, Gormus B, Marx PA. 2004. Classic AIDS in a sooty mangabey after an 18-year natural infection. *J Virol* **78:** 8902–8908.

Mandl JN, Barry AP, Vanderford TH, Kozyr N, Chavan R, Klucking S, Barrat FJ, Coffman RL, Staprans SI, Feinberg MB. 2008. Divergent TLR7 and TLR9 signaling and type I interferon production distinguish pathogenic and nonpathogenic AIDS virus infections. *Nat Med* **14:** 1077–1087.

Mattapallil JJ, Douek DC, Buckler-White A, Montefiori D, Letvin NL, Nabel GJ, Roederer M. 2006. Vaccination preserves CD4 memory T cells during acute simian immunodeficiency virus challenge. *J Exp Med* **203:** 1533–1541.

McCune JM. 2001. The dynamics of CD4$^+$ T-cell depletion in HIV disease. *Nature* **410:** 974–979.

Cite this article as *Cold Spring Harb Perspect Med* doi: 10.1101/cshperspect.a007153

Milush JM, Reeves JD, Gordon SN, Zhou D, Muthukumar A, Kosub DA, Chacko E, Giavedoni LD, Ibegbu CC, Cole KS, et al. 2007. Virally induced CD4$^+$ T cell depletion is not sufficient to induce AIDS in a natural host. *J Immunol* **179:** 3047–3056.

Milush JM, Mir KD, Sundaravaradan V, Gordon SN, Engram J, Cano CA, Reeves JD, Anton E, O'Neill E, Butler E, et al. 2011. Lack of clinical AIDS in SIV-infected sooty mangabeys with significant CD4$^+$ T cell loss is associated with double-negative T cells. *J Clin Invest* **121:** 1102–1110.

Muller MC, Barre-Sinoussi F. 2003. SIVagm: Genetic and biological features associated with replication. *Front Biosci* **8:** d1170–d1185.

Murphey-Corb M, Martin LN, Rangan SR, Baskin GB, Gormus BJ, Wolf RH, Andes WA, West M, Montelaro RC. 1986. Isolation of an HTLV-III-related retrovirus from macaques with simian AIDS and its possible origin in asymptomatic mangabeys. *Nature* **321:** 435–437.

Novembre FJ, Saucier M, Anderson DC, Klumpp SA, O'Neil SP, Brown CR II, Hart CE, Guenthner PC, Swenson RB, McClure HM. 1997. Development of AIDS in a chimpanzee infected with human immunodeficiency virus type 1. *J Virol* **71:** 4086–4091.

Nowak MA, Lloyd AL, Vasquez GM, Wiltrout TA, Wahl LM, Bischofberger N, Williams J, Kinter A, Fauci AS, Hirsch VM, et al. 1997. Viral dynamics of primary viremia and antiretroviral therapy in simian immunodeficiency virus infection. *J Virol* **71:** 7518–7525.

Okoye A, Meier-Schellersheim M, Brenchley JM, Hagen SI, Walker JM, Rohankhedkar M, Lum R, Edgar JB, Planer SL, Legasse A, et al. 2007. Progressive CD4$^+$ central memory T cell decline results in CD4$^+$ effector memory insufficiency and overt disease in chronic SIV infection. *J Exp Med* **204:** 2171–2185.

Onanga R, Souquière S, Makuwa M, Mouinga-Ondeme A, Simon F, Apetrei C, Roques P. 2006. Primary simian immunodeficiency virus SIVmnd-2 infection in mandrills (*Mandrillus sphinx*). *J Virol* **80:** 3301–3309.

O'Neil SP, Novembre FJ, Hill AB, Suwyn C, Hart CE, Evans-Strickfaden T, Anderson DC, deRosayro J, Herndon JG, Saucier M, et al. 2000. Progressive infection in a subset of HIV-1-positive chimpanzees. *J Infect Dis* **182:** 1051–1062.

Paiardini M. 2010. Th17 cells in natural SIV hosts. *Curr Opin HIVAIDS* **5:** 166–172.

Paiardini M, Cervasi B, Dunham R, Sumpter B, Radziewicz H, Silvestri G. 2004. Cell-cycle dysregulation in the immunopathogenesis of AIDS. *Immunol Res* **29:** 253–268.

Paiardini M, Cervasi B, Sumpter B, McClure HM, Sodora DL, Magnani M, Staprans SI, Piedimonte G, Silvestri G. 2006. Perturbations of cell cycle control in T cells contribute to the different outcomes of simian immunodeficiency virus infection in rhesus macaques and sooty mangabeys. *J Virol* **80:** 634–642.

Paiardini M, Pandrea I, Apetrei C, Silvestri G. 2009a. Lessons learned from the natural hosts of HIV-related viruses. *Annu Rev Med* **60:** 485–495.

Paiardini M, Pandrea I, Apetrei C, Silvestri G. 2009b. Lessons learned from the natural hosts of HIV-related viruses. *Ann Rev Med* **60:** 485–495.

Paiardini M, Cervasi B, Reyes-Aviles E, Micci L, Ortiz AM, Chahroudi A, Vinton C, Gordon SN, Bosinger SE, Francella N, et al. 2011. Low levels of SIV infection in sooty mangabey central-memory CD4$^+$ T-cells is associated with limited CCR5 expression. *Nat Med* (in press).

Pandrea I, Onanga R, Kornfeld C, Rouquet P, Bourry O, Clifford S, Telfer PT, Abernethy K, White LT, Ngari P, et al. 2003. High levels of SIVmnd-1 replication in chronically infected *Mandrillus sphinx*. *Virology* **317:** 119–127.

Pandrea I, Silvestri G, Onanga R, Veazey RS, Marx PA, Hirsch V, Apetrei C. 2006a. Simian immunodeficiency viruses replication dynamics in African non-human primate hosts: Common patterns and species-specific differences. *J Med Primatol* **35:** 194–201.

Pandrea I, Apetrei C, Dufour J, Dillon N, Barbercheck J, Metzger M, Jacquelin B, Bohm R, Marx PA, Barre-Sinoussi F, et al. 2006b. Simian immunodeficiency virus SIVagm.sab infection of Caribbean African green monkeys: A new model for the study of SIV pathogenesis in natural hosts. *J Virol* **80:** 4858–4867.

Pandrea IV, Gautam R, Ribeiro RM, Brenchley JM, Butler IF, Pattison M, Rasmussen T, Marx PA, Silvestri G, Lackner AA, et al. 2007a. Acute loss of intestinal CD4$^+$ T cells is not predictive of simian immunodeficiency virus virulence. *J Immunol* **179:** 3035–3046.

Pandrea I, Apetrei C, Gordon S, Barbercheck J, Dufour J, Bohm R, Sumpter B, Roques P, Marx PA, Hirsch VM, et al. 2007b. Paucity of CD4$^+$CCR5$^+$ T cells is a typical feature of natural SIV hosts. *Blood* **109:** 1069–1076.

Pandrea I, Ribeiro RM, Gautam R, Gaufin T, Pattison M, Barnes M, Monjure C, Stoulig C, Dufour J, Cyprian W, et al. 2008a. Simian immunodeficiency virus SIVagm dynamics in African green monkeys. *J Virol* **82:** 3713–3724.

Pandrea I, Gaufin T, Brenchley JM, Gautam R, Monjure C, Gautam A, Coleman C, Lackner AA, Ribeiro RM, Douek DC, et al. 2008b. Cutting edge: Experimentally induced immune activation in natural hosts of SIV induces significant increases in viral replication and CD4$^+$ T cell depletion. *J Immunol* **181:** 6687–6691.

Pandrea I, Silvestri G, Apetrei C. 2009. AIDS in african non-human primate hosts of SIVs: A new paradigm of SIV infection. *Curr HIV Res* **7:** 57–72.

Pandrea I, Apetrei C. 2010. Where the wild things are: Pathogenesis of SIV infection in African nonhuman primate hosts. *Curr HIV/AIDS Rep* **7:** 28–36.

Peeters M, Chaix ML, Delaporte E. 2008. Genetic diversity and phylogeographic distribution of SIV: How to understand the origin of HIV. *Med Sci (Paris)* **24:** 621–628.

Perelson AS, Kirschner DE, De Boer R. 1993. Dynamics of HIV infection of CD4$^+$ T cells. *Math Biosci* **114:** 81–125.

Perelson AS, Essunger P, Cao Y, Vesanen M, Hurley A, Saksela K, Markowitz M, Ho DD. 1997. Decay characteristics of HIV-1-infected compartments during combination therapy. *Nature* **387:** 188–191.

Picker LJ. 2006. Immunopathogenesis of acute AIDS virus infection. *Curr Opin Immunol* **18:** 399–405.

Rey-Cuillé MA, Berthier JL, Bomsel-Demontoy MC, Chaduc Y, Montagnier L, Hovanessian AG, Chakrabarti LA. 1998. Simian immunodeficiency virus replicates to high levels in sooty mangabeys without inducing disease. *J Virol* **72:** 3872–3886.

Riddick NE, Hermann EA, Loftin LM, Elliott ST, Wey WC, Cervasi B, Taaffe J, Engram JC, Li B, Else JG, et al. 2010. A novel CCR5 mutation common in sooty mangabeys reveals SIVsmm infection of CCR5-null natural hosts and efficient alternative coreceptor use in vivo. *PLoS Pathog* **6:** e1001064.

Rotger M, Dalmau J, Rauch A, McLaren P, Bosinger S, Martinez R, Sandler NG, Roque A, Liebner J, Battegay M, et al. 2011. Comparative transcriptomics of extreme phenotypes of human HIV-1 infection and SIV infection in sooty mangabey and rhesus macaque. *J Clin Invest* **121:** 2391–2400.

Sauter D, Schindler M, Specht A, Landford WN, Münch J, Kim KA, Votteler J, Schubert U, Bibollet-Ruche F, Keele BF, et al. 2009. Tetherin-driven adaptation of Vpu and Nef function and the evolution of pandemic and nonpandemic HIV-1 strains. *Cell Host Microbe* **6:** 409–421.

Schmitz JE, Kuroda MJ, Santra S, Sasseville VG, Simon MA, Lifton MA, Racz P, Tenner-Racz K, Dalesandro M, Scallon BJ, et al. 1999. Control of viremia in simian immunodeficiency virus infection by CD8$^+$ lymphocytes. *Science* **283:** 857–860.

Schmitz JE, Zahn RC, Brown CR, Rett MD, Li M, Tang H, Pryputniewicz S, Byrum RA, Kaur A, Montefiori DC, et al. 2009. Inhibition of adaptive immune responses leads to a fatal clinical outcome in SIV-infected pigtailed macaques but not vervet African green monkeys. *PLoS Pathog* **5:** e1000691.

Sharp PM, Hahn BH. 2010. The evolution of HIV-1 and the origin of AIDS. *Philos Trans R Soc Lond B Biol Sci* **365:** 2487–2494.

* Sharp PM, Hahn BH. 2011. Origins of HIV and the AIDS pandemic. *Cold Spring Harb Perspect Med* **1:** a006841.

Silvestri G, Sodora DL, Koup RA, Paiardini M, O'Neil SP, McClure HM, Staprans SI, Feinberg MB. 2003. Nonpathogenic SIV infection of sooty mangabeys is characterized by limited bystander immunopathology despite chronic high-level viremia. *Immunity* **18:** 441–452.

Sodora DL, Allan JS, Apetrei C, Brenchley JM, Douek DC, Else JG, Estes JD, Hahn BH, Hirsch VM, Kaur A, et al. 2009. Toward an AIDS vaccine: Lessons from natural simian immunodeficiency virus infections of African nonhuman primate hosts. *Nat Med* **15:** 861–865.

Sumpter B, Dunham R, Gordon S, Engram J, Hennessy M, Kinter A, Paiardini M, Cervasi B, Klatt N, McClure H, et al. 2007. Correlates of preserved CD4$^+$ T cell homeostasis during natural, nonpathogenic simian immunodeficiency virus infection of sooty mangabeys: Implications for AIDS pathogenesis. *J Immunol* **178:** 1680–1691.

Wei X, Ghosh SK, Taylor ME, Johnson VA, Emini EA, Deutsch P, Lifson JD, Bonhoeffer S, Nowak MA, Hahn BH, et al. 1995. Viral dynamics in human immunodeficiency virus type 1 infection. *Nature* **373:** 117–122.

Worobey M, Telfer P, Souquière S, Hunter M, Coleman CA, Metzger MJ, Reed P, Makuwa M, Hearn G, Honarvar S, et al. 2010. Island biogeography reveals the deep history of SIV. *Science* **329:** 1487.

HIV-1 Antiretroviral Drug Therapy

Eric J. Arts[1] and Daria J. Hazuda[2]

[1]Ugandan CFAR Laboratories, Division of Infectious Diseases, Department of Medicine, Case Western Reserve University, Cleveland, Ohio 44106

[2]Merck Research Laboratories, West Point, Pennsylvania 19486

Correspondence: eja3@case.edu; daria_hazuda@merck.com

The most significant advance in the medical management of HIV-1 infection has been the treatment of patients with antiviral drugs, which can suppress HIV-1 replication to undetectable levels. The discovery of HIV-1 as the causative agent of AIDS together with an ever-increasing understanding of the virus replication cycle have been instrumental in this effort by providing researchers with the knowledge and tools required to prosecute drug discovery efforts focused on targeted inhibition with specific pharmacological agents. To date, an arsenal of 24 Food and Drug Administration (FDA)-approved drugs are available for treatment of HIV-1 infections. These drugs are distributed into six distinct classes based on their molecular mechanism and resistance profiles: (1) nucleoside-analog reverse transcriptase inhibitors (NNRTIs), (2) non–nucleoside reverse transcriptase inhibitors (NNRTIs), (3) integrase inhibitors, (4) protease inhibitors (PIs), (5) fusion inhibitors, and (6) coreceptor antagonists. In this article, we will review the basic principles of antiretroviral drug therapy, the mode of drug action, and the factors leading to treatment failure (i.e., drug resistance).

BASIC PRINCIPLES OF ANTIRETROVIRAL THERAPY

Before 1996, few antiretroviral treatment options for HIV-1 infection existed. The clinical management of HIV-1 largely consisted of prophylaxis against common opportunistic pathogens and managing AIDS-related illnesses. The treatment of HIV-1 infection was revolutionized in the mid-1990s by the development of inhibitors of the reverse transcriptase and protease, two of three essential enzymes of HIV-1, and the introduction of drug regimens that combined these agents to enhance the overall efficacy and durability of therapy. A timeline of antiretroviral drug development and approval for human use is described in Figure 1.

Since the first HIV-1 specific antiviral drugs were given as monotherapy in the early 1990s, the standard of HIV-1 care evolved to include the administration of a cocktail or combination of antiretroviral agents (ARVs). The advent of combination therapy, also known as HAART, for the treatment of HIV-1 infection was seminal in reducing the morbidity and mortality associated with HIV-1 infection and AIDS (Collier et al. 1996; D'Aquila et al. 1996; Staszewski et al. 1996). Combination antiretroviral therapy dramatically suppresses viral replication and reduces the plasma HIV-1 viral load (vLoad) to below the limits of detection of the

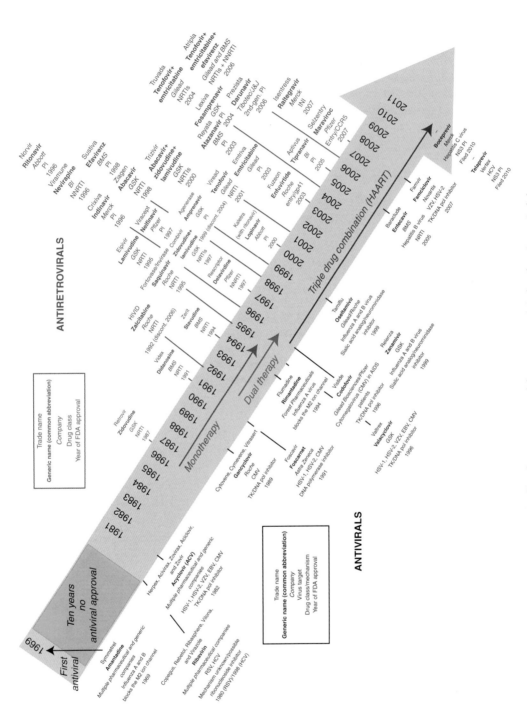

Figure 1. Timeline for FDA approval for current antiviral and antiretroviral drugs.

Cite this article as *Cold Spring Harb Perspect Med* doi: 10.1101/cshperspect.a007161

most sensitive clinical assays (<50 RNA copies/ mL) resulting in a significant reconstitution of the immune system (Autran et al. 1997; Komanduri et al. 1998; Lederman et al. 1998;) as measured by an increase in circulating $CD4^+$ T-lymphocytes. Importantly, combination therapy using three antiretroviral agents directed against at least two distinct molecular targets is the underlying basis for forestalling the evolution drug resistance.

In an untreated individual, on average there are 10^4–10^5 or more HIV-1 particles per mL of plasma, which turn over at a rate of $\sim 10^{10}$/d (Ho et al. 1995; Wei et al. 1995; Perelson et al. 1996). Owing to the error-prone reverse transcription process, it is estimated that one mutation is introduced for every 1000–10,000 nucleotides synthesized (Mansky and Temin 1995; O'Neil et al. 2002; Abram et al. 2010). As the HIV-1 genome is $\sim 10,000$ nucleotides in length, one to 10 mutations may be generated in each viral genome with every replication cycle. With this enormous potential for generating genetic diversity, HIV-1 variants with reduced susceptibility to any one or two drugs will often preexist in the viral quasispecies before initiating therapy (Coffin 1995). The success of HAART results in part from using drug combinations that decrease the probability of selecting virus clones (from an intrapatient HIV-1 population) bearing multiple mutations and conferring resistance to a three-antiretroviral-drug regimen.

Given the rate of HIV-1 turnover and the size of the virus population, mathematical modeling studies have suggested that any combinations in which at least three mutations are required should provide durable inhibition (Frost and McLean 1994; Coffin 1995; Nowak et al. 1997; Stengel 2008). In the simplest interpretation of these models, three drug combinations should be more advantageous than two drug regimens, and in fact, this was the precedent established in early clinical trials of combination antiretroviral therapy. However, this interpretation assumes that all drugs have equal activity, that they require the same number of mutations to engender resistance, and that resistance mutations impact viral replication

capacity or viral fitness to a similar degree. Trial and error with early antiretroviral agents helped to establish the basic principles for effective drug combinations in HAART. Since these early days, therapies have evolved, with the introduction of newer drugs with greater potency and higher barriers to the development of resistance. Moreover, some antiretroviral agents have been shown to select for mutations which are either incompatible with or engender hypersensitivity to other antiretroviral drugs, suggesting certain ARVs may offer an advantage with respect to resistance barrier when used in the context of specific combinations (Larder et al. 1995; Kempf et al. 1997; Hsu et al. 1998). Therefore, whether HIV-1 treatment can be simplified to two or even one potent drug(s) remains an open question that can only be answered with future clinical studies.

In 2010, HIV-1 treatment guidelines in the United States and European Union recommend the initiation of HAART with three fully active antiretroviral agents when CD4 cells in peripheral blood decline to 350 per cubic mm, a stage at which viral levels can often reach 10,000–100,000 copies per mL (as measured by RNA in the blood) (see http://aidsinfo.nih.gov/ Guidelines/). With proper adherence, HAART can suppress viral replication for decades, dramatically increasing the life expectancy of the HIV-infected individual. However, HAART alone cannot eliminate HIV-1 infection. HIV-1 is a chronic infection for which there is currently no cure—the prospect of maintaining therapy for the lifetime of a patient presents major challenges. The potential for persistent viral replication in compartments and reservoirs may continue to drive pathogenic disease processes (Finzi et al. 1997, 1999). The effect of therapy can be impaired by nonadherence, poor drug tolerability, and drug interactions among antiretroviral agents and other medications that decrease optimal drug levels. Each of these can lead to virologic failure and the evolution of drug resistance.

For all antiretroviral drug classes, drug resistance has been documented in patients failing therapy as well as in therapy-naïve patients infected with transmitted, drug-resistant viruses.

Resistance testing is therefore recommended before initiating HAART in therapy-naïve patients as well as when reoptimizing antiretroviral therapy after treatment failure. Given the number of agents and distinct classes of antiretroviral drugs available today, most patients, even those with a history of failure, can be successfully treated. However, as the virus continues to evolve and escape, with even the most effective therapies, new HIV-1 treatments will always be needed.

THE HIV REPLICATION CYCLE AND DRUG TARGETS

Antiretroviral agents for the treatment of HIV-1 are a relatively new addition to the armamentarium of antiviral drugs. In the 1960s, amantadine and rimantidine were the first approved antiviral drugs for treatment of a human influenza virus infection (Davies et al. 1964; Wingfield et al. 1969), but more than 20 years passed before the elucidation of their mechanism of action (Hay et al. 1985). With the advent of modern molecular biology, such serendipitous approaches to antiviral drug discovery have been largely replaced by mechanistic-based approaches, which include (1) high throughput compound screens with virus-specific replication or enzymatic assays, (2) optimization of inhibitors using lead compounds based on homologous enzymes or targets, and/or (3) rational drug design modeled on the structures of viral proteins. These methods together with advances in the corresponding enabling technologies greatly accelerated the development of antiretroviral drugs in the early 1990s. The highly divergent evolution of HIV-1 genes from the human host provided the basis

for implementing targeted screening efforts and/or designing and optimizing inhibitors with minimal off-target activities, thus capitalizing on these technological advances. A full timeline in the development of antiviral and antiretroviral drugs for human use is described in Figure 1.

Whereas the HIV-1 life cycle presents many potential opportunities for therapeutic intervention, only a few have been exploited. The replication scheme of HIV-1 is shown in Figure 2, marked with the steps blocked by approved inhibitors (numbers in panel 2A). A timing of the retroviral lifecycle is described in panel B based on the specific time window of inhibition by a specific drug class. In panel 2C, the inhibitors in development (normal text) or FDA approved (italic/bold text) are listed by inhibition of a specific retroviral replication event. The first step in the HIV-1 replication cycle, viral entry (Doms and Wilen 2011), is the target for several classes of antiretroviral agents: attachment inhibitors, chemokine receptor antagonists, and fusion inhibitors. The HIV-1 envelope gp120/gp41 has affinity for the CD4 receptor and directs HIV-1 to $CD4^+$ immune cells (Dalgleish et al. 1984; Klatzmann et al. 1984). Interaction of the gp120 subunit of the HIV-1 envelope with CD4 is followed by binding to an additional coreceptor, either the CC chemokine receptor CCR5 or the CXC chemokine receptor CXCR4 (Alkhatib et al. 1996; Deng et al. 1996; Doranz et al. 1996; Feng et al. 1996). The disposition of these coreceptors on the surface of lymphocytes and monocyte/macrophages, and coreceptor recognition by the viral envelope, are major determinants of tropism for different cell types. These sequential receptor-binding

---

Figure 2. Identifying distinct steps in HIV-1 life cycle as potential or current target for antiretroviral drugs. (*A*) Schematic of the HIV-1 life cycle in a susceptible $CD4^+$ cell. (*B*) Time frame for antiretroviral drug action during a single-cycle HIV-1 replication assay. In this experiment, HIV-1 inhibitors are added following a synchronized inhibition. The addition of drug following the HIV-1 replication step targeted by the drug will result in a lack of inhibition. The time window of drug inhibition provides an estimate for the time required for these replication steps. For example, T30 or enfuvirtide (T20) only inhibits within 1–2 h of infection, whereas lamivudine (3TC) inhibits within a 2- to 10-h time frame, which coincides with reverse transcription. (*C*) Preclinical, abandoned (normal text), or FDA-approved (bold italic text) inhibitors are listed in relation to specificity of action and drug target.

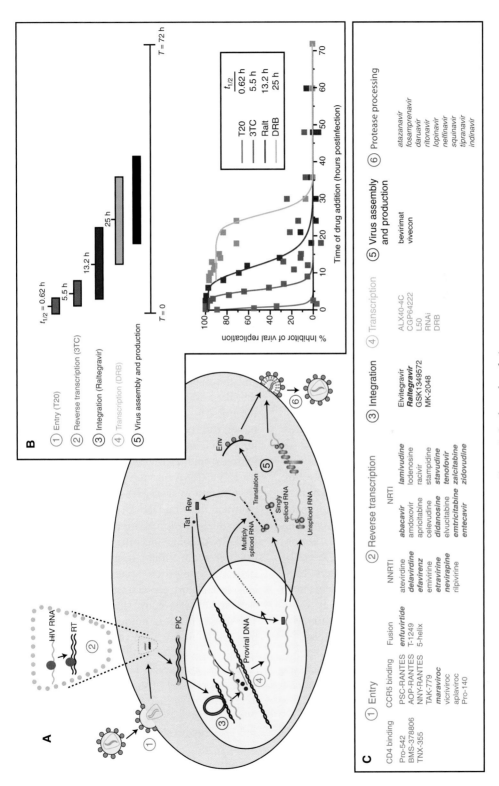

Figure 2. *See legend on facing page.*

events trigger conformational changes in the HIV-1 envelope, exposing a hydrophobic domain on gp41 that mediates fusion with the cellular membrane. The entire entry process is completed within 1 h of virus contact with the cell (Fig. 2B). Gp120 and CD4 are targets for small-molecule and antibody-based attachment inhibitors BMS-378806 and TNX-355, each of which have shown some clinical promise, although neither is approved for use in HIV-1 patients (Reimann et al. 1997; Lin et al. 2003; Kuritzkes et al. 2004). BMS-378806 binds to a pocket on gp120 important for binding CD4 and alters the conformation of the envelope protein such that it cannot recognize CD4 (Lin et al. 2003). TNX-355 is a humanized anti-CD4 monoclonal antibody that binds to CD4 and inhibits HIV-1 envelope docking, but does not inhibit CD4 function in immunological contexts (Reimann et al. 1997). Gp41 and the coreceptor CCR5 are the targets for the two approved entry agents that will be discussed in more detail below: the peptide-based fusion inhibitor, fuzeon, and the small-molecule CCR5 chemokine receptor antagonist, maraviroc.

Viral entry and fusion of the HIV-1 envelope with the host cell membrane allow for uncoating of the viral core and initiate a slow dissolution process that maintains protection of the viral RNA genome while permitting access to deoxyribonucleoside triphosphates (dNTPs) necessary for reverse transcription and proviral DNA synthesis (Fig. 1). Reverse transcription is a process extending over the next 10 h of infection (Fig. 2A,B). Reverse transcriptase (RT) was the first HIV-1 enzyme to be exploited for antiretroviral drug discovery (Fig. 1). RT is a multifunctional enzyme with RNA-dependent DNA polymerase, RNase-H, and DNA-dependent DNA polymerase activities, all of which are required to convert the single-stranded HIV-1 viral RNA into double-stranded DNA (Hughes and Hu 2011). RT is the target for two distinct classes of antiretroviral agents: the NRTIs (Fig. 2C), which are analogs of native nucleoside substrates, and the NNRTIs (Fig. 2C), which bind to a noncatalytic allosteric pocket on the enzyme. Together,

the 12 licensed agents in these two classes account for the nearly half of all approved antiretroviral drugs. Although the NRTIs and NNRTIs differ with respect to their site of interaction on the enzyme and molecular mechanism, both affect the DNA polymerization activity of the enzyme and block the generation of full-length viral DNA.

The completion of reverse transcription is required to form the viral preintegration complex, or PIC. The PIC, comprised of viral as well as cellular components, is transported to the nucleus where the second essential HIV-1 enzyme, integrase, catalyzes the integration of the viral DNA with the host DNA (Craigie and Bushman 2011). Integrase orchestrates three sequence-specific events required for integration, assembly with the viral DNA, endonucleolytic processing of the 3′ ends of the viral DNA, and strand transfer or joining of the viral and cellular DNA. In the context of HIV-1 infection, the process occurs in a stepwise manner, with the rate-limiting event being strand transfer and the stable integration of the viral genome into the human chromosome occurring within the first 15–20 h of infection (Fig. 2B). The newest class of approved ARVs, integrase inhibitors (INIs or InSTIs) (Fig. 2C), specifically inhibit strand transfer and block integration of the HIV-1 DNA into the cellular DNA.

Integration of the HIV-1 DNA is required to maintain the viral DNA in the infected cell and is essential for expression of HIV-1 mRNA and viral RNA. Following integration, the cellular machinery can initiate transcription; however, transcript elongation requires binding of the HIV-1 regulatory protein Tat to the HIV-1 RNA element (TAR) (Karn and Stoltzfus 2011). This mechanism is unique to HIV-1 and is thus considered a highly desirable therapeutic target. A variety of candidate small-molecule inhibitors of either HIV transcription, or more specifically, the Tat–TAR interaction, have been identified during the last 15 yrs (Fig. 2A,C, section 4) (Hsu et al. 1991; Cupelli and Hsu 1995; Hamy et al. 1997; Hwang et al. 2003). Unfortunately, none of these compounds were sufficiently potent and/or selective to progress beyond phase I clinical trials.

Recent reports describe a new cyclic Tat peptidomimetic that binds to TAR with high affinity and shows broad and potent HIV-1 inhibition (Davidson et al. 2009; Lalonde 2011). Surprisingly, this drug inhibits both HIV-1 reverse transcription and Tat-mediated mRNA transcription (Lalonde 2011).

The assembly and maturation of HIV-1 on the inner plasma membrane is also an active area for drug discovery. Inhibitors such as betulinic acid have been shown to block HIV-1 maturation by interacting with the viral capsid (Fig. 2A,C, section 5) (Fujioka et al. 1994; Li et al. 2003). Although a promising new mechanism of action, insufficient antiviral activity precluded the development beyond early phase clinical trials (Smith et al. 2007).

The context of the HIV-1 life cycle, the final class of approved ARVs, is the HIV-1 PIs. PIs block proteolysis of the viral polyprotein, a step required for the production of infectious viral particles (Sundquist and Kräusslich 2011). PIs are among the most potent agents developed to date, but are large, peptidelike compounds that generally require the coadministration of a "boosting" agent to inhibit their metabolism and enhance drug levels. Therefore, PI-containing regimens contain a fourth drug, albeit one that does not directly contribute to overall antiviral activity. To date, ritonavir (RTV) is the only boosting agent or pharmacokinetic enhancer (PKE) available for use (Kempf et al. 1997; Hsu et al. 1998), although other compounds are in early stages of clinical development.

This description of the HIV-1 replication cycle (Fig. 2) provides a cursory overview of the most advanced antiretroviral drug targets with a focus on the approved agents that will be covered in more detail below. However, it should be noted that nearly all viral processes that are distinct from the cellular life cycle are potentially suitable for screening/designing inhibitors. Enhancing or modulating the activities of cellular restriction factors (Malim and Bieniasz 2011) could also potentially provide an approach to inhibiting HIV-1 replication and/or modulate pathogenesis and transmission, but this topic is not covered further here.

NUCLEOSIDE/NUCLEOTIDE REVERSE TRANSCRIPTASE INHIBITORS

NRTIs were the first class of drugs to be approved by the FDA (Fig. 1) (Young 1988). NRTIs are administered as prodrugs, which require host cell entry and phosphorylation (Mitsuya et al. 1985; Furman et al. 1986; Mitsuya and Broder 1986; St Clair et al. 1987; Hart et al. 1992) by cellular kinases before enacting an antiviral effect (Fig. 3). Lack of a $3'$-hydroxyl group at the sugar ($2'$-deoxyribosyl) moiety of the NRTIs prevents the formation of a $3'$-$5'$-phosphodiester bond between the NRTIs and incoming $5'$-nucleoside triphosphates, resulting in termination of the growing viral DNA chain. Chain termination can occur during RNA-dependent DNA or DNA-dependent DNA synthesis, inhibiting production of either the ($-$) or ($+$) strands of the HIV-1 proviral DNA (Cheng et al. 1987; Balzarini et al. 1989; Richman 2001). Currently, there are eight FDA-approved NRTIs: abacavir (ABC, Ziagen), didanosine (ddI, Videx), emtricitabine (FTC, Emtriva), lamivudine (3TC, Epivir), stavudine (d4T, Zerit), zalcitabine (ddC, Hivid), zidovudine (AZT, Retrovir), and Tenofovir disoprovil fumarate (TDF, Viread), a nucleotide RT inhibitor (Fig. 3).

As with all antiretroviral therapies, treatment with any of these agents often results in the emergence of HIV-1 strains with reduced drug susceptibility. Resistance to NRTIs is mediated by two mechanisms: ATP-dependent pyrophosphorolysis, which is the removal of NRTIs from the $3'$ end of the nascent chain, and reversal of chain termination (Arion et al. 1998; Meyer et al. 1999; Boyer et al. 2001) and increased discrimination between the native deoxyribonucleotide substrate and the inhibitor. NRTI mutations occur in RT and are classified as nucleoside/nucleotide associated mutations (NAMs) or thymidine analog mutations (TAMs). TAMs promote pyrophosphorolysis and are involved in the excision of AZT and d4T (Arion et al. 1998; Meyer et al. 2002; Naeger et al. 2002). TAM amino acid changes in HIV-1 RT include two distinct pathways: the TAM1 pathway (M41L, L210W, T215Y,

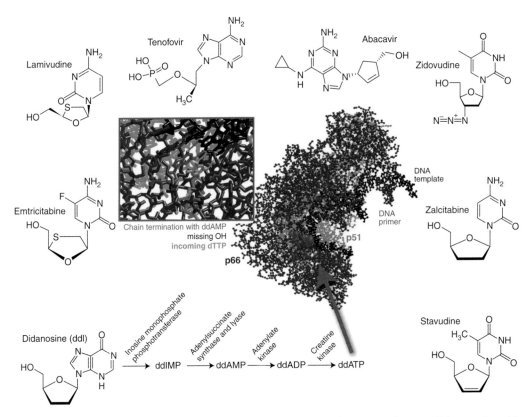

Figure 3. Nucleos(t)ide reverse transcriptase inhibitors and X-ray crystal structure of HIV-1 RT in complex with DNA primer/template chain terminated with ddAMP and with an incoming dTTP. The cartoon of the crystal structure data was adapted from coordinates deposited by Huang et al. (1998) (1RTD).

and occasionally D67N) and the TAM2 pathway (D67N, K70R, T215F, and 219E/Q) (Larder and Kemp 1989; Boucher et al. 1992; Kellam et al. 1992; Harrigan et al. 1996; Bacheler et al. 2001; Marcelin et al. 2004; Yahi et al. 2005).

A second mechanism of NRTI resistance is the prevention of NRTI incorporation into the nascent chain. Mutations associated with this mechanism include the M184V/I and the K65R. The M184V mutation emerges with 3TC or FTC therapy (Schinazi et al. 1993; Quan et al. 1996), whereas treatment with Tenofovir, ddC, ddI, d4T, and ABC can select K65R (Wainberg et al. 1999; Margot et al. 2002; Garcia-Lerma et al. 2003; Shehu-Xhilaga et al. 2005). In general, K65R rarely emerges in patients receiving any AZT-containing regimen because this mutation is phenotypically antagonistic to the TAMs (Parikh et al. 2006;

White et al. 2006). M184V restores Tenofovir susceptibility in the presence of K65R (Deval et al. 2004), thus K65R viruses are also infrequent in patients on Tenofivir who fail 3TC or emtricitabine (FTC) with M184V.

Many primary and secondary NRTI mutations (or combinations of these) have been shown to decrease RT function and viral replicative fitness (Quinones-Mateu and Arts 2002, 2006). Although several studies have suggested a potential for a clinical benefit associated with reduced replicative fitness of NRTI-resistant variants, it is important to note that additional mutations can accumulate in the presence of ongoing treatment resulting in higher levels of resistance. The loss in replicative fitness owing to drug resistance mutations (in the absence of drug) can also be compensated by accumulating secondary mutations.

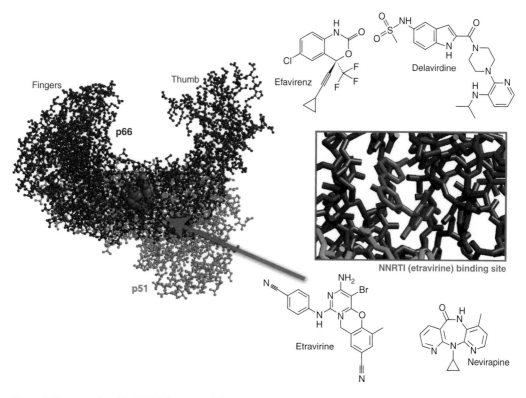

Figure 4. Non–nucleoside RT inhibitors and the X-ray crystal structure of HIV-1 RT complexed with etravirine (Lansdon et al. 2010) (3MEE).

NON–NUCLEOSIDE REVERSE TRANSCRIPTASE INHIBITORS

NNRTIs inhibit HIV-1 RT by binding and inducing the formation of a hydrophobic pocket proximal to, but not overlapping the active site (Fig. 4) (Kohlstaedt et al. 1992; Tantillo et al. 1994). The binding of NNRTIs changes the spatial conformation of the substrate-binding site and reduces polymerase activity (Kohlstaedt et al. 1992; Spence et al. 1995). The NNRTI-binding pocket only exists in the presence of NNRTIs (Rodgers et al. 1995; Hsiou et al. 1996) and consists of hydrophobic residues (Y181, Y188, F227, W229, and Y232), and hydrophilic residues such as K101, K103, S105, D192, and E224 of the p66 subunit and E138 of the p51 subunit (Fig. 4) (Sluis-Cremer et al. 2004). Unlike NRTIs, these non/uncompetitive inhibitors do not inhibit the RT of other lentiviruses such as HIV-2 and simian immunodeficiency

virus (SIV) (Kohlstaedt et al. 1992; Witvrouw et al. 1999). Currently, there are four approved NNRTIs: etravirine, delavirdine, efavirenz, and nevirapine, and several in development, including rilpivirine in phase 3 (Fig. 4).

NNRTI resistance generally results from amino acid substitutions such as L100, K101, K103, E138, V179, Y181, and Y188 in the NNRTI-binding pocket of RT (Tantillo et al. 1994). The most common NNRTI mutations are K103N and Y181C (Bacheler et al. 2000, 2001; Demeter et al. 2000; Dykes et al. 2001). As with NRTI resistance, complex patterns of NNRTI-resistant mutations can arise and alternative pathways have been observed in nonsubtype B infected individuals (Brenner et al. 2003; Spira et al. 2003; Gao et al. 2004). Most NNRTI mutations engender some level of cross resistance among different NNRTIs, especially in the context of additional secondary mutations (Antinori et al. 2002).

In contrast to the significant reductions in replicative fitness observed with resistance to other drug classes, with NNRTIs, single nucleotide changes can result in high-level resistance with only a slight loss of replicative fitness (Deeks 2001; Dykes et al. 2001; Imamichi et al. 2001). A lower genetic barrier, minimal impact on replicative fitness, and the slow reversion of these mutations in patients in the absence of drug contribute to transmission and stability of NNRTI-resistant HIV-1 in the population. Interestingly, the majority of NNRTI-resistance mutations selected under NNRTI treatment are commonly found as wild-type sequence in HIV-1 group O and HIV-2. HIV-1 group O can actually be subdivided into lineages based on a C181 or Y181 amino acid in RT (Tebit et al. 2010). Furthermore, nearly all primate lentiviruses can be phylogenetically classified into different lineages based on signature sequences in NNRTI-binding pocket and linked to a Cys, Ile, or Tyr at position 181, i.e., the primary codon-conferring resistance to NNRTIs (Tebit et al. 2010). Given the intrinsic resistance in most primate lentiviruses, aside from HIV-1 group M, it is not surprising that acquired resistance to NNRTIs has the least fitness impact.

INTEGRASE INHIBITORS

Integrase was the most recent HIV-1 enzyme to be successfully targeted for drug development (Espeseth et al. 2000; Hazuda et al. 2004a,b). Raltegravir (RAL), MK-0518 was FDA approved in 2007, and other integrase inhibitors, including Elvitegravir (EVG), GS-9137 are progressing through clinical development (Fig. 5) (Sato et al. 2006; Shimura et al. 2008). As mentioned above, integrase catalyzes $3'$ end processing and viral DNA and strand transfer. All integrase inhibitors in development target the strand transfer reaction and are thus referred to as either INIs or more specifically, integrase strand transfer inhibitors (InSTIs) (Espeseth et al. 2000; Hazuda et al. 2004a,b; McColl and Chen 2010). The selective effect on strand transfer is a result of a now well-defined mechanism of action in which the inhibitor (1) binds only to the specific complex between integrase and the viral DNA and (2) interacts with the two essential magnesium metal ion cofactors in the integrase active site and also the DNA (Fig. 5). Therefore, all InSTIs are comprised of two essential components: a metal-binding pharmacophore, which sequestors the active site magnesiums, and a hydrophobic group, which interacts with the viral DNA as well as the enzyme in the complex (Grobler et al. 2002). InSTIs are therefore the only ARV class that interacts with two essential elements of the virus, the integrase enzyme as well as the viral DNA, which is the substrate for integration.

The recent cocrystallization of the foamy virus integrase DNA complex or intasome with both RAL and EVG (Hare et al. 2010) corroborates the biochemical mechanism and provides a structural basis for understanding the unique breadth of antiviral activity that has been observed for InSTIs across all HIV-1 subtypes as well as other retroviruses, such as HIV-2 and XMRV (Fig. 5) (Maignan et al. 1998; Damond et al. 2008; Shimura et al. 2008; Van Baelen et al. 2008; Garrido et al. 2010; Singh et al. 2010). In the cocrystal structure, the general architecture and amino acids within the active site of the foamy virus intasome are highly conserved with other retroviral integrases, as are the immediate surrounding interactions with InSTIs. The common mechanism of action and conserved binding mode for InSTIs also has important implications for understanding resistance to the class. Mutations that engender resistance to InSTIs almost always map within the integrase active site near the amino acid residues that coordinate the essential magnesium cofactors (Hazuda et al. 2004a; Hare et al. 2010). Thus, these mutations have deleterious effects on enzymatic function and viral replicative capacity (Marinello et al. 2008; Quercia et al. 2009). In clinical studies, resistance to Raltegravir is associated with three independent pathways or sets of mutations in the integrase gene, as defined by primary or signature mutations at Y143, N155, or Q148 (Fransen et al. 2009). These primary mutations are generally observed with specific secondary mutations; for N155(H) these include E92Q,

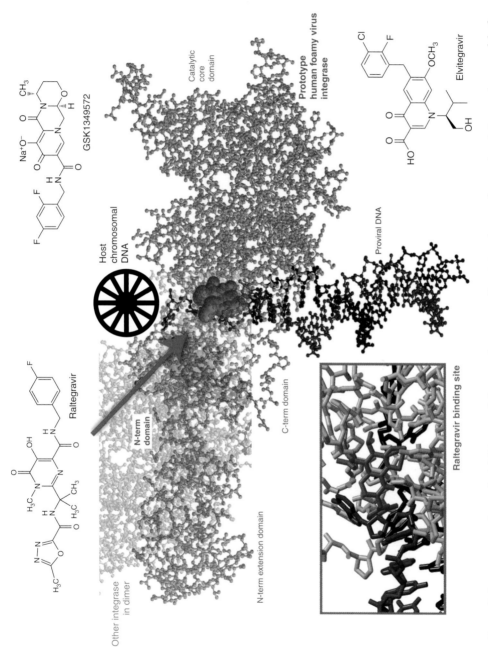

Figure 5. Integrase strand transfer inhibitors and the crystal structure of prototype human foamy virus integrase (as a model of HIV-1 IN) complexed to dsDNA and Raltegravir (Hare et al. 2010) (3OYH). N-term, amino-terminal; C-term, carboxy-terminal.

V151L, T97A, G163R, and L74M, whereas for Q148(K/R/H), G140S/A and E138K are common. Significant cross resistance is observed among the InSTIs almost regardless of the primary/secondary mutation sets (Goethals et al. 2008; Marinello et al. 2008). Although cross resistance is prevalent, different agents appear to preferentially select different patterns of mutations (Hazuda et al. 2004a).

PROTEASE INHIBITORS

The HIV-1 protease is the enzyme responsible for the cleavage of the viral gag and gag-pol polyprotein precursors during virion maturation (Park and Morrow 1993; Miller 2001). Ten PIs are currently approved: amprenavir (APV, Agenerase), atazanavir (ATZ, Reyataz), darunavir (TMC114, Prezista), fosamprenavir (Lexiva), indinavir (IDV, Crixivan), lopinavir (LPV), nelfinavir (NFV, Viracept), ritonavir (RTV, Norvir), saquinavir (SQV, Fortovase/ Invirase), and tipranavir (TPV, Aptivus) (Fig. 6).

Because of its vital role in the life cycle of HIV-1 and relatively small size (11 kDa), it was initially expected that resistance to protease inhibitors would be rare. However, the protease gene has great plasticity, with polymorphisms observed in 49 of the 99 codons, and more than 20 substitutions known to be associated with resistance (Shafer et al. 2000). The emergence of protease inhibitor resistance likely requires the stepwise accumulation of primary and compensatory mutations (Molla et al. 1996a) and each PI usually selects for certain signature primary mutations and a characteristic pattern of compensatory mutations. Unlike NNRTIs, primary drug-resistant substitutions are rarely observed in the viral populations in protease inhibitor-naïve individuals (Kozal et al. 1996).

All PIs share relatively similar chemical structures (Fig. 6) and cross resistance is commonly observed. For most PIs, primary resistance mutations cluster near the active site of the enzyme, at positions located at the substrate/inhibitor binding site (e.g., D30N, G48V, I50V, V82A, or I84V, among others). These amino acid changes usually have a deleterious effect on the replicative fitness (Nijhuis et al. 2001; Quinones-Mateu and Arts 2002; Quinones-Mateu et al. 2008). In addition to mutations in the protease gene, changes located within eight major protease cleavage sites (i.e., gag and pol genes), have been associated with resistance to protease inhibitors (Doyon et al. 1996; Zhang et al. 1997; Clavel et al. 2000; Miller 2001; Nijhuis et al. 2001). Cleavage site mutants are better substrates for the mutated protease, and thus partially compensate for the resistance-associated loss of viral fitness (Doyon et al. 1996; Mammano et al. 1998; Zennou et al. 1998; Clavel et al. 2000; Nijhuis et al. 2001). With PI resistance, HIV-1 appears to follows a "stepwise" pathway to overcome drug selection: (1) acquisition of primary resistance mutations in the protease gene, (2) selection of secondary/compensatory protease mutations to repair the enzymatic function and rescue viral fitness, and (3) selection of mutations in the major cleavage sites of the gag and gag-pol polyprotein precursors that restore protein processing and increase production of the HIV-1 protease itself (Condra et al. 1995; Molla et al. 1996b; Doyon et al. 1998; Berkhout 1999; Nijhuis et al. 2001).

ENTRY INHIBITORS

HIV-1 entry exploits several host proteins for a set of intricate events leading to membrane fusion and virus core release into the cytoplasm (Fig. 7). HIV-1 entry inhibitors can be subdivided into distinct classes based on disruption/inhibition of distinct targets/steps in the process.

Fusion Inhibitors

The crystal structure of the gp41 ectodomain and of the ectodomain partnered with an inhibitory peptide (C34) revealed that the fusion-active conformation of gp41 was a six-helix bundle in which three N helices form an interior, trimeric coiled coil onto which three anti-parallel C helices pack (Doms and Wilen 2011). Peptide fusion inhibitors were designed based on the discovery that two homologous domains

    Cite this article as *Cold Spring Harb Perspect Med* doi: 10.1101/cshperspect.a007161

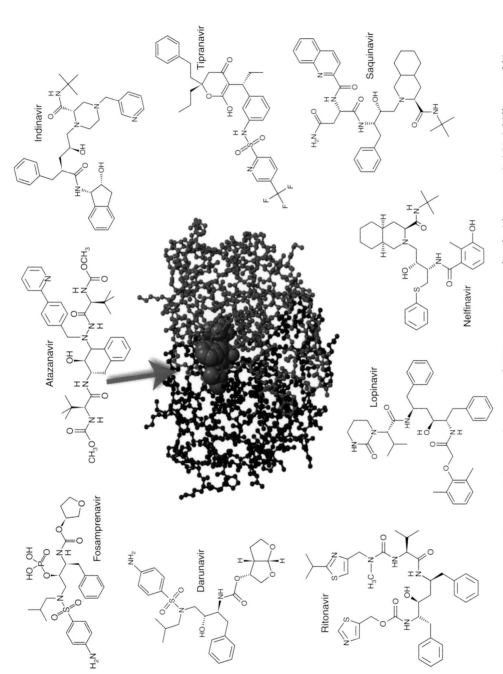

Figure 6. Protease inhibitors and the crystal structure of HIV-1 protease complexed with atazanavir (CA Schiffer, unpubl.). (3EKY).

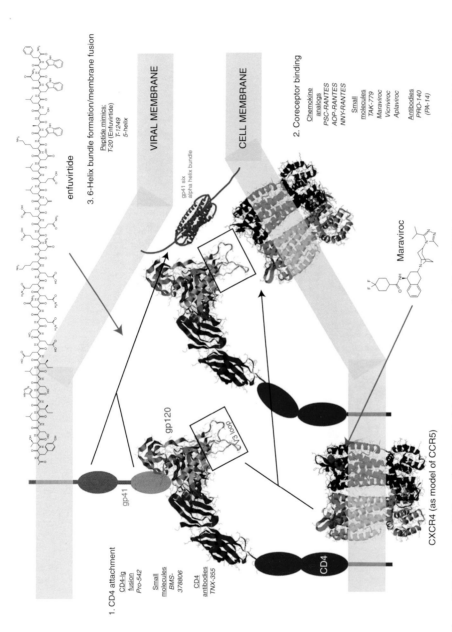

Figure 7. Structure predictions of various viral-host components involved in the HIV-1 entry process and the inhibitors. Section 1 describes the components involved for initial CD4 attachment, specifically the D1 domain of CD4 and the C4 domain of gp120. The gp120 structure is shown as an overlay of two structures (2NY2 and 3HI1) (Zhou et al. 2007; Chen et al. 2009). Inhibitors of this CD4 process are listed. Interactions between gp120 and CXCR4 are described in section 2. A rough model of maraviroc (MVC) binding to CCR5 in Figure 7 is based on data from a recently published structure of CXCR4 six complexed to a small molecule, IT1t (Wu et al. 2010). The final step in the entry process involves the formation of the gp41 six alpha-helix bundle, which can be blocked by T20 (enfuvirtide). The structure for HIV-1 gp41 six alpha-helix bundle is based on that of SIV gp41 (Malashkevich et al. 1998) (2SIV).

in the viral gp41 protein must interact with each other to promote fusion, and that mimicry of one of these domains by a heterologous protein can bind and disrupt the intramolecular interactions of the virus protein. Alpha-helical peptides homologous to the leucine zipper domain of gp41 had significant antiviral activity against HIV-1, and this activity depended on their ordered solution structure (Wild et al. 1993, 1994). Rational design of helical inhibitors ultimately produced a molecule (T-20, enfuvirtide) with potent antiviral activity in vivo (Fig. 7) (Kilby et al. 1998; Lalezari et al. 2003).

Resistance to early alpha-helical inhibitors was shown to be mediated by mutations in the amino-terminal heptad repeat region of gp41 (Rimsky et al. 1998), which provide further evidence for binding of these peptides to the virus. Monotherapy with enfuvirtide resulted in viral load rebounds after 14 days with resistance which mapped to determinants in the HR1 domain (G36D, I37T, V38A, V38M, N42T, N42D, N43K) (Wei et al. 2002). Mutations that confer resistance to enfuvirtide result in reduced replication capacity/replicative fitness presumably because mutations that reduce enfuvirtide binding also reduce the efficiency of six-helix bundle formation and overall fusion rates (Reeves et al. 2004, 2005). These mutations do not confer cross resistance to other entry inhibitors (attachment inhibitors or coreceptor inhibitors) (Ray et al. 2005) but can sensitize viruses to neutralization by monoclonal antibodies that target the gp41 domain by prolonging the exposure of fusion intermediates that are specifically sensitive to these antibodies (Reeves et al. 2005). Adaptation to enfuvirtide has even resulted in viruses that require enfuvirtide for fusion (Baldwin et al. 2004).

Resistance mutations in gp41 decrease fusion efficiency and reduce viral fitness (Labrosse et al. 2003). Nonetheless, studies of baseline susceptibility to enfuvirtide suggested that large variations in intrinsic susceptibility existed in diverse HIV-1 isolates, and that these variations mapped to regions outside the enfuvirtide-binding site (Derdeyn et al. 2000). Sequences associated with the V3 loop were correlated with intrinsic enfuvirtide susceptibility, suggesting

that interactions with the coreceptor were important determinants of susceptibility of a drug that inhibits virus fusion. A seminal observation in the understanding of entry inhibitor susceptibility was the discovery that efficiency of the fusion process was the principal modulator of intrinsic enfuvirtide susceptibility (Reeves et al. 2002). Mutations in the coreceptor-binding site that reduced gp120 affinity for CCR5 resulted in viruses with reduced fusion kinetics (Reeves et al. 2004; Biscone et al. 2006). Engagement of CD4 by gp120 initiates a process of structural rearrangement in the envelope glycoprotein resulting in fusion. Completion of this process requires engagement of the coreceptor molecule, but enfuvirtide susceptibility is limited to the time between CD4 engagement and six-helix bundle formation. Any decrease in the rate of this entry process (e.g., reducing the levels of coreceptor expression) also increases susceptibility of the virus to inhibition by enfuvirtide. Consistent with this, ENF is synergistic with compounds that inhibit CD4 or coreceptor engagement (Tremblay et al. 2000; Nagashima et al. 2001).

Small-Molecule CCR5 Antagonists

Small-molecule CCR5 antagonists bind to hydrophobic pockets within the transmembrane helices of CCR5 (Dragic et al. 2000; Tsamis et al. 2003). This site does not overlap the binding sites of either CCR5 agonists or HIV-1 envelope. Instead, drug binding induces and stabilizes a receptor conformation that is not recognized by either. Thus, these molecules are considered allosteric inhibitors. Ideally, a small-molecule inhibitor of CCR5 would block binding by HIV-1 envelope but continue to bind native chemokines and allow signal transduction. Most small-molecule inhibitors, however, are pure antagonists of the receptor. Oral administration of small-molecule antagonists has been shown to inhibit viral replication in macaque models (Veazey et al. 2003) and to prevent vaginal transmission (Veazey et al. 2005). Thus far, three antagonists (VCV, MVC, and Aplaviroc) have been shown to inhibit virus replication in humans (Dorr et al. 2005). The compound MVC was approved for therapeutic use by the FDA in 2007 (Fig. 7).

MVC binds a hydrophobic transmembrane cavity of CCR5. Binding alters the conformation of the second extracellular loop of the receptor and prevents interaction with the V3 stem loop of gp120 (Dragic et al. 2000; Kondru et al. 2008). A rough model of MVC binding to CCR5 in Figure 7 is based on a recently published structure of CXCR4 complexed to a small molecule, IT1t (Wu et al. 2010). CXCR4 also serves as a coreceptor for HIV-1 but attempts at development of CXCR4 antagonists (e.g., AMD3100) fail in clinical studies (Hendrix et al. 2004). Because MVC binds to a host cell protein, resistance to MVC is unlike that of other ARVs. Potential resistance mechanisms include (1) tropism switching (utilization of CXCR4 instead of CCR5 for entry), (2) increased affinity for the coreceptor, (3) utilization of inhibitor-bound receptor for entry, and (4) faster rate of entry. Tropism switching has been a concern in the therapeutic administration of this class as primary infection with, or early emergence of CXCR4 tropic virus, although rare, typically leads to faster disease progression. Thus, the selection of CXCR4 tropic virus owing to CCR5 antagonist treatment could have a negative impact on HIV-1 pathogenesis.

Small-molecule CCR5 inhibitors have been used to select for drug resistance in peripheral blood mononuclear cell cultures (PBMC), which express CCR5 and CXCR4, as well as a variety of other chemokine receptors that could potentially substitute for HIV-1 coreceptors. In these experiments, inhibitor-resistant viruses continue to require CCR5 for entry (Trkola et al. 2002; Marozsan et al. 2005; Baba et al. 2007; Westby et al. 2007). Furthermore, evaluation of coreceptor tropism of viruses from patients who failed MVC therapy during clinical trials has suggested that tropism change occurred only when X4 tropic viruses were preexisting in the patient quasispecies before initiating treatment with MVC (Westby et al. 2006). Thus, it appears that de novo mutations conferring altered coreceptor usage is not the favored pathway for resistance in vitro or in vivo. It

should be noted that in some treatment failures, the use of CCR5 was maintained even in the presence of MVC. These "resistant" HIV-1 isolates did not display the same shift in drug susceptibility, typically characterized by an increase in IC_{50} values, but were capable of using both the free and inhibitor-bound CCR5 for entry (Trkola et al. 2002; Tsibris et al. 2008). In such cases, resistance is reported as MPI (or maximum percent inhibition) for saturating concentrations of drug.

Although it is still early with respect to the clinical experience for CCR5 antagonists, there are documented cases of treatment failures that are not accounted for by either CXCR4 tropism switch or resistance owing to increased MPI. Recent studies suggest discrepancies in the sensitivity to CCR5 antagonists may be assay dependent. Susceptibility to entry CCR5 antagonists can be affected by cell type, state of cellular activation, and number of virus replication cycles (Kuhmann et al. 2004; Marozsan et al. 2005; Lobritz et al. 2007; Westby et al. 2007). Also, different primary HIV-1 isolates can vary in sensitivity by as much as 100-fold in IC_{50} values (Torre et al. 2000; Dorr et al. 2005; Lobritz et al. 2007), and this difference is much more demonstrable with infection assays using replication-competent primary HIV-1 isolates as compared with defective viruses limited to single-cycle replication. These complexities make it quite challenging to detect resistance at the time of treatment failure with routine resistance testing assays. Given the issues, the use of CCR5 antagonists in clinical practice is somewhat more complex than other classes of ARV agents.

CONCLUSIONS

The breadth and depth of the HIV-1 therapy pipeline may arguably be among the most successful for treating any single human disease, infection, or disorder as illustrated by the number of antiretroviral agents and unique drug classes available. In reviewing the history of ARV drug development, however, there are some key lessons and parallels that need to be kept in mind as we consider the development of small-molecule prevention strategies for HIV-1 and evolving treatment strategies for other viral infections, including hepatitis C virus (HCV). The road to successful HIV-1 treatment was hard, and in the early days many patients were inadequately treated with suboptimal regimens that rapidly led to failure and drug resistance. Although it is unknown whether the prevention of HIV-1 transmission will require the same number of agents, the inherent plasticity of HIV-1 would suggest erring on the side of caution and focusing early on combination products that would mitigate this risk. In the case of HCV, the breadth of genetic diversity appears to be greater than that observed in an HIV-1 infected individual. Anti-HCV drugs in the most advanced stages for approval inhibit a small number of targets (e.g., the NS5b polymerase and NS3 protease) and each class appears to share significant cross resistance; when tested as single agents, the emergence of HCV drug resistance is rapid. The success of HAART should provide the benchmark for HCV drug development and a roadmap for the development of novel prevention strategies in HIV-1 to avoid potential risk to both the individual patient and the population by preventing the acquisition and transmission of drug resistance.

REFERENCES

*Reference is also in this collection.

Abram ME, Ferris AL, Shao W, Alvord WG, Hughes SH. 2010. Nature, position, and frequency of mutations made in a single cycle of HIV-1 replication. *J Virol* **84:** 9864–9878.

Alkhatib G, Combadiere C, Broder CC, Feng Y, Kennedy PE, Murphy PM, Berger EA. 1996. CC CKR5: A RANTES, MIP-1a, MIP-1b receptor as a fusion cofactor for macrophage-tropic HIV-1. *Science* **272:** 1955–1958.

Antinori A, Zaccarelli M, Cingolani A, Forbici F, Rizzo MG, Trotta MP, Di Giambenedetto S, Narciso P, Ammassari A, Girardi E, et al. 2002. Cross-resistance among nonnucleoside reverse transcriptase inhibitors limits recycling efavirenz after nevirapine failure. *AIDS Res Hum Retroviruses* **18:** 835–838.

Arion D, Kaushik N, McCormick S, Borkow G, Parniak MA. 1998. Phenotypic mechanism of HIV-1 resistance to 3′-azido-3′-deoxythymidine (AZT): Increased polymerization processivity and enhanced sensitivity to pyrophosphate of the mutant viral reverse transcriptase. *Biochemistry* **37:** 15908–15917.

Autran B, Carcelain G, Li TS, Blanc C, Mathez D, Tubiana R, Katlama C, Debre P, Leibowitch J. 1997. Positive effects of combined antiretroviral therapy on $CD4^{+}$ T cell homeostasis and function in advanced HIV disease. *Science* **277:** 112–116.

Baba M, Miyake H, Wang X, Okamoto M, Takashima K. 2007. Isolation and characterization of human immunodeficiency virus type 1 resistant to the small-molecule CCR5 antagonist TAK-652. *Antimicrob Agents Chemother* **51:** 707–715.

Bacheler LT, Anton ED, Kudish P, Baker D, Bunville J, Krakowski K, Bolling L, Aujay M, Wang XV, Ellis D, et al. 2000. Human immunodeficiency virus type 1 mutations selected in patients failing efavirenz combination therapy. *Antimicrob Agents Chemother* **44:** 2475–2484.

Bacheler L, Jeffrey S, Hanna G, D'Aquila R, Wallace L, Logue K, Cordova B, Hertogs K, Larder B, Buckery R, et al. 2001. Genotypic correlates of phenotypic resistance to efavirenz in virus isolates from patients failing nonnucleoside reverse transcriptase inhibitor therapy. *J Virol* **75:** 4999–5008.

Baldwin CE, Sanders RW, Deng Y, Jurriaans S, Lange JM, Lu M, Berkhout B. 2004. Emergence of a drug-dependent human immunodeficiency virus type 1 variant during therapy with the T20 fusion inhibitor. *J Virol* **78:** 12428–12437.

Balzarini J, Herdewijn P, De Clercq E. 1989. Differential patterns of intracellular metabolism of 2′,3′-didehydro-2′,3′-dideoxythymidine and 3′-azido-2′,3′-dideoxythymidine, two potent anti-human immunodeficiency virus compounds. *J Biol Chem* **264:** 6127–6133.

Berkhout B. 1999. HIV-1 evolution under pressure of protease inhibitors: Climbing the stairs of viral fitness. *J Biomed Sci* **6:** 298–305.

Biscone MJ, Miamidian JL, Muchiri JM, Baik SS, Lee FH, Doms RW, Reeves JD. 2006. Functional impact of HIV coreceptor-binding site mutations. *Virology* **351:** 226–236.

Boucher CA, O'Sullivan E, Mulder JW, Ramautarsing C, Kellam P, Darby G, Lange JM, Goudsmit J, Larder BA. 1992. Ordered appearance of zidovudine resistance mutations during treatment of 18 human immunodeficiency virus-positive subjects. *J Infect Dis* **165:** 105–110.

Boyer PL, Sarafianos SG, Arnold E, Hughes SH. 2001. Selective excision of AZTMP by drug-resistant human immunodeficiency virus reverse transcriptase. *J Virol* **75:** 4832–4842.

Brenner B, Turner D, Oliveira M, Moisi D, Detorio M, Carobene M, Marlink RG, Schapiro J, Roger M, Wainberg MA. 2003. A V106M mutation in HIV-1 clade C viruses exposed to efavirenz confers cross-resistance to nonnucleoside reverse transcriptase inhibitors. *AIDS* **17:** F1–F5.

Chen L, Kwon YD, Zhou T, Wu X, O'Dell S, Cavacini L, Hessell AJ, Pancera M, Tang M, Xu L, et al. 2009. Structural basis of immune evasion at the site of CD4 attachment on HIV-1 gp120. *Science* **326:** 1123–1127.

Cheng YC, Dutschman GE, Bastow KF, Sarngadharan MG, Ting RY. 1987. Human immunodeficiency virus reverse transcriptase. General properties and its interactions with nucleoside triphosphate analogs. *J Mol Biol* **262:** 2187–2189.

Clavel F, Race E, Mammano F. 2000. HIV drug resistance and viral fitness. *Adv Pharmacol* **49:** 41–66.

Coffin JM. 1995. HIV population dynamics in vivo: Implications for genetic variation, pathogenesis, and therapy. *Science* **267:** 483–489.

Collier AC, Coombs RW, Schoenfeld DA, Bassett RL, Timpone J, Baruch A, Jones M, Facey K, Whitacre C, McAuliffe VJ, et al. 1996. Treatment of human immunodeficiency virus infection with saquinavir, zidovudine, and zalcitabine. AIDS Clinical Trials Group. *N Engl J Med* **334:** 1011–1017.

Condra JH, Schleif WA, Blahy OM, Gabryelski LJ, Graham DJ, Quintero JC, Rhodes A, Robbins HL, Roth E, Shivaprakash M. 1995. In vivo emergence of HIV-1 variants resistant to multiple protease inhibitors. *Nature (London)* **374:** 569–571.

* Craigie R, Bushman FD. 2011. HIV DNA integration. *Cold Spring Harb Perspect Med* doi: 10.1101/cshperspect. a006890.

Cupelli LA, Hsu MC. 1995. The human immunodeficiency virus type 1 Tat antagonist, Ro 5-3335, predominantly inhibits transcription initiation from the viral promoter. *J Virol* **69:** 2640–2643.

D'Aquila RT, Hughes MD, Johnson VA, Fischl MA, Sommadossi JP, Liou SH, Timpone J, Myers M, Basgoz N, Niu M, et al. 1996. Nevirapine, zidovudine, and didanosine compared with zidovudine and didanosine in patients with HIV-1 infection. A randomized, double-blind, placebo-controlled trial. National Institute of Allergy and Infectious Diseases AIDS Clinical Trials Group Protocol 241 Investigators. *Ann Intern Med* **124:** 1019–1030.

Dalgleish AG, Beverley PC, Clapham PR, Crawford DH, Greaves MF, Weiss RA. 1984. The CD4 (T4) antigen is an essential component of the receptor for the AIDS retrovirus. *Nature* **312:** 763–767.

Damond F, Lariven S, Roquebert B, Males S, Peytavin G, Morau G, Toledano D, Descamps D, Brun-Vezinet F, Matheron S. 2008. Virological and immunological response to HAART regimen containing integrase inhibitors in HIV-2-infected patients. *AIDS* **22:** 665–666.

Davidson A, Leeper TC, Athanassiou Z, Patora-Komisarska K, Karn J, Robinson JA, Varani G. 2009. Simultaneous recognition of HIV-1 TAR RNA bulge and loop sequences by cyclic peptide mimics of Tat protein. *Proc Natl Acad Sci* **106:** 11931–11936.

Davies WL, Grunert RR, Haff RF, Mcgahen JW, Neumayer EM, Paulshock M, Watts JC, Wood TR, Hermann EC, Hoffmann CE. 1964. Antiviral activity of 1-adamantanamine (amantadine). *Science* **144:** 862–863.

Deeks SG. 2001. International perspectives on antiretroviral resistance. Nonnucleoside reverse transcriptase inhibitor resistance. *J Acquir Immune Defic Syndr* **26:** S25–S33.

Demeter LM, Shafer RW, Meehan PM, Holden-Wiltse J, Fischl MA, Freimuth WW, Para MF, Reichman RC. 2000. Delavirdine susceptibilities and associated reverse transcriptase mutations in human immunodeficiency virus type 1 isolates from patients in a phase I/II trial of delavirdine monotherapy (ACTG 260). *Antimicrob Agents Chemother* **44:** 794–797.

Deng H, Liu R, Ellmeier W, Choe S, Unutmaz D, Burkhart M, Di Marzio P, Marmon S, Sutton RE, Hill CM, et al.

1996. Identification of a major co-receptor for primary isolates of HIV-1. *Nature (London)* **381:** 661–666.

Derdeyn CA, Decker JM, Sfakianos JN, Wu X, O'Brien WA, Ratner L, Kappes JC, Shaw GM, Hunter E. 2000. Sensitivity of human immunodeficiency virus type 1 to the fusion inhibitor T-20 is modulated by coreceptor specificity defined by the V3 loop of gp120. *J Virol* **74:** 8358–8367.

Deval J, White KL, Miller MD, Parkin NT, Courcambeck J, Halfon P, Selmi B, Boretto J, Canard B. 2004. Mechanistic basis for reduced viral and enzymatic fitness of HIV-1 reverse transcriptase containing both K65R and M184V mutations. *J Biol Chem* **279:** 509–516.

Doranz BJ, Rucker J, Yi Y, Smyth RJ, Samson M, Peiper SC, Parmentier M, Collman RG, Doms RW. 1996. A dual-tropic primary HIV-1 isolate that uses fusin and the beta-chemokine receptors CKR-5, CKR-3, and CKR-2b as fusion cofactors. *Cell* **85:** 1149–1158.

Dorr P, Westby M, Dobbs S, Griffin P, Irvine B, Macartney M, Mori J, Rickett G, Smith-Burchnell C, Napier C, et al. 2005. Maraviroc (UK-427,857), a potent, orally bioavailable, and selective small-molecule inhibitor of chemokine receptor CCR5 with broad-spectrum anti-human immunodeficiency virus type 1 activity. *Antimicrob Agents Chemother* **49:** 4721–4732.

Doyon L, Croteau G, Thibeault D, Poulin F, Pilote L, Lamarre D. 1996. Second locus involved in human immunodeficiency virus type 1 resistance to protease inhibitors. *J Virol* **70:** 3763–3769.

Doyon L, Payant C, Brakier-Gingras L, Lamarre D. 1998. Novel Gag-Pol frameshift site in human immunodeficiency virus type 1 variants resistant to protease inhibitors. *J Virol* **72:** 6146–6150.

Dragic T, Trkola A, Thompson DA, Cormier EG, Kajumo FA, Maxwell E, Lin SW, Ying W, Smith SO, Sakmar TP, et al. 2000. A binding pocket for a small molecule inhibitor of HIV-1 entry within the transmembrane helices of CCR5. *Proc Natl Acad Sci* **97:** 5639–5644.

Dykes C, Fox K, Lloyd A, Chiulli M, Morse E, Demeter LM. 2001. Impact of clinical reverse transcriptase sequences on the replication capacity of HIV-1 drug-resistant mutants. *Virology* **285:** 193–203.

Espeseth AS, Felock P, Wolfe A, Witmer M, Grobler J, Anthony N, Egbertson M, Melamed JY, Young S, Hamill T, et al. 2000. HIV-1 integrase inhibitors that compete with the target DNA substrate define a unique strand transfer conformation for integrase. *Proc Natl Acad Sci* **97:** 11244–11249.

Feng Y, Broder CC, Kennedy PE, Berger EA. 1996. HIV-1 entry cofactor: Functional cDNA cloning of a seven-transmembrane, G protein-coupled receptor. *Science* **272:** 872–877.

Finzi D, Hermankova M, Pierson T, Carruth LM, Buck C, Chaisson RE, Quinn TC, Chadwick K, Margolick J, Brookmeyer R, et al. 1997. Identification of a reservoir for HIV-1 in patients on highly active antiretroviral therapy. *Science* **278:** 1295–1300.

Finzi D, Blankson J, Siliciano JD, Margolick JB, Chadwick K, Pierson T, Smith K, Lisziewicz J, Lori F, Flexner C, et al. 1999. Latent infection of CD4+ T cells provides a mechanism for lifelong persistence of HIV-1, even in patients on effective combination therapy. *Nat Med* **5:** 512–517.

Fransen S, Gupta S, Danovich R, Hazuda D, Miller M, Witmer M, Petropoulos CJ, Huang W. 2009. Loss of Raltegravir susceptibility of HIV-1 is conferred by multiple non-overlapping genetic pathways. *J Virol* **83:** 11440–11446.

Frost SD, McLean AR. 1994. Quasispecies dynamics and the emergence of drug resistance during zidovudine therapy of HIV infection. *AIDS* **8:** 323–332.

Fujioka T, Kashiwada Y, Kilkuskie RE, Cosentino LM, Ballas LM, Jiang JB, Janzen WP, Chen IS, Lee KH. 1994. Anti-AIDS agents, 11. Betulinic acid and platanic acid as anti-HIV principles from Syzigium claviflorum, and the anti-HIV activity of structurally related triterpenoids. *J Nat Prod* **57:** 243–247.

Furman PA, Fyfe JA, St Clair MH, Weinhold K, Rideout JL, Freeman GA, Lehrman SN, Bolognesi DP, Broder S, Mitsuya H. 1986. Phosphorylation of 3′-azido-3′-deoxythymidine and selective interaction of the 5′-triphosphate with human immunodeficiency virus reverse transcriptase. *Proc Natl Acad Sci* **83:** 8333–8337.

Gao Y, Paxinos E, Galovich J, Troyer R, Baird H, Abreha M, Kityo C, Mugyenyi P, Petropoulos C, Arts EJ. 2004. Characterization of a subtype D human immunodeficiency virus type 1 isolate that was obtained from an untreated individual and that is highly resistant to nonnucleoside reverse transcriptase inhibitors. *J Virol* **78:** 5390–5401.

Garcia-Lerma JG, MacInnes H, Bennett D, Reid P, Nidtha S, Weinstock H, Kaplan JE, Heneine W. 2003. A novel genetic pathway of human immunodeficiency virus type 1 resistance to stavudine mediated by the K65R mutation. *J Virol* **77:** 5685–5693.

Garrido C, Geretti A, Zahonero N, Booth C, Strang A, Soriano V, De Mendoza C, et al. 2010. Integrase variability and susceptibility to HIV integrase inhibitors: Impact of subtypes, antiretroviral experience and duration of HIV infection. *J Antimicrob Chemother* **65:** 320–326.

Goethals O, Clayton R, Van Ginderen M, Vereycken I, Wagemans E, Geluykens P, Dockx K, Strijbos R, Smits V, Vos A, et al. 2008. Resistance mutations in HIV type 1 integrase selected with elvitegravir confer reduced susceptibility to a wide range of integrase inhibitors. *J Virol* **82:** 10366–10374.

Grobler JA, Stillmock K, Hu B, Witmer M, Felock P, Espeseth AS, Wolfe A, Egbertson M, Bourgeois M, Melamed J, et al. 2002. Diketo acid inhibitor mechanism and HIV-1 integrase: Implications for metal binding in the active site of phosphotransferase enzymes. *Proc Natl Acad Sci* **99:** 6661–6666.

Hamy F, Felder ER, Heizmann G, Lazdins J, Aboul-ela F, Varani G, Karn J, Klimkait T. 1997. An inhibitor of the Tat/TAR RNA interaction that effectively suppresses HIV-1 replication. *Proc Natl Acad Sci* **94:** 3548–3553.

Hare S, Vos AM, Clayton RF, Thuring JW, Cummings MD, Cherepanov P. 2010. Molecular mechanisms of retroviral integrase inhibition and the evolution of viral resistance. *Proc Natl Acad Sci* **107:** 20057–20062.

Harrigan PR, Kinghorn I, Bloor S, Kemp SD, Najera I, Kohli A, Larder BA. 1996. Significance of amino acid variation at human immunodeficiency virus type 1 reverse transcriptase residue 210 for zidovudine susceptibility. *J Virol* **70:** 5930–5934.

Hart GJ, Orr DC, Penn CR, Figueiredo HT, Gray NM, Boehme RE, Cameron JM. 1992. Effects of (−)-2′-deoxy-3′-thiacytidine (3TC) 5′-triphosphate on human immunodeficiency virus reverse transcriptase and mammalian DNA polymerases alpha, beta, and gamma. *Antimicrob Agents Chemother* **36:** 1688–1694.

Hay AJ, Wolstenholme AJ, Skehel JJ, Smith MH. 1985. The molecular basis of the specific anti-influenza action of amantadine. *EMBO J* **4:** 3021–3024.

Hazuda DJ, Anthony NJ, Gomez RP, Jolly SM, Wai JS, Zhuang L, Fisher TE, Embrey M, Guare JP Jr, Egbertson MS, et al. 2004a. A naphthyridine carboxamide provides evidence for discordant resistance between mechanistically identical inhibitors of HIV-1 integrase. *Proc Natl Acad Sci* **101:** 11233–11238.

Hazuda DJ, Young SD, Guare JP, Anthony NJ, Gomez RP, Wai JS, Vacca JP, Handt L, Motzel SL, Klein HJ, et al. 2004b. Integrase inhibitors and cellular immunity suppress retroviral replication in rhesus macaques. *Science* **305:** 528–532.

Hendrix CW, Collier AC, Lederman MM, Schols D, Pollard RB, Brown S, Jackson JB, Coombs RW, Glesby MJ, Flexner CW, et al. 2004. Safety, pharmacokinetics, and antiviral activity of AMD3100, a selective CXCR4 receptor inhibitor, in HIV-1 infection. *J Acquir Immune Defic Syndr* **37:** 1253–1262.

Ho DD, Neumann AU, Perelson AS, Chen W, Leonard JM, Markowitz M. 1995. Rapid turnover of plasma virions and CD4 lymphocytes in HIV-1 infection. *Nature (London)* **373:** 123–126.

Hsiou Y, Ding J, Das K, Clark AD Jr, Hughes SH, Arnold E. 1996. Structure of unliganded HIV-1 reverse transcriptase at 2.7 Å resolution: Implications of conformational changes for polymerization and inhibition mechanisms. *Structure* **4:** 853–860.

Hsu MC, Schutt AD, Holly M, Slice LW, Sherman MI, Richman DD, Potash MJ, Volsky DJ. 1991. Inhibition of HIV replication in acute and chronic infections in vitro by a Tat antagonist. *Science* **254:** 1799–1802.

Hsu A, Granneman GR, Cao G, Carothers L, El-Shourbagy T, Baroldi P, Erdman K, Brown F, Sun E, Leonard JM. 1998. Pharmacokinetic interactions between two human immunodeficiency virus protease inhibitors, ritonavir and saquinavir. *Clin Pharmacol Ther* **63:** 453–464.

* Hu W-S, Hughes SH. 2011. HIV-1 reverse transcription. *Cold Spring Harb Perspect Med* doi: 10/1101/cshperspect.a006882.

Huang H, Chopra R, Verdine GL, Harrison SC. 1998. Structure of a covalently trapped catalytic complex of HIV-1 reverse transcriptase: Implications for drug resistance. *Science* **282:** 1669–1675.

Hwang S, Tamilarasu N, Kibler K, Cao H, Ali A, Ping YH, Jeang KT, Rana TM. 2003. Discovery of a small molecule Tat-trans-activation-responsive RNA antagonist that potently inhibits human immunodeficiency virus-1 replication. *J Biol Chem* **278:** 39092–39103.

Imamichi T, Murphy MA, Imamichi H, Lane HC. 2001. Amino acid deletion at codon 67 and Thr-to-Gly change at codon 69 of human immunodeficiency virus type 1 reverse transcriptase confer novel drug resistance profiles. *J Virol* **75:** 3988–3992.

* Karn J, Stoltzfus CM. 2011. Transcriptional and posttranscriptional regulation of HIV-1 gene expression. *Cold Spring Harb Perspect Med* doi: 10.1101/cshperspect.a006916.

Kellam P, Boucher CA, Larder BA. 1992. Fifth mutation in human immunodeficiency virus type 1 reverse transcriptase contributes to the development of high-level resistance to zidovudine. *Proc Natl Acad Sci* **89:** 1934–1938.

Kempf DJ, Marsh KC, Kumar G, Rodrigues AD, Denissen JF, McDonald E, Kukulka MJ, Hsu A, Granneman GR, Baroldi PA, et al. 1997. Pharmacokinetic enhancement of inhibitors of the human immunodeficiency virus protease by coadministration with ritonavir. *Antimicrob Agents Chemother* **41:** 654–660.

Kilby JM, Hopkins S, Venetta TM, DiMassimo B, Cloud GA, Lee JY, Alldredge L, Hunter E, Lambert D, Bolognesi D, et al. 1998. Potent suppression of HIV-1 replication in humans by T-20, a peptide inhibitor of gp41-mediated virus entry. *Nat Med* **4:** 1302–1307.

Klatzmann D, Champagne E, Chamaret S, Gruest J, Guetard D, Hercend T, Gluckman JC, Montagnier L. 1984. T-lymphocyte T4 molecule behaves as the receptor for human retrovirus LAV. *Nature* **312:** 767–768.

Kohlstaedt LA, Wang J, Friedman JM, Rice PA, Steitz TA. 1992. Crystal structure at 3.5 A resolution of HIV-1 reverse transcriptase complexed with an inhibitor. *Science* **256:** 1783–1790.

Komanduri KV, Viswanathan MN, Wieder ED, Schmidt DK, Bredt BM, Jacobson MA, McCune JM. 1998. Restoration of cytomegalovirus-specific CD4+ T-lymphocyte responses after ganciclovir and highly active antiretroviral therapy in individuals infected with HIV-1. *Nat Med* **4:** 953–956.

Kondru R, Zhang J, Ji C, Mirzadegan T, Rotstein D, Sankuratri S, Dioszegi M. 2008. Molecular interactions of CCR5 with major classes of small-molecule anti-HIV CCR5 antagonists. *Mol Pharmacol* **73:** 789–800.

Kozal MJ, Shah N, Shen N, Yang R, Fucini R, Merigan TC, Richman DD, Morris D, Hubbel E, Chee M, et al. 1996. Extensive polymorphisms observed in HIV-1 clade B protease gene using high-density oligonucleotide arrays. *Nature Med* **2:** 753–759.

Kuhmann SE, Pugach P, Kunstman KJ, Taylor J, Stanfield RL, Snyder A, Strizki JM, Riley J, Baroudy BM, Wilson IA, et al. 2004. Genetic and phenotypic analyses of human immunodeficiency virus type 1 escape from a small-molecule CCR5 inhibitor. *J Virol* **78:** 2790–2807.

Kuritzkes DR, Jacobson J, Powderly WG, Godofsky E, DeJesus E, Haas F, Reimann KA, Larson JL, Yarbough PO, Curt V Jr, et al. 2004. Antiretroviral activity of the anti-CD4 monoclonal antibody TNX-355 in patients infected with HIV type 1. *J Infect Dis* **189:** 286–291.

Labrosse B, Labernardiere JL, Dam E, Trouplin V, Skrabal K, Clavel F, Mammano F. 2003. Baseline susceptibility of primary human immunodeficiency virus type 1 to entry inhibitors. *J Virol* **77:** 1610–1613.

Lalezari JP, Henry K, O'Hearn M, Montaner JS, Piliero PJ, Trottier B, Walmsley S, Cohen C, Kuritzkes DR, Eron JJ Jr, et al. 2003. Enfuvirtide, an HIV-1 fusion inhibitor, for drug-resistant HIV infection in North and South America. *N Engl J Med* **348:** 2175–2185.

Lalonde M, Lobritz M, Ratcliff A, Chaminian M, Athanassiou Z, Tyagi M, Karn J, Robinson JA, Varani G, Arts EJ. 2011. Inhibition of both HIV-1 reverse transcription and gene expression by a cyclic peptide that binds the Tat-transactivating response element (TAR) RNA. *PLoS Pathog* **7:** e1002038.

Lansdon EB, Brendza KM, Hung M, Wang R, Mukund S, Jin D, Birkus G, Kutty N, Liu X. 2010. Crystal structures of HIV-1 reverse transcriptase with etravirine (TMC125) and rilpivirine (TMC278): Implications for drug design. *J Med Chem* **53:** 4295–4299.

Larder BA, Kemp SD. 1989. Multiple mutations in HIV-1 reverse transcriptase confer high-level resistance to zidovudine (AZT). *Science* **246:** 1155–1158.

Larder BA, Kemp SD, Harrigan PR. 1995. Potential mechanism for sustained antiretroviral efficacy of AZT-3TC combination therapy. *Science* **269:** 696–699.

Lederman MM, Connick E, Landay A, Kuritzkes DR, Spritzler J, St Clair M, Kotzin BL, Fox L, Chiozzi MH, Leonard JM, et al. 1998. Immunologic responses associated with 12 weeks of combination antiretroviral therapy consisting of zidovudine, lamivudine, and ritonavir: Results of AIDS Clinical Trials Group Protocol 315. *J Infect Dis* **178:** 70–79.

Li F, Goila-Gaur R, Salzwedel K, Kilgore NR, Reddick M, Matallana C, Castillo A, Zoumplis D, Martin DE, Orenstein JM, et al. 2003. PA-457: A potent HIV inhibitor that disrupts core condensation by targeting a late step in Gag processing. *Proc Natl Acad Sci* **100:** 13555–13560.

Lin PF, Blair W, Wang T, Spicer T, Guo Q, Zhou N, Gong YF, Wang HG, Rose R, Yamanaka G, et al. 2003. A small molecule HIV-1 inhibitor that targets the HIV-1 envelope and inhibits CD4 receptor binding. *Proc Natl Acad Sci* **100:** 11013–11018.

Lobritz MA, Marozsan AJ, Troyer RM, Arts EJ. 2007. Natural variation in the V3 crown of human immunodeficiency virus type 1 affects replicative fitness and entry inhibitor sensitivity. *J Virol* **81:** 8258–8269.

Maignan S, Guilloteau JP, Zhou-Liu Q, Clement-Mella C, Mikol V. 1998. Crystal structures of the catalytic domain of HIV-1 integrase free and complexed with its metal cofactor: High level of similarity of the active site with other viral integrases. *J Mol Biol.* **282:** 359–68.

Malashkevich VN, Chan DC, Chutkowski CT, Kim PS. 1998. Crystal structure of the simian immunodeficiency virus (SIV) gp41 core: Conserved helical interactions underlie the broad inhibitory activity of gp41 peptides. *Proc Natl Acad Sci* **95:** 9134–9139.

* Malim MH, Bieniasz PD. 2011. HIV restriction factors and mechanisms of evasion. *Cold Spring Harb Perspect Med* doi: 10.1101/cshperspect.a006940.

Mammano F, Petit C, Clavel F. 1998. Resistance-associated loss of viral fitness in human immunodeficiency virus type 1: Phenotypic analysis of protease and gag coevolution in protease inhibitor-treated patients. *J Virol* **72:** 7632–7637.

Mansky LM, Temin HM. 1995. Lower in vivo mutation rate of human immunodeficiency virus type 1 than that predicted from the fidelity of purified reverse transcriptase. *J Virol* **69:** 5087–5094.

Marcelin AG, Delaugerre C, Wirden M, Viegas P, Simon A, Katlama C, Calvez V. 2004. Thymidine analogue reverse transcriptase inhibitors resistance mutations profiles and association to other nucleoside reverse transcriptase inhibitors resistance mutations observed in the context of virological failure. *J Med Virol* **72:** 162–165.

Marinello J, Marchand C, Mott B, Bain A, Thomas CJ, Pommier Y. 2008. Comparison of raltegravir and elvitegravir on HIV-1 integrase catalytic reactions and on a series of drug-resistant integrase mutants. *Biochemistry* **47:** 9345–54.

Margot NA, Isaacson E, McGowan I, Cheng AK, Schooley RT, Miller MD. 2002. Genotypic and phenotypic analyses of HIV-1 in antiretroviral-experienced patients treated with tenofovir DF. *AIDS* **16:** 1227–1235.

Marozsan AJ, Moore DM, Lobritz MA, Fraundorf E, Abraha A, Reeves JD, Arts EJ. 2005. Differences in the fitness of two diverse wild-type human immunodeficiency virus type 1 isolates are related to the efficiency of cell binding and entry. *J Virol* **79:** 7121–7134.

McColl DJ, Chen X. 2010. Strand transfer inhibitors of HIV-1 integrase: Bringing IN a new era of antiretroviral therapy. *Antiviral Res* **85:** 101–118.

Meyer PR, Matsuura SE, Mian AM, So AG, Scott WA. 1999. A mechanism of AZT resistance: An increase in nucleotide-dependent primer unblocking by mutant HIV-1 reverse transcriptase. *Mol Cell Biol* **4:** 35–43.

Meyer PR, Matsuura SE, Tolun AA, Pfeifer I, So AG, Mellors JW, Scott WA. 2002. Effects of specific zidovudine resistance mutations and substrate structure on nucleotide-dependent primer unblocking by human immunodeficiency virus type 1 reverse transcriptase. *Antimicrob Agents Chemother* **46:** 1540–1545.

Miller V. 2001. International perspectives on antiretroviral resistance. Resistance to protease inhibitors. *J Acquir Immune Defic Syndr* **26** (Suppl 1): S34–S50.

Mitsuya H, Broder S. 1986. Inhibition of the in vitro infectivity and cytopathic effect of human T-lymphotrophic virus type III/lymphadenopathy-associated virus (HTLV-III/LAV) by 2′,3′-dideoxynucleosides. *Proc Natl Acad Sci* **83:** 1911–1915.

Mitsuya H, Weinhold KJ, Furman PA, St Clair MH, Lehrman SN, Gallo RC, Bolognesi D, Barry DW, Broder S. 1985. 3′-Azido-3′-deoxythymidine (BW A509U): An antiviral agent that inhibits the infectivity and cytopathic effect of human T-lymphotropic virus type III/lymphadenopathy-associated virus in vitro. *Proc Natl Acad Sci* **82:** 7096–7100.

Molla A, Korneyeva M, Gao Q, Vasavanonda S, Schipper PJ, Mo HM, Markowitz M, Chernyavskiy T, Niu P, Lyons N, et al. 1996a. Ordered accumulation of mutations in HIV protease confers resistance to ritonavir. *Nat Med* **2:** 760–766.

Molla A, Korneyeva M, Gao Q, Vasavanonda S, Schipper PJ, Mo HM, Markowitz M, Chernyavskiy T, Niu P, Lyons N, et al. 1996b. Ordered accumulation of mutations in HIV protease confers resistance to ritonavir. *Nat Med* **2:** 760–766.

Naeger LK, Margot NA, Miller MD. 2002. ATP-dependent removal of nucleoside reverse transcriptase inhibitors by human immunodeficiency virus type 1 reverse transcriptase. *Antimicrob Agents Chemother* **46:** 2179–2184.

Nagashima KA, Thompson DA, Rosenfield SI, Maddon PJ, Dragic T, Olson WC. 2001. Human immunodeficiency virus type 1 entry inhibitors PRO 542 and T-20 are potently synergistic in blocking virus-cell and cell-cell fusion. *J Infect Dis* **183:** 1121–1125.

Nijhuis M, Deeks S, Boucher C. 2001. Implications of antiretroviral resistance on viral fitness. *Curr Opin Infect Dis* **14:** 23–28.

Nowak MA, Bonhoeffer S, Shaw GM, May RM. 1997. Antiviral drug treatment: Dynamics of resistance in free virus and infected cell populations. *J Theor Biol* **184:** 203–217.

O'Neil PK, Sun G, Yu H, Ron Y, Dougherty JP, Preston BD. 2002. Mutational analysis of HIV-1 long terminal repeats to explore the relative contribution of reverse transcriptase and RNA polymerase II to viral mutagenesis. *J Biol Chem* **277:** 38053–38061.

Parikh UM, Bacheler L, Koontz D, Mellors JW. 2006. The K65R mutation in human immunodeficiency virus type 1 reverse transcriptase exhibits bidirectional phenotypic antagonism with thymidine analog mutations. *J Virol* **80:** 4971–4977.

Park J, Morrow CD. 1993. Mutations in the protease gene of human immunodeficiency virus type 1 affect release and stability of virus particles. *Virology* **194:** 843–850.

Perelson AS, Neumann AU, Markowitz M, Leonard JM, Ho DD. 1996. HIV-1 dynamics in vivo: Virion clearance rate, infected cell life-span, and viral generation time. 1582–1586.

Quan Y, Gu Z, Li X, Li Z, Morrow CD, Wainberg MA. 1996. Endogenous reverse transcription assays reveal high-level resistance to the triphosphate of (−)2′-dideoxy-3′-thiacytidine by mutated M184V human immunodeficiency virus type 1. *J Virol* **70:** 5642–5645.

Quercia R, Dam E, Perez-Bercoff D, Clavel F. 2009. Selective-advantage profile of human immunodeficiency virus type 1 integrase mutants explains in vivo evolution of raltegravir resistance genotypes. *Virol J* **83:** 10245–10249.

Quinones-Mateu ME, Arts EJ. 2002. Fitness of drug resistant HIV-1: Methodology and clinical implications. *Drug Resist Updat* **5:** 224–233.

Quinones-Mateu ME, Arts EJ. 2006. Virus fitness: Concept, quantification, and application to HIV population dynamics. *Curr Top Microbiol Immunol* **299:** 83–140.

Quinones-Mateu ME, Moore-Dudley DM, Jegede O, Weber J, Arts J. 2008. Viral drug resistance and fitness. *Adv Pharmacol* **56:** 257–296.

Ray PE, Soler-Garcia AA, Xu L, Soderland C, Blumenthal R, Puri A. 2005. Fusion of HIV-1 envelope-expressing cells to human glomerular endothelial cells through an CXCR4-mediated mechanism. *Pediatr Nephrol* **20:** 1401–1409.

Reeves JD, Gallo SA, Ahmad N, Miamidian JL, Harvey PE, Sharron M, Pohlmann S, Sfakianos JN, Derdeyn CA, Blumenthal R, et al. 2002. Sensitivity of HIV-1 to entry inhibitors correlates with envelope/coreceptor affinity, receptor density, and fusion kinetics. *Proc Natl Acad Sci* **99:** 16249–16254.

Reeves JD, Miamidian JL, Biscone MJ, Lee FH, Ahmad N, Pierson TC, Doms RW. 2004. Impact of mutations in the coreceptor binding site on human immunodeficiency virus type 1 fusion, infection, and entry inhibitor sensitivity. *J Virol* **78:** 5476–5485.

Reeves JD, Lee FH, Miamidian JL, Jabara CB, Juntilla MM, Doms RW. 2005. Enfuvirtide resistance mutations: Impact on human immunodeficiency virus envelope function, entry inhibitor sensitivity, and virus neutralization. *J Virol* 79: 4991–4999.

Reimann KA, Lin W, Bixler S, Browning B, Ehrenfels BN, Lucci J, Miatkowski K, Olson D, Parish TH, Rosa MD, et al. 1997. A humanized form of a CD4-specific monoclonal antibody exhibits decreased antigenicity and prolonged plasma half-life in rhesus monkeys while retaining its unique biological and antiviral properties. *AIDS Res Hum Retroviruses* 13: 933–943.

Richman DD. 2001. HIV chemotherapy. *Nature* 410: 995–1001.

Rimsky LT, Shugars DC, Matthews TJ. 1998. Determinants of human immunodeficiency virus type 1 resistance to gp41-derived inhibitory peptides. *J Virol* 72: 986–993.

Rodgers DW, Gamblin SJ, Harris BA, Ray S, Culp JS, Hellmig B, Woolf DJ, Debouck C, Harrison SC. 1995. The structure of unliganded reverse transcriptase from the human immunodeficiency virus type 1. *Proc Natl Acad Sci* 92: 1222–1226.

Sato M, Motomura T, Aramaki H, Matsuda T, Yamashita M, Ito Y, Kawakami H, Matsuzaki Y, Watanabe W, Yamataka K, et al. 2006. Novel HIV-1 integrase inhibitors derived from quinolone antibiotics. *J Med Chem* 49: 1506–1508.

Schinazi RF, Lloyd RM Jr, Nguyen MH, Cannon DL, McMillan A, Ilksoy N, Chu CK, Liotta DC, Bazmi HZ, Mellors JW. 1993. Characterization of human immunodeficiency viruses resistant to oxathiolane-cytosine nucleosides. *Antimicrob Agents Chemother* 37: 875–881.

Shafer RW, Dupnik K, Winters MA, Eshleman SH. 2000. A guide to HIV-1 reverse transcriptase and protease sequencing for drug resistance studies. In *HIV sequence compendium 2000* (ed. Kuiken CL, et al.). Los Alamos National Laboratory, Los Alamos, NM.

Shehu-Xhilaga M, Tachedjian G, Crowe SM, Kedzierska K. 2005. Antiretroviral compounds: Mechanisms underlying failure of HAART to eradicate HIV-1. *Curr Med Chem* 12: 1705–1719.

Shimura K, Kodama E, Sakagami Y, Matsuzaki Y, Watanabe W, Yamataka K, Watanabe Y, Ohata Y, Doi S, Sato M, et al. 2008. Broad antiretroviral activity and resistance profile of the novel human immunodeficiency virus integrase inhibitor elvitegravir (JTK-303/GS-9137). *J Virol* 82: 764–774.

Singh I, Gorzynski J, Drobysheva D, Bassit L, Schinazi R. 2010. Raltegravir is a potent inhibitor of XMRV, a virus implicated in prostate cancer and chronic fatigue syndrome. *PLoS One.* 5: e9948. doi: 10.1371/journal.pone.0009948.

Sluis-Cremer N, Temiz NA, Bahar I. 2004. Conformational changes in HIV-1 reverse transcriptase induced by nonnucleoside reverse transcriptase inhibitor binding. *Curr HIV Res* 2: 323–332.

Smith PF, Ogundele A, Forrest A, Wilton J, Salzwedel K, Doto J, Allaway GP, Martin DE. 2007. Phase I and II study of the safety, virologic effect, and pharmacokinetics/pharmacodynamics of single-dose 3-o-(3′,3′-dimethylsuccinyl)betulinic acid (bevirimat) against human immunodeficiency virus infection. *Antimicrob Agents Chemother* 51: 3574–3581.

Spence RA, Kati WM, Anderson KS, Johnson KA. 1995. Mechanism of inhibition of HIV-1 reverse transcriptase by nonnucleoside inhibitors. *Science* 267: 988–993.

Spira S, Wainberg MA, Loemba H, Turner D, Brenner BG. 2003. Impact of clade diversity on HIV-1 virulence, antiretroviral drug sensitivity and drug resistance. *J Antimicrob Chemother* 51: 229–240.

Staszewski S, Miller V, Rehmet S, Stark T, De CJ, De BM, Peeters M, Andries K, Moeremans M, De RM, et al. 1996. Virological and immunological analysis of a triple combination pilot study with loviride, lamivudine and zidovudine in HIV-1-infected patients. *AIDS* 10: F1–F7.

St Clair MH, Richards CA, Spector T, Weinhold KJ, Miller WH, Langlois AJ, Furman PA. 1987. 3′-Azido-3′-deoxythymidine triphosphate as an inhibitor and substrate of purified human immunodeficiency virus reverse transcriptase. *Antimicrob Agents Chemother* 31: 1972–1977.

Stengel RF. 2008. Mutation and control of the human immunodeficiency virus. *Math Biosci* 213: 93–102.

* Sundquist WI, Kräusslich H-G. 2011. Assembly, budding, and maturation. *Cold Spring Harb Perspect Med* doi: 10.1101/cshperspect.1006924.

Tantillo C, Ding J, Jacobo-Molina A, Nanni RG, Boyer PL, Hughes SH, Pauwels R, Andries K, Janssen PA, Arnold EA. et al. 1994. Locations of anti-AIDS drug binding sites and resistance mutations in the three-dimensional structure of HIV-1 reverse transcriptase. Implications for mechanisms of drug inhibition and resistance. *J Mol Biol* 243: 369–387.

Tebit DM, Lobritz M, Lalonde M, Immonen T, Singh K, Sarafianos S, Herchenroder O, Krausslich HG, Arts EJ. 2010. Divergent evolution in reverse transcriptase (RT) of HIV-1 group O and M lineages: Impact on structure, fitness, and sensitivity to nonnucleoside RT inhibitors. *J Virol* 84: 9817–9830.

Torre VS, Marozsan AJ, Albright JL, Collins KR, Hartley O, Offord RE, Quinones-Mateu ME, Arts EJ. 2000. Variable sensitivity of CCR5-tropic human immunodeficiency virus type 1 isolates to inhibition by RANTES analogs. *J Virol* 74: 4868–4876.

Tremblay CL, Kollmann C, Giguel F, Chou TC, Hirsch MS. 2000. Strong in vitro synergy between the fusion inhibitor T-20 and the CXCR4 blocker AMD-3100. *J Acquir Immune Defic Syndr* 25: 99–102.

Trkola A, Kuhmann SE, Strizki JM, Maxwell E, Ketas T, Morgan T, Pugach P, Xu S, Wojcik L, Tagat J, et al. 2002. HIV-1 escape from a small molecule, CCR5-specific entry inhibitor does not involve CXCR4 use. *Proc Natl Acad Sci* 99: 395–400.

Tsamis F, Gavrilov S, Kajumo F, Seibert C, Kuhmann S, Ketas T, Trkola A, Palani A, Clader JW, Tagat JR, et al. 2003. Analysis of the mechanism by which the small-molecule CCR5 antagonists SCH-351125 and SCH-350581 inhibit human immunodeficiency virus type 1 entry. *J Virol* 77: 5201–5208.

Tsibris AM, Sagar M, Gulick RM, Su Z, Hughes M, Greaves W, Subramanian M, Flexner C, Giguel F, Leopold KE, et al. 2008. In vivo emergence of vicriviroc resistance in a human immunodeficiency virus type 1 subtype C-infected subject. *J Virol* 82: 8210–8214.

Van Baelen K, Van Eygen V, Rondelez E, Stuyver LJ. 2008. Clade-specific HIV-1 integrase polymorphisms do not reduce raltegravir elvitegravir phenotypic susceptibility. *AIDS* **22:** 1877–80.

Veazey RS, Klasse PJ, Ketas TJ, Reeves JD, Piatak M Jr, Kunstman K, Kuhmann SE, Marx PA, Lifson JD, Dufour J, et al. 2003. Use of a small molecule CCR5 inhibitor in macaques to treat simian immunodeficiency virus infection or prevent simian-human immunodeficiency virus infection. *J Exp Med* **198:** 1551–1562.

Veazey RS, Klasse PJ, Schader SM, Hu Q, Ketas TJ, Lu M, Marx PA, Dufour J, Colonno RJ, Shattock RJ, et al. 2005. Protection of macaques from vaginal SHIV challenge by vaginally delivered inhibitors of virus-cell fusion. *Nature (London)* **438:** 99–102.

Wainberg MA, Miller MD, Quan Y, Salomon H, Mulato AS, Lamy PD, Margot NA, Anton KE, Cherrington JM. 1999. In vitro selection and characterization of HIV-1 with reduced susceptibility to PMPA. *Antiviral Therapy* **4:** 87–94.

Wei X, Ghosh SK, Taylor ME, Johnson VA, Emini EA, Deutsch P, Lifson JD, Bonhoeffer S, Nowak MA, Hahn BH. 1995. Viral dynamics in human immunodeficiency virus type 1 infection. *Nature (London)* **373:** 117–122.

Wei X, Decker JM, Liu H, Zhang Z, Arani RB, Kilby JM, Saag MS, Wu X, Shaw GM, Kappes JC. 2002. Emergence of resistant human immunodeficiency virus type 1 in patients receiving fusion inhibitor (T-20) monotherapy. *Antimicrob Agents Chemother* **46:** 1896–1905.

Westby M, Lewis M, Whitcomb J, Youle M, Pozniak AL, James IT, Jenkins TM, Perros M, van der Ryst E. 2006. Emergence of CXCR4-using human immunodeficiency virus type 1 (HIV-1) variants in a minority of HIV-1-infected patients following treatment with the CCR5 antagonist maraviroc is from a pretreatment CXCR4-using virus reservoir. *J Virol* **80:** 4909–4920.

Westby M, Smith-Burchnell C, Mori J, Lewis M, Mosley M, Stockdale M, Dorr P, Ciaramella G, Perros M. 2007. Reduced maximal inhibition in phenotypic susceptibility assays indicates that viral strains resistant to the CCR5 antagonist maraviroc utilize inhibitor-bound receptor for entry. *J Virol* **81:** 2359–2371.

White KL, Chen JM, Feng JY, Margot NA, Ly JK, Ray AS, Macarthur HL, McDermott MJ, Swaminathan S, Miller MD. 2006. The K65R reverse transcriptase mutation in HIV-1 reverses the excision phenotype of zidovudine resistance mutations. *Antiviral Therapy* **11:** 155–163.

Wild C, Greenwell T, Matthews T. 1993. A synthetic peptide from HIV-1 gp41 is a potent inhibitor of virus-mediated cell-cell fusion. *AIDS Res Hum Retroviruses* **9:** 1051–1053.

Wild CT, Shugars DC, Greenwell TK, McDanal CB, Matthews TJ. 1994. Peptides corresponding to a predictive alpha-helical domain of human immunodeficiency virus type 1 gp41 are potent inhibitors of virus infection. *Proc Natl Acad Sci* **91:** 9770–9774.

* Wilen CB, Tilton JC, Doms RW. 2011. HIV: Cell binding and entry. *Cold Spring Harb Perspect Med* doi: 10.1101/cshperspect.a006866.

Wingfield WL, Pollack D, Grunert RR. 1969. Therapeutic efficacy of amantadine HCl and rimantadine HCl in naturally occurring influenza A2 respiratory illness in man. *N Engl J Med* **281:** 579–584.

Witvrouw M, Pannecouque C, Van Laethem K, Desmyter J, De Clercq E, Vandamme AM. 1999. Activity of non-nucleoside reverse transcriptase inhibitors against HIV-2 and SIV. *AIDS* **13:** 1477–1483.

Wu B, Chien EY, Mol CD, Fenalti G, Liu W, Katritch V, Abagyan R, Brooun A, Wells P, Bi FC, et al. 2010. Structures of the CXCR4 chemokine GPCR with small-molecule and cyclic peptide antagonists. *Science* **330:** 1066–1071.

Yahi N, Fantini J, Henry M, Tourres C, Tamalet C. 2005. Structural analysis of reverse transcriptase mutations at codon 215 explains the predominance of T215Y over T215F in HIV-1 variants selected under antiretroviral therapy. *J Biomed Sci* **12:** 701–710.

Young FE. 1988. The role of the FDA in the effort against AIDS. *Public Health Rep* **103:** 242–245.

Zennou V, Mammano F, Paulous S, Mathez D, Clavel F. 1998. Loss of viral fitness associated with multiple Gag and Gag-Pol processing defects in human immunodeficiency virus type 1 variants selected for resistance to protease inhibitors in vivo. *J Virol* **72:** 3300–3306.

Zhang YM, Imamichi H, Imamichi T, Lane HC, Falloon J, Vasudevachari MB, Salzman NP. 1997. Drug resistance during indinavir therapy is caused by mutations in the protease gene and in its Gag substrate cleavage sites. *J Virol* **71:** 6662–6670.

Zhou T, Xu L, Dey B, Hessell AJ, Van RD, Xiang SH, Yang X, Zhang MY, Zwick MB, Arthos J, et al. 2007. Structural definition of a conserved neutralization epitope on HIV-1 gp120. *Nature* **445:** 732–737.

Novel Cell and Gene Therapies for HIV

James A. Hoxie[1] and Carl H. June[2]

[1]Division of Hematology/Oncology, Department of Medicine, University of Pennsylvania School of Medicine, Philadelphia, Pennsylvania 19104

[2]Abramson Family Cancer Research Institute and Department of Pathology and Laboratory Medicine, University of Pennsylvania School of Medicine, Philadelphia, Pennsylvania 19104

Correspondence: hoxie@mail.med.upenn.edu

Highly active antiretroviral therapy dramatically improves survival in HIV-infected patients. However, persistence of HIV in reservoirs has necessitated lifelong treatment that can be complicated by cumulative toxicities, incomplete immune restoration, and the emergence of drug-resistant escape mutants. Cell and gene therapies offer the promise of preventing progressive HIV infection by interfering with HIV replication in the absence of chronic antiviral therapy. Individuals homozygous for a deletion in the CCR5 gene (CCR5Δ32) are largely resistant to infection from R5-topic HIV-1 strains, which are most commonly transmitted. A recent report that an HIV-infected patient with relapsed acute myelogenous leukemia was effectively cured from HIV infection after transplantation of hematopoietic stem/progenitor cells (HSC) from a CCR5Δ32 homozygous donor has generated renewed interest in developing treatment strategies that target viral reservoirs and generate HIV resistance in a patient's own cells. Although the development of cell-based and gene transfer therapies has been slow, progress in a number of areas is evident. Advances in the fields of gene-targeting strategies, T-cell-based approaches, and HSCs have been encouraging, and a series of ongoing and planned trials to establish proof of concept for strategies that could lead to successful cell and gene therapies for HIV are under way. The eventual goal of these studies is to eliminate latent viral reservoirs and the need for lifelong antiretroviral therapy.

In the era of highly active antiretroviral therapy (HAART), HIV replication can be suppressed to low or undetectable levels (Kaufmann et al. 2000; Maartens et al. 2007; Kitahata et al. 2009) with a corresponding but variable increase in CD4 T-cell counts (Valdez et al. 2002; Moore et al. 2005; Geng and Deeks 2009). Survival has improved dramatically, and HAART has turned HIV infection into a chronic, manageable illness. However, survival remains reduced by at least 10 years compared with the general population, and there are long-term complications (Lewden et al. 2005; Lohse et al. 2007) including accelerated cardiovascular disease, liver and renal failure, neurocognitive dysfunction, and HIV-associated malignancies (Cheung et al. 2005; Lekakis and Ikonomidis 2010; Nunez 2010; Xia et al. 2011). Ongoing immune activation and inflammation can occur and may contribute to increased morbidity and mortality (Giorgi et al. 1999; Rodger et al. 2009). Many patients, particularly those who began HAART in the setting of advanced immunodeficiency, do not achieve complete CD4 reconstitution (≥ 500 cells/ mm^3) (Valdez et al. 2002; Kelley et al. 2009), which has been linked to increased morbidity (Lewden et al. 2007; Baker et al. 2008). Moreover, even when viral replication is suppressed for many years, replicating HIV typically reappears in plasma once HAART is discontinued, indicating that patients are not cured and that long-lived viral reservoirs persist (Chun and Fauci 1999; Blankson et al. 2002). Given the limitations and the potential complications of HAART, there has been renewed interest in novel approaches to control or preferably cure HIV infection (Richman et al. 2009).

With advances in understanding the molecular basis for HIV replication and mechanisms for host control, a number of investigators are focusing on cell-based and gene therapy either as stand-alone approaches or as adjuvants to pharmacological drug regimens. There have been several excellent reviews of this topic (Poluri et al. 2003; Strayer et al. 2005; Edelstein et al. 2007; Rossi et al. 2007). These approaches have potential advantages compared to conventional drugs. In particular, the self-renewing properties of cell and genetic strategies at least in principal can obviate the need for lifelong antiretroviral medications. As will be discussed, cell-based approaches involving hematopoietic stem cell transplantation have the potential to be curative (Hütter et al. 2009; Allers et al. 2010; Deeks and McCune 2010). This review will consider novel therapies that focus on modifying host cells to resist HIV infection with approaches directed at viral or cellular targets. One general consideration is that although nucleic acid-based antivirals can be designed to have high specificity for HIV-1 targets at the gene, RNA, or protein level, as with pharmacologic approaches, viral escape is a major concern. In contrast, cellular targets are far less prone to mutational escape, although with the exception of CCR5, the side effects of down-regulating cellular targets for the long term are unknown.

HISTORY

Cell and gene therapy strategies have been proposed from the earliest days of the HIV epidemic. Friedmann and Anderson have reviewed the conception and early history of gene transfer therapy (Friedmann and Roblin 1972; Anderson 1984; Friedmann 1992). Baltimore first proposed a gene therapy approach for treating HIV/AIDS, in which a retrovirus could be engineered to antagonize HIV by expressing dominant negative RNA or protein inhibitors (Baltimore 1988).

Allogeneic and Xenogeneic Transplantation

Xenotransplantation was proposed as a cell-based therapy, with the rationale that HIV-1-infected humans could be reconstituted with nonhuman primate bone marrow, an approach that would take advantage of the intrinsic resistance of nonhuman species to infection with HIV-1. In 1995, a 38-yr-old patient infected with HIV for more than 15 years underwent a controversial experiment following approval by the U.S. Food and Drug Administration. His bone marrow was suppressed by sublethal doses of radiation and chemotherapeutic drugs, after which he received an infusion of bone marrow cells from a baboon. This experiment occurred before the advent of HAART therapy, at a time when the standard treatment approaches were failing to control HIV, and the AIDS activist community was increasingly convinced that radical approaches should be explored. The patient survived and improved for reasons that are unclear, as the baboon cells failed to engraft and were not detectable in his bone marrow beyond the first month posttransplant (Michaels et al. 2004).

Allogeneic and syngeneic (twin) hematopoietic stem cell transplantation (HSCT) were proposed as a treatment for AIDS, based on the positive experiences with allogeneic HSCT in children with severe combined immunodeficiency (Cavazzana-Calvo and Fischer 2007). Patients with HIV/AIDS-associated malignancies for whom HSCT was standard therapy provided an opportunity to determine the regenerative potential of bone marrow and

Cite this article as *Cold Spring Harb Perspect Med* doi: 10.1101/cshperspect.a007179

mature lymphocytes in the setting of HIV infection. The first report of allogeneic HSCT described two men with AIDS and Kaposi's sarcoma who were given infusions of partially compatible bone marrow from uninfected family donors (Hassett et al. 1983). No clinical benefit occurred, and engraftment of the donor bone marrow was brief, probably because no conditioning regimen was given to the patients. In retrospect, it was not realized at that time that even though AIDS patients have severe immunosuppression, they are still able to mount potent rejection responses to organ and bone marrow grafts. Lane and colleagues conducted the first study using infusion of bone marrow from a healthy identical twin to a patient with AIDS. The patient was given a single infusion of bone marrow without conditioning, followed by repeated infusions of syngeneic peripheral blood lymphocytes from the donor (Lane et al. 1984). Transient clinical benefit was observed with increased CD4 counts that peaked three months after stem cell infusion. These early experiments were conducted in the pre-HAART days in the presence of high levels HIV viremia, and the results may have been more durable with long-term immune reconstitution had effective antiretroviral therapy been available. Consistent with this notion, Holland and coworkers reported the case of a patient with HIV and lymphoma who received an allogeneic HSCT with concomitant zidovudine therapy. The patient engrafted with donor marrow, but died from progressive lymphoma 47 days after HSCT. An autopsy revealed no evidence for residual HIV-1 by culture or polymerase chain reaction (PCR) of numerous tissues, suggesting that myeloablative chemoradiotherapy combined with antiretroviral therapy had a substantial impact on latent viral reservoirs in host cells (Holland et al. 1989).

PROTEIN-BASED INHIBITORS

A number of criteria have been proposed for developing genetic inhibitors of HIV for human clinical trials, including (1) potency of inhibition, (2) lack of immunogenicity, (3) lack of toxicity, (4) the stage in the viral life cycle that they target, and (5) their potential for selecting resistant viruses (Dropulic and June 2006). Mathematical modeling has predicted that postintegration inhibitors lead to the persistence of cells carrying an integrated provirus, resulting in an accumulation of HIV-1-infected cells that could ultimately counteract their antiviral effect. In contrast, inhibitors that act before integration, even those with lower potency, are predicted to exert a systemic antiviral effect with the expansion of transduced cells capable of resisting HIV infection (von Laer et al. 2006).

Protein-based inhibitors can be directed toward cellular or viral targets and have included dominant negative inhibitors, intrabodies, intrakines, fusion inhibitors, and zinc finger nucleases (ZFNs). Each type of protein inhibitor is described in detail below. These inhibitors are typically expressed from a viral vector long terminal repeat (LTR), but in several instances are produced from strong constitutive promoters inserted within the bodies of the viral vectors. Strictly speaking, ZFNs are protein-based inhibitors that have a fundamentally different mechanism than the other inhibitors in this class, and so are discussed in a separate section.

Dominant Negative Inhibitory Proteins

The first protein used in a gene therapy trial for treating HIV infection was a mutant form of the HIV Rev protein called M10 (Woffendin et al. 1996). Rev M10 is believed to work by blocking the export of singly spliced and unspliced HIV RNA from the nucleus to the cytoplasm thereby preventing viral assembly and subsequent transmission (Malim et al. 1992). This mutant protein remains one of the most potent inhibitors of HIV replication and has been evaluated in human clinical trials (Ranga et al. 1998; Morgan et al. 2005; Podsakoff et al. 2005). Although there was no long-term benefit, a modest survival advantage of M10-expressing CD4 cells was observed.

Intrabodies and Intrakines

The expression of intracellular antibodies (termed "intrabodies"), typically as single-chain Fv fragments that target viral or cellular proteins, and chemokines (termed "intrakines") have also proven to be very potent inhibitors of HIV replication in vitro (Yang et al. 1997; Schroers et al. 2002; Poluri et al. 2003; Lobato and Rabbitts 2004; Lo and Marasco 2008; Zhang et al. 2009). These proteins bind to viral or cellular target proteins leading to their intracellular degradation. Notable cellular targets for this approach have included the HIV coreceptors CCR5 and CXCR4 (Yang et al. 1997; Schroers et al. 2002; Zhang et al. 2009).

Fusion Inhibitors

von Laer and colleagues developed a novel protein-based fusion inhibitor, termed C46, which binds to HIV gp41 at the cell surface and blocks viral entry (Hildinger et al. 2001; Egelhofer et al. 2004; Perez et al. 2005). C46 is comprised of amino acids from the second heptad repeat of gp41. As with other protein-based inhibitors, it can be expressed constitutively by retroviral vectors and applied to a gene therapy setting. A membrane-anchored form of C46 (maC46) was shown to confer HIV resistance to hematopoietic stem cells in nonhuman primates (Trobridge et al. 2009). In a comparative study, the maC46 fusion inhibitor was compared to an HIV-1 tat/rev-specific small hairpin (sh) RNA (Lee et al. 2002) and an RNA antisense gene specific for the HIV-1 envelope glycoprotein (Dropulic et al. 1996). Notably, maC46 proved to be the most potent in conferring a selection advantage to transduced cells following HIV-1 inoculation in vitro and in humanized mice in vivo (Kimpel et al. 2010). The maC46 transgene was tested in a phase I trial in 10 patients with late-stage HIV infection (van Lunzen et al. 2007). The infusions of transduced T cells were well tolerated, and a significant albeit transient increase of CD4 counts was observed after infusion.

Inhibitors of Viral Restriction Factor, TRIM5α

An early event in the viral life cycle that is well suited to protein-based inhibition is uncoating of the viral capsid following entry, which is highly restricted for HIV-1 in rhesus macaques by the host protein tripartite motif (TRIM)5-α (Stremlau et al. 2004; Nisole et al. 2005; Malin and Bieniasz 2011). Under permissive conditions the HIV and SIV capsid proteins have evolved to escape binding by TRIM5-α, which otherwise binds and leads to degradation of the capsid and a failure of reverse transcription. It has been shown in vitro that in human cells expressing either TRIM5-α from a nonpermissive primate species or a human TRIM5-α that was rendered competent for binding to the HIV capsid, these proteins act in a dominant negative manner to inhibit replication (Yap et al. 2005; Stremlau et al. 2006). These approaches represent promising interventions for future protein-based gene therapy.

Although protein-based approaches can have potent antiviral effects in vitro, a principal limitation is not only the need to maintain sufficient levels of expression from the vector, but to avoid immunogenicity in vivo, as immune responses could confound persisting antiviral activity. Thus, for protein inhibitors that are advanced to clinical trials both their potency and immunogenicity will need to be assessed.

NUCLEIC ACID–BASED INHIBITORS

A number of RNA-based approaches, including antisense RNAs, aptamers, decoys, ribozymes, and si/shRNAs have been described. An advantage of these inhibitors is that in contrast to protein-based inhibitors, they do not elicit adaptive T-cell responses and are unlikely to be immunogenic. However, RNA approaches can potentially have off-target toxicity because of activation of innate immune responses and competition with endogenous RNA functions (Lares et al. 2010).

Antisense RNA

Short and long antisense RNA transgenes that pair with HIV transcripts to form duplexes that are nonfunctional have been effective in blocking HIV replication in hematopoietic cells. The first demonstration of this principle came from studies using adeno-associated virus vectors to deliver a short anti-U5 region antisense

RNA (Chatterjee et al. 1992). More recently, Levine and coworkers infused patients with autologous CD4 T cells that were genetically modified by a conditionally replicating lentiviral vector expressing a long antisense RNA to the HIV envelope mRNA (Levine et al. 2006). In a follow-up study (clinicaltrials.gov NCT00295477), Tebas and coworkers showed a decrease in viral load in the majority (88% [7/8]) of these patients after antiretroviral therapy was discontinued, with one patient maintaining undetectable HIV RNA for 104 days (Tebas et al. 2010). Although the mechanism by which these antisense transcripts inhibited HIV replication is not clear, it could have triggered extensive adenosine deamination of the HIV/antisense duplex, resulting in nuclear retention of transcripts or the generation of multiple viral disabling mutations (Lu et al. 2004). A further analysis of transduced cells from patients in this trial using 454 pyrosequencing showed no evidence for abnormal expansion of cells because of vector-mediated insertional activation of proto-oncogenes, suggesting that no adverse events were apparent at the molecular level (Wang et al. 2009).

Aptamers

RNA aptamers are RNA molecules with structural features that facilitate high-affinity interactions with targeted ligands. They are amenable to rapid selection in vitro and can be designed to bind to virtually any protein of choice (Nimjee et al. 2005). There are now a number of highly effective aptamers available for testing in gene therapy settings (Symensma et al. 1996; Joshi et al. 2003; Held et al. 2006, 2007; Kissel et al. 2007), although to date none have been tested in clinical trials for HIV infection.

RNA Decoys

The HIV Tat protein binds to a structure on the 5′ UTR of mRNAs termed the TAR motif, and is required for efficient transcription (Arya et al. 1985; Sodroski et al. 1985). Short RNA oligonucleotides corresponding to the TAR sequence, termed TAR decoys, bind to Tat and block its

interaction with the authentic TAR region to inhibit HIV gene expression and replication (Lisziewicz et al. 1993; Bohjanen et al. 1997). RNA decoys have also been developed to prevent Rev from binding to the Rev response element (RRE), which is required to transport unspliced and singly spliced mRNAs from the nucleus (Lee et al. 1994). A clinical trial to evaluate the safety and feasibility of an RRE decoy was conducted using bone marrow from four HIV-1-infected pediatric subjects (Kohn et al. 1999). The approach was safe; however, expression of the RNA decoy was likely insufficient for antiviral effects to be seen.

Ribozymes

Ribozymes are antisense RNAs that not only bind but enzymatically cleave targeted mRNAs. The first demonstration that ribozymes were effective in inhibiting HIV replication was published in 1990 (Sarver et al. 1990). Since then, many ribozyme-based antivirals have been developed, and some have been evaluated in clinical trials including those targeting HIV genes *tat* and *rev*, and the U5 region of the viral LTR. Ribozymes in these trials were either expressed from the retroviral LTRs as long, capped, polyadenylated transcripts from the retroviral LTR promoter (Bauer et al. 1997; Ngok et al. 2004), or as a discrete, chimeric Pol III tRNA-ribozyme transcript (Leavitt et al. 1994; Li et al. 2005). Four of these trials involved retroviral vector delivery of decoy or ribozyme genes into autologous hematopoietic progenitor cells isolated from HIV-1-infected individuals (Kohn et al. 1999; Amado et al. 2004; Mitsuyasu et al. 2009; DiGiusto et al. 2010). Following retroviral transduction, cells were reinfused either in the absence of bone marrow conditioning, or in one case with bone marrow conditioning to treat an AIDS-related lymphoma (DiGiusto et al. 2010). Although these trials showed that ribozymes and decoys could be safely introduced into mobilized stem cells or peripheral blood mononuclear cell cultures (PBMCs) and reinfused in patients, no significant anti-HIV effects were observed. However, in a recent phase II clinical trial, Mitsuyasu et al. (2009) showed that an anti-tat ribozyme could be safely introduced in autologous stem cells from a gammaretroviral vector. This vector (termed OZ1) expressed a hammerhead ribozyme that targeted the overlapping reading frames of the viral *vpr* and *tat* genes in unspliced and spliced viral transcripts, respectively. Although no significant differences were observed in the viral load between OZ1-treated and placebo groups at the time of study's primary end point at 12 months, they observed a significantly lower viral load at later time points, and OZ1-treated patients had higher CD4 cell numbers at all time points. This study is notable because it is the first randomized, double-blinded phase II clinical trial for HIV-1 infected patients to test gene-modified HSC, and showed the feasibility of conducting complex cell and gene transfer trials that will likely be used in future protocols to eradicate HIV infection.

RNA Interference

RNA interference (RNAi) is a regulatory mechanism of most eukaryotic cells that uses small double-stranded RNA (dsRNA) molecules as triggers to direct homology-dependent control of gene activity (Hannon and Rossi 2004). Known as small interfering RNAs (siRNA) these ~21- to 22-bp-long dsRNA molecules have characteristic two-nucleotide $3'$ overhangs that allows them to be recognized by the enzymatic machinery of RNAi leading to homology-dependent degradation of the target mRNA. To date, preclinical studies indicate that this is the most potent RNA-based inhibitory mechanism available for therapeutic application. Virtually all of the HIV-encoded mRNAs have been shown to be susceptible to RNAi down-regulation in cell lines, including tat, rev, gag, pol, nef, vif, env, vpr, and the LTR (Rossi et al. 2007).

A substantial challenge for clinical applications of RNAi inhibitors is the high mutation rate of HIV, which readily generates escape mutants. Ideally, conserved sequences on transcripts encoding gene products required for critical functions in the viral life cycle and that cannot tolerate mutations will need to be targeted (Zhang et al. 2007). An alternative approach to relying solely on RNAi is to combine a single shRNA with other antiviral genes, thereby providing synergistic inhibition. This approach has been successfully performed by coexpressing an anti-tat/rev shRNA, a nucleolar localizing TAR decoy, and an anti-CCR5 ribozyme in a single vector backbone (Li et al. 2005). A pilot trial testing this triple combination lentiviral vector transduced into stem cells from patients with HIV and non-Hodgkin's lymphoma was recently reported (DiGiusto et al. 2010). Engraftment occurred by 11 d, with low but persisting levels of gene marking observed in several cell lineages for up to 24 mo. In preclinical studies, this triple combination vector was compared to the protein-based maC46 fusion inhibitor and to a long RNA antisense vector, and was found to be less efficient than C46 in providing a selective advantage in HIV-1-infected cultures of primary T cells (Kimpel et al. 2010).

CCR5 AS AN ANTI-HIV TARGET AND ERADICATION STRATEGIES

The identification of CCR5 as the major coreceptor for transmitted HIV-1 isolates provided an explanation for the previously noted resistance of individuals who were frequently exposed to HIV, but remained uninfected, and were found to be homozygous for the CCR5Δ32 allele (Samson et al. 1996). The mutation is caused by a 32-bp deletion resulting in a frameshift mutation that truncates CCR5 and prevents its expression on the cell surface (Huang et al. 1996; Samson et al. 1996; Zimmerman et al. 1997). Given that individuals lacking CCR5 are entirely healthy, this observation provided a strong impetus for the development of drugs that target the virus-CCR5 interaction, one of which is now FDA approved (Gulick et al. 2008).

The "Berlin Patient"

Although CCR5 small molecule inhibitors are proving to be clinically useful (Gulick et al. 2008), their activity did not predict the remarkable outcome of a patient who has apparently been cured of HIV infection following an allogeneic HSC transplant from a homozygous CCR5Δ32 donor in an extraordinary experiment conducted by Gero Hütter and colleagues in Berlin (Hütter et al. 2009). The patient, infected with HIV for 10 years and well controlled on antiretroviral therapy (ART), developed acute myelogenous leukemia unrelated to his HIV. When ART was interrupted during antileukemia chemotherapy, HIV plasma RNA not surprisingly increased to a high level, and was again controlled when ART was restarted. Despite achieving a remission, the patient's leukemia relapsed, requiring that an allogeneic HSC transplant be performed as lifesaving therapy. In the German database for unrelated donors, one individual was identified who was MHC compatible and also homozygous for the CCR5Δ32 mutation in *CCR5*. The patient received a HSC transplant from this donor, with ART being discontinued at the time of stem cell infusion. Strikingly, 3½ years posttransplant and in the continued absence of antiretroviral therapy, no HIV has been detected using the most sensitive molecular assays for viral RNA or DNA in plasma, lymphoid tissues, and peripheral blood CD4 cells (Allers et al. 2010; Hutter and Thiel 2011). For a variety of reasons, including the low frequency of CCR5Δ32 homozygotes in the general population and the logistics and feasibility of identifying suitable donors, similar transplants in HIV-infected patients are unlikely to be performed anytime soon, leaving open many questions (discussed further below) as to how HIV was eradicated in this patient (Deeks and McCune 2010). However, given this proof of concept that HIV infection can be apparently cured, the challenge to the field is to develop more generally applicable approaches that (1) do not require intensive myeloablative chemotherapy and (2) can be performed using autologous rather than allogeneic cells.

HIV Latency

The establishment of HIV latency and the failure of antiviral therapies to eradicate persisting, long-lived viral reservoirs is a central issue in the field (discussed by Siliciano and Greene 2011). Whereas the clinical result from the Berlin patient suggests that donor CCR5-negative HSCs and engraftment of HIV-resistant T cells can ultimately suppress or prevent HIV replication and eradicate HIV reservoirs, other mechanisms likely contributed (Deeks and McCune 2010; Cannon and June 2011). It is possible or even likely that the myeloablative therapy could have provided a synergistic effect. However, there are numerous reports of HIV-infected individuals undergoing allogeneic HSC or bone marrow transplants from donors not selected to be CCR5-negative, demonstrating that HIV-1 was not eliminated by the particular chemotherapy regimens used (Hassett et al. 1983; Angelucci et al. 1990; Bardini et al. 1991; Schlegel et al. 2000; Kang et al. 2002; Avettand-Fenoel et al. 2007; Wolf et al. 2007; Woolfrey et al. 2008; Kamp et al. 2010; Polizzotto et al. 2010), with one possible exception (Holland et al. 1989). Of note, the Berlin patient's myeloablative therapy consisted of amsacrine, fludarabine, cytarabine, and cyclophosphamide, a regimen that has not previously been reported in patients with HIV infection. He was also treated with total body irradiation (TBI), both to enhance the antileukemic chemotherapy and for immunosuppression to facilitate engraftment; although historically, TBI used in conjunction with myeloblative therapy has not eradicated HIV (Giri et al. 1992; Tomonari et al. 2005; Polizzotto et al. 2010). Of potential importance, the Berlin patient was also treated with several agents to prevent graft versus host disease, including antithymocyte globulin (ATG), cyclosporine, and mycophenolate mofetil, all of which have potent cytolytic or suppressive effects on T cells. Given the role of CD4/CCR5 memory T cells as a long-lived reservoir for HIV (Han et al. 2007), it is likely that these agents directly targeted this reservoir. ATG, in particular, is not only a potent immunosuppressive drug, but contains polyclonal antibodies directed against all known lymphocyte subsets (Rebellato et al. 1994). The effects of ATG in the setting are largely unknown, because this potent immunosuppressive agent has only rarely been given to patients with chronic HIV-1 infection (Wolf et al. 2007). Finally, it is possible that innate or acquired immunity delivered by the donor immune system may have contributed to the elimination of residual HIV reservoirs. The Berlin patient developed graft versus host disease, and an allogeneic immune response directed against host lymphocytes could have had a purging effect on latent HIV reservoirs. Allogeneic immune effects are among the most powerful known immune responses, and are capable of eradicating host lymphocytes and stem cells as well as malignant hematopoetic cells in the recipient (Barrett and Malkovska 1996; McSweeney et al. 2001). In summary, although an allogeneic transplant from a normal donor cannot in itself explain the remarkable clinical outcome reported by Hütter and colleagues, it is possible that transplantation with HIV-resistant cells, myeloablative therapy, anti-T-cell therapy, and alloimmune responses could all have contributed to the observed long-term control of HIV (Hütter et al. 2009; Allers et al. 2010; Hutter and Thiel 2011).

GENE THERAPY STRATEGIES TO REDUCE CCR5 EXPRESSION

As noted previously, there are a number of gene therapy approaches to inhibit CCR5 expression with a goal of mimicking the CCR5 null phenotype of a ccr5Δ32 homozygote (Nazari and Joshi 2008). These can act at the RNA level with RNA interference (Anderson et al. 2003; Qin et al. 2003; Anderson and Akkina 2005) or ribozymes (Bai et al. 2000, 2001; Cordelier et al. 2004), or at the protein level with CCR5-targeted intrabodies (Cordelier et al. 2004; Swan et al. 2006) or intrakines (Yang et al. 1997; Schroers et al. 2002). Improvements in humanized mouse models that support the generation of human T cells in vivo (Denton and Garcia 2009) are permitting analyses of their relative efficacies. Recent reports have highlighted the potential of RNA interference to down-regulate CCR5 expression, including the possibility of exploiting in vivo delivery of siRNAs through the use of T-cell-targeted

nanoparticles (Kumar et al. 2008; Kim et al. 2009). Alternatively, RNA interference can be achieved through the stable expression of shRNAs targeting CCR5 from lentiviral vectors. Transduction of such vectors into human CD34[+] HSC allowed HIV resistance to be conferred on both macrophages derived in vitro from the transduced cells (Liang et al. 2010) as well as T-cell progeny that differentiated in vivo in a bone marrow/liver/thymus (BLT) mouse model (Shimizu et al. 2010). A targeted strategy to deliver lentiviral vectors expressing an anti-CCR5 shRNA specifically to CCR5[+] cells in vivo was also shown using a PBMC transplanted mouse (Anderson et al. 2009) and in nonhuman primates after stem cell transplants (An et al. 2007).

Aside from these encouraging results in humanized mouse and nonhuman primate models, the utility of genetic approaches to treat HIV infection will ultimately require evaluation in patients. Given the chronic but typically manageable nature of HIV infection with ART, the clinical and ethical criteria for patient selection in gene therapy trials will need to be carefully considered. Notably, patients who develop AIDS-related malignancies, particularly lymphomas, a subset of which will require autologous HSC transplants, represent a unique cohort in which gene therapy approaches can be evaluated, because their HSCs are mobilized and harvested before chemotherapy. This provides both an opportunity to engineer HSCs to be resistant to HIV infection and to increase the chances of engraftment of the modified cells following a myeloablative conditioning regimen (DiGiusto et al. 2010). Moreover, as noted above, anti-lymphoma chemotherapy in this setting may have the added benefit of targeting HIV reservoirs.

GENE-EDITING STRATEGIES TO DISRUPT CCR5

All of the strategies described above have the challenge of achieving and maintaining sufficient levels of anti-CCR5 activity to decrease CCR5 expression to levels that are insufficient for HIV infection. In addition, in the case of lentiviral vector-based approaches, there remains the theoretical risk of these integrating vectors causing insertional mutagenesis in host DNA. More recently, gene-editing approaches to achieve permanent CCR5 gene disruption have been described using ZFNs that do not require continuous expression of a therapeutic gene.

ZFNs are engineered fusion proteins that contain two linked domains: a DNA-binding zinc finger protein and the endonuclease domain of a type-1 restriction enzyme (Fig. 1) (Urnov et al. 2010). ZFNs have been designed to target many genes, including CCR5 and CXCR4, and are currently being evaluated in preclinical and clinical settings as HIV therapeutics (Lombardo et al. 2007; Perez et al. 2008; Holt et al. 2010; Wilen et al. 2011). The zinc finger protein domain is an artificial array of zinc finger peptides that confer sequence-specific DNA-binding properties. This occurs because three to four residues toward the tip of each zinc finger peptide make contact with three or four base pairs of DNA. Altering the contact residues changes the specificity of the zinc finger for a DNA sequence. Moreover, linking multiple zinc fingers extends the length of the DNA sequence that is recognized, and because ZFNs are dimerized via their endonuclease domain, this further extends the length and the specificity of a targeted sequence. In this way, a ZFN pair engineered to bind to the sense and the antisense regions of a gene can target a sequence that is theoretically unique in the human genome, although the potential for off-target effects is always a consideration.

On binding, ZFNs act as designer restriction enzymes, cutting both strands of DNA at the bound target sequence. Following cleavage, the double-stranded break is repaired in

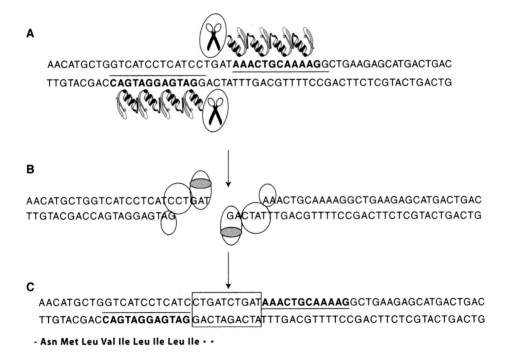

Figure 1. Zinc finger nucleases. (*A*) ZFNs bind CCR5 target sequence (underlined), and create a five-base-pair double-stranded staggered cut. (*B*) The double-stranded break is repaired by the error-prone nonhomologous end-joining (NHEJ) repair pathway. (*C*) The NHEJ results in a variety of insertions or deletions that disrupt the open reading frame (ORF). In the case of the CCR5 ZFNs, the most common repair leads to a 5-pb duplication (boxed). This insertion introduces two adjacent stop codons in the ORF that result in premature termination. (Figure adapted from Cannon and June 2011; reprinted, with permission, from the author.)

mammalian cells, predominantly by the error-prone nonhomologous end-joining (NHEJ) pathway, which typically generates a series of small deletions or additions at the break site with disruption of the open reading frame (ORF). ZFNs that target the coding sequence for the CCR5 amino terminus have been described (Perez et al. 2008) in which a 5-bp duplication at the target site is commonly observed (Perez et al. 2008; Holt et al. 2010). This insertion changes the translational reading frame and so introduces two adjacent stop codons that result in premature termination (Fig. 1).

A key feature of ZFN gene editing is that the ZFNs are only required to be expressed during a short time window; once the double-stranded break is created, the host NHEJ repair pathway creates a permanent gene knockout that is propagated with cell division. ZFNs can be delivered to a wide variety of human cells using standard gene delivery techniques. Notably, vectors that transiently express ZFNs are effective and may even be preferable because they eliminate the risk of immunogenicity that is inherent in protein-based gene therapy approaches. Vectors currently in use include adenoviral and nonintegrating lentiviral vectors. Nucleofection of plasmid DNA or in vitro-transcribed mRNA has also been used successfully, albeit with less efficiency (Lombardo et al. 2007; Perez et al. 2008; Holt et al. 2010).

ZFNs can be directed to the mature target cells that HIV-1 infects, or HSCs that give rise to these cells. Delivery of ZFNs to peripheral T cells has been especially effective using adenovirus vectors. In preclinical studies, Perez and colleagues (Perez et al. 2008) reported that adenovirus vector delivery of a CCR5-targeted ZFN pair led to disruption of ~50% of CCR5 alleles in populations of primary human CD4$^+$ T cells. They further showed that the ZFN treatment generated HIV-resistant primary CD4$^+$ T cells that expanded stably in HIV-infected cultures for several weeks, resulting in enrichment of ZFN-generated CCR5-modified cells in the population on long-term exposure to virus (>50 d). In addition, when the cells were transplanted into an immunodeficient NOD/LtSz-scid IL2R λ null (NSG) mouse and followed by infection with a CCR5-tropic strain of HIV-1, the ZFN-modified T cells preferentially expanded, so that the proportion of modified cells present at the end of the experiment was greater than two- to threefold higher in the HIV-infected mice. Thus, a major finding from this work was that HIV infection itself can play an important role in selecting CCR5-negative cells. Even when present at low frequency, these cells are able to replenish and stabilize T-cell populations, whereas nonedited CCR5$^+$ cells are destroyed by HIV.

Disrupting the CCR5 gene in HSCs could create a more potent antiviral effect than in peripheral T cells, given the added advantage that CCR5-negative cells would be generated in all hematopoietic lineages that HIV-1 infects, including macrophages as well as T cells. However, manipulating HSCs is technically challenging, as these cells are difficult to maintain in culture without losing viability or undergoing differentiation. An additional problem is that CCR5 is not expressed in HSCs, and therefore it is not possible to positively select for CCR5-disrupted cells. An early report described using nonintegrating lentiviral vectors to deliver ZFNs to CD34$^+$ HSCs, but the efficiency was poor (Lombardo et al. 2007). However, Holt and colleagues recently showed that nucleofection of human HSCs with ZFN plasmid DNA was particularly effective with an average disruption rate of 17% of CCR5 alleles (Holt et al. 2010). Importantly, these ZFN-modified cells retained their "stemness" and were subsequently able to engraft NSG mice with the same efficiency as were nonmodified HSCs. In addition, the frequency of CCR5-disrupted genes in HSCs before engraftment was maintained in mature cells following differentiation in vivo and persisted through a secondary transplantation, clearly indicating that in SCID mice the ZFNs were able to modify HSCs and that these cells retained the capacity to regenerate all hematopoietic lineages. Moreover, similar to the study of Perez and colleagues, when mice transplanted with ZFN-modified HSCs were challenged with a

CCR5-tropic strain of HIV-1, CD4 cells that persisted in gut mucosa showed a potent selection for CCR5-negative, ZFN-modified cells (Holt et al. 2010).

CLINICAL TRIALS TESTING CELL AND GENE TRANSFER APPROACHES FOR CHRONIC HIV INFECTION

Cell-delivered gene therapy for HIV/AIDS has the potential to generate an immune system that is resistant to HIV that could theoretically contribute to the eradication of HIV reservoirs. As described in earlier sections, anti-HIV gene(s) can be introduced into HSC and/or T lymphocytes to provide a population of cells that is protected. Both HSC- and T-cell-directed approaches have shown promise. Although the potential benefits of HSC are substantial, there are two major challenges. First, the efficiencies of gene transfer with HSCs have been lower than with T-cell-directed therapies (Rossi et al. 2007; June et al. 2009), particularly regarding the ability to genetically modify long-term repopulating cells. Second, HSC appear to be more susceptible than mature T cells to genotoxicity from integrating viral vectors (Newrzela et al. 2008; Montini et al. 2009; Matrai et al. 2010). In contrast to the challenges in evaluating efficacy of anti-HIV gene therapy in HSCs, an attractive feature of gene therapy in T cells is that it is straightforward to determine therapeutic effects. Supervised treatment interruptions, if carefully performed, are safe and can provide definitive information on the antiviral efficacy of the vector by measuring changes in viral load or CD4 cell counts after the interruption. Recently, a subpopulation of postthymic T cells with extensive self-renewal capacity has been described (Zhang et al. 2005; Stemberger et al. 2007; Turtle et al. 2009), and it is possible that the generation of populations of autologous, HIV-resistant HSCs and T cells will be an efficient approach to replicate the encouraging findings with allogeneic HSC using CCR5Δ32 donor cells.

A variety of the anti-HIV strategies discussed above has been tested in tissue culture and in animal models, and some have progressed to clinical trials (Table 1). The first direct test of the safety and feasibility of CCR5 knockout T cells is currently being evaluated in a phase I clinical trial sponsored by Sangamo Biosciences (California), in conjunction with investigators at the University of Pennsylvania. Details of the protocol design can be found at clinicaltrials.gov NCT00842634. An overview of the clinical trial is shown in Figure 2.

CONCLUDING REMARKS: TOWARD HIV ERADICATION

Combination ART has dramatically improved the survival of patients with HIV infection and reduced the frequency of opportunistic infections and cancers. However, its use can be associated with significant long-term toxicities, and persisting inflammation and immune activation even in the face ART can impact clinical outcomes and survival (Butler et al. 2011). In addition, even with improvements in pharmacologic formulations of anti-HIV drugs, HIV infection continues to represent a significant economic burden, with the discounted, lifetime medical care cost of HIV-1 infection in the United States estimated at $303,100 (2005 dollars) per person with 73% of this cost attributed to ARV drugs (Schackman et al. 2006). These realities provide an additional rationale for developing eradication therapies with the goal of long-term, drug-free control of HIV-1, a concept that before the Berlin patient, was unthinkable. As described in this review, the toolbox of strategies to render hematopoietic and lymphoid systems resistant to HIV infection continues to expand and to improve. Key challenges facing this emerging field are numerous and include the development of robust assays to measure the impact of therapies on latent HIV-1 reservoirs. Another issue is whether pleuripotent stem cells or progenitor cells constitute a potential reservoir, as suggested by a recent study (Carter et al. 2010). If confirmed, this study has major implications for therapies that contemplate the use of gene-modified HSCs. Another barrier to implementing gene modification strategies is a possible requirement for myeloablative therapy to

Table 1. Completed and ongoing cell and gene transfer trials for HIV

Protocol description	Phase	Status	Payload	Cellular vehicle	Transfer vector	Reference(s)
T cells						
A randomized study of HIV-specific T-cell gene therapy in subjects with undetectable plasma viremia on combination antiretroviral therapy	2	Completed	CD4 receptor coupled with the CD3 signaling chain ζ	Autologous CD4+ and CD8+ T cells; (3) repeat doses	Murine retrovirus	Deeks et al. 2002
Evaluation of safety, tolerability, and persistence of escalating and repeat doses of genetically modified syngeneic CD8+ or CD4+/CD8+ cells	1	Completed	CD4 receptor coupled with the CD3 signaling chain ζ	Syngeneic CD8+ or CD4+/CD8+ cells; single or multiple doses	Murine retrovirus	Walker et al. 2000
Evaluation of safety, tolerability, and tissue trafficking of a single dose of genetically modified autologous CD4+ and CD8+ cells	1	Completed	CD4 receptor coupled with the CD3 signaling chain ζ	Autologous CD4+ and CD8+ cells; single dose	Murine retrovirus	Mitsuyasu et al. 2000
A clinical trial of CD4ζ gene-modified T-cell infusion with and without IL-2 in HIV-infected participants	1/2	Completed	CD4 receptor coupled with the CD3 signaling chain ζ	Autologous CD4ζ cells, randomized comparison to cells only or IL-2 only	Murine retrovirus	Aronson et al. 2008
Evaluation of safety and tolerability of a single infusion of autologous CD4+ T cells modified with a dominant negative anti-HIV gene	1	Completed	Rev M10	Autologous CD4+ cells; single dose	Gold particles	Woffendin et al. 1996
Evaluation of safety and tolerability of a single infusion of autologous retrovirally modified CD4+ T cells to express a dominant negative anti-HIV gene	1	Completed	Rev M10	Autologous CD4+ cells; single dose	Murine retrovirus	Ranga et al. 1998
A marker study of therapeutically transduced CD4+ peripheral blood lymphocytes in HIV discordant identical twins	1	Completed	Anti-HIV-1 tat ribozyme (Rz2)	Syngeneic CD4+ cells; single dose	Murine retrovirus	Macpherson et al. 2005

Description	Phase	Status	Gene/Intervention	Cells	Vector	Reference
Evaluation of safety and tolerability of multiple infusions of syngeneic CD4+ lymphocytes modified with anti-HIV genes	1	Completed	Trans-dominant rev and/or trans-dominant rev with TAR antisense	Syngeneic CD4+ cells; two doses	Murine retrovirus	Morgan et al. 2005
Evaluation of safety and tolerability of ribozyme gene therapy of HIV-1 infection	1	Completed	Anti-HIV-1 ribozyme to the U5 leader sequence and pol	Autologous CD4+ cells; single dose	Murine retrovirus	Wong-Staal et al. 1998
Evaluation of safety and tolerability of a single dose of autologous T cells transduced with VRX496 in HIV-positive patient subjects	1	Completed	Anti-HIV-1 antisense against the envelope gene	Autologous CD4+ T cells; single dose	HIV-derived lentivirus, conditionally replicating	Levine et al. 2006
Evaluation of safety, tolerability, and antiviral effects of autologous CD4+ T cells expressing the HIV fusion inhibitor M87	1	Completed	gp41 fusion peptide inhibitor	Autologous CD4+ T cells	Murine retrovirus	van Lunzen et al. 2007
An open-label, multicenter study to evaluate the safety, tolerability, and biological activity of repeated doses of autologous T cells transduced with VRX496 in HIV-positive subjects	2	Ongoing	Anti-HIV-1 antisense against the envelope gene (VRX496)	Autologous CD4+ T cells; (4 or 8) repeat doses	HIV-derived lentivirus, conditionally replicating	ClinicalTrials.gov NCT00131560
An open-label, single center study to evaluate the tolerability, trafficking, and therapeutic effects of repeated doses of autologous T cells transduced with VRX496 in HIV-infected subjects	1/2	Ongoing	Anti-HIV-1 antisense against the envelope gene (VRX496)	Autologous CD4+ T cells; (6) repeat doses	HIV-derived lentivirus, conditionally replicating	Collman et al. 2009; ClinicalTrials.gov NCT00295477
Autologous T cells genetically modified at the CCR5 gene by zinc finger nucleases SB-728 for HIV (zinc-finger)	1	Ongoing	CCR5 modfied T cells (SB 728 T)	Autologous CD4+ T cells	Ad5/35 ZFN	ClinicalTrials.gov NCT00842634
Study of autologous T cells genetically modified at the CCR5 gene by zinc finger nucleases	1/2	Ongoing	CCR5-modfied T cells (SB 728 T)	Autologous CD4+ T cells	Ad5/35 ZFN	ClinicalTrials.gov NCT01252641
Phase 1 dose escalation study of autologous T cells genetically modified at the CCR5 gene by zinc finger nucleases in HIV-infected patients	1	Ongoing	CCR5-modfied T cells (SB 728 T)	Autologous CD4+ T cells	Ad5/35 ZFN	ClinicalTrials.gov NCT01044654

Continued

Table 1. *Continued*

Protocol description	Phase	Status	Payload	Cellular vehicle	Transfer vector	Reference(s)
Redirected high-affinity gag-specific autologous T cells for HIV gene therapy	1	Ongoing	HLA-A2 gag-specific CTLs (wild-type and high-affinity TCR)	Autologous CD8$^+$ T cells	HIV-derived SIN lentivirus	ClinicalTrials.gov NCT00991224
HSCs						
Nonmyeloablative conditioning followed by transplantation of genetically modified HLA-matched peripheral blood progenitor cells for hematologic malignancies in patients with AIDS	1	Completed	*Trans*-dominant Rev	Autologous CD34$^+$ cells isolated from mobilized peripheral blood	Murine retrovirus	Kang et al. 2002; Hayakawa et al. 2009
Evaluation of retroviral-mediated transfer of a *rev*-responsive element decoy gene into CD34$^+$ cells from the bone marrow of HIV-1-infected children	1	Completed	RRE-decoy	Autologous CD34$^+$ bone marrow cells	Murine retrovirus	Kohn et al. 1999
Evaluation of safety, tolerability, and persistence of transplantation with autologous bone marrow transduced with a retroviral vector expressing dominant negative Rev or a control gene	1	Completed	Dominant negative RevM10	Autologous CD34$^+$ bone marrow cells	Murine retrovirus	Podsakoff et al. 2005

Cite this article as *Cold Spring Harb Perspect Med* doi: 10.1101/cshperspect.a007179

Tat and *tat/rev* ribozyme in autologous CD34+ cells and control vector in patients with and without ablation	1	Completed	Rev/tat ribozyme	Autologous CD34+ bone marrow cells	Murine retrovirus	Michienzi et al. 2003; Look for Rossi
Evaluation of safety and tolerability of autologous CD34+ hematopoietic progenitor cells transduced with an anti-HIV ribozyme	1	Completed	Anti-HIV-1 tat ribozyme (Rz2)	Autologous CD34+ cells isolated from mobilized peripheral blood	Murine retrovirus	Amado et al. 1999, 2004
A randomized, double-blind, controlled trial to evaluate the safety and efficacy of autologous CD34+ hematopoietic progenitor cells transduced with placebo or an anti-HIV-1 ribozyme (OZ1) in patients with HIV-1 infection	2	Completed	Anti-HIV ribozyme OZ-1	Autologous CD34+ cells isolated from mobilized peripheral blood	Murine retrovirus	Mitsuyasu et al. 2009
A pilot study of safety and feasibility of stem cell therapy for AIDS lymphoma using stem cells treated with a lentivirus vector encoding multiple anti-HIV RNAs	1	Completed	Triple combination vector coexpressing an anti-*tat/rev* shRNA, a nucleolar localizing TAR decoy, and an anti-CCR5 ribozyme in a single vector backbone	Autologous CD34+ HSCs	HIV-derived lentivirus	DiGiusto et al. 2010

CCR5 knockout in autologous T cells can be achieved by infection with an Ad5/35 vector (SB-728) expressing left and right ZFNs, linked by a self-cleaving protease 2A sequence. In an ongoing phase I clinical trial, the CCR5-modified T cells are expanded ex vivo using antibodies to CD3 and CD28 for ~10 days and adoptively transferred to the patient. The study end points are listed.

TAR, *trans*-activation response; TCR, T-cell receptor; RRE, Rev response element.

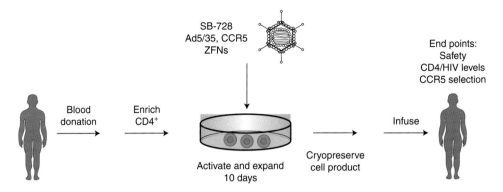

Figure 2. First clinical application of CCR5 ZFNs (protocol NCT00842634; see text).

facilitate engraftment of gene-modified HSCs. However, there is progress in the area of inducible selection systems for HSCs, for example, using modified methylguanine methyltransferase (Milsom and Williams 2007; Trobridge et al. 2009; Beard et al. 2010), and if proven safe, these approaches may circumvent this issue. Given the remarkable proof of concept that HIV infection can be eradicated (Hütter et al. 2009; Allers et al. 2010; Hutter and Thiel 2011), this field will be an exciting one for renewed research efforts.

REFERENCES

*Reference is also in this collection.

Allers K, Hütter G, Hofmann J, Loddenkemper C, Rieger K, Thiel E, Schneider T. 2010. Evidence for the cure of HIV infection by CCR5Δ32/Δ32 stem cell transplantation. *Blood* **117:** 2791–2799.

Amado RG, Mitsuyasu RT, Symonds G, Rosenblatt JD, Zack J, Sun LQ, Miller M, Ely J, Gerlach W. 1999. A phase I trial of autologous CD34+ hematopoietic progenitor cells transduced with an anti-HIV ribozyme. *Human Gene Ther* **10:** 2255–2270.

Amado RG, Mitsuyasu RT, Rosenblatt JD, Ngok FK, Bakker A, Cole S, Chorn N, Lin LS, Bristol G, Boyd MP, et al. 2004. Anti-human immunodeficiency virus hematopoietic progenitor cell-delivered ribozyme in a phase I study: Myeloid and lymphoid reconstitution in human immunodeficiency virus type-1-infected patients. *Hum Gene Ther* **15:** 251–262.

An DS, Donahue RE, Kamata M, Poon B, Metzger M, Mao SH, Bonifacino A, Krouse AE, Darlix JL, Baltimore D, et al. 2007. Stable reduction of CCR5 by RNAi through hematopoietic stem cell transplant in non-human primates. *Proc Natl Acad Sci* **104:** 13110–13115.

Anderson WF. 1984. Prospects for human gene therapy. *Science* **226:** 401–409.

Anderson J, Akkina R. 2005. HIV-1 resistance conferred by siRNA cosuppression of CXCR 4 and CCR 5 coreceptors by a bispecific lentiviral vector. *AIDS Res Theory* **2:** 1–12.

Anderson J, Banerjea A, Akkina R. 2003. Bispecific short hairpin siRNA constructs targeted to CD4, CXCR4, and CCR5 confer HIV-1 resistance. *Oligonucleotides* **13:** 303–312.

Anderson JS, Walker J, Nolta JA, Bauer G. 2009. Specific transduction of HIV-susceptible cells for CCR5 knockdown and resistance to HIV infection: A novel method for targeted gene therapy and intracellular immunization. *J Acquir Immune Defic Syndr* **52:** 152–161.

Angelucci E, Lucarelli G, Baronciani D, Durazzi SM, Galimberti M, Maddaloni D, Polchi P. 1990. Bone marrow transplantation in an HIV positive thalassemic child following therapy with azidothymidine. *Haematologica* **75:** 285–287.

Aronson N, Benstein W, Levine B, Jagodzinski L, Plesa G, Cox J, Darden J, Polonis V, Gibbs B, Flaks H, et al. 2008. A clinical trial of CD4 ζ gene-modified T cell infusion with and without IL-2 in HIV-infected participants. In *AIDS 2008 XVII International AIDS Conference*, 3–8 August 2008, Abstract TUPDA104, p. 328. Mexico City.

Arya SK, Guo C, Josephs SF, Wong-Staal F. 1985. Transactivator gene of human T-lymphotropic virus type III (HTLV-III). *Science* **229:** 69.

Avettand-Fenoel V, Mahlaoui N, Chaix ML, Milliancourt C, Burgard M, Cavazzana-Calvo M, Rouzioux C, Blanche S. 2007. Failure of bone marrow transplantation to eradicate HIV reservoir despite efficient HAART. *AIDS* **21:** 776–777.

Bai J, Gorantla S, Banda N, Cagnon L, Rossi J, Akkina R. 2000. Characterization of anti-CCR5 ribozyme-transduced CD34+ hematopoietic progenitor cells in vitro and in a SCID-hu mouse model in vivo. *Mol Ther* **1:** 244–254.

Bai J, Rossi J, Akkina R. 2001. Multivalent anti-CCR5 ribozymes for stem cell-based HIV type 1 gene therapy. *AIDS Res Hum Retrov* **17:** 385–399.

Baker JV, Peng G, Rapkin J, Abrams DI, Silverberg MJ, MacArthur RD, Cavert WP, Henry WK, Neaton JD Terry Beirn Community Programs for Clinical Research on AIDS (CPCRA). 2008. CD4+ count and risk of non-AIDS diseases following initial treatment for HIV infection. *AIDS* 22: 841–848.

Baltimore D. 1988. Gene therapy. Intracellular immunization. *Nature* 335: 395–396.

Bardini G, Re MC, Rosti G, Belardinelli AR. 1991. HIV infection and bone-marrow transplantation. *Lancet* 337: 1163–1164.

Barrett A, Malkovska V. 1996. Graft-versus-leukaemia: Understanding and using the alloimmune response to treat haematological malignancies. *Brit J Haematol* 93: 754–761.

Bauer G, Valdez P, Kearns K, Bahner I, Wen SF, Zaia JA, Kohn DB. 1997. Inhibition of human immunodeficiency virus-1 (HIV-1) replication after transduction of granulocyte colony-stimulating factor-mobilized CD34$^+$ cells from HIV-1-infected donors using retroviral vectors containing anti-HIV-1 genes. *Blood* 89: 2259–2267.

Beard BC, Trobridge GD, Ironside C, McCune JS, Adair JE, Kiem HP. 2010. Efficient and stable MGMT-mediated selection of long-term repopulating stem cells in nonhuman primates. *J Clin Invest* 120: 2345–2354.

Blankson JN, Persaud D, Siliciano RF. 2002. The challenge of viral reservoirs in HIV-1 infection. *Annu Rev Med* 53: 557–593.

Bohjanen PR, Liu Y, Garcia-Blanco MA. 1997. TAR RNA decoys inhibit tat-activated HIV-1 transcription after preinitiation complex formation. *Nucleic Acids Res* 25: 4481–4486.

Butler SL, Valdez H, Westby M, Perros M, June CH, Jacobson JM, Levy Y, Cooper DA, Douek D, Lederman MM, Tebas P. 2011. Disease-modifying therapeutic concepts for HIV in the era of highly active antiretroviral therapy. *J Acquir Immune Defic Syndr* (in press).

Cannon P, June C. 2011. Chemokine receptor 5 knockout strategies. *Curr Opin HIV AIDS* 6: 74–79.

Carter CC, Onafuwa-Nuga A, McNamara LA, Riddell JIV, Bixby D, Savona MR, Collins KL. 2010. HIV-1 infects multipotent progenitor cells causing cell death and establishing latent cellular reservoirs. *Nat Med* 16: 446–451.

Cavazzana-Calvo M, Fischer A. 2007. Gene therapy for severe combined immunodeficiency: Are we there yet? *J Clin Invest* 117: 1456–1465.

Chatterjee S, Johnson PR, Wong KKJr. 1992. Dual-target inhibition of HIV-1 in vitro by means of an adeno-associated virus antisense vector. *Science* 258: 1485–1488.

Cheung MC, Pantanowitz L, Dezube BJ. 2005. AIDS-related malignancies: Emerging challenges in the era of highly active antiretroviral therapy. *Oncologist* 10: 412.

Chun TW, Fauci AS. 1999. Latent reservoirs of HIV: Obstacles to the eradication of virus. *Proc Natl Acad Sci* 96: 10958–10961.

Collman R, Shaheen F, Boyer J, Binder G, Zifchak L, Aberra F, McGarrity G, Levine B, Tebas P, June C, et al. 2009. Safety, antiviral effects, and quantitative measurement of modified CD4 T cells trafficking to gut lymphoid tissue in a phase I/II open-label clinical trial evaluating multiple infusions of lentiviral vector-modified CD4 T cells expressing long env antisense. In *16th Conference on Retroviruses and Opportunistic Infections Abstract #83, Feb 9, 2009.* Montreal, Canada.

Cordelier P, Kulkowsky JW, Ko C, Matskevitch AA, McKee HJ, Rossi JJ, Bouhamdan M, Pomerantz RJ, Kari G, Strayer DS. 2004. Protecting from R5-tropic HIV: Individual and combined effectiveness of a hammerhead ribozyme and a single-chain Fv antibody that targets CCR5. *Gene Ther* 11: 1627–1637.

Deeks SG, McCune JM. 2010. Can HIV be cured with stem cell therapy? *Nat Biotechnol* 28: 807–810.

Deeks SG, Wagner B, Anton PA, Mitsuyasu RT, Scadden DT, Huang C, Macken C, Richman DD, Christopherson C, June CH, et al. 2002. A phase II randomized study of HIV-specific T-cell gene therapy in subjects with undetectable plasma viremia on combination anti-retroviral therapy. *Mol Ther* 5: 788–797.

Denton P, Garcia J. 2009. Novel humanized murine models for HIV research. *Curr HIV/AIDS Rep* 6: 13–19.

DiGiusto DL, Krishnan A, Li L, Li H, Li S, Rao A, Mi S, Yam P, Stinson S, Kalos M, et al. 2010. RNA-based gene therapy for HIV with Lentiviral vector–modified CD34$^+$ cells in patients undergoing transplantation for AIDS-related lymphoma. *Sci Transl Med* 2: 36–43.

Dropulic B, June CH. 2006. Gene-based immunotherapy for human immunodeficiency virus infection and acquired immunodeficiency syndrome. *Hum Gene Ther* 17: 577–588.

Dropulić B, Heermánková M, Pitha PM. 1996. A conditionally replicating HIV-1 vector interferes with wild-type HIV-1 replication and spread. *Proc Natl Acad Sci* 93: 11103–11108.

Edelstein ML, Abedi MR, Wixon J. 2007. Gene therapy clinical trials worldwide to 2007—An update. *J Gene Med* 9: 833–842.

Egelhofer M, Brandenburg G, Martinius H, Schult-Dietrich P, Melikyan G, Kunert R, Baum C, Choi I, Alexandrov A, von Laer D, et al. 2004. Inhibition of human immunodeficiency virus type 1 entry in cells expressing gp41-derived peptides. *J Virol* 78: 568–575.

Friedmann T. 1992. A brief history of gene therapy. *Nat Genet* 2: 93–98.

Friedmann T, Roblin R. 1972. Gene therapy for human genetic disease? *Science* 175: 949–955.

Geng EH, Deeks SG. 2009. CD4$^+$T cell recovery with antiretroviral therapy: More than the sum of the parts. *Clin Infect Dis* 48: 362–364.

Giorgi JV, Hultin LE, McKeating JA, Johnson TD, Owens B, Jacobson LP, Shih R, Lewis J, Wiley DJ, Phair JP, et al. 1999. Shorter survival in advanced human immunodeficiency virus type 1 infection is more closely associated with T lymphocyte activation than with plasma virus burden or virus chemokine coreceptor usage. *J Infect Dis* 179: 859–870.

Giri N, Vowels MR, Ziegler JB. 1992. Failure of allogeneic bone marrow transplantation to benefit HIV infection. *J Paediatr Child Health* 28: 331–333.

Gulick RM, Lalezari J, Goodrich J, Clumeck N, DeJesus E, Horban A, Nadler J, Clotet B, Karlsson A, Wohlfeiler M, et al. 2008. Maraviroc for previously treated patients with R5 HIV-1 infection. *N Engl J Med* 359: 1429–1441.

Han Y, Wind-Rotolo M, Yang HC, Siliciano JD, Siliciano RF. 2007. Experimental approaches to the study of HIV-1 latency. *Nat Rev Microbiol* 5: 95–106.

Hannon G, Rossi J. 2004. Unlocking the potential of the human genome with RNA interference. *Nature* **431:** 371–378.

Hassett J, Zaroulis C, Greenberg ML, Siegal FP, et al. 1983. Bone-marrow transplantation in AIDS. *N Engl J Med* **309:** 665–665.

Hayakawa J, Washington K, Uchida N, Phang O, Kang EM, Hsieh MM, Tisdale JF. 2009. Long-term vector integration site analysis following retroviral mediated gene transfer to hematopoietic stem cells for the treatment of HIV infection. *PLoS One* **4:** e4211. doi: 10.1371/journal.pone.0004211.

Held DM, Kissel JD, Patterson JT, Nickens DG, Burke DH. 2006. HIV-1 inactivation by nucleic acid aptamers. *Front Biosci* **11:** 89–112.

Held DM, Kissel JD, Thacker SJ, Michalowski D, Saran D, Ji J, Hardy RW, Rossi JJ, Burke DH. 2007. Cross-clade inhibition of recombinant human immunodeficiency virus type 1 (HIV-1), HIV-2, and simian immunodeficiency virus SIVcpz reverse transcriptases by RNA pseudoknot aptamers. *J Virol* **81:** 5375–5384.

Hildinger M, Dittmar MT, Schult-Dietrich P, Fehse B, Schnierle BS, Thaler S, Stiegler G, Welker R, von Laer D. 2001. Membrane-anchored peptide inhibits human immunodeficiency virus entry. *J Virol* **75:** 3038–3042.

Holland HK, Saral R, Rossi JJ, Donnenberg AD, Burns WH, Beschorner WE, Farzadegan H, Jones RJ, Quinnan GV, Vogelsang GB. 1989. Allogeneic bone marrow transplantation, zidovudine, and human immunodeficiency virus type 1 (HIV-1) infection. Studies in a patient with non-Hodgkin lymphoma. *Ann Intern Med* **111:** 973–981.

Holt N, Wang J, Kim K, Friedman G, Wang X, Taupin V, Crooks GM, Kohn DB, Gregory PD, Holmes MC, et al. 2010. Human hematopoietic stem/progenitor cells modified by zinc-finger nucleases targeted to CCR5 control HIV-1 in vivo. *NatBiotechnol* **28:** 839–847.

Huang Y, Paxton WA, Wolinsky SM, Neumann AU, Zhang L, He T, Kang S, Ceradini D, Jin Z, Yazdanbakhsh K, et al. 1996. The role of a mutant CCR5 allele in HIV-1 transmission and disease progression. *Nat Med* **2:** 1240–1243.

Hutter G, Thiel E. 2011. Allogeneic transplantation of CCR5-deficient progenitor cells in a patient with HIV infection: An update after 3 years and the search for patient no. 2. *AIDS* **25:** 273–274.

Hütter G, Nowak D, Mossner M, Ganepola S, Müssig A, Allers K, Schneider T, Hofmann J, Kücherer C, Blau O, et al. 2009. Long-term control of HIV by CCR5 Δ32/Δ32 stem-cell transplantation. *N Engl J Med* **360:** 692–696.

Joshi PJ, Fisher TS, Prasad VR. 2003. Anti-HIV inhibitors based on nucleic acids: Emergence of aptamers as potent antivirals. *Curr Drug Targets Infect Disord* **3:** 383–400.

June CH, Blazar BR, Riley JL. 2009. Engineering lymphocyte subsets: Tools, trials and tribulations. *Nat Rev Immunol* **9:** 704–716.

Kamp C, Wolf T, Bravo IG, Kraus B, Krause B, Neumann B, Winskowsky G, Thielen A, Werner A, Schnierle BS. 2010. Decreased HIV diversity after allogeneic stem cell transplantation of an HIV-1 infected patient: A case report. *Virol J* **7:** 55.

Kang EM, de Witte M, Malech H, Morgan RA, Phang S, Carter C, Leitman SF, Childs R, Barrett AJ, Little R, et al.

2002. Nonmyeloablative conditioning followed by transplantation of genetically modified HLA-matched peripheral blood progenitor cells for hematologic malignancies in patients with acquired immunodeficiency syndrome. *Blood* **99:** 698–701.

Kaufmann GR, Zaunders JJ, Cunningham P, Kelleher AD, Grey P, Smith D, Carr A, Cooper DA. 2000. Rapid restoration of CD4 T cell subsets in subjects receiving antiretroviral therapy during primary HIV-1 infection. *AIDS* **14:** 2643–2651.

Kelley CF, Kitchen CM, Hunt PW, Rodriguez B, Hecht FM, Kitahata M, Crane HM, Willig J, Mugavero M, Saag M, et al. 2009. Incomplete peripheral CD4$^+$ cell count restoration in HIV-infected patients receiving long-term antiretroviral treatment. *Clin Infect Dis* **48:** 787–794.

Kim SS, Peer D, Kumar P, Subramanya S, Wu H, Asthana D, Habiro K, Yang YG, Manjunath N, Shimaoka M, et al. 2009. RNAi-mediated CCR5 silencing by LFA-1-targeted nanoparticles prevents HIV infection in BLT mice. *Mol Ther* **18:** 370–376.

Kimpel J, Braun SE, Qiu G, Wong FE, Conolle M, Schmitz JE, Brendel C, Humeau LM, Dropulic B, Rossi JJ, et al. 2010. Survival of the fittest: Positive selection of CD4$^+$ T cells expressing a membrane-bound fusion inhibitor following HIV-1 infection. *PLoS One* **5:** 378–384.

Kissel JD, Held DM, Hardy RW, Burke DH. 2007. Active site binding and sequence requirements for inhibition of HIV-1 reverse transcriptase by the RT1 family of single-stranded DNA aptamers. *Nucleic Acids Res* **35:** 5039–5050.

Kitahata MM, Gange SJ, Abraham AG, Merriman B, Saag MS, Justice AC, Hogg RS, Deeks SG, Eron JJ, Brooks JT, et al. 2009. Effect of early versus deferred antiretroviral therapy for HIV on survival. *N Engl J Med* **360:** 1815–1826.

Kohn DB, Bauer G, Rice CR, Rothschild JC, Carbonaro DA, Valdez P, Hao Q, Zhou C, Bahner I, Kearns K, et al. 1999. A clinical trial of retroviral-mediated transfer of a rev-responsive element decoy gene into CD34$^+$ cells from the bone marrow of human immunodeficiency virus-1-infected children. *Blood* **94:** 368–371.

Kumar P, Ban HS, Kim SS, Wu H, Pearson T, Greiner DL, Laouar A, Yao J, Haridas V, Habiro K, et al. 2008. T cell-specific siRNA delivery suppresses HIV-1 infection in humanized mice. *Cell* **134:** 577–586.

Lane HC, Masur H, Longo DL, Klein HG, Rook AH, Quinnan GVJr, Steis RG, Macher A, Whalen G, Edgar LC. 1984. Partial immune reconstitution in a patient with the acquired immunodeficiency syndrome. *N Engl J Med* **311:** 1099–1103.

Lares MR, Rossi JJ, Ouellet DL. 2010. RNAi and small interfering RNAs in human disease therapeutic applications. *Trends Biotechnol* **28:** 570–579.

Leavitt MC, Yu M, Yamada O, Kraus G, Looney D, Poeschla E, Wong-Staal F. 1994. Transfer of an anti-HIV-1 ribozyme gene into primary human lymphocytes. *Hum Gene Ther* **5:** 1115–1120.

Lee SW, Gallardo HF, Gilboa E, Smith C. 1994. Inhibition of human immunodeficiency virus type 1 in human T cells by a potent Rev response element decoy consisting of the 13-nucleotide minimal Rev-binding domain. *J Virol* **68:** 8254–8264.

Lee NS, Dohjima T, Bauer G, Li H, Li MJ, Ehsani A, Salva-terra P, Rossi J. 2002. Expression of small interfering RNAs targeted against HIV-1 rev transcripts in human cells. *Nat Biotechnol* **20:** 500–505.

Lekakis J, Ikonomidis I. 2010. Cardiovascular complications of AIDS. *Curr Opin Crit Care* **16:** 408–412.

Levine BL, Humeau LM, Boyer J, MacGregor RR, Rebello T, Lu X, Binder GK, Slepushkin V, Lemiale F, Mascola JR, et al. 2006. Gene transfer in humans using a conditionally replicating lentiviral vector. *Proc Natl Acad Sci* **103:** 17372–17377.

Lewden C, Salmon D, Morlat P, Bévilacqua S, Jougla E, Bon-net F, Héripret L, Costagliola D, May T, Chène G. Mortal-ity 2000 Study Group. 2005. Causes of death among human immunodeficiency virus (HIV)-infected adults in the era of potent antiretroviral therapy: emerging role of hepatitis and cancers, persistent role of AIDS. *Int J Epidemiol* **34:** 121–130.

Lewden C, Chene G, Morlat P, Raffi F, Dupon M, Dellamon-ica P, Pellegrin JL, Katlama C, Dabis F, Leport C, Agence Nationale de Recherches sur le Sida et les Hepatites Vir-ales. 2007. HIV-infected adults with a CD4 cell count greater than 500 cells/mm^3 on long-term combination antiretroviral therapy reach same mortality rates as the general population. *J Acquir Immune Defic Syndr* **46:** 72–77.

Li MJ, Kim J, Li S, Zaia J, Yee JK, Anderson J, Akkina R, Rossi JJ. 2005. Long-term inhibition of HIV-1 infection in pri-mary hematopoietic cells by lentiviral vector delivery of a triple combination of anti-HIV shRNA, anti-CCR5 ribo-zyme, and a nucleolar-localizing TAR decoy. *Mol Ther* **12:** 900–909.

Liang M, Kamata M, Chen KN, Pariente N, An DS, Chen IS. 2010. Inhibition of HIV-1 infection by a unique short hairpin RNA to chemokine receptor 5 delivered into macrophages through hematopoietic progenitor cell transduction. *J Gene Med* **12:** 255–265.

Lisziewicz J, Sun D, Smythe J, Lusso P, Lori F, Louie A, Mark-ham P, Rossi J, Reitz M, Gallo RC. 1993. Inhibition of human immunodeficiency virus type 1 replication by regulated expression of a polymeric Tat activation response RNA decoy as a strategy for gene therapy in AIDS. *Proc Natl Acad Sci* **90:** 8000–8004.

Lo AS, Zhu Q, Marasco WA. 2008. Intracellular antibodies (intrabodies) and their therapeutic potential. In *Thera-peutic antibodies* (ed. Chernajovsky Y, Nissim A), Vol. 181, pp. 343–373. Springer-Verlag, Berlin.

Lobato M, Rabbitts T. 2004. Intracellular antibodies as spe-cific reagents for functional ablation: Future therapeutic molecules. *Curr Mol Med* **4:** 519–528.

Lohse N, Hansen AB, Pedersen G, Kronborg G, Gerstoft J, Sørensen HT, Vaeth M, Obel N. 2007. Survival of persons with and without HIV infection in Denmark, 1995–2005. *Ann Intern Med* **146:** 87–95.

Lombardo A, Genovese P, Beausejour CM, Colleoni S, Lee YL, Kim KA, Ando D, Urnov FD, Galli C, Gregory PD, et al. 2007. Gene editing in human stem cells using zinc finger nucleases and integrase-defective lentiviral vector delivery. *Nat Biotechnol* **25:** 1298–1306.

Lu X, Yu Q, Binder GK, Chen Z, Slepushkina T, Rossi J, Dropulic B. 2004. Antisense-mediated inhibition of human immunodeficiency virus (HIV) replication by

use of an HIV type 1-based vector results in severely atte-nuated mutants incapable of developing resistance. *J Virol* **78:** 7079–7088.

Maartens G, Boulle A, Mocroft A, Phillips AN, Gatell J, Ledergerber B, Fisher M, Clumeck N, Losso M, Lazzarin A, et al. 2007. Normalisation of CD4 counts in patients with HIV-1 infection and maximum virological suppres-sion who are taking combination antiretroviral therapy: An observational cohort study. Commentary. *Lancet* **370:** 407–413.

Macpherson JL, Boyd MP, Arndt AJ, Todd AV, Fanning GC, Ely JA, Elliott F, Knop A, Raponi M, Murray J, et al. 2005. Long-term survival and concomitant gene expres-sion of ribozyme-transduced CD4$^+$ T-lymphocytes in HIV-infected patients. *J.Gene Med* **7:** 552–564.

* Malim MH, Bieniasz PD. 2011. HIV restriction factors and mechanisms of evasion. *Cold Spring Harb Pespect Med* doi: 10.1101/cshperspect.a006940.

Malim MH, Freimuth WW, Liu J, Boyle TJ, Lyerly HK, Cullen BR, Nabel GJ, et al. 1992. Stable expression of transdominant Rev protein in human T cells inhibits human immunodeficiency virus replication. *J Exp Med* **176:** 1197–1201.

Mátrai JM, Chuah MK, VandenDriessche T. 2010. Recent advances in lentiviral vector development and applica-tions. *Mol Ther* **18:** 477–490.

McSweeney P, Niederwieser D, Shizuru JA, Sandmaier BM, Molina AJ, Maloney DG, Chauncey TR, Gooley TA, Hegenbart U, Nash RA, et al. 2001. Hematopoietic cell transplantation in older patients with hematologic malignancies: Replacing high-dose cytotoxic therapy with graft-versus-tumor effects. *Blood* **97:** 3390–3400.

Michaels MG, Kaufman C, Volberding PA, Gupta P, Switzer WM, Heneine W, Sandstrom P, Kaplan L, Swift P, Damon L, et al. 2004. Baboon bone-marrow xenotransplant in a patient with advanced HIV disease: Case report and 8-year follow-up. *Transplantation* **78:** 1582–1589.

Michienzi A, Castanotto D, Lee N, LI S, Zaia JA, Rossi JJ. 2003. RNA-mediated inhibition of HIV in a gene therapy setting. *Ann NY Acad Sci* **1002:** 63–71.

Mitsuyasu R, Merigan T, Carr A, Zack JA, Winters MA, Workman C, Bloch M, Lalezari J, Becker S, Thornton L, et al. 2009. Phase 2 gene therapy trial of an anti-HIV ribo-zyme in autologous CD34$^+$ cells. *Nat Med* **15:** 285–292.

Milsom MD, Williams DA. 2007. Live and let die: In vivo selection of gene-modified hematopoietic stem cells via MGMT-mediated chemoprotection. *DNA Repair* **6:** 1210–1221.

Mitsuyasu RT, Anton PA, Deeks SG, Scadden DT, Connick E, Downs MT, Bakker A, Roberts MR, June CH, Jalali S, et al. 2000. Prolonged survival and tissue trafficking following adoptive transfer of CD4ζ gene-modified autologous CD4$^+$ and CD8$^+$ T cells in human immuno-deficiency virus-infected subjects. *Blood* **96:** 785–793.

Montini E, Cesana D, Schmidt M, Sanvito F, Bartholomae CC, Ranzani M, Benedicenti F, Sergi Sergi L, Ambrosi A, Ponzoni M, et al. 2009. The genotoxic potential of ret-roviral vectors is strongly modulated by vector design and integration site selection in a mouse model of HSC gene therapy. *J Clin Invest* **119:** 964–975.

Moore R, Keruly J, Gebo K, Lucas G. 2005. An improvement in virologic response to highly active antiretroviral therapy in clinical practice from 1996 through 2002. *J Acquir Immune Defic Syndr* **39:** 195–198.

Morgan RA, Walker R, Carter CS, Natarajan V, Tavel JA, Bechtel C, Herpin B, Muul L, Zheng Z, Jagannatha S, et al. 2005. Preferential survival of CD4[+] T lymphocytes engineered with anti-human immunodeficiency virus (HIV) genes in HIV-infected individuals. *Hum Gene Ther* **16:** 1065–1074.

Nazari R, Joshi S. 2008. CCR5 as target for HIV-1 gene therapy. *Curr Gene Ther* **8:** 264–272.

Newrzela S, Cornils K, Li Z, Baum C, Brugman MH, Hartmann M, Meyer J, Hartmann S, Hansmann ML, Fehse B, et al. 2008. Resistance of mature T cells to oncogene transformation. *Blood* **112:** 2278–2286.

Ngok FK, Mitsuyasu RT, Macpherson JL, Boyd MP, Symonds GP, Amado RG. 2004. Clinical gene therapy research utilizing ribozymes: Application to the treatment of HIV/AIDS. *Methods Mol Biol* **252:** 581–598.

Nimjee SM, Rusconi CP, Sullenger BA. 2005. Aptamers: An emerging class of therapeutics. *Medicine* **56:** 555–583.

Nisole S, Stoye JP, Saïb A. 2005. TRIM family proteins: Retroviral restriction and antiviral defence. *Nat Rev Microbiol* **3:** 799–808.

Nunez M. 2010. Clinical syndromes and consequences of antiretroviral-related hepatotoxicity. *Hepatology* **52:** 1143–1155.

Perez EE, Riley JL, Carroll RG, von Laer D, June CH. 2005. Suppression of HIV-1 infection in primary CD4 T cells transduced with a self-inactivating lentiviral vector encoding a membrane expressed gp41-derived fusion inhibitor. *Clin Immunol* **115:** 26–32.

Perez EE, Wang J, Miller JC, Jouvenot Y, Kim KA, Liu O, Wang N, Lee G, Bartsevich VV, Lee YL, et al. 2008. Establishment of HIV-1 resistance in CD4[+] T cells by genome editing using zinc-finger nucleases. *Nat Biotechnol* **26:** 808–816.

Podsakoff GM, Engel BC, Carbonaro DA, Choi C, Smogorzewska EM, Bauer G, Selander D, Csik S, Wilson K, Betts MR, et al. 2005. Selective survival of peripheral blood lymphocytes in children with HIV-1 following delivery of an anti-HIV gene to bone marrow CD34[+] cells. *Mol Ther* **12:** 77–86.

Polizzotto MN, Skinner M, Cole-Sinclair MF, Opat SS, Spencer A, Avery S. 2010. Allo-SCT for hematological malignancies in the setting of HIV. *Bone Marrow Transpl* **45:** 584–586.

Poluri A, van Maanen M, Sutton RE. 2003. Genetic therapy for HIV/AIDS. *Expert Opin Biol Ther* **3:** 951–963.

Qin XF, An DS, Chen IS, Baltimore D. 2003. Inhibiting HIV-1 infection in human T cells by lentiviral-mediated delivery of small interfering RNA against CCR5. *Proc Natl Acad Sci* **100:** 183–188.

Ranga U, Woffendin C, Verma S, Xu L, June CH, Bishop DK, Nabel GJ. 1998. Retroviral delivery of an antiviral gene in HIV-infected individuals. *Proc Natl Acad Sci* **95:** 1201–1206.

Rebellato LM, Gross U, Verbanac KM, Thomas JM. 1994. A comprehensive definition of the major antibody specificities in polyclonal rabbit antithymocyte globulin. *Transplantation* **57:** 685–694.

Richman DD, Margolis DM, Delaney M, Greene WC, Hazuda D, Pomerantz RJ. 2009. The challenge of finding a cure for HIV infection. *Science* **323:** 1304–1307.

Rodger AJ, Fox Z, Lundgren JD, Kuller LH, Boesecke C, Gey D, Skoutelis A, Goetz MB, Phillips AN. INSIGHT Strategies for Management of Antiretroviral Therapy (SMART) Study Group. 2009. Activation and coagulation biomarkers are independent predictors for the development of opportunistic disease in patients with HIV infection. *J Infect Dis* **200:** 973–983.

Rossi JJ, June CH, Kohn DB, et al. 2007. Genetic therapies for HIV/AIDS. *Nat Biotechnol* **25:** 1444–1454.

Samson M, Libert F, Doranz BJ, Rucker J, Liesnard C, Farber CM, Saragosti S, Lapoumeroulie C, Cognaux J, Forceille C, et al. 1996. Resistance to HIV-1 infection in caucasian individuals bearing mutant alleles of the CCR-5 chemokine receptor gene [see comments]. *Nature* **382:** 722–725.

Sarver N, Cantin EM, Chang PS, Zaia JA, Ladne PA, Stephens DA, Rossi JJ. 1990. Ribozymes as potential anti-HIV-1 therapeutic agents. *Science* **247:** 1222–1225.

Schackman BR, Gebo KA, Walensky RP, Losina E, Muccio T, Sax PE, Weinstein MC, Seage GR III, Moore RD, Freedberg KA. 2006. The lifetime cost of current human immunodeficiency virus care in the United States. *Med Care* **44:** 990–997.

Schlegel P, Beatty P, Halvorsen R, McCune J. 2000. Successful allogeneic bone marrow transplant in an HIV-1-positive man with chronic myelogenous leukemia. *J Acquir Immune Defic Syndr* **24:** 289–290.

Schroers R, Davis CM, Wagner HJ, Chen SY. 2002. Lentiviral transduction of human T-lymphocytes with a RANTES intrakine inhibits human immunodeficiency virus type 1 infection. *Gene Ther* **9:** 889–897.

Shimizu S, Hong P, Arumugam B, Pokomo L, Boyer J, Koizumi N, Kittipongdaja P, Chen A, Bristol G, Galic Z, et al. 2010. A highly efficient short hairpin RNA potently down-regulates CCR5 expression in systemic lymphoid organs in the hu-BLT mouse model. *Blood* **115:** 1534–1544.

* Siliciano RF, Greene WC. 2011. HIV latency. *Cold Spring Harb Perspect Med* doi: 10.1101/cshperspect.a007096.

Sodroski J, Rosen C, Goh WC, Haseltine W. 1985. A transcriptional activator protein encoded by the x-lor region of the human T-cell leukemia virus. *Science* **228:** 1430.

Stemberger C, Huster KM, Koffler M, Anderl F, Schiemann M, Wagner H, Busch DH. 2007. A single naïve CD8[+] T cell precursor can develop into diverse effector and memory subsets. *Immunity* **27:** 985–997.

Strayer DS, Akkina R, Bunnell BA, Dropulic B, Planelles V, Pomerantz RJ, Rossi JJ, Zaia JA. 2005. Current status of gene therapy strategies to treat HIV/AIDS. *Mol Ther* **11:** 823–842.

Stremlau M, Owens CM, Perron MJ, Kiessling M, Autissier P, Sodroski J. 2004. The cytoplasmic body component TRIM5α restricts HIV-1 infection in Old World monkeys. *Nature* **427:** 848–853.

Stremlau M, Perron M, Lee M, Li Y, Song B, Javanbakht H, Diaz-Griffero F, Anderson DJ, Sundquist WI, Sodroski J. 2006. Specific recognition and accelerated uncoating of retroviral capsids by the TRIM5α restriction factor. *Proc Natl Acad Sci* **103:** 5514–5519.

Cite this article as *Cold Spring Harb Perspect Med* doi: 10.1101/cshperspect.a007179

Swan CH, Bühler B, Steinberger P, Tschan MP, Barbas CF III, Torbett BE. 2006. T-cell protection and enrichment through lentiviral CCR5 intrabody gene delivery. *Gene Ther* **13**: 1480–1492.

Symensma TL, Giver L, Zapp M, Takle GB, Ellington AD. 1996. RNA aptamers selected to bind human immunodeficiency virus type 1 Rev in vitro are Rev responsive in vivo. *J Virol* **70**: 179–187.

Tebas P, Stein D, Zifchak L, Seda A, Binder G, Aberra F, Collman R, McGarrity G, Levine B, June C, et al. 2010. Prolonged control of viremia after transfer of autologous CD4 T cells genetically modified with a lentiviral vector expressing long antisense to HIV env (VRX496). In *17th Conference of Retroviruses and Opportunistic Infections*. San Francisco.

Tomonari A, Takahashi S, Shimohakamada Y, Ooi J, Takasugi K, Ohno N, Konuma T, Uchimaru K, Tojo A, Odawara T, et al. 2005. Unrelated cord blood transplantation for a human immunodeficiency virus-1-seropositive patient with acute lymphoblastic leukemia. *Bone Marrow Transpl* **36**: 261–262.

Trobridge GD, Wu RA, Beard BC, Chiu SY, Muñoz NM, von Laer D, Rossi JJ, Kiem HP. 2009. Protection of stem cell-derived lymphocytes in a primate AIDS gene therapy model after in vivo selection. *PLoS One* **4**: e7693. doi: 10.1371/journal.pone.0007693.

Turtle CJ, Swanson HM, Fujii N, Estey EH, Riddell SR. 2009. A distinct subset of self-renewing human memory CD8[+] T cells survives cytotoxic chemotherapy. *Immunity* **31**: 834–844.

Urnov FD, Rebar EJ, Holmes MC, Zhang HS, Gregory PD. 2010. Genome editing with engineered zinc finger nucleases. *Nat Rev Genet* **11**: 636–646.

Valdez H, Connick E, Smith KY, Lederman MM, Bosch RJ, Kim RS, St Clair M, Kuritzkes DR, Kessler H, Fox L, AIDS Clinical Trials Group Protocol 375 Team, et al. 2002. Limited immune restoration after 3 years' suppression of HIV-1 replication in patients with moderately advanced disease. *AIDS* **16**: 1859–1866.

van Lunzen J, Glaunsinger T, Stahmer I, von Baehr V, Baum C, Schilz A, Kuehlcke K, Naundorf S, Martinius H, Hermann F, et al. 2007. Transfer of autologous gene-modified T cells in HIV-infected patients with advanced immunodeficiency and drug-resistant virus. *Mol Ther* **15**: 1024–1033.

von Laer D, Hasselmann S, Hasselmann K.. 2006. Impact of gene-modified T cells on HIV infection dynamics. *J Theor Biol* **238**: 60–77.

Walker RE, Bechtel CM, Natarajan V, Baseler M, Hege KM, Metcalf JA, Stevens R, Hazen A, Blaese RM, Chen CC, et al. 2000. Long-term in vivo survival of receptor-modified syngeneic T cells in patients with human immunodeficiency virus infection. *Blood* **96**: 467–474.

Wang GP, Levine BL, Binder GK, Berry CC, Malani N, McGarrity G, Tebas P, June CH, Bushman FD. 2009. Analysis of lentiviral vector integration in HIV[+] study subjects receiving autologous infusions of gene modified CD4[+] T cells. *Mol Ther* **17**: 844–850.

Wilen CB, Wang J, Tilton JC, Miller JC, Kim KA, Rebar EJ, Sherrill-Mix SA, Patro SC, Secreto AJ, Jordan AP, et al. 2011. Engineering HIV-resistant human CD4[+] T cells with CXCR4-specific zinc-finger nucleases. *PLoS Pathog* **7**: e1002020. doi: 10.1371/journal.ppat.1002020.

Woffendin C, Ranga U, Yang Z, Xu L, Nabel GJ. 1996. Expression of a protective gene prolongs survival of T cells in human immunodeficiency virus-infected patients. *Proc Natl Acad Sci* **93**: 2889–2894.

Wolf T, Rickerts V, Staszewski S, Kriener S, Wassmann B, Bug G, Bickel M, Gute P, Brodt HR, Martin H. 2007. First case of successful allogeneic stem cell transplantation in an HIV-patient who acquired severe aplastic anemia. *Haematologica* **92**: e56–e58.

Wong-Staal F, Poeschla EM, Looney DJ. 1998. A controlled, phase 1 clinical trial to evaluate the safety and effects in HIV-1 infected humans of autologous lymphocytes transduced with a ribozyme that cleaves HIV-1 RNA. *Hum Gene Ther* **9**: 2407–2425.

Woolfrey AE, Malhotra U, Harrington RD, McNevin J, Manley TJ, Riddell SR, Coombs RW, Appelbaum FR, Corey L, Storb R. 2008. Generation of HIV-1-specific CD8[+] cell responses following allogeneic hematopoietic cell transplantation. *Blood* **112**: 3484–3487.

Xia C, Luo D, Yu X, Jiang S, Liu S. 2011. HIV-associated dementia in the era of highly active antiretroviral therapy (HAART). *Microbes Infect* **13**: 419–425.

Yang AG, Bai X, Huang XF, Yao C, Chen S. 1997. Phenotypic knockout of HIV type 1 chemokine coreceptor CCR-5 by intrakines as potential therapeutic approach for HIV-1 infection. *Proc Natl Acad Sci* **94**: 11567.

Yap MW, Nisole S, Stoye JP. 2005. A single amino acid change in the SPRY domain of human Trim5 αleads to HIV-1 restriction. *Curr Biol* **15**: 73–78.

Zhang Y, Joe G, Hexner E, Zhu J, Emerson SG. 2005. Host-reactive CD8[+] memory stem cells in graft-versus-host disease. *Nat Med* **11**: 1299–1305.

Zhang J, Wu YO, Xiao L, Li K, Chen LL, Sirois P. 2007. Therapeutic potential of RNA interference against cellular targets of HIV infection. *Mol Biotechnol* **37**: 225–236.

Zhang JC, Sun L, Nie QH, Huang CX, Jia ZS, Wang JP, Lian JQ, Li XH, Wang PZ, Zhang Y, et al. 2009. Downregulation of CXCR4 expression by SDF-KDEL in CD34[+] hematopoietic stem cells: An anti-human immunodeficiency virus strategy. *J Virol Methods* **161**: 30–37.

Zimmerman PA, Buckler-White A, Alkhatib G, Spalding T, Kubofcik J, Combadiere C, Weissman D, Cohen O, Rubbert A, Lam G, et al. 1997. Inherited resistance to HIV-1 conferred by an inactivating mutation in CC chemokine receptor 5: Studies in populations with contrasting clinical phenotypes, defined racial background, and quantified risk. *Mol Med* **3**: 23–36.

The HIV-1 Epidemic: Low- to Middle-Income Countries

Yiming Shao[1] and Carolyn Williamson[2]

[1]The State Key Laboratory for Infectious Disease Control and Prevention, National Center for AIDS/STD Control and Prevention, Chinese Center for Disease Control and Prevention, Changping District, Beijing 102206, China

[2]Institute of Infectious Diseases and Molecular Medicine, Division of Medical Virology, University of Cape Town and National Health Laboratory Service, 7925 South Africa

Correspondence: carolyn.williamson@uct.ac.za

Low- to middle-income countries bear the overwhelming burden of the human immunodeficiency virus type 1 (HIV-1) epidemic in terms of the numbers of their citizens living with HIV/AIDS (acquired immunodeficiency syndrome), the high degrees of viral diversity often involving multiple HIV-1 clades circulating within their populations, and the social and economic factors that compromise current control measures. Distinct epidemics have emerged in different geographical areas. These epidemics differ in their severity, the population groups they affect, their associated risk behaviors, and the viral strains that drive them. In addition to inflicting great human cost, the high burden of HIV infection has a major impact on the social and economic development of many low- to middle-income countries. Furthermore, the high degrees of viral diversity associated with multiclade HIV epidemics impacts viral diagnosis and pathogenicity and treatment and poses daunting challenges for effective vaccine development.

Although the first cases of acquired immunodeficiency syndrome (AIDS) were identified in North America in 1981, molecular epidemiology studies have placed the geographical origin of HIV-1 in west-central equatorial Africa (Keele et al. 2006). Today, Africa still shoulders the greatest burden of the epidemic, harboring >68% of all infections despite accommodating only 13% of the world's population (UNAIDS 2010). The last five years have seen the epidemic reaching a peak in Africa, Latin America, and the Caribbean, and most parts of Asia. However, rates of transmission are still rising in Eastern Europe and Central Asia (UNAIDS 2010). In this work, we describe what is known about the emergence of HIV-1 in low- to middle-income countries, focusing specifically on the countries in sub-Saharan Africa, Asia, Latin America, and the Caribbean that are most affected by the epidemic. We provide an overview of the current status of the epidemic, its molecular epidemiology, and the impact of HIV diversity on disease progression, treatment, and vaccine development.

EPIDEMIOLOGICAL HISTORY

The first cases of AIDS in Africa were described in the early 1980s where it affected heterosexual populations in a band of equatorial countries including Zaire (now Democratic Republic of Congo), Rwanda, and Uganda (Piot et al. 1984; van de Perre et al. 1984; Serwadda et al. 1985). It was initially known as "slim disease" because it was associated with diarrhea and weight loss. The spread of HIV-1 to southern African countries occurred later where detection in the general population of, for example, South Africa was only observed in the late 1980s (Gouws et al. 2010).

Whereas the epidemic in Africa has always been driven by heterosexual transmission, in other regions such as Latin America and the Caribbean, the HIV epidemic was also associated with injection drug use (IDU) and men who have sex with men (MSM). In these regions, the HIV epidemic started in the late 1970s and early 1980s. By the year 2000, the Caribbean had the second highest prevalence of HIV-1-infected adult population outside sub-Saharan Africa (PAHO, WHO, UNAIDS 2001).

In Asia, HIV/AIDS was first reported in Thailand, India, and China around the mid-to late 1980s (Phanuphak et al. 1985; Zeng et al. 1986; Simoes et al. 1987). Since the early 1990s, HIV prevalences in excess of 10% were documented among female sex workers (FSW) in Thailand and India, as well as among IDUs in countries near the "golden triangle" region (where the borders of Thailand, Myanmar, and Laos meet) (Ma et al. 1990; Estebanez et al. 1993; Dore et al. 1996; Crofts et al. 1998; WHO 1998). The epidemic then spread to more countries and populations in Asia through IDU, heterosexual transmission, and, at a later time, through MSM contacts, as well as mother-to-child transmission.

CURRENT STATUS

The brunt of the epidemic globally is carried by the low- to middle-income countries, which provide the virus with a fertile environment to spread owing to a combination of factors including poor socioeconomic conditions, lack of access to health care, economic or political displacement of communities, and, to a certain extent, cultural practices and gender inequalities. The highest HIV prevalence in the adult population aged 15–49 yr is within sub-Saharan Africa (5%), followed by the Caribbean (1%), Eastern Europe and Central Asia (0.8%), Central and South America (0.5%), South and Southeast Asia (0.3%), and East Asia (0.1%) (Fig. 1 and Table 1) (UNAIDS 2010). Staggeringly, 91% of the world's total HIV-1-infected population, numbering 30 million people (UNAIDS 2010), resided in low- to middle-income countries in 2009. Being hit the hardest by the epidemic is sub-Saharan Africa, which harbors about three-quarters of the world's HIV-1-infected individuals from low- and middle-income countries and more than two-thirds of the world's total (Table 1). Other regions with high HIV/AIDS burdens include South and Southeast Asia, Eastern Europe and Central Asia, and Central and South America, which, respectively, account for approximately 12%, 4%, and 4% of the world's HIV-1-infected individuals (Table 1).

Low- to middle-income countries also carry the heaviest burden of drug use–related HIV infections. The three regions with both the highest numbers and the highest prevalence of HIV-positive IDUs are all low- to middle-income regions: Latin America, with 580,000 HIV-infected IDUs (a prevalence of 29%), Eastern Europe with 940,000 HIV-infected IDUs (a prevalence of 27%), and East and Southeast Asia with 661,000 HIV-infected IDUs (a prevalence of 17%) (Table 2). These associations suggest that the fight against drug trafficking and abuse in these regions is inextricably linked with the fight against HIV.

Trends in Africa

The epidemic in sub-Saharan Africa is devastating. In 2009, approximately 22.5 million Africans were living with HIV-1, with the worst affected region being southern Africa where about one-third of the world's HIV-infected

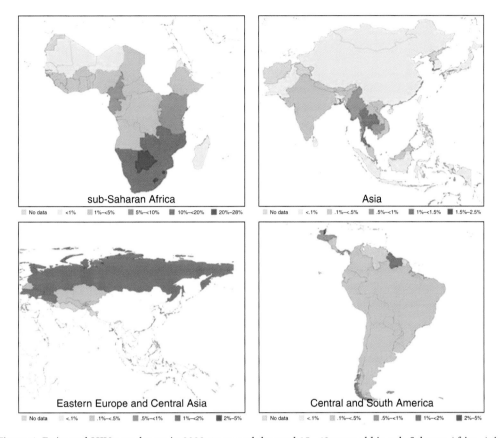

Figure 1. Estimated HIV prevalences in 2009 among adults aged 15–49 years old in sub-Saharan Africa, Asia, Eastern Europe, and Central Asia. (Source: UNAIDS.)

people reside (UNAIDS 2010). There are huge country-to-country variations in the severity of the epidemic. Whereas most southern African countries (Zambia, Zimbabwe, Malawi, Mozambique, Namibia, Botswana, Lesotho, South Africa, and Swaziland) have an adult (15–49 yr) prevalence of >10%, those in East and Central Africa (Uganda, Kenya, Tanzania, Cameroon, and Rwanda) have less severe epidemics, with adult HIV prevalences ranging between 5% and 10%. Most West African countries have adult prevalences below 1%. Although the worst affected country globally is Swaziland, where approximately 26% of adults are infected with the virus, South Africa is the country with the greatest number of individuals living with HIV: an estimated 5.6 million. In the countries worst affected by the epidemic, life expectancy has been reduced by

as much as 20 years and there has been a concomitant reduction in income and an increase in household poverty (UNAIDS 2010).

The social, economic, and political factors responsible for the widespread epidemic in southern Africa include poverty, poor social and health infrastructure, lack of education, political and social instability, the low status of women, and sexual violence, as well as lack of political commitment, slow control response, and ineffective preventive measures during critical periods of HIV epidemic in the region. There is no single factor that accounts for the high HIV prevalences observed in certain countries in Africa. Behavioral factors are likely to have contributed, including the number of casual sexual partners, condom usage, age of sexual debut, intergenerational sex, concurrency, and sexual networks (Gregson et al. 2002; Abdool

Table 1. HIV/AIDS statistics for selected regions in 2009

Regions	Estimated people living with HIV (percent of world total)	People newly infected with HIV	Estimated percent adult prevalence (15–49)	Estimated AIDS-related deaths (percent of world total)
Sub-Saharan Africa	22.5 million (68%)	1.8 million (69%)	5.0	1.3 million (72%)
South and Southeast Asia	4.1 million (12%)	270,000 (10%)	0.3	260,000 (14%)
East Asia	770,000 (2.3%)	82,000 (3%)	0.1	36,000 (2.0%)
Eastern Europe and Central Asia	1.4 million (4.2%)	130,000 (5%)	0.8	76,000 (4.2%)
Caribbean	240,000 (0.7%)	17,000 (0.7%)	1.0	12,000 (0.7%)
Central and South America	1.4 million (4.2%)	92,000 (3.5%)	0.5	58,000 (3.2%)
Subtotal	30.41 million (92%)	2.39 million (92%)		1.74 million (97%)
World total	33.3 million	2.6 million		1.8 million

Data from UNAIDS (2010), with modification.

Karim et al 2010; UNAIDS 2010; Ott et al. 2011; Steffenson et al. 2011). Differences in male circumcision practices may also play a role as circumcision has been associated with a ~60% reduction in risk of infection, and lack of circumcision in some communities may put these men at increased risk of infection (Auvert et al. 2005; Gray et al. 2007). The epidemic in Africa is fueled by genetically diverse viral genotypes, which may differ in terms of pathogenicity and rates of transmission (Kaleebu et al. 2002). Early HIV infection is associated with high viral loads, and in a region in southern Africa, nearly 40% of transmissions were attributed to this phase of infection (Powers et al. 2011). Lastly, host genetics affects susceptibility to infection (Pereyra et al. 2010). It is likely that a combination of these factors has contributed to the varying intensity of the epidemic in Africa; however, the relative contribution that each of these factors makes is unknown.

The Africa HIV/AIDS epidemic appears to have peaked in the mid-1990s, and over the last 10 years there has been a decrease in the number of new infections (estimated to be 1.8 million in 2009 compared with 2.2 million in 1999) (UNAIDS 2010). Although this decrease is thought to be attributable, in part, to the natural course of the HIV epidemic, it is probably also attributable to significant changes in human behavior. Specifically, in 13 sub-Saharan African countries, the age of sexual debut and condom usage has significantly increased, and the number of sexual partners has significantly decreased (UNAIDS 2010). Despite declining infection rates, the massive increase in access to antiretrovirals is expected to result in people living longer with HIV and consequently, to an increase in the number of HIV-1-infected individuals.

In Africa, women are disproportionally affected by the epidemic and accordingly account for 60% of infections. Women are eight times more likely to become infected than their male counterparts (Powers et al. 2008). Although part of this discrepancy is attributable to women being biologically more susceptible to infection, women are also more vulnerable to infection owing to both gender-associated violence and sociocultural factors that restrict their economic empowerment and access to health care. Along with large numbers of HIV-1-infected women come large numbers of infected babies: There are currently 2.3 million children living with HIV in sub-Saharan Africa.

Table 2. Regional and global estimates of the number of IDUs and the number of HIV-positive IDUs in 2007

Region	Estimated number of IDUs		Estimated number of HIV-positive IDUs		Estimated HIV prevalence of IDUs
	Median number	%	Median number	%	%
Latin America	2,018,000	13	580,500	19	29
Eastern Europe	3,476,500	22	940,000	31	27
East and Southeast Asia	3,957,500	25	661,000	22	17
Canada and USA	2,270,500	14	347,000	12	15
South Asia	569,500	3.6	74,500	2.5	13
Caribbean	186,000	1.2	24,000	0.8	13
Sub-Saharan Africa	1,778,500	11	221,000	7.4	12
Central Asia	247,500	1.6	29,000	1.0	12
Western Europe	1,044,000	6.6	114,000	3.8	11
Middle East and North Africa	121,000	0.8	3500	0.1	2.9
Pacific Island states and territories	19,500	0.1	500	0.02	2.6
Australia and New Zealand	173,500	1.1	2500	0.1	1.4
Extrapolated global estimates	15,862,000	100.00	2,997,500	100.00	19

Data from Mathers et al. (2008), with modification.

Trends in Asia

In many Asian countries, although new HIV infections peaked around the year 2000 and have declined in the FSW and IDU populations, the initially hidden HIV epidemic in the MSM population has experienced a steady increase (Chan et al. 1998; Colby et al. 2003; Girault et al. 2004; UNAIDS 2010). Consequently, very high HIV prevalences and incidences in MSM have now been reached in many Asian countries, such as Cambodia (9%), Myanmar (29%), India (5%–25%), China (1%–9%, incidence 3.6/100 py), and Thailand (17%–31%, incidence 5.7/100 py) (Ruan et al. 2007; UNAIDS 2010; van Griensven et al. 2010). In China, the proportion of MSM in the annual reported HIV cases increased from 12% in 2007 to 33% in 2009 (Mi et al. 2010).

Because of cultural and social pressures, many MSM in Asia are married or otherwise engaged with female partners so as to avoid social stigma and discrimination. For example, approximately one-quarter of the Chinese MSM are married (Tang et al. 2008), and ∼30% of these individuals have sex with a steady female partner (Ruan et al. 2007). In a Bangkok survey of 1121 MSM, 22% reported having sex with both men and women during

the six months before the survey. Of these men, 36% had engaged in unprotected sex and 17% were HIV positive (van Griensven et al. 2005). This characteristic of Asian MSM behavior has blurred the line between what, in other parts of the world, are usually distinct HIV-1 epidemics within the MSM community and the general population. MSM in Asia is serving as a bridging group to effectively transmit HIV from high-risk groups to the general population.

IDU is a major route of HIV-1 transmission in this region. Afghanistan and Myanmar are the world's largest opiate- and heroin-producing areas. Whereas one-third of the heroin produced in Afghanistan reaches Europe, one-quarter goes to Central Asia and the Russian Federation. The rest is trafficked to other South and Southeast Asia countries. Most of the heroin produced in Myanmar supplies the local and regional markets, including China, South and Southeast Asia, and Oceania.

Trends in Latin America and the Caribbean

The HIV prevalence is generally around 1% in the Caribbean with the notable exception of Cuba, which has a very low prevalence of 0.1% (UNAIDS 2010). In Latin America and the

Caribbean, the HIV cases are mostly distributed within discreet MSM, IDU, and FSW networks. The current HIV prevalence in the MSM community remains extremely high in countries such as Peru (9.8%–22.3%), Argentina (14%), Uruguay (22%), Ecuador (15%), and Columbia (10%–25%) (UNAIDS 2008; Cáceres and Mendoza 2009). The world's largest cocaine-producing area is located in the Andean region, covering Columbia, Peru, and Bolivia, and HIV transmission follows the drug-trafficking routes. High HIV prevalences were found in IDUs in Peru (9.8%–22.3%), Brazil (8%–65%), Argentina (14%–80%), and Uruguay (24%–76%) (Hacker et al. 2005; Cáceres and Mendoza 2009).

MOLECULAR EPIDEMIOLOGY

During the 100 years since HIV-1 was first transmitted to humans, its evolution has resulted in a highly divergent epidemic. The pandemic today is caused by HIV group M viruses, and currently circulating viruses can be categorized into nine distinct clusters, referred to as subtypes or clades (subtypes A–D, F–H, J, and K), two of which have evolved further into distinct "sub-subtype" lineages known as A1 and A2 for subtype A viruses, and F1 and F2 for subtype F viruses. In addition, lineages have evolved comprising viruses that have been derived through genetic recombination between viruses of different subtypes, which arose when individuals were simultaneously infected with viruses belonging to two different subtypes. Whereas countless unique recombinant viruses have emerged in areas where two or more subtypes cocirculate, when recombinant viruses are discovered to have spread to form their own subepidemics, they are referred to as circulating recombinant forms. There are currently 48 circulating recombinant forms, or CRFs named CRF01 through CRF48 (Los Alamos HIV Sequence Database 2010).

The global spread of HIV has yielded a situation today where the different HIV subtypes and CRFs often have quite distinct geographical distributions. These distributions, along with the numbers of people infected with the different HIV-1M lineages, are illustrated in Figure 2. Equatorial West Africa, the site where HIV has presumably been evolving in humans for the longest period, is also unsurprisingly at the global epicenter of HIV diversity, and here an extraordinary pool of diverse forms exists (Carr et al. 2010).

It is apparent that the global epidemic has been seeded by only a handful of these lineages (Fig. 2). Analysis of HIV-1 sequences is an extremely effective tool for tracking this historical dissemination. Also, HIV evolves rapidly enough that the timescales of viral spread can be traced using the genomic sequences of viruses sampled from the epidemic. Outside of central equatorial Africa, founder events where viruses have moved into new transmission networks and rapidly spread have played a major role in shaping the current diversity of the epidemic (Tebit et al. 2010). The genetic bottlenecks that occur when an entire subepidemic is initiated by a single virus being introduced have resulted in relatively low viral diversities in some regions. There are multiple examples of founder effects, with subtype C being the most successful, globally having founded epidemics in South Africa, India, Ethiopia, and South America (Hermelaar et al. 2006). Other founder effects are seen in Thailand, with subtype B IDU epidemic and CRF01_AE heterosexual epidemic (Korber et al. 2000), in Russia and Eastern Europe with subtype A IDU epidemic (Bobkov et al. 2001), and in China with CRF07_B'C IDU epidemic (Shao et al. 1999; Piyasirisilp et al. 2000; Su et al. 2000; Tee et al. 2008).

Globally, subtype C is now the most successful of the HIV-1M lineages and today accounts for >50% of infections, whereas subtypes A and B each account for over 10% of worldwide HIV infections. Subtypes D and G, CRF01_AE, and CRF02_AG account for only between 2% and 6% each. Subtypes F, H, J, K, other CRFs, and all other unclassified recombinant forms individually make only a minor contribution to the global HIV population (<1% each) but together account for the remaining 15% of worldwide HIV infections (Hemelaar et al. 2006). The regional distribution of

Figure 2. Phylogenetic trees illustrating the diversity of HIV-1 in Africa, Asia, and Latin America with the number of sequences included within each phylogeny roughly proportional to the percentage of subtypes contributing to each epidemic (as described by Hemelaar et al. 2006). The southern African epidemic is illustrated separately from the rest of Africa. Each sequence in the African phylogenies represents 250,000 infections, each sequence in the Asian phylogeny represents 200,000 infections, and each sequence in the Latin American phylogeny represents 26,000 infections. HIV-1 gp160 sequences from Los Alamos HIV database were used (http://www.hiv.lanl.gov/) and branch-length scale is in expected nucleotide substitutions per site. The figure was generated by Nobubelo Ngandu, University of Cape Town.

different subtypes and CRFs is illustrated in Figure 3.

Tracking the Spread of HIV-1 in Africa

The oldest HIV-1M sequences that have been analyzed were sampled in 1959 and 1960 from individuals in Central Africa (specifically in Kinshasa in the Democratic Republic of Congo [or DRC]) (Worobey et al. 2008) and appear to be quite closely related to ancestral subtype A and D viruses. Although all HIV-1M subtypes have been identified in the DRC and Cameroon, subtype A dominates in the DRC, and CRF02_AG dominates in Cameroon (Carr et al. 2010). This AG recombinant, and subtypes A1 and G, are very successful in Western Africa where they today collectively dominate the epidemics in Nigeria, Ghana, Senegal, Mali, and Cote d'Ivoire (Hemelaar et al. 2006).

Outside of West Africa, East Africa has the oldest epidemic that is driven primarily by infections with subtypes A and D and their derived recombinants. Subtype A was likely to have entered East Africa some time shortly after 1950, with subtype D being introduced about ten years later (Vidal et al. 2000; Gray et al. 2009). Today, subtype A dominates in Kenya (Dowling et al. 2002), whereas both subtypes A and D have emerged in Uganda (Herbeck et al. 2007). Tanzania, which lies between the subtype A/D epidemic on its northern border and the subtype C epidemic on its southern border, is a melting pot of subtype A, C, and, to a lesser extent, D infections and various recombinant forms. Subtype C accounts for almost all infections in southern Africa.

The Spread of HIV to South America

In South America, subtype B was initially introduced into MSM transmission networks from Haiti and North America (see Vermund and Leigh-Brown 2011), with subtypes C and F1, and recombinants BC and BF1, becoming more dominant later. Although initially it was thought that the South American subtype C epidemic was founded by viruses from Burundi, a country that does not share social, cultural, or economic relations with any South American country, it was later found to be linked to a small subtype C epidemic in the United Kingdom, a country which shares strong ties with Brazil (de Oliveira et al. 2010). Subtype F was thought to be introduced into South America in the late 1970s, and a BF1 recombinant rapidly spread, becoming the dominant virus in Argentina (Aulicino et al. 2007).

The Spread of HIV to Asia and the Pacific Rim

In Asia, there are four major HIV lineages, including subtypes C and B, CRF01_AE, and CRF07 and 08, which are B/C recombinants (Figs. 2 and 4). Subtype C accounts for close to 97% of infections in India, whereas in Southeast Asia, CRF01_AE accounts for more than 90% of the infections in Thailand, Vietnam, Cambodia, and Indonesia. In China, CRF07/08_B'C, CRF01_AE, and subtype B' viruses are in circulation, whereas in West Asia (mainly Russia and former Soviet states) subtypes A, B, C, and various recombinants all account for similar numbers of infections. Exceptional epidemics in the Asian region occur in the Philippines, where subtype B dominates, and in Pakistan through Central Asia, where subtype A is dominant (Fig. 4).

The Chinese national HIV molecular epidemiology surveys provided an opportunity to view the evolution of the HIV-1 subtypes throughout the country. They showed that the initial epidemic in IDUs in the Yunnan Province bordering Myanmar was dominated by a well-defined lineage of subtype B viruses (known as either the Thai B lineage or subtype B') (Ma et al.1990; Shao et al. 1994). This subtype B' was also responsible for a severe epidemic among commercial plasma donors in central China in the mid-1990s, triggered by unsafe practices at that time (Li et al. 1997; Cui et al. 2004). In 1992, subtype C viruses, most likely originating from northern India, were detected in IDUs in the Yunnan (Luo et al.1995; Yu et al. 1998; Zhao et al. 1999). In the early 1990s, the Chinese subtype C and subtype B' lineages recombined with one another to produce CRF07_B'C and 08_B'C (Shao et al. 1999;

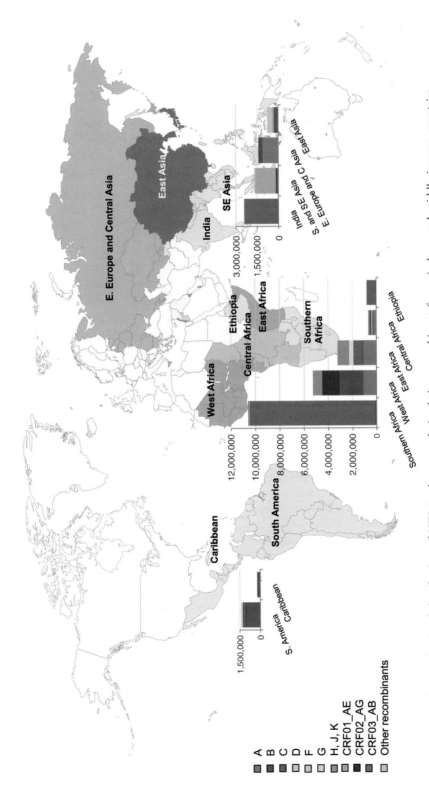

Figure 3. Regional distribution of HIV-1 subtypes and circulating recombinant forms in low- and middle-income countries. Regions comprising different countries are colored in different shades of gray. The size of the bar is proportional to the size of the epidemic (UNAIDS 2010) with the proportion of the subtypes contributing to each epidemic (as calculated by Hemelaar et al. 2006) illustrated as a percentage of the bar.

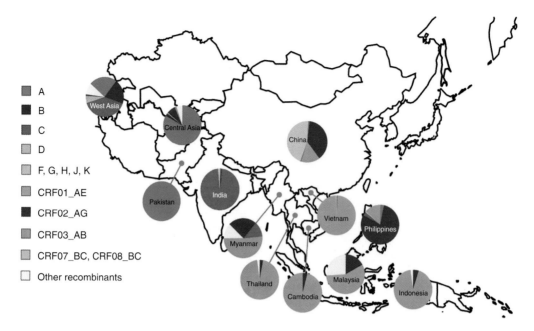

Figure 4. Regional distribution of HIV-1 subtypes and recombinants in Asia. The data of Malaysia, Indonesia, Philippines, Pakistan, Central Asia (including Afghanistan, Kazakstan, and Uzbekistan), and West Asia (including Saudi Arabia, Iran, and Israel) are from the Los Alamos HIV databases (http://www.hiv.lanl.gov/). The data of China, Japan, India, Myanmar, Vietnam, and Thailand are from former studies in the region and estimates based on the unpublished data of research projects (Y Shao, pers. comm.).

Piyasirisilp et al. 2000; Su et al. 2000; Tee et al. 2008). These recombinants were also transmitted among IDUs from Yunnan northward to Sichuan, Qinghai, and Xinjiang region, and eastward to Guangxi Province (Shao et al. 1999; Piyasirisilp et al. 2000; Su et al. 2000). Although CRF01_AE was initially spread in South and Eastern China primarily through heterosexual contacts (Xing et al. 2002; Zhang et al. 2006), it later became one of the dominant HIV-1 strains circulating within the Chinese MSM population living with HIV (Zhang et al. 2007). Within China today, CRF01_AE, CRF07_B′C, and subtype B are the major lineages, with subtype B decreasing in prominence within this epidemic (Teng et al. 2011).

HIV PATHOGENESIS IN LOW- TO MIDDLE-INCOME COUNTRIES

It is difficult to directly compare disease progression profiles between lower- and higher-income countries because of the confounding influences of country-to-country differences in both the subtype makeup of circulating viruses, host population differences related to age distributions, gender-dependent infection biases, modes of transmission, and host genetics. Additional confounding factors are the increased frequency in lower-income countries of mixed infections of HIV with other diseases such as tuberculosis, together with decreased access to basic health care such as antibiotics for treating opportunistic infections.

There have been only a few studies that have sought to determine whether there exist subtype-dependent differences in rates of disease progression. In East Africa, where subtypes A and D are cocirculating, studies have shown that individuals infected with subtype D viruses tended to progress to AIDS faster than those infected with subtype A viruses (Kaleebu et al. 2002; Kiwanuka et al. 2008). For example, a study by Kiwanuka et al. (2008), involving 350 Ugandan HIV-1 seroconverters infected with subtype A (15%), D (59%), and recombinant forms

(21%), found that whereas individuals infected with recombinant viruses had a median time to AIDS of 5.6 years, those infected with subtype D and A viruses had a median time to AIDS of 6.5 and 8 years, respectively. Progression to AIDS has been associated with coreceptor switch from CCR5 (R5) to CXCR4 (X4), and one possible explanation for subtype D viruses being more virulent than subtype A viruses is that subtype D viruses are more likely to use CXCR4 as a coreceptor for cellular entry.

In regions of the world where the incidence of HIV is high, individuals are at increased risk of being infected with multiple divergent HIV-1 strains (i.e., diverse members of either the same or different subtypes) (Herbinger et al. 2006). Both coinfection and superinfection have been associated with higher viral loads and increased rates of disease progression (Gottlieb et al. 2004; Grobler et al. 2004). Similarly, in Tanzania where up to 24% of HIV-positive individuals may be infected with multiple divergent HIV lineages, multiple infections were associated with increased viral loads (Saathoff et al. 2010).

HIV GENETIC DIVERSITY AND DRUG RESISTANCE

The development of highly effective antiretroviral drugs over the past two decades has saved millions of lives; however, the continuing evolution of resistance to these drugs among circulating HIV variants is of concern and presents a new global challenge. In low- to middle-income countries, alteration of drug combinations to combat multidrug resistance is often not an available treatment option. It is therefore important in these countries that effective HIV drug resistance (HIVDR) surveillance infrastructure is set up, to combat the emergence of HIV lineages that are resistant to multiple antiretrovirals (Bennett et al. 2008).

Surveillance of Transmitted HIVDR in Antiretroviral Treatment-Naïve HIV-1-Infected Individuals

With increasing numbers of people on antiretroviral therapy, there is also an increased probability of viruses both developing drug

resistance mutations and these mutant viruses being transmitted. Individuals infected with viruses carrying multiple drug resistance mutations may have reduced survival prospects, and for this reason, estimates of the rate at which drug resistance mutations are transmitted are a key target metric of the World Health Organization (WHO) HIVDR threshold survey (Bennett et al. 2008). This survey has revealed that fewer than 5% of HIV transmissions in lower-income countries such as Malawi (Kamoto et al. 2008), Tanzania, Ethiopia, Swaziland, South Africa, Thailand, and Vietnam involve the transmission of viruses carrying known drug resistance mutations, whereas specific areas of Brazil and China have transmitted drug resistance mutation rates between 5% and 15% (Booth et al. 2007; Inocencio et al. 2009; Liao et al. 2010). The prevalence of transmitted HIV drug resistance (TDR) in developing countries is lower than that in developed countries, where in some countries, such as the United States, rates of transmitted drug resistance within newly diagnosed antiretroviral naïve HIV-1-infected individuals are in the range of 15% (Wheeler et al. 2010). This difference reflects the fact that individuals in higher-income countries have had a longer history of antiretroviral therapy that initially involved single or dual drug treatments (as opposed to the triple drug treatments that are currently preferred).

TDR viruses from higher-income countries may also be spreading to the lower- to middle-income countries. TDR viruses with multidrug resistances to the second-line drugs had been detected in treatment-naïve MSM in Beijing in 2006, before China's national treatment program starting the second-line drug treatment. All of these MSMs with multidrug resistances carried the North American prototype B HIV-1, not the local prevalent B′, CRF01_AE and CRF07_B′C viruses (Zhang et al. 2007).

HIVDR in Different Subtypes or CRFs

The probability of drug resistance mutations arising during antiretroviral treatment increases primarily with the duration of the treatment, and there appears to be no difference between lower- and higher-income countries with respect to the rate at which resistance mutations arise in individuals treated with the same drugs (reviewed by Ferradini et al. 2006; Hamers et al. 2008; Kouanfack et al. 2009).

Although many of the drug resistance mutations have been identified in subtype B viruses (the best studied group in this regard) (Kantor et al. 2005), there also exist numerous examples of subtype-specific differences in the spectrum of resistance mutations that naturally arise in response to common ARVs. The reason for this is that evolution of drug resistance often involves more than just a single mutation, and usually a pathway to drug resistance is required whereby accessory mutations enable the emergence of primary resistance mutation by compensating for the damaging effects that resistant mutations may have on viral fitness. For example, subtype C and CRF07_B′C viruses are predisposed to rapidly acquire the V106M resistance mutation, a well-known efavirenz drug resistance mutation, as the genetic barrier to these viruses acquiring this mutation is substantially lower (need only one nucleotide change) compared with viruses belonging to other subtypes and CRFs (Liao et al. 2007). Because of the similar preexisting background mutations, subtype C viruses more rapidly developed K65R mutation than other subtypes when exposed to tenofovir (TDF) (Brenner et al. 2006; Doualla-Bell et al. 2006). In some cases the primary drug resistance mutations themselves are already naturally present at relatively high frequencies. For example, the M46I nelfinavir resistance mutation is naturally present in 0.6%–1.0% of CRF01_AE isolates, and this drug should therefore not be widely used on people infected with such viruses (Shafer et al. 2008).

CONSEQUENCES OF HIV GENETIC DIVERSITY ON VACCINE DEVELOPMENT

A globally effective vaccine would need to protect against the huge variation experienced in different regions of the world, where viruses may differ by as much as 30% in their envelope protein sequences (Fig. 2) (Gaschen et al. 2002).

Even a vaccine designed to cover a specific region (e.g., a subtype C-specific vaccine in southern Africa) would need to be effective against viruses that are up to 20% different from one another.

The induction of antiviral neutralizing antibodies is usually the primary aim of vaccines in that this provides the first line of defense blocking incoming virus particles from entering susceptible cells. In early HIV infection, these neutralizing antibodies, directed at the viral Env protein, are very specific and generally only recognize the autologous viruses within that person. However, after many years of infection, some people develop antibodies that recognize a broader range of heterologous viruses, including those that are very different from those which initially generated the antibody response. The production of such broadly acting antibodies is referred to as a broadly cross-reactive neutralizing response (see Overbaugh and Morris 2011). These cross-neutralizing antibodies tend to recognize a diverse array of subtypes, although there is an association between neutralization phenotype and the subtype that induced these antibodies, suggesting that neutralizing epitopes are more conserved within a subtype (Seaman et al. 2010). One of the biggest challenges of current HIV vaccine research is the design of an immunogen which will elicit neutralizing antibodies in vaccinated individuals, which will protect them from infection with the entire range of circulating HIV-1M lineages (Fig. 2) (Kwong et al. 2011).

As an alternative to producing vaccines that will provide antibody-based protection from infection, much effort is being focused on designing vaccines that induce the cellular immune system to attack and kill HIV-infected cells. During normal HIV infections it is the cytotoxic T lymphocytes (CTLs) of the cellular immune system that, at least temporarily, hold the virus in check during the prolonged progression to AIDS. CTLs identify infected cells through the recognition of short linear peptides (usually nine amino acids long) derived from degraded HIV proteins that are presented on the surface of infected cells by human leukocyte antigen (HLA) class I molecules.

CTL responses tend to recognize peptides matched to the subtype that elicited the response more frequently than to peptides derived from divergent viruses, suggesting that genetic diversity is likely to impact on effectiveness of vaccines (Korber et al. 2009). One strategy for producing a broadly protective CTL-based vaccine is to induce CTL responses with vaccine-derived peptides that differ as little as possible from those of circulating viruses. Probably the best developed version of this strategy to date uses synthetic proteins, called mosaic immunogens, that have been designed in silico to mimic HIV proteins and both maximize CTL epitope coverage and minimize the genetic distance between the immunogens and circulating viruses (Korber et al. 2009). Mosaic immunogens have been found to elicit CTL responses that are of greater magnitude and breadth than those elicited by naturally derived immunogens (Barouch et al. 2010; Santra et al. 2010; Koup and Douek 2011).

CONCLUDING REMARKS

Low- to middle-income countries continue to be hardest hit by the epidemic, and the health burden imposed by HIV is further exacerbated by poor socioeconomic conditions, the high prevalence of opportunistic infections, poor access to health care, and widespread drug abuse. Further challenges include the stigmatization, and even criminalization, of MSM and IDU in some African and Asian countries, respectively, which makes prevention in these groups difficult. The epidemics in Eastern Europe and Central Asia continue to increase, mainly driven through IDU, whereas in Africa, Latin America, the Caribbean, and most parts of Asia, the epidemics appear to have stabilized. However, within these stabilized epidemics are communities which have frighteningly high epidemic levels, such as urban government antenatal clinics in South Africa where as many as 60% of pregnant women were HIV-1 positive (Abdool Karim et al. 2010). The HIV epidemic in low- to middle-income countries can only be effectively controlled by combined measures, through social and economic approaches targeting the roots of the epidemic, and with various biomedical prevention, intervention, and treatment.

Molecular epidemiology is an effective tool for tracking the global spread of HIV. The genetic diversity of HIV epidemics in lower-income countries is greater than that in higher-income countries. This huge diversity, together with high transmission rates and risk of multiple HIV infections, and the occurrence and transmission of drug resistance viruses, will negatively impact pathogenesis and the sustainability of the lifelong antiviral treatment. It is also a major concern that divergent viral lineages, together with new recombinant viruses, will undermine vaccine development efforts. A vaccine against HIV-1 would be the most effective method of controlling this pandemic. Low- to middle-income countries can make a major contribution to these efforts as their high incidences of infection make them ideal locations for prevention and treatment researches, as well as for phase III HIV vaccine trials. It is hoped that closer global cooperation from upstream basic research to downstream clinical trials will greatly speed better intervention and treatment strategies as well as the ultimate production of a successful AIDS vaccine (Kaleebu et al. 2008; Kent et al. 2010).

ACKNOWLEDGMENTS

We thank Min Wei for helping look for background materials, Lingjie Liao, Xiang He, and Tao Teng for searching and arranging the references, Yi Feng and Qifne Sun for generating the tables and some of the figures, Nobubelo Ngandu for generating phylogenetic trees to illustrate global diversity, Debbie Stewart for her help with figures and database searches, and Darren Martin for review and comment.

REFERENCES

*Reference is also in this collection.

Abdool Karim Q, Kharsany AB, Frohlich JA, Werner L, Mashego M, Mlotshwa M, Madlala BT, Ntombela F, Abdool Karim SS. 2010. Stabilizing HIV prevalence masks high HIV incidence rates amongst rural and urban women in KwaZulu-Natal, South Africa. *Int J Epidemiol* doi: 10.1093/ije/dyq176.

Aulicino PC, Holmes EC, Rocco C, Mangano A, Sen L. 2007. Extremely rapid spread of human immunodeficiency virus type 1 BF recombinants in Argentina. *J Virol* **81:** 427–429.

Auvert B, Taljaard D, Lagarde E, Sobngwi-Tambekou J, Sitta R, Puren A. 2005. Randomized, controlled intervention trial of male circumcision for reduction of HIV infection risk: The ANRS 1265 trial. *PLoS Med* **2:** e298. doi: 10.1371/journal.pmed.0020298.

Barouch DH, O'Brien KL, Simmons NL, King SL, Abbink P, Maxfield LF, Sun YH, La Porte A, Riggs AM, Lynch DM, et al. 2010. Mosaic HIV-1 vaccines expand the breadth and depth of cellular immune responses in rhesus monkeys. *Nat Med* **16:** 319–323.

Bennett DE, Bertagnolio S, Sutherland D, Gilks CF. 2008. The World Health Organization's global strategy for prevention and assessment of HIV drug resistance. *Antivir Ther* **13:** 1–13.

Bobkov A, Kazennova E, Khanina T, Bobkova M, Selimova L, Kravchenko A, Pokrovsky V, Weber J. 2001. An HIV type 1 subtype A strain of low genetic diversity continues to spread among injecting drug users in Russia: Study of the new local outbreaks in Moscow and Irkutsk. *AIDS Res Hum Retroviruses* **17:** 257–261.

Booth CL, Geretti AM. 2007. Prevalence and determinants of transmitted antiretroviral drug resistance in HIV-1 infection. *J Antimicrob Chemother* **59:** 1047–1056.

Brenner BG, Oliveira M, Doualla-Bell F, Moisi DD, Ntemgwa M, Frankel F, Essex M, Wainberg MA. 2006. HIV-1 subtype C viruses rapidly develop K65R resistance to tenofovir in cell culture. *AIDS* **20:** F9–F13.

Cáceres CF, Mendoza W. 2009. The national response to the HIV/AIDS epidemic in Peru: Accomplishments and gaps—A review. *J Acquir Immune Defic Syndr* **51:** S60–S66.

Carr JK, Wolfe ND, Torimiro JN, Tamoufe U, Mpoudi-Ngole E, Eyzaguirre L, Birx DL, McCutchan FE, Burke DS. 2010. HIV-1 recombinants with multiple parental strains in low-prevalence, remote regions of Cameroon: Evolutionary relics? *Retrovirology* **7:** 39.

Chan R, Kavi AR, Carl G, Khan S, Oetomo D, Tan ML, Brown T. 1998. HIV and men who have sex with men: Perspectives from selected Asian countries. *AIDS* **12:** 59–68.

China Ministry of Health, UNAIDS, WHO. 2009. Estimates for the HIV/AIDS epidemic in China. http://www.unaids.org.cn/en/index/Document_view.asp?id=413.

Cui W, Xing H, Wang Zh, Huang H, Li H, Ma P, Xue X, Wei M, Zhu X, Shao Y. 2004. Study on subtype and sequence of the C2-V3 region of env gene among HIV-1 strains in Henan Province. *Chin J AIDS STD* **10:** 403–406.

Colby DJ. 2003. HIV knowledge and risk factors among men who have sex with men in Ho Chi Minh City, Vietnam. *J Acquir Immune Defic Syndr* **32:** 80–85.

Crofts N, Reid G, Deany P. 1998. Injecting drug use and HIV infection in Asia. *AIDS* **12:** 69–78.

De Oliveira T, Pillay D, Gifford RJ. UK Collaborative Group on HIV Drug Resistance. 2010. The HIV-1 subtype C epidemic in South America is linked to the United Kingdom. *PLoS One* **19:** 9311. doi: 10.1371/journal.pone.0009311.

Dore GJ, Kaldor JM, Ungchusak K, Mertens TE. 1996. Epidemiology of HIV and AIDS in the Asia-Pacific region. *Med J Aust* **165:** 494–498.

Doualla-Bell F, Avalos A, Brenner B, Gaolathe T, Mine M, Gaseitsiwe S, Oliveira M, Moisi D, Ndwapi N, Moffat H, et al. 2006. High prevalence of the K65R mutation in human immunodeficiency virus type 1 subtype C isolates from infected patients in Botswana treated with didanosine-based regimens. *Antimicrob Agents Chemother* **50:** 4182–4185.

Dowling WE, Kim B, Mason CJ, Wasunna KM, Alam U, Elson L, Birx DL, Robb ML, McCutchan FE, Carr JK. 2002. Forty-one near full-length HIV-1 sequences from Kenya reveal an epidemic of subtype A and A-containing recombinants. *AIDS* **16:** 1809–1820.

Estebanez P, Fitch K, Najera R. 1993. HIV and female sex workers. *Bull World Health Organ* **71:** 397–412.

Ferradini L, Jeannin A, Pinoges L, Izopet J, Odhiambo D, Mankhambo L, Karungi G, Szumilin E, Balandine S, Fedida G, et al. 2006. Scaling up of highly active antiretroviral therapy in a rural district of Malawi: An effectiveness assessment. *Lancet* **367:** 1335–1342.

Gaschen B, Taylor J, Yusim K, Foley B, Gao F, Lang D, Novitsky V, Haynes B, Hahn BH, Bhattacharya T, et al. 2002. Diversity considerations in HIV-1 vaccine selection. *Science* **296:** 2354–2360.

Girault P, Saidel T, Song N, de Lind Van Wijngaarden JW, Dallabetta G, Stuer F, Mills S, Or V, Grosjean P, Glaziou P, et al. 2004. HIV, STIs, and sexual behaviors among men who have sex with men in Phnom Penh, Cambodia. *AIDS Educ Prev* **16:** 31–44.

Gottlieb GS, Nickle DC, Jensen MA, Wong KG, Grobler J, Li F, Liu SL, Rademeyer C, Learn GH, Karim SS, et al. 2004. Dual HIV-1 infection associated with rapid disease progression. *Lancet* **363:** 619–622.

Gouws E, Abdool Karim Q. 2010. HIV infection in South Africa: The evolving epidemic. In *HIV/AIDS in South Africa*, 2nd ed., pp. 55–73. Cambridge University Press, Cambridge.

Gray RH, Kigozi G, Serwadda D, Makumbi F, Watya S, Nalugoda F, Kiwanuka N, Moulton LH, Chaudhary MA, Chen MZ, et al. 2007. Male circumcision for HIV prevention in men in Rakai, Uganda: A randomised trial. *Lancet* **369:** 657–666.

Gray RR, Tatem AJ, Lamers S, Hou W, Laeyendecker O, Serwadda D, Sewankambo N, Gray RH, Wawer M, Quinn TC, et al. 2009. Spatial phylodynamics of HIV-1 epidemic emergence in East Africa. *AIDS* **23:** F9–F17.

Gregson S, Nyamukapa CA, Garnett GP, Mason PR, Zhuwau T, Caraël M, Chandiwana SK, Anderson RM. 2002. Sexual mixing patterns and sex-differentials in teenage exposure to HIV infection in rural Zimbabwe. *Lancet* **359:** 1896–1903.

Grobler J, Gray CM, Rademeyer C, Seoighe C, Ramjee G, Karim SA, Morris L, Williamson C. 2004. Incidence of HIV-1 dual infection and its association with increased viral load set point in a cohort of HIV-1 subtype C-infected female sex workers. *J Infect Dis* **190:** 1355–1359.

Gupta RK, Chrystie IL, O'Shea S, Mullen JE, Kulasegaram R, Tong CY. 2005. K65R and Y181C are less prevalent in

HAART-experienced HIV-1 subtype A patients. *AIDS* **19:** 1916–1919.

Hacker MA, Malta M, Enriquez M, Bastos FI. 2005. Human immunodeficiency virus, AIDS, and drug consumption in South America and the Caribbean: Epidemiological evidence and initiatives to curb the epidemic. *Rev Panam Salud Publica* **18:** 303–313.

Hamers RL, Derdelinckx I, van Vugt M, Stevens W, Rinke de Wit TF, Schuurman R. 2008. The status of HIV-1 resistance to antiretroviral drugs in sub-Saharan Africa. *Antivir Ther* **13:** 625–639.

Hemelaar J, Gouws E, Ghys PD, Osmanov S. 2006. Global and regional distribution of HIV-1 genetic subtypes and recombinants in 2004. *AIDS* **20:** W13–W23.

Herbeck JT, Lyagoba F, Moore SW, Shindo N, Biryahwaho B, Kaleebu P, Mullins JI. 2007. Prevalence and genetic diversity of HIV type 1 subtypes A and D in women attending antenatal clinics in Uganda. *AIDS Res Hum Retroviruses* **23:** 755–760.

Herbinger KH, Gerhardt M, Piyasirisilp S, Mloka D, Arroyo MA, Hoffmann O, Maboko L, Birx DL, Mmbando D, McCutchan FE, et al. 2006. Frequency of HIV type 1 dual infection and HIV diversity: Analysis of low- and high-risk populations in Mbeya Region, Tanzania. *AIDS Res Hum Retroviruses* **22:** 599–606.

Inocencio LA, Pereira AA, Sucupira MC, Fernandez JC, Jorge CP, Souza DF, Fink HT, Diaz RS, Becker IM, Suffert TA, et al. 2009. Brazilian network for HIV drug resistance surveillance: A survey of individuals recently diagnosed with HIV. *J Int AIDS Soc* **12:** 20.

Kaleebu P, French N, Mahe C, Yirrell D, Watera C, Lyagoba F, Nakiyingi J, Rutebemberwa A, Morgan D, Weber J, et al. 2002. Effect of human immunodeficiency virus (HIV) type 1 envelope subtypes A and D on disease progression in a large cohort of HIV-1-positive persons in Uganda. *J Infect Dis* **185:** 1244–1250.

Kaleebu P, Abimiku A, El-Halabi S, Koulla-Shiro S, Mamotte N, Mboup S, Mugerwa R, Nkengasong J, Toure-Kane C, Tucker T, et al. 2008. African AIDS vaccine programme for a coordinated and collaborative vaccine development effort on the continent. *PLoS Med* **5:** e236. doi: 10.1371/journal.pmed.0050236.

Kamoto K, Aberle-Grasse J. 2008. Surveillance of transmitted HIV drug resistance with the World Health Organization threshold survey method in Lilongwe, Malawi. *Antivir Ther* **13:** 83–87.

Kantor R, Katzenstein DA, Efron B, Carvalho AP, Wynhoven B, Cane P, Clarke J, Sirivichayakul S, Soares MA, Snoeck J, et al. 2005. Impact of HIV-1 subtype and antiretroviral therapy on protease and reverse transcriptase genotype: Results of a global collaboration. *PLoS Med* **2:** e112. doi: 10.1371/journal.pmed.0020112.

Keele BF, Van Heuverswyn F, Li Y, Bailes E, Takehisa J, Santiago ML, Bibollet-Ruche F, Chen Y, Wain LV, Liegeois F, et al. 2006. Chimpanzee reservoirs of pandemic and non-pandemic HIV-1. *Science* **313:** 523–526.

Kent SJ, Cooper DA, Chhi Vun M, Shao Y, Zhang L, Ganguly N, Bela B, Tamashiro H, Ditangco R, Rerks-Ngarm S, et al. 2010. AIDS vaccine for Asia network (AVAN): Expanding the regional role in developing HIV vaccines. *PLoS Med* **7:** e1000331. doi: 10.1371/journal.pmed.1000331.

Kiwanuka N, Laeyendecker O, Robb M, Kigozi G, Arroyo M, McCutchan F, Eller LA, Eller M, Makumbi F, Birx D, et al. 2008. Effect of human immunodeficiency virus type 1 (HIV-1) subtype on disease progression in persons from Rakai, Uganda, with incident HIV-1 infection. *J Infect Dis* **197:** 707–713.

Korber B, Muldoon M, Theiler J, Gao F, Gupta R, Lapedes A, Hahn BH, Wolinsky S, Bhattacharya T. 2000. Timing the ancestor of the HIV-1 pandemic strains. *Science* **288:** 1789–1796.

Korber BT, Letvin NL, Haynes BF. 2009. T-cell vaccine strategies for human immunodeficiency virus, the virus with a thousand faces. *J Virol* **83:** 8300–8314.

Kouanfack C, Montavon C, Laurent C, Aghokeng A, Kenfack A, Bourgeois A, Koulla-Shiro S, Mpoudi-Ngole E, Peeters M, Delaporte E. 2009. Low levels of antiretroviral-resistant HIV infection in a routine clinic in Cameroon that uses the World Health Organization (WHO) public health approach to monitor antiretroviral treatment and adequacy with the WHO recommendation for second-line treatment. *Clin Infect Dis* **48:** 1318–1322.

* Koup RA, Douek DC. 2011. Vaccine design for CTL responses. *Cold Spring Harb Perspect Med* doi: 10.1101/cshperspect.a007252.

* Kwong PD, Mascola JR, Nabel GJ. 2011. Rational design of vaccines to elicit broadly neutralizing antibodies to HIV-1. *Cold Spring Harb Perspect Med* doi: 10.1101/cshperspect.a007278.

Li Y, Shao Y, Luo X, Su L, Zhang L, Chen J, Fang Y, Yaun J, Zhang Y, Chen H, et al. 1997. Subtype and sequence analysis of the C2-V3 region of gp120 genes among HIV-1 strains in Hubei province. *Chin J Epidemiol* **18:** 217–219.

Li Y, Uenishi R, Hase S, Liao H, Li XJ, Tsuchiura T, Tee KK, Pybus OG, Takebe Y. 2010. Explosive HIV-1 subtype B′ epidemics in Asia driven by geographic and risk group founder events. *Virology* **402:** 223–227.

Liao L, Xing H, Li X, Ruan Y, Zhang Y, Qin G, Shao Y. 2007. Genotypic analysis of the protease and reverse transcriptase of HIV type 1 isolates from recently infected injecting drug users in western China. *AIDS Res Hum Retroviruses* **23:** 1062–1065.

Liao L, Xing H, Shang H, Li J, Zhong P, Kang L, Cheng H, Si X, Jiang S, Li X, et al. 2010. The prevalence of transmitted antiretroviral drug resistance in treatment-naive HIV-infected individuals in China. *J Acquir Immune Defic Syndr* **53:** S10–S14.

Los Alamos HIV Sequence Database. 2010. Overview of the subtypes of primate immunodeficiency viruses. http://www.hiv.lanl.gov/content/sequence/HelpDocs/subtypes.html.

Luo CC, Tian C, Hu DJ, Kai M, Dondero T, Zheng X. 1995. HIV-1 subtype C in China. *Lancet* **345:** 1051–1052.

Ma Y, Li Z, Zhao S. 1990. HIV infected people were first identified in intravenous drug users in China. *Chin J Epidemiol* **11:** 184–185.

Mathers BM, Degenhardt L, Phillips B, Wiessing L, Hickman M, Strathdee SA, Wodak A, Panda S, Tyndall M, Toufik A, et al. 2008. Global epidemiology of injecting drug use and HIV among people who inject drugs: A systematic review. *Lancet* **372:** 1733–1745.

Mi G, Wu Z. 2010. A review of studies on sero-sorting among HIV positive men who have sex with men. *Chin J AIDS STD* **16:** 201–203.

Ott MQ, Bärnighausen T, Tanser F, Lurie MN, Newell ML. 2011. Age-gaps in sexual partnerships: Seeing beyond "sugar daddies." *AIDS* **25:** 861–863.

* Overbaugh J, Morris L. 2011. The antibody response against HIV-1. *Cold Spring Harb Perspect Med* doi: 10.1011/cshperspect.a007039.

PAHO, WHO, UNAIDS. 2001. Monitoring the AIDS Epidemic (MAP). HIV and AIDS in the Americas: An epidemic with many faces. http://www.who.int/hiv/strategic/pubrio00/en/index.html.

Pereyra F, Jia X, McLaren PJ, Telenti A, de Bakker PI, Walker BD et al.. 2010. The major genetic determinants of HIV-1 control affect HLA class I peptide presentation. International HIV controllers study. *Science* **330:** 1551–1557.

Phanuphak P, Locharernkul C, Panmuong W, Wilde H. 1985. A report of three cases of AIDS in Thailand. *Asian Pac J Allergy Immunol* **3:** 195–199.

Piot P, Quinn TC, Taelman H, Feinsod FM, Minlangu KB, Wobin O, Mbendi N, Mazebo P, Ndangi K, Stevens W, et al.1984. Acquired immunodeficiency syndrome in a heterosexual population in Zaire. *Lancet* **2:** 65–69.

Piyasirisilp S, McCutchan FE, Carr JK, Sanders-Buell E, Liu W, Chen J, Wagner R, Wolf H, Shao Y, Lai S, et al. 2000. A recent outbreak of human immunodeficiency virus type 1 infection in southern China was initiated by two highly homogeneous, geographically separated strains, circulating recombinant form AE and a novel BC recombinant. *J Virol* **74:** 11286–11295.

Powers KA, Ghani AC, Miller WC, Hoffman IF, Pettifor AE, Kamanga G, Martinson FE, Cohen MS. 2011. The role of acute and early HIV infection in the spread of HIV and implications for transmission prevention strategies in Lilongwe, Malawi: A modelling study. *Lancet* **378:** 256–268.

Powers KA, Poole C, Pettifor AE, Cohen MS. 2008. Rethinking the heterosexual infectivity of HIV-1: A systematic review and meta-analysis. *Lancet Infect Dis* **8:** 553–563.

Ruan Y, Li D, Li X, Qian HZ, Shi W, Zhang X, Yang Z, Zhang X, Wang C, Liu Y, et al. 2007. Relationship between syphilis and HIV infections among men who have sex with men in Beijing, China. *Sex Transm Dis* **34:** 592–597.

Saathoff E, Pritsch M, Geldmacher C, Hoffmann O, Koehler RN, Maboko L, Maganga L, Geis S, McCutchan FE, Kijak GH, et al. 2010. Viral and host factors associated with the HIV-1 viral load setpoint in adults from Mbeya Region, Tanzania. *J Acquir Immune Defic Syndr* **54:** 324–330.

Santra S, Liao HX, Zhang R, Muldoon M, Watson S, Fischer W, Theiler J, Szinger J, Balachandran H, Buzby A, et al. 2010. Mosaic vaccines elicit CD8$^+$ T lymphocyte responses that confer enhanced immune coverage of diverse HIV strains in monkeys. *Nat Med* **16:** 324–328.

Seaman MS, Janes H, Hawkins N, Grandpre LE, Devoy C, Giri A, Coffey RT, Harris L, Wood B, Daniels MG, et al. 2010. Tiered categorization of a diverse panel of HIV-1 Env pseudoviruses for assessment of neutralizing antibodies. *J Virol* **84:** 1439–1452.

Serwadda D, Mugerwa RD, Sewankambo NK, Lwegaba A, Carswell JW, Kirya GB, Bayley AC, Downing RG, Tedder RS, Clayden SA, et al. 1985. Slim disease: A new disease in Uganda and its association with HTLV-III infection. *Lancet* **2:** 849–852.

Shafer RW, Rhee SY, Bennett DE. 2008. Consensus drug resistance mutations for epidemiological surveillance: Basic principles and potential controversies. *Antivir Ther* **13:** 59–68.

Shao Y, Zhao Q, Wang B, Chen Zh, Su L, Zeng Y, Wolf H. 1994. Sequence analysis of HIV env genes among HIV infected drug injecting users in Dehong epidemic area of Yunnan Province, China. *Chin J Virol* **10:** 291–299 (in Chinese).

Shao Y, Zhao F, Yang W, XC Gong. 1999. The identification of recombinant HIV-1 strains in IDUs in southwest and northwest China. *Chin J Exp Clin Virol* **13:** 109–112 (in Chinese).

Simoes EA, Babu PG, John TJ, Nirmala S, Solomon S, Lakshminarayana CS, Quinn TC. 1987. Evidence for HTLV-III infection in prostitutes in Tamil Nadu. (India). *Indian J Med Res* **85:** 335–338.

Steffenson AE, Pettifor AE, Seage GR III, Rees HV, Cleary PD. 2011. Concurrent sexual partnerships and human immunodeficiency virus risk among South African youth. *Sex Transm Dis* **38:** 459–466.

Su L, Graf M, Zhang Y, von Briesen H, Xing H, Kostler J, Melzl H, Wolf H, Shao YM, Wagner R. 2000. Characterization of a virtually full-length human immunodeficiency virus type 1 genome of a prevalent intersubtype (C/B) recombinant strain in China. *J Virol* **74:** 11367–11376.

Tang WM, Ding JP, Yan HJ, Xuan XP, Yang HT, Zhao JK. 2008. Sexual behaviors and HIV/syphilis infection among men who have sex with men: A meta analysis of data collected between 2001 and 2006 in the Chinese mainland. *Chin J AIDS STD* **14:** 471–488.

Taylor BS, Sobieszczyk ME, McCutchan FE, Hammer SM. 2008. The challenge of HIV-1 subtype diversity. *N Engl J Med* **358:** 1590–1602.

Tebit DM, Arts EJ. 2010. Tracking a century of global expansion and evolution of HIV to drive understanding and to combat disease. *Lancet Infect Dis* **1:** 45–56.

Tee KK, Pybus OG, Li XJ, Han X, Shang H, Kamarulzaman A, Takebe Y. 2008. Temporal and spatial dynamics of human immunodeficiency virus type 1 circulating recombinant forms 08_BC and 07_BC in Asia. *J Virol* **82:** 9206–9215.

Teng T, Shao Y. 2011. Scientific approaches to AIDS prevention and control in China. *Adv Dent Res* **23:** 10–12.

Tully DC, Wood C. 2010. Chronology and evolution of the HIV-1 subtype C epidemic in Ethiopia. *AIDS* **24:** 1577–1582.

Xing H, Pan L, Su L, Fan X, Feng Y, Qiang L, Shao Y. 2002. Molecular epidemiological study of a HIV-1 strain of subtype E in China between 1996 and 1998. *Chin J STD/AIDS Prev Cont* **8:** 200–203.

Xu F, Kilmarx PH, Supawitkul S, Yanpaisarn S, Limpakarnjanarat K, Manopaiboon C, Korattana S, Mastro TD, StLouis ME. 2000. HIV-1 seroprevalence, risk factors, and preventive behaviors among women in northern Thailand. *J Acquir Immune Defic Syndr* **25:** 353–359.

UNAIDS. 2008. UNAIDS report on the global HIV epidemic: 2008. http://www.unaids.org/GlobalReport/.

UNAIDS. 2010. UNAIDS report on the global HIV epidemic: 2010. http://www.unaids.org/GlobalReport/.

UNODC. 2010. The global heroin market. *World Drug Report 2010*, pp. 37–63.

Van de Perre P, Rouvroy D, Lepage P, Bogaerts J, Kestelyn P, Kayihigi J, Hekker AC, Butzler JP, Clumeck NA. 1984. Acquired immunodeficiency syndrome in Rwanda. *Lancet* **2:** 62–65.

van Griensven F, Thanprasertsuk S, Jommaroeng R, Mansergh G, Naorat S, Jenkins RA, Ungchusak K, Phanuphak P, Tappero JW, Bangkok MSM Study Group. 2005. Evidence of a previously undocumented epidemic of HIV infection among men who have sex with men in Bangkok, Thailand. *AIDS* **19:** 521–526.

van Griensven F, Varangrat A, Wimonsate W, Tanpradech S, Kladsawad K, Chemnasiri T, Suksripanich O, Phanuphak P, Mock P, Kanggarnrua K, et al. 2010. Trends in HIV prevalence, estimated HIV incidence and risk behavior among men who have sex with men in Bangkok, Thailand, 2003-2007. *J Acquir Immune Defic Syndr* **53:** 234–249.

* Vermund ST, Leigh-Brown A. 2011. The HIV epidemic: High-income countries. *Cold Spring Harb Perspect Med* doi: 10.1101/cshperspect.a007195.

Vidal N, Peeters M, Mulanga-Kabeya C, Nzilambi N, Robertson D, Ilunga W, Sema H, Tshimanga K, Bongo B, Delaporte E. 2000. Unprecedented degree of human immunodeficiency virus type 1 (HIV-1) group M genetic diversity in the Democratic Republic of Congo suggests that the HIV-1 pandemic originated in Central Africa. *J Virol* **74:** 10498–10507.

Wheeler WH, Ziebell RA, Zabina H, Pieniazek D, Prejean J, et al. 2010. Prevalence of transmitted drug resistance associated mutations and HIV-1 subtypes in new HIV-1 diagnoses, U.S.-2006. *AIDS* **24:** 1203–1212.

Xin R, He X, Xing H, Sun F, Ni M, Zhang Y, Meng Z, Feng Y, Liu S, Wei J, et al. 2009. Genetic and temporal dynamics of human immunodeficiency virus type 1 CRF07_BC in Xinjiang, China. *J Gen Virol* **90:** 1757–1761.

WHO. 1998. Fifty years of WHO in the Western Pacific Region, Chapter 26. Sexually transmitted diseases, including HIV/AIDS, pp. 245–259. http://www.wpro.who.int/NR/rdonlyres/3E44D634-C0F9-480C-B6C2-1C418AE8F3A0/0/chapter26.pdf.

Worobey M, Gemmel M, Teuwen DE, Haselkorn T, Kunstman K, Bunce M, Muyembe JJ, Kabongo JM, Kalengayi RM, Van Marck E, et al. 2008. Direct evidence of extensive diversity of HIV-1 in Kinshasa by 1960. *J Nature* **455:** 661–664.

Yu XF, Chen J, Shao Y, Beyrer C, Lai S. 1998. Two subtypes of HIV-1 among injection-drug users in southern China. *Lancet* **351:** 1250.

Zeng Y, Fan J, Zhang Q, Wang P, Tang D, Zhon S, Zheng X, Liu D. 1986. Detection of antibody to LAV/HTLV-III in sera from hemophiliacs in China. *AIDS Res* **2:** S147–S149.

Zhang Y, Lu L, Ba L, Liu L, Yang L, Jia M, Wang H, Fang Q, Shi Y, Yan W, et al. 2006. Dominance of HIV-1 subtype CRF01_AE in sexually acquired cases leads to a new epidemic in Yunnan province of China. *PLoS Med* **3:** e443. doi: 10.1371/journal.pmed.0030443.

Zhang X, Li Sh, Li X, Li X, Xu J, Li D, Ruan Y, Xing H, Zhang X, Shao Y. 2007. Characterization of HIV-1 subtypes and viral antiretroviral drug resistance in men who have sex with men in Beijing, China. *AIDS* **21:** S59–S66.

Zhao F, Shao Y, Duan Y, Chen X, Zhao Q, Su L, Guan Y, Zeng Y, Wolf H. 1999. Sequence analysis of human immunodeficiency virus type 1 tat gene among the long term HIV infected non-progressor in Yunnan province. *Chin J Exp Clin Virol* **13:** 37–40.

The HIV Epidemic: High-Income Countries

Sten H. Vermund[1] and Andrew J. Leigh-Brown[2]

[1]Institute for Global Health and Department of Pediatrics, Vanderbilt University School of Medicine, Nashville, Tennessee 37203

[2]Institute of Evolutionary Biology, University of Edinburgh, Edinburgh, Scotland EH9 3JT, United Kingdom

Correspondence: sten.vermund@vanderbilt.edu

The HIV epidemic in higher-income nations is driven by receptive anal intercourse, injection drug use through needle/syringe sharing, and, less efficiently, vaginal intercourse. Alcohol and noninjecting drug use increase sexual HIV vulnerability. Appropriate diagnostic screening has nearly eliminated blood/blood product-related transmissions and, with antiretroviral therapy, has reduced mother-to-child transmission radically. Affected subgroups have changed over time (e.g., increasing numbers of Black and minority ethnic men who have sex with men). Molecular phylogenetic approaches have established historical links between HIV strains from central Africa to those in the United States and thence to Europe. However, Europe did not just receive virus from the United States, as it was also imported from Africa directly. Initial introductions led to epidemics in different risk groups in Western Europe distinguished by viral clades/sequences, and likewise, more recent explosive epidemics linked to injection drug use in Eastern Europe are associated with specific strains. Recent developments in phylodynamic approaches have made it possible to obtain estimates of sequence evolution rates and network parameters for epidemics.

The social and molecular epidemiology of the HIV epidemic in higher-income countries has had characteristics quite distinct from the situation in the most highly affected regions of sub-Saharan Africa (Kilmarx 2009). Since the recognition of AIDS in 1981, high-income nations have had their own characteristic epidemic trends, transmission dynamics, affected subgroups, and recent trends. We focus on the higher-income nations of Western Europe, North America, and many nations of Oceania, including Australia and New Zealand. We acknowledge that one must also consider the proximity of Eastern Europe, Caribbean, and the rest of Oceania, acknowledging that most of them are high- or low- middle-income nations, not high income; neither are they as resource-limited as those reviewed by Williamson and Shao 2011 (Haiti and a few others are exceptions). We seek to summarize a complex matrix of biological phenomena in a complex disease that is steeped in human behavior and stigma, fear, and prejudice (Remien and Mellins 2007; Mahajan et al. 2008).

HIV/AIDS shares transmission characteristics with other sexual and blood-borne agents. Higher sexual mixing rates and lack of condom use are conspicuous risk factors (Vermund et al. 2009). Reuse of syringes and needles, receipt of contaminated blood or blood products, or occupational needle sticks in a health care setting all are associated with HIV infection. Mother-to-child transmission prepartum, intrapartum, or postpartum via breast milk feeding is an additional major transmission route (Fowler et al. 2010). Since the advent of antiretroviral therapy and routine screening of women in pregnancy, the number of infants infected with HIV in high-income nations has plummeted, although cases still occur (Birkhead et al. 2010; Lampe et al. 2010; Whitmore et al. 2010, 2011).

HIV is the most dangerous of the sexual and blood-borne diseases in its epidemic potential and its virulence. Our inability to cure HIV infection adds to the complexity of its chronic management. In addition, HIV seems to be spread more like hepatitis B virus (HBV) than any other infectious agent (sex, blood, or perinatal), although it is not as infectious as HBV (Barth et al. 2010; Dwyre et al. 2011). Classical sexually transmitted diseases (SDIs) are rarely blood-borne, although sexual and perinatal HIV transmission occurs as with HBV, herpes simplex type 2 (HSV-2), syphilis, gonorrhea, and chlamydia. Hepatitis C virus (HCV) is transmitted overwhelmingly through blood-borne routes, with sexual and perinatal cases occurring rarely. Human T-lymphotropic virus type 1 (HTLV-1) is transmitted most often via breast milk; HTLV-2 transmission has been reported via needle exchange. Although HIV is surely reminiscent of other infectious agents, its unique $CD4^+$ T-lymphocyte tropism helps dictate its exact transmission routes and frequencies (see Shaw and Hunter 2011).

Sadly, the early history of the HIV epidemic is one of missed opportunities, over and over, in country after country. This continues to the present day (Mahy et al. 2009). The failure to see the now-obvious parallels in risk with HBV led to a failure to protect the blood supply in the early 1980s during an intense transmission period, resulting in the preventable infection of thousands of blood recipients, particularly men with hemophilia. The failure to aggressively advocate sexual risk reduction, including delayed coital debut among adolescents, reduced numbers of sexual partners,

and use of condoms, led to unknown numbers of same sex and heterosexual HIV transmissions. Even in 2011, members of the U.S. Congress demonize such organizations as Planned Parenthood that advocate pregnancy and HIV/STD prevention, empowering the poor and the young to access their unmet needs for these services. The failure to promptly, consistently, and extensively provide clean needles and syringes for IDUs (and for health care in resource-limited settings) led to an estimated 4394 to 9666 preventable infections in the U.S. alone (33% of the incidence in IDUs) from 1987 to 1995, costing society US$244 to US$538 million (Lurie and Drucker 1997). The U.S. Federal Government did not lift its ban on funding for needle/syringe exchange programs (N/SEP) until 2009, despite overwhelming evidence indicating its effectiveness in preventing HIV transmission. Although beyond the scope of this work, the failure of many high-income nations to liberate public health professionals and community activists to address the problem using evidence-based solutions, rather than using HIV/AIDS as a politicized point of ideological conflict, costs, and continues to cost, lives (Mathers et al. 2010).

Here we present the characteristics of viral evolution as seen in high-income nations, primarily in the northern hemisphere, including issues of viral diversity and transmission. We further present a brief review of major risk exposures for HIV and their importance in the epidemic patterns of higher-income nations.

OVERVIEW OF VIRUS EVOLUTION AND RATES OF CHANGE

The rapidity of HIV evolution is such a prominent feature of the virus that it has attracted extensive attention since the development of thermostable polymerases allowed it to be easily studied (Meyerhans et al. 1989; Balfe et al. 1990; Wolfs et al. 1990). This required a revolution in the thinking of virologists, many of whom had previously been familiar with the evolution of DNA viruses, which diverge slowly and often in parallel with their hosts (Sambrook et al. 1980).[3] These differences gradually became apparent more widely, and both the data and methodology developed (Crandall et al. 1999) to reveal the conflicting forces that act on the HIV population to force the highest rates of amino acid substitution known (under selection by drugs, cytotoxic T-lymphocytes [CTLs], or by neutralizing antibody) (Frost et al. 2000, 2005) (see Carrington and Alter 2011; Lackner et al. 2011; Overbaugh and Morris 2011; Swanstrom and Coffin 2011; Walker and McMichael 2011) and yet allow chance effects to play a significant role in viral evolution (Leigh Brown 1997; Leigh Brown and Richman 1997). The impact on HIV evolution of the complexity of the processes and forces involved, and their interactions, have been reviewed effectively elsewhere (Rambaut et al. 2004), so only recent developments will be looked at further here.

Powerful, statistically rigorous techniques for estimation of molecular evolutionary parameters are now available and have been applied to HIV sequences from infected populations to estimate both the rate of evolutionary change and dates of divergence (Drummond and Rambaut 2007). These studies, however, have a long history: Estimates of the rate of synonymous nucleotide substitution made 20 years ago (Balfe et al. 1990; Wolfs et al. 1990) of the order of 5×10^{-3} were not very different from those made recently ($2-6 \times 10^{-3}$) (Korber et al. 2000; Robbins et al. 2003; Lewis et al. 2008). What has changed is that such estimates can now be made on large and complex data sets instead of being restricted to relatively uniform well-defined outbreaks. The first demonstration of the power of the new approaches was the timing of the origin of the pandemic HIV "M" strain to the early 20th century (Korber et al. 2000) (see also Sharp and Hahn 2011). More recently they have allowed estimates to be made of the origins of diverse epidemics in different countries and risk groups (Lemey et al. 2003; Salemi et al. 2008). Finally, in combination with very large data sets, these approaches have allowed a bridge to be made across to infectious disease epidemiology, as under certain assumptions, the viral evolutionary history can be used to estimate critical epidemiological parameters that are not accessible from other routes (Volz et al. 2009).

Major Clades

Across the globe, comparisons between viruses isolated in the 1980s from different populations of HIV-infected individuals frequently showed much greater differences than were seen within populations. These became termed "subtypes," although, as there is no clear serological distinction, the generic term "clade" is more appropriate. National epidemics were initially only associated with one clade, with the exception of Thailand, where two major forms were recognized early on. The distinction between "Thai A" (now called CRF01) and "Thai B" strains (Ou et al. 1993) gave rise to the name of the HIV clade that had spread across the United States and Western Europe in the 1980s: "subtype B." Likewise, East Africa was another locus where two subtypes became established and cocirculated: the "A" and "D" clades (see Sharp and Hahn 2011; Williamson and Shao 2011). In addition, owing to the recombinogenic nature of HIV replication, recombinants between these clades arise wherever cocirculation occurs, and several have given rise to epidemics themselves; these have become termed circulating recombinant forms, or "CRFs" (Robertson et al. 2000), whose names identify the parent strains. Classification systems are frequently superseded by later knowledge, and this is no exception, as the best known "recombinant," the Thai "A" strain or CRF01/AE, evidence for whose hypothesized second parent "E" has always been lacking, now appears rather to be a divergent form of subtype A (Anderson et al. 2000). Similarly, it appears that CRF02/AG is a parental, rather than recombinant form (Abecasis et al. 2007). In recent years the genetic complexity of the global epidemic has increased

---

[3]Summarized succinctly by one of the leading figures in the molecular characterization of HIV in the 1980s, Simon Wain-Hobson: "The problem of HIV evolution ... is one of population genetics" (in conversation with ALB at the Institut Pasteur, August 1988).

substantially, with more than 40 CRFs now recognized (Kuiken et al. 2010), instead of the original four (Robertson et al. 2000).

Genetic diversity of HIV has always been greatest in west central Africa. Here all major clades (A–K, with the exception of the "B" and missing "E" clades) have been found, and indeed, especially among strains from the 1980s, many intermediate and unclassifiable forms (Kalish et al. 2004), as expected for the region where the pandemic originated (Rambaut et al. 2001; see Sharp and Hahn 2011). The lack of major biological differences among HIV clades has led to the view that their distinctiveness has arisen primarily through the stochastic process of founder events, via the chance appearance of virus from a particular origin in a highly susceptible population (Rambaut et al. 2004). Dating the divergence events using molecular clock–based approaches suggests the divergence of the major clades, which took place here, occurred around the middle of the 20th century (Korber et al. 2000), but seeding of the northern hemisphere has often been from other parts of Africa, and after distinct clades emerged.

Virus Diversity in the Northern Hemisphere

North America

Given what is known of the pathogenesis of HIV disease, the origin of HIV infection in the northern hemisphere had to have predated the original description of AIDS in 1981 by several years. Indeed, serosurveillance studies showed that the prevalence of HIV infection among men who have sex with men (MSM), which was 20%–40% by the early 1980s, was already between 4% and 7% in San Francisco, New York, and Los Angeles in the late 1970s (Coutinho et al. 1989). Nevertheless, analysis of nucleotide sequences from strains isolated in the 1980s (Robbins et al. 2003) confirmed the earlier estimate (Korber et al. 2000), that the common ancestor of the HIV strains found in the United States preceded that period by approximately 10 years. The final piece in this puzzle was found when a number of sequences

of HIV from Haitian immigrants to the United States who were HIV-infected in the 1980s were compared to United States sequences (Gilbert et al. 2007). These belonged to the same (B) clade but all branched off earlier. Using similar dating approaches, the origin of the U.S. clade is confirmed at 1969, whereas the common ancestor of the Haitian strains predated this by a few years at 1966 (Gilbert et al. 2007). This is the closest that can currently be reached to the reconstruction of the origin of the northern hemisphere epidemic and the B subtype. The B subtype was not found among early sequences from Africa (it was introduced later to South African MSM); the most closely related sequences from Africa are from the D clade, within which, when early African sequences are included, the B clade nests (Kalish et al. 2004). The date for this B-D divergence is around 1954 (Gilbert et al. 2007), consistent with the placing of a sequence fragment from a sample from Kinshasa in 1957 very near the split, but in the D clade (Korber et al. 2000). It has been speculated that the return in the 1960s of expatriate Haitians from the diaspora of the 1950s to Francophone West Africa was responsible. It is known that AIDS was recognized in Haiti much earlier than anywhere else, outside the three U.S. cities (Los Angeles, New York, and San Francisco), and that initially the infection was predominantly in men, only becoming a generalized epidemic over the next decade (Pape and Johnson 1993).

The reconstruction, through molecular phylogenetics, of the connections which linked a then obscure syndrome among a number of homosexual men in U.S. cities to central Africa, where the infection had not at the time been recognized, is the most striking of many examples where sequencing of viral strains has yielded important information for epidemiology. The pattern of introduction in Europe, however, was more complex.

Europe 1980–1990

Identification of AIDS in Europe followed its recognition in the United States. Once recognized, retrospective investigation identified a few cases that could be traced back to the 1970s, but among MSM many of the earliest cases had direct recent links to the United States (Pinching 1984), and all MSM at that time were infected with a virus later characterized as the B clade. These introductions of the virus probably continued: It has been estimated by reconstructing viral lineages among MSM in the United Kingdom that at least six lineages are represented in present day populations, most, but not all, of which had an origin in the early 1980s (Hue et al. 2005). These were already distinct B clade lineages which represented independent introductions into the United Kingdom. Another introduction that appeared to be independent was into the injecting drug user population of northern Europe (Lukashov et al. 1996), among whom the virus was remarkably similar, whether in Amsterdam, Dublin, or Edinburgh, reflecting another founder introduction to a highly susceptible population (Leigh Brown et al. 1997; Op de Coul et al. 2001). The distinctiveness of this strain from that in MSM (Kuiken and Goudsmit 1994; Holmes et al. 1995; Kuiken et al. 1996) contrasted with that of hemophiliacs in Edinburgh, infected from locally prepared

factor VIII in the early 1980s, whose virus could be linked directly to that found in Scottish MSM (Brown et al. 1997). Interestingly, the distinction is less apparent in IDUs from southern Europe from this period, whose sequences overlapped more with those from MSM (Lukashov et al. 1996). The epidemics also behaved differently in these two areas, stabilizing in the north while leading to the highest prevalence in Europe in Spain, Italy, and Portugal (Hamers et al. 1997).

Europe did not just receive virus from the United States, however; it was also imported from Africa directly (Fig. 1). Molecular epidemiological evidence suggests strongly that since the 1960s, HIV-1 and HIV-2 have been transported from Africa to North America/Caribbean and Europe, and have been transported, too, between the European and North American continents (Fig. 1). Cases were recognized almost as early among individuals having links to Africa (Alizon et al. 1986), and these led to local epidemics of non-B clades. One hospital in Paris recorded eight clades, including CRF01AE by 1995 (Simon et al. 1996), but although several clades were recorded in the United Kingdom, in relative terms, the prevalence of

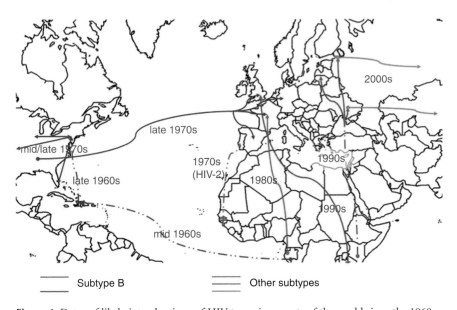

Figure 1. Dates of likely introductions of HIV to various parts of the world since the 1960s.

non-B strains at that time was low (Arnold et al. 1995). Patterns of strain prevalence reflected past colonial links and frequency of contemporary interchange, strikingly so in the case of Portugal, the only country outside West Africa to record a significant epidemic of HIV-2 (De Cock and Brun-Vezinet 1989), reflecting the location of its former colony Guinea-Bissau at the epicenter of the HIV-2 epidemic. The molecular distinctions between European epidemics extended to the HIV-1 clades, however, with clade G, associated with West Africa, being well represented in Portugal but not elsewhere in Western Europe. However, it gave rise to an iatrogenic outbreak in southern Russia in 1988 (in Elista, 300 km south of Volgograd) (Bobkov et al. 1994), and the devastating iatrogenic outbreak in Romanian children (Hersh et al. 1991b) was associated with clade F (Thomson et al. 2002).

Europe 1990–2005

Those initial introductions into Western Europe led to epidemics in different risk groups in different regions often distinguished by viral clade, thus in France, transmission among IDUs and MSM remained of the B subtype, whereas heterosexual transmission was more commonly of a variety of non-B subtypes, as was the case in Switzerland (Op de Coul et al. 2001). Non-B subtypes were present but rare in the UK until 1995, after which a substantial increase in immigration from southern Africa, particularly Zimbabwe, led to a dramatic rise in subtypes C and A (Parry et al. 2001). By 2005, the prevalence of non-B clades taken together equaled the B clade, and recently, in both France and the United Kingdom, there has been noticeable, and nonreciprocal, crossover of non-B clades into MSM (de Oliveira et al. 2010; Fox et al. 2010). Similarly, Swiss investigators have confirmed changes in B clade viral transmission over time (Kouyos RD et al. 2010).

The breakup of the Soviet Union and the Eastern European bloc at the end of the 1980s, the subsequent economic decline in some of these countries, and the freeing of travel in the 1990s was a combination that had devastating

consequences regarding HIV transmission. A very large population of injection drug users adopting highly unsafe practices (Dehne et al. 1999) was the focus of massive, explosive epidemics (Aceijas et al. 2004). Virus obtained in 1995 from Donetsk in eastern Ukraine and Krasnodar in southern Russia, about 350 km distant, were very similar, with about 2% divergence in the C2-V3 region, and belonged to subtype A but had some unique features (Bobkov et al. 1997) that allowed it to be easily tracked as it spread from Ukraine to St. Petersburg and Estonia in the 1990s and then in the 21st century, on to other parts of Russia, including Moscow and Irkutsk (Siberia) (Bobkov et al. 2001), as well as neighbors, including Kazakhstan (Bobkov et al. 2004) and the Baltic states (Zetterberg et al. 2004). Soon after the original outbreak in Ukraine, there was an explosive outbreak in the Former Soviet Union enclave of Kaliningrad. This was characterized by a quite distinct A/B recombinant (Liitsola et al. 1998), which detailed analysis suggests was derived from subtype B and A variants found among IDUs in Ukraine (Liitsola et al. 2000).

DRUG-RESISTANT HIV AND ITS TRANSMISSION

From very early days of antiretroviral therapy, the transmission of drug-resistant virus was already being recorded. Clear evidence was presented in three different countries that before 1995, mutations at amino acid 215 in reverse transcriptase were present in untreated individuals (Perrin et al. 1994; de Ronde et al. 1996; Quigg et al. 1997). Systematic surveys of the presence of such mutations have followed, initially linked to studies of acute infection (Little et al. 1999, 2002). Later studies extended this to include individuals in chronic infection, which could be studied in larger numbers (Weinstock et al. 2004; Cane et al. 2005; Wensing et al. 2005; Chaix et al. 2009; Vercauteren et al. 2009). The overall picture appears to indicate an increase in transmission of drug resistance in the late 1990s as more individuals went on therapy (Leigh Brown et al. 2003), which was followed by a decrease, as more individuals on therapy became completely suppressed, but the exact timing was difficult to determine from chronic infection studies and may have varied between countries. In addition, there were differences in the classes of drug that were most strongly affected, with nonnucleoside resistance consistently appearing at a higher frequency than nucleoside or protease inhibitor resistance, reflecting differences in transmission fitness of nucleoside reverse transcriptase inhibitor (NNRTI) resistance-associated mutations (Leigh Brown et al. 2003). Most countries in the developed world have developed systematic surveillance for transmitted resistance to minimize its impact on individualized antiretroviral therapy.

HIV Transmission Networks and Dynamics Revealed by Molecular Epidemiology

Analyses of nucleotide sequence variation have been used for almost 20 years to describe the relationships between viral strains within infected communities (Balfe et al. 1990; Kuiken and Goudsmit 1994; Leigh Brown et al. 1997). Recently a combination of greatly enhanced data sets and improvements in methodology have allowed much more detail to be added to the depictions of the structure of the transmission networks deduced from sequence data. This has shown that some risk groups (e.g., injection drug users) can give rise to tight clusters where many individuals may share very similar sequences (Brenner et al. 2008; Yerly et al. 2009). In addition, for the first time it has become possible to investigate the dynamics of the epidemic based on such data. Although many studies of sexual behavior have shown that individuals can vary greatly in their number of sexual partners over time and this concept has underpinned many epidemiological models of HIV transmission, however, contact tracing has been shown to be poor at recovering the HIV transmission network relative to other infections (Yirrell et al. 1998; Resik et al. 2007). The development of the field of molecular phylodynamics, used first to estimate dates of the origin of the zoonotic transmission of

HIV, provided new routes to obtain such parameter estimates, which can be applied to small-scale intensive studies of specific outbreaks, as well as to obtain global estimates relating to long-term spread.

By analyzing the transmission network revealed by partial *pol* gene sequences obtained during routine clinical care, it was shown that the cluster size distribution of MSM attending a single large clinic in London had a long right tail, implying many individuals were associated with large clusters: In fact, of individuals in clusters >2, 25% were found in clusters ≥10 (Lewis et al. 2008). Within these clusters, 25% of intervals between transmissions, inferred from the dated transmission network, were 6 months or less, implying a significant role for acute infection in driving the epidemic (Cohen et al. 2011).

Extending this work to non-B subtypes, primarily associated with heterosexual infection in the United Kingdom, revealed that although clustering could be detected, it was much less extensive than among MSM, and the mean intertransmission interval was substantially longer. In fact, there were hardly any intervals ≤6 mo, indicating important differences between the epidemics in the two risk groups (Hughes et al. 2009).

Example of Phylodynamic Modeling from the United Kingdom

The phylodynamic approach has been extended to a population survey of viral sequences from the entire U.K. HIV epidemic among MSM (Leigh Brown et al. 2011). Partial *pol* gene sequences from approximately two-thirds of the MSM under care in the United Kingdom in 2007 have been analyzed. Of those linked to any other, 29% were linked to only 1, 41% linked to 2–10, and 29% linked to ≥10. In striking contrast, a recent update on transmission of non-B subtypes in the United Kingdom has revealed that only one large cluster is attributable to heterosexual transmission, all others being "crossovers" to IDU or MSM (S Hue et al., unpubl.). It has thus been possible to obtain estimates of network parameters for the entire epidemic in the United Kingdom. The scale of sequence data coverage greatly exceeds that generally achievable in epidemiological surveys of sexual contact. This provides an opportunity to make critical inferences on the impact of clustering for the epidemic, and its importance for intervention strategies. It is well understood that under certain conditions there is no "epidemic threshold," such that a randomly distributed (i.e., untargeted) intervention, will be unable to stop the epidemic (Keeling and Eames 2005). The estimates of epidemiological parameters based on data from routinely performed HIV genotyping tests therefore provide critical information for the successful and efficient implementation of transmission interventions including vaccination (when available), and for both antiretroviral treatment for prevention among infectous seropositives and preexposure prophylaxis among seronegatives (Burns et al. 2010; Kurth et al. 2011).

RISK EXPOSURES: MSM

The AIDS epidemic was recognized first among MSM in Los Angeles, New York, and San Francisco. Clinicians were alarmed to see gay men who presented with low CD4$^+$ T-lymphocytes and opportunistic infections like *Pneumocystis jirovecii* (*P. carinii* is the form seen in animals, now distinguished from the former human form) or malignancies like Kaposi's sarcoma (KS) (Centers for Disease Control 1981; Gottlieb et al. 1981). *P. jirovecii* pneumonia was a disease of the profoundly immunosuppressed, as with persons on cancer treatment or with profound malnutrition. KS in high-income countries was previously recognized as a disease of elderly men in the Mediterranean basin that was far more indolent than the form that was later associated with HIV infection (Di Lorenzo 2008). Both were previously unknown in apparently healthy young American men. A rush on the rarely used drug pentamidine for young men without coexisting cancer from the Centers for Disease Control and Prevention in 1981 was a sentinel event that alerted public health officials to the new

P. jirovecii outbreak (Selik et al. 1984). In quick succession, MSM with the same opportunistic infections (OIs) or malignancies (OMs) were reported from around the world.

From 1981 through the present (2011), MSM have remained the most afflicted group in high-income nations. For example, about half of all cases in the United States in 2009 are estimated to have occurred in MSM (El-Sadr et al. 2010). STD rates remain high for MSM in major U.S. cities, as documented for gonorrhea and chlamydia by the CDC STD Surveillance Network in 2009 (Fig. 2). Trends have been dramatically upward for gonorrhea in the CDC Gonococcal Isolate Surveillance Project from 1992 to 2009 (Fig. 3). Despite the worrisome trends among MSM as a whole, the proportion of all reported HIV/AIDS cases that have occurred among White MSM in the United States has dropped markedly from the 80+% range in the early 1980s to 25% in 2010 (Fig. 4). At the same time, the proportion of cases among Black MSM in the United States has risen from just a couple percent in the early 1980s to about 24% in 2010 (El-Sadr et al. 2010). Both the prevalence and incidence rates for HIV/AIDS among Black MSM are now higher (Fig. 4) than any other subgroup of Americans, including IDUs. In Canada, Mexico, Western Europe, Australia, New Zealand, and higher-income island nations of the Caribbean, MSM continue to be the principal driver of the HIV/AIDS epidemic. This is also true of lower-income nations of Latin America and represents a growing proportion of cases in urban Asia (Baral et al. 2007; van Griensven and de Lind van Wijngaarden 2010).

The epidemic spread exceedingly among MSM, reaching peaks of greater than 50% prevalence of young MSM in the high-prevalence neighborhoods of three cities where it was first recognized (Los Angeles, New York, and San Francisco) by 1984. It is estimated that MSM incidence rates in the United States rose to 15% per annum in the early 1980s in high-prevalence communities. Rates rose among MSM in Western Europe, Australia, Canada, and other higher-income, non-Muslim countries, although not quite as quickly and gener-

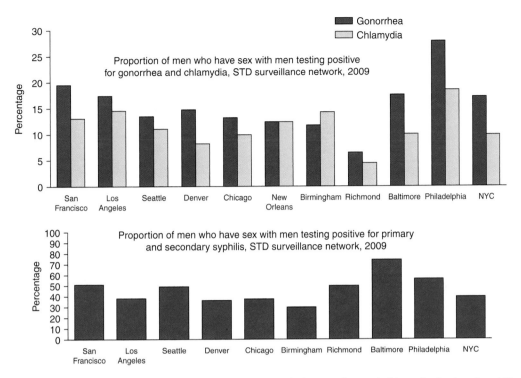

Figure 2. Proportion of men who have sex with men who had gonorrhea and chlamydia (*top*) and syphilis (*bottom*) in the STD (sexually transmitted disease) Surveillance Program of the Centers for Disease Control and Prevention (CDC), United States, 2009.

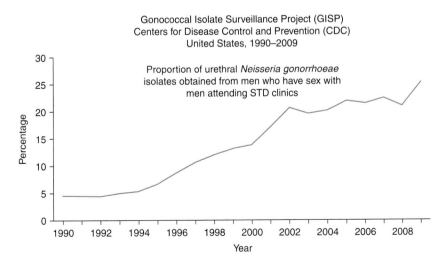

Figure 3. Proportion of urethral *Neisseria gonorrhoeae* isolates obtained from men who have sex with men attending sexually transmitted disease clinics from the Gonococcal Isolate Surveillance Project (GISP), Centers for Disease Control and Prevention (CDC), United States, 1990–2009.

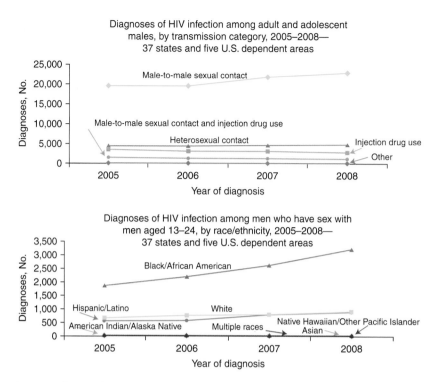

Figure 4. Diagnoses of HIV infection among adolescents and adult males by transmission category (*top*; all ages) and race/ethnicity (*bottom*; ages 13–24 years), Centers for Disease Control and Prevention (CDC), 37 states and five dependent areas of the United States, 2005–2008.

ally not as high as in the United States. In the mid-1980s, incidence rates declined in the Western world, likely due to both reduced risk behaviors and to a saturation effect of infection in the high-risk pool (i.e., the probability of becoming infected dropped owing to the uninfected persons not sexually mixing with infected, higher-risk persons). Evidence of reduced sexual risk taking by MSM in the late 1980s is found in declines in rectal gonorrhea and syphilis rates among MSM beginning in the mid-1980s. However, rates of STDs have rebounded among MSM, particularly among the young (Fig. 4). In 2009 surveillance, it is likely that such phenomena as "prevention fatigue," young MSM without a memory of AIDS in the pretreatment era, and minority MSM who are not reached by prevention messages targeting LGBTI communities (lesbian, gay, bisexual, transgender, intersex) are contributing to the worrisome trends in nations as

diverse as the United States and China (Wolitski et al. 2001; Rietmeijer et al. 2003; Johnson et al. 2008; Scheer et al. 2008; Xu et al. 2010; Hao et al. 2011).

Both behavior and biology are relevant to the MSM epidemic. Multiple sexual partners, use of recreational drugs and alcohol proximate to sex, and the practice of unprotected anal intercourse all increase HIV risk markedly. Rectal exposure to HIV is more likely to infect than vaginal exposure (Powers et al. 2008; Boily et al. 2009; Baggaley et al. 2010). This is likely owing to the large surface area and thin epithelial layer of the rectum, its capacity for fluid absorption, trauma during anal sex, and the high numbers of immunological target cells in the gastrointestinal tract (Dandekar 2007; Brenchley and Douek 2008). Antiretroviral drugs taken orally by HIV-uninfected MSM as preexposure prophylaxis (PrEP) can protect against rectal infection from HIV, but only when these drugs are adhered to and systemic plasma levels are detectable (Grant et al. 2010). This was predicted by nonhuman primate challenge studies (Garcia-Lerma et al. 2008). It is not yet known whether topical PrEP in the form of rectal microbicides will be effective biologically and, if so, whether they will be used enough to make a difference in the epidemic (McGowan 2011). Circumcision is not likely to benefit MSM, as the principal risk derives from receptive anal intercourse (Millett et al. 2008; Vermund and Qian 2008).

RISK EXPOSURES: INJECTION DRUG USE, NEEDLE SHARING, AND USE OF OTHER DRUGS/ALCOHOL

Both illicit and licit drug and alcohol use are associated with HIV risk in several ways (Samet et al. 2007). Opioids like heroin and stimulants like cocaine and methamphetamines may be injected using needles and syringes that have been shared with other drug users. This sharing of injection "works" commonly shares blood from one user to another. This injection route is highly efficient in spreading HIV, as was seen in both North America and Western Europe (Booth et al. 2006, 2008). Needle/syringe exchange programs (N/SEP) are effective at reducing HIV transmission among IDUs, but have been politically charged (Des Jarlais 2000; Downing et al. 2005; Shaw 2006; Tempalski et al. 2007). Some politicians and policymakers have argued that N/SEP should be banned because they will increase IDU use, although there is no evidence of this. Others have argued for the bans because they perceive N/SEP as an inadequate response to a large need of comprehensive drug abuse prevention and treatment programs; support of N/SEP, they have argued, will discourage funding for the more comprehensive response.

Given the policy debates, it is not surprising that some countries were far more aggressive and successful in controlling the epidemic than others (Mathers et al. 2010). Australia is one model of IDU risk reduction efficiency. Making clean needles and syringes available widely, from any pharmacy as well as special services in neighborhoods with high drug use rates, is credited for maintaining IDU HIV prevalence <2% for the past 30 years (Burrows 1998). In addition, the Australians have mainstreamed methadone use into the world of the general practitioner, reducing stigma and increasing access. Similarly, aggressive public health responses in Australia, Canada, and several Western European countries may have blunted the magnitude of HIV in their IDUs using N/SEP, and opioid substitution therapy. Even "medically supervised injecting centres" may have played a role, a topic of current study (Fischer et al. 2002).

Other nations were not as fortunate as Australia in having policies suitable to control the epidemic among IDUs (Lurie and Drucker 1997; Drucker et al. 1998). In the United States, for example, federal funding for N/SEP was banned by Congress until the early Obama administration (Millett et al. 2010), a ban with the active or the tacit approval of Reagan, G.H.W. Bush, Clinton, and G.W. Bush administrations. Funding for opioid substitution therapy did not meet the demand and waiting lists to join such programs were common. Doctors and venues licensed to distribute methadone are highly restricted, in direct contrast to the Australian approach. During that time, IDU incidence was high until community, local, and state initiatives introduced N/SEP programs that proved highly effective in reducing HIV incidence in American IDUs, as had been achieved by such programs earlier in Europe (Kwon et al. 2009; Jones et al. 2010; Palmateer et al. 2010).

Russia has had even worse policies and laws than the United States, banning opioid substitution therapy (both methadone and buprenorphine) altogether (as of 2011). Russia also has very limited N/SEP, and continues to experience some of the highest HIV seroconversion rates in the world among its IDUs (Aceijas et al. 2007; Sarang et al. 2008). The rest of Europe is a patchwork of both progressive and dysfunctional public policies vis-à-vis risk reduction for IDUs, depending in large part on whether the given nation is strongly connected to the European Community or whether they are more influenced by Russia. In summary, countries that have taken a more punitive, less pragmatic public health approach have been beleaguered by ongoing HIV spread among IDUs.

There has been a substantial increase in the 1980s and subsequent decrease of incidence and prevalence among IDUs in most high-income nations. Modeling suggests that the greater the volume of "dead space" in a used syringe, the higher the risk of residual blood infecting another drug user, as derived from ecological data on local IDU prevalence and local syringe usage patterns (Bobashev and Zule 2010). Evidence from the 1990s suggested that use of clean needles and syringes had dropped the likelihood of HIV transmission and acquisition substantially (Heimer 2008). The principal challenge is reaching all those IDUs who need N/SEP, drug treatment, and ancillary services, as well as to stem the tide of new users recruited among the young. Unfortunately, IDU itself seems to be expanding worldwide, partly owing to poor addiction and HIV care for many IDUs already infected (Wolfe et al. 2010).

Already highly endemic in Asia where most of the world's poppies are grown, the epidemic has been a principal driver of the epidemic in Iran, Pakistan, eastern India, China, Burma, Thailand, Cambodia, and Vietnam. In Africa, new reports of IDU in such unexpected venues as South Africa and Tanzania suggest the need for new research and action in that already HIV-beleaguered continent. Eastern Europe has seen some success in risk and incidence reduction, as in Estonia and Ukraine, but Russia continues to see robust HIV incidence among its IDUs owing to its failure to embrace opiate agonist therapy and vigorous N/SEP efforts (Sarang et al. 2007; Bobrova et al. 2008; Elovich and Drucker 2008; Krupitsky et al. 2010). That IDUs often cannot access clean needles and "works" when they need to in many countries of Eastern Europe and even in the United States, HIV continues to spread and cause preventable disease.

Non-IDU drug use and alcohol abuse were recognized early on as cofactors for high-risk sexual activity. "Poppers" are recreational stimulants that are popular among some MSM communities and used directly as a sexual enhancement. Disinhibition and failure to use condoms during sex have been associated particularly with stimulant and alcohol use (Khan et al. 2009; Parry et al. 2009; Hagan et al. 2011; Qian et al. 2011). There is also a body of literature citing the nonspecific immune activation resulting from drug use, from antigenic stimulation from contaminants, and immunosuppressive effects of alcohol use (Barve et al. 2002; Cabral 2006; Hahn and Samet 2010). Such immunostimulation could increase risk of acquisition of HIV by activated immune cells, whereas immunosuppression might further aggravate the $CD4^+$ cell depletions caused by HIV. Prevention efforts to reduce HIV risk among noninjection drug and alcohol users have shown mixed results (Strauss and Rindskopf 2009; Strauss et al. 2009; Crawford and Vlahov 2010).

RISK EXPOSURES: ADMINISTRATION OF BLOOD AND BLOOD PRODUCTS, AND IATROGENIC EXPOSURES

Early in the epidemic, HIV-contaminated blood and blood products were a major contributor to the high incidence of infection in the Western world. The first cases reported in 1982 were men with hemophilia (Centers for Disease Control 1982; Evatt et al. 1985). Many nations began to screen for risk based on personally reported risk characteristics (such as MSM or IDU), but this was an imperfect way to tag the at-risk blood donations. A few countries added a surrogate for HIV risk, such as hepatitis B antigen or antibody, which improved the sensitivity of the screening in the pre-HIV test period before 1985 (Busch et al. 1997; Jackson et al. 2003). As concentrated factor VIII was a notable contributor to risk, heat-treated products were introduced, although for more than a year, the old products were still marketed, despite the known risk. When the first commercially available HIV antibody test was licensed in 1985, blood banking practices changed radically in most high-income nations. Sadly, politics interfered with policy in some settings (e.g., France) or economic consideration delayed screening, such that thousands of persons were still being infected, even in richer nations, where delays in introducing HIV screening resulted in transmissions. Today in 2011, many countries like the United States use both antibody screening and antigen screening using nucleic acid tests. This is to detect those persons who might be in the period between infection and seropositivity. Given the high viral load in these infected pre-seroconverters, risk of transmission is very high. Antibody testing of the blood supply is cost-effective, although antigen testing with PCR of seronegative blood products is not (AuBuchon et al. 1997). The new generation antibody-antigen dual serology tests will surely lead to reconsideration of PCR testing, as more persons infected with HIV in the window period will now be detected with serology.

Sharing needles and/or syringes is an exceedingly efficient way to transmit HIV, and this is not merely a phenomenon of IDU. It is also seen in dysfunctional health care settings such as orphanages in Romania, and hospitals in Libya (Hersh et al. 1991a; Rosenthal 2006). It remains a global disgrace that health care workers still reuse injection equipment in some settings; even national vaccine programs have been implicated when syringes/needles have run short owing to logistical supply chain problems or owing to graft (persons redirected syringes and needles for personal gain, but reusing needles/syringes for vaccine administration). Hence, WHO and other organizations have taken a strong interest in single-use syringes to avoid reuse of needles in the health care setting (Kane et al. 1999; Simonsen et al. 1999; Ekwueme et al. 2002; Sikora et al. 2010).

RISK EXPOSURES: HETEROSEXUAL

Heterosexual HIV spread has been a reality in the HIV epidemic from the very beginning of the recognition of the epidemic in high-income nations, although far less prevalent than in Africa. In the early days of statistics dominated by MSM, and then by IDUs, it was psychologically tempting to think that heterosexuals would be spared, although no other STD or blood-borne pathogen is transmitted only among MSM and not via male-female contact. In fact, the proportion of reported HIV/AIDS cases has risen steadily in high-income countries since the start of the epidemic in the West (Burchell et al. 2008; Kramer et al. 2008; Adimora et al. 2009; Mercer et al. 2009; Toussova et al. 2009). Although expanding more slowly than in Africa or among MSM, the large population at risk results in a substantial proportion of the persons infected with HIV attributable to heterosexual contact, one-fifth to one-third of the epidemic cases in non-IDUs in most high-income countries (Malebranche 2008; Rothenberg 2009; Mah and Halperin 2010).

The magnitude of the heterosexual epidemic in the northern hemisphere and higher-income nations has not reached anywhere near the magnitude seen in southern Africa. There are many possible explanations, all of them speculative and hard to demonstrate definitively. Viral infectiousness may be higher with the C clade of virus prevalence in southern Africa than with the B clade most common in the Americas, Europe, and Australia (Novitsky et al. 2010). A second speculation is that host genetics differ in Africans than in Caucasians such that African host susceptibility is higher (Pereyra et al. 2010). A third idea is that sexual partner mixing rates are higher in Africa where absentee worker husbands and intergenerational sex both drive risky behavior (Konde-Lule et al. 1997; Pickering et al. 1997; Yirrell et al. 1998; Gregson et al. 2002; Ekanem et al. 2005; Latora et al. 2006; Hallett et al. 2007; Helleringer and Kohler 2007; Katz and Low-Beer 2008; Doherty 2011; Fritz et al. 2011; Steffenson et al. 2011). A fourth explanation is that African tribal and ethnic groups that do not ritually circumcise their men are at especially high risk of high HIV transmission. There is little scientific consensus to date as to the relative importance of these and other factors.

Male circumcision is powerfully associated with reduced risk for HIV acquisition in both epidemiologic and experimental studies (Bongaarts et al. 1989; Weiss et al. 2000; Auvert et al. 2005; Bailey et al. 2007; Gray et al. 2007; Sahasrabuddhe and Vermund 2007; Mills et al.

2008; Siegfried et al. 2009; Weiss et al. 2010). Protection was more than 50% in three independent clinical trials from South Africa, Kenya, and Uganda. Although American men are circumcised at high rates, European men are not, such that the dearth of a marked heterosexual epidemic in both North America and Europe suggests, at least, that circumcision status has not been a dominant factor in predicting heterosexual spread in the northern hemisphere.

RISK EXPOSURES: MOTHER-TO-CHILD TRANSMISSION

The steady increase of HIV among women in the West was the consequence of expanded transmission among MSM, some of whom were bisexual, and IDUs who were most often heterosexual and were sometimes women themselves. As the numbers of women with HIV infection rose, so too did the numbers of babies born who were exposed to HIV (Stringer and Vermund 1999). Fully one in four in North America and somewhat fewer (about one in eight) in Europe of exposed infants were infected in utero and intrapartum in the pre-ART era (Stringer and Vermund 1999). Viral load of the mother correlated with transmission risk. Breastfeeding was avoided early on in the epidemic once it was apparent that breast milk of an HIV-infected nursing mother could infect her infant. Once zidovudine (and then nevirapine) was demonstrated to be safe and effective in the prevention of mother-to-child transmission (PMTCT), monotherapy was instituted with tremendous impact, reducing transmission to the infant by half or more (Lindegren et al. 1999; Mofenson and McIntyre 2000; Mofenson 2003).

Without intervention, about 5% of infected mothers will be expected to transmit HIV in utero to the fetus, about 20% will transmit intrapartum, and about 15% will transmit through the breast milk, for an aggregate of 40% risk of infection to the infant (Mofenson and Fowler 1999). Prematurity increases this risk, as does high maternal viral load (VL), low maternal $CD4^+$ cell count, and vaginal (vs. cesarean)

delivery (Goldenberg et al. 2002; Fowler et al. 2007). Discovery of a risk factor does not necessarily translate directly into a remediable action. For example, although bacterial vaginosis is a risk factor for preterm birth, and preterm birth is a risk factor for HIV infection from mother to child, use of antibiotics during pregnancy to reduce bacterial vaginosis did not reduce either prematurity or HIV infection in African women in a clinical trial (HPTN 024), beyond nevirapine (NVP) alone (Goldenberg et al. 2006a,b; Taha et al. 2006).

Problems with drug resistance were noted in PMTCT, as with ART therapy in general (Fogel et al. 2011a,b). Happily, most ART was demonstrated to be tolerable and safe in pregnancy (efavirenz is an exception), and combination ART (cART) is now the rule for HIV-infected pregnant women in high-income countries. Pediatric HIV has ceased to be a major public health problem in higher-income countries as routine "opt-out" testing for all pregnant women maximizes case detection, and use of cART reduces transmission to a minimum, with replacement "formula" feeding (Jamieson et al. 2007; Committee on Pediatric AIDS 2008; Gazzard et al. 2008). In Africa, in contrast, much progress is needed, although our tools for program evaluation and breastfeeding management have improved (Stringer et al. 2005, 2008, 2010; Reithinger et al. 2007; Mofenson 2010). The impact of opt-out testing in pregnancy, high coverage with cART, and replacement feeding for infants is encouraging in high-income nations. The peak of 1700 reported cases of pediatric AIDS occurred in 1992; subsequently, cases have decreased to <50 new cases of AIDS annually (a >96% reduction) and <300 annual perinatal HIV transmissions in 2005; despite that, the number of HIV-infected women continues to rise (Fowler et al. 2010; Lansky et al. 2010). Given stability of HIV incidence and the frequency of pregnancy in HIV-infected women, the programs to prevent perinatal transmission must be maintained and access assured (Wade et al. 2004; Volmink et al. 2007; Peters et al. 2008; McDonald et al. 2009; Birkhead et al. 2010; Yang et al. 2010).

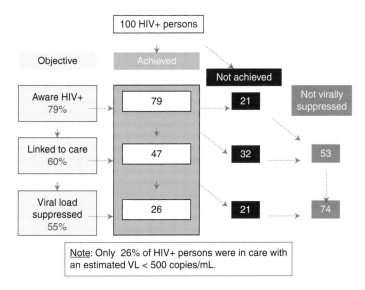

Figure 5. This is a model of how we estimate that only 26 of each 100 HIV-infected people in the United States are virally suppressed such that they would be expected to have a very slow disease progression and would be minimally infectious to others. Based on Centers for Disease Control and Prevention (CDC), United States, 2009 estimates of the proportion of HIV-infected persons in the United States who know their HIV-seropositive status (79%), the proportion of those persons who are linked to HIV care (60%), and the proportion of them who are virally suppressed (55%), this is a cascade model of the overall number of 100 HIV-infected persons who are currently immunologically suppressed (only 26). (From Burns et al. 2010.)

PREVENTION OF HIV IN HIGH-INCOME NATIONS

Other works in this collection discuss preexposure prophylaxis and behavioral change, and we discussion prevention only very briefly here. Behavior change remains a backbone of interventions in higher-income nations, although results have been somewhat disappointing as with Project EXPLORE/HIVNET 015 (Chesney et al. 2003; Koblin et al. 2003, 2004; Colfax et al. 2004, 2005; Salomon et al. 2009). Combination interventions are now being explored (Barrow et al. 2008; Corsi and Booth 2008; Buchbinder 2009; Rotheram-Borus et al. 2009; Burns et al. 2010; Cohen et al. 2010; Crawford and Vlahov 2010; DeGruttola et al. 2010; El-Bassel et al. 2010; Read 2010; Reynolds and Quinn 2010), as are behavioral approaches with a better base in evidence (Collins et al. 2006; Harshbarger et al. 2006; Lyles et al. 2006; Wingood and DiClemente 2006; Margaret Dolcini et al. 2010). One strategy has been a conspicuous failure, namely "abstinence only" education, discovered to be associated with higher pregnancy rates than more comprehensive approaches that also highlighted STD prevention with condom use (Ott and Santelli 2007; Kohler et al. 2008; Trenholm et al. 2008). It is thought that the more pragmatic sexual education approaches in Europe and Australia, with structural changes like widespread provision of condom dispensing machines in bathrooms, may explain why the heterosexual epidemic among heterosexuals has been lower than in the United States where comprehensive sex education and condom advertising and distribution have been comparatively curtailed (Dworkin and Ehrhardt 2007). Structural interventions and behavior interventions can support biomedical advances in PrEP (both oral and topical [microbicides]), cART for prevention by reducing viral load among infectious persons (Fig. 5). Multicomponent interventions show promise and must be studied (Vermund et al. 2010; Kurth et al. 2011).

REFERENCES

Reference is also in this collection.

Abecasis AB, Lemey P, Vidal N, de Oliveira T, Peeters M, Camacho R, Shapiro B, Rambaut A, Vandamme AM. 2007. Recombination confounds the early evolutionary history of human immunodeficiency virus type 1: Subtype G is a circulating recombinant form. *J Virol* **81**: 8543–8551.

Aceijas C, Stimson GV, Hickman M, Rhodes T. 2004. Global overview of injecting drug use and HIV infection among injecting drug users. *AIDS* **18**: 2295–2303.

Aceijas C, Hickman M, Donoghoe MC, Burrows D, Stuikyte R. 2007. Access and coverage of needle and syringe programmes (NSP) in Central and Eastern Europe and Central Asia. *Addiction* **102**: 1244–1250.

Adimora AA, Schoenbach VJ, Floris-Moore MA. 2009. Ending the epidemic of heterosexual HIV transmission among African Americans. *Am J Prev Med* **37**: 468–471.

Alizon M, Wain-Hobson S, Montagnier L, Sonigo P. 1986. Genetic variability of the AIDS virus: Nucleotide sequence analysis of two isolates from African patients. *Cell* **46**: 63–74.

Anderson JP, Rodrigo AG, Learn GH, Madan A, Delahunty C, Coon M, Girard M, Osmanov S, Hood L, Mullins JI. 2000. Testing the hypothesis of a recombinant origin of human immunodeficiency virus type 1 subtype E. *J Virol* **74**: 10752–10765.

Arnold C, Barlow KL, Parry JV, Clewley JP. 1995. At least five HIV-1 sequence subtypes (A, B, C, D, A/E) occur in England. *AIDS Res Hum Retroviruses* **11**: 427–429.

AuBuchon JP, Birkmeyer JD, Busch MP. 1997. Cost-effectiveness of expanded human immunodeficiency virus-testing protocols for donated blood. *Transfusion* **37**: 45–51.

Auvert B, Taljaard D, Lagarde E, Sobngwi-Tambekou J, Sitta R, Puren A. 2005. Randomized, controlled intervention trial of male circumcision for reduction of HIV infection risk: The ANRS 1265 trial. *PLoS Med* **2**: e298. doi: 10.1371/journal.pmed.0020298.

Baggaley RF, White RG, Boily MC. 2010. HIV transmission risk through anal intercourse: Systematic review, meta-analysis and implications for HIV prevention. *Int J Epidemiol* **39**: 1048–1063.

Bailey RC, Moses S, Parker CB, Agot K, Maclean I, Krieger JN, Williams CF, Campbell RT, Ndinya-Achola JO. 2007. Male circumcision for HIV prevention in young men in Kisumu, Kenya: A randomised controlled trial. *Lancet* **369**: 643–656.

Balfe P, Simmonds P, Ludlam CA, Bishop JO, Brown AJ. 1990. Concurrent evolution of human immunodeficiency virus type 1 in patients infected from the same source: Rate of sequence change and low frequency of inactivating mutations. *J Virol* **64**: 6221–6233.

Baral S, Sifakis F, Cleghorn F, Beyrer C. 2007. Elevated risk for HIV infection among men who have sex with men in low- and middle-income countries 2000–2006: A systematic review. *PLoS Med* **4**: e339. doi: 10.1371/journal.pmed.0040339.

Barrow RY, Berkel C, Brooks LC, Groseclose SL, Johnson DB, Valentine JA. 2008. Traditional sexually transmitted disease prevention and control strategies: Tailoring for African American communities. *Sex Transm Dis* **35**: S30–S39.

Barth RE, Huijgen Q, Taljaard J, Hoepelman AI. 2010. Hepatitis B/C and HIV in sub-Saharan Africa: An association between highly prevalent infectious diseases. A systematic review and meta-analysis. *Int J Infect Dis* **14**: e1024–e1031.

Barve SS, Kelkar SV, Gobejishvilli L, Joshi-Barve S, McClain CJ. 2002. Mechanisms of alcohol-mediated CD4$^+$ T lymphocyte death: Relevance to HIV and HCV pathogenesis. *Front Biosci* **7**: d1689–d1696.

Birkhead GS, Pulver WP, Warren BL, Hackel S, Rodriguez D, Smith L. 2010. Acquiring human immunodeficiency virus during pregnancy and mother-to-child transmission in New York: 2002–2006. *Obstet Gynecol* **115**: 1247–1255.

Bobashev GV, Zule WA. Modeling the effect of high dead-space syringes on the human immunodeficiency virus (HIV) epidemic among injecting drug users. 2010. *Addiction* **105**: 1439–1447.

Bobkov A, Cheingsong-Popov R, Garaev M, Rzhaninova A, Kaleebu P, Beddows S, Bachmann MH, Mullins JI, Louwagie J, Janssens W, et al. 1994. Identification of an env G subtype and heterogeneity of HIV-1 strains in the Russian Federation and Belarus. *AIDS* **8**: 1649–1655.

Bobkov A, Cheingsong-Popov R, Selimova L, Ladnaya N, Kazennova E, Kravchenko A, Fedotov E, Saukhat S, Zverev S, Pokrovsky V, et al. 1997. An HIV type 1 epidemic among injecting drug users in the former Soviet Union caused by a homogeneous subtype A strain. *AIDS Res Hum Retroviruses* **13**: 1195–1201.

Bobkov A, Kazennova E, Khanina T, Bobkova M, Selimova L, Kravchenko A, Pokrovsky V, Weber J. 2001. An HIV type 1 subtype A strain of low genetic diversity continues to spread among injecting drug users in Russia: Study of the new local outbreaks in Moscow and Irkutsk. *AIDS Res Hum Retroviruses* **17**: 257–261.

Bobkov AF, Kazennova EV, Sukhanova AL, Bobkova MR, Pokrovsky VV, Zeman VV, Kovtunenko NG, Erasilova IB. 2004. An HIV type 1 subtype A outbreak among injecting drug users in Kazakhstan. *AIDS Res Hum Retroviruses* **20**: 1134–1136.

Bobrova N, Rughnikov U, Neifeld E, Rhodes T, Alcorn R, Kirichenko S, Power R. 2008. Challenges in providing drug user treatment services in Russia: Providers' views. *Subst Use Misuse* **43**: 1770–1784.

Boily MC, Baggaley RF, Wang L, Masse B, White RG, Hayes RJ, Alary M. 2009. Heterosexual risk of HIV-1 infection per sexual act: Systematic review and meta-analysis of observational studies. *Lancet Infect Dis* **9**: 118–129.

Bongaarts J, Reining P, Way P, Conant F. 1989. The relationship between male circumcision and HIV infection in African populations. *AIDS* **3**: 373–377.

Booth RE, Kwiatkowski CF, Brewster JT, Sinitsyna L, Dvoryak S. 2006. Predictors of HIV sero-status among drug injectors at three Ukraine sites. *AIDS* **20**: 2217–2223.

Booth RE, Lehman WE, Kwiatkowski CF, Brewster JT, Sinitsyna L, Dvoryak S. 2008. Stimulant injectors in Ukraine: The next wave of the epidemic? *AIDS Behav* **12**: 652–661.

Brenchley JM, Douek DC. 2008. HIV infection and the gastrointestinal immune system. *Mucosal Immunol* **1:** 23–30.

Brenner BG, Roger M, Moisi DD, Oliveira M, Hardy I, Turgel R, Charest H, Routy JP, Wainberg MA. 2008. Transmission networks of drug resistance acquired in primary/early stage HIV infection. *AIDS* **22:** 2509–2515.

Buchbinder S. 2009. The epidemiology of new HIV infections and interventions to limit HIV transmission. *Top HIV Med* **17:** 37–43.

Burchell AN, Calzavara LM, Orekhovsky V, Ladnaya NN. 2008. Characterization of an emerging heterosexual HIV epidemic in Russia. *Sex Transm Dis* **35:** 807–813.

Burns DN, Dieffenbach CW, Vermund SH. 2010. Rethinking prevention of HIV type 1 infection. *Clin Infect Dis* **51:** 725–731.

Burrows D. 1998. Injecting equipment provision in Australia: The state of play. *Subst Use Misuse* **33:** 1113–1127.

Busch MP, Dodd RY, Lackritz EM, AuBuchon JP, Birkmeyer JD, Petersen LR. 1997. Value and cost-effectiveness of screening blood donors for antibody to hepatitis B core antigen as a way of detecting window-phase human immunodeficiency virus type 1 infections. The HIV Blood Donor Study Group. *Transfusion* **37:** 1003–1011.

Cabral GA. 2006. Drugs of abuse, immune modulation, and AIDS. *J Neuroimmune Pharmacol* **1:** 280–295.

Cane P, Chrystie I, Dunn D, Evans B, Geretti AM, Green H, Phillips A, Pillay D, Porter K, Pozniak A, et al. 2005. Time trends in primary resistance to HIV drugs in the United Kingdom: Multicentre observational study. *BMJ* **331:** 1368.

* Carrington M, Alter G. 2011. Innate immune control of HIV disease. *Cold Spring Harbor Perspect Med* doi: 10.1101/cshperspect.a007070.

Centers for Disease Control. 1981. Kaposi's sarcoma and *Pneumocystis* pneumonia among homosexual men—New York City and California. *MMWR Morb Mortal Wkly Rep* **30:** 305–308.

Centers for Disease Control. 1982. Update on acquired immune deficiency syndrome (AIDS)—United States. *MMWR Morb Mortal Wkly Rep* **31:** 507–508, 513–504.

Chaix ML, Descamps D, Wirden M, Bocket L, Delaugerre C, Tamalet C, Schneider V, Izopet J, Masquelier B, Rouzioux C, et al. 2009. Stable frequency of HIV-1 transmitted drug resistance in patients at the time of primary infection over 1996–2006 in France. *AIDS* **23:** 717–724.

Chesney MA, Koblin BA, Barresi PJ, Husnik MJ, Celum CL, Colfax G, Mayer K, McKirnan D, Judson FN, Huang Y, et al. 2003. An individually tailored intervention for HIV prevention: Baseline data from the EXPLORE Study. *Am J Public Health* **93:** 933–938.

Cohen MS, Gay CL, Busch MP, Hecht FM. 2010. The detection of acute HIV infection. *J Infect Dis* **202** Suppl 2: S270–S277.

Cohen MS, Haynes BF, McMichael AJ, Shaw GM. 2011. Medical Progress: Acute HIV-1 infection. *N Engl J Med* **364:** 1943–1954.

Colfax G, Vittinghoff E, Husnik MJ, McKirnan D, Buchbinder S, Koblin B, Celum C, Chesney M, Huang Y, Mayer K, et al. 2004. Substance use and sexual risk: A participant- and episode-level analysis among a cohort of men who have sex with men. *Am J Epidemiol* **159:** 1002–1012.

Colfax G, Coates TJ, Husnik MJ, Huang Y, Buchbinder S, Koblin B, Chesney M, Vittinghoff E. 2005. Longitudinal patterns of methamphetamine, popper (amyl nitrite), and cocaine use and high-risk sexual behavior among a cohort of San Francisco men who have sex with men. *J Urban Health* **82:** 62–70.

Collins C, Harshbarger C, Sawyer R, Hamdallah M. 2006. The diffusion of effective behavioral interventions project: Development, implementation, and lessons learned. *AIDS Educ Prev* **18:** 5–20.

Committee on Pediatric AIDS. 2008. HIV testing and prophylaxis to prevent mother-to-child transmission in the United States. *Pediatrics* **122:** 1127–1134.

Corsi KF, Booth RE. 2008. HIV sex risk behaviors among heterosexual methamphetamine users: Literature review from 2000 to present. *Curr Drug Abuse Rev* **1:** 292–296.

Coutinho RA, van Griensven GJ, Moss A. 1989. Effects of preventive efforts among homosexual men. *AIDS* **3** Suppl 1: S53–S56.

Crandall KA, Vasco DA, Posada D, Imamichi H. 1999. Advances in understanding the evolution of HIV. *AIDS* **13** Suppl A: S39–S47.

Crawford ND, Vlahov D. 2010. Progress in HIV reduction and prevention among injection and noninjection drug users. *J Acquir Immune Defic Syndr* **55** Suppl 2: S84–S87.

Dandekar S. 2007. Pathogenesis of HIV in the gastrointestinal tract. *Curr HIV/AIDS Rep* **4:** 10–15.

De Cock KM, Brun-Vezinet F. 1989. Epidemiology of HIV-2 infection. *AIDS* **3** Suppl 1: S89–S95.

DeGruttola V, Smith DM, Little SJ, Miller V. 2010. Developing and evaluating comprehensive HIV infection control strategies: Issues and challenges. *Clin Infect Dis* **50** Suppl 3: S102–S107.

Dehne KL, Khodakevich L, Hamers FF, Schwartlander B. 1999. The HIV/AIDS epidemic in eastern Europe: Recent patterns and trends and their implications for policymaking. *AIDS* **13:** 741–749.

Des Jarlais DC. 2000. Research, politics, and needle exchange. *Am J Public Health* **90:** 1392–1394.

de Oliveira T, Pillay D, Gifford RJ. 2010. The HIV-1 subtype C epidemic in South America is linked to the United Kingdom. *PLoS One* **5:** e9311. doi: 10.1371/journal.pone.0009311.

de Ronde A, Schuurman R, Goudsmit J, van den Hoek A, Boucher C. 1996. First case of new infection with zidovudine-resistant HIV-1 among prospectively studied intravenous drug users and homosexual men in Amsterdam, The Netherlands. *AIDS* **10:** 231–232.

Di Lorenzo G. 2008. Update on classic Kaposi sarcoma therapy: New look at an old disease. *Crit Rev Oncol Hematol* **68:** 242–249.

Doherty IA. 2011. Sexual networks and sexually transmitted infections: Innovations and findings. *Curr Opin Infect Dis* **24:** 70–77.

Downing M, Riess TH, Vernon K, Mulia N, Hollinquest M, McKnight C, Jarlais DC, Edlin BR. 2005. What's community got to do with it? Implementation models of syringe exchange programs. *AIDS Educ Prev* **17:** 68–78.

Drucker E, Lurie P, Wodak A, Alcabes P. 1998. Measuring harm reduction: The effects of needle and syringe exchange programs and methadone maintenance on the ecology of HIV. *AIDS* **12** Suppl A: S217–S230.

Drummond AJ, Rambaut A. 2007. BEAST: Bayesian evolutionary analysis by sampling trees. *BMC Evol Biol* **7**: 214.

Dworkin SL, Ehrhardt AA. 2007. Going beyond "ABC" to include "GEM": Critical reflections on progress in the HIV/AIDS epidemic. *Am J Public Health* **97**: 13–18.

Dwyre DM, Fernando LP, Holland PV. 2011. Hepatitis B, hepatitis C and HIV transfusion-transmitted infections in the 21st century. *Vox Sang* **100**: 92–98.

Ekanem EE, Afolabi BM, Nuga AO, Adebajo SB. 2005. Sexual behaviour, HIV-related knowledge and condom use by intra-city commercial bus drivers and motor park attendants in Lagos, Nigeria. *Afr J Reprod Health* **9**: 78–87.

Ekwueme DU, Weniger BG, Chen RT. 2002. Model-based estimates of risks of disease transmission and economic costs of seven injection devices in sub-Saharan Africa. *Bull World Health Organ* **80**: 859–870.

El-Bassel N, Gilbert L, Witte S, Wu E, Hunt T, Remien RH. 2010. Couple-based HIV prevention in the United States: Advantages, gaps, and future directions. *J Acquir Immune Defic Syndr* **55** Suppl 2: S98–S101.

Elovich R, Drucker E. 2008. On drug treatment and social control: Russian narcology's great leap backwards. *Harm Reduct J* **5**: 23.

El-Sadr WM, Mayer KH, Hodder SL. 2010. AIDS in America—Forgotten but not gone. *N Engl J Med* **362**: 967–970.

Evatt BL, Gomperts ED, McDougal JS, Ramsey RB. 1985. Coincidental appearance of LAV/HTLV-III antibodies in hemophiliacs and the onset of the AIDS epidemic. *N Engl J Med* **312**: 483–486.

Fischer B, Rehm J, Kirst M, Casas M, Hall W, Krausz M, Metrebian N, Reggers J, Uchtenhagen A, van den Brink W, et al. 2002. Heroin-assisted treatment as a response to the public health problem of opiate dependence. *Eur J Public Health* **12**: 228–234.

Fogel J, Hoover DR, Sun J, Mofenson LM, Fowler MG, Taylor AW, Kumwenda N, Taha TE, Eshleman SH. 2011a. Analysis of nevirapine resistance in HIV-infected infants who received extended nevirapine or nevirapine/zidovudine prophylaxis. *AIDS* **25**: 911–917.

Fogel J, Li Q, Taha TE, Hoover DR, Kumwenda NI, Mofenson LM, Kumwenda JJ, Fowler MG, Thigpen MC, Eshleman SH. 2011b. Initiation of antiretroviral treatment in women after delivery can induce multiclass drug resistance in breastfeeding HIV-infected infants. *Clin Infect Dis* **52**: 1069–1076.

Fowler MG, Lampe MA, Jamieson DJ, Kourtis AP, Rogers MF. 2007. Reducing the risk of mother-to-child human immunodeficiency virus transmission: Past successes, current progress and challenges, and future directions. *Am J Obstet Gynecol* **197**: S3–S9.

Fowler MG, Gable AR, Lampe MA, Etima M, Owor M. 2010. Perinatal HIV and its prevention: Progress toward an HIV-free generation. *Clin Perinatol* **37**: 699–719, vii.

Fox J, Castro H, Kaye S, McClure M, Weber JN, Fidler S. 2010. Epidemiology of non-B clade forms of HIV-1 in men who have sex with men in the UK. *AIDS* **24**: 2397–2401.

Fritz K, McFarland W, Wyrod R, Chasakara C, Makumbe K, Chirowodza A, Mashoko C, Kellogg T, Woelk G. 2011. Evaluation of a peer network-based sexual risk reduction intervention for men in beer halls in Zimbabwe: Results from a randomized controlled trial. *AIDS Behav.* doi: 10.1007/s10461-011-9922-1.

Frost SD, Nijhuis M, Schuurman R, Boucher CA, Brown AJ. 2000. Evolution of lamivudine resistance in human immunodeficiency virus type 1-infected individuals: The relative roles of drift and selection. *J Virol* **74**: 6262–6268.

Frost SD, Wrin T, Smith DM, Kosakovsky Pond SL, Liu Y, Paxinos E, Chappey C, Galovich J, Beauchaine J, Petropoulos CJ, et al. 2005. Neutralizing antibody responses drive the evolution of human immunodeficiency virus type 1 envelope during recent HIV infection. *Proc Natl Acad Sci* **102**: 18514–18519.

Garcia-Lerma JG, Otten RA, Qari SH, Jackson E, Cong ME, Masciotra S, Luo W, Kim C, Adams DR, Monsour M, et al. 2008. Prevention of rectal SHIV transmission in macaques by daily or intermittent prophylaxis with emtricitabine and tenofovir. *PLoS Med* **5**: e28. doi: 10.1371/journal.pmed.0050028.

Gazzard B, Clumeck N, d'Arminio Monforte A, Lundgren JD. 2008. Indicator disease-guided testing for HIV— The next step for Europe? *HIV Med* **9** Suppl 2: 34–40.

Gilbert MT, Rambaut A, Wlasiuk G, Spira TJ, Pitchenik AE, Worobey M. 2007. The emergence of HIV/AIDS in the Americas and beyond. *Proc Natl Acad Sci* **104**: 18566–18570.

Goldenberg RL, Stringer JS, Sinkala M, Vermund SH. 2002. Perinatal HIV transmission: Developing country considerations. *J Matern Fetal Neonatal Med* **12**: 149–158.

Goldenberg RL, Mudenda V, Read JS, Brown ER, Sinkala M, Kamiza S, Martinson F, Kaaya E, Hoffman I, Fawzi W, et al. 2006a. HPTN 024 study: Histologic chorioamnionitis, antibiotics and adverse infant outcomes in a predominantly HIV-1-infected African population. *Am J Obstet Gynecol* **195**: 1065–1074.

Goldenberg RL, Mwatha A, Read JS, Adeniyi-Jones S, Sinkala M, Msmanga G, Martinson F, Hoffman I, Fawzi W, Valentine M, et al. 2006b. The HPTN 024 Study: The efficacy of antibiotics to prevent chorioamnionitis and preterm birth. *Am J Obstet Gynecol* **194**: 650–661.

Gottlieb MS, Schroff R, Schanker HM, Weisman JD, Fan PT, Wolf RA, Saxon A. 1981. *Pneumocystis carinii* pneumonia and mucosal candidiasis in previously healthy homosexual men: Evidence of a new acquired cellular immunodeficiency. *N Engl J Med* **305**: 1425–1431.

Grant RM, Lama JR, Anderson PL, McMahan V, Liu AY, Vargas L, Goicochea P, Casapia M, Guanira-Carranza JV, Ramirez-Cardich ME, et al. 2010. Preexposure chemoprophylaxis for HIV prevention in men who have sex with men. *N Engl J Med* **363**: 2587–2599.

Gray RH, Kigozi G, Serwadda D, Makumbi F, Watya S, Nalugoda F, Kiwanuka N, Moulton LH, Chaudhary MA, Chen MZ, et al. 2007. Male circumcision for HIV prevention in men in Rakai, Uganda: A randomised trial. *Lancet* **369**: 657–666.

Gregson S, Nyamukapa CA, Garnett GP, Mason PR, Zhuwau T, Carael M, Chandiwana SK, Anderson RM. 2002. Sexual mixing patterns and sex-differentials in teenage exposure to HIV infection in rural Zimbabwe. *Lancet* **359:** 1896–1903.

Hagan H, Perlman DC, Des Jarlais DC. 2011. Sexual risk and HIV infection among drug users in New York City: A pilot study. *Subst Use Misuse* **46:** 201–207.

Hahn JA, Samet JH. 2010. Alcohol and HIV disease progression: Weighing the evidence. *Curr HIV/AIDS Rep* **7:** 226–233.

Hallett TB, Gregson S, Lewis JJ, Lopman BA, Garnett GP. 2007. Behaviour change in generalised HIV epidemics: impact of reducing cross-generational sex and delaying age at sexual debut. *Sex Transm Infect* **83** (Suppl 1): i50–i54.

Hamers FF, Batter V, Downs AM, Alix J, Cazein F, Brunet JB. 1997. The HIV epidemic associated with injecting drug use in Europe: Geographic and time trends. *AIDS* **11:** 1365–1374.

Hao C, Yan H, Yang H, Huan X, Guan W, Xu X, Zhang M, Tang W, Wang N, Gu J, et al. 2011. The incidence of syphilis, HIV and HCV and associated factors in a cohort of men who have sex with men in Nanjing, China. *Sex Transm Infect* **87:** 199–201.

Harshbarger C, Simmons G, Coelho H, Sloop K, Collins C. 2006. An empirical assessment of implementation, adaptation, and tailoring: The evaluation of CDC's National Diffusion of VOICES/VOCES. *AIDS Educ Prev* **18:** 184–197.

Heimer R. 2008. Community coverage and HIV prevention: Assessing metrics for estimating HIV incidence through syringe exchange. *Int J Drug Policy* **19** (Suppl 1): S65–S73.

Helleringer S, Kohler HP. 2007. Sexual network structure and the spread of HIV in Africa: Evidence from Likoma Island, Malawi. *AIDS* **21:** 2323–2332.

Hersh BS, Popovici F, Apetrei RC, Zolotusca L, Beldescu N, Calomfirescu A, Jezek Z, Oxtoby MJ, Gromyko A, Heymann DL. 1991a. Acquired immunodeficiency syndrome in Romania. *Lancet* **338:** 645–649.

Hersh BS, Popovici F, Zolotusca L, Beldescu N, Oxtoby MJ, Gayle HD. 1991b. The epidemiology of HIV and AIDS in Romania. *AIDS* **5** (Suppl 2): S87–S92.

Holmes EC, Zhang LQ, Robertson P, Cleland A, Harvey E, Simmonds P, Leigh Brown AJ. 1995. The molecular epidemiology of human immunodeficiency virus type 1 in Edinburgh. *J Infect Dis* **171:** 45–53.

Hue S, Pillay D, Clewley JP, Pybus OG. 2005. Genetic analysis reveals the complex structure of HIV-1 transmission within defined risk groups. *Proc Natl Acad Sci* **102:** 4425–4429.

Hughes GJ, Fearnhill E, Dunn D, Lycett SJ, Rambaut A, Leigh Brown AJ. 2009. Molecular phylodynamics of the heterosexual HIV epidemic in the United Kingdom. *PLoS Pathog* **5:** e1000590. doi: 10.1371/journal.ppat.1000590.

Jackson BR, Busch MP, Stramer SL, AuBuchon JP. 2003. The cost-effectiveness of NAT for HIV, HCV, and HBV in whole-blood donations. *Transfusion* **43:** 721–729.

Jamieson DJ, Clark J, Kourtis AP, Taylor AW, Lampe MA, Fowler MG, Mofenson LM. 2007. Recommendations for human immunodeficiency virus screening, prophylaxis, and treatment for pregnant women in the United States. *Am J Obstet Gynecol* **197:** S26–S32.

Johnson CV, Mimiaga MJ, Bradford J. 2008. Health care issues among lesbian, gay, bisexual, transgender and intersex (LGBTI) populations in the United States: Introduction. *J Homosex* **54:** 213–224.

Jones JH, Handcock MS. 2003. An assessment of preferential attachment as a mechanism for human sexual network formation. *Proc Biol Sci* **270:** 1123–1128.

Jones L, Pickering L, Sumnall H, McVeigh J, Bellis MA. 2010. Optimal provision of needle and syringe programmes for injecting drug users: A systematic review. *Int J Drug Policy* **21:** 335–342.

Kalish ML, Robbins KE, Pieniazek D, Schaefer A, Nzilambi N, Quinn TC, St Louis ME, Youngpairoj AS, Phillips J, Jaffe HW, et al. 2004. Recombinant viruses and early global HIV-1 epidemic. *Emerg Infect Dis* **10:** 1227–1234.

Kane A, Lloyd J, Zaffran M, Simonsen L, Kane M. 1999. Transmission of hepatitis B, hepatitis C and human immunodeficiency viruses through unsafe injections in the developing world: Model-based regional estimates. *Bull World Health Organ* **77:** 801–807.

Katz I, Low-Beer D. 2008. Why has HIV stabilized in South Africa, yet not declined further? Age and sexual behavior patterns among youth. *Sex Transm Dis* **35:** 837–842.

Keeling MJ, Eames KT. 2005. Networks and epidemic models. *J R Soc Interface* **2:** 295–307.

Khan MR, Bolyard M, Sandoval M, Mateu-Gelabert P, Krauss B, Aral SO, Friedman SR. 2009. Social and behavioral correlates of sexually transmitted infection- and HIV-discordant sexual partnerships in Bushwick, Brooklyn, New York. *J Acquir Immune Defic Syndr* **51:** 470–485.

Kilmarx PH. 2009. Global epidemiology of HIV. *Curr Opin HIVAIDS* **4:** 240–246.

Koblin BA, Chesney MA, Husnik MJ, Bozeman S, Celum CL, Buchbinder S, Mayer K, McKirnan D, Judson FN, Huang Y, et al. 2003. High-risk behaviors among men who have sex with men in 6 US cities: Baseline data from the EXPLORE Study. *Am J Public Health* **93:** 926–932.

Koblin B, Chesney M, Coates T. 2004. Effects of a behavioural intervention to reduce acquisition of HIV infection among men who have sex with men: The EXPLORE randomised controlled study. *Lancet* **364:** 41–50.

Kohler PK, Manhart LE, Lafferty WE. 2008. Abstinence-only and comprehensive sex education and the initiation of sexual activity and teen pregnancy. *J Adolesc Health* **42:** 344–351.

Konde-Lule JK, Sewankambo N, Morris M. 1997. Adolescent sexual networking and HIV transmission in rural Uganda. *Health Transit Rev* **7** (Suppl): 89–100.

Korber B, Muldoon M, Theiler J, Gao F, Gupta R, Lapedes A, Hahn BH, Wolinsky S, Bhattacharya T. 2000. Timing the ancestor of the HIV-1 pandemic strains. *Science* **288:** 1789–1796.

Kouyos RD, von W, Yerly S, Boni J, Taffe P, Shah C, Burgisser P, Klimkait T, Weber R, Hirschel B, et al. 2010. Molecular epidemiology reveals long-term changes in HIV type 1 subtype B transmission in Switzerland. *J Infect Dis* **201:** 1488–1497.

Kramer MA, van Veen MG, de Coul EL, Geskus RB, Coutinho RA, van de Laar MJ, Prins M. 2008. Migrants travelling to their country of origin: A bridge population for HIV transmission? *Sex Transm Infect* **84:** 554–555.

Krupitsky E, Woody GE, Zvartau E, O'Brien CP. 2010. Addiction treatment in Russia. *Lancet* **376:** 1145.

Kuiken CL, Goudsmit J. 1994. Silent mutation pattern in V3 sequences distinguishes virus according to risk group in Europe. *AIDS Res Hum Retroviruses* **10:** 319–320.

Kuiken CL, Cornelissen MT, Zorgdrager F, Hartman S, Gibbs AJ, Goudsmit J. 1996. Consistent risk group-associated differences in human immunodeficiency virus type 1 vpr, vpu and V3 sequences despite independent evolution. *J Gen Virol* **77** (Pt 4)**:** 783–792.

Kuiken C, Foley B, Leitner T, Apetrei RC, Hahn BH, Mizrachi I, Mullins JI, Rambaut A, Wolinsky S, Korber B, ed. 2010. *HIV Sequence Compendium 2010.* Theoretical Biology and Biophysics Group, Los Alamos National Laboratory, NM.

Kurth AE, Celum C, Baeten JM, Vermund SH, Wasserheit JN. 2011. Combination HIV prevention: Significance, challenges, and opportunities. *Curr HIV/AIDS Rep* **8:** 62–72.

Kwon JA, Iversen J, Maher L, Law MG, Wilson DP. 2009. The impact of needle and syringe programs on HIV and HCV transmissions in injecting drug users in Australia: A model-based analysis. *J Acquir Immune Defic Syndr* **51:** 462–469.

* Lackner AA, Lederman MM, Rodriguez B. 2011. HIV pathogenesis—The host. *Cold Spring Harb Perspect Med* doi: 10.1101/cshperspect.a007005.

Lampe MA, Nesheim S, Shouse RL, Borkowf CB, Minasandram V, Little K, Kilmarx PH, Whitmore S, Taylor A, Valleroy L. 2010. Racial/ethnic disparities among children with diagnoses of perinatal HIV infection—34 states, 2004–2007. *MMWR Morb Mortal Wkly Rep* **59:** 97–101.

Lansky A, Brooks JT, DiNenno E, Heffelfinger J, Hall HI, Mermin J. 2010. Epidemiology of HIV in the United States. *J Acquir Immune Defic Syndr* **55** Suppl 2**:** S64–S68.

Latora V, Nyamba A, Simpore J, Sylvette B, Diane S, Sylvere B, Musumeci S. 2006. Network of sexual contacts and sexually transmitted HIV infection in Burkina Faso. *J Med Virol* **78:** 724–729.

Leigh Brown AJ. 1997. Analysis of HIV-1 env gene sequences reveals evidence for a low effective number in the viral population. *Proc Natl Acad Sci* **94:** 1862–1865.

Leigh Brown AJ, Richman DD. 1997. HIV-1: Gambling on the evolution of drug resistance? *Nat Med* **3:** 268–271.

Leigh Brown AJ, Lobidel D, Wade CM, Rebus S, Phillips AN, Brettle RP, France AJ, Leen CS, McMenamin J, McMillan A, et al. 1997. The molecular epidemiology of human immunodeficiency virus type 1 in six cities in Britain and Ireland. *Virology* **235:** 166–177.

Leigh Brown AJ, Frost SD, Mathews WC, Dawson K, Hellmann NS, Daar ES, Richman DD, Little SJ. 2003. Transmission fitness of drug-resistant human immunodeficiency virus and the prevalence of resistance in the antiretroviral-treated population. *J Infect Dis* **187:** 683–686.

Leigh Brown AJ, Lycett SJ, Weinert L, Hughes GJ, Fearnhill E, Dunn DT. UK HIV Drug Resistance Collaboration. 2011. Transmission network parameters estimated from HIV sequences for a nation-wide epidemic. *J Infect Dis* (in press).

Lemey P, Pybus OG, Wang B, Saksena NK, Salemi M, Vandamme AM. 2003. Tracing the origin and history of the HIV-2 epidemic. *Proc Natl Acad Sci* **100:** 6588–6592.

Lewis F, Hughes GJ, Rambaut A, Pozniak A, Leigh Brown AJ. 2008. Episodic sexual transmission of HIV revealed by molecular phylodynamics. *PLoS Med* **5:** e50. doi: 10.1371/journal.pmed.0050050.

Liitsola K, Tashkinova I, Laukkanen T, Korovina G, Smolskaja T, Momot O, Mashkilleyson N, Chaplinskas S, Brummer-Korvenkontio H, Vanhatalo J, et al. 1998. HIV-1 genetic subtype A/B recombinant strain causing an explosive epidemic in injecting drug users in Kaliningrad. *AIDS* **12:** 1907–1919.

Liitsola K, Holm K, Bobkov A, Pokrovsky V, Smolskaya T, Leinikki P, Osmanov S, Salminen M. 2000. An AB recombinant and its parental HIV type 1 strains in the area of the former Soviet Union: Low requirements for sequence identity in recombination. UNAIDS Virus Isolation Network. *AIDS Res Hum Retroviruses* **16:** 1047–1053.

Lindegren ML, Byers RHJr, Thomas P, Davis SF, Caldwell B, Rogers M, Gwinn M, Ward JW, Fleming PL. 1999. Trends in perinatal transmission of HIV/AIDS in the United States. *JAMA* **282:** 531–538.

Little SJ, Daar ES, D'Aquila RT, Keiser PH, Connick E, Whitcomb JM, Hellmann NS, Petropoulos CJ, Sutton L, Pitt JA, et al. 1999. Reduced antiretroviral drug susceptibility among patients with primary HIV infection. *JAMA* **282:** 1142–1149.

Little SJ, Holte S, Routy JP, Daar ES, Markowitz M, Collier AC, Koup RA, Mellors JW, Connick E, Conway B, et al. 2002. Antiretroviral-drug resistance among patients recently infected with HIV. *N Engl J Med* **347:** 385–394.

Lukashov VV, Kuiken CL, Vlahov D, Coutinho RA, Goudsmit J. 1996. Evidence for HIV type 1 strains of U.S. intravenous drug users as founders of AIDS epidemic among intravenous drug users in northern Europe. *AIDS Res Hum Retroviruses* **12:** 1179–1183.

Lurie P, Drucker E. 1997. An opportunity lost: HIV infections associated with lack of a national needle-exchange programme in the USA. *Lancet* **349:** 604–608.

Lyles CM, Crepaz N, Herbst JH, Kay LS. 2006. Evidence-based HIV behavioral prevention from the perspective of the CDC's HIV/AIDS Prevention Research Synthesis Team. *AIDS Educ Prev* **18:** 21–31.

Mah TL, Halperin DT. 2010. Concurrent sexual partnerships and the HIV epidemics in Africa: Evidence to move forward. *AIDS Behav* **14:** 11–16; dicussion 34–17.

Mahajan AP, Sayles JN, Patel VA, Remien RH, Sawires SR, Ortiz DJ, Szekeres G, Coates TJ. 2008. Stigma in the HIV/AIDS epidemic: A review of the literature and recommendations for the way forward. *AIDS* **22** Suppl 2**:** S67–S79.

Mahy M, Warner-Smith M, Stanecki KA, Ghys PD. 2009. Measuring the impact of the global response to the

AIDS epidemic: Challenges and future directions. *J Acquir Immune Defic Syndr* 52 Suppl 2: S152–S159.

Malebranche DJ. 2008. Bisexually active Black men in the United States and HIV: Acknowledging more than the "down low". *Arch Sex Behav* 37: 810–816.

Margaret Dolcini M, Gandelman AA, Vogan SA, Kong C, Leak TN, King AJ, Desantis L, O'Leary A. 2010. Translating HIV interventions into practice: Community-based organizations' experiences with the diffusion of effective behavioral interventions (DEBIs). *Soc Sci Med* 71: 1839–1846.

Mathers BM, Degenhardt L, Ali H, Wiessing L, Hickman M, Mattick RP, Myers B, Ambekar A, Strathdee SA. 2010. HIV prevention, treatment, and care services for people who inject drugs: A systematic review of global, regional, and national coverage. *Lancet* 375: 1014–1028.

McDonald AM, Zurynski YA, Wand HC, Giles ML, Elliott EJ, Ziegler JB, Kaldor JM. 2009. Perinatal exposure to HIV among children born in Australia, 1982–2006. *Med J Aust* 190: 416–420.

McGowan I. 2011. Rectal microbicides: Can we make them and will people use them? *AIDS Behav* 15 (Suppl 1): S66–S71.

Mercer CH, Copas AJ, Sonnenberg P, Johnson AM, McManus S, Erens B, Cassell JA. 2009. Who has sex with whom? Characteristics of heterosexual partnerships reported in a national probability survey and implications for STI risk. *Int J Epidemiol* 38: 206–214.

Meyerhans A, Cheynier R, Albert J, Seth M, Kwok S, Sninsky J, Morfeldt-Manson L, Asjo B, Wain-Hobson S. 1989. Temporal fluctuations in HIV quasispecies in vivo are not reflected by sequential HIV isolations. *Cell* 58: 901–910.

Millett GA, Flores SA, Marks G, Reed JB, Herbst JH. 2008. Circumcision status and risk of HIV and sexually transmitted infections among men who have sex with men: A meta-analysis. *JAMA* 300: 1674–1684.

Millett GA, Crowley JS, Koh H, Valdiserri RO, Frieden T, Dieffenbach CW, Fenton KA, Benjamin R, Whitescarver J, Mermin J, et al. 2010. A way forward: The National HIV/AIDS Strategy and reducing HIV incidence in the United States. *J Acquir Immune Defic Syndr* 55 Suppl 2: S144–S147.

Mills E, Cooper C, Anema A, Guyatt G. 2008. Male circumcision for the prevention of heterosexually acquired HIV infection: A meta-analysis of randomized trials involving 11,050 men. *HIV Med* 9: 332–335.

Mofenson LM. 2003. Advances in the prevention of vertical transmission of human immunodeficiency virus. *Semin Pediatr Infect Dis* 14: 295–308.

Mofenson LM. 2010. Antiretroviral drugs to prevent breast-feeding HIV transmission. *Antivir Ther* 15: 537–553.

Mofenson LM, Fowler MG. 1999. Interruption of materno-fetal transmission. *AIDS* 13 (Suppl A): S205–S214.

Mofenson LM, McIntyre JA. 2000. Advances and research directions in the prevention of mother-to-child HIV-1 transmission. *Lancet* 355: 2237–2244.

Novitsky V, Wang R, Bussmann H, Lockman S, Baum M, Shapiro R, Thior I, Wester C, Wester CW, Ogwu A, et al. 2010. HIV-1 subtype C-infected individuals maintaining high viral load as potential targets for the "test-and-treat" approach to reduce HIV transmission. *PLoS One* 5: e10148. doi: 10.1371/journal.pone.0010148.

Op de Coul EL, Prins M, Cornelissen M, van der Schoot A, Boufassa F, Brettle RP, Hernandez-Aguado L, Schiffer V, McMenamin J, Rezza G, et al. 2001. Using phylogenetic analysis to trace HIV-1 migration among western European injecting drug users seroconverting from 1984 to 1997. *AIDS* 15: 257–266.

Ott MA, Santelli JS. 2007. Abstinence and abstinence-only education. *Curr Opin Obstet Gynecol* 19: 446–452.

Ou CY, Takebe Y, Weniger BG, Luo CC, Kalish ML, Auwanit W, Yamazaki S, Gayle HD, Young NL, Schochetman G. 1993. Independent introduction of two major HIV-1 genotypes into distinct high-risk populations in Thailand. *Lancet* 341: 1171–1174.

* Overbaugh J, Morris L. 2011. The antibody response against HIV-1. *Cold Spring Harbor Perspect Med* doi: 10.1101/cshperspect.a007039.

Palmateer N, Kimber J, Hickman M, Hutchinson S, Rhodes T, Goldberg D. 2010. Evidence for the effectiveness of sterile injecting equipment provision in preventing hepatitis C and human immunodeficiency virus transmission among injecting drug users: A review of reviews. *Addiction* 105: 844–859.

Pape J, Johnson WDJr. 1993. AIDS in Haiti 1982–1992. *Clin Infect Dis* 17 (Suppl 2): S341–S345.

Parry JV, Murphy G, Barlow KL, Lewis K, Rogers PA, Belda FJ, Nicoll A, McGarrigle C, Cliffe S, Mortimer PP, et al. 2001. National surveillance of HIV-1 subtypes for England and Wales: Design, methods, and initial findings. *J Acquir Immune Defic Syndr* 26: 381–388.

Parry CD, Carney T, Petersen P, Dewing S, Needle R. 2009. HIV-risk behavior among injecting or non-injecting drug users in Cape Town, Pretoria, and Durban, South Africa. *Subst Use Misuse* 44: 886–904.

Pereyra F, Jia X, McLaren PJ, Telenti A, de Bakker PI, Walker BD, Ripke S, Brumme CJ, Pulit SL, Carrington M, et al. 2010. The major genetic determinants of HIV-1 control affect HLA class I peptide presentation. *Science* 330: 1551–1557.

Perrin L, Yerly S, Rakik A, Kinloch S, Hirschel B. 1994. Transmission of 215 mutants in primary HIV infection and analysis after 6 months of zidovudine. *AIDS* 8: pS3.

Peters VB, Liu KL, Robinson LG, Dominguez KL, Abrams EJ, Gill BS, Thomas PA. 2008. Trends in perinatal HIV prevention in New York City, 1994–2003. *Am J Public Health* 98: 1857–1864.

Pickering H, Okongo M, Ojwiya A, Yirrell D, Whitworth J. 1997. Sexual networks in Uganda: Mixing patterns between a trading town, its rural hinterland and a nearby fishing village. *Int J STD AIDS* 8: 495–500.

Pinching AJ. 1984. The acquired immune deficiency syndrome. *Clin Exp Immunol* 56: 1–13.

Powers KA, Poole C, Pettifor AE, Cohen MS. 2008. Rethinking the heterosexual infectivity of HIV-1: A systematic review and meta-analysis. *Lancet Infect Dis* 8: 553–563.

Qian HZ, Stinnette SE, Rebeiro PF, Kipp AM, Shepherd BE, Samenow CP, Jenkins CA, No P, McGowan CC, Hulgan T, et al. 2011. The relationship between injection and non-injection drug use and HIV disease progression. *J Subst Abuse Treat* 41: 14–20.

Quigg M, Rebus S, France AJ, McMenamin J, Darby G, Leigh Brown AJ. 1997. Mutations associated with zidovudine resistance in HIV-1 among recent seroconvertors. *AIDS* **11:** 835–836.

Rambaut A, Robertson DL, Pybus OG, Peeters M, Holmes EC. 2001. Human immunodeficiency virus. Phylogeny and the origin of HIV-1. *Nature* **410:** 1047–1048.

Rambaut A, Posada D, Crandall KA, Holmes EC. 2004. The causes and consequences of HIV evolution. *Nat Rev Genet* **5:** 52–61.

Read JS. 2010. Prevention of mother-to-child transmission of HIV: Antiretroviral strategies. *Clin Perinatol* **37:** 765–776, viii.

Reithinger R, Megazzini K, Durako SJ, Harris DR, Vermund SH. 2007. Monitoring and evaluation of programmes to prevent mother to child transmission of HIV in Africa. *BMJ* **334:** 1143–1146.

Remien RH, Mellins CA. 2007. Long-term psychosocial challenges for people living with HIV: Let's not forget the individual in our global response to the pandemic. *AIDS* **21** (Suppl 5)**:** S55–S63.

Resik S, Lemey P, Ping LH, Kouri V, Joanes J, Perez J, Vandamme AM, Swanstrom R. 2007. Limitations to contact tracing and phylogenetic analysis in establishing HIV type 1 transmission networks in Cuba. *AIDS Res Hum Retroviruses* **23:** 347–356.

Reynolds SJ, Quinn TC. 2010. Setting the stage: Current state of affairs and major challenges. *Clin Infect Dis* **50** (Suppl 3)**:** S71–S76.

Rietmeijer CA, Patnaik JL, Judson FN, Douglas JMJr. 2003. Increases in gonorrhea and sexual risk behaviors among men who have sex with men: A 12-year trend analysis at the Denver Metro Health Clinic. *Sex Transm Dis* **30:** 562–567.

Robbins KE, Lemey P, Pybus OG, Jaffe HW, Youngpairoj AS, Brown TM, Salemi M, Vandamme AM, Kalish ML. 2003. U.S. human immunodeficiency virus type 1 epidemic: Date of origin, population history, and characterization of early strains. *J Virol* **77:** 6359–6366.

Robertson DL, Anderson JP, Bradac JA, Carr JK, Foley B, Funkhouser RK, Gao F, Hahn BH, Kalish ML, Kuiken C, et al. 2000. HIV-1 nomenclature proposal. *Science* **288:** 55–56.

Rosenthal E. 2006. HIV injustice in Libya—Scapegoating foreign medical professionals. *N Engl J Med* **355:** 2505–2508.

Rothenberg R. 2009. HIV transmission networks. *Curr Opin HIV AIDS* **4:** 260–265.

Rotheram-Borus MJ, Swendeman D, Chovnick G. 2009. The past, present, and future of HIV prevention: Integrating behavioral, biomedical, and structural intervention strategies for the next generation of HIV prevention. *Annu Rev Clin Psychol* **5:** 143–167.

Sahasrabuddhe VV, Vermund SH. 2007. The future of HIV prevention: Control of sexually transmitted infections and circumcision interventions. *Infect Dis Clin North Am* **21:** 241–257, xi.

Salemi M, de Oliveira T, Ciccozzi M, Rezza G, Goodenow MM. 2008. High-resolution molecular epidemiology and evolutionary history of HIV-1 subtypes in Albania. *PLoS One* **3:** e1390. doi: 10.1371/journal.pone.0001390.

Salomon EA, Mimiaga MJ, Husnik MJ, Welles SL, Manseau MW, Montenegro AB, Safren SA, Koblin BA, Chesney MA, Mayer KH. 2009. Depressive symptoms, utilization of mental health care, substance use and sexual risk among young men who have sex with men in EXPLORE: Implications for age-specific interventions. *AIDS Behav* **13:** 811–821.

Sambrook J, Sleigh M, Engler JA, Broker TR. 1980. The evolution of the adenoviral genome. *Ann NY Acad Sci* **354:** 426–452.

Samet JH, Walley AY, Bridden C. 2007. Illicit drugs, alcohol, and addiction in human immunodeficiency virus. *Panminerva Med* **49:** 67–77.

Sarang A, Stuikyte R, Bykov R. 2007. Implementation of harm reduction in Central and Eastern Europe and Central Asia. *Int J Drug Policy* **18:** 129–135.

Sarang A, Rhodes T, Platt L. 2008. Access to syringes in three Russian cities: Implications for syringe distribution and coverage. *Int J Drug Policy* **19** (Suppl 1)**:** S25–S36.

Scheer S, Kellogg T, Klausner JD, Schwarcz S, Colfax G, Bernstein K, Louie B, Dilley JW, Hecht J, Truong HM, et al. 2008. HIV is hyperendemic among men who have sex with men in San Francisco: 10-year trends in HIV incidence, HIV prevalence, sexually transmitted infections and sexual risk behaviour. *Sex Transm Infect* **84:** 493–498.

Selik RM, Haverkos HW, Curran JW. 1984. Acquired immune deficiency syndrome (AIDS) trends in the United States, 1978–1982. *Am J Med* **76:** 493–500.

* Sharp PM, Hahn BH. 2011. Origins of HIV and the AIDS pandemic. *Cold Spring Harb Perspec Med* doi: 10.1101/cshperspect.a006841.

Shaw SJ. 2006. Public citizens, marginalized communities: The struggle for syringe exchange in Springfield, Massachusetts. *Med Anthropol* **25:** 31–63.

* Shaw GM, Hunter E. 2011. HIV transmission. *Cold Spring Harb Perspect Med* doi: 10.1011/cshperspect.a006965.

Siegfried N, Muller M, Deeks JJ, Volmink J. 2009. Male circumcision for prevention of heterosexual acquisition of HIV in men. *Cochrane Database Syst Rev* CD003362.

Sikora C, Chandran AU, Joffe AM, Johnson D, Johnson M. 2010. Population risk of syringe reuse: Estimating the probability of transmitting bloodborne disease. *Infect Control Hosp Epidemiol* **31:** 748–754.

Simon F, Loussert-Ajaka I, Damond F, Saragosti S, Barin F, Brun-Vezinet F. 1996. HIV type 1 diversity in northern Paris, France. *AIDS Res Hum Retroviruses* **12:** 1427–1433.

Simonsen L, Kane A, Lloyd J, Zaffran M, Kane M. 1999. Unsafe injections in the developing world and transmission of bloodborne pathogens: A review. *Bull World Health Organ* **77:** 789–800.

Steffenson AE, Pettifor AE, Seage GR 3rd, Rees HV, Cleary PD. 2011. Concurrent sexual partnerships and human immunodeficiency virus risk among South African youth. *Sex Transm Dis.* **38:** 459–466.

Strauss SM, Rindskopf DM. 2009. Screening patients in busy hospital-based HIV care centers for hazardous and harmful drinking patterns: The identification of an optimal screening tool. *J Int Assoc Physicians AIDS Care (Chic)* **8:** 347–353.

Strauss SM, Tiburcio NJ, Munoz-Plaza C, Gwadz M, Lunievicz J, Osborne A, Padilla D, McCarty-Arias M, Norman R. 2009. HIV care providers' implementation of routine alcohol reduction support for their patients. *AIDS Patient Care STDS* **23:** 211–218.

Stringer JS, Vermund SH. 1999. Prevention of mother-to-child transmission of HIV-1. *Curr Opin Obstet Gynecol* **11:** 427–434.

Stringer JS, Sinkala M, Maclean CC, Levy J, Kankasa C, Degroot A, Stringer EM, Acosta EP, Goldenberg RL, Vermund SH. 2005. Effectiveness of a city-wide program to prevent mother-to-child HIV transmission in Lusaka, Zambia. *AIDS* **19:** 1309–1315.

Stringer EM, Chi BH, Chintu N, Creek TL, Ekouevi DK, Coetzee D, Tih P, Boulle A, Dabis F, Shaffer N, et al. 2008. Monitoring effectiveness of programmes to prevent mother-to-child HIV transmission in lower-income countries. *Bull World Health Organ* **86:** 57–62.

Stringer EM, Ekouevi DK, Coetzee D, Tih PM, Creek TL, Stinson K, Giganti MJ, Welty TK, Chintu N, Chi BH, et al. 2010. Coverage of nevirapine-based services to prevent mother-to-child HIV transmission in 4 African countries. *JAMA* **304:** 293–302.

* Swanstrom R, Coffin J. 2011. HIV-1 pathogenesis: The virus. *Cold Spring Harb Perspect Biol.* doi: 10.1101/cshperspect.a007443.

Taha TE, Brown ER, Hoffman IF, Fawzi W, Read JS, Sinkala M, Martinson FE, Kafulafula G, Msamanga G, Emel L, et al. 2006. A phase III clinical trial of antibiotics to reduce chorioamnionitis-related perinatal HIV-1 transmission. *AIDS* **20:** 1313–1321.

Tempalski B, Flom PL, Friedman SR, Des Jarlais DC, Friedman JJ, McKnight C, Friedman R. 2007. Social and political factors predicting the presence of syringe exchange programs in 96 US metropolitan areas. *Am J Public Health* **97:** 437–447.

Thomson MM, Perez-Alvarez L, Najera R. 2002. Molecular epidemiology of HIV-1 genetic forms and its significance for vaccine development and therapy. *Lancet Infect Dis* **2:** 461–471.

Toussova O, Shcherbakova I, Volkova G, Niccolai L, Heimer R, Kozlov A. 2009. Potential bridges of heterosexual HIV transmission from drug users to the general population in St. Petersburg, Russia: Is it easy to be a young female? *J Urban Health* **86** (Suppl 1): 121–130.

Trenholm C, Devaney B, Fortson K, Clark M, Bridgespan LQ, Wheeler J. 2008. Impacts of abstinence education on teen sexual activity, risk of pregnancy, and risk of sexually transmitted diseases. *J Policy Anal Manage* **27:** 255–276.

van Griensven F, de Lind van Wijngaarden JW. 2010. A review of the epidemiology of HIV infection and prevention responses among MSM in Asia. *AIDS* **24:** (Suppl 3): S30–S40.

Vercauteren J, Wensing AM, van de Vijver DA, Albert J, Balotta C, Hamouda O, Kucherer C, Struck D, Schmit JC, Asjo B, et al. 2009. Transmission of drug-resistant HIV-1 is stabilizing in Europe. *J Infect Dis* **200:** 1503–1508.

Vermund SH, Allen KL, Karim QA. 2009. HIV-prevention science at a crossroads: Advances in reducing sexual risk. *Curr Opin HIV AIDS* **4:** 266–273.

Vermund SH, Hodder SL, Justman JE, Koblin BA, Mastro TD, Mayer KH, Wheeler DP, El-Sadr WM. 2010. Addressing research priorities for prevention of HIV infection in the United States. *Clin Infect Dis* **50** Suppl 3: S149–155.

Vermund SH, Qian HZ. 2008. Circumcision and HIV prevention among men who have sex with men: No final word. *JAMA* **300:** 1698–1700.

Volmink J, Siegfried NL, van der Merwe L, Brocklehurst P. 2007. Antiretrovirals for reducing the risk of mother-to-child transmission of HIV infection. *Cochrane Database Syst Rev* CD003510.

Volz EM, Kosakovsky Pond SL, Ward MJ, Leigh Brown AJ, Frost SD. 2009. Phylodynamics of infectious disease epidemics. *Genetics* **183:** 1421–1430.

Wade NA, Zielinski MA, Butsashvili M, McNutt LA, Warren BL, Glaros R, Cheku B, Pulver W, Pass K, Fox K, et al. 2004. Decline in perinatal HIV transmission in New York State (1997–2000). *J Acquir Immune Defic Syndr* **36:** 1075–1082.

* Walker B, McMichael A. 2011. The T-cell response to HIV. *Cold Spring Harbor Perspect Med* doi: 10.1101/cshperspect.a007054.

Weinstock HS, Zaidi I, Heneine W, Bennett D, Garcia-Lerma JG, Douglas JMJr, LaLota M, Dickinson G, Schwarcz S, Torian L, et al. 2004. The epidemiology of antiretroviral drug resistance among drug-naive HIV-1-infected persons in 10 US cities. *J Infect Dis* **189:** 2174–2180.

Weiss HA, Quigley MA, Hayes RJ. 2000. Male circumcision and risk of HIV infection in sub-Saharan Africa: A systematic review and meta-analysis. *AIDS* **14:** 2361–2370.

Weiss HA, Dickson KE, Agot K, Hankins CA. 2010. Male circumcision for HIV prevention: Current research and programmatic issues. *AIDS* **24** (Suppl 4): S61–S69.

Wensing AM, van de Vijver DA, Angarano G, Asjo B, Balotta C, Boeri E, Camacho R, Chaix ML, Costagliola D, De Luca A, et al. 2005. Prevalence of drug-resistant HIV-1 variants in untreated individuals in Europe: Implications for clinical management. *J Infect Dis* **192:** 958–966.

Whitmore SK, Patel-Larson A, Espinoza L, Ruffo NM, Rao S. 2010. Missed opportunities to prevent perinatal human immunodeficiency virus transmission in 15 jurisdictions in the United States during 2005–2008. *Women Health* **50:** 414–425.

Whitmore SK, Zhang X, Taylor AW, Blair JM. 2011. Estimated number of infants born to HIV-infected women in the United States and five dependent areas, 2006. *J Acquir Immune Defic Syndr.* **57:** 218–222.

* Williamson C, Shao Y. 2011. The HIV epidemic: Low- to middle-income countries. *Cold Spring Harb Perspect Med* doi: 10.1101/cshperspect.a007187.

Wingood GM, DiClemente RJ. 2006. Enhancing adoption of evidence-based HIV interventions: Promotion of a suite of HIV prevention interventions for African American women. *AIDS Educ Prev* **18:** 161–170.

Wolfe D, Carrieri MP, Shepard D. 2010. Treatment and care for injecting drug users with HIV infection: A review of barriers and ways forward. *Lancet* **376:** 355–366.

Wolfs TF, de Jong JJ, Van den Berg H, Tijnagel JM, Krone WJ, Goudsmit J. 1990. Evolution of sequences encoding the principal neutralization epitope of human immunodeficiency virus 1 is host dependent, rapid, and continuous. *Proc Natl Acad Sci* **87:** 9938–9942.

Wolitski RJ, Valdiserri RO, Denning PH, Levine WC. 2001. Are we headed for a resurgence of the HIV epidemic among men who have sex with men? *Am J Public Health* **91:** 883–888.

Xu JJ, Zhang M, Brown K, Reilly K, Wang H, Hu Q, Ding H, Chu Z, Bice T, Shang H. 2010. Syphilis and HIV seroconversion among a 12-month prospective cohort of men who have sex with men in Shenyang, China. *Sex Transm Dis* **37:** 432–439.

Yang Q, Boulos D, Yan P, Zhang F, Remis RS, Schanzer D, Archibald CP. 2010. Estimates of the number of prevalent and incident human immunodeficiency virus (HIV) infections in Canada, 2008. *Can J Public Health* **101:** 486–490.

Yerly S, Junier T, Gayet-Ageron A, Amari EB, von Wyl V, Gunthard HF, Hirschel B, Zdobnov E, Kaiser L. 2009. The impact of transmission clusters on primary drug resistance in newly diagnosed HIV-1 infection. *AIDS* **23:** 1415–1423.

Yirrell DL, Pickering H, Palmarini G, Hamilton L, Rutemberwa A, Biryahwaho B, Whitworth J, Brown AJ. 1998. Molecular epidemiological analysis of HIV in sexual networks in Uganda. *AIDS* **12:** 285–290.

Zetterberg V, Ustina V, Liitsola K, Zilmer K, Kalikova N, Sevastianova K, Brummer-Korvenkontio H, Leinikki P, Salminen MO. 2004. Two viral strains and a possible novel recombinant are responsible for the explosive injecting drug use-associated HIV type 1 epidemic in Estonia. *AIDS Res Hum Retroviruses* **20:** 1148–1156.

Host Genes Important to HIV Replication and Evolution

Amalio Telenti[1] and Welkin E. Johnson[2]

[1]Institute of Microbiology, University Hospital and University of Lausanne, 1011 Lausanne, Switzerland

[2]New England Primate Research Center, Department of Microbiology and Molecular Genetics, Harvard Medical School, Southborough, Massachusetts 01772

Correspondence: Amalio.Telenti@chuv.ch

Recent years have seen a significant increase in understanding of the host genetic and genomic determinants of susceptibility to HIV-1 infection and disease progression, driven in large part by candidate gene studies, genome-wide association studies, genome-wide transcriptome analyses, and large-scale in vitro genome screens. These studies have identified common variants in some host loci that clearly influence disease progression, characterized the scale and dynamics of gene and protein expression changes in response to infection, and provided the first comprehensive catalogs of genes and pathways involved in viral replication. Experimental models of AIDS and studies in natural hosts of primate lentiviruses have complemented and in some cases extended these findings. As the relevant technology continues to progress, the expectation is that such studies will increase in depth (e.g., to include host whole exome and whole genome sequencing) and in breadth (in particular, by integrating multiple data types).

Host genetics has been of considerable interest to the field of HIV/AIDS since the identification of the role of CCR5Δ32 (Dean et al. 1996; Huang et al. 1996; Liu et al. 1996) in resistance to infection and of human leukocyte antigen (HLA) alleles in disease progression (Kaslow et al. 1996). Further observations confirm that there is a significant component of heredity in the susceptibility to HIV-1. Identical twins infected with the same viral strain progressed at a similar pace, whereas their fraternal twin had a different clinical course (Draenert et al. 2006). In vitro, the study of cells from large pedigrees—immortalized B lymphocytes from multigeneration families—allowed the identification of a host genetic contribution to >50% of the observed differences in cell susceptibility to transduction with a vesicular stomatitis virus (VSV)-pseudotyped HIV-1 vector (Loeuillet et al. 2008). In the larger context of infection by simian immunodeficiency viruses, host genetic variation influences transmission between species as well as replication and pathogenesis within individuals of the same species (Kirmaier et al. 2010). Host genetic variation includes fixed differences between species (divergence) and variation within populations (polymorphism). Genetic barriers to viral transmission and spread can arise owing to variation in genes required for optimal viral replication (e.g., receptors, transcription factors, chaperones, etc.) as well as variation in genes that actively thwart viral replication and pathogenesis (e.g., innate and adaptive immune effectors such as antibody and cytotoxic T lymphocyte (CTL) responses, and restriction factor loci such as *TRIM5α*, the *APOBEC3* cluster, and *tetherin*). The influence of host selection on viral evolution manifests as changes in primary sequence (escape and reversion variants) in the viral genome structure, and over the long term, in the acquisition of accessory functions. In this article, we begin with an overview of the current understanding of genes and gene variants influencing susceptibility to HIV-1 in humans, and susceptibility to HIV/simian immunodeficiency virus (SIV) in nonhuman primate models of infection. This includes data from candidate gene studies, genome-wide association studies, and other genome-wide screens. We then analyze notable data on host genome pressure on the viral genome structure. Finally, we present a global view on the evolutionary genomics of susceptibility to HIV-1 and other retroviruses.

KEY ADVANCES

Human Genomics

Genome-Wide Association Studies

The HIV field was one of the first to embrace the opportunities offered by new technologies that allowed genome-wide association studies (GWASs). This approach, which assesses 500,000 to 1 million genetic variants (single nucleotide polymorphism, SNP) for each individual, had advantages and limitations (Telenti and Goldstein 2006). First, it allowed for the first time the non–a priori analysis, at genome-wide scale, of possible genetic determinants of a trait. This would allow discovery, unbiased validation of all previously reported variants, and scoring the contribution of the genetic variant in the context of other possible genome influences. However, the limitations of the approach include the need for sufficient power, requiring large numbers of subjects, and the identification

of very precise study traits (Evangelou et al. 2011). In addition, only common variation (present in >5% in a given population) is investigated in GWASs. The need for large populations (power) is determined by the statistical requirements that impose a very strict threshold of significance (i.e., defined by $P < 5 \times 10^{-8}$ owing to correction for multiple comparisons) and by the limited contribution of any given genetic variant to the population phenotype. The study traits need to be defined by strict criteria to avoid heterogeneity in the phenotype— an important consideration because the field uses study outcomes, such as time to AIDS or death, which represent composite end points. Finally, the genotyping arrays used in GWASs target common variants that tag other variants; the actual causal variant or functional polymorphism is unlikely to be interrogated directly. Current arrays are generally adequate to capture common variation in Caucasians and Asians, although less effective in individuals of African ancestry.

Eight GWASs have been published during the period 2007–2010 (Table 1). The first study (Fellay et al. 2007) investigated the genomic determinants of viral set point after seroconversion—the relatively steady state of viral replication in the 3 years following a documented infection. As a secondary end point, the study evaluated disease progression, as defined by time to CD4$^+$ T-cell count less than 350 cells/ μL, or else initiation of treatment. This study identified three variants in chromosome 6: a variant in *HCP5* that tagged the protective allele *HLA-B*5707*, a variant upstream of *HLA-C* associated with difference in HLA-C expression levels, and a variant in *ZNRD1* that may be merely associated with other influences from the major histocompatibility complex (*MHC*) locus, or contribute directly to pathogenesis (Ballana et al. 2010). Extension of this work by the same group of researchers (Fellay et al. 2009) confirmed the various associations and ruled out major variants elsewhere in the genome with the exception of those in the *CCR5-CCR2* locus (Fig. 1). The overall variance explained by the genome-wide significant hits was close to 20%—much larger than what has

Table 1. Genome-wide association studies, 2007–2010

Study/year	Trait	N	Population	Genome-wide significant hits[a]	Comment
Fellay et al. 2007	Viral load set point, disease progression	486 (seroconverters)	Caucasian	rs2395029 (HCP5/HLA-B*57:01), rs9264942 (HLA-C), rs9261174 (ZNRD1)	
Dalmasso et al. 2008	Plasma HIV-RNA levels and cellular HIV-DNA levels	605 (seroconverters)	Caucasian	rs10484554, rs2523619, and rs2395029 (HLA-C, HLA-B locus)	
Limou et al. 2009	HIV nonprogression	275 (HIV-1+ nonprogressors), and 1352 (negative controls)	Caucasian	rs2395029 (HCP5/HLA-B*57:01)	Subset study (Limou et al. 2010) used to validate candidate rs2234358 (CXCR6) in three independent cohort studies (n = 1028)
Le Clerc et al. 2009	Rapid disease progression	85 (HIV-1+ rapid progressors), and 2049 (negative controls)	Caucasian	—	
Fellay et al. 2009	Viral load set point, disease progression	2554 (seroconverters and seroprevalent)	Caucasian	rs2395029, rs9264942, rs259919, rs9468692, rs9266409 (MHC locus), and rs333(CCR5Δ32)	Failed to validate previously reported candidate genes with the exception of CCR5/CCR2 variants
Herbeck et al. 2010	Disease progression	156 (rapid, moderate, and nonprogressors)	Caucasian	—	Among the 25 top-ranking variants, rs17762192 (PROX1) was validated in an independent replication cohort of 590 seroconverters
Pelak et al. 2010	Viral set point	515 (seroconverters)	African American	HLA-B*57:03 (rs2523608)	In contrast with the tagged HLA, rs2523608 does not reach genome-wide significance
International HIV Controllers Study 2010	HIV nonprogression	974 (HIV controllers), 2648 (HIV progressors)	Caucasian, African American, and Hispanics	313 SNPs in the MHC locus captured by rs9264942, rs2395029, rs4418214, and rs3131018 as independent markers	Arg97, Cys67, Gly62, and Glu63, all in HLA-B; Ser77 in HLA-A; and Met304 in HLA-C collectively explain 20% of the observed variance in Caucasians

[a]Varies between $p < 10^{-7}$ and $p < 10^{-8}$ depending on the study design.

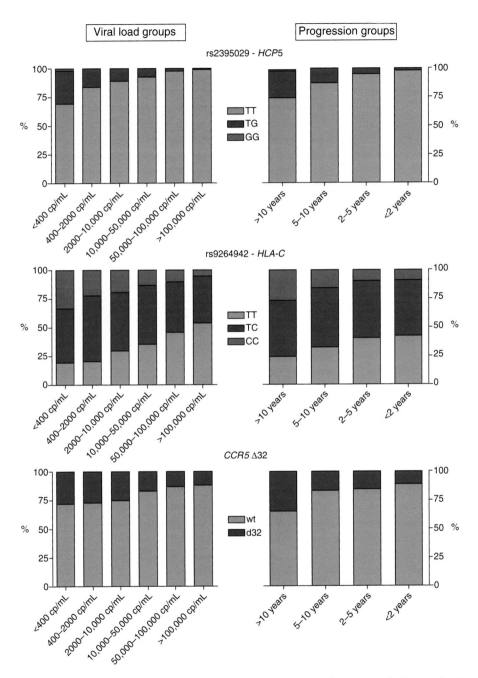

Figure 1. Distribution of the protective alleles according to viral or clinical phenotypes. The bar graphs show the allelic distribution of three variants that have a genome-wide significant association with HIV-1 set point (left-hand-side graphs) and disease progression (right-hand-side graphs) in a population of 2362 HIV-infected individuals. The data on *HCP5*, a perfect tag of HLA*B57:01, illustrate the nature of the association between protective alleles and long-term nonprogression: although 31%–37% of elite controllers carry HLA*B57:01, only 6%–22% of HLA*B57:01 carriers are elite controllers. (Adapted, with permission, from Fellay et al. 2009; reprinted, with permission, from *PLoS Genetics* © 2009.)

been observed in GWASs of other diseases or in studies of anthropomorphic diversity (height, weight) but still a small percentage of the total. Although the overwhelming message is one of confirmation of the critical importance of the *HLA-B/HLA-C* locus, some studies have reported additional candidate loci (Table 1).

A significant step forward in the understanding of the complexity of signals emerging from the *MHC* locus came from the International HIV Controllers Study (2010). This large consortium compared the genetic data of 974 subjects defined as viral controllers and 2648 progressors. After an initial step that confirmed the critical importance of the MHC region in the trait of nonprogression, the study proceeded to mapping putative causal variants within the *HLA* locus. This step is complex because this gene-rich region has high levels of genetic diversity and a complex pattern of linkage disequilibrium. The linkage disequilibrium reflects the measurement of nonrandom association between two or more alleles that occur together on a chromosome. The International HIV Controllers Study developed analytical tools that specified unique residues in the HLA-B groove as putative causal variants. These included Gly62, Glu63, Cys67, and Arg97, all in HLA-B. In addition, the study identified Ser77 in HLA-A and Met304 in HLA-C. With the exception of Met304 in the transmembrane domain of HLA-C, these residues are all located in the MHC class I peptide binding groove, underscoring that the conformational presentation of class I restricted epitopes to T cells play a key role in host control. Chimpanzees can be infected with HIV-1; however, most do not develop AIDS. The contemporary MHC class I repertoire of chimpanzees targets analogous conserved domains of HIV-1/SIV$_{cpz}$ to those targeted by human protective alleles *HLA-B*57* and *B*27*, in particular, of the Gag protein (de Groot et al. 2010). The functional characteristics of the chimpanzee *MCH*-repertoire may be the result of a selective sweep caused by lentiviruses (de Groot et al. 2010). The consistent identification of HLA-B*57, HLA-B*27, and HLA-B*51 in host genome studies is attributed to fitness cost of mutating in the targeted viral

epitopes and to the patterns of viral escape from CD8$^+$ T-cell recognition. Differences in prevalence of the restricting HLA allele in different human groups consequently lead to HIV-1 adaptation to HLA at a population level (Kawashima et al. 2009).

Following the completion of the GWASs mentioned above, the field is planning a joint meta-analysis of all available data sets combining information presented in Table 1 and new GWASs. This approach can address issues of insufficient power and thus facilitate the identification of common variants that would be associated with smaller effects, or more rare variants included in the typing arrays. Follow-up of GWASs includes resequencing and functional analysis of the main hits and putative causal variants. This has been performed for *HCP5* (Yoon et al. 2010), *HLA-C* (Thomas et al. 2009), and *ZNRD1* (Ballana et al. 2010) hits. Of particular interest is the recent work that identifies the −35 SNP in *HLA-C* as a marker of variation at the binding of microRNA Hsa-miR-148a to its target site within the 3′ untranslated region of HLA-C (Kulkarni et al. 2011). This mechanism of posttranscriptional regulation results in relatively low surface expression of alleles that bind this microRNA and high expression of HLA-C alleles. The exact role of the individual HLA-B amino acids identified by the International HIV Controllers Study on epitope presentation, T-cell receptor binding, and CTL activity is the subject of intensive research in the field.

Vaccine Genomics

A particularly attractive use of GWASs is in the understanding of differences between individuals in the response to immunogens. Although vaccine genomics is in its infancy, some initial applications have been reported. A GWAS assessed determinants of HIV-specific T-cell responses to the MRKAd5 HIV-1 gag/pol/nef vaccine (Fellay et al. 2011) tested in 831 subjects of the Step HIV-1 vaccine trial, as measured by IFN-γ ELISpot assays. No genetic variant reached genome-wide significance, but polymorphisms located in the MHC showed the strongest association with response to the HIV-1 Gag protein. HLA-B alleles known to associate with differences in HIV-1 control were found to be responsible. The authors concluded that the host immunogenetic background needs to be considered in the analysis of immune responses to T-cell vaccines. Increasingly, vaccination studies will include host genetic analysis, informed consent, and DNA storage for later analysis.

Vacccine genomics can also address the selective pressure from vaccine-induced T-cell responses on HIV-1 infection in humans. Rolland et al. (2011) analyzed HIV-1 from 68 infected volunteers in the STEP trial to identify signatures distinguishing vaccine from placebo recipients. Gag amino acid 84, a site encompassed by several epitopes contained in the vaccine and restricted by HLA alleles common in the study cohort, was identified by this approach. Viral genome regions excluded from the vaccine components did not carry distinctive signatures of selective pressure from vaccine-induced T-cell responses on HIV-1 infection in humans.

Advanced Genome Analyses

Although the platforms and analytical tools for genome-wide genotyping are well established, it is clear that the GWAS approach will not capture some aspects of the host influence on the pathogenesis of HIV infection. Additional types of data include the transcriptome and proteome of the infected cell or individual. Further techniques use evolutionary and comparative genetic tools for the identification of host genes involved in genetic conflicts with lentiviral or retroviral pathogens, or large-scale functional genomics using loss-of-function (siRNA) and gain-of-function screens (Bushman et al. 2009; Telenti 2009).

Expression analyses have generated a collection of data from in vitro and in vivo studies on cell lines, whole blood, and cellular subsets. Many publications were based on earlier technologies interrogating small gene subsets, were limited by the number of samples, and rarely captured the dynamic nature of the transcriptome (Giri et al. 2006). Globally, this body of literature has described a number of features

of the infectious process that massively modulates the antiviral defense systems (the interferon response, including the antiretroviral intrinsic cellular defense apparatus), as well as genes involved in the cell cycle and degradation/proteasome pathway (Rotger et al. 2010). Particularly relevant is the observation that elite controllers have $CD4^+$ T-cell transcriptome profiles that are similar to those from individuals receiving effective treatment (Rotger et al. 2010). The transcriptome profile of a subset of elite controllers is indistinguishable from that of the uninfected individuals—at least in HLA-DR-$CD4^+$ T-cell subsets that were the object of a recent study (Vigneault et al. 2011).

Increasingly, studies are successfully using expression data to single out genes for functional analyses. Notable examples include observations on SOCS1 (Rotger et al. 2011), BATF (Quigley et al. 2010), and CXCR6 (Paust et al. 2010) that originated in transcriptome analyses. The suppressor of cytokine signaling 1 (SOCS1) suppresses interferon signaling by direct binding to phosphorylated type I interferon and active JAK kinase and by orchestrating the events leading to proteasomal degradation of a number of target proteins. Although viruses such as HTLV-1 may use induction of SOCS1 to evade the antiviral effects of interferon signaling, it is differentially up-regulated in the nonpathogenic primate models (Bosinger et al. 2009; Jacquelin et al. 2009; Lederer et al. 2009) and in HIV-1 infected humans that tolerate high levels of viral replication and do not progress (Rotger et al. 2011). The basic leucine transcription factor, ATF-like (BATF), a transcription factor in the AP-1 family, was identified during the study of exhausted $CD8^+$ T cells. PD-1 coordinately regulates a program of exhaustion genes in humans and mice, including up-regulation of BATF. Silencing BATF in T cells from individuals with chronic viremia rescues HIV-specific T-cell function (Quigley et al. 2010). The third example, CXCR6, emerged from the study of natural killer (NK) cells in mice (Paust et al. 2010). Hepatic NK cells, but not splenic or naive NK cells, develop specific memory of vaccines containing antigens from HIV-1 and other viruses (Paust et al. 2010).

NK cell memory depends on CXCR6, a chemokine receptor on hepatic NK cells that is required for the persistence of NK memory.

Additional efforts are also directed at comparative transcriptomics, which is the cross-species analysis of expression patterns of human and nonhuman primates during infection and disease (Rotger et al. 2011). HIV-infected subjects with rapid disease progression have gene expression patterns in CD4$^+$ and CD8$^+$ T cells similar to that in pathogenic SIV infection of rhesus macaque (Fig. 2). In contrast, humans that do not progress despite prolonged and extreme levels of viral replication share transcriptional features with the sooty mangabey model of natural infection, including a common profile of regulation of a set of genes that includes *CASP1, CD38, LAG3, TNFSF13B, SOCS1,* and *EEF1D* (Rotger et al. 2011).

Next-generation sequencing offers an unprecedented opportunity to jointly analyze cellular and viral transcriptional activity. This approach served to show that at peak infection of SupT1 cells—a T-cell line—one transcript in 143 is of viral origin (0.7%), including a small component of antisense viral transcription (Lefebvre et al. 2011). Deep sequencing also showed very active transcription of repetitive elements and endogenous retroviruses: Approximately 0.4% of all cellular transcripts are of such origin whether the cell is infected or not (Lefebvre et al. 2011).

Proteomic studies analyzed 2000–3200 proteins and identified 15%–21% to be differentially expressed on infection (Chan et al 2007; Ringrose et al. 2008), including changes in the abundance of proteins with known interactions with HIV-1 viral proteins. The NCBI HIV-1 Human Protein Interaction Database (http://www.ncbi.nlm.nih.gov/RefSeq/HIVInteractions/) summarizes >3000 interactions with almost 1500 human genes (Fu et al. 2009). A complete library of viral-host protein coimmunoprecipitation relevant to the early innate immune responses to HIV is being built by the HINT consortium (HIV Networks Team, www.hint.org) and will be publicly available.

Genome-wide siRNA and shRNA screens have generated a comprehensive view of genes required for efficient viral replication (Brass et al. 2008; Konig et al. 2008; Zhou et al. 2008; Yeung et al. 2009). Individually, each published screen has identified a few hundred such candidate genes. However, there is a limited overlap

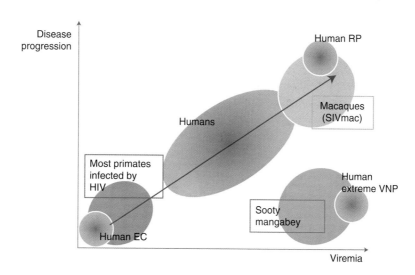

Figure 2. Schematic representation of the parallelism between human and nonhuman primate (NHP) models of HIV/SIV pathogenesis. EC, elite controllers; RP, rapid progressors; VNP, extreme viremic nonprogressors. (Adapted, with permission, from Guido Silvestri 2010.)

across studies that has been attributed to differences in study design and to the type and source of the si/shRNA library (Bushman et al. 2009). The studies were also not specifically designed to identify restriction factors; no hits resulted in increasing levels of viral replication. Despite these limitations, meta-analysis of the data convincingly identified sets of genes and pathways that are common to two or more studies. A pattern emerged among the 34 shared genes that involve the nuclear pore machinery, the mediator complex, a number of key kinases, and components of the NF-κB complex. Studies under way address those various technical issues by the systematic comparison of multiple siRNA libraries, the development of experimental approaches that can identify restriction factors, the integration of these data with other sources of experimental data, notably with gene expression (Fellay et al. 2010), and by the use of primary cells.

One gain-of-function screen used a cDNA library representing 15,000 unique genes in an infectious HIV-1 system (Nguyen et al. 2007). A more recent overexpression screening approach aimed at characterizing the antiviral activity of more than 380 interferon-stimulated genes (ISGs) against a panel of viral pathogens (Schoggins et al. 2011). Broadly antiviral effectors included *IRF1*, *C6orf150/MB21D1*, *HPSE*, *RIG-I/DDX58*, *MDA5/IFIH1*, and *IFITM3*. In addition, anti-HIV activity was proposed for *MX2*, *IFITM2*, *CD74*, *TNFRDF10A*, *IRF7*, and *UNC84B*. Several ISGs, including *ADAR*, *FAM46C*, *LY6E*, and *MCOLN2*, enhanced the replication of certain viruses, highlighting the complexity in the type I interferon responses.

Primate Genetics

Nonhuman primates (NHP), whether naturally or experimentally infected with simian immunodeficiency viruses (SIVs), display phenotypic variation on multiple levels, including differences in relative susceptibility to infection, variability in both acute and long-term viral replication levels, differing rates of disease progression, and differences in degree of pathogenesis (see Klatt et al. 2011; Lifson and Haigwood

2011). NHP populations, including both wild populations and captive-bred colonies, comprise genetically variable, outbred individuals, and it is reasonable to assume that variation in virological phenotypes reflects, in part, host genetic variation. Phenotypic variation in SIV-infected NHPs provides a considerable but largely unexplored opportunity to examine the influence of host genetics on primate immunodeficiency virus replication and disease. In the case of rhesus macaques, the most commonly used NHP in AIDS research, the availability of whole genome sequence (WGS) data has facilitated discovery and cataloging of SNPs and copy-number variants that may prove useful for genetic and genomic analyses (Malhi et al. 2007; Lee et al. 2008). WGS data are also available for chimpanzees and a variety of other NHPs representing all main primate lineages (www.ensembl.org). Thus, comparative studies of the different SIVs and their respective primate hosts have the potential to identify and characterize genes that govern the transmission of viruses within and between populations.

Relevance to HIV infection and human disease exists on multiple fronts. For many gene products with well-established roles in HIV-1 infection or replication, there is evidence for an analogous or similar role for other SIV, and in such cases, the existence of animal models can confirm and even extend understanding of the biological relevance of such interactions (Fig. 2). Humans are frequently exposed to retroviruses of other, nonhuman primates (Wolfe et al. 2004), and SIV from chimpanzees and sooty mangabeys have made the jump several times (see Hahn et al. 2000; Apetrei et al. 2004; Sharp and Hahn 2011). Thus, in addition to reconstructing the natural history of the primate lentiviruses, comparative studies can reveal the role of genetics in determining which retroviruses jump from primates into humans, and allow us to ask whether adaptation to the human genetic landscape is a prerequisite for emergence of new viruses. As a practical matter, identification and molecular analysis of genes that interfere with cross-species transmission are also helping pave the way toward improved animal models of HIV infection and AIDS

(Hatziioannou et al. 2006; Kamada et al. 2006; Ambrose et al. 2007; Igarashi et al. 2007; Hatziioannou et al. 2009). Finally, understanding the role of genetics in promoting the non-pathogenic outcome in natural hosts of SIV could some day suggest pharmacological strategies for uncoupling HIV infection from disease in humans.

Studies in Macaques (Macaca sp.)

In a striking parallel to the emergence of HIV-1 and HIV-2 in humans, the first SIV was isolated as an emerging pathogen in colonies of captive Asian macaques in the early 1980s (Daniel et al. 1985; Mansfield et al. 1995; Gardner 2003; Apetrei et al. 2004). The SIV-infected macaque has since served as the primary model for preclinical AIDS research, in part because there is still no practical, small animal model that faithfully recapitulates HIV infection and AIDS in all essential parameters. Among the practical difficulties encountered in the SIV/macaque model are animal-to-animal variability in susceptibility to infection, viral replication levels, and disease progression (not unlike the variation observed among HIV-infected humans). Although this variability can confound small-scale animal studies (for practical reasons, studies in NHPs are often underpowered), it also means that nonhuman primate models can be used to investigate the influence of host genetics on viral replication, emergence, and pathogenesis.

As a model system for understanding the influence of host genetic variation on virological outcomes, SIV-infected macaques have several advantages over analysis of human cohorts. Initial infection is established either by a cloned viral isolate of known sequence, or by a biological isolate that can be genetically defined by sequencing. Unlike naturally occurring infection in humans, most macaque cohorts consist of animals infected in parallel with identical or very similar viruses. Both the time and route of infection are known precisely. Because most or all animals in a study are under the care of a specific team of investigators following a set of standardized protocols, sample collection,

evaluation, and care of animals in SIV cohorts are inherently more uniform. Furthermore, end points (such as definition of AIDS) are consistently applied. Finally, and most importantly, hypotheses can be tested or confirmed in either prospectively or retrospectively genotyped animals. Familial relationships are also often known for captive-bred animals, raising the possibility that approaches incorporating pedigree analysis might ultimately be applied to macaque cohorts. The primary disadvantage of macaque cohorts as subjects for genetic association studies is the relatively small size of any given study (typically <50 animals). Nonetheless, given the extensive use of the SIV/macaque model for the past 25 years and the widespread use of a limited number of closely related SIV strains, it should be feasible to assemble sufficiently large study cohorts by combining samples from multiple SIV studies. To date, GWASs have not been reported for SIV/macaque cohorts. However, studies focusing on specific genes/loci serve to illustrate the potential contributions of SIV cohorts to our understanding of host genetics and AIDS.

The MHC Locus in Primate Models

Perhaps the most significant contributions of SIV animal models to understanding the role of host genetics have been in elucidating the influence of MHC class I genes on lentiviral replication and disease progression, and the potential for vaccine-induced CTL responses. The first robust correlations between specific MHC genotypes and epitope-specific viral escape from CTL emerged from studies of SIV-infected rhesus macaques (Evans et al. 1999; Allen et al. 2000). For example, Evans et al. documented emergence of epitope-specific escape variants by tracking SIV sequences in MHC-defined macaques (Evans et al. 1999). Similarly, Allen et al. reported rapid emergence of CTL-escape variants in a *Mamu-*A01*-restricted Tat epitope during the first 8 wk of infection, corresponding to the primary onset of acute-phase virus-specific CTL responses (Allen et al. 2000). Such studies were corroborated by monoclonal-antibody-based depletion of $CD8^+$ T cells in SIV-infected macaques, which led to transient increases in viral replication levels and provided confirmation that cellular immune responses have a major influence on viral replication (Schmitz et al. 1999). Later studies also revealed significant correlations between specific MHC alleles and control of SIV replication levels; most notably, *MHC B*-locus alleles *Mamu-B*08* and *B*17*, which are associated with elite control of SIVmac infection in rhesus macaques (Yant et al. 2006; Loffredo et al. 2007a,b).

More recently, O'Connor and colleagues took advantage of a unique population of animals, the Mauritian cynomolgus macaques, to analyze the impact of *MHC* diversity on viral replication levels in vivo (O'Connor et al. 2010). Because Mauritian cynomolgus macaques display a limited number of *MHC* haplotypes, *MHC*-homozygosity is fairly common—a situation enabling direct comparison of viral replication in homozygous and heterozygous individuals, many with shared *MHC* haplotypes. In this study, *MHC*-homozygous animals infected with SIVmac239 had chronic-phase plasma viral RNA levels 80 times higher on average than *MHC*-heterozygous animals analyzed in parallel, suggesting that the increased breadth of potential virus-specific $CD8^+$ T-lymphocyte responses in heterozygous individuals on average gave rise to significantly enhanced control of viral replication. These results provided strong experimental confirmation of prior studies describing heterozygous advantage in human HIV/AIDS cohorts (Carrington et al. 1999; Tang et al. 1999). As a practical matter, evidence of heterozygous advantage in these animals also lends credence to the hypothesis that an effective HIV-1 vaccine will be one that induces a broad range of virus-specific immune responses.

MHC class I molecules also interact with killer immunoglobulin-like receptors (KIR), influencing NK cell responses to HIV infection (Martin et al. 2002, 2007). Several candidate MHC-KIR interactions have been reported for rhesus macaques based on in vitro binding assays (Rosner et al. 2011). Colantonio and colleagues made the serendipitous discovery that a recombinant soluble MHC tetramer folded around certain SIV-derived peptides could bind

to the surface of lymphocytes from a subset of uninfected animals with no prior exposure to SIV or SIV antigens (Colantonio et al. 2011). This observation led to the identification of a specific MHC-KIR interaction (Mamu-A1*00201/Mamu-KIR3DL05) and the first functional demonstration of ligand-mediated NK cell inhibition in primary macaque cells (Colantonio et al. 2011). Further investigation also revealed that the particular SIV peptide/epitope bound by the Mamu-A1*00201 tetramer influenced the ligand/KIR interaction. Although the impact of MHC-bound peptide on MHC-KIR interaction has been reported (Malnati et al. 1995; Peruzzi et al. 1996; Mandelboim et al. 1997; Rajagopalan et al. 1997; Zappacosta et al. 1997; Hansasuta et al. 2004; Thananchai et al. 2007), the analysis by Colantonio et al. pointed toward specific involvement of residues in the KIR molecule. By taking advantage of polymorphic variants of the rhesus macaque *KIR3DL05* gene that differed in peptide selectivity, they pinpointed residues in the third loop of the KIR D1 domain that influence the peptide dependency of the MHC-KIR interaction (Colantonio et al. 2011). Structural models of MHC-KIR interaction place this loop in close proximity to the MHC-peptide surface (Boyington et al. 2000; Fan et al. 2001; Sharma et al. 2009). Importantly, genetic characterization of the rhesus macaque *KIR* locus and the development of reagents specific for macaques will permit incorporation of *KIR* genetics into animal models of AIDS and preclinical vaccine research.

Restriction Factors

The discovery of the host restriction factor TRIM5α illustrates the benefits of a comparative approach to the study of HIV and AIDS. The antiviral activity of TRIM5α was first identified by Stremlau and colleagues by screening a rhesus macaque cDNA library for genes that conferred resistance to HIV-1 infection of human cells (Stremlau et al. 2004). Similarly, an unusual Trim5 ortholog was identified as the source of a genetic block to HIV-1 infection in cells from South American owl monkeys (*Aotus sp.*)

(Nisole et al. 2004; Sayah et al. 2004). It is now widely assumed that TRIM5 is a major modulator of cross-species transmission of retroviruses, and as a practical consequence, *TRIM5* poses a significant genetic barrier to development of a NHP model of HIV-1 infection (Hatziioannou et al. 2006; Kamada et al. 2006; Ambrose et al. 2007; Igarashi et al. 2007; Hatziioannou et al. 2009).

Two independent association studies in SIV-infected rhesus macaques have shown the ability of TRIM5 to suppress lentiviral replication levels in vivo (Kirmaier et al. 2010; Lim et al. 2010a). It is noteworthy that both studies uncovered significant associations with relatively few animals ($n < 100$ animals in both cases). This differs from *TRIM5* studies in human HIV/AIDS cohorts, where reported associations have been modest (Goldschmidt et al. 2006; Javanbakht et al. 2006; Speelmon et al. 2006). The Lim et al. study (Lim et al. 2010a), which focused on a cohort of SIVmac251-infected rhesus macaques, most closely resembles the situation in human cohorts. Just as HIV-1 is only weakly susceptible to human TRIM5, SIVmac strains are relatively resistant to rhesus macaque TRIM5. The enhanced ability to detect a significant effect in macaques was owed, in part, to the presence at high frequency of functionally distinct TRIM5 alleles in the rhesus macaque *TRIM5* locus (Newman et al. 2006; Wilson et al. 2008). Lim et al. also revealed that complete viral resistance to TRIM5 is not required for pathogenesis—animals with restrictive alleles displayed lower but significant levels of SIVmac251 replication and developed AIDS—a fact that could not have been appreciated from tissue culture experiments alone.

Goldstein and colleagues first reported evidence for variation in inherent susceptibility of T lymphocytes from naïve, uninfected rhesus macaques to infection with SIV strain SIVsmE543-3 (Goldstein et al. 2000). They further showed that susceptibility of an animal's cells to infection in tissue culture correlated with in vivo susceptibility to SIVsmE543-3 infection. These results argued for the existence of an intrinsic cellular block to infection unrelated to virus-specific adaptive immunity. More

recently, Kirmaier et al. found that this inherent resistance to SIVsmE543-3 was due to rhesus macaque TRIM5, and that variation in susceptibility of animals to SIVsmE543 infection correlated with allelic variation in the rhesus macaque *TRIM5* gene (Kirmaier et al. 2010). Compared to SIVmac251, which is well adapted to rhesus macaques as a host, the impact of rhesus TRIM5 polymorphism on SIVsmE543 was far more dramatic (2–3 log differences in viral loads). The greater susceptibility of SIVsmE543-3 to multiple alleles of rhesus TRIM5 likely reflects its derivation by brief passage of a sooty mangabey virus (SIVsm) through only two rhesus macaques (Hirsch et al. 1997).

Restriction of HIV-1 by the most common alleles of rhesus macaque TRIM5 also poses a barrier to development of simian-tropic strains of HIV-1 (Ambrose et al. 2007). However, additional genetic barriers remain, including those imposed by the *APOBEC3* and *BST2/Tetherin* loci. Although intraspecies surveys of NHP *BST2/Tetherin* and *APOBEC3* loci have been limited, some reports suggest that these genes may also be polymorphic in rhesus macaques (Weiler et al. 2006; Jia et al. 2009; McNatt et al. 2009). Whether allelic variants of tetherin and *APOBEC3* have an impact on viral infection or pathogenesis in the macaque model remains to be seen (a weak correlation between APOBEC3G variation and SIVmac replication levels in vivo has been reported [Weiler et al. 2006]). In contrast to rhesus macaques, pig-tailed macaques (species *Macaca nemestrina*) uniformly carry a single allele of *TRIM5* ($TRIM5^{CypA}$) that does not restrict HIV-1 infection in tissue culture assays (Liao et al. 2007), raising the possibility that this species may provide a more permissive host for developing an experimental model of HIV-1 infection (see Igarashi et al. 2007; Hatziioannou et al. 2009; Lifson and Haigwood 2011).

Although SIV-infected macaque cohorts have not been routinely subjected to gene association or GWASs, the impact of host genes and genetic variation on SIV replication and disease is disclosed by adaptive countermeasures acquired by viruses during replication in vivo. Kirmaier et al. identified specific amino acid alterations in the SIVsmE543 CA protein that emerged in vivo in several animals bearing suppressive *TRIM5* genotypes. Such changes also appeared during the emergence of SIV in rhesus macaque colonies in the 1970s (Kirmaier et al. 2010). Among its many functions, the SIVmac Nef protein clearly prevents viral inhibition by rhesus macaque tetherin in tissue culture assays (Jia et al. 2009; Zhang et al. 2009). Compelling evidence in support of a similar role in vivo came from retrospective analysis of animals infected with a *nef*-deleted variant of SIV-mac239. Replication of the SIVmac239 nef mutants was initially attenuated in vivo, and strains that eventually grew out in these animals had acquired novel anti-tetherin activity through adaptive changes in the viral Env protein (Serra-Moreno et al. 2011). Taken together, the emergence of adaptations to overcome CTL and restriction factors in animals with defined genotypes provides evidence that expression of these host genes (e.g., *MHC*, *TRIM5*, and *tetherin*) has biological relevance, by inhibiting viral replication in vivo, consistent with their observed mechanisms of action in the laboratory.

Chemokine Receptors and Chemokines

Mutations analogous to the Δ32 base-pair deletion in the CCR5 coding sequence in humans have also been found in two closely related nonhuman primates that serve as natural hosts of SIV infection (Chen et al. 1998; Palacios et al. 1998; Riddick et al. 2010). At least two distinct mutations are found in the *CCR5* gene of sooty mangabeys (*Cercocebus atys*), the natural hosts of SIVsm (Riddick et al. 2010). Both are deletion mutations, encompassing 2 and 24 nucleotides of coding sequence, respectively. In one colony of captive sooty mangabeys, the frequencies of the Δ2 and Δ24 alleles were reported to be 26% and 3%, respectively (Riddick et al. 2010). The Δ24 variant is also present at high frequency in red-capped mangabeys (*Cercocebus torquatus*), and the naturally occurring SIV that is endemic to these animals (SIVrcm) can use a different molecule (CCR2b) as a coreceptor for viral entry (Chen et al. 1998).

Only as recently as 2004, researchers came to recognize that copy-number variation (CNV) is a major source of human genetic diversity (Iafrate et al. 2004; Sebat et al. 2004). CNV, which can encompass expansions and contractions of large segments of chromosomal DNA (and the genes contained therein), can result in phenotypic diversity. Nonhuman primates with relevance to AIDS and AIDS research, including chimpanzees and Indian origin rhesus macaques, also display significant levels of genome-wide CNV (Perry et al. 2006; Lee et al. 2008). In 2005, Gonzalez et al. reported an inverse correlation between copy number of *CCL3L1*, which encodes a ligand for CCR5, and susceptibility to HIV-1 infection (Gonzalez et al. 2005). Interestingly, a similar link between *CCL3L1* copy number and disease progression in SIV-mac-infected macaques has also been described (Degenhardt et al. 2009). The correlations have intuitive appeal, as ligands of CCR5 have been shown experimentally to inhibit HIV-1 replication (Cocchi et al. 1995; Menten et al. 1999, 2002; Nibbs et al. 1999; Xin et al. 1999). However, the link between *CCL3L1* copy number and HIV-1 infection has been called into question by subsequent, independent studies, which have failed to reproduce the correlation (Bhattacharya et al. 2009; Urban et al. 2009). Likewise, after correcting for known influence of the MHC class I and *TRIM5* loci on SIV infection, Lim et al. did not find a significant correlation between *CCL3L1* copy number and replication of SIVmac in rhesus macaques (Lim et al. 2010b). At present, the original claims remain controversial and additional work may be needed to resolve the discrepancies; the fact that *CCL3L1* is copy number variable in rhesus macaques raises the possibility that some issues may be addressable experimentally.

Joint Viral-Host Genome Analysis

A particularly attractive application of genomics is in the analysis of the reciprocal genetic signals resulting from the interaction between the host and the pathogen. The HIV-1 genome is conducive to such analyses because the expected plasticity and mutability would effectively reflect the genetic signals of escape. However, there are constraints to viral escape, which reflect RNA and protein structural requirements (Watts et al. 2009) that may translate into loss of fitness. Although the HIV-1 genome is considered to be highly variable, 77% of amino acid positions are conserved, whereas 10% of the genome is under positive selection. This class of sites defines critical residues in host-pathogen interaction, whether resulting from CTL or other host-selective pressures. Although half of the sites under positive selection in the HIV-1 genome are mapped to CTL epitopes, there is considerable interest in identifying the nature of the pressures that are driving evolution in nonepitope sites. These may reflect pressure from KIR or host restriction factors. Figure 3 depicts the superposed signals and influences that may account for conservation and variation in the viral genome.

Studies of monozygotic twins infected with the same viral strain (Draenert et al. 2006) showed the power of the host genome to control and drive viral diversity. The initial CD8+ T-cell response targeted 17 epitopes, 15 of which were identical in each twin. Three years after infection, 14 of 15 initial responses were still detectable, whereas of four responses that declined in both twins, three showed mutations

at the same residue. The antibody responses cross-neutralized the other twin's virus, with similar changes in the pattern of evolution in the envelope gene. These results indicate a considerable concordance of cellular and humoral immune responses and HIV evolution in the same genetic environment (Draenert et al. 2006). To discover a larger number of host HLA-viral genome mutual associations at the population level, Moore et al. (2002) and Bhattacharya et al. (2007) searched for signatures of selection driven by specific HLA alleles across the viral genome. The challenge here is to expand the analysis to identify all possible driving forces, which may include mechanisms in addition to CTL pressure. Eventually, there is a need to conduct a full discovery effort that considers both the viral and the corresponding host at the genome level.

NEW RESEARCH AREAS

Next-Generation Sequencing

The field of host genomics is facing a change in paradigm (Fellay et al. 2010). GWASs served to understand the role of common variants in HIV-1 disease. However, the extent of variance explained—around 20% of viral load or disease

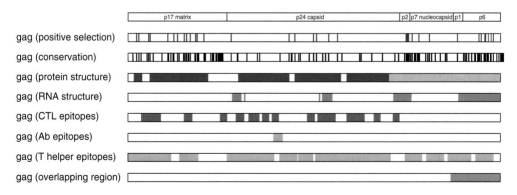

Figure 3. Multilayer representation of HIV-1 clade B Gag. The various information layers align the sites under positive selective pressure (red), conservation scores (<90% conserved, black), the structured domains at the protein (dark blue), and viral RNA level (light blue) (Watts et al. 2009), the position of CTL (dark green), antibody (light green), and T helper epitopes (turquoise) compiled in the Los Alamos HIV database, and the Gag region overlapping with the viral protease (purple).

progression—indicates that other factors are yet to be discovered. New technology in genome sequencing now allows the study of rare variation, whether through the capture and sequencing of the whole exome (the 2% of the genome that is protein coding), or increasingly through the deep sequencing of the whole genome or transcriptome (RNA-seq). These sequencing and resequencing tools associated with effective decoding bioinformatic support are also becoming affordable. A number of projects are under way in the HIV field that use next-generation sequencing for the study of extremes of disease (elite controllers vs. rapid progressors), and of unique populations that are resistant to infection (noninfected exposed hemophiliacs). Next-generation sequencing also offers an unprecedented opportunity to jointly analyze cellular and viral transcriptional activity (Lefebvre et al. 2011).

Evolutionary Genomics

The close association of retroviruses with the human, primate, and mammalian genome supports the notion that there has been significant coevolution, which results in signals of selection in both genomes. The availability of complete genomes for dozens of mammals and, as of 2010, nine primates representing all main primate and prosimian groups permits screening for signals of genetic conflict genome-wide. The analyses can extend from the comparative assessment of the coding regions of single individuals, to the analyses of whole genome signatures and the inclusion of multiple individuals. When applied to human diversity, genome evolutionary studies mark regions that likely reflect the impact of pathogens and population bottlenecks. These genes and genetic regions can inform next-generation sequencing projects (King et al. 2010), as analysis of the latter will need to be supported by priors—functional or other—because they test a very large sampling space. That is, genome sequencing generates very large numbers of variants that cannot be prioritized based solely on statistics. A new range of tools are now available for the identification of evolutionary signals during

human population diversification (Grossman et al. 2010).

Paleovirology and Protein Reconstruction

The infection of the germline can lead to viral genes becoming inherited as host alleles. A broad range of retroviral and nonretroviral virus groups are now recognized as part of modern genomes (Johnson 2010; Katzourakis et al. 2010). The long-lasting association of retroviral genomes with the mammal genome, as well as exogenous retro/lentiviral infections, could influence contemporary susceptibility to HIV-1 in humans. Paleovirology is a novel field of research that uses genome data to reconstruct extinct viruses and the ancestral state of restriction factors (Emerman et al. 2010). Specific examples include the reconstruction of the core protein of PtERV, a 4-million-year-old endogenous retrovirus identified in the chimpanzee and gorilla genomes (Kaiser et al 2007). The resurrection of this virus could show that the human variant of TRIM5α actively prevented infection by this virus. A second example of reconstruction, this time of 20-million-year-old TRIM5α, suggested that restriction of HIV-1 has decreased during evolution leading to humans (Goldschmidt et al. 2008). Soll et al. investigated two chimpanzee endogenous retrovirus-1 and -2 (CERV1 and CERV2) relatives of modern murine leukemia viruses (MLVs) that are present in the genomes of Old World primates, but absent from the human genome (Soll et al. 2010). Using CERV2 Env-pseudotyped MLV vectors, Soll et al. identified copper transport protein 1 (CTR1) as a receptor that was presumably used by CERV2 during its ancient exogenous replication in primates. CTR1-inactivating mutations may represent an evolutionary barrier to the acquisition of CERV2 resistance in primates. In addition, the reconstruction of the first examples of endogenous lentiviruses, PSIV in lemurs (Gilbert et al. 2009) and RELIK in rabbits (Katzourakis et al. 2007; Keckesova et al. 2009), opens the door to the study of the activity of restriction factors on these ancient elements (Rahm et al. 2011).

Data Integration, Network, and Systems Biology

There is an urgent need to design experiments that will allow the incorporation of multiple types of genomic information with function. An excellent example from immunology is the identification of regulatory networks that control the transcriptional response of mouse primary dendritic cells to toll-like receptor agonists (Amit et al. 2009). This network model identified 24 core regulators and 76 fine tuners that explain how pathogen-sensing pathways achieve specificity. This study established a broadly applicable approach to dissecting the regulatory networks controlling transcriptional responses in mammalian cells. A second paradigmatic use of systems biology is the study of heterogeneity in cell populations during viral infection (Snijder et al. 2009). Much of the variation in viral infection, endocytosis, and membrane lipid composition is determined by the adaptation of cells to their population context. Perturbation screens, combined with quantitative modeling of single cells, revealed the molecular networks that underlie the heterogeneous patterns in cell populations that likely mediate collective behavior. Although developing system and network approaches in in vitro models is reductionist, the next challenge is to set the conditions for HIV-1 systems biology in vivo that can create series of perturbations, implement the iterative acquisition of high throughput data, and model the observed variation.

CONCLUSIONS

The field of host genomics is at a crossroads thanks to the availability of new technology and sequences of human and primate genomes. Some steps have been completed, such as clarifying the role of common variants in HIV viral load and disease progression, and the description of transcription in infected individuals. The next studies will be technology driven, but will also be designed to capture and integrate multiple types of data. These include the study of the genetics of resistance to infection and disease mediated by rare human variants, the use of deep RNA sequencing to improve the quantification of transcriptome changes in specific cell populations, and the capture of the dynamic processes in gene expression. At the postgenome level, there is a need for completing functional analysis of human variants or of candidate regulatory genes identified through the novel approaches. The integration of data and system biology will generate new lists of candidates for such functional studies.

Primate genetics is also progressing, led by the identification of important variants that stratify the susceptibility to infection in the macaque model. Live-attenuated SIV mutants and chimeric viruses (SIV-HIV hybrids, or SHIVs) first revealed the presence of dominant-acting genetic barriers to lentiviral replication in nonhuman primates, some of which have been identified. However, despite more than 25 years of work, cohorts of SIV-infected macaques have yet to be analyzed by large-scale gene association or GWASs.

Understanding of the host genome should inform the study of the viral genome, and vice versa. The plasticity of the viral genome provides several kinds of information: from the structural requirements for viral function, to the faithful reflection of host pressures exerted during cross-species transmission, adaptation to a new host, and spread in a genetically diverse population. Research priorities include the mapping of all sites in the viral genome that are the likely result of host-selective pressure, followed by the identification of the host factors that exert those pressures, and their characterization. A general framework needs to be designed that will address whether the role of the respective host factors is to act as cross-species and interindividual barriers of transmission, to exert control of viral replication in the infected individual, or to limit pathogenicity. Progress in the HIV-1 field will benefit from improved understanding of the evolution of the innate immunity against retroviruses, and this should clarify the genetic conflicts between retrovirus and the host.

ACKNOWLEDGMENTS

Work in the laboratory of A.T. is supported by the Swiss National Science Foundation. Work in the laboratory of W.E.J. is supported by the

National Institutes of Health. We thank Jacques Fellay for comments and for Figure 1, and Joke Snoeck for data on Figure 3.

REFERENCES

*Reference is also in this collection.

Allen TM, O'Connor DH, Jing P, Dzuris JL, Mothe BR, Vogel TU, Dunphy E, Liebl ME, Emerson C, Wilson N, et al. 2000. Tat-specific cytotoxic T lymphocytes select for SIV escape variants during resolution of primary viraemia. *Nature* **407:** 386–390.

Ambrose Z, KewalRamani VN, Bieniasz PD, Hatziioannou T. 2007. HIV/AIDS: In search of an animal model. *Trends Biotechnol* **25:** 333–337.

Amit I, Garber M, Chevrier N, Leite AP, Donner Y, Eisenhaure T, Guttman M, Grenier JK, Li W, Zuk O, et al. 2009. Unbiased reconstruction of a mammalian transcriptional network mediating pathogen responses. *Science* **326:** 257–263.

Apetrei C, Robertson DL, Marx PA. 2004. The history of SIVS and AIDS: Epidemiology, phylogeny and biology of isolates from naturally SIV infected non-human primates (NHP) in Africa. *Front Biosci* **9:** 225–254.

Ballana E, Senserrich J, Pauls E, Faner R, Mercader JM, Uyttebroeck F, Palou E, Mena MP, Grau E, Clotet B, et al. 2010. ZNRD1 (zinc ribbon domain-containing 1) is a host cellular factor that influences HIV-1 replication and disease progression. *Clin Infect Dis* **50:** 1022–1032.

Bhattacharya T, Daniels M, Heckerman D, Foley B, Frahm N, Kadie C, Carlson J, Yusim K, McMahon B, Gaschen B, et al. 2007. Founder effects in the assessment of HIV polymorphisms and HLA allele associations. *Science* **315:** 1583–1586.

Bhattacharya T, Stanton J, Kim EY, Kunstman KJ, Phair JP, Jacobson LP, Wolinsky SM. 2009. CCL3L1 and HIV/AIDS susceptibility. *Nat Med* **15:** 1112–1115.

Bosinger SE, Li Q, Gordon SN, Klatt NR, Duan L, Xu L, Francella N, Sidahmed A, Smith AJ, Cramer EM, et al. 2009. Global genomic analysis reveals rapid control of a robust innate response in SIV-infected sooty mangabeys. *J Clin Invest* **119:** 3556–3572.

Boyington JC, Motyka SA, Schuck P, Brooks AG, Sun PD. 2000. Crystal structure of an NK cell immunoglobulin-like receptor in complex with its class I MHC ligand. *Nature* **405:** 537–543.

Brass AL, Dykxhoorn DM, Benita Y, Yan N, Engelman A, Xavier RJ, Lieberman J, Elledge SJ. 2008. Identification of host proteins required for HIV infection through a functional genomic screen. *Science* **319:** 921–926.

Bushman FD, Malani N, Fernandes J, D'Orso I, Cagney G, Diamond TL, Zhou H, Hazuda DJ, Espeseth AS, Konig R, et al. 2009. Host cell factors in HIV replication: Meta-analysis of genome-wide studies. *PLoS Pathog* **5:** e1000437. doi: 10.1371/journal.ppat.1000437.

Carrington M, Nelson GW, Martin MP, Kissner T, Vlahov D, Goedert JJ, Kaslow R, Buchbinder S, Hoots K, O'Brien SJ. 1999. HLA and HIV-1: Heterozygote advantage and B*35-Cw*04 disadvantage. *Science* **283:** 1748–1752.

Chan EY, Qian WJ, Diamond DL, Liu T, Gritsenko MA, Monroe ME, Camp DG, Smith RD, Katze MG. 2007. Quantitative analysis of human immunodeficiency virus type 1-infected CD4$^+$ cell proteome: Dysregulated cell cycle progression and nuclear transport coincide with robust virus production. *J Virol* **81:** 7571–7583.

Chen Z, Kwon D, Jin Z, Monard S, Telfer P, Jones MS, Lu CY, Aguilar RF, Ho DD, Marx PA. 1998. Natural infection of a homozygous Δ24 CCR5 red-capped mangabey with an R2b-tropic simian immunodeficiency virus. *J Exp Med* **188:** 2057–2065.

Cocchi F, DeVico AL, Garzino-Demo A, Arya SK, Gallo RC, Lusso P. 1995. Identification of RANTES, MIP-1α, and MIP-1β as the major HIV-suppressive factors produced by CD8$^+$ T cells. *Science* **270:** 1811–1815.

Colantonio AD, Bimber BN, Neidermyer WJ, Reeves RK, Alter G, Altfeld M, Johnson RP, Carrington M, O'Connor DH, Evans DT. 2011. KIR polymorphisms modulate peptide-dependent binding to an MHC class I ligand with a Bw6 motif. *PLoS Pathog* **7:** e1001316. doi:10.1371/journal.ppat.1001316 .

Dalmasso C, Carpentier W, Meyer L, Rouzioux C, Goujard C, Chaix ML, Lambotte O, Avettand-Fenoel V, Le CS, de Senneville LD, et al. 2008. Distinct genetic loci control plasma HIV-RNA and cellular HIV-DNA levels in HIV-1 infection: The ANRS genome wide association 01 study. *PLoS One* **3:** e3907. doi:10.1371/journal.pone.0003907.

Daniel MD, Letvin NL, King NW, Kannagi M, Sehgal PK, Hunt RD, Kanki PJ, Essex M, Desrosiers RC. 1985. Isolation of T-cell tropic HTLV-III-like retrovirus from macaques. *Science* **228:** 1201–1204.

Dean M, Carrington M, Winkler C, Huttley GA, Smith MW, Allikmets R, Goedert JJ, Buchbinder SP, Vittinghoff E, Gomperts E, et al. 1996. Genetic restriction of HIV-1 infection and progression to AIDS by a deletion allele of the CKR5 structural gene. Hemophilia growth and development study, multicenter AIDS cohort study, multicenter hemophilia cohort study, San Francisco City cohort, ALIVE study. *Science* **273:** 1856–1862.

Degenhardt JD, de Candia P, Chabot A, Schwartz S, Henderson L, Ling B, Hunter M, Jiang Z, Palermo RE, Katze M, et al. 2009. Copy number variation of *CCL3*-like genes affects rate of progression to simian-AIDS in Rhesus Macaques (*Macaca mulatta*). *PLoS Genet* **5:** e1000346. doi: 10.1371/journal.pgen.1000346.

de Groot NG, Heijmans CM, Zoet YM, de Ru AH, Verreck FA, van Veelen PA, Drijfhout JW, Doxiadis GG, Remarque EJ, Doxiadis II, et al. 2010. AIDS-protective HLA-B*27/B*57 and chimpanzee MHC class I molecules target analogous conserved areas of HIV-1/SIVcpz. *Proc Natl Acad Sci* **107:** 15175–15180.

Draenert R, Allen TM, Liu Y, Wrin T, Chappey C, Verrill CL, Sirera G, Eldridge RL, Lahaie MP, Ruiz L, et al. 2006. Constraints on HIV-1 evolution and immunodominance revealed in monozygotic adult twins infected with the same virus. *J Exp Med* **203:** 529–539.

Emerman M, Malik HS. 2010. Paleovirology—Modern consequences of ancient viruses. *PLoS Biol* **8:** e1000301. doi:10.1371/journal.pbio.1000301.

Evangelou E, Fellay J, Colombo S, Martinez-Picado J, Obel N, Goldstein DB, Telenti A, Ioannidis JP. 2011. Impact of phenotype definition on genome-wide association signals: Empirical evaluation in HIV-1 infection. *Am J Epidemiol* 173: 1336–1342.

Evans DT, O'Connor DH, Jing P, Dzuris JL, Sidney J, da Silva J, Allen TM, Horton H, Venham JE, Rudersdorf RA, et al. 1999. Virus-specific cytotoxic T-lymphocyte responses select for amino-acid variation in simian immunodeficiency virus Env and Nef. *Nat Med* 5: 1270–1276.

Fan QR, Long EO, Wiley DC. 2001. Crystal structure of the human natural killer cell inhibitory receptor KIR2DL1-HLA-Cw4 complex. *Nat Immunol* 2: 452–460.

Fellay J, Shianna KV, Ge D, Colombo S, Ledergerber B, Weale M, Zhang K, Gumbs C, Castagna A, Cossarizza A, et al. 2007. A whole-genome association study of major determinants for host control of HIV-1. *Science* 317: 944–947.

Fellay J, Ge D, Shianna KV, Colombo S, Ledergerber B, Cirulli ET, Urban TJ, Zhang K, Gumbs CE, Smith JP, et al. 2009. Common genetic variation and the control of HIV-1 in humans. *PLoS Genet* 5: e1000791. doi: 10.1371/journal.pgen.1000791.

Fellay J, Shianna KV, Telenti A, Goldstein DB. 2010. Host genetics and HIV-1: The final phase? *PLoS Pathog* 6: e1001033. doi:10.1371/journal. ppat.1001033.

Fellay J, Frahm N, Shianna KV, Cirulli ET, Casimiro DR, Robertson MN, Haynes BF, Geraghty DE, McElrath MJ, Goldstein DB. 2011. Host genetic determinants of T cell responses to the MRKAd5 HIV-1 gag/pol/nef vaccine in the step trial. *J Infect Dis* 203: 773–779.

Fu W, Sanders-Beer BE, Katz KS, Maglott DR, Pruitt KD, Ptak RG. 2009. Human immunodeficiency virus type 1, human protein interaction database at NCBI. *Nucleic Acids Res* 37: D417–D422.

Gardner MB. 2003. Simian AIDS: An historical perspective. *J Med Primatol* 32: 180–186.

Gilbert C, Maxfield DG, Goodman SM, Feschotte C. 2009. Parallel germline infiltration of a lentivirus in two Malagasy lemurs. *PLoS Genet* 5: e1000425. doi: 10.1371/journal.pgen.1000425.

Giri MS, Nebozhyn M, Showe L, Montaner LJ. 2006. Microarray data on gene modulation by HIV-1 in immune cells: 2000–2006. *J Leukoc Biol* 80: 1031–1043.

Goldschmidt V, Bleiber G, May M, Martinez R, Ortiz M, Telenti A. 2006. Role of common human TRIM5α variants in HIV-1 disease progression. *Retrovirology* 3: 54.

Goldschmidt V, Ciuffi A, Ortiz M, Brawand D, Munoz M, Kaessmann H, Telenti A. 2008. Antiretroviral activity of ancestral TRIM5α. *J Virol* 82: 2089–2096.

Goldstein S, Brown CR, Dehghani H, Lifson JD, Hirsch VM. 2000. Intrinsic susceptibility of rhesus macaque peripheral CD4⁺ T cells to simian immunodeficiency virus in vitro is predictive of in vivo viral replication. *J Virol* 74: 9388–9395.

Gonzalez E, Kulkarni H, Bolivar H, Mangano A, Sanchez R, Catano G, Nibbs RJ, Freedman BI, Quinones MP, Bamshad MJ, et al. 2005. The influence of CCL3L1 gene-containing segmental duplications on HIV-1/AIDS susceptibility. *Science* 307: 1434–1440.

Grossman SR, Shylakhter I, Karlsson EK, Byrne EH, Morales S, Frieden G, Hostetter E, Angelino E, Garber M, Zuk O, et al. 2010. A composite of multiple signals distinguishes causal variants in regions of positive selection. *Science* 327: 883–886.

Hahn BH, Shaw GM, De Cock KM, Sharp PM. 2000. AIDS as a zoonosis: Scientific and public health implications. *Science* 287: 607–614.

Hansasuta P, Dong T, Thananchai H, Weekes M, Willberg C, Aldemir H, Rowland-Jones S, Braud VM. 2004. Recognition of HLA-A3 and HLA-A11 by KIR3DL2 is peptide-specific. *Eur J Immunol* 34: 1673–1679.

Hatziioannou T, Princiotta M, Piatak MJr, Yuan F, Zhang F, Lifson JD, Bieniasz PD. 2006. Generation of simian-tropic HIV-1 by restriction factor evasion. *Science* 314: 95.

Hatziioannou T, Ambrose Z, Chung NP, Piatak MJr, Yuan F, Trubey CM, Coalter V, Kiser R, Schneider D, Smedley J, et al. 2009. A macaque model of HIV-1 infection. *Proc Natl Acad Sci* 106: 4425–4429.

Herbeck JT, Gottlieb GS, Winkler CA, Nelson GW, An P, Maust BS, Wong KG, Troyer JL, Goedert JJ, Kessing BD, et al. 2010. Multistage genomewide association study identifies a locus at 1q41 associated with rate of HIV-1 disease progression to clinical AIDS. *J Infect Dis* 201: 618–626.

Hirsch V, Adger-Johnson D, Campbell B, Goldstein S, Brown C, Elkins WR, Montefiori DC. 1997. A molecularly cloned, pathogenic, neutralization-resistant simian immunodeficiency virus, SIVsmE543–3. *J Virol* 71: 1608–1620.

Huang Y, Paxton WA, Wolinsky SM, Neumann AU, Zhang L, He T, Kang S, Ceradini D, Jin Z, Yazdanbakhsh K, et al. 1996. The role of a mutant CCR5 allele in HIV-1 transmission and disease progression. *Nat Med* 2: 1240–1243.

Iafrate AJ, Feuk L, Rivera MN, Listewnik ML, Donahoe PK, Qi Y, Scherer SW, Lee C. 2004. Detection of large-scale variation in the human genome. *Nat Genet* 36: 949–951.

Igarashi T, Iyengar R, Byrum RA, Buckler-White A, Dewar RL, Buckler CE, Lane HC, Kamada K, Adachi A, Martin MA. 2007. Human immunodeficiency virus type 1 derivative with 7% simian immunodeficiency virus genetic content is able to establish infections in pig-tailed macaques. *J Virol* 81: 11549–11552.

International HIV Controllers Study. 2010. The major genetic determinants of HIV-1 control affect HLA class I peptide presentation. *Science* 330: 1551–1557.

Jacquelin B, Mayau V, Targat B, Liovat AS, Kunkel D, Petitjean G, Dillies MA, Roques P, Butor C, Silvestri G, et al. 2009. Nonpathogenic SIV infection of African green monkeys induces a strong but rapidly controlled type I IFN response. *J Clin Invest* 119: 3544–3555.

Javanbakht H, An P, Gold B, Petersen DC, O'Huigin C, Nelson GW, O'Brien SJ, Kirk GD, Detels R, Buchbinder S, et al. 2006. Effects of human TRIM5α polymorphisms on antiretroviral function and susceptibility to human immunodeficiency virus infection. *Virology* 354: 15–27.

Jia B, Serra-Moreno R, Neidermyer W, Rahmberg A, Mackey J, Fofana IB, Johnson WE, Westmoreland S, Evans DT. 2009. Species-specific activity of SIV Nef and HIV-1 Vpu in overcoming restriction by tetherin/BST2. *PLoS Pathog* 5: e1000429. doi: 10.1371/journal.ppat.1000429.

Johnson WE. 2010. Endless forms most viral. *PLoS Genet* 6: e1001210. doi: 10.1371/journal.pgen.1001210.

Kaiser SM, Malik H, Emerman M. 2007. Restriction of an extinct retrovirus by the human TRIM5 α antiviral protein. *Science* **316:** 1756–1758.

Kamada K, Igarashi T, Martin MA, Khamsri B, Hatcho K, Yamashita T, Fujita M, Uchiyama T, Adachi A. 2006. Generation of HIV-1 derivatives that productively infect macaque monkey lymphoid cells. *Proc Natl Acad Sci* **103:** 16959–16964.

Kaslow RA, Carrington M, Apple R, Park L, Munoz A, Saah AJ, Goedert JJ, Winkler C, O'Brien SJ, Rinaldo C, et al. 1996. Influence of combinations of human major histocompatibility complex genes on the course of HIV-1 infection. *NatMed* **2:** 405–411.

Katzourakis A, Gifford RJ. 2010. Endogenous viral elements in animal genomes. *PLoS Genet* **6:** e1001191. doi: 10.1371/journal.pgen.1001191.

Katzourakis A, Tristem M, Pybus OG, Gifford RJ. 2007. Discovery and analysis of the first endogenous lentivirus. *Proc Natl Acad Sci* **104:** 6261–6265.

Kawashima Y, Pfafferott K, Frater J, Matthews P, Payne R, Addo M, Gatanaga H, Fujiwara M, Hachiya A, Koizumi H, et al. 2009. Adaptation of HIV-1 to human leukocyte antigen class I. *Nature* **458:** 641–645.

Keckesova Z, Ylinen LM, Towers GJ, Gifford RJ, Katzourakis A. 2009. Identification of a RELIK orthologue in the European hare (*Lepus europaeus*) reveals a minimum age of 12 million years for the lagomorph lentiviruses. *Virology* **384:** 7–11.

King CR, Rathouz PJ, Nicolae DL. 2010. An evolutionary framework for association testing in resequencing studies. *PLoS Genet* **6:** e1001202. doi: 10.1371/journal.pgen.1001202.

Kirmaier A, Wu F, Newman RM, Hall LR, Morgan JS, O'Connor S, Marx PA, Meythaler M, Goldstein S, Buckler-White A, et al. 2010. *TRIM5* suppresses cross-species transmission of a primate immunodeficiency virus and selects for emergence of resistant variants in the new species. *PLoS Biol* **8:** e1000462. doi: 10.1371/journal.pbio.1000462.

* Klatt NR, Silvestri G, Hirsch V. 2011. Nonpathogenic simian immunodeficiency virus infections. *Cold Spring Harb Perspect Med* doi: 10.1101/cshperspect.a007153.

Konig R, Zhou Y, Elleder D, Diamond TL, Bonamy GM, Irelan JT, Chiang CY, Tu BP, De Jesus PD, Lilley CE, et al. 2008. Global analysis of host-pathogen interactions that regulate early-stage HIV-1 replication. *Cell* **135:** 49–60.

Kulkarni SS, Savan R, Qi Y, Gao X, Yuki Y, Bass SE, Martin MP, Hunt P, Deeks S, Telenti A, et al. 2011. Differential microRNA regulation of HLA-C expression and it association with HIV control. *Nature* **472:** 495–498.

Le Clerc S, Limou S, Coulonges C, Carpentier W, Dina C, Taing L, Delaneau O, Labib T, Sladek R, Deveau C, et al. 2009. Genomewide association study of a rapid progression cohort identifies new susceptibility alleles for AIDS (ANRS Genomewide Association Study 03). *J Infect Dis* **200:** 1194–1201.

Lederer S, Favre D, Walters KA, Proll S, Kanwar B, Kasakow Z, Baskin CR, Palermo R, McCune JM, Katze MG. 2009. Transcriptional profiling in pathogenic and non-pathogenic SIV infections reveals significant distinctions in kinetics and tissue compartmentalization. *PLoS Pathog* **5:** e1000296. doi: 10.1371/journal.ppat.1000296.

Lee AS, Gutierrez-Arcelus M, Perry GH, Vallender EJ, Johnson WE, Miller GM, Korbel JO, Lee C. 2008. Analysis of copy number variation in the rhesus macaque genome identifies candidate loci for evolutionary and human disease studies. *Hum Mol Genet* **17:** 1127–1136.

Lefebvre G, Desfarges S, Uyttebroeck F, Muñoz M, Beerenwinkel N, Rougemont J, Telenti A, Ciuffi A. 2011. Analysis of HIV-1 expression level and sense of transcription by high-throughput sequencing of the infected cell. *J Virol* **85:** 6205–6211.

Liao CH, Kuang YQ, Liu HL, Zheng YT, Su B. 2007. A novel fusion gene, TRIM5-Cyclophilin A in the pig-tailed macaque determines its susceptibility to HIV-1 infection. *AIDS* **21** (Suppl 8): S19–S26.

* Lifson JD, Haigwood NL. 2011. Lessons in nonhuman primate models for AIDS vaccine research: From minefields to milestones. *Cold Spring Harb Perspect Med* doi: 10.1101/cshperspect.a007310.

Lim SY, Rogers T, Chan T, Whitney JB, Kim J, Sodroski J, Letvin NL. 2010a. TRIM5α modulates v control in Rhesus Monkeys. *PLoS Pathog* **6:** e1000738. doi: 10.1371/journal.ppat.1000738.

Lim SY, Chan T, Gelman RS, Whitney JB, O'Brien KL, Barouch DH, Goldstein DB, Haynes BF, Letvin NL. 2010b. Contributions of Mamu-A*01 status and TRIM5 allele expression, but not CCL3L copy number variation, to the control of SIVmac251 replication in Indian-origin rhesus monkeys. *PLoS Genet* **6:** e1000997. doi: 10.1371/journal.pgen.1000997.

Limou S, Le Clerc S, Coulonges C, Carpentier W, Dina C, Delaneau O, Labib T, Taing L, Sladek R, Deveau C, et al. 2009. Genomewide association study of an AIDS-nonprogression cohort emphasizes the role played by HLA genes (ANRS Genomewide Association Study 02). *J Infect Dis* **199:** 419–426.

Limou S, Coulonges C, Herbeck JT, van Manen D, An P, Le Clerc S, Delaneau O, Diop G, Taing L, Montes M, et al. 2010. Multiple-cohort genetic association study reveals CXCR6 as a new chemokine receptor involved in long-term nonprogression to AIDS. *J Infect Dis* **202:** 908–915.

Liu R, Paxton WA, Choe S, Ceradini D, Martin SR, Horuk R, MacDonald ME, Stuhlmann H, Koup RA, Landau NR. 1996. Homozygous defect in HIV-1 coreceptor accounts for resistance of some multiply-exposed individuals to HIV-1 infection. *Cell* **86:** 367–377.

Loeuillet C, Deutsch S, Ciuffi A, Robyr D, Taffe P, Munoz M, Beckmann JS, Antonarakis SE, Telenti A. 2008. In vitro whole-genome analysis identifies a susceptibility locus for HIV-1. *PLoS Biol* **6:** e32. doi: 10.1371/journal.pbio.0060032.

Loffredo JT, Friedrich TC, Leon EJ, Stephany JJ, Rodrigues DS, Spencer SP, Bean AT, Beal DR, Burwitz BJ, Rudersdorf RA, et al. 2007a. CD8+ T cells from SIV elite controller macaques recognize Mamu-B*08-bound epitopes and select for widespread viral variation. *PLoS One* **2:** e1152. doi: 10.1371/journal.pone.0001152.

Loffredo JT, Maxwell J, Qi Y, Glidden CE, Borchardt GJ, Soma T, Bean AT, Beal DR, Wilson NA, Rehrauer WM, et al. 2007b. Mamu-B*08-positive macaques control simian immunodeficiency virus replication. *J Virol* **81:** 8827–8832.

Cite this article as *Cold Spring Harb Perspect Med* doi: 10.1101/cshperspect.a007203

Malhi RS, Sickler B, Lin D, Satkoski J, Tito RY, George D, Kanthaswamy S, Smith DG. 2007. MamuSNP: A resource for Rhesus Macaque (*Macaca mulatta*) genomics. *PLoS ONE* **2**: e438. doi: 10.1371/journal.pone.0000438.

Malnati MS, Peruzzi M, Parker KC, Biddison WE, Ciccone E, Moretta A, Long EO. 1995. Peptide specificity in the recognition of MHC class I by natural killer cell clones. *Science* **267**: 1016–1018.

Mandelboim O, Wilson SB, Vales-Gomez M, Reyburn HT, Strominger JL. 1997. Self and viral peptides can initiate lysis by autologous natural killer cells. *Proc Natl Acad Sci* **94**: 4604–4609.

Mansfield KG, Lerch NW, Gardner MB, Lackner AA. 1995. Origins of simian immunodeficiency virus infection in macaques at the New England Regional Primate Research Center. *J Med Primatol* **24**: 116–122.

Martin MP, Gao X, Lee JH, Nelson GW, Detels R, Goedert JJ, Buchbinder S, Hoots K, Vlahov D, Trowsdale J, et al. 2002. Epistatic interaction between KIR3DS1 and HLA-B delays the progression to AIDS. *Nat Genet* **31**: 429–434.

Martin MP, Qi Y, Gao X, Yamada E, Martin JN, Pereyra F, Colombo S, Brown EE, Shupert WL, Phair J, et al. 2007. Innate partnership of HLA-B and KIR3DL1 subtypes against HIV-1. *Nat Genet* **39**: 733–740.

McNatt MW, Zang T, Hatziioannou T, Bartlett M, Fofana IB, Johnson WE, Neil SJ, Bieniasz PD. 2009. Species-specific activity of HIV-1 Vpu and positive selection of tetherin transmembrane domain variants. *PLoS Pathog* **5**: e1000300. doi: 10.1371/journal.ppat.1000300.

Menten P, Struyf S, Schutyser E, Wuyts A, De Clercq E, Schols D, Proost P, Van Damme J. 1999. The LD78β isoform of MIP-1α is the most potent CCR5 agonist and HIV-1-inhibiting chemokine. *J Clin Invest* **104**: R1–R5.

Menten P, Wuyts A, Van Damme J. 2002. Macrophage inflammatory protein-1. *Cytokine Growth Factor Rev* **13**: 455–481.

Moore CB, John M, James IR, Christiansen FT, Witt CS, Mallal SA. 2002. Evidence of HIV-1 adaptation to HLA-restricted immune responses at a population level. *Science* **296**: 1439–1443.

Newman RM, Hall L, Connole M, Chen GL, Sato S, Yuste E, Diehl W, Hunter E, Kaur A, Miller GM, et al. 2006. Balancing selection and the evolution of functional polymorphism in Old World monkey TRIM5α. *Proc Natl Acad Sci* **103**: 19134–19139.

Nguyen DG, Yin H, Zhou Y, Wolff KC, Kuhen KL, Caldwell JS. 2007. Identification of novel therapeutic targets for HIV infection through functional genomic cDNA screening. *Virology* **362**: 16–25.

Nibbs RJ, Yang J, Landau NR, Mao JH, Graham GJ. 1999. LD78β, a non-allelic variant of human MIP-1α (LD78α), has enhanced receptor interactions and potent HIV suppressive activity. *J Biol Chem* **274**: 17478–17483.

Nisole S, Lynch C, Stoye JP, Yap MW. 2004. A Trim5-cyclophilin A fusion protein found in owl monkey kidney cells can restrict HIV-1. *Proc Natl Acad Sci* **101**: 13324–13328.

O'Connor SL, Lhost JJ, Becker EA, Detmer AM, Johnson RC, Macnair CE, Wiseman RW, Karl JA, Greene JM, Burwitz BJ, et al. 2010. MHC heterozygote advantage in simian immunodeficiency virus-infected Mauritian cynomolgus macaques. *Sci Trans Med* **2**: 22ra18.

Palacios E, Digilio L, McClure HM, Chen Z, Marx PA, Goldsmith MA, Grant RM. 1998. Parallel evolution of CCR5-null phenotypes in humans and in a natural host of simian immunodeficiency virus. *Curr Biol* **8**: 943–946.

Paust S, Gill HS, Wang BZ, Flynn MP, Moseman EA, Senman B, Szczepanik M, Telenti A, Askenase PW, Compans RW, et al. 2010. Critical role for the chemokine receptor CXCR6 in NK cell-mediated antigen-specific memory of haptens and viruses. *Nat Immunol* **11**: 1127–1135.

Pelak K, Goldstein DB, Walley NM, Fellay J, Ge D, Shianna KV, Gumbs C, Gao X, Maia JM, Cronin KD, et al. 2010. Host determinants of HIV-1 control in African Americans. *J Infect Dis* **201**: 1141–1149.

Perry GH, Tchinda J, McGrath SD, Zhang J, Picker SR, Caceres AM, Iafrate AJ, Tyler-Smith C, Scherer SW, Eichler EE, et al. 2006. Hotspots for copy number variation in chimpanzees and humans. *Proc Natl Acad Sci* **103**: 8006–8011.

Peruzzi M, Parker KC, Long EO, Malnati MS. 1996. Peptide sequence requirements for the recognition of HLA-B*2705 by specific natural killer cells. *J Immunol* **157**: 3350–3356.

Quigley M, Pereyra F, Nilsson B, Porichis F, Fonseca C, Eichbaum Q, Julg B, Jesneck JL, Brosnahan K, Imam S, et al. 2010. Transcriptional analysis of HIV-specific CD8+ T cells shows that PD-1 inhibits T cell function by upregulating BATF. *Nat Med* **16**: 1147–1151.

Rahm N, Yap M, Snoeck J, Zoete V, Munoz M, Radespiel U, Zimmermann E, Michielin O, Stoye JP, Ciuffi A, et al. 2011. Unique spectrum of activity of prosimian TRIM5α against exogenous and endogenous retroviruses. *J Virol* **85**: 4173–4183.

Rajagopalan S, Long EO. 1997. The direct binding of a p58 killer cell inhibitory receptor to human histocompatibility leukocyte antigen (HLA)-Cw4 exhibits peptide selectivity. *J Exp Med* **185**: 1523–1528.

Riddick NE, Hermann EA, Loftin LM, Elliott ST, Wey WC, Cervasi B, Taaffe J, Engram JC, Li B, Else JG, et al. 2010. A novel CCR5 mutation common in sooty mangabeys reveals SIVsmm infection of CCR5-null natural hosts and efficient alternative coreceptor use in vivo. *PLoS Pathog* **6**: e1001064. doi: 10.1371/journal.ppat.1001064.

Ringrose JH, Jeeninga RE, Berkhout B, Speijer D. 2008. Proteomic studies reveal coordinated changes in T-cell expression patterns upon infection with human immunodeficiency virus type 1. *J Virol* **82**: 4320–4330.

Rolland M, Tovanabutra S, deCamp AC, Frahm N, Gilbert PB, Sanders-Buell E, Heath L, Magaret CA, Bose M, Bradfield A, et al. 2011. Genetic impact of vaccination on breakthrough HIV-1 sequences from the STEP trial. *Nat Med* **17**: 366–371.

Rosner C, Kruse PH, Hermes M, Otto N, Walter L. 2011. Rhesus macaque inhibitory and activating KIR3D interact with Mamu-A-encoded ligands. *J Immunol* **186**: 2156–2163.

Rotger M, Dang KK, Fellay J, Heinzen EL, Feng S, Descombes P, Shianna KV, Ge D, Gunthard HF, Goldstein DB, et al. 2010. Genome-wide mRNA expression correlates of viral control in CD4+ T-cells from HIV-1-

infected individuals. *PLoS Pathog* **6:** e1000781. doi: 10.1371/journal.ppat.1000781.

Rotger M, Dalmau J, Rauch A, McLaren P, Bosinger SE, Marttinez R, Sandler NG, Roque A, Liebner J, Battegay M, et al. 2011. Comparative transcriptome analysis of extreme phenotypes of human HIV-1 infection and sooty mangabey and rhesus macaque models of SIV infection. *J Clin Invest* **121:** 2391–2400.

Sayah DM, Luban J. 2004. Selection for loss of Ref1 activity in human cells releases human immunodeficiency virus type 1 from cyclophilin A dependence during infection. *J Virol* **78:** 12066–12070.

Schmitz JE, Kuroda MJ, Santra S, Sasseville VG, Simon MA, Lifton MA, Racz P, Tenner-Racz K, Dalesandro M, Scallon BJ, et al. 1999. Control of viremia in simian immunodeficiency virus infection by CD8+ lymphocytes. *Science* **283:** 857–860.

Schoggins JW, Wilson SJ, Panis M, Murphy MY, Jones CT, Bieniasz P, Rice CM. 2011. A diverse range of gene products are effectors of the type I interferon antiviral response. *Nature* **472:** 481–485.

Sebat J, Lakshmi B, Troge J, Alexander J, Young J, Lundin P, Maner S, Massa H, Walker M, Chi M, et al. 2004. Large-scale copy number polymorphism in the human genome. *Science* **305:** 525–528.

Serra-Moreno R, Jia B, Breed M, Alvarez X, Evans DT. 2011. Compensatory changes in the cytoplasmic tail of gp41 confer resistance to tetherin/BST-2 in a pathogenic nef-deleted SIV. *Cell Host Microbe* **9:** 46–57.

Sharma D, Bastard K, Guethlein LA, Norman PJ, Yawata N, Yawata M, Pando M, Thananchai H, Dong T, Rowland-Jones S, et al. 2009. Dimorphic motifs in D0 and D1+D2 domains of killer cell Ig-like receptor 3DL1 combine to form receptors with high, moderate, and no avidity for the complex of a peptide derived from HIV and HLA-A*2402. *J Immunol* **183:** 4569–4582.

* Sharp PM, Hahn BH. 2011. Origins of HIV and the AIDs pandemic. *Cold Spring Harb Perspect Med* doi: 10.1101/cshperspect.a006841.

Silvestri G. 2010. Pathogenic vs. nonpathogenic retrovirus infections. Presented at the Conference on Retroviruses and Opportunistic Infections. San Francisco, February 16–19, 2010.

Snijder B, Sacher R, Ramo P, Damm EM, Liberali P, Pelkmans L. 2009. Population context determines cell-to-cell variability in endocytosis and virus infection. *Nature* **461:** 520–523.

Soll SJ, Neil SJ, Bieniasz PD. 2010. Identification of a receptor for an extinct virus. *Proc Natl Acad Sci* **107:** 19496–19501.

Speelmon EC, Livingston-Rosanoff D, Li SS, Vu Q, Bui J, Geraghty DE, Zhao LP, McElrath MJ. 2006. Genetic association of the antiviral restriction factor TRIM5α with human immunodeficiency virus type 1 infection. *J Virol* **80:** 2463–2471.

Stremlau M, Owens CM, Perron MJ, Kiessling M, Autissier P, Sodroski J. 2004. The cytoplasmic body component TRIM5α restricts HIV-1 infection in Old World monkeys. *Nature* **427:** 848–853.

Tang J, Costello C, Keet IP, Rivers C, Leblanc S, Karita E, Allen S, Kaslow RA. 1999. HLA class I homozygosity accelerates disease progression in human immunodefi-

ciency virus type 1 infection. *AIDS Res Hum Retroviruses* **15:** 317–324.

Telenti A. 2009. HIV-1 host interactions—Integration of large scale datasets. *F1000 Biol Rep* **1:** 71.

Telenti A, Goldstein DB. 2006. Genomics meets HIV. *Nat Rev Microbiol* **4:** 9–18.

Thananchai H, Gillespie G, Martin MP, Bashirova A, Yawata N, Yawata M, Easterbrook P, McVicar DW, Maenaka K, Parham P, et al. 2007. Cutting edge: Allele-specific and peptide-dependent interactions between KIR3DL1 and HLA-A and HLA-B. *J Immunol* **178:** 33–37.

Thomas R, Apps R, Qi Y, Gao X, Male V, O'hUigin C, O'Connor G, Ge D, Fellay J, Martin JN, et al. 2009. HLA-C cell surface expression and control of HIV/AIDS correlate with a variant upstream of HLA-C. *Nat Genet* **41:** 1290–1294.

Urban TJ, Weintrob AC, Fellay J, Colombo S, Shianna KV, Gumbs C, Rotger M, Pelak K, Dang KK, Detels R, et al. 2009. CCL3L1 and HIV/AIDS susceptibility. *Nat Med* **15:** 1110–1112.

Vigneault F, Woods M, Buzon MJ, Li C, Pereyra F, Crosby SD, Rychert J, Church G, Martinez-Picado J, Rosenberg ES, et al. 2011. Transcriptional profiling of CD4 T cells identifiesd subgroups of HIV-1 elite controllers. *J Virol* **85:** 3015–3019.

Watts JM, Dang KK, Gorelick RJ, Leonard CW, Bess JWJr, Swanstrom R, Burch CL, Weeks KM. 2009. Architecture and secondary structure of an entire HIV-1 RNA genome. *Nature* **460:** 711–716.

Weiler A, May GE, Qi Y, Wilson N, Watkins DI. 2006. Polymorphisms in eight host genes associated with control of HIV replication do not mediate elite control of viral replication in SIV-infected Indian rhesus macaques. *Immunogenetics* **58:** 1003–1009.

Wilson SJ, Webb BL, Maplanka C, Newman RM, Verschoor EJ, Heeney JL, Towers GJ. 2008. Rhesus macaque TRIM5 alleles have divergent antiretroviral specificities. *J Virol* **82:** 7243–7247.

Wolfe ND, Switzer WM, Carr JK, Bhullar VB, Shanmugam V, Tamoufe U, Prosser AT, Torimiro JN, Wright A, Mpoudi-Ngole E, et al. 2004. Naturally acquired simian retrovirus infections in central African hunters. *Lancet* **363:** 932–937.

Xin X, Shioda T, Kato A, Liu H, Sakai Y, Nagai Y. 1999. Enhanced anti-HIV-1 activity of CC-chemokine LD78β, a non-allelic variant of MIP-1α/LD78α. *FEBS Lett* **457:** 219–222.

Yant LJ, Friedrich TC, Johnson RC, May GE, Maness NJ, Enz AM, Lifson JD, O'Connor DH, Carrington M, Watkins DI. 2006. The high-frequency major histocompatibility complex class I allele Mamu-B*17 is associated with control of simian immunodeficiency virus SIVmac239 replication. *J Virol* **80:** 5074–5077.

Yeung ML, Houzet L, Yedavalli VS, Jeang KT. 2009. A genome-wide short hairpin RNA screening of jurkat T-cells for human proteins contributing to productive HIV-1 replication. *J Biol Chem* **284:** 19463–19473.

Yoon W, Ma BJ, Fellay J, Huang W, Xia SM, Zhang R, Shianna KV, Liao HX, Haynes BF, Goldstein DB. 2010. A polymorphism in the HCP5 gene associated with HLA-B*5701 does not restrict HIV-1 in vitro. *AIDS* **24:** 155–157.

Zappacosta F, Borrego F, Brooks AG, Parker KC, Coligan JE. 1997. Peptides isolated from HLA-Cw*0304 confer different degrees of protection from natural killer cell-mediated lysis. *Proc Natl Acad Sci* **94:** 6313–6318.

Zhang F, Wilson SJ, Landford WC, Virgen B, Gregory D, Johnson MC, Munch J, Kirchhoff F, Bieniasz PD, Hatziioannou T. 2009. Nef proteins from simian immunodeficiency viruses are tetherin antagonists. *Cell Host Microbe* **6:** 54–67.

Zhou H, Xu M, Huang Q, Gates AT, Zhang XD, Castle JC, Stec E, Ferrer M, Strulovici B, Hazuda DJ, et al. 2008. Genome-scale RNAi screen for host factors required for HIV replication. *Cell Host Microbe* **4:** 495–504.

Vaccine Design for CD8 T Lymphocyte Responses

Richard A. Koup and Daniel C. Douek

Immunology Laboratory, Vaccine Research Center, National Institute of Allergy and Infectious Diseases, National Institutes of Health, Bethesda, Maryland 20892

Correspondence: rkoup@mail.nih.gov

Vaccines are arguably the most powerful medical intervention in the fight against infectious diseases. The enormity of the global human immunodeficiency virus type 1 (HIV)/acquired immunodeficiency syndrome (AIDS) pandemic makes the development of an AIDS vaccine a scientific and humanitarian priority. Research on vaccines that induce T-cell immunity has dominated much of the recent development effort, mostly because of disappointing efforts to induce neutralizing antibodies through vaccination. Whereas T cells are known to limit HIV and other virus infections after infection, their role in protection against initial infection is much less clear. In this article, we will review the rationale behind a T-cell-based vaccine approach, provide an overview of the methods and platforms that are being applied, and discuss the impact of recent vaccine trial results on the future direction of T-cell vaccine research.

Ongoing efforts to develop effective vaccines against HIV are partly based on the principle that the specific antiviral CD8 T lymphocyte (CTL) response is crucial for immune control of viral replication. This certainly applies to many chronic persistent infections with viruses such as hepatitis B virus (HBV), hepatitis C virus (HCV), cytomegalovirus (CMV), and Epstein-Barr virus (EBV). The same appears to be the case for HIV infection, with a substantial body of evidence suggesting that HIV-specific CD8 T-cell responses suppress HIV replication in vivo. Aside from the temporal association of an increase in CD8 T-cell responses with a decrease in viral load in acute infection (Borrow et al. 1994; Koup et al. 1994), the targeting of particular epitopes restricted by certain human leukocyte antigen (HLA) alleles, such as HLA-B*5701, is consistently associated with low levels of virus load (Goulder and Watkins 2008; Hunt and Carrington 2008). In addition, CD8 T-cell depletion in simian immunodeficiency virus (SIV)-infected macaques is associated with an increase in viral load that is likely because of loss of SIV-specific T-cell responses (Jin et al. 1999; Schmitz et al. 1999). However, whereas the majority of T-cell-based vaccines tested in the macaque model have resulted in variably reduced viral load after SIV challenge (Shiver et al. 2002; Liu et al. 2009), the SIV-specific T-cell responses they elicit are insufficient in terms of frequency alone to define outcome (Casimiro et al. 2005; Moniuszko et al. 2005). Furthermore, it is not apparent what distinguishes the immunity afforded by a macaque CMV-based vaccine that profoundly controls SIV replication from those that merely blunt viral load (Hansen et al. 2009; Hansen et al. 2011). What is clear is that simple quantitative correlates of virus control have proved elusive (Ogg et al. 1998; Betts et al. 2001; Edwards et al. 2002; Addo et al. 2003), whereas qualitative aspects of the HIV-specific CD8 T-cell response seem to play a critical role in the efficacy of antiviral control (Betts et al. 2006).

T-CELL CHARACTERISTICS ASSOCIATED WITH VIRUS CONTROL

Qualitative aspects of immune control have generally been gleaned from observational studies in long-term nonprogressors, elite controllers, and HIV-2-infected nonprogressors, and have revealed a multitude of characteristics, which all appear to contribute to virus control (Fig. 1). First, it is likely that CD4 T cells will need to play an important role as effector cells per se, or in giving help to CD8 T cells (Rosenberg et al. 1997). Parenthetically, one should bear in mind that CD4 T-cell help is likely critical to the development of Env-specific high-affinity neutralizing antibodies. The

phenotypes of CD8$^+$ T cells that correlate with lower viral loads in chronic HIV infection are either central memory cells (Burgers et al. 2009) or effector memory cells (Hess et al. 2004; Addo et al. 2007) that do not express exhaustion markers such as PD-1 (Day et al. 2006; Petrovas et al. 2006; Trautmann et al. 2006). In terms of functional capacity, virus control has been associated with so-called polyfunctional CD8 T cells that secrete multiple cytokines (a property that is related to the sensitivity of antigen recognition) (Betts et al. 2006) as well as proliferative capacity (Day et al. 2007) and the ability to kill HIV-infected target cells (Yang et al. 1996; Migueles et al. 2008; Hersperger et al. 2010) or suppress HIV replication in vitro (Blackbourn et al. 1996; Yang et al. 1997; Spentzou et al. 2010). However, the qualitative properties of CD8$^+$ T-cell populations are also clearly impacted on by viral replication itself, thus rendering it difficult to disentangle cause from effect when interpreting associations between low viral load and particular phenotypic or functional profiles. Nevertheless, the rationale for what a T-cell-based vaccine should look like has been largely driven by data from individuals chronically infected with HIV.

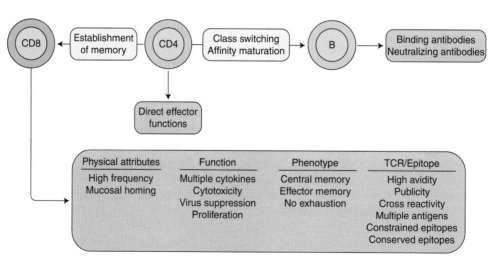

Figure 1. Attributes of CD8 T cells associated with virus control in infected individuals. These characteristics are thought important to emulate in the response elicited by a vaccine. The role of CD8 T cells should be viewed in the light of the roles of CD4 T cells and B cells.

Cite this article as *Cold Spring Harb Perspect Med* doi: 10.1101/cshperspect.a007252

It should be noted that different vaccine modalities are available that can induce different patterns of CD4, CD8, and antibody responses (Table 1). Despite the availability of such an armamentarium with which to tailor the character of the vaccine-induced immune response, a pragmatic approach has dominated the T-cell vaccine field, which has concentrated on vaccines to stimulate CD8 T cells. The origin of this bias is partially historical, and partially based on the general knowledge that CD8 T cells are efficient mediators of viral clearance, and would therefore be an appropriate component of a T-cell-based vaccine designed to lower viral load, if not clear HIV infection after challenge. It is for this reason that much of the subsequent discussion of actual vaccine approaches in this article will focus on methods that have been used to stimulate CD8, rather than CD4, T-cell responses through vaccination.

DNA VACCINES

Induction of major histocompatibility complex (MHC)-I-restricted CD8 T-cell responses is best accomplished when the vaccine antigen is produced endogenously in which presentation of peptides to major histocompatability class one molecules is more efficient than occurs via cross-presentation of exogenous proteins. In vivo injection of plasmid DNA is one of the most direct, although arguably not the most efficient, methods for accomplishing

endogenous expression of foreign proteins. Whereas early studies in mice demonstrated the potential of unmodified plasmid DNA as a vaccine modality (Fynan et al. 1993; Ulmer et al. 1993), early clinical trials in humans showed a general lack of potency of this approach, especially in stimulating class-I-restricted responses. This probably relates to the fact that after intramuscular injection, most DNA was being taken up and expressed by muscle cells, and that directing the uptake to professional antigen-presenting cells was required to improve immunogenicity (Wolff et al. 1990; Dupuis et al. 2000). Despite these early disappointing results, further refinements in the composition (promoter and codon use), manufacture, and purification (maintaining supercoiled structures) of plasmid DNA led to improvements in overall immunogenicity (Gao et al. 2003; Barouch et al. 2005; Pillai et al. 2008; Cai et al. 2009). Despite these improvements even the best preparations required several immunizations to achieve reasonable levels of immunity.

Multiple efforts have been made to further improve the immunogenicity of plasmid DNA. Among these have been attempts to improve delivery by use of needle-free delivery systems or incorporation of the DNA into various particle, metal, or lipid formulations in hopes of improving uptake into antigen-presenting cells (Klavinskis et al. 1997; Catanzaro et al. 2007; Helson et al. 2008). Although these approaches have been somewhat successful, another approach that has gained popularity lately is the application of pulsed electrical currents to the region of immunization, termed in vivo electroporation (Luxembourg et al. 2007). A number of devices have been developed and tested that differ in their degrees of complexity, ease of use, and comfort to the vaccinee. Generally the devices have been shown to improve the immunogenicity of plasmid DNA, and could help to make plasmid DNA a viable platform for routine vaccination, either by decreasing the number of inoculations or the dose of a vaccine that is required to generate protection. Because neither DNA nor these devices have yet been licensed, their development would

Table 1. Vaccine modalities and the immunity that they elicit

Modality	CD4 T cells	CD8 T cells	Antibodies
Whole killed HIV	+++	−	+++
Whole attenuated HIV	++	++	++
Live vector viruses	+++	+++	+++
Live bacterial vectors	+++	+	+++
Pseudovirions	++	−	+++
Replicons	++	++	++
DNA plasmids	++	+	+
Viral proteins	++	−	+++
HIV peptides	++	++	−

likely require a combination license application to the Food and Drug Administration.

Another approach that has been tried to improve the immunogenicity of DNA vaccines is to include cytokines or chemokines (often termed molecular adjuvants), either in *trans* during or soon after the vaccination, or in *cis* by encoding for them within the plasmid DNA. The list of molecular adjuvants that have been evaluated include those designed to increase inflammation, skew the response toward Th1 or Th2, lead to proliferation of responding cells, attract appropriate T cells and anaphase-promoting complexes to the site of vaccination, or improve the induction of long-term memory (Abdulhaqq and Weiner 2008). Although some molecular adjuvants have shown a modest increase in the frequency of vaccine-induced T cells, others have shown a profound effect (notably IL12 and IL15) especially when combined with electroporation (Chong et al. 2007; Hirao et al. 2008). In certain studies the responses induced in conjunction with molecular adjuvants were shown to improve aspects of protection against retroviral challenge; in some these were associated with improvements in T-cell responses whereas in others the impact appears to have been on aspects of the antibody response (Lai et al. 2007). A few of these approaches have been or are planned to be advanced into phase II trials.

Where DNA vaccines probably show the greatest promise is in prime-boost combination with other platforms. In the long history of DNA vaccine testing, it is generally accepted that DNA acts better as a prime than as a boost when combined with other modalities. This may be related to the observation that DNA immunization tends to stimulate a CD4-biased response, thereby providing the necessary T-cell help for an antibody response when proteins are used as the boost, or CD8 T-cell responses when viral vectors serve as the boost (Tritel et al. 2003). The only HIV vaccine currently in phase IIb testing is a DNA prime, recombinant adenovirus type 5 boost combination, and several other earlier phase trials use DNA as a prime for various different boosts. Even in prime-boost combination, multiple immunizations with DNA are often required to achieve a maximum boost. In comparison, vector–vector or vector–protein combinations often require fewer immunizations to achieve adequate priming. One must therefore balance the impact of preexisting immunity and induction of vector-specific immune responses that are inherent to the use of a vector prime, to the possible increased number of vaccinations that may be required with DNA priming, when coming up with the optimum regimen for a given vaccine target. The lack of vector-specific immunity to DNA remains a compelling consideration when developing vaccines for worldwide distribution.

VIRAL VECTORS

Mass and targeted vaccination with vaccinia virus is responsible for the eradication of smallpox, making it arguably the most successful vaccine ever. Because of the extensive clinical experience with vaccinia, it is not surprising that this was one of the first viruses to be used in the development of recombinant vectors expressing foreign viral antigens as potential vaccine platforms. Indeed, the first documented HIV vaccine trial used a recombinant vaccinia virus expressing HIV genes (Zagury et al. 1988). In the subsequent 25 years, innumerable pox-based vectors expressing HIV antigens have been made and tested in human trials with mixed results (Pantaleo et al. 2010). Initial attempts used vaccinia, which is a live virus vaccine. This led to episodes of local and disseminated vaccinia infections and deaths when the vaccine was given (even in a presumably inactivated form) as immunotherapy to individuals already infected with HIV (Picard et al. 1991; Zagury 1991). The field then began to search for more attenuated pox viruses to use as vectors, and a number of alternatives were developed, including modified vaccinia ankara (MVA), fowlpox, ALVAC, and NyVac, most of which have been tested for efficacy in nonhuman primates, and for safety and immunogenicity in humans (Pantaleo et al. 2010). Not surprisingly, there is a general rule that with increasing attenuation comes decreasing immunogenicity, so by a standard measure of immunogenicity, these attenuated vectors tended to perform worse than parental vaccinia vectors, although with a possible increased margin of safety. Another characteristic of pox vectors is that unlike some viral vectors that express only the insert of choice, pox vectors have a large and complex genome and express many proteins in addition to the vaccine insert. As a result, the immunogenicity of the insert is often diluted by potentially more immunodominant responses directed to vector-specific antigens (Smith et al. 2005).

Despite these many potential pitfalls, a recent trial of an ALVAC-based vaccine in combination with protein boost showed moderate (31%) efficacy against acquisition of HIV infection in a low-risk heterosexual cohort in Thailand (the RV144 trial, Table 2) (Rerks-Ngarm et al. 2009). That efficacy may have been higher in the first 6 months after full vaccination, but then waned over time. Of interest, virtually no CD8 T-cell responses were induced by this viral vector-based vaccine, and only CD4 T-cell and antibody responses appear to have been generated. Therefore, despite the use of a viral vector, the vaccine performed as would be expected of a protein-based vaccine. Intensive studies are ongoing to dissect any potential immune correlate of protection, but early indicators suggest it will not be a CD8 response.

Adenoviruses make up the majority of the other widely tested viral vectors (Lasaro and Ertl 2009). Adenovirus-based vaccines are easy to engineer, manufacture, and test. Adenovirus vectors can be made replication incompetent while expressing only the vaccine insert of interest, thereby alleviating the problems of safety and epitope competition that hamper

pox-vector-based platforms. The fact that only the insert gene is expressed may be one factor responsible for the general ability of recombinant Ad vectors to stimulate high-frequency CD4 and CD8 T-cell responses. Levels of preexisting immunity vary among the five major serogroups of human adenoviruses, and can greatly influence the immunogenicity of the vector. The preexisting immunity that appears to be most important is the level of neutralizing antibody to the vector, as there is broad cross-reactivity among T-cell responses to the different serogroups. The impact of preexisting immunity can be at least partially alleviated by priming with either DNA or a low seroprevalent adenovirus.

Clinical trials of HIV vaccines based on adenovirus type 5 (rAd5) alone or in combination with DNA priming are ongoing (Table 2; Catanzaro et al. 2006; Koup et al. 2010). The rAd5 vaccine tested by Merck in the Step and Phambili trials contained HIV Gag, Pol, and Nef, but no Env, and therefore represents a true T-cell vaccine in that it stimulated neither virion binding nor neutralizing antibodies. Protection was intended to be afforded by stimulation of CD8 T-cell responses that would contain viral replication on infection, a contention supported by preclinical nonhuman primate testing (Shiver et al. 2002). Whereas the vaccine was shown to stimulate strong T-cell responses, the vaccine failed either to protect volunteers from acquisition of infection or to reduce viral loads after infection. In fact, there was an apparent increased risk of HIV infection in individuals with preexisting immunity to Ad5, raising the specter that Ad-specific CD4 T cells were being induced at the mucosa, thereby leading to enhanced risk of infection (Buchbinder

Table 2. Vaccines tested in human efficacy trials

Product	Trial	Antigens	Immunity	Population
VaxGen rgp120	Vax003/Vax004	Env	Ab, CD4	MSM and IVDU
Merck rAd5	Step/Phambili	Gag/Pol/Nef	CD8, CD4	MSM
Sanofi Alvac/rgp120	RV144	Env + Gag	Ab, CD4	General
VRC DNA/rAd5	HVTN505	Env/Gag/Pol/Nef	Ab, CD8, CD4	MSM

MSM, men who have sex with men; IVDU, intravenous drug user.

et al. 2008; McElrath et al. 2008). Subsequent analyses have failed to bear out this hypothesis. Efficacy testing of a vaccine developed at the Vaccine Research Center based on a DNA prime, rAd5 boost expressing HIV Gag, Pol, Nef, and Env is in progress (Koup et al. 2010). In addition, lower seroprevalent Ad-based vaccines are in preclinical testing in nonhuman primates and phase I testing in humans (Liu et al. 2009; Barouch 2010).

In addition to pox and Ad-based vectors, there is a long list of other viruses that have been tried as vaccine vectors with varying levels of success (Robert-Guroff 2007). These include adeno-associated virus (AAV), vesticular stomatitis virus (VSV), lymphocytic choriomeningitis virus (LCMV), herpes simplex virus (HSV), semliki forest virus (SFV), Venezuelan equine encephalitis virus (VEE), and others. Some of these have been tested for immunogenicity in humans, and have mostly proven to be less immunogenic than Ad-based platforms. Some are being reengineered to improve immunogenicity whereas others have been dropped from development. At this time it is difficult to predict which, if any, of these alternative vector designs may ultimately prove highly immunogenic and move forward in clinical development.

All of the viral vectors described so far are either replication incompetent or highly attenuated. However, some vaccine efforts are relying on the development of fully replication-competent viral vectors. The two most prominent in this class are replication-competent adenovirus and CMV. The goal with these approaches is to induce sustained effector memory T-cell responses that may prove more effective at blocking or controlling HIV infection than the central memory responses that are induced by most replication-defective vectors (Peng et al. 2005). Of note is the fact that preclinical testing of a rhesus-based CMV vector in nonhuman primates has shown about 30%–50% efficacy in rapid and profound control of SIV infection in monkeys (Hansen et al. 2009, 2011). This efficacy is correlated with effector memory CD8 T cells and not with antibody responses. Although the exact mechanism of protection is far from defined in these

studies, and a variety of safety issues still need to be addressed, the ability to achieve profound control of SIV infection in the absence of antibodies warrants further investigation and development of this platform.

OTHER APPROACHES

DNA and viral vectors offer the advantage of having the vaccine antigen expressed within the host cell, essentially assuring some antigen processing and MHC class I expression leading to the induction of CD8 T-cell responses. However, other vaccine strategies have been used to specifically target HIV antigens to MHC class I molecules. One way has been to combine soluble antigens with adjuvants that directly target and/or activate dendritic cells (DCs). There are many subsets of DCs, each with their own unique features and locations, but all of which are intimately involved in the stimulation of adaptive immune responses, and many of which have the unique capacity for efficient cross-presentation of soluble antigens to the class I pathway (Steinman 2008; Ueno et al. 2011). By understanding the location and expression of surface antigens and toll-like receptors (TLRs) on each subset, vaccine strategies can be tailored to stimulate the response of choice. Of importance in these strategies is an understanding of the movement of DCs after targeting by the vaccine antigen. Linkage of the antigen to the TLR ligand may be necessary because, once stimulated, DCs will migrate away from the depot of antigen to regional draining lymph nodes where T-cell responses will be induced.

Many combinations of protein and TLR ligand have undergone preclinical testing with varying effects on CD4 and CD8 T-cell responses, and one DC targeting approach has advanced to a phase I clinical trial. This involves linking HIV Gag antigen to an antibody to DEC205 (a surface marker on DCs) combined with TLR stimulation of DCs using synthetic double-stranded RNA poly:IC (Cheong et al. 2010; Nchinda et al. 2010; Tewari et al. 2010). Even with the combination of DEC205 targeting and TLR stimulation of DCs, the majority of the T-cell response stimulated in response to the Gag antigen is mediated by CD4 rather than CD8 T cells. A variety of other DC targeting strategies are undergoing preclinical evaluation.

Bacteria have also been proposed as vectors capable of stimulating CD8 T-cell responses. Among the candidates are bacteria that replicate inside monocytes or macrophages (listeria moncytogenes, Bacillus Calmette-Guerin [BCG]) and those that are easily engineered to secrete proteins via the type III pathway (Salmonella, Shigella) (Garmory et al. 2003). Among these, BCG has the obvious advantage of a long history of use as a clinical vaccine in humans, and is certainly the furthest along in clinical development. Still, recombinant BCG and other mycobacterial vectors are probably still years away from clinical testing in humans.

Probably the simplest approach to inducing a class-I-restricted response is to vaccinate with peptides that can bind directly to class I molecules, thereby bypassing the need for intracellular processing and presentation. Different peptide preparations have been tested extensively, and various attempts have been made to improve the in vivo immunogenicity of this approach. Despite these efforts, standard vaccine delivery of peptides appears to have limited utility in stimulating CD8 T-cell responses, with the exception of one approach. Intravenous infusion of peripheral blood mononuclear cells coated ex vivo with peptides has shown great potency in preclinical animal studies (De Rose et al. 2008). Whether this approach will prove useful in human testing, especially considering the impracticality of such an approach for mass vaccination efforts, remains to be seen.

ANTIGENS

HIV expresses nine structural (Gag, Env) and nonstructural (Pol, Nef, Tat, Rev, Vpr, Vpu, Vif) proteins, some of which (Gag, Pol, Env) are further processed by viral or host proteases into more proteins. Several different criteria can be applied to decide which of these proteins to include in a T-cell-based vaccine: Sequence conservation, level of protein expression, and timing of protein expression are three obvious ones. The polymerase polyprotein (inclusive of

protease, reverse transcriptase, and integrase) is certainly the most conserved across HIV strains and would therefore seem like a natural choice (Korber et al. 2007). The problem lies in the fact that these proteins are transcribed from a long Gag/Pol RNA through an inefficient frame shift mechanism, which leads to very low protein expression in infected cells. The efficiency with which CTL targeting these antigens can recognize and kill HIV-infected cells has therefore been called into question (Chen et al. 2009).

From the standpoint of the level of immune recognition, Gag and Env are the clear winners. Gag is reasonably well conserved across HIV strains, whereas Env has a mixture of highly conserved and highly variable regions, the latter being mostly the result of immune pressure from antibodies. In addition, both are structural proteins so they are expressed to high levels at the same time during the viral life cycle. The rest of the HIV proteins are produced from multiply spliced RNA species (Nef, Tat, Rev, Vpr, Vpu, Vif), are therefore expressed at lower levels than the structural proteins. Although these would therefore appear to be poor vaccine targets, the potential for low-level constitutive expression, or very early expression during the viral life cycle, has led to the inclusion of some of them within various T-cell vaccines.

Some have evaluated T-cell responses during chronic infection as an indicator of which antigens to include in a vaccine. From a pure frequency standpoint, Gag, Pol, and Env are among the dominant responses, followed by Nef and then the other nonstructural proteins (Betts et al. 2001). This hierarchy may be somewhat different during acute infection (Goulder et al. 2001; Lichterfeld et al. 2004). As another strategy, some have assessed associations between CTL responses and HIV plasma RNA during chronic infection as an indicator of what antigens to include in a vaccine. When these types of evaluations are performed, it is clear that CTL responses to Gag are most strongly associated with virus control during chronic infection (Zuniga et al. 2006; Rolland et al. 2008). These types of analyses are probably measuring a complex combination of factors encompassing the efficiency of antiviral activity to an antigen and ease with which it can escape from that immune response (among other things). How this might predict protection in a prophylactic vaccine setting, in which the response would be present before virus infection, is not clear. In fact, at least one HIV antigen (Env) that ranks very poorly in this type of assessment was clearly the antigen that was responsible for protection in the only human trial of an HIV vaccine to show efficacy (Rerks-Ngarm et al. 2009). One must therefore be careful when using data from chronic infections in deciding what antigens to include in a prophylactic vaccine.

Rather than relying on antigens known to be encoded by HIV, some are investigating unconventional antigens. It is known that there are cryptic start sites in the HIV genome that could lead to protein production from alternative reading frames. Recent evidence suggests that HIV-infected individuals and SIV-infected monkeys often make T-cell responses to some of these cryptic epitopes, raising the question of whether they should or could be included in a vaccine (Bansal et al. 2010; Maness et al. 2010a,b). Even less conventional antigens have also been considered. Evidence suggests that transcriptional control of human endogenous retroviruses (HERV) is compromised in HIV infection, leading to protein production and the induction of T-cell responses to endogenous retroviral antigens (Garrison et al. 2007). Because these HERV proteins are very conserved, a vaccine approach based on expression of HERV antigens has been proposed.

LOCATION

The mucosal surfaces are not only the major route of transmission of HIV infection in both men and women but also the anatomical site of the greatest depletion of CD4 T cells during the course of the disease owing to the high availability of $CCR5^+$ $CD4^+$ T cells for viral replication (Kotler et al. 1984; Guadalupe et al. 2003; Brenchley et al. 2004; Mehandru et al. 2004; Li et al. 2005; Mattapallil et al. 2005). The depletion of gastrointestinal CD4 T cells and the structural disruption of these mucosal surfaces

are both massive and rapid, and therefore an effective HIV vaccine must interfere with virus replication during the narrow temporal window between the moment of mucosal transmission and the establishment of disseminated infection (Douek et al. 2006; Haase 2011). Recent data suggest that the majority of HIV infections among men who have sex with men and heterosexual HIV infections in women occur through a transmission event in which a single viral variant is responsible for establishing the initial disseminated infection (Keele et al. 2008; Abrahams et al. 2009; Haaland et al. 2009; Li et al. 2010). Thus, suppression of the initial transmission event is theoretically simple but the targeting of that suppression event presents a considerable problem of localization. This may require a high frequency of HIV-specific immunity in a very localized region, at the site of exposure, rather than systemic immunity. Although systemic immunization may elicit antigen-specific responses at mucosal surfaces (Belyakov et al. 1998; Baig et al. 2002; Pal et al. 2006) such approaches have not been overwhelmingly successful in significantly blocking or attenuating infection after mucosal challenge. However, the targeting of the mucosal surface itself with local immunization may elicit high-frequency local responses that can confer protection against mucosal virus challenge (Belyakov et al. 1998; Barnett et al. 2008). In this light, it is important to note that recent studies suggest that the ratio of SIV-specific CD8 T cells to SIV-infected CD4 T cells at the site of primary infection is critical in determining the degree of control of viral replication (Li et al. 2009). Thus, these data highlight a narrow window of opportunity when preexisting virus-specific memory T-cell populations may endow the vaccinated host with a heightened ability to attenuate local virus replication at the mucosal surfaces.

BREADTH, EPITOPE ESCAPE, AND THE IMPACT OF THE HOST RESPONSE

In terms of the number of epitopes targeted, there is some debate over the optimal breadth of recognition that is required for virus control in vaccinated individuals. Studies from HIV-infected people in Africa and from vaccinated rhesus macaques suggest that the more epitopes targeted, in Gag rather than Env, the better (Zuniga et al. 2006; Kiepiela et al. 2007; Rolland et al. 2008). However, it would be clearly advantageous to target epitopes that are both conserved across different HIV subtypes (Goulder et al. 1996; Turnbull et al. 2006) and also sequence limited by viral fitness constraints (Martinez-Picado et al. 2006; Schneidewind et al. 2007). As mentioned above, the striking association between the expression of particular MHC alleles and virologic outcome speaks to a role for CD8 T cells in the control of viral replication (Kaslow et al. 1996; Hendel et al. 1999; Migueles et al. 2000; Gao et al. 2001), but the mechanisms underlying such protective effects remain poorly understood save for the presentation by particular MHC alleles of particular epitopes that are conserved owing to fitness constraints (Wang et al. 2009) and diversity in the T-cell receptor repertoire that targets such epitopes (Simons et al. 2008; Geldmacher et al. 2009). Although TCR diversity may seem an obvious advantage when targeting a virus that undergoes mutational epitope escape, studies in SIV-infected rhesus macaques have shown that the beneficial virologic outcome conferred by expression of the MHC allele Mamu A*01 may be mediated by the use of particular TCRs that target an immunodominant epitope (Price et al. 2009). Such TCRs are termed public because they are common to the epitope-specific response of more than one infected animal and as such represent an example of extreme bias in TCR usage.

However, the associations described above are far from clear-cut and still raise the question of whether responses of considerably greater magnitude than those elicited by current vaccine modalities would beneficially affect outcome. Furthermore, viral escape from the immune response by mutation presents a significant problem for any vaccine strategy. Indeed, studies in acute HIV infection suggest that CD8 T-cell responses to immunodominant epitopes are associated with control of viral replication but only during the acute phase. As the infected host enters the chronic phase, these epitopes are found to have escaped and the emerging HIV-specific CD8 T cells, which target a new set of epitopes, may not be as effective in the control of virus (Goulder et al. 2001; Leslie et al. 2004; Goonetilleke et al. 2009).

How does one increase the breadth of a vaccine to cover strain diversity? Historically this has been accomplished by including multiple strains of a virus in a vaccine. The seasonal influenza vaccine and polio vaccine are two classic examples. This approach is also being applied to HIV in which vaccines based on three or more strains of HIV are being tested. This approach clearly increases the likelihood that the T-cell response will recognize more than one strain of HIV (Seaman et al. 2005), however, the cost and complexity of a vaccine increases significantly with each new strain that is added, and it is difficult to determine the impact of vaccine valency in actual protection in human clinical trials. HIV sequence alignments and knowledge of where T-cell epitopes reside within those alignments have been used to generate HIV antigens that encompass the most prevalent HIV strain sequences within just one or a few constructs. These approaches are based on linking sequences back to their common ancestor (center of tree approach), using consensus sequences, or creating mosaics of multiple strains in a single reading frame (Nickle et al. 2003; Mullins et al. 2004; Fischer et al. 2007). The latter approach (mosaic inserts) has progressed the furthest, both in terms of theoretical coverage (Fischer et al. 2007) and breadth of the T-cell response generated in mouse and nonhuman primate testing (Kong et al. 2009; Barouch et al. 2010; Santra et al. 2010). Although these approaches increase the breadth of the T-cell response to a vaccine, their impact on breadth of protection across multiple strains remains to be determined.

CONCLUDING REMARKS

T-cell vaccine approaches to HIV have dominated much of the vaccine research agenda over the last decade. The enthusiasm for a T-cell approach was driven by the inability to stimulate broad neutralizing antibodies through vaccination combined with a plethora of data from HIV-infected individuals indicating that CD8 T cells are instrumental in viral control. A vaccine representing a pure T-cell approach (that is, one which contained no envelope antigen) underwent efficacy testing in humans and failed to protect from acquisition of infection (STEP Trial) (Buchbinder et al. 2008). More recently, the RV144 trial in Thailand showed a moderate efficacy of 31% protection from acquisition (P = 0.04) with a vaccine that elicited envelope-specific antibodies and CD4 T cells but no CD8 T-cell responses (Rerks-Ngarm et al. 2009). Although these results may at first seem somewhat damning with respect to CD8 T-cell-based vaccines, many approaches remain to be tested. These newer modalities are designed to stimulate CD8 T cells, which differ in quality, quantity, phenotype, breadth, and location from the vaccines tested previously. Whether any of these approaches will prove beneficial remains to be seen.

It is important to look critically at the concept of a pure T-cell vaccine in the context of virus-specific adaptive immunity as a whole. The result of the RV144 trial suggests that the single-minded pursuit of solely eliciting CD8 T cells as the antiviral effectors may be of scientific interest, but may not be the best approach in practice. The combination of modalities that stimulate CD4 and CD8 T cells as well as antibodies, for which multiple approaches are available, appears to be the logical direction to follow. Indeed, it appears that the field is taking precisely this direction.

ACKNOWLEDGMENTS

The authors thank Dr. Barney Graham for his suggestions and helpful input.

REFERENCES

Abdulhaqq SA, Weiner DB. 2008. DNA vaccines: Developing new strategies to enhance immune responses. *Immunol Res* **42**: 219–232.

Abrahams MR, Anderson JA, Giorgi EE, Seoighe C, Mlisana K, Ping LH, Athreya GS, Treurnicht FK, Keele BF, Wood N, et al. 2009. Quantitating the multiplicity of infection with human immunodeficiency virus type 1 subtype C reveals a non-poisson distribution of transmitted variants. *J Virol* **83**: 3556–3567.

Addo MM, Yu XG, Rathod A, Cohen D, Eldridge RL, Strick D, Johnston MN, Corcoran C, Wurcel AG, Fitzpatrick CA, et al. 2003. Comprehensive epitope analysis of human immunodeficiency virus type 1 (HIV-1)-specific T-cell responses directed against the entire expressed HIV-1 genome demonstrate broadly directed responses, but no correlation to viral load. *J Virol* **77**: 2081–2092.

Addo MM, Draenert R, Rathod A, Verrill CL, Davis BT, Gandhi RT, Robbins GK, Basgoz NO, Stone DR, Cohen DE, et al. 2007. Fully differentiated HIV-1 specific CD8$^+$ T effector cells are more frequently detectable in controlled than in progressive HIV-1 infection. *PLoS One* **2**: e321. doi: 10.1371/journal.pone.0000321.

Baig J, Levy DB, McKay PF, Schmitz JE, Santra S, Subbramanian RA, Kuroda MJ, Lifton MA, Gorgone DA, Wyatt LS, et al. 2002. Elicitation of simian immunodeficiency virus-specific cytotoxic T lymphocytes in mucosal compartments of rhesus monkeys by systemic vaccination. *J Virol* **76**: 11484–11490.

Bansal A, Carlson J, Yan J, Akinsiku OT, Schaefer M, Sabbaj S, Bet A, Levy DN, Heath S, Tang J, et al. 2010. CD8 T cell response and evolutionary pressure to HIV-1 cryptic epitopes derived from antisense transcription. *J Exp Med* **207**: 51–59.

Barnett SW, Srivastava IK, Kan E, Zhou F, Goodsell A, Cristillo AD, Ferrai MG, Weiss DE, Letvin NL, Montefiori D, et al. 2008. Protection of macaques against vaginal SHIV challenge by systemic or mucosal and systemic vaccinations with HIV-envelope. *AIDS* **22**: 339–348.

Barouch DH. 2010. Novel adenovirus vector-based vaccines for HIV-1. *Curr Opin HIV AIDS* **5**: 386–390.

Barouch DH, Yang ZY, Kong WP, Korioth-Schmitz B, Sumida SM, Truitt DM, Kishko MG, Arthur JC, Miura A, Mascola JR, et al. 2005. A human T-cell leukemia virus type 1 regulatory element enhances the immunogenicity of human immunodeficiency virus type 1 DNA vaccines in mice and nonhuman primates. *J Virol* **79**: 8828–8834.

Barouch DH, O'Brien KL, Simmons NL, King SL, Abbink P, Maxfield LF, Sun YH, La Porte A, Riggs AM, Lynch DM, et al. 2010. Mosaic HIV-1 vaccines expand the breadth and depth of cellular immune responses in rhesus monkeys. *Nat Med* **16**: 319–323.

Belyakov IM, Derby MA, Ahlers JD, Kelsall BL, Earl P, Moss B, Strober W, Berzofsky JA. 1998. Mucosal immunization with HIV-1 peptide vaccine induces mucosal and systemic cytotoxic T lymphocytes and protective immunity in mice against intrarectal recombinant HIV-vaccinia challenge. *Proc Natl Acad Sci* **95**: 1709–1714.

Betts MR, Ambrozak DR, Douek DC, Bonhoeffer S, Brenchley JM, Casazza JP, Koup RA, Picker LJ. 2001. Analysis of total human immunodeficiency virus (HIV)-specific CD4$^+$ and CD8$^+$ T-cell responses: Relationship to viral load in untreated HIV infection. *J Virol* **75**: 11983–11991.

Betts MR, Nason MC, West SM, De Rosa SC, Migueles SA, Abraham J, Lederman MM, Benito JM, Goepfert PA, Connors M, et al. 2006. HIV nonprogressors preferentially maintain highly functional HIV-specific CD8$^+$ T cells. *Blood* **107**: 4781–4789.

Blackbourn DJ, Mackewicz CE, Barker E, Hunt TK, Herndier B, Haase AT, Levy JA. 1996. Suppression of HIV replication by lymphoid tissue CD8$^+$ cells correlates with the clinical state of HIV-infected individuals. *Proc Natl Acad Sci* **93:** 13125–13130.

Borrow P, Lewicki H, Hahn BH, Shaw GM, Oldstone MB. 1994. Virus-specific CD8$^+$ cytotoxic T-lymphocyte activity associated with control of viremia in primary human immunodeficiency virus type 1 infection. *J Virol* **68:** 6103–6110.

Brenchley JM, Schacker TW, Ruff LE, Price DA, Taylor JH, Beilman GJ, Nguyen PL, Khoruts A, Larson M, Haase AT, et al. 2004. CD4$^+$ T cell depletion during all stages of HIV disease occurs predominantly in the gastrointestinal tract. *J Exp Med* **200:** 749–759.

Buchbinder SP, Mehrotra DV, Duerr A, Fitzgerald DW, Mogg R, Li D, Gilbert PB, Lama JR, Marmor M, Del Rio C, et al. 2008. Efficacy assessment of a cell-mediated immunity HIV-1 vaccine (the Step Study): A double-blind, randomised, placebo-controlled, test-of-concept trial. *Lancet* **372:** 1881–1893.

Burgers WA, Riou C, Mlotshwa M, Maenetje P, de Assis Rosa D, Brenchley J, Mlisana K, Douek DC, Koup R, Roederer M, et al. 2009. Association of HIV-specific and total CD8$^+$ T memory phenotypes in subtype C HIV-1 infection with viral set point. *J Immunol* **182:** 4751–4761.

Cai Y, Rodriguez S, Hebel H. 2009. DNA vaccine manufacture: scale and quality. *Expert Rev Vaccines* **8:** 1277–1291.

Casimiro DR, Wang F, Schleif WA, Liang X, Zhang ZQ, Tobery TW, Davies ME, McDermott AB, O'Connor DH, Fridman A, et al. 2005. Attenuation of simian immunodeficiency virus SIVmac239 infection by prophylactic immunization with DNA and recombinant adenoviral vaccine vectors expressing Gag. *J Virol* **79:** 15547–15555.

Catanzaro AT, Koup RA, Roederer M, Bailer RT, Enama ME, Moodie Z, Gu L, Martin JE, Novik L, Chakrabarti BK, et al. 2006. Phase 1 safety and immunogenicity evaluation of a multiclade HIV-1 candidate vaccine delivered by a replication-defective recombinant adenovirus vector. *J Infect Dis* **194:** 1638–1649.

Catanzaro AT, Roederer M, Koup RA, Bailer RT, Enama ME, Nason MC, Martin JE, Rucker S, Andrews CA, Gomez PL, et al. 2007. Phase I clinical evaluation of a six-plasmid multiclade HIV-1 DNA candidate vaccine. *Vaccine* **25:** 4085–4092.

Chen H, Piechocka-Trocha A, Miura T, Brockman MA, Julg BD, Baker BM, Rothchild AC, Block BL, Schneidewind A, Koibuchi T, et al. 2009. Differential neutralization of human immunodeficiency virus (HIV) replication in autologous CD4 T cells by HIV-specific cytotoxic T lymphocytes. *J Virol* **83:** 3138–3149.

Cheong C, Choi JH, Vitale L, He LZ, Trumpfheller C, Bozzacco L, Do Y, Nchinda G, Park SH, Dandamudi DB, et al. 2010. Improved cellular and humoral immune responses in vivo following targeting of HIV Gag to dendritic cells within human anti-human DEC205 monoclonal antibody. *Blood* **116:** 3828–3838.

Chong SY, Egan MA, Kutzler MA, Megati S, Masood A, Roopchard V, Garcia-Hand D, Montefiori DC, Quiroz J, Rosati M, et al. 2007. Comparative ability of plasmid IL-12 and IL-15 to enhance cellular and humoral

immune responses elicited by a SIVgag plasmid DNA vaccine and alter disease progression following SHIV(89.6P) challenge in rhesus macaques. *Vaccine* **25:** 4967–4982.

Day CL, Kaufmann DE, Kiepiela P, Brown JA, Moodley ES, Reddy S, Mackey EW, Miller JD, Leslie AJ, DePierres C, et al. 2006. PD-1 expression on HIV-specific T cells is associated with T-cell exhaustion and disease progression. *Nature* **443:** 350–354.

Day CL, Kiepiela P, Leslie AJ, van der Stok M, Nair K, Ismail N, Honeyborne I, Crawford H, Coovadia HM, Goulder PJ, et al. 2007. Proliferative capacity of epitope-specific CD8 T-cell responses is inversely related to viral load in chronic human immunodeficiency virus type 1 infection. *J Virol* **81:** 434–438.

De Rose R, Fernandez CS, Smith MZ, Batten CJ, Alcantara S, Peut V, Rollman E, Loh L, Mason RD, Wilson K, et al. 2008. Control of viremia and prevention of AIDS following immunotherapy of SIV-infected macaques with peptide-pulsed blood. *PLoS Pathog* **4:** e1000055.

Douek DC, Kwong PD, Nabel GJ. 2006. The rational design of an AIDS vaccine. *Cell* **124:** 677–681.

Dupuis M, Denis-Mize K, Woo C, Goldbeck C, Selby MJ, Chen M, Otten GR, Ulmer JB, Donnelly JJ, Ott G, et al. 2000. Distribution of DNA vaccines determines their immunogenicity after intramuscular injection in mice. *J Immunol* **165:** 2850–2858.

Edwards BH, Bansal A, Sabbaj S, Bakari J, Mulligan MJ, Goepfert PA. 2002. Magnitude of functional CD8$^+$ T-cell responses to the gag protein of human immunodeficiency virus type 1 correlates inversely with viral load in plasma. *J Virol* **76:** 2298–2305.

Fischer W, Perkins S, Theiler J, Bhattacharya T, Yusim K, Funkhouser R, Kuiken C, Haynes B, Letvin NL, Walker BD, et al. 2007. Polyvalent vaccines for optimal coverage of potential T-cell epitopes in global HIV-1 variants. *Nat Med* **13:** 100–106.

Fynan EF, Webster RG, Fuller DH, Haynes JR, Santoro JC, Robinson HL. 1993. DNA vaccines: Protective immunizations by parenteral, mucosal, and gene-gun inoculations. *Proc Natl Acad Sci* **90:** 11478–11482.

Gao X, Nelson GW, Karacki P, Martin MP, Phair J, Kaslow R, Goedert JJ, Buchbinder S, Hoots K, Vlahov D, et al. 2001. Effect of a single amino acid change in MHC class I molecules on the rate of progression to AIDS. *N Engl J Med* **344:** 1668–1675.

Gao F, Li Y, Decker JM, Peyerl FW, Bibollet-Ruche F, Rodenburg CM, Chen Y, Shaw DR, Allen S, Musonda R, et al. 2003. Codon usage optimization of HIV type 1 subtype C gag, pol, env, and nef genes: In vitro expression and immune responses in DNA-vaccinated mice. *AIDS Res Hum Retroviruses* **19:** 817–823.

Garmory HS, Leary SE, Griffin KF, Williamson ED, Brown KA, Titball RW. 2003. The use of live attenuated bacteria as a delivery system for heterologous antigens. *J Drug Target* **11:** 471–479.

Garrison KE, Jones RB, Meiklejohn DA, Anwar N, Ndhlovu LC, Chapman JM, Erickson AL, Agrawal A, Spotts G, Hecht FM, et al. 2007. T cell responses to human endogenous retroviruses in HIV-1 infection. *PLoS Pathog* **3:** e165.

Geldmacher C, Metzler IS, Tovanabutra S, Asher TE, Gostick E, Ambrozak DR, Petrovas C, Schuetz A,

Ngwenyama N, Kijak G, et al. 2009. Minor viral and host genetic polymorphisms can dramatically impact the biologic outcome of an epitope-specific CD8 T-cell response. *Blood* **114:** 1553–1562.

Goonetilleke N, Liu MK, Salazar-Gonzalez JF, Ferrari G, Giorgi E, Ganusov VV, Keele BF, Learn GH, Turnbull EL, Salazar MG, et al. 2009. The first T cell response to transmitted/founder virus contributes to the control of acute viremia in HIV-1 infection. *J Exp Med* **206:** 1253–1272.

Goulder PJ, Watkins DI. 2008. Impact of MHC class I diversity on immune control of immunodeficiency virus replication. *Nat Rev Immunol* **8:** 619–630.

Goulder PJ, Bunce M, Krausa P, McIntyre K, Crowley S, Morgan B, Edwards A, Giangrande P, Phillips RE, McMichael AJ. 1996. Novel, cross-restricted, conserved, and immunodominant cytotoxic T lymphocyte epitopes in slow progressors in HIV type 1 infection. *AIDS Res Hum Retroviruses* **12:** 1691–1698.

Goulder PJ, Altfeld MA, Rosenberg ES, Nguyen T, Tang Y, Eldridge RL, Addo MM, He S, Mukherjee JS, Phillips MN, et al. 2001. Substantial differences in specificity of HIV-specific cytotoxic T cells in acute and chronic HIV infection. *J Exp Med* **193:** 181–194.

Guadalupe M, Reay E, Sankaran S, Prindiville T, Flamm J, McNeil A, Dandekar S. 2003. Severe CD4⁺ T-cell depletion in gut lymphoid tissue during primary human immunodeficiency virus type 1 infection and substantial delay in restoration following highly active antiretroviral therapy. *J Virol* **77:** 11708–11717.

Haaland RE, Hawkins PA, Salazar-Gonzalez J, Johnson A, Tichacek A, Karita E, Manigart O, Mulenga J, Keele BF, Shaw GM, et al. 2009. Inflammatory genital infections mitigate a severe genetic bottleneck in heterosexual transmission of subtype A and C HIV-1. *PLoS Pathog* **5:** e1000274.

Haase AT. 2011. Early events in sexual transmission of HIV and SIV and opportunities for interventions. *Annu Rev Med* **62:** 127–139.

Hansen SG, Vieville C, Whizin N, Coyne-Johnson L, Siess DC, Drummond DD, Legasse AW, Axthelm MK, Oswald K, Trubey CM, et al. 2009. Effector memory T cell responses are associated with protection of rhesus monkeys from mucosal simian immunodeficiency virus challenge. *Nat Med* **15:** 293–299.

Hansen SG, Ford JC, Lewis MS, Ventura AB, Hughes CM, Coyne-Johnson L, Whizin N, Oswald K, Shoemaker R, Swanson T, et al. 2011. Profound early control of highly pathogenic SIV by an effector memory T-cell vaccine. *Nature* **473:** 523–527.

Helson R, Olszewska W, Singh M, Megede JZ, Melero JA, O'Hagan D, Openshaw PJ. 2008. Polylactide-co-glycolide (PLG) microparticles modify the immune response to DNA vaccination. *Vaccine* **26:** 753–761.

Hendel H, Caillat-Zucman S, Lebuanec H, Carrington M, O'Brien S, Andrieu JM, Schachter F, Zagury D, Rappaport J, Winkler C, et al. 1999. New class I and II HLA alleles strongly associated with opposite patterns of progression to AIDS. *J Immunol* **162:** 6942–6946.

Hersperger AR, Pereyra F, Nason M, Demers K, Sheth P, Shin LY, Kovacs CM, Rodriguez B, Sieg SF, Teixeira-Johnson L, et al. 2010. Perforin expression directly ex vivo by HIV-specific CD8 T-cells is a correlate of HIV elite control. *PLoS Pathog* **6:** e1000917.

Hess C, Altfeld M, Thomas SY, Addo MM, Rosenberg ES, Allen TM, Draenert R, Eldrige RL, van Lunzen J, Stellbrink HJ, et al. 2004. HIV-1 specific CD8⁺ T cells with an effector phenotype and control of viral replication. *Lancet* **363:** 863–866.

Hirao LA, Wu L, Khan AS, Hokey DA, Yan J, Dai A, Betts MR, Draghia-Akli R, Weiner DB. 2008. Combined effects of IL-12 and electroporation enhances the potency of DNA vaccination in macaques. *Vaccine* **26:** 3112–3120.

Hunt PW, Carrington M. 2008. Host genetic determinants of HIV pathogenesis: An immunologic perspective. *Curr Opin HIV AIDS* **3:** 342–348.

Jin X, Bauer DE, Tuttleton SE, Lewin S, Gettie A, Blanchard J, Irwin CE, Safrit JT, Mittler J, Weinberger L, et al. 1999. Dramatic rise in plasma viremia after CD8⁺ T cell depletion in simian immunodeficiency virus-infected macaques. *J Exp Med* **189:** 991–998.

Kaslow RA, Carrington M, Apple R, Park L, Munoz A, Saah AJ, Goedert JJ, Winkler C, O'Brien SJ, Rinaldo C, et al. 1996. Influence of combinations of human major histocompatibility complex genes on the course of HIV-1 infection. *Nat Med* **2:** 405–411.

Keele BF, Giorgi EE, Salazar-Gonzalez JF, Decker JM, Pham KT, Salazar MG, Sun C, Grayson T, Wang S, Li H, et al. 2008. Identification and characterization of transmitted and early founder virus envelopes in primary HIV-1 infection. *Proc Natl Acad Sci* **105:** 7552–7557.

Kiepiela P, Ngumbela K, Thobakgale C, Ramduth D, Honeyborne I, Moodley E, Reddy S, de Pierres C, Mncube Z, Mkhwanazi N, et al. 2007. CD8⁺ T-cell responses to different HIV proteins have discordant associations with viral load. *Nat Med* **13:** 46–53.

Klavinskis LS, Gao L, Barnfield C, Lehner T, Parker S. 1997. Mucosal immunization with DNA-liposome complexes. *Vaccine* **15:** 818–820.

Kong WP, Wu L, Wallstrom TC, Fischer W, Yang ZY, Ko SY, Letvin NL, Haynes BF, Hahn BH, Korber B, et al. 2009. Expanded breadth of the T-cell response to mosaic human immunodeficiency virus type 1 envelope DNA vaccination. *J Virol* **83:** 2201–2215.

Korber BTM, Brander C, Haynes BF, Koup RA, Moore JP, Walker BD, Watkins DI, ed. 2007. *HIV molecular immunology.* Los Alamos National Laboratory, Los Alamos, NM.

Kotler DP, Gaetz HP, Lange M, Klein EB, Holt PR. 1984. Enteropathy associated with the acquired immunodeficiency syndrome. *Ann Intern Med* **101:** 421–428.

Koup RA, Safrit JT, Cao Y, Andrews CA, McLeod G, Borkowsky W, Farthing C, Ho DD. 1994. Temporal association of cellular immune responses with the initial control of viremia in primary human immunodeficiency virus type 1 syndrome. *J Virol* **68:** 4650–4655.

Koup RA, Roederer M, Lamoreaux L, Fischer J, Novik L, Nason MC, Larkin BD, Enama ME, Ledgerwood JE, Bailer RT, et al. 2010. Priming immunization with DNA augments immunogenicity of recombinant adenoviral vectors for both HIV-1 specific antibody and T-cell responses. *PLoS One* **5:** e9015.

Lai L, Vodros D, Kozlowski PA, Montefiori DC, Wilson RL, Akerstrom VL, Chennareddi L, Yu T, Kannanganat S,

Ofielu L, et al. 2007. GM-CSF DNA: An adjuvant for higher avidity IgG, rectal IgA, and increased protection against the acute phase of a SHIV-89.6P challenge by a DNA/MVA immunodeficiency virus vaccine. *Virology* **369:** 153–167.

Lasaro MO, Ertl HC. 2009. New insights on adenovirus as vaccine vectors. *Mol Ther* **17:** 1333–1339.

Leslie AJ, Pfafferott KJ, Chetty P, Draenert R, Addo MM, Feeney M, Tang Y, Holmes EC, Allen T, Prado JG, et al. 2004. HIV evolution: CTL escape mutation and reversion after transmission. *Nat Med* **10:** 282–289.

Li Q, Duan L, Estes JD, Ma ZM, Rourke T, Wang Y, Reilly C, Carlis J, Miller CJ, Haase AT. 2005. Peak SIV replication in resting memory CD4$^+$ T cells depletes gut lamina propria CD4$^+$ T cells. *Nature* **434:** 1148–1152.

Li Q, Skinner PJ, Ha SJ, Duan L, Mattila TL, Hage A, White C, Barber DL, O'Mara L, Southern PJ, et al. 2009. Visualizing antigen-specific and infected cells in situ predicts outcomes in early viral infection. *Science* **323:** 1726–1729.

Li H, Bar KJ, Wang S, Decker JM, Chen Y, Sun C, Salazar-Gonzalez JF, Salazar MG, Learn GH, Morgan CJ, et al. 2010. High multiplicity infection by HIV-1 in men who have sex with men. *PLoS Pathog* **6:** e1000890.

Lichterfeld M, Yu XG, Cohen D, Addo MM, Malenfant J, Perkins B, Pae E, Johnston MN, Strick D, Allen TM, et al. 2004. HIV-1 Nef is preferentially recognized by CD8 T cells in primary HIV-1 infection despite a relatively high degree of genetic diversity. *AIDS* **18:** 1383–1392.

Liu J, O'Brien KL, Lynch DM, Simmons NL, La Porte A, Riggs AM, Abbink P, Coffey RT, Grandpre LE, Seaman MS, et al. 2009. Immune control of an SIV challenge by a T-cell-based vaccine in rhesus monkeys. *Nature* **457:** 87–91.

Luxembourg A, Evans CF, Hannaman D. 2007. Electroporation-based DNA immunisation: Translation to the clinic. *Expert Opin Biol Ther* **7:** 1647–1664.

Maness NJ, Walsh AD, Piaskowski SM, Furlott J, Kolar HL, Bean AT, Wilson NA, Watkins DI. 2010a. CD8$^+$ T cell recognition of cryptic epitopes is a ubiquitous feature of AIDS virus infection. *J Virol* **84:** 11569–11574.

Maness NJ, Wilson NA, Reed JS, Piaskowski SM, Sacha JB, Walsh AD, Thoryk E, Heidecker GJ, Citron MP, Liang X, et al. 2010b. Robust, vaccine-induced CD8$^+$ T lymphocyte response against an out-of-frame epitope. *J Immunol* **184:** 67–72.

Martinez-Picado J, Prado JG, Fry EE, Pfafferott K, Leslie A, Chetty S, Thobakgale C, Honeyborne I, Crawford H, Matthews P, et al. 2006. Fitness cost of escape mutations in p24 Gag in association with control of human immunodeficiency virus type 1. *J Virol* **80:** 3617–3623.

Mattapallil JJ, Douek DC, Hill B, Nishimura Y, Martin M, Roederer M. 2005. Massive infection and loss of memory CD4$^+$ T cells in multiple tissues during acute SIV infection. *Nature* **434:** 1093–1097.

McElrath MJ, De Rosa SC, Moodie Z, Dubey S, Kierstead L, Janes H, Defawe OD, Carter DK, Hural J, Akondy R, et al. 2008. HIV-1 vaccine-induced immunity in the test-of-concept Step Study: A case-cohort analysis. *Lancet* **372:** 1894–1905.

Mehandru S, Poles MA, Tenner-Racz K, Horowitz A, Hurley A, Hogan C, Boden D, Racz P, Markowitz M. 2004. Primary HIV-1 infection is associated with preferential depletion of CD4$^+$ T lymphocytes from effector sites in the gastrointestinal tract. *J Exp Med* **200:** 761–770.

Migueles SA, Sabbaghian MS, Shupert WL, Bettinotti MP, Marincola FM, Martino L, Hallahan CW, Selig SM, Schwartz D, Sullivan J, et al. 2000. HLA B*5701 is highly associated with restriction of virus replication in a subgroup of HIV-infected long term nonprogressors. *Proc Natl Acad Sci* **97:** 2709–2714.

Migueles SA, Osborne CM, Royce C, Compton AA, Joshi RP, Weeks KA, Rood JE, Berkley AM, Sacha JB, Cogliano-Shutta NA, et al. 2008. Lytic granule loading of CD8$^+$ T cells is required for HIV-infected cell elimination associated with immune control. *Immunity* **29:** 1009–1021.

Moniuszko M, Bogdan D, Pal R, Venzon D, Stevceva L, Nacsa J, Tryniszewska E, Edghill-Smith Y, Wolinsky SM, Franchini G. 2005. Correlation between viral RNA levels but not immune responses in plasma and tissues of macaques with long-standing SIVmac251 infection. *Virology* **333:** 159–168.

Mullins JI, Nickle DC, Heath L, Rodrigo AG, Learn GH. 2004. Immunogen sequence: The fourth tier of AIDS vaccine design. *Expert Rev Vaccines* **3**(4 Suppl): S151–S159.

Nchinda G, Amadu D, Trumpfheller C, Mizenina O, Uberla K, Steinman RM. 2010. Dendritic cell targeted HIV gag protein vaccine provides help to a DNA vaccine including mobilization of protective CD8$^+$ T cells. *Proc Natl Acad Sci* **107:** 4281–4286.

Nickle DC, Jensen MA, Gottlieb GS, Shriner D, Learn GH, Rodrigo AG, Mullins JI. 2003. Consensus and ancestral state HIV vaccines. *Science* **299:** 1515–1518.

Ogg GS, Jin X, Bonhoeffer S, Dunbar PR, Nowak MA, Monard S, Segal JP, Cao Y, Rowland-Jones SL, Cerundolo V, et al. 1998. Quantitation of HIV-1-specific cytotoxic T lymphocytes and plasma load of viral RNA. *Science* **279:** 2103–2106.

Pal R, Venzon D, Santra S, Kalyanaraman VS, Montefiori DC, Hocker L, Hudacik L, Rose N, Nacsa J, Edghill-Smith Y, et al. 2006. Systemic immunization with an ALVAC-HIV-1/protein boost vaccine strategy protects rhesus macaques from CD4$^+$ T-cell loss and reduces both systemic and mucosal simian-human immunodeficiency virus SHIVKU2 RNA levels. *J Virol* **80:** 3732–3742.

Pantaleo G, Esteban M, Jacobs B, Tartaglia J. 2010. Poxvirus vector-based HIV vaccines. *Curr Opin HIV AIDS* **5:** 391–396.

Peng B, Wang LR, Gomez-Roman VR, Davis-Warren A, Montefiori DC, Kalyanaraman VS, Venzon D, Zhao J, Kan E, Rowell TJ, et al. 2005. Replicating rather than nonreplicating adenovirus-human immunodeficiency virus recombinant vaccines are better at eliciting potent cellular immunity and priming high-titer antibodies. *J Virol* **79:** 10200–10209.

Petrovas C, Casazza JP, Brenchley JM, Price DA, Gostick E, Adams WC, Precopio ML, Schacker T, Roederer M, Douek DC, et al. 2006. PD-1 is a regulator of virus-specific CD8$^+$ T cell survival in HIV infection. *J Exp Med* **203:** 2281–2292.

Picard O, Lebas J, Imbert JC, Bigel P, Zagury D. 1991. Complication of intramuscular/subcutaneous immune therapy in severely immune-compromised individuals. *J Acquir Immune Defic Syndr* **4:** 641–643.

Pillai VB, Hellerstein M, Yu T, Amara RR, Robinson HL. 2008. Comparative studies on in vitro expression and in vivo immunogenicity of supercoiled and open circular forms of plasmid DNA vaccines. *Vaccine* **26:** 1136–1141.

Price DA, Asher TE, Wilson NA, Nason MC, Brenchley JM, Metzler IS, Venturi V, Gostick E, Chattopadhyay PK, Roederer M, et al. 2009. Public clonotype usage identifies protective Gag-specific CD8$^+$ T cell responses in SIV infection. *J Exp Med* **206:** 923–936.

Rerks-Ngarm S, Pitisuttithum P, Nitayaphan S, Kaewkungwal J, Chiu J, Paris R, Premsri N, Namwat C, de Souza M, Adams E, et al. 2009. Vaccination with ALVAC and AIDSVAX to prevent HIV-1 infection in Thailand. *N Engl J Med* **361:** 2209–2220.

Robert-Guroff M. 2007. Replicating and non-replicating viral vectors for vaccine development. *Curr Opin Biotechnol* **18:** 546–556.

Rolland M, Heckerman D, Deng W, Rousseau CM, Coovadia H, Bishop K, Goulder PJ, Walker BD, Brander C, Mullins JI. 2008. Broad and Gag-biased HIV-1 epitope repertoires are associated with lower viral loads. *PLoS One* **3:** e1424.

Rosenberg ES, Billingsley JM, Caliendo AM, Boswell SL, Sax PE, Kalams SA, Walker BD. 1997. Vigorous HIV-1-specific CD4$^+$ T cell responses associated with control of viremia. *Science* **278:** 1447–1450.

Santra S, Liao HX, Zhang R, Muldoon M, Watson S, Fischer W, Theiler J, Szinger J, Balachandran H, Buzby A, et al. 2010. Mosaic vaccines elicit CD8$^+$ T lymphocyte responses that confer enhanced immune coverage of diverse HIV strains in monkeys. *Nat Med* **16:** 324–328.

Schmitz JE, Kuroda MJ, Santra S, Sasseville VG, Simon MA, Lifton MA, Racz P, Tenner-Racz K, Dalesandro M, Scallon BJ, et al. 1999. Control of viremia in simian immunodeficiency virus infection by CD8$^+$ lymphocytes. *Science* **283:** 857–860.

Schneidewind A, Brockman MA, Yang R, Adam RI, Li B, Le Gall S, Rinaldo CR, Craggs SL, Allgaier RL, Power KA, et al. 2007. Escape from the dominant HLA-B27-restricted cytotoxic T-lymphocyte response in Gag is associated with a dramatic reduction in human immunodeficiency virus type 1 replication. *J Virol* **81:** 12382–12393.

Seaman MS, Xu L, Beaudry K, Martin KL, Beddall MH, Miura A, Sambor A, Chakrabarti BK, Huang Y, Bailer R, et al. 2005. Multiclade human immunodeficiency virus type 1 envelope immunogens elicit broad cellular and humoral immunity in rhesus monkeys. *J Virol* **79:** 2956–2963.

Shiver JW, Fu TM, Chen L, Casimiro DR, Davies ME, Evans RK, Zhang ZQ, Simon AJ, Trigona WL, Dubey SA, et al. 2002. Replication-incompetent adenoviral vaccine vector elicits effective anti-immunodeficiency-virus immunity. *Nature* **415:** 331–335.

Simons BC, Vancompernolle SE, Smith RM, Wei J, Barnett L, Lorey SL, Meyer-Olson D, Kalams SA. 2008. Despite biased TRBV gene usage against a dominant HLA B57-restricted epitope, TCR diversity can provide recognition of circulating epitope variants. *J Immunol* **181:** 5137–5146.

Smith CL, Mirza F, Pasquetto V, Tscharke DC, Palmowski MJ, Dunbar PR, Sette A, Harris AL, Cerundolo V. 2005. Immunodominance of poxviral-specific CTL in a human trial of recombinant-modified vaccinia Ankara. *J Immunol* **175:** 8431–8437.

Spentzou A, Bergin P, Gill D, Cheeseman H, Ashraf A, Kaltsidis H, Cashin-Cox M, Anjarwalla I, Steel A, Higgs C, et al. 2010. Viral inhibition assay: A CD8 T cell neutralization assay for use in clinical trials of HIV-1 vaccine candidates. *J Infect Dis* **201:** 720–729.

Steinman RM. 2008. Dendritic cells in vivo: A key target for a new vaccine science. *Immunity* **29:** 319–324.

Tewari K, Flynn BJ, Boscardin SB, Kastenmueller K, Salazar AM, Anderson CA, Soundarapandian V, Ahumada A, Keler T, Hoffman SL, et al. 2010. Poly(I:C) is an effective adjuvant for antibody and multi-functional CD4$^+$ T cell responses to Plasmodium falciparum circumsporozoite protein (CSP) and alphaDEC-CSP in non human primates. *Vaccine* **28:** 7256–7266.

Trautmann L, Janbazian L, Chomont N, Said EA, Gimmig S, Bessette B, Boulassel MR, Delwart E, Sepulveda H, Balderas RS, et al. 2006. Upregulation of PD-1 expression on HIV-specific CD8$^+$ T cells leads to reversible immune dysfunction. *Nat Med* **12:** 1198–1202.

Tritel M, Stoddard AM, Flynn BJ, Darrah PA, Wu CY, Wille U, Shah JA, Huang Y, Xu L, Betts MR, et al. 2003. Prime-boost vaccination with HIV-1 Gag protein and cytosine phosphate guanosine oligodeoxynucleotide, followed by adenovirus, induces sustained and robust humoral and cellular immune responses. *J Immunol* **171:** 2538–2547.

Turnbull EL, Lopes AR, Jones NA, Cornforth D, Newton P, Aldam D, Pellegrino P, Turner J, Williams I, Wilson CM, et al. 2006. HIV-1 epitope-specific CD8$^+$ T cell responses strongly associated with delayed disease progression cross-recognize epitope variants efficiently. *J Immunol* **176:** 6130–6146.

Ueno H, Klechevsky E, Schmitt N, Ni L, Flamar AL, Zurawski S, Zurawski G, Palucka K, Banchereau J, Oh S. 2011. Targeting human dendritic cell subsets for improved vaccines. *Semin Immunol* **23:** 21–27.

Ulmer JB, Donnelly JJ, Parker SE, Rhodes GH, Felgner PL, Dwarki VJ, Gromkowski SH, Deck RR, DeWitt CM, Friedman A, et al. 1993. Heterologous protection against influenza by injection of DNA encoding a viral protein. *Science* **259:** 1745–1749.

Wang YE, Li B, Carlson JM, Streeck H, Gladden AD, Goodman R, Schneidewind A, Power KA, Toth I, Frahm N, et al. 2009. Protective HLA class I alleles that restrict acute-phase CD8$^+$ T-cell responses are associated with viral escape mutations located in highly conserved regions of human immunodeficiency virus type 1. *J Virol* **83:** 1845–1855.

Wolff JA, Malone RW, Williams P, Chong W, Acsadi G, Jani A, Felgner PL. 1990. Direct gene transfer into mouse muscle in vivo. *Science* **247:** 1465–1468.

Yang OO, Kalams SA, Rosenzweig M, Trocha A, Jones N, Koziel M, Walker BD, Johnson RP. 1996. Efficient lysis of human immunodeficiency virus type 1-infected cells by cytotoxic T lymphocytes. *J Virol* **70:** 5799–5806.

Yang OO, Kalams SA, Trocha A, Cao H, Luster A, Johnson RP, Walker BD. 1997. Suppression of human immunodeficiency virus type 1 replication by CD8+ cells: Evidence for HLA class I-restricted triggering of cytolytic and noncytolytic mechanisms. *J Virol* **71:** 3120–3128.

Zagury D. 1991. Anti-HIV cellular immunotherapy in AIDS. *Lancet* **338:** 694–695.

Zagury D, Bernard J, Cheynier R, Desportes I, Leonard R, Fouchard M, Reveil B, Ittele D, Lurhuma Z, Mbayo K, et al. 1988. A group specific anamnestic immune reaction against HIV-1 induced by a candidate vaccine against AIDS. *Nature* **332:** 728–731.

Zuniga R, Lucchetti A, Galvan P, Sanchez S, Sanchez C, Hernandez A, Sanchez H, Frahm N, Linde CH, Hewitt HS, et al. 2006. Relative dominance of Gag p24-specific cytotoxic T lymphocytes is associated with human immunodeficiency virus control. *J Virol* **80:** 3122–3125.

Rational Design of Vaccines to Elicit Broadly Neutralizing Antibodies to HIV-1

Peter D. Kwong, John R. Mascola, and Gary J. Nabel

Vaccine Research Center, National Institute of Allergy and Infectious Diseases, National Institutes of Health, Bethesda, Maryland 20892

Correspondence: gnabel@nih.gov

The development of a highly effective AIDS vaccine will likely depend on success in designing immunogens that elicit broadly neutralizing antibodies to naturally circulating strains of HIV-1. Although the antibodies induced after natural infection with HIV-1 are often directed to strain-specific or nonneutralizing determinants, it is now evident that 10%–25% of HIV-infected individuals generate neutralizing antibody responses of considerable breadth. In the past, only four broadly neutralizing monoclonal antibodies had been defined, but more than a dozen monoclonal antibodies of substantial breadth have more recently been isolated. An understanding of their recognition sites, the structural basis of their interaction with the HIV Env, and their development pathways provides new opportunities to design vaccine candidates that will elicit broadly protective antibodies against this virus.

For the majority of licensed vaccines, neutralizing antibodies have provided the best correlate of vaccine efficacy. Although a variety of immune mechanisms may contribute to protection, immunity is in part caused by inactivation of the infecting virus that aborts productive replication. In the case of HIV, it has been difficult to define such antibodies because the virus has evolved a multitude of mechanisms to evade humoral immunity. Because of its error-prone DNA-dependent RNA polymerase and its ability to undergo RNA recombination, the virus has generated unprecedented diversity (Korber et al. 2000). The number of common determinants shared by naturally circulating strains is therefore diminished. In addition, HIV envelope glycoprotein (Env) displays a low spike density on the virion surface (Klein et al. 2009; Klein and Bjorkman 2010), potentially reducing the efficiency of cross-linking and the advantage of antibody avidity that enhances the neutralization of many viruses. Its high carbohydrate content further masks critical structures that may be sensitive to neutralization (Wyatt et al. 1998). Finally, other mechanisms, including conformational flexibility, strain-specific amino acid variability, and decoy forms of the HIV Env, such as the free monomer (Douek et al. 2006), stimulate nonneutralizing antibody responses to irrelevant viral structures. Thus, the definition of serotypes that has proven a successful approach for many vaccines has not been available to guide the design of broadly neutralizing antibody immunogens. These challenges have prompted efforts to understand the immunobiology of HIV-1 Env, with an emphasis on understanding the structural basis for HIV-1 Env neutralization. In addition, the definition and characterization of monoclonal antibodies that mediate such broad neutralization, as well as the structural basis for its interaction with HIV envelope, have provided opportunities for the design of HIV-1 vaccines that stimulate the production of antibodies that are directed against specific conserved regions of the virus.

Rational design of immunogens that elicit broadly reactive neutralizing antibodies is facilitated by the identification of HIV-infected individuals with broadly neutralizing sera, from which individual monoclonal antibodies can be isolated. Two strategies have led to the identification of such antibodies. First, individual B cells have been isolated, grown in microcultures, and the secreted antibodies have been tested for neutralization. Antibodies that neutralize diverse HIV-1 viruses were identified, and the immunoglobulin genes from the cells of interest were cloned and expressed. The neutralization specificity was then confirmed for the cloned expressed IgG genes. A second approach built on knowledge of structure to design resurfaced and stabilized HIV Env cores that were used as probes to select individual B cells targeted to a specific site. The immunoglobulin genes from these B cells were then rescued by PCR amplification. The neutralization breadths of expressed antibodies were defined against a panel of genetically diverse circulated viruses. Finally, targeted approaches to other specific regions of the virus, including the membrane-proximal region (MPR), CD4-induced (CD4i), and Env glycans have all provided specific targets for which immunogens can be specifically designed. Taken together, these approaches have enabled the design of probes that allow detection of antibodies to specific viral structures at the same time that they serve as prototype immunogens to elicit these responses. Nonetheless, impediments remain to the elicitation of such antibodies, including the ability to overcome the elimination of autoreactive B cells and to stimulate the relevant necessary somatic mutations that give rise to

antibodies of the appropriate specificity. Finally, elucidation of the critical structures that confer relevant antigenicity while defining the determinants required for immunogenicity represents a key scientific question whose solution will facilitate the success of this rational vaccine design strategy.

ROLE OF ANTIBODIES IN PROTECTION

The design of immunogens able to elicit neutralizing antibodies (NAb) remains a major goal of HIV-1 vaccine development. Most licensed viral vaccines induce antibodies that neutralize the infecting virus, thereby protecting against infection or disease. Although the specific immune responses required to protect humans against HIV-1 infection are not known, studies of lentiviral infection in nonhuman primates (NHPs) have shown that passive infusion of antibodies can prevent infection. Specifically, antibodies that neutralize HIV-1 have been shown to prevent infection by a chimeric simian-human immunodeficiency virus (SHIV) containing the *env* gene of HIV-1. Early SHIV challenge studies used intravenous inoculation of the challenge virus. In this setting, high concentrations of NAbs were required to block infection, and nonneutralizing antibodies were unable to protect against infection (Mascola et al. 1999; Shibata et al. 1999). Subsequent studies, using a single oral or vaginal inoculation sufficient to infect 100% of control animals, also showed protection by NAb (Baba et al. 2000; Mascola et al. 2000; Hofmann-Lehmann et al. 2001; Parren et al. 2001). The protection against SHIV infection has been most directly associated with the neutralization potency of the infused antibodies (Nishimura et al. 2002; Mascola 2003). However, recent studies have also suggested a role of Fc-mediated antibody effector functions in conferring protection. The neutralizing monoclonal antibody (mAb) IgG1 b12 showed a diminished protective effect if the Fc region of the IgG was altered to knock out complement binding and ADCC activity (Hessel et al. 2007). Importantly, recent passive transfer studies have employed low-dose mucosal inoculation that requires multiple challenges to infect all control animals (Hessell et al. 2009a,b). This model may be more physiologically relevant to the relatively inefficient HIV-1 infection in humans. In the low-dose NHP model, approximately 10-fold less antibody was required to mediate protection against infection compared to prior studies with high-dose inoculations; serum antibody titers sufficient to mediate 90% virus neutralization at 1:5 serum dilution were associated with protection. Hence, vaccines may not need to achieve extraordinarily high levels of HIV-1 NAbs. However, the antibody response will likely need to be durable, and NAbs will have to cross-react with a genetically diverse spectrum of HIV-1 strains.

CLINICAL TRIALS OF CANDIDATE HIV VACCINES

To protect against HIV-1, antibodies must bind the viral surface envelope glycoprotein, a homotrimer composed of the gp120 surface unit and the gp41 *trans*-membrane domain. The earliest phase I vaccine trials included recombinant envelope products, gp120 or gp160, formulated in various adjuvants. Live recombinant vectors such as *vaccinia* and canary pox and nucleic acid–based vaccines have also been tested, usually including gene inserts expressing the Env glycoprotein. Recombinant protein vaccines used alone, or as a boost to vaccine vectors, generally elicited high titers of anti-Env antibodies. Initial immunogenicity studies showed that vaccine-elicited antibodies could neutralize HIV-1 in vitro, but it was soon realized that the viral neutralization was limited to prototype laboratory-adapted HIV-1 strains and did not extend to primary HIV-1 isolates (Wrin and Nunberg 1994; Mascola et al. 1996). These neutralization data generated considerable debate regarding the rationale for efficacy testing of protein-based vaccines. Despite this uncertainty, two phase III gp120 vaccine trials were conducted, each with a bivalent formulation of two strains of gp120 formulated in Alum. The VAX004 and VAX003 studies were initiated in 1998 and 1999, respectively, and the results reported in 2003. These gp120 vaccines showed no significant impact on acquisition of HIV-1 infection and had no impact on plasma viremia or peripheral CD4$^+$ T-cell counts (Flynn et al. 2005; Gilbert et al. 2005).

The failure of these gp120 vaccines was generally viewed as evidence that a successful vaccine would need to induce more potent NAbs that can neutralize circulating strains of HIV-1. Prior to the release of the VAX003 and VAX004 results, an additional phase III trial was planned. The RV144 study included priming immunizations with an avipox vector (ALVAC) and boosting with the same bivalent gp120s used in the VAX003 study. The results of this study, released in 2009, showed that volunteers in the vaccine arm of the study acquired 31% fewer HIV-1 infections than those in the placebo arm (Rerks-Ngarm et al. 2009). This modest efficacy, although not deemed adequate for licensure, was the first indication that a vaccine could protect against HIV-1 infection. The immunologic data from the RV144 vaccine trial are still being analyzed. Immune correlates analysis will include a set of in vitro antibody assays, including traditional virus neutralization, and assays of antibody binding, ADCC, and additional measures of Fc-mediated antibody effector functions. Whether any of these measures of vaccine-elicited antibodies will correlate with protection is yet to be determined. It is also unclear why this ALVAC/gp120 trial produced a modest protective effect whereas the gp120-only VAX003 and VAX004 studies did not. Differences in the studies include the ALVAC prime in the RV144 study and substantial differences in the risk factors and routes of HIV-1 infection for the populations studied. In summary, human efficacy trials indicate that vaccine-elicited protection against HIV-1 infection is achievable, but the specific antibody responses that may contribute to protection have not been elucidated. Despite these limitations in our knowledge of immune correlates, the modest 31% protection observed in the RV144 study suggests much room for improvement. One means of achieving improved vaccine efficacy may be through the elicitation of more potent and cross-reactive NAbs.

HUMORAL IMMUNE RESPONSE DURING HIV-1 INFECTION

Most vaccines seek to mimic the immune response generated during natural infection. HIV-1, however, shows extensive genetic variability owing to error-prone reverse transcription of the viral genome and a high tolerance for mutations that mediate viral immune evasion to antibody and CD8 T-cell responses. As a consequence of chronic persistent replication in the face of immune pressure, the HIV-1 epidemic is comprised of diverse genetic variants and a poorly understood level of antigenic diversity. A critical feature of HIV-1 appears to be its general resistance to NAbs. This resistance is manifest during the early phase of HIV-1 infection, in which NAb responses to the infecting virus may not appear until several months after infection. Once such autologous NAbs arise, the virus quickly escapes, leading to an ongoing cycle of adaptive antibody responses and further viral escape (Albert et al. 1990; Montefiori et al. 1991; Richman et al. 2003; Wei et al. 2003; Flynn et al. 2005; Gilbert et al. 2005). Despite this impressive immune evasion capability, recent evidence suggests that there are some vulnerabilities in the protective armor of HIV-1 that could be exploited by rationally designed antibody-based vaccines.

The advent of high-throughput neutralization assays using recombinant Env pseudoviruses has permitted the screening of sera from relatively large cohorts of HIV-1-infected donors. These data reveal that between 10% and 25% of HIV-1-infected subjects make NAb that cross-react with a substantial portion of diverse HIV-1 strains (Dhillon et al. 2007; Li et al. 2007; Doria-Rose et al. 2009; Sather et al. 2009; Stamatatos et al. 2009). A smaller subset of these sera is able to neutralize the large majority of circulating HIV-1 isolates (Li et al. 2007; Simek et al. 2009). The observation that some individuals make broadly cross-reactive neutralizing antibodies has spurred a renewed effort to isolate neutralizing mAbs to define vulnerable epitopes on the viral Env that could serve as targets of vaccine design. Until 2009, only a handful of known neutralizing mAbs were able to neutralize primary isolates of HIV-1, and even these antibodies displayed limitations in overall potency or breadth of reactivity (reviewed in Zolla-Pazner 2004; Pantophlet and Burton 2006; Mascola and Montefiori 2010). The effort to isolate new mAbs has been bolstered by several recent technological advances. Multiparameter flow cytometry can be used to identify and sort individual HIV-1 Env-specific memory B cells (Scheid et al. 2009). The antibody heavy and light chain can then be genetically recovered from the cDNA of single B cells and the full IgG expressed. An alternate methodology involves the screening of thousands of unselected individual memory B cells, each stimulated to secrete IgG. The supernatants of these B-cell cultures are screened for HIV-1 neutralization using a high-throughput microneutralization assay (Walker et al. 2009). The application of these new B-cell technologies has resulted in the isolation of several new HIV-1 mAbs that are more potent and broadly reactive than were prior antibodies. For example, the PG9 and PG16 mAbs are directed to a previously undefined site within the V2-V3 region of the HIV-1 viral spike (Walker et al. 2009). Another example is the VRC01, VRC02, and VRC03 mAbs that bind to a functionally conserved region of gp120 that interacts with the host cell receptor CD4 (Wu et al. 2010; Zhou et al. 2010). The fairly rapid and efficient isolation of these new mAbs suggests that numerous additional neutralizing mAbs will be forthcoming. Because each category of new broadly neutralizing mAbs potentially identifies a highly conserved target of HIV-1 neutralization, the structural analysis of these antibodies bound to HIV-1 Env can provide valuable insights for vaccine design.

STRUCTURAL VIROLOGY

Structural biology provides atomic-level details about the three-dimensional organization and chemical structure of proteins. Such details facilitate an understanding of mechanism and—combined with recent advances of in silico protein design—provide a means by which to manipulate and to optimize Env-based immunogens (Nabel et al. 2011). Overall, the host-derived envelope that surrounds HIV-1 sequesters most of its proteins from antibody-mediated recognition. Only two HIV-1 proteins protrude through this protective membrane: the HIV-1 gp120 envelope glycoprotein and its gp41 *trans*-membrane partner. Three gp120s associate noncovalently with three gp41s to make up the viral spike (reviewed in Wyatt and Sodroski 1998).

A combination of X-ray crystallography and electron microscopy has served to illuminate many of the details of the viral spike (Fig. 1). HIV-1 uses a two-receptor mechanism involving the host receptor CD4 and a coreceptor (generally CCR5 or CXCR4), along with considerable Env-based conformational change, to fuse viral and target cell membranes and to enter host cells (Wyatt and Sodroski 1998). In the first or "unliganded" state, the HIV-1 viral spike covers most of its surface with host-derived N-linked glycan (Wyatt et al. 1998). These glycans appear as "self" to the humoral immune system and therefore are mostly immunologically silent. In the unliganded state, only a few nonglycosylated sites are available for antibody-based recognition. After CD4 engages gp120 at the cell surface, the HIV-1 viral spike undergoes substantial structural rearrangements: in the CD4-bound conformation of gp120, a four-stranded bridging sheet forms and the V3 loop is sprung, and in gp41, the previously occluded N-heptad repeat becomes available for T-20 recognition (reviewed in Wyatt and Sodroski 1998). In the CD4-bound state, much more of the glycan-free viral spike surface is uncovered and potentially available for recognition by neutralizing antibody, such as V3-directed antibodies or CD4-induced (CD4i) antibodies that bind to the bridging sheet. However, the close proximity of viral and target cell membrane sterically occludes access of immunoglobulins to these epitopes (Labrijn et al. 2003). CD4 engagement induces the formation of a high-affinity site for coreceptor binding (Wu et al. 1996). Although details remain to be deciphered, coreceptor binding facilitates additional conformational changes. A transition intermediate has been hypothesized to form, in which gp41 is fully

extended, with a fusion peptide thrown into the target cell membrane and the carboxy-terminal *trans*-membrane region embedded in the viral membrane. This transient intermediate resolves itself by forming a highly stable six-helix bundle, the postfusion state, in which fusion peptide and *trans*-membrane regions of gp41 are juxtaposed.

SITES OF HIV-1 VULNERABILITY

The assembled viral spike is highly protected from antibody-mediated recognition. Nonetheless, as described above, a number of broadly neutralizing antibodies have recently been discovered. The breadth of these antibodies indicates that they recognize conserved Env regions of functional importance; conversely, the susceptibility of these regions to antibody-mediated recognition indicates that they form sites of HIV-1 vulnerability to the humoral immune response.

One such site of Env vulnerability is the initial site of attachment for the CD4 receptor (Fig. 2A,B) (Zhou et al. 2007; Chen et al. 2009). This site, on the outer domain of gp120, consists of roughly two-thirds of the surface HIV-1 uses to fully engage the CD4 receptor. Although the site is itself glycan-free, N-linked glycosylation and variable loops (e.g., V5) surround much of it. Moreover, most of the surrounding surface that is glycan-free is highly susceptible to conformational change; antibodies that bind to these regions induce conformations in the neighboring V1/V2 variable regions that are incompatible with the functional viral spike. Precise targeting by antibodies onto the initial site of CD4 attachment thus appears to be required for effective neutralization at this site (Chen et al. 2009).

The first monoclonal antibody found to bind effectively to this site, the b12 antibody, was identified from a phage library (Burton et al. 1994) and found to use a heavy-chain-only means of recognition (Zhou et al. 2007), which is generally not observed in naturally elicited antibodies. The b12 antibody nevertheless displays reasonable recognition of the site (Zhou et al. 2007), although extension outside the

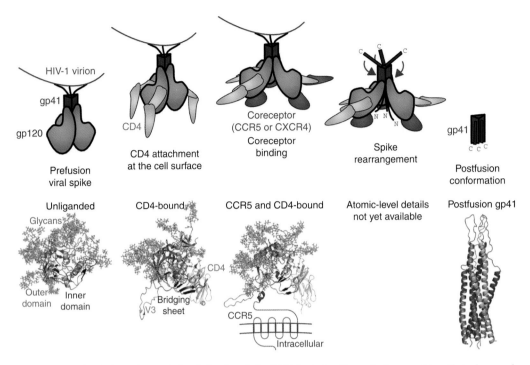

Figure 1. Mechanistic and atomic-level details of the HIV-1 viral spike, a fusion machine that also evades humoral detection. (*Top* row) Entry schematic. The viral spike is composed of three gp120 envelope glycoproteins (cyan) and three gp41 *trans*-membrane (red). In the prefusion conformation, the outer surface of the spike is covered with N-linked glycan, which is seen as "self" by the humoral immune system therefore virtually invisible to potentially neutralizing antibody. At the cell surface, binding to CD4 (yellow) induces large structural rearrangements, which include the formation of a binding site for a second requisite coreceptor (purple). Coreceptor binding induces a transient intermediate, with the amino-terminal fusion peptide of gp41 thrown into the target cell membrane while the carboxy-terminal gp41-*trans*-membrane region is buried in the viral membrane. Subsequent spike rearrangements resolve into a stable postfusion conformation, with the fusion peptide and *trans*-membrane regions of gp41 in close proximity. (*Bottom* row) Atomic-level structures, with polypeptide backbones shown in ribbon representation and N-linked glycans in stick representation. The unliganded conformation of HIV-1 gp120 in its viral spike conformation is unknown, but an atomic model has been solved for an SIV core (Chen et al. 2005), in which the N-linked glycans (cyan) cluster onto one face of gp120. The CD4-bound conformation of gp120 (Kwong et al. 1998; Huang et al. 2005) contains a four-stranded bridging sheet (blue) and a protruding V3 region (orange), both of which interact with CCR5 (Rizzuto et al. 1998), although only the amino-terminal region of the bound CCR5 structure has been determined (Huang et al. 2007). The gp41 conformations of HIV-1 in prefusion and intermediate stages is unknown, but the postfusion conformation (Chan et al. 1997; Weissenhorn et al. 1998) reveals a stable six-helix bundle, with structural similarity to other type 1 viral fusion machines.

target site, especially around the CD4-binding loop, allowed for antigenic variation, especially with non-B clade isolates (Wu et al. 2009b). As a consequence, the b12 antibody only neutralizes ~35% of circulating isolates.

Other antibodies have been isolated more recently that also target this site. Antibody HJ16 is a natural human antibody isolated by direct assessment of neutralization coupled to single B-cell antibody sequencing (Corti et al. 2010). The atomic-level structure of this antibody in complex with gp120 has not yet been determined, but initial mapping suggests that antibody HJ16 strays into the "loop D" region, outside of the site of vulnerability (Corti et al. 2010); overall HJ16 neutralizes ~30% of

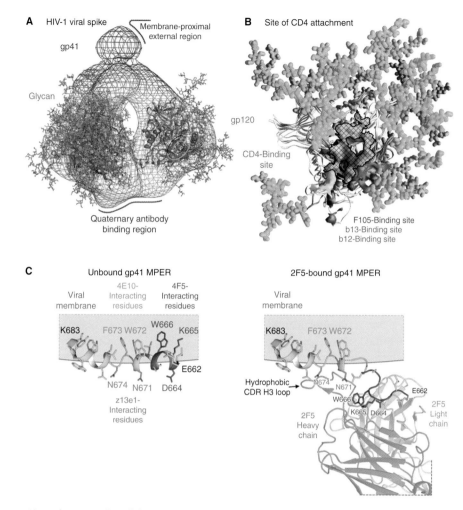

Figure 2. Sites of HIV-1 vulnerability to neutralizing antibody. (*A*) Electron tomogram of the HIV-1 viral spike (Liu et al. 2008), with docked atomic-level structure of gp120 (Pancera et al. 2010a) and with sites of vulnerability to antibody-mediated neutralization identified. (*B*) Initial site of CD4 attachment (Zhou et al. 2007; Chen et al. 2009). The site of vulnerability to antibody is shown as a cross-hatched yellow surface, with regions that induce conformational change and are binding sites for antibody F105 (blue) and antibody b13 (purple) or regions that extend outside of the site such as that recognized by antibody b12 (red). Neighboring N-linked glycan is shown in cyan. (*C*) MPER and membrane context. The MPER contains a number of highly conserved tryptophans, which are important for its entry function. (*Left*) Model of MPER in membrane from NMR/EPR measurements. The structure of MPER residues 662–683 is shown in the context of a DPC micelle (Sun et al. 2008), with residues required for recognition by neutralizing antibodies 2F5, z13e1, and 4E10 colored in red, green, and cyan, respectively. (*Right*) Model of MPER bound by broadly neutralizing antibody 2F5 as inferred from the crystal structure of the 2F5-epitope in complex with its gp41-MPER epitope (Ofek et al. 2004). The 2F5 antibody (partially shown with heavy chain in blue and light chain in gray) extracts its epitope from a helical conformation and induces an extended loop (Song et al. 2009). This has been modeled with the epitope as defined in the crystal structure (red) connected through a schematic dashed yellow line into the carboxy-terminal portion of the MPER (the structure is not known of the complete MPER when bound by the antibody nor the relative orientations of the amino- and carboxy-terminal portions of the MPER in this context). Virus neutralization by antibodies 2F5, z13e1, and 4E10, furthermore, also appears to require interactions with the surrounding membrane, likely mediated by extended CDR H3 loops, all of which contain hydrophobic motifs capable of interacting with membrane (Alam et al. 2009; Julien et al 2010; Ofek et al. 2010a,b; Scherer et al. 2010).

circulating isolates (Corti et al. 2010). Antibody VRC01 and related antibodies VRC02 and VRC03 appear to target the site of vulnerability more precisely. These antibodies show neutralization breadths of up to ~90% of circulating isolates (Wu et al. 2010). The structure of VRC01 in complex with gp120 reveals an extraordinary mimicry between VRC01 heavy chain and CD4 receptor (Zhou et al. 2010).

Another site of vulnerability on the viral spike shows quaternary structural constraints, and maps to the second and third variable regions of gp120 (variable loops V2 and V3) (Fig. 2C). The structure of this portion of the Env has not yet been determined, and its functional importance is also not clear, but may relate to HIV-1 recognition of the $\alpha 4\beta 7$-integrin, the gut-homing receptor for HIV-1 used by primary isolates during early stages of infection (Arthos et al. 2008). A number of antibodies have been identified that recognize this region, including the human monoclonal antibodies 2909, PG9, and PG16 and a number of rhesus antibodies (Gorny et al. 2005; Walker et al. 2009; Robinson et al. 2010). Structural analysis indicates that all of these antibodies use extended heavy chain third-complementarity determining regions, which are anionic and tyrosine sulfated (Pancera et al. 2010b; Pejchal et al. 2010; Changela et al. 2011). Despite this similarity, these antibodies vary dramatically in their neutralization breadth, with 2909 being extremely strain-specific and PG9/PG16 able to neutralize 70%–80% of current circulating isolates (Honnen et al. 2007; Walker et al. 2009). This divergence in breadth appears related to the specific immunotype of the quaternary site of vulnerability recognized, with strain-specific variants recognizing rare variants of the site, and more broadly neutralizing antibodies recognizing more common variants of the site (Wu et al. 2011).

Another critical function the HIV-1 envelope performs relates to fusion of viral and target cell membranes, which is required for virus entry. Virtually all of the functionally conserved surfaces required for fusion are occluded in the functional viral spike and available for antibody-mediated neutralization only as transient intermediates in the entry process. Even then, access is limited. For example, entry requires the N-heptad repeats of gp41 to snap back on themselves, and small molecule mimics of the carboxy-terminal heptad repeat (e.g., T-20 or Fuzeon) are effective therapeutics (Baldwin et al. 2003). Nonetheless, antibodies that recognize this potential site of vulnerability such as D5 or HK20 have weak potency and limited breadths of neutralization, properties attributed to steric occlusion at the viral membrane-target cell membrane interface (Luftig et al. 2006; Gustchina et al. 2010; Sabin et al. 2010). One area of vulnerability that appears less sterically occluded is the membrane-proximal external region (MPER) (Fig. 2C). The precise role that this region plays in viral entry is unclear, but alteration of hydrophobic residues in the MPER leads to loss of fusion capabilities (Munoz-Barroso et al. 1999; Salzwedel et al. 1999). Human antibodies 2F5, Z13e, and 4E10 have been found to recognize the MPER and to show reasonable neutralization breadths and potencies (Muster et al. 1993, 1994; Trkola et al. 1995; Stiegler et al. 2001; Zwick et al. 2001; Binley et al. 2004). Interestingly, this site of HIV-1 vulnerability includes not only the HIV-1 Env (i.e., specific amino acids in the MPER), but also the neighboring or surrounding lipid membrane. Antibodies that recognize the MPER thus require a hydrophobic membrane-binding component to neutralize virus (Ofek et al. 2004; Sun et al. 2008; Alam et al. 2009; Julien et al. 2010; Ofek et al. 2010b). Such membrane "corecognition" appears to lead to self-recognition, and the MPER-directed antibodies are generally self-reactive (Haynes et al. 2005; Alam et al. 2007). Such self-reactivity may impede antibody development (B-cell deletion or anergy), and few MPER-directed neutralizing antibodies are observed in sera from HIV-1-infected individuals (Walker et al. 2009).

A number of other sites of vulnerability can be inferred from antibodies that neutralize with different specificities. A conserved cluster of high-mannose glycans (around residues 295, 332, and 392 of gp120) is recognized by the 2G12 antibody (Sanders et al. 2002). This antibody has a highly unusual structure that involves variable-domain swapping (Calarese et al. 2003), and most sera do not competitively inhibit 2G12 binding, suggesting that antibodies against this site are very rare. A recent analysis of the elite neutralizers from the protocol G screen of sera from almost 2000 HIV-1-infected individuals, however, indicates that some of the best neutralizers recognize this face of the Env trimer (Simek et al. 2009; Walker et al. 2009, 2010a). Other antibodies directed at the V3 loop (up to 10%–30% breadth) or of CD4-induced specificity (up to 10% breadth) indicate that these regions form partial sites of vulnerability (Xiang et al. 2003; Zolla-Pazner and Cardozo 2010).

IMMUNOGEN DESIGN

CD4bs Immunogens

The molecular interactions of broadly neutralizing antibodies to the CD4bs have suggested at least four strategies to elicit these antibodies (Fig. 3). First, trimeric forms of HIV-1 Env have been generated by inclusion of the gp41 trimerization sites after deletion of the trans-membrane domain.The trimeric protein can be further stabilized by addition of trimerization sequences from such proteins as the fibritin protein from phage λ. Such trimers can be further stabilized with site-specific mutations previously shown to fix the core structure (Yang et al. 2002). In these prototypes, the variable V1-V3 domains are often removed to minimize immune responses to irrelevant strain-specific structures.

An alternative approach is to develop immunogens based on a monomer structure (Fig. 3, second panel). Such proteins have been derived from stabilized core Env proteins further altered based on an understanding of structure (Zhou et al. 2007, 2010; Wu et al. 2010). Bioinformatic design has suggested mutations that replace HIV residues on the nonneutralizing inner domain of gp120 with SIV Env residues, minimizing serologic cross-reactivity with HIV-1. The surface of the conformationally stabilized Env core protein was modified and masked further with glycans. Such probes have been used to

analyze antisera for the presence of broadly neutralizing antibodies and served also as prototype immunogens to elicit antibodies to this site.

A third approach focuses on generating a subdomain of the HIV-1 Env, the outer domain that contains the initial site of CD4 attachment (Fig. 3, third panel). In this protein, a large portion of the inner domain that elicits nonneutralizing antibodies is eliminated. The immune response is therefore directed to the relevant site of initial CD4 binding. In addition, by removing parts of the inner domain required for CD4 binding, potential inhibition by CD4 attachment is avoided. In theory, modified forms of the outer domain will allow targeting of the immune response to the most relevant conserved region of binding (e.g., the critical β15 loop that interacts with both the b12 and VRC01 broadly neutralizing antibodies). Although a soluble form of the outer domain that contains the β15 loop did not bind to b12 with high affinity in early studies (Yang et al. 2004), inclusion of a trans-membrane domain (Wu et al. 2009a) or further site-directed mutagenesis based on the VRC01/Env structure has stabilized this interaction and increased b12 or VRC01 binding. Additional mutations in the outer domain region have been designed to preserve high-affinity binding and are currently under evaluation both as probes of serum neutralizing antibody activity and as immunogens. A fourth approach to CD4bs immunogen development employs scaffolds based on informatics and epitope transplantion (Fig. 3, lower panel) (e.g., adding the β15 loop to an unrelated structure that presents the epitope naturally). Scaffolds have been identified that bind CD4bs antibodies and are the subject of continued investigation.

Glycan and Quaternary Immunogens

At least two types of antibodies recognize carbohydrate determinants on HIV-1 Env. Among these antibodies are those like the prototypic 2G12 mAb that recognize high-mannose structures on the outer domain (Buchacher et al. 1994; Trkola et al. 1995; Calarese et al. 2003). Although 2G12 interacts with these glycans

Trimers

Monomeric gp120 (core)

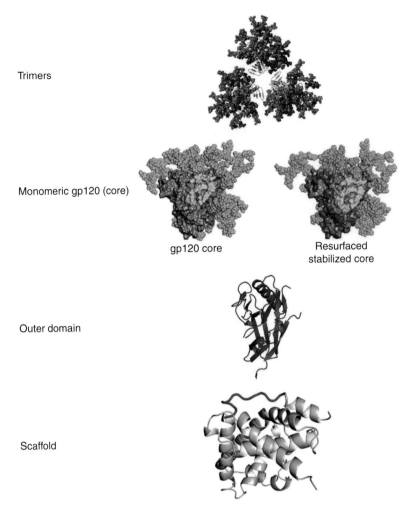

gp120 core

Resurfaced
stabilized core

Outer domain

Scaffold

Figure 3. Alternative forms of HIV Env serve as prototype immunogens for neutralizing antibody vaccines. Different forms of the HIV Env can be used to elicit neutralizing antibody responses. They range from the most complex form, the HIV trimer that most closely resembles the form found on the viral spike (*upper* row), monomeric forms, which include the gp120 core or resurfaced stabilized cores (*second* row), a region of the core that is composed primarily of the outer domain (OD) which includes the CD4-binding site (*third* row), or selected subdomains, such as the CD4-binding β-15 loop or MPER attached to a heterologous stable scaffold (*bottom* row). Modification of these prototypes by deletion of variable regions, removal, or addition of glycans, stabilization with disulfide bonds or addition of space filling mutations can serve to alter immunogenicity and elicit antibodies of the desired specificity.

through an unconventional arrangement of its antigen-binding sites, additional antibodies directed to this region have been described recently that display exceptionally high affinity and breadth (Walker et al. 2010a). There is concern that it will be difficult to generate antibodies to carbohydrate structures added by endogenous glycosylation machinery, and even

if it is possible, there is concern that such antibodies may react with such carbohydrates on host proteins. Nonetheless, several groups have attempted to develop immunogens by chemical conjugation with nonhuman glycans or by selection in yeast (Agrawal-Gamse et al. 2011). Although some structures have been defined that react with 2G12, it has not yet been possible to elicit such neutralizing antibodies with these immunogens.

Another class of exceptionally neutralizing antibodies has been directed to glycans, and possibly peptide sequences in the V1/V2 region, with additional interactions dependent on the V3 region (Walker et al. 2009). Such immunoglobulins have been termed quaternary antibodies, although it is not certain that they are directed to complex conformational determinants from different parts of Env. Although the structure of at least one such antibody has been elucidated, the molecular details of its interaction with Env remain hypothetical. Efforts at immunogen design have focused thus far on membrane-bound trimers, which show the strongest binding to these antibodies (Walker et al. 2009); however, occasional monomeric gp120 derivatives have been identified and are also under investigation. Although this class of antibody represents ~25% of the antibodies in the sera from subjects with broadly neutralizing responses (Walker et al. 2010b), it has not yet been possible to elicit these antibodies by vaccination in animal models or humans.

MPER, V3, and CD4i Immunogens

At least two broadly neutralizing mAbs, 2F5 and 4E10, recognize the highly conserved MPER region of Env. Both show reasonable breadth of neutralization, although their potency is generally low. Structures of these antibodies complexed to their cognate peptides have been determined and suggest that hydrophobic patches are required for stable interactions needed for neutralization. In the case of 4E10, efforts to develop immunogens based on stabilized peptides have allowed the definition of vaccine candidates with appropriate antigenic profiles but they do not elicit neutralizing antibodies. In the case of 2F5, a variety of approaches have elicited antibodies that react with 2F5 peptides but these antibodies do not neutralize diverse viral strains. An understanding of the structural interactions of 2F5 with its viral target has suggested that the structure of the 2F5 peptide in the context of gp120 as

well as its interaction with the viral membrane through a hydrophobic patch are required for neutralization (Ofek et al. 2010b). Based on these findings, Ofek and colleagues have generated scaffolds that present the constrained 2F5 peptide which can elicit antibodies that interact similar to the 2F5 ab, suggesting a potential strategy to elicit such neutralizing antibodies (Ofek et al. 2010a). At the same time, it is difficult to elicit antibodies that retain the hydrophobic patch required for neutralization because these antibodies are usually polyreactive (Haynes et al. 2005) and therefore likely to undergo clonal deletion because of their autoreactivity.

The V3 and CD4i regions have also received considerable attention with respect to immunogen design and have been reviewed elsewhere. It is possible to elicit neutralizing antibodies to these V3 epitopes using modified murine retroviral gp or HIV-1 Env proteins (Chakrabarti et al. 2002; Zolla-Pazner et al. 2008). Because the V3 region is not exposed on most naturally circulating viral isolates, it is unlikely that these responses would be effective in the absence of an activity that would increase its exposure. Thus far, no such agonistic antibodies have been identified and this target therefore poses a considerable challenge. Similarly, CD4i antibodies can also be readily induced after immunization with HIV Env stabilized core proteins (Dey et al. 2009). Although such antibodies are seen frequently in HIV-infected individuals (Decker et al. 2005; Gray et al. 2007), they do not mediate neutralization, although it remains possible that they may contribute to protection through ADCC function.

Fundamental B-Cell Biology and the Antibody Response

To elicit a robust neutralizing antibody response to HIV-1, an understanding of B-cell biology is required. Immunogens must engage the appropriate naïve B-cell receptors to induce antibodies of the appropriate specificity. In addition, autoreactive B cells will be eliminated by clonal deletion, and the process of immunization likely must drive somatic mutations that are required for affinity maturation and development of high-affinity antibodies with the appropriate specificity. Critical to the success of rational vaccine design is the ability to take advantage of these factors and address the basic aspects of B-cell development that control antibody specificity and synthesis. For example, immunogens will need to engage the low-affinity germline precursors in a way that facilitates the development of high-affinity antibodies. In this regard, it is important to understand the impact of adjuvants and/or delivery matrices on the generation of antibody diversity and production. At the same time, such delivery agents must have the necessary safety and immunogenicity profiles required for widespread use. Such adjuvants may include alum, saponin-based emulsions, ASO1A and B, ASO2, MF59, nanoparticles, and multimeric viral carriers, such as Qβ. The choice of animal models for testing is also critical, because not all species show similar degrees of somatic mutation, have similar genomic precursors, nor have the ability to make long CDR3 regions as found in humans. Candidates would be optimally tested in relevant humanized mouse models and NHPs before progression into phase I human clinical trials. When possible, it is also desirable to test relevant immunogens for protection in NHP challenge studies and proof of concept and to determine effective preventive antibody levels.

CONCLUDING REMARKS

HIV-1 has evolved multiple mechanisms to evade the neutralizing antibody response. Specifically, Env evades host recognition by virtue of its sequence diversity, limited exposure to the immune system because of carbohydrate masking, and conformational flexibility. HIV-1 has been resistant to classification by traditional serotyping, suggesting that standard approaches to vaccine development are unlikely to succeed. Recent progress in the definition of broadly neutralizing antibodies, the elucidation of the structures of these antibodies complexed to HIV-1, and the utilization of structural biology to define the relationship between antigenicity and immunogenicity in vaccine candidates has catalyzed a resurgence in this area of research. An understanding of the biology of HIV-1, the human immune response to the virus, and the application of structural biology to immunogen design have provided new opportunities to advance the goal of identifying vaccines that elicit broadly neutralizing antibodies that prevent or contain infection and/or progression to AIDS.

ACKNOWLEDGMENTS

We thank Ati Tislerics for manuscript editing, Gilad Ofek for assistance with Figure 2C, and Brenda Hartman and Jonathan Stuckey for assistance with graphic arts.

REFERENCES

Agrawal-Gamse C, Luallen RJ, Liu B, Fu H, Lee FH, Geng Y, Doms RW. 2011. Yeast-elicited cross-reactive antibodies to HIV Env glycans efficiently neutralize virions expressing exclusively high-mannose N-linked glycans. *J Virol* **85:** 470–480.

Alam SM, McAdams M, Boren D, Rak M, Scearce RM, Gao F, Camacho ZT, Gewirth D, Kelsoe G, Chen P, et al. 2007. The role of antibody polyspecificity and lipid reactivity in binding of broadly neutralizing anti-HIV-1 envelope human monoclonal antibodies 2F5 and 4E10 to glycoprotein 41 membrane proximal envelope epitopes. *J Immunol* **178:** 4424–4435.

Alam SM, Morelli M, Dennison SM, Liao HX, Zhang R, Xia SM, Rits-Volloch S, Sun L, Harrison SC, Haynes BF, et al. 2009. Role of HIV membrane in neutralization by two broadly neutralizing antibodies. *Proc Natl Acad Sci* **106:** 20234–20239.

Albert J, Abrahamsson B, Nagy K, Aurelius E, Gaines H, Nystrom G, Fenyo EM. 1990. Rapid development of isolate-specific neutralizing antibodies after primary HIV-1 infection and consequent emergence of virus variants which resist neutralization by autologous sera. *AIDS* **4:** 107–112.

Arthos J, Cicala C, Martinelli E, Macleod K, Van Ryk D, Wei D, Xiao Z, Veenstra TD, Conrad TP, Lempicki RA, et al. 2008. HIV-1 envelope protein binds to and signals through integrin $\alpha_4\beta_7$, the gut mucosal homing receptor for peripheral T cells. *Nat Immunol* **9:** 301–309.

Baba TW, Liska V, Hofmann-Lehmann R, Vlasak J, Xu W, Ayehunie S, Cavacini LA, Posner MR, Katinger H, Stiegler G, et al. 2000. Human neutralizing monoclonal antibodies of the IgG1 subtype protect against mucosal simian-human immunodeficiency virus infection. *Nat Med* **6:** 200–206.

Baldwin CE, Sanders RW, Berkhout B. 2003. Inhibiting HIV-1 entry with fusion inhibitors. *Curr Med Chem* **10:** 1633–1642.

Binley JM, Wrin T, Korber B, Zwick MB, Wang M, Chappey C, Stiegler G, Kunert R, Zolla-Pazner S, Katinger H, et al. 2004. Comprehensive cross-clade neutralization analysis of a panel of anti-human immunodeficiency virus type 1 monoclonal antibodies. *J Virol* **78:** 13232–13252.

Buchacher A, Predl R, Strutzenberger K, Steinfellner W, Trkola A, Purtscher M, Gruber G, Tauer C, Steindl F, Jungbauer A, et al. 1994. Generation of human monoclonal antibodies against HIV-1 proteins; electrofusion and Epstein–Barr virus transformation for peripheral blood lymphocyte immortalization. *AIDS Res Hum Retroviruses* **10:** 359–369.

Burton DR, Pyati J, Koduri R, Sharp SJ, Thornton GB, Parren PW, Sawyer LS, Hendry RM, Dunlop N, Nara PL, et al. 1994. Efficient neutralization of primary isolates of HIV-1 by a recombinant human monoclonal antibody. *Science* **266:** 1024–1027.

Calarese DA, Scanlan CN, Zwick MB, Deechongkit S, Mimura Y, Kunert R, Zhu P, Wormald MR, Stanfield RL, Roux KH, et al. 2003. Antibody domain exchange is an immunological solution to carbohydrate cluster recognition. *Science* **300:** 2065–2071.

Chakrabarti BK, Kong WP, Wu B-Y, Yang Z-Y, Friborg J Jr, Ling X, King SR, Montefiori DC, Nabel GJ. 2002. Modifications of the human immunodeficiency virus envelope glycoprotein enhance immunogenicity for genetic immunization. *J Virol* **76:** 5357–5368.

Changela A, Wu X, Yang Y, Zhang B, Zhu J, Nardone GA, O'Dell S, Pancera M, Gorny MK, Phogat S, Robinson JE, Stamatatos L, Zolla-Pazner S, Mascola JR, Kwong PD. 2011. Crystal structure of human antibody 2909 reveals conserved features of quaternary-specific antibodies that potently neutralize HIV-1. *J Virol* **85:** 2524–2535.

Chan DC, Fass D, Berger JM, Kim PS. 1997. Core structure of gp41 from the HIV envelope glycoprotein. *Cell* **89:** 263–273.

Chen B, Vogan EM, Gong H, Skehel JJ, Wiley DC, Harrison SC. 2005. Structure of an unliganded simian immunodeficiency virus gp120 core. *Nature* **433:** 834–841.

Chen L, Kwon YD, Zhou T, Wu X, O'Dell S, Cavacini L, Hessell AJ, Pancera M, Tang M, Xu L, et al. 2009. Structural basis of immune evasion at the site of CD4 attachment on HIV-1 gp120. *Science* **326:** 1123–1127.

Corti D, Langedijk JP, Hinz A, Seaman MS, Vanzetta F, Fernandez-Rodriguez BM, Silacci C, Pinna D, Jarrossay D, Balla-Jhagjhoorsingh S, et al. 2010. Analysis of memory B cell responses and isolation of novel monoclonal antibodies with neutralizing breadth from HIV-1-infected individuals. *PLoS ONE* **5:** e8805.

Decker JM, Bibollet-Ruche F, Wei X, Wang S, Levy DN, Wang W, Delaporte E, Peeters M, Derdeyn CA, Allen S, et al. 2005. Antigenic conservation and immunogenicity of the HIV coreceptor binding site. *J Exp Med* **201:** 1407–1419.

Dey B, Svehla K, Xu L, Wycuff D, Zhou T, Voss G, Phogat A, Chakrabarti BK, Li Y, Shaw G, et al. 2009. Structure-based stabilization of HIV-1 gp120 enhances humoral immune responses to the induced co-receptor binding site. *PLoS Pathog* **5:** e1000445.

Dhillon AK, Donners H, Pantophlet R, Johnson WE, Decker JM, Shaw GM, Lee FH, Richman DD, Doms RW, Vanham

G, et al. 2007. Dissecting the neutralizing antibody specificities of broadly neutralizing sera from human immunodeficiency virus type 1–infected donors. *J Virol* **81:** 6548–6562.

Doria-Rose NA, Klein RM, Manion MM, O'Dell S, Phogat A, Chakrabarti B, Hallahan CW, Migueles SA, Wrammert J, Ahmed R, et al. 2009. Frequency and phenotype of human immunodeficiency virus envelope-specific B cells from patients with broadly cross-neutralizing antibodies. *J Virol* **83:** 188–199.

Douek DC, Kwong PD, Nabel GJ. 2006. The rational design of an AIDS vaccine. *Cell* **124:** 677–681.

Flynn NM, Forthal DN, Harro CD, Judson FN, Mayer KH, Para MF. 2005. Placebo-controlled phase 3 trial of a recombinant glycoprotein 120 vaccine to prevent HIV-1 infection. *J Infect Dis* **191:** 654–665.

Gilbert PB, Peterson ML, Follmann D, Hudgens MG, Francis DP, Gurwith M, Heyward WL, Jobes DV, Popovic V, Self SG, et al. 2005. Correlation between immunologic responses to a recombinant glycoprotein 120 vaccine and incidence of HIV-1 infection in a phase 3 HIV-1 preventive vaccine trial. *J Infect Dis* **191:** 666–677.

Gorny MK, Stamatatos L, Volsky B, Revesz K, Williams C, Wang XH, Cohen S, Staudinger R, Zolla-Pazner S. 2005. Identification of a new quaternary neutralizing epitope on human immunodeficiency virus type 1 virus particles. *J Virol* **79:** 5232–5237.

Gray ES, Moore PL, Choge IA, Decker JM, Bibollet-Ruche F, Li H, Leseka N, Treurnicht F, Mlisana K, Shaw GM, et al. 2007. Neutralizing antibody responses in acute human immunodeficiency virus type 1 subtype C infection. *J Virol* **81:** 6187–6196.

Gustchina E, Li M, Louis JM, Anderson DE, Lloyd J, Frisch C, Bewley CA, Gustchina A, Wlodawer A, Clore GM. 2010. Structural basis of HIV-1 neutralization by affinity matured Fabs directed against the internal trimeric coiled-coil of gp41. *PLoS Pathog* **6:** e1001182.

Haynes BF, Fleming J, St Clair WE, Katinger H, Stiegler G, Kunert R, Robinson J, Scearce RM, Plonk K, Staats HF, et al. 2005. Cardiolipin polyspecific autoreactivity in two broadly neutralizing HIV-1 antibodies. *Science* **308:** 1906–1908.

Hessell AJ, Hangartner L, Hunter M, Havenith CE, Beurskens FJ, Bakker JM, Lanigan CM, Landucci G, Forthal DN, Parren PW, et al. 2007. Fc receptor but not complement binding is important in antibody protection against HIV. *Nature* **449:** 101–104.

Hessell AJ, Poignard P, Hunter M, Hangartner L, Tehrani DM, Bleeker WK, Parren PW, Marx PA, Burton DR. 2009a. Effective, low-titer antibody protection against low-dose repeated mucosal SHIV challenge in macaques. *Nat Med* **15:** 951–954.

Hessell AJ, Rakasz EG, Poignard P, Hangartner L, Landucci G, Forthal DN, Koff WC, Watkins DI, Burton DR. 2009b. Broadly neutralizing human anti-HIV antibody 2G12 is effective in protection against mucosal SHIV challenge even at low serum neutralizing titers. *PLoS Pathog* **5:** e1000433.

Hofmann-Lehmann R, Vlasak J, Rasmussen RA, Smith BA, Baba TW, Liska V, Ferrantelli F, Montefiori DC, McClure HM, Anderson DC, et al. 2001. Postnatal passive immunization of neonatal macaques with a triple combination of human monoclonal antibodies against oral simian-human immunodeficiency virus challenge. *J Virol* 75: 7470–7480.

Honnen WJ, Krachmarov C, Kayman SC, Gorny MK, Zolla-Pazner S, Pinter A. 2007. Type-specific epitopes targeted by monoclonal antibodies with exceptionally potent neutralizing activities for selected strains of human immunodeficiency virus type 1 map to a common region of the V2 domain of gp120 and differ only at single positions from the clade B consensus sequence. *J Virol* 81: 1424–1432.

Huang CC, Tang M, Zhang MY, Majeed S, Montabana E, Stanfield RL, Dimitrov DS, Korber B, Sodroski J, Wilson IA, et al. 2005. Structure of a V3-containing HIV-1 gp120 core. *Science* 310: 1025–1028.

Huang CC, Lam SN, Acharya P, Tang M, Xiang SH, Hussan SS, Stanfield RL, Robinson J, Sodroski J, Wilson IA, et al. 2007. Structures of the CCR5 N terminus and of a tyrosine-sulfated antibody with HIV-1 gp120 and CD4. *Science* 317: 1930–1934.

Julien JP, Huarte N, Maeso R, Taneva SG, Cunningham A, Nieva JL, Pai EF. 2010. Ablation of the complementarity-determining region H3 apex of the anti-HIV-1 broadly neutralizing antibody 2F5 abrogates neutralizing capacity without affecting core epitope binding. *J Virol* 84: 4136–4147.

Klein JS, Bjorkman PJ. 2010. Few and far between: How HIV may be evading antibody avidity. *PLoS Pathog* 6: e1000908.

Klein JS, Gnanapragasam PN, Galimidi RP, Foglesong CP, West AP Jr, Bjorkman PJ. 2009. Examination of the contributions of size and avidity to the neutralization mechanisms of the anti-HIV antibodies b12 and 4E10. *Proc Natl Acad Sci* 106: 7385–7390.

Korber B, Muldoon M, Theiler J, Gao F, Gupta R, Lapedes A, Hahn BH, Wolinsky S, Bhattacharya T. 2000. Timing the ancestor of the HIV-1 pandemic strains. *Science* 288: 1789–1796.

Kwong PD, Wyatt R, Robinson J, Sweet RW, Sodroski J, Hendrickson WA. 1998. Structure of an HIV gp120 envelope glycoprotein in complex with the CD4 receptor and a neutralizing human antibody. *Nature* 393: 648–659.

Labrijn AF, Poignard P, Raja A, Zwick MB, Delgado K, Franti M, Binley J, Vivona V, Grundner C, Huang CC, et al. 2003. Access of antibody molecules to the conserved co-receptor binding site on glycoprotein gp120 is sterically restricted on primary human immunodeficiency virus type 1. *J Virol* 77: 10557–10565.

Li Y, Migueles SA, Welcher B, Svehla K, Phogat A, Louder MK, Wu X, Shaw GM, Connors M, Wyatt RT, et al. 2007. Broad HIV-1 neutralization mediated by CD4-binding site antibodies. *Nat Med* 13: 1032–1034.

Liu J, Bartesaghi A, Borgnia MJ, Sapiro G, Subramaniam S. 2008. Molecular architecture of native HIV-1 gp120 trimers. *Nature* 455: 109–113.

Luftig MA, Mattu M, Di Giovine P, Geleziunas R, Hrin R, Barbato G, Bianchi E, Miller MD, Pessi A, Carfi A. 2006. Structural basis for HIV-1 neutralization by a gp41 fusion intermediate-directed antibody. *Nat Struct Mol Biol* 13: 740–747.

Mascola JR. 2003. Defining the protective antibody response for HIV-1. *Curr Mol Med* 3: 209–216.

Mascola JR, Montefiori DC. 2010. The role of antibodies in HIV vaccines. *Annu Rev Immunol* 28: 413–444.

Mascola JR, Snyder SW, Weislow OS, Belay SM, Belshe RB, Schwartz DH, Clements ML, Dolin R, Graham BS, Gorse GJ, et al. 1996. Immunization with envelope subunit vaccine products elicits neutralizing antibodies against laboratory-adapted but not primary isolates of human immunodeficiency virus type 1. *J Infect Dis* 173: 340–348.

Mascola JR, Lewis MG, Stiegler G, Harris D, VanCott TC, Dayes D, Louder MK, Brown CR, Sapan CV, Frankel SS, et al. 1999. Protection of macaques against pathogenic simian/human immunodeficiency virus 89.6PD by passive transfer of neutralizing antibodies. *J Virol* 73: 4009–4018.

Mascola JR, Stiegler G, VanCott TC, Katinger H, Carpenter CB, Hanson CE, Beary H, Hayes D, Frankel SS, Birx DL, et al. 2000. Protection of macaques against vaginal transmission of a pathogenic HIV-1/SIV chimeric virus by passive infusion of neutralizing antibodies. *Nat Med* 6: 207–210.

Montefiori DC, Zhou IY, Barnes B, Lake D, Hersh EM, Masuho Y, Lefkowitz LB Jr. 1991. Homotypic antibody responses to fresh clinical isolates of human immunodeficiency virus. *Virology* 182: 635–643.

Munoz-Barroso I, Salzwedel K, Hunter E, Blumenthal R. 1999. Role of the membrane-proximal domain in the initial stages of human immunodeficiency virus type 1 envelope glycoprotein-mediated membrane fusion. *J Virol* 73: 6089–6092.

Muster T, Steindl F, Purtscher M, Trkola A, Klima A, Himmler G, Ruker F, Katinger H. 1993. A conserved neutralizing epitope on gp41 of human immunodeficiency virus type 1. *J Virol* 67: 6642–6647.

Muster T, Guinea R, Trkola A, Purtscher M, Klima A, Steindl F, Palese P, Katinger H. 1994. Cross-neutralizing activity against divergent human immunodeficiency virus type 1 isolates induced by the gp41 sequence ELDKWAS. *J Virol* 68: 4031–4034.

Nabel GJ, Kwong PD, Mascola JR. 2011. Progress in the rational design of an AIDS vaccine. *Philos Trans R Soc Lond B Biol Sci* (in press).

Nishimura Y, Igarashi T, Haigwood N, Sadjadpour R, Plishka RJ, Buckler-White A, Shibata R, Martin MA. 2002. Determination of a statistically valid neutralization titer in plasma that confers protection against simian-human immunodeficiency virus challenge following passive transfer of high-titered neutralizing antibodies. *J Virol* 76: 2123–2130.

Ofek G, Tang M, Sambor A, Katinger H, Mascola JR, Wyatt R, Kwong PD. 2004. Structure and mechanistic analysis of the anti-human immunodeficiency virus type 1 antibody 2F5 in complex with its gp41 epitope. *J Virol* 78: 10724–10737.

Ofek G, Guenaga FJ, Schief WR, Skinner J, Baker D, Wyatt R, Kwong PD. 2010a. Elicitation of structure-specific antibodies by epitope scaffolds. *Proc Natl Acad Sci* 107: 17880–17887.

Ofek G, McKee K, Yang Y, Yang ZY, Skinner J, Guenaga FJ, Wyatt R, Zwick MB, Nabel GJ, Mascola JR, et al. 2010b. Relationship between antibody 2F5 neutralization of HIV-1 and hydrophobicity of its heavy chain third

 Cite this article as *Cold Spring Harb Perspect Med* doi: 10.1101/cshperspect.a007278

complementarity–determining region. *J Virol* **84**: 2955–2962.

Pancera M, Majeed S, Ban YE, Chen L, Huang CC, Kong L, Kwon YD, Stuckey J, Zhou T, Robinson JE, et al. 2010a. Structure of HIV-1 gp120 with gp41-interactive region reveals layered envelope architecture and basis of conformational mobility. *Proc Natl Acad Sci* **107**: 1166–1171.

Pancera M, McLellan JS, Wu X, Zhu J, Changela A, Schmidt SD, Yang Y, Zhou T, Phogat S, Mascola JR, et al. 2010b. Crystal structure of PG16 and chimeric dissection with somatically related PG9: structure–function analysis of two quaternary-specific antibodies that effectively neutralize HIV-1. *J Virol* **84**: 8098–8110.

Pantophlet R, Burton DR. 2006. GP120: Target for neutralizing HIV-1 antibodies. *Annu Rev Immunol* **24**: 739–769.

Parren PW, Marx PA, Hessell AJ, Luckay A, Harouse J, Cheng-Mayer C, Moore JP, Burton DR. 2001. Antibody protects macaques against vaginal challenge with a pathogenic R5 simian/human immunodeficiency virus at serum levels giving complete neutralization in vitro. *J Virol* **75**: 8340–8347.

Pejchal R, Walker LM, Stanfield RL, Phogat SK, Koff WC, Poignard P, Burton DR, Wilson IA. 2010. Structure and function of broadly reactive antibody PG16 reveal an H3 subdomain that mediates potent neutralization of HIV-1. *Proc Natl Acad Sci* **107**: 11483–11488.

Rerks-Ngarm S, Pitisuttithum P, Nitayaphan S, Kaewkungwal J, Chiu J, Paris R, Premsri N, Namwat C, de Souza M, Adams E, et al. 2009. Vaccination with ALVAC and AIDSVAX to prevent HIV-1 infection in Thailand. *N Engl J Med* **361**: 2209–2220.

Richman DD, Wrin T, Little SJ, Petropoulos CJ. 2003. Rapid evolution of the neutralizing antibody response to HIV type 1 infection. *Proc Natl Acad Sci* **100**: 4144–4149.

Rizzuto CD, Wyatt R, Hernández-Ramos N, Sun Y, Kwong PD, Hendrickson WA, Sodroski J. 1998. A conserved HIV gp120 glycoprotein structure involved in chemokine receptor binding. *Science* **280**: 1949–1953.

Robinson JE, Franco K, Elliott DH, Maher MJ, Reyna A, Montefiori DC, Zolla-Pazner S, Gorny MK, Kraft Z, Stamatatos L. 2010. Quaternary epitope specificities of anti-HIV-1 neutralizing antibodies generated in rhesus macaques infected by the simian/human immunodeficiency virus SHIVSF162P4. *J Virol* **84**: 3443–3453.

Sabin C, Corti D, Buzon V, Seaman MS, Lutje HD, Hinz A, Vanzetta F, Agatic G, Silacci C, Mainetti L, et al. 2010. Crystal structure and size-dependent neutralization properties of HK20, a human monoclonal antibody binding to the highly conserved heptad repeat 1 of gp41. *PLoS Pathog* **6**: e1001195.

Salzwedel K, West JT, Hunter E. 1999. A conserved tryptophan-rich motif in the membrane-proximal region of the human immunodeficiency virus type 1 gp41 ectodomain is important for Env-mediated fusion and virus infectivity. *J Virol* **73**: 2469–2480.

Sanders RW, Venturi M, Schiffner L, Kalyanaraman R, Katinger H, Lloyd KO, Kwong PD, Moore JP. 2002. The mannose-dependent epitope for neutralizing antibody 2G12 on human immunodeficiency virus type 1 glycoprotein gp120. *J Virol* **76**: 7293–7305.

Sather DN, Armann J, Ching LK, Mavrantoni A, Sellhorn G, Caldwell Z, Yu X, Wood B, Self S, Kalams S, et al. 2009.

Factors associated with the development of cross-reactive neutralizing antibodies during human immunodeficiency virus type 1 infection. *J Virol* **83**: 757–769.

Scheid JF, Mouquet H, Feldhahn N, Seaman MS, Velinzon K, Pietzsch J, Ott RG, Anthony RM, Zebroski H, Hurley A, et al. 2009. Broad diversity of neutralizing antibodies isolated from memory B cells in HIV-infected individuals. *Nature* **458**: 636–640.

Scherer EM, Leaman DP, Zwick MB, McMichael AJ, Burton DR. 2010. Aromatic residues at the edge of the antibody combining site facilitate viral glycoprotein recognition through membrane interactions. *Proc Natl Acad Sci* **107**: 1529–1534.

Shibata R, Igarashi T, Haigwood N, Buckler-White A, Ogert R, Ross W, Willey R, Cho MW, Martin MA. 1999. Neutralizing antibody directed against the HIV-1 envelope glycoprotein can completely block HIV-1/SIV chimeric virus infections of macaque monkeys. *Nat Med* **5**: 204–210.

Simek MD, Rida W, Priddy FH, Pung P, Carrow E, Laufer DS, Lehrman JK, Boaz M, Tarragona-Fiol T, Miiro G, et al. 2009. Human immunodeficiency virus type 1 elite neutralizers: Individuals with broad and potent neutralizing activity identified by using a high-throughput neutralization assay together with an analytical selection algorithm. *J Virol* **83**: 7337–7348.

Song L, Sun ZY, Coleman KE, Zwick MB, Gach JS, Wang JH, Reinherz EL, Wagner G, Kim M. 2009. Broadly neutralizing anti-HIV-1 antibodies disrupt a hinge-related function of gp41 at the membrane interface. *Proc Natl Acad Sci* **106**: 9057–9062.

Stamatatos L, Morris L, Burton DR, Mascola JR. 2009. Neutralizing antibodies generated during natural HIV-1 infection: Good news for an HIV-1 vaccine? *Nat Med* **15**: 866–870.

Stiegler G, Kunert R, Purtscher M, Wolbank S, Voglauer R, Steindl F, Katinger H. 2001. A potent cross-clade neutralizing human monoclonal antibody against a novel epitope on gp41 of human immunodeficiency virus type 1. *AIDS Res Hum Retroviruses* **17**: 1757–1765.

Sun ZY, Oh KJ, Kim M, Yu J, Brusic V, Song L, Qiao Z, Wang JH, Wagner G, Reinherz EL. 2008. HIV-1 broadly neutralizing antibody extracts its epitope from a kinked gp41 ectodomain region on the viral membrane. *Immunity* **28**: 52–63.

Trkola A, Pomales AB, Yuan H, Korber B, Maddon PJ, Allaway GP, Katinger H, Barbas CFIII, Burton DR, Ho DD. 1995. Cross-clade neutralization of primary isolates of human immunodeficiency virus type 1 by human monoclonal antibodies and tetrameric CD4-IgG. *J Virol* **69**: 6609–6617.

Walker LM, Phogat SK, Chan-Hui PY, Wagner D, Phung P, Goss JL, Wrin T, Simek MD, Fling S, Mitcham JL, et al. 2009. Broad and potent neutralizing antibodies from an African donor reveal a new HIV-1 vaccine target. *Science* **326**: 285–289.

Walker L, Chan-Hui P, Ramos A, Simek M, Falkowska E, Doores K, Hammond P, Wrin T, Mosley B, Olsen O, et al. 2010a. High through-put functional screening of activated B cells from 4 African elite neutralizers yields a panel of novel broadly neutralizing antibodies [abstract]. *AIDS Res Hum Retroviruses* **26**: A-149–A-150.

Walker LM, Simek MD, Priddy F, Gach JS, Wagner D, Zwick MB, Phogat SK, Poignard P, Burton DR. 2010b. A limited number of antibody specificities mediate broad and potent serum neutralization in selected HIV-1 infected individuals. *PLoS Pathog* **6:** e1001028.

Wei X, Decker JM, Wang S, Hui H, Kappes JC, Wu X, Salazar-Gonzalez JF, Salazar MG, Kilby JM, Saag MS, Komarova NL, Nowak MA, Hahn BH, Kwong PD, Shaw GM. 2003. Antibody neutralization and escape by HIV-1. *Nature* **422:** 307–312.

Weissenhorn W, Carfi A, Lee KH, Skehel JJ, Wiley DC. 1998. Crystal structure of the Ebola virus membrane fusion subunit, GP2, from the envelope glycoprotein ectodomain. *Mol Cell* **2:** 605–616.

Wrin T, Nunberg JH. 1994. HIV-1MN recombinant gp120 vaccine serum, which fails to neutralize primary isolates of HIV-1, does not antagonize neutralization by antibodies from infected individuals. *AIDS* **8:** 1622–1623.

Wu L, Gerard NP, Wyatt R, Choe H, Parolin C, Ruffing N, Borsetti A, Cardoso AA, Desjardin E, Newman W, et al. 1996. CD4-induced interaction of primary HIV-1 gp120 glycoproteins with the chemokine receptor CCR-5. *Nature* **384:** 179–183.

Wu L, Zhou T, Yang ZY, Svehla K, O'Dell S, Louder MK, Xu L, Mascola JR, Burton DR, Hoxie JA, et al. 2009a. Enhanced exposure of the CD4-binding site to neutralizing antibodies by structural design of a membrane-anchored human immunodeficiency virus type 1 gp120 domain. *J Virol* **83:** 5077–5086.

Wu X, Zhou T, O'Dell S, Wyatt RT, Kwong PD, Mascola JR. 2009b. Mechanism of human immunodeficiency virus type 1 resistance to monoclonal antibody b12 that effectively targets the site of CD4 attachment. *J Virol* **83:** 10892–10907.

Wu X, Yang ZY, Li Y, Hogerkorp CM, Schief WR, Seaman MS, Zhou T, Schmidt SD, Wu L, Xu L, et al. 2010. Rational design of envelope identifies broadly neutralizing human monoclonal antibodies to HIV-1. *Science* **329:** 856–861.

Wu X, Changela A, O'Dell S, Schmidt SD, Pancera M, Yang Y, Zhang B, Gorny MK, Phogat S, Robinson JE, Stamatatos L, Zolla-Pazner S, Kwong PD, Mascola JR. 2011. Immunotypes of a quaternary site of HIV-1 vulnerability and their recognition by antibodies. *J Virol* **85:** 4578–4585.

Wyatt R, Sodroski J. 1998. The HIV-1 envelope glycoproteins: Fusogens, antigens, and immunogens. *Science* **280:** 1884–1888.

Wyatt R, Kwong PD, Desjardins E, Sweet RW, Robinson J, Hendrickson WA, Sodroski JG. 1998. The antigenic structure of the HIV gp120 envelope glycoprotein. *Nature* **393:** 705–711.

Xiang SH, Wang L, Abreu M, Huang CC, Kwong PD, Rosenberg E, Robinson JE, Sodroski J. 2003. Epitope mapping and characterization of a novel CD4-induced human monoclonal antibody capable of neutralizing primary HIV-1 strains. *Virology* **315:** 124–134.

Yang X, Lee J, Mahony EM, Kwong PD, Wyatt R, Sodroski J. 2002. Highly stable trimers formed by human immunodeficiency virus type 1 envelope glycoproteins fused with the trimeric motif of T4 bacteriophage fibritin. *J Virol* **76:** 4634–4642.

Yang X, Tomov V, Kurteva S, Wang L, Ren X, Gorny MK, Zolla-Pazner S, Sodroski J. 2004. Characterization of the outer domain of the gp120 glycoprotein from human immunodeficiency virus type 1. *J Virol* **78:** 12975–12986.

Zhou T, Xu L, Dey B, Hessell AJ, Van Ryk D, Xiang SH, Yang X, Zhang MY, Zwick MB, Arthos J, et al. 2007. Structural definition of a conserved neutralization epitope on HIV-1 gp120. *Nature* **445:** 732–737.

Zhou T, Georgiev I, Wu X, Yang ZY, Dai K, Finzi A, Kwon YD, Scheid JF, Shi W, Xu L, et al. 2010. Structural basis for broad and potent neutralization of HIV-1 by antibody VRC01. *Science* **329:** 811–817.

Zolla-Pazner S. 2004. Identifying epitopes of HIV-1 that induce protective antibodies. *Nat Rev Immunol* **4:** 199–210.

Zolla-Pazner S, Cardozo T. 2010. Structure–function relationships of HIV-1 envelope sequence-variable regions refocus vaccine design. *Nat Rev Immunol* **10:** 527–535.

Zolla-Pazner S, Cohen SS, Krachmarov C, Wang S, Pinter A, Lu S. 2008. Focusing the immune response on the V3 loop, a neutralizing epitope of the HIV-1 gp120 envelope. *Virology* **372:** 233–246.

Zwick MB, Labrijn AF, Wang M, Spenlehauer C, Saphire EO, Binley JM, Moore JP, Stiegler G, Katinger H, Burton DR, et al. 2001. Broadly neutralizing antibodies targeted to the membrane-proximal external region of human immunodeficiency virus type 1 glycoprotein gp41. *J Virol* **75:** 10892–10905.

Lessons in Nonhuman Primate Models for AIDS Vaccine Research: From Minefields to Milestones

Jeffrey D. Lifson[1] and Nancy L. Haigwood[2]

[1]AIDS and Cancer Virus Program, SAIC Frederick, Inc., National Cancer Institute, Frederick, Maryland 21702

[2]Oregon National Primate Research Center and Vaccine & Gene Therapy Institute, Oregon Health & Science University, Beaverton, Oregon 97006

Correspondence: haigwoon@ohsu.edu

Nonhuman primate (NHP) disease models for AIDS have made important contributions to the search for effective vaccines for AIDS. Viral diversity, persistence, capacity for immune evasion, and safety considerations have limited development of conventional approaches using killed or attenuated vaccines, necessitating the development of novel approaches. Here we highlight the knowledge gained and lessons learned in testing vaccine concepts in different virus/NHP host combinations.

In the early years of the AIDS pandemic, the search for an animal model for HIV-1 infection focused on experimental infection of chimpanzees (*Pan troglodytes*), resulting in productive infections and new information about transmission (Fultz et al. 1986a,b). However, disease in chimpanzees occurred rarely and only after >10 years of infection (Novembre et al. 1997). During the early 1980s, outbreaks of immunodeficiency-associated diseases occurred in Asian macaques (*Macaca* species; Figs. 1 and 2) at multiple primate centers. The animals succumbed to neoplasms and opportunistic infections, paralleling the newly described human disease now known as AIDS (Gottlieb et al. 1981; Masur et al. 1981; Siegal et al. 1981). Whereas some cases were associated with a D-type retrovirus (Daniel et al. 1984; Marx et al. 1984; Stromberg et al. 1984), others

were linked to novel simian lentiviruses (Daniel et al. 1985; Letvin et al. 1985; Benveniste et al. 1986; Murphey-Corb et al. 1986) related to the newly discovered etiologic agent for human AIDS (Barre-Sinoussi et al. 1983; Gallo et al. 1984; Popovic et al. 1984). The source of the lentiviral infections in Asian macaques was cross-species transmission via documented or presumed exposure in captivity to African nonhuman primates (NHPs) infected in the wild (Apetrei et al. 2005).

Disease in Asian macaques—*M. mulatta* (*rhesus*), *M. nemestrina* (*pigtailed*), and *M. fascicularis* (*cynomolgus*)—had striking similarities with human AIDS, including acute and then progressive loss of CD4$^+$ T cells followed by clinical immunodeficiency, opportunistic infections, and neoplasms (reviewed in Hirsch and Lifson 2000). The causative viruses—simian

Figure 1. Many monkeys, more viruses; right choice brings truth—wrong choice ... delusion. (Original artwork by Joel Ito, Oregon National Primate Research Center.)

immunodeficiency viruses (SIVs)—were shown to transmit the infection and reproduce the characteristic disease course. The molecular cloning of these SIVs allowed the development of chimeric viral constructs with HIV genes

spliced into SIV backbones, termed simian-human immunodeficiency viruses (SHIVs; Li et al. 1992) intended to address certain questions in NHP models. SHIVs bearing HIV Envelope protein (Env) allowed testing of HIV Env-based vaccines or other Env-targeted interventions. The ability to infect NHP with a known amount of a characterized SIV or SHIV stock, via a defined route, and to obtain blood and tissue samples at specific times after inoculation represented powerful experimental advantages. The recapitulation of key aspects of the pathogenesis of human HIV-1 infection, combined with the compressed time course of disease progression, and the ability to perform interventions make experimental SIV or SHIV infection a valuable tool for addressing many important questions in AIDS research.

This work focuses on the AIDS vaccine research using NHP models that has guided vaccine development. NHP models allow analyses of (1) the transmission of viruses across mucosal barriers, (2) the definition of relevant challenges via different routes, (3) the timing of viral dissemination, (4) the establishment

Figure 2. Representative image of *Macaca mulatta*. (Original artwork courtesy of Joel Ito, Oregon National Primate Research Center.)

of viral reservoirs in the newly infected host, (5) the quality and magnitude of innate and adaptive immune responses, and (6) demonstration of the protective effects of passive antibody. Comparative immunogenicity and challenge studies have allowed comparisons of vaccines. The outcomes of selected vaccine trials with NHP are summarized in Table 1. NHP models have provided invaluable information for HIV vaccine development; this work aims to provide an understanding of the utility and the limits of this information, critical to interpretation of the results and effective utilization of the models.

NHP Models

The first described SIVs were isolated from infected rhesus or pigtailed macaques (Daniel et al. 1985; Benveniste et al. 1986; Murphey-Corb et al. 1986). Following these studies, additional SIVs were identified and transmitted to other macaque species (Apetrei et al. 2004; Gautam et al. 2009; Souquiere et al. 2009). In addition, molecular chimeras of these viruses have been engineered, allowing multiple viruses with diverse properties to be used in different species and subspecies of macaque. Each combination of virus and route of administration in a different macaque species arguably constitutes a distinct model. Some viruses are molecular clones, some are "swarms" or complex quasispecies mixtures of related viruses, the exact composition and diversity of which may vary with the passage history and production method used, even for different virus preparations designated by the same isolate nomenclature. Each of these individual models has different strengths and limitations. One of the key lessons from the experience of using NHP models for AIDS research is the importance of choosing a model that is appropriate to the question posed. This issue has been at the heart of controversies surrounding the interpretation of vaccine experiments in NHP. This work highlights the key features of some different experimental models and points out areas in which there is room for continued development.

Advances in Understanding Mucosal Transmission in NHP

Most HIV infections involve sexual transmission via mucosal routes. Whereas initial experiments in macaques and chimpanzees used intravenous inoculation, subsequent studies modeled mucosal transmission via rectal (Keele et al. 2009), vaginal (Stone et al. 2010), and penile routes (Ma et al. 2011) or oropharyngeal transmission in neonates (Abel et al. 2006). In the absence of overt mucosal lesions, attempted mucosal transmission with infected cells rather than cell-free virus has been unsuccessful (Sodora et al. 1998), although a recent report claimed successful vaginal transmission using infected cells (Salle et al. 2010). Mechanistic details of mucosal transmission, including how virus applied to a mucosal surface reaches initial target cells, the identity of these initial target cells, and the mechanisms of local amplification at mucosal sites and pathways of dissemination leading to systemic infection remain only partially understood and somewhat controversial, but NHP models allow experimental investigation (Haase 2010). Whereas Langerhans cells can take up virus from vaginal inocula (Miller and Hu 1999), $CD4^+$ T cells are the first cells productively infected at mucosal sites (Li et al. 2009b). Far from being simply a physical barrier, the female genital mucosa is a complex, dynamic tissue in which mere exposure to viral inocula can trigger complex responses, including innate immune responses with elements that both hinder and facilitate the establishment of infection, in part by recruiting additional target cells (Li et al. 2009a). Infection can occur across either the vaginal or cervical mucosae, influenced by the stage of menstrual cycle and vaginal epithelial thickness, with sites of preexisting inflammation predisposing to local infection. Whereas local innate and adaptive immune responses influence mucosal infection, it appears that in both naïve and suboptimally vaccinated hosts, the virus-specific mucosal T-cell responses are too little, and too late, to significantly alter the course of infection (Reynolds et al. 2005). Understanding how deposition of

Table 1. Selected vaccine trials in NHPs and their outcomes

Vaccine	Species	Challenge virus	Route[a]	Outcome	Reference(s)
A. Protein only vaccines					
Gp120 protein (HIV-IIIB)	*Pan troglodytes*	HIV-1 HXB2 (IIIB)	IV	2/2 protected	Berman et al. 1990
Gp120 protein (HIV-SF2)	*P. troglodytes*	HIV-SF2	IV	1/2 infected virus control in 1/2?	el-Amad et al. 1995
Gp130 protein (SIVmac1A11)	*Macaca mulatta*	SIVmac251	IV	0/4 protected	Giavedoni et al. 1993
B. Vaccinia (Vac) prime, subunit boost vaccines					
Vac-Env gp160 (SIVmne) prime + gp160 protein (SIVmne) boost	*Macaca nemestrina*	SIVmne clone	IV	4/4 protected from infection	Hu et al. 1992
Vac-Env gp160 (SIVmne) prime + gp160 protein (SIVmne) boost	*M. nemestrina*	SIVmne clone or swarm	IR	4/4 protected from swarm by IR	Polacino et al. 1999a
Vac-Env + Vac-Gag/Pol + Gag/Pol/Env particles; Vac-Gag/Pol + Gag/Pol particles	*M. nemestrina*	SIVmne clone or swarm	IV	Only Gag/Pol/Env (gp160) was fully protective	Polacino et al. 1999b
Vac-Env gp160 or gp130 (SIVmac239) + gp160 or gp130 protein (SIVmac239)	*M. mulatta*	SIVmac251	IV	0/4 protected; virus measured by p27 lower in virus primed	Giavedoni et al. 1993; Ahmad et al. 1994
[Vac-Env gp160 (SIVmne) + vac-Gag (SIVmne)] prime + [VLPs (SIVmne) boost or DNA boost with all SHIV genes]	*M. nemestrina*	SHIV-89.6P	IR	Lower PVL and protection from CD4 loss in DNA + Vaccinia groups	Doria-Rose et al. 2003
C. Live attenuated SIV					
SIVmac239Δnef	*M. mulatta*	SIVmac251	IV	4/4 protected	Daniel et al. 1992
SIVmac239Δnef	*M. mulatta*	SIVsmE660	IV	10/10 infected; 2 $\log_{10}$ difference in PVL	Reynolds et al. 2005
SIVmac239Δnef, -vpr, NRE	*M. mulatta* newborns	SIVmac251	None	All rapidly developed AIDS	Baba et al. 1995
D. Adenovirus vaccines with and without boosting					
Ad5 Gag (SIVmac239) (replication defective) alone or primed with DNA	*M. mulatta*	SHIV-89.6P	IR	Reduction in viremia (2 $\log_{10}$) and protection from CD4 loss	Shiver et al. 2002
DNA prime + Ad5 Gag (SIVmac239) (replication incompetent)	*M. mulatta*	SIVmac239	IR	Temporary reduction in early viremia (0.5 $\log_{10}$)	Casimiro et al. 2010

[a]IV, intravenous inoculation; IR, atraumatic intrarectal inoculation.

Cite this article as *Cold Spring Harb Perspect Med* doi: 10.1101/cshperspect.a007310

a viral inoculum on a mucosal surface leads to a rampant, disseminated, systemic infection of lymphoid tissues within 10–14 days is a critical area of active research.

Early NHP mucosal transmission studies used large inocula to maximize the chances of infection. More recently, researchers have challenged using repeated exposure with lower titered inocula, either through a specified number of challenges, or until all control animals become infected. Such studies are more complex and resource intensive but better represent typical human mucosal exposures. In the majority of human heterosexual mucosal HIV infections, the initial disseminated systemic infection is established by a single viral variant 80% of the time, indicative of infection with a single viral particle, with only a few variants typically involved in the remaining cases (Keele et al. 2008). This information stimulated efforts to improve macaque mucosal transmission models by characterizing the genetic diversity of challenge stocks, the identity of variants present in challenge stocks, and the number of variants involved in establishing new infections in naïve hosts by different routes and doses to best mimic HIV mucosal transmission (Keele et al. 2009; Liu et al. 2010; Stone et al. 2010; Ma et al. 2011).

A Concise History of NHP Vaccine Approaches

Following the cloning of the HIV-1 genome (Alizon et al. 1984; Hahn et al. 1984; Luciw et al. 1984), investigators were quick to design recombinant protein vaccines intended to elicit responses to Env, since it was shown that neutralizing antibodies (NAbs), which can block virus infection in vitro, are directed exclusively to Env. Recombinant Envs were produced in diverse expression systems and shown to elicit NAbs against laboratory-adapted HIV-1 isolates (Lasky et al. 1986; Arthur et al. 1987; Palker et al. 1988). Broader responses were associated with Envs produced in mammalian cells (Haigwood et al. 1992), and the glycoprotein was shown to bind antibodies from HIV-positive serum that could neutralize divergent strains

of HIV-1 (Steimer et al. 1991). However, the discovery that primary HIV-1 isolates were far more difficult to neutralize than laboratory-adapted stocks caused significant concern (Berman et al. 1992; Matthews 1994). Mammalian cell-produced monomeric gp120 protected a handful of chimpanzees from infection with one lab-adapted HIV-1 strain (Berman et al. 1990), although only partially protecting against another such strain (el-Amad et al. 1995). A SIV Env gp130 monomer provided only a hint of control and no protection against SIVmac251 infection in rhesus macaques (Ahmad et al. 1994), and did not protect in another, heterologous SIV challenge system (Stott et al. 1998).

We now know that Env is a highly variable glycosylated protein trimer that efficiently disguises its conserved domains, which are only transiently exposed during binding to CD4 and the chemokine coreceptors in the process of membrane fusion. It is thus not surprising that even mammalian cell-produced recombinant Env proteins do not readily display the determinants required for the generation of NAbs that can bind native Env trimers on virions to block infection effectively. Approaches to modifying Env for better presentation of key neutralization determinants have been reviewed (Haynes and Montefiori 2006). Recombinant protein vaccines typically show partial protection from SHIV challenge (Earl et al. 2001), but not from SIV challenge, thus fueling the debate about the relative merits of SHIV versus SIV challenge models. A major conclusion from the early Env vaccine work was that clinically relevant NAbs, although likely to be important in a protective vaccine, were difficult to generate and maintain, likely requiring more authentic trimeric forms of Env.

Because most licensed viral vaccines are attenuated or inactivated versions of the pathogen, efforts subsequently focused on these approaches for HIV-1, recognizing the safety issues inherent in an attenuated HIV. Live attenuated vaccines (LAV) were pursued in SIV models, with impressive protection against high-dose, highly pathogenic, intravenous challenge (Daniel et al. 1992). Progression to AIDS in some LAV-immunized newborn macaques underscored safety concerns (Baba et al. 1995) even as other studies demonstrated an inverse relationship between extent of attenuation and protective efficacy (Wyand et al. 1996). However, the underlying mechanism(s) accounting for LAV protection remain unclear and a topic of continuing investigation (Wyand et al. 1999; Koff et al. 2006). The contribution of persistent low-level viral replication and associated ongoing immune stimulation to the protection afforded by live attenuated SIV vaccines is an important question that remains to be clarified.

The other traditional approach to vaccination, using inactivated virions or infected cells as immunogens, protected macaques against challenge with homologous SIV (Desrosiers et al. 1989; Murphey-Corb et al. 1989) or heterologous SIV (Johnson et al. 1992). However, follow-up experiments showed the mechanism for protection in these early studies was xenoreactivity to HLA antigens in the immunogens produced from human cells. Protection was seen only when matched HLA antigens were present on the SIV challenge virus. Indeed, immunization with purified HLA antigens alone conferred such protection, when the challenge virus was grown in HLA matched cells (Arthur et al. 1995). Studies controlling for confounding effects of host cell proteins incorporated into virions demonstrated that conventional inactivation methods can destroy the conformation of envelope glycoproteins, resulting in inferior immunogens (Cranage et al. 1995). Novel approaches that inactivate retroviral infectivity by preferential covalent modification of internal virion proteins required for infection while not affecting envelope glycoproteins on the virion surface (Arthur et al. 1998; Rossio et al. 1998) provided noninfectious virions with native envelope glycoproteins for use as immunogens, alone or in prime boost regimens, conferring partial protection against SIV or SHIV challenge (Lifson et al. 2002, 2004). Further developments in this "native whole virion immunogen" strategy include vaccination with proviral genomes that can produce such particles in vivo (Wang et al. 2000) and "single cycle" virus vaccines

that infected target cells, but produce noninfectious progeny virions and can reduce virus loads after SIV challenge (Evans et al. 2005; Jia et al. 2009). Although not fully effective alone against high-dose SIV challenge, these one-round vaccines showed significant improvement against repeated low-dose challenge (Alpert et al. 2010). Vaccines that present authentic virion structures in vivo may be useful as boosting agents and continue to be explored (Poon et al. 2005).

Subsequent efforts aimed at developing recombinant viral vectors expressing HIV or SIV genes that are targets of cytotoxic T cells, in an approach aimed at controlling rather than preventing infection. Prime boost or combination approaches employed more than one means of antigen presentation and showed better protection in macaques after SIV challenge compared with recombinant protein-only vaccines. Vaccinia emerged early as a preferred recombinant viral vector, eliciting strong T-cell responses to Env in humans. Responses were higher in vaccinia-naïve individuals (Cooney et al. 1991) and were boosted by the addition of recombinant protein to the regimen (Cooney et al. 1993). Because of safety concerns with replication competent vaccinia strains, increased emphasis was placed on more attenuated variants such as Modified Vaccinia Ankara (MVA), which were immunogenic in macaques (Barouch et al. 2001b). Fowlpox vectors such as ALVAC were used to evade preexisting immunity to vaccinia, but by themselves showed only modest activity in macaques (Pal et al. 2002). Once again, there were conflicting data with nominally the same approach tested in different SIV/macaque models. Env gp160-based vaccinia virus prime and subunit boost immunization provided "sterilizing" immunity in *M. nemestrina* challenged intravenously with SIVmne (Hu et al. 1992). However, whereas similar vaccines based on SIVmac239 were immunogenic, they failed to protect in *M. mulatta* challenged with the related virus swarm SIVmac251 (Ahmad et al. 1994; Table 1). Differential outcomes like these were some of the factors that led some investigators to attempt to standardize studies by utilizing a common species, the rhesus macaque, and a common SIV challenge virus, SIVmac239, and to develop a common SHIV challenge, SHIV-89.6P (Uberla 2005). This approach facilitated comparisons between studies, but adoption of a limited number of standardized models, without a compelling rationale for their superiority or relevance also carried significant risk.

In 1993, DNA vaccination emerged as a promising new tool, utilizing mammalian expression vectors as vaccines with impressive results in mice (Ulmer et al. 1993). However, such vaccines proved poorly immunogenic in macaques (Barouch et al. 2001a). The addition of cytokine genes to the DNA vaccines increased responses (Barouch et al. 2000, 2002), and the use of electroporation to enhance DNA uptake dramatically improved immunogenicity (Otten et al. 2004). A multitude of combination, or prime boost, experiments were performed with SIV or SHIV challenge to explore the DNA vaccines in combination with poxvirus vectors and proteins (Pal et al. 2006), with varying degrees of success (Doria-Rose et al. 2003; Dale et al. 2004; Mossman et al. 2004; Rosati et al. 2005; summarized in Table 1).

Interest in cellular immunity increased following the negative clinical trial results with the antibody targeting VaxGen vaccine and findings suggesting that T cells might be responsible for protection in multiply exposed yet uninfected sex workers in Africa (Rowland-Jones et al. 1998; Kaul et al. 2001). The daunting challenges in developing immunogens capable of inducing broadly neutralizing antibodies also contributed to the shift in emphasis. The development of recombinant adenovirus vectors to induce/enhance T-cell responses, when used alone or in combination with DNA or proteins, was vigorously pursued by a number of groups due to their impressive immunogenicity in model systems including NHP (Barouch and Nabel 2005; Robert-Guroff 2007).

Significant viral control after intravenous challenge with SHIV-89.6P was observed following vaccination with Ad5-Gag(SIV) when used alone or as a boost to a DNA prime (Shiver et al. 2002). These findings were used to support the approach of the STEP clinical trial, although

the vaccine components and immunization regimens were not an exact match. Subsequently, a similar vaccine experiment using DNA prime/Ad5-Gag(SIV) boost was performed in NHP using SIVmac239 as the challenge virus. The effects on viremia in this experiment were modest and transient, and limited to the subset of animals expressing the MHC Class I allele MamuA*01, with evidence for viral escape at 6 months postchallenge (Casimiro et al. 2005). A very recent NHP study explicitly designed to simulate the STEP trial as closely as possible yielded negative results, matching the clinical results (D Watkins, pers. comm.).

The results of the STEP trial forced the field to take a hard look at the T-cell-only vaccine hypothesis, adenovirus vectors, and the NHP results (Watkins et al. 2008). Replication of competent adenovirus vectors had been developed in parallel with the Ad5 vectors. Although DNA/MVA vaccines had shown robust virus control after SHIV-89.6P challenge (Amara et al. 2001), a vaccine based on DNA priming (with or without cytokines IL-12 or IL-15), replicating Ad5-SIV boost, and a recombinant gp140 protein plus SIV Nef, showed essentially no virus control after SIVmac251 challenge (Demberg et al. 2008). These results suggest that comparative studies in different models are not only important, but that they remain critical until one or more of the models is validated as convincingly replicating vaccine efficacy demonstrated in human trials. Importantly, different NHP models may be better suited for evaluation of different vaccine approaches.

Despite the generation of adenovirus-specific immunity, which limited the use of the vector to one or two immunizations, results using Ad5 were encouraging with respect to the ability to generate strong T-cell responses (Santra et al. 2005), and in combination with DNA, strong B- and T-cell responses (Seaman et al. 2005) in macaques. The vectors were also immunogenic in humans in Phase I trials, although human responses were not as robust as those in macaque, and limited by preexisting antivector immunity (Koup et al. 2010). Combination vaccine experiments to test Ad vectors with other vectors continue, although the sobering lack of efficacy in the STEP trial and relatively wide-spread preexisting immunity to Ad5 (O'Brien et al. 2009) have diminished enthusiasm for this vector. To circumvent preexisting antivector immunity issues, efforts are currently directed to developing vaccine vectors based on rare human (Ko et al. 2009; McVey et al. 2010) or chimpanzee adenovirus serotypes.

The conventional anamnestic expansion-dependent central memory T-cell type of vaccine responses induced by these vaccines typically were unable to provide truly effective, sustained control of viral replication (Reynolds et al. 2005). To test the hypothesis that a qualitatively different type of immune response might be more effective, persistent rhesus cytomegalovirus-based vectors expressing most of the SIV genome were explored and elicited strong, broad, persistent CD4+ and CD8+ effector memory T-cell responses that were associated with stringent control of replication following repeated titered intrarectal challenge with SIV-mac239 (Hansen et al. 2009, 2011). If these results can be translated by developing safe human cytomegalovirus vectors, or other vectors that generate persistent effector T-cell responses, this could be a major advance toward developing an effective vaccine.

IMMUNE RESPONSES

Antiviral immune responses operative in primate lentivirus infection can be considered as (1) intrinsic (Malim and Bieniasz 2011), (2) innate (Carrington and Alter 2011), or (3) adaptive, comprising both humoral (Overbaugh and Morris 2011) and cellular (Walker and McMichael 2011) immunity. For each response, there is evidence that lentiviruses evolve countermeasures to overcome host mechanisms. These responses in NHP are discussed in turn below.

"Intrinsic" Immune Responses

The intrinsic immune responses that may restrict lentiviral infection include the interferon inducible APOBEC3 protein system (Simon et al. 2005; Goila-Gaur and Strebel 2008) and tetherin (CD317, BST-2 [Neil et al. 2008; Jia et al. 2009; McNatt et al. 2009; Zhang et al. 2009]), which are covered by Malim and Bieniasz (2011). Another intrinsic immune mechanism particularly relevant to vaccines studies is the TRIM5α protein, which interferes with primate lentiviral replication after entry (Stremlau et al. 2004). Importantly, genetic polymorphisms have recently been described in the TRIM5α sequences of rhesus macaques that represent a key new variable to control in future NHP studies (Newman et al. 2006; Newman and Johnson 2007; Kirmaier et al. 2010; Lim et al. 2010). Depending on the sequence of the target SIV capsid protein, TRIM5α polymorphisms can have profound influence on permissiveness for viral replication. The sequence of SIVmac239 appears to be refractory to these effects, whereas those of SIVsmE543 and SIVsmE040 appear to be quite sensitive. Other viruses and viral swarms, including SIVmac251 and particularly SIVsmE660, may show an intermediate range of sensitivities, perhaps reflecting a mixture of sensitive and resistant capsid sequences in the viral stocks. Because these polymorphisms can exert profound effects on viral replication for susceptible viruses, studies conducted with such viruses must be stringently controlled for TRIM5α genotypes, much like controlling for MHC I alleles associated with spontaneous control of viral replication (Goulder and Watkins 2008). If not, control of viral replication could be erroneously attributed to an experimental intervention rather than the uneven distribution of the TRIM5α genotypes among groups.

Innate Immune Responses

In NHP models, innate immune responses can be broadly considered as those associated with (1) cells that secrete soluble immune active mediators or (2) cells that are either dedicated components of the innate immune system (e.g., NK cells) or that bridge the innate and adaptive immune systems (dendritic cells). The complexity of these interactions is exemplified by the response to vaginal inoculation of rhesus macaques with SIV. In response to such inoculations, vaginal epithelial cells produce chemokines, including CCL20, that attract plasmacytoid dendritic cells (PDC) and T cells (Li et al. 2009a). The PDCs produce abundant amounts of the antiviral cytokine interferon-α. However, the net effect of the cytokine responses is to facilitate viral replication by recruiting activated CD4$^+$ T cells to the site of local inflammation, providing additional targets for infection. This local amplification at mucosal portals of viral entry may facilitate systemic dissemination and is a potential target for intervention (Li et al. 2009a; Haase 2010). There are important differences in NK cell populations between humans and different NHP species (for NK cell function, see Reeves et al. 2010; Siliciano and Greene 2011 and references). There are also differences in the numbers and apparent trafficking of these NK populations during primary and chronic SIV infection. A role for NK cells in controlling viral replication in SIV infection remains to be convincingly demonstrated, although this is well established for HIV in humans by genetic association studies demonstrating that pairing of certain KIR genotypes and MHC I alleles can affect disease progression (Martin et al. 2002, 2007; Martin and Carrington 2005; Qi et al. 2006). Current efforts to apply pyrosequencing methods to characterize these loci in experimentally important NHP species should help to determine whether such associations exist for infected NHP (Wiseman et al. 2009).

Adaptive Immune Responses

The MHC loci of monkeys and humans differ. Among the NHP species commonly used for AIDS research, the MHC of the Indian rhesus macaque has been most extensively studied (Wiseman et al. 2009). In contrast to humans, rhesus macaques lack a MHC I C locus, but have greater numbers of alleles for their A and B loci than do humans (Bontrop and Watkins 2005; Goulder and Watkins 2008). The full impact of this difference is incompletely understood.

B-CELL RESPONSES

The daunting diversity of HIV presents a challenge for humoral immune protection. For highly variable pathogens like HIV, broadly

neutralizing antibodies directed to native Env oligomers on the virion surface block infection in vitro and can protect macaques from infection in passive transfer studies but are difficult to raise by vaccination. Depletion of B cells using anti-CD20 antibodies in NHP models has been pivotal in demonstrating the role of antibodies in limiting infection in vivo. Even when antibodies failed to fully block infection, some clinical benefits accrued (summarized below). When given prophylactically, high doses of both polyclonal and monoclonal NAbs can prevent or limit infection (Haigwood and Stamatatos 2003; Mascola 2003). The effectiveness of prophylactic passive IgG or mAbs was observed in several primate HIV, SIV, and SHIV studies (Prince et al. 1991; Putkonen et al. 1991; Conley et al. 1996; Shibata et al. 1999) and SCID-hu mouse models (Gauduin et al. 1995, 1997). These animal models have allowed assessment of the importance of dose, timing, and specificity of the antibody preparations. NAbs are effective in preventing the establishment of infection in vivo when present at high concentrations at the time of viral challenge or a few hours later, including challenge at mucosal surfaces (Baba et al. 2000; Mascola et al. 2000). Protection of newborn macaques with passively transferred serum containing SIV NAbs was fully effective in blocking oral infection (Van Rompay et al. 1998). In newborn macaques exposed orally to SHIV-SF162p3, substerilizing levels of IgG matched to the challenge virus resulted in better virus control and rapid development of de novo NAbs (Ng et al. 2010). Vaccinated macaque dams transferred anti-SIV antibodies to their infants, which were subsequently protected from oral/conjunctival challenge (Van Rompay et al. 2003). Whereas HIV immune globulin (HIVIG) IgG can directly limit the infectivity of HIV-1 and SHIV in vivo (Igarashi et al. 1999), non-neutralizing IgG was not effective, implicating NAbs as the active component in polyclonal HIVIG. Challenge studies in macaques with the R5-tropic SHIV$_{DH12}$, using HIVIG purified from chimpanzee-derived IgG, were matched for HIV$_{DH12}$ and protected only at doses of HIVIG that neutralized virus to 100% in vitro, in the range of 200 mg/kg (Shibata et al. 1999).

A summary of many of the passive immunization studies using SHIV challenge is shown (Table 2), with testing of both polyclonal serum (HIVIG from HIV+ sera and CHIVIG from the sera of infected chimpanzees) as well as human neutralizing mAbs. All of these Ab preparations targeted native Env. At least four lessons have been learned: (1) Antibodies protect in a dose-dependent manner, with higher doses protecting more animals; (2) relatively high doses of mAbs, or mAbs plus polyclonal HIVIG or CHIVIG, are needed for protection against viral challenge; (3) lower levels of antibodies are better able to protect when the challenge dose is reduced; and (4) intravenously administered IgG can protect against mucosal challenge, presumably through transudation. In SIV models, NAbs delivered within the first 24 hours of infection can block infection or, at lower doses, affect control of viral replication and the development of de novo responses (Haigwood et al. 2004). Passive immunization with NAbs has progressed to proof of concept studies in which a NAb is expressed from a transgene delivered by recombinant adenovirus-associated virus (AAV), resulting in sustained high plasma NAb levels in vivo and protection from infection (Johnson et al. 2009). Related approaches are under consideration for feasibility studies in humans.

Although NAb function has correlated with protection in these studies, other antibody-mediated effector functions, such as antibody-dependent cellular cytotoxicity (ADCC) or antibody-dependent cell-mediated viral inhibition (ADCVI), might also be critical in vivo. ADCC, measuring the ability of antibody and Fcγ receptor (FcγR)-bearing effector cells to kill target cells expressing HIV antigens, correlated inversely with disease progression in the Multicenter AIDS Cohort Study (Baum et al. 1996). Non-neutralizing IgG with ADCC activity was ineffective in blocking SIV infection in newborn macaques (Florese et al. 2009). ADCVI, which occurs when antibody forms a bridge between an infected target cell and FcγR-bearing effector cell, thereby limiting viral

production, can also limit HIV, SIV, and SHIV infection by direct killing. During acute HIV infection, ADCVI antibodies develop weeks to months earlier than do NAbs and can inhibit both autologous and heterologous strains of HIV-1. Moreover, there is an inverse correlation between ADCVI antibody activity and plasma viremia during the acute phase of HIV infection (Forthal et al. 2001). Serum from rhesus macaques that does not neutralize PBMC-passaged, uncloned SIV_{Mac251} can have potent ADCVI activity (Forthal et al. 2006). Such non-neutralizing serum protected newborn macaques from oral challenge with SIV_{Mac251} (Van Rompay et al. 1998). Engineered antibodies have provided a direct demonstration of a role for FcγR-mediated antibody functions in preventing lentivirus infection (Hessell et al. 2007). Different forms of the neutralizing mAb IgG1b12, with equivalent ELISA and neutralizing activity, provided different levels of protection against vaginal challenge with SHIV162p3 depending on whether they had a functional FcγR-binding domain. Thus, ADCVI, ADCC, or related FcγR-mediated activities might be involved in both modulating established infection and in preventing new infection.

T-Cell Responses

Responses by SIV-specific CD8 T cells are believed to contribute to control of viral replication, although the extent and precise mechanisms remain controversial. SIV-specific $CD8^+$ T-cell responses become measurable around the time of the decline from peak viremia in primary infection (Veazey et al. 2001). Depletion of $CD8^+$ cells by administration of a monoclonal antibody during chronic infection results in a transient increase in plasma viremia, coinciding with the period of depletion of $CD8^+$ cells (Schmitz et al. 1999). However, the available mAbs directed against CD8α also deplete some NK cells, and depletion of $CD8^+$ lymphocytes can result in proliferation of $CD4^+$ T cells, potentially providing additional targets for viral infection and increasing viremia via this mechanism (Okoye et al. 2009). Perhaps the clearest evidence of $CD8^+$ T-cell control of viral replication is provided by evidence of immune escape

Table 2. Effects of dose and route of challenge on passive protection in SHIV infection of rhesus macaques

Virus	Dose[a]	Route[b]	Treatment (mg/kg)	Protection	Lessons learned[c]	Reference
SHIV-89.6	40 (AID_{50})	IV	HIVIG (400); mAb 2F5 (15); mAb 2G12 (15)	3/6	High doses of HIVIG plus NmAbs partially protect from IV challenge.	Mascola et al. 1999
SHIV-89.6	100 ($TCID_{50}$)	IVag	HIVIG (400); mAb 2F5 (15); mAb 2G12 (15)	4/5	The same combination and dose partially protects vs. high-dose mucosal challenge.	Mascola et al. 2000
SHIV-DH12	100 ($TCID_{50}$)	IV	CHIVIG (230)	1/1	High levels of polyclonal IgG protect vs. high-dose IV challenge.	Shibata et al. 1999
SHIV-DH12	100 ($TCID_{50}$)	IV	CHIVIG (25)	0/2	10-fold lower levels failed to protect, demonstrating a dose response.	
SHIV-DH12	10 ($TCID_{50}$)	IV	CHIVIG (25)	1/1	This 10-fold lower dose could protect against a 10-fold lower IV challenge.	
SHIV-vpu CCR5	10 (AID_{50})	IV	mAb F105 (10); mAb 2F5 (10); mAb **2G12** (10)	4/4 adults	Neutralizing mAbs alone protects vs. low-dose IV challenge.	Baba et al. 2000
SHIV-vpu CCR5	10 (AID_{50})	oral	mAb F105 (10); mAb 2F5 (10); mAb 2G12 (10)	4/4 infants	NmAbs—IgGs—can fully protect infants vs. oral challenge.	Baba et al. 2000
SHIV-SF162P3	500 ($TCID_{50}$)	IVag	mAb 2G12 (40)	3/5	A single NmAb partially protects vs. high-dose mucosal challenge.	Hessell et al. 2009
SHIV-SF162P4	300 ($TCID_{50}$)	IVag	mAb b12 (25)	4/4	NmAb directed to the CD4bs protects vs. a lower dose mucosal challenge.	Parren et al. 2001
SHIV-SF162P4	300 ($TCID_{50}$)	IVag	mAb b12 (5)	2/4	A fivefold lower dose is less effective in protection, same dose and mucosal route.	Parren et al. 2001

Continued

Table 2. *Continued*

Virus	Dose[a]	Route[b]	Treatment (mg/kg)	Protection	Lessons learned[c]	Reference
SHIV-SF162P4	300 (TCID$_{50}$)	IVag	mAb b12 (1)	0/4	A fivefold still lower dose is ineffective against the same dose and mucosal route.	
SHIV-BaL	2000 (TCID$_{50}$)	IR	mAb 2F5 (50)	5/6	Partially protects vs. IR challenge with gp41 NmAb 2F5.	Hessell et al. 2010
SHIV-BaL	2000 (TCID$_{50}$)	IR	mAb 4E10 (50)	5/6	Partially protects vs. the same IR challenge with gp41 NmAb 4E10.	

[a]AID$_{50}$, animal infectious dose that infects 50% of test animals in a titration experiment; TCID$_{50}$, tissue-culture infectious dose that infects 50% of test wells using Reed–Muench analysis.

[b]IV, intravenous inoculation; IVag, atraumatic intravaginal inoculation; IR, atraumatic intrarectal inoculation; HIVIG, HIV immune globulin, purified IgG; CHIVIG, purified IgG from HIV-infected chimpanzee.

[c]NmAbs, neutralizing monoclonal antibodies; CD4bs, CD4 binding site.

by mutation of viral sequences affecting CTL epitopes. Examples have been reported for both early escape for epitopes that are not structurally constrained, such as the MamuA*01 restricted epitope tat SL8, and later escape of structurally constrained epitopes such as the MamuA*01 restricted epitope gag CM9, which typically emerges later in infection, in combination with compensatory mutations outside the epitope itself, partially restoring replicative fitness of the mutant virus (Goulder and Watkins 2004). Similar evidence based on in vivo selection implicating MHC-I restricted activity of CD8[+] T cells comes from a study of a SIV-mac239 variant engineered to be incapable of the MHIC-I downregulation via Nef, but intact for other Nef functions. The SIVmac239 variant underwent extensive mutation in vivo in infected macaques to restore the capacity to down-regulate MHC-I, associated with a relative loss of control of viral replication, supporting the importance of MHC-I restricted killing by CD8[+] T cells (Swigut et al. 2004).

CD8[+] T cells could potentially contribute to control of viral replication via both cytolytic and noncytolytic mechanisms, including production of antiviral cytokines or chemokines, including MIP-1β (Cocchi et al. 1995; Gauduin

1998). The ability of CD8[+] T cells to efficiently load lytic granules and up-regulate granzyme B in response to stimulation by HIV-1 infected autologous CD4[+] T cells, and to eliminate infected cells in co-cultures, correlated with elite controller status in HIV-infected humans (Migueles et al. 2008). Preliminary observations in SIV-infected rhesus macaques suggest a similar correlation between these activities and the extent of control of viral replication (SA Migueles, in prep.). Recent progress with systems for adoptive transfer of ex vivo selected and expanded autologous cells (Berger et al. 2008; Bolton et al. 2010; Minang et al. 2010) and characterization of Mauritian cynomolgus macaques with limited MHC heterogeneity (Burwitz et al. 2009) show promise for enabling adoptive transfer studies to address the in vivo antiviral activities of T-cell populations with different functional properties. Studies in Mauritian cynomolgous have also demonstrated an improved capacity of responses from MHC heterozygotes to control SIV replication in vivo (O'Connor et al. 2010), paralleling similar observations in humans (Carrington et al. 1999).

The role of virus-specific CD4[+] T cells in antiviral immune responses is incompletely understood. In addition to the "helper" function traditionally ascribed to CD4$^+$ T cells, they may mediate other functions. Animals chronically infected with live attenuated SIVmac239Δnef virus were found to develop abundant populations of SIV antigen-specific effector memory CD4$^+$ T cells, capable of up-regulating perforin expression and degranulation upon stimulation with SIV antigens (Gauduin et al. 2006), although any direct role such cells play in the robust protection mediated by SIVmac239Δnef immunization remains to be demonstrated. The extent to which the prominent responses by SIV-specific CD4$^+$ T cells contribute to the impressive protection afforded by vaccines based on recombinant rhesus CMV vectors (Hansen et al. 2009), and the mechanisms responsible, remain tantalizing areas for future research.

CONCLUSIONS

NHP models have been successfully used to evaluate the safety, immunogenicity, and protective efficacy of different candidate AIDS vaccine approaches and to conduct proof of concept studies for novel vaccine concepts. The available data suggest the potential predictive value of certain NHP models for some vaccines. In some models, most notably those involving X4-tropic SHIV challenges, lack of predictive value has been demonstrated. In others, NHP models represent arguably the only feasible, relevant approach to rigorously investigate mechanistic questions related to vaccine efficacy, such as the early events that lead to a rampant systemic infection. Despite their many unique advantages, NHP models have been controversial, in part because of a lack of understanding about the differences in their biology and the extent to which they mimic human disease. Recognition of the range of different models available, along with better informed selection of appropriate models for particular experiments, coupled with appropriately informed interpretation of the results of these studies, should not only improve the utility of the studies but also increase our understanding of the immunopathology of disease. Continuing development and refinement of NHP models will be critical for developing an effective vaccine for the prevention of HIV infection and AIDS.

ACKNOWLEDGMENTS

This article is dedicated to Harvey Crystal, MD, dedicated clinician and inspired teacher (J.D.L.) and to Marshall Hall Edgell, PhD, whose insight and creativity inspired two generations of scientists (N.L.H.). Preparation of this material was supported in part with federal funds from the National Cancer Institute, National Institutes of Health, under contract HHSN261200800001E (J.D.L.), and from the National Center for Research Resources, and under grant P51-RR000163 (N.L.H.). The authors thank Jeremy Smedley, DVM, for helpful insight and contributions to early drafts of this manuscript.

REFERENCES

*Reference is also in this collection.

Abel K, Pahar B, Van Rompay KK, Fritts L, Sin C, Schmidt K, Colon R, McChesney M, Marthas ML. 2006. Rapid virus dissemination in infant macaques after oral simian immunodeficiency virus exposure in the presence of local innate immune responses. *J Virol* **80:** 6357–6367.

Ahmad S, Lohman B, Marthas M, Giavedoni L, el-Amad Z, Haigwood NL, Scandella CJ, Gardner MB, Luciw PA, Yilma T. 1994. Reduced virus load in rhesus macaques immunized with recombinant gp160 and challenged with simian immunodeficiency virus. *AIDS Res Hum Retroviruses* **10:** 195–204.

Alizon M, Sonigo P, Barre-Sinoussi F, Chermann JC, Tiollais P, Montagnier L, Wain-Hobson S. 1984. Molecular cloning of lymphadenopathy-associated virus. *Nature* **312:** 757–760.

Alpert MD, Rahmberg AR, Neidermyer W, Ng SK, Carville A, Camp JV, Wilson RL, Piatak M Jr, Mansfield KG, Li W, et al. 2010. Envelope-modified single-cycle simian immunodeficiency virus selectively enhances antibody responses and partially protects against repeated, low-dose vaginal challenge. *J Virol* **84:** 10748–10764.

Amara RR, Villinger F, Altman JD, Lydy SL, O'Neil SP, Staprans SI, Montefiori DC, Xu Y, Herndon JG, Wyatt LS, et al. 2001. Control of a mucosal challenge and prevention of AIDS by a multiprotein DNA/MVA vaccine. *Science* **292:** 69–74.

Apetrei C, Gormus B, Pandrea I, Metzger M, ten Haaft P, Martin LN, Bohm R, Alvarez X, Koopman G, Murphey-Corb M, et al. 2004. Direct inoculation of simian immunodeficiency virus from sooty mangabeys in black mangabeys (*Lophocebus aterrimus*): First evidence of AIDS in a heterologous African species and different pathologic outcomes of experimental infection. *J Virol* **78:** 11506–11518.

Apetrei C, Kaur A, Lerche NW, Metzger M, Pandrea I, Hardcastle J, Falkenstein S, Bohm R, Koehler J, Traina-Dorge V, et al. 2005. Molecular epidemiology of simian immunodeficiency virus SIVsm in U.S. primate centers unravels the origin of SIVmac and SIVstm. *J Virol* **79:** 8991–9005.

Arthur LO, Pyle SW, Nara PL, Bess JW Jr, Gonda MA, Kelliher JC, Gilden RV, Robey WG, Bolognesi DP, Gallo RC, et al. 1987. Serological responses in chimpanzees inoculated with human immunodeficiency virus glycoprotein (gp120) subunit vaccine. *Proc Natl Acad Sci* **84:** 8583–8587.

Arthur LO, Bess JW Jr, Urban RG, Strominger JL, Morton WR, Mann DL, Henderson LE, Benveniste RE. 1995. Macaques immunized with HLA-DR are protected from challenge with simian immunodeficiency virus. *J Virol* **69:** 3117–3124.

Arthur LO, Bess JW Jr, Chertova EN, Rossio JL, Esser MT, Benveniste RE, Henderson LE, Lifson JD. 1998. Chemical inactivation of retroviral infectivity by targeting nucleocapsid protein zinc fingers: A candidate SIV vaccine. *AIDS Res Hum Retroviruses* **14 Suppl 3:** S311–319.

Baba TW, Jeong YS, Pennick D, Bronson R, Greene MF, Ruprecht RM. 1995. Pathogenicity of live, attenuated SIV after mucosal infection of neonatal macaques [see comments]. *Science* **267:** 1820–1825.

Baba TW, Liska V, Hofmann-Lehmann R, Vlasak J, Xu W, Ayehunie S, Cavacini LA, Posner MR, Katinger H, Stiegler G, et al. 2000. Human neutralizing monoclonal antibodies of the IgG1 subtype protect against mucosal simian-human immunodeficiency virus infection. *Nat Med* **6:** 200–206.

Barouch DH, Nabel GJ. 2005. Adenovirus vector-based vaccines for human immunodeficiency virus type 1. *Hum Gene Ther* **16:** 149–156.

Barouch DH, Santra S, Schmitz JE, Kuroda MJ, Fu TM, Wagner W, Bilska M, Craiu A, Zheng XX, Krivulka GR, et al. 2000. Control of viremia and prevention of clinical AIDS in rhesus monkeys by cytokine-augmented DNA vaccination. *Science* **290:** 486–492.

Barouch DH, Craiu A, Santra S, Egan MA, Schmitz JE, Kuroda MJ, Fu TM, Nam JH, Wyatt LS, Lifton MA, et al. 2001a. Elicitation of high-frequency cytotoxic T-lymphocyte responses against both dominant and subdominant simian-human immunodeficiency virus epitopes by DNA vaccination of rhesus monkeys. *J Virol* **75:** 2462–2467.

Barouch DH, Santra S, Kuroda MJ, Schmitz JE, Plishka R, Buckler-White A, Gaitan AE, Zin R, Nam JH, Wyatt LS, et al. 2001b. Reduction of simian-human immunodeficiency virus 89.6p viremia in rhesus monkeys by recombinant modified vaccinia virus ankara vaccination. *J Virol* **75:** 5151–5158.

Barouch DH, Santra S, Tenner-Racz K, Racz P, Kuroda MJ, Schmitz JE, Jackson SS, Lifton MA, Freed DC, Perry HC, et al. 2002. Potent CD4+ T cell responses elicited by a bicistronic HIV-1 DNA vaccine expressing gp120 and GM-CSF. *J Immunol* **168:** 562–568.

Barre-Sinoussi F, Chermann JC, Rey F, Nugeyre MT, Chamaret S, Gruest J, Dauguet C, Axler-Blin C, Vezinet-Brun F, Rouzioux C, et al. 1983. Isolation of a T-lymphotropic retrovirus from a patient at risk for acquired immune deficiency syndrome (AIDS). *Science* **220:** 868–871.

Baum LL, Cassutt KJ, Knigge K, Khattri R, Margolick J, Rinaldo C, Kleeberger CA, Nishanian P, Henrard DR, Phair J. 1996. HIV-1 gp120-specific antibody-dependent cell-mediated cytotoxicity correlates with rate of disease progression. *J Immunol* **157:** 2168–2173.

Benveniste RE, Arthur LO, Tsai CC, Sowder R, Copeland TD, Henderson LE, Oroszlan S. 1986. Isolation of a lentivirus from a macaque with lymphoma: Comparison with HTLV-III/LAV and other lentiviruses. *J Virol* **60:** 483–490.

Berger C, Jensen MC, Lansdorp PM, Gough M, Elliott C, Riddell SR. 2008. Adoptive transfer of effector CD8+ T cells derived from central memory cells establishes persistent T cell memory in primates. *J Clin Invest* **118:** 294–305.

Berman PW, Gregory TJ, Riddle L, Nakamura GR, Champe MA, Porter JP, Wurm FM, Hershberg RD, Cobb EK, Eichberg JW. 1990. Protection of chimpanzees from infection by HIV-1 after vaccination with recombinant glycoprotein gp120 but not gp160. *Nature* **345:** 622–625.

Berman PW, Matthews TJ, Riddle L, Champe M, Hobbs MR, Nakamura GR, Mercer J, Eastman DJ, Lucas C, Langlois AJ, et al. 1992. Neutralization of multiple laboratory and clinical isolates of human immunodeficiency virus type 1 (HIV-1) by antisera raised against gp120 from the MN isolate of HIV-1. *J Virol* **66:** 4464–4469.

Bolton DL, Minang JT, Trivett MT, Song K, Tuscher JJ, Li Y, Piatak M Jr, O'Connor D, Lifson JD, Roederer M, et al. 2010. Trafficking, persistence, and activation state of adoptively transferred allogeneic and autologous Simian Immunodeficiency Virus-specific CD8(+) T cell clones during acute and chronic infection of rhesus macaques. *J Immunol* **184:** 303–314.

Bontrop RE, Watkins DI. 2005. MHC polymorphism: AIDS susceptibility in non-human primates. *Trends Immunol* **26:** 227–233.

Burwitz BJ, Pendley CJ, Greene JM, Detmer AM, Lhost JJ, Karl JA, Piaskowski SM, Rudersdorf RA, Wallace LT, Bimber BN, et al. 2009. Mauritian cynomolgus macaques share two exceptionally common major histocompatibility complex class I alleles that restrict simian immunodeficiency virus-specific CD8+ T cells. *J Virol* **83:** 6011–6019.

* Carrington M, Alter G. 2011. Innate immune control of HIV. *Cold Spring Harb Perspect Med* doi: 10.1101/cshperspect.a007070.

Carrington M, Nelson GW, Martin MP, Kissner T, Vlahov D, Goedert JJ, Kaslow R, Buchbinder S, Hoots K, O'Brien SJ. 1999. HLA and HIV-1: Heterozygote advantage and B*35-Cw*04 disadvantage. *Science* **283:** 1748–1752.

Casimiro DR, Wang F, Schleif WA, Liang X, Zhang ZQ, Tobery TW, Davies ME, McDermott AB, O'Connor DH, Fridman A, et al. 2005. Attenuation of simian immunodeficiency virus SIVmac239 infection by prophylactic immunization with DNA and recombinant adenoviral vaccine vectors expressing Gag. *J Virol* **79:** 15547–15555.

Casimiro DR, Cox K, Tang A, Sykes KJ, Wang F, Bett A, Schleif WA, Liang X, Flynn J, et al. 2010. Efficiency of multivalent adenovirus-based vaccine against simian immunodeficiency virus challenge. *J Virol* **84:** 2996–3003.

Cocchi F, DeVico AL, Garzino-Demo A, Arya SK, Gallo RC, Lusso P. 1995. Identification of RANTES, MIP-1a, and MIP-1b as the major HIV-suppressive factors produced by CD8+ T cells. *Science* **270:** 1811–1815.

Conley AJ, Kessler JA II, Boots LJ, McKenna PM, Schleif WA, Emini EA, Mark GE III, Katinger H, Cobb EK, Lunceford SM, et al. 1996. The consequence of passive administration of an anti-human immunodeficiency virus type 1 neutralizing monoclonal antibody before challenge of chimpanzees with a primary virus isolate. *J Virol* **70:** 6751–6758.

Cooney EL, Collier AC, Greenberg PD, Coombs RW, Zarling J, Arditti DE, Hoffman MC, Hu SL, Corey L. 1991. Safety of and immunological response to a recombinant vaccinia virus vaccine expressing HIV envelope glycoprotein. *Lancet* **337:** 567–572.

Cooney EL, McElrath MJ, Corey L, Hu SL, Collier AC, Arditti D, Hoffman M, Coombs RW, Smith GE, Greenberg PD. 1993. Enhanced immunity to human immunodeficiency virus (HIV) envelope elicited by a combined vaccine regimen consisting of priming with a vaccinia recombinant expressing HIV envelope and boosting with gp160 protein. *Proc Natl Acad Sci* **90:** 1882–1886.

Cranage MP, McBride BW, Rud EW. 1995. The simian immunodeficiency virus transmembrane protein is poorly immunogenic in inactivated virus vaccine. *Vaccine* **13:** 895–900.

Dale CJ, De Rose R, Stratov I, Chea S, Montefiori DC, Thomson S, Ramshaw IA, Coupar BE, Boyle DB, Law M, et al. 2004. Efficacy of DNA and fowlpox virus priming/boosting vaccines for simian/human immunodeficiency virus. *J Virol* **78:** 13819–13828.

Daniel MD, King NW, Letvin NL, Hunt RD, Sehgal PK, Desrosiers RC. 1984. A new type D retrovirus isolated from macaques with an immunodeficiency syndrome. *Science* **223:** 602–605.

Daniel MD, Letvin NL, King NW, Kannagi M, Sehgal PK, Hunt RD, Kanki PJ, Essex M, Desrosiers RC. 1985. Isolation of T-cell tropic HTLV-III-like retrovirus from macaques. *Science* **228:** 1201–1204.

Daniel MD, Kirchhoff F, Czajak SC, Sehgal PK, Desrosiers RC. 1992. Protective effects of a live attenuated SIV vaccine with a deletion in the nef gene. *Science* **228:** 1201–1204.

Demberg T, Boyer JD, Malkevich N, Patterson LJ, Venzon D, Summers EL, Kalisz I, Kalyanaraman VS, Lee EM, Weiner DB, et al. 2008. Sequential priming with simian immunodeficiency virus (SIV) DNA vaccines, with or without encoded cytokines, and a replicating adenovirus-SIV recombinant followed by protein boosting does not control a pathogenic SIVmac251 mucosal challenge. *J Virol* **82:** 10911–10921.

Desrosiers RC, Wyand MS, Kodama T, Ringler DJ, Arthur LO, Sehgal PK, Letvin NL, King NW, Daniel MD. 1989. Vaccine protection against simian immunodeficiency virus infection. *Proc Natl Acad Sci* **86:** 6353–6357.

Doria-Rose NA, Ohlen C, Polacino P, Pierce CC, Hensel MT, Kuller L, Mulvania T, Anderson D, Greenberg PD, Hu SL, et al. 2003. Multigene DNA priming-boosting vaccines

protect macaques from acute CD4(+)-T-cell depletion after simian–human immunodeficiency virus SHIV89.6P mucosal challenge. *J Virol* **77:** 11563–11577.

Earl PL, Sugiura W, Montefiori DC, Broder CC, Lee SA, Wild C, Lifson J, Moss B. 2001. Immunogenicity and protective efficacy of oligomeric human immunodeficiency virus type 1 gp140. *J Virol* **75:** 645–653.

Eel-Amad Z, Murthy KK, Higgins K, Cobb EK, Haigwood NL, Levy JA, Steimer KS. 1995. Resistance of chimpanzees immunized with recombinant gp120SF2 to challenge by HIV-1SF2. *Aids* **9:** 1313–1322.

Evans DT, Bricker JE, Sanford HB, Lang S, Carville A, Richardson BA, Piatak M Jr, Lifson JD, Mansfield KG, Desrosiers RC. 2005. Immunization of macaques with single-cycle simian immunodeficiency virus (SIV) stimulates diverse virus-specific immune responses and reduces viral loads after challenge with SIVmac239. *J Virol* **79:** 7707–7720.

Florese RH, Demberg T, Xiao P, Kuller L, Larsen K, Summers LE, Venzon D, Cafaro A, Ensoli B, Robert-Guroff M. 2009. Contribution of nonneutralizing vaccine-elicited antibody activities to improved protective efficacy in rhesus macaques immunized with Tat/Env compared with multigenic vaccines. *J Immunol* **182:** 3718–3727.

Forthal DN, Landucci G, Daar ES. 2001. Antibody from patients with acute human immunodeficiency virus (HIV) infection inhibits primary strains of HIV type 1 in the presence of natural-killer effector cells. *J Virol* **75:** 6953–6961.

Forthal DN, Landucci G, Cole KS, Marthas M, Becerra JC, Van Rompay K. 2006. Rhesus macaque polyclonal and monoclonal antibodies inhibit simian immunodeficiency virus in the presence of human or autologous rhesus effector cells. *J Virol* **80:** 9217–9225.

Fultz PN, McClure HM, Daugharty H, Brodie A, McGrath CR, Swenson B, Francis DP. 1986a. Vaginal transmission of human immunodeficiency virus (HIV) to a chimpanzee. *J Infect Dis* **154:** 896–900.

Fultz PN, McClure HM, Swenson RB, McGrath CR, Brodie A, Getchell JP, Jensen FC, Anderson DC, Broderson JR, Francis DP. 1986b. Persistent infection of chimpanzees with human T-lymphotropic virus type III/lymphadenopathy-associated virus: A potential model for acquired immunodeficiency syndrome. *J Virol* **58:** 116–124.

Gallo RC, Salahuddin SZ, Popovic M, Shearer GM, Kaplan M, Haynes BF, Palker TJ, Redfield R, Oleske J, Safai B, et al. 1984. Frequent detection and isolation of cytopathic retroviruses (HTLV-III) from patients with AIDS and at risk for AIDS. *Science* **224:** 500–503.

Gauduin M-C, Safrit JT, Weir R, Fung MSC, Koup RA. 1995. Pre- and postexposure protection against human immunodeficiency virus type 1 infection mediated by a monoclonal antibody. *J Infect Dis* **171:** 1203–1209.

Gauduin M-C, Parren PWHI, Weir R, Barbas CFI, Burton DR, Koup RA. 1997. Passive immunization with a potent neutralizing human monoclonal antibody protects HU-PBL-SCID mice against challenge by primary isolates of human immunodeficiency virus type 1. *Nat Med* **3:** 1389–1393.

Gauduin MC, Glickman RL, Means R, Johnson RP. 1998. Inhibition of simian immunodeficiency virus (SIV) replication by CD8$^+$ T lymphocytes from macaques immunized with live attenuated SIV. *J Virol* **72:** 6315–6324.

Gauduin MC, Yu Y, Barabasz A, Carville A, Piatak M, Lifson JD, Desrosiers RC, Johnson RP. 2006. Induction of a virus-specific effector-memory CD4+ T cell response by attenuated SIV infection. *J Exp Med* **203:** 2661–2672.

Gautam R, Gaufin T, Butler I, Gautam A, Barnes M, Mandell D, Pattison M, Tatum C, Macfarland J, Monjure C, et al. 2009. Simian immunodeficiency virus SIVrcm, a unique CCR2-tropic virus, selectively depletes memory CD4+ T cells in pigtailed macaques through expanded coreceptor usage *in vivo. J Virol* **83:** 7894–7908.

Giavedoni LD, Planelles V, Haigwood NL, Ahmad S, Kluge JD, Marthas ML, Gardner MB, Luciw PA, Yilma TD. 1993. Immune response of rhesus macaques to recombinant simian immunodeficiency virus gp130 does not protect from challenge infection. *J Virol* **67:** 577–583.

Goila-Gaur R, Strebel K. 2008. HIV-1 Vif, APOBEC, and intrinsic immunity. *Retrovirology* **5:** 51.

Gottlieb MS, Schroff R, Schanker HM, Weisman JD, Fan PT, Wolf RA, Saxon A. 1981. *Pneumocystis carinii* pneumonia and mucosal candidiasis in previously healthy homosexual men: Evidence of a new acquired cellular immunodeficiency. *N Engl J Med* **305:** 1425–1431.

Goulder PJ, Watkins DI. 2004. HIV and SIV CTL escape: Implications for vaccine design. *Nat Rev Immunol* **4:** 630–640.

Goulder PJ, Watkins DI. 2008. Impact of MHC class I diversity on immune control of immunodeficiency virus replication. *Nat Rev Immunol* **8:** 619–630.

Haase AT. 2010. Targeting early infection to prevent HIV-1 mucosal transmission. *Nature* **464:** 217–223.

Hahn BH, Shaw GM, Arya SK, Popovic M, Gallo RC, Wong-Staal F. 1984. Molecular cloning and characterization of the HTLV-III virus associated with AIDS. *Nature* **312:** 166–169.

Haigwood NL, Stamatatos L. 2003. Role of neutralizing antibodies in HIV infection. *Aids* **17** Suppl 4: S67–71.

Haigwood NL, Nara PL, Brooks E, Van Nest GA, Ott G, Higgins KW, Dunlop N, Scandella CJ, Eichberg JW, Steimer KS. 1992. Native but not denatured recombinant human immunodeficiency virus type 1 gp120 generates broad-spectrum neutralizing antibodies in baboons. *J Virol* **66:** 172–182.

Haigwood NL, Montefiori DC, Sutton WF, McClure J, Watson AJ, Voss G, Hirsch VM, Richardson BA, Letvin NL, Hu SL, et al. 2004. Passive immunotherapy in simian immunodeficiency virus-infected macaques accelerates the development of neutralizing antibodies. *J Virol* **78:** 5983–5995.

Hansen SG, Vieville C, Whizin N, Coyne-Johnson L, Siess DC, Drummond DD, Legasse AW, Axthelm MK, Oswald K, Trubey CM, et al. 2009. Effector memory T cell responses are associated with protection of rhesus monkeys from mucosal simian immunodeficiency virus challenge. *Nat Med* **15:** 293–299.

Hansen SG, Ford JC, Lewis MS, Ventura AB, Hughes CM, Coyne-Johnson L, Whizin N, Oswald K, Shoemaker R, Swanson T, et al. 2011. Profound early control of highly pathogenic SIV by an effector memory T-cell vaccine. *Nature* **473:** 523–527.

Haynes BF, Montefiori DC. 2006. Aiming to induce broadly reactive neutralizing antibody responses with HIV-1 vaccine candidates. *Expert Rev Vaccines* **5:** 347–363.

Hessell AJ, Hangartner L, Hunter M, Havenith CE, Beurskens FJ, Bakker JM, Lanigan CM, Landucci G, Forthal DN, Parren PW, et al. 2007. Fc receptor but not complement binding is important in antibody protection against HIV. *Nature* **449:** 101–104.

Hessell AJ, Rakasz EG, Poignard P, Hangartner L, Landucci G, Forthal DN, Koff WC, Watkins DI, Burton DR. 2009. Broadly neutralizing human anti-HIV antibody 2G12 is effective in protection against mucosal SHIV challenge even at low serum neutralizing titers. *PLoS Pathog* **5:** e1000433.

Hessell AJ, Rakasz EG, Tehrani DM, Huber M, Weisgrau KL, Landucci G, Forthal DN, Koff WC, Poignard P, Watkins DI, et al. 2010. Broadly neutralizing monoclonal antibodies 2F5 and 4E10 directed against the human immunodeficiency virus type 1 gp41 membrane-proximal external region protect against mucosal challenge by simian-human immunodeficiency virus SHIVBa-L. *J Virol* **84:** 1302–1313.

Hirsch VM, Lifson JD. 2000. Simian immunodeficiency virus infection of monkeys as a model system for the study of AIDS pathogenesis, treatment, and prevention. *Adv Pharmacol* **49:** 437–477.

Hu SL, Abrams K, Barber GN, Moran P, Zarling JM, Langlois AJ, Kuller L, Morton WR, Benveniste RE. 1992. Protection of macaques against SIV infection by subunit vaccines of SIV envelope glycoprotein gp160. *Science* **255:** 456–459.

Igarashi T, Brown C, Azadegan A, Haigwood N, Dimitrov D, Martin MA, Shibata R. 1999. Human immunodeficiency virus type 1 neutralizing antibodies accelerate clearance of cell-free virions from blood plasma [see comments]. *Nat Med* **5:** 211–216.

Jia B, Serra-Moreno R, Neidermyer W, Rahmberg A, Mackey J, Fofana IB, Johnson WE, Westmoreland S, Evans DT. 2009. Species-specific activity of SIV Nef and HIV-1 Vpu in overcoming restriction by tetherin/BST2. *PLoS Pathog* **5:** e1000429.

Johnson PR, Montefiori DC, Goldstein S, Hamm TE, Zhou J, Kitov S, Haigwood NL, Misher L, London WT, Gerin JL, et al. 1992. Inactivated whole-virus vaccine derived from a proviral DNA clone of simian immunodeficiency virus induces high levels of neutralizing antibodies and confers protection against heterologous challenge. *Proc Natl Acad Sci* **89:** 2175–2179.

Johnson PR, Schnepp BC, Zhang J, Connell MJ, Greene SM, Yuste E, Desrosiers RC, Clark KR. 2009. Vector-mediated gene transfer engenders long-lived neutralizing activity and protection against SIV infection in monkeys. *Nat Med* **15:** 901–906.

Kaul R, Rowland-Jones SL, Kimani J, Dong T, Yang HB, Kiama P, Rostron T, Njagi E, Bwayo JJ, MacDonald KS, et al. 2001. Late seroconversion in HIV-resistant Nairobi prostitutes despite pre-existing HIV-specific CD8+ responses. *J Clin Invest* **107:** 341–349.

Keele BF, Giorgi EE, Salazar-Gonzalez JF, Decker JM, Pham KT, Salazar MG, Sun C, Grayson T, Wang S, Li H, et al. 2008. Identification and characterization of transmitted and early founder virus envelopes in primary HIV-1 infection. *Proc Natl Acad Sci* **105**: 7552–7557.

Keele BF, Li H, Learn GH, Hraber P, Giorgi EE, Grayson T, Sun C, Chen Y, Yeh WW, Letvin NL, et al. 2009. Low-dose rectal inoculation of rhesus macaques by SIVsmE660 or SIVmac251 recapitulates human mucosal infection by HIV-1. *J Exp Med* **206**: 1117–1134.

Kirmaier A, Wu F, Newman RM, Hall LR, Morgan JS, O'Connor S, Marx PA, Meythaler M, Goldstein S, Buckler-White A, et al. 2010. TRIM5 suppresses cross-species transmission of a primate immunodeficiency virus and selects for emergence of resistant variants in the new species. *PLoS Biol* **8**: e1000462.

Ko SY, Cheng C, Kong WP, Wang L, Kanekiyo M, Einfeld D, King CR, Gall JG, Nabel GJ. 2009. Enhanced induction of intestinal cellular immunity by oral priming with enteric adenovirus 41 vectors. *J Virol* **83**: 748–756.

Koff WC, Johnson PR, Watkins DI, Burton DR, Lifson JD, Hasenkrug KJ, McDermott AB, Schultz A, Zamb TJ, Boyle R, et al. 2006. HIV vaccine design: Insights from live attenuated SIV vaccines. *Nat Immunol* **7**: 19–23.

Koup RA, Roederer M, Lamoreaux L, Fischer J, Novik L, Nason MC, Larkin BD, Enama ME, Ledgerwood JE, Bailer RT, et al. 2010. Priming immunization with DNA augments immunogenicity of recombinant adenoviral vectors for both HIV-1 specific antibody and T-cell responses. *PLoS One* **5**: e9015.

Lasky LA, Groopman JE, Fennie CW, Benz PM, Capon DJ, Dowbenko DJ, Nakamura GR, Nunes WM, Renz ME, Berman PW. 1986. Neutralization of the AIDS retrovirus by antibodies to a recombinant envelope glycoprotein. *Science* **233**: 209–212.

Letvin NL, Daniel MD, Sehgal PK, Desrosiers RC, Hunt RD, Waldron LM, MacKey JJ, Schmidt DK, Chalifoux LV, King NW. 1985. Induction of AIDS-like disease in macaque monkeys with T-cell tropic retrovirus STLV-III. *Science* **230**: 71–73.

Li L, Lord CI, Haseltine W, Letvin NL, Sodroski J. 1992. Infection of cynomolgus monkeys with a chimeric HIV-1/SIVmac virus that expresses the HIV-1 envelope glycoproteins. *J Acquir Immune Defic Syndr* **5**: 639–646.

Li Q, Estes JD, Schlievert PM, Duan L, Brosnahan AJ, Southern PJ, Reilly CS, Peterson ML, Schultz-Darken N, Brunner KG, et al. 2009a. Glycerol monolaurate prevents mucosal SIV transmission. *Nature* **458**: 1034–1038.

Li Q, Skinner PJ, Duan L, Haase AT. 2009b. A technique to simultaneously visualize virus-specific CD8+ T cells and virus-infected cells in situ. *J Vis Exp* **30**: 1561.

Lifson JD, Piatak M Jr, Rossio JL, Bess J Jr, Chertova E, Schneider D, Kiser R, Coalter V, Poore B, Imming R, et al. 2002. Whole inactivated SIV virion vaccines with functional envelope glycoproteins: Safety, immunogenicity, and activity against intrarectal challenge. *J Med Primatol* **31**: 205–216.

Lifson JD, Rossio JL, Piatak M Jr, Bess J Jr, Chertova E, Schneider DK, Coalter VJ, Poore B, Kiser RF, Imming RJ, et al. 2004. Evaluation of the safety, immunogenicity, and protective efficacy of whole inactivated simian immunodeficiency virus (SIV) vaccines with conformationally and functionally intact envelope glycoproteins. *AIDS Res Hum Retroviruses* **20**: 772–787.

Lim SY, Rogers T, Chan T, Whitney JB, Kim J, Sodroski J, Letvin NL. 2010. TRIM5alpha modulates immunodeficiency virus control in rhesus monkeys. *PLoS Pathog* **6**: e1000738.

Liu J, Keele BF, Li H, Keating S, Norris PJ, Carville A, Mansfield KG, Tomaras GD, Haynes BF, Kolodkin-Gal D, et al. 2010. Low-dose mucosal simian immunodeficiency virus infection restricts early replication kinetics and transmitted virus variants in rhesus monkeys. *J Virol* **84**: 10406–10412.

Luciw PA, Potter SJ, Steimer K, Dina D, Levy JA. 1984. Molecular cloning of AIDS-associated retrovirus. *Nature* **312**: 760–763.

Ma Z, Keele BF, Qureshi H, Stone M, Desilva V, Fritts L, Lifton JD, Miller CJ. 2011. SIVmac251 is inefficiently transmitted to rhesus macaques by penile inoculation with a single SIVenv variant found in ramp-up phase plasma. *AIDS Res Hum Retroviruses* (in press).

* Malim MH, Bieniasz PD. 2011. HIV restriction factors and mechanisms of evasion. *Cold Spring Harb Perspect Med* doi: 10.1101/cshperspect.a006940.

Martin MP, Carrington M. 2005. Immunogenetics of viral infections. *Curr Opin Immunol* **17**: 510–516.

Martin MP, Gao X, Lee JH, Nelson GW, Detels R, Goedert JJ, Buchbinder S, Hoots K, Vlahov D, Trowsdale J, et al. 2002. Epistatic interaction between KIR3DS1 and HLA-B delays the progression to AIDS. *Nat Genet* **31**: 429–434.

Martin MP, Qi Y, Gao X, Yamada E, Martin JN, Pereyra F, Colombo S, Brown EE, Shupert WL, Phair J, et al. 2007. Innate partnership of HLA-B and KIR3DL1 subtypes against HIV-1. *Nat Genet* **39**: 733–740.

Marx PA, Maul DH, Osborn KG, Lerche NW, Moody P, Lowenstine LJ, Henrickson RV, Arthur LO, Gilden RV, Gravell M, et al. 1984. Simian AIDS: Isolation of a type D retrovirus and transmission of the disease. *Science* **223**: 1083–1086.

Mascola JR. 2003. Defining the protective antibody response for HIV-1. *Curr Mol Med* **3**: 209–216.

Mascola JR, Lewis MG, Stiegler G, Harris D, VanCott TC, Hayes D, Louder MK, Brown CR, Sapan CV, Frankel SS, et al. 1999. Protection of macaques against pathogenic simian/human immunodeficiency virus 89.6PD by passive transfer of neutralizing antibodies. *J Virol* **73**: 4009–4018.

Mascola JR, Stiegler G, VanCott TC, Katinger H, Carpenter CB, Hanson CE, Beary H, Hayes D, Frankel SS, Birx DL, et al. 2000. Protection of macaques against vaginal transmission of a pathogenic HIV-1/SIV chimeric virus by passive infusion of neutralizing antibodies. *Nat Med* **6**: 207–210.

Masur H, Michelis MA, Greene JB, Onorato I, Stouwe RA, Holzman RS, Wormser G, Brettman L, Lange M, Murray HW, et al. 1981. An outbreak of community-acquired *Pneumocystis carinii* pneumonia: Initial manifestation of cellular immune dysfunction. *N Engl J Med* **305**: 1431–1438.

Matthews TJ. 1994. Dilemma of neutralization resistance of HIV-1 field isolates and vaccine development. *AIDS Res Hum Retroviruses* **10**: 631–632.

McNatt MW, Zang T, Hatziioannou T, Bartlett M, Fofana IB, Johnson WE, Neil SJ, Bieniasz PD. 2009. Species-specific activity of HIV-1 Vpu and positive selection of tetherin transmembrane domain variants. *PLoS Pathog* **5**: e1000300.

McVey D, Zuber M, Ettyreddy D, Reiter CD, Brough DE, Nabel GJ, Richter King C, Gall JG. 2010. Characterization of human adenovirus 35 and derivation of complex vectors. *Virol J* **7**: 276.

Migueles SA, Osborne CM, Royce C, Compton AA, Joshi RP, Weeks KA, Rood JE, Berkley AM, Sacha JB, Cogliano-Shutta NA, et al. 2008. Lytic granule loading of CD8+ T cells is required for HIV-infected cell elimination associated with immune control. *Immunity* **29**: 1009–1021.

Miller CJ, Hu J. 1999. T cell-tropic simian immunodeficiency virus (SIV) and simian-human immunodeficiency viruses are readily transmitted by vaginal inoculation of rhesus macaques, and Langerhans' cells of the female genital tract are infected with SIV. *J Infect Dis* 179 Suppl **3**: S413–S417.

Minang JT, Trivett MT, Bolton DL, Trubey CM, Estes JD, Li Y, Smedley J, Pung R, Rosati M, Jalah R, et al. 2010. Distribution, persistence, and efficacy of adoptively transferred central and effector memory-derived autologous simian immunodeficiency virus-specific CD8+ T cell clones in rhesus macaques during acute infection. *J Immunol* **184**: 315–326.

Mossman SP, Pierce CC, Watson AJ, Robertson MN, Montefiori DC, Kuller L, Richardson BA, Bradshaw JD, Munn RJ, Hu SL, et al. 2004. Protective immunity to SIV challenge elicited by vaccination of macaques with multigenic DNA vaccines producing virus-like particles. *AIDS Res Hum Retroviruses* **20**: 425–434.

Murphey-Corb M, Martin LN, Rangan SR, Baskin GB, Gormus BJ, Wolf RH, Andes WA, West M, Montelaro RC. 1986. Isolation of an HTLV-III-related retrovirus from macaques with simian AIDS and its possible origin in asymptomatic mangabeys. *Nature* **321**: 435–437.

Murphey-Corb M, Martin LN, Davison-Fairburn B, Montelaro RC, Miller M, West M, Ohkawa S, Baskin GB, Zhang J-Y, Putney SD, et al. 1989. A formalin-inactivated whole SIV vaccine confers protection in macaques. *Science* **246**: 1293–1297.

Neil SJ, Zang T, Bieniasz PD. 2008. Tetherin inhibits retrovirus release and is antagonized by HIV-1 Vpu. *Nature* **451**: 425–430.

Newman RM, Johnson WE. 2007. A brief history of TRIM5α. *AIDS Rev* **9**: 114–125.

Newman RM, Hall L, Connole M, Chen GL, Sato S, Yuste E, Diehl W, Hunter E, Kaur A, Miller GM, et al. 2006. Balancing selection and the evolution of functional polymorphism in Old World monkey TRIM5alpha. *Proc Natl Acad Sci* **103**: 19134–19139.

Ng CT, Jaworski JP, Jayaraman P, Sutton WF, Delio P, Kuller L, Anderson D, Landucci G, Richardson BA, Burton DR, et al. 2010. Passive neutralizing antibody controls SHIV viremia and enhances B cell responses in infant macaques. *Nat Med* **16**: 1117–1119.

Novembre FJ, Saucier M, Anderson DC, Klumpp SA, O'Neil SP, Brown CR, 2nd Hart, CE, Guenthner PC, Swenson RB, McClure HM. 1997. Development of AIDS in a chimpanzee infected with human immunodeficiency virus type 1. *J Virol* **71**: 4086–4091.

O'Brien KL, Liu J, King SL, Sun YH, Schmitz JE, Lifton MA, Hutnick NA, Betts MR, Dubey SA, Goudsmit J, et al. 2009. Adenovirus-specific immunity after immunization with an Ad5 HIV-1 vaccine candidate in humans. *Nat Med* **15**: 873–875.

O'Connor SL, Lhost JJ, Becker EA, Detmer AM, Johnson RC, Macnair CE, Wiseman RW, Karl JA, Greene JM, Burwitz BJ, et al. 2010. MHC heterozygote advantage in simian immunodeficiency virus-infected Mauritian cynomolgus macaques. *Sci Transl Med* **2**: 22ra18.

Okoye A, Park H, Rohankhedkar M, Coyne-Johnson L, Lum R, Walker JM, Planer SL, Legasse AW, Sylwester AW, Piatak M Jr, et al. 2009. Profound CD4+/CCR5+ T cell expansion is induced by CD8+ lymphocyte depletion but does not account for accelerated SIV pathogenesis. *J Exp Med* **206**: 1575–1588.

Otten G, Schaefer M, Doe B, Liu H, Srivastava I, zur Megede J, O'Hagan D, Donnelly J, Widera G, Rabussay D, et al. 2004. Enhancement of DNA vaccine potency in rhesus macaques by electroporation. *Vaccine* **22**: 2489–2493.

* Overbaugh J, Morris L. 2011. The antibody response against HIV-1. *Cold Spring Harb Perspect Med* doi: 10.1101/cshperspect.a007039.

Pal R, Venzon D, Letvin NL, Santra S, Montefiori DC, Miller NR, Tryniszewska E, Lewis MG, VanCott TC, Hirsch V, et al. 2002. ALVAC-SIV-gag-pol-env-based vaccination and macaque major histocompatibility complex class I (A*01) delay simian immunodeficiency virus SIVmac-induced immunodeficiency. *J Virol* **76**: 292–302.

Pal R, Wang S, Kalyanaraman VS, Nair BC, Whitney S, Keen T, Hocker L, Hudacik L, Rose N, Mboudjeka I, et al. 2006. Immunization of rhesus macaques with a polyvalent DNA prime/protein boost human immunodeficiency virus type 1 vaccine elicits protective antibody response against simian human immunodeficiency virus of R5 phenotype. *Virology* **348**: 341–353.

Palker TJ, Clark ME, Langlois AJ, Matthews TJ, Weinhold KJ, Randall RR, Bolognesi DP, Haynes BF. 1988. Type-specific neutralization of the human immunodeficiency virus with antibodies to env-encoded synthetic peptides. *Proc Natl Acad Sci* **85**: 1932–1936.

Parren PW, Marx PA, Hessell AJ, Luckay A, Harouse J, Cheng-Mayer C, Moore JP, Burton DR. 2001. Antibody protects macaques against vaginal challenge with a pathogenic R5 simian/human immunodeficiency virus at serum levels giving complete neutralization in vitro. *J Virol* **75**: 8340–8347.

Polacino P, Stallard V, Montefiori DC, Brown CR, Richardson BA, Morton WR, Benveniste RE, Hu SL. 1999a. Protection of macaques against intrarectal infection by a combination immunization regimen with recombinant simian immunodeficiency virus SIVmne gp160 vaccines. *J Virol* **73**: 3134–3146.

Polacino PS, Stallard V, Klaniecki JE, Pennathur S, Montefiori DC, Langlois AJ, Richardson BA, Morton WR, Benveniste RE, Hu SL. 1999b. Role of immune responses against the envelope and the core antigens of simian immunodeficiency virus SIVmne in protection against us cloned and uncloned virus challenge in Macaques. *J Virol* **73**: 8201–8215.

Poon B, Safrit JT, McClure H, Kitchen C, Hsu JF, Gudeman V, Petropoulos C, Wrin T, Chen IS, Grovit-Ferbas K. 2005.

Induction of humoral immune responses following vaccination with envelope-containing, formaldehyde-treated, thermally inactivated human immunodeficiency virus type 1. *J Virol* **79:** 4927–4935.

Popovic M, Sarngadharan MG, Read E, Gallo RC. 1984. Detection, isolation, and continuous production of cytopathic retroviruses (HTLV-III) from patients with AIDS and pre-AIDS. *Science* **224:** 497–500.

Prince AM, Reesink H, Pascual D, Horowitz B, Hewlett I, Murthy KK, Cobb KE, Eichberg J. 1991. Prevention of HIV infection by passive immunization with HIV immunoglobulin. *AIDS Res Hum Retrovir* **7:** 971–973.

Putkonen P, Thorstensson R, Ghavamzadeh L, Albert J, Hild K, Biberfeld G, Norrby E. 1991. Prevention of HIV-2 and SIVsm infection by passive immunization in cynomolgous monkeys. *Nature* **352:** 436–438.

Qi Y, Martin MP, Gao X, Jacobson L, Goedert JJ, Buchbinder S, Kirk GD, O'Brien SJ, Trowsdale J, Carrington M. 2006. KIR/HLA pleiotropism: Protection against both HIV and opportunistic infections. *PLoS Pathog* **2:** e79.

Reeves RK, Evans TI, Gillis J, Johnson RP. 2010. Simian immunodeficiency virus infection induces expansion of alpha4beta7+ and cytotoxic CD56+ NK cells. *J Virol* **84:** 8959–8963.

Reynolds MR, Rakasz E, Skinner PJ, White C, Abel K, Ma ZM, Compton L, Napoe G, Wilson N, Miller CJ, et al. 2005. CD8+ T-lymphocyte response to major immunodominant epitopes after vaginal exposure to simian immunodeficiency virus: too late and too little. *J Virol* **79:** 9228–9235.

Robert-Guroff M. 2007. Replicating and non-replicating viral vectors for vaccine development. *Curr Opin Biotechnol* **18:** 546–556.

Rosati M, von Gegerfelt A, Roth P, Alicea C, Valentin A, Robert-Guroff M, Venzon D, Montefiori DC, Markham P, Felber BK, et al. 2005. DNA vaccines expressing different forms of simian immunodeficiency virus antigens decrease viremia upon SIVmac251 challenge. *J Virol* **79:** 8480–8492.

Rossio JL, Esser MT, Suryanarayana K, Schneider DR, Bess JW, Vasquez GM, Wiltrout TA, Chertova F, Grimes MK, Sattentau Q, et al. 1998. Inactivation of human immunodeficiency virus type 1 infectivity with preservation of conformational and functional integrity of virion surface proteins. *J Virol* **72:** 7992–8001.

Rowland-Jones SL, Dong T, Fowke KR, Kimani J, Krausa P, Newell H, Blanchard T, Ariyoshi K, Oyugi J, Ngugi E, et al. 1998. Cytotoxic T cell responses to multiple conserved HIV epitopes in HIV-resistant prostitutes in Nairobi. *J Clin Invest* **102:** 1758–1765.

Salle B, Brochard P, Bourry O, Mannioui A, Andrieu T, Prevot S, Dejucq-Rainsford N, Dereuddre-Bosquet N, Le Grand R. 2010. Infection of macaques after vaginal exposure to cell-associated simian immunodeficiency virus. *J Infect Dis* **202:** 337–344.

Santra S, Seaman MS, Xu L, Barouch DH, Lord CI, Lifton MA, Gorgone DA, Beaudry KR, Svehla K, Welcher B, et al. 2005. Replication-defective adenovirus serotype 5 vectors elicit durable cellular and humoral immune responses in nonhuman primates. *J Virol* **79:** 6516–6522.

Schmitz JE, Kuroda MJ, Santra S, Sasseville VG, Simon MA, Lifton MA, Racz P, Tenner-Racz K, Dalesandro M, Scallon BJ, et al. 1999. Control of viremia in simian immunodeficiency virus infection by CD8+ lymphocytes. *Science* **283:** 857–860.

Seaman MS, Xu L, Beaudry K, Martin KL, Beddall MH, Miura A, Sambor A, Chakrabarti BK, Huang Y, Bailer R, et al. 2005. Multiclade human immunodeficiency virus type 1 envelope immunogens elicit broad cellular and humoral immunity in rhesus monkeys. *J Virol* **79:** 2956–2963.

Shibata R, Igarashi T, Haigwood N, Buckler-White A, Ogert R, Ross W, Willey R, Cho MW, Martin MA. 1999. Neutralizing antibody directed against the HIV-1 envelope glycoprotein can completely block HIV-1/SIV chimeric virus infections of macaque monkeys [see comments]. *Nat Med* **5:** 204–210.

Shiver JW, Fu TM, Chen L, Casimiro DR, Davies ME, Evans RK, Zhang ZQ, Simon AJ, Trigona WL, Dubey SA, et al. 2002. Replication-incompetent adenoviral vaccine vector elicits effective anti-immunodeficiency-virus immunity. *Nature* **415:** 331–335.

Siegal FP, Lopez C, Hammer GS, Brown AE, Kornfeld SJ, Gold J, Hassett J, Hirschman SZ, Cunningham-Rundles C, Adelsberg BR, et al. 1981. Severe acquired immunodeficiency in male homosexuals, manifested by chronic perianal ulcerative herpes simplex lesions. *N Engl J Med* **305:** 1439–1444.

* Siliciano RF, Greene WC. 2011. HIV latency. *Cold Spring Harb Perspect Med* doi: 10.1101/cshperspect.a007096.

Simon V, Zennou V, Murray D, Huang Y, Ho DD, Bieniasz PD. 2005. Natural variation in Vif: differential impact on APOBEC3G/3F and a potential role in HIV-1 diversification. *PLoS Pathog* **1:** e6.

Sodora DL, Gettie A, Miller CJ, Marx PA. 1998. Vaginal transmission of SIV: Assessing infectivity and hormonal influences in macaques inoculated with cell-free and cell-associated viral stocks. *AIDS Res Hum Retroviruses* **14 Suppl 1:** S119–S123.

Souquiere S, Onanga R, Makuwa M, Pandrea I, Ngari P, Rouquet P, Bourry O, Kazanji M, Apetrei C, Simon F, et al. 2009. Simian immunodeficiency virus types 1 and 2 (SIV mnd 1 and 2) have different pathogenic potentials in rhesus macaques upon experimental cross-species transmission. *J Gen Virol* **90:** 488–499.

Steimer KS, Scandella CJ, Skiles PV, Haigwood NL. 1991. Neutralization of divergent HIV-1 isolates by conformation-dependent human antibodies to Gp120. *Science* **254:** 105–108.

Stone M, Keele BF, Ma ZM, Bailes E, Dutra J, Hahn BH, Shaw GM, Miller CJ. 2010. A limited number of simian immunodeficiency virus (SIV) env variants are transmitted to rhesus macaques vaginally inoculated with SIV-mac251. *J Virol* **84:** 7083–7095.

Stott EJ, Almond N, Kent K, Walker B, Hull R, Rose J, Silvera P, Sangster R, Corcoran T, Lines J, et al. 1998. Evaluation of a candidate human immunodeficiency virus type 1 (HIV-1) vaccine in macaques: Effect of vaccination with HIV-1 gp120 on subsequent challenge with heterologous simian immunodeficiency virus-HIV-1 chimeric virus. *J Gen Virol* **79** (Pt 3)**:** 423–432.

Stremlau M, Owens CM, Perron MJ, Kiessling M, Autissier P, Sodroski J. 2004. The cytoplasmic body component

TRIM5alpha restricts HIV-1 infection in Old World monkeys. *Nature* **427:** 848–853.

Stromberg K, Benveniste RE, Arthur LO, Rabin H, Giddens WE Jr, Ochs HD, Morton WR, Tsai CC. 1984. Characterization of exogenous type D retrovirus from a fibroma of a macaque with simian AIDS and fibromatosis. *Science* **224:** 289–282.

Swigut T, Alexander L, Morgan J, Lifson J, Mansfield KG, Lang S, Johnson RP, Skowronski J, Desrosiers R. 2004. Impact of Nef-mediated downregulation of major histocompatibility complex class I on immune response to simian immunodeficiency virus. *J Virol* **78:** 13335–13344.

Uberla K. 2005. Efficacy of AIDS vaccine strategies in non-human primates. *Med Microbiol Immunol* **194:** 201–206.

Ulmer JB, Donnelly JJ, Parker SE, Rhodes GH, Felgner PL, Dwarki VJ, Gromkowski SH, Deck RR, EdWitt CM, Friedman A, et al. 1993. Heterologous protection against influenza by injection of DNA encoding a viral protein. *Science* **259:** 1745–1749.

Van Rompay KK, Berardi CJ, Dillard-Telm S, Tarara RP, Canfield DR, Valverde CR, Montefiori DC, Cole KS, Montelaro RC, Miller CJ, et al. 1998. Passive immunization of newborn rhesus macaques prevents oral simian immunodeficiency virus infection. *J Infect Dis* **177:** 1247–1259.

Van Rompay KK, Greenier JL, Cole KS, Earl P, Moss B, Steckbeck JD, Pahar B, Rourke T, Montelaro RC, Canfield DR, et al. 2003. Immunization of newborn rhesus macaques with simian immunodeficiency virus (SIV) vaccines prolongs survival after oral challenge with virulent SIV-mac251. *J Virol* **77:** 179–190.

Veazey RS, Gauduin MC, Mansfield KG, Tham IC, Altman JD, Lifson JD, Lackner AA, Johnson RP. 2001. Emergence and kinetics of simian immunodeficiency virus-specific CD8(+) T cells in the intestines of macaques during primary infection. *J Virol* **75:** 10515–10519.

* Walker B, McMichael A. 2011. The T-cell response to HIV. *Cold Spring Harb Perspect Med* doi: 10.1101/cshperspect.a007054.

Wang SW, Kozlowski PA, Schmelz G, Manson K, Wyand MS, Glickman R, Montefiori D, Lifson JD, Johnson RP, Neutra MR, et al. 2000. Effective induction of simian immunodeficiency virus-specific systemic and mucosal immune responses in primates by vaccination with proviral DNA producing intact but noninfectious virions. *J Virol* **74:** 10514–10522.

Watkins DI, Burton DR, Kallas EG, Moore JP, Koff WC. 2008. Nonhuman primate models and the failure of the Merck HIV-1 vaccine in humans. *Nat Med* **14:** 617–621.

Wiseman RW, Karl JA, Bimber BN, O'Leary CE, Lank SM, Tuscher JJ, Detmer AM, Bouffard P, Levenkova N, Turcotte CL, et al. 2009. Major histocompatibility complex genotyping with massively parallel pyrosequencing. *Nat Med* **15:** 1322–1326.

Wyand MS, Manson KH, Garcia-Moll M, Montefiori D, Desrosiers RC. 1996. Vaccine protection by a triple deletion mutant of simian immunodeficiency virus. *J Virol* **70:** 3724–3733.

Wyand MS, Manson K, Montefiori DC, Lifson JD, Johnson RP, Desrosiers RC. 1999. Protection by live, attenuated simian immunodeficiency virus against heterologous challenge. *J Virol* **73:** 8356–8363.

Zhang F, Wilson SJ, Landford WC, Virgen B, Gregory D, Johnson MC, Munch J, Kirchhoff F, Bieniasz PD, Hatziioannou T. 2009. Nef proteins from simian immunodeficiency viruses are tetherin antagonists. *Cell Host Microbe* **6:** 54–67.

Human Immunodeficiency Virus Vaccine Trials

Robert J. O'Connell[1,3], Jerome H. Kim[1,3], Lawrence Corey[2,3], and Nelson L. Michael[1,3]

[1]U.S. Military HIV Research Program, Walter Reed Army Institute of Research, Silver Spring, Maryland 20910

[2]Fred Hutchinson Cancer Research Center, Seattle, Washington 98109-1024

Correspondence: nmichael@hivresearch.org

More than 2 million AIDS-related deaths occurred globally in 2008, and more than 33 million people are living with HIV/AIDS. Despite promising advances in prevention, an estimated 2.7 million new HIV infections occurred in that year, so that for every two patients placed on combination antiretroviral treatment, five people became infected. The pandemic poses a formidable challenge to the development, progress, and stability of global society 30 years after it was recognized. Experimental preventive HIV-1 vaccines have been administered to more than 44,000 human volunteers in more than 187 separate trials since 1987. Only five candidate vaccine strategies have been advanced to efficacy testing. The recombinant glycoprotein (rgp)120 subunit vaccines, AIDSVAX B/B and AIDSVAX B/E, and the Merck Adenovirus serotype (Ad)5 viral-vector expressing HIV-1 Gag, Pol, and Nef failed to show a reduction in infection rate or lowering of postinfection viral set point. Most recently, a phase III trial that tested a heterologous prime-boost vaccine combination of ALVAC-HIV vCP1521 and bivalent rgp120 (AIDSVAX B/E) showed 31% efficacy in protection from infection among community-risk Thai participants. A fifth efficacy trial testing a DNA/recombinant(r) Ad5 prime-boost combination is currently under way. We review the clinical trials of HIV vaccines that have provided insight into human immunogenicity or efficacy in preventing HIV-1 infection.

CHALLENGES FOR HIV-1 VACCINES

Several factors make development of a vaccine protective against HIV-1 infection a formidable scientific and technological challenge (Douek et al. 2006; Barouch 2008). Extraordinary viral diversity is perhaps the most intractable obstacle to vaccine development. Envelope amino acid sequence diversity among the nine subtypes (A, B, C, D, F, G, H, J, and K) and more than 35 circulating recombinant forms can vary up to 20% within a particular subtype and 35% between subtypes (Walker and Korber 2001; Gaschen et al. 2002). Extremely rapid and error-prone replication yields a large number of mutant genomes, some of which are able to escape immune control (Richman et al. 2003; Goulder and Watkins 2004; Mascola and Montefiori 2010). Another major obstacle is the lack of clear immune correlates of protection in humans (Pantaleo and Koup 2004; Plotkin 2008). As natural immune responses

---

[3]The views expressed here belong solely to the investigators and are not to be construed to reflect the views of the Department of the Army, nor the Department of Defense.

against HIV fail to prevent infection or eradicate the virus, HIV-1 vaccine development cannot emulate the disease-free immune state. Candidate vaccine immunogenicity can be characterized, but these responses cannot be rationally weighted for further evaluation in the absence of correlates of protection. Broadly neutralizing antibodies (NAb) do occur rarely in HIV-1-infected individuals (Simek et al. 2009; Zhou et al. 2010), and passive administration of high doses of monoclonal antibodies affords protection to simian human immunodeficiency virus (SHIV) infection in nonhuman primates (Baba et al. 2000; Mascola et al. 2000; Hessell et al. 2010). However, immunogens that elicit such antibodies have been elusive for many reasons including tolerance control and immunoregulation (Johnson and Desrosiers 2002; Haynes et al. 2005), sequestration of the epitope in the lipid membrane (Sun et al. 2008), and exposure of epitopes only transiently during viral entry (Frey et al. 2008). Cell mediated immunity also develops in most infected individuals in the form of cytotoxic T-lymphocyte (CTL) cell activity which suppresses HIV replication and produces β-chemokines but fails to eradicate infection (Cocchi et al. 1995; D'Souza and Harden 1996; Mackewicz et al. 1996). Finally, long-lived latent tissue reservoirs are established very early in infection, greatly complicating eradication of infection (Chun et al. 1997, 1998).

GENERAL APPROACHES TO VACCINE DEVELOPMENT

Prophylactic vaccines against HIV-1 have attempted to accomplish one of two goals: prevent establishment of infection through generation of NAb, or generation of T-cell responses that result in attenuation of pathogenesis once infection occurs (Douek et al. 2006; McMichael 2006). The latter approach has been referred to as "T-cell" vaccination (Korber et al. 2009). Whole inactivated HIV-1 vaccines (WIVs) have not been seriously considered for human vaccination largely owing to concerns that inactivation might be incomplete, especially when viral aggregates occur. Additional

complicating factors include a loss of antigenicity seen with conventional virus inactivation strategies, relatively modest neutralization induced by WIVs using alternative inactivated approaches (Poon et al. 2005), and marginal protective efficacy in rhesus macaques (Lifson et al. 2004). Although live simian immunodeficiency virus (SIV) vaccine attenuated by deletion of the *nef* gene has shown protection against SIV infection (Daniel et al. 1992; Joag et al. 1998; Wyand et al. 1999), safety concerns preclude its development in humans. First, in vivo *nef* repair and evolution occurs in SIV-infected macaques (Whatmore et al. 1995), and pathogenicity is at least partially retained both in macaques infected with SIVΔ*nef* (Baba et al. 1995; Cohen 1997) and in humans infected with *nef*-deleted HIV-1(Mariani et al. 1996).

These considerations have directed most vaccine efforts toward newer strategies that employ synthetic envelope protein subunits or HIV-1 protein expression via recombinant viral vectors with HIV-specific inserts, or naked DNA. Heterologous prime-boost approaches are frequently used because of the early observation that such regimens often strengthen and broaden HIV-specific immune responses (Cooney et al. 1993; Excler and Plotkin 1997; Ranasinghe and Ramshaw 2009). Prime-boost strategies are not new to medical science: Knowledge that naturally occurring immune responses may be boosted has existed since Robert Koch showed that microbe-derived antigen provoked an immune response at injection sites in tuberculosis patients (Burke 1993). More than 100 NHP and human clinical trials have evaluated prime-boost HIV-1 vaccine strategies (Paris et al. 2010). In this article, we focus on HIV preventive vaccine strategies that have progressed to at least phase II human testing (Table 1).

MODELING HIV VACCINE DEVELOPMENT

Studies evaluating chronically HIV-1-infected individuals have led to many important insights that inform vaccine development. For a complete discussion of HIV-1 pathogenesis, please refer to Lackner et al. (2011).

Table 1. Completed phase II and III human HIV-1 vaccine trials

Category	Study protocol	Candidate vaccine	Phase	Volunteers	Location	Year published	Result
Pox-protein	RV144	ALVAC-HIV vCP1521/AIDSVAX MN-CM244 rgp120 (CRF01_AE, B)	III	16,403	Thailand	2009	31% efficacy
Pox-protein	HVTN203	ALVAC vCP1452/MN-GNE8 rgp120 (B)	II	330	US	2007	Interferon-γ ELISpot in 16% of volunteers
Pox-protein	HIVNET 026	ALVAC vCP1452/MN rgp120 (B)	II	200	Multinational	2007	Not immunogenic
Pox-protein	AVEG 202/HIVNET 014	ALVAC-HIV vCP205/SF2 rgp120 (B)	II	420	US	2001	CD8$^+$ T cells in 33% of volunteers
DNA-pox	IAVI 010	DNA-HIVA/MVA-HIVA (A)	IIa	115	East Africa	2007	Rare interferon-γ ELISpot responses
DNA-pox	HVTN 205	GeoVax JS7 DNA/MVA HIV62 (B)	II	225	US, Peru, RSA	–	Ongoing
DNA-Ad5	RV 172	DNA (VRC-HIVDNA016-00-VP)/rAd5(VRC-HIVADV014-00-VP (A, B, and C)	I/IIa	324	East Africa	2010	Interferon-γ ELISpot in 63% of volunteers
DNA-Ad5	HVTN 204	DNA (VRC-HIVDNA016-00-VP)/rAd5(VRC-HIVADV014-00-VP (A, B, and C)	IIa	480	Americas, RSA	–	Interferon-γ ELISpot in >60% of volunteers
DNA-Ad5	HVTN 505	DNA (VRC-HIVDNA016-00-VP)/rAd5(VRC-HIVADV014-00-VP (A, B, and C)	IIb	1350	US	–	Ongoing
DNA-Ad5	HVTN 502/Merck 023	MRKAd5 HIV-1 gag/pol/nef (B)	IIb	3000	US	2008	No efficacy; transient infection risk
DNA-Ad5	HVTN 503	MRKAd5 HIV-1 gag/pol/nef	IIb	3000	RSA	–	No efficacy
Pox-protein	ACTG 326; PACTG 326	ALVAC vCP1452/AIDSVAX B/B	I/II	48	US	2005	Safe, poorly immunogenic
Peptide	ANRS VAC 18	LIPO-5	II	156	France	2010	CD8$^+$ responses in >60%
Protein-protein	AVEG 201	rgp120/HIV-1 SF-2/MN rgp120	II	296	US	2000	NAb in 87%, DTH to gp120 in 59% of volunteers
AAV	IAVI A002	tgAAC09	II	84	RSA, Uganda, Zambia	2010	Interferon-γ ELISpot in 25% of volunteers
Protein	VAX 003	AIDSVAX B/E	III	2500	Thailand	2006	No efficacy
Protein	VAX 004	AIDSVAX B/B	III	5400	US	2005	No efficacy

Shown are completed clinical HIV vaccine trials grouped by vaccine types (Category).

Pox, recombinant poxvirus-vectored vaccine; DNA, deoxyribonucleic acid; Ad5, recombinant adenovirus 5-vectored vaccine; US, United States; RSA, Republic of South Africa; Nab, neutralizing antibodies; DTH, delayed type hypersensitivity.

Acute Infection

HIV infection acquired sexually begins in CD4$^+$ T cells or macrophages in vaginal or rectal mucosa and remains confined to mucosa or regional lymphoid tissue for a few days during the "eclipse phase" before exponential replication of virus in plasma and establishment of the reservoir (McMichael et al. 2010). Unfortunately, immune responses elicited by HIV-1 infection fail to prevent infection. However, many insights from these well-characterized adaptive and innate responses offer hope that optimized vaccine-induced responses may be protective. Studies of acutely HIV-1 infected humans have shown that the majority of sexually acquired infections are caused by a single transmitted/founder (T/F) virus (Keele et al. 2008; Abrahams et al. 2009; Salazar-Gonzalez et al. 2009). Unlike viruses circulating in chronically infected humans, T/F viruses are more likely to be CCR5-tropic and are less macrophage-tropic (Salazar-Gonzalez et al. 2009). Initial viral uniformity could make the virus more easily neutralized if an effective immune response were present at the time of exposure. Adaptive cellular (CD8$^+$) immune responses drive both viral suppression and diversity through escape mutants (Goonetilleke et al. 2009; Treurnicht et al. 2010). Similarly, HIV-specific antibody responses, which typically mature over time, significantly shape the generation of neutralization escape mutants but fail to neutralize contemporaneous strains (Richman et al. 2003; Wei et al. 2003; Moore et al. 2009).

Nonhuman Primate Models

Nonhuman primate (NHP) models have provided important insights into HIV-1 vaccine development. Four models utilizing different viruses or host species have been used: HIV-1/chimpanzee, HIV-2/macaque, SIV/macaque, and SHIV/macaque. The SIV/macaque and SHIV/macaque models are currently most commonly used because of low levels of HIV replication, prolonged time to progression, and high cost of HIV-2 and chimpanzee models (Franchini et al. 2004). NHP experiments

inform vaccine development through elegant experiments aimed at elucidation of SIV or SHIV pathogenesis or vaccine performance. An SIV-macaque study showed increased survival, reduction in viremia, and preservation of central memory CD4$^+$ T lymphocytes following delivery of a plasmid DNA prime followed by type 5 Adenovirus vector boost vaccine (Letvin et al. 2006). This study, along with human immunogenicity data from phase I studies, raised enthusiasm for an ongoing phase IIB efficacy study of these vaccines (ClinicalTrials.gov, NCT00865566), in the HIV Vaccine Trial Network (HVTN) 505 protocol. A similar approach using a SHIV/macaque model (Shiver et al. 2002) failed to predict the lack of efficacy in a phase IIB evaluation of the Merck Ad5 gag/pol/nef vaccine (Buchbinder et al. 2008). This vaccine failed to show postinfection viremic control in a SIV challenge model (Casimiro et al. 2005). Unfortunately, there are no animal models that accurately predict efficacy in humans. High-dose challenge NHP models are potentially confounded by high SIV challenge doses used to achieve 100% infection rates after a single exposure in placebo animals (McDermott et al. 2004). Repeated mucosal challenges with a lower dose of virus (10–50 TCID$_{50}$) may more accurately approach human exposure conditions, but such studies remain to be correlated with human clinical trial outcomes.

DNA AND PROTEIN SUBUNIT VACCINE TRIALS

Protein Subunit Alone

Initial clinical trials of candidate HIV vaccines in the late 1980s and early 1990s attempted to follow the template for hepatitis B vaccine: Use recombinant subunits or synthetic peptide fragments to elicit neutralizing antibodies against viral antigens expressed on the virion surface. For HIV-1, these are the Env proteins gp120 and gp41. These antigens typically elicit strong binding antibody, limited neutralizing antibody (Nab) and CD4$^+$, but not CD8$^+$, T-cell responses (Pantophlet and Burton 2006).

Following a phase I/II trial that showed safety and immunogenicity (Belshe et al. 1994), the first phase III efficacy trial of a prophylactic HIV-1 vaccine (VAX 004) investigated a recombinant HIV-1 envelope glycoprotein subunit (rgp120) derived from MN, a laboratory-grown strain, and a second envelope derived from a clade B primary isolate (GNE8) (AIDSVAX B/B') in alum adjuvant. Study participants were men who have sex with men and women at high risk for heterosexual transmission of HIV-1 from 61 sites in the United States, Puerto Rico, Canada, or the Netherlands (Harro et al. 2004; Flynn et al. 2005). Participants received vaccine or placebo at months 0, 1, 6, 12, 18, 24, and 30. Infection rates among the 3598 vaccinees and 1805 placebo recipients were similar at 6.7% and 7.0%, respectively. There were no differences in postinfection secondary end points including viral load, $CD4^+$ T-cell count, time interval to initiation of antiretroviral therapy, or genetic characteristics of the infecting virus. Vaccine induced binding antibody responses were inversely correlated with risk of infection. For all eight antibody variables, the mean responses tended to be slightly higher in uninfected vaccinees than in infected vaccinees, suggesting that such antibody responses either (1) caused both increased (low responders) and decreased (high responders) risk of HIV acquisition or (2) represented a correlate as opposed to causative mechanism for enhanced HIV-1 acquisition. There was some suggestion that there were both higher antibody titers and protection in African-American subjects, but incorporation of a correction for multiple sampling diminished this finding. Neutralizing antibodies did not correlate with infection incidence (Gilbert et al. 2005).

Following the demonstration of safety and immunogenicity in a phase I/II trial in Bangkok, Thailand (Pitisuttithum et al. 2004), a second phase III efficacy trial of a vaccine candidate was undertaken (VAX003). This study represented the first efficacy trial conducted in a developing country, and it was the first to exclusively study an intravenous drug user (IDU) population. The vaccine contained two different rgp120 antigens, one from subtype B MN, and the other from a primary isolate CRF_AE (A244) produced in Chinese hamster ovary (CHO) cells in alum adjuvant. A total of 2546 IDUs were enrolled between March 1999 and August 2000 and received vaccine or placebo at months 0, 1, 6, 12, 18, 24, and 36. Adverse events were rare and occurred with equal frequency among vaccine and placebo recipients. Vaccine efficacy was estimated at 0.1% (95% CI, −30.8% to 23.8%), and no effect was observed on secondary end points (Pitisuttithum et al. 2006; Pitisuttithum 2008).

Among 2099 uninfected subjects in phase I and II trials of Env-based subunit AIDS vaccines, 23 were diagnosed with intercurrent HIV-1 infection. There were no significant differences in secondary end points, including virus load, CD4 lymphocyte count, or V3 loop amino acid sequence (Graham et al. 1998).

Further nonprespecified analyses of Vax 004 antibody-directed cell-mediated viral inhibition (ADCVI) showed an inverse correlation between ADCVI levels and HIV acquisition. This effect was influenced by Fc-γ IIa and IIIa polymorphisms (Forthal et al. 2007). More recently, Gilbert et al. reported low levels of neutralizing antibody against Tier 2 isolates in Vax 004 (Gilbert et al. 2010); however, when the same data are analyzed using two different assays (Monogram and TZM-bl), there appeared to be an inverse correlation between breadth of (low-level) Tier 2 neutralization and infection (B Korber, unpubl.).

DNA Prime Protein Subunit Boost

Strategies employing DNA prime followed by protein boost have only been studied in a single published phase I human trial. A polyvalent DNA prime vaccine containing five plasmids each encoding a codon-optimized protein including gp120 sequences from subtypes A, B, C, and E, as well as a sixth plasmid encoding a subtype C *gag* gene was performed. Protein boost components included equal parts of five gp120 proteins matching those used in DNA prime components formulated with QS-21 adjuvant and excipient cyclodextrin (Wang et al. 2008). DNA vaccination was administered in two different doses intramuscularly (IM), as well as in one dose intradermally (ID). The vaccine strategy elicited cross-subtype HIV-specific T cell responses as well as high titer serum antibody responses against HIV-1 viruses with diverse genetic backgrounds. These results were tempered by observations of delayed type hypersensitivity in 43% of subjects following protein vaccination, and two cases of vasculitis temporally related to inoculation with recombinant Env protein + QS21 adjuvant (Kennedy et al. 2008).

Early naked DNA plasmid vaccines given alone were poorly immunogenic and responses lacked durability (MacGregor et al. 1998, 2000, 2002; Wang et al. 1998; Roy et al. 2000). Such observations have led to pursuit of alternative strategies to enhance immune responses (Abdulhaqq and Weiner 2008), and contemporary strategies typically use DNA vaccines as a "prime" with a heterologous boost. Electroporation is a promising approach that has been used to enhance in vivo transfection rates in nonhuman primates (Rosati et al. 2008; Patel et al. 2010) and more recently in several phase I HIV-1 preventive DNA vaccine human trials (Vasan et al. 2009; clinicaltrials.gov NCT00991354 and NCT01260727).

PEPTIDE VACCINE TRIALS

Several synthetic peptide constructs have been investigated as potential preventive HIV-1 vaccines. Sequences from the envelope V3 loop induce Nab in some laboratory animals but were tolerogenic in macaques and chimpanzees. An octameric V3 multiple antigen peptide formulated in alum was found to be safe but generated neither consistent nor robust lymphoproliferative responses in human volunteers (Kelleher et al. 1997). Peptides do not usually induce a class I-restricted CD8$^+$ response in vivo, but some lipopeptides elicit such a response (Deres et al. 1989; Deprez et al. 1996). Building on this concept, a phase II human trial evaluated HIV-LIPO-5 vaccine (five long peptides, Gag17−35, 253−284, Pol325−355, Nef66−97, and 116−145, containing multiple

CD8$^+$ and CD4$^+$ epitopes, coupled to a palmitoyl tail) in 132 volunteers. Vaccination, which was given IM at one of three doses at weeks 0, 4, 12, and 24, elicited CD8$^+$ responses as measured by IFN-γ ELISpot in two-thirds of vaccine recipients regardless of dose, and CD4$^+$ responses as measured by PBMC lymphoproliferation in approximately half of vaccine recipients regardless of dose (Salmon-Céron et al. 2010). Future studies will likely incorporate lipopeptides in prime-boost regimens.

ADENOVIRAL VECTOR VACCINE TRIALS

DNA and adenovirus constructs have been some of the most extensively tested in the NHP model. Enthusiasm for Merck DNA–Ad5 was generated by the comparative study by Shiver et al. in which macaques that received either DNA–Ad5 or Ad5 alone, and were challenged with a homologous SIV with an HIV-derived envelope (SHIV)89.6P virus, controlled viral replication better than monkeys that received rMVA (Shiver et al. 2002). A postinfection viral load reduction was not seen when SIVmac239 was the challenge virus (Casimiro et al. 2005). The results of the Merck Step trial (discussed in the following paragraphs) have led to a reevaluation of the NHP challenge model, in particular, questions about dose (low vs. high dose), route (intravenous vs. mucosal), SHIV versus SIV challenge, virus challenge stock (single clone vs. "swarm," heterologous vs. homologous), as well as end points (Fauci et al. 2008; Watkins et al. 2008). Recently, statistically significant differences in control of viral replication after challenge have been observed in the macaque model using a heterologous challenge with SIVsmE660 after vaccination with DNA–Ad5 that expresses all HIV proteins except Env from SIVmac239 (Wilson et al. 2009). DNA-Ad5 SIV-based vaccines protected against an SIVsmE660 acquisition after low-dose intrarectal challenge (Letvin et al. 2011). Adenovirus serotypes that are less commonly associated with human disease have also recently been developed to overcome problems with preexisting immunity, and Ad26 and 35 vaccines are now in a Phase I clinical trial

(Baden et al. 2009). Results from NHP challenge studies are encouraging—particularly the strategy of priming with Ad5 and boosting with Ad26, which was associated with a 2.4 log reduction in set-point viral load and improved survival compared with control animals (Liu et al. 2009).

The most extensively tested of the adenoviral vector vaccine candidates in human trials is Merck's replication-defective Ad5, containing clade B *gag*, *pol*, and *nef* gene inserts. This candidate was advanced to a Phase IIb efficacy evaluation that was stopped at a scheduled interim analysis when it was apparent that there was no effect on postinfection viral RNA level (Buchbinder et al. 2008), despite demonstration of apparent immunogenicity in Phase I clinical trials and evidence of viral load control in the SHIV/NHP model (Shiver et al. 2002; Casimiro et al. 2003; Wilson et al. 2006; Priddy et al. 2008; Harro et al. 2009; Asmuth et al. 2010). Although mechanisms underlying failure of this vaccine remain elusive, one explanation is that this homologous vector vaccine, which elicits frequent IFN-γ ELISpot responses (77% of vaccinees overall) when given in three doses, generates a limited breadth of antigen-specific responses (McElrath et al. 2008). Likewise, the absolute level of IFN-γ ELISpot may have been too low, or given the inferior IFN-γ ELISpot generation by the ALVAC + AIDSVAX B/E combination, the ELISpot assay may not measure relevant responses. Intense study evaluated initial vaccine-related enhancement of infection among a subgroup of anti-Ad5 positive uncircumcised men who have sex with men (MSM) (Buchbinder and Duerr 2009; D'Souza and Frahm 2010). Extended follow-up analyses (Benlahrech et al. 2009; Masek-Hammerman et al. 2010) have indicated that increased risk of acquisition decreased over time. The significance and potential mechanism of vaccination-enhanced acquisition among uncircumcised men is unclear, therefore a cautious approach has been taken with Ad5-vectored *gag/pol/nef* vaccines, especially in uncircumcised MSM (Duerr et al. 2010).

The NIH/NIAID Vaccine Research Center has also developed a DNA–rAd5 prime-boost regimen, which has been extensively tested for safety and immunogenicity, and a Phase IIb efficacy, test-of-concept trial is ongoing in the United States (HVTN 505) among circumcised men who have sex with men and lack preexisting antivector-NAbs. The first DNA–rAd5 combination consisted of four separate DNA plasmids (VRC-HIVDNA009–00-VP): One plasmid with gene inserts for *gag* (clade B strain HXB2), *pol* (clade B NL4–3), and *nef* (clade B, PV22) with mutations in *gag* and *pol* to prevent enzymatic activity, creating a fusion protein product that does not produce pseudoparticles. *Env* genes from clades A (92RW020), B (HXB2), and C (97ZA012) were truncated immediately downstream from the transmembrane domain of gp41, along with other deletions, to create three gp145-expressing plasmids. For the rAd5 boost (VRC-HIVADV014–00-VP), four separate vectors were produced with the same inserts (clade/strain matched) expressing a Gag/Pol polyprotein along with gp140 versions of Env. Nef was not included in Ad5 as the vectors were not stable. Separate Phase I dose-ranging trials of each of these products showed that the plasmid DNA was similarly immunogenic at the 4 ($n = 20$) or 8 mg ($n = 15$) doses with 100% and 93% with $CD4^+$ responses by ICS, respectively, and 35% and 33% with $CD8^+$ responses, respectively, with both T-cell subsets exhibiting more Env-specific responses. The rAd5-vectored HIV-immunogens generated ELISpot responses in 22 out of 30 patients (73%), with 28 out of 30 patients having a detectable $CD4^+$ response and 18 out of 30 patients a $CD8^+$ response by ICS at week 4 (peak immunogenicity) after a single IM dose, and mostly Env-specific. Although no NAb were detected in either the DNA or rAd5 trials, Env-specific binding antibody responses measured using ELISA were detected in 50% and 71% of rAd5 and DNA recipients, respectively (Catanzaro et al. 2006, 2007; Graham et al. 2006).

These constructs were subsequently tested in three Phase IIa trials, IAVI V001, HVTN 204, and RV172 (Kibuuka et al. 2010). In the RV172 study, the VRC DNA (VRC-HIVDNA-016–00-VP) tested was a six-plasmid mixture encoding HIV *env* from subtypes A, B, and C, and subtype B *gag*, *pol*, and *nef*, and VRC-HIVADV014–00-VP. Volunteers were randomized to receive either 10^{10} ($n = 24$) or 10^{11} ($n = 24$) particle units (PUs) of rAd5 on day 0 only; 4 mg of DNA at 0, 1, and 2 months, followed by rAd5 at either 10^{10} ($n = 114$) or 10^{11} PUs ($n = 24$) boosting at 6 months. HIV-specific T-cell responses were detected in 63% of vaccinees. ELISpot responses for DNA prime with low-dose (63%) or high-dose (60%) rAd5 were similar—positive responses were predominantly to Env peptides, followed by Pol or Gag, regardless of the immunization regimen. The high-dose rAd5 boost had the highest frequency of responders to all three antigens tested (Env, Gag, or Pol), whereas responses were approximately equal for the other immunization groups for two antigens (20%–26%). Preexisting Ad5-NAbs did not appear to affect the frequency or magnitude of T-cell responses in the prime-boost vaccinees. Response rates in participants that received rAd5 alone were lower, but not statistically significant.

Rare Serotype Ad Vectors

Antivector immunity against adenovirus vectors varies by population and serotype. Ad5 and Ad35 seropositivity was detected in 60% and 7%, respectively, in individuals at risk for AIDS, respectively, living in the Netherlands, and 90% vs. 20% among those living in sub-Saharan Africa (Kostense et al. 2004). Vector neutralization threatens to attenuate immunogenicity, prompting construction of vectors using Ad serotypes that have lower frequencies of natural infection in humans, such as Ad36 and Ad26 (Barouch 2010).

POXVIRUS AND PROTEIN SUBUNIT PRIME-BOOST VACCINE TRIALS

Poxviruses have properties that make them excellent expression systems. Characteristics such as large capacity for integration of foreign DNA (more than 25,000 base pairs for vaccinia virus [VV]) and cytoplasmic gene expression are possessed by members of the poxviridae

family. Successful recombinant gene expression using poxvirus was first shown in 1982 and has been used for such diverse activities as analysis of protein structure/function relationships, protein processing and intracellular trafficking, antigen presentation, and the determinants of cellular and humoral immunity, and live recombinant vaccines. Meaningful differences among poxviridae include host specificity and susceptibility to antivector host immunity (Paoletti 1996; Carroll and Moss 1997).

Vaccinia Vectors

Modified Vaccinia Ankara (MVA) was initially developed near the end of the smallpox eradication program by the technique of passaging virus about 500 times on primary chicken embryo fibroblasts (Franchini et al. 2004). Multiple resultant genetic deletions have rendered MVA unable to replicate in most mammalian cells, which is a likely explanation for its excellent safety record: It was assessed as a smallpox vaccine in over 120,000 recipients in Germany without significant adverse reactions. Its large genome renders it amenable to genetic manipulation, making it an ideal candidate vector for vaccine development (Moss 1991; Moss et al. 1996; Blanchard et al. 1998; Sutter and Staib 2003).

Macaque-SIV studies of DNA followed by MVA-vectored boost vaccination showed induction of CTL responses, primarily to a single SIV gag epitope (Hanke et al. 1999; Allen et al. 2000). Given alone IM, MVA-based vaccines have been shown to be safe and immunogenic in human clinical trials (Cebere et al. 2006; Dorrell et al. 2007; Peters et al. 2007; Jaoko et al. 2008). Antivector immunity is an important consideration: Many older adults as well as most members of the U.S. military have been inoculated with vaccinia and thus have at least some degree of antivaccinia immunity.

The first report of immunogenicity data for using prime-boost DNA/MVA vaccine approach evaluated pTHr–HIVA plasmid DNA and MVA–HIVA prime-boost combination, after initial Phase I safety testing (Mwau et al. 2004; Cebere et al. 2006). The HIVA immunogen contained consensus clade A Gag p24/p17 sequences and a string of CTL epitopes inserted into plasmid DNA, and the same immunogens as a transgene insert in recombinant MVA (rMVA). Safety was first shown in several small trials (IAVI 001, IAVI 003 and IAVI 005), followed by Phase I/II studies. Only one antibody response to p24/p17 was detected in any of the participants in these studies (Mwau et al. 2004; Cebere et al. 2006; Jaoko et al. 2008). In the IAVI 006 study, 119 volunteers were randomized into several arms and received pTHr–HIVA DNA in a dose of 0, 0.5, or 2.0 mg in two doses IM, followed by two doses of 5×10^7 rMVA after 4 or 16 weeks. Using a fresh peripheral blood mononuclear cell (PBMC) IFN-γ ELISpot assay, less than 15% of vaccine recipients had transient HIV-specific Gag responses, with no significant effect of DNA priming observed (Hanke et al. 2007; Guimaraes-Walker et al. 2008). In IAVI 016, 24 volunteers received either two doses of rMVA alone at the higher dose of 2.5×10^8 plaque-forming units (pfu) or two doses of pTHr–HIVA DNA at a higher dose of 4 mg (Casimiro et al. 2003; Goonetilleke et al. 2006; Peters et al. 2007; Baden et al. 2009). In this study, immunomonitoring was conducted by ex vivo ELISpot, a "cultured" ELISpot using a 10-d, peptide-stimulated PBMC expansion step, and polyfunctional or multicolor flow cytometry. No antigen-specific responses were detected by the validated ex vivo ELISpot in the rMVA-only group, while four of eight in the DNA–rMVA group had responses. In the "cultured" ELISpot assay, five of eight in the rMVA-only group had responses, while all of those who received DNA–rMVA had Gag-specific responses. This modified ELISpot format detected predominantly CD4+ T-cell responses, with two vaccinees having CD8+ responses, which were of greater magnitude and breadth in those who received the DNA–rMVA combination compared with the two doses of rMVA. Larger trials at these higher doses are ongoing (IAVI 2010).

Another DNA–MVA prime-boost Phase I study, HIVIS 02, has been conducted. This study utilized plasmid DNA constructs developed at the Karolinska Institute, and the Walter Reed Army Institute of Research/

National Institutes of Health (WRAIR/NIH)-produced recombinant MVA-CMDR (Chiang Mai Double Recombinant), which has inserts based on CRF01_AE isolates from Thailand that express gp150 (CM235), gag and pol, with a deleted integrase and nonfunctional reverse transcriptase (CM240). MVA-CMDR was shown to be safe and immunogenic in a separate Phase I trial (Currier and de Souza 2009). A total of 40 healthy HIV-negative participants were randomized into four groups ($n = 10$ per group) and injected with seven DNA plasmids expressing subtypes A, B, and C gp160 *env*; B *rev*; subtype A and B p17/p24 *gag*, and clade B *rt-mut*. The DNA priming dose was administered ID or IM with or without recombinant granulocyte-macrophage colony-stimulating factor (GM-CSF) using the Biojector 2000. A single boost of MVA-CMDR was given 6 months after the last DNA injection at either 10^7 pfu ID or 10^8 pfu IM. Eleven of 37 (30%) had positive ELISpot after DNA alone, whereas significant boosting was observed after MVA-CMDR with 34 of 37 (92%) vaccinees responding (32 to Gag and 24 to Env). One mg of DNA administered ID was as effective as 4 mg of DNA IM as a prime for the MVA-CMDR boost. In addition, 68% had an IL-2 ELISpot response and 92% a HIV-1-specific lymphocyte proliferation assay (LPA) response. These results were confirmed in a second study conducted in Tanzania (Aboud et al. 2010). A comparison of these DNA vectors delivered by Biojector and electroporation with subsequent rMVA boosting will evaluate the role of DNA delivery route (clinicaltrials.gov NCT01260727).

The GeoVax clade B DNA–MVA was tested in HVTN 065. Volunteers received IM DNA and MVA that expressed clade B Gag, Pol, and Env, with DNA given at 0 and 2 months and boosted at months 4 and 6 with MVA. Both low- (0.3 mg of DNA and 107 of the 50% tissue culture infective dose [TCID50] of MVA) and high-dose (3 mg of DNA and 108 TCID50 of MVA) formulations were tested. Preliminary results showed similar T-cell responses in low- and high-dose groups. In the high-dose group, 75% of volunteers had CD4 and 37% had

CD8 T-cell responses. Cosecretion of IL-2 and IFN-g was observed in 82% of CD4 and 67% of CD8 responses at the peak immunogenicity time point (Robinson et al. 2008). Additional DNA/MVA candidates based on subtype C (TBC-M4 MVA) and recombinant B′/C antigens (ADVAX DNA and ADMVA) have completed Phase I testing, but none in combination. ADVAX DNA given alone at 0, 1, and 3 months showed ELISpot responses in four of 12 high-dose recipients, whereas the MVA constructs (given IM at 0, 1, and 6 months) were more immunogenic, with ELISpot positivity in 60%–100% of vaccine recipients at the highest dose given (Ramanathan et al. 2009; Vasan et al. 2010a,b).

Despite initially disappointing results with the DNA–MVA prime-boost approach (Guimaraes-Walker et al. 2008), more recent studies are promising. Effects of MVA boost appear consistent with differences in immunogenicity probably because of DNA priming and dose (Hanke et al. 2007). One concern with using MVA as a vector is the presence of preexisting immunity to vaccinia, which may affect the magnitude and quality of immune responses, as shown with recombinant Ad5 vaccine vector (Priddy et al. 2008; Harro et al. 2009). However, data from the HIVIS-02 study indicate that only the magnitude of the response was attenuated, whereas the frequency of responders was maintained in those with preexisting vaccinia immunity (Gudmundsdotter et al. 2009; Barouch 2010).

NYVAC was derived from VC-2, a plaque-cloned isolate of the COPENHAGEN vaccinia strain. NYVAC construction entailed deletion of 18 open reading frames encoding for replication competency in mammalian cells and virulence factors (Tartaglia et al. 1992).

The EuroVacc 02 Phase I trial evaluated recombinant DNA and NYVAC, which express matched immunogens containing *gag, pol, env*, and *nef* sequences from a Chinese clade B and C recombinant virus isolate, 97CN54 (CRF70_B/C′). Forty volunteers received DNA C or placebo on day 0 and at week 4, followed by NYVAC C boosting (20 received DNA/ NYVAC) at weeks 20 and 24. A total of 90%

showed IFN-γ ELISpot responses compared with 33% who received NYVAC C alone. T-cell responses were of relatively high magnitude, with CD4$^+$ (16 of 18 assessed) more frequent than CD8$^+$ responses (eight of 16). Responses were skewed toward Env (91% of vaccinees), with 48% responding to Gag, Pol, or Nef. Responses after NYVAC boost compared with NYVAC alone suggest that DNA priming was occurring, even when no immune response was detected after the second dose of DNA (Harari et al. 2008). Induced T cells were typically polyfunctional, with 60% exhibiting two to three functions, and both CD4$^+$ and CD8$^+$ T cells typically expressing IFN-γ, IL-2, and TNF-γ. Prime-boost regimens with DNA and poxvirus vectors appear to induce polyfunctional T-cell responses that are biased toward Env-specific T-cell responses, with a predominance of CD4$^+$ over CD8$^+$ T-cell responses.

Canarypox Vectors

ALVAC is a recombinant canarypox virus (CPV)-based vector that functions as an immunization vehicle by expressing gene products in the absence of productive viral replication (Tartaglia et al. 1992). Although there are no licensed ALVAC-based human vaccines, commercial production feasibility is clearly shown by current marketing of five licensed veterinary canarypox-based vaccines: Recombitek (Canine distemper), Purevax (rabies[cats]), Recombitek WNV (West Nile [equine]), Eurifel (feline leukemia), and Proteqflu (influenza[horses]) (Merial 2010). Because CPV is a bird pathogen, canarypox recombinant vectors infect but are unable to replicate in humans, predicting that they will not disseminate in vaccine recipients nor be transmitted to unvaccinated contacts (Taylor et al. 1988; Baxby and Paoletti 1992). In both guinea pig and macaque models, gp160 MN/LAI-2 significantly boosted antibody responses primed with ALVAC-HIV (Excler and Plotkin 1997; Jaoko et al. 2008). ALVAC-HIV-1 expressing homologous vaccine native Env proteins protected macaques from high-dose mucosal challenge, and this protection was better when gp120 antigen was included in the vaccine (Pialoux et al. 1995; Excler and Plotkin 1997). ALVAC-SIV given to neonatal macaques protected against low dose SIV challenge in milk; interestingly, an MVA-SIV vaccine developed better ELISpot responses but did not protect (Van Rompay et al. 2005).

Multiple HIV-1 inserts have been inserted into ALVAC vectors, yielding extensive safety and immunogenicity data in humans (Gilbert et al. 2003; de Bruyn et al. 2004). Phase I human trials with recombinant canarypox vectors have shown induction of CD8$^+$ CTL (notably these are after in vitro stimulation; when direct ex vivo IFN-g or CD107 expression is measured, CD8$^+$ CTL are not identified). However, only 15%–30% of subjects have such responses at any given time postvaccination (Belshe et al. 1998, 2001; Clements-Mann et al. 1998; Evans et al. 1999; Salmon-Céron et al. 1999). Three recombinant ALVAC HIV-1 vaccines have advanced to Phase II studies.

ALVAC-HIV vCP205 expresses four gene products: Gag p55 protein from the HIV-1 LAI strain, p15 protein, a portion of the protease gene from the HIV-1 LAI strain, a portion of gp120 from the HIV-1 MN strain, and the anchoring transmembrane region of gp41 from the HIV-1 LAI strain (Paris et al. 2010). This recombinant vector was studied in a phase II trial conducted in the US among 60 study participants at lower risk for HIV-1 infection and 375 individuals at higher risk for HIV-1 infection. ALVAC-HIV vCP205 was given with or without an SF-2 rgp120 boost at 0, 1, 3, and 6 months (Belshe et al. 2001). More than 90% of combination regimen recipients developed Nab responses against homologous TCLA virus, and approximately one-third of those who received vCP205-containing regimens developed anti-HIV CTL responses, without differences in immune responses among the higher and lower risk groups.

More complex ALVAC vectors, vCP1433 and vCP1452, were designed to provide better gene expression and more durable CD8$^+$ CTL responses. Both express *gag*, protease, *nef*, and *pol* genes, and a part of the *env* gene expressing the gp120 and anchoring region of gp41. The ALVAC vCP1452 vector was modified by insertion of two vaccinia virus-coding sequences (E3L and K3L) to enhance expression efficiency in ALVAC-infected human cells. In large, phase II studies, HVTN 203 and HIVNET 026, vCP1452 was boosted with the bivalent rgp120-containing sequence from the clade B primary isolate GNE8 as well as MN. Cellular immunogenicity in this study was not deemed sufficient to proceed with a planned Phase III immune correlates trial (Russell et al. 2007).

Two Phase I/II trials of vCP1521, which is similar to vCP205 except for a different *env*, 92Th023 gp120 (CRF01_AE), substituted for MN, in combination with three different subunit boosts have been conducted in Thailand. In RV132 (vCP1521 + oligomeric rgp160 92TH023/polyphosphazene or rgp120 CM235/SF2/MF59), the majority of recipients (68%–93%) developed either SF2, TH023 or CM235 lymphoproliferative responses to gp120 antigens. In RV132, NAb to the TCLA subtype E strain NPO3 was found in 84% and 89% of recipients of vCP1521 + gp160 TH023 or gp120 CM235/SF2, respectively. In total, 96% and 100% of vCP1521 + gp160 TH023 and vCP1521 + gp120 CM235/SF2 prime-boost recipients had NAb against any subtype E-adapted HIV-1, respectively. Neutralization of homologous and heterologous laboratory-adapted strains of HIV-1 was observed for the majority of vaccine recipients in both prime-boost arms. Cross-neutralization of SF2 by recipients of oligomeric gp160 (27%) was also observed. The NAb response observed for subjects boosted with bivalent gp120 B/CRF01_AE was similar to a previous clinical trial of these antigens among Thai volunteers (Pitisuttithum et al. 2003). In that study, 84% and 96% of volunteers who received both vaccine antigens had NAb to NPO3 and SF2, respectively. Some data suggest that oligomeric antigens may be more likely to induce relevant NAb, although this was not seen in relation to the heterologous CM244 strain in a PBMC-based neutralization assay (Fouts et al. 1997; VanCott et al. 1997). HIV-specific CD8$^+$ CTLs were detected in 11% and 25% of prime-boost recipients in the gp160 TH023 + gp120

CM235/SF2 arms of RV132, respectively (Thongcharoen et al. 2007).

In RV135, 58% and 67% of persons receiving vCP1521 plus high- or low-dose rgp120 (MN/A244) in alum developed A244-specific proliferative responses, whereas 71% had a NAb response against subtype E-adapted HIV-1. HIV-specific CD8[+] CTLs were identified in 23% of vaccinees receiving either boost (Nitayaphan et al. 2004). Significant ADCC activity was also observed in over 85% of vaccine recipients to either clade B or CRF01_AE gp120 (Karnasuta et al. 2005). The vCP1521 prime and high-dose 300 μg of MN and 300 μg of A244 boost was chosen for use in a Phase IIb efficacy trial, RV144.

RV144—Proof of Concept for Protective Efficacy

As neither the oligomeric gp160 92TH023 or the bivalent gp120 B/E vaccine were available for further testing, the vaccine combination in RV135, which had passed the predetermined immunogenicity criteria for advancement, was selected for efficacy testing. The Thai "Phase III" trial, RV144, provided the first evidence that an HIV vaccine could provide protective efficacy (Rerks-Ngarm et al. 2009). The modified intent-to-treat analysis (excluding those randomized, but HIV-infected at the first vaccination visit) showed 31.2% efficacy (95% CI: 1.1,52.1; $p = 0.04$) after 42 months of follow-up after vaccination with ALVAC vCP1521 at 0, 1, 3, and 6 months with boosting with AIDSVAX B/E rgp120 at 3 and 6 months. Although not included in the prespecified analysis plan, vaccine efficacy appeared to be higher (60%, 95% CI 22,80) at 12 months postvaccination, suggesting an early, but nondurable, vaccine effect (Michael 2009). In a subset of HIV-negative volunteers who completed all immunizations, 20% had CD8[+] ELISpot responses to Gag or Env. Intracellular cytokine staining showed predominantly CD4[+] responses favoring Env over Gag (34% vs. 1.4%). Lymphoproliferation to gp120 was approximately 87% for MN and 90% for A244. Few Nab were identified. ELISA binding against gp120 A244, gp120 MN, and

p24 were positive in 98.6%, 98.6%, and 52.1% of vaccinees at 12 months, respectively. The most striking parallel to declining immunogenicity over time was a 10-fold drop in binding antibody to gp120 B and AE over the 6 months from the last vaccination.

THE WAY FORWARD FROM RV144

Identifying correlates of protection would revolutionize HIV-1 vaccine development. To this end, more than 30 investigators are collaborating to intensely characterize immune factors associated with vaccine efficacy (Kim et al. 2010). Studies include measures of Nab, ADCC, blocking of monoclonal antibody binding, immunoglobulin glycosylation and binding kinetics, phagocytosis, transcriptional profiling, immune phenotyping, and antigen specific cellular immune responses, cellular immunogenicity, host genetics testing, and other approaches to attempt to identify potential immune correlates or mechanisms of protection (Koup et al. 2010). Discovery of a correlate of protection from RV144 would provide a rational basis for further improvement of this regimen and newer vaccine approaches. In the absence of such a discovery, HIV vaccine development will have to build on RV144 in a more empirical fashion. This goal can be approached in two parallel, mutually supportive clinical development pathways (Fig. 1).

Product Development Pathway

This pathway takes the most direct extension from RV144, minimizes variable changes, and seeks the shortest pathway to licensure of a public health tool including the use of vaccines tailored for regional distribution of HIV subtypes in populations at highest risk for HIV infection. Because vaccine efficacy in RV144 appeared to wane from 60% at 12 months to 31% at 3.5 years, schedules incorporating booster doses will be evaluated in immunogenicity studies as a means to incrementally improve and sustain a vaccine efficacy >50% at 24 months. Efficacy studies are being discussed in populations at high risk of HIV infection including heterosex-

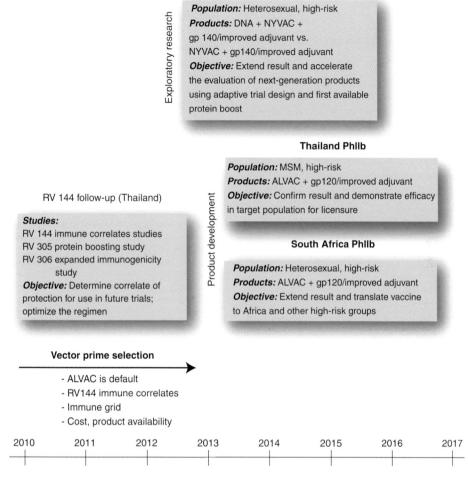

South Africa PhIIb

Population: Heterosexual, high-risk
Products: DNA + NYVAC + gp 140/improved adjuvant vs. NYVAC + gp140/improved adjuvant
Objective: Extend result and accelerate the evaluation of next-generation products using adaptive trial design and first available protein boost

Thailand PhIIb

Population: MSM, high-risk
Products: ALVAC + gp120/improved adjuvant
Objective: Confirm result and demonstrate efficacy in target population for licensure

South Africa PhIIb

Population: Heterosexual, high-risk
Products: ALVAC + gp120/improved adjuvant
Objective: Extend result and translate vaccine to Africa and other high-risk groups

RV 144 follow-up (Thailand)

Studies:
RV 144 immune correlates studies
RV 305 protein boosting study
RV 306 expanded immunogenicity study
Objective: Determine correlate of protection for use in future trials; optimize the regimen

Exploratory research

Product development

Vector prime selection

- ALVAC is default
- RV144 immune correlates
- Immune grid
- Cost, product availability

2010   2011   2012   2013   2014   2015   2016   2017

Figure 1. Global clinical AIDS vaccine development. Depicted are three distinct, but interrelated, spheres of activity placed along a timeline. RV144 follow-up is ongoing, and seeks to identify a correlate of immune protection by interrogating RV144 samples or samples generated by follow-up studies. Exploratory research activities will include adaptive design clinical trials, which aim to rapidly identify next generation products that should proceed to efficacy testing. Product development activities seek to extend the RV144 proof of concept by studying pox-protein prime-boost strategies in high-risk populations from Africa and Thailand with different routes of exposure in pursuit of licensure to realize rapid public health benefit.

uals in southern Africa and MSM in Thailand that could begin in 2014. These studies would build on RV144 by keeping vaccine constituents and schedule constant. Given that RV144 was conducted in a population at risk for primarily CRF01_AE infection, this variable would also be held constant in subsequent studies in Thailand. However, both the ALVAC inserts and the recombinant gp120 protein to be tested

in southern Africa would need to be substituted with subtype C strains that match the prevalent regional circulating HIV variants. Any change in vaccine, population, or subtype will impact comparability of the studies. Finally, the gp120 recombinant protein boost for subsequent Thai and southern Africa studies has yet to be determined. This discussion is theoretically more straightforward for Thailand as AIDSVAX

B/E is the established precedent but is less intuitive for studies elsewhere.

Exploratory Research Pathway: Pox-Protein and Beyond

Parallel exploration of novel pox vectors and gp120 proteins/adjuvants to improve protective efficacy is warranted through iterative clinical research. Poxvirus vectors, such as NYVAC, merit consideration for research track exploration as they appear to generate greater cellular and humoral immune responses than ALVAC. Although the relevance of those evoked immune responses to protective efficacy is unknown, products that evoke qualitative and quantitative differences in immunogenicity compared to RV144 should be evaluated. The rgp120 used in RV144 was formulated with alum, a relatively weak adjuvant. Studies using more potent adjuvants such as MF59, are warranted as well as rgp120s that generate novel immunogenicity profiles in phase I clinical studies. The use of accelerated clinical study designs that efficiently eliminate poorly performing vaccine candidates should be part of this exploratory research pathway (Koup et al. 2010; Corey et al. 2011).

RV144 and the subsequent immunogenicity and efficacy studies that are being currently discussed are limited by the need the need to match HIV subtypes to specific regional epidemics. Increasing the valency of HIV vaccine candidates to capture a broader range of global HIV subtype diversity would provide the possibility of a globally effective preventive vaccine. Such an advance could enable a concerted HIV elimination strategy similar to the smallpox eradication campaign (Breman and Arita 1980). These factors have prompted development of strategies aimed at generating responses broad enough to afford vaccine recipients universal protection. One promising approach involves the use of polyvalent "mosaic" inserts which cover the range of HIV T-cell epitopes across HIV subtypes (Fischer et al. 2007; Abdulhaqq and Weiner 2008; Barouch et al. 2010). Replication-incompetent Ad26 vectored mosaic HIV-1 Gag, Pol, and Env antigens augmented breadth

and depth of cellular immune responses in rhesus monkeys (Barouch et al. 2010), and both Ad26 and Ad35 vectored HIV-1 vaccine candidates, the former in combination with MVA mosaics are moving forward into phase I human clinical trials.

REFERENCES

*Reference is also in this collection.

Abdulhaqq SA, Weiner DB. 2008. DNA vaccines: Developing new strategies to enhance immune responses. *Immunol Res* **42**: 219–232.

Abrahams MR, Anderson JA, Giorgi EE, Seoighe C, Mlisana K, Ping LH, Athreya GS, Treurnicht FK, Keele BF, Wood N, et al. 2009. Quantitating the multiplicity of infection with human immunodeficiency virus type 1 subtype C reveals a non-Poisson distribution of transmitted variants. *J Virol* **83**: 3556–3567.

Allen TM, Vogel TU, Fuller DH, Mothé BR, Steffen S, Boyson JE, Shipley T, Fuller J, Hanke T, Sette A, et al. 2000. Induction of AIDS virus-specific CTL activity in fresh, unstimulated peripheral blood lymphocytes from rhesus macaques vaccinated with a DNA prime/modified vaccinia virus Ankara boost regimen. *J Immunol* **164**: 4968–4978.

Almeida JR, Price DA, Papagno L, Arkoub ZA, Sauce D, Bornstein E, Asher TE, Samri A, Schnuriger A, Theodorou I, et al. 2007. Superior control of HIV-1 replication by CD8[+] T cells is reflected by their avidity, polyfunctionality, and clonal turnover. *J Exp Med* **204**: 2473–2485.

Asmuth DM, Brown EL, DiNubile MJ, Sun X, del Rio C, Harro C, Keefer MC, Kublin JG, Dubey SA, Kierstead LS, et al. 2010. Comparative cell-mediated immunogenicity of DNA/DNA, DNA/adenovirus type 5 (Ad5), or Ad5/Ad5 HIV-1 clade B gag vaccine prime-boost regimens. *J Infect Dis* **201**: 132–141.

Baba TW, Jeong YS, Pennick D, Bronson R, Greene MF, Ruprecht RM. 1995. Pathogenicity of live, attenuated SIV after mucosal infection of neonatal macaques. *Science* **267**: 1820–1825.

Baba TW, Liska V, Hofmann-Lehmann R, Vlasak J, Xu W, Ayehunie S, Cavacini LA, Posner MR, Katinger H, Stiegler G, et al. 2000. Human neutralizing monoclonal antibodies of the IgG1 subtype protect against mucosal simian-human immunodeficiency virus infection. *Nat Med* **6**: 200–206.

Baden LR, Dolin R, O'Brien KL, Abbink P, La Porte A, Seaman MS, Choi E, Tucker R, Weithens M, Pau MG, et al. 2009. First-in-human Phase 1 safety and immunogenicity of an adenovirus serotype 26 HIV-1 vaccine vector. *AIDS Vaccine 2009.* Paris, France.

Barouch DH. 2008. Challenges in the development of an HIV-1 vaccine. *Nature* **455**: 613–619.

Barouch DH. 2010. Novel adenovirus vector-based vaccines for HIV-1. *Curr Opin HIV AIDS* **5**: 386–390.

Barouch DH, O'Brien KL, Simmons NL, King SL, Abbink P, Maxfield LF, Sun YH, La Porte A, Riggs AM, Lynch DM, et al. 2010. Mosaic HIV-1 vaccines expand the breadth and depth of cellular immune responses in rhesus monkeys. *Nat Med* **16:** 319–323.

Baxby D, Paoletti E. 1992. Potential use of non-replicating vectors as recombinant vaccines. *Vaccine* **10:** 8–9.

Belshe RB, Graham BS, Keefer MC, Gorse GJ, Wright P, Dolin R, Matthews T, Weinhold K, Bolognesi DP, Sposto R, et al. 1994. Neutralizing antibodies to HIV-1 in seronegative volunteers immunized with recombinant gp120 from the MN strain of HIV-1. NIAID AIDS Vaccine Clinical Trials Network. *JAMA* **272:** 475–480.

Belshe RB, Gorse GJ, Mulligan MJ, Evans TG, Keefer MC, Excler JL, Duliege AM, Tartaglia J, Cox WI, McNamara J, et al. 1998. Induction of immune responses to HIV-1 by canarypox virus (ALVAC) HIV-1 and gp120 SF-2 recombinant vaccines in uninfected volunteers. NIAID AIDS Vaccine Evaluation Group. *AIDS* **12:** 2407–2415.

Belshe RB, Stevens C, Gorse GJ, Buchbinder S, Weinhold K, Sheppard H, Stablein D, Self S, McNamara J, Frey S, et al. 2001. Safety and immunogenicity of a canarypox-vectored human immunodeficiency virus Type 1 vaccine with or without gp120: A phase 2 study in higher- and lower-risk volunteers. *J Infect Dis* **183:** 1343–1352.

Benlahrech A, Harris J, Meiser A, Papagatsias T, Hornig J, Hayes P, Lieber A, Athanasopoulos T, Bachy V, Csomor E, et al. 2009. Adenovirus vector vaccination induces expansion of memory CD4 T cells with a mucosal homing phenotype that are readily susceptible to HIV-1. *Proc Natl Acad Sci* **106:** 19940–19945.

Betts MR, Nason MC, West SM, De Rosa SC, Migueles SA, Abraham J, Lederman MM, Benito JM, Goepfert PA, Connors M, et al. 2006. HIV nonprogressors preferentially maintain highly functional HIV-specific CD8$^+$ T cells. *Blood* **107:** 4781–4789.

Bhattacharya T, Daniels M, Heckerman D, Foley B, Frahm N, Kadie C, Carlson J, Yusim K, McMahon B, Gaschen B, et al. 2007. Founder effects in the assessment of HIV polymorphisms and HLA allele associations. *Science* **315:** 1583–1586.

Blanchard TJ, Alcami A, Andrea P, Smith GL. 1998. Modified vaccinia virus Ankara undergoes limited replication in human cells and lacks several immunomodulatory proteins: Implications for use as a human vaccine. *J Gen Virol* **79:** 1159–1167.

Breman JG, Arita I. 1980. The confirmation and maintenance of smallpox eradication. *N Engl J Med* **303:** 1263–1273.

Brumme ZL, Brumme CJ, Heckerman D, Korber BT, Daniels M, Carlson J, Kadie C, Bhattacharya T, Chui C, Szinger J, et al. 2007. Evidence of differential HLA class I-mediated viral evolution in functional and accessory/regulatory genes of HIV-1. *PLoS Pathog* **3:** e94.

Buchbinder S, Duerr A. 2009. Clinical outcomes from the Step study. *AIDS Vaccine* 2009. Paris, France.

Buchbinder SP, Mehrotra DV, Duerr A, Fitzgerald DW, Mogg R, Li D, Gilbert PB, Lama JR, Marmor M, Del Rio C, et al. 2008. Efficacy assessment of a cell-mediated immunity HIV-1 vaccine (the Step Study): A double-blind, randomised, placebo-controlled, test-of-concept trial. *Lancet* **372:** 1881–1893.

Burke DS. 1993. Vaccine therapy for HIV: A historical review of the treatment of infectious diseases by active specific immunization with microbe-derived antigens. *Vaccine* **11:** 883–891.

Carrington M, O'Brien SJ. 2003. The influence of HLA genotype on AIDS. *Annu Rev Med* **54:** 535–551.

Carrington M, Nelson GW, Martin MP, Kissner T, Vlahov D, Goedert JJ, Kaslow R, Buchbinder S, Hoots K, O'Brien SJ. 1999. HLA and HIV-1: Heterozygote advantage and B*35-Cw*04 disadvantage. *Science* **283:** 1748–1752.

Carroll MW, Moss B. 1997. Poxviruses as expression vectors. *Curr Opin Biotechnol* **8:** 573–577.

Casimiro DR, Chen L, Fu TM, Evans RK, Caulfield MJ, Davies ME, Tang A, Chen M, Huang L, Harris V, et al. 2003. Comparative immunogenicity in rhesus monkeys of DNA plasmid, recombinant vaccinia virus, and replication-defective adenovirus vectors expressing a human immunodeficiency virus type 1 gag gene. *J Virol* **77:** 6305–6313.

Casimiro DR, Wang F, Schleif WA, Liang X, Zhang ZQ, Tobery TW, Davies ME, McDermott AB, O'Connor DH, Fridman A, et al. 2005. Attenuation of simian immunodeficiency virus SIVmac239 infection by prophylactic immunization with dna and recombinant adenoviral vaccine vectors expressing Gag. *J Virol* **79:** 15547–15555.

Catanzaro AT, Koup RA, Roederer M, Bailer RT, Enama ME, Moodie Z, Gu L, Martin JE, Novik L, Chakrabarti BK, et al. 2006. Phase 1 safety and immunogenicity evaluation of a multiclade HIV-1 candidate vaccine delivered by a replication-defective recombinant adenovirus vector. *J Infect Dis* **194:** 1638–1649.

CDC 1999. Ten great public health achievements—United States, 1900–1999. *MMWR Morb Mortal Wkly Rep* **48:** 241–243.

Catanzaro AT, Roederer M, Koup RA, Bailer RT, Enama ME, Nason MC, Martin JE, Rucker S, Andrews CA, Gomez PL, et al. 2007. Phase I clinical evaluation of a six-plasmid multiclade HIV-1 DNA candidate vaccine. *Vaccine* **25:** 4085–4092.

Cebere I, Dorrell L, McShane H, Simmons A, McCormack S, Schmidt C, Smith C, Brooks M, Roberts JE, Darwin SC, et al. 2006. Phase I clinical trial safety of DNA- and modified virus Ankara-vectored human immunodeficiency virus type 1 (HIV-1) vaccines administered alone and in a prime-boost regime to healthy HIV-1-uninfected volunteers. *Vaccine* **24:** 417–425.

Chun TW, Carruth L, Finzi D, Shen X, DiGiuseppe JA, Taylor H, Hermankova M, Chadwick K, Margolick J, Quinn TC, et al. 1997. Quantification of latent tissue reservoirs and total body viral load in HIV-1 infection. *Nature* **387:** 183–188.

Chun TW, Engel D, Berrey MM, Shea T, Corey L, Fauci AS. 1998. Early establishment of a pool of latently infected, resting CD4$^+$ T cells during primary HIV-1 infection. *Proc Natl Acad Sci* **95:** 8869–8873.

Clements-Mann ML, Weinhold K, Matthews TJ, Graham BS, Gorse GJ, Keefer MC, McElrath MJ, Hsieh RH, Mestecky J, Zolla-Pazner S, et al. 1998. Immune responses to human immunodeficiency virus (HIV) type 1 induced by canarypox expressing HIV-1MN gp120, HIV-1SF2 recombinant gp120, or both vaccines in seronegative adults. NIAID AIDS Vaccine Evaluation Group. *J Infect Dis* **177:** 1230–1246.

Cocchi F, DeVico AL, Garzino-Demo A, Arya SK, Gallo RC, Lusso P. 1995. Identification of RANTES, MIP-1 α, and MIP-1 β as the major HIV-suppressive factors produced by CD8+ T cells. *Science* **270:** 1811–1815.

Cohen J. 1997. Weakened SIV vaccine still kills. *Science* **278:** 24–25.

Cooney EL, McElrath MJ, Corey L, Hu SL, Collier AC, Arditti D, Hoffman M, Coombs RW, Smith GE, Greenberg PD. 1993. Enhanced immunity to human immunodeficiency virus (HIV) envelope elicited by a combined vaccine regimen consisting of priming with a vaccinia recombinant expressing HIV envelope and boosting with gp160 protein. *Proc Natl Acad Sci* **90:** 1882–1886.

Corey L, Nabel GJ, Dieffenbach C, Gilbert P, Haynes BF, Johnston M, Kublin J, Lane HC, Pantaleo G, Picker LJ, et al. 2011. HIV-1 vaccines and adaptive trial designs. *Sci Transl Med* **3:** 79ps13.

Currier J, de Souza M. 2009. Characterization of cell-mediated immune responses generated by a recombinant modified vaccinia Ankara (rMVA)-HIV-1 in a Phase I vaccine Trial. *AIDS Vaccine 2009.* Paris, France.

D'Souza MP, Frahm N. 2010. Adenovirus 5 serotype vector-specific immunity and HIV-1 infection: A tale of T cells and antibodies. *AIDS* **24:** 803–809.

D'Souza MP, Harden VA. 1996. Chemokines and HIV-1 second receptors. Confluence of two fields generates optimism in AIDS research. *Nat Med* **2:** 1293–1300.

Daniel MD, Kirchhoff F, Czajak SC, Sehgal PK, Desrosiers RC. 1992. Protective effects of a live attenuated SIV vaccine with a deletion in the *nef* gene. *Science* **258:** 1938–1941.

de Bruyn G, Rossini AJ, Chiu YL, Holman D, Elizaga ML, Frey SE, Burke D, Evans TG, Corey L, Keefer MC. 2004. Safety profile of recombinant canarypox HIV vaccines. *Vaccine* **22:** 704–713.

Deprez B, Sauzet JP, Boutillon C, Martinon F, Tartar A, Sergheraert C, Guillet JG, Gomard E, Gras-Masse H. 1996. Comparative efficiency of simple lipopeptide constructs for in vivo induction of virus-specific CTL. *Vaccine* **14:** 375–382.

Deres K, Schild H, Wiesmuller KH, Jung G, Rammensee HG. 1989. In vivo priming of virus-specific cytotoxic T lymphocytes with synthetic lipopeptide vaccine. *Nature* **342:** 561–564.

Dorrell L, Williams P, Suttill A, Brown D, Roberts J, Conlon C, Hanke T, McMichael A. 2007. Safety and tolerability of recombinant modified vaccinia virus Ankara expressing an HIV-1 gag/multiepitope immunogen (MVA.HIVA) in HIV-1-infected persons receiving combination antiretroviral therapy. *Vaccine* **25:** 3277–3283.

Douek DC, Kwong PD, Nabel GJ, et al. 2006. The rational design of an AIDS vaccine. *Cell* **124:** 677–681.

Duerr A, Huang Y, Moodie Z, Lawrence D, Robertson M, Buchbinder S. 2010. Analysis of the relative risk of HIV acquisition among Step study participants with extended follow-up. *AIDS Res Human Retroviruses* **26:** A-1-A-184. Oral Abstract OA03.05.

Evans TG, Keefer MC, Weinhold KJ, Wolff M, Montefiori D, Gorse GJ, Graham BS, McElrath MJ, Clements-Mann ML, Mulligan MJ, et al. 1999. A canarypox vaccine expressing multiple human immunodeficiency virus type 1 genes given alone or with rgp120 elicits broad and durable CD8+ cytotoxic T lymphocyte responses in seronegative volunteers. *J Infect Dis* **180:** 290–298.

Excler JL, Plotkin S. 1997. The prime-boost concept applied to HIV preventive vaccines. *AIDS* **11:** S127–S137.

Fauci AS, Johnston MI, Dieffenbach CW, Burton DR, Hammer SM, Hoxie JA, Martin M, Overbaugh J, Watkins DI, Mahmoud A, et al. 2008. HIV vaccine research: The way forward. *Science* **321:** 530–532.

Fellay J, Shianna KV, Ge D, Colombo S, Ledergerber B, Weale M, Zhang K, Gumbs C, Castagna A, Cossarizza A, et al. 2007. A whole-genome association study of major determinants for host control of HIV-1. *Science* **317:** 944–947.

Fischer W, Perkins S, Theiler J, Bhattacharya T, Yusim K, Funkhouser R, Kuiken C, Haynes B, Letvin NL, Walker BD, et al. 2007. Polyvalent vaccines for optimal coverage of potential T-cell epitopes in global HIV-1 variants. *Nat Med* **13:** 100–106.

Flynn NM, Forthal DN, Harro CD, Judson FN, Mayer KH, Para MF. 2005. Placebo-controlled phase 3 trial of a recombinant glycoprotein 120 vaccine to prevent HIV-1 infection. *J Infect Dis* **191:** 654–665.

Forthal DN, Gilbert PB, Landucci G, Phan T. 2007. Recombinant gp120 vaccine-induced antibodies inhibit clinical strains of HIV-1 in the presence of Fc receptor-bearing effector cells and correlate inversely with HIV infection rate. *J Immunol* **178:** 6596–6603.

Fouts TR, Binley JM, Trkola A, Robinson JE, Moore JP. 1997. Neutralization of the human immunodeficiency virus type 1 primary isolate JR-FL by human monoclonal antibodies correlates with antibody binding to the oligomeric form of the envelope glycoprotein complex. *J Virol* **71:** 2779–2785.

Franchini G, Gurunathan S, Baglyos L, Plotkin S, Tartaglia J. 2004. Poxvirus-based vaccine candidates for HIV: Two decades of experience with special emphasis on canarypox vectors. *Expert Rev Vaccines* **3:** S75–88.

Frey G, Peng H, Rits-Volloch S, Morelli M, Cheng Y, Chen B. 2008. A fusion-intermediate state of HIV-1 gp41 targeted by broadly neutralizing antibodies. *Proc Natl Acad Sci* **105:** 3739–3744.

Gao X, Nelson GW, Karacki P, Martin MP, Phair J, Kaslow R, Goedert JJ, Buchbinder S, Hoots K, Vlahov D, et al. 2001. Effect of a single amino acid change in MHC class I molecules on the rate of progression to AIDS. *N Engl J Med* **344:** 1668–1675.

Gaschen B, Taylor J, Yusim K, Foley B, Gao F, Lang D, Novitsky V, Haynes B, Hahn BH, Bhattacharya T, et al. 2002. Diversity considerations in HIV-1 vaccine selection. *Science* **296:** 2354–2360.

Gilbert PB, Chiu YL, Allen M, Lawrence DN, Chapdu C, Israel H, Holman D, Keefer MC, Wolff M, Frey SE, et al. 2003. Long-term safety analysis of preventive HIV-1 vaccines evaluated in AIDS vaccine evaluation group NIAID-sponsored Phase I and II clinical trials. *Vaccine* **21–22:** 2933–2947.

Gilbert PB, Peterson ML, Follmann D, Hudgens MG, Francis DP, Gurwith M, Heyward WL, Jobes DV, Popovic V, Self SG, et al. 2005. Correlation between immunologic responses to a recombinant glycoprotein 120 vaccine and incidence of HIV-1 infection in a phase 3 HIV-1 preventive vaccine trial. *J Infect Dis* **191:** 666–677.

Gilbert P, Wang M, Wrin T, Petropoulos C, Gurwith M, Sinangil F, D'Souza P, Rodriguez-Chavez IR, DeCamp A, Giganti M, et al. 2010. Magnitude and breadth of a nonprotective neutralizing antibody response in an efficacy trial of a candidate HIV-1 gp120 vaccine. *J Infect Dis* **202**: 595–605.

Goonetilleke N, Moore S, Dally L, Winstone N, Cebere I, Mahmoud A, Pinheiro S, Gillespie G, Brown D, Loach V, et al. 2006. Induction of multifunctional human immunodeficiency virus type 1 (HIV-1)-specific T cells capable of proliferation in healthy subjects by using a prime-boost regimen of DNA- and modified vaccinia virus Ankara-vectored vaccines expressing HIV-1 Gag coupled to CD8$^+$ T-cell epitopes. *J Virol* **80**: 4717–4728.

Goonetilleke N, Liu MK, Salazar-Gonzalez JF, Ferrari G, Giorgi E, Ganusov VV, Keele BF, Learn GH, Turnbull EL, Salazar MG, et al. 2009. The first T cell response to transmitted/founder virus contributes to the control of acute viremia in HIV-1 infection. *J Exp Med* **206**: 1253–1272.

Goulder PJ, Brander C, Tang Y, Tremblay C, Colbert RA, Addo MM, Rosenberg ES, Nguyen T, Allen R, Trocha A, et al. 2001. Evolution and transmission of stable CTL escape mutations in HIV infection. *Nature* **412**: 334–338.

Goulder PJ, Watkins DI. 2004. HIV and SIV CTL escape: Implications for vaccine design. *Nat Rev Immunol* **4**: 630–640.

Graham BS, McElrath MJ, Connor RI, Schwartz DH, Gorse GJ, Keefer MC, Mulligan MJ, Matthews TJ, Wolinsky SM, Montefiori DC, et al. 1998. Analysis of intercurrent human immunodeficiency virus type 1 infections in phase I and II trials of candidate AIDS vaccines. AIDS vaccine evaluation group, and the correlates of HIV immune protection group. *J Infect Dis* **177**: 310–319.

Graham BS, Koup RA, Roederer M, Bailer RT, Enama ME, Moodie Z, Martin JE, McCluskey MM, Chakrabarti BK, Lamoreaux L, et al. 2006. Phase 1 safety and immunogenicity evaluation of a multiclade HIV-1 DNA candidate vaccine. *J Infect Dis* **194**: 1650–1660.

Gudmundsdotter L, Nilsson C, Brave A, Hejdeman B, Earl P, Moss B, Robb M, Cox J, Michael N, Marovich M, et al. 2009. Recombinant Modified Vaccinia Ankara (MVA) effectively boosts DNA-primed HIV-specific immune responses in humans despite pre-existing vaccinia immunity. *Vaccine* **27**: 4468–4474.

Guimarães-Walker A, Mackie N, McCormack S, Hanke T, Schmidt C, Gilmour J, Barin B, McMichael A, Weber J, Legg K, et al. 2008. Lessons from IAVI-006, a phase I clinical trial to evaluate the safety and immunogenicity of the pTHr.HIVA DNA and MVA.HIVA vaccines in a prime-boost strategy to induce HIV-1 specific T-cell responses in healthy volunteers. *Vaccine* **26**: 6671–6677.

Hanke T, Samuel RV, Blanchard TJ, Neumann VC, Allen TM, Boyson JE, Sharpe SA, Cook N, Smith GL, Watkins DI, et al. 1999. Effective induction of simian immunodeficiency virus-specific cytotoxic T lymphocytes in macaques by using a multiepitope gene and DNA prime-modified vaccinia virus Ankara boost vaccination regimen. *J Virol* **73**: 7524–7532.

Hanke T, Goonetilleke N, McMichael AJ, Dorrell L. 2007. Clinical experience with plasmid DNA- and modified vaccinia virus Ankara-vectored human immunodeficiency virus type 1 clade A vaccine focusing on T-cell induction. *J Gen Virol* **88**: 1–12.

Harari A, Bart PA, Stohr W, Tapia G, Garcia M, Medjitna-Rais E, Burnet S, Cellerai C, Erlwein O, Barber T, et al. 2008. An HIV-1 clade C DNA prime, NYVAC boost vaccine regimen induces reliable, polyfunctional, and long-lasting T cell responses. *J Exp Med* **205**: 63–77.

Harro CD, Judson FN, Gorse GJ, Mayer KH, Kostman JR, Brown SJ, Koblin B, Marmor M, Bartholow BN, Popovic V, et al. 2004. Recruitment and baseline epidemiologic profile of participants in the first phase 3 HIV vaccine efficacy trial. *J Acquir Immune Defic Syndr* **37**: 1385–1392.

Harro CD, Robertson MN, Lally MA, O'Neill LD, Edupuganti S, Goepfert PA, Mulligan MJ, Priddy FH, Dubey SA, Kierstead LS, et al. 2009. Safety and immunogenicity of adenovirus-vectored near-consensus HIV type 1 clade B gag vaccines in healthy adults. *AIDS Res Hum Retroviruses* **25**: 103–114.

Haynes BF, Fleming J, St Clair EW, Katinger H, Stiegler G, Kunert R, Robinson J, Scearce RM, Plonk K, Staats HF, et al. 2005. Cardiolipin polyspecific autoreactivity in two broadly neutralizing HIV-1 antibodies. *Science* **308**: 1906–1908.

Hessell AJ, Rakasz EG, Tehrani DM, Huber M, Weisgrau KL, Landucci G, Forthal N, Koff WC, Poignard P, Watkins DI, et al. 2010. Broadly neutralizing monoclonal antibodies 2F5 and 4E10 directed against the human immunodeficiency virus type 1 gp41 membrane-proximal external region protect against mucosal challenge by simian-human immunodeficiency virus SHIVBa-L. *J Virol* **84**: 1302–1313.

IAVI. 2010. Database of AIDS Vaccine Candidates in Clinical Trials. Retrieved 27 October, 2010, from http://www.iavireport.org/trials-db/Pages/default.aspx.

Jaoko W, Nakwagala FN, Anzala O, Manyonyi GO, Birungi J, Nanvubya A, Bashir F, Bhatt K, Ogutu H, Wakasiaka S, et al. 2008. Safety and immunogenicity of recombinant low-dosage HIV-1 A vaccine candidates vectored by plasmid pTHr DNA or modified vaccinia virus Ankara (MVA) in humans in East Africa. *Vaccine* **26**: 2788–2795.

Joag SV, Liu ZQ, Stephens EB, Smith MS, Kumar A, Li Z, Wang C, Sheffer D, Jia F, Foresman L, et al. 1998. Oral immunization of macaques with attenuated vaccine virus induces protection against vaginally transmitted AIDS. *J Virol* **72**: 9069–9078.

Johnson WE, Desrosiers RC. 2002. Viral persistence: HIV's strategies of immune system evasion. *Annu Rev Med* **53**: 499–518.

Kannanganat S, Kapogiannis BG, Ibegbu C, Chennareddi L, Goepfert P, Robinson HL, Lennox J, Amara R. 2007. Human immunodeficiency virus type 1 controllers but not noncontrollers maintain CD4 T cells coexpressing three cytokines. *J Virol* **81**: 12071–12076.

Karnasuta C, Paris RM, Cox JH, Nitayaphan S, Pitisuttithum P, Thongcharoen P, Brown AE, Gurunathan S, Tartaglia J, Heyward WL, et al. 2005. Antibody-dependent cell-mediated cytotoxic responses in participants enrolled in a phase I/II ALVAC-HIV/AIDSVAX B/E prime-boost HIV-1 vaccine trial in Thailand. *Vaccine* **23**: 2522–2529.

Kaslow RA, Carrington M, Apple R, Park L, Munoz A, Saah AJ, Goedert JJ, Winkler C, O'Brien SJ, Rinaldo C, et al. 1996. Influence of combinations of human major histocompatibility complex genes on the course of HIV-1 infection. *Nat Med* **2:** 405–411.

Keele BF, Giorgi EE, Salazar-Gonzalez JF, Decker JM, Pham KT, Salazar MG, Sun C, Grayson T, Wang S, Li H, et al. 2008. Identification and characterization of transmitted and early founder virus envelopes in primary HIV-1 infection. *Proc Natl Acad Sci* **105:** 7552–7557.

Kelleher AD, Emery S, Cunningham P, Duncombe C, Carr A, Golding H, Forde S, Hudson J, Roggensack M, Forrest BD, et al. 1997. Safety and immunogenicity of UBI HIV-1MN octameric V3 peptide vaccine administered by subcutaneous injection. *AIDS Res Hum Retroviruses* **13:** 29–32.

Kennedy JS, Co M, Green S, Longtine K, Longtine J, O'Neill MA, Adams JP, Rothman AL, Yu Q, Johnson-Leva R, et al. 2008. The safety and tolerability of an HIV-1 DNA prime-protein boost vaccine (DP6-001) in healthy adult volunteers. *Vaccine* **26:** 4420–4424.

Kibuuka H, Kimutai R, Maboko L, Sawe F, Schunk MS, Kroidl A, Shaffer D, Eller LA, Kibaya R, Eller MA, et al. 2010. A phase 1/2 study of a multiclade HIV-1 DNA plasmid prime and recombinant adenovirus serotype 5 boost vaccine in HIV-Uninfected East Africans (RV 172). *J Infect Dis* **201:** 600–607.

Kiepiela P, Leslie AJ, Honeyborne I, Ramduth D, Thobakgale C, Chetty S, Rathnavalu P, Moore C, Pfafferott KJ, Hilton L, et al. 2004. Dominant influence of HLA-B in mediating the potential co-evolution of HIV and HLA. *Nature* **432:** 769–775.

Kim JH, Rerks-Ngarm S, Excler JL, Michael NL. 2010. HIV vaccines: Lessons learned and the way forward. *Curr Opin HIV AIDS* **5:** 428–434.

Korber BT, Letvin NL, Haynes BF. 2009. T-cell vaccine strategies for human immunodeficiency virus, the virus with a thousand faces. *J Virol* **83:** 8300–8314.

Kostense S, Koudstaal W, Sprangers M, Weverling GJ, Penders G, Helmus N, Vogels R, Bakker M, Berkhout B, Havenga M, et al. 2004. Adenovirus types 5 and 35 seroprevalence in AIDS risk groups supports type 35 as a vaccine vector. *AIDS* **18:** 1213–1216.

Koup RA, Graham BS, Douek DC. 2010. The quest for a T cell-based immune correlate of protection against HIV: A story of trials and errors. *Nat Rev Immunol* **11:** 65–70.

* Lackner AA, Lederman MM, Rodriguez B. 2011. HIV pathogenesis—The host. *Cold Spring Harb Perspect Med* doi: 10.1101/cshperspect.a007005.

Letvin NL, Mascola JR, Sun Y, Gorgone DA, Buzby AP, Xu L, Yang ZY, Chakrabarti B, Rao SS, Schmitz JE, et al. 2006. Preserved CD4$^+$ central memory T cells and survival in vaccinated SIV-challenged monkeys. *Science* **312:** 1530–1533.

Letvin NL, Rao SS, Montefiori DC, Seaman MS, Sun Y, Lim SY, Asmal M, Gelman RS, Shen L, Whitney JB, et al. 2011. Immune and genetic correlates of vaccine protection against mucosal infection by SIV in monkeys. *Sci Transl Med* **3:** 81ra36.

Lifson JD, Rossio JL, Piatak M Jr, Bess JJr, Chertova E, Schneider DK, Coalter VJ, Poore B, Kiser RF, Imming RJ, et al. 2004. Evaluation of the safety, immunogenicity, and protective efficacy of whole inactivated simian immunodeficiency virus (SIV) vaccines with conformationally and functionally intact envelope glycoproteins. *AIDS Res Hum Retroviruses* **20:** 772–787.

Limou S, Le Clerc S, Coulonges C, Carpentier W, Dina C, Delaneau O, Labib T, Taing L, Sladek R, Deveau C, et al. 2009. Genomewide association study of an AIDS-nonprogression cohort emphasizes the role played by HLA genes (ANRS Genomewide Association Study 02). *J Infect Dis* **199:** 419–426.

Liu J, O'Brien KL, Lynch DM, Simmons NL, La Porte A, Riggs AM, Abbink P, Coffey RT, Grandpre LE, Seaman LS, et al. 2009. Immune control of an SIV challenge by a T-cell-based vaccine in rhesus monkeys. *Nature* **457:** 87–91.

MacGregor RR, Boyer JD, Ugen KE, Lacy KE, Gluckman SJ, Bagarazzi ML, Chattergoon MA, Baine Y, Higgins TJ, Ciccarelli RB, et al. 1998. First human trial of a DNA-based vaccine for treatment of human immunodeficiency virus type 1 infection: Safety and host response. *J Infect Dis* **178:** 92–100.

MacGregor RR, Boyer JD, Ciccarelli RB, Ginsberg RS, Weiner DB. 2000. Safety and immune responses to a DNA-based human immunodeficiency virus (HIV) type I env/rev vaccine in HIV-infected recipients: Follow-up data. *J Infect Dis* **181:** 406.

MacGregor RR, Ginsberg R, Ugen KE, Baine Y, Kang CU, Tu XM, Higgins T, Weiner DB, Boyer JD. 2002. T-cell responses induced in normal volunteers immunized with a DNA-based vaccine containing HIV-1 env and rev. *AIDS* **16:** 2137–2143.

Mackewicz CE, Barker E, Levy JA. 1996. Role of β-chemokines in suppressing HIV replication. *Science* **274:** 1393–1395.

Mariani R, Kirchhoff F, Greenough TC, Sullivan JL, Desrosiers RC, Skowronski J. 1996. High frequency of defective nef alleles in a long-term survivor with nonprogressive human immunodeficiency virus type 1 infection. *J Virol* **70:** 7752–7764.

Markel H. 2004. Taking shots: The modern miracle of vaccines. *Medscape Pediatrics* **6.** www.medscape.com/view article/481059.

Mascola JR, Montefiori DC. 2010. The role of antibodies in HIV vaccines. *Annu Rev Immunol* **28:** 413–444.

Mascola JR, Stiegler G, VanCott TC, Katinger H, Carpenter CB, Hanson CE, Beary H, Hayes D, Frankel SS, Birx DL, et al. 2000. Protection of macaques against vaginal transmission of a pathogenic HIV-1/SIV chimeric virus by passive infusion of neutralizing antibodies. *Nat Med* **6:** 207–210.

Masek-Hammerman K, Li H, Liu J, Abbink P, La Porte A, O'Brien KL, Whitney JB, Carville A, Mansfield KG, Barouch DH. 2010. Mucosal trafficking of vector-specific CD4$^+$ T lymphocytes following vaccination of rhesus monkeys with adenovirus serotype 5. *J Virol* **84:** 9810–9816.

McDermott AB, Mitchen J, Piaskowski S, De Souza I, Yant LJ, Stephany J, Furlott J, Watkins DI. 2004. Repeated low-dose mucosal simian immunodeficiency virus SIV-mac239 challenge results in the same viral and immunological kinetics as high-dose challenge: A model for the evaluation of vaccine efficacy in nonhuman primates. *J Virol* **78**: 3140–3144.

McElrath MJ, De Rosa SC, Moodie Z, Dubey S, Kierstead L, Janes H, Defawe OD, Carter DK, Hural J, Akondy R, et al. 2008. HIV-1 vaccine-induced immunity in the test-of-concept Step Study: A case-cohort analysis. *Lancet* **372**: 1894–1905.

McMichael AJ. 2006. HIV vaccines. *Annu Rev Immunol* **24**: 227–255.

McMichael AJ, Borrow P, Tomaras GD, Goonetilleke N, Haynes BF. 2010. The immune response during acute HIV-1 infection: Clues for vaccine development. *Nat Rev Immunol* **10**: 11–23.

Merial 2010. *ProteqFlu* Harlow, UK, Merial Animal Health, Ltd. 2010.

Michael NL. 2009. Primary and sub-group analyses of the Thai Phase III HIV Vaccine Trial. *AIDS Vaccine*. Paris, France.

Moore CB, John M, James IR, Christiansen FT, Witt CS, Mallal SA. 2002. Evidence of HIV-1 adaptation to HLA-restricted immune responses at a population level. *Science* **296**: 1439–1443.

Moore PL, Ranchobe N, Lambson BE, Gray ES, Cave E, Abrahams MR, Bandawe G, Mlisana K, Abdool Karim SS, Williamson C, et al. 2009. Limited neutralizing antibody specificities drive neutralization escape in early HIV-1 subtype C infection. *PLoS Pathog* **5**: e1000598.

Moss B. 1991. Vaccinia virus: A tool for research and vaccine development. *Science* **252**: 1662–1667.

Moss B, Carroll MW, Wyatt LS, Bennink JR, Hirsch VM, Goldstein S, Elkins WR, Fuerst TR, Lifson JD, Piatak M, et al. 1996. Host range restricted, non-replicating vaccinia virus vectors as vaccine candidates. *Adv Exp Med Biol* **397**: 7–13.

Mwau M, Cebere I, Sutton J, Chikoti P, Winstone N, Wee EG, Beattie T, Chen YH, Dorrell L, McShane H, et al. 2004. A human immunodeficiency virus 1 (HIV-1) clade A vaccine in clinical trials: Stimulation of HIV-specific T-cell responses by DNA and recombinant modified vaccinia virus Ankara (MVA) vaccines in humans. *J Gen Virol* **85**: 911–919.

Nitayaphan S, Pitisuttithum P, Karnasuta C, Eamsila C, de Souza M, Morgan P, Polonis V, Benenson M, VanCott T, Ratto-Kim S, et al. 2004. Safety and immunogenicity of an HIV subtype B and E prime-boost vaccine combination in HIV-negative Thai adults. *J Infect Dis* **190**: 702–706.

Pantaleo G, Koup RA. 2004. Correlates of immune protection in HIV-1 infection: What we know, what we don't know, what we should know. *Nat Med* **10**: 806–810.

Pantophlet R, Burton DR. 2006. GP120: Target for neutralizing HIV-1 antibodies. *Annu Rev Immunol* **24**: 739–769.

Paoletti E. 1996. Applications of pox virus vectors to vaccination: An update. *Proc Natl Acad Sci* **93**: 11349–11353.

Paris RM, Kim JH, Robb ML, Michael NL. 2010. Prime-boost immunization with poxvirus or adenovirus vectors as a strategy to develop a protective vaccine for HIV-1. *Expert Rev Vaccines* **9**: 1055–1069.

Patel V, Valentin A, Kulkarni V, Rosati M, Bergamaschi C, Jalah R, Alicea C, Minang JT, Trivett MT, Ohlen C,

et al. 2010. Long-lasting humoral and cellular immune responses and mucosal dissemination after intramuscular DNA immunization. *Vaccine* **28**: 4827–4836.

Pelak K, Goldstein DB, Walley NM, Fellay J, Ge D, Shianna KV, Gumbs C, Gao X, Maia JM, Cronin KD, et al. 2010. Host determinants of HIV-1 control in African Americans. *J Infect Dis* **201**: 1141–1149.

Pereyra F, Addo MM, Kaufmann DE, Liu Y, Miura T, Rathod A, Baker B, Trocha A, Rosenberg R, Mackey E, et al. 2008. Genetic and immunologic heterogeneity among persons who control HIV infection in the absence of therapy. *J Infect Dis* **197**: 563–571.

Pereyra F, Jia X, et al. 2010. The major genetic determinants of HIV-1 control affect HLA class I peptide presentation. *Science* **330**: 1551–1557.

Peters BS, Jaoko W, Vardas E, Panayotakopoulos G, Fast P, Schmidt C, Gilmour J, Bogoshi M, Omosa-Manyonyi G, Dally L, et al. 2007. Studies of a prophylactic HIV-1 vaccine candidate based on modified vaccinia virus Ankara (MVA) with and without DNA priming: Effects of dosage and route on safety and immunogenicity. *Vaccine* **25**: 2120–2127.

Pialoux G, Excler JL, Riviere Y, Gonzalez-Canali G, Feuillie V, Coulaud P, Gluckman JC, Matthews TJ, Meignier B, Kieny MP, et al. 1995. A prime-boost approach to HIV preventive vaccine using a recombinant canarypox virus expressing glycoprotein 160 (MN) followed by a recombinant glycoprotein 160 (MN/LAI). The AGIS Group, and l'Agence Nationale de Recherche sur le SIDA. *AIDS Res Hum Retroviruses* **11**: 373–381.

Pitisuttithum P. 2008. HIV vaccine research in Thailand: Lessons learned. *Expert Rev Vaccines* **7**: 311–317.

Pitisuttithum P, Nitayaphan S, Thongcharoen P, Khamboonruang C, Kim J, de Souza M, Chuenchitra T, Garner RP, Thapinta D, Polonis V, et al. 2003. Safety and immunogenicity of combinations of recombinant subtype E and B human immunodeficiency virus type 1 envelope glycoprotein 120 vaccines in healthy Thai adults. *J Infect Dis* **188**: 219–227.

Pitisuttithum P, Berman PW, Phonrat B, Suntharasamai P, Raktham S, Srisuwanvilai LO, Hirunras K, Kitayaporn D, Kaewkangwal J, Migasena S, et al. 2004. Phase I/II study of a candidate vaccine designed against the B and E subtypes of HIV-1. *J Acquir Immune Defic Syndr* **37**: 1160–1165.

Pitisuttithum P, Gilbert P, Gurwith M, Heyward W, Martin M, van Griensven F, Hu D, Tappero JW, Choopanya K. 2006. Randomized, double-blind, placebo-controlled efficacy trial of a bivalent recombinant glycoprotein 120 HIV-1 vaccine among injection drug users in Bangkok, Thailand. *J Infect Dis* **194**: 1661–1671.

Plotkin SA. 2008. Vaccines: Correlates of vaccine-induced immunity. *Clin Infect Dis* **47**: 401–409.

Poon B, Hsu JF, Gudeman V, Chen IS, Grovit-Ferbas K. 2005. Formaldehyde-treated, heat-inactivated virions with increased human immunodeficiency virus type 1 env can be used to induce high-titer neutralizing antibody responses. *J Virol* **79**: 10210–10217.

Poropatich K, Sullivan DJ Jr. 2010. HIV-1 long-term non-progressors: The viral, genetic and immunological basis for disease non-progression. *J Gen Virol*.

Potter SJ, Lacabaratz C, Lambotte O, Perez-Patrigeon S, Vingert B, Sinet M, Colle JH, Urrutia A, Scott-Algara E, Boufassa F, et al. 2007. Preserved central memory and activated effector memory CD4$^+$ T-cell subsets in human immunodeficiency virus controllers: An ANRS EP36 study. *J Virol* **81:** 13904–13915.

Priddy FH, Brown D, Kublin J, Monahan K, Wright DP, Lalezari J, Santiago S, Marmor M, Lally M, Novak RM, et al. 2008. Safety and immunogenicity of a replication-incompetent adenovirus type 5 HIV-1 clade B gag/pol/nef vaccine in healthy adults. *Clin Infect Dis* **46:** 1769–1781.

Ramanathan VD, Kumar M, Mahalingam J, Sathyamoorthy P, Narayanan PR, Solomon S, Panicali D, Chakrabarty S, Cox J, Sayeed E, et al. 2009. A Phase 1 study to evaluate the safety and immunogenicity of a recombinant HIV type 1 subtype C-modified vaccinia Ankara virus vaccine candidate in Indian volunteers. *AIDS Res Hum Retroviruses* **25:** 1107–1116.

Ranasinghe C, Ramshaw IA. 2009. Genetic heterologous prime-boost vaccination strategies for improved systemic and mucosal immunity. *Expert Rev Vaccines* **8:** 1171–1181.

Rerks-Ngarm S, Pitisuttithum P, Nitayaphan S, Kaewkungwal J, Chiu J, Paris R, Premsri N, Namwat C, de Souza M, Adams E, et al. 2009. Vaccination with ALVAC and AIDSVAX to prevent HIV-1 infection in Thailand. *N Engl J Med* **361:** 2209–2220.

Richman DD, Wrin T, Little SJ, Petropoulos CJ. 2003. Rapid evolution of the neutralizing antibody response to HIV type 1 infection. *Proc Natl Acad Sci* **100:** 4144–4149.

Robinson H, Goepfert P, Hay P, Team HP. 2008. GeoVax clade B DNA/MVA HIV/AIDS vaccine is well tolerated and immunogenic when administered to healthy seronegative adults. Conference on Retroviruses and Opportunistic Infections, Boston, MA.

Rosati M, Valentin A, Jalah R, Patel V, von Gegerfelt A, Bergamaschi C, Alicea C, Weiss D, Treece J, Pal R, et al. 2008. Increased immune responses in rhesus macaques by DNA vaccination combined with electroporation. *Vaccine* **26:** 5223–5229.

Roy MJ, Wu MS, Barr LJ, Fuller JT, Tussey LG, Speller S, Culp J, Burkholder JK, Swain WF, Dixon RM, et al. 2000. Induction of antigen-specific CD8$^+$ T cells, T helper cells, and protective levels of antibody in humans by particle-mediated administration of a hepatitis B virus DNA vaccine. *Vaccine* **19:** 764–778.

Russell ND, Graham BS, Keefer MC, McElrath MJ, Self SG, Weinhold KJ, Montefiori DC, Ferrari G, Horton H, Tomaras GD, et al. 2007. Phase 2 study of an HIV-1 canarypox vaccine (vCP1452) alone and in combination with rgp120: Negative results fail to trigger a phase 3 correlates trial. *J Acquir Immune Defic Syndr* **44:** 203–212.

Salazar-Gonzalez JF, Salazar MG, Keele BF, Learn GH, Giorgi EE, Li H, Decker JM, Wang S, Baalwa J, Kraus MH, et al. 2009. Genetic identity, biological phenotype, and evolutionary pathways of transmitted/founder viruses in acute and early HIV-1 infection. *J Exp Med* **206:** 1273–1289.

Salmon-Céron D, Excler JL, Finkielsztejn L, Autran B, Gluckman JC, Sicard D, Matthews TJ, Meignier B, Valentin C, El Habib R, et al. 1999. Safety and immunogenicity of a live recombinant canarypox virus expressing HIV type 1 gp120 MN MN tm/gag/protease LAI (ALVAC-HIV, vCP205) followed by a p24E–V3 MN synthetic peptide (CLTB-36) administered in healthy volunteers at low risk for HIV infection. AGIS Group and L'Agence Nationale de Recherches sur Le Sida. *AIDS Res Hum Retroviruses* **15:** 633–645.

Salmon-Céron D, Durier C, Desaint C, Cuzin L, Surenaud M, Hamouda NB, Lelièvre JD, Bonnet B, Pialoux G, Poizot-Martin I, et al. 2010. Immunogenicity and safety of an HIV-1 lipopeptide vaccine in healthy adults: A phase 2 placebo-controlled ANRS trial. *AIDS* **24:** 2211–2223.

Sandström E, Nilsson C, Hejdeman B, Bråve A, Bratt G, Robb M, Cox J, Vancott T, Marovich M, Stout R, et al. 2008. Broad immunogenicity of a multigene, multiclade HIV-1 DNA vaccine boosted with heterologous HIV-1 recombinant modified vaccinia virus Ankara. *J Infect Dis* **198:** 1482–1490.

Serwanga J, Shafer LA, Pimego E, Auma B, Watera C, Rowland S, Yirrell D, Pala P, Grosskurth H, Whitworth J, et al. 2009. Host HLA B*allele-associated multi-clade Gag T-cell recognition correlates with slow HIV-1 disease progression in antiretroviral therapy-naive Ugandans. *PLoS One* **4:** e4188.

Shiver JW, Fu TM, Chen L, Casimiro DR, Davies ME, Evans RK, Zhang ZQ, Simon AJ, Trigona WL, Dubey SA, et al. 2002. Replication-incompetent adenoviral vaccine vector elicits effective anti-immunodeficiency-virus immunity. *Nature* **415:** 331–335.

Simek MD, Rida W, Priddy FH, Pung P, Carrow E, Laufer DS, Lehrman JK, Boaz M, Tarragona-Fiol T, Miiro G, et al. 2009. Human immunodeficiency virus type 1 elite neutralizers: Individuals with broad and potent neutralizing activity identified by using a high-throughput neutralization assay together with an analytical selection algorithm. *J Virol* **83:** 7337–7348.

Singh P, Kaur G, Sharma G, Mehra NK. 2008. Immunogenetic basis of HIV-1 infection, transmission and disease progression. *Vaccine* **26:** 2966–2980.

Spearman P. 2006. Current progress in the development of HIV vaccines. *Curr Pharm Des* **12:** 1147–1167.

Sun ZY, Oh KJ, Kim M, Yu J, Brusic V, Song L, Qiao Z, Wang JH, Wagner G, Reinherz EL. 2008. HIV-1 broadly neutralizing antibody extracts its epitope from a kinked gp41 ectodomain region on the viral membrane. *Immunity* **28:** 52–63.

Sutter G, Staib C. 2003. Vaccinia vectors as candidate vaccines: The development of modified vaccinia virus Ankara for antigen delivery. *Curr Drug Targets Infect Disord* **3:** 263–271.

Tartaglia J, Cox WI, Taylor J, Perkus M, Riviere M, Meignier B, Paoletti E. 1992. Highly attenuated poxvirus vectors. *AIDS Res Hum Retroviruses* **8:** 1445–1447.

Taylor J, Weinberg R, Languet B, Desmettre P, Paoletti E. 1988. Recombinant fowlpox virus inducing protective immunity in non-avian species. *Vaccine* **6:** 497–503.

Thongcharoen P, Suriyanon V, Paris RM, Khamboonruang C, de Souza MS, Ratto-Kim S, Karnasuta C, Polonis VR, Baglyos L, Habib RE, et al. 2007. A phase 1/2 comparative vaccine trial of the safety and immunogenicity of a CRF01_AE (subtype E) candidate vaccine: ALVAC-HIV (vCP1521) prime with oligomeric gp160 (92TH023/LAI-

DID) or bivalent gp120 (CM235/SF2) boost. *J Acquir Immune Defic Syndr* **46**: 48–55.

TIHC. 2010. The major genetic determinants of HIV-1 control affect HLA class I peptide presentation. *Science* doi: 101126/science1195271.

Tartaglia J, Cox WI, et al. 1992. Highly attenuated poxvirus vectors. *AIDS Res Hum Retroviruses* **8**: 1445–1447.

Treurnicht FK, Seoighe C, Martin DP, Wood N, Abrahams MR, Rosa Dde A, Bredell H, Woodman Z, Hide W, Mlisana K, et al. 2010. Adaptive changes in HIV-1 subtype C proteins during early infection are driven by changes in HLA-associated immune pressure. *Virology* **396**: 213–225.

Vaccari M, Poonam P, Franchini G. 2010. Phase III HIV vaccine trial in Thailand: A step toward a protective vaccine for HIV. *Expert Rev Vaccines* **9**: 997–1005.

Van Rompay KK, Abel K, Lawson JR, Singh RP, Schmidt KA, Evans T, Earl P, Harvey D, Franchini G, Tartaglia J, et al. 2005. Attenuated poxvirus-based simian immunodeficiency virus (SIV) vaccines given in infancy partially protect infant and juvenile macaques against repeated oral challenge with virulent SIV. *J Acquir Immune Defic Syndr* **38**: 124–134.

VanCott TC, Mascola JR, Kaminski RW, Kalyanaraman V, Hallberg PL, Burnett PR, Ulrich JT, Rechtman DJ, Birx DL. 1997. Antibodies with specificity to native gp120 and neutralization activity against primary human immunodeficiency virus type 1 isolates elicited by immunization with oligomeric gp160. *J Virol* **71**: 4319–4330.

Vasan S, Hurley A, Schlesinger SJ, Hannaman D, Gardiner DF, Dugin DP, Boente-Carrera M, Vittorino R, Caskey M, Andersen J, et al. 2009. In vivo electroporation enhances the immunogenicity of ADVAX, a DNA-based HIV-1 vaccine candidate, in healthy volunteers. *Retrovirology* **6**: O31.

Vasan S, Schlesinger SJ, Chen Z, Hurley A, Lombardo A, Than S, Adesanya P, Bunce C, Boaz M, Boyle R, et al. 2010a. Phase 1 safety and immunogenicity evaluation of ADMVA, a multigenic, modified vaccinia Ankara-HIV-1 B'/C candidate vaccine. *PLoS One* **5**: e8816.

Vasan S, Schlesinger SJ, Huang Y, Hurley A, Lombardo A, Chen Z, Than S, Adesanya P, Bunce C, Boaz M, et al. 2010b. Phase 1 safety and immunogenicity evaluation of ADVAX, a multigenic, DNA-based clade C/B' HIV-1 candidate vaccine. *PLoS One* **5**: e8617.

Walker BD, Korber BT. 2001. Immune control of HIV: The obstacles of HLA and viral diversity. *Nat Immunol* **2**: 473–475.

Wang R, Doolan DL, Le TP, Hedstrom RC, Coonan KM, Charoenvit Y, Jones TR, Hobart P, Margalith M, Ng J, et al. 1998. Induction of antigen-specific cytotoxic T lymphocytes in humans by a malaria DNA vaccine. *Science* **282**: 476–480.

Wang S, Kennedy JS, West K, Montefiori DC, Coley S, Lawrence J, Shen S, Green S, Rothman AL, Ennis FA, et al. 2008. Cross-subtype antibody and cellular immune responses induced by a polyvalent DNA prime-protein boost HIV-1 vaccine in healthy human volunteers. *Vaccine* **26**: 3947–3957.

Watkins DI, Burton DR, Kallas EG, Moore JP, Koff WC. 2008. Nonhuman primate models and the failure of the Merck HIV-1 vaccine in humans. *Nat Med* **14**: 617–621.

Wei X, Decker JM, Wang S, Hui H, Kappes JC, Wu X, Salazar-Gonzalez JF, Salazar MG, Kilby JM, Saag MS, et al. 2003. Antibody neutralization and escape by HIV-1. *Nature* **422**: 307–312.

Whatmore AM, Cook N, Hall GA, Sharpe S, Rud EW, Cranage MP. 1995. Repair and evolution of nef in vivo modulates simian immunodeficiency virus virulence. *J Virol* **69**: 5117–5123.

Wilson NA, Reed J, Napoe GS, Piaskowski S, Szymanski A, Furlott J, Gonzalez EJ, Yant LJ, Maness NJ, May GE, et al. 2006. Vaccine-induced cellular immune responses reduce plasma viral concentrations after repeated low-dose challenge with pathogenic simian immunodeficiency virus SIVmac239. *J Virol* **80**: 5875–5885.

Wilson NA, Keele BF, Reed JS, Piaskowski SM, MacNair CE, Bett AJ, Liang X, Wang F, Thoryk E, et al. 2009. Vaccine-induced cellular responses control simian immunodeficiency virus replication after heterologous challenge. *J Virol* **83**: 6508–6521.

Wyand MS, Manson K, Montefiori DC, Lifson JD, Johnson RP, Desrosiers RC. 1999. Protection by live, attenuated simian immunodeficiency virus vaccines against heterologous challenge. *J Virol* **73**: 8356–8363.

Zhou T, Georgiev I, Wu X, Yang ZY, Dai K, Finzi A, Kwon YD, Scheid JF, Shi W, Xu L, et al. 2010. Structural basis for broad and potent neutralization of HIV-1 by antibody VRC01. *Science* **329**: 811–817.

Microbicides: Topical Prevention against HIV

Robin J. Shattock[1] and Zeda Rosenberg[2]

[1]Centre for Infection and Immunity, Division of Clinical Sciences, St George's, University of London, London SW17 0RE, United Kingdom

[2]International Partnership for Microbicides, Silver Spring, Maryland 20910

Correspondence: shattock@sgul.ac.uk

Microbicides represent a potential intervention strategy for preventing HIV transmission. Vaginal microbicides would meet the need for a discreet method that women could use to protect themselves against HIV. Although early-generation microbicides failed to demonstrate efficacy, newer candidates are based on more potent antiretroviral (ARV) products. Positive data from the CAPRISA 004 trial of tenofovir gel support use in women and represent a turning point for the field. This article reviews current progress in development of ARV-based microbicides. We discuss the consensus on selection criteria, the potential for drug resistance, rationale for drug combinations, and the use of pharmacokinetic (PK)/pharmacodynamic (PD) assessment in product development. The urgent need for continued progress in development of formulations for sustained delivery is emphasized. Finally, as the boundaries between different prevention technologies become increasingly blurred, consideration is given to the potential synergy of diverse approaches across the prevention landscape.

An effective microbicide may be one of the best ways to address a central gap in current HIV prevention strategies: lack of a discreet method that women can use to protect themselves from infection. Recently, the World Health Organization reported that AIDS is the leading cause of death among women of reproductive age globally, and particularly in sub-Saharan Africa (World Health Organization 2009). Methods available to prevent HIV include condoms, male circumcision, and behavioral interventions, but data indicate that they are insufficient to protect women. Among women in sub-Saharan Africa, one of the highest-risk factors for acquiring HIV is being in a stable long-term relationship where condom use is low (Shattock and Solomon 2004).

Condoms are impractical for women who want to conceive children or who cannot persuade their partners to use them. Next to an effective vaccine, microbicides (topical preexposure prophylaxis [PrEP]) and oral PrEP have the greatest potential to provide women with protection they can control. Both could be configured to protect men and women from transmission of HIV during unprotected anal intercourse.

Microbicides are topical PrEP products, such as gels, capsules, tablets, films, and intravaginal rings (IVR). They are designed to be applied either around the time of coitus, used on a daily basis (gels and films), or to deliver product over a prolonged period of time (IVR). The premise is inhibition or blockade of the earliest steps in the infection process at the vaginal or rectal mucosa. Because microbicides are topical, higher local drug concentrations can be delivered to virally exposed surfaces without significant systemic exposure, thereby reducing the risk of long-term toxicity in healthy but at-risk individuals.

GENERAL PRINCIPLES FOR PRIORITIZING MICROBICIDE DEVELOPMENT

Characteristics important to microbicide development are:

- Safety: Microbicides are designed for use by healthy individuals and should not demonstrate any localized toxicity. Avoiding any potential impact on epithelial surfaces and natural innate barriers that might tip the balance in favor of infection is particularly important. Long-term systemic toxicity associated with frequency and duration of product is equally important. Impact on fertility and/or fetal abnormalities is a crucial consideration for women.

- Efficacy: Any product must have a significant degree of efficacy in real-world situations. The level of efficacy required for adoption of a microbicide into different national programs remains the subject of much debate. Controversy also continues about the level of protection necessary to avoid negatively impacting condom use or discouraging other safe sex practices.

- Cost: A microbicide must be affordable to at-risk populations. Biological interventions may be effective, but if they are too expensive for mass distribution, they will never achieve widespread use.

- Acceptability: A microbicide must be acceptable for use in conjunction with sex. A product that demonstrates efficacy in a clinical trial with ongoing adherence counseling may be less acceptable in the real world and may not be adopted by at-risk populations.

- Appropriate drug delivery: For any microbicide to be effective, sufficient drug levels must be maintained in the appropriate compartments of the genital tract or rectum during exposure to virus. This feature has different implications for coitally dependent products than for those designed for sustained delivery.

- Long-term efficacy: As the field moves toward use of ARVs, the potential circulation within a given community of viruses resistant to the drug or drugs contained in a microbicide must be considered.

- Potential for resistance and impact on therapy: As ARV drugs used in treatment are increasingly included in microbicides, it will be important to ensure that their use for prevention does not induce drug resistance that might limit therapeutic options for individuals who become infected during product use.

- Prioritization of best-in-class products: It is not feasible to test every promising ARV-based microbicide in large clinical efficacy trials. Best-in-class products must be prioritized to maximize progress and prevent duplication. Prioritization should include in vitro activity (potency and breadth), stage of product development (including manufacturing processes), stability under diverse environmental conditions in multiple topical dosage forms, and positive prior clinical experience.

THE EARLY HISTORY OF MICROBICIDE DEVELOPMENT

Microbicides were originally conceived as products that could offer broad protection against all or most sexually transmitted infections. The expectation was that they could be generated using simple nondrug-based compounds and could be provided over the counter (OTC) without a prescription.

Initial approaches focused on identification of existing OTC products with antiviral activity and were encouraged by early observations that surfactant-containing spermicides could disrupt the integrity of HIV in vitro

(Malkovsky et al. 1988). Based on the potential for rapid introduction of existing products with in vitro activity against HIV, a large Phase III trial was performed using an OTC nonoxynol 9 spermicide, COL-1492 (Van Damme et al. 2002). It quickly became apparent, however, that surfactant-based products were insufficient to prevent HIV infection and also perturbed protective epithelial barriers when used frequently, potentially leading to higher rates of infection (Hillier et al. 2005). This early setback led to discontinuation of other strategies using surfactant-based products. An alternative approach was proposed, based on the observation that low pH ($<$4.5) could inactivate HIV (O'Connor et al. 1995). Based on these observations, it was hypothesized that acid-buffering gels could be used to lower vaginal pH and inactivate HIV (Mayer et al. 2001). Although shown to be safe in clinical trials, this is not sufficiently potent to prevent infection (Ramjee et al. 2010).

Researchers also discovered that long chain polyanionic compounds could prevent HIV infection in vitro. These compounds prevented HIV entry into target cells in vitro through nonspecific inhibition of viral binding to target cells (Fletcher and Shattock 2008). However, a range of polyanionic gels failed to show significant efficacy against HIV in clinical trials (Ramjee et al. 2010). Although a number of reasons for their lack of efficacy have been proposed, most researchers recognize they had insufficient potency. Disappointing results brought to a close this early chapter of microbicide development, opening a new phase of development focused on use of potent HIV-specific ARV drugs that interdict key stages in HIV replication: viral entry, reverse transcription, integration, and maturation.

MATCHING MICROBICIDES TO MECHANISMS OF TRANSMISSION

HIV transmission is relatively infrequent when considering the risk of infection per coital act (Dosekun and Fox 2010). However, risk of infection can be greatly influenced by the infectiousness of a sexual partner and the susceptibility of the exposed individual (Dosekun and Fox 2010). Recent genetic studies indicate that HIV infection is established by a single isolate in most women ($>$80%) infected by the vaginal route (Keele et al. 2008; Salazar-Gonzalez et al. 2009; Keele 2010). The balance between exposure and successful or nonsuccessful infection is likely to be relatively small. Therefore, anything that can reduce the infectious dose during any exposure and/or increase an exposed individual's resistance to infection may have a significant impact on transmission rates. Likewise, the balance could be tipped in favor of infection if the normal flora are disturbed, innate protective factors are reduced, or protective epithelium is damaged or inflamed (Haase 2011).

ARV DRUGS TO PREVENT INITIAL EVENTS IN MUCOSAL TRANSMISSION

As noted above, current microbicide development is focused on the use of ARVs that interdict key stages in HIV replication: viral entry, reverse transcription, integration, and maturation (Fig. 1A). Viral entry into target cells represents the first point at which microbicides could interrupt initial transmission events (Fig. 1B). Although much is known about the viral life cycle, far less is known about initial events in the mucosa of the genital tract and rectum that must be prevented or aborted to ensure protection from HIV infection (Fig. 2). Microbicides have a relatively short window of opportunity for blocking infection. Increasing evidence from animal models suggests that infection is established relatively quickly at the mucosa after exposure to HIV. In a nonhuman primate (NHP) model, a 30- to 60-min exposure to an infectious inoculum is sufficient to establish infection (Shattock and Moore 2003). That is most likely the time frame required for viral attachment to target cells and the optimal window for prevention of initial infection by entry inhibitors. Transmitted/founder viruses recently have been found to primarily target CD4 central and effector memory populations within mucosal tissue, which express high levels of CCR5. Transmitted

A

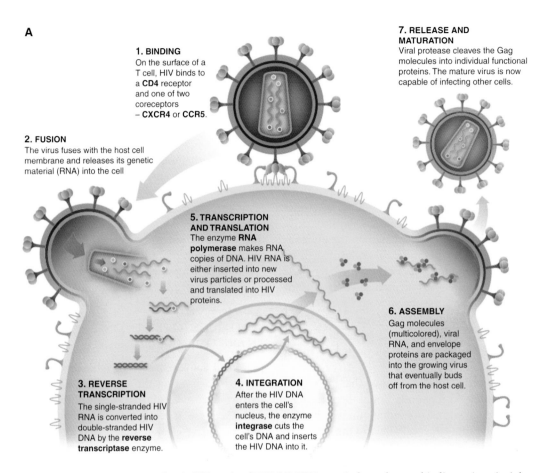

1. BINDING
On the surface of a
T cell, HIV binds to
a **CD4** receptor
and one of two
coreceptors
– **CXCR4** or **CCR5**.

**7. RELEASE AND
MATURATION**
Viral protease cleaves the Gag
molecules into individual functional
proteins. The mature virus is now
capable of infecting other cells.

2. FUSION
The virus fuses with the host cell
membrane and releases its genetic
material (RNA) into the cell

**5. TRANSCRIPTION
AND TRANSLATION**
The enzyme **RNA
polymerase** makes RNA
copies of DNA. HIV RNA is
either inserted into new
virus particles or processed
and translated into HIV
proteins.

6. ASSEMBLY
Gag molecules
(multicolored), viral
RNA, and envelope
proteins are packaged
into the growing virus
that eventually buds
off from the host cell.

**3. REVERSE
TRANSCRIPTION**
The single-stranded HIV
RNA is converted into
double-stranded HIV
DNA by the **reverse
transcriptase** enzyme.

4. INTEGRATION
After the HIV DNA
enters the cell's
nucleus, the enzyme
integrase cuts the
cell's DNA and inserts
the HIV DNA into it.

Figure 1. ARV drugs targeting the viral life cycle of HIV. (*A*) HIV entry is dependent on binding to its principle receptor, CD4, which triggers conformational change in the viral envelope (gp120) and subsequent engagement with one or two possible coreceptors (CCR5 or CXCR4), triggering subsequent fusion between the viral and cellular membranes. In most cases, virus using the CCR5 coreceptor (R5 virus) is the predominant strain mediating initial infection. Coreceptor engagement, in turn, causes conformational changes in gp41, leading to insertion of the fusion peptide into the cell membrane. Following membrane fusion, the viral core enters the cytoplasm of target cells and the process of reverse transcription begins. Following reverse transcription and generation of full-length double stranded proviral DNA, the viral enzyme integrase ensures provirus incorporation into the host cell DNA. Once incorporated into a target cell, the proviral DNA will persist for the lifetime of that cell and any potential progeny. Activation of host transcription factors known to bind to the long terminal repeat of the HIV genome initiates viral replication, which leads to assembly and budding of new virus. As part of this process, the viral enzyme protease cleaves two precursor viral polyproteins (Pr55 and Pr160), generating respective functional proteins essential for the maturation required to generate infectious virions. (*B*) Viral entry into target cells represents the first point at which microbicides could interrupt initial transmission events either binding to the viral envelope or blocking interaction with CD4 or coreceptor. Reverse transcription represents the second point at which a microbicide could potentially intervene through the activity of nucleoside/nucleotide reverse transcriptase inhibitors (NRTIs) or nonnucleotide reverse transcriptase inhibitors (NNRTIs). Integration, a third point of drug intervention, can be prevented by integrase inhibitors. Finally, protease inhibitors can block viral maturation leading to the release of defective noninfectious virions. (*Continued on next page.*)

Cite this article as *Cold Spring Harb Perspect Med* doi: 10.1101/cshperspect.a007385

B

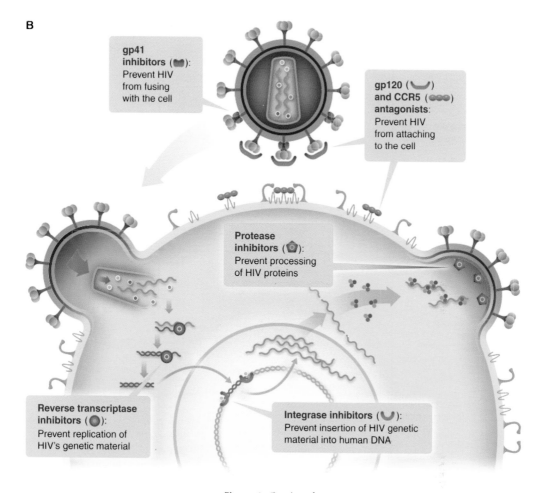

gp41 inhibitors ( ): Prevent HIV from fusing with the cell

gp120 ( ) and CCR5 ( ) antagonists: Prevent HIV from attaching to the cell

Protease inhibitors ( ): Prevent processing of HIV proteins

Reverse transcriptase inhibitors ( ): Prevent replication of HIV's genetic material

Integrase inhibitors ( ): Prevent insertion of HIV genetic material into human DNA

Figure 1. *Continued*

founder viruses appear to show low macrophage tropism, suggesting that infection of antigen-presenting cells may not be critical to establishment of an initial infection (Salazar-Gonzalez et al. 2009).

Tracking of labeled viruses in explant models and NHP studies has shown that the virus can penetrate superficial layers of stratified epithelium and come in contact with potentially susceptible T cells and Langerhans cells within these epithelial surfaces (Haase 2011). In NHP studies, infection has often been associated with areas of disruption in epithelial integrity and where stromal papillae are nearest to the tissue surface, providing the greatest chance of interaction with target cells (Haase 2011). The

columnar epithelium in the endocervix and the rectum remains an important potential site of infection and may be more susceptible to damage because they are only a single cell deep.

Detailed studies in NHPs show that infection is established rapidly (16–72 h) within local mucosal tissue, forming an initial foci of infection (Haase 2011). Reverse transcriptase and integrase inhibitors that target preintegration steps in the viral lifecycle may readily prevent establishment of such foci. In order for infection to take hold, local propagation of virus requires the influx of additional activated CD4 cells (Haase 2011). Local expansion of these infected cells is thought to produce new

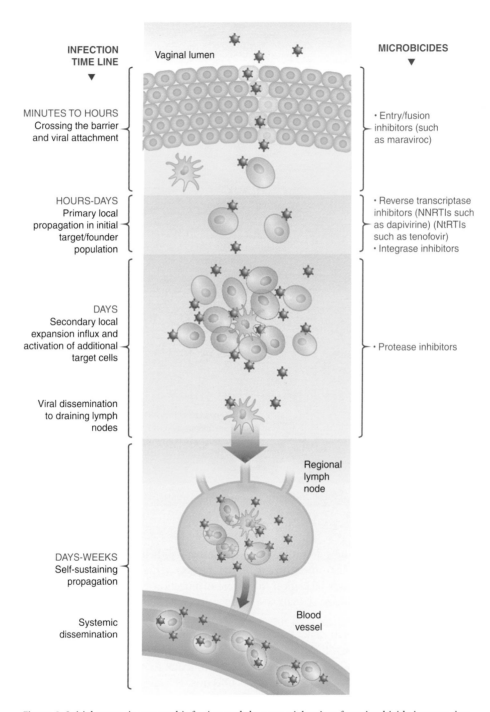

INFECTION TIME LINE ▼

Vaginal lumen

MICROBICIDES ▼

MINUTES TO HOURS
Crossing the barrier
and viral attachment

• Entry/fusion
inhibitors (such
as maraviroc)

HOURS-DAYS
Primary local
propagation in initial
target/founder
population

• Reverse transcriptase
inhibitors (NNRTIs such
as dapivirine) (NtRTIs
such as tenofovir)
• Integrase inhibitors

DAYS
Secondary local
expansion influx and
activation of additional
target cells

• Protease inhibitors

Viral dissemination
to draining lymph
nodes

Regional
lymph
node

DAYS-WEEKS
Self-sustaining
propagation

Systemic
dissemination

Blood
vessel

Figure 2. Initial events in mucosal infection and the potential points for microbicide intervention.

virus sufficient for dissemination into draining lymph nodes, where infected cells can be detected within 24–72 h of exposure (Haase 2011). This local expansion is thought to be critical to establishing productive infection, and blockade of this step may abort an infection. Here, local expansion of an infected founder population might be inhibited by protease inhibitors able to prevent production of virus from the initial foci of infection. Virus dissemination to draining lymph nodes leads to self-sustaining propagation and latent reservoirs refractory to topical microbicides. Systemic dissemination then occurs in the days to weeks following initial exposure (Haase 2011).

Success with microbicides requires delivery of appropriate drugs to these target sites. Therefore, drugs need to be formulated and chosen with an understanding about PK following topical application. Likewise, drugs being used for oral PrEP need to be chosen and evaluated to ensure adequate penetration into the mucosal portals of entry so that sufficient drug is at the right place at the right time. Formulation considerations are discussed in more detail below.

TARGETING OF VIRAL ENTRY

Because binding of viral gp120 to CD4 is essential for productive infection, targeting this step likely would significantly impact infection. Drugs that can effectively block gp120–CD4 interaction, however, are few and, to date, none of those identified has been evaluated in humans as microbicides. One compound, cyclotriazadisulfonamide CADA, has been shown to down-regulate CD4 expression. If it proves safe, CADA may represent an interesting new strategy (Vermeire et al. 2008). Because CD4 is known to have important functional characteristics in driving T cell responses, the impact of sustained suppression on immunity is unknown. An alternative approach would be to target gp120 and prevent either its interaction with CD4 or subsequent downstream conformational change. A small protein mimic of CD4 has been identified and is known to effectively block HIV infection in vitro (Van

Herrewege et al. 2008). Manufacture and scale up of such an approach, however, represents a significant hurdle. An alternative approach is a family of small-molecule drugs known to bind to gp120 and prevent subsequent conformational change. This includes BMS-806, which has been shown to inhibit infection in vitro and prevent infection of NHPs (Si et al. 2004; Veazey et al. 2005). A derivative of BMS-806, DS003, with a similar mechanism of action has been put into formal microbicide development. Both compounds are likely to bind to gp120 and prevent CD4 from inducing structural changes in gp120 that drive the fusion protein gp41 to form six helix bundles and cause viral and cellular membranes to fuse (Si et al. 2004). A parallel approach is use of synthetic peptides based on the gp41 sequence, which can bind to gp41 trimers and prevent formation of the six-helix bundle configuration necessary for juxtaposition of the viral and cellular membranes (Fig. 1B) (Shattock and Moore 2003). Two of these peptides—C52L and T1249— have been shown to protect Rhesus macaques from SHIV challenge (Si et al. 2004; Veazey et al. 2008). This approach is biologically plausible but peptides are expensive and creating formulations that are stable in a mucosal environment remains challenging.

INHIBITING CORECEPTOR INTERACTION

In contrast to the approaches above, targeting coreceptors currently is the most plausible strategy for preventing viral fusion. HIV uses two principal coreceptors for fusion in vivo: CXCR4 and CCR5. Epidemiological evidence shows that virus using CCR5 (known as R5 virus) are responsible for >90% of infections worldwide. Therefore, targeting CCR5 is a highly plausible intervention strategy. Drugs that target CCR5 include RANTES protein analogs (Lederman et al. 2004; Gaertner et al. 2008), and small module inhibitors such as CMPD167 (Veazey et al. 2005) and maraviroc (Veazey et al. 2010). Several CCR5 antagonists have been shown to protect against vaginal SHIV challenge in NHP studies (Lederman et al. 2004; Veazey et al. 2005, 2010). Maraviroc is the only CCR5 inhibitor shown to be safe and effective in humans when used as a therapeutic drug. CCR5 antagonists, and specifically maraviroc, are thought to bind to the *trans*-membrane domains of CCR5 and modify the conformation of its extracellular domains, thereby inhibiting gp120 binding to CCR5 (Fig. 3) (Kondru et al. 2008; Hu et al. 2010; Garcia-Perez et al. 2011). Recently, maraviroc also has been shown to prevent SHIV infection in NHP models when formulated as a microbicide (Veazey et al. 2010). Wide clinical experience and established manufacturing processes make maraviroc the most attractive candidate for microbicide development in this class of compounds.

TARGETING REVERSE TRANSCRIPTASE

NRTIs

Reverse transcriptase inhibitors and, in particular, nucleoside reverse transcriptase inhibitors (NRTIs) were the first class of drugs developed for HIV therapy (Broder 2010). They were also shown to have utility in postexposure prophylaxis and in prevention of mother-to-child transmission (PMTCT) (Wiznia et al. 1996). NRTIs block reverse transcription by competing with the natural nucleoside counterparts for incorporation into newly forming HIV DNA. Once successfully incorporated, termination of the elongating DNA chain ensues, and DNA synthesis is interrupted (Fig. 4). NRTIs'

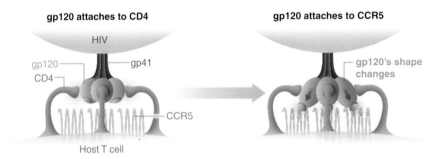

1. HIV ATTACHMENT
The HIV life cycle begins when a protein on the virus—**gp120**—binds to a **CD4** receptor on the cell. This in turn induces conformational change in gp120, facilitating its binding to a **CCR5** or **CXCR4** coreceptor and triggering viral fusion with the cell.

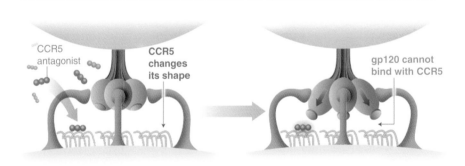

2. CCR5 ANTAGONISTS
CCR5 antagonists like **maraviroc** change the shape of CCR5, making it impossible for gp120 to bind to the CCR5 receptor, thereby inhibiting HIV attachment.

Figure 3. How CCR5 antagonists like maraviroc work.

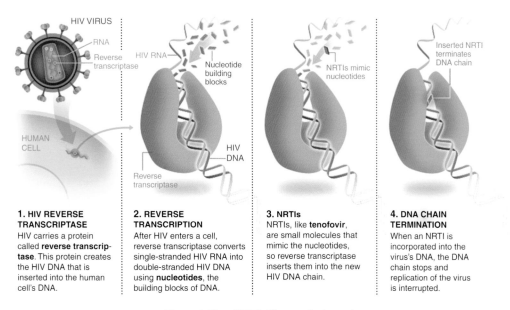

Figure 4. How NRTIs like tenofovir work.

potential use in microbicides was first demonstrated in a pivotal NHP vaginal challenge study in which 1% tenofovir gel was shown to prevent acquisition of SIV infection (Parikh et al. 2009). Given the promising data in macaques with NRTIs, tenofovir gel was evaluated for safety both preclinically and clinically, and a group of South African investigators carried out the first efficacy trial of an ARV-containing microbicide (Abdool Karim et al. 2010). The CAP-RISA 004 trial demonstrated for the first time that an ARV-based microbicide could protect women from HIV infection. A number of interesting observations stemmed from this study. First, over a 36-month period, 1% tenofovir gel was shown to provide a 39% reduction in HIV transmission. Interestingly, at 1 year, the figure was 50%, and in those using the drug with >80% compliance, it was shown to be 53% effective (Abdool Karim et al. 2010).

Tenofovir has characteristics that may make it particularly suitable for use as a microbicide. It is the only licensed nucleotide RT inhibitor, which requires only two phosphorylations to become active, whereas nucleoside inhibitors such as AZT need three phosphorylations. More importantly, tenofovir has an extremely long tissue half-life, meaning that after the

microbicide is applied, drug may remain active in tissue for prolonged periods (Rohan et al. 2010). This may provide a wider window of protection than other short-lived NRTIs, and in NHP studies, some protection was evident 3 days postapplication (Dobard et al. 2010), but the window remains to be determined in humans. Seven other NRTIs are licensed and already in use as therapeutic drugs and may be used in future combinations with tenofovir. Tenofovir + emtricitabine (FTC) has already shown activity in NHP studies and is in clinical development (Parikh et al. 2009).

NNRTIs

The second class of reverse transcriptase inhibitors is NNRTIs, drugs that are active through noncompetitive inhibition of the reverse transcriptase enzyme (Fig. 5). Several NNRTIs are widely used in treatment. Nevirapine has been used for PMTCT, providing biological plausibility for its use in prevention (Chasela et al. 2010). NNRTIs that have undergone preclinical evaluation as microbicides are dapivirine, UC781, MIV-150, and MC1220 (Fletcher et al. 2005; Fernandez-Romero et al. 2007; Fletcher and Shattock 2008; Caron et al. 2010). All of

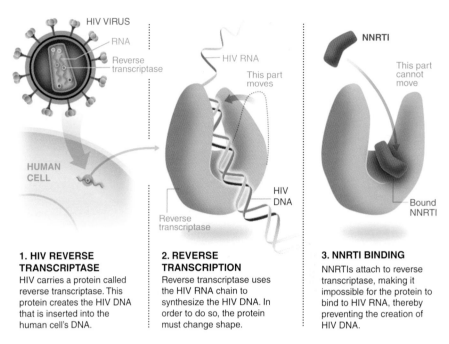

1. HIV REVERSE TRANSCRIPTASE
HIV carries a protein called reverse transcriptase. This protein creates the HIV DNA that is inserted into the human cell's DNA.

2. REVERSE TRANSCRIPTION
Reverse transcriptase uses the HIV RNA chain to synthesize the HIV DNA. In order to do so, the protein must change shape.

3. NNRTI BINDING
NNRTIs attach to reverse transcriptase, making it impossible for the protein to bind to HIV RNA, thereby preventing the creation of HIV DNA.

Figure 5. How NNRTIs like dapivirine work.

these drugs are known to show an extremely high-level affinity for binding to the RT enzyme; MIV-150 and MC1220 have been shown to prevent infection in NHP studies after vaginal challenge (Crostarosa et al. 2009; Caron et al. 2010; Kenney et al. 2011). Dapivirine is the most advanced of these candidates in terms of clinical development. Its poor oral bioavailability could potentially provide an advantage as a microbicide because topical application can achieve high levels of localized concentration with minimal systemic exposure. This feature may be important when considering potential long-term safety and/or resistance, as will be discussed later. Dapivirine shows extremely high activity in a range of preclinical models (Fletcher et al. 2009) by inference; it has the potential to be active in preventing HIV transmission when delivered as a microbicide.

TARGETING VIRAL INTEGRATION

Raltegravir, the only integrase inhibitor currently in therapeutic use (Powderly 2010), is undergoing preclinical study for potential use as a microbicide. Little published data currently

exist on its activity in relevant preclinical microbicide models, but inhibition of viral integration remains an important target for prevention of initial events essential for viral transmission. Indeed, one integrase inhibitor has been shown to prevent SHIV challenge in a pilot NHP study (Dobard et al. 2010). The microbicide field will continue to watch the development of integrase inhibitors and their potential use in microbicide strategies.

TARGETING VIRAL MATURATION

Ten licensed protease inhibitors currently are in use for therapy (De Clercq 2010). These drugs have the highest barrier to resistance when used for therapy and represent interesting potential candidates as microbicides. Whether drugs that block postintegration events would be able to prevent transmission remains unclear. However, infection of mucosal tissue is hypothesized to be initiated by a small foci of infected cells and establishment of infection is hypothesized to require virus dissemination after such initial events (Fig. 2) (Haase 2011). It is quite plausible that protease inhibitors may be able to block dissemination from any initial focus for a sufficient time before the localized infection is eliminated or dies through lack of onward spread. Protease inhibitors are being assessed for microbicide activity in preclinical in vitro models and NHP studies. Should they show promise, they likely will be used in future microbicide formulations, most probably in combination with other drugs.

NEW RESEARCH AREAS

Development of Formulation Strategies

The initial concept of a microbicide was based on coitally dependent application of a gel formulation (i.e., dosing close to each act of sexual intercourse). This mode of delivery remains an important strategy, as it may be preferable to many women. The advantage of this approach is that it provides a bolus of drug directly before potential viral exposure. The disadvantages are related to compliance. Women must be able to anticipate when they may have sex and be able to apply a product discreetly in advance. For many women in the developing world, this may be impractical, and it would be complicated by the need to ensure that a microbicide is always readily available and could still be reliably used if sex took place in the context of drugs, alcohol, or abuse. Experience indicates that it is often difficult to ensure compliance with coitally dependent microbicides in the context of clinical trials (Turner et al. 2009; Greene et al. 2010).

The timing of nevirapine in its proven strategy for preventing mother-to-child HIV transmission led the CAPRISA 004 team to design a clinical trial with the BAT-24 "before and after sex" dosing regimen (Abdool Karim et al. 2010). BAT-24 required administration of tenofovir gel any time in the 12 hours preceding intercourse and a second dose of gel in the 12 hours after intercourse for no more than two doses in any 24-h period (Fig. 6). This regimen was envisaged to be more flexible and to increase the chance of having two drug doses available around the time of transmission—one before and one after a sex act. BAT-24 with tenofovir

gel was the first microbicide regimen to demonstrate protection from HIV. How critical each dose and the timing around sex were for optimal protection is unclear. Nevertheless, the results from CAPRISA 004 represent a significant step forward. A second trial with the BAT-24 dosing regimen is planned to begin in early 2011.

A third approach has been development of trials around daily dosing with tenofovir gel, as is currently being assessed in the VOICE trial by the Microbicide Trials Network (http://www.mtnstopshiv.org/news/studies/mtn003/backgrounder). With this regimen, subjects are asked to apply the product once a day independent of any sex acts. It is hoped that compliance with daily dosing would provide steady-state drug levels within the optimal dosing concentration required for protection (Fig. 6). This approach is anticipated to lead to greater compliance because it can be incorporated as a routine act in daily life.

A fourth approach is sustained drug delivery, which is designed around IVRs already being used to deliver hormone replacement therapy and hormonal contraception. Silicone elastomer-impregnated rings can release hydrophobic drugs for prolonged periods. Recent studies have shown that an IVR can deliver the NNRTI dapivirine under steady release conditions for a minimum of a month (Nel et al. 2009; Malcolm et al. 2010). Studies in Africa have shown that such rings are highly acceptable in at-risk populations (Smith et al. 2008; Woodsong et al. 2010). The advantage of an IVR is that it can be worn safely for a month or longer. The premise is that IVRs can deliver optimal concentrations of drug required for protection (Fig. 6). However, one of the tradeoffs against the benefits of an IVR is that it may not release drug at levels equal to what is achievable with daily or coitally related dosing with gels, films, etc. Nevertheless, compliance with the ring may be higher because there is no requirement to plan for each sexual act or to use the product on a daily basis.

Ultimately, development of multiple formulations would give women the choice of an approach that best fits their circumstances.

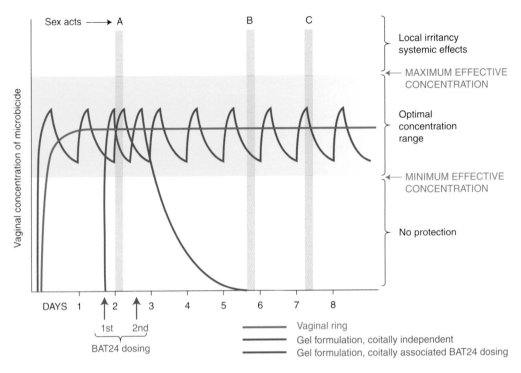

Figure 6. Microbicide dosing schemes: coitally dependent and independent.

Ongoing efforts to increase the range of formulations include development of rapid-dissolving vaginal tablets and small dissolvable films for vaginal application (Romano et al. 2008). Once a drug candidate has shown efficacy in clinical trials, other formulations of the same product may be evaluated to provide greater choice to women.

PHARMACOKINETICS (PK) AND PHARMACODYNAMICS (PD) IN CLINICAL ASSESSMENT

Phase III clinical trials are expensive, time-consuming, and require a large number of participants to determine efficacy. Such trials cost between $50–$100 million. Furthermore, as successive interventions are introduced, incidence of infection within communities will decrease and the window for performing placebo-controlled trials will close. This will mean that testing of later-generation products will become more complex because reduced incidence and provision of an increasing

number of prevention options will require even larger trials. The move to ARV-based microbicides has meant that candidates can be assessed using traditional drug development tools, including PK (measurement of drug absorption, retention, and distribution over time) and in parallel PD (assessment of drug activity within biological compartments). These tools are becoming increasingly important in microbicide development to accelerate prioritization of candidates, combinations, and formulations prior to Phase III. Once the concentration required to provide protection within mucosal tissue is known, different formulations can be assessed against this potential biological parameter.

More recent efforts have also focused on establishment of potential PD markers. The basic principle is to take biopsies from mucosal tissue following drug application and determine whether the tissue is resistant to infection when challenged with live virus in the laboratory. In the first PD study of a rectal microbicide in humans, two different doses of the NNRTI

UC781 and a placebo gel showed different PD properties. The highest dose tested was shown to significantly reduce viral replication in tissue biopsies challenged after drug application (Anton et al. 2009). This technology now is being applied to development of vaginal microbicides, which is more complex because only one to two biopsy samples can be taken from a woman's vagina, in contrast to the 10–30 biopsies that can safely be taken from the colon and rectum. It remains to be seen whether ex vivo challenge of single cervical biopsy from women postdosing with product would be sufficient to effectively assess microbicide PD. An alternative approach to establish a marker of PD has been the analysis of anti-HIV activity in cervical fluid following vaginal application of microbicide formulations (Keller et al. 2010; Herold et al. 2011). PK and PD are emerging tools that may be critical for rapid assessment and introduction of new formulations of a product once it has shown efficacy in a clinical trial. Furthermore, these studies may be important in assessing ongoing drug combinations designed to increase the efficacy of products once single-ARV microbicides have been proven effective.

RESISTANCE TO ARV MICROBICIDES

When ARV-based microbicides are introduced, the potential impact on current and future therapeutic strategies must be considered. This consequence is particularly important because many drugs used as microbicides are components of products already in use for front-line and second-line treatment. The long-term implications would be serious if a microbicide significantly reduced treatment because it induced resistance. Such concerns are not without precedence, as treatment with nevirapine for PMTCT has been associated with induction of resistance to that drug, although the impact on subsequent therapy remains controversial (Lockman et al. 2010; Palumbo et al. 2010). The potential for resistance following topical use of a drug that does not deliver significant systemic levels, however, remains theoretical and can only be tested in a clinical setting (IPM 2008; Wilson et al. 2008). Interestingly,

no evidence of resistance to tenofovir was seen in women in CAPRISA 004 who became infected during the trial. It remains unclear, however, how compliant with the BAT 24 regimen the subjects who became infected were and, therefore, what impact the dosing frequency had on induction of resistance.

Resistance can occur under two potential scenarios. The first is transmission of resistant isolates. In any community where treatment is widespread, resistant isolates will exist. An individual using a microbicide based on a single ARV potentially would be susceptible to transmission of such a resistant isolate because it might not be blocked by the drug to which it is resistant. It should be noted, however, that resistance generally is relative and not absolute. Therefore, viruses resistant to an individual drug can still be inhibited if the drug is applied in sufficient concentration. The advantage of topical dosing with ARVs is that they can be safely applied at concentrations many times higher than can be achieved with oral delivery. Whether resistant virus would survive the high concentrations within the mucosal environment is unclear. Nevertheless, that needs to be closely monitored.

The second scenario that could lead to resistance involves the continued use of an ARV-based microbicide by an individual following HIV infection. Here active replication within the individual could give rise to selection of resistance mutations to the drug contained in the microbicide. Again, the risk is theoretical and the likelihood is unknown. However, the resistance scenarios have led to the notion that ARV-based microbicides should not be available over the counter after they are licensed. Access to such products would require demonstration of seronegative status and potentially frequent monitoring to ensure that seroconversion has not occurred. If an individual using an ARV-based microbicide were to become infected, she or he immediately would be advised to stop using the product. This approach is being implemented in all clinical trials. How frequently individuals should be tested once a microbicide is licensed is unclear, but it is likely that monitoring for resistance will be done during Phase IV roll out of any ARV-based microbicide. If resistance becomes an issue, increased focus will be placed on development of combination microbicides aimed at limiting this risk.

COMBINATION MICROBICIDES

Development and efficacy testing of single-drug candidates is the current focus of microbicide clinical trials, and microbicide combinations are already being assessed as next-generation products. There are three principal reasons for combination-based microbicides. The first is to increase the breadth of activity. Any single ARV microbicide would be circumvented by a virus that is naturally resistant to the drug or exploits a mechanism that allows it to bypass the drug's activity. Second, ARV combinations may be additive or synergistic, thus allowing lower concentrations of drugs to have more potent activity against HIV. Third, combining drugs that interfere with different stages in the transmission process and the viral lifecycle would provide greater chances of protection. This is supported by the observed superiority of highly active ART in prevention of mother–child transmission (PMTCT) (Sturt et al. 2010).

Clinical trials to test microbicide drug combinations have their own challenges. Combination studies traditionally require testing individual drugs in separate arms against the combination. Here the emphasis is on demonstrating that the combination is more effective than either drug alone. Although safety and PK trials of combination microbicides are feasible, there may not be sufficient scope, funding, and/or subjects to provide continual iterations of drug combinations as microbicides are introduced. It is more likely that single-drug microbicides will be licensed initially, and that ongoing studies then will assess whether combining other drugs with these candidates can increase their potency. The first drugs to show efficacy in clinical trials likely will dominate subsequent microbicide drug development, with combinations focused on the first licensed products. Innovative strategies to bridge from PK/PD to efficacy will be needed to ensure that optimal microbicide combinations are available.

MICROBICIDES IN THE CONTEXT OF OTHER PREVENTION TECHNOLOGIES

Early models of the impact of an effective microbicide on transmission rates anticipate that it would be bi-directional, that is, preventing infection from men to women and from women to men (Watts et al. 2002). Whether ARV-based microbicides would directly reduce transmission from women to men is unclear. Nevertheless, reducing the incidence of infection in women would, by implication, reduce the number of women who could transmit HIV to additional male partners.

Concern has been raised that microbicides might lead to reduced use of condoms. Models of the impact of microbicide introduction, based on different levels of efficacy and the potential impact of such a strategy on current condom use, suggest that the effects are likely to be small if a microbicide is >50% effective and consistent condom use in the population is <70% (Foss et al. 2003). Recent results from the iPrEX trial of oral PrEP showed 44% (MIT) efficacy overall and up to 50% efficacy in participants who were >50% compliant (Grant et al. 2010). It is unclear whether combining oral PrEP with microbicide might provide even higher levels of protection.

The major challenge for PrEP with both oral agents and topical microbicides will be the ongoing issue of compliance. Individuals are often poor judges of potential risk when it comes to different partners. This means that if a microbicide is used intermittently based on perception of risk, rather than routinely, then exposure is likely to occur in the absence of a microbicide. In that scenario, combining PrEP (topical or oral) with a suboptimal vaccine could have potential. Recent data from the RV-144 vaccine trial in Thailand suggest that vaccine candidates may be successful in low-risk cohorts (Rerks-Ngarm et al. 2009). Microbicide use could turn high-risk cohorts into low-risk cohorts in which superimposition of a suboptimal vaccine could further reduce transmission rates (Excler et al. 2010). The potential exists for use of microbicides in the context of a vaccine, but assessing potential synergies between the two approaches is a significant challenge for clinical trial design (Excler et al. 2010).

CONCLUSIONS

With the adoption of highly active ARV-based products for development, the microbicide field has benefitted from a highly successful therapeutic field. Partnerships between pharmaceutical companies and public–private partnerships, together with support from government agencies, have resulted in accelerated development of ARV-based microbicides. Researchers are optimistic that efficacy of microbicides can be increased through different dosing regimens and formulation strategies. The groundbreaking results from the CAPRISA 004 trial show that microbicides can be effective.

A number of Phase III trials of microbicides are currently underway or about to be initiated, including research on tenofovir gel and a dapivirine IVR. The challenge is to build on such early proof-of-concept, and develop approaches that are even more effective, together with a wider range of products that provide women with optimal choice in HIV prevention. It is our fervent hope that in the near future microbicides are introduced as part of a comprehensive, multicomponent prevention strategy that continues to evolve as additional options become available.

ACKNOWLEDGMENTS

We greatly acknowledge the editorial help of Pamela N. Norick, Judy Orvos, and Sarah Harman. Illustrations were produced by 5W Infographics based on original diagrams from the authors. Z.F.R. is the Chief Executive Officer of the International Partnership for Microbicides.

REFERENCES

Abdool Karim Q, Abdool Karim SS, Frohlich JA, Grobler AC, Baxter C, Mansoor LE, Kharsany AB, Sibeko S, Mlisana KP, Omar Z, et al. 2010. Effectiveness and safety of tenofovir gel, an antiretroviral microbicide, for the prevention of HIV infection in women. *Science* **329:** 1168–1174.

Anton P, Tanner K, Cho D, Johnson E, Cumberland B, Zhou Y, Mauck C, McGowan I. 2009. Strong suppression of HIV-1 infection of colorectal explants following in vivo rectal application of UC781 gel: A novel endpoint in a phase 1 trial. In *16th Conference on Retroviruses and Opportunistic Infections.* Montreal, Canada.

Broder S. 2010. Twenty-five years of translational medicine in antiretroviral therapy: Promises to keep. *Sci Transl Med* **2:** 39ps33.

Caron M, Besson G, Etenna SL, Mintsa-Ndong A, Mourtas S, Radaelli A, Morghen Cde G, Loddo R, La Colla P, Antimisiaris SG, et al. 2010. Protective properties of non-nucleoside reverse transcriptase inhibitor (MC1220) incorporated into liposome against intravaginal challenge of Rhesus macaques with RT-SHIV. *Virology* **405:** 225–233.

Chasela CS, Hudgens MG, Jamieson DJ, Kayira D, Hosseinipour MC, Kourtis AP, Martinson F, Tegha G, Knight RJ, Ahmed YI, et al. 2010. Maternal or infant antiretroviral drugs to reduce HIV-1 transmission. *N Engl J Med* **362:** 2271–2281.

Crostarosa F, Aravantinou M, Akpogheneta OJ, Jasny E, Shaw A, Kenney J, Piatak M, Lifson JD, Teitelbaum A, Hu L, et al. 2009. A macaque model to study vaginal HSV-2/immunodeficiency virus co-infection and the impact of HSV-2 on microbicide efficacy. *PLoS One* **4:** e8060.

De Clercq E. 2010. Antiretroviral drugs. *Curr Opin Pharmacol* **10:** 507–515.

Dobard C, Sharma S, Holder A, Kuklenyik Z, Hanson D, Smith J, Otten RA, Novembre FJ, Garcia-Lerma1 G, Heneine W. 2010a. Protection by TFV gel against vaginal SHIV infection in macaques three days after gel application and its relationship to tissue drug levels. In *Microbicides 2010.* Pittsburgh.

Dobard CW, Sharma S, Martin A, Hazuda D, Hanson D, Smith J, Otten RA, Novembre F, Garcia-Lerma G, Heneine W. 2010b. Protection against repeated vaginal SHIV exposures in macaques by a topical gel with an integrase inhibitor. In *Microbicides 2010.* Pittsburgh.

Dosekun O, Fox J. 2010. An overview of the relative risks of different sexual behaviours on HIV transmission. *Curr Opin HIV AIDS* **5:** 291–297.

Excler JL, Rida W, Priddy F, Gilmour J, McDermott AB, Kamali A, Anzala O, Mutua G, Sanders EJ, Koff W, et al. 2010. AIDS vaccines and preexposure prophylaxis: Is synergy possible? *AIDS Res Hum Retrov* doi: 101089/aid20100206.

Fernández-Romero JA, Thorn M, Turville SG, Titchen K, Sudol K, Li J, Miller T, Robbiani M, Maguire RA, Buckheit RW Jr, et al. 2007. Carrageenan/MIV-150 (PC-815), a combination microbicide. *Sex Transm Dis* **34:** 9–14.

Fletcher PS, Shattock RJ. 2008. PRO-2000, an antimicrobial gel for the potential prevention of HIV infection. *Curr Opin Invest Drugs* **9:** 189–200.

Fletcher P, Kiselyeva Y, Wallace G, Romano J, Griffin G, Margolis L, Shattock R. 2005. The nonnucleoside reverse transcriptase inhibitor UC-781 inhibits human immunodeficiency virus type 1 infection of human cervical tissue and dissemination by migratory cells. *J Virol* **79:** 11179–11186.

Fletcher P, Harman S, Azijn H, Armanasco N, Manlow P, Perumal D, de Bethune MP, Nuttall J, Romano J, Shattock R. 2009. Inhibition of human immunodeficiency virus type 1 infection by the candidate microbicide dapivirine, a nonnucleoside reverse transcriptase inhibitor. *Antimicrob Agents Chemother* **53:** 487–495.

Foss AM, Vickerman PT, Heise L, Watts CH. 2003. Shifts in condom use following microbicide introduction: Should we be concerned? *AIDS* **17:** 1227–1237.

Gaertner H, Cerini F, Escola JM, Kuenzi G, Melotti A, Offord R, Rossitto-Borlat I, Nedellec R, Salkowitz J, Gorochov G, et al. 2008. Highly potent, fully recombinant anti-HIV chemokines: Reengineering a low-cost microbicide. *Proc Natl Acad Sci* **105:** 17706–17711.

Garcia-Perez J, Rueda P, Staropoli I, Kellenberger E, Alcami J, Arenzana-Seisdedos F, Lagane B. 2011. New insights into the mechanisms whereby low molecular weight CCR5 ligands inhibit HIV-1 infection. *J Biol Chem* **286:** 4978–4990.

Grant RM, Lama JR, Anderson PL, McMahan V, Liu AY, Vargas L, Goicochea P, Casapía M, Guanira-Carranza JV, Ramirez-Cardich ME, et al. 2010. Preexposure chemoprophylaxis for HIV prevention in men who have sex with men. *N Engl J Med* **363:** 2587–2599.

Greene E, Batona G, Hallad J, Johnson S, Neema S, Tolley EE. 2010. Acceptability and adherence of a candidate microbicide gel among high-risk women in Africa and India. *Cult Health Sex* **12:** 739–754.

Haase AT. 2011. Early events in sexual transmission of HIV and SIV and opportunities for interventions. *Annu Rev Med* **62:** 127–139.

Herold BC, Mesquita PM, Madan RP, Keller MJ. 2011. Female genital tract secretions and semen impact the development of microbicides for the prevention of HIV and other sexually transmitted infections. *Am J Reprod Immunol* **65:** 325–333.

Hillier SL, Moench T, Shattock R, Black R, Reichelderfer P, Veronese F. 2005. In vitro and in vivo: The story of nonoxynol 9. *J Acq Immune Def Synd* **39:** 1–8.

Hu Q, Huang X, Shattock RJ. 2010. C-C chemokine receptor type 5 (CCR5) utilization of transmitted and early founder human immunodeficiency virus type 1 envelopes and sensitivity to small-molecule CCR5 inhibitors. *J Gen Virol* **91:** 2965–2973.

IPM. 2008. Consensus Statement by IPM and the Executive Committee of the IPM Scientific Advisory Board. The Use of Antiretroviral Drugs (ARVs) in Microbicides and the Potential for the Development of Drug-resistant Strains of HIV. http://www.ipmglobal.org/publications-media/ipm-consensus-statement-resistance.

Keele BF. 2010. Identifying and characterizing recently transmitted viruses. *Curr Opin HIV AIDS* **5:** 327–334.

Keele BF, Giorgi EE, Salazar-Gonzalez JF, Decker JM, Pham KT, Salazar MG, Sun C, Grayson T, Wang S, Li H, et al. 2008. Identification and characterization of transmitted and early founder virus envelopes in primary HIV-1 infection. *Proc Natl Acad Sci* **105:** 7552–7557.

Keller MJ, Mesquita PM, Torres NM, Cho S, Shust G, Madan RP, Cohen HW, Petrie J, Ford T, Soto-Torres L, et al. 2010. Postcoital bioavailability and antiviral activity of 0.5%

PRO 2000 gel: Implications for future microbicide clinical trials. *PLoS One* **5:** e8781.

Kenney J, Aravantinou M, Singer R, Hsu M, Rodriguez A, Kizima L, Abraham CJ, Menon R, Seidor S, Chudolij A, et al. 2011. An antiretroviral/zinc combination gel provides 24 hours of complete protection against vaginal SHIV infection in macaques. *PLoS One* **6:** e15835.

Kondru R, Zhang J, Ji C, Mirzadegan T, Rotstein D, Sankuratri S, Dioszegi M. 2008. Molecular interactions of CCR5 with major classes of small-molecule anti-HIV CCR5 antagonists. *Mol Pharmacol* **73:** 789–800.

Lederman MM, Veazey RS, Offord R, Mosier DE, Dufour J, Mefford M, Piatak MJr, Lifson JD, Salkowitz JR, Rodriguez B, et al. 2004. Prevention of vaginal SHIV transmission in rhesus macaques through inhibition of CCR5. *Science* **306:** 485–487.

Lockman S, Hughes MD, McIntyre J, Zheng Y, Chipato T, Conradie F, Sawe F, Asmelash A, Hosseinipour MC, Mohapi L, et al. 2010. Antiretroviral therapies in women after single-dose nevirapine exposure. *N Engl J Med* **363:** 1499–1509.

Malcolm RK, Edwards KL, Kiser P, Romano J, Smith TJ. 2010. Advances in microbicide vaginal rings. *Antiviral Res* **88:** S30–S39.

Malkovsky M, Newell A, Dalgleish AG. 1988. Inactivation of HIV by nonoxynol-9. *Lancet* **1:** 645.

Mayer KH, Peipert J, Fleming T, Fullem A, Moench T, Cu-Uvin S, Bentley M, Chesney M, Rosenberg Z. 2001. Safety and tolerability of BufferGel, a novel vaginal microbicide, in women in the United States. *Clin Infect Dis* **32:** 476–482.

Nel A, Smythe S, Young K, Malcolm K, McCoy C, Rosenberg Z, Romano J. 2009. Safety and pharmacokinetics of dapivirine delivery from matrix and reservoir intravaginal rings to HIV-negative women. *J Acq Immune Def Synd* **51:** 416–423.

O'Connor TJ, Kinchington D, Kangro HO, Jeffries DJ. 1995. The activity of candidate virucidal agents, low pH and genital secretions against HIV-1 in vitro. *Int J STD AIDS* **6:** 267–272.

Palumbo P, Lindsey JC, Hughes MD, Cotton MF, Bobat R, Meyers T, Bwakura-Dangarembizi M, Chi BH, Musoke P, Kamthunzi P, et al. 2010. Antiretroviral treatment for children with peripartum nevirapine exposure. *N Engl J Med* **363:** 1510–1520.

Parikh UM, Dobard C, Sharma S, Cong ME, Jia H, Martin A, Pau CP, Hanson DL, Guenthner P, Smith J, et al. 2009. Complete protection from repeated vaginal simian-human immunodeficiency virus exposures in macaques by a topical gel containing tenofovir alone or with emtricitabine. *J Virol* **83:** 10358–10365.

Powderly WG. 2010. Integrase inhibitors in the treatment of HIV-1 infection. *J Antimicrob Chemother* **65:** 2485–2488.

Ramjee G, Kamali A, McCormack S. 2010. The last decade of microbicide clinical trials in Africa: From hypothesis to facts. *AIDS* **24:** S40–S49.

Rerks-Ngarm S, Pitisuttithum P, Nitayaphan S, Kaewkungwal J, Chiu J, Paris R, Premsri N, Namwat C, de Souza M, Adams E, et al. 2009. Vaccination with ALVAC and AIDSVAX to prevent HIV-1 infection in Thailand. *N Engl J Med* **361:** 2209–2220.

Rohan LC, Moncla BJ, Kunjara Na Ayudhya RP, Cost M, Huang Y, Gai F, Billitto N, Lynam JD, Pryke K, Graebing P, et al. 2010. In vitro and ex vivo testing of tenofovir shows it is effective as an HIV-1 microbicide. *PLoS One* **5:** e9310.

Romano J, Malcolm RK, Garg S, Rohan LC, Kaptur PE. 2008. Microbicide delivery: Formulation technologies and strategies. *Curr Opin HIVAIDS* **3:** 558–566.

Salazar-Gonzalez JF, Salazar MG, Keele BF, Learn GH, Giorgi EE, Li H, Decker JM, Wang S, Baalwa J, Kraus MH, et al. 2009. Genetic identity, biological phenotype, and evolutionary pathways of transmitted/founder viruses in acute and early HIV-1 infection. *J Exp Med* **206:** 1273–1289.

Shattock RJ, Moore JP. 2003. Inhibiting sexual transmission of HIV-1 infection. *Nat Rev Microbiol* **1:** 25–34.

Shattock R, Solomon S. 2004. Microbicides—aids to safer sex. *Lancet* **363:** 1002–1003.

Si Z, Madani N, Cox JM, Chruma JJ, Klein JC, Schön A, Phan N, Wang L, Biorn AC, Cocklin S, et al. 2004. Small-molecule inhibitors of HIV-1 entry block receptor-induced conformational changes in the viral envelope glycoproteins. *Proc Natl Acad Sci* **101:** 5036–5041.

Smith DJ, Wakasiaka S, Hoang TD, Bwayo JJ, Del Rio C, Priddy FH. 2008. An evaluation of intravaginal rings as a potential HIV prevention device in urban Kenya: Behaviors and attitudes that might influence uptake within a high-risk population. *J Womens Health* **17:** 1025–1034.

Sturt AS, Dokubo EK, Sint TT. 2010. Antiretroviral therapy (ART) for treating HIV infection in ART-eligible pregnant women. *Cochrane Database Syst Rev* **3:** CD008440.

Turner AN, De Kock AE, Meehan-Ritter A, Blanchard K, Sebola MH, Hoosen AA, Coetzee N, Ellertson C. 2009. Many vaginal microbicide trial participants acknowledged they had misreported sensitive sexual behavior in face-to-face interviews. *J Clin Epidemiol* **62:** 759–765.

Van Damme L, Ramjee G, Alary M, Vuylsteke B, Chandeying V, Rees H, Sirivongrangson P, Mukenge-Tshibaka L, Ettiègne-Traoré V, Uaheowitchai C, et al. 2002. Effectiveness of COL-1492, a nonoxynol-9 vaginal gel, on HIV-1 transmission in female sex workers: A randomised controlled trial. *Lancet* **360:** 971–977.

Van Herrewege Y, Morellato L, Descours A, Aerts L, Michiels J, Heyndrickx L, Martin L, Vanham G. 2008. CD4 mimetic miniproteins: Potent anti-HIV compounds with promising activity as microbicides. *J Antimicrob Chemother* **61:** 818–826.

Veazey RS, Klasse PJ, Schader SM, Hu Q, Ketas TJ, Lu M, Marx PA, Dufour J, Colonno RJ, Shattock RJ, et al. 2005. Protection of macaques from vaginal SHIV challenge by vaginally delivered inhibitors of virus-cell fusion. *Nature* **438:** 99–102.

Veazey RS, Ketas TA, Klasse PJ, Davison DK, Singletary M, Green LC, Greenberg ML, Moore JP. 2008. Tropism-independent protection of macaques against vaginal transmission of three SHIVs by the HIV-1 fusion inhibitor T-1249. *Proc Natl Acad Sci* **105:** 10531–10536.

Veazey RS, Ketas TJ, Dufour J, Moroney-Rasmussen T, Green LC, Klasse PJ, Moore JP. 2010. Protection of rhesus macaques from vaginal infection by vaginally delivered maraviroc, an inhibitor of HIV-1 entry via the CCR5 co-receptor. *J Infect Dis* **202:** 739–744.

Vermeire K, Brouwers J, Van Herrewege Y, Le Grand R, Vanham G, Augustijns P, Bell TW, Schols D. 2008. CADA, a potential anti-HIV microbicide that specifically targets the cellular CD4 receptor. *Curr HIV Res* **6:** 246–256.

Watts C, Vickerman P, Terris-Prestholt F. 2002. *The public health benefits of microbicides in lower-income countries.* Rockefeller Foundation, New York.

Wilson DP, Coplan PM, Wainberg MA, Blower SM. 2008. The paradoxical effects of using antiretroviral-based microbicides to control HIV epidemics. *Proc Natl Acad Sci* **105:** 9835–9840.

Wiznia AA, Crane M, Lambert G, Sansary J, Harris A, Solomon L. 1996. Zidovudine use to reduce perinatal HIV type 1 transmission in an urban medical center. *JAMA* **275:** 1504–1506.

Woodsong C, Masenga G, Rees H, Bekker LG, Ganesh S, Young K, Romano J, Nel A. 2010. Safety and acceptability of vaginal ring as microbicide delivery method in African women. In *Microbicides 2010.* Pittsburgh.

World Health Organization. 2009. *Women and health: Today's evidence, tomorrow's agenda.* World Health Organization. http://www.who.int/gender/women_health_report/en/index.html.

HIV Prevention by Oral Preexposure Prophylaxis

Walid Heneine[1] and Angela Kashuba[2]

[1]Laboratory Branch, Division of HIV/AIDS Prevention, National Center for HIV, Hepatitis, STD, and Prevention, Centers for Disease Control and Prevention, Atlanta, Georgia 30333

[2]Division of Pharmacotherapy and Experimental Therapeutics, Eshelman School of Pharmacy, and UNC Center for AIDS Research, University of North Carolina at Chapel Hill, Chapel Hill, North Carolina 27599

Correspondence: wheneine@cdc.gov; akashuba@unc.edu

The impressive advances in antiretroviral (ARV) therapy of chronic human immunodeficiency virus (HIV) infections during the last decade and the availability of potent ARV drugs have fueled interest in using chemoprophylaxis as a novel HIV prevention strategy. Preexposure prophylaxis (PrEP) refers to the use of ARV drugs in HIV-negative persons to prevent HIV infection. The rationale for PrEP builds on the success of ARV prophylaxis in preventing mother-to-child transmission of HIV and on a large body of animal studies that show the efficacy of PrEP against mucosal and parenteral infection. We focus on oral administration of ARV drugs for prevention of HIV infection. Identifying an effective prophylactic pill that individuals can take outside the setting of sexual intercourse precludes the necessity to disclose such use to their partners, thereby empowering those who might not be in a position to negotiate with their partners. Several human clinical trials evaluating the efficacy of daily regimens of the HIV reverse-transcriptase (RT) inhibitors tenofovir disoproxil fumarate (TDF) or Truvada (TDF and emtricitabine [FTC]) are under way among high-risk populations. The results of one trial among men who have sex with men showed that daily Truvada was safe and effective, providing the first support for oral PrEP as a prevention strategy. Here we outline the preclinical and clinical research on oral PrEP, pharmacologic considerations, and future directions and challenges.

The prevention of human immunodeficiency virus (HIV) infection remains a critical public health priority. It is estimated that 2.7 million new HIV infections have occurred worldwide in 2008 at a rate that continues to outpace the rate at which HIV-infected persons enter treatment (UNAIDS 2009). While work on HIV vaccine discovery continues to progress, prevention research has focused in recent years on a variety of new biomedical strategies for preventing infection, such as male circumcision, topical gels containing antiretroviral (ARV) drugs, or preexposure prophylaxis (PrEP) by oral ARV drugs (Padian et al. 2008). PrEP entails providing HIV-negative individuals with oral ARV drugs to prevent HIV acquisition. Since 1995, the continually impressive advances in ARV therapy of HIV-infected individuals and the availability of potent ARV drugs with known safety and potency profiles have fueled interest in using ARV drugs for HIV prevention. The rationale for PrEP is further supported by the fact that ARV drugs provided to pregnant women with HIV infection were shown to dramatically reduce the risk of perinatal transmission and protect treated breastfed infants of HIV-infected mothers (Hu 2000). Mathematical models estimate that over the next 10 years, an effective PrEP program could prevent 2.7–3.2 million new HIV-1 infections in sub-Saharan Africa (Li 2009). This potentially significant public health benefit requires very high PrEP efficacy, which might be lost or substantially reduced with a PrEP efficacy of <50%. Therefore, identifying highly effective PrEP modalities is critical. In this work we provide an overview of oral PrEP for HIV prevention, discuss ARV drug selection and pharmacology, animal studies, current clinical trials, and implementation planning.

EARLY EVENTS IN HIV TRANSMISSION AND IMPLICATIONS FOR PrEP

The current understanding of mucosal HIV transmission suggests that HIV first replicates at a low level at the mucosal point of entry in the new host. PrEP can be designed to exploit this brief period of virus vulnerability and block HIV from establishing itself as a persistent infection. Early infection events have been largely derived from monkey model studies of vaginal infection with the simian immunodeficiency virus (SIV) (Haase 2005). These studies have consistently shown that, after penetration of SIV into the cervicovaginal epithelium, infection in cervicovaginal tissues during the first 1–3 d is limited to extremely small numbers of productively infected cells in rare foci (Hu et al. 2000; Miller et al. 2005; Li et al. 2009). This small local founder population of infected cells expands in the following days, possibly by accretion of new infections around the initial clusters (Li et al. 2009). Continuous expansion at the point of entry and dissemination of both virus and infected cells through lymphatic drainage and the bloodstream establishes a sustainable infection in secondary lymphoid organs (Miller et al. 2005). In <2 wk, a very small founder population of productively infected cells at the portal of entry progresses

to systemic infection with a burst of virus production and depletion of gut $CD4^+$ T cells. At this point, a robust virus-specific immune response can contain viral replication only to a certain degree (Reynolds et al. 2005). Thus, the first days of infection at the mucosa when replication is limited to small clusters of infected cells are the periods of maximum virus vulnerability and represent a window of opportunity for intervention. Mucosal HIV infection can be conceivably prevented by a rapid and efficient host immune response or by limiting the size of founder populations of infected cells to a theoretical threshold under which infection cannot be established. HIV vaccines have so far been unable to elicit highly protective immune responses. Anti-inflammatory agents that interfere with innate host responses and limit expansion of founder populations have shown promising results in macaques, although the existence of occult infections is still a possibility (Li et al. 2009). However, by inhibiting key steps in HIV replication such as entry or reverse transcription, ARV drugs delivered by oral PrEP may conceivably block the establishment of founder populations of infected cells, or prevent their expansion leading to a dead-end infection. As described below, data from animal models and also from a recent human trial support the promise of this prevention strategy.

ARV DRUGS FOR ORAL PrEP

Beyond mechanical barriers, there are only two biological strategies to prevent HIV infection at the moment of exposure: modification of a host defense (such as a vaccine stimulating neutralizing antibodies) and/or the use of ARV therapy to saturate the cells receptive to HIV infection and replication. The significance of finding a drug, or drug combination, that people can control and use intermittently to protect them against HIV acquisition is immense. In order to choose an effective ARV regimen for PrEP, investigators must choose the right drugs, which target the right sites of infection in the right concentration for the right amount of time.

The Right Drug

Drug candidates for oral PrEP have largely been selected from currently approved drugs for treatment of individuals infected with HIV-1 because development of drugs exclusively for HIV prevention has been limited. There are currently >30 drugs or drug combinations that have been approved for treatment of HIV (Department of Health and Human Services Panel on Antiretroviral Guidelines for Adults and Adolescents January 10, 2011), and a number of desirable drug characteristics for PrEP overlap with those for treatment: good tolerability, low pill burden, infrequent dosing, and resistance profiles with minimal cross resistance. Preintegration drugs (chemokine receptor antagonists, nucleoside and nonnucleoside analog reverse-transcriptase inhibitors, and integrase inhibitors) are currently thought to be more suitable than postintegration drugs (protease inhibitors, maturation inhibitors) for prevention, although direct evidence to support this assumption is lacking.

One pharmacokinetic property considered important for PrEP drugs targeting sexual transmission includes the ability to rapidly reach and accumulate in genital and rectal tissues. Antiretrovirals differ greatly in their ability to penetrate mucosal tissues or secretions (Cohen et al. 2007; Dumond et al. 2007, 2009; Jones et al. 2009; Brown et al. 2011). In general, highly protein-bound compounds do not gain access to these secondary compartments because of their affinity for plasma proteins such as albumin and α_1-acid glycoprotein. Drugs such as the protease inhibitors, which are 95%–99% bound to plasma proteins, generally achieve female genital tract concentrations <50% those in the plasma (Nicol and Kashuba 2010). In contrast, most nucleoside-analog reverse-transcriptase inhibitors (NRTIs) have a low degree of protein binding (<0.7%–49%) and achieve concentrations two- to sixfold higher in mucosal tissue than in plasma.

However, plasma protein binding is not the only predictor of ARV exposure. For example, maraviroc, a cellular entry inhibitor that shows 75% protein binding in blood plasma, has high penetration into cervicovaginal fluid (CVF) and vaginal tissue. After 7 d of dosing at 300 mg of maraviroc twice daily, the areas under the concentration–time curve (AUCs) in CVF and vaginal tissue are 2.7 and 1.9 times higher, respectively, than blood plasma (Dumond et al. 2009). Raltegravir, an integrase inhibitor that is 83% protein bound in plasma, has also been shown to penetrate well in the female genital tract: The concentrations of the drug in CVF after multiple dosing are up to twofold higher than those in plasma (Jones et al. 2009). Figure 1 summarizes all available data of antiretroviral penetration into cervicovaginal fluid, cervical and vaginal tissues, and rectal tissues.

One additional factor relating ARV pharmacokinetics to efficacy at mucosal surfaces is protein binding within the mucosal secretions themselves. The concentrations of albumin and α_1-acid glycoprotein in cervicovaginal fluid are <1% of the values in plasma (Salas Herrera et al. 1991). Although the protein binding of drugs in genital secretions has not been extensively evaluated, maraviroc has recently been shown to have 10-fold less protein binding in CVF than in plasma (7.5% vs. 75%) (Dumond et al. 2009). This phenomenon must be considered in pharmacokinetic–pharmacodynamic analysis of ARV prevention strategies, as the free drug concentration represents the fraction of drug available to be active against HIV infection. Therefore, even though total drug concentration may be lower than blood plasma in mucosal secretions, the free drug concentration may be similar to, or higher than, these plasma concentrations.

The extent of penetration of drugs into rectal tissues also has implications for HIV transmission. Data in these tissues are sparse, but are available for some nucleoside-analog reverse-transcriptase inhibitors (NRTI), nonnucleoside-analog reverse-transcriptase inhibitors (NNRTI), protease inhibitors (PI), and entry inhibitors. Taken orally, tenofovir and emtricitabine concentrate in rectal tissues, achieving concentrations 33- and fourfold greater than plasma. Exposures ($AUC_{12\,h}$) of maraviroc are ~30 times higher in rectal tissues than in plasma (Brown et al. 2011). Darunavir, etravirine, and ritonavir have exposures three-, seven-, and 13-fold higher than plasma.

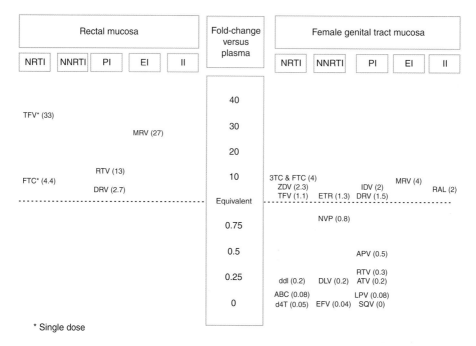

Figure 1. Penetration of oral antiretrovirals into mucosal surfaces. Blood plasma AUC ratios of tissue or mucosal secretions under steady-state conditions relative to blood plasma are reported unless otherwise marked. The dotted line represents drug concentrations in tissue or secretions equivalent to blood plasma. Drugs falling above the line concentrate at mucosal surfaces. Drugs falling below the line achieve concentrations at mucosal surfaces lower than blood plasma. The number in parentheses represents the ratio of mucosal surface AUC to blood plasma AUC. NRTI, nucleoside-analog reverse-transcriptase inhibitors; NNRTI, nonnucleoside-analog reverse-transcriptase inhibitors; PI, protease inhibitors; EI, entry inhibitors; II, integrase inhibitors TFV (tenofovir); FTC (emtricitabine); 3TC (lamivudine); ZDV (zidovudine); ddI (didanosine); ABC (abacavir); d4T (stavudine); ETR (etravirine); NVP (nevirapine); DLV (delavirdine); EFV (efavirenz); IDV (indinavir); RTV (ritonavir); DRV (darunavir); APV (amprenavir); LPV (lopinavir); ATV (atazanavir); SQV (saquinavir); MRV (maraviroc); RAL (raltegravir).

The nucleoside/tide analogs require cellular uptake and phosphorylation to be active against the RT enzyme. Intracellular concentrations of tenofovir diphosphate and emtricitabine triphosphate have been recently evaluated in cervical, vaginal, and rectal tissues after a single dose (Patterson et al. 2010). The accumulation of tenofovir diphosphate and emtricitabine triphosphate was wide-ranging depending on tissue type. In rectal tissue, the exposure of tenofovir diphosphate was 100-fold higher than in vaginal or cervical tissue. Yet in vaginal and cervical tissue, the exposure of FTC was 10-fold higher than in rectal tissue. In all tissues, emtricitabine triphosphate was not detectable beyond 2 d postdose. These results provide one plausible pharmacologic explanation for the recent disappointing results from the FEM-PrEP trial with Truvada (Matassa 2011) (see below). This study was stopped prematurely for futility. FEM PrEP used the same daily dosing of Truvada, which showed 44% efficacy in preventing HIV transmission in men who have sex with men, but enrolled only high-risk women. These results show the need for a better understanding between ARV exposure in mucosal tissues and protection from HIV infection.

The Right Site

Over the past few years the HIV transmission event has become increasingly well understood (Berger et al. 1999; Keele et al. 2008; Keele and

Derdeyn 2009). This information is critical for designing effective strategies for HIV prevention. The most current evidence suggests the initial target of the transmitted/founder virus at mucosal sites is a $CD4^+$ T cell expressing high levels of CCR5 and $\alpha_4\beta_7$ receptors (Chun et al. 1998). A strategy targeting these cells within vulnerable mucosal tissues (vaginal, cervical, and colorectal) may assist in selecting certain drugs for HIV prevention strategies. For example, orally administered drugs that reach these tissues in high concentrations for an extended period of time and target these specific cells would be preferentially selected for further development.

The Right Concentration

The target drug exposure required for preventing HIV infection at human mucosal surfaces is unknown. Therefore, the working assumption is that higher drug exposures are better at conferring protection. Although human trials will ultimately determine whether tissue concentrations achieved with oral drug dosing can prevent HIV transmission, preclinical research (animal studies and human tissue culture experiments) can provide valuable information and can guide the next steps in the research and practice of PrEP. These models can evaluate potential cell populations/subpopulations that are not adequately protected by current PrEP regimens, can help to define better drug regimens that include other drug classes such as entry inhibitors, and can evaluate the effect of drug-resistant viruses on protective efficacy (Veazey et al. 2005).

Performing early dose-ranging tissue concentration studies in phase I pharmacokinetic investigations, coupled with a preclinical understanding of concentration targets required to prevent HIV infection, allows for an optimal selection of drugs, doses, and dosing frequency to be implemented in later stage clinical trials. These studies should evaluate not only drug penetration in rectal and vaginal tissues, but also the degree of drug exposure in cells that are primary targets during early mucosal infection such as activated/resting T cells, dendritic

cells, or Langerhans cells. As an example, the threshold for concentrations of TFV-DP and FTC-TP that results in tissue protection has not been absolutely defined in macaques or tissue explants. These data, coupled with the knowledge of how drugs behave in tissues with certain dosing strategies, can be used to determine the minimal dose and dose frequency required for protection, which can then be used to optimize phases II and III clinical trial design.

One such dose-ranging tissue concentration study currently ongoing is HPTN 066 within the HIV Prevention Trials Network: This multi-site phase I study is scheduled to be completed in 2011. It involves four different treatment regimens in HIV-negative men and women (arm 1: 300 mg TDF/200 mg FTC weekly; arm 2: 300 mg TDF/200 mg FTC twice weekly; arm 3: 600 mg TDF/400 mg FTC twice weekly; and arm 4: 300 mg TDF/200 mg FTC daily). Sampling of rectal, seminal, and vaginal fluids, plasma, cells, and tissues will be performed to assess the dose proportionality of intracellular phosphorylated metabolites of TFV and FTC and to quantify their intraindividual variability. The information from this study will fill a large gap in the knowledge of intracellular kinetics of these medications at multiple mucosal surfaces and will help identify the time periods during which specific dosing strategies confer protection, once target concentrations are identified (using in vitro, animal, or clinical study data). Quantification of the intra-individual variability will enable future studies to use drug concentrations as a determinant of adherence to medication regimens—a critical component to interpret the findings of HIV prevention studies, which remains suboptimal.

The Right Amount of Time

Infection of HIV in mucosal tissues occurs quickly. These data are reviewed in detail in Shaw and Hunter (2011). Knowing that an initial round of HIV replication can occur in mucosal tissue within 24 h of exposure suggests that antiretroviral drugs need to be at the site of infection before, or very shortly after, HIV exposure. Additionally, data demonstrating recoverable infectious virus up to 8 d after inoculation of cervical tissue cultures suggest that significant drug exposure may need to be available for days after exposure (Collins et al. 2000). Particular emphasis is being placed on longer-acting drug formulations for next-generation PrEP agents (vaginal rings, intramuscular depot injections), which will require less frequent dosing and could be taken independently of virus exposure.

PRECLINICAL RESEARCH IN ANIMAL MODELS

The potential use of antiretroviral drugs for HIV prophylaxis has been studied extensively in nonhuman primate models of mucosal and parenteral SIV or SHIV (SIV/HIV chimera) transmission and, more recently, in humanized mouse models (Table 1). Early work with subcutaneous TFV in macaques showed the first proof-of-concept data on the efficacy of ARV prophylaxis against intravenous virus inoculation (Tsai et al. 1995). Subsequent work showed that postexposure prophylaxis with TFV can protect against intravenous SIV inoculation and helped define the optimal timing for initiating ARV therapy and the need for a 4 wk treatment to achieve protection (Tsai et al. 1998). Indications that ARV drugs administered before exposure could also prevent oral SIV infection came from studies that used different doses of TFV (Van Rompay et al. 1998, 2006). More recently, repeat low-dose macaque models of mucosal transmission have been developed and used to assess PrEP efficacy of different ARV regimens and modalities (Otten et al. 2005; Subbarao et al. 2006). These models closely mimic human transmission of HIV in many aspects, including the use of a lower and more physiologic virus inoculum than that used in conventional single high-dose challenge models. In addition, the SHIV challenge contains an R5-tropic HIV-1_{SF162} envelope similar to naturally transmitted viruses. Virus exposures are repeated to mimic high-risk human exposures, thereby providing the opportunity to measure protection against multiple transmission

events in each animal (Subbarao et al. 2006; Garcia-Lerma et al. 2008; Keele and Derdeyn 2009). Using such a model of rectal infection to assess the efficacy of TDF, FTC, or TDF/FTC combination at human equivalent dosing, it was found that daily TDF provided little protection, whereas FTC reduced risk by 3.8-fold (Subbarao et al. 2006, Garcia-Lerma et al. 2008, 2010). In contrast, TDF/FTC combination was more protective and provided a nearly eightfold lower risk of infection; a higher FTC/TFV dose afforded full protection (Garcia-Lerma et al. 2008). These experiments showed a dose-response relationship and suggested that TDF/FTC may be more effective than either TDF or FTC alone against rectal infection. Data on PrEP efficacy against vaginal challenges in macaques are not available. These studies are important because similar to what is observed in humans, oral Truvada in macaques achieves different drug exposures in vaginal tissues than in rectal tissues. However, data from a humanized mouse model showed that a high dose of TDF/FTC combination administered intraperitoneally protected mice against a vaginal HIV infection (Denton et al. 2008). Recent findings also showed that oral PrEP with either raltegravir or maraviroc protected humanized mice from vaginal HIV infection (Neff et al. 2011), although drug concentrations were not measured.

Several important observations of potential relevance to humans have been made from the analysis of PrEP breakthrough infections in macaques. First, drug resistance can emerge if ARV therapy continues after PrEP fails. In one macaque study, two of six animals infected during daily PrEP with FTC or Truvada showed selection for drug-resistant viruses (Garcia-Lerma et al. 2008). In both macaques, the M184V mutation associated with FTC resistance was selected, thus reiterating the importance of closely monitoring PrEP failures to minimize drug-resistance emergence. Second, PrEP breakthroughs during FTC and Truvada treatment had lower acute viremias than control animals. A reduction in viremia during PrEP might conceivably contribute to a decrease in HIV-1 transmissibility at the population level

Table 1. Efficacy of preexposure prophylaxis modalities in animal models of mucosal and intravenous infection

Reference	Animal	Drugs and dose	Route of drug administration	Virus exposure and dose	Interventions	Main findings
Tsai et al. 1995	Long-tailed macaques	TFV, 20 mg/kg	Subcutaneous	Single intravenous exposure to SIV_{mne} (10^3 $TCID_{50}$)	TFV initiated 48 h before exposure and continued for 4 wk	Full protection
Van Rompay et al. 1998	Rhesus macaques	TFV, 30 mg/kg	Subcutaneous	Oral, SIVmac251 (10^5 $TCID_{50}$)	Two doses given 4 h before and 24 h after	Full protection
Van Rompay et al. 2006	Rhesus macaques	TDF, 0.01–0.02 mg/kg	Oral	Multiple oral exposures to SIVmac251 (10^4 $TCID_{50}$)	TDF initiated 1 d before exposure and maintained during continuous virus inoculations; one additional dose after the last virus exposure	No protection; low doses of TDF
Van Rompay et al. 2006	Rhesus macaques	TDF, 10 mg/kg	Oral	Multiple oral exposures to SIVmac251 (10^5 $TCID_{50}$)	Repeated cycles of daily TDF initiated 1–2 d before exposure	Partial prophylactic efficacy; infection associated with low systemic drug exposures
Subbarao et al. 2006	Rhesus macaques	TDF, 22 mg/kg	Oral	Repeated low-dose atraumatic rectal exposures to $SHIV_{162p3}$ (10 $TCID_{50}$)	Daily or weekly TDF	All controls animals infected after 1.5 exposures; 3 of 4 TDF-treated animals infected after 6–7 exposures
García-Lerma et al. 2008	Rhesus macaques	FTC, 20 mg/kg; TDF, 22 mg/kg; TFV, 20 mg/kg	Oral (Truvada) or subcutaneous (FTC, TFV)	Repeated low-dose atraumatic rectal exposures to $SHIV_{162p3}$ (10 $TCID_{50}$)	Daily FTC (subcutaneous); daily oral Truvada; daily or intermittent FTC/TFV (subcutaneous)	Risk of infection reduced with human equivalent doses of FTC (fourfold) and Truvada (eightfold); no infection with FTC and high TFV doses (daily or intermittent)

Continued

Table 1. *Continued*

Reference	Animal	Drugs and dose	Route of drug administration	Virus exposure and dose	Interventions	Main findings
Denton et al. 2008	Humanized mice	FTC, 3.5 mg; TDF, 5.2 mg	Intraperitoneal	Single atraumatic intravaginal exposure to HIV-1$_{JR-CSF}$ (10^5 TCID$_{50}$)	Daily FTC/TDF initiated 48 h before exposure and continued for 7 d	Full protection
García-Lerma et al. 2010	Rhesus macaques	FTC, 20 mg/kg; TDF, 20 mg/kg	Oral	Repeated low-dose atraumatic rectal exposures to SHIV$_{162p3}$ (10 TCID$_{50}$)	Event-driven and exposure-independent intermittent Truvada dosing	High to moderate efficacy with 2 weekly human equivalent doses of Truvada given at different intervals
Neff et al. 2010	Humanized mice	Raltegravair (164 mg/kg) or Maraviroc (62 mg/kg)	Oral	Single high-dose HIV-1 (3000 TCID)	Daily given 4 d before and 3 d after challenge	Full protection, human equivalent dosing

TCID, tissue culture infectious dose.

Cite this article as *Cold Spring Harb Perspect Med* doi: 10.1101/cshperspect.a007419

and could add to the overall effectiveness of PrEP. Attenuated acute viremia might also reduce early CD4[+] T cell depletion, help to preserve immune function, and attenuate the course of HIV infection (Mehandru et al. 2004).

Animal models have also been used to explore the efficacy of intermittent drug dosing with TFV or Truvada. Intermittent PrEP can reduce the risks of drug toxicities, increase adherence, minimize drug-resistance emergence, and be more cost effective. Both FTC-TP and TFV-DP have long (40 to >100 h) intracellular half-lives in humans and can potentially achieve extended prophylactic activity when administered intermittently (Wang et al. 2004; Hawkins et al. 2005; Pruvost et al. 2005). Intermittent PrEP regimens of TDF or Truvada can be designed to be exposure driven or to follow a fixed schedule. Studies in macaques showing protection from oral or rectal SIV/SHIV exposures by a two-dose subcutaneous regimen containing TFV or TFV/FTC have provided the first proof-of-concept evidence for intermittent PrEP (Van Rompay et al. 1998; Garcia-Lerma et al. 2008). However, the high drug doses and subcutaneous drug delivery might have overestimated efficacy in both studies. More recent work using human equivalent doses of Truvada showed that macaques can be protected from rectal SHIV infection by several PrEP modalities, including a single oral dose given 1–7 d before exposure, followed by a second dose 2 h after exposure (Garcia-Lerma et al. 2010). Exposure-driven prophylactic modalities initiated around the time of exposure also maintained protection. These studies showed that intermittent PrEP, particularly with long-acting ARV drugs, can be highly effective and have a wide window of protection. They strengthen the possibility of developing feasible, cost-effective strategies to prevent HIV transmission in humans.

HUMAN CLINICAL TRIALS

A number of challenges currently exist for human HIV prevention trials. Unlike antiretroviral treatment efficacy (with HIV RNA concentrations and CD4[+] T cell counts), there is currently no surrogate marker to use in place of new HIV infections. Therefore, prevention trials must enroll thousands of participants, followed for several years, at significant expense, in order to document efficacy. Additionally, the regulatory environment for demonstrating safe and effective marketable prevention strategies is not well defined.

However, a number of clinical studies have been initiated in multiple at-risk populations with standard dosing of TDF with or without FTC. As of February 2011, nine oral PrEP trials enrolling >22,000 participants are at varying stages of completion (Table 2). The at-risk populations in these studies are heterosexual men and women, men who have sex with men (MSM), and intravenous drug users (IDUs). All trials are using TDF-containing products, either alone or in combination with FTC, and all are evaluating daily use with the exception of one pilot study of intermittent PrEP.

The first phase II safety study of daily TDF among 936 high-risk women in Ghana, Nigeria, and Cameroon for up to 12 mo saw no differences in adverse events or grade-3 or -4 laboratory abnormalities between placebo and TDF users (Peterson al. 2007).

The second clinical study to be published evaluated daily use of the fixed-dose combination of TDF + emtricitabine (Truvada) in men who have sex with men. A total of 2499 gay and bisexual men, other MSM, and transgender women at high risk of HIV infection participated in the six-country, four-continent preexposure prophylaxis initiative (iPrEx study). In this study, daily use of Truvada reduced HIV acquisition by 44% (Grant et al. 2010). Based on drug concentrations in plasma and cryopreserved peripheral blood mononuclear cell cultures (PBMCs), a substantial number of subjects only appeared to be taking their drug sporadically. In those subjects who were ≥90% adherent by pill counts, drug exposure, and self-report, Truvada conferred 68% protection against HIV acquisition (Celum 2011). Truvada was safe and generally well tolerated, with higher rates of nausea and weight loss during the first 4 wk of treatment compared with placebo. The overall rate of side effects was

Table 2. Summary of clinical trials investigating oral antiretrovirals for HIV prevention and expected dates of results publication

2010	2011	2012 +
iPrEx	**FEM-PrEP**	**Partners PrEP**
Phase III trial of once-daily oral TDF/FTC (Brazil, Ecuador, Peru, South Africa, Thailand, United States) *Showed that once-daily TDF/FTC reduced risk of HIV infection in gay men, transgender women, and other men who have sex with men an average of 43.8%.*	Phase III trial of a once-daily dose of TDF/FTC (Kenya, South Africa, Tanzania) *Study's data review committee determined that the trial would not be able to answer the question of whether the study drug decreased risk of HIV infection among HIV-negative women at risk via sexual transmission. The study will be discontinued.*	Phase III trial of once-daily oral TDF and once-daily oral TDF/FTC (Kenya, Uganda)
CDC 4323	**CDC 4940 (TDF2)**	**VOICE (MTN-003)**
Phase II trial of once-daily oral TDF (United States) *The trial reported no serious adverse events and preliminary data show PrEP use did not have a significant effect on HIV risk behavior. Additional data expected in 2011.*	Phase II trial of once-daily TDF/FTC (Botswana)	Phase IIb trial of once-daily oral TDF, once-daily oral TDF/FTC, and 1% tenofovir gel (South Africa, Uganda, Zimbabwe). This study has fully enrolled.
	IAVI E001 and E002 in Kenya and Uganda	**CDC 4370**
	This study is evaluating the safety and acceptability of intermittent and daily PrEP regimens using TDF/FTC. This study has completed.	Phase II/III trial of once-daily oral TDF (Thailand)
		ATN 082
		An exploratory mixed-methods study comparing behavioral interventions alone and combined with daily TDF/FTC

very similar in both the FTC-TDF and placebo groups, and severe side effects were rare.

Most recently, however, a trial in women using daily dosing of Truvada (FEM-PrEP) was halted for futility: It was determined that substantial HIV prevention (i.e., >30%) could not be attained with daily use of Truvada (Matassa 2011; Roehr 2011). As of May 2011, the study is undergoing orderly closure, and sample and data analysis will determine whether this result was because of a biological process (e.g., increased genital tract inflammation in this population), adherence, or differential drug exposure in genital tract tissues (as discussed above).

In contrast to FEM-PrEP, interim results from two studies, the Partners PrEP trial and TDF2 (CDC 4940), have demonstrated clear efficacy in reducing HIV acquisition. In the Partners PrEP, both daily TDF and FTC/TDF reduced HIV risk by 62% and 73%, respectively, in both men and women, and the effects in women and men were statistically similar (Baeten 2011). The study is continuing with two active arms only to gather additional comparative information. Likewise, results of the TDF2 trial among heterosexual participants in Botswana showed that daily FTC/TDF provided an overall protective efficacy of 62.6% (Thigpen 2011). Limiting analysis to participants on study medication when infected, the protective efficacy was 77.9%.

Looking beyond tenofovir and emtricitabine, next-generation PrEP drugs are being evaluated, including maraviroc (a CCR5 receptor antagonist) and TMC278 (a long-acting NNRTI).

Finally, intermittent PrEP is being considered as an attractive strategy. Daily dosing may not be practical for individuals who are only occasionally exposed to high-risk encounters, which can subsequently result in decreased rates of adherence to medication regimens (prophylaxis fatigue) and unnecessary systemic toxicity. A preliminary clinical study in Kenya and Uganda is currently under way among 150 serodiscordant couples to evaluate the safety and acceptability of iPrEP (http://www.iavi.org). In this investigation, subjects will take standard doses of oral TDF/FTC once daily or intermittently (defined as twice weekly plus coitus-related dosing). Samples for blood plasma and intracellular concentrations will be obtained.

FUTURE DIRECTIONS AND CHALLENGES

Based on the results of iPrEX, the Centers for Disease Control and Prevention has issued interim guidance for PrEP use among MSMs and has begun with other U.S. Public Health Service (PHS) agencies to develop PHS guidelines on the use of PrEP for MSM at high risk for HIV infection (Centers for Disease Control 2011). However, until the safety and efficacy of PrEP is determined in trials now under way with populations at high risk for HIV acquisition by other routes of transmission, PrEP should be considered only for MSM. The iPrEX trial results that showed a substantially higher efficacy (~68%) among adherent participants with detectable ARV drugs provide strong evidence that support for adherence to the ARV regimen must be a routine component of any PrEP program. To minimize the risk for drug resistance, PrEP should not be started in persons with signs or symptoms of acute viral infection unless

HIV-uninfected status is confirmed. Despite the indications of biologic effectiveness, the implementation of PrEP will need to overcome many challenges to provide a meaningful benefit at the population level. Mathematical models have suggested that the effectiveness of PrEP may be offset by low uptake, suboptimal adherence, and risk compensation, which refers to increases in HIV risk behavior among PrEP users on the assumption that they are protected against HIV infection (Abbas et al. 2007; Desai et al. 2008; Paltiel et al. 2009). As was the case during the implementation of ARV for treating HIV infections, plans for PrEP implementation will likely include multiple components on optimal drug delivery, safety screening, behavioral intervention, integration of PrEP as part of comprehensive care, and monitoring the impact of PrEP at the population level (Underhill et al. 2011). The promise of PrEP also raises important research questions. These range from the development and evaluation of next-generation PrEP agents and modalities including episodic dosing, new drug classes and combinations, and long-acting formulations, to the assessment of risks of PrEP-induced drug resistance (Supervie et al. 2010).

REFERENCES

*Reference is also in this collection.

Abbas UL, Anderson RM, Mellors JW. 2007. Potential impact of antiretroviral chemoprophylaxis on HIV-1 transmission in resource-limited settings. *PLoS One* **2:** e875. doi: 10.1371/journal.pone.0000875.

Baeten J, Colem C. 2011. Antiretroviral pre-exposure prophylaxis for HIV-1 prevention among heterosexual African men and women: The partners PrEP study. In *The 6th International AIDS Society Conference on HIV Pathogenesis, Treatment and Prevention.* Rome.

Berger EA, Murphy PM, Farber JM. 1999. Chemokine receptors as HIV-1 coreceptors: Roles in viral entry, tropism, and disease. *Annu Rev Immunol* **17:** 657–700.

Brown KC, Patterson KB, Malone SA, Shaheen NJ, Prince HM, Dumond JB, Spacek MB, Heidt PE, Cohen MS, Kashuba AD. 2011. Single and multiple dose pharmacokinetics of maraviroc in saliva, semen, and rectal tissue of healthy HIV-negative men. *J Infect Dis* **203:** 1484–1490.

Celum C. 2011. Drugs for prevention-topical and systemic PrEP. In *18th Conference on Retroviruses and Opportunistic Infections,* Boston.

Centers for Disease Control 2011. Interim guidance: Preexposure prophylaxis for the prevention of HIV infection in men who have sex with men. *MMWR Morb Mortal Wkly Rep* **60**: 65–68.

Chun TW, Engel D, Berrey MM, Shea T, Corey L, Fauci AS. 1998. Early establishment of a pool of latently infected, resting CD4$^+$ T cells during primary HIV-1 infection. *Proc Natl Acad Sci* **95**: 8869–8873.

Cohen MS, Gay C, Kashuba AD, Blower S, Paxton L. 2007. Narrative review: Antiretroviral therapy to prevent the sexual transmission of HIV-1. *Ann Intern Med* **146**: 591–601.

Collins KB, Patterson BK, Naus GJ, Landers DV, Gupta P. 2000. Development of an in vitro organ culture model to study transmission of HIV-1 in the female genital tract. *Nat Med* **6**: 475–479.

Denton PW, Estes JD, Sun Z, Othieno FA, Wei BL, Wege AK, Powell DA, Payne D, Haase AT, Garcia JV. 2008. Antiretroviral pre-exposure prophylaxis prevents vaginal transmission of HIV-1 in humanized BLT mice. *PLoS Med* **5**: e16. doi: 10.1371/journal.pmed.0050016.

Department of Health and Human Services Panel on Antiretroviral Guidelines for Adults and Adolescents (January 10, 2011). Guidelines for the use of antiretroviral agents in HIV-1-infected adults and adolescents, pp. 1–166.

Desai K, Sansom SL, Ackers ML, Stewart SR, Hall HI, Hu DJ, Sanders R, Scotton CR, Soorapanth S, Boily MC, et al. 2008. Modeling the impact of HIV chemoprophylaxis strategies among men who have sex with men in the United States: HIV infections prevented and cost-effectiveness. *AIDS* **22**: 1829–1839.

Dumond JB, Yeh RF, Patterson KB, Corbett AH, Jung BH, Rezk NL, Bridges AS, Stewart PW, Cohen MS, Kashuba AD. 2007. Antiretroviral drug exposure in the female genital tract: Implications for oral pre- and post-exposure prophylaxis. *AIDS* **21**: 1899–1907.

Dumond JB, Patterson KB, Pecha AL, Werner RE, Andrews E, Damle B, Tressler R, Worsley J, Kashuba AD. 2009. Maraviroc concentrates in the cervicovaginal fluid and vaginal tissue of HIV-negative women. *J Acquir Immune Defic Syndr* **51**: 546–553.

Garcia-Lerma JG, Otten RA, Qari SH, Jackson E, Cong ME, Masciotra S, Luo W, Kim C, Adams DR, Monsour M, et al. 2008. Prevention of rectal SHIV transmission in macaques by daily or intermittent prophylaxis with emtricitabine and tenofovir. *PLoS Med* **5**: e28. doi: 10.1371/journal.pmed.0050028.

Garcia-Lerma JG, Cong ME, Mitchell J, Youngpairoj AS, Zheng Q, Masciotra S, Martin A, Kuklenyik Z, Holder A, Lipscomb J, et al. 2010. Intermittent prophylaxis with oral truvada protects macaques from rectal SHIV infection. *Sci Transl Med* **2**: 14ra4. doi: 10.1126/scitranslmed.3000391.

Grant RM, Lama JR, Anderson PL, McMahan V, Liu AY, Vargas L, Goicochea P, Casapia M, Guanira-Carranza JV, Ramirez-Cardich ME, et al. 2010. Preexposure chemoprophylaxis for HIV prevention in men who have sex with men. *N Engl J Med* **363**: 2587–2599.

Haase A. 2005. Perils at the mucosal front lines for HIV and SIV and their hosts. *Nature Rev Immunol* **5**: 783–792.

Hawkins T, Veikley W, St Claire RL III, Guyer B, Clark N, Kearney BP. 2005. Intracellular pharmacokinetics of tenofovir diphosphate, carbovir triphosphate, and

lamivudine triphosphate in patients receiving triple-nucleoside regimens. *J Acquir Immune Defic Syndr* **39**: 406–411.

Hu J, Gardner MB, Miller CJ. 2000. Simian immunodeficiency virus rapidly penetrates the cervicovaginal mucosa after intravaginal inoculation and infects intraepithelial dendritic cells. *J Virol* **74**: 6087–6095.

Jones AE, Talameh JA, Patterson KB, Rezk N, Prince H, Kashuba ADM. 2009. First-dose and steady-state pharmacokinetics of raltegravir in the genital tract of HIV negative women. *In 10th International Workshop on Clinical Pharmacology of HIV Therapy.* Amsterdam.

Keele BF, Derdeyn CA. 2009. Genetic and antigenic features of the transmitted virus. *Curr Opin HIV AIDS* **4**: 352–357.

Keele BF, Giorgi EE, Salazar-Gonzalez JF, Decker JM, Pham KT, Salazar MG, Sun C, Grayson T, Wang S, Li H, et al. 2008. Identification and characterization of transmitted and early founder virus envelopes in primary HIV-1 infection. *Proc Natl Acad Sci* **105**: 7552–7557.

Li Q. 2009. Glycerol monolaurate prevents mucosal SIV transmission. *Nature Rev Immunol* **458**: 1034–1038.

Li Q, Estes JD, Schlievert PM, Duan L, Brosnahan AJ, Southern PJ, Reilly CS, Peterson ML, Schultz-Darken N, Brunner KG, et al. 2009. Glycerol monolaurate prevents mucosal SIV transmission. *Nature* **458**: 1034–1038.

Matassa M. 2011. FHI statement on the FEM-PrEP HIV prevention study: FHI to initiate orderly closure of FEM-PrEP. http://www.fhi.org/en/AboutFHI/Media/Releases/FEM-PrEP_statement041811.htm.

Mehandru S, Poles MA, Tenner-Racz K, Horowitz A, Hurley A, Hogan C, Boden D, Racz P, Markowitz M. 2004. Primary HIV-1 infection is associated with preferential depletion of CD4$^+$ T lymphocytes from effector sites in the gastrointestinal tract. *J Exp Med* **200**: 761–770.

Miller CJ, Li Q, Abel K, Kim EY, Ma ZM. 2005. Propagation and dissemination of infection after vaginal transmission of simian immunodeficiency virus. *J Virol* **79**: 9217–9227.

Neff CP, Ndolo T, Tandon A, Habu Y, Akkina R. 2010. Oral pre-exposure prophylaxis by anti-retrovirals raltegravir and maraviroc protects against HIV-1 vaginal transmission in a humanized mouse model. *PLoS One* **5**: e15257. doi: 10.1371/journal.pone.0015257.

Nicol MR, Kashuba AD. 2010. Pharmacologic opportunities for HIV prevention. *Clin Pharmacol Ther* **88**: 598–609.

Otten RA, Adams DR, Kim CN, Jackson E, Pullium JK, Lee K, Grohskopf LA, Monsour M, Butera S, Folks TM. 2005. Multiple vaginal exposures to low doses of R5 simian-human immunodeficiency virus: Strategy to study HIV preclinical interventions in nonhuman primates. *J Infect Dis* **191**: 164–173.

Padian NS, Buve A, Balkus J, Serwadda D, Cates W Jr. 2008. Biomedical interventions to prevent HIV infection: Evidence, challenges, and way forward. *Lancet* **372**: 585–599.

Paltiel AD, Freedberg KA, Scott CA, Schackman BR, Losina E, Wang B, Seage GR III, Sloan CE, Sax PE, Walensky RP. 2009. HIV preexposure prophylaxis in the United States: Impact on lifetime infection risk, clinical outcomes, and cost-effectiveness. *Clin Infect Dis* **48**: 806–815.

Patterson K, Prince H, Kraft E, Jones A, Paul S, Shaheen N, Spacek M, Heidt P, Reddy S, Rooney J, et al. 2010. Exposure of extracellular and intracellular tenofovir and emtricitabine in mucosal tissues after a single fixed dose of TDF/FTC: Implications for pre-exposure HIV prophylaxis (PrEP). XVIII International AIDS Conference, Vienna, Austria.

Peterson L, Taylor D, Roddy R, Belai G, Phillips P, Nanda K, Grant R, Clarke EE, Doh AS, Ridzon R, et al. 2007. Tenofovir disoproxil fumarate for prevention of HIV infection in women: A phase 2, double-blind, randomized, placebo-controlled trial. *PLoS Clin Trials* **2:** e27. doi: 10.1371/journal.pctr.0020027.

Pruvost A, Negredo E, Benech H, Theodoro F, Puig J, Grau E, Garcia E, Molto J, Grassi J, Clotet B. 2005. Measurement of intracellular didanosine and tenofovir phosphorylated metabolites and possible interaction of the two drugs in human immunodeficiency virus-infected patients. *Antimicrob Agents Chemother* **49:** 1907–1914.

Reynolds MR, Rakasz E, Skinner PJ, White C, Abel K, Ma Z-M, Compton L, Napoé G, Wilson N, et al. 2005. $CD8^+$ T-lymphocyte response to major immunodominant epitopes after vaginal exposure to simian immunodeficiency virus: Too late and too little. *J Virol* **79:** 9228–9235.

Roehr B. 2011. HIV prevention trial in women is abandoned after drugs show no impact on infection rates. *BMJ* **342:** d2613. doi: 10.1136/bmj.d2613.

Salas Herrera IG, Pearson RM, Turner P 1991. Quantitation of albumin and α-1-acid glycoprotein in human cervical mucus. *Hum Exp Toxicol* **10:** 137–139.

* Shaw GM, Hunter E. 2011. HIV transmission. *Cold Spring Harb Perspect Biol* doi: 10.1101/cshperspect.a006965.

Subbarao S, Otten RA, Ramos A, Kim C, Jackson E, Monsour M, Adams DR, Bashirian S, Johnson J, Soriano V, et al. 2006. Chemoprophylaxis with tenofovir disoproxil fumarate provided partial protection against infection with simian human immunodeficiency virus in macaques given multiple virus challenges. *J Infect Dis* **194:** 904–911.

Supervie V, García-Lerma JG, Heneine W, Blower S. 2010. HIV, transmitted drug resistance, and the paradox of pre-exposure prophylaxis. *Proc Natl Acad Sci* **107:** 12381–12386.

Thigpen MC, Kebaabetswe PM, Smith DK, Segolodi TM, Soud FA, Chillag K, Chirwa LA, Kasonde M, Mutanhaurwa R, Henderson FL. 2011. Daily oral antiretroviral use for the prevention of HIV infection in heterosexually active young adults in Botswana: Results from the TDF2 study. In *The 6th Annual International AIDS Society Conference on HIV Pathogenesis, Treatment and Prevention*. Rome.

Tsai CC, Follis KE, Sabo A, Beck TW, Grant RF, Bischofberger N, Benveniste RE, Black R. 1995. Prevention of SIV infection in macaques by (*R*)-9-(2-phosphonylmethoxypropyl)adenine. *Science* **270:** 1197–1199.

Tsai CC, Emau P, Follis KE, Beck TW, Benveniste RE, Bischofberger N, Lifson JD, Morton WR. 1998. Effectiveness of postinoculation (*R*)-9-(2-phosphonylmethoxypropyl) adenine treatment for prevention of persistent simian immunodeficiency virus SIV_{mne} infection depends critically on timing of initiation and duration of treatment. *J Virol* **72:** 4265–4273.

UNAIDS. 2009. Global Facts & Figures. Joint United Nations Programme on HIV/AIDS.

Underhill K, Operario D, Skeer M, Mimiaga M, Mayer K. 2011. Packaging PrEP to prevent HIV: An integrated framework to plan for pre-exposure prophylaxis implementation in clinical practice. *J Acquir Immune Defic Syndr* **55:** 8–13.

Van Rompay KK, Berardi CJ, Aguirre NL, Bischofberger N, Lietman PS, Pedersen NC, Marthas ML. 1998. Two doses of PMPA protect newborn macaques against oral simian immunodeficiency virus infection. *Aids* **12:** F79–F83.

Van Rompay KK, Kearney BP, Sexton JJ, Colon R, Lawson JR, Blackwood EJ, Lee WA, Bischofberger N, Marthas ML. 2006. Evaluation of oral tenofovir disoproxil fumarate and topical tenofovir GS-7340 to protect infant macaques against repeated oral challenges with virulent simian immunodeficiency virus. *J Acquir Immune Defic Syndr* **43:** 6–14.

Veazey RS, Springer MS, Marx PA, Dufour J, Klasse PJ, Moore JP. 2005. Protection of macaques from vaginal SHIV challenge by an orally delivered CCR5 inhibitor. *Nat Med* **11:** 1293–1294.

Wang LH, Begley J, St Claire RL III, Harris J, Wakeford C, Rousseau FS. 2004. Pharmacokinetic and pharmacodynamic characteristics of emtricitabine support its once daily dosing for the treatment of HIV infection. *AIDS Res Hum Retroviruses* **20:** 1173–1182.

Behavioral and Biomedical Combination Strategies for HIV Prevention

Linda-Gail Bekker[1,2], Chris Beyrer[3], and Thomas C. Quinn[4]

[1]The Desmond Tutu HIV Centre, Institute of Infectious Disease and Molecular Medicine, University of Cape Town, Cape Town 7925, South Africa

[2]Department of Medicine, University of Cape Town, Cape Town 7925, South Africa

[3]Department of Epidemiology, Johns Hopkins Bloomberg School of Public Health, Baltimore, Maryland 21205

[4]Section on International HIV/STD Research, National Institute of Allergy and Infectious Diseases, National Institutes of Health, Bethesda, Maryland 20892

Correspondence: Linda-gail.bekker@hiv-research.org.za

Around 2.5 million people become infected with HIV each year. This extraordinary toll on human life and public health worldwide will only be reversed with effective prevention. What's more, in the next few years, it is likely at least, that no single prevention strategy will be sufficient to contain the spread of the disease. There is a need for combination prevention as there is for combination treatment, including biomedical, behavioral, and structural interventions. Expanded HIV prevention must be grounded in a systematic analysis of the epidemic's dynamics in local contexts. Although 85% of HIV is transmitted sexually, effective combinations of prevention have been shown for people who inject drugs. Combination prevention should be based on scientifically derived evidence, with input and engagement from local communities that fosters the successful integration of care and treatment.

History has taught us that the way to eradicate a global viral epidemic is to design, mass produce, and then systematically vaccinate the population at risk with an effective prophylactic vaccine (Fifty-Fifth World Assembly 2008; http://www.who.int/mediacentre/news/releases/releasewha01/en/index.html). Past experience has also shown that the path to an effective AIDS vaccine may be long and complicated (Nabel 2001; Barouche 2008). Although the modest RV144 or "Thai" vaccine trial efficacy results in 2009 provided the first hope that a prophylactic HIV vaccine may be possible (Rerks-Ngarm et al. 2009), with almost 60 million men, women, and children having been infected and more than 25 million attributable deaths, 30 years of this epidemic has taken a monstrous toll. More worrying still, modeling exercises have indicated that staggering numbers of new infections may occur given current infection rates. The world may be facing 20–60 million new HIV infections in the 15–20 years it may take to develop and evaluate a highly efficacious prophylactic vaccine. Such alarming projections emphasize the urgency of finding effective alternative approaches to prevention more immediately (Legakos and Gable 2008).

King Holmes has proposed a synergistic combination of sociobehavioral and medical interventions and coined the phrase highly active retroviral prevention (HARP). He further stated "only fools today still advocate a single method of HIV prevention" (Vandenbruaene 2007). Coates and colleagues captured this concept previously (Fig. 1) (Coates et al. 2008).

The emphasis in prevention research in 2011 has therefore shifted to the design and evaluation of combination prevention packages. The justification for this is summarized in Table 1.

A challenge for researchers in the HIV prevention field is to design combination packages that are feasible, effective, affordable, community- and population-specific, and acceptable. This article examines the available biomedical and behavioral interventions and the evidence of their suitability for inclusion in combination prevention packages.

KNOW YOUR EPIDEMIC, KNOW YOUR RESPONSE

It is likely that prevention packages will not be a "one size fits all endeavor" and ideal menus will need to be tailored for specific behaviors, regions, and risk categories. This requires that physicians and public health workers know their local populations and the basis for transmission and thus tailor responses accordingly.

The rapidly changing face of the epidemic calls not only for increased epidemiologic HIV transmission and behavioral surveillance but also for nuanced investigation that accounts for preference, social, cultural, and gender contexts. To this end, UNAIDS has launched a program entitled "know your epidemic, know your response" which encourages country-led investigation of the relevant drivers and risk behaviors (UNAIDS 2007; Wilson and Halperin 2008).

Which Level of Evidence?

Padian and colleagues (2010) have examined all randomized controlled trials with HIV transmission as an endpoint performed in the last 2 decades (summarized and updated in Table 2) and asked what types of evidence will be considered in the selection of interventions in prevention packages of the future. Public health agencies have long recognized the shortfalls posed by relying solely on randomized controlled trial (RCT) data and have consequently developed guidance that also considers a variety of other data sources (Padian et al. 2010). HIV prevention researchers may need to adopt similar strategies. Furthermore, a clear understanding of how partially efficacious interventions may be combined for synergistic or additive effect remains a challenge (Piot et al. 2008).

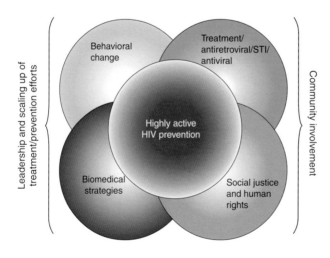

Figure 1. Highly active HIV prevention, a term coined by King Holmes, University of Washington School of Medicine, Seattle, WA. (STI) Sexually transmitted infections (Coates and Gable 2008).

Table 1. Reasons for a shift to prevention packages

Concept	Reason	Reference
1.	It is clear from results of recent trials that there will be no single "magic bullet" for HIV prevention in the near term.	Padian et al. 2010
2.	Interventions with modest levels of effect might lead to more substantial efficacy overall if combined.	Auerbach and Coates 2000
3.	Current biomedical interventions are affected by human behavioral factors (adherence, risk compensation) and will rely on sociobehavioral interventions to strengthen effectiveness.	Padian et al. 2008
4.	Biomedical interventions are being offered in addition to an already relatively robust prevention package, including regular HIV testing, risk reduction counseling, latex condoms, and postexposure chemoprophylaxis.	Grant et al. 2010; Karim et al. 2010
5.	In most of the recently reported randomized controlled HIV prevention trials, there is some reduction in reported sexual risk over time and a lower than expected HIV incidence in the trial population overall, even when the modality under investigation has not shown any positive effect.	Padian et al. 2010

BEHAVIORAL STRATEGIES

Efforts to modify sexual and drug-using human behaviors to reduce HIV risk have fallen short in the last 2 decades. Behavioral strategies have been defined as interventions to "motivate behavioral change in individuals and social units by use of a range of educational, motivational, peer-led, skill-building approaches as well as community normative approaches" (Coates and Gable 2008). A list of some available behavioral strategies is given in Table 3. The first successful examples of behavior change that led to decreased HIV transmission incidence were reported in men who have sex with men (MSM) (Winkelstein et al. 1988; Kippax and Race 2003). Subsequently, a number of countries have attributed decreases in HIV incidence to changes in sexual behavior, including Brazil, Cote d'Ivoire, Kenya, Uganda, Malawi, Tanzania, Zimbabwe, Burkino Faso, Namibia, and Swaziland (Stoneburner and Low-Beer 2004; Slutkin et al. 2006).

Behavioral Intervention Research

The HIV literature is full of behavioral interventional and observational studies in a variety of settings and target groups, most of which have not objectively altered HIV transmission or acquisition rates. Seven randomized trials of behavioral interventions, described by Padian and colleagues, summarized in Table 4, showed neither benefit nor harm (Padian et al. 2010). Project EXPLORE is the only interventional study for HIV behavior with an HIV infection endpoint. Using a counseling intervention to reduce HIV incidence, the average follow-up was 3.25 yr and HIV acquisition was reduced by 18.2% in the experimental arm, but this effect compared with controls was insignificant. On more careful examination, there were more dramatic effects on HIV incidence in the first year (39%), but this effect was lost over time. There is a recurring theme in the literature that behavioral change is hard to maintain (Koblin et al. 2004; Karim et al. 2010).

Table 2. Results of 40 randomized controlled trials reporting on 42 interventions to prevent sexual transmission of HIV

| Type of intervention | HIV prevention efficacy | | | |
	Positive effect (significantly reduced HIV incidence compared with control)	Adverse effect (significantly increased HIV incidence compared with control)	No effect (either way)	Total
Behavioral	–	–	7	7
Microfinance	–	–	1	1
Diaphragm	–	–	1	1
Vaginal microbicides	1	1	11	13
Preexposure prophylaxis	1	–	2	3
Male circumcision	3	–	1	4
Sexually transmitted infection (STI) treatment	1	–	8	9
Vaccine	1	–	3	4
Antiretroviral therapy (ART) in discordancy	1	–	–	1
Total	8	1	34	43

Data adapted from Padian 2010.

Behavioral Intervention Programming

Auerbach and Coates (2008) state "behavioral strategies are necessary but not sufficient to reduce HIV transmission, but are essential in a comprehensive HIV prevention strategy. Behavioral strategies themselves need to be combinations of approaches at multiple levels of influence." The list of possible behavioral interventions that have been identified over this time is long (Table 3), and inclusion of any component in a combination prevention package would depend on the target population and the risk activity.

The slogan, the "ABC of HIV prevention" was apparently first coined as part of a prevention campaign in Botswana in the late 1990s (Fig. 2). It was well known by then that individuals could reduce their risk of becoming HIV infected through sexual transmission; however, an "inappropriate and ineffective" emphasis on "abstinence only" in prevention programming was counterproductive in many settings where HIV risk occurred (Coates et al. 2008; Collins et al. 2008).

The Need for Combination Prevention

Most forms of prevention need continual behavior modification to be effective. This effect was highlighted in two efficacy trials of preexposure prophylaxis (PrEP), one oral (Grant et al. 2010) and one topical (Karim et al. 2010). Both showed modest efficacy overall but improved efficacy in those participants who were most adherent. Future combination prevention packages are likely to contain one or more of these

Table 3. A list of some behavioral strategies

Sexual debut delay
Sexual partner reduction
Consistent condom usage
HIV counseling and testing
Sexual abstinence
Monogamy
Biomedical intervention uptake and consistent usage
Adherence to harm reduction strategies
Decreased substance use

Data adapted from Coates and Gable 2008.

Table 4. Randomized controlled trials assessing behavioral HIV prevention interventions

| Author | Citation | Brief description of participants and study | Risk-taking behavior decreased during the study? | |
			Intervention	Control
Kamali et al. 2003	Syndromic management of sexually transmitted infections and behaviour change interventions on transmission of HIV-1 in rural Uganda: A community randomised trial.	**Adults in rural Uganda** were randomized to syndromic STI management, behavioral interventions in combination with syndromic STI management and routine care and community development services. The primary outcome was HIV-1 incidence. Secondary outcomes were incidence of STIs and markers of behavioral change.	Yes	Yes
Koblin et al. 2004	Effects of a behavioural intervention to reduce acquisition of HIV infection among men who have sex with men: The EXPLORE randomised controlled study.	**Men that have sex with men**. The experimental intervention consisted of 10 one-on-one counseling sessions followed by maintenance sessions every 3 mo. Outcomes include HIV incidence and assessment of behavioral change, including occurrence of unprotected receptive anal intercourse with HIV-positive and unknown-status partners.	Yes	Yes
Ross et al. 2007	Biological and behavioral impact of an adolescent sexual health intervention in Tanzania: A community-randomized trial.	**Youth in Tanzania.** The intervention had four components: Community activities; teacher-led, peer-assisted sexual health education in years 5–7 of primary school; training and supervision of health workers to provide "youth-friendly" sexual health services; and peer condom social marketing versus standard activities. Impacts on HIV incidence, STI symptoms, as well as knowledge, reported attitudes, and other sexual health and behavioral outcomes were measured.	NA[a]	NA[a]
Corbett et al. 2007	HIV incidence during a cluster-randomized trial of two strategies providing voluntary counseling and testing at the workplace, Zimbabwe.	**Business employees in Zimbabwe**. Comparison of voluntary counseling and testing (VCT) when counseling and rapid testing were available onsite versus using prepaid vouchers for an external provider (which was the standard VCT). The main measured outcomes were rate of HIV incidence and VCT uptake.	NR	NR

Continued

Table 4. *Continued*

Author	Citation	Brief description of participants and study	Risk-taking behavior decreased during the study?	
			Intervention	Control
Jewkes et al. 2008	Impact of stepping stones on incidence of HIV and HSV-2 and sexual behaviour in rural South Africa: Cluster randomized controlled trial.	**Youth (15–26 yrs) in South Africa.** The intervention was Stepping Stones, a 50-h program that aims to improve sexual health by using participatory learning approaches to build knowledge, risk awareness, and communication skills and to stimulate critical reflection versus a 3-h intervention on HIV and safer sex. HIV and HSV-2 incidence, unwanted pregnancy, reported sexual practices, depression, and substance misuse were measured.	Yes	Yes
Patterson et al. 2008	Efficacy of a brief behavioral intervention to promote condom use among female sex workers in Tijuana and Ciudad Juarez, Mexico.	**Female sex workers living in Tijuana, Mexico** were provided with a 30-min behavioral intervention or a didactic control condition. At baseline and 6 mo, women underwent interviews and testing for HIV, syphilis, gonorrhea, and chlamydia.	Yes	Yes
Cowan et al. 2009	The Regai Dzive Shiri project: The results of a cluster randomized trial of a multi-component HIV prevention intervention for young people in rural Zimbabwe.	Youth–Community-based HIV prevention intervention for **adolescents** based in 30 communities **in rural Zimbabwe.** HIV and STI incidence, pregnancy, attitude, and self-reported sexual behavior were measured.	NR	NR

Data adapted from Padian et al. 2010.

NA, not available; NR, not recorded.

[a]Changes in behavior are not reported over time for this cohort as many of the participants were not sexually active at study enrollment.

biomedical interventions: vaginal microbicides providing a female-initiated methodology and oral PrEP as an alternative for those at risk for HIV exposures other than via the vaginal mucosa. Research is currently under way on rectal microbicides (Microbicide Trials Network 2011). The challenge will be to design prevention studies that embrace this integrated model of combined interventions. In addition, it will be important to monitor at the community level what happens when the combination intervention is scaled up (Piot et al. 2008).

CONDOMS

Intact latex and polyurethrane condoms have been shown in vitro to be impenetrable to particles the size of sexually transmitted pathogens (Lyle et al. 1997). When male condoms are used consistently, their effectiveness in reducing HIV transmission can be as high as 95% (Pinkerton et al. 1997). In most countries with generalized epidemics, condoms are actively promoted for all sexually active individuals as part of a comprehensive prevention approach despite

Figure 2. The "ABC" HIV prevention campaign. Billboard in Botswana. Sexual HIV risk could be avoided altogether by avoiding any sexual activities that could cause transmission of HIV (*A*, Abstain) or through avoiding sexual intercourse other than with a mutually faithful uninfected partner (*B*, Be faithful) or through the correct and consistent use of condoms (*C*, Condomise).

the social, economic, and psychological factors that limit their consistent use.

Several longitudinal cohort studies of serodiscordant couples estimated the effectiveness of male condoms for prevention of HIV transmission at around 85% (Weller et al. 2002). A study in MSM showed 76% efficacy in HIV prevention if used consistently (Golden 2006). The type of lubricant used, rather than strength of condom, was reported to be more important in safe usage of condoms in MSM (Harding et al. 2000). One of the most cited examples of how campaigns and policy can markedly increase male condom uptake is the much cited "100% condom use policy" in Thailand involving, in particular, the military and sex workers. Following an increase in condom usage from just over 10% in 1989 to >90% by 1993 in sex workers, incidence of new sexually transmitted infections (STIs) was seven times lower and HIV incidence was 50% lower (UNAIDS 2000). Similar success has been described in other female sex worker populations from other parts of the world (Weir et al. 1998).

Condom use and effectiveness at the population level is not well established. Demographic and health surveys from Latin America and Africa have reported increased usage of male condoms in recent years (Kerrigan et al. 2006; Shisana et al. 2009; Rehle et al. 2010). Recent encouraging reports suggest that the

use of mass media and creative condom social marketing to change attitudes and increase use have been successful (Foss et al. 2007).

The Female Condom

Female condoms were developed to provide a female-controlled biomedical HIV prevention method. Also made from polyurethrane or latex, they provide a physical barrier to HIV particles and prevent exposure to genital secretions containing HIV (Drew et al. 1990). No specific trials have been conducted to assess the efficacy of the female condom to prevent HIV infection, although impact on STI rates have been reported (French et al. 2003). Although the male partner is not unaware of the female condom, it does provide a prevention strategy that is less dependent on male willingness. Cost and mechanical difficulties have been cited as reasons for limited uptake (Galvao et al. 2005), but newer designs are easier to use, can be reused (Thomsen et al. 2006), and are made of materials that improve sensation with the promise of enhancing acceptability, uptake, and cost (Coffey et al. 2006). There are increasing reports of female condoms being used for HIV prevention in anal sex (http://doh.dc.gov/doh/cwp/view,a,1371,q,602647,dohNav_GID,1839,dohNav,|33815|,,.asp).

The perception exists that condom use indicates high-risk sexual partnerships associated with infidelity and infection. More than 2/3 of new infections occur in sub-Saharan Africa (SSA) between heterosexual cohabiting partners, and data from Rwanda and Zambia suggest that 60%–95% of new infections occur between married couples living together (Dunkle et al. 2008). In this setting, consistent condom use is challenging and couples-based HIV testing and counseling may have a greater role (The Voluntary HIV-1 Counseling and Testing Study Group 2000).

Additionally, there is a gap between supply and demand of condoms, especially in the developing world. Ideally, condoms should be free in these environments as research suggests even extremely low prices are a barrier to use (UNAIDS 2009). The United Nations Population Fund (UNFPA) estimated that at least 13.1 billion condoms are required annually to reduce the spread of HIV (UNFPA 2007). In 2008, <15% of this target was distributed globally.

HIV COUNSELING AND TESTING

HIV counseling and testing (HCT) services are important entry points for prevention and care. Studies from different countries have shown that individuals take precautions to protect their partners once they know they are HIV positive (Sweat et al. 2000; Allen et al. 2003), and modeling studies have found HCT to offer substantial clinical benefits and to be cost effective even in settings where linkage and access to care is limited (Walensky et al. 2009).

The past decade has seen a rapid global scale-up of HCT (WHO 2009; Kranzer et al. 2010). HCT uptake is associated with a range of sociodemographic factors and identifying characteristics of individuals who have never been tested is important to develop services targeted at first-time testers and thus to achieve universal access to HCT (Khumalo-Sakutukwa et al. 2008; Helleringer et al. 2009).

Sexually active individuals in high HIV prevalence settings are at continuous risk of infection and should therefore be tested at regular intervals. The World Health Organization (WHO) recommends annual testing in these settings, and a recent study from South Africa found annual screening to be very cost effective even in the Western Cape, the province with the lowest rates of HIV infection in South Africa (Walensky et al. 2009).

Evidence for HCT Role in Prevention

Studies have shown that many infected persons decrease high-risk sexual or needle-sharing behaviors once they are aware of their positive HIV status. The majority of this research is from high-income countries, with the strongest evidence for behavior change within discordant couples who also received counseling. Most studies that have assessed the effect of HCT on sexual behavior have focused on the change in behavior over periods of less than a year (Denison et al. 2008).

The challenge is to increase testing coverage and identify those who are positive for care (Kranzer et al. 2010). Strategies to increase testing have included national campaigns, provider-initiated counseling such as has been implemented in Botswana (Bateganya et al. 2007), couples counseling services, and community-level campaigns such as Project Accept (Fig. 3) (Khumalo-Sakutukwa et al. 2008). These alternative strategies not only increase coverage but ensure inaccessible populations such as men, the working population, and asymptomatic HIV-infected individuals are also tested (Matovu and Makumbi 2007). The role of incentives to increase testing coverage is also being investigated in a number of settings and populations (http://www.cgdev.org/content/publications/detail/1424161).

PREVENTION WITH POSITIVES

Traditional prevention has been thought of as protecting individuals from becoming HIV infected. Positive prevention embraces the concept that individuals who have tested positive may be helped to avoid spreading the infection further. Positive prevention also recognizes that infected individuals may want to remain sexually active and may wish to have children, both of which can be done with minimized harm to others (PEPFAR Prevention 2008).

Figure 3. Testing coverage can be markedly increased by using a variety of nontraditional venues and outlets such as household campaigns, high-risk venues, work environments, and community-based mobile testing units such as The Tutu Tester is a mobile testing service operating in Cape Town, South Africa (van Schaik et al. 2010).

Research in Positive Prevention

Antiretroviral therapy (ART) has dramatically reduced the morbidity and mortality of HIV infection through sustained reduction in HIV viral replication (De Cock et al. 2009). This reduction in HIV viral load (plasma HIV ribonucleic acid [RNA] levels) reduces infectiousness in the infected individual and, as a result, susceptibility for the noninfected partner (Musicco et al. 1994).

Evidence for Reduced HIV Transmission

Viral load is the single greatest risk factor for all transmission modes. ART can reduce the plasma and genital HIV viral load in the infected individual to undetectable levels (Granich et al. 2010). In a study of 415 HIV serodiscordant couples in Uganda, 21.7% of the initially uninfected partners became infected over 30 months of follow-up, translating to a transmission rate of approximately 12 infections per 100 person years (Fig. 4) (Quinn et al. 2000). No transmission events occurred in those couples in which the infected partner had a plasma HIV-1 RNA level of less than 1500 copies/mL, and the transmission risk increased as plasma HIV-1 RNA levels increased. For every 10-fold increase in viral load, there was a greater than twofold risk of transmission. This was similarly shown in HIV-serodiscordant couples in Zambia (Fideli et al. 2001). Plasma HIV-1 RNA levels generally correlate positively with the concentration of HIV in genital secretions, rectal mucosa, and saliva, although inflammation can stimulate local replication (Cu-Uvin et al. 2000; Lampinen et al. 2000; Shugars et al. 2000). Other studies have shown that transmission events may be observed at a very low plasma HIV-1 RNA level, suggesting that plasma viral load is not the only determinant of transmission (Vernazza et al. 2000; Tovanabutra et al. 2002).

Clinical Research in Discordant Couples

The outcomes of two retrospective clinical studies that showed the benefit of ART on HIV transmission (Musicco et al. 1994; Castilla et al. 2005) have been corroborated by the recent release of early results from a randomized trial, known as HPTN 052. The deferred treatment study arm was prematurely halted after a scheduled interim review by an independent Data and Safety Monitoring Board (DSMB) that concluded that initiation of ART by HIV-infected individuals substantially protected their HIV-uninfected sexual partners from acquiring HIV infection, with a 96% reduction in risk of HIV transmission. The study enrolled 1763, mostly heterosexual discordant couples in which the infected index case was ART-naive and had a CD4 T-cell count of 350–550 cells/mm^3. Treatment was commenced at 250 cells/mm^3 in the control or "treatment deferment" arm (Cohen et al. 2011).

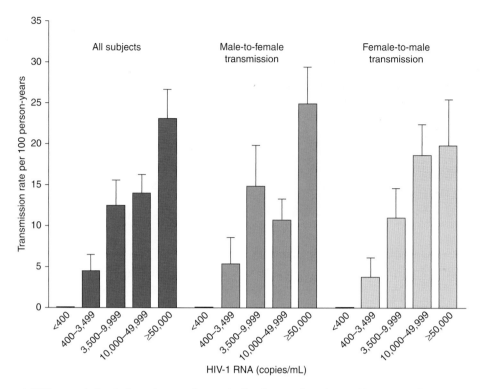

Figure 4. HIV transmission is dependent on plasma viral load. Mean (+SE) rate of heterosexual transmission of HIV-1 among 415 couples, according to the sex and the serum HIV-1 RNA level of the HIV-1-positive partner (Quinn et al. 2000). At baseline, among the 415 couples, 228 male partners and 187 female partners were HIV-1-positive. The limit of detection of the assay was 400 HIV-1 RNA copies per milliliter. For partners with fewer than 400 HIV-1 RNA copies per milliliter, there were zero transmissions.

Mathematical Models and Population-Level Impact

Extrapolating from this result, reduction of viral load within a population would likely lower the rate of heterosexual transmission within that population. Substantial reduction in the number of anticipated HIV cases in concentrated epidemics of injection drug users and MSM have been reported in at least two population-based studies of HIV incidence before and after the availability of ART (Katz et al. 2002; Fang et al. 2004; Porco et al. 2004). Mathematical models have been used to predict the ability of HIV treatment to reduce HIV incidence and prevalence. In a model generated by Gray et al. from the Ugandan transmission study data, ART would be predicted to reduce incident HIV by 80% (Gray et al. 2001). Conversely, others

have argued that ART could not reduce HIV prevalence in resource-constrained regions (Baggaley et al. 2006). More recently, this has led to the concept of "Test and Treat," modeled by Granich et al. (2009), which espouses universal HIV testing with immediate commencement of ART, regardless of clinical or immune status. This controversial model, based on the South African epidemic, with annual testing, heterosexual transmission, and a number of other assumptions, reported that immediate ART initiation could reduce HIV incidence by 95% over a 10-year period. To be considered, of course, are the cost and operational challenges as well as the risk of drug resistance and toxicity (Padian et al. 2008).

However, this shift from a focus on downstream therapeutic application of ART to more upstream preventive benefits of eliminating

HIV transmission has received considerable interest. A number of community-based feasibility studies are under way or planned in the United States and Southern Africa, and it is envisaged that definitive randomized trials will be designed and executed in the next 5–10 years.

MALE CIRCUMCISION

Practiced since at least the Sixth Dynasty (2345–2181 B.C.), approximately 30%–34% of adult men globally are circumcised (Padian et al. 2008). Male circumcision was first proposed in 1986 as an intervention to reduce risk of HIV acquisition (Fink 1986). Ecological and observational studies had shown that, in regions where HIV transmission is predominantly heterosexual, the prevalence of HIV and of male circumcision are inversely correlated (Bailey et al. 2001). The prevalence of HIV has been shown to be significantly higher in uncircumcised than in circumcised men in more than 30 cross-sectional studies, and a number of prospective studies have shown a protective effect, ranging from 48% to 88% (Bongaarts et al. 1989; Moses et al. 1990; Caldwell and Caldwell 1996; Gray et al. 2000; Reynolds et al. 2004; Buchbinder et al. 2005). A meta-analysis of studies from SSA reported an adjusted relative risk of 0.42 (95% CI 0.34–0.54) in all circumcised men, with a stronger adjusted relative risk of 0.29 (0.20–0.41) in circumcised men who were at higher risk of acquiring HIV (Weiss et al. 2000).

RCTs

To definitively document the protective associations of circumcision, three RCTs were designed and undertaken in three sub-Saharan countries: South Africa, Kenya, and Uganda. Design and results are presented (Table 5). A total of 11,054 HIV-negative men aged between 15 and 49 yr were randomized and similar to the observational data estimates, the summary rate ratio was 0.42 or a protective effect of 58% (Padian et al. 2010). Unlike many other biomedical interventions, male circumcision is a one-time procedure for which adherence issues are limited to refraining from intercourse during healing (Padian et al. 2010). As a result of these three randomized controlled clinical trials, WHO and UNAIDS (2007) have now made strong recommendations to roll out male circumcision with all possible urgency. Most recent long-term follow-up from these studies indicate that efficacy does not decrease with time, suggesting that the long-term efficacy of the intervention outweighs any risk compensation should this phenomenon be occurring (Kong et al. 2011).

MC to Prevent Male-to-Male HIV Transmission

Observational studies of MC to reduce HIV transmission between MSM have shown inconsistent results perhaps because men may adopt both receptive and insertive sexual roles. In a cohort study of HIV-negative MSM, no association between circumcision status and HIV acquisition was shown (Grulich et al. 2001). It is unclear what role MC would have in bisexual men.

MC to Prevent Male-to-Female HIV Transmission

A previous observational study in HIV-discordant couples in Rakai suggested a lower rate of male-to-female HIV transmission from circumcised HIV-infected men, particularly if their viral load was below 50,000 copies per mL (Gray et al. 2000). A fourth prospective randomized controlled trial was conducted in discordant couples and examined HIV transmission to female partners of HIV-infected men in Rakai, Uganda (Table 4). The study was stopped prematurely for futility; however, HIV acquisition was increased in the subgroup of female partners of men who resumed sexual activity early before complete wound healing (relative risk: 2.92, 95% CI 1.02–8.46, $P = 0.06$).

Biological Explanation

There are a number of biological studies that suggest plausible mechanisms for the protection offered by removal of the male foreskin. These are summarized in Table 6.

Table 5. Randomized trials of male circumcision (MC) to prevent HIV transmission

Country	Funding	Study population	Design/question/ method	Outcome	Reference
South Africa	Agence Nationale de Recherche sur le Sida (ANRS)/ National Institute for Communicable Diseases (South Africa)	3274 18- to 24-yr-old men in a semiurban, informal settlement called Orange Farm	Does circumcision reduce male risk of HIV infection by female partners? Study visits at months 3, 12, 21 postrandomization; circumcision performed using the sleeve method. Both the intervention and the control group received an enhanced prevention package beyond the local standard of care.	Stopped early. MC reduced the risk of HIV infection by 60%–61%.	Auvert et al. 2005
Uganda	National Institutes of Health/Johns Hopkins University, Rakai Health Sciences Project	Approximately 5000 15- to 49-yr-old men in rural Uganda (Rakai District)	Does circumcision reduce male risk of HIV infection by female partners? Four visits over 2 yr of follow-up; circumcision performed using the sleeve method. Both the intervention and the control group received an enhanced prevention package beyond the local standard of care.	Stopped early. MC reduced the risk of HIV infection by 48%.	Gray et al. 2007
Kenya	National Institutes of Health and Canadian Institute of Health Research/ University of Nairobi, University of Manitoba	2784 18- to 24-yr-old urban HIV-negative men	Does circumcision reduce male risk of HIV infection by female partners? Six study visits (months 1, 3, 6, 12, 18, 24) over 2 yr; circumcision performed by forceps-guided method. Both the intervention and the control group received an enhanced prevention package beyond the local standard of care.	Stopped early. MC reduced the risk of HIV infection by 53%.	Bailey et al. 2007

Table 5. *Continued*

Country	Funding	Study population	Design/question/ method	Outcome	Reference
Uganda	Bill and Melinda Gates Foundation/ Johns Hopkins University, Rakai Health Sciences Project	200 men had been enrolled concurrently with female partners and had couples HCT. In some, the women were enrolled separately	Is circumcision safe for HIV-positive men? How does it affect rates of acquisition of STIs? Does circumcision reduce female risk of infection by HIV-positive, circumcised male partners? Four visits over 2 yr of follow-up; sleeve method.	Trial suspended: DSMB review determined that the study lacked statistical power to answer its study question.	Wawer et al. 2009

Data from AVAC 2007 and Padian et al. 2010.

Other Benefits

The Rakai MC circumcision study conducted in 2005 also reported that men circumcised at the beginning of the study had a 50% reduction in rates of genital ulcer disease (GUD), attributable to herpes, syphilis, and chancroid, a reduction in acquisition of genital herpes (27%), and a 30%–35% reduction in rates of human papillomavirus (HPV), including the types that cause cancer. Benefits also extended to the female partners of the circumcised men: a 50% reduction in rates of GUD, as well as a dramatic reduction in trichomoniasis, HPV infection, and bacterial vaginosis (Bailey et al. 2001; Weis et al. 2006). HIV-infected men may also experience less GUD (Schneider et al. 2010).

Current Implementation Programs

Although the efficacy of MC on reducing individual risk is clear, the population-level effectiveness of this procedure in reducing HIV transmission will depend heavily on the acceptability of male circumcision programs in specific populations (Westercamp and Bailey 2007). Data on its acceptability among adults show this is likely to be highly context-specific and influenced by local cultural norms and practices (Eaton and Kalichman 2009).

Table 6. Some biological reasons for the protection against HIV acquisition in men undergoing male circumcision

Explanation	Reference(s)
1. The inner mucosal surface of the human foreskin, exposed on erection, has nine times higher density of HIV target cells (Langerhans' cells, $CD4^+$ T cells, and macrophages) than does cervical tissue.	Patterson et al. 2002
2. The foreskin's inner surface lacks the protective layer of squamous epithelial cell HIV target cells found on the outer surface and the glans.	McCoombe and Short 2006
3. In explant culture, several times more HIV is taken up by Langerhans' cells and $CD4^+$ T cells in foreskin than in cervical tissue.	Patterson et al. 2002
4. A foreskin is associated with increased incidence of ulcerative sexually transmitted infections; the number of HIV target cells in the prepuce is increased in the setting of recent STIs.	Weis et al. 2006; Bailey et al. 2001; Donoval et al. 2006
5. Susceptibility of the foreskin to abrasions.	

STI INTERVENTIONS

Longitudinal studies have shown substantial relative risks for HIV infection associated with various STIs with syphilis, chancroid, and genital herpes having larger effects on susceptibility than gonorrhea, chlamydia, and trichomonas (Stamm et al. 1988; Rottingen et al. 2001). These ulcerative diseases appear to create an entry point for the virus by disrupting the genital epithelial barrier leading to a greater susceptibility (Ghys et al. 1997). In addition, studies have shown that HIV viral shedding in the genital tract is substantially increased with a sexually transmitted coinfection, and this replication is reduced after treatment of the STI (Eron et al. 1996).

As a result, efforts to ensure prompt diagnosis and treatment of STIs along with behavioral risk reduction have been part of HIV prevention programming since the 1980s. In 1989, Pepin and colleagues suggested that the interaction of HIV and STIs may present an opportunity for intervention (Pepin et al. 1989). Empirical evidence for this intervention has included uncontrolled intervention studies among sex workers and community-based RCTs in general populations (Laga et al. 1994).

Clinical Trials of STI Treatment

Eight of the nine RCTs of STI treatment for HIV prevention showed no effect, although one additional study found a significant reduction on HIV incidence in a subgroup of men who attended program meetings (Table 7) (Gregson et al. 2007). Four community-randomized trials have been conducted to assess the effect on HIV transmission and HIV acquisition through reduction of the incidence of the most common curable STIs. Of all four study outcomes, only the Mwanza trial reported significant reduction (38%) in HIV incidence. Many possible reasons for this discrepancy have been cited but most compelling are the differences between the stage of the epidemic in Uganda and Tanzania when the studies were performed. The epidemic in Uganda was more established (HIV prevalence 16% and stable) with lower risk behavior and

lower rates of curable STIs. In contrast, the HIV prevalence in Mwanza was 4% and rising with much greater rates of STIs (Grosskurth et al. 2000).

These data would suggest that STI treatment interventions can have an impact where treatable STIs are prevalent and where HIV incidence is very high in the general populations. However, even if the HIV epidemic has matured in the general adult population, adolescents as they sexually debut may initially have low HIV prevalence and constitute a population where STI control may be very important. It is still recommended that STI treatment should be an essential component of HIV control programs in communities in which the burden of STIs is substantial (Grosskurth et al. 2000; Hayes et al. 2010).

Herpes Simplex 2 and HIV Transmission

In SSA, HSV-2 infections have a two- to three-fold increased effect on HIV acquisition in the general population (Freeman et al. 2006). Initial proof of concept, randomized trials of suppressive treatment with valiciclovir reported reduced HIV shedding in genital secretions of coinfected individuals, suggesting potential for reduced HIV transmission risk (Nagot et al. 2007; Zuckerman et al. 2007). Subsequently, in three RCTs, antivirals for herpes simplex virus (HSV) suppression were insufficiently potent to alleviate persistent genital inflammation in HIV-negative HSV2-positive persons, and the reduction in HIV levels in HIV-positive persons was insufficient to reduce HIV transmission (Table 7) (Hayes et al. 2010).

HIV PREVENTION AMONG INJECTING DRUG USERS

HIV prevention for people who inject drugs (commonly referred to as injecting drug users, or IDUs) presents a difficult paradox. There is abundant evidence for the efficacy of a number of interventions for this population, and clear and compelling data has emerged on the efficacy of combinations of these interventions in

Table 7. The randomized trials of treatment of STIs to reduce HIV transmission

Intervention	Country/region	Target population and annual HIV incidence (ppy)	Efficacy/outcome	Reference
Individual syndromic STI[a] treatment to reduce HIV incidence (CRCT)	Mwanza, Tanzania	General population; 0.9%	38% reduction in HIV incidence	Grosskurth et al. 2000
STI therapy[b] to reduce HIV incidence (everyone treated every 10 mo) (CRCT)	Rakai, Uganda	General population; 1.5 ppy	Nil	Wawer et al. 2009
Individual RCT of intensive, microscopy-assisted STI[a] screening and treatment to reduce HIV incidence		FSW; 7.6 ppy	Nil	Ghys et al. 1997
Individual syndromic STI[b] to reduce HIV incidence (CRCT)	Masaka, (rural) Uganda	General population; 0.8 ppy	Nil	Kamali et al. 2003
Treatable STI[a] periodic presumptive therapy Individual RCT	Kenya	FSW; 3.2 ppy	Nil	Kaul et al. 2004
Individual syndromic STI treatment[a] to reduce HIV incidence (CRCT)	Manicaland, Zimbabwe	General population; 1.5 ppy	Nil; subgroup of men who attended program meetings (IRR 0.48; $p = 0.04$)	Gregson et al. 2007
HSV-2 suppression[c]	Tanzania	HSV-2-positive women; 4.1 ppy	Nil	Watson-Jones et al. 2008
HSV-2 suppression[c]	Africa; Peru and United States	WSM; MSM HSV2 seropositive; 3.3 ppy	Nil (some benefit in subset of women who took >90% of doses)	Celum et al. 2008
HSV-2 suppression[c]	Africa	HIV/HSV-2 positive; 2.7 ppy	Nil	Celum et al. 2010

ppy, per hundred person years or annual percentage; CRCT, community randomized control trial; FSW, female sex workers; HSV-2, herpes simplex virus 2; IRR, incidence rate ratio; WSM; women who have sex with men.
[a]Treatable STIs: Chancroid, syphilis, gonorrhea, chlamydial infections, and trichomonas.
[b]Single-dose oral antibiotic.
[c]Acyclovir treatment.

achieving control of HIV spread via this route (Degenhardt et al. 2010). Yet IDU remain the least served of any risk group globally for prevention, treatment, and care (Wolfe et al. 2010). Epidemics driven by IDU risks and by risk-enhancing structural and policy environments continue to expand in 2010 (Beyrer et al. 2010). These policy failures include punitive and repressive drug laws, criminalization of drug dependency and possession, and the continued resistance

to the provision of evidence-based drug treatment, including methadone maintenance therapy in many states and regions (Strathdee et al. 2010).

HIV spread among IDUs has been driven largely, but not exclusively, by injecting use of heroin. Cocaine, methamphetamine, and combinations are also important substances associated with injecting risks. Heroin predominates in Eastern Europe and Central Asia, North, South, and Southeast Asia, and Western Europe, encompassing the major populations at risk for HIV through injecting (Mathers et al. 2008). The most recent global estimate, from the reference group to the United Nations on HIV and injecting drug use was that some 15.9 million persons (range from 11.0 to 21.2 million) worldwide were IDU in 2007 (Beyrer 2010).

Evidence for Efficacy

The literature on HIV prevention for this population is large and growing (Degenhardt et al. 2010). The most compelling recent data suggest, as with prevention of sexual transmission, that no single intervention alone can reduce HIV risks enough to control injecting-driven epidemics. Encouragingly, however, recent modeling studies show that combination approaches to HIV prevention for this population can be synergistic in effect and have real impact on HIV risks at individual, couple, network, and population levels of spread (Degenhardt et al. 2010).

The components of effective prevention services for IDU include individual and higher-level interventions (Table 8). An essential component is access to safe injecting equipment. Because the primary risk for HIV acquisition and transmission among drug users is the reuse of contaminated injecting equipment, multiple approaches to reducing equipment reuse,

termed needle and syringe exchange programs (NSP), have been developed. The provision of equipment for people who inject has proven politically challenging in many contexts, because this has been seen (based on no empirical evidence) as "encouraging" injecting. Indeed, the U.S. federal ban on funding for such programs, lifted in 2010 by the Obama administration, was based on this unsound premise (Beyrer et al. 2010). A recent global review and modeling exercise of the evidence for efficacy suggests that with high coverage, NSP can reduce HIV incidence at population levels by 20% over 5 years, but the reduction is too modest to control HIV spread (Degenhardt et al. 2010).

A second critical component of HIV prevention for IDU is drug treatment. The first agent shown to have efficacy in reduction of HIV transmission among drug users was methadone (Metzger et al. 1999). Because methadone is an oral administered liquid and an opiate agonist, opioid-dependent patients can be maintained on the agent and reduce dramatically their injecting drug use. This simple "substitution" therapy, as it has come to be known, was shown by Metzger and colleagues (1999) to markedly reduce HIV infection rates among IDU in Philadelphia in the 1990s. Newer agents are also now available, but there have been significant obstacles to the

Table 8. Structural interventions associated with prevention in injection drug use

Level	Physical environment	Social environment	Economic environment	Policy environment
Micro	Unsafe drug use, injecting, and sex work locations Injecting in public places Detention centers	Social and peer group risk norms Policing and crackdowns Lack of health and welfare services	Cost of living and health services Cost of prevention services Lack of income-generation opportunities	Access to clean needles and syringes Policy and programs for distribution Housing access
Macro	Drug traffic and distribution routes Population mobility	Sex inequalities and risk Stigmatization and marginalization of IDU Inadequate public advocacy	Reduced public spending Increase in informal income generation Economic uncertainty	Laws governing human rights Laws governing drug possession Public health policy and harm-reduction services

Data adapted from Degenhardt et al. 2010.

 Cite this article as *Cold Spring Harb Perspect Med* doi: 10.1101/cshperspect.a007435

widespread use of these agents. Methadone was strongly opposed by the Soviet Union when it was first introduced, and opioid substitution therapy (OST) remains illegal in Russia (Beyrer et al. 2010).

Although NSP and OST in combination can reduce HIV risks, recent modeling work by Hallett et al., reported by Degenhardt et al. (2010), showed that a third element is essential for individuals and for epidemics: access to antiretroviral therapies (ARVs). ARV access for IDUs alone had roughly the same impact on HIV incidence as the combination of OST and NSP, and was significantly higher when ARVs were available to HIV-infected IDU at higher (<350 CD4s) levels. They found a dramatic synergistic impact of provision of NSP, OST, and ARV on reducing HIV incidence over time, with a 39% reduction in population levels of HIV infection over 5 years with the combination approach (Degenhardt et al. 2010). This model assumed quite modest levels of efficacy for each component at the individual level (60% for OST, 40% for NSP, and 90% for ARVs when initiated at the higher CD4 level) (Degenhardt et al. 2010).

Challenges and Opportunities for Implementation

Although it is tremendously encouraging to show the synergistic effects of combined preventive interventions on HIV incidence at population levels among IDUs, the realities of access to care for this population are sobering. Wolfe et al. reviewed access to care for IDUs in selected high-burden countries and found that among all populations at risk for HIV infection, IDUs remained the least served (Wolfe et al. 2010). An even more telling finding was that in China and Vietnam, the number of drug users in detention is three times and 33 times higher, respectively, than those in treatment. Incarceration is not an evidence-based approach to HIV prevention, but rather a well-described risk for HIV infection among drug users (Beyrer et al. 2010).

Strathdee et al. used the risk-environment framework to investigate another aspect of IDU risks that poses real challenges—the social,

policy, and legal environments that can reduce, or *drive*, HIV risks (Strathdee et al. 2010). They found that structural aspects of risk environments had substantial impacts on HIV risks and disease spread. As IDU risks emerge in new settings, as was happening in East and South Africa in 2010, these challenges are likely to continue to undermine our responses (Beyrer et al. 2010).

STRUCTURAL APPROACHES TO HIV PREVENTION

Structural factors in HIV epidemics are defined as "physical, social, political, cultural, organizational, community, economic, legal, or policy aspects of the environment that facilitate or obstruct efforts to avoid HIV infection" (Gupta et al. 2008). Structural interventions are programs or policies that seek to address these factors and prevent HIV acquisition through altering the context and mechanisms by which the behavior occurs rather than targeting the behavior itself (Gupta et al. 2008).

Evidence Linking Structural Factors to HIV

The literature cites both indirect associations, such as increased HIV vulnerability among orphaned or homeless children (Hallman 2005), and more direct associations, such as studies on migration linking South African mine workers and their high-risk work conditions plus separation from family to unprotected sex with prostitutes (Lurie et al. 2003). Associations, however, are not always clear-cut: Although poverty has been linked to HIV vulnerability, in a study of household wealth and HIV prevalence in population-based surveys in SSA, HIV prevalence was highest in the wealthier sector of the population in six of the eight surveys studied (Sumartojo et al. 2000; Hallman 2005).

Examples of Structural Approaches in HIV Prevention

Structural approaches may be single policies or programs (revoking a discriminatory law, e.g., the South African sodomy law in 1994) or transformational processes (community mobilization for MC in KwaZulu Natal following King Goodwill's pronouncement that Zulu men should be circumcised). Their common purpose is to change the social, economic, political, or environmental factors that determine HIV risk and vulnerability in specific settings (Gupta et al. 2008). The policy and legal shifts that enabled NSP and OST to operate for IDUs is an example of a successful structural intervention (Drucker et al. 1998).

Within the social framework, school retention has been linked to a lower HIV acquisition risk. This effect may be caused by reducing child prostitution, child labor, and antisocial behavior as well as opportunities for HIV and sex education (Hallman 2006). A higher level of education is associated with safer sexual practices and a later sexual debut (Prata et al. 2005), yet school fees are a major barrier to school attendance in the developing world.

One of the ways economic frameworks create a risk environment for HIV is through women's financial dependency on men and the lack of opportunities for both sexes in the developing world. Microcredit programs, for example, may reduce women's HIV vulnerability by strengthening their economic options (Hargreaves et al. 2002; Hall 2006).

Designing Structural Interventions and Their Evaluation

This effort requires an analysis of social, political, economic, and environmental factors in a given setting that increase risk or vulnerability. Different factors may have more than one causal pathway from the structural factor to the behavior(s) that need to change to reduce risk (Over 1998).

A review of 24 IDU harm reduction programs in 1989 found prevention efficacy for such programs without associated adverse events. Further structural interventions to reduce some legal limitations, increase community and social mobilization, and upscale such programs holds great potential for reducing HIV vulnerabilities among IDUs (Table 7) (Degenhardt et al. 2010). Some successful structural interventions have occurred at the country or district level. The Mbeya region in Tanzania is one of the most affected regions in the country but reported a 7% decrease in HIV prevalence over a few years. The intervention included a single regional HIV plan with political support, regional AIDS coordination, and involvement of business and nongovernmental organizations (NGOs) (Vogel 2007).

It is possible that individuals will find it difficult to change their behavior in relation to HIV risk until they have a stronger social, political, economic, and environmental framework in which to live their lives. As a result, structural approaches that are context-specific and evidence-based should be a part of an overall HIV prevention package that includes behavioral and biomedical approaches (Gupta et al. 2008).

CONCLUSION

The aim of the Sixth Millennium Development Goal is to halt, and reverse, the spread of HIV by 2015. This article describes an impressive array of evidence-based devices (condoms, harm reduction, MC) that can be implemented along with information, skills, and services. Concerted HIV prevention efforts from countries as diverse as Thailand, Australia, and Senegal have resulted in maintenance of low seroprevalence rates (Winkelstein et al. 1987; Kippax et al. 2003). Other studies conducted in high-risk populations have shown that HIV prevention can work, even in the most challenging settings. Yet, despite this, UNAIDS tells us that only 60% of sex workers, 46% of injection drug users, and 40% of MSM were reached with HIV prevention programs in 2008 (UNAIDS 2008). The positive results of biomedical interventions in 2010 and those expected over the next several years give promise that a number of other interventions can be added to the menu. The era is one of HARP-targeted, strategic, and creative combinations of behavioral, biomedical, and structural interventions. These programs will require universal access, widescale implementation, careful monitoring, and evaluation, financial, and technical resources and robust commitment at regional and country levels (Coates et al. 2008; Merson et al. 2008). We may then begin to see a substantial impact on the spread of HIV globally.

REFERENCES

Allen S, Meinzen-Derr J, Kautzman M, Zulu I, Trask S. 2003. Sexual behavior of HIV discordant couples after HIV counseling and testing. *Aids* **17:** 733–740.

Auerbach J, Coates T. 2000. HIV prevention research: Accomplishments and challenges for the third decade of AIDS. *Am J Public Health* **90:** 1029–1032.

Auvert B, Taljaard D, Lagarde E, Sobngwi-Tambekou J, Sitta R, Puren A. 2005. Randomized, controlled intervention trial of male circumcision for reduction of HIV infection risk: The ANRS 1265 trial. *PLoS Med* **2:** e298. doi: 10.1371/journal.pmed.0020298.

AVAC. 2007. A new way to protect against HIV? *Understanding the results of male circumcision studies for HIV prevention*. AIDS Vaccine Advocacy Coalition's Anticipating and Understanding Results series, September 2007, updated ed.

Baggaley RF, Garnett GP, Ferguson NM. 2006. Modelling the impact of antiretroviral use in resource-poor settings. *PLoS Med* **3:** e124. doi: 10.1371/journal.pmed. 0030124.

Bailey RC, Plummer FA, Moses S. 2001. Male circumcision and HIV prevention: Current knowledge and future research directions. *Lancet Infect Dis* **1:** 223–231.

Bailey RC, Moses S, Parker CB, Agot K, Maclean I, Krieger JN, Williams CF, Campbell RT, Ndinya-Achola JO. 2007. Male circumcision for HIV prevention in young men in Kisumu, Kenya: A randomised controlled trial. *Lancet* **369:** 643–656.

Barouche D. 2008. Challenges in the development of an HIV-1 vaccine. *Nature* **455:** 613–617.

Bateganya MH, Abdulwadud OA, Kiene SM. 2007. Home-based HIV voluntary counseling and testing in developing countries. *Cochrane DB Syst Rev* **4:** CD006493.

Beyrer C, Malinowska-Sempruch K, Kamarulzaman A, Kazatchkine M, Sidibe M, Strathdee SA. 2010. Time to act: A call for comprehensive responses to HIV in people who use drugs. *Lancet* **376:** 551–563.

Bongaarts J, Reining P, Way P, Conant F. 1989. The relationship between male circumcision and HIV infection in African populations. *AIDS* **3:** 373–377.

Buchbinder SP, Vittinghoff E, Heagerty PJ, Celum CL, Seage GR III, Judson FN, McKirnan D, Mayer KH, Koblin BA. 2005. Sexual risk, nitrite inhalant use, and lack of circumcision associated with HIV seroconversion in men who have sex with men in the United States. *J Acquir Immune Defic Syndr* **39:** 82–89.

Caldwell JC, Caldwell P. 1996. The African AIDS epidemic. *Sci Am* **274:** 62–63, 66–68.

Castilla J, Del Romero J, Hernando V, Marincovich B, Garcia S, Rodriguez C. 2005. Effectiveness of highly active antiretroviral therapy in reducing heterosexual transmission of HIV. *J Acquir Immune Defic Syndr* **40:** 96–101.

Celum C, Wald A, Hughes J, Sanchez J, Reid S, Delany-Moretlwe S, Cowan F, Casapia M, Ortiz A, Fuchs J, et al. 2008. Effect of aciclovir on HIV-1 acquisition in herpes simplex virus 2 seropositive women and men who have sex with men: A randomised, double-blind, placebo-controlled trial. *Lancet* **371:** 2109–2119.

Celum C, Wald A, Lingappa JR, Magaret AS, Wang RS, Mugo N, Mujugira A, Baeten JM, Mullins JI, Hughes JP, et al. 2010. Acyclovir and transmission of HIV-1 from persons infected with HIV-1 and HSV-2. *N Engl J Med* **362:** 427–439.

Coates T, Richter L, Caceres C. 2008. Behavioural strategies to reduce HIV transmission: How to make them work better. *Lancet* **372:** 669–684.

Coffey PS, Kilbourne-Brook M, Austin G, Seamans Y, Cohen J. 2006. Short-term acceptability of the PATH woman's condom among couples at three sites. *Contraception* **73:** 588–593.

Cohen J. 2005. HIV/AIDS: Prevention cocktails: Combining tools to stop HIV's spread. *Science* **309:** 1002–1005.

Cohen M, Hoffman I, Royce R, Kazembe P, Dyer JR, Costello Daly C, Zimba D, Vernazza PL, Maida M, et al. 1997. Reduction of concentration of HIV-1 in semen after treatment of urethritis: Implications for prevention of sexual transmission of HIV-1. *Lancet* **349:** 1868–1873.

Cohen MS, Chen YQ, McCauley M, Gamble T, Hosseinipour MC, Kumarasamy N, Hakim JG, Kumwenda J, Grinsztejn B, Pilotto JH, et al. 2011. Prevention of HIV-1 infection with early antiretroviral therapy. *N Engl J Med* **365:** 493–505.

Cohen MS, Gay C, Kashuba ADM, Blower S, Paxton L. 2007. Narrative Review: Antiretroviral therapy for prevention of the sexual transmission of HIV-1. *Ann Intern Med* **146:** 591–601.

Collins C, Coates TJ, Curran J. 2008. Moving beyond the alphabet soup of HIV prevention. *AIDS* **22** (Suppl. 2): S5–S8.

Corbett EL, Makamure B, Cheung YB, Dauya E, Matambo R, Bandason T, Munyati SS, Mason PR, Butterworth AE, Hayes RJ. 2007. HIV incidence during a cluster-randomized trial of two strategies providing voluntary counselling and testing at the workplace, Zimbabwe. *AIDS* **21:** 483–489.

Cowan FM, Pascoe SJS, Langhaug LF, Dirawo J, Mavhu W, Chidiya S, Jaffar S, Mbizuo M, Stephenson JM, Johnson AM, et al. 2009. The Regai Dzive Shiri Project: The results of a cluster randomized trial of a multi-component HIV prevention intervention for young people in rural Zimbabwe. 18th International Society for Sexually Transmitted Disease Research, London.

Cu-Uvin S, Caliendo AM, Reinert S, Chang A, Juliano-Remollino C, Flanigan TP, Mayer KH, Carpenter CC. 2000. Effect of highly active antiretroviral therapy on cervicovaginal HIV-1 RNA. *AIDS* **14:** 415–421.

De Cock KM, Crowley SP, Lo Y.-R, Granich RM, Williams BG. 2009. Preventing HIV transmission with antiretrovirals. *B World Health Organ* **87:** 488–488A.

Degenhardt L, Mathers B, Vickerman P, Rhodes T, Latkin C, Hickman M. 2010. Prevention for people who inject drugs: Why individual, structural, and combination approaches are required. *Lancet* **376:** 285–301.

Denison JA, O'Reilly KR, Schmid GP, Kennedy CE, Sweat MD. 2008. HIV voluntary counseling and testing and behavioral risk reduction in developing countries: A meta-analysis, 1990–2005. *Aids Behav* **12:** 363–373.

Donoval BA, Landay AL, Moses S, Agot K, Ndinya-Achola JO, Nyagaya EA, MacLean I, Bailey RC. 2006. HIV-1 target cells in foreskins of African men with varying histories of sexually transmitted infections. *Am J Clin Pathol* **125:** 386–391.

Drew WL, Blair M, Miner RC, Conant M. 1990. Evaluation of the virus permeability of a new condom for women. *Sex Transm Dis* **17:** 110–112.

Drucker E, Lurie P, Wodak A, Alcabes P. 1998. Measuring harm reduction: The effects of needle and syringe exchange programs and methadone maintenance on the ecology of HIV. *AIDS* **12:** S217–S230.

Dunkle KL, Stephenson R, Karita E, Chomba E, Kayitenkore K, Vwalika C, Greenberg L, Allen S. 2008. New heterosexually transmitted HIV infections in married or cohabiting couples in urban Zambia and Rwanda: An analysis of survey and clinical data. *Lancet* **371:** 2183–2191.

Eaton L, Kalichman SC. Behavioral aspects of male circumcision for the prevention of HIV infection. *Curr HIV/AIDS Rep* **6:** 187–193.

Eron JJ Jr, Gilliam B, Fiscus S, Dyer J, Cohen MS. 1996. HIV-1 shedding and chlamydial urethritis. *JAMA* **275:** 36.

Fang CT, Hsu HM, Twu SJ, Chen MY, Chang YY, Hwang JS, Wang JD, Chuang CY. 2004. Decreased HIV transmission after a policy of providing free access to highly active antiretroviral therapy in Taiwan. *J Infect Dis* **190:** 879–885.

Fideli US, Allen SA, Musonda R, Trask S, Hahn BH, Weiss H, Mulenga J, Kasolo F, Vermund SH, Aldrovandi GM. 2001. Virologic and immunologic determinants of heterosexual transmission of human immunodeficiency virus type 1 in Africa. *AIDS Res Hum Retrov* **17:** 901–910.

Fink AJ. 1986. In defense of circumcision. *Pediatrics* **77:** 265–267.

Foss AM, Hossain M, Vickerman PT, Watts CH. 2007. A systematic review of published evidence on intervention impact on condom use in sub-Saharan Africa and Asia. *Sex Transm Infect* **83:** 510–516.

Freeman EE, Weiss HA, Glynn JR, Cross PL, Whitworth JA, Hayes RJ. 2006. Herpes simplex virus 2 infection increases HIV acquisition in men and women: Systematic review and meta-analysis of longitudinal studies. *AIDS* **20:** 73–83.

French PP, Latka M, Gollub EL, Rogers C, Hoover DR, Stein ZA. 2003. Use-effectiveness of the female versus male condom in preventing sexually transmitted disease in women. *Sex Transm Dis* **30:** 433–439.

Galvao LW, Oliveira LC, Diaz J, Kime D-J, Marchif N, van Damg J, Castilhoc RF, Chenh M, Maurizio M. 2005. Effectiveness of female and male condoms in preventing exposure to semen during vaginal intercourse: A randomized trial. *Contraception* **71:** 130–136.

Ghys P, Fransen K, Diallo O, Ettiègne-Traoré V, Coulibaly IM, Yeboué KM, Kalish ML, Maurice C, Whitaker JP, Greenberg AE, et al. 1997. The associations between cervicovaginal HIV shedding, sexually transmitted diseases, and immunosuppression in sex workers in Abidjan, Cote d'Ivoire. *AIDS* **11:** F85–F93.

Golden M. 2006. HIV serosorting among men who have sex with men: Implications for prevention. 13th Conference on Retroviruses and Opportunistic Infections, Denver, CO, abstract 163.

Granich R, Gilks C, Dye C, DeCock K, Williams B. 2009. Universal voluntary HIV testing with immediate antiretroviral therapy as a strategy for elimination of HIV transmission: A mathematical model. *Lancet* **373:** 48–57.

Granich R, Crowley S, Vitoria M, Smyth C, Kahn JG, Bennett R, Lo YR, Souteyrand Y, Williams B. 2010. Highly active antiretroviral treatment for the prevention of HIV transmission. *J Int AIDS Soc* **13:** 1.

Grant RM, Lama JR, Anderson PL, McMahan V, Liu AY, Vargas L, Goicochea P, Casapía M, Guanira-Carranza JV, Ramirez-Cardich ME, et al. 2010. Preexposure chemoprophylaxis for HIV prevention in men who have sex with men. *N Engl J Med* **363:** 2587–2599.

Gray RH, Kiwanuka N, Quinn TC, Sewankambo NK, Serwadda D, Mangen FW, Lutalo T, Nalugoda F, Kelly R, Meehan M, et al. 2000. Male circumcision and HIV acquisition and transmission: Cohort studies in Rakai, Uganda. Rakai Project Team. *AIDS* **14:** 2371–2381.

Gray RH, Wawer MJ, Brookmeyer R, Sewankambo NK, Serwadda D, Wabwire-Mangen F, Lutalo T, Li X, vanCott T, Quinn TC, Rakai Project Team, et al. 2001. Probability of HIV-1 transmission per coital act in monogamous, heterosexual, HIV-1-discordant couples in Rakai, Uganda. *Lancet* **357:** 1149–1153.

Gray RH, Kigozi G, Serwadda D, Makumbi F, Watya S, Nalugoda F, Kiwanuka N, Moulton LH, Chaudhary MA, Chen MZ, et al. 2007. Male circumcision for HIV prevention in men in Rakai, Uganda: A randomised trial. *Lancet* **369:** 657–666.

Gregson S, Adamson S, Papaya S, Mundondo J, Nyamukapa CA, Mason PR, Garnett GP, Chandiwana SK, Foster G, Anderson RM. 2007. Impact and process evaluation of integrated community and clinic-based HIV-1 control: A cluster-randomised trial in eastern Zimbabwe. *PLoS Med* **4:** e102. doi: 10.1371/journal.pmed.0040102.

Grosskurth H, Gray R, Hayes R, Mabey D, Wawer M. 2000. Control of sexually transmitted diseases for HIV-1 prevention: Understanding the implications of the Mwanza and Rakai trials. *Lance* **355:** 1981–1987.

Grulich AE, Hendry O, Clark E, Kippax S, Kaldor JM. 2001. Circumcision and male-to-male sexual transmission of HIV. *AIDS* **15:** 1188–1189.

Gupta GR, Parkhurst J, Ogden J, Aggleton P, Mahal A. 2008. Structural approaches to HIV prevention. *Lancet* **372:** 765–775.

Hall J. 2006. Microfinance brief: Tap and reposition youth (TRY) program. Population Council, New York.

Hallman K. 2005. Gendered socioeconomic conditions and HIV risk behaviours among young people in South Africa. *Afr J AIDS Res* **4:** 37–50.

Hallman K. 2006. HIV vulnerability of non-enrolled and urban poor girls in Kwa-Zulu Natal, South Africa. Population Council, New York.

Harding R, Golombok S, Sheldon J. 2000. A clinical trial of a thicker versus a standard condom for gay men. 13th International AIDS Conference, Durban. Abstract WePpC1395.

 Cite this article as *Cold Spring Harb Perspect Med* doi: 10.1101/cshperspect.a007435

Hargreaves J, Atsbeha T, Gear J, Kim J, Mzamani Makhubele B, Mashaba K, Morison L, Motsei M, Peters C, Porter J, et al. 2002. Social interventions for HIV/AIDS: Intervention with microfinance for AIDS gender equity. *IMAGE Study Evaluation Monograph.* RADAR and Small Enterprise Foundation Publishers.

Hayes R, Watson-Jones D, Celum C, van de Wijgert J, Wasserheit J. 2010. Treatment of sexually transmitted infections for HIV prevention: End of the road or new beginning? *AIDS* 24: S15–S26.

Helleringer S, Kohler HP, Frimpong JA, Mkandawire J. 2009. Increasing uptake of HIV testing and counseling among the poorest in sub-Saharan countries through home-based service provision. *J Acquir Immune Defic Syndr* 51: 185–193.

Jewkes R, Nduna M, Levin J, Jama N, Dunkle K, Puren A, Duvvury N. 2008. Impact of stepping stones on incidence of HIV and HSV-2 and sexual behaviour in rural South Africa: Cluster randomised controlled trial. *BMJ* 337: a506. doi: 10.1136/bmj.a506.

Jürgens R, Csete J, Amon J, Baral S, Beyrer C. 2010. People who use drugs, HIV and human rights. *Lancet* 376: 355–366.

Kamali A, Quigley M, Nakiyingi J, Kinsman J, Kengeya-Kayondo J, Gopal R, Ojwiya A, Hughes P, Carpenter LM, Whitworth J, et al. 2003. Syndromic management of sexually-transmitted infections and behaviour change interventions on transmission of HIV-1 in rural Uganda: A community randomised trial. *Lancet* 361: 645–652.

Karim QA, Karim SSA, Frohlich JA, Grobler AC, Baxter C, Mansoor LE, Kharsany AB, Sibeko S Mlisana KP, Omar Z, et al. 2010. Effectiveness and safety of tenofovir gel, an antiretroviral microbicide, for the prevention of HIV infection in women. *Science* 329: 1168–1174.

Katz MH, Schwarcz SK, Kellogg TA, Klausner JD, Dilley JW, Gibson S, McFarland W, et al. 2002. Impact of highly active antiretroviral treatment on HIV seroincidence among men who have sex with men: San Francisco. *Am J Public Health* 92: 388–394.

Kaul R, Kimani J, Nagelkerke NJ, Fonck K, Ngugi EN, Keli F, MacDonald KS, Maclean IW, Bwayo JJ, Temmerman M, et al. 2004. Monthly antibiotic chemoprophylaxis and incidence of sexually transmitted infections and HIV-1 infection in Kenyan sex workers: A randomized controlled study. *JAMA* 291: 2555–2562.

Kerrigan D, Moreno L, Rosario S, Gomez B, Jerez H, Barrington C, Weiss E, Sweat M. 2006. Environmental–structural interventions to reduce HIV/STI risk among female sex workers in the Dominican Republic. *Am J Public Health* 96: 120–125.

Khumalo-Sakutukwa G, Morin SF, Fritz K, Charlebois ED, van Rooyen H, Chingono A, Modiba P, Mrumbi K, Visrutaratna S, Singh B, et al. 2008. Project accept (HPTN 043): A community-based intervention to reduce HIV incidence in populations at risk for HIV in sub-Saharan Africa and Thailand. *J Acquir Immune Defic Syndr* 49: 422–431.

Kippax S, Race K. 2003. Sustaining safe practice: Twenty years on. *Soc Sci Med* 57: 1–12.

Koblin B, Chesney M, Coates TJ, for the EXPLORE Study Team. 2004. Effects of a behavioural intervention to reduce acquisition of HIV infection among men who have sex with men: The EXPLORE randomised controlled study. *Lancet* 364: 41–50.

Kong X, Kigozi G, Ssempija V, Serwadda D, Nalugoda F, Makumbi F, Lutako T, Watya S, Wawer M, Gray R. 2011. Longer-term effects of male circumcision on HIV incidence and risk behaviors during post-trial surveillance in Rakai, Uganda. 18th Conference on Retroviruses and Opportunistic Infections (CROI 2011), Boston, February 27-March 2. Abstract 36.

Kranzer K, Zeinecker J, Ginsberg P, Orrell C, Kalaw NN, Lawn SD, Bekker L-G, Wood R. 2010. Linkage to HIV care and antiretroviral therapy in Cape Town, South Africa. *PLoS One* 5: e13801. doi: 10.1371/journal.pone.0013801.

Laga M, Alary M, Nzila N, Manoka AT, Tuliza M, Behets F, Goeman J, St Louis M, Piot P. 1994. Condom promotion, sexually transmitted diseases treatment, and declining incidence of HIV-1 infection in female Zairian sex workers. *Lancet* 344: 246–248.

Lampinen TM, Critchlow CW, Kuypers JM, Hurt CS, Nelson PJ, Hawes SE, Coombs RW, Holmes KK, Kiviat NB. 2000. Association of antiretroviral therapy with detection of HIV-1 RNA and DNA in the anorectal mucosa of homosexual men. *AIDS* 14: F69–F75.

Legakos S, Gable A. 2008. Challenges to HIV prevention—Seeking effective measures in the absence of a vaccine. *N Engl J Med* 358: 1543–1545.

Lurie MN, Williams B, Zuma K, Mkaya-Mwamburi D, Garnett G, Sturm AW, Sweat MD, Gittelsohn J, Abdool Karim SS. 2003. The impact of migration on HIV-1 transmission in South Africa. *Sex Transm Dis* 30: 149–156.

Lyle CD, Routson LB, Seaborn GB, Dixon LG, Bushar HF, Cyr WH. 1997. An in vitro evaluation of condoms as barriers to a small virus. *Sex Transm Dis* 24: 161–164.

Mathers B, Degenhardt L, Phillips B, Wiessing L, Hickman M, Strathdee S, Wodak A, Panda S, Tyndall M, Toufik A, et al. 2008. Global epidemiology of injecting drug use and HIV among people who inject drugs: A systematic review. *Lancet* 372: 1733–1745.

Matovu JK, Makumbi FE. 2007. Expanding access to voluntary HIV counselling and testing in sub-Saharan Africa: Alternative approaches for improving uptake, 2001–2007. *Trop Med Int Health* 12: 1315–1322.

McCoombe SG, Short RV. 2006. Potential HIV-1 target cells in the human penis. *AIDS* 20: 1491–1495.

Merson M, Padian N, Coates TJ, Gupta GR, Bertozzi SM, Piot P, Mane P, Barton M. Lancet HIV Prevention Series authors. 2008. Combination HIV prevention. *Lancet* 372: 1805–1806.

Metzger DS, Navaline H, Woody GE. 1999. Drug abuse treatment as AIDS prevention. *Public Health* 113: 97–106.

Microbicide Trials Network. 2011. Researchers reformulate tenofovir vaginal gel for rectal use. http://www.mtnstopshiv.org/node/2934..

Moses S, Bradley JE, Nagelkerke NJ, Ronald AR, Ndinya-Achola JO, Plummer FA. 1990. Geographical patterns of male circumcision practices in Africa: Association with HIV seroprevalence. *Int J Epidemiol* 19: 693–697.

Musicco M, Lazzarin A, Nicolosi A, Gasparini M, Costigliola P, Arici C, Saracco A. 1994. Antiretroviral treatment of men infected with human immunodeficiency virus type 1 reduces the incidence of heterosexual transmission. Italian Study Group on HIV Heterosexual Transmission. *Arch Intern Med* **154:** 1971–1976.

Myron S, Cohen YQ, Chen M, McCauley T, Gamble MC, Hosseinipour N, Kumarasamy JG, Hakim J, Kumwenda J, Grinsztejn B, et al. 2011. Prevention of HIV-1 infection with early antiretroviral therapy. *N Engl J Med* **365:** 493–505.

Nabel GJ. 2001. Challenges and opportunities for development of an AIDS vaccine. *Nature* **410:** 1002–1007.

Nagot N, Ouedraogo A, Foulongne V, Konaté I, Weiss HA, Vergne L, Defer M-C, Djagbaré D, Sanon A, Andonaba J-B, et al. 2007. Reduction of HIV-1 RNA levels with therapy to suppress herpes simplex virus. *N Engl J Med* **356:** 790–799.

Over M. 1998. The effect of societal variables on urban rates of HIV infection in developing countries: An exploratory analysis. In *Confronting AIDS: The evidence from the developing world* (ed. Ainsworth M, et al.), pp. 39–51. World Bank, Washington, DC.

Padian N, Buve A, Balkus J, Serwadda D, Cates W. 2008. Biomedical intervention to prevent HIV infection: Evidence, challenges, and way forward. *Lancet* **372:** 585–599.

Padian N, McCoy S, Balkus J, Wasserheit J. 2010. Weighing the gold in the gold standard: Challenges in HIV prevention research. *AIDS* **24:** 621–635.

Patterson BK, Landay A, Siegel JN, Flener Z, Pessis D, Chaviano A, Bailey RC. 2002. Susceptibility to human immunodeficiency virus-1 infection of human foreskin and cervical tissue grown in explant culture. *Am J Pathol* **161:** 867–873.

Patterson TL, Mausbach B, Lozada R, Staines-Orozco H, Semple SJ, Fraga-Vallejo M, Orozovich P, Abramovitz D, de la Torre A, Amaro H, et al. 2008. Efficacy of a brief behavioral intervention to promote condom use among female sex workers in Tijuana and Ciudad Juarez, Mexico. *Am J Public Health* **98:** 2051–2057.

PEPFAR. 2008. Prevention for positives. http://www.pepfar.gov/press/92934.htm (accessed April 24, 2008).

Pepin J, Plummer FA, Brunham RC. 1989. The interaction of HIV infection and other sexually transmitted diseases: An opportunity for intervention. *AIDS* **3:** 3–9.

Pinkerton SD, Abramson PR. 1997. Effectiveness of condoms in preventing HIV transmission. *Soc Sci Med* **44:** 1303–1312.

Piot P, Bartos M, Larson H, Zewdie D, Mane P. 2008. Coming to terms with complexity: A call to action for HIV prevention. *Lancet* **372:** 845–859.

Porco TC, Martin JN, Page-Shafer KA, Cheng A, Charlebois E, Grant RM, Osmond DH. 2004. Decline in HIV infectivity following the introduction of highly active antiretroviral therapy. *AIDS* **18:** 81–88.

Prata N, Vahidnia F, Fraser A. 2005. Gender and relationship differences in condom use among 15–24-year-olds in Angola. *Int Fam Plan Perspec* **31:** 192–199.

Quinn TC, Wawer MJ, Sewankambo N, Serwadda D, Li C, Wabwire-Mangen F, Meehan MO, Lutalo T, Gray RH. 2000. A study in rural Uganda of heterosexual transmission of human immunodeficiency virus. *N Engl J Med* **343:** 364.

Rehle T, Hallett T, Shisana O, Pillay-van Wyk V, Zuma K, Carrara H, Jooste S. 2010. A decline in new HIV infections in South Africa: Estimating HIV incidence from three national HIV surveys in 2002, 2005 and 2008. *PLoS One* **5:** e11094.

Rerks-Ngarm S, Pitisuttithum P, Nitayaphan S, Kaewkungwal J, Chiu J, Paris R, Nakorn P, Namwat C, de Souza M, Adams E, et al. 2009. Vaccination with ALVAC and AIDSVAX to prevent HIV-1 infection in Thailand. *N Engl J Med* **361:** 2209–2220.

Reynolds SJ, Shepherd ME, Risbud AR, Gangakhedkar RR, Brookmeyer RS, Divekar AD, Mehendale SM, Bollinger RC. 2004. Male circumcision and risk of HIV-1 and other sexually transmitted infections in India. *Lancet* **363:** 1039–1040.

Ross DA, Changalucha J, Obasi A, Todd J, Plummer ML, Cleopas-Mazige B, Anemona A, Everett D, Weiss HA, Mabey DC, et al. 2007. Biological and behavioral impact of an adolescent sexual health intervention in Tanzania: A community-randomized trial. *AIDS* **21:** 1943–1955.

Rottingen J-A, Cameron D, Garnett G. 2001. A systematic review of the epidemiologic interactions between classic sexually transmitted diseases and HIV. How much really is known? *Sex Transm Dis* **28:** 579–597.

Schneider JA, Lakshmi V, Dandona R, Kumar GA, Sudha T, Dandona L. 2010. Population-based seroprevalence of HSV-2 and syphilis in Andhra Pradesh state of India. *N Engl J Med* **362:** 427–439.

Shisana O, Rehele T, Simbayi L, Zuma K, Jooste S. 2009. South African national prevalence, incidence, behaviour and communication survey, 2008. In *A turning tide among teenagers?*, HSRC, Cape Town, South Africa.

Shugars DC, Slade GD, Patton LL, Fiscus SA. 2000. Oral and systemic factors associated with increased levels of human immunodeficiency virus type 1 RNA in saliva. *Oral Surg Oral Med Oral Pathol Oral Radiol Endod* **89:** 432–440.

Slutkin G, Okware S, Naamara W, Sutherland D, Flanagan D, Carael M, Blas E, Delay P, Tarantola D. 2006. How Uganda reversed its HIV epidemic. *AIDS Behav* **10:** 351–360.

Stamm WE, Handsfield HH, Rompalo AM, Ashley RL, Roberts PL, Corey L. 1988. The association between genital ulcer disease and acquisition of HIV infection in homosexual men. *JAMA* **260:** 1429–1433.

Stoneburner RL, Low-Beer D. 2004. Population-level HIV declines and behavioral risk avoidance in Uganda. *Science* **304:** 714–718.

Strathdee S, Hallett T, Bobrova N, Rhodes T, Booth R, Abdool R, Hankins CA. 2010. HIV and the risk environment among people who inject drugs: Past, present, and projections for the future. *Lancet* **376:** 268–284.

Sumartojo E, Doll L, Holtgrave D, Gayle H, Merson M. 2000. Enriching the mix: Incorporating structural factors into HIV prevention. *AIDS* **14:** S1.

Sweat M, Gregorich S, Sangiwa G, Furlonge C, Balmer D, Kamenga C, Grinstead O, Coates T. 2000. Cost-effectiveness of voluntary HIV-1 counselling and testing in reducing sexual transmission of HIV-1 in Kenya and Tanzania. *Lancet* **356:** 113–121.

The Voluntary HIV-1 Counseling Testing Efficacy Study Group. 2000. Efficacy of voluntary HIV-1 counselling and testing in individuals and couples in Kenya, Tanzania, and Trinidad: A randomised trial. *Lancet* **356**: 103–112.

Thomsen SC, Ombidi W, Toroitich-Ruto C, Wong EL, Tucker HO, Homan R, Kingola N, Luchters S. 2006. A prospective study assessing the effects of introducing the female condom in a sex worker population in Mombasa, Kenya. *Sex Transm Infect* **82**: 397–402.

Tovanabutra S, Robison V, Wongtrakul J, Sennum S, Suriyanon V, Kingkeow D, Kawichai S, Tanan P, Duerr A, Nelson KE, et al. 2002. Male viral load and heterosexual transmission of HIV-1 subtype E in northern Thailand. *J Acquir Immune Defic Syndr* **29**: 275–283.

UNAIDS. 2000. Evaluation of the 100% condom programme in Thailand, case study. Geneva: Joint United Nations Programme on HIV/AIDS.

UNAIDS. 2006. Report on the global AIDS epidemic. Geneva.

UNAIDS. 2007. AIDS epidemic update. Geneva: Joint United Nations Programme on HIV/AIDS.

UNAIDS. 2008. Report for the global AIDS epidemic. UNAIDS, Geneva.

UNAIDS. 2009. Condoms and HIV prevention: Position statement by UNAIDS, UNFPA and WHO. Geneva.

UNAIDS/WHO. 2004. UNAIDS/WHO policy statement on HIV testing. Geneva, Switzerland.

UNAIDS and WHO. 2007. New data on male circumcision and HIV prevention: Policy and programme implications; WHO/UNAIDS technical consultation male circumcision and HIV prevention: Research implications for policy and programming Montreux. Geneva: Joint United Nations Programme on HIV/AIDS and World Health Organization.

UNFPA. 2007. Donor support for contraceptives and condoms for STI/HIV prevention. Geneva.

Vandenbruaene M. 2007. King Kennard Holmes—Chair of the Department of Global Health of the University of Washington. *Lancet Infect Dis* **7**: 516–520.

van Schaik N, Kranzer K, Wood R, Bekker LG. 2010. Earlier HIV diagnosis—are mobile services the answer? *S Afr Med J* **100**: 671–674.

Vernazza PL, Troiani L, Flepp MJ, Cone RW, Schock J, Roth F, Boggian K, Cohen MS, Fiscus SA, Eron JJ. 2000. Potent antiretroviral treatment of HIV-infection results in suppression of the seminal shedding of HIV. The Swiss HIV Cohort Study. *AIDS* **14**: 117–121.

Vogel UF. 2007. Towards universal access to prevention, treatment and care: Experiences and challenges from the Mbeya region in Tanzania—A case study. In *UNAIDS best practices collection*. Joint United Nations Programme on HIV/AIDS, Geneva.

Walensky RP, Wood R, Fofana MO, Martinson NA, Losina E, April MD, Bassett IV, Morris BL, Freedberg KA, Paltiel AD, et al. 2009. The clinical impact and cost-effectiveness of routine, voluntary HIV screening in South Africa. *J Acquir Immune Defic Syndr* **56**: 26–35.

Watson-Jones D, Weiss HA, Rusizoka M, Changalucha J, Baisley K, Mugeye K, Tanton C, Ross D, Everett D, Clayton T, et al. 2008. Effect of herpes simplex suppression on incidence of HIV among women in Tanzania. *N Engl J Med* **358**: 1560–1571.

Wawer MJ, Makumbi F, Kigozi G, Serwadda D, Watya S, Nalugoda F, Buwembo D, Ssempijja V, Kiwanuka N, Moulton LH, et al. 2009. Circumcision in HIV-infected men and its effect on HIV transmission to female partners in Rakai, Uganda: A randomised controlled trial. *Lancet* **374**: 229–237.

Weir SS, Fox LJ, De Moya A, Gomez B, Guerrero G, Hassiq SE. 1998. Measuring condom use among sex workers in the Dominican Republic. *Int J STD AIDS* 223–226.

Weiss HA, Quigley MA, Hayes RJ. 2000. Male circumcision and risk of HIV infection in sub-Saharan Africa: A systematic review and meta-analysis. *AIDS* **14**: 2361–2370.

Weiss HA, Thomas SL, Munabi SK, Hayes RJ. 2006. Male circumcision and risk of syphilis, chancroid, and genital herpes: A systematic review and meta-analysis. *Sex Transm Infect* **82**: 101–109.

Weller S, Davis K2002. Condom effectiveness in reducing heterosexual HIV transmission. *Cochrane DB Syst Rev* **1**: CD003255.

Westercamp N, Bailey RC. 2007. Acceptability of male circumcision for prevention of HIV/AIDS in sub-Saharan Africa: A review. *AIDS Behav* **11**: 341–355.

WHA 5515. 2002. Small pox eradication-destruction of stocks. Fifty-Fifth World Assembly Agenda item 13,16;18 May 2002.

WHO. 2009. Towards universal access: scaling up priority HIV/AIDS interventions in the health sector: Progress report 2009. Geneva, Switzerland.

Wilson D, Halperin D. 2008. "Know your epidemic, know your response": A useful approach, if we get it right. *Lancet* **372**: 423–426.

Winkelstein W Jr, Samuel M, Padian NS, Wiley JA, Lang W, Anderson RE, Levy JA. 1987. The San Francisco Men's Health Study: III. Reduction in human immunodeficiency virus transmission among homosexual/bisexual men, 1982–86. *Am J Public Health* **77**: 685–689.

Winkelstein W Jr, Wiley JA, Padian NS, M Samuel M, Shiboski S, Ascher MS, Levy JA, et al. 1988. The San Francisco Men's Health Study: Continued decline in HIV seroconversion rates among homosexual/bisexual men. *Am J Public Health* **78**: 1472–1474.

Wolfe D, Carrieri P, Shepard D, Walker D. 2010. Treatment and care for HIV-infected injecting drug users: A review of barriers and ways forward. *Lancet* **376**: 355–366.

Zuckerman RA, Lucchetti A, Whittington WL, Sanchez J, Coombs RW, Zuñiga R, Magaret AS, Wald A, Corey L, Celum C, et al. 2007. Herpes simplex virus (HSV) suppression with valacyclovir reduces rectal and blood plasma HIV-1 levels in HIV-1/HSV-2-seropositive men: A randomized, double-blind, placebo-controlled crossover trial. *J Infect Dis* **196**: 1500–1508.

Index